1896	**Bequerel** discovers radioactivity.
1897	**Thomson** identifies cathode rays as negative corpuscles (electrons).
1900	**Planck** introduces the quantum idea.
1905	**Einstein** introduces the light corpuscle (photon) concept.
1905	**Einstein** advances the special theory of relativity.
1911	**Rutherford** reveals the nuclear atom.
1913	**Bohr** gives a quantum theory of the hydrogen atom.
1915	**Einstein** advances the general theory of relativity.
1923	**Compton's** experiments confirm the existence of the photon.
1924	**de Broglie** advances the wave theory of matter.
1925	**Goudsmit** and **Uhlenbeck** establish the spin of the electron.
1925	**Pauli** states the exclusion principle.
1926	**Schrödinger** develops the wave theory of quantum mechanics.
1927	**Davisson** and **Germer** and **Thomson** verify the wave nature of electrons.
1927	**Heisenberg** proposes the uncertainty principle.
1928	**Dirac** blends relativity and quantum mechanics in a theory of the electron.
1929	**Hubble** discovers the expanding universe.
1932	**Anderson** discovers antimatter in the form of the positron.
1932	**Chadwick** discovers the neutron.
1932	**Heisenberg** gives the neutron–proton explanation of nuclear structure.
1934	**Fermi** proposes a theory of the annihilation and creation of matter.
1938	**Meitner** and **Frisch** interpret results of **Hahn** and **Strassmann** as nuclear fission.
1939	**Bohr** and **Wheeler** give a detailed theory of nuclear fission.
1942	**Fermi** builds and operates the first nuclear reactor.
1945	**Oppenheimer's** Los Alamos team creates a nuclear explosion.
1947	**Bardeen**, **Brattain**, and **Shockley** develop the transistor.
1956	**Reines** and **Cowan** identify the antineutrino.
1957	**Feynman** and **Gell-Mann** account for all weak interactions with a "left-handed" neutrino.
1960	**Maiman** invents the laser.
1965	**Penzias** and **Wilson** discover background radiation in the universe left over from the Big Bang.
1967	**Bell** and **Hewish** discover pulsars, which are neutron stars.
1968	**Wheeler** names black holes.
1969	**Gell-Mann** suggests quarks as the building blocks of nucleons.
1977	**Lederman** and his team discover the bottom quark.
1981	**Binning** and **Rohrer** invent the scanning tunneling microscope.
1987	**Bednorz** and **Müller** discover high-temperature superconductivity.
1995	**Cornell** and **Wieman** create a "Bose–Einstein condensate" at 20 billionths of a degree.
2000	**Pogge** and **Martini** provide evidence for supermassive black holes in other galaxies.

Unfold to explore how **The Physics Place Web site** makes learning easy and fun!

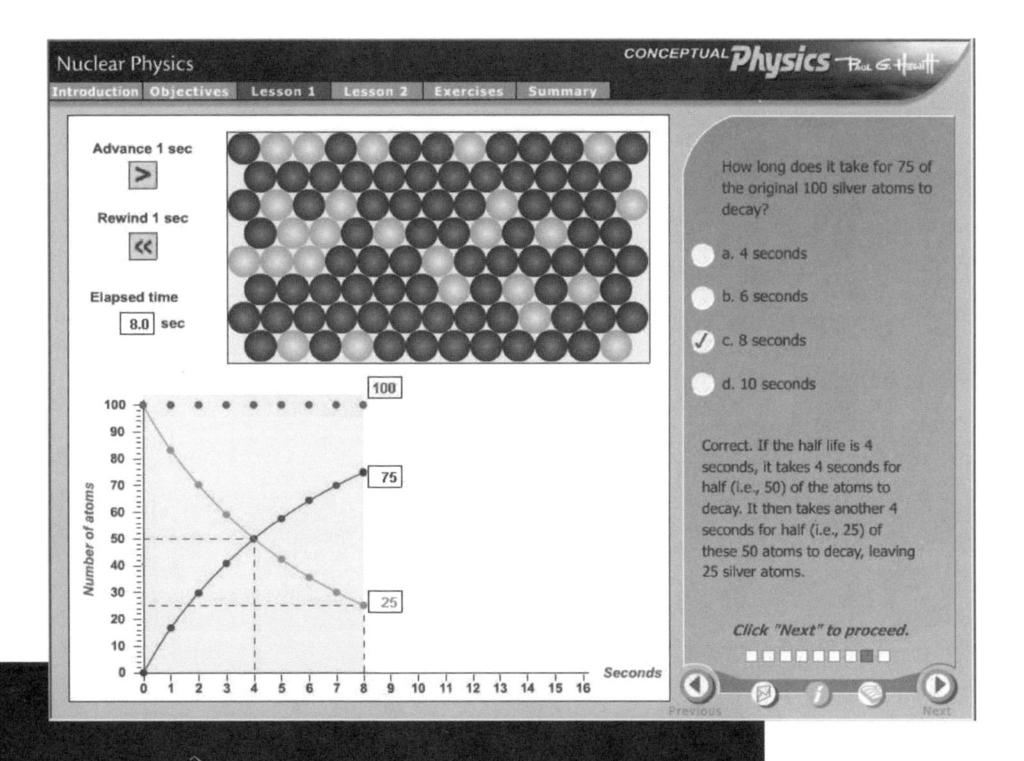

This media grid provides an at-a-glance, chapter-by-chapter overview of how the text and media work together to create a complete learning package. The grid features a wealth of tutorials, videos, and new interactive figures available for use with ***Conceptual Physics Media Update,* Tenth Edition**. In addition, every chapter includes assignable online quizzes, self-assessment tools, and hundreds of simulations and animations with ActivPhysics Online.

Paul G. Hewitt

ewton's Third aw of Motion	6 Momentum	7 Energy	8 Rotational Motion	9 Gravit
IGURE IN *CPX*: .10, 5.11, 5.16, p. 81, .22, 5.26 **STRATIONS:** nteraction eaction on sses tion on Rifle esentation: How to tract Vectors ddition of Vectors	**INTERACTIVE FIGURE IN *CPX*:** • 6.11, 6.13, 6.14, 6.15, 6.18 **TUTORIAL:** • Momentum and Collisions **VIDEO DEMONSTRATIONS:** • Definition of Momentum • Changing Momentum—Follow Through • Decreasing Momentum Over a Short Time	**INTERACTIVE FIGURE IN *CPX*:** • 7.9, 7.12, 7.15, 7.16, 7.17 **TUTORIAL:** • Energy **VIDEO DEMONSTRATIONS:** • Bowling Ball and Conservation of Energy • Conservation of Energy: Numerical Example • Machines: Pulleys	**INTERACTIVE FIGURE IN *CPX*:** • 8.1, 8.2, 8.3, 8.18, 8.20, 8.53 **TUTORIAL:** • Rotational Motion **VIDEO DEMONSTRATIONS:** • Rotational Speed • Rotational Inertia Using Weighted Pipes • Rotational Inertia Using a Hammer • Rotational Inertia with a Weighted Rod • Difference Between Torque and Weight • Why a Ball Rolls Down a Hill • Locating the Center of Gravity • Toppling • Centripetal Force • Simulated Gravity • Conservation of Angular Momentum Using a Rotating Platform	**INTERACTIVE FIGURE IN *CP*** • 9.6, 9.9 **TUTORIAL:** • Gravity **VIDEO DEMONSTRATIONS:** • Von Jolly's Method of Measuring the Attraction Between Two Masses • Inverse-Square Law • Weight and Weightlessn • Apparent Weightlessnes • Gravitational Field Inside Hollow Planet • The Weight of an Object a Hollow Planet but Not Center • Discovery of Neptune

ermodynamics	19 Vibrations and Waves	20 Sound	21 Musical Sounds	22 Electrost
IGURE IN *CPX*: **STRATION:** ocess	**INTERACTIVE FIGURE IN *CPX*:** • 19.3, 19.7, 19.9, 19.13, 19.14, 19.16, 19.17 **TUTORIAL:** • Waves and Sound **VIDEO DEMONSTRATIONS:** • Longitudinal vs. Transverse Waves • Doppler Effect	**INTERACTIVE FIGURE IN *CPX*:** • 20.12, 20.15, 20.16 **VIDEO DEMONSTRATIONS:** • Refraction of Sound • Resonance • Resonance and Bridges • Interference and Beats		**INTERACTIVE FIGURE IN *CP*** • 22.1, 22.2, 22.7, 22.8, 22.12, 22.18 **TUTORIAL:** • Electrostatics **VIDEO DEMONSTRATIONS:** • Electric Potential • Van de Graaff Generator

ight Quanta	32 The Atom and the Quantum	33 The Atomic Nucleus and Radioactivity	34 Nuclear Fission and Fusion	35 Special Th of Relati
IGURE IN *CPX*:	**INTERACTIVE FIGURE IN *CPX*:** • 32.13, 32.15 **VIDEO DEMONSTRATION:** • Electron Waves	**INTERACTIVE FIGURE IN *CPX*:** • 33.2, 33.3, 33.4, 33.9 **TUTORIAL:** • Nuclear Physics **VIDEO DEMONSTRATIONS:** • Radioactive Decay • Half-Life • Carbon Dating	**INTERACTIVE FIGURE IN *CPX*:** • 34.10, 34.11, 34.16, 34.18 **TUTORIAL:** • Nuclear Physics **VIDEO DEMONSTRATIONS:** • Nuclear Fission • Plutonium • Controlling Nuclear Fusion	**INTERACTIVE FIGURE IN *CPX*** • 35.4, 35.5, 35.9, 35.10, 35.2 **VIDEO DEMONSTRATIONS:** • The Twin Trip Animation • Space and Time Travel

10 Projectile and Satellite Motion

INTERACTIVE FIGURE IN *CPX:*
- 10.4, 10.5, 10.9, 10.11, 10.13, 10.21, 10.25, 10.33

TUTORIALS:
- Projectile Motion
- Orbital Motion and Kepler's Laws

VIDEO DEMONSTRATIONS:
- Projectile Motion Demonstration
- More Projectile Motion
- Circular Orbits

11 The Atomic Nature of Matter

VIDEO DEMONSTRATIONS:
- Evidence for Atoms
- Atoms Are Recyclable

12 Solids

INTERACTIVE FIGURE IN *CPX:*
- 12.16

VIDEO DEMONSTRATIONS:
- Surface Area vs. Volume
- Scaling

13 Liquids

INTERACTIVE FIGURE IN *CPX:*
- 13.9, 13.14

VIDEO DEMONSTRATIONS:
- Dam Keeps Water in Place, Water Keeps Dam in Place
- Buoyancy
- Archimedes' Principle
- Flotation

23 Electric Current

INTERACTIVE FIGURE IN *CPX:*
- 23.17, 23.18

TUTORIAL:
- Circuits

VIDEO DEMONSTRATIONS:
- Ohm's Law
- Caution on Handling Electrical Wires
- Birds and High Voltage Wires
- Alternating Current
- Electric Circuits

24 Magnetism

INTERACTIVE FIGURE IN *CPX:*
- 24.2, 24.7, 24.15, 24.19

TUTORIAL:
- Magnetism

VIDEO DEMONSTRATIONS:
- Oersted's Discovery
- Magnetic Forces on Current-Carrying Wires

25 Electromagnetic Induction

INTERACTIVE FIGURE IN *CPX:*
- 25.3, 25.6, 25.7

VIDEO DEMONSTRATIONS:
- Faraday's Law
- Application of E&M Induction

26 Properties of Light

INTERACTIVE FIGURE IN *CPX:*
- 26.2, 26.3, 26.4, 26.12, 26.13

VIDEO DEMONSTRATION:
- Light and Transparent Materials

36 General Theory of Relativity

INTERACTIVE FIGURE IN *CPX:*
- 36.3, 36.4, 36.10

D Appendix

TUTORIAL:
- Vectors

Media for *Conceptual Physics Media Update,* Tenth Edition, b

1 About Science

2 Newton's First Law of Motion — Inertia

VIDEO DEMONSTRATIONS:
- Newton's Law of Inertia
- The Old Tablecloth Trick
- Toilet Paper Roll
- Inertia of a Ball
- Inertia of a Cylinder
- Inertia of an Anvil
- Definition of a Newton

3 Linear Motion

INTERACTIVE FIGURE IN *CPX*:
- 3.6, 3.7, 3.8

VIDEO DEMONSTRATIONS:
- Definition of Speed
- Average Speed
- Velocity
- Changing Velocity
- Definition of Acceleration
- Numerical Example of Acceleration
- Free Fall: How Fast?
- $v = gt$
- Free Fall: How Far?
- Falling and Air Resistance

4 Newton's Second Law of Motion

INTERACTIVE FIGURE IN *CPX*:
- 4.9, 4.12, 4.15

TUTORIAL:
- Newton's Second Law

VIDEO DEMONSTRATIONS:
- Force Causes Acceleration
- Friction Second
- Newton's Second Law
- Free-Fall Acceleration Explained
- Falling and Air Resistance
- Air Resistance and Falling Objects

5

INTERACTIVE
- 5.1, 5.8, 5.9, Tug of War

TUTORIAL:
- Vectors

VIDEO DEMO
- Forces and
- Action and Different M
- Action; Rea and Bullet
- Vector Rep Add and Su
- Geometric

14 Gases and Plasmas

VIDEO DEMONSTRATIONS:
- Air Has Weight
- Air Is Matter: Pouring Air from One Glass to Another
- Air Has Pressure
- Buoyancy of Air

15 Temperature, Heat, and Expansion

VIDEO DEMONSTRATIONS:
- Low Temperatures with Liquid Nitrogen
- How a Thermostat Works

16 Heat Transfer

INTERACTIVE FIGURE IN *CPX*:
- 16.12, 16.20

VIDEO DEMONSTRATIONS:
- The Secret to Walking on Hot Coals
- Air Is a Poor Conductor

17 Change of Phase

INTERACTIVE FIGURE IN *CPX*:
- 17.16

VIDEO DEMONSTRATIONS:
- Condensation Is a Warming Process
- Boiling Is a Cooling Process
- Pressure Cooker and Boiling and Freezing at the Same Time

18 T

INTERACTIVE
- 18.10, 18.12

VIDEO DEMON
- Adiabatic F

27 Color

INTERACTIVE FIGURE IN *CPX*:
- 27.9, 27.10, 27.18

VIDEO DEMONSTRATIONS:
- Yellow-Green Peak of Sunlight
- Colored Shadows
- Why the Sky Is Blue and Why the Sunset Is Red

28 Reflection and Refraction

INTERACTIVE FIGURE IN *CPX*:
- 28.6, 28.14, 28.36, 28.46

VIDEO DEMONSTRATIONS:
- Image Formation in a Mirror
- Model of Refraction
- The Rainbow

29 Light Waves

INTERACTIVE FIGURE IN *CPX*:
- 29.3, 29.4, 29.8, 29.18, 29.19, 29.35

VIDEO DEMONSTRATION:
- Soap Bubble Interference
- Polarized Light and 3-D Viewing

30 Light Emission

INTERACTIVE FIGURE IN *CPX*:
- 30.2, 30.5, 30.7, 30.8, 30.10

31

INTERACTIVE
- 31.1, 31.6

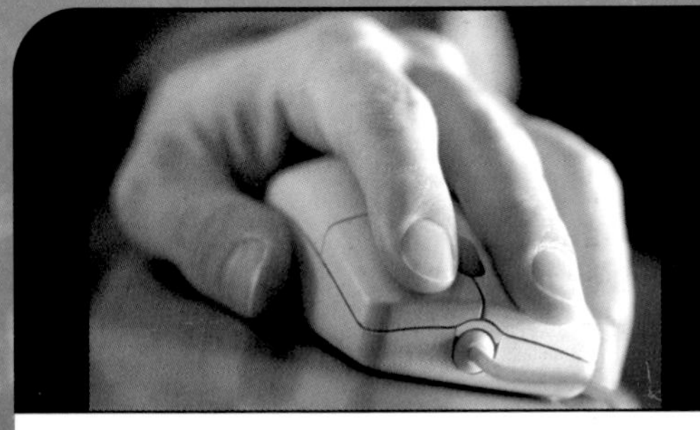

Your steps to success.

STEP 1: Register

All you need to get started is a valid email address and the access code below.To register, simply:

1. Go to www.physicsplace.com
2. Click the appropriate book cover.
   Cover must match the textbook edition being used for your class.
3. Click **"Register"** under **"First-Time User?"**
4. Leave **"No, I Am a New User"** selected.
5. Using a coin, scratch off the silver coating below to reveal your access code.
   Do not use a knife or other sharp object, which can damage the code.
6. Enter your access code in lowercase or uppercase, without the dashes.
7. Follow the on-screen instructions to complete registration.
   During registration, you will establish a personal login name and password to use for logging into the website. You will also be sent a registration confirmation email that contains your login name and password.

Your Access Code is:

Note: If there is no silver foil covering the access code, it may already have been redeemed, and therefore may no longer be valid. In that case, you can purchase access online using a major credit card. To do so, go to www.physicsplace.com, click the cover of your textbook, click **"Buy Now"**, and follow the on-screen instructions.

STEP 2: Log in

1. Go to www.physicsplace.com and click the appropriate book cover.
2. Under **"Established User?"** enter the login name and password that you created during registration. *If unsure of this information, refer to your registration confirmation email.*
3. Click **"Log In"**.

STEP 3: (Optional) Join a class

Instructors have the option of creating an online class for you to use with this website. If your instructor decides to do this, you'll need to complete the following steps using the Class ID your instructor provides you. By "joining a class," you enable your instructor to view the scored results of your work on the website in his or her online gradebook.

To join a class:

1. Log into the website. For instructions, see "STEP 2: Log in."
2. Click **"Join a Class"** near the top right.
3. Enter your instructor's **"Class ID"** and then click **"Next"**.
4. At the Confirm Class page you will see your instructor's name and class information. If this information is correct, click **"Next"**.
5. Click **"Enter Class Now"** from the Class Confirmation page.

- *To confirm your enrollment in the class, check for your instructor and class name at the top right of the page. You will be sent a class enrollment confirmation email.*
- *As you complete activities on the website from now through the class end date, your results will post to your instructor's gradebook, in addition to appearing in your personal view of the Results Reporter.*

To log into the class later, follow the instructions under "STEP 2: Log in."

Got technical questions?

Customer Technical Support: To obtain support, please visit us online anytime at http://247.aw.com where you can search our knowledgebase for common solutions, view product alerts, and review all options for additional assistance.

SITE REQUIREMENTS

For the latest updates on Site Requirements, go to www.physicsplace.com, choose your text cover, and click Site Reqs.

WINDOWS
OS: Windows 2000, XP
Resolution: 1024 x 768
Plugins: Latest Version of Flash/QuickTime/Shockwave (as needed)
Browsers: Internet Explorer 6.0; Firefox 1.0

MACINTOSH
OS: 10.2.4, 10.3.2
Resolution: 1024 x 768
Plugins: Latest Version of Flash/QuickTime/Shockwave (as needed)
Browsers: Firefox 1.0; Safari 1.3

Internet Connection: 56k modem minimum

Register and log in

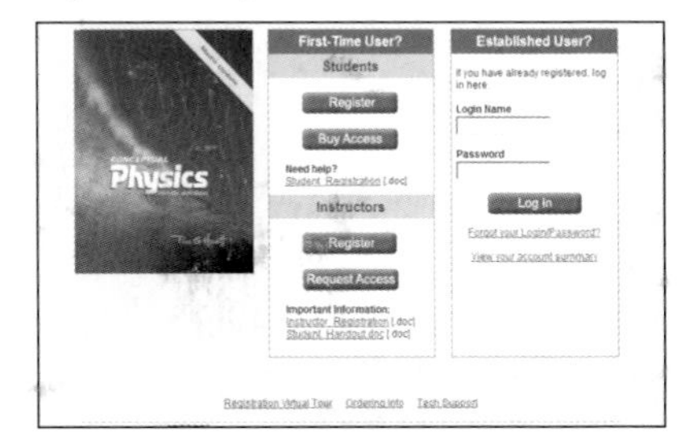

Join a class

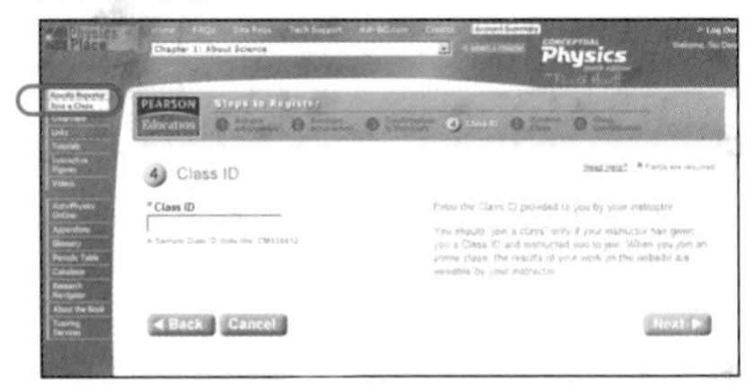

Important: Please read the Subscription and End-User License agreement, accessible from the book website's login page, before using *The Physics Place* website. By using the website, you indicate that you have read, understood, and accepted the terms of this agreement.

Conceptual Physics

tenth edition

Conceptual

Physics

tenth edition

written and illustrated by

City College of San Francisco

San Francisco Boston New York
Capetown Hong Kong London Madrid Mexico City
Montreal Munich Paris Singapore Sydney Tokyo Toronto

Publisher: Jim Smith
Assistant Editor and Media Producer: Ashley Eklund
Editorial Manager: Laura Kenney
Development Manager: Michael Gillespie
Executive Marketing Manager: Scott Dustan
Managing Editor: Corinne Benson
Senior Production Supervisor: Nancy Tabor
Production Service and Composition: Aptara, Inc.
Text Design: Carolyn Deacy
Cover Design: Yvo Riezebos Design
Cover Production: Seventeenth Street Studios
Manufacturing Buyer: Jeff Sargent
Photo Researcher: Ira Kleinberg
Cover Printer: Phoenix Color Corporation
Text Printer and Binder: Courier Kendallville
Cover Image: Wave and surfer, Photolibrary.com/AMANA AMERICA INC.
IMA USA INC.; particle tracks, Lawrence Berkeley National Laboratory,
University of California

ISBN-13: 978-0-321-54833-7 (Student copy)
ISBN-10: 0-321-54833-7 (Student copy)
ISBN-13: 978-0-321-54830-6 (Professional copy)
ISBN-10: 0-321-54830-2 (Professional copy)

www.aw-bc.com

1 2 3 4 5 6 7 8 9 10—CRK—10 09 08 07

To Lillian Lee Hewitt

Contents in Brief

Contents in Detail

PART THREE

Heat 289

PART FOUR

Sound 361

PART FIVE

Electricity and Magnetism 409

PART SIX

Light 495

PART EIGHT

Relativity 685

The Conceptual Physics Photo Album

Conceptual Physics is a very personal book, reflected in the many photographs of family and friends that grace its pages. Foremost to providing suggestions and feedback to this and previous editions is Ken Ford, former CEO of the American Institute of Physics, to whom the Eighth Edition was dedicated. Ken's hobby of gliding is appropriately shown on page 391, and his dedication to teaching Germantown Academy high-school students in Pennsylvania is seen on page 686. The First Edition of *Conceptual Physical Science* was dedicated to resourceful Charlie Spiegel, shown on page 503. Although Charlie passed away in 1996, his personal touch carries over to this book. Assisting in the production of this edition is my wife Lillian, to whom this book is dedicated. She is shown on pages 2, 143, 316, and 382. Lillian holds our colorful pet conure, Sneezlee, on page 521. Her dad Wai Tsan Lee is on page 461, and her niece and nephew, Allison and Erik Wong, demonstrate thermodynamics on page 353.

Part openers with the cartoon style blurbs about physics are of family and close friends. The book opener on page 1 is of my great nephew Evan Suchocki (pronounced Su-hock′-ee, with silent c) holding a pet chickie while sitting on my lap. Part 1 on page 21 shows Debbie and Natalie Limogan, children of my dear San Francisco friends, Hideko and Herman Limogan. The little boy in the middle is Genichiro Nakada. Part 2 on page 209 is Andres Riveros Mendoza, son of David Riveros, my co-author of a physics textbook in Spanish, *Fisica 1 & 2: Las Reglas de la Naturaleza*, published in Mexico City, Mexico. Part 3 opens on page 289 with Terrence Jones, son of my niece Corine Jones, who is now grown up and created the computer font for the *Next-Time Questions*. Part 4, page 361, is my grandson Alexander Hewitt, and Part 5, page 409, is my granddaughter Megan, daughter of Leslie and Bob Abrams. Part 6, page 495, is Lillian's nephew, Christopher Lee. Grandchildren Alexander and Grace Hewitt open Part 7 on page 619. Grace alone begins Part 8 on page 685.

To celebrate this Tenth Edition, chapter-opening photographs are of teacher friends and colleagues, mostly in their classrooms demonstrating physics typical of the chapter material. Their names appear with their photos.

City College of San Francisco friends and colleagues open Chapters 3, 4, 7, 13, 21, and 23. On page 101 we see Will Maynez with the air track he designed and built, and again burning a peanut on page 324. Dave Wall is on page 595.

Physics teacher friends from high schools include Chicago's finest, close friend Marshall Ellenstein, who swings the water-filled bucket on page 146 and walks barefoot on broken glass on page 265. Marshall, a longtime contributor to *Conceptual Physics*, has recently converted videos made of my lectures in 1982 to a 3-disc DVD set, *Conceptual Physics Alive!—The San Francisco Years* (which predate the DVDs of 34 lectures in Hawaii). Page 122 shows dear friend and dedicated San Mateo physics teacher, Pablo Robinson, who risks his body for science sandwiched between beds of nails. Pablo, shown again on page 515, is the author of the lab manual that accompanies this book. An old black and white photo of Pablo's children, David and Kristin, is shown on page 159. Pablo's wife Ellyn, author of *Biotechnology: Science for the New Millennium*, EMC-Paradigm Publishing, 2006, is on page 290. Dean Baird, co-author of the *Conceptual Physical Science* lab manuals, is on page 325.

Family photos begin with the touching photo on page 81 of son Paul and his daughter Grace. Another photo on page 88 linking touching to Newton's third law is of my brother Steve with his daughter Gretchen at their coffee farm in Costa Rica. Gretchen is shown again, grown up, on page 319. Steve's son Travis is seen on page 155, and his oldest daughter Stephanie on page 232. My son Paul is again shown on pages 309 and 346. His lovely wife, Ludmila, holds crossed Polaroids on page 573, and their dog Hanz pants on page 326. The endearing girl on page 217 is my daughter Leslie, now a mom, teacher, and earth-science co-author of our *Conceptual Physical Science* textbooks. This photo of Leslie, now colorized, has been a trademark of *Conceptual Physics* since the Third Edition. A more recent photo of her with husband Bob is on page 595. Their children, Megan and Emily, along with son Paul's children, make up the colorful set of photos on page 520. Photos of my late son James are on pages 151, 400, and 550. He left me my first grandson, Manuel, seen on pages 238 and 388. Manuel's grandmom, my wife

Millie, before passing away in early 2004, bravely holds her hand above the active pressure cooker on page 309. Brother Dave (no, not a twin) and his wife Barbara demonstrate atmospheric pressure on page 272. Their son Dave is on page 451, and grandson John Perry Hewitt is on page 282. Sister Marjorie Hewitt Suchocki, an author and theologian at Claremont School of Theology, illustrates reflection on page 533. Marjorie's son, John Suchocki, author of the third edition (2007) of *Conceptual Chemistry*, Benjamin Cummings, and my chemistry co-author of the *Conceptual Physical Science* textbooks, walks fearlessly across hot coals on page 306 (for emphasis, David Willey does the same on page 337). Nephew John is also a talented vocalist and guitarist known as John Andrew in his popular CDs, seen with his guitar on page 480. The group listening to music on page 404 are part of John's and Tracy's wedding party; from left to right, Butch Orr, my niece Cathy Candler, bride and groom, niece Joan Lucas, sister Marjorie, Tracy's parents Sharon and David Hopwood, teacher friends Kellie Dippel and Mark Werkmeister, and myself.

Personal friends who were my former students begin with Tenny Lim, who draws her bow on page 114 as she has been doing since the Sixth Edition. Tenny is now a "rocket scientist" at Jet Propulsion Lab in Pasadena. Another rocket-scientist friend is Helen Yan, who develops satellites for Lockheed Martin in Sunnyvale in addition to stints at part-time physics teaching. Her hand lettering, now in computer font, adorns the Next-Time Questions for this edition. On page 315 Helen poses with the box she first posed with for the Fifth Edition when she was my teaching assistant. Another former student is Alexei Cogan, who demonstrates center of gravity on page 143. The karate gal on page 96 is former CCSF student Cassy Cosme. This is a colorized black and white photo that graced three editions of this book before it went full color in the Seventh Edition. On page 151 student Cliff Braun is at the far left of my son James in Figure 8.51, with nephew Robert Baruffaldi at the far right.

Two dear friends who go back to my own school days are Howie Brand on page 91 and Dan Johnson on page 285. Other cherished friends are Paul Ryan, who drags his finger through molten lead on page 337, and Tim Gardner, demonstrating Bernoulli's principle on page 278. My friend and mentor from sign painting days, Burl Grey, is on page 22, and is discussed on pages 30 and 31. Lifelong friend Ernie Brown, who designed the cover logos for all my conceptual books, is seen on page 750—not in a photo, but in a cartoon—for Ernie is a cartoonist and my cartooning mentor of earlier years. Physics buddy John Hubisz opens Chapter 12, page 229, and appears in the entropy photo on page 356. Friends Larry and Tammy Tunison wear radiation badges on page 656. Suzanne Lyons, co-author of *Conceptual Science*, poses with her children Tristan and Simone on page 528. Phil Wolf, co-author of the *Problem Solving in Conceptual Physics* book that accompanies this edition, is shown on page 600. Helping create that book is Diane Riendeau, shown on page 362.

My dear Hawaii friends include Walter Steiger, page 634, Jean and George Curtis, pages 477 and 582, Richard Crowe, page 720, Praful Shah, page 161, and the Hu family of Honolulu, beginning with Meidor, page 430, who took many photos for previous editions that carry over here. Mom Ping Hu is on page 137, with uncle Chiu Man Wu on page 326, and his daughter, Andrea, on page 129.

The inclusion of these people who are so dear to me makes *Conceptual Physics* all the more my labor of love.

To the Student

You know you can't enjoy a game unless you know its rules; whether it's a ball game, a computer game, or simply a party game. Likewise, you can't fully appreciate your surroundings until you understand the rules of nature. Physics is the study of these rules, which show how everything in nature is beautifully connected. So the main reason to study physics is to enhance the way you see the physical world. You'll see the mathematical structure of physics in frequent equations, but more than being recipes for computation, you'll see the equations as *guides to thinking*.

I enjoy physics, and you will too — because you'll understand it. If you get hooked and take a follow-up course, *then* you can focus on mathematical problems. Go for comprehension of concepts now, and if computation follows, it will be with understanding.

Enjoy your physics!

Paul G. Hewitt

To the Instructor

The sequence of chapters in this Tenth Edition is identical to that of the previous edition. In addition to extensive edits in every chapter, there are many new photographs, new chapter-end material, and other new features described below.

As with the previous edition, Chapter 1, "About Science," begins your course on a high note with coverage on early measurements of the Earth and distances to the Moon and the Sun.

Part One, "Mechanics," begins with Chapter 2, which, as in the previous edition, presents a brief historical overview of Aristotle and Galileo, progressing to Newton's first law and to mechanical equilibrium. The high tone of Chapter 1 is maintained as forces are treated before velocity and acceleration. Students get their first taste of physics via a very comprehensible treatment of parallel force vectors. They enter a comfortable part of physics before being introduced to kinematics.

Chapter 3, "Linear Motion," is the only chapter in Part One that is devoid of physics laws. Kinematics has no laws, only definitions, mainly for *speed, velocity,* and *acceleration*—likely the least exciting concepts that your course has to offer. Too often kinematics becomes a pedagogical "black hole" of instruction—too much time for too little physics. Being more math than physics, the kinematics equations can appear to the student as the most intimidating in the book. Although the experienced eye doesn't see them as such, this is how *students* first see them:

$$\varsigma = \varsigma_o + \delta\ni$$
$$\zeta = \varsigma_o\ni + \tfrac{1}{2}\,\delta\ni^2$$
$$\varsigma^2 = {\varsigma_o}^2 + 2\delta\,\zeta$$
$$\varsigma_\alpha = \tfrac{1}{2}\,(\,\varsigma_o + \varsigma\,)$$

If you wish to reduce class size, display these equations on the first day and announce that class effort for much of the term will be making sense of them. Don't we do much the same with the standard symbols?

Ask any college graduate these two questions: What is the acceleration of an object in free fall? What keeps Earth's interior hot? You'll see where their education was focused, for many more will correctly answer the first question than the second. Traditionally, physics courses have been top-heavy in kinematics with little or no coverage of modern physics. Radioactive decay almost never gets the attention given to falling bodies. So my recommendation is to pass quickly through Chapter 3, making the distinction between velocity and acceleration, and then to move on to Chapter 4, "Newton's Second Law of Motion," where the concepts of velocity and acceleration find their application.

Chapter 5 continues with Newton's third law. The end of the chapter treats the parallelogram rule for combining vectors—first force vectors and then velocity vectors. It also introduces vector components. More on vectors is found in Appendix D, and especially in the *Practicing Physics* book.

Chapter 6, "Momentum," is a logical extension of Newton's third law. One reason I prefer teaching it before energy is that students find mv much simpler and easier to grasp than $\frac{1}{2}mv^2$. Another reason for treating momentum first is that the vectors of the previous chapter are employed with momentum but not with energy.

Chapter 7, "Energy," is a longer chapter, rich with everyday examples and current energy concerns. Energy is central to mechanics, so this chapter has the greatest number of exercises (70) in the chapter-end material. Work, energy, and power also get generous coverage in the *Practicing Physics* book.

After Chapters 8 and 9 (on rotational mechanics and gravity), mechanics culminates with Chapter 10 (on projectile motion and satellite motion). Students are fascinated to learn that any projectile moving fast enough can become an Earth satellite. Moving faster, it can become a satellite of the Sun. Projectile motion and satellite motion belong together.

Part Two, "Properties of Matter," begins with a new chapter on atoms, with much of the historical treatment of the previous edition moved forward to Chapter 32 in an expanded treatment of atoms and quanta.

Parts Three through Eight continue, like earlier parts, with enriched examples of current technology. The chapter with the fewest changes is Chapter 36, "General Theory of Relativity," which now acknowledges the recent finding that the universe is flat.

This edition retains the boxes with short essays on such topics as energy and technology, railroad train wheels, magnetic strips on credit cards, and magnetically levitated trains. Also featured are boxes on pseudoscience, crystal power, the placebo effect, water dowsing, magnetic therapy, electromagnetic waves surrounding power lines, and the phobia about food irradiation and anything nuclear. To the person who works in the arena of science, who knows about the care, checking, and cross-checking that go into understanding something, these fads and misconceptions are laughable. But to those who don't work in the science arena, including even your best students, pseudoscience can seem compelling when purveyors clothe their wares in the language of science while skillfully sidestepping the tenets of science. It is my hope that these boxes may help to stem this rising tide.

A new feature of this edition is "One-Step Calculations," sets of simple "plug-and-chug" problems requiring only single-step solutions. They appear in more equation-oriented chapters. Students become familiar with the equations by substitution of given numerical values. More math-physics challenges are found in the problem sets. These are preceded by qualitative exercises, expanded by an average of ten additional new ones per chapter throughout the book.

New to this edition are the insightful boxes in many of the margins. Every page of every introductory textbook ought to have information that perks up the brain. The Insight boxes add to that.

The most striking new feature of this edition is the student supplement *Problem Solving in Conceptual Physics,* co-authored with Phil Wolf. While problem solving is not the main thrust of a conceptual course, Phil and I nevertheless love solving problems. In a novel and student-friendly way, our supplement features problems that are more physics than math, nicely extending *Conceptual Physics*—even to courses oriented to problem solving. We think that many professors will enjoy the options offered by this student supplement to the textbook. Problem solutions are posted on the website in the Instructor's Resource area.

Supporting this edition is the *Instructor's Manual,* with suggested lectures, demonstrations, and answers to all chapter-end material. The *Next-Time*

Questions book has a greater number of questions, which are now in a horizontal format to make them more compatible with computer monitors and PowerPoint® displays. Although the printed book is still in black and white, the electronic version is in color. The student *Practicing Physics* book (with answers to the odd-numbered exercises and problems herein) has several new practice pages and is now available electronically. The *Laboratory Manual* has been improved. Still available is the set of *Transparencies* with an accompanying instruction guide. The *Test Bank* (both in print and in computerized format) has been revised and has many contributions by Herb Gottlieb. Perhaps most important, however, is the ambitious range of teaching and learning media developed to support this new edition.

To help with your in-class presentations, we have built a new instructor supplement called *The Conceptual Physics Lecture Launcher*. This CD-ROM provides a wealth of presentation tools to help support your fun and dynamic lectures. It includes more than 100 clips from my favorite video demonstrations, more than 130 interactive applets developed specifically to help you illustrate particularly tricky concepts, and chapter-by-chapter weekly in-class quizzes in PPT for use with Classroom Response Systems (easy-to-use wireless polling systems that allow you to pose questions in class, have each student vote, and then display and discuss results in real time). *The Conceptual Physics Lecture Launcher* also provides all the line images from the book (in high resolution) and the *Instructor's Manual* in convenient, editable Word format.

For out-of-class help for your students, a critically acclaimed website, which can be found at http://www.physicsplace.com, now provides even more study resources. The Physics Place is the most educationally advanced, most highly rated by students, and most widely used website available for students taking this course. The enhanced website now provides more of the students' favorite interactive online tutorials (covering topics that many of you requested), and a new library of Interactive Figures (key figures from each chapter in the book that are better understood through interactive experimentation owing to reasons of scale, geometry, time evolution, or multiple representation). Quizzes, flash cards, and a wealth of other chapter-specific study aids are also provided.

All of these innovative, targeted, and effective online learning media are easily integrated in your course using a new online gradebook (allowing you to "assign" the tutorials, quizzes, and other activities as out-of-class homework or projects that are automatically graded and recorded), simple icons throughout the text (highlighting for you and your students key tutorials, Interactive Figures, and other online resources), and *The Conceptual Physics Lecture Launcher* CD-ROM. A new Online Resources section at the Physics Place summarizes the media available to you and your students, chapter by chapter, week by week.

For more information on the support ancillaries, check out http://www.aw-bc.com/physics or contact your Addison Wesley representative, or contact me, Pghewitt@aol.com.

Acknowledgments

I am enormously grateful to Ken Ford for checking this edition for accuracy and for his many insightful suggestions. Many years ago, I admired Ken's own books, one of which, *Basic Physics,* first inspired me to write *Conceptual Physics*. Today I am honored that he has given so much of his time and energy to assist in making this edition my best edition ever. Errors invariably crop up after manuscript is submitted, so I take full responsibility for any errors that have survived his scrutiny.

For extensive feedback, I'm thankful to Diane Riendeau. For valued suggestions, I thank my friends Dean Baird, Howie Brand, George Curtis, Marshall Ellenstein, Mona El Tawil-Nassar, Jim Hicks, John Hubisz, Dan Johnson, Fred Myers, Kenn Sherey, Chuck Stone, Pablo Robinson, and Phil Wolf. I'm grateful for suggestions from Matthew Griffiths, Paul Hammer, Francisco Izaguirre, Les Sawyer, Dan Sulke, Richard W. Tarara, and Lawrence Weinstein. I am grateful to the resourcefulness of my Exploratorium friends and colleagues: Judith Brand, Paul Doherty, Ron Hipschman, and Modesto Tamez. For photos, I thank my brother Dave Hewitt, my son Paul Hewitt, Keith Bardin, Burl Grey, Lillian Lee Hewitt, Will Maynez, Fred Myers, Milo Patterson, Jay Pasachoff, and David Willey. For strengthening the test bank, I thank Herb Gottlieb.

I remain grateful to the authors of books that initially served as influences and references many years ago: Theodore Ashford, *From Atoms to Stars;* Albert Baez, *The New College Physics: A Spiral Approach;* John N. Cooper and Alpheus W. Smith, *Elements of Physics;* Richard P. Feynman, *The Feynman Lectures on Physics;* Kenneth Ford, *Basic Physics;* Eric Rogers, *Physics for the Inquiring Mind;* Alexander Taffel, *Physics: Its Methods and Meanings;* UNESCO, *700 Science Experiments for Everyone;* and Harvey E. White, *Descriptive College Physics*. For this and the previous edition, I'm thankful to Bob Park, whose book *Voodoo Science* motivated me to include the boxes on pseudoscience.

For the *Problem Solving in Conceptual Physics* ancillary, co-authored with Phil Wolf, we both thank Tsing Bardin, Howie Brand, George Curtis, Ken Ford, Herb Gottlieb, Jim Hicks, David Housden, Chelcie Liu, Fred Myers, Stan Schiocchio, Diane Riendeau, and David Williamson for valuable feedback.

I am particularly grateful to my wife, Lillian Lee Hewitt, for her assistance in all phases of book-and-ancillary preparation. I'm grateful to my niece Gretchen Hewitt Rojas for keyboarding help. Thanks go to my lifelong friend Ernie Brown for designing the physics logo and for chapter headers in the new problems book.

For their dedication to this edition, I am grateful to the staff at Addison Wesley in San Francisco. I am especially thankful to Liana Allday and Editor-in-Chief Adam Black. I'm grateful to Ira Kleinberg for securing the many new photographs. A note of appreciation is due Claire Masson for the cyberspace components of this and the previous edition. I thank David Vasquez, my dear friend of many years, for his insightful tutorials. And I thank the production folks at Techbooks GTS for their patience with my last-minute changes. I've been blessed with a first-rate team!

Paul G. Hewitt
St. Petersburg, Florida

Conceptual

Physics

tenth edition

Wow, Great Uncle Paul! Before this chickie exhausted its inner space resources and poked out of its shell, it must have thought it was at its last moments. But what seemed like its end was a new beginning. Are we like chickies, ready to poke through to a new environment and a new understanding of our place in the universe?

CHAPTER 1

About Science

The circular spots of light surrounding Lillian are images of the Sun, cast through small openings between leaves above. During a partial eclipse, the spots are crescents.

First of all, science is the body of knowledge that describes the order within nature and the causes of that order. Second, science is an ongoing human activity that represents the collective efforts, findings, and wisdom of the human race, an activity that is dedicated to gathering knowledge about the world and organizing and condensing it into testable laws and theories. Science had its beginnings before recorded history, when people first discovered regularities and relationships in nature, such as star patterns in the night sky and weather patterns—when the rainy season started or when the days grew longer. From these regularities, people learned to make predictions that gave them some control over their surroundings.

Science made great headway in Greece in the third and fourth centuries BC. It spread throughout the Mediterranean world. Scientific advance came to a near halt in Europe when the Roman Empire fell in the fifth century AD. Barbarian hordes destroyed almost everything in their paths as they overran Europe and ushered in what came to be known as the Dark Ages. During this time, the Chinese and Polynesians were charting the stars and the planets and Arab nations were developing mathematics and learning about the production of glass, paper, metals, and various chemicals. Greek science was reintroduced to Europe by Islamic influences that penetrated into Spain during the tenth, eleventh, and twelfth centuries. Universities emerged in Europe in the thirteenth century, and the introduction of gunpowder changed the social and political structure of Europe in the fourteenth century. The fifteenth century saw art and science beautifully blended by Leonardo da Vinci. Scientific thought was furthered in the sixteenth century with the advent of the printing press.

The sixteenth-century Polish astronomer Nicolaus Copernicus caused great controversy when he published a book proposing that the Sun is stationary and that Earth revolves around the Sun. These ideas conflicted with the popular view that Earth was the center of the universe. They also conflicted with Church teachings and were banned for 200 years. The Italian physicist Galileo Galilei was arrested for popularizing the Copernican theory and for his other contributions to scientific thought. Yet, a century later, those who advocated Copernican ideas were accepted.

This kind of cycle happens age after age. In the early 1800s, geologists met with violent condemnation because they differed with the Genesis account of creation. Later in the same century, geology was accepted, but theories of evolution were condemned and the teaching of them was forbidden. Every age has its groups of intellectual rebels who are condemned and sometimes persecuted at the time but who later seem harmless and often essential to the elevation of human conditions. "At every crossway on the road that leads to the future, each progressive spirit is opposed by a thousand men appointed to guard the past."[1]

Scientific Measurements

Measurements are a hallmark of good science. How much you know about something is often related to how well you can measure it. This was well put by the famous physicist Lord Kelvin in the nineteenth century: "I often say that when you can measure something and express it in numbers, you know something about it. When you cannot measure it, when you cannot express it in numbers, your knowledge is of a meager and unsatisfactory kind. It may be the beginning of knowledge, but you have scarcely in your thoughts advanced to the stage of science, whatever it may be." Scientific measurements are not something new but go back to ancient times. In the third century BC, for example, fairly accurate measurements were made of the sizes of Earth, Moon, and Sun, as well as of the distances between them.

Size of Earth

The size of Earth was first measured in Egypt by the geographer and mathematician Eratosthenes about 235 BC.[2] Eratosthenes calculated the circumference of Earth in the following way. He knew that the Sun is highest in the sky at noon on June 22, the summer solstice. At this time, a vertical stick casts its shortest shadow. If the Sun is directly overhead, a vertical stick casts no shadow at all, which occurs at the summer solstice in Syene, a city south of Alexandria (where the Aswan Dam stands today). Eratosthenes learned that the Sun was directly overhead at the summer solstice in Syene from library information, which reported that, at this particular time, sunlight shines directly down a deep well in Syene and is reflected back up again. Eratosthenes reasoned that, if the Sun's rays were extended into Earth at this point, they would pass through the center. Likewise, a vertical line extended into Earth at Alexandria (or anywhere else) would also pass through Earth's center.

At noon on June 22, Eratosthenes measured the shadow cast by a vertical pillar in Alexandria and found it to be 1/8 the height of the pillar (Figure 1.1). This corresponds to a 7.2° angle between the Sun's rays and the vertical pillar. Since 7.2° is 7.2/360, or 1/50 of a circle, Eratosthenes reasoned that the distance between Alexandria and Syene must be 1/50 the circumference of Earth. Thus the circumference of Earth becomes 50 times the distance between these two cities. This distance, quite flat and frequently traveled, was measured by

[1]From Count Maurice Maeterlinck's "Our Social Duty."

[2]Eratosthenes was second chief librarian at the University of Alexandria in Egypt, founded by Alexander the Great. Eratosthenes, one of the foremost scholars of his time, wrote on philosophical, scientific, and literary matters. As a mathematician, he invented a method for finding prime numbers. His reputation among his contemporaries was immense—Archimedes dedicated a book to him. As a geographer, he wrote *Geography*, the first book to give geography a mathematical basis and to treat Earth as a globe divided into frigid, temperate, and torrid zones. It long remained a standard work, and it was used a century later by Julius Caesar. Eratosthenes spent most of his life in Alexandria, and he died there in 195 BC.

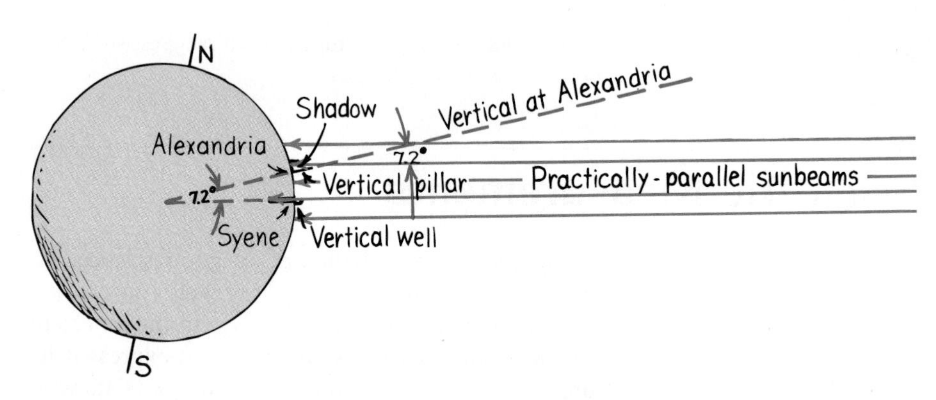

FIGURE 1.1
When the Sun is directly overhead at Syene, it is not directly overhead in Alexandria, 800 km north. When the Sun's rays shine directly down a vertical well in Syene, they cast a shadow of a vertical pillar in Alexandria. The verticals at both locations extend to the center of Earth, and they make the same angle that the Sun's rays make with the pillar at Alexandria. Eratosthenes measured this angle to be 1/50 of a complete circle. Therefore, the distance between Alexandria and Syene is 1/50 Earth's circumference. (Equivalently, the shadow cast by the pillar is 1/8 the height of the pillar, which means the distance between the locations is 1/8 Earth's radius.)

surveyors to be 5000 stadia (800 kilometers). So Eratosthenes calculated Earth's circumference to be 50×5000 stadia $= 250{,}000$ stadia. This is within 5% of the currently accepted value of Earth's circumference.

We get the same result by bypassing degrees altogether and comparing the length of the shadow cast by the pillar to the height of the pillar. Geometrical reasoning shows, to a close approximation, that the ratio *shadow length/pillar height* is the same as the ratio *distance between Alexandria and Syene/Earth's radius*. So just as the pillar is 8 times greater than its shadow, the radius of Earth must be 8 times greater than the distance between Alexandria and Syene.

Since the circumference of a circle is 2π times its radius ($C = 2\pi r$), Earth's radius is simply its circumference divided by 2π. In modern units, Earth's radius is 6370 kilometers and its circumference is 40,000 km.

Size of the Moon

Aristarchus was perhaps the first to suggest that Earth spins on a daily axis, which accounted for the daily motion of the stars. He also hypothesized that Earth moves around the Sun in a yearly orbit, and that the other planets do likewise.[3] He correctly measured the Moon's diameter and its distance from

[3]Aristarchus was unsure of his heliocentric hypothesis, likely because Earth's unequal seasons seemed not to support the idea that Earth circles the Sun. More important, it was noted that the Moon's distance from Earth varies—clear evidence that the Moon does not perfectly circle Earth. If the Moon does not follow a circular path about Earth, it was hard to argue that Earth follows a circular path about the Sun. The explanation, the elliptical paths of planets, was not discovered until centuries later by Johannes Kepler. In the meantime, the epicycles proposed by other astronomers accounted for these discrepancies. It is interesting to speculate about the course of astronomy if the Moon didn't exist. Its irregular orbit would not have contributed to the early discrediting of the heliocentric theory, which may have taken hold centuries earlier.

Earth. He did all this in about 240 BC, seventeen centuries before his findings became fully accepted.

Aristarchus compared the size of the Moon with the size of Earth by watching an eclipse of the Moon. Earth, like any body in sunlight, casts a shadow. An eclipse of the Moon is simply the event wherein the Moon passes into this shadow. Aristarchus carefully studied this event and found that the width of Earth's shadow out at the Moon was 2.5 Moon diameters. This would seem to indicate that the Moon's diameter is 2.5 times smaller than Earth's. But because of the huge size of the Sun, Earth's shadow tapers, as evidenced during a solar eclipse. (Figure 1.2 shows this in exaggerated scale.) At that time, Earth intercepts the Moon's shadow—but just barely. The Moon's shadow tapers almost to a point at Earth's surface, evidence that the taper of the Moon's shadow at this distance is one Moon diameter. So during a lunar eclipse Earth's shadow, covering the same distance, must also taper one Moon diameter. Taking the tapering of the Sun's rays into account, Earth's diameter must be (2.5 + 1) times the Moon's diameter. In this way, Aristarchus showed that the Moon's diameter is 1/3.5 that of Earth's. The

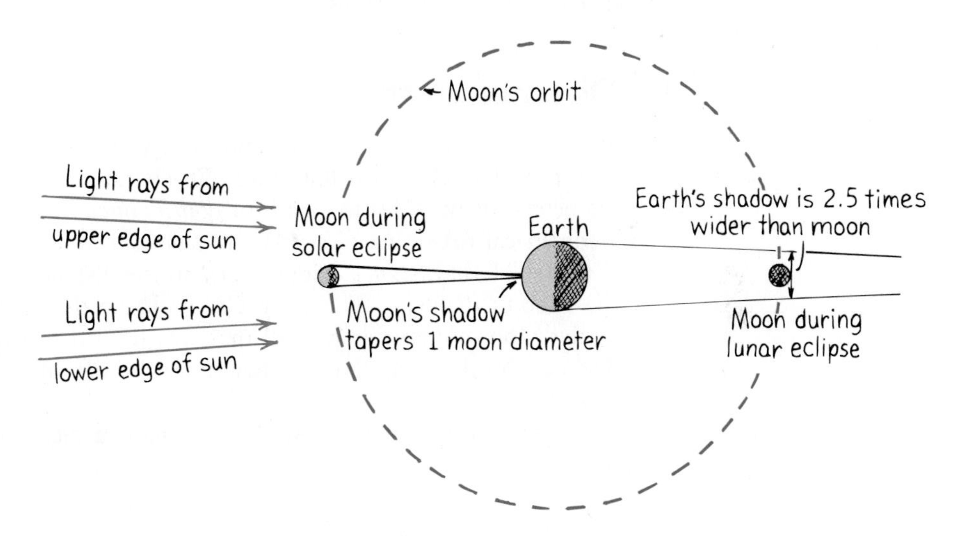

FIGURE 1.2
During a lunar eclipse, Earth's shadow is observed to be 2.5 times as wide as the Moon's diameter. Because of the Sun's large size, Earth's shadow must taper. The amount of taper is evident during a solar eclipse, where the Moon's shadow tapers a whole Moon diameter from Moon to Earth. So Earth's shadow tapers the same amount in the same distance. Therefore, Earth's diameter must be 3.5 Moon diameters.

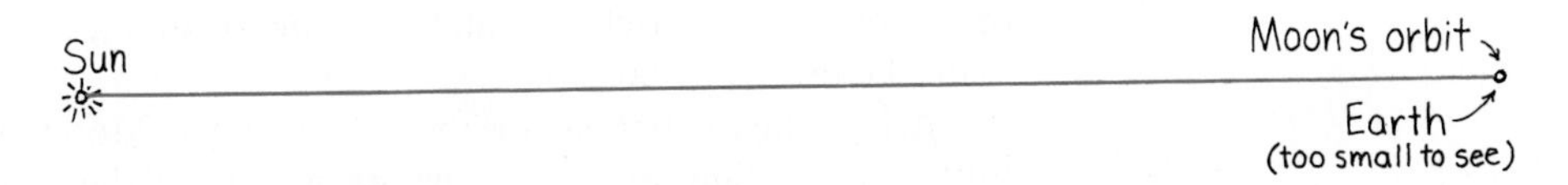

FIGURE 1.3
Correct scale of solar and lunar eclipses, which shows why eclipses are rare. (They are even rarer because the Moon's orbit is tilted about 5° from the plane of Earth's orbit about the Sun.)

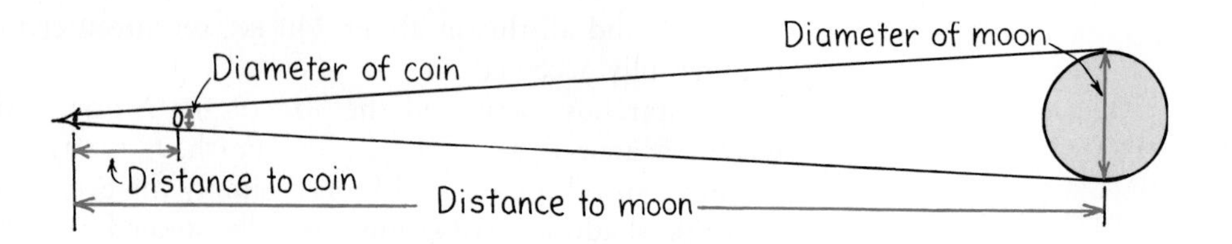

$$\frac{\text{Coin diameter}}{\text{Coin distance}} = \frac{\text{Moon diameter}}{\text{Moon distance}} = \frac{1}{110}$$

FIGURE 1.4
An exercise in ratios: When the coin barely "eclipses" the Moon, then the diameter of the coin to the distance between you and the coin is equal to the diameter of the Moon to the distance between you and the Moon (not to scale here). Measurements give a ratio of 1/110 for both.

presently accepted diameter of the Moon is 3640 km, within 5% of the value calculated by Aristarchus.

Distance to the Moon

Tape a small coin, such as a dime, to a window and view it with one eye so that it just blocks out the full Moon. This occurs when your eye is about 110 coin diameters away. Then the ratio of *coin diameter/coin distance* is about 1/110. Geometrical reasoning of similar triangles shows this is also the ratio of *Moon diameter/Moon distance* (Figure 1.4). So the distance to the Moon is 110 times the Moon's diameter. The early Greeks knew this. Aristarchus's measurement of the Moon's diameter was all that was needed to calculate the Earth–Moon distance. So the early Greeks knew both the size of the Moon and its distance from Earth.

With this information, Aristarchus made a measurement of the Earth–Sun distance.

Distance to the Sun

If you were to repeat the coin-on-the-window-and-Moon exercise for the Sun (which would be dangerous to do because of the Sun's brightness), guess what: The ratio *Sun diameter/Sun distance* is also 1/110. This is because the Sun and Moon both have the same size to the eye. They both taper the same angle (about 0.5°). So, although the ratio of diameter to distance was known to the early Greeks, diameter alone or distance alone would have to be determined by some other means. Aristarchus found a means of doing this and made a rough estimate. Here's what he did.

Aristarchus watched for the phase of the Moon when it was *exactly* half full, with the Sun still visible in the sky. Then the sunlight must be falling on the Moon at right angles to his line of sight. This meant that the lines between Earth and the Moon, between Earth and the Sun, and between the Moon and the Sun form a right triangle (Figure 1.5).

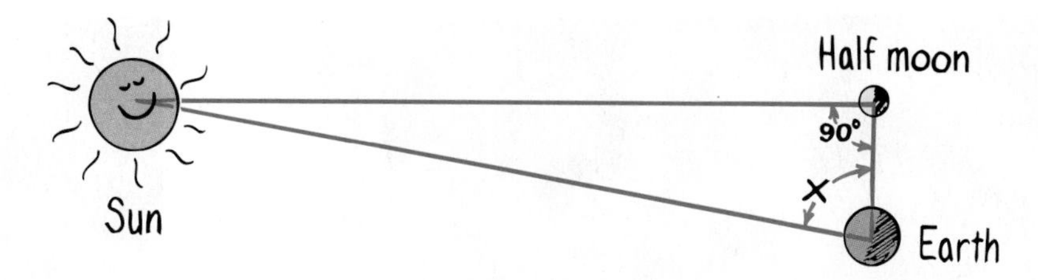

FIGURE 1.5
When the Moon appears exactly half full, the Sun, Moon, and Earth form a right triangle (not to scale). The hypotenuse is the Earth–Sun distance. By simple trigonometry, the hypotenuse of a right triangle can be found if you know the value of either nonright angle and the length of one side. The Earth–Moon distance is a known side. Measure angle *X* and you can calculate the Earth–Sun distance.

A rule of trigonometry states that, if you know all the angles in a right triangle plus the length of any one of its sides, you can calculate the length of any other side. Aristarchus knew the distance from Earth to the Moon. At the time of the half Moon he also knew one of the angles, 90°. All he had to do was measure the second angle between the line of sight to the Moon and the line of sight to the Sun. Then the third angle, a very small one, is 180° minus the sum of the first two angles (the sum of the angles in any triangle = 180°).

Measuring the angle between the lines of sight to the Moon and Sun is difficult to do without a modern transit. For one thing, both the Sun and Moon are not points, but are relatively big. He had to sight on their centers (or either edge) and measure the angle between—quite large, almost a right angle itself! By modern-day measure, his measurement was very crude. He measured 87°, while the true value was 89.8°. He figured the Sun to be about 20 times the Moon's distance, when in fact it is about 400 times as distant. So although his method was ingenious, his measurements were not. Perhaps Aristarchus found it difficult to believe the Sun was so far away, and he erred on the nearer side. We don't know.

Today we know the Sun to be an average of 150,000,000 kilometers away. It is somewhat closer in December (147,000,000 km), and somewhat farther in June (152,000,000 km).

Size of the Sun

Once the distance to the Sun is known, the 1/110 ratio of diameter/distance enables a measurement of the Sun's diameter. Another way to measure the 1/110 ratio, besides the method of Figure 1.4, is to measure the diameter of the Sun's image cast through a pinhole opening. You should try this. Poke a hole in a sheet of opaque cardboard and let sunlight shine on it. The round image that is cast on a surface below is actually an image of the Sun. You'll see that the size of the image does not depend on the size of the pinhole but, rather, on how far away the pinhole is from the image. Bigger holes make brighter images, not bigger ones. Of course, if the hole is very big, no image is formed. Careful measurement will show the ratio of image size to pinhole distance is 1/110—the same as the ratio of *Sun diameter/Sun–Earth distance* (Figure 1.6).

FIGURE 1.6
The round spot of light cast by the pinhole is an image of the Sun. Its *diameter/distance* ratio is the same as the *Sun diameter/Sun's distance* ratio, 1/110. The Sun's diameter is 1/110 its distance from Earth.

Interestingly, at the time of a partial solar eclipse, the image cast by the pinhole will be a crescent shape—the same as that of the partially covered Sun. This provides an interesting way to view a partial eclipse without looking at the Sun.

Have you noticed that the spots of sunlight you see on the ground beneath trees are perfectly round when the Sun is overhead and spread into ellipses when the Sun is low in the sky? These are pinhole images of the Sun, where light

FIGURE 1.7
Renoir accurately painted the spots of sunlight on his subjects' clothing and surroundings—images of the Sun cast by relatively small openings in the leaves above.

FIGURE 1.8
The crescent-shaped spots of sunlight are images of the Sun when it is partially eclipsed.

shines through openings in the leaves that are small compared with the distance to the ground below. A round spot 10 cm in diameter is cast by an opening that is 110×10 cm above ground. Tall trees make large images; short trees make small images. And, at the time of a partial solar eclipse, the images are crescents (Figure 1.8).

Mathematics—The Language of Science

Science and human conditions advanced dramatically after science and mathematics became integrated some four centuries ago. When the ideas of science are expressed in mathematical terms, they are unambiguous. The equations of science provide compact expressions of relationships between concepts. They don't have the multiple meanings that so often confuse the discussion of ideas expressed in common language. When findings in nature are expressed mathematically, they are easier to verify or to disprove by experiment. The mathematical structure of physics will be evident in the many equations you will encounter throughout this book. The equations are guides to thinking that show the connections between concepts in nature. The methods of mathematics and experimentation led to enormous success in science.[4]

[4]We distinguish between the mathematical structure of physics and the practice of mathematical problem solving—the focus of most nonconceptual courses. Note the relatively small number of problems at the ends of the chapters in this book, compared with the number of exercises. The focus is on comprehension comfortably before computation. Additional problems for this edition are in the *Problem Solving in Conceptual Physics* booklet.

Scientific Methods

There is no *one* scientific method. But there are common features in the way scientists do their work. This all dates back to the Italian physicist Galileo Galilei (1564–1642) and the English philosopher Francis Bacon (1561–1626). They broke free from the methods of the Greeks, who worked "upward or downward," depending on the circumstances, reaching conclusions about the physical world by reasoning from arbitrary assumptions (axioms). The modern scientist works "upward," first examining the way the world actually works and then building a structure to explain findings.

Although no cookbook description of the **scientific method** is really adequate, some or all of the following steps are likely to be found in the way most scientists carry out their work.

1. Recognize a question or a puzzle—such as an unexplained fact.
2. Make an educated guess—a **hypothesis**—that might resolve the puzzle.
3. Predict consequences of the hypothesis.
4. Perform experiments or make calculations to test the predictions.
5. Formulate the simplest general rule that organizes the three main ingredients: hypothesis, predicted effects, and experimental findings.

Although these steps are appealing, much progress in science has come from trial and error, experimentation without hypotheses, or just plain accidental discovery by a well-prepared mind. The success of science rests more on an attitude common to scientists than on a particular method. This attitude is one of inquiry, experimentation, and humility—that is, a willingness to admit error.

The Scientific Attitude

It is common to think of a fact as something that is unchanging and absolute. But, in science, a **fact** is generally a close agreement by competent observers who make a series of observations about the same phenomenon. For example, where it was once a fact that the universe is unchanging and permanent, today it is a fact that the universe is expanding and evolving. A scientific hypothesis, on the other hand, is an educated guess that is only presumed to be factual until supported by experiment. When a hypothesis has been tested over and over again and has not been contradicted, it may become known as a **law** or *principle*.

If a scientist finds evidence that contradicts a hypothesis, law, or principle, then, in the scientific spirit, it must be changed or abandoned—regardless of the reputation or authority of the persons advocating it (unless the contradicting evidence, upon testing, turns out to be wrong—which sometimes happens). For example, the greatly respected Greek philosopher Aristotle (384–322 BC) claimed that an object falls at a speed proportional to its weight. This idea was held to be true for nearly 2000 years because of Aristotle's compelling authority. Galileo allegedly showed the falseness of Aristotle's claim with one experiment—demonstrating that heavy and light objects dropped from the Leaning Tower of

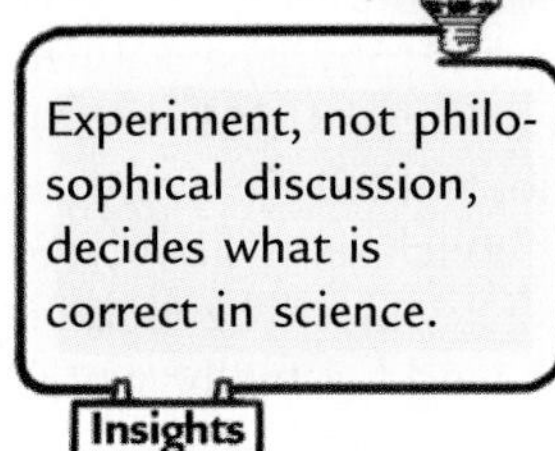
Experiment, not philosophical discussion, decides what is correct in science.
Insights

Pisa fell at nearly equal speeds. In the scientific spirit, a single verifiable experiment to the contrary outweighs any authority, regardless of reputation or the number of followers or advocates. In modern science, argument by appeal to authority has little value.[5]

Scientists must accept their experimental findings even when they would like them to be different. They must strive to distinguish between what they see and what they wish to see, for scientists, like most people, have a vast capacity for fooling themselves.[6] People have always tended to adopt general rules, beliefs, creeds, ideas, and hypotheses without thoroughly questioning their validity and to retain them long after they have been shown to be meaningless, false, or at least questionable. The most widespread assumptions are often the least questioned. Most often, when an idea is adopted, particular attention is given to cases that seem to support it, while cases that seem to refute it are distorted, belittled, or ignored.

Scientists use the word *theory* in a way that differs from its usage in everyday speech. In everyday speech, a theory is no different from a hypothesis—a supposition that has not been verified. A scientific **theory,** on the other hand, is a synthesis of a large body of information that encompasses well-tested and verified hypotheses about certain aspects of the natural world. Physicists, for example, speak of the quark theory of the atomic nucleus, chemists speak of the theory of metallic bonding in metals, and biologists speak of the cell theory.

Facts are revisable data about the world.
Theories interpret facts.

The theories of science are not fixed; rather, they undergo change. Scientific theories evolve as they go through stages of redefinition and refinement. During the past hundred years, for example, the theory of the atom has been repeatedly refined as new evidence on atomic behavior has been gathered. Similarly, chemists have refined their view of the way molecules bond together, and biologists have refined the cell theory. The refinement of theories is a strength of science, not a weakness. Many people feel that it is a sign of weakness to change their minds. Competent scientists must be experts at changing their minds. They change their minds, however, only when confronted with solid experimental evidence or when a conceptually simpler hypothesis forces them to a new point of view. More important than defending beliefs is improving them. Better hypotheses are made by those who are honest in the face of experimental evidence.

Away from their profession, scientists are inherently no more honest or ethical than most other people. But in their profession they work in an arena that puts a high premium on honesty. The cardinal rule in science is that all hypotheses must be testable—they must be susceptible, at least in principle, to being shown to be *wrong*. In science, it is more important that there be a means of proving an idea wrong than that there be a means of proving it right. This is a major factor that distinguishes science from nonscience. At first this may seem strange, for when we wonder about most things, we concern ourselves with ways of finding out whether they are true. Scientific hypotheses are different. In fact, if you want to distinguish whether a hypothesis is scientific or not, look to see if there is a test

[5]But appeal to *beauty* has value in science. More than one experimental result in modern times has contradicted a lovely theory, which, upon further investigation, proved to be wrong. This has bolstered scientists' faith that the ultimately correct description of nature involves conciseness of expression and economy of concepts—a combination that deserves to be called beautiful.

[6]In your education it is not enough to be aware that other people may try to fool you; it is more important to be aware of your own tendency to fool yourself.

for proving it wrong. If there is no test for its possible wrongness, then the hypothesis is not scientific. Albert Einstein put it well when he stated, "No number of experiments can prove me right; a single experiment can prove me wrong."

Consider the biologist Charles Darwin's hypothesis that life forms evolve from simpler to more complex forms. This could be proved wrong if paleontologists were to find that more complex forms of life appeared before their simpler counterparts. Einstein hypothesized that light is bent by gravity. This might be proved wrong if starlight that grazed the Sun and could be seen during a solar eclipse were undeflected from its normal path. As it turns out, less complex life forms are found to precede their more complex counterparts and starlight is found to bend as it passes close to the Sun, which support the claims. If and when a hypothesis or scientific claim is confirmed, it is regarded as useful and as a stepping-stone to additional knowledge.

Consider the hypothesis "The alignment of planets in the sky determines the best time for making decisions." Many people believe it, but this hypothesis is not scientific. It cannot be proven wrong, nor can it be proven right. It is *speculation*. Likewise, the hypothesis "Intelligent life exists on other planets somewhere in the universe" is not scientific. Although it can be proven correct by the verification of a single instance of intelligent life existing elsewhere in the universe, there is no way to prove it wrong if no intelligent life is ever found. If we searched the far reaches of the universe for eons and found no life, that would not prove that it doesn't exist "around the next corner." A hypothesis that is capable of being proved right but not capable of being proved wrong is not a scientific hypothesis. Many such statements are quite reasonable and useful, but they lie outside the domain of science.

CHECK YOURSELF

Which of these is a scientific hypothesis?

a. Atoms are the smallest particles of matter that exist.

b. Space is permeated with an essence that is undetectable.

c. Albert Einstein was the greatest physicist of the twentieth century.

CHECK YOUR ANSWER

Only *a* is scientific, because there is a test for falseness. The statement is not only *capable* of being proved wrong, but it in fact *has* been proved wrong. Statement *b* has no test for possible wrongness and is therefore unscientific. Likewise for any principle or concept for which there is no means, procedure, or test whereby it can be shown to be wrong (if it is wrong). Some pseudoscientists and other pretenders of knowledge will not even consider a test for the possible wrongness of their statements. Statement *c* is an assertion that has no test for possible wrongness. If Einstein was not the greatest physicist, how could we know? It is important to note that, because the name Einstein is generally held in high esteem, it is a favorite of pseudoscientists. So we should not be surprised that the name of Einstein, like that of Jesus or of any other highly respected person, is cited often by charlatans who wish to bring respect to themselves and their points of view. In all fields, it is prudent to be skeptical of those who wish to credit themselves by calling upon the authority of others.

None of us has the time, energy, or resources to test every idea, so most of the time we take somebody's word. How do we know whose word to take? To reduce the likelihood of error, scientists accept only the word of those whose ideas, theories, and findings are testable—if not in practice, at least in principle. Speculations that cannot be tested are regarded as "unscientific." This has the long-run effect of compelling honesty—findings widely publicized among fellow scientists are generally subjected to further testing. Sooner or later, mistakes (and deception) are found out; wishful thinking is exposed. A discredited scientist does not get a second chance in the community of scientists. The penalty for fraud is professional excommunication. Honesty, so important to the progress of science, thus becomes a matter of self-interest to scientists. There is relatively little bluffing in a game in which all bets are called. In fields of study where right and wrong are not so easily established, the pressure to be honest is considerably less.

The ideas and concepts most important to our everyday life are often unscientific; their correctness or incorrectness cannot be determined in the laboratory. Interestingly enough, it seems that people honestly believe their own ideas about things to be correct, and almost everyone is acquainted with people who hold completely opposite views—so the ideas of some (or all) must be incorrect. How do you know whether or not *you* are one of those holding erroneous beliefs? There is a test. Before you can be reasonably convinced that you are right about a particular idea, you should be sure that you understand the objections and the positions of your most articulate antagonists. You should find out whether your views are supported by sound knowledge of opposing ideas or by your *misconceptions* of opposing ideas. You make this distinction by seeing whether or not you can state the objections and positions of your opposition to *their* satisfaction. Even if you can successfully do this, you cannot be absolutely certain of being right about your own ideas, but the probability of being right is considerably higher if you pass this test.

CHECK YOURSELF

Suppose that, in a disagreement between two people, A and B, you note that person A only states and restates one point of view, whereas person B clearly states both her own position and that of person A. Who is more likely to be correct? *(Think before you read the answer below!)*

We each need a knowledge filter to tell the difference between what is valid and what only pretends to be valid. The best knowledge filter ever invented is science.

Insights

CHECK YOUR ANSWER

Who knows for sure? Person B may have the cleverness of a lawyer who can state various points of view and still be incorrect. We can't be sure about the "other guy." The test for correctness or incorrectness suggested here is not a test of others, but of and for *you.* It can aid your personal development. As you attempt to articulate the ideas of your antagonists, be prepared, like scientists who are prepared to change their minds, to discover evidence contrary to your own ideas—evidence that may alter your views. Intellectual growth often comes about in this way.

PSEUDOSCIENCE

In prescientific times, any attempt to harness nature meant forcing nature against her will. Nature had to be subjugated, usually with some form of magic or by means that were above nature—that is, supernatural. Science does just the opposite, and it works within nature's laws. The methods of science have largely displaced reliance on the supernatural—but not entirely. The old ways persist, full force in primitive cultures, and they survive in technologically advanced cultures too, sometimes disguised as science. This is fake science—**pseudoscience.** The hallmark of a pseudoscience is that it lacks the key ingredients of evidence and having a test for wrongness. In the realm of pseudoscience, skepticism and tests for possible wrongness are downplayed or flatly ignored.

There are various ways to view cause-and-effect relations in the universe. Mysticism is one view, appropriate perhaps in religion but not applicable to science. Astrology is an ancient belief system that supposes there is a mystical correspondence between individuals and the universe as a whole—that human affairs are influenced by the positions and movements of planets and other celestial bodies. This nonscientific view can be quite appealing. However insignificant we may feel at times, astrologers assure us that we are intimately connected to the workings of the cosmos, which has been created for humans—particularly the humans belonging to one's own tribe, community, or religious group. Astrology as ancient magic is one thing, but astrology in the guise of science is another. When it poses as a science related to astronomy, then it becomes pseudoscience. Some astrologers present their craft in a scientific guise. When they use up-to-date astronomical information and computers that chart the movements of heavenly bodies, astrologers are operating in the realm of science. But when they use these data to concoct astrological revelations, they have crossed over into full-fledged pseudoscience.

Pseudoscience, like science, makes predictions. The predictions of a dowser, who locates underground water with a dowsing stick, have a very high rate of success—nearly 100%. Whenever the dowser goes through his or her ritual and points to a spot on the ground, the well digger is sure to find water. Dowsing works. Of course, the dowser can hardly miss, because there is groundwater within 100 meters of the surface at nearly every spot on Earth. (The real test of a dowser would be finding a place where water wouldn't be found!)

A shaman who studies the oscillations of a pendulum suspended over the abdomen of a pregnant woman can predict the sex of the fetus with an accuracy of 50%. This means that, if he tries his magic many times on many fetuses, half his predictions will be right and half will be wrong—the predictability of ordinary guessing. In comparison, determining the sex of unborns by scientific means gives a 95% success rate via sonograms, and 100% by amniocentesis. The best that can be said for the shaman is that the 50% success rate is a lot better than that of astrologers, palm readers, or other pseudoscientists who predict the future.

An example of a pseudoscience that has zero success is provided by energy-multiplying machines. These machines, which are alleged to deliver more energy than they take in, are, we are told, "still on the drawing boards and needing funds for development." They are touted by quacks who sell shares to an ignorant public who succumb to the pie-in-the-sky promises of success. This is junk science. Pseudoscientists are everywhere, are usually successful in recruiting apprentices for money or labor, and can be very convincing even to seemingly reasonable people. Their books greatly outnumber books on science in bookstores. Junk science is thriving.

Four centuries ago, most humans were dominated by superstition, devils, demons, disease, and magic in their short and difficult lives. Only through enormous effort did they gain scientific knowledge and overthrow superstition. We have come a long way in comprehending nature and freeing ourselves from ignorance. We should rejoice in what we've learned. We no longer have to die whenever an infectious disease strikes, to live in fear of demons, and to fear torture by moral authorities. Life was cruel in medieval times. Today we have no need to pretend that superstition is anything but superstition, or that junk notions are anything but junk notions—whether voiced by street-corner quacks, by loose thinkers who write promise-heavy health books, by hucksters who sell magnetic therapy, or by demagogues who inflict fear.

Yet there is cause for alarm when the superstitions that people once fought to erase are back in force, enchanting a growing number of people. James Randi reports in his book *Flim-Flam!* that more than twenty thousand practicing astrologers in the United States serve millions of credulous believers. Science writer Martin Gardner reports that a greater percentage of Americans today believe in astrology and occult phenomena than did citizens of medieval Europe. Few newspapers print a daily science column, but nearly all provide daily horoscopes. Although goods and medicines around us have improved with scientific advances, much human thinking has not.

Many believe that the human condition is slipping backward because of growing technology. More likely, however, we'll slip backward because science and technology will bow to the irrationality, superstitions, and demagoguery of the past. Watch for their spokespeople. Pseudoscience is a huge and lucrative business.

Although the notion of being familiar with counter points of view seems reasonable to most thinking people, just the opposite—shielding ourselves and others from opposing ideas—has been more widely practiced. We have been taught to discredit unpopular ideas without understanding them in proper context. With the 20/20 vision of hindsight, we can see that many of the "deep truths" that were the cornerstones of whole civilizations were shallow reflections of the prevailing ignorance of the time. Many of the problems that plagued societies stemmed from this ignorance and the resulting misconceptions; much of what was held to be true simply wasn't true. This is not confined to the past. Every scientific advance is by necessity incomplete and partly inaccurate, for the discoverer sees with the blinders of the day, and can only discard a part of that blockage.

Science, Art, and Religion

The search for order and meaning in the world around us has taken different forms: One is science, another is art, and another is religion. Although the roots of all three go back thousands of years, the traditions of science are relatively recent. More important, the domains of science, art, and religion are different, although they often overlap. Science is principally engaged with discovering and recording natural phenomena, the arts are concerned with personal interpretation and creative expression, and religion addresses the source, purpose, and meaning of it all.

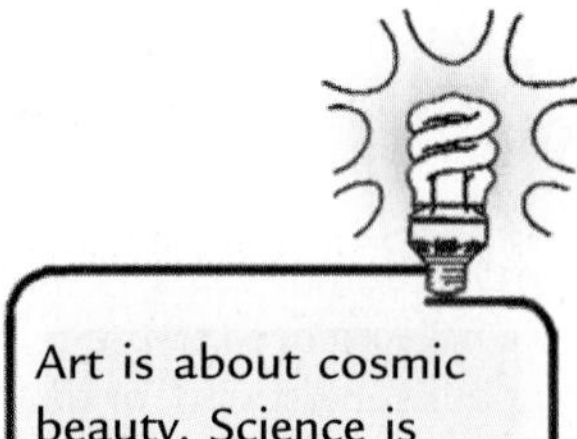

Art is about cosmic beauty. Science is about cosmic order. Religion is about cosmic purpose.
Insights

Science and the arts are comparable. In the art of literature, we discover what is possible in human experience. We can learn about emotions ranging from anguish to love, even if we haven't experienced them. The arts do not necessarily give us those experiences, but they describe them to us and suggest what may be possible for us. Science tells us what is possible in nature. Scientific knowledge helps us to predict possibilities in nature even before those possibilities have been experienced. It provides us with a way of connecting things, of seeing relationships between and among them, and of making sense of the great variety of natural events around us. Science broadens our perspective of nature. A knowledge of both the arts and the sciences makes for a wholeness that affects the way we view the world and the decisions we make about it and ourselves. A truly educated person is knowledgeable in both the arts and the sciences.

Science and religion have similarities also, but they are basically different—principally because their domains are different. The domain of science is natural order; the domain of religion is nature's purpose. Religious beliefs and practices usually involve faith in, and worship of, a supreme being and the creation of human community—not the practices of science. In this respect, science and religion are as different as apples and oranges: They are two different yet complementary fields of human activity.

When we study the nature of light later in this book, we will treat light first as a wave and then as a particle. To the person who knows a little bit about science, waves and particles are contradictory; light can be only one or the other, and we have to choose between them. But to the enlightened person, waves and particles complement each other and provide a deeper understanding of light. In a similar way, it is mainly people who are either uninformed or misinformed about the deeper natures of both science and religion who feel that they must choose between believing in religion and believing in science. Unless one has a shallow understanding of either or both, there is no contradiction in being religious and being scientific in one's thinking.[7]

Many people are troubled about not knowing the answers to religious and philosophical questions. Some avoid uncertainty by uncritically accepting almost any comforting answer. An important message in science, however, is that uncertainty is acceptable. For example, in Chapter 31 you'll learn that it is not possible to know with certainty both the momentum and position of an electron in an atom. The more you know about one, the less you can know about the other. Uncertainty is a part of the scientific process. It's okay not to know the answers to fundamental questions. Why are apples gravitationally attracted to Earth? Why do electrons repel one another? Why do magnets interact with other magnets? Why does energy have mass? At the deepest level, scientists don't know the answers to these questions—at least not yet. Scientists in general are comfortable with not knowing. We know a lot about where we are, but nothing really about *why* we are. It's okay not to know the answers to such religious questions—especially if we keep an open mind and heart with which to keep exploring.

The belief that there is only one truth and that oneself is in possession of it seems to me the deepest root of all the evil that is in the world. —*Max Born*

Insights

Science and Technology

Science and technology are also different from each other. Science is concerned with gathering knowledge and organizing it. Technology lets humans use that knowledge for practical purposes, and it provides the tools needed by scientists in their further explorations.

Technology is a double-edged sword that can be both helpful and harmful. We have the technology, for example, to extract fossil fuels from the ground and then to burn the fossil fuels for the production of energy. Energy production from fossil fuels has benefited our society in countless ways. On the flip side, the burning of fossil fuels endangers the environment. It is tempting to blame technology itself for problems such as pollution, resource depletion, and even overpopulation. These problems, however, are not the fault of technology any more than a shotgun wound is the fault of the shotgun. It is humans who use the technology, and humans who are responsible for how it is used.

Remarkably, we already possess the technology to solve many environmental problems. This twenty-first century will likely see a switch from fossil fuels to more sustainable energy sources, such as photovoltaics, solar thermal electric generation, or biomass conversion. Whereas the paper on which this book is printed came from trees, paper will soon come from fast-growing weeds, and less of it may be needed if small, easy-to-read computer screens gain popularity. We are more and more recycling waste products. In some parts of the world, progress is being made on stemming the human population explosion that aggravates almost every problem faced by humans today. The greatest obstacle to solving today's problems lies more with social inertia than with a

[7]Of course, this doesn't apply to certain religious extremists who steadfastly assert that one cannot embrace both their brand of religion and science.

RISK ASSESSMENT

The numerous benefits of technology are paired with risks. When the benefits of a technological innovation are seen to outweigh its risks, the technology is accepted and applied. X rays, for example, continue to be used for medical diagnosis despite their potential for causing cancer. But when the risks of a technology are perceived to outweigh its benefits, it should be used very sparingly or not at all.

Risk can vary for different groups. Aspirin is useful for adults, but for young children it can cause a potentially lethal condition known as *Reye's syndrome*. Dumping raw sewage into the local river may pose little risk for a town located upstream, but for towns downstream the untreated sewage is a health hazard. Similarly, storing radioactive wastes underground may pose little risk for us today, but for future generations the risks of such storage are greater if there is leakage into groundwater. Technologies involving different risks for different people, as well as differing benefits, raise questions that are often hotly debated. Which medications should be sold to the general public over the counter and how should they be labeled? Should food be irradiated in order to put an end to food poisoning, which kills more than 5000 Americans each year? The risks to all members of society need consideration when public policies are decided.

The risks of technology are not always immediately apparent. No one fully realized the dangers of combustion products when petroleum was selected as the fuel of choice for automobiles early in the last century. From the hindsight of 20/20 vision, alcohols from biomass would have been a superior choice environmentally, but they were banned by the prohibition movements of the day. Because we are now more aware of the environmental costs of fossil-fuel combustion, biomass fuels are making a slow comeback. An awareness of both the short-term risks and the long-term risks of a technology is crucial.

People seem to have a hard time accepting the impossibility of zero risk. Airplanes cannot be made perfectly safe. Processed foods cannot be rendered completely free of toxicity, for all foods are toxic to some degree. You cannot go to the beach without risking skin cancer, no matter how much sunscreen you apply. You cannot avoid radioactivity, for it's in the air you breathe and the foods you eat, and it has been that way since before humans first walked Earth. Even the cleanest rain contains radioactive carbon-14, not to mention the same in our bodies. Between each heartbeat in each human body, there have always been about 10,000 naturally occurring radioactive decays. You might hide yourself in the hills, eat the most natural of foods, practice obsessive hygiene, and still die from cancer caused by the radioactivity emanating from your own body. The probability of eventual death is 100%. Nobody is exempt.

Science helps to determine the most probable. As the tools of science improve, then assessment of the most probable gets closer to being on target. Acceptance of risk, on the other hand, is a societal issue. Placing zero risk as a societal goal is not only impractical but selfish. Any society striving toward a policy of zero risk would consume its present and future economic resources. Isn't it more noble to accept nonzero risk and to minimize risk as much as possible within the limits of practicality? A society that accepts no risks receives no benefits.

lack of technology. Technology is our tool. What we do with this tool is up to us. The promise of technology is a cleaner and healthier world. Wise applications of technology *can* lead to a better world.

Physics—The Basic Science

Science, once called *natural philosophy,* encompasses the study of living things and nonliving things, the life sciences and the physical sciences. The life sciences include biology, zoology, and botany. The physical sciences include geology, astronomy, chemistry, and physics.

Physics is more than a part of the physical sciences. It is the *basic* science. It's about the nature of basic things such as motion, forces, energy, matter, heat, sound, light, and the structure of atoms. Chemistry is about how matter is put together, how atoms combine to form molecules, and how the molecules combine to make up the many kinds of matter around us. Biology is more complex and involves matter that is alive. So underneath biology is chemistry, and underneath chemistry is physics. The concepts of physics reach up to these more complicated sciences. That's why physics is the most basic science.

An understanding of science begins with an understanding of physics. The following chapters present physics conceptually so that you can enjoy understanding it.

CHECK YOURSELF

Which of the following activities involves the utmost human expression of passion, talent, and intelligence?

a. painting and sculpture
b. literature
c. music
d. religion
e. science

In Perspective

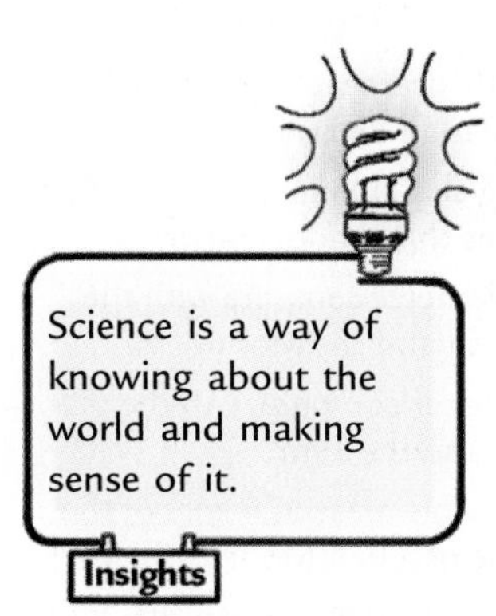

Science is a way of knowing about the world and making sense of it.
Insights

Only a few centuries ago the most talented and most skilled artists, architects, and artisans of the world directed their genius and effort to the construction of the great cathedrals, synagogues, temples, and mosques. Some of these architectural structures took centuries to build, which means that nobody witnessed both the beginning and the end of construction. Even the architects and early builders who lived to a ripe old age never saw the finished results of their labors. Entire lifetimes were spent in the shadows of construction that must have seemed without beginning or end. This enormous focus of human energy was inspired by a vision that went beyond worldly concerns—a vision of the cosmos. To the people of that time, the structures they erected were their "spaceships of faith," firmly anchored but pointing to the cosmos.

Today the efforts of many of our most skilled scientists, engineers, artists, and artisans are directed to building the spaceships that already orbit Earth and

CHECK YOUR ANSWER

All of them! The human value of science, however, is the least understood by most individuals in our society. The reasons are varied, ranging from the common notion that science is incomprehensible to people of average ability to the extreme view that science is a dehumanizing force in our society. Most of the misconceptions about science probably stem from the confusion between the *abuses* of science and science itself.

Science is an enchanting human activity shared by a wide variety of people who, with present-day tools and know-how, are reaching further and finding out more about themselves and their environment than people in the past were ever able to do. The more you know about science, the more passionate you feel toward your surroundings. There is physics in everything you see, hear, smell, taste, and touch!

others that will voyage beyond. The time required to build these spaceships is extremely brief compared with the time spent building the stone and marble structures of the past. Many people working on today's spaceships were alive before the first jetliner carried passengers. Where will younger lives lead in a comparable time?

We seem to be at the dawn of a major change in human growth, for as little Evan suggests in the photo that precedes the beginning of this chapter, we may be like the hatching chicken who has exhausted the resources of its inner-egg environment and is about to break through to a whole new range of possibilities. Earth is our cradle and has served us well. But cradles, however comfortable, are one day outgrown. So with the inspiration that in many ways is similar to the inspiration of those who built the early cathedrals, synagogues, temples, and mosques, we aim for the cosmos.

We live in an exciting time!

Summary of Terms

Fact A phenomenon about which competent observers who have made a series of observations are in agreement.

Hypothesis An educated guess; a reasonable explanation of an observation or experimental result that is not fully accepted as factual until tested over and over again by experiment.

Law A general hypothesis or statement about the relationship of natural quantities that has been tested over and over again and has not been contradicted. Also known as a *principle.*

Pseudoscience Fake science that pretends to be real science.

Scientific method Principles and procedures for the systematic pursuit of knowledge involving the recognition and formulation of a problem, the collection of data through observation and experiment, and the formulation and testing of hypotheses.

Theory A synthesis of a large body of information that encompasses well-tested and verified hypotheses about certain aspects of the natural world.

Suggested Reading

Bodanis, David. *$E = mc^2$: A Biography of the World's Most Famous Equation*. New York: Berkeley Publishing Group, 2002.

Bronowski, Jacob. *Science and Human Values*. New York: Harper & Row, 1965.

Bryson, Bill. *A Short History of Nearly Everything*. New York: Broadway Books, 2003.

Cole, K. C. *First You Build a Cloud.* New York: Morrow, 1999.

Feynman, Richard P. *Surely You're Joking, Mr. Feynman.* New York: Norton, 1986.

Gleick, James. *Genius—The Life and Science of Richard Feynman*. New York: Pantheon Books, 1992.

Sagan, Carl. *The Demon-Haunted World*. New York: Random House, 1995.

Review Questions

1. Briefly, what is science?
2. Throughout the ages, what has been the general reaction to new ideas about established "truths"?

Scientific Measurements

3. When the Sun was directly overhead in Syene, why was it not directly overhead in Alexandria?
4. The Earth, like everything else illuminated by the Sun, casts a shadow. Why does this shadow taper?
5. How does the Moon's diameter compare with the distance between the Earth and the Moon?
6. How does the Sun's diameter compare with the distance between the Earth and the Sun?
7. Why did Aristarchus make his measurements of the Sun's distance at the time of a half Moon?
8. What are the circular spots of light seen on the ground beneath a tree on a sunny day?

Mathematics—The Language of Science

9. What is the role of equations in this course?

Scientific Methods

10. Outline the steps of the classic scientific method.

The Scientific Attitude

11. Distinguish among a scientific fact, a hypothesis, a law, and a theory.
12. In daily life, people are often praised for maintaining some particular point of view, for the "courage of

their convictions." A change of mind is seen as a sign of weakness. How is this different in science?

13. What is the test for whether a hypothesis is scientific or not?

14. In daily life, we see many cases of people who are caught misrepresenting things and who soon thereafter are excused and accepted by their contemporaries. How is this different in science?

15. What test can you perform to increase the chance in your own mind that you are right about a particular idea?

Science, Art, and Religion

16. Why are students of the arts encouraged to learn about science and science students encouraged to learn about the arts?

17. Why do many people believe they must choose between science and religion?

18. Psychological comfort is a benefit of having solid answers to religious questions. What benefit accompanies a position of not knowing the answers?

Science and Technology

19. Clearly distinguish between science and technology.

Physics—The Basic Science

20. Why is physics considered to be the basic science?

Project

1. Poke a hole in a piece of cardboard and hold the cardboard horizontally in the sunlight. Note the image of the Sun that is cast below. To convince yourself that the round spot of light is an image of the round Sun, try holes of different shapes. A square or triangular hole will still cast a round image when the distance to the image is large compared with the size of the hole. When the Sun's rays and the image surface are perpendicular, the image is a circle; when the Sun's rays make an angle with the image surface, the image is a "stretched-out" circle, an ellipse. Let the solar image fall upon a coin, say a dime. Position the cardboard so the image just covers the coin. This is a convenient way to measure the diameter of the image—the same as the diameter of the easy-to-measure coin. Then measure the distance between the cardboard and the coin. Your ratio of image size to image distance should be about 1/110. This is the ratio of solar diameter to solar distance to the Earth. Using the information that the Sun is 150,000,000 kilometers distant, calculate the diameter of the Sun.

2. Choose a particular day in the very near future—and during that day carry a small notebook with you and record every time you come in contact with modern technology. After your recording is done, write a short page or two discussing your dependencies on your list of technologies. Make a note of how you'd be affected if each suddenly vanished, and how you'd cope with the loss.

Exercises

1. What is the penalty for scientific fraud in the science community?

2. Which of the following are scientific hypotheses?

(a) Chlorophyll makes grass green. (b) The Earth rotates about its axis because living things need an alternation of light and darkness. (c) Tides are caused by the Moon.

3. In answer to the question, "When a plant grows, where does the material come from?" Aristotle hypothesized by logic that all material came from the soil. Do you consider his hypothesis to be correct, incorrect, or partially correct? What experiments do you propose to support your choice?

4. The great philosopher and mathematician Bertrand Russell (1872–1970) wrote about ideas in the early part of his life that he rejected in the latter part of his life. Do you see this as a sign of weakness or as a sign of strength in Bertrand Russell? (Do you speculate that your present ideas about the world around you will change as you learn and experience more, or do you speculate that further knowledge and experience will solidify your present understanding?)

5. Bertrand Russell wrote, "I think we must retain the belief that scientific knowledge is one of the glories of man. I will not maintain that knowledge can never do harm. I think such general propositions can almost always be refuted by well-chosen examples. What I will maintain—and maintain vigorously—is that knowledge is very much more often useful than harmful and that fear of knowledge is very much more often harmful than useful." Think of examples to support this statement.

6. When you step from the shade into the sunlight, the Sun's heat is as evident as the heat from hot coals in a fireplace in an otherwise cold room. You feel the Sun's heat not because of its high temperature (higher temperatures can be found in some welder's torches), but because the Sun is big. Which do you estimate is larger, the Sun's radius or the distance between the Moon and the Earth? Check your answer in the list of physical data on the inside back cover. Do you find your answer surprising?

7. The shadow cast by a vertical pillar in Alexandria at noon during the summer solstice is found to be 1/8 the height of the pillar. The distance between Alexandria and Syene is 1/8 the Earth's radius. Is there a geometric connection between these two 1-to-8 ratios?
8. If Earth were smaller than it is, would the shadow of the vertical pillar in Alexandria have been longer or shorter at noon during the summer solstice?
9. What is probably being misunderstood by a person who says, "But that's only a scientific theory"?
10. Scientists call a theory that unites many ideas in a simple way "beautiful." Are unity and simplicity among the criteria of beauty outside of science? Support your answer.

PART ONE

Mechanics

Burl Grey, who first introduced the author to the concept of tension, shows a 2-lb bag producing a tension of 9 N.

CHAPTER 2

Newton's First Law of Motion—Inertia

More than 2000 years ago, ancient Greek scientists were familiar with some of the ideas in physics that we study today. They had a good understanding of some of the properties of light, but they were confused about motion. One of the first to study motion seriously was Aristotle, the most outstanding philosopher-scientist of his time in ancient Greece. Aristotle attempted to clarify motion by classification.

Aristotle on Motion

Aristotle divided motion into two main classes: *natural motion* and *violent motion*. We shall briefly consider each, not as study material, but only as a background to present-day ideas about motion.

Aristotle asserted that natural motion proceeds from the "nature" of an object, dependent on what combination of the four elements (earth, water, air, and fire) the object contains. In his view, every object in the universe has a proper place, determined by this "nature"; any object not in its proper place will "strive" to get there. Being of earth, an unsupported lump of clay properly falls to the ground; being of the air, an unimpeded puff of smoke properly rises; being a mixture of earth and air but predominantly earth, a feather properly falls to the ground, but not as rapidly as a lump of clay. He stated that heavier objects would strive harder. Hence, argued Aristotle, objects should fall at speeds proportional to their weights: the heavier the object, the faster it should fall.

Natural motion could be either straight up or straight down, as in the case of all things on Earth, or it could be circular, as in the case of celestial objects. Unlike up-and-down motion, circular motion has no beginning or end, repeating itself without deviation. Aristotle believed that different rules apply in the heavens, and he asserted that celestial bodies are perfect spheres made of a perfect and unchanging substance, which he called *quintessence*.[1] (The only celestial object with any detectable variation on its face was the Moon. Medieval Christians, still under the sway of Aristotle's teaching, explained that lunar

[1]Quintessence is the *fifth* essence, the other four being earth, water, air, and fire.

ARISTOTLE (384–322 BC)

Greek philosopher, scientist, and educator, Aristotle was the son of a physician who personally served the king of Macedonia. At 17, he entered the Academy of Plato, where he worked and studied for 20 years until Plato's death. He then became the tutor of young Alexander the Great. Eight years later, he formed his own school. Aristotle's aim was to systematize existing knowledge, just as Euclid had systematized geometry. Aristotle made critical observations, collected specimens, and gathered together, summarized, and classified almost all existing knowledge of the physical world. His systematic approach became the method from which Western science later arose. After his death, his voluminous notebooks were preserved in caves near his home and were later sold to the library at Alexandria. Scholarly activity ceased in most of Europe through the Dark Ages, and the works of Aristotle were forgotten and lost in the scholarship that continued in the Byzantine and Islamic empires. Various texts were reintroduced to Europe during the eleventh and twelfth centuries and translated into Latin. The Church, the dominant political and cultural force in Western Europe, first prohibited the works of Aristotle and then accepted and incorporated them into Christian doctrine.

imperfections were due to the closeness of the Moon and its contamination by the corrupted Earth.)

Violent motion, Aristotle's other class of motion, resulted from pushing or pulling forces. Violent motion was imposed motion. A person pushing a cart or lifting a heavy weight imposed motion, as did someone hurling a stone or winning a tug of war. The wind imposed motion on ships. Floodwaters imposed it on boulders and tree trunks. The essential thing about violent motion was that it was externally caused and was imparted to objects; they moved not of themselves, not by their "nature," but because of pushes or pulls.

The concept of violent motion had its difficulties, for the pushes and pulls responsible for it were not always evident. For example, a bowstring moved an arrow until the arrow left the bow; after that, further explanation of the arrow's motion seemed to require some other pushing agent. Aristotle imagined, therefore, that a parting of the air by the moving arrow resulted in a squeezing effect on the rear of the arrow as the air rushed back to prevent a vacuum from forming. The arrow was propelled through the air as a bar of soap is propelled in the bathtub when you squeeze one end of it.

To sum up, Aristotle taught that all motions are due to the nature of the moving object, or due to a sustained push or pull. Provided that an object is in its proper place, it will not move unless subjected to a force. Except for celestial objects, the normal state is one of rest.

Aristotle's statements about motion were a beginning in scientific thought, and, although he did not consider them to be the final words on the subject, his followers for nearly 2000 years regarded his views as beyond question. Implicit in the thinking of ancient, medieval, and early Renaissance times was the notion that the normal state of objects is one of rest. Since it was evident to most thinkers until the sixteenth century that Earth must be in its proper place, and since a force capable of moving Earth was inconceivable, it seemed quite clear to them that Earth does not move.

CHECK YOURSELF

Isn't it common sense to think of Earth in its proper place, and that a force to move it *is* inconceivable, as Aristotle held, and that Earth *is* at rest in this universe?

Copernicus and the Moving Earth

Nicolaus Copernicus (1473–1543)

It was in this intellectual climate that the Polish astronomer Nicolaus Copernicus (1474–1543) formulated his theory of the moving Earth. Copernicus reasoned that the simplest way to account for the observed motions of the Sun, Moon, and planets through the sky was to assume that Earth (and other planets) circle around the Sun. For years he worked without making his thoughts public—for two reasons. The first was that he feared persecution; a theory so completely different from common opinion would surely be taken as an attack on established order. The second reason was that he had grave doubts about it himself; he could not reconcile the idea of a moving Earth with the prevailing ideas of motion. Finally, in the last days of his life, at the urging of close friends, he sent his *De Revolutionibus* to the printer. The first copy of his famous exposition reached him on the day he died—May 24, 1543.

Most of us know about the reaction of the medieval Church to the idea that Earth travels around the Sun. Because Aristotle's views had become so formidably a part of Church doctrine, to contradict them was to question the Church itself. For many Church leaders, the idea of a moving Earth threatened not only their authority but the very foundations of faith and civilization as well. For better or for worse, this new idea was to overturn their conception of the cosmos—although eventually the Church embraced it.

Galileo and the Leaning Tower

FIGURE 2.1
Galileo's famous demonstration.

It was Galileo, the foremost scientist of the early seventeenth century, who gave credence to the Copernican view of a moving Earth. He accomplished this by discrediting the Aristotelian ideas about motion. Although he was not the first to point out difficulties in Aristotle's views, Galileo was the first to provide conclusive refutation through observation and experiment.

Galileo easily demolished Aristotle's falling-body hypothesis. Galileo is said to have dropped objects of various weights from the top of the Leaning Tower of Pisa and to have compared their falls. Contrary to Aristotle's assertion,

CHECK YOUR ANSWER

Aristotle's views were logical and consistent with everyday observations. So, unless you become familiar with the physics to follow in this book, Aristotle's views about motion *do* make common sense. But, as you acquire new information about nature's rules, you'll likely find your common sense progressing beyond Aristotelian thinking.

GALILEO GALILEI (1564–1642)

Galileo was born in Pisa, Italy, in the same year Shakespeare was born and Michelangelo died. He studied medicine at the University of Pisa and then changed to mathematics. He developed an early interest in motion and was soon at odds with his contemporaries, who held to Aristotelian ideas on falling bodies. He left Pisa to teach at the University of Padua and became an advocate of the new Copernican theory of the solar system. He was one of the first to build a telescope, and he was the first to direct it to the nighttime sky and discover mountains on the Moon and the moons of Jupiter. Because he published his findings in Italian instead of in the Latin expected of so reputable a scholar, and because of the recent invention of the printing press, his ideas reached a wide readership. He soon ran afoul of the Church, and he was warned not to teach, and not to hold to, Copernican views. He restrained himself publicly for nearly 15 years and then defiantly published his observations and conclusions, which were counter to Church doctrine. The outcome was a trial in which he was found guilty, and he was forced to renounce his discoveries. By then an old man, broken in health and spirit, he was sentenced to perpetual house arrest. Nevertheless, he completed his studies on motion, and his writings were smuggled from Italy and published in Holland. Earlier, he had damaged his eyes looking at the Sun through a telescope, which led to blindness at the age of 74. He died 4 years later.

Galileo found that a stone twice as heavy as another did not fall twice as fast. Except for the small effect of air resistance, he found that objects of various weights, when released at the same time, fell together and hit the ground at the same time. On one occasion, Galileo allegedly attracted a large crowd to witness the dropping of two objects of different weight from the top of the tower. Legend has it that many observers of this demonstration who saw the objects hit the ground together scoffed at the young Galileo and continued to hold fast to their Aristotelian teachings.

Galileo's Inclined Planes

> Galileo was concerned with *how* things move rather than *why* they move. He showed that experiment rather than logic is the best test of knowledge.
>
> **Insights**

Aristotle was an astute observer of nature, and he dealt with problems around him rather than with abstract cases that did not occur in his environment. Motion always involved a resistive medium such as air or water. He believed a vacuum to be impossible and therefore did not give serious consideration to motion in the absence of an interacting medium. That's why it was basic to Aristotle that an object requires a push or pull to keep it moving. And it was this basic principle that Galileo denied when he stated that, if there is no interference with a moving object, it will keep moving in a straight line forever; no push, pull, or force of any kind is necessary.

Galileo tested this hypothesis by experimenting with the motion of various objects on plane surfaces tilted at various angles. He noted that balls rolling on downward-sloping planes picked up speed, while balls rolling on upward-sloping planes lost speed. From this he reasoned that balls rolling along a horizontal plane would neither speed up nor slow down. The ball would finally come to rest not because of its "nature" but because of friction. This idea was supported

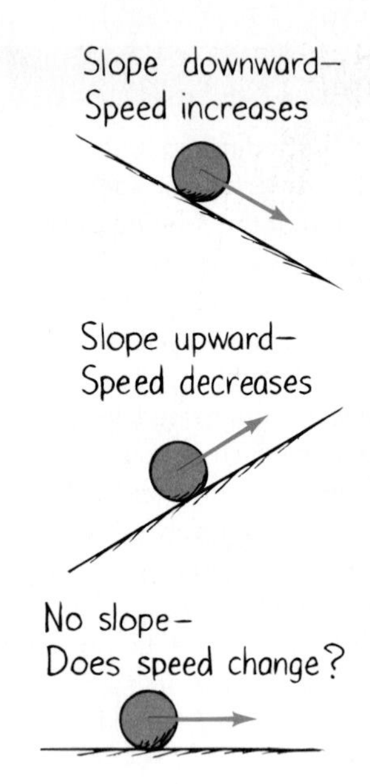

FIGURE 2.2
Motion of balls on various planes.

by Galileo's observation of motion along smoother surfaces: When there was less friction, the motion of objects persisted for a longer time; the less the friction, the more the motion approached constant speed. He reasoned that, in the absence of friction or other opposing forces, a horizontally moving object would continue moving indefinitely.

This assertion was supported by a different experiment and another line of reasoning. Galileo placed two of his inclined planes facing each other. He observed that a ball released from a position of rest at the top of a downward-sloping plane rolled down and then up the slope of the upward-sloping plane until it almost reached its initial height. He reasoned that only friction prevented it from rising to exactly the same height, for the smoother the planes, the more nearly the ball rose to the same height. Then he reduced the angle of the upward-sloping plane. Again the ball rose to the same height, but it had to go farther. Additional reductions of the angle yielded similar results; to reach the same height, the ball had to go farther each time. He then asked the question, "If I have a long horizontal plane, how far must the ball go to reach the same height?" The obvious answer is "Forever—it will never reach its initial height."[2]

Galileo analyzed this in still another way. Because the downward motion of the ball from the first plane is the same for all cases, the speed of the ball when it begins moving up the second plane is the same for all cases. If it moves up a steep slope, it loses its speed rapidly. On a lesser slope, it loses its speed more slowly and rolls for a longer time. The less the upward slope, the more slowly it loses its speed. In the extreme case in which there is no slope at all—that is, when the plane is horizontal—the ball should not lose any speed. In the absence of retarding forces, the tendency of the ball is to move forever without slowing down. We call this property of an object to resist changes in motion *inertia.*

Galileo's concept of inertia discredited the Aristotelian theory of motion. Aristotle did not recognize the idea of inertia because he failed to imagine what motion would be like without friction. In his experience, all motion was subject to resistance, and he made this fact central to his theory of motion. Aristotle's

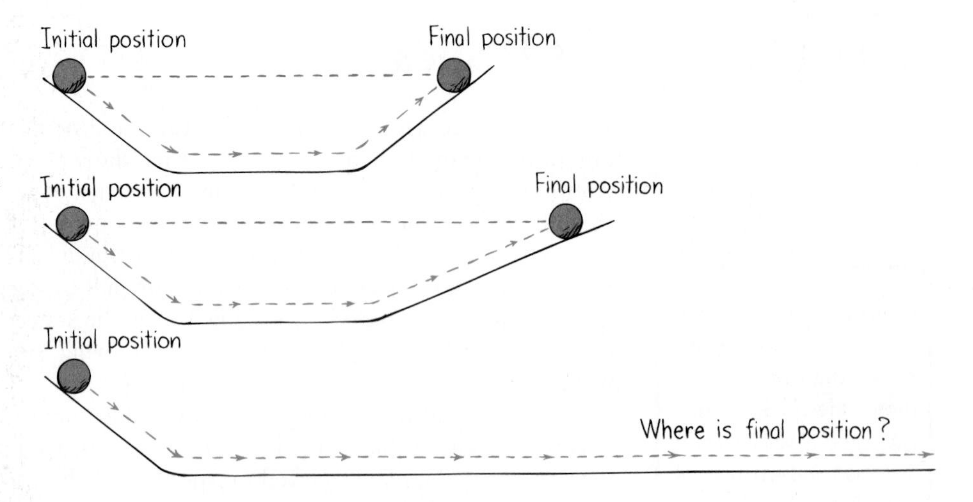

FIGURE 2.3
A ball rolling down an incline on the left tends to roll up to its initial height on the right. The ball must roll a greater distance as the angle of incline on the right is reduced.

[2]From Galileo's *Dialogues Concerning the Two New Sciences.*

failure to recognize friction for what it is—namely, a force like any other—impeded the progress of physics for nearly 2000 years, until the time of Galileo. An application of Galileo's concept of inertia would show that no force is required to keep Earth moving forward. The way was open for Isaac Newton to synthesize a new vision of the universe.

CHECK YOURSELF

Would it be correct to say that inertia is the *reason* a moving object continues in motion when no force acts upon it?

In 1642, several months after Galileo died, Isaac Newton was born. By the time Newton was 23, he developed his famous laws of motion, which completed the overthrow of the Aristotelian ideas that had dominated the thinking of the best minds for nearly two millennia. In this chapter, we will consider the first of Newton's laws. It is a restatement of the concept of inertia as proposed earlier by Galileo. (Newton's three laws of motion first appeared in one of the most important books of all time, Newton's *Principia.*)

Newton's First Law of Motion

Newton's Law of Inertia
The Old Tablecloth Trick
Toilet Paper Roll
Inertia of a Ball
Inertia of a Cylinder
Inertia of an Anvil

Aristotle's idea that a moving object must be propelled by a steady force was completely turned around by Galileo, who stated that, in the *absence* of a force, a moving object will continue moving. The tendency of things to resist changes in motion was what Galileo called *inertia.* Newton refined Galileo's idea and made it his first law, appropriately called the **law of inertia.** From Newton's *Principia* (translated from the original Latin):

> **Every object continues in its state of rest, or of uniform motion in a straight line, unless it is compelled to change that state by forces impressed upon it.**

The key word in this law is *continues:* An object *continues* to do whatever it happens to be doing unless a force is exerted upon it. If it is at rest, it *continues* in a state of rest. This is nicely demonstrated when a tablecloth is skillfully whipped from under dishes on a tabletop, leaving the dishes in their initial state of rest. This property of objects to resist changes in motion is called **inertia.**

CHECK YOUR ANSWER

In the strict sense, no. We don't know the reason for objects persisting in their motion when no forces act upon them. We refer to the property of material objects to behave in this predictable way as *inertia*. We understand many things and have labels and names for these things. There are many things we do not understand, and we have labels and names for these things also. Education consists not so much in acquiring new names and labels but in learning which phenomena we understand and which we don't.

FIGURE 2.4
Examples of inertia.

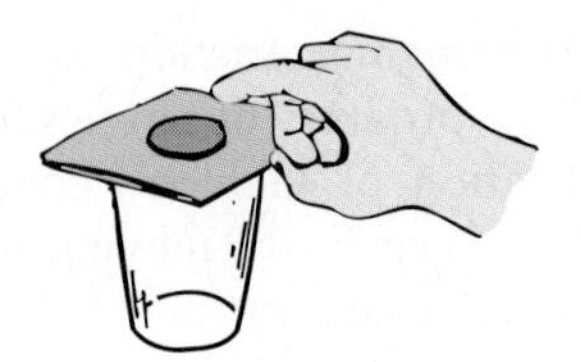

Why will the coin drop into the glass when a force accelerates the card?

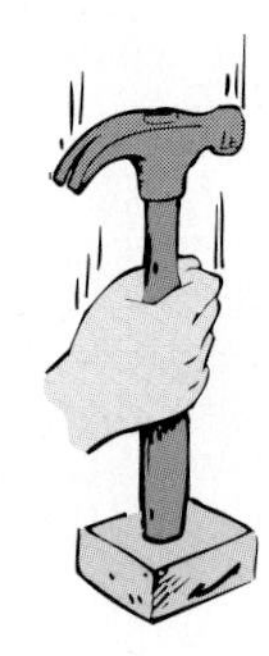

Why does the downward motion and sudden stop of the hammer tighten the hammerhead?

Why is it that a slow continuous increase in the downward force breaks the string above the massive ball, but a sudden increase breaks the lower string?

If an object is moving, it *continues* to move without turning or changing its speed. This is evident in space probes that continually move in outer space. Changes in motion must be imposed against the tendency of an object to retain its state of motion. In the absence of net forces, a moving object tends to move along a straight-line path indefinitely.

CHECK YOURSELF

A hockey puck sliding across the ice finally comes to rest. How would Aristotle have interpreted this behavior? How would Galileo and Newton have interpreted it? How would you interpret it? *(Think before you read the answers below!)*

Net Force

Changes in motion are produced by a force or combination of forces (in the next chapter we'll refer to changes in motion as *acceleration*). A **force,** in the simplest sense, is a push or a pull. Its source may be gravitational, electrical, magnetic, or simply muscular effort. When more than a single force acts on an object, we consider the ***net* force.** For example, if you and a friend pull in the

CHECK YOUR ANSWERS

Aristotle would probably say that the puck slides to a stop because it seeks its proper and natural state, one of rest. Galileo and Newton would probably say that, once in motion, the puck would continue in motion, and that what prevents continued motion is not its nature or its proper rest state but the friction the puck encounters. This friction is small compared with the friction between the puck and a wooden floor, which is why the puck slides so much farther on ice. Only you can answer the last question.

ISAAC NEWTON (1642–1727)

Isaac Newton was born prematurely, and barely survived, on Christmas Day, 1642, the same year that Galileo died. Newton's birthplace was his mother's farmhouse in Woolsthorpe, Lincolnshire, England. His father died several months before his birth, and he grew up under the care of his mother and grandmother. As a child, he showed no particular signs of brightness, and, at the age of $14\frac{1}{2}$, he was taken out of school to work on his mother's farm. As a farmer, he was a failure, preferring to read books he borrowed from a neighboring druggist. An uncle, who sensed the scholarly potential in young Isaac, prompted him to study at the University of Cambridge, which he did for 5 years, graduating without particular distinction.

A plague swept through England, and Newton retreated to his mother's farm—this time to continue his studies. At the farm, at the age of 23 and 24, he laid the foundations for the work that was to make him immortal. Seeing an apple fall to the ground led him to consider the force of gravity extending to the Moon and beyond, and he formulated the law of universal gravitation and applied it to solving the centuries-old mysteries of planetary motion and ocean tides; he invented the calculus, an indispensable mathematical tool in science; he extended Galileo's work and formulated the three fundamental laws of motion; and he formulated a theory of the nature of light and showed, with prisms, that white light is composed of all colors of the rainbow. It was his experiments with prisms that first made him famous.

When the plague subsided, Newton returned to Cambridge and soon established a reputation for himself as a first-rate mathematician. His mathematics teacher resigned in his favor, and Newton was appointed the Lucasian professor of mathematics. He held this post for 28 years. In 1672, he was elected to the Royal Society, where he exhibited the world's first reflector telescope. It can still be seen, preserved at the library of the Royal Society in London with the inscription: "The first reflecting telescope, invented by Sir Isaac Newton, and made with his own hands."

It wasn't until Newton was 42 that he began to write what is generally acknowledged as the greatest scientific book ever written, the *Principia Mathematica Philosophiae Naturalis.* He wrote the work in Latin and completed it in 18 months. It appeared in print in 1687, but it wasn't printed in English until 1729, 2 years after his death. When asked how he was able to make so many discoveries, Newton replied that he found his solutions to problems were not by sudden insight but by continually thinking very long and hard about them until he worked them out.

At the age of 46, his energies turned somewhat from science when he was elected a member of Parliament. He attended the sessions in Parliament for 2 years and never gave a speech. One day he rose and the House fell silent to hear the great man. Newton's "speech" was very brief; he simply requested that a window be closed because of a draft.

A further turn from his work in science was his appointment as warden, and then as master, of the mint. Newton resigned his professorship and directed his efforts toward greatly improving the workings of the mint, to the dismay of counterfeiters who flourished at that time. He maintained his membership in the Royal Society and was elected president, then was re-elected each year for the rest of his life. At the age of 62, he wrote *Opticks,* which summarized his work on light. Nine years later, he wrote a second edition to his *Principia.*

Although Newton's hair turned gray at 30, it remained full, long, and wavy all his life, and, unlike others in his time, he did not wear a wig. He was a modest man, overly sensitive to criticism, and he never married. He remained healthy in body and mind into old age. At 80, he still had all his teeth, his eyesight and hearing were sharp, and his mind was alert. In his lifetime, he was regarded by his countrymen as the greatest scientist who ever lived. In 1705, he was knighted by Queen Anne. Newton died at the age of 85 and was buried in Westminister Abbey along with England's kings and heroes.

Newton showed that the universe operated according to natural laws that were neither capricious nor malevolent—a knowledge that provided hope and inspiration to scientists, writers, artists, philosophers, and people of all walks of life and that ushered in the Age of Reason. The ideas and insights of Isaac Newton truly changed the world and elevated the human condition.

PERSONAL ESSAY

When I was in high school, my counselor advised me not to enroll in science and math classes and instead to focus on what seemed to be my gift for art. I took this advice. I was then interested in drawing comic strips and in boxing, neither of which earned me much success. After a stint in the army, I tried my luck at sign painting, and the cold Boston winters drove me south to warmer Miami, Florida. There, at age 26, I got a job painting billboards and met my intellectual mentor, Burl Grey. Like me, Burl had never studied physics in high school. But he was passionate about science in general, and he shared his passion with many questions as we painted together.

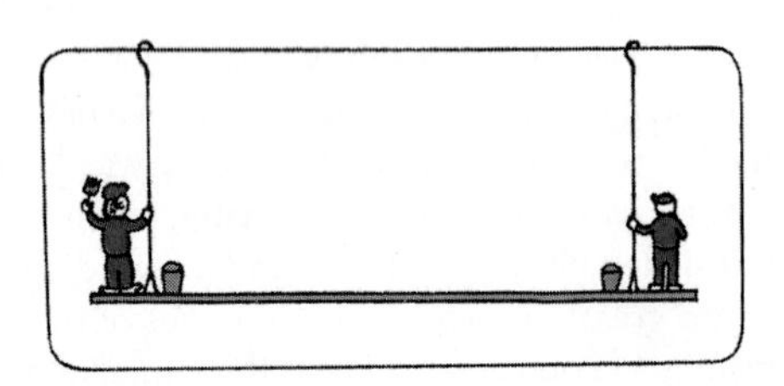

I remember Burl asking me about the tensions in the ropes that held up the scaffold we were on. The scaffold was simply a heavy horizontal plank suspended by a pair of ropes. Burl twanged the rope nearest his end of the scaffold and asked me to do the same with mine. He was comparing the tensions in both ropes—to determine which was greater. Burl was heavier than I was, and he guessed that the tension in his rope was greater. Like a more tightly stretched guitar string, the rope with greater tension twangs at a higher pitch. The finding that Burl's rope had a higher pitch seemed reasonable because his rope supported more of the load.

When I walked toward Burl to borrow one of his brushes, he asked if the tensions in the ropes had changed. Did tension in his rope increase as I moved closer? We agreed that it should have, because even more of the load was supported by Burl's rope. How about my rope? Would its tension decrease? We agreed that it would, for it would be supporting less of the total load. I was unaware at the time that I was discussing physics.

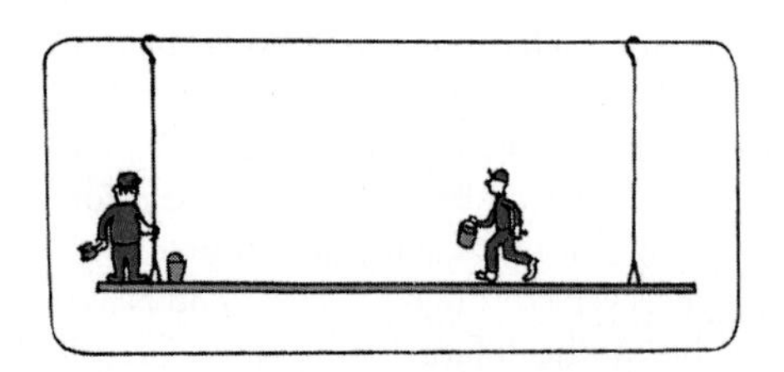

Burl and I used exaggeration to bolster our reasoning (just as physicists do). If we both stood at an extreme end of the scaffold and leaned outward, it was easy to imagine the opposite end of the staging rising like the end of a seesaw—with the opposite rope going limp. Then there would be no tension in that rope. We then reasoned the

same direction with equal forces on an object, the forces combine to produce a net force twice as great as your single force. If each of you pull with equal forces in *opposite* directions, the net force is zero. The equal but oppositely directed forces cancel each other. One of the forces can be considered to be the negative of the other, and they add algebraically to zero, with a resulting net force of zero.

Figure 2.5 shows how forces combine to produce a net force. A pair of 5-newton forces in the same direction produce a net force of 10 newtons. If the 5-newton forces are in opposite directions, the net force is zero. If 10 newtons of force is exerted to the right and 5 newtons to the left, the net force is 5 newtons to the right. The forces are shown by arrows. A quantity such as force that has both magnitude and direction is called a *vector quantity*. Vector quantities can be represented by arrows whose length and direction show the magnitude and direction of the quantity. (More about vectors in Chapter 4.)

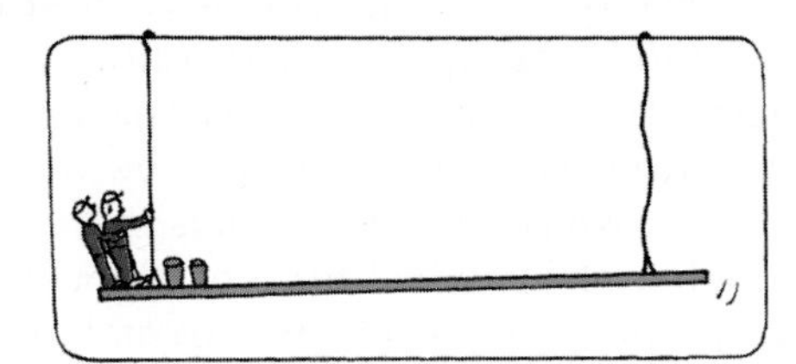

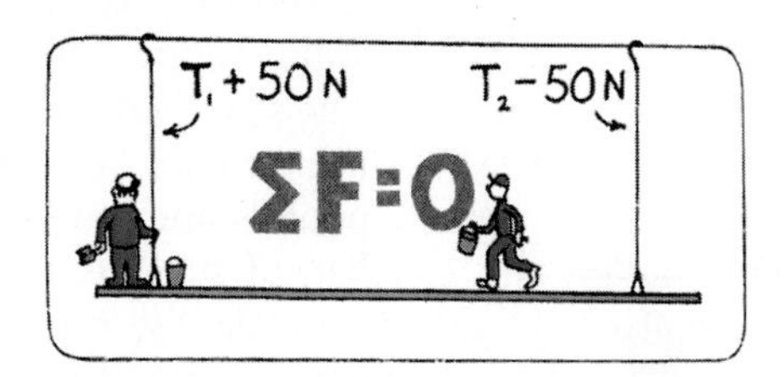

tension in my rope would gradually decrease as I walked toward Burl. It was fun posing such questions and seeing if we could answer them.

A question that we couldn't answer was whether or not the decrease in tension in my rope when I walked away from it would be *exactly* compensated by a tension increase in Burl's rope. For example, if my rope underwent a decrease of 50 newtons, would Burl's rope gain 50 newtons? (We talked pounds back then, but here we use the scientific unit of force, the *newton*—abbreviated N.) Would the gain be *exactly* 50 N? And, if so, would this be a grand coincidence? I didn't know the answer until more than a year later, when Burl's stimulation resulted in my leaving full-time painting and going to college to learn more about science.[3]

At college, I learned that any object at rest, such as the sign-painting scaffold that supported us, is said to be in equilibrium. That is, all the forces that act on it balance to zero. So the sum of the upward forces supplied by the supporting ropes indeed do add up to our weights plus the weight of the scaffold. A 50-N loss in one would be accompanied by a 50-N gain in the other.

I tell this true story to make the point that one's thinking is very different when there is a rule to guide it. Now when I look at any motionless object I know immediately that all the forces acting on it cancel out. We view nature differently when we know its rules. Without the rules of physics, we tend to be superstitious and to see magic where there is none. Quite wonderfully, everything is connected to everything else by a surprisingly small number of rules, and in a beautifully simple way. The rules of nature are what the study of physics is about.

[3]I am forever indebted to Burl Grey for the stimulation he provided, for when I continued with formal education, it was with enthusiasm. I lost touch with Burl for 40 years. A student in my class at the Exploratorium in San Francisco, Jayson Wechter, who was a private detective, located him in 1998 and put us in contact. Friendship renewed, we once again continue in spirited conversations.

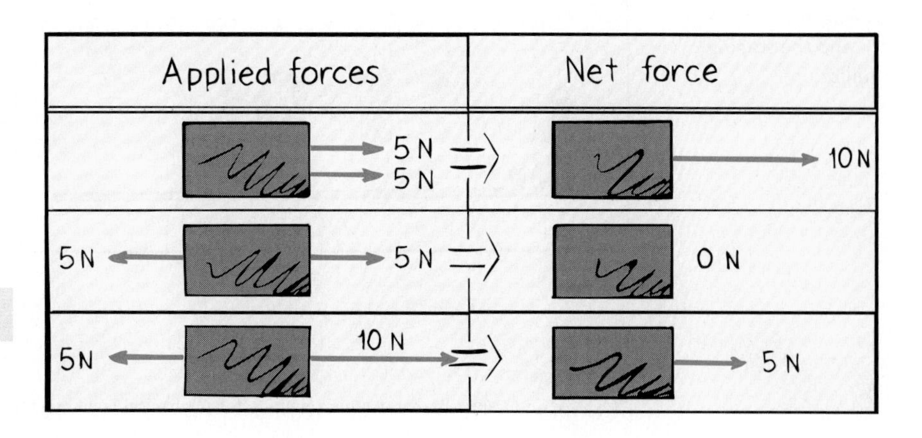

FIGURE 2.5
Net force. Whether the forces are expressed in newtons or in pounds, the result is the same.

The Equilibrium Rule

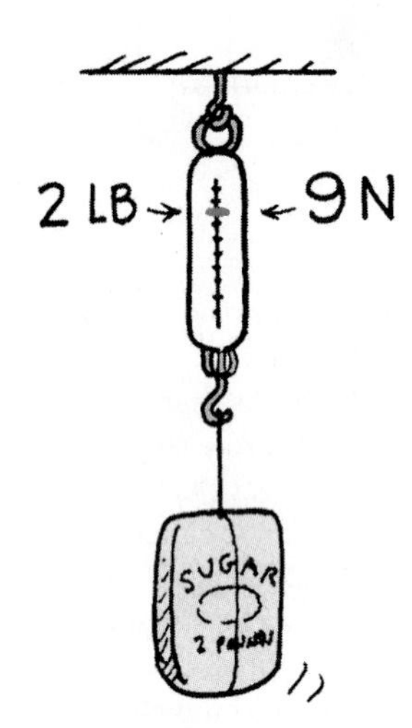

FIGURE 2.6
The upward tension in the string has the same magnitude as the weight of the bag, so the net force on the bag is zero.

If you tie a string around a 2-pound bag of sugar and hang it on a weighing scale (Figure 2.6), a spring in the scale stretches until the scale reads 2 pounds. The stretched spring is under a "stretching force" called *tension*. The same scale in a science lab is likely calibrated to read the same force as 9 newtons. Both pounds and newtons are units of weight, which in turn are units of *force*. The bag of sugar is attracted to Earth with a gravitational force of 2 pounds—or, equivalently, 9 newtons. Hang twice as much sugar from the scale and the reading will be 18 newtons.

Note that there are two forces acting on the bag of sugar—tension force acting upward and weight acting downward. The two forces on the bag are equal and opposite, and they cancel to zero. Hence, the bag remains at rest. In accord with Newton's first law, no net force acts on the bag. We can look at Newton's first law in a different light—*mechanical equilibrium*.

When the net force on something is zero, we say that something is in **mechanical equilibrium.**[4] In mathematical notation, the equilibrium rule is

$$\Sigma F = 0$$

The symbol Σ stands for "the vector sum of" and F stands for "forces." For a suspended object at rest, like the bag of sugar, the rule says that the forces acting upward on the object must be balanced by other forces acting downward to make the vector sum equal zero. (Vector quantities take direction into account, so if upward forces are +, downward ones are −, and, when added, they actually subtract.)

In Figure 2.7, we see the forces involved for Burl and Hewitt on their sign-painting scaffold. The sum of the upward tensions is equal to the sum of their weights plus the weight of the scaffold. Note how the magnitudes of the two upward vectors equal the magnitude of the three downward vectors. Net force on the scaffold is zero, so we say it is in mechanical equilibrium.

FIGURE 2.7
The sum of the upward vectors equals the sum of the downward vectors. $\Sigma F = 0$ and the staging is in equilibrium.

[4]We'll see, in Chapter 8, that another condition for mechanical equilibrium is that the net *torque* equal zero.

PRACTICING PHYSICS

1. When Burl stands alone in the exact middle of his staging, the left scale reads 500 N. Fill in the reading on the right scale. The total weight of Burl and the staging must be __________ N.

2. Burl stands farther from the left. Fill in the reading on the right scale.

3. In a silly mood, Burl dangles from the right end. Fill in the reading on the right scale.

CHECK YOURSELF

Consider the gymnast hanging from the rings.

1. If she hangs with her weight evenly divided between the two rings, how would scale readings in both supporting ropes compare with her weight?
2. Suppose she hangs with slightly more of her weight supported by the left ring. How will the right scale read?

CHECK YOUR ANSWERS

(Are you reading this before you have formulated reasoned answers in your thinking? If so, do you also exercise your body by watching others do push-ups? Exercise your thinking: When you encounter the many Check Yourself questions as above throughout this book, think before you check the footnoted answers!)

1. The reading on each scale will be half her weight. The sum of the readings on both scales then equals her weight.
2. When more of her weight is supported by the left ring, the reading on the right reduces to less than half her weight. No matter how she hangs, the sum of the scale readings equals her weight. For example, if one scale reads two-thirds her weight, the other scale will read one-third her weight. Get it?

Everything not undergoing changes in motion is in mechanical equilibrium. That's because $\sum F = 0$.

Insights

PRACTICING PHYSICS ANSWERS

Do your answers illustrate the equilibrium rule? In Question 1, the right rope must be under **500 N** of tension because Burl is in the middle and both ropes support his weight equally. Since the sum of upward tensions is 1000 N, the total weight of Burl and the scaffold must be **1000 N.** Let's call the upward tension forces +1000 N. Then the downward weights are −1000 N. What happens when you add +1000 N and −1000 N? The answer is that they equal zero. So we see that $\Sigma F = 0$.

For Question 2, did you get the correct answer of **830 N**? Reasoning: We know from Question 1 that the sum of the rope tensions equals 1000 N, and since the left rope has a tension of 170 N, the other rope must make up the difference—that 1000 N − 170 N = 830 N. Get it? If so, great. If not, talk about it with your friends until you do. Then read further.

The answer to Question 3 is **1000 N.** Do you see that this illustrates $\Sigma F = 0$?

Support Force

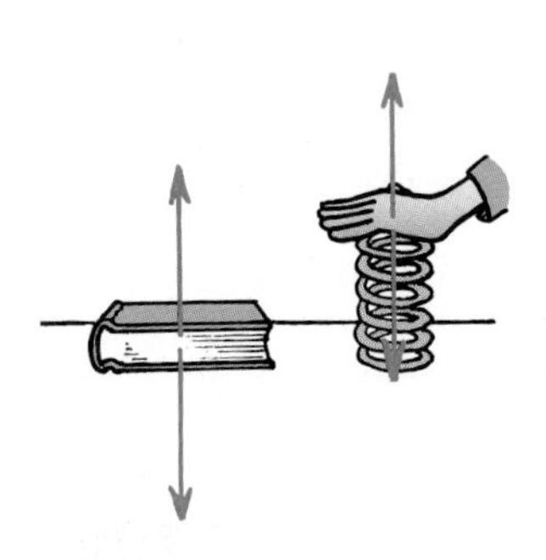

FIGURE 2.8
(Left) The table pushes up on the book with as much force as the downward force of gravity on the book. (Right) The spring pushes up on your hand with as much force as you exert to push down on the spring.

Consider a book lying at rest on a table. It is in equilibrium. What forces act on the book? One force is that due to gravity—the *weight* of the book. Since the book is in equilibrium, there must be another force acting on the book to produce a net force of zero—an upward force opposite to the force of gravity. The table exerts this upward force. We call this the upward *support force.* This upward support force, often called the *normal force,* must equal the weight of the book.[5] If we call the upward force positive, then the downward weight is negative, and the two add to become zero. The net force on the book is zero. Another way to say the same thing is $\Sigma F = 0$.

To understand better that the table pushes up on the book, compare the case of compressing a spring (Figure 2.8). If you push the spring down, you can feel the spring pushing up on your hand. Similarly, the book lying on the table compresses atoms in the table, which behave like microscopic springs. The weight of the book squeezes downward on the atoms, and they squeeze upward on the book. In this way, the compressed atoms produce the support force.

When you step on a bathroom scale, two forces act on the scale. One is your downward push on the scale—the result of gravity pulling on you—and the other is the upward support force of the floor. These forces squeeze a mechanism (in effect, a spring) within the scale that is calibrated to show the magnitude of the support force (Figure 2.9). It is this support force that shows your weight. When you weigh yourself on a bathroom scale at rest, the support force and the force of gravity pulling you down have the same magnitude. Hence we can say that your weight is the force of gravity acting on you.

CHECK YOURSELF

1. What is the net force on a bathroom scale when a 150-pound person stands on it?
2. Suppose you stand on two bathroom scales with your weight evenly divided between the two scales. What will each scale read? What happens when you stand with more of your weight on one foot than the other?

[5]This force acts at right angles to the surface. When we say "normal to," we are saying "at right angles to," which is why this force is called a normal force.

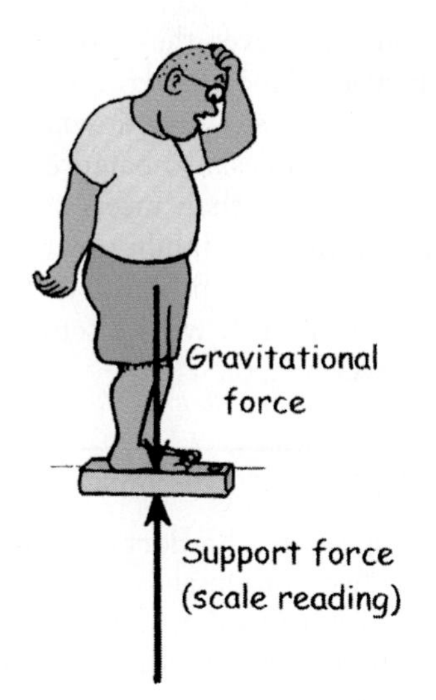

FIGURE 2.9
The upward support is as much as the downward gravitational force.

Equilibrium of Moving Things

FIGURE 2.10
When the push on the crate is as great as the force of friction between the crate and the floor, the net force on the crate is zero and it slides at an unchanging speed.

Rest is only one form of equilibrium. An object moving at constant speed in a straight-line path is also in equilibrium. Equilibrium is a state of no change. A bowling ball rolling at constant speed in a straight line is in equilibrium—until it hits the pins. Whether at rest (static equilibrium) or steadily rolling in a straight-line path (dynamic equilibrium), $\Sigma F = 0$.

It follows from Newton's first law that an object under the influence of only one force cannot be in equilibrium. Net force couldn't be zero. Only when two or more forces act on it can it be in equilibrium. We can test whether or not something is in equilibrium by noting whether or not it undergoes changes in its state of motion.

Consider a crate being pushed horizontally across a factory floor. If it moves at a steady speed in a straight-line path, it is in dynamic equilibrium. This tells us that more than one force acts on the crate. Another force exists—likely the force of friction between the crate and the floor. The fact that the net force on the crate equals zero means that the force of friction must be equal and opposite to our pushing force.

The **equilibrium rule,** $\Sigma F = 0$, provides a reasoned way to view all things at rest—balancing rocks, objects in your room, or the steel beams in bridges

CHECK YOUR ANSWERS

1. Zero, as evidenced by the scale remaining at rest. The scale reads the *support force*, which has the same magnitude as weight—not the net force.
2. The reading on each scale is half your weight. Then the sum of the scale readings will balance your weight and the net force on you will be zero. If you lean more on one scale than the other, more than half your weight will be read on that scale but less on the other, so they will still add up to your weight. Like the example of the gymnast hanging by the rings, if one scale reads two-thirds your weight, the other scale will read one-third your weight.

or in building construction. Whatever their configuration, if in static equilibrium, all acting forces always balance to zero. The same is true of objects that move steadily, not speeding up, slowing down, or changing direction. For dynamic equilibrium, all acting forces also balance to zero. The equilibrium rule is one that allows you to see more than meets the eye of the casual observer. It's nice to know the reasons for the stability of things in our everyday world.

There are different forms of equilibrium. In Chapter 8, we'll talk about rotational equilibrium, and, in Part 4, we'll discuss thermal equilibrium associated with heat. Physics is everywhere.

CHECK YOURSELF

An airplane flies at constant speed in a horizontal straight path. In other words, the flying plane is in equilibrium. Two horizontal forces act on the plane. One is the thrust of the propellers that push it forward, and the other is the force of air resistance that acts in the opposite direction. Which force is greater?

The Moving Earth

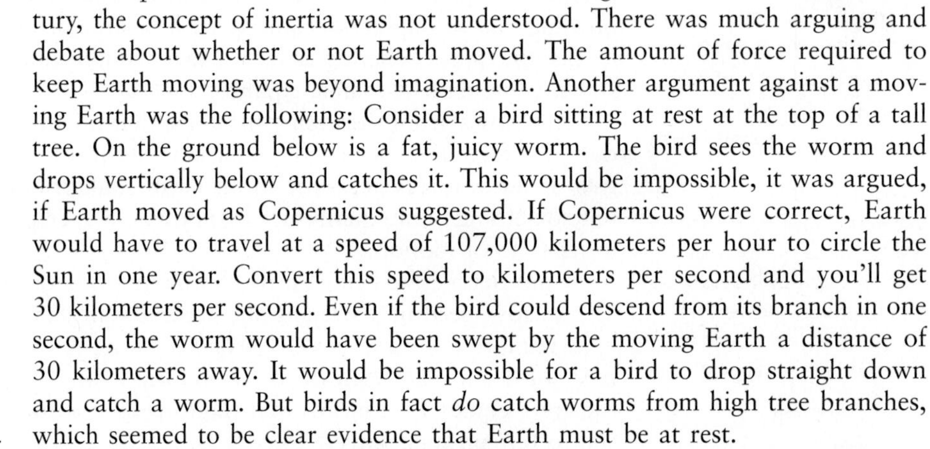

When Copernicus announced the idea of a moving Earth in the sixteenth century, the concept of inertia was not understood. There was much arguing and debate about whether or not Earth moved. The amount of force required to keep Earth moving was beyond imagination. Another argument against a moving Earth was the following: Consider a bird sitting at rest at the top of a tall tree. On the ground below is a fat, juicy worm. The bird sees the worm and drops vertically below and catches it. This would be impossible, it was argued, if Earth moved as Copernicus suggested. If Copernicus were correct, Earth would have to travel at a speed of 107,000 kilometers per hour to circle the Sun in one year. Convert this speed to kilometers per second and you'll get 30 kilometers per second. Even if the bird could descend from its branch in one second, the worm would have been swept by the moving Earth a distance of 30 kilometers away. It would be impossible for a bird to drop straight down and catch a worm. But birds in fact *do* catch worms from high tree branches, which seemed to be clear evidence that Earth must be at rest.

FIGURE 2.11
Can the bird drop down and catch the worm if Earth moves at 30 km/s?

Can you refute this argument? You can if you invoke the idea of inertia. You see, not only is Earth moving at 30 kilometers per second, but so are the tree, the branch of the tree, the bird that sits on it, the worm below, and even the air in between. All are moving at 30 kilometers per second. Things in motion remain in motion if no unbalanced forces are acting upon them. So, when the bird drops from the branch, its initial sideways motion of 30 kilometers per

CHECK YOUR ANSWER

Both forces have the same magnitude. Call the forward force exerted by the propellers positive. Then the air resistance is negative. Since the plane is in dynamic equilibrium, can you see that the two forces combine to equal zero? Hence it neither gains or loses speed.

second remains unchanged. It catches the worm, quite unaffected by the motion of its total environment.

Stand next to a wall. Jump up so that your feet are no longer in contact with the floor. Does the 30-kilometer-per-second wall slam into you? It doesn't, because you are also traveling at 30 kilometers per second—before, during, and after your jump. The 30 kilometers per second is the speed of Earth relative to the Sun, not the speed of the wall relative to you.

People four hundred years ago had difficulty with ideas like these, not only because they failed to acknowledge the concept of inertia but because they were not accustomed to moving in high-speed vehicles. Slow, bumpy rides in horse-drawn carriages did not lend themselves to experiments that would reveal the effect of inertia. Today we flip a coin in a high-speed car, bus, or plane, and we catch the vertically moving coin as we would if the vehicle were at rest. We see evidence for the law of inertia when the horizontal motion of the coin before, during, and after the catch is the same. The coin keeps up with us. The vertical force of gravity affects only the vertical motion of the coin.

FIGURE 2.12
When you flip a coin in a high-speed airplane, it behaves as if the airplane were at rest. The coin keeps up with you—inertia in action!

Our notions of motion today are very different from those of our ancestors. Aristotle did not recognize the idea of inertia because he did not see that all moving things follow the same rules. He imagined that rules for motion in the heavens were very different from the rules of motion on Earth. He saw vertical motion as natural but horizontal motion as unnatural, requiring a sustained force. Galileo and Newton, on the other hand, saw that all moving things follow the same rules. To them, moving things require *no* force to keep moving if there are no opposing forces, such as friction. We can only wonder how differently science might have progressed if Aristotle had recognized the unity of all kinds of motion.

Summary of Terms

Newton's first law of motion (the law of inertia) Every object continues in its state of rest, or of uniform motion in a straight line, unless compelled to change that state by forces impressed upon it.

Inertia The property of things to resist changes in motion.

Force In the simplest sense, a push or a pull.

Net force The vector sum of forces that act on an object.

Mechanical equilibrium The state of an object or system of objects for which there are no changes in motion. In accord with Newton's first law, if at rest, the state of rest persists. If moving, motion continues without change.

Equilibrium rule For any object or system of objects in equilibrium, the sum of the forces acting equals zero. In equation form, $\Sigma F = 0$.

Review Questions

Each chapter in this book concludes with a set of review questions, exercises, and, for some chapters, problems. In some chapters, there is a set of single-step numerical problems that are meant for acquainting you with equations in the chapter— ***One-Step Calculations.*** *The* ***Review Questions*** *are designed to help you comprehend ideas and catch the essentials of the chapter material. You'll notice that answers to the questions can be found within the chapters. The* ***Exercises*** *stress thinking rather than mere recall of information, and they call for an understanding of the definitions, principles, and relationships of the chapter material. In many cases, the intention of particular exercises is to help you apply the ideas of physics to familiar situations. Unless you cover only a few chapters in your course, you will likely be expected to tackle only a few exercises for each chapter. Answers should be in complete sentences, with an explanation or sketches when applicable. The large number of exercises is to allow your instructor a wide choice of assignments.* ***Problems*** *go further than One-Step Calculations and feature concepts that are more clearly understood with more challenging computations. Additional problems are in the supplement* ***Problem Solving in Conceptual Physics.***

Aristotle on Motion

1. Contrast Aristotle's ideas of natural motion and violent motion.
2. What class of motion, natural or violent, did Aristotle attribute to motion of the Moon?

3. What state of motion did Aristotle attribute to the Earth?

Copernicus and the Moving Earth

4. What relationship between the Sun and Earth did Copernicus formulate?

Galileo and the Leaning Tower

5. What did Galileo discover in his legendary experiment on the Leaning Tower of Pisa?

Galileo's Inclined Planes

6. What did Galileo discover about moving bodies and force in his experiments with inclined planes?
7. What does it mean to say that a moving object has inertia? Give an example.
8. Is inertia the *reason* for moving objects maintaining motion, or the *name* given to this property?

Newton's First Law of Motion

9. Cite Newton's first law of motion.

Net Force

10. What is the net force on a cart that is pulled to the right with 100 pounds and to the left with 30 pounds?
11. Why do we say that force is a vector quantity?

The Equilibrium Rule

12. Can force be expressed in units of pounds and also in units of newtons?
13. What is the net force on an object that is pulled with 80 newtons to the right and 80 newtons to the left?
14. What is the net force on a bag pulled down by gravity with 18 newtons and pulled upward by a rope with 18 newtons?
15. What does it mean to say something is in mechanical equilibrium?
16. State the equilibrium rule in symbolic notation.

Support Force

17. Consider a book that weighs 15 N at rest on a flat table. How many newtons of support force does the table provide? What is the net force on the book in this case?
18. When you stand at rest on a bathroom scale, how does your weight compare with the support force by the scale?

Equilibrium of Moving Things

19. A bowling ball at rest is in equilibrium. Is the ball in equilibrium when it moves at constant speed in a straight-line path?
20. What is the test for whether or not a moving object is in equilibrium?
21. If you push on a crate with a force of 100 N and it slides at constant velocity, how much is the friction acting on the crate?

The Moving Earth

22. What concept was missing in people's minds in the sixteenth century when they couldn't believe the Earth was moving?
23. A bird sitting in a tree is traveling at 30 km/s relative to the faraway Sun. When the bird drops to the ground below, does it still go 30 km/s, or does this speed become zero?
24. Stand next to a wall that travels at 30 km/s relative to the Sun. With your feet on the ground, you also travel the same 30 km/s. Do you maintain this speed when your feet leave the ground? What concept supports your answer?
25. What did Aristotle fail to recognize about the rules of nature for objects on Earth and in the heavens?

Project

Ask a friend to drive a small nail into a piece of wood placed on top of a pile of books on your head. Why doesn't this hurt you?

Exercises

Please do not be intimidated by the large number of exercises in this book. If your course work is to cover many chapters, your instructor will likely assign only a few exercises from each.

1. A ball rolling along a floor doesn't continue rolling indefinitely. Is it because it is seeking a place of rest or because some force is acting upon it? If the latter, identify the force.
2. Copernicus postulated that the Earth moves around the Sun (rather than the other way around), but he was troubled about the idea. What concepts of mechanics was he missing (concepts later introduced by Galileo and Newton) that would have eased his doubts?
3. What Aristotelian idea did Galileo discredit in his fabled Leaning Tower demonstration?
4. What Aristotelian idea did Galileo demolish with his experiments with inclined planes?
5. Was it Galileo or Newton who first proposed the concept of inertia?

6. Asteroids have been moving though space for billions of years. What keeps them moving?
7. A space probe may be carried by a rocket into outer space. What keeps the probe moving after the rocket no longer pushes it?
8. In answer to the question "What keeps the Earth moving around the Sun?", a friend asserts that inertia keeps it moving. Correct your friend's erroneous assertion.
9. Your friend says that inertia is a force that keeps things in their place, either at rest or in motion. Do you agree? Why or why not?
10. Your other friend says that bureaucratic organizations have a lot of inertia. Is this akin to Newton's first law of inertia?
11. Consider a ball at rest in the middle of a toy wagon. When the wagon is pulled forward, the ball rolls against the back of the wagon. Interpret this observation in terms of Newton's first law.
12. In tearing a paper towel or plastic bag from a roll, why is a sharp jerk more effective than a slow pull?
13. If you're in a car at rest that gets hit from behind, you can suffer a serious neck injury called whiplash. What does whiplash have to do with Newton's first law?
14. In terms of Newton's first law (the law of inertia), how does a car headrest help to guard against whiplash in a rear-end collision?
15. Why do you lurch forward in a bus that suddenly slows? Why do you lurch backward when it picks up speed? What law applies here?
16. Suppose that you're in a moving car and the motor stops running. You step on the brakes and slow the car to half speed. If you release your foot from the brakes, will the car speed up a bit, or will it continue at half speed and slow due to friction? Defend your answer.
17. Each bone in the chain of bones forming your spine is separated from its neighbors by disks of elastic tissue. What happens, then, when you jump heavily onto your feet from an elevated position? (Hint: think about the hammerhead in Figure 2.4.) Can you think of a reason why you are a little taller in the morning than at night?
18. When you push a cart, it moves. When you stop pushing, it comes to rest. Does this violate Newton's law of inertia? Defend your answer.
19. Start a ball rolling down a bowling alley and you'll find that it moves slightly slower with time. Does this violate Newton's law of inertia? Defend your answer.
20. Consider a pair of forces, one having a magnitude of 20 N and the other a magnitude of 12 N. What maximum net force is possible for these two forces? What is the minimum net force possible?
21. When any object is in mechanical equilibrium, what can be correctly said about all the forces that act on it? Must the net force necessarily be zero?
22. A monkey hangs stationary at the end of a vertical vine. What two forces act on the monkey? Which, if either, is greater?
23. Can an object be in mechanical equilibrium when only a single force acts on it? Explain.
24. When a ball is tossed straight up, it momentarily comes to a stop at the top of its path. Is it in equilibrium during this brief moment? Why or why not?
25. A hockey puck slides across the ice at a constant speed. Is it in equilibrium? Why or why not?
26. The sketch shows a painter's staging in mechanical equilibrium. The person in the middle weighs 250 N, and the tensions in each rope are 200 N. What is the weight of the staging?

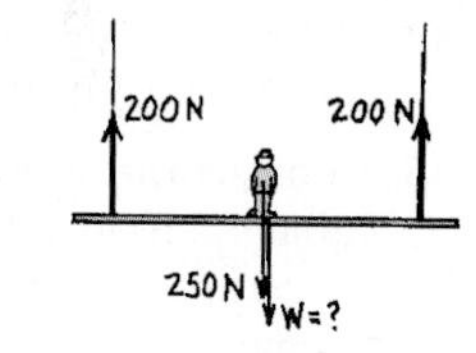

27. A different staging that weighs 300 N supports two painters, one 250 N and the other 300 N. The reading in the left scale is 400 N. What is the reading in the right-hand scale?

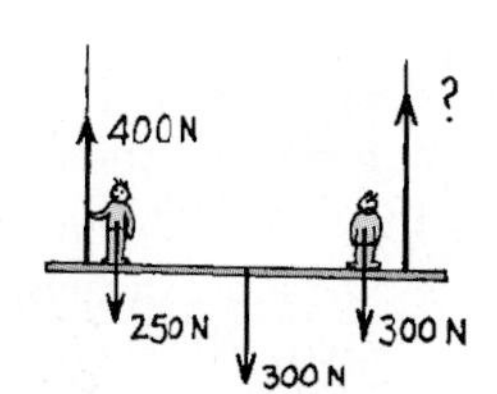

28. Nellie Newton hangs at rest from the ends of the rope as shown. How does the reading on the scale compare with her weight?

29. Harry the painter swings year after year from his bosun's chair. His weight is 500 N and the rope, unknown to him, has a breaking point of 300 N. Why doesn't the rope break when he is supported as shown at the left? One day, Harry is painting near a flagpole, and, for a change, he ties the free end of the rope to the flagpole instead of to his chair, as shown at the right. Why did Harry end up taking his vacation early?

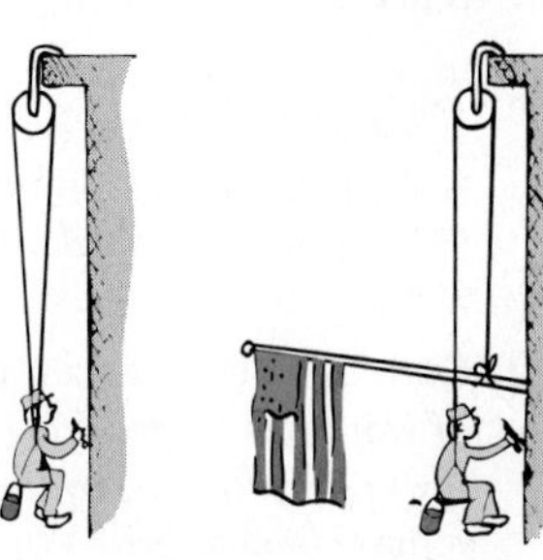

30. Can you say that no force acts on a body at rest? Or is it correct to say that no *net* force acts on it? Defend your answer.

31. For the pulley system shown, what is the upper limit of weight the strong man can lift?

32. If the strong man in the previous exercise exerts a downward force of 800 N on the rope, how much upward force is exerted on the block?

33. A force of gravity pulls downward on a book on a table. What force prevents the book from accelerating downward?

34. How many significant forces act on a book at rest on a table? Identify the forces.

35. Consider the normal force on a book at rest on a tabletop. If the table is tilted so that the surface forms an inclined plane, will the magnitude of the normal force change? If so, how?

36. When you push downward on a book at rest on a table, you feel an upward force. Does this force depend on friction? Defend your answer.

37. As you stand on a floor, does the floor exert an upward force against your feet? How much force does it exert? Why are you not moved upward by this force?

38. Place a heavy book on a table and the table pushes up on the book. Why doesn't this upward push cause the book to rise from the table?

39. An empty jug of weight W rests on a table. What is the support force exerted on the jug by the table? What is the support force when water of weight w is poured into the jug?

40. If you pull horizontally on a crate with a force of 200 N, it slides across the floor in dynamic equilibrium. How much friction is acting on the crate?

41. In order to slide a heavy cabinet across the floor at constant speed, you exert a horizontal force of 600 N. Is the force of friction between the cabinet and the floor greater than, less than, or equal to 600 N? Defend your answer.

42. Consider a crate at rest on a factory floor. As a pair of workmen begin lifting it, does the support force on the crate provided by the floor increase, decrease, or remain unchanged? What happens to the support force on the workmen's feet?

43. Two people each pull with 300 N on a rope in a tug of war. What is the net force on the rope? How much force is exerted on each person by the rope?

44. Two forces act on a parachutist falling in air: weight and air drag. If the fall is steady, with no gain or loss of speed, then the parachutist is in dynamic equilibrium. How do the magnitudes of weight and air drag compare?

45. Before the time of Galileo and Newton, some learned scholars thought that a stone dropped from the top of a tall mast of a moving ship would fall vertically and hit the deck behind the mast by a distance equal to how far the ship had moved forward while the stone was falling. In light of your understanding of Newton's first law, what do you think about this?

46. Because the Earth rotates once every 24 hours, the west wall in your room moves in a direction toward you at a linear speed that is probably more than 1000 km per hour (the exact speed depends on your latitude). When you stand facing the wall, you are carried along at the same speed, so you don't notice it. But when you jump upward, with your feet no longer in contact with the floor, why doesn't the high-speed wall slam into you?

47. A child learns in school that the Earth is traveling faster than 100,000 kilometers per hour around the Sun and, in a frightened tone, asks why we aren't swept off. What is your explanation?

48. If you toss a coin straight upward while riding in a train, where does the coin land when the motion of the train is uniform along a straight-line track? When the train slows while the coin is in the air? When the train is turning?

49. The smokestack of a stationary toy train consists of a vertical spring gun that shoots a steel ball a meter or so straight into the air—so straight that the ball always falls back into the smokestack. Suppose the train moves at constant speed along the straight track. Do you think the ball will still return to the smokestack if shot from the moving train? What if the train gains speed along the straight track? What if it moves at a constant speed on a circular track? Why do your answers differ?

50. Consider an airplane that flies due east on a trip, then returns flying due west. Flying in one direction, the plane flies with the rotation of the Earth, and in the opposite direction, against the Earth's rotation. But, in the absence of winds, the times of flight are equal either way. Why is this so?

CHAPTER 3

Linear Motion

Chelcie Liu asks his students to check their neighbors and predict which ball will reach the end of the equal-length tracks first.

More than 2000 years ago, the ancient Greek scientists were familiar with some of the ideas of physics that we study today. They had a good understanding of some of the properties of light, but they were confused about motion. Great progress in understanding motion occurred with Galileo and his study of balls on inclined planes, as discussed in the previous chapter. In this chapter, we will learn rules of motion that involve three concepts: *speed, velocity,* and *acceleration*. Mastering these concepts would be nice, but it will be enough if you only become familiar with them and become able to distinguish among them. You will gain more familiarity with these concepts in the following chapters. Here we'll consider only the simplest form of motion—that along a straight-line path: *linear motion*.

Motion Is Relative

Everything moves—even things that appear to be at rest. They move relative to the Sun and stars. As you're reading this, you're moving at about 107,000 kilometers per hour relative to the Sun, and you're moving even faster relative to the center of our galaxy. When we discuss the motion of something, we describe the motion relative to something else. If you walk down the aisle of a moving bus, your speed relative to the floor of the bus is likely quite different from your speed relative to the road. When we say a racing car reaches a speed of 300 kilometers per hour, we mean relative to the track. Unless stated otherwise, when we discuss the speeds of things in our environment, we mean relative to the surface of Earth. Motion is relative.

Speed

Definition of Speed
Average Speed

Before the time of Galileo, people described moving things as simply "slow" or "fast." Such descriptions were vague. Galileo is credited with being the first to measure speed by considering the distance covered and the time it takes. He defined **speed** as the distance covered per unit of time.

$$\text{Speed} = \frac{\text{distance}}{\text{time}}$$

A cyclist who covers 30 meters in a time of 2 seconds, for example, has a speed of 15 meters per second.

Any combination of distance and time units is legitimate for measuring speed; for motor vehicles (or long distances), the units kilometers per hour (km/h) or miles per hour (mi/h or mph) are commonly used. For shorter distances, meters per second (m/s) is more useful. The slash symbol (/) is read as *per* and means "divided by." Throughout this book, we'll primarily use meters per second (m/s). Table 3.1 shows some comparative speeds in different units.[1]

FIGURE 3.1
When sitting on a chair, your speed is zero relative to Earth but 30 km/s relative to the Sun.

Instantaneous Speed

Things in motion often have variations in speed. A car, for example, may travel along a street at 50 km/h, slow to 0 km/h at a red light, and speed up to only 30 km/h because of traffic. You can tell the speed of the car at any instant by looking at its speedometer. The speed at any instant is the *instantaneous speed.* A car traveling at 50 km/h usually goes at that speed for less than one hour. If it did go at that speed for a full hour, it would cover 50 km. If it continued at that speed for half an hour, it would cover half that distance: 25 km. If it continued for only one minute, it would cover less than 1 km.

TABLE 3.1
Approximate Speeds in Different Units

mi/h	km/h	m/s
12 mi/h =	20 km/h =	6 m/s
25 mi/h =	40 km/h =	11 m/s
37 mi/h =	60 km/h =	17 m/s
50 mi/h =	80 km/h =	22 m/s
62 mi/h =	100 km/h =	28 m/s
75 mi/h =	120 km/h =	33 m/s
100 mi/h =	160 km/h =	44 m/s

Average Speed

In planning a trip by car, the driver often wants to know the time of travel. The driver is concerned with the *average speed* for the trip. Average speed is defined as

$$\text{Average speed} = \frac{\text{total distance covered}}{\text{time interval}}$$

Average speed can be calculated rather easily. For example, if we drive a distance of 80 kilometers in a time of 1 hour, we say our average speed is 80 kilometers per hour. Likewise, if we travel 320 kilometers in 4 hours,

$$\text{Average speed} = \frac{\text{total distance covered}}{\text{time interval}} = \frac{320 \text{ km}}{4 \text{ h}} = 80 \text{ km/h}$$

FIGURE 3.2
A speedometer gives readings in both miles per hour and kilometers per hour.

We see that, when a distance in kilometers (km) is divided by a time in hours (h), the answer is in kilometers per hour (km/h).

Since average speed is the whole distance covered divided by the total time of travel, it doesn't indicate the different speeds and variations that may have taken place during shorter time intervals. On most trips, we experience a variety of speeds, so the average speed is often quite different from the instantaneous speed.

If we know average speed and time of travel, distance traveled is easy to find. A simple rearrangement of the definition above gives

$$\text{Total distance covered} = \text{average speed} \times \text{time}$$

[1]Conversion is based on 1 h = 3600 s, 1 mi = 1609.344 m.

If you get a traffic ticket for speeding, is it because of your *instantaneous speed* or your *average speed?*

Insights

If your average speed is 80 kilometers per hour on a 4-hour trip, for example, you cover a total distance of 320 kilometers.

CHECK YOURSELF

1. What is the average speed of a cheetah that sprints 100 meters in 4 seconds? How about if it sprints 50 m in 2 s?
2. If a car moves with an average speed of 60 km/h for an hour, it will travel a distance of 60 km.
   a. How far would it travel if it moved at this rate for 4 h?
   b. For 10 h?
3. In addition to the speedometer on the dashboard of every car is an odometer, which records the distance traveled. If the initial reading is set at zero at the beginning of a trip and the reading is 40 km one-half hour later, what has been your average speed?
4. Would it be possible to attain this average speed and never go faster than 80 km/h?

Velocity

Velocity
Changing Velocity

When we know both the speed and the direction of an object, we know its **velocity.** For example, if a car travels at 60 km/h, we know its speed. But if we say it moves at 60 km/h to the north, we specify its *velocity*. Speed is a description of how fast; velocity is how fast *and* in what direction. A quantity such

CHECK YOUR ANSWERS

(Are you reading this before you have reasoned answers in your mind? As mentioned in the previous chapter, when you encounter Check Yourself questions throughout this book, check your ***thinking*** *before you read the footnoted answers. You'll not only learn more, you'll enjoy learning more.)*

1. In both cases the answer is 25 m/s:

$$\text{Average speed} = \frac{\text{distance covered}}{\text{time interval}} = \frac{100 \text{ meters}}{4 \text{ seconds}} = \frac{50 \text{ meters}}{2 \text{ seconds}} = 25 \text{ m/s}$$

2. The distance traveled is the average speed × time of travel, so
   a. Distance = 60 km/h × 4 h = 240 km
   b. Distance = 60 km/h × 10 hr = 600 km
3. $\text{Average speed} = \dfrac{\text{total distance covered}}{\text{time interval}} = \dfrac{40 \text{ km}}{0.5 \text{ h}} = 80 \text{ km/h}$
4. No, not if the trip starts from rest and ends at rest. There are times in which the instantaneous speeds are less than 80 km/h, so the driver must drive at speeds of greater than 80 km/h during one or more time intervals in order to average 80 km/h. In practice, average speeds are usually much lower than high instantaneous speeds.

as velocity that specifies direction as well as magnitude is called a **vector quantity.** Recall from Chapter 2 that force is a vector quantity, requiring both magnitude and direction for its description. Likewise, velocity is a vector quantity. In contrast, quantities that require only magnitude for a description are called *scalar quantities*. Speed is a scalar quantity.

Constant Velocity

Constant speed means steady speed. Something with constant speed doesn't speed up or slow down. Constant velocity, on the other hand, means both constant speed *and* constant direction. Constant direction is a straight line—the object's path doesn't curve. So constant velocity means motion in a straight line at a constant speed.

Changing Velocity

If either the speed or the direction changes (or if both change), then the velocity changes. A car on a curved track, for example, may have a constant speed, but, because its direction is changing, its velocity is not constant. We'll see in the next section that it is *accelerating*.

FIGURE 3.3
The car on the circular track may have a constant speed, but its velocity is changing every instant. Why?

CHECK YOURSELF

1. "She moves at a constant speed in a constant direction." Rephrase the same sentence in fewer words.
2. The speedometer of a car moving to the east reads 100 km/h. It passes another car that moves to the west at 100 km/h. Do both cars have the same speed? Do they have the same velocity?
3. During a certain period of time, the speedometer of a car reads a constant 60 km/h. Does this indicate a constant speed? A constant velocity?

Acceleration

THE Physics Place
Definition of Acceleration
Numerical Example of Acceleration

We can change the velocity of something by changing its speed, by changing its direction, or by changing both its speed *and* its direction. How quickly velocity changes is **acceleration:**

$$\text{Acceleration} = \frac{\text{change in velocity}}{\text{time interval}}$$

CHECK YOUR ANSWERS

1. "She moves at a constant velocity."
2. Both cars have the same speed, but they have opposite velocities because they are moving in opposite directions.
3. The constant speedometer reading indicates a constant speed but not a constant velocity, because the car may not be moving along a straight-line path, in which case it is accelerating.

FIGURE 3.4
We say that a body undergoes acceleration when there is a *change* in its state of motion.

We are familiar with acceleration in an automobile. When the driver depresses the gas pedal (appropriately called the accelerator), the passengers then experience acceleration (or "pickup," as it is sometimes called) as they are pressed against their seats. The key idea that defines acceleration is *change*. Suppose we are driving and, in 1 second, we steadily increase our velocity from 30 kilometers per hour to 35 kilometers per hour, and then to 40 kilometers per hour in the next second, to 45 in the next second, and so on. We change our velocity by 5 kilometers per hour each second. This change in velocity is what we mean by acceleration.

$$\text{Acceleration} = \frac{\text{change in velocity}}{\text{time interval}} = \frac{5\text{ km/h}}{1\text{ s}} = 5\text{ km/h·s}$$

In this case, the acceleration is 5 kilometers per hour second (abbreviated as 5 km/h·s). Note that a unit for time enters twice: once for the unit of velocity and again for the interval of time in which the velocity is changing. Also note that acceleration is not just the total change in velocity; it is the *time rate of change,* or *change per second,* in velocity.

CHECK YOURSELF

1. A particular car can go from rest to 90 km/h in 10 s. What is its acceleration?
2. In 2.5 s, a car increases its speed from 60 km/h to 65 km/h while a bicycle goes from rest to 5 km/h. Which undergoes the greater acceleration? What is the acceleration of each?

The term *acceleration* applies to decreases as well as to increases in velocity. We say the brakes of a car, for example, produce large retarding accelerations; that is, there is a large decrease per second in the velocity of the car. We often call this *deceleration.* We experience deceleration when the driver of a bus or car applies the brakes and we tend to lurch forward.

We accelerate whenever we move in a curved path, even if we are moving at constant speed, because our direction is changing—hence, our velocity is changing. We experience this acceleration as we tend to lurch toward the outer

CHECK YOUR ANSWERS

1. Its acceleration is 9 km/h·s. Strictly speaking, this would be its average acceleration, for there may have been some variation in its rate of picking up speed.
2. The accelerations of both the car and the bicycle are the same: 2 km/h·s.

$$\text{Acceleration}_{\text{car}} = \frac{\text{change of velocity}}{\text{time interval}} = \frac{65\text{ km/h} - 60\text{ km/h}}{2.5\text{ s}} = \frac{5\text{ km/h}}{2.5\text{ s}} = 2\text{ km/h·s}$$

$$\text{Acceleration}_{\text{bike}} = \frac{\text{change of velocity}}{\text{time interval}} = \frac{5\text{ km/h} - 0\text{ km/h}}{2.5\text{ s}} = \frac{5\text{ km/h}}{2.5\text{ s}} = 2\text{ km/h·s}$$

Although the velocities are quite different, the rates of change of velocity are the same. Hence, the accelerations are equal.

FIGURE 3.5
Rapid deceleration is sensed by the driver, who lurches forward (in accord with Newton's first law).

There are three controls that change velocity in an automobile: the gas pedal (accelerator), the brakes, and the steering wheel.

part of the curve. We distinguish speed and velocity for this reason and define *acceleration* as the rate at which velocity changes, thereby encompassing changes both in speed and in direction.

Anyone who has stood in a crowded bus has experienced the difference between velocity and acceleration. Except for the effects of a bumpy road, you can stand with no extra effort inside a bus that moves at constant velocity, no matter how fast it is going. You can flip a coin and catch it exactly as if the bus were at rest. It is only when the bus accelerates—speeds up, slows down, or turns—that you experience difficulty standing.

In much of this book, we will be concerned only with motion along a straight line. When straight-line motion is being considered, it is common to use *speed* and *velocity* interchangeably. When direction doesn't change, acceleration may be expressed as the rate at which *speed* changes.

$$\text{Acceleration (along a straight line)} = \frac{\text{change in speed}}{\text{time interval}}$$

CHECK YOURSELF

1. What is the acceleration of a race car that whizzes past you at a constant velocity of 400 km/h?
2. Which has the greater acceleration, an airplane that goes from 1000 km/h to 1005 km/h in 10 seconds or a skateboard that goes from zero to 5 km/h in 1 second?

CHECK YOUR ANSWERS

1. Zero, because its velocity doesn't change.
2. Both gain 5 km/h, but the skateboard does so in one-tenth the time. The skateboard therefore has the greater acceleration—in fact, ten times greater. A little figuring will show that the acceleration of the airplane is 0.5 km/h·s, while acceleration of the slower-moving skateboard is 5 km/h·s. Velocity and acceleration are very different concepts. Distinguishing between them is very important.

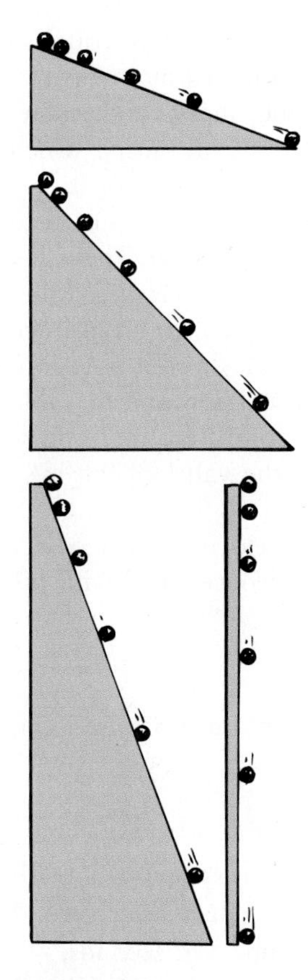

FIGURE 3.6

Interactive Figure

The greater the slope of the incline, the greater the acceleration of the ball. What is its acceleration if the incline is vertical?

Acceleration on Galileo's Inclined Planes

Galileo developed the concept of acceleration in his experiments on inclined planes. His main interest was falling objects, and, because he lacked suitable timing devices, he used inclined planes effectively to slow accelerated motion and to investigate it more carefully.

Galileo found that a ball rolling down an inclined plane picks up the same amount of speed in successive seconds; that is, the ball rolls with unchanging acceleration. For example, a ball rolling down a plane inclined at a certain angle might be found to pick up a speed of 2 meters per second for each second it rolls. This gain per second is its acceleration. Its instantaneous velocity at 1-second intervals, at this acceleration, is then 0, 2, 4, 6, 8, 10, and so forth, meters per second. We can see that the instantaneous speed or velocity of the ball at any given time after being released from rest is simply equal to its acceleration multiplied by the time:[2]

$$\text{Velocity acquired} = \text{acceleration} \times \text{time}$$

If we substitute the acceleration of the ball in this relationship (2 meters per second squared), we can see that, at the end of 1 second, the ball is traveling at 2 meters per second; at the end of 2 seconds, it is traveling at 4 meters per second; at the end of 10 seconds, it is traveling at 20 meters per second; and so on. The instantaneous speed or velocity at any time is simply equal to the acceleration multiplied by the number of seconds it has been accelerating.

Galileo found greater accelerations for steeper inclines. The ball attains its maximum acceleration when the incline is tipped vertically. Then the acceleration is the same as that of a falling object (Figure 3.6). Regardless of the weight or size of the object, Galileo discovered that, when air resistance is small enough to be neglected, all objects fall with the same unchanging acceleration.

Free Fall

Free Fall: How Fast?

How Fast

Things fall because of the force of gravity. When a falling object is free of all restraints—no friction, with the air or otherwise—and falls under the influence of gravity alone, the object is in a state of **free fall.** (We'll consider the effects of air resistance on falling objects in Chapter 4.) Table 3.2 shows the instantaneous speed of a freely falling object at 1-second intervals. The important thing to note in these numbers is the way in which the speed changes. *During each second of fall, the object gains a speed of 10 meters per second.* This gain

[2]Note that this relationship follows from the definition of acceleration. From $a = v/t$, simple rearrangement (multiplying both sides of the equation by t) gives $v = at$.

TABLE 3.2
Free-Fall from Rest

Time of Fall (seconds)	*Velocity Acquired (meters/second)*
0	0
1	10
2	20
3	30
4	40
5	50
.	.
.	.
.	.
t	$10t$

per second is the acceleration. Free-fall acceleration is approximately equal to 10 meters per second each second, or, in shorthand notation, 10 m/s^2 (read as 10 meters per second squared). Note that the unit of time, the second, enters twice—once for the unit of speed and again for the time interval during which the speed changes.

In the case of freely falling objects, it is customary to use the letter g to represent the acceleration (because the acceleration is due to *gravity*). The value of g is very different on the surface of the Moon and on the surfaces of other planets. Here on Earth, g varies slightly in different locations, with an average value equal to 9.8 meters per second each second, or, in shorter notation, 9.8 m/s^2. We round this off to 10 m/s^2 in our present discussion and in Table 3.2 to establish the ideas involved more clearly; multiples of 10 are more obvious than multiples of 9.8. Where accuracy is important, the value of 9.8 m/s^2 should be used.

Note in Table 3.2 that the instantaneous speed or velocity of an object falling from rest is consistent with the equation that Galileo deduced with his inclined planes:

$$\text{Velocity acquired} = \text{acceleration} \times \text{time}$$

The instantaneous velocity v of an object falling from rest[3] after a time t can be expressed in shorthand notation as

$$v = gt$$

THE Physics Place
$v = gt$

To see that this equation makes good sense, take a moment to check it with Table 3.2. Note that the instantaneous velocity or speed in meters per second is simply the acceleration $g = 10$ m/s^2 multiplied by the time t in seconds.

Free-fall acceleration is clearer when we consider a falling object equipped with a speedometer (Figure 3.7). Suppose a rock is dropped from a high cliff and you witness it with a telescope. If you focus the telescope on the speedometer, you'd note increasing speed as time progresses. By how much? The answer is, by 10 m/s each succeeding second.

CHECK YOURSELF

What would the speedometer reading on the falling rock shown in Figure 3.7 be 5 s after it drops from rest? How about 6 s after it is dropped? 6.5 s after it is dropped?

CHECK YOUR ANSWERS

The speedometer readings would be 50 m/s, 60 m/s, and 65 m/s, respectively. You can reason this from Table 3.2 or use the equation $v = gt$, where g is 10 m/s^2.

Free Fall: How Far?

[3]If, instead of being dropped from rest, the object is thrown downward at speed v_o, the speed v after any elapsed time t is $v = v_o + gt$. We will not be concerned with this added complication here; we will instead learn as much as we can from the simplest cases. That will be a lot!

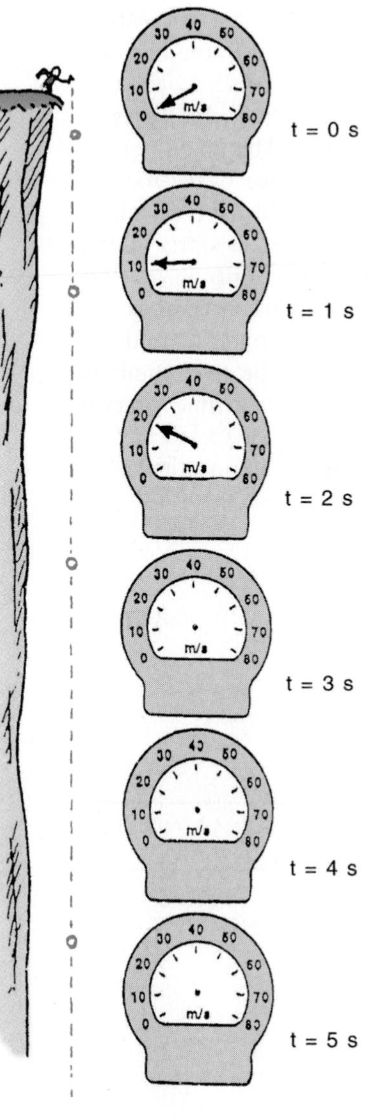

FIGURE 3.7
Pretend that a falling rock is equipped with a speedometer. In each succeeding second of fall, you'd find the rock's speed increasing by the same amount: 10 m/s. Sketch in the missing speedometer needle at t = 3 s, 4 s, and 5 s. (Table 3.2 shows the speeds we would read at various seconds of fall.)

So far, we have been considering objects moving straight downward in the direction of the pull of gravity. How about an object thrown straight upward? Once released, it continues to move upward for a time and then comes back down. At its highest point, when it is changing its direction of motion from upward to downward, its instantaneous speed is zero. Then it starts downward *just as if it had been dropped from rest at that height.*

During the upward part of this motion, the object slows as it rises. It should come as no surprise that it slows at the rate of 10 meters per second each second—the same acceleration it experiences on the way down. So, as Figure 3.8 shows, the instantaneous speed at points of equal elevation in the path is the same whether the object is moving upward or downward. The velocities are opposite, of course, because they are in opposite directions. Note that the downward velocities have a negative sign, indicating the downward direction (it is customary to call *up* positive, and *down* negative.) Whether moving upward or downward, the acceleration is 10 m/s^2 the whole time.

CHECK YOURSELF

A ball is thrown straight upward and leaves your hand at 20 m/s. What predictions can you make about the ball? (Please think about this *before* reading the suggested predictions!)

How Far

How *far* an object falls is altogether different from how *fast* it falls. With his inclined planes, Galileo found that the distance a uniformly accelerating object travels is proportional to the *square of the time.* The details of this relationship are in Appendix B. We will state here only the results. The distance traveled by a uniformly accelerating object starting from rest is

$$\text{Distance traveled} = 1/2\ (\text{acceleration} \times \text{time} \times \text{time})$$

This relationship applies to the distance something falls. We can express it, for the case of a freely falling object, in shorthand notation as

$$d = 1/2\ gt^2$$

CHECK YOUR PREDICTIONS

There are several. One prediction is that it will slow to 10 m/s one second after it leaves your hand and will come to a momentary stop two seconds after leaving your hand, when it reaches the top of its path. This is because it loses 10 m/s each second going up. Another prediction is that one second later, three seconds total, it will be moving downward at 10 m/s. In another second, it will return to its starting point and be moving at 20 m/s. So the time each way is two seconds, and its total time in flight takes four seconds. We'll now treat how far it travels up and down.

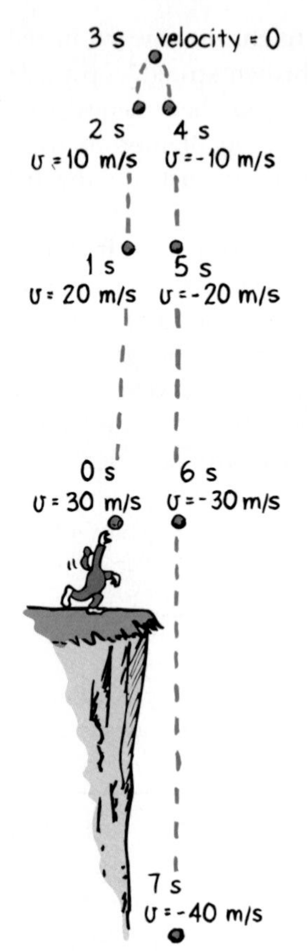

FIGURE 3.8

Interactive Figure

The rate at which the velocity changes each second is the same.

TABLE 3.3

Distance Fallen in Free Fall

Time of Fall (seconds)	*Distance Fallen (meters)*
0	0
1	5
2	20
3	45
4	80
5	125
.	.
.	.
.	.
t	$\frac{1}{2}10\,t^2$

in which d is the distance something falls when the time of fall in seconds is substituted for t and squared.[4] If we use 10 m/s^2 for the value of g, the distance fallen for various times will be as shown in Table 3.3.

Note that an object falls a distance of only 5 meters during the first second of fall, although its speed is then 10 meters per second. This may be confusing, for we may think that the object should fall a distance of 10 meters. But for it to fall 10 meters in its first second of fall, it would have to fall at an *average* speed of 10 meters per second for the entire second. It starts its fall at 0 meters per second, and its speed is 10 meters per second only in the last instant of the 1-second interval. Its average speed during this interval is the average of its initial and final speeds, 0 and 10 meters per second. To find the average value of these or any two numbers, we simply add the two numbers and divide by 2. This equals 5 meters per second, which, over a time interval of 1 second, gives a distance of 5 meters. As the object continues to fall in succeeding seconds, it will fall through ever-increasing distances because its speed is continuously increasing.

CHECK YOURSELF

A cat steps off a ledge and drops to the ground in 1/2 second.

a. What is its speed on striking the ground?

b. What is its average speed during the 1/2 second?

c. How high is the ledge from the ground?

CHECK YOUR ANSWERS

a. Speed: $v = gt = 10\text{ m/s}^2 \times 1/2\text{ s} = 5\text{ m/s}$.

b. Average speed: $\bar{v} = \dfrac{\text{initial } v + \text{final } v}{2} = \dfrac{0\text{ m/s} + 5\text{m/s}}{2} = 2.5\text{ m/s}$.

We put a bar over the symbol to denote average speed: $\bar{v}$.

c. Distance: $d = vt = 2.5\text{ m/s} \times 1/2\text{ s} = 1.25\text{ m}$.

Or equivalently,

$$d = 1/2\,gt^2 = 1/2 \times 10\text{ m/s}^2 = (1/2\text{ s})^2 = 1/2 \times 10\text{ m/s}^2 \times 1/4\text{ s}^2 = 1.25\text{ m}$$

Notice that we can find the distance by either of these equivalent relationships.

[4] d = *average velocity* × *time*

$$d = \frac{\textit{initial velocity} + \textit{final velocity}}{2} \times \textit{time}$$

$$d = \frac{0 + gt}{2} \times t$$

$$d = 1/2\,gt^2$$

(See Appendix B for further explanation.)

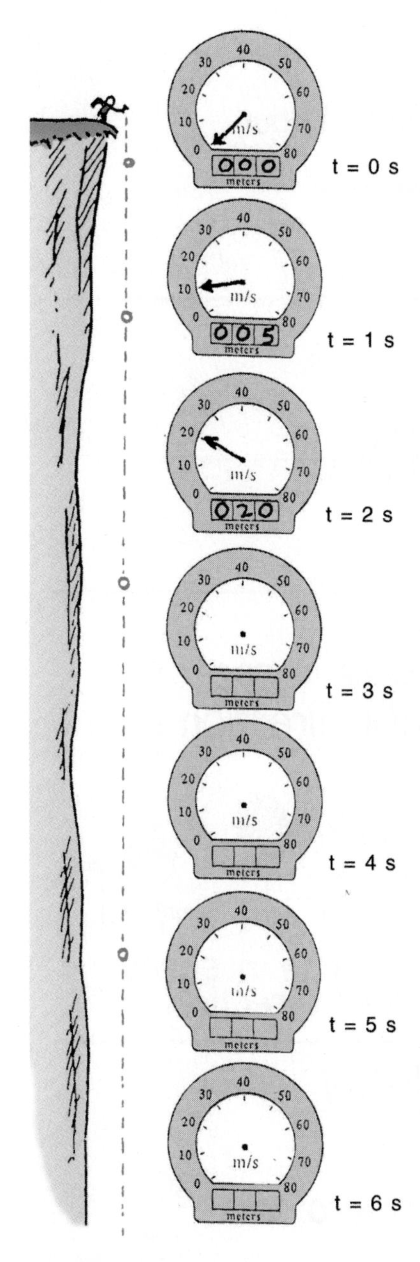

FIGURE 3.9
Pretend that a falling rock is equipped with a speedometer and an odometer. Speed readings increase by 10 m/s and distance readings by $\frac{1}{2}gt^2$. Can you complete the speedometer positions and odometer readings?

It is a common observation that many objects fall with unequal accelerations. A leaf, a feather, or a sheet of paper may flutter to the ground slowly. The fact that air resistance is responsible for these different accelerations can be shown very nicely with a closed glass tube containing light and heavy objects—a feather and a coin, for example. In the presence of air, the feather and coin fall with quite different accelerations. But, if the air in the tube is removed by a vacuum pump and the tube is quickly inverted, the feather and coin fall with the same acceleration (Figure 3.10). Although air resistance appreciably alters the motion of things like falling feathers, the motion of heavier objects like stones and baseballs at ordinary low speeds is not appreciably affected by the air. The relationships $v = gt$ and $d = 1/2\, gt^2$ can be used to a very good approximation for most objects falling in air from an initial position of rest.

FIGURE 3.10
A feather and a coin fall at equal accelerations in a vacuum.

How Quickly "How Fast" Changes

Much of the confusion that arises in analyzing the motion of falling objects comes about because it is easy to get "how fast" mixed up with "how far." When we wish to specify how fast something is falling, we are talking about *speed* or *velocity,* which is expressed as $v = gt$. When we wish to specify how far something falls, we are talking about *distance,* which is expressed as $d = 1/2\, gt^2$. Speed or velocity (how fast) and distance (how far) are entirely different from each other.

A most confusing concept, and probably the most difficult encountered in this book, is "how quickly does how fast change"—acceleration. What makes acceleration so complex is that it is *a rate of a rate*. It is often confused with velocity, which is itself a rate (the rate of change of position). Acceleration is not velocity, nor is it even a change in velocity. Acceleration is the rate at which velocity itself changes.

Please remember that it took people nearly 2000 years from the time of Aristotle to reach a clear understanding of motion, so be patient with yourself if you find that you require a few hours to achieve as much!

Velocity = {speed *and* direction}

San Francisco

E

Velocity = 300 km/h, east

Acceleration = {Rate of change in velocity} *due to* {change in speed and/or direction}

Change in speed but *not* direction

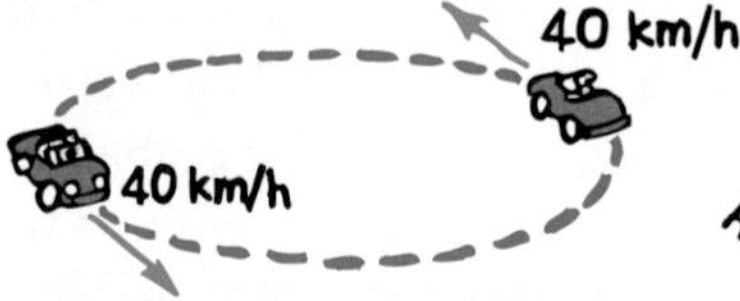

Change in direction but *not* speed

Change in speed *and* direction

Acceleration = change in velocity / time

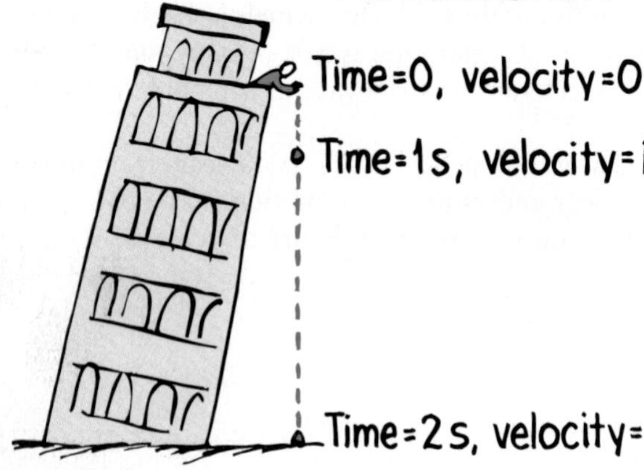

$$\text{Acceleration} = \frac{20 \text{ m/s}}{2 \text{ s}}$$

$$a = 10 \frac{\text{m/s}}{\text{s}}$$

$$a = 10 \text{ m/ss}$$

$$a = 10 \text{ m/s}^2$$

FIGURE 3.11
Motion analysis.

HANG TIME

Some athletes and dancers have great jumping ability. Leaping straight up, they seem to "hang in the air," defying gravity. Ask your friends to estimate the "hang time" of the great jumpers—the time a jumper is airborne with feet off the ground. They may say two or three seconds. But, surprisingly, the hang time of the greatest jumpers is almost always less than 1 second! A longer time is one of many illusions we have about nature.

A related illusion is the vertical height a human can jump. Most of your classmates probably cannot jump higher than 0.5 meter. They can step over a 0.5-meter fence, but, in doing so, their body rises only slightly. The height of the barrier is different than the height a jumper's "center of gravity" rises. Many people can leap over a 1-meter fence, but only rarely does anybody raise the "center of gravity" of their body 1 meter. Even basketball star Michael Jordan in his prime couldn't raise his body 1.25 meters high, although he could easily reach considerably above the more-than-3-meter-high basket.

Jumping ability is best measured by a standing vertical jump. Stand facing a wall with feet flat on the floor and arms extended upward. Make a mark on the wall at the top of your reach. Then make your jump and, at the peak, make another mark. The distance between these two marks measures your vertical leap. If it's more than 0.6 meters (2 feet), you're exceptional.

Here's the physics. When you leap upward, jumping force is applied only while your feet make contact with the ground. The greater the force, the greater your launch speed and the higher the jump. When your feet leave the ground, your upward speed immediately decreases at the steady rate of g—10 m/s^2. At the top of your jump, your upward speed decreases to zero. Then you begin to fall, gaining speed at exactly the same rate, g. If you land as you took off, upright with legs extended, then time rising equals time falling; hang time is time up plus time down. While airborne, no amount of leg or arm pumping or other bodily motions can change your hang time.

The relationship between time up or down and vertical height is given by

$$d = \frac{1}{2}gt^2$$

If we know the vertical height d, we can rearrange this expression to read

$$t = \sqrt{\frac{2d}{g}}$$

American basketball star Spud Webb recorded a vertical standing jump of 1.25 meters back in 1986.[5] This was the world record at that time. Let's use his jumping height of 1.25 meters for d, and use the more precise value of 9.8 m/s^2 for g. Solving for t, half the hang time, we get

$$t = \sqrt{\frac{2d}{g}} = \sqrt{\frac{2(1.25\text{ m})}{9.8\text{ m/s}^2}} = 0.50\text{ s}$$

Double this (because this is the time for one way of an up-and-down round trip) and we see that Spud's record-breaking hang time is 1 second.

We're talking here of vertical motion. How about running jumps? Hang time depends only on the jumper's vertical speed at launch. While airborne, the jumper's horizontal speed remains constant while the vertical speed undergoes acceleration. Interesting physics!

[5]The value of 1.25 m for *d* represents the maximum height of the jumper's center of gravity, not the height of the bar. The height gained by the jumper's center of gravity is what's important in determining jumping ability. We'll learn about center of gravity in Chapter 8.

Summary of Terms

Speed How fast something moves; the distance traveled per unit of time.

Velocity The speed of an object and a specification of its direction of motion.

Vector quantity Quantity in physics that has both magnitude and direction.

Acceleration The rate at which velocity changes with time; the change in velocity may be in magnitude, or direction, or both.

Free Fall Motion under the influence of gravity only.

Summary of Formulas

$$\text{Speed} = \frac{\text{distance}}{\text{time}}$$

$$\text{Average speed} = \frac{\text{total distance covered}}{\text{time interval}}$$

$$\text{Acceleration} = \frac{\text{change of velocity}}{\text{time interval}}$$

$$\text{Acceleration (along a straight line)} = \frac{\text{change in speed}}{\text{time interval}}$$

Velocity acquired in free fall, from rest; $v = gt$

Distance fallen in free fall, from rest; $d = 1/2\, gt^2$

Review Questions

Motion Is Relative

1. As you read this, how fast are you moving relative to the chair you are sitting on? Relative to the Sun?

Speed

2. What two units of measurement are necessary for describing speed?

Instantaneous Speed

3. What kind of speed is registered by an automobile speedometer—average speed or instantaneous speed?

Average Speed

4. Distinguish between instantaneous speed and average speed.
5. What is the average speed in kilometers per hour for a horse that gallops a distance of 15 km in a time of 30 min?
6. How far does a horse travel if it gallops at an average speed of 25 km/h for 30 min?

Velocity

7. Distinguish between speed and velocity.

Constant Velocity

8. If a car moves with a constant velocity, does it also move with a constant speed?

Changing Velocity

9. If a car is moving at 90 km/h and it rounds a corner, also at 90 km/h, does it maintain a constant speed? A constant velocity? Defend your answer.

Acceleration

10. Distinguish between velocity and acceleration.
11. What is the acceleration of a car that increases its velocity from 0 to 100 km/h in 10 s?
12. What is the acceleration of a car that maintains a constant velocity of 100 km/h for 10 s? (Why do some of your classmates who correctly answer the previous question get this question wrong?)
13. When are you most aware of motion in a moving vehicle—when it is moving steadily in a straight line or when it is accelerating? If a car moved with absolutely constant velocity (no bumps at all), would you be aware of motion?
14. Acceleration is generally defined as the time rate of change of velocity. When can it be defined as the time rate of change of speed?

Acceleration on Galileo's Inclined Planes

15. What did Galileo discover about the amount of speed a ball gained each second when rolling down an inclined plane? What did this say about the ball's acceleration?
16. What relationship did Galileo discover for the velocity acquired on an incline?
17. What relationship did Galileo discover about a ball's acceleration and the steepness of an incline? What acceleration occurs when the plane is vertical?

Free Fall

How Fast

18. What exactly is meant by a "freely falling" object?
19. What is the gain in speed per second for a freely falling object?
20. What is the velocity acquired by a freely falling object 5 s after being dropped from a rest position? What is the velocity 6 s after?
21. The acceleration of free fall is about 10 m/s^2. Why does the seconds unit appear twice?
22. When an object is thrown upward, how much speed does it lose each second?

How Far

23. What relationship between distance traveled and time did Galileo discover for accelerating objects?
24. What is the distance fallen for a freely falling object 1 s after being dropped from a rest position? What is it 4 s after?
25. What is the effect of air resistance on the acceleration of falling objects? What is the acceleration with no air resistance?

How Quickly "How Fast" Changes

26. Consider these measurements: 10 m, 10 m/s, and 10 m/s^2. Which is a measure of distance, which of speed, and which of acceleration?

Projects

1. Grandma is interested in your educational progress. Like most grandmothers, she likely has little science background and may be mathematically challenged. Write a letter to Grandma, without using equations, and explain to her the difference between velocity and acceleration. Tell her why some of your classmates confuse the two, and state some examples that clear up the confusion.
2. Stand flatfooted next to a wall and make a mark at the highest point you can reach. Then jump vertically and mark this highest point. The distance between the marks is your vertical jumping distance. Use this to calculate your hang time.

One-Step Calculations

These are "plug-in-the-number" type activities to familiarize you with the equations that link the concepts of physics. They are mainly one-step substitutions and are less challenging than the Problems.

$$\textbf{Speed} = \frac{\textbf{distance}}{\textbf{time}}$$

1. Calculate your walking speed when you step 1 meter in 0.5 second.
2. Calculate the speed of a bowling ball that travels 4 meters in 2 seconds.

$$\textbf{Average speed} = \frac{\textbf{total distance covered}}{\textbf{time interval}}$$

3. Calculate your average speed if you run 50 meters in 10 seconds.
4. Calculate the average speed of a tennis ball that travels the full length of the court, 24 meters, in 0.5 second.
5. Calculate the average speed of a cheetah that runs 140 meters in 5 seconds.
6. Calculate the average speed (in km/h) of Larry, who runs to the store 4 kilometers away in 30 minutes.

Distance = average speed × time

7. Calculate the distance (in km) that Larry runs if he maintains an average speed of 8 km/h for 1 hour.
8. Calculate the distance you will travel if you maintain an average speed of 10 m/s for 40 seconds.
9. Calculate the distance you will travel if you maintain an average speed of 10 km/h for one-half hour.

$$\textbf{Acceleration} = \frac{\textbf{change of velocity}}{\textbf{time interval}}$$

10. Calculate the acceleration of a car (in km/h·s) that can go from rest to 100 km/h in 10 s.
11. Calculate the acceleration of a bus that goes from 10 km/h to a speed of 50 km/h in 10 seconds.
12. Calculate the acceleration of a ball that starts from rest, rolls down a ramp, and gains a speed of 25 m/s in 5 seconds.
13. On a distant planet, a freely falling object gains speed at a steady rate of 20 m/s during each second of fall. Calculate its acceleration.

Instantaneous speed = acceleration × time

14. Calculate the instantaneous speed (in m/s) at the 10-second mark for a car that accelerates at 2 m/s^2 from a position of rest.
15. Calculate the speed (in m/s) of a skateboarder who accelerates from rest for 3 seconds down a ramp at an acceleration of 5 m/s^2.

Velocity acquired in free fall, from rest;
$v = gt$ **(where** $g = 10\ m/s^2$**)**

16. Calculate the instantaneous speed of an apple that falls freely from a rest position and accelerates at 10 m/s^2 for 1.5 seconds.
17. An object is dropped from rest and falls freely. After 7 seconds, calculate its instantaneous speed.
18. A skydiver steps from a high-flying helicopter. In the absence of air resistance, how fast would she be falling at the end of a 12-second jump?
19. On a distant planet, a freely falling object has an acceleration of 20 m/s^2. Calculate the speed that an object dropped from rest on this planet acquires in 1.5 seconds.

Distance fallen in free fall, from rest; $d = 1/2\ gt^2$

20. An apple drops from a tree and hits the ground in 1.5 seconds. Calculate how far it falls.
21. Calculate the vertical distance an object dropped from rest covers in 12 seconds of free fall.
22. On a distant planet a freely falling object has an acceleration of 20 m/s^2. Calculate the vertical distance an object dropped from rest on this planet covers in 1.5 seconds.

Exercises

1. What is the impact speed when a car moving at 100 km/h bumps into the rear of another car traveling in the same direction at 98 km/h?
2. Harry Hotshot can paddle a canoe in still water at 8 km/h. How successful will he be at canoeing upstream in a river that flows at 8 km/h?

3. Is a fine for speeding based on one's average speed or one's instantaneous speed? Explain.
4. One airplane travels due north at 300 km/h while another travels due south at 300 km/h. Are their speeds the same? Are their velocities the same? Explain.
5. Light travels in a straight line at a constant speed of 300,000,000 m/s. What is the acceleration of light?
6. Can an automobile with a velocity toward the north simultaneously have an acceleration toward the south? Explain.
7. You're in a car traveling at some specified speed limit. You see a car moving at the same speed coming toward you. How fast is the car approaching you, compared with the speed limit?
8. For straight-line motion, how does a speedometer indicate whether or not acceleration is occurring?
9. Can an object reverse its direction of travel while maintaining a constant acceleration? If so, give an example. If not, provide an explanation.
10. You are driving north on a highway. Then, without changing speed, you round a curve and drive east. (a) Does your velocity change? (b) Do you accelerate? Explain.
11. Correct your friend who says, "The dragster rounded the curve at a constant velocity of 100 km/h."
12. Harry says acceleration is how fast you go. Carol says acceleration is how fast you get fast. They look to you for confirmation. Who's correct?
13. Starting from rest, one car accelerates to a speed of 50 km/h, and another car accelerates to a speed of 60 km/h. Can you say which car underwent the greater acceleration? Why or why not?
14. Cite an instance in which your speed could be zero while your acceleration is nonzero.
15. Cite an example of something with a constant speed that also has a varying velocity. Can you cite an example of something with a constant velocity and a varying speed? Defend your answers.
16. Cite an example of something that undergoes acceleration while moving at constant speed. Can you also give an example of something that accelerates while traveling at constant velocity? Explain.
17. (a) Can an object be moving when its acceleration is zero? If so, give an example. (b) Can an object be accelerating when its speed is zero? If so, give an example.
18. Can you cite an example in which the acceleration of a body is opposite in direction to its velocity? If so, what is your example?
19. On which of these hills does the ball roll down with increasing speed and decreasing acceleration along the path? (Use this example if you wish to explain to someone the difference between speed and acceleration.)

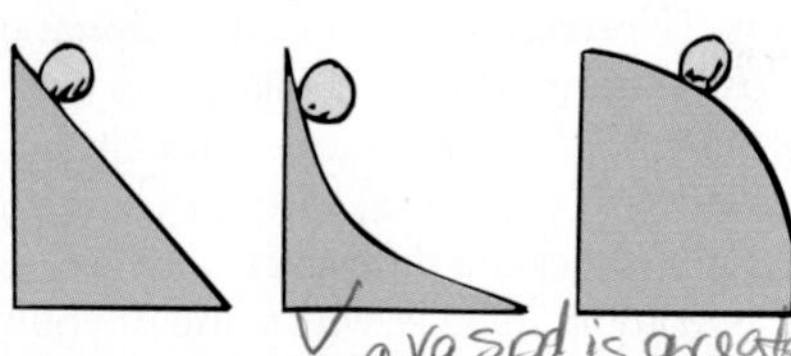

20. Suppose that the three balls shown in Exercise 19 start simultaneously from the tops of the hills. Which one reaches the bottom first? Explain.
21. What is the acceleration of a car that moves at a steady velocity of 100 km/h for 100 seconds? Explain your answer.
22. Which is greater, an acceleration from 25 km/h to 30 km/h or one from 96 km/h to 100 km/h if both occur during the same time?
23. Galileo experimented with balls rolling on inclined planes with angles ranging from 0° to 90°. What range of accelerations corresponds to this range in angles (from what acceleration to what)?
24. Be picky and correct your friend who says, "In free fall, air resistance is more effective in slowing a feather than a coin."
25. Suppose that a freely falling object were somehow equipped with a speedometer. By how much would its reading in speed increase with each second of fall?
26. Suppose that the freely falling object in the preceding exercise were also equipped with an odometer. Would the readings of distance fallen each second indicate equal or different falling distances for successive seconds?
27. For a freely falling object dropped from rest, what is the acceleration at the end of the fifth second of fall? At the end of the tenth second of fall? Defend your answers.
28. If air resistance can be neglected, how does the acceleration of a ball that has been tossed straight upward compare with its acceleration if simply dropped?
29. When a ballplayer throws a ball straight up, by how much does the speed of the ball decrease each second while ascending? In the absence of air resistance, by how much does it increase each second while descending? How much time is required for rising compared to falling?
30. Someone standing at the edge of a cliff (as in Figure 3.8) throws a ball nearly straight up at a certain speed and another ball nearly straight down with the same initial speed. If air resistance is negligible, which ball will have the greater speed when it strikes the ground below?
31. Answer the previous question for the case where air resistance is *not* negligible—where air drag affects motion.

32. If you drop an object, its acceleration toward the ground is 10 m/s^2. If you throw it down instead, would its acceleration after throwing be greater than 10 m/s^2? Why or why not?

33. In the preceding exercise, can you think of a reason why the acceleration of the object thrown downward through the air might be appreciably less than 10 m/s^2?

34. While rolling balls down an inclined plane, Galileo observes that the ball rolls 1 cubit (the distance from elbow to fingertip) as he counts to ten. How far will the ball have rolled from its starting point when he has counted to twenty?

35. Consider a vertically launched projectile when air drag is negligible. When is the acceleration due to gravity greater? When ascending, at the top, or when descending? Defend your answer.

36. If it were not for air resistance, why would it be dangerous to go outdoors on rainy days?

37. Extend Tables 3.2 and 3.3 to include times of fall of 6 to 10 seconds, assuming no air resistance.

38. As speed increases for an object in free fall, does acceleration increase also?

39. A ball tossed upward will return to the same point with the same initial speed when air resistance is negligible. When air resistance is not negligible, how does the return speed compare with its initial speed?

40. Two balls are released simultaneously from rest at the left end of equal-length tracks A and B as shown. Which ball reaches the end of its track first?

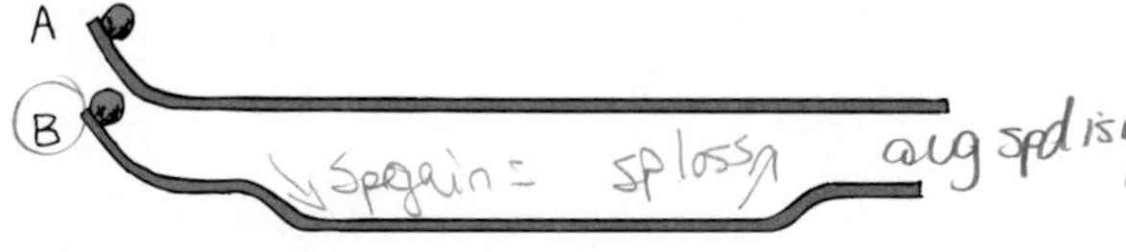

41. Refer to the pair of tracks in exercise 40. (a) On which track is the average speed greater? (b) Why is the speed of the ball at the end of the tracks the same?

42. In this chapter, we studied idealized cases of balls rolling down smooth planes and objects falling with no air resistance. Suppose a classmate complains that all this attention focused on idealized cases is valueless because idealized cases simply don't occur in the everyday world. How would you respond to this complaint? How do you suppose the author of this book would respond?

43. Why does a stream of water get narrower as it falls from a faucet?

44. A person's hang time would be considerably greater on the Moon. Why?

45. Make up two multiple-choice questions that would check a classmate's understanding of the distinction between velocity and acceleration.

Problems

1. The ocean's level is currently rising at about 1.5 mm per year. At this rate, in how many years will sea level be 3 meters higher than now?

2. What is the acceleration of a vehicle that changes its velocity from 100 km/h to a dead stop in 10 s?

3. A ball is thrown straight up with an initial speed of 30 m/s. How high does it go, and how long is it in the air (neglecting air resistance)?

4. A ball is thrown with enough speed straight up so that it is in the air several seconds. (a) What is the velocity of the ball when it reaches its highest point? (b) What is its velocity 1 s before it reaches its highest point? (c) What is the change in its velocity during this 1-s interval? (d) What is its velocity 1 s after it reaches its highest point? (e) What is the change in velocity during this 1-s interval? (f) What is the change in velocity during the 2-s interval? (Careful!) (g) What is the acceleration of the ball during any of these time intervals and at the moment the ball has zero velocity?

5. What is the instantaneous velocity of a freely falling object 10 s after it is released from a position of rest? What is its average velocity during this 10-s interval? How far will it fall during this time?

6. A car takes 10 s to go from $v = 0$ m/s to $v = 25$ m/s at approximately constant acceleration. If you wish to find the distance traveled using the equation $d = \frac{1}{2}at^2$, what value should you use for a?

7. A reconnaissance plane flies 600 km away from its base at 200 km/h, then flies back to its base at 300 km/h. What is its average speed?

8. A car travels along a certain road at an average speed of 40 km/h and returns along the same road at an average speed of 60 km/h. Calculate the average speed for the round trip. (Don't say 50 km/h!)

9. If there were no air drag, how fast would drops fall from a cloud 1 kilometer above the Earth's surface? (Be glad that raindrops experience air drag in fall!)

10. Surprisingly, very few athletes can jump more than 2 feet (0.6 m) straight up. Use $d = \frac{1}{2}gt^2$ and solve for the time one spends moving upward in a 2-foot vertical jump. Then double it for the "hang time"—the time one's feet are off the ground.

CHAPTER 4

Newton's Second Law of Motion

Efrain Lopez shows that when the forces balance to zero, no acceleration occurs.

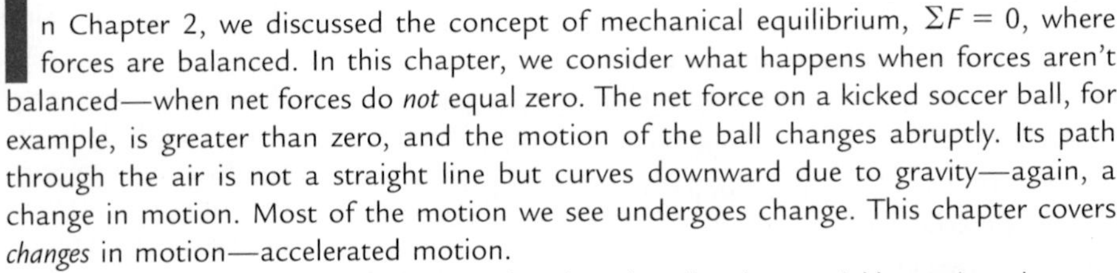

In Chapter 2, we discussed the concept of mechanical equilibrium, $\Sigma F = 0$, where forces are balanced. In this chapter, we consider what happens when forces aren't balanced—when net forces do *not* equal zero. The net force on a kicked soccer ball, for example, is greater than zero, and the motion of the ball changes abruptly. Its path through the air is not a straight line but curves downward due to gravity—again, a change in motion. Most of the motion we see undergoes change. This chapter covers *changes* in motion—accelerated motion.

In Chapter 3, we learned that acceleration describes how quickly motion changes. Specifically, it is the change in velocity during a certain time interval. Recall the definition of acceleration:

$$\text{acceleration} = \frac{\text{change in velocity}}{\text{time interval}}$$

We will now focus on the *cause* of acceleration: *force.*

Force Causes Acceleration

Consider a hockey puck at rest on ice. Apply a **force,** and it starts to move—it accelerates. When the hockey stick is no longer pushing it, the puck moves at constant velocity. Apply another force by striking the puck again, and again the motion changes. Applied force produces acceleration.

FIGURE 4.1
Kick the ball and it accelerates.

Most often, the applied force is not the only force acting on an object. Other forces may act as well. Recall, from Chapter 2, that the combination of forces acting on an object is the *net force*. Acceleration depends on the *net force*. To increase the acceleration of an object, you must increase the net force acting on it. If you double the net force on an object, its acceleration doubles; if you triple the net force, its acceleration triples; and so on. This makes good sense. We say an object's acceleration is directly proportional to the net force acting on it. We write

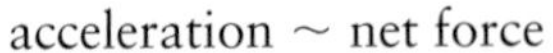

$$\text{acceleration} \sim \text{net force}$$

The symbol ~ stands for "is directly proportional to." That means any change in one is the same amount of change in the other.

CHECK YOURSELF

1. You push on a crate that sits on a smooth floor, and it accelerates. If you apply four times the net force, how much greater will be the acceleration?
2. If you push with the same increased force on the same crate, but it slides on a very rough floor, how will the acceleration compare with pushing the crate on a smooth floor? *(Think before you read the answer below!)*

Friction

When surfaces slide or tend to slide over one another, a force of **friction** acts. When you apply a force to an object, friction usually reduces the net force and the resulting acceleration. Friction is caused by the irregularities in the surfaces in mutual contact, and it depends on the kinds of material and how much they are pressed together. Even surfaces that appear to be very smooth have microscopic irregularities that obstruct motion. Atoms cling together at many points of contact. When one object slides against another, it must either rise over the irregular bumps or else scrape atoms off. Either way requires force.

The direction of the friction force is always in a direction opposing motion. An object sliding *down* an incline experiences friction directed *up* the incline; an object that slides to the *right* experiences friction toward the *left*. Thus, if an object is to move at constant velocity, a force equal to the opposing force of friction must be applied so that the two forces exactly cancel each other. The zero net force then results in zero acceleration and constant velocity.

Force of hand accelerates the brick

Twice as much force produces twice as much acceleration

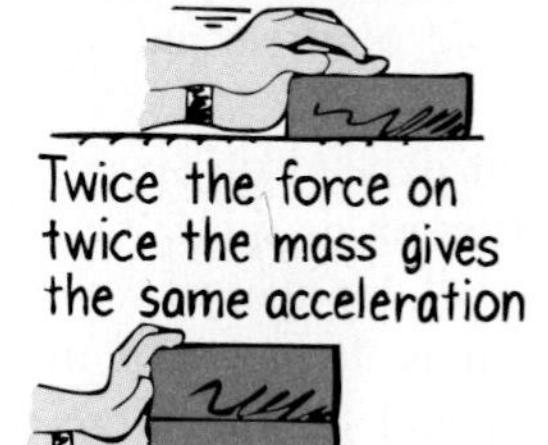

Twice the force on twice the mass gives the same acceleration

FIGURE 4.2
Acceleration is directly proportional to force.

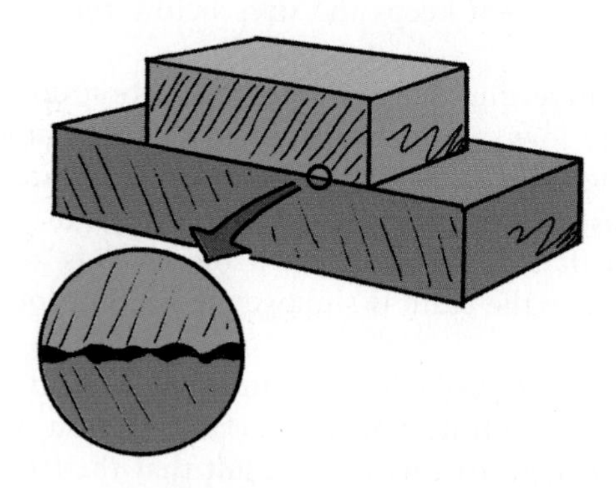

FIGURE 4.3
Friction results from the mutual contact of irregularities in the surfaces of sliding objects. Even surfaces that appear to be smooth have irregular surfaces when viewed at the microscopic level.

CHECK YOUR ANSWERS

1. It will have four times as much acceleration.
2. It will have less acceleration because friction will reduce the net force.

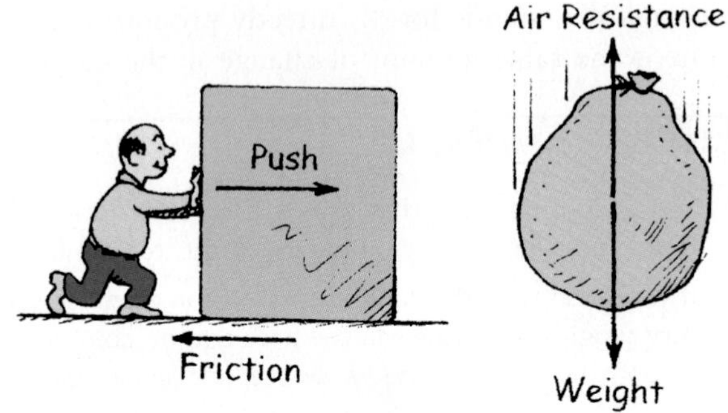

FIGURE 4.4
The direction of the force of friction always opposes the direction of motion. (Left) Push the crate to the right, and friction acts toward the left. (Right) The sack falls downward, and air friction (air resistance) acts upward. (What is the acceleration of the sack when air resistance equals the sack's weight?)

No friction exists on a crate that sits at rest on a level floor. But, if you push the crate horizontally, you'll disturb the contact surfaces and friction is produced. How much? If the crate is still at rest, then the friction that opposes motion is just enough to cancel your push. If you push horizontally with, say, 70 newtons, friction builds up to become 70 newtons. If you push harder—say, 100 newtons—and the crate is on the verge of sliding, the friction between the crate and floor opposes your push with 100 newtons. If 100 newtons is the most the surfaces can muster, then, when you push a bit harder, the clinging gives way and the crate slides.[1]

Interestingly, the friction of sliding is somewhat less than the friction that builds up before sliding takes place. Physicists and engineers distinguish between *static friction* and *sliding friction*. For given surfaces, static friction is somewhat greater than sliding friction. If you push on a crate, it takes more force to get it going than it takes to keep it sliding. Before the time of antilock brake systems, slamming on the brakes of a car was quite problematic. When tires lock, they slide, providing less friction than if they are made to roll to a stop. A rolling tire does not slide along the road surface, and friction is static friction, with more grab than sliding friction. But once the tires start to slide, the frictional force is reduced—not a good thing. An antilock brake system keeps the tires below the threshold of breaking loose into a slide.

It's also interesting that the force of friction does not depend on speed. A car skidding at low speed has approximately the same friction as the same car skidding at high speed. If the friction force of a crate that slides against a floor is 90 newtons at low speed, to a close approximation it is 90 newtons at a greater speed. It may be more when the crate is at rest and on the verge of sliding, but, once the crate is sliding, the friction force remains approximately the same.

More interesting still, friction does not depend on the area of contact. If you slide the crate on its smallest surface, all you do is concentrate the same weight on a smaller area with the result that the friction is the same. So those extra wide tires you see on some cars provide no more friction than narrower tires. The wider tire simply spreads the weight of the car over more surface

[1]Even though it may not seem so yet, most of the concepts in physics are not really complicated. But friction is different. Unlike most concepts in physics, it is a very complicated phenomenon. The findings are empirical (gained from a wide range of experiments) and the predictions are approximate (also based on experiment).

FIGURE 4.5
Friction between the tire and the ground is nearly the same whether the tire is wide or narrow. The purpose of the greater contact area is to reduce heating and wear.

area to reduce heating and wear. Similarly, the friction between a truck and the ground is the same whether the truck has four tires or eighteen! More tires spread the load over more ground area and reduces the pressure per tire. Interestingly, stopping distance when brakes are applied is not affected by the number of tires. But the wear that tires experience very much depends on the number of tires.

Friction is not restricted to solids sliding over one another. Friction occurs also in liquids and gases, both of which are called *fluids* (because they flow). Fluid friction occurs as an object pushes aside the fluid it is moving through. Have you ever attempted a 100-m dash through waist-deep water? The friction of fluids is appreciable, even at low speeds. So unlike the friction between solid surfaces, fluid friction depends on speed. A very common form of fluid friction for something moving through air is *air resistance,* also called *air drag.* You usually aren't aware of air resistance when walking or jogging, but you notice it at the higher speeds when riding a bicycle or when skiing downhill. Air resistance increases with increasing speed. The falling sack shown in Figure 4.4 will reach a constant velocity when air resistance balances the sack's weight.

CHECK YOURSELF

What net force does a sliding crate experience when you exert a force of 110 N and friction between the crate and the floor is 100 N?

Mass and Weight

FIGURE 4.6
An anvil in outer space—between the Earth and the Moon, for example—may be weightless, but it is not massless.

The acceleration imparted to an object depends not only on applied forces and friction forces, but on the inertia of the object. How much inertia an object possesses depends on the amount of matter in the object—the more matter, the more inertia. In speaking of how much matter something has, we use the term *mass*. The greater the mass of an object, the greater its inertia. **Mass** is a measure of the inertia of a material object.

Mass corresponds to our intuitive notion of **weight.** We casually say that something has a lot of matter if it weighs a lot. But there is a difference between mass and weight. We can define each as follows:

> **Mass:** ***The quantity of matter in an object. It is also the measure of the inertia or sluggishness that an object exhibits in response to any effort made to start it, stop it, or change its state of motion in any way.***
>
> **Weight:** ***The force upon an object due to gravity.***

In the absence of acceleration, mass and weight are directly proportional to each other.[2] If the mass of an object is doubled, its weight is also doubled;

CHECK YOUR ANSWER

10 N in the direction of your push (110 N − 100 N).

[2]Weight and mass are directly proportional; weight = mg, where g is the constant of proportionality and has the value 9.8 N/kg. Equivalently, g is the acceleration due to gravity, 9.8 m/s^2 (the units N/kg are equivalent to m/s^2). In Chapter 9, we'll extend the definition of weight as the force that an object exerts on a supporting surface.

FIGURE 4.7
The astronaut in space finds that it is just as difficult to shake the "weightless" anvil as it would be on Earth. If the anvil were more massive than the astronaut, which would shake more—the anvil or the astronaut?

FIGURE 4.8
Why will a slow, continuous increase in downward force break the string above the massive ball, while a sudden increase would break the lower string?

if the mass is halved, the weight is halved. Because of this, mass and weight are often interchanged. Also, mass and weight are sometimes confused because it is customary to measure the quantity of matter in things (mass) by their gravitational attraction to the Earth (weight). But mass is more fundamental than weight; it is a fundamental quantity that completely escapes the notice of most people.

There are times when weight corresponds to our unconscious notion of inertia. For example, if you are trying to determine which of two small objects is the heavier one, you might shake them back and forth in your hands or move them in some way instead of lifting them. In doing so, you are judging which of the two is more difficult to get moving, feeling which of the two is most resistant to a change in motion. You are really comparing the inertias of the objects.

In the United States, the quantity of matter in an object is commonly described by the gravitational pull between it and the Earth, or its *weight,* usually expressed in *pounds*. In most of the world, however, the measure of matter is commonly expressed in a mass unit, the **kilogram.** At the surface of the Earth, a brick with a mass of 1 kilogram weighs 2.2 pounds. In metric units, the unit of force is the **newton,** which is equal to a little less than a quarter-pound (like the weight of a quarter-pound hamburger *after* it is cooked). A 1-kilogram brick weighs about 10 newtons (more precisely, 9.8 N).[3] Away from the Earth's surface, where the influence of gravity is less, a 1-kilogram brick weighs less. It would also weigh less on the surface of planets with less gravity than Earth. On the Moon's surface, for example, where the gravitational force on things is only one-sixth as strong as on Earth, a 1-kilogram brick weighs about 1.6 newtons (or 0.36 pounds). On planets with stronger gravity, it would weigh more, but the mass of the brick is the same everywhere. The brick offers the same resistance to speeding up or slowing down regardless of whether it's on Earth, on the Moon, or on any other body attracting it. In a drifting spaceship, where a scale with a brick on it reads zero, the brick still has mass. Even though it doesn't press down on the scale, the brick has the same resistance to a change in motion as it has on Earth. Just as much force would have to be exerted by an astronaut in the spaceship to shake it back and forth as would be required to shake it back and forth while on Earth. You'd have to provide the same amount of push to accelerate a huge truck to a given speed on a level surface on the Moon as on Earth. The difficulty of *lifting* it against gravity (weight), however, is something else. Mass and weight are different from each other (Figures 4.6 and 4.7).

A nice demonstration that distinguishes mass and weight is the massive ball suspended on the string, shown in Figure 4.8. The top string breaks when the lower string is pulled with a gradual increase in force, but the bottom string breaks when the lower string is jerked. Which of these cases illustrates the weight of the ball, and which illustrates the mass of the ball? Note that only the top string bears the weight of the ball. So, when the lower string is gradually pulled, the tension supplied by the pull is transmitted to the top string. The total tension in the top string is caused by the pull plus the weight of the ball. The top string breaks when the breaking point is reached. But, when the

[3]So 2.2 lb equals 9.8 N, or 1 N is approximately equal to 0.22 lb—about the weight of an apple. In the metric system, it is customary to specify quantities of matter in units of mass (in grams or kilograms) and rarely in units of weight (in newtons). In the United States and countries that use the British system of units, however, quantities of matter are customarily specified in units of weight (in pounds). (The British unit of mass, the *slug,* is not well known.) See Appendix I for more about systems of measurement.

bottom string is jerked, the mass of the ball—its tendency to remain at rest—is responsible for the bottom string breaking.

It is also easy to confuse mass and **volume.** When we think of a massive object, we often think of a big object. An object's size (volume), however, is not necessarily a good way to judge its mass. Which is easier to get moving: a car battery or an empty cardboard box of the same size? So, we find that mass is neither weight nor volume.

CHECK YOURSELF

1. Does a 2-kg iron brick have twice as much *inertia* as a 1-kg iron brick? Twice as much *mass*? Twice as much *volume*? Twice as much *weight*?
2. Would it be easier to lift a cement truck on the Earth's surface or to lift it on the Moon's surface?
3. Ask a friend to drive a small nail into a piece of wood placed on top of a pile of books on your head. Why doesn't this hurt you?

Mass Resists Acceleration

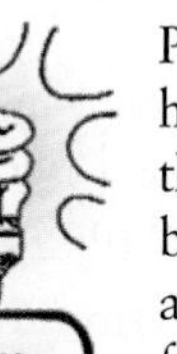

When two things are directly proportional to each other, as one increases, the other increases also. However, when two things are inversely proportional to each other, as one increases, the other decreases.

Insights

Push your friend on a skateboard and your friend accelerates. Now push equally hard on an elephant on a skateboard and the acceleration is much less. You'll see that the amount of acceleration depends not only on the force but on the mass being pushed. The same force applied to twice the mass produces half the acceleration; for three times the mass, one-third the acceleration. We say that, for a given force, the acceleration produced is inversely proportional to the mass. That is,

$$\text{Acceleration} \sim \frac{1}{\text{mass}}$$

CHECK YOUR ANSWERS

1. The answers to all parts are *yes*. A 2-kg iron brick has twice as many iron atoms and therefore twice the amount of matter and mass. In the same location, it also has twice the weight. And, since both bricks have the same density (the same ratio of mass to volume), the 2-kg brick also has twice the volume.
2. A cement truck would be easier to lift on the Moon because the gravitational force is less on the Moon. When you *lift* an object, you are contending with the force of gravity (its weight). Although its mass is the same on the Earth, on the Moon, or anywhere, its weight is only 1/6 as much on the Moon, so only 1/6 as much effort is required to lift it there. To move it horizontally, however, you are not pushing against gravity. When mass is the only factor, equal forces will produce equal accelerations, whether the object is on the Earth or the Moon.
3. The relatively large mass of the books and block atop your head resists being moved. The force that is successful in driving the nail will not be as successful in accelerating the massive books and block, which don't move very much when the nail is struck. Can you see the similarity between this and the suspended massive ball demonstration, where the supporting string doesn't break when the bottom string is jerked?

FIGURE 4.9
Interactive Figure
The greater the mass, the greater the force must be for a given acceleration.

FIGURE 4.10
An enormous force is required to accelerate this three-story-high earth mover when it carries a typical 350-ton load.

By inversely we mean that the two values change in opposite directions. As the denominator increases, the whole quantity decreases. For example, the quantity 1/100 is less than 1/10.

Newton's Second Law of Motion

Newton's Second Law

Newton was the first to discover the relationship among three basic physical concepts—acceleration, force, and mass. He proposed one of the most important rules of nature, his second law of motion. **Newton's second law** states

> ***The acceleration of an object is directly proportional to the net force acting on the object, is in the direction of the net force, and is inversely proportional to the mass of the object.***

In summarized form, this is

$$\text{Acceleration} \sim \frac{\text{net force}}{\text{mass}}$$

We use the wiggly line $\sim$ as a symbol meaning "is proportional to." We say that acceleration a is directly proportional to the overall net force F and inversely proportional to the mass m. By this we mean that, if F increases, a increases by the same factor (if F doubles, a doubles); but if m increases, a decreases by the same factor (if m doubles, a is cut in half).

By using consistent units, such as newtons (N) for force, kilograms (kg) for mass, and meters per second squared (m/s^2) for acceleration, the proportionality may be expressed as an exact equation:

$$\text{acceleration} = \frac{\text{net force}}{\text{mass}}$$

In its briefest form, where a is acceleration, F_{net} is net force, and m is mass, it becomes

$$a = \frac{F_{\text{net}}}{m}$$

An object is accelerated in the direction of the force acting on it. Applied in the direction of the object's motion, a force will increase the object's speed.

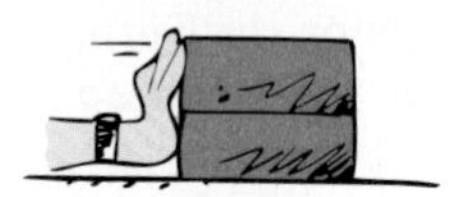

The same force accelerates 2 bricks 1/2 as much

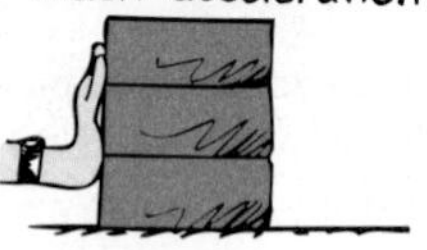

FIGURE 4.11
Acceleration is inversely proportional to mass.

Applied in the opposite direction, it will decrease the speed of the object. Applied at right angles, it will deflect the object. Any other direction of application will result in a combination of speed change and deflection. *The acceleration of an object is always in the direction of the net force.*

CHECK YOURSELF

1. In the previous chapter, acceleration was defined to be the time rate of change in velocity; that is, a = (change in v)/time. In this chapter, are we saying that acceleration is instead the ratio of force to mass—that is, $a = F/m$? Which is it?
2. A jumbo jet cruises at a constant velocity of 1000 km/h when the thrusting force of its engines is a constant 100,000 N. What is the acceleration of the jet? What is the force of air resistance on the jet?

When Acceleration Is *g*—Free Fall

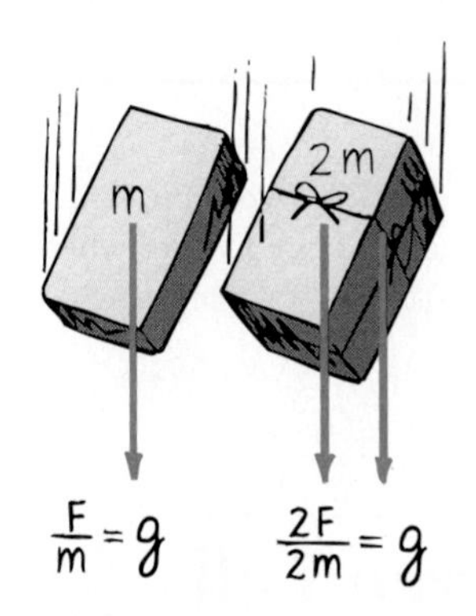

FIGURE 4.12
Interactive Figure
The ratio of weight (F) to mass (m) is the same for all objects in the same locality; hence, their accelerations are the same in the absence of air resistance.

Although Galileo founded both the concept of inertia and the concept of acceleration, and although he was the first to measure the acceleration of falling objects, he could not explain why objects of various masses fall with equal accelerations. Newton's second law provides the explanation.

We know that a falling object accelerates toward Earth because of the gravitational force of attraction between the object and Earth. When the force of gravity is the only force—that is, when friction (such as air resistance) is negligible—we say that the object is in a state of **free fall.**

The greater the mass of an object, the greater is the gravitational force of attraction between it and the Earth. The double brick in Figure 4.12, for example, has twice the gravitational attraction of the single brick. Why, then, as Aristotle supposed, doesn't the double brick fall twice as fast? The answer is that the acceleration of an object depends not only on the force—in this case, the

Free-Fall Acceleration Explained

Only a single force acts on something in free fall—the force of gravity.

CHECK YOUR ANSWERS

1. Acceleration is *defined* as the time rate of change of velocity and is *produced by* a force. How much force/mass (the cause) determines the rate change in v/time (the effect). So, whereas we defined acceleration in Chapter 3, in this chapter we define the terms that produce acceleration.
2. The acceleration is zero because the velocity is constant. Since the acceleration is zero, it follows from Newton's second law that the net force is zero, which means that the force of air drag must just equal the thrusting force of 100,000 N and must act in the opposite direction. So the air drag on the jet is 100,000 N. (Note that we don't need to know the velocity of the jet to answer this question. We need only to know that it is constant, our clue that acceleration—and therefore net force—is zero.)

---

weight—but also on the object's resistance to motion, its inertia. Whereas a force produces an acceleration, inertia is a *resistance* to acceleration. So twice the force exerted on twice the inertia produces the same acceleration as half the force exerted on half the inertia. Both accelerate equally. The acceleration due to gravity is symbolized by g. We use the symbol g, rather a, to denote that acceleration is due to gravity alone.

The ratio of weight to mass for freely falling objects equals a constant—g. This is similar to the constant ratio of circumference to diameter for circles, which equals the constant π. The ratio of weight to mass is the same both for heavy objects and for light objects, just as the ratio of circumference to diameter is the same both for large circles and for small circles (Figure 4.13).

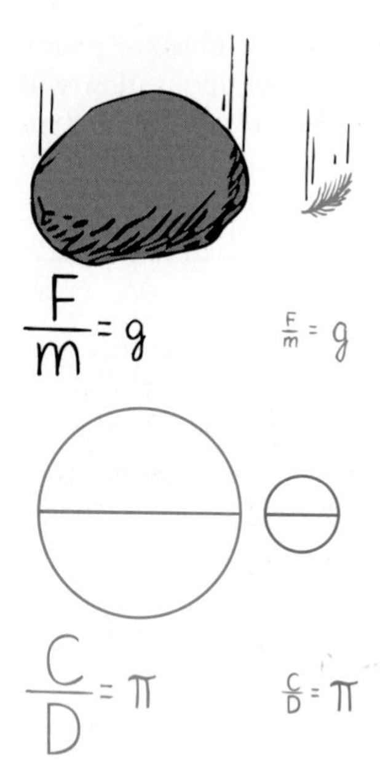

FIGURE 4.13
The ratio of weight (F) to mass (m) is the same for the large rock and the small feather; similarly, the ratio of circumference (C) to diameter (D) is the same for the large and the small circle.

We now understand that the acceleration of free fall is independent of an object's mass. A boulder 100 times more massive than a pebble falls with the same acceleration as the pebble because, although the force on the boulder (its weight) is 100 times greater than the force on the pebble, its resistance to a change in motion (its mass) is 100 times that of the pebble. The greater force offsets the equally greater mass.

CHECK YOURSELF

In a vacuum, a coin and a feather fall at the same rate, side by side. Would it be correct to say that equal forces of gravity act on both the coin and the feather when in a vacuum?

When Acceleration Is Less Than *g*—Nonfree Fall

Objects falling in a vacuum are one thing, but what of the practical cases of objects falling in air? Although a feather and a coin will fall equally fast in a vacuum, they fall quite differently in air. How do Newton's laws apply to objects falling in air? The answer is that Newton's laws apply for *all* objects, whether freely falling or falling in the presence of resistive forces. The accelerations, however, are quite different for the two cases. The important thing to keep in mind is the idea of *net force*. In a vacuum or in cases in which air resistance can be neglected, the net force is the weight because it is the only force. In the presence

When Galileo tried to explain why all objects fall with equal accelerations, wouldn't he have loved to know the rule $a = F/m$?

Insights

The Physics Place
Falling and Air Resistance

CHECK YOUR ANSWER

No, no, no, a thousand times no! These objects accelerate equally not because the forces of gravity on them are equal, but because the *ratios* of their weights to their masses are equal. Although air resistance is not present in a vacuum, gravity is. (You'd know this if you stuck your hand into a vacuum chamber and the truck shown in Figure 4.10 rolled over it!) If you answered *yes* to this question, let this be a warning to be more careful when you think physics!

FIGURE 4.14
When weight mg is greater than air resistance R, the falling sack accelerates. At higher speeds, R increases. When $R = mg$, acceleration reaches zero, and the sack reaches its terminal velocity.

of air resistance, however, the net force is less than the weight—it is the weight minus air drag, the force arising from air resistance.[4]

The force of air drag experienced by a falling object depends on two things. First, it depends on the frontal area of the falling object—that is, on the amount of air the object must plow through as it falls. Second, it depends on the speed of the falling object; the greater the speed, the greater the number of air molecules an object encounters per second and the greater the force of molecular impact. Air drag depends on the size and the speed of a falling object.

In some cases, air drag greatly affects falling; in other cases, it doesn't. Air drag is important for a falling feather. Because a feather has so much area compared with its small weight, it doesn't have to fall very fast before the upward-acting air resistance cancels the downward-acting weight. The net force on the feather is then zero and acceleration terminates. When acceleration terminates, we say that the object has reached its **terminal speed.** If we are concerned with direction, down for falling objects, we say the object has reached its **terminal velocity.** The same idea applies to all objects falling in air. Consider skydiving. As a falling skydiver gains speed, air drag may finally build up until it equals the weight of the skydiver. If and when this happens, the *net force* becomes zero and the skydiver no longer accelerates; she has reached her terminal velocity. For a feather, terminal velocity is a few centimeters per second, whereas, for a skydiver, it is about 200 kilometers per hour. A skydiver may vary this speed by varying position. Head or feet first is a way of encountering less air and thus less air drag and attaining maximum terminal velocity. A smaller terminal velocity is attained by spreading oneself out like a flying squirrel. Minimum terminal velocity is attained when the parachute is opened.

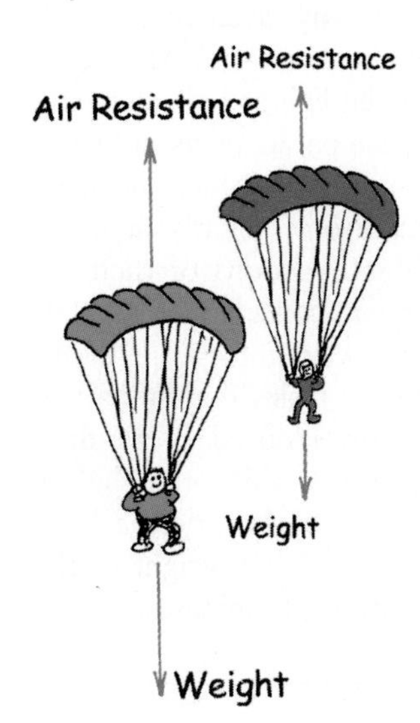

FIGURE 4.15
The heavier parachutist must fall faster than the lighter parachutist for air resistance to cancel his greater weight.

Consider a man and woman parachuting together from the same altitude (Figure 4.15). Suppose that the man is twice as heavy as the woman and that their same-sized parachutes are initially opened. Having parachutes of the same size means that, at equal speeds, the air resistance is the same on both of them. Who reaches the ground first—the heavy man or the lighter woman? The answer is that the person who falls fastest gets to the ground first—that is, the person with the greatest terminal speed. At first we might think that, because the parachutes are the same, the terminal speeds for each would be the same and, therefore, that both would reach the ground at the same time. This doesn't happen, however, because air drag depends on speed. Greater speed means greater force of air impact. The woman will reach her terminal speed when the air drag against her parachute equals her weight. When this occurs, the air drag against the parachute of the man will not yet equal his weight. He must fall faster than she does for the air drag to match his greater weight.[5] Terminal velocity is greater for the heavier person, with the result that the heavier person reaches the ground first.

---

[4]In mathematical notation,

$$a = \frac{F_{\text{net}}}{m} = \frac{mg - R}{m}$$

where mg is the weight and R is the air resistance. Note that when $R = mg$, $a = 0$; then, with no acceleration, the object falls at constant velocity. With elementary algebra, we can go another step and get

$$a = \frac{F_{\text{net}}}{m} = \frac{mg - R}{m} = g - \frac{R}{m}$$

We see that the acceleration a will always be less than g if air resistance R impedes falling. Only when $R = 0$ does $a = g$.

[5]Terminal speed for the twice-as-heavy man will be about 41 percent greater than the woman's terminal speed, because the retarding force of air resistance is proportional to speed squared. ($v_{\text{man}}^2/v_{\text{woman}}^2 = 1.41^2 = 2$.)

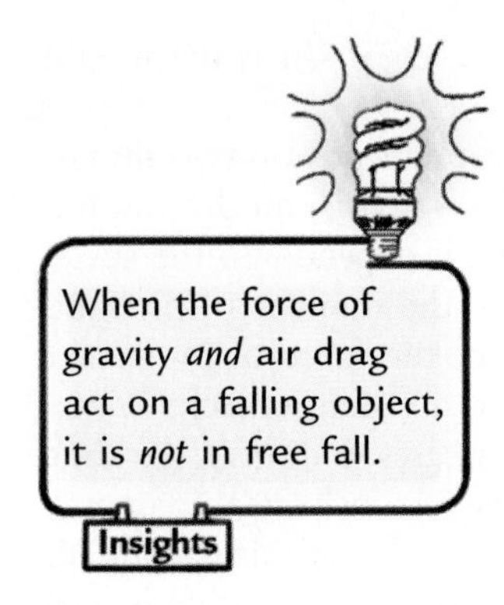

When the force of gravity *and* air drag act on a falling object, it is *not* in free fall.
Insights

CHECK YOURSELF

A skydiver jumps from a high-flying helicopter. As she falls faster and faster through the air, does her acceleration increase, decrease, or remain the same?

Consider a pair of tennis balls, one a regular hollow ball and the other filled with iron pellets. Although they are the same size, the iron-filled ball is considerably heavier than the regular ball. If you hold them above your head and drop them simultaneously, you'll see that they strike the ground at about the same time. But if you drop them from a greater height—say, from the top of a building—you'll note the heavier ball strikes the ground first. Why? In the first case, the balls do not gain much speed in their short fall. The air drag they encounter is small compared with their weights, even for the regular ball. The tiny difference in their arrival time is not noticed. But, when they are dropped from a greater height, the greater speeds of fall are met with greater air resistance. At any given speed, each ball encounters the same air resistance because each has the same size. This same air resistance may be a lot compared with the weight of the lighter ball, but only a little compared with the weight of the heavier ball (like the parachutists in Figure 4.15). For example, 1 N of air drag acting on a 2-N object will reduce its acceleration by half, but 1 N of air drag on a 200-N object will only slightly diminish its acceleration. So, even with equal air resistances, the accelerations of each are different. There is a moral to be learned here. Whenever you consider the acceleration of something, use the equation of Newton's second law to guide your thinking: The acceleration is equal to the ratio of *net* force to the mass. For the falling tennis balls, the net force on the hollow ball is appreciably reduced as air drag builds up, while the net force on the iron-filled ball is comparably only slightly reduced. Acceleration decreases as net force decreases, which, in turn, decreases as air drag increases. If and when the air drag builds up to equal the weight of the falling object, then the net force becomes zero and acceleration terminates.

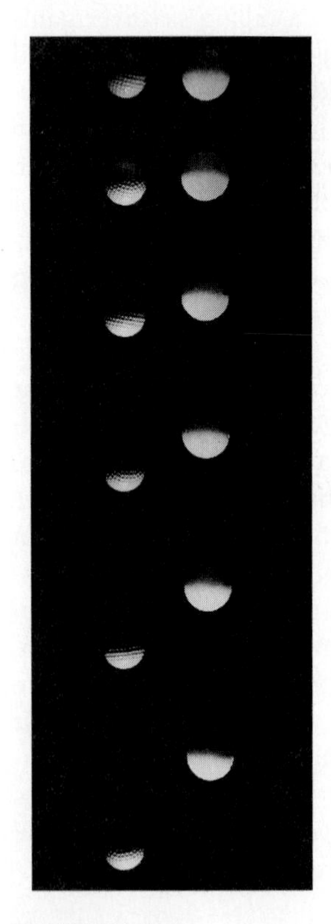

FIGURE 4.16
A stroboscopic study of a golf ball (left) and a Styrofoam ball (right) falling in air. The air resistance is negligible for the heavier golf ball, and its acceleration is nearly equal to *g*. Air resistance is not negligible for the lighter Styrofoam ball, which reaches its terminal velocity sooner.

CHECK YOUR ANSWER

Acceleration decreases because the net force on her decreases. Net force is equal to her weight minus her air resistance, and, because air resistance increases with increasing speed, net force and hence acceleration decrease. By Newton's second law,

$$a = \frac{F_{net}}{m} = \frac{mg - R}{m}$$

where mg is her weight and R is the air resistance she encounters. As R increases, a decreases. Note that, if she falls fast enough so that $R = mg$, $a = 0$; then, with no acceleration, she falls at a constant speed.

Summary of Terms

Friction The resistive force that opposes the motion or attempted motion of an object either past another object with which it is in contact or through a fluid.

Mass The quantity of matter in an object. More specifically, it is the measure of the inertia or sluggishness that an object exhibits in response to any effort made to start it, stop it, deflect it, or change in any way its state of motion.

Weight The force due to gravity on an object.

Kilogram The fundamental SI unit of mass. One kilogram (symbol kg) is the mass of 1 liter (1 L) of water at 4°C.

Newton The SI unit of force. One newton (symbol N) is the force that will give an object of mass 1 kg an acceleration of $1 m/s^2$.

Volume The quantity of space an object occupies.

Newton's second law The acceleration of an object is directly proportional to the net force acting on the object, is in the direction of the net force, and is inversely proportional to the mass of the object.

Force Any influence that can cause an object to be accelerated, measured in newtons (or in pounds, in the British system).

Free fall Motion under the influence of gravitational pull only.

Terminal speed The speed at which the acceleration of a falling object terminates because air resistance balances its weight. When direction is specified, then we speak of **terminal velocity.**

Review Questions

Force Causes Acceleration

1. Is acceleration proportional to net force, or does acceleration equal net force?

Friction

2. How does friction affect the net force on an object?

3. How great is the force of friction compared with your push on a crate that doesn't move on a level floor?

4. As you increase your push, will friction on the crate increase also?

5. Once the crate is sliding, how hard do you push to keep it moving at constant velocity?

6. Which is normally greater, static friction or sliding friction on the same object?

7. How does the force of friction vary with speed?

8. Slide a block on its widest surface, then tip the block so it slides on its narrowest surface. In which case is friction greater?

9. Does fluid friction vary with speed and area of contact?

Mass and Weight

10. What relationship does mass have with inertia?

11. What relationship does mass have with weight?

12. Which is more fundamental, *mass* or *weight*? Which varies with location?

13. Fill in the blanks: Shake something to and fro and you're measuring its ________. Lift it against gravity and you're measuring its _______.

14. Fill in the blanks: The Standard International unit for mass is the _______. The Standard International unit for force is the ________.

15. What is the approximate weight of a quarter-pound hamburger after it is cooked?

16. What is the weight of a 1-kilogram brick?

17. In the string-pull illustration in Figure 4.8, a gradual pull of the lower string results in the top string breaking. Does this illustrate the ball's weight or its mass?

18. In the string-pull illustration in Figure 4.8, a sharp jerk on the bottom string results in the bottom string breaking. Does this illustrate the ball's weight or its mass?

19. Clearly distinguish among *mass, weight*, and *volume*.

Mass Resists Acceleration

20. Is acceleration *directly* proportional to mass, or is it *inversely* proportional to mass? Give an example.

Newton's Second Law of Motion

21. State Newton's second law of motion.

22. If we say that one quantity is *directly proportional* to another quantity, does this mean they are *equal* to each other? Explain briefly, using mass and weight as an example.

23. If the net force acting on a sliding block is somehow tripled, by how much does the acceleration increase?

24. If the mass of a sliding block is tripled while a constant net force is applied, by how much does the acceleration decrease?

25. If the mass of a sliding block is somehow tripled at the same time the net force on it is tripled, how does the resulting acceleration compare with the original acceleration?

26. How does the direction of acceleration compare with the direction of the net force that produces it?

When Acceleration Is g—Free Fall

27. What is meant by *free fall*?
28. The ratio of circumference to diameter for all circles is π. What is the ratio of force to mass for freely falling bodies?
29. Why doesn't a heavy object accelerate more than a light object when both are freely falling?

When Acceleration Is Less Than g—Nonfree Fall

30. What is the net force that acts on a 10-N freely falling object?
31. What is the net force that acts on a 10-N falling object when it encounters 4 N of air resistance? 10 N of air resistance?
32. What two principal factors affect the force of air resistance on a falling object?
33. What is the acceleration of a falling object that has reached its terminal velocity?
34. Why does a heavy parachutist fall faster than a lighter parachutist who wears a parachute of the same size?
35. If two objects having the same size fall through air at different speeds, which encounters the greater air resistance?

Projects

1. Write a letter to Grandma, similar to the one of Project 1 in Chapter 3. Tell her that Galileo introduced the concepts of acceleration and inertia, and was familiar with forces, but didn't see the connection among these three concepts. Tell her how Isaac Newton did see the connection and how it explains why heavy and light objects in free fall gain the same speed in the same time. In this letter, it's okay to use an equation or two, as long as you make it clear to Grandma that an equation is a shorthand notation of ideas you've explained.
2. Drop a sheet of paper and a coin at the same time. Which reaches the ground first? Why? Now crumple the paper into a small, tight wad and again drop it with the coin. Explain the difference observed. Will they fall together if dropped from a second-, third-, or fourth-story window? Try it and explain your observations.
3. Drop a book and a sheet of paper, and note that the book has a greater acceleration—g. Place the paper beneath the book so that it is forced against the book as both fall, so both fall at g. How do the accelerations compare if you place the paper on top of the raised book and then drop both? You may be surprised, so try it and see. Then explain your observation.
4. Drop two balls of different weight from the same height, and, at small speeds, they practically fall together. Will they roll together down the same inclined plane? If each is suspended from an equal length of string, making a pair of pendulums, and displaced through the same angle, will they swing back and forth in unison? Try it and see; then explain using Newton's laws.
5. The net force acting on an object and the resulting acceleration are always in the same direction. You can demonstrate this with a spool. If the spool is pulled horizontally to the right, in which direction will it roll?

One-Step Calculations

Make these simple one-step calculations and familiarize youself with the equations that link the concepts of force, mass, and acceleration.

Weight = mg

1. Calculate the weight of a person having a mass of 50 kg in newtons.
2. Calculate the weight of a 2000-kg elephant in newtons.
3. Calculate the weight of a 2.5-kg melon in newtons. What is its weight in pounds?
4. An apple weighs about 1 N. What is its mass in kilograms? What is its weight in pounds?
5. Susie Small finds that she weighs 300 N. Calculate her mass.

Acceleration: $a = F_{net} / m$

6. Calculate the acceleration of a 2000-kg, single-engine airplane just before takeoff when the thrust of its engine is 500 N.
7. Calculate the acceleration of a 300,000-kg jumbo jet just before takeoff when the thrust on the aircraft is 120,000 N.
8. (a) Calculate the acceleration of a 2-kg block on a horizontal friction-free air table when you exert a horizontal net force of 20 N. (b) What acceleration occurs if the friction force is 4 N?

Force = ma

9. Calculate the horizontal force that must be applied to a 1-kg puck to make it accelerate on a horizontal

friction-free air table with the same acceleration it would have if it were dropped and fell freely.

10. Calculate the horizontal force that must be applied to produce an acceleration of 1.8 g for a 1.2-kg puck on a horizontal friction-free air table.

Exercises

1. Can the velocity of an object reverse direction while maintaining a constant acceleration? If so, give an example; if not, provide an explanation.

2. On a long alley, a bowling ball slows down as it rolls. Is any horizontal force acting on the ball? How do you know?

3. Is it possible to move in a curved path in the absence of a force? Defend your answer.

4. An astronaut tosses a rock on the Moon. What force(s) act(s) on the rock during its curved path?

5. Since an object weighs less on the surface of the Moon than on Earth's surface, does it have less inertia on the Moon's surface?

6. Which contains more apples, a 1-pound bag of apples on Earth or a 1-pound bag of apples on the Moon? Which contains more apples, a 1-kilogram bag of apples on Earth or a 1-kilogram bag of apples on the Moon?

7. A 400-kg bear grasping a vertical tree slides down at constant velocity. What is the friction force that acts on the bear?

8. A crate remains at rest on a factory floor while you push on it with a horizontal force *F*. How big is the friction force exerted on the crate by the floor? Explain.

9. In an orbiting space shuttle, you are handed two identical boxes, one filled with sand and the other filled with feathers. How can you determine which is which without opening the boxes?

10. Your empty hand is not hurt when it bangs lightly against a wall. Why does it hurt if you're carrying a heavy load? Which of Newton's laws is most applicable here?

11. Why is a massive cleaver more effective for chopping vegetables than an equally sharp knife?

12. Does the mass of an astronaut change when he or she is visiting the International Space Station? Defend your answer.

13. When a junked car is crushed into a compact cube, does its mass change? Its weight? Explain.

14. Gravity on the surface of the Moon is only 1/6 as strong as gravity on the Earth. What is the weight of a 10-kg object on the Moon and on the Earth? What is its mass on each?

15. Does a dieting person more accurately lose mass or lose weight?

16. What weight change occurs when your mass increases?

17. What is your own mass in kilograms? Your weight in newtons?

18. A grocery bag can withstand 300 N of force before it rips apart. How many kilograms of apples can it safely hold?

19. Consider a heavy crate resting on the bed of a flatbed of a truck. When the truck accelerates, the crate also accelerates and remains in place. Identify the force that accelerates the crate.

20. Explain how Newton's first law of motion can be considered to be a consequence of Newton's second law.

21. A rocket becomes progressively easier to accelerate as it travels through space. Why is this so? (Hint: About 90% of the mass of a newly launched rocket is fuel.)

22. Which requires less fuel, launching a rocket from the Moon or from the Earth? Defend your answer.

23. Aristotle claimed that the speed of a falling object depends on its weight. We now know that objects in free fall, whatever their weights, undergo the same gain in speed. Why does weight not affect acceleration?

24. When blocking in football, a defending lineman often attempts to get his body under the body of his opponent and push upward. What effect does this have on the friction force between the opposing lineman's feet and the ground?

25. A race car travels along a raceway at a constant velocity of 200 km/h. What horizontal forces act, and what is the net force acting on the car?

26. To pull a wagon across a lawn with constant velocity, you have to exert a steady force. Reconcile this fact with Newton's first law, which says that motion with constant velocity requires no force.

27. Three identical blocks are pulled, as shown, on a horizontal frictionless surface. If tension in the rope held by the hand is 30 N, what is the tension in the other ropes?

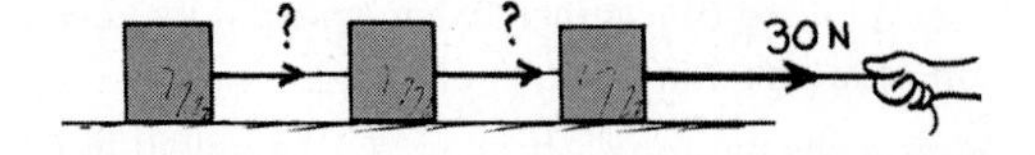

28. Free fall is motion in which gravity is the only force acting. (a) Is a skydiver who has reached terminal speed in free fall? (b) Is a satellite above the atmosphere that circles Earth in free fall?

29. When a coin is tossed upward, what happens to its velocity while ascending? Its acceleration? (Neglect air resistance.)

30. How much force acts on a tossed coin when it is halfway to its maximum height? How much force acts on it when it reaches its peak? (Neglect air resistance.)

31. Sketch the path of a ball tossed vertically into the air. (Neglect air resistance.) Draw the ball halfway to the top, at the top, and halfway down to its starting point. Draw a force vector on the ball in all three positions. Is the vector the same or different in the three locations? Is the acceleration the same or different in the three locations?

32. As you leap upward in a standing jump, how does the force that you exert on the ground compare with your weight?

33. When you jump vertically off the ground, what is your acceleration when you reach your highest point?

34. What is the acceleration of a rock at the top of its trajectory when it has been thrown straight upward? (Is your answer consistent with Newton's second law?)

35. A common saying goes, "It's not the fall that hurts you; it's the sudden stop." Translate this into Newton's laws of motion.

36. A friend says that, as long as a car is at rest, no forces act on it. What do you say if you're in the mood to correct the statement of your friend?

37. When your car moves along the highway at constant velocity, the net force on it is zero. Why, then, do you have to keep running your engine?

38. A "shooting star" is usually a grain of sand from outer space that burns up and gives off light as it enters the atmosphere. What exactly causes this burning?

39. What is the net force on a 1-N apple when you hold it at rest above your head? What is the net force on it after you release it?

40. Does a stick of dynamite contain force?

41. A parachutist, after opening her parachute, finds herself gently floating downward, no longer gaining speed. She feels the upward pull of the harness, while gravity pulls her down. Which of these two forces is greater? Or are they equal in magnitude?

42. Does a falling object increase in speed if its acceleration of fall decreases?

43. What is the net force acting on a 1-kg ball in free fall?

$F_{net} = mg = 9.8$ N

44. What is the net force acting on a falling 1-kg ball if it encounters 2 N of air resistance?

$F_{net} = 9.8 - 2 = 7.8$ N.

45. A friend says that, before the falling ball in the previous exercise reaches terminal velocity, it *gains* speed while acceleration *decreases*. Do you agree or disagree with your friend? Defend your answer.

46. Why will a sheet of paper fall more slowly than one that is wadded into a ball?

47. Upon which will air resistance be greater—a sheet of falling paper or the same paper wadded into a ball that falls at a faster terminal speed? (Careful!)

48. Hold a Ping-Pong ball and a golf ball at arm's length and drop them simultaneously. You'll see them hit the floor at about the same time. But, if you drop them off the top of a high ladder, you'll see the golf ball hit first. What is your explanation?

49. How does the force of gravity on a raindrop compare with the air drag it encounters when it falls at constant velocity?

50. When a car is moving in reverse, backing from a driveway, the driver applies the brakes. In what direction is the car's acceleration?

51. When a parachutist opens her parachute, in what direction does she accelerate?

52. How does the terminal speed of a parachutist before opening a parachute compare to terminal speed after? Why is there a difference?

53. How does the gravitational force on a falling body compare with the air resistance it encounters before it reaches terminal velocity? After reaching terminal velocity?

54. Why is it that a cat that accidentally falls from the top of a 50-story building hits a safety net below no faster than if it fell from the twentieth story?

Once terminal spd. is attained, falling extra dist. does not affect speed.

55. Under what conditions would a metal sphere dropping through a viscous liquid be in equilibrium?

56. When and if Galileo dropped two balls from the top of the Leaning Tower of Pisa, air resistance was not really negligible. Assuming that both balls were of the same size, one made of wood and one of metal, which ball struck the ground first? Why?

57. If you drop a pair of tennis balls simultaneously from the top of a building, they will strike the ground at the same time. If you fill one of the balls with lead pellets and then drop them together, which one will hit the ground first? Which one will experience greater air resistance? Defend your answers.

58. In the absence of air resistance, if a ball is thrown vertically upward with a certain initial speed, on returning to its original level it will have the same speed. When air resistance is a factor, will the ball be moving faster, the same, or more slowly than its throwing speed when it gets back to the same level? Why? (Physicists often use a "principle of exaggeration" to help them analyze a problem. Consider the exaggerated case of a feather, not a ball, because the effect of air resistance on the feather is more pronounced and therefore easier to visualize.)

59. If a ball is thrown vertically into the air in the presence of air resistance, would you expect the time during which it rises to be longer or shorter than the time during which it falls? (Again use the "principle of exaggeration.")

60. Make up two multiple-choice questions that would check a classmate's understanding of the distinction between mass and weight.

Problems

1. What is the greatest acceleration a runner can muster if the friction between her shoes and the pavement is 90% of her weight?

2. What is the acceleration of a 40-kg block of cement when pulled sideways with a net force of 200 N?

3. What is the acceleration of a 20-kg pail of cement that is pulled upward (not sideways!) with a force of 300 N?

4. If a mass of 1 kg is accelerated 1 m/s^2 by a force of 1 N, what would be the acceleration of 2 kg acted on by a force of 2 N?

5. How much acceleration does a 747 jumbo jet of mass 30,000 kg experience in takeoff when the thrust for each of four engines is 30,000 N?

6. Two boxes are seen to accelerate at the same rate when a force F is applied to the first and $4F$ is applied to the second. What is the mass ratio of the boxes?

7. A firefighter of mass 80 kg slides down a vertical pole with an acceleration of 4 m/s^2. What is the friction force that acts on the firefighter?

8. What will be the acceleration of a skydiver when air resistance builds up to be equal to half her weight?

9. Sprinting near the end of a race, a runner with a mass of 60 kg accelerates from a speed of 6 m/s to a speed of 7 m/s in 2 s. (a) What is the runner's average acceleration during this time? (b) To gain speed, the runner produces a backward force on the ground so that the ground pushes the runner forward, providing the force necessary for acceleration. Calculate this average force.

10. Before going into orbit, an astronaut has a mass of 55 kg. While in orbit, a measurement determines that a force of 100 N causes her to move with an acceleration of 1.90 m/s^2. To regain her original weight, should she go on a diet or start eating more candy?

Remember, review questions provide you with a self check of whether or not you grasp the central ideas of the chapter. The exercises and problems are extra "pushups" for you to try after you have at least a fair understanding of the chapter and can handle the review questions.

CHAPTER 5

Newton's Third Law of Motion

Darlene Librero pulls with one finger; Paul Doherty pulls with both hands. Question they ask of their Exploratorium class: "Who exerts more force on the scale?"

If you drop a sheet of tissue paper in front of the heavyweight boxing champion of the world and challenge him to hit it in midair with a force of only 50 pounds (222 N)—sorry, the champ can't do it. In fact, his best punch couldn't even come close. Why is this? We'll see in this chapter that the tissue has insufficient inertia for a 50-pound interaction with the champ's fist.

Forces and Interactions

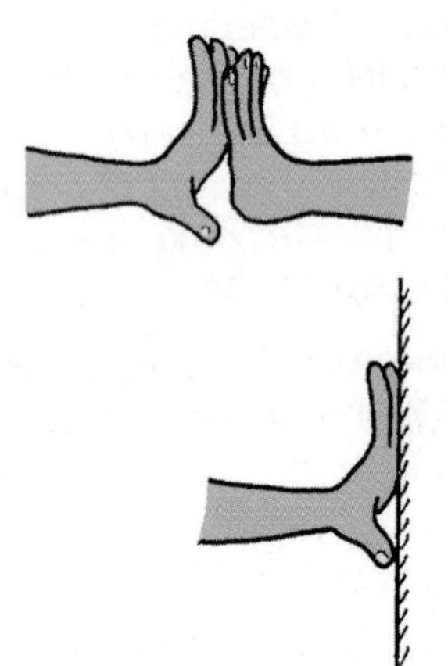

FIGURE 5.1
Interactive Figure
You can feel your fingers being pushed by your friend's fingers. You also feel the same amount of force when you push on a wall and it pushes back on you. As a point of fact, you can't push on the wall *unless* it pushes back on you!

So far we've treated force in its simplest sense—as a push or pull. Yet no push or pull ever occurs alone. Every force is part of an *interaction* between one thing and another. When you push on a wall with your fingers, more is happening than your pushing on the wall. You're interacting with the wall, which also pushes back on you. This is evident in your bent fingers, as illustrated in Figure 5.1. There is a pair of forces involved: your push on the wall and the wall's push back on you. These forces are equal in magnitude (have the same strength) and opposite in direction, and they constitute a single interaction. In fact, you can't push on the wall *unless* the wall pushes back.[1]

Consider a boxer's fist hitting a massive punching bag. The fist hits the bag (and dents it) while the bag hits back on the fist (and stops its motion). A pair of forces is involved in hitting the bag. The force pair can be quite large. But what of hitting a piece of tissue paper, as discussed earlier? The boxer's fist can only exert as much force on the tissue paper as the tissue paper can exert on the fist. Furthermore, the fist can't exert any force at all unless what is being hit exerts the same amount of force back. An interaction requires a *pair* of forces acting on *two* separate objects.

Other examples: You pull on a cart and it accelerates. But, in doing so, the cart pulls back on you, as evidenced perhaps by the tightening of the rope wrapped around your hand. A hammer hits a stake and drives it into the

[1]We tend to think that only living things are capable of pushing and pulling. But inanimate things can do the same. So please don't be troubled about the idea of the inanimate wall pushing on you. It does, just as another person leaning against you would.

FIGURE 5.2
When you lean against a wall, you exert a force on the wall. The wall simultaneously exerts an equal and opposite force on you. Hence you don't topple over.

ground. In doing so, the stake exerts an equal amount of force on the hammer, which brings the hammer to an abrupt halt. One thing interacts with another—you with the cart, or the hammer with the stake.

Which exerts the force and which receives the force? Isaac Newton's response was that neither force has to be identified as "exerter" or "receiver"; he concluded that both objects must be treated equally. For example, when you pull the cart, the cart pulls on you. This pair of forces, your pull on the cart and the cart's pull on you, make up the single interaction between you and the cart. In the interaction between the hammer and the stake, the hammer exerts a force against the stake but is itself brought to a halt in the process. Such observations led Newton to his third law of motion.

FIGURE 5.3
He can hit the massive bag with considerable force. But with the same punch he can exert only a tiny force on the tissue paper in midair.

FIGURE 5.4
In the interaction between the hammer and the stake, each exerts the same amount of force on the other.

THE Physics Place
Forces and Interactions

Newton's Third Law of Motion

Newton's **third law** states:

> **Whenever one object exerts a force on a second object, the second object exerts an equal and opposite force on the first.**

We can call one force the *action force* and the other the *reaction force.* Then we can express Newton's third law in the form:

> **To every action there is always an opposed equal reaction.**

It doesn't matter which force we call *action* and which we call *reaction.* The important thing is that they are co-parts of a single interaction, and that neither force exists without the other.

When you walk, you interact with the floor. You push against the floor, and the floor pushes against you. The pair of forces occurs at the same time (they are *simultaneous*). Likewise, the tires of a car push against the road while the road pushes back on the tires—the tires and road simultaneously push against each other. In swimming, you interact with the water, pushing the water backward, while the water simultaneously pushes you forward—you and the water push against each other. The reaction forces are what account for our motion in these examples. These forces depend on friction; a person or car on ice, for example, may be unable to exert the action force to produce the needed reaction force. Neither force exists without the other.

Push your fingers together and notice that, as you push harder, discoloration is the same for both. Aha, they both experience the same magnitude of force!

Insights

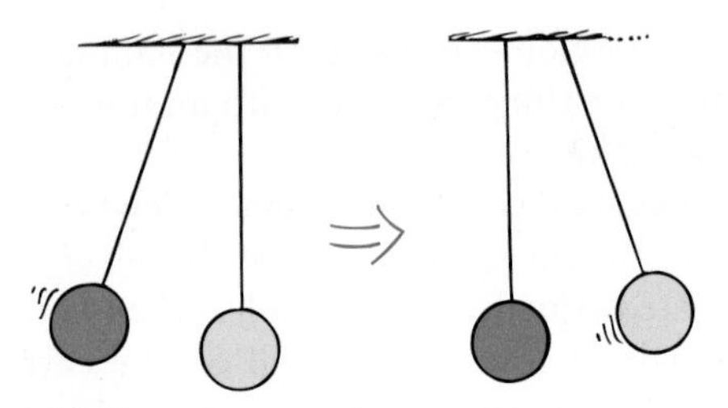

FIGURE 5.5
The impact forces between the blue ball and the yellow ball move the yellow ball and stop the blue ball.

FIGURE 5.6
In the interaction between the car and the truck, is the force of impact the same on each? Is the damage the same?

CHECK YOURSELF

Does a speeding missile possess force?

Action: tire pushes on road
Reaction: road pushes on tire

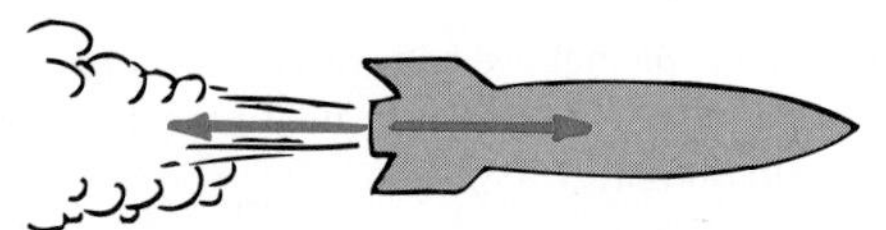
Action: rocket pushes on gas
Reaction: gas pushes on rocket
Action: man pulls on spring
Reaction: spring pulls on man
Action: earth pulls on ball
Reaction: ball pulls on earth

FIGURE 5.7
Action and reaction forces. Note that, when action is "*A* exerts force on *B*," the reaction is then simply "*B* exerts force on *A*."

CHECK YOUR ANSWER

No, a force is not something an object *has*, like mass, but is part of an interaction between one object and another. A speeding missile may possess the capability of exerting a force on another object when interaction occurs, but it does not possess force as a thing in itself. As we will see in the following chapters, a speeding missile possesses momentum and kinetic energy.

Defining Your System

FIGURE 5.8
Interactive Figure
A force acts on the orange, and the orange accelerates to the right.

An interesting question often arises: Since action and reaction forces are equal and opposite, why don't they cancel to zero? To answer this question, we must consider the *system* involved. Consider, for example, a system consisting of a single orange, Figure 5.8. The dashed line surrounding the orange encloses and defines the system. The vector that pokes outside the dashed line represents an external force on the system. The system accelerates in accord with Newton's second law. In Figure 5.9, we see that this force is provided by an apple, which doesn't change our analysis. The apple is outside the system. The fact that the orange simultaneously exerts a force on the apple, which is external to the system, may affect the apple (another system), but not the orange. You can't cancel a force on the orange with a force on the apple. So, in this case, the action and reaction forces don't cancel.

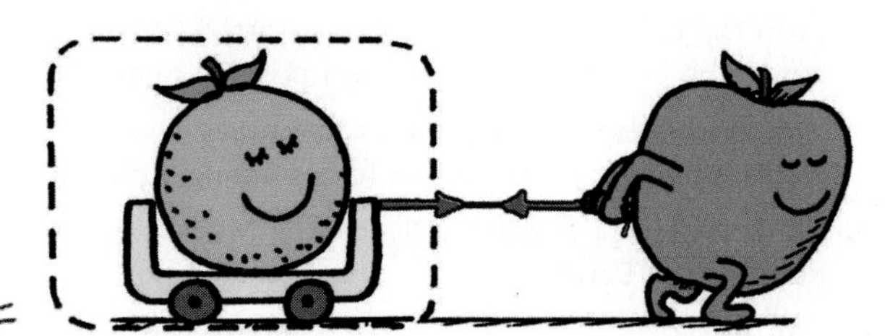

FIGURE 5.9
Interactive Figure
The force on the orange, provided by the apple, is not cancelled by the reaction force on the apple. The orange still accelerates.

A system may be as tiny as an atom or as large as the universe.

Insights

Now let's consider a larger system, enclosing *both* the orange and the apple. We see the system bounded by the dashed line in Figure 5.10. Notice that the force pair is *internal* to the orange–apple system. Then these forces *do* cancel each other. They play no role in accelerating the system. A force external to the system is needed for acceleration. That's where friction with the floor comes into play (Figure 5.11). When the apple pushes against the floor, the floor simultaneously pushes on the apple—an external force on the system. The system accelerates to the right.

FIGURE 5.10 Interactive Figure
In the larger system of orange + apple, action and reaction forces are internal and cancel. If these are the only horizontal forces, with no external force, no acceleration of the system occurs.

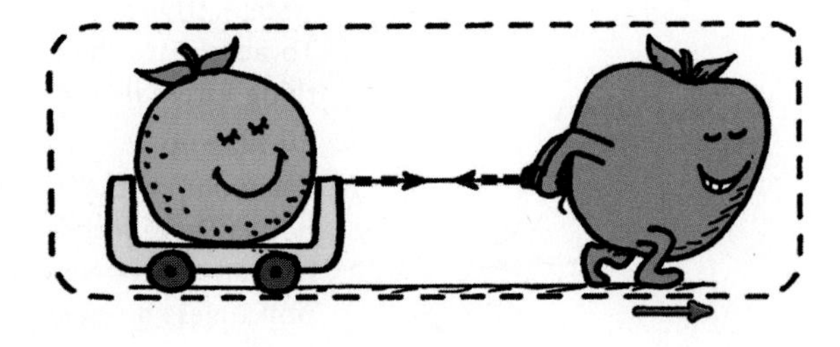

FIGURE 5.11 Interactive Figure
An external horizontal force occurs when the floor pushes on the apple (reaction to the apple's push on the floor). The orange–apple system accelerates.

Inside a football are trillions and trillions of interatomic forces at play. They hold the ball together, but they play no role in accelerating the ball. Although every one of the interatomic forces is part of an action–reaction pair within the ball, they combine to zero, no matter how many of them there are. A force

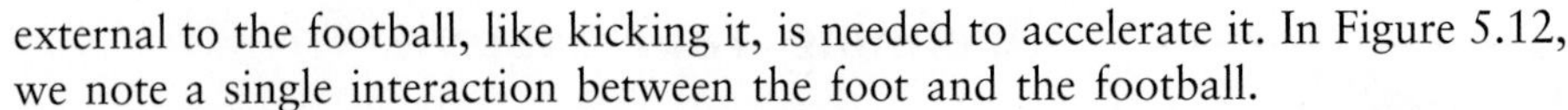

external to the football, like kicking it, is needed to accelerate it. In Figure 5.12, we note a single interaction between the foot and the football.

FIGURE 5.12
A acts on B, and B accelerates.

The football in Figure 5.13, however, does not accelerate. In this case, there are two interactions occurring—two forces acting on the football. If they are simultaneous, equal and opposite, then the net force is zero. Do the two opposing kicks make up an action–reaction pair? No, for they act on the same object, not on different objects. They may be equal and opposite, but, unless they act on different objects, they are not an action–reaction pair. Get it?

If this is confusing, it may be well to note that Newton had difficulties with the third law himself. (See insightful examples of Newton's third law on pages 21 and 22 in the *Concept Development Practice Book*.)

CHECK YOURSELF

1. On a cold, rainy day, you find yourself in a car with a dead battery. You must push the car to move it and get it started. Why can't you move the car by remaining comfortably inside and pushing against the dashboard?
2. Why does a book sitting on a table never accelerate "spontaneously" in response to the trillions of interatomic forces acting within it?
3. We know that the Earth pulls on the Moon. Does it follow that the Moon also pulls on the Earth?
4. Can you identify the action and reaction forces in the case of an object falling in a vacuum?

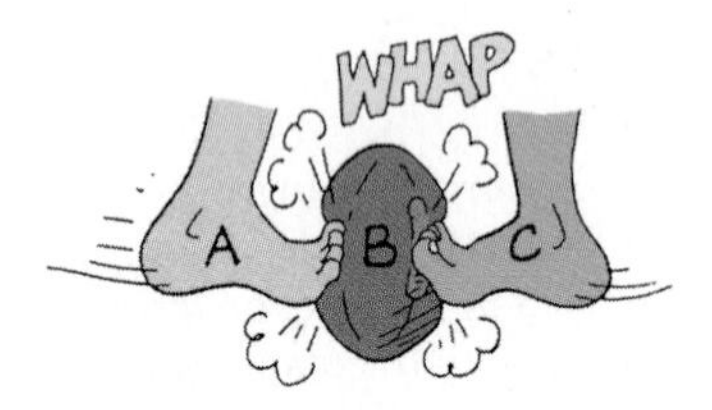

FIGURE 5.13
Both A and C act on B. They can cancel each other, so B does not accelerate.

CHECK YOUR ANSWERS

1. In this case, the system to be accelerated is the car. If you remain inside and push on the dashboard, the force pair you produce acts and reacts within the system. These forces cancel out as far as any motion of the car is concerned. To accelerate the car, there must be an interaction between the car and something external—for example, you on the outside pushing against the road and on the car.
2. Every one of these interatomic forces is part of an action-reaction pair within the book. These forces add up to zero, no matter how many of them there are. This is what makes Newton's *first* law apply to the book. The book has zero acceleration unless an *external* force acts on it.
3. Yes, both pulls make up an action-reaction pair of forces associated with the gravitational interaction between Earth and Moon. We can say that (1) Earth pulls on Moon and (2) Moon likewise pulls on Earth; but it is more insightful to think of this as a single interaction—both Earth and Moon simultaneously pulling on each other, each with the *same* amount of force.
4. To identify a pair of action-reaction forces in any situation, first identify the pair of interacting objects involved—Body A and Body B. Body A, the falling object, is interacting (gravitationally) with Body B, the whole Earth. So the Earth pulls downward on the object (call it action), while the object pulls upward on the Earth (reaction).

You can't push or pull on something unless that something simultaneously pushes or pulls on you. That's the law!

Insights

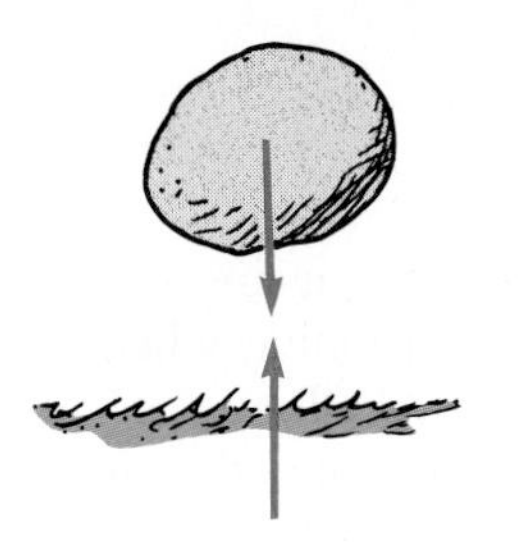

FIGURE 5.14
The Earth is pulled up by the boulder with just as much force as the boulder is pulled downward by the Earth.

THE Physics Place
Action and Reaction on Different Masses
Action and Reaction on Rifle and Bullet

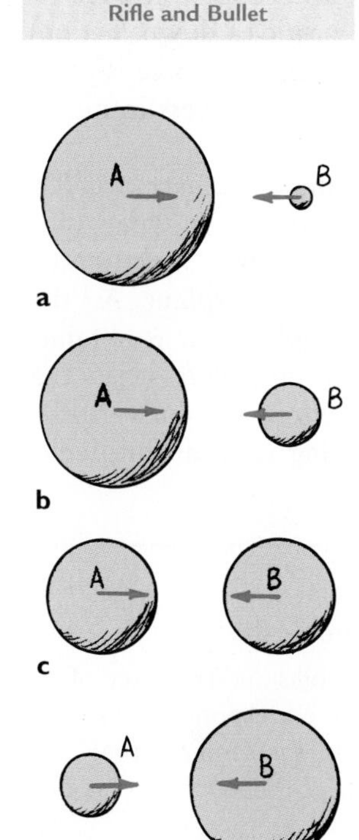

FIGURE 5.15
Which falls toward the other, A or B? Do the accelerations of each relate to their relative masses?

Action and Reaction on Different Masses

As strange as it may first seem, a falling object pulls upward on Earth with as much force as Earth pulls downward on it. The resulting acceleration of the falling object is evident, while the upward acceleration of Earth is too small to detect. So strictly speaking, when you step off a curb, the street rises ever so slightly to meet you.

We can see that the Earth accelerates slightly in response to a falling object by considering the exaggerated examples of two planetary bodies, *a* through *e* in Figure 5.15. The forces between bodies A and B are equal in magnitude and oppositely directed in *each* case. If acceleration of planet A is unnoticeable in *a*, then it is more noticeable in *b*, where the difference between the masses is less extreme. In *c*, where both bodies have equal mass, acceleration of object A is as evident as it is for B. Continuing, we see that the acceleration of A becomes even more evident in *d* and even more so in *e*.

The role of different masses is evident in a fired cannon. When a cannon is fired, there is an interaction between the cannon and the cannonball (Figure 5.16). A pair of forces acts on both cannon and cannonball. The force exerted on the cannonball is as great as the reaction force exerted on the cannon; hence, the cannon recoils. Since the forces are equal in magnitude, why doesn't the cannon recoil with the same speed as the cannonball? In analyzing changes in motion, Newton's second law reminds us that we must also consider the masses involved. Suppose we let F represent both the action and reaction force, m the mass of the cannonball, and $\mathcal{M}$ the mass of the much more massive cannon. The accelerations of the cannonball and the cannon are then found by comparing the ratio of force to mass. The accelerations are:

$$\text{Cannonball: } \frac{F}{m} = \Large a$$

$$\text{Cannon: } \frac{F}{\Large m} = a$$

This shows why the change in velocity of the cannonball is so large compared with the change in velocity of the cannon. A given force exerted on a small mass produces a large acceleration, while the same force exerted on a large mass produces a small acceleration.

Going back to the example of the falling object, if we used similarly exaggerated symbols to represent the acceleration of the Earth reacting to a falling object, the symbol m for the Earth's mass would be astronomical in size. The force F, the weight of the falling object, divided by this large mass would result in a microscopic a to represent the acceleration of the Earth toward the falling object.

FIGURE 5.16
Interactive Figure
The force exerted against the recoiling cannon is just as great as the force that drives the cannonball inside the barrel. Why, then, does the cannonball accelerate more than the cannon?

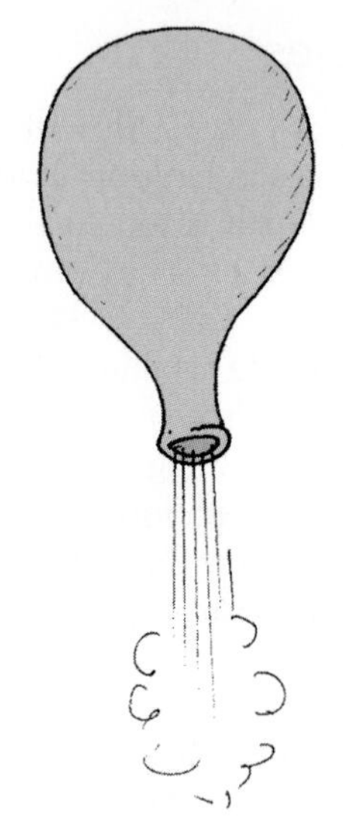

FIGURE 5.17
The balloon recoils from the escaping air, and it moves upward.

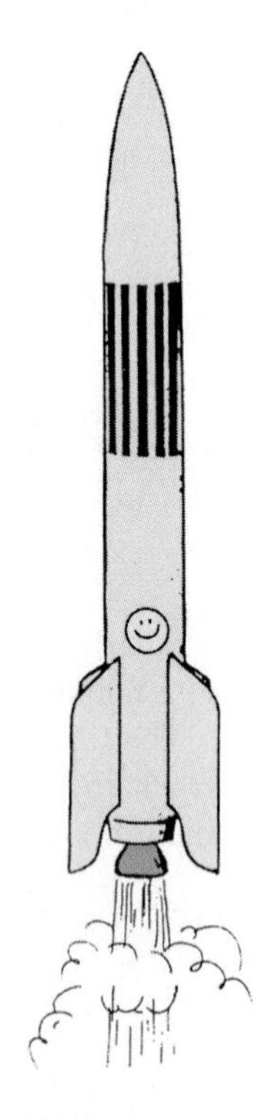

FIGURE 5.18
The rocket recoils from the "molecular cannonballs" it fires, and it moves upward.

We can extend the idea of a cannon recoiling from the ball it fires to understanding rocket propulsion. Consider an inflated balloon recoiling when air is expelled (Figure 5.17). If the air is expelled downward, the balloon accelerates upward. The same principle applies to a rocket, which continually "recoils" from the ejected exhaust gas. Each molecule of exhaust gas is like a tiny cannonball shot from the rocket (Figure 5.18).

A common misconception is that a rocket is propelled by the impact of exhaust gases against the atmosphere. In fact, before the advent of rockets, it was generally thought that sending a rocket to the Moon was impossible. Why? Because there is no air above Earth's atmosphere for the rocket to push against. But this is like saying a cannon wouldn't recoil unless the cannonball had air to push against. Not true! Both the rocket and recoiling cannon accelerate because of the reaction forces exerted by the material they fire—not because of any pushes on the air. In fact, a rocket operates better above the atmosphere where there is no air resistance.

Using Newton's third law, we can understand how a helicopter gets its lifting force. The whirling blades are shaped to force air particles down (action), and the air forces the blades up (reaction). This upward reaction force is called *lift*. When lift equals the weight of the aircraft, the helicopter hovers in midair. When lift is greater, the helicopter climbs upward.

This is true for birds and airplanes. Birds fly by pushing air downward. The air in turn pushes the bird upward. When the bird is soaring, the wings must be shaped so that moving air particles are deflected downward. Slightly tilted wings that deflect oncoming air downward produce lift on an airplane. Air that is pushed downward continuously maintains lift. This supply of air is obtained by the forward motion of the aircraft, which results from propellers or jets that push air backward. The air, in turn, pushes the propellers or jets forward. We will learn in Chapter 14 that the curved surface of a wing is an airfoil, which enhances the lifting force.

CHECK YOURSELF

1. A car accelerates along a road. Identify the force that moves the car.
2. A high-speed bus and an innocent bug have a head-on collision. The force of impact splatters the poor bug over the windshield. Is the corresponding force that the bug exerts against the windshield greater, less, or the same? Is the resulting deceleration of the bus greater than, less than, or the same as that of the bug?

FIGURE 5.19
Geese fly in a V formation because air pushed downward at the tips of their wings swirls upward, creating an updraft that is strongest off to the side of the bird. A trailing bird gets added lift by positioning itself in this updraft, pushes air downward, and creates another updraft for the next bird, and so on. The result is a flock flying in a V formation.

PRACTICING PHYSICS

Tug of War Interactive Figure

Perform a tug-of-war between boys and girls. Do it on a polished floor that's somewhat slippery, with boys wearing socks and girls wearing rubber-soled shoes. Who will surely win, and why? (Hint: Who wins a tug-of-war, those that pull harder on the rope or those who push harder against the floor?)

FIGURE 5.20
You cannot touch without being touched—Newton's third law.

We see Newton's third law at work everywhere. A fish pushes the water backward with its fins, and the water pushes the fish forward. When the wind pushes against the branches of a tree and the branches push back on the wind, we have whistling sounds. Forces are interactions between different things. Every contact requires at least a twoness; there is no way that an object can exert a force on nothing. Forces, whether large shoves or slight nudges, always occur in pairs, each of which is opposite to the other. Thus, we cannot touch without being touched.

CHECK YOUR ANSWERS

1. It is the road that pushes the car along. Really! Only the road provides the horizontal force to move the car forward. How does it do this? The rotating tires of the car push back on the road (action). The road simultaneously pushes forward on the tires (reaction). How about that!
2. The magnitudes of both forces are the same, for they constitute an action-reaction force pair that makes up the interaction between the bus and the bug. The accelerations, however, are very different because the masses are different. The bug undergoes an enormous and lethal deceleration, while the bus undergoes a very tiny deceleration—so tiny that the very slight slowing of the bus is unnoticed by its passengers. But if the bug were more massive—as massive as another bus, for example—the slowing down would unfortunately be very apparent. (Can you see the wonder of physics here? Although so much is *different* for the bug and the bus, the amount of force each encounters is the *same*. Amazing!)

Summary of Newton's Three Laws

Newton's first law, the law of inertia: An object at rest tends to remain at rest; an object in motion tends to remain in motion at constant speed along a straight-line path. This property of objects to resist change in motion is called *inertia.* Mass is a measure of inertia. Objects will undergo changes in motion only in the presence of a net force.

Newton's second law, the law of acceleration: When a net force acts on an object, the object will accelerate. The acceleration is directly proportional to the net force and inversely proportional to the mass. Symbolically, $a = F/m$. Acceleration is always in the direction of the net force. When objects fall in a vacuum, the net force is simply the weight, and the acceleration is g (the symbol g denotes that acceleration is due to gravity alone). When objects fall in air, the net force is equal to the weight minus the force of air resistance, and the acceleration is less than g. If and when the force of air resistance equals the weight of a falling object, acceleration terminates, and the object falls at constant speed (called *terminal speed*).

Newton's third law, the law of action–reaction: Whenever one object exerts a force on a second object, the second object exerts an equal and opposite force on the first. Forces occur in pairs, one action and the other reaction, which together constitute the interaction between one object and the other. Action and reaction always occur simultaneously and act on different objects. Neither force exists without the other.

Isaac Newton's three laws of motion are rules of nature that enable us to see how beautifully so many things connect with one another. We see these rules at play in our everyday environment.

Vectors

Vectors

FIGURE 5.21
This vector, scaled so that 1 cm equals 20 N, represents a force of 60 N to the right.

We have learned that any quantity that requires both magnitude and direction for a complete description is a **vector quantity.** Examples of vector quantities include force, velocity, and acceleration. By contrast, a quantity that can be described by magnitude only, not involving direction, is called a **scalar quantity.** Mass, volume, and speed are scalar quantities.

A vector quantity is nicely represented by an arrow. When the length of the arrow is scaled to represent the quantity's magnitude, and the direction of the arrow shows the direction of the quantity, we refer to the arrow as a **vector.**

Adding vectors that act along parallel directions is simple enough: If they act in the same direction, they add; if they act in opposite directions, they subtract. The sum of two or more vectors is called their **resultant.** To find the resultant of two vectors that don't act in exactly the same or opposite direction, we use the *parallelogram rule.*[2] Construct a parallelogram wherein the

The valentine vector says, "I was only a scalar until you came along and gave me direction."

[2] A parallelogram is a four-sided figure with opposite sides parallel to each other. Usually, you determine the length of the diagonal by measurement; but, in the special case in which the two vectors **X** and **Y** are perpendicular, you can apply the Pythagorean Theorem, $\mathbf{R}^2 = \mathbf{X}^2 + \mathbf{Y}^2$, to find the resultant: $\mathbf{R} = \sqrt{(\mathbf{X}^2 + \mathbf{Y}^2)}$.

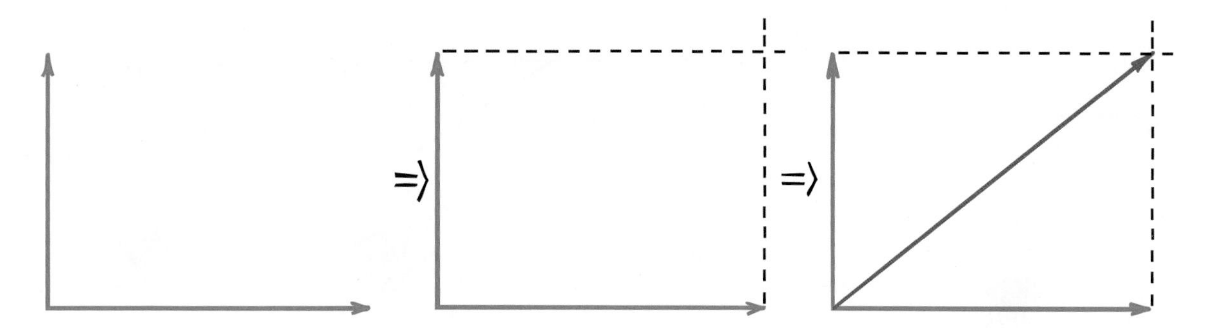

FIGURE 5.22 Interactive Figure

The pair of vectors at right angles to each other make two sides of a rectangle, the diagonal of which is their resultant.

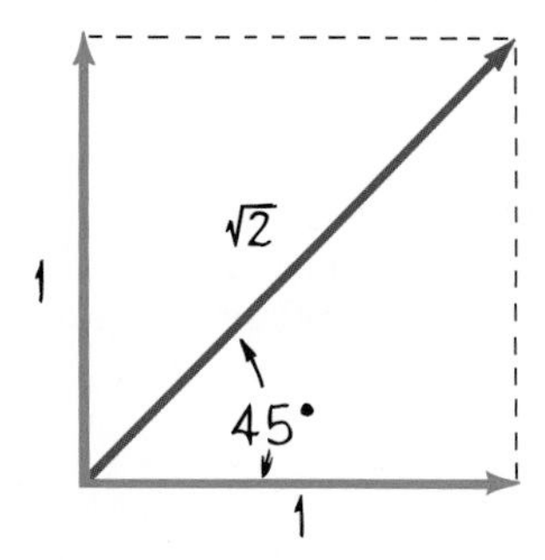

FIGURE 5.23
When a pair of equal-length vectors at right angles to each other are added, they form a square. The diagonal of the square is the resultant, $\sqrt{2}$ times the length of either side.

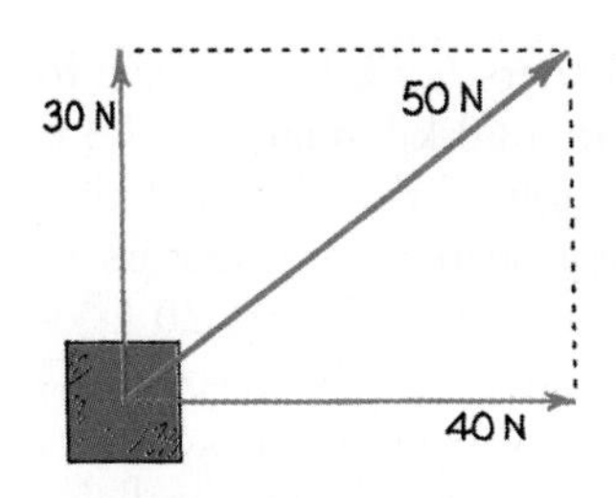

FIGURE 5.24
The resultant of the 30-N and 40-N forces is 50 N.

two vectors are adjacent sides—the diagonal of the parallelogram shows the resultant. In Figure 5.22, the parallelograms are rectangles.

In the special case of two vectors that are equal in magnitude and perpendicular to each other, the parallelogram is a square (Figure 5.23). Since for any square the length of a diagonal is $\sqrt{2}$, or 1.41, times one of the sides, the resultant is $\sqrt{2}$ times one of the vectors. For example, the resultant of two equal vectors of magnitude 100 acting at a right angle to each other is 141.

Force Vectors

Figure 5.24 shows a pair of forces acting on a box. One is 30 newtons and the other is 40 newtons. Simple measurement shows the resultant of this pair of forces is 50 newtons.

Figure 5.25 shows Nellie Newton hanging at rest from a clothesline. Note that the clothesline acts like a pair of ropes that make different angles with the vertical. Which side has the greater tension? Investigation will show there are three forces acting on Nellie: her weight, a tension in the left-hand side of the rope, and a tension in the right-hand side of the rope. Because of the different angles, different rope tensions will occur in each side. Figure 5.25 shows a step-by-step solution. Because Nellie hangs in equilibrium, her weight must be supported by two rope tensions, which must add vectorially to be equal and opposite to her weight. The parallelogram rule shows that the tension in the right-hand rope is greater than the tension in the left-hand rope. If you measure the vectors, you'll see that tension in the right rope is about twice the tension in the left rope. Both rope tensions combine to support her weight.

More about force vectors can be found in Appendix D at the end of this book and on pages 23–30 in the *Practicing Physics* book.

Velocity Vectors

Recall, from Chapter 3, the difference between speed and velocity—speed is a measure of "how fast"; velocity is a measure of both how fast and "in which direction." If the speedometer in a car reads 100 kilometers per hour, you

FIGURE 5.25
Nellie Newton hangs motionless by one hand from a clothesline. If the line is on the verge of breaking, which side is most likely to break?

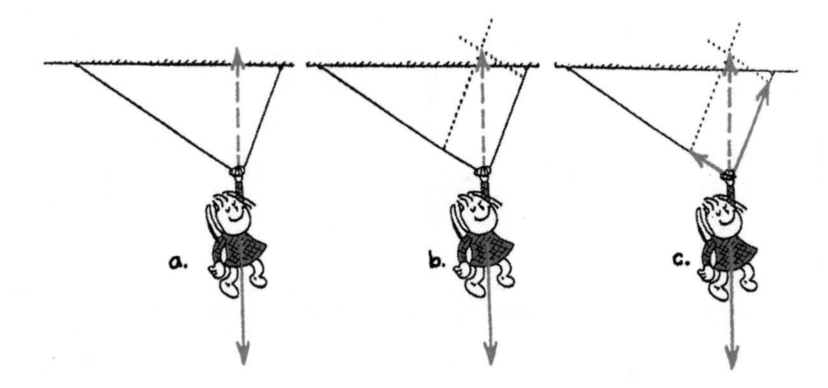

FIGURE 5.26 Interactive Figure
(a) Nellie's weight is shown by the downward vertical vector. An equal and opposite vector is needed for equilibrium, shown by the dashed vector. (b) This dashed vector is the diagonal of a parallelogram defined by the dotted lines. (c) Both rope tensions are shown by the constructed vectors. Tension is greater in the right rope, the one most likely to break.

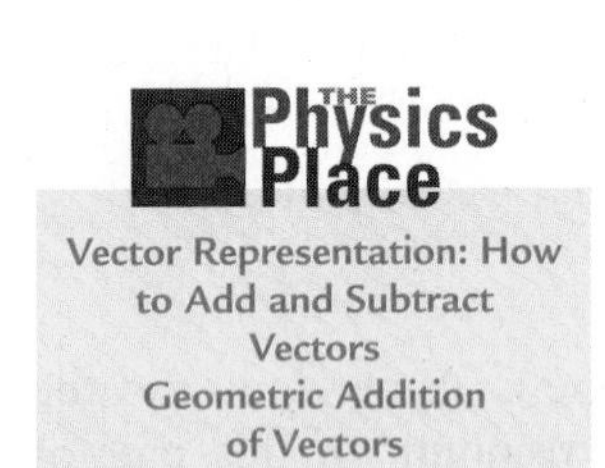

THE Physics Place
Vector Representation: How to Add and Subtract Vectors
Geometric Addition of Vectors

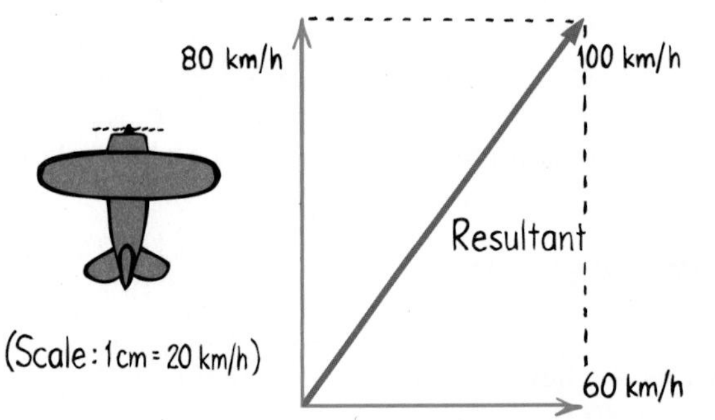

FIGURE 5.27
The 60-km/h crosswind blows the 80-km/h aircraft off course at 100 km/h.

The pair of 6-unit and 8-unit vectors at right angles to each other say, "We may be a six and an eight, but together we're a perfect ten."

Insights

know your speed. If there is also a compass on the dashboard, indicating that the car is moving due north, for example, you know your velocity—100 kilometers per hour north. To know your velocity is to know your speed and your direction.

Consider an airplane flying due north at 80 kilometers per hour relative to the surrounding air. Suppose that the plane is caught in a 60-kilometer-per-hour crosswind (wind blowing at right angles to the direction of the airplane) that blows it off its intended course. This example is represented with vectors in Figure 5.27 with velocity vectors scaled so that 1 centimeter represents 20 kilometers per hour. Thus, the 80-kilometer-per-hour velocity of the airplane is shown by the 4-centimeter vector and the 60-kilometer-per-hour crosswind is shown by the 3-centimeter vector. The diagonal of the constructed parallelogram (a rectangle, in this case) measures 5 cm, which represents 100 km/h. So the airplane moves at 100 km/h relative to the ground, in a direction between north and northeast.

PRACTICING PHYSICS

Here we see a top view of an airplane being blown off course by wind in various directions. Using a pencil and the parallelogram rule, sketch the vectors that show the resulting velocities for each case. In which case does the airplane travel fastest across the ground? Slowest?

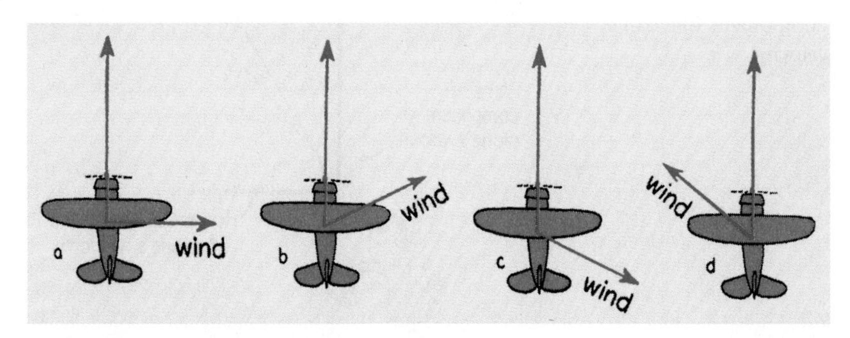

CHECK YOURSELF

Consider a motorboat that normally travels 10 km/h in still water. If the boat heads directly across the river, which also flows at a rate of 10 km/h, what will be its velocity relative to the shore?

PRACTICING PHYSICS

Here we see top views of three motorboats crossing a river. All have the same speed relative to the water, and all experience the same water flow. Construct resultant vectors showing the speed and direction of the boats. Then answer these questions:

(a) Which boat takes the shortest path to the opposite shore?

(b) Which boat reaches the opposite shore first?

(c) Which boat provides the fastest ride?

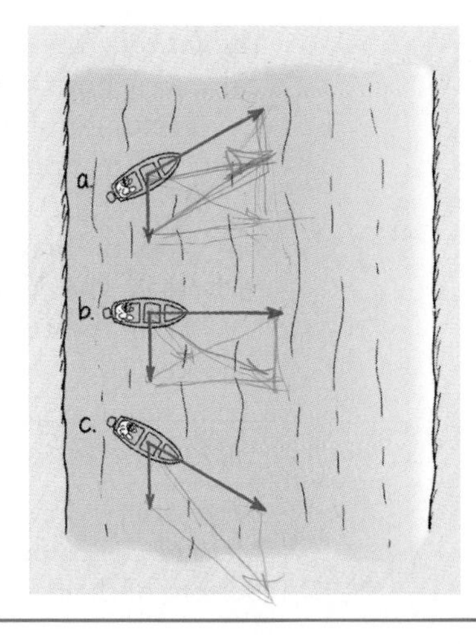

Components of Vectors

Just as two vectors at right angles can be combined into one resultant vector, any vector can be resolved into two *component* vectors perpendicular to each other. These two vectors are known as the *components* of the given vector they replace. The process of determining the components of a vector is called

CHECK YOUR ANSWER

When the boat heads cross-stream (at right angles to the river flow), its velocity is 14.1 km/h, 45 degrees downstream (in accord with the diagram in Figure 5.23).

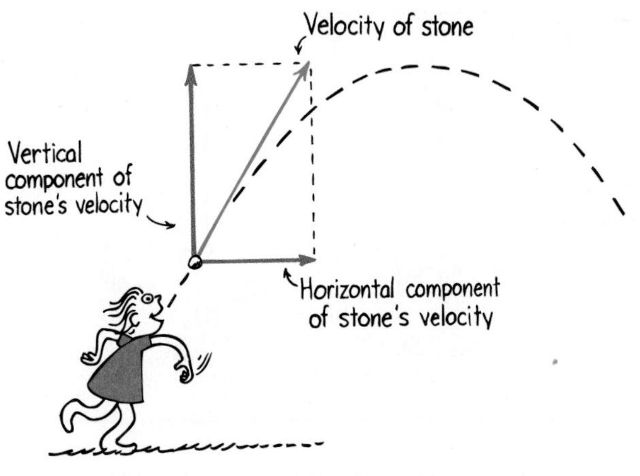

FIGURE 5.28
The horizontal and vertical components of a ball's velocity.

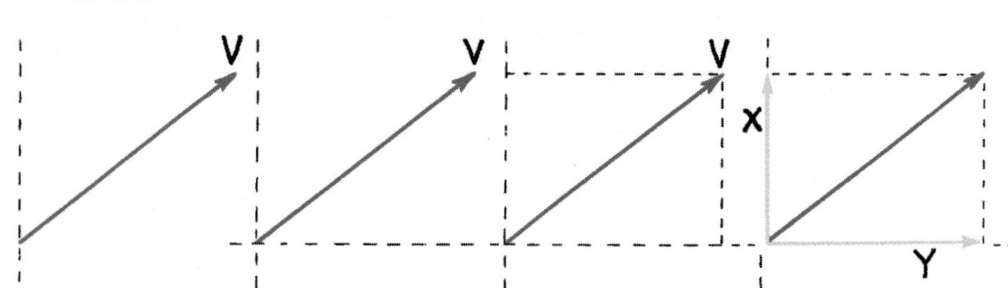

FIGURE 5.29
Construction of the vertical and horizontal components of a vector.

resolution. Any vector drawn on a piece of paper can be resolved into a vertical and a horizontal component.

Vector resolution is illustrated in Figure 5.29. A vector **V** is drawn in the proper direction to represent a vector quantity. Then vertical and horizontal lines (*axes*) are drawn at the tail of the vector. Next, a rectangle is drawn that has **V** as its diagonal. The sides of this rectangle are the desired components, vectors **X** and **Y**. In reverse, note that the vector sum of vectors **X** and **Y** is **V**.

We'll return to vector components when we treat projectile motion in Chapter 10.

EXERCISE

With a ruler, draw the horizontal and vertical components of the two vectors shown. Measure the components and compare your findings with the answers given at the bottom of the page.

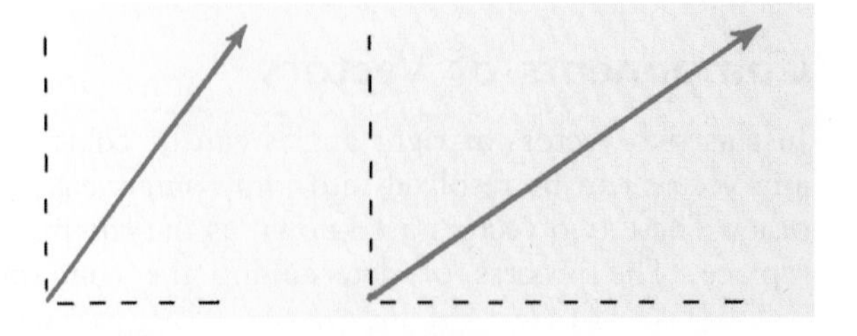

ANSWERS

Left vector: The horizontal component is 2 cm; the vertical component is 2.6 cm. Right vector: The horizontal component is 3.8 cm; the vertical component is 2.6 cm.

Summary of Terms

Newton's third law Whenever one object exerts a force on a second object, the second object exerts an equal and opposite force on the first.

Vector quantity A quantity that has both magnitude and direction. Examples are force, velocity, and acceleration.

Scalar quantity A quantity that has magnitude but not direction. Examples are mass, volume, and speed.

Vector An arrow drawn to scale used to represent a vector quantity.

Resultant The net result of a combination of two or more vectors.

Review Questions

Forces and Interactions

1. When you push against a wall with your fingers, they bend because they experience a force. Identify this force.
2. A boxer can hit a heavy bag with great force. Why can't he hit a piece of tissue paper in midair with the same amount of force?
3. How many forces are required for an interaction?

Newton's Third Law of Motion

4. State Newton's third law of motion.
5. Consider hitting a baseball with a bat. If we call the force on the bat against the ball the *action* force, identify the *reaction* force.
6. Consider the apple and the orange (Figure 5.9). If the system is considered to be only the orange, is there a net force on the system when the apple pulls?
7. If the system is considered to be the apple and the orange together (Figure 5.10), is there a net force on the system when the apple pulls (ignoring friction with the floor)?
8. To produce a net force on a system, must there be an externally applied net force?
9. Consider the system of a single football. If you kick it, is there a net force to accelerate the system? If a friend kicks it at the same time with an equal and opposite force, is there a net force to accelerate the system?

Action and Reaction on Different Masses

10. The Earth pulls down on you with a gravitational force that you call your weight. Do you pull up on the Earth with the same amount of force?
11. If the forces that act on a cannonball and the recoiling cannon from which it is fired are equal in magnitude, why do the cannonball and cannon have very different accelerations?
12. Identify the force that propels a rocket.
13. How does a helicopter get its lifting force?
14. Can you physically touch a person without that person touching you with the same amount of force?

Summary of Newton's Three Laws

15. Fill in the blanks: Newton's first law is often called the law of ________; Newton's second law is the law of ________; and Newton's third law is the law of ________ and ________.
16. Which of the three laws defines the concept of force interaction?

Vectors

17. Cite three examples of a vector quantity and three examples of a scalar quantity.
18. Why is speed considered a scalar and velocity a vector?
19. According to the parallelogram rule, what quantity is represented by the diagonal of a constructed parallelogram?
20. Consider Nellie hanging at rest in Figure 5.25. If the ropes were vertical, with no angle involved, what would be the tension in each rope?
21. When Nellie's ropes make an angle, what quantity must be equal and opposite to her weight?
22. When a pair of vectors are at right angles, is the resultant always greater than either of the vectors separately?

Project

Hold your hand like a flat wing outside the window of a moving automobile. Then slightly tilt the front edge upward and notice the lifting effect. Can you see Newton's laws at work here?

One-Step Calculations

1. Calculate the resultant of the pair of velocities 100 km/h north and 75 km/h south. Calculate the resultant if both of the velocities are directed north.

Resultant of Two Vectors at Right Angles to Each Other; $R = \sqrt{(X^2 + Y^2)}$

2. Calculate the magnitude of the resultant of a pair of 100-km/h velocity vectors that are at right angles to each other.

3. Calculate the resultant of a horizontal vector with a magnitude of 4 units and a vertical vector with a magnitude of 3 units.
4. Calculate the resultant velocity of an airplane that normally flies at 200 km/h if it encounters a 50-km/h wind from the side (at a right angle to the airplane).

Exercises

1. Reconcile the fact that friction acts in a direction to oppose motion even though you rely on friction to propel you forward when walking.
2. The photo shows Steve Hewitt and daughter Gretchen. Is Gretchen touching her dad, or is dad touching her? Explain.

3. When you rub your hands together, can you push harder on one hand than the other?
4. For each of the following interactions, identify action and reaction forces. (a) A hammer hits a nail. (b) Earth gravity pulls down on a book. (c) A helicopter blade pushes air downward.
5. You hold an apple over your head. (a) Identify all the forces acting on the apple and their reaction forces. (b) When you drop the apple, identify all the forces acting on it as it falls and the corresponding reaction forces. Neglect air drag.
6. Identify the action–reaction pairs of forces for the following situations: (a) You step off a curb. (b) You pat your tutor on the back. (c) A wave hits a rocky shore.
7. Consider a baseball player batting a ball. (a) Identify the action–reaction pairs when the ball is being hit, and (b) while the ball is in flight.
8. What physics is involved for a passenger feeling pushed backward into the seat of an airplane when it accelerates along the runway in takeoff?
9. If you drop a rubber ball on the floor, it bounces back up. What force acts on the ball to provide the bounce?
10. When you kick a football, what action and reaction forces are involved? Which force, if any, is greater?
11. Is it true that, when you drop from a branch to the ground below, you pull upward on Earth? If so, then why is the acceleration of Earth not noticed?
12. Within a book on a table, there are billions of forces pushing and pulling on all the molecules. Why is it that these forces never by chance add up to a net force in one direction, causing the book to accelerate "spontaneously" across the table?
13. Two 100-N weights are attached to a spring scale as shown. Does the scale read 0, 100, or 200 N, or does it give some other reading? (Hint: Would it read any differently if one of the ropes were tied to the wall instead of to the hanging 100-N weight?)

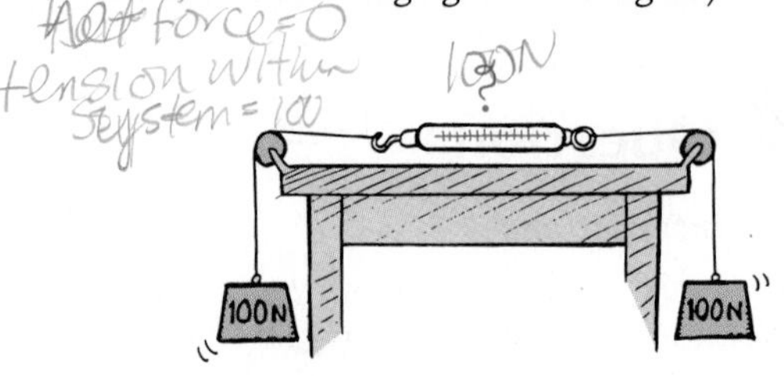

14. When the athlete holds the barbell overhead, the reaction force is the weight of the barbell on his hand. How does this force vary for the case in which the barbell is accelerated upward? Downward?
15. Consider the two forces acting on the person who stands still—namely, the downward pull of gravity and the upward support of the floor. Are these forces equal and opposite? Do they form an action–reaction pair? Why or why not?

16. Why can you exert greater force on the pedals of a bicycle if you pull up on the handlebars?
17. Does a baseball bat slow down when it hits a ball? Defend your answer.
18. Why does a rope climber pull downward on the rope to move upward?
19. You push a heavy car by hand. The car, in turn, pushes back with an opposite but equal force on you. Doesn't this mean that the forces cancel one another, making acceleration impossible? Why or why not?
20. A farmer urges his horse to pull a wagon. The horse refuses, saying that to try would be futile, for it would flout Newton's third law. The horse concludes that she can't exert a greater force on the wagon than the wagon exerts on her, and, therefore, that she won't be able to accelerate the wagon. What is your explanation to convince the horse to pull?

21. The strong man will push the two initially stationary freight cars of equal mass apart before he himself drops straight to the ground. Is it possible for him to give either of the cars a greater speed than the other? Why or why not?

22. Suppose that two carts, one twice as massive as the other, fly apart when the compressed spring that joins them is released. How fast does the heavier cart roll compared with the lighter cart?

23. If you exert a horizontal force of 200 N to slide a crate across a factory floor at constant velocity, how much friction is exerted by the floor on the crate? Is the force of friction equal and oppositely directed to your 200-N push? If the force of friction isn't the reaction force to your push, what is?

24. If a Mack truck and Honda Civic have a head-on collision, upon which vehicle is the impact force greater? Which vehicle experiences the greater deceleration? Explain your answers.

25. Ken and Joanne are astronauts floating some distance apart in space. They are joined by a safety cord whose ends are tied around their waists. If Ken starts pulling on the cord, will he pull Joanne toward him, or will he pull himself toward Joanne, or will both astronauts move? Explain.

26. Which team wins in a tug-of-war—the team that pulls harder on the rope, or the team that pushes harder against the ground? Explain.

27. In a tug-of-war between Sam and Maddy, each pulls on the rope with a force of 250 N. What is the tension in the rope? If both remain motionless, what horizontal force does each exert against the ground?

28. Consider a tug-of-war on a smooth floor between boys wearing socks and girls wearing rubber-soled shoes. Why do the girls win?

29. Two people of equal mass attempt a tug-of-war with a 12-m rope while standing on frictionless ice. When they pull on the rope, each of them slides toward the other. How do their accelerations compare, and how far does each person slide before they meet?

30. What aspect of physics was not known by the writer of this newspaper editorial that ridiculed early experiments by Robert H. Goddard on rocket propulsion above the Earth's atmosphere? "Professor Goddard . . . does not know the relation of action to reaction, and of the need to have something better than a vacuum against which to react . . . seems to lack the knowledge ladled out daily in high schools."

31. Which of the following are scalar quantities, which are vector quantities, and which are neither? (a) velocity; (b) age; (c) speed; (d) acceleration; (e) temperature.

32. What can you correctly say about two vectors that add together to equal zero?

33. Can a pair of vectors with unequal magnitudes ever add to zero? Can three unequal vectors add to zero? Defend your answers.

34. When can a nonzero vector have a zero horizontal component?

35. When, if ever, can a vector quantity be added to a scalar quantity?

36. Which is more likely to break—a hammock stretched tightly between a pair of trees or one that sags more when you sit on it?

37. A heavy bird sits on a clothesline. Will the tension in the clothesline be greater if the line sags a lot or if it sags a little?

38. The rope supports a lantern that weighs 50 N. Is the tension in the rope less than, equal to, or more than 50 N? Use the parallelogram rule to defend your answer.

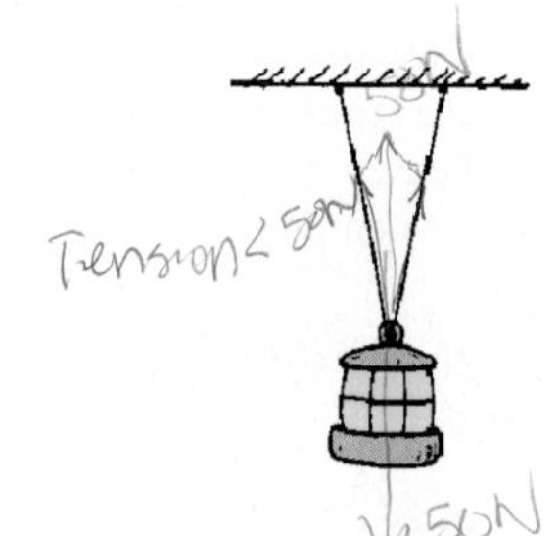

39. The rope is repositioned as shown, and still supports the 50-N lantern. Is the tension in the rope less than, equal to, or more than 50 N? Use the parallelogram rule to defend your answer.

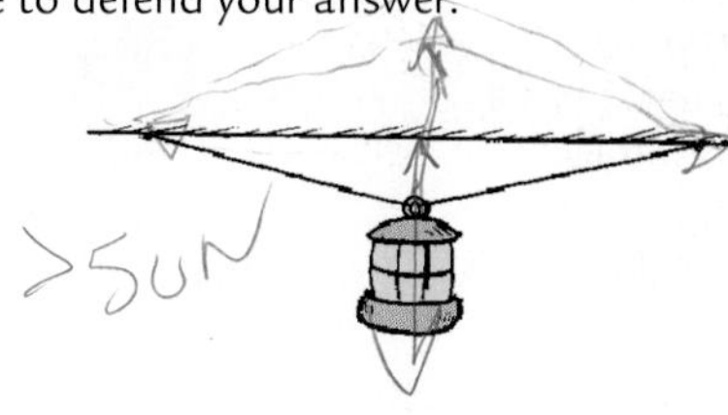

40. Why does vertically falling rain make slanted streaks on the side windows of a moving automobile? If the streaks make an angle of 45°, what does this tell you about the relative speed of the car and the falling rain?

41. A balloon floats motionless in the air. A balloonist begins climbing the supporting cable. In which direction does the balloon move as the balloonist climbs? Defend your answer.

42. Consider a stone at rest on the ground. There are two interactions that involve the stone. One is between the stone and the Earth: Earth pulls down on the stone (its weight) and the stone pulls up on the Earth. What is the other interaction?

43. A stone is shown at rest on the ground. (a) The vector shows the weight of the stone. Complete the vector diagram showing another vector that results in zero net force on the stone. (b) What is the conventional name of the vector you have drawn?

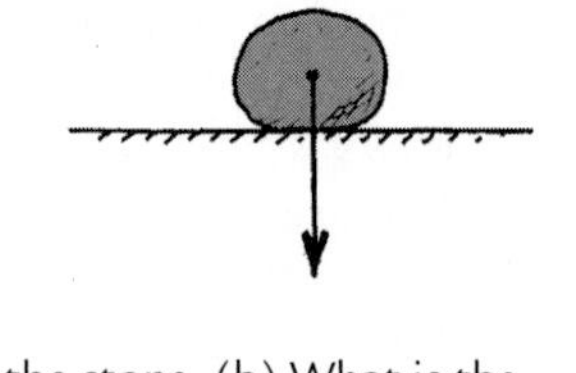

44. Here a stone is suspended at rest by a string. (a) Draw force vectors for all the forces that act on the stone. (b) Should your vectors have a zero resultant? (c) Why, or why not?

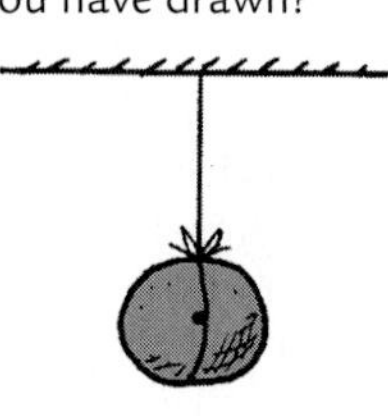

45. Here the same stone is being accelerated vertically upward. (a) Draw force vectors to some suitable scale showing relative forces acting on the stone. (b) Which is the longer vector, and why?

46. Suppose the string in the preceding exercise breaks and the stone slows in its upward motion. Draw a force vector diagram of the stone when it reaches the top of its path.

47. What is the acceleration of the stone of Exercise 46 at the top of its path?

48. Here the stone is sliding down a friction-free incline. (a) Identify the forces that act on it, and draw appropriate force vectors. (b) By the parallelogram rule, construct the resultant force on the stone (carefully showing that it has a direction parallel to the incline—the same direction as the stone's acceleration).

49. Here the stone is at rest, interacting with both the surface of the incline and the block. (a) Identify all the forces that act on the stone, and draw appropriate force vectors. (b) Show that the net force on the stone is zero. (Hint 1: There are two normal forces on the stone. Hint 2: Be sure the vectors you draw are for forces that act *on* the stone, not *by* the stone on the surfaces.)

50. In drawing a diagram of forces that act on a sprinter, which of these should *not* be drawn: Weight, *mg*; force the sprinter exerts on the ground; tension in the sprinter's lower legs?

Problems

1. A boxer punches a sheet of paper in midair and brings it from rest up to a speed of 25 m/s in 0.05 s. If the mass of the paper is 0.003 kg, what force does the boxer exert on it?

2. If you stand next to a wall on a frictionless skateboard and push the wall with a force of 40 N, how hard does the wall push on you? If your mass is 80 kg, what's your acceleration?

3. If raindrops fall vertically at a speed of 3 m/s and you are running at 4 m/s, how fast do they hit your face?

4. Forces of 3.0 N and 4.0 N act at right angles on a block of mass 2.0 kg. How much acceleration occurs?

5. Consider an airplane that normally has an airspeed of 100 km/h in a 100-km/h crosswind blowing from west to east. Calculate its ground velocity when its nose is pointed north in the crosswind.

6. A canoe is paddled at 4 km/h directly across a river that flows at 3 km/h, as shown in the figure. (a) What is the resultant speed of the canoe relative to the shore? (b) In approximately what direction should the canoe be paddled to reach a destination directly across the river?

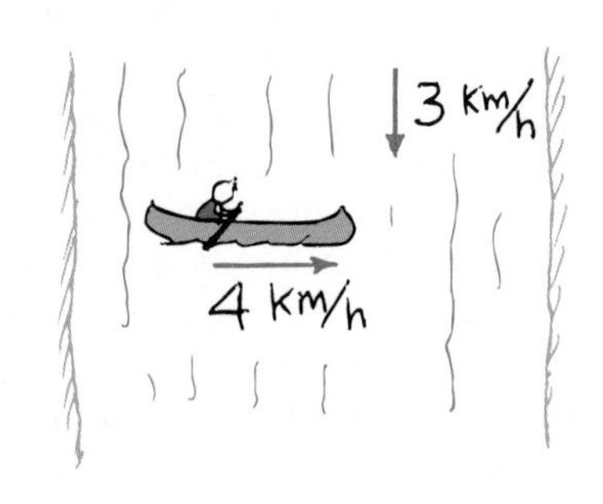

CHAPTER 6

Momentum

Howie Brand demonstrates the different results when a dart bounces from a wooden block rather than sticking to it.

In Chapter 2, we introduced Galileo's concept of inertia and, in Chapter 4, we showed how it was incorporated into Newton's first law of motion. We discussed inertia in terms of objects at rest and objects in motion. In this chapter, we will concern ourselves only with the inertia of moving objects. When we combine the ideas of inertia and motion, we are dealing with momentum. *Momentum* is a property of moving things.

Momentum

We all know that a heavy truck is harder to stop than a small car moving at the same speed. We state this fact by saying that the truck has more momentum than the car. By **momentum** we mean inertia in motion. More specifically, momentum is defined as the product of the mass of an object and its velocity; that is,

$$\text{Momentum} = \text{mass} \times \text{velocity}$$

Or, in shorthand notation,

$$\text{Momentum} = mv$$

When direction is not an important factor, we can say:

$$\text{Momentum} = \text{mass} \times \text{speed},$$

which we still abbreviate mv.

We can see from the definition that a moving object can have a large momentum if either its mass or its velocity is large or if both its mass and its velocity are large. The truck has more momentum than the car moving at the same speed because it has a greater mass. We can see that a huge ship moving at a small speed can have a large momentum, as can a small bullet moving at a high speed. And, of course, a huge object moving at a high speed, such as a massive truck rolling down a steep hill with no brakes, has a huge momentum, whereas the same truck at rest has no momentum at all—because the v term in mv is zero.

FIGURE 6.1
The boulder, unfortunately, has more momentum than the runner.

FIGURE 6.2
Why are the engines of a supertanker normally cut off 25 km from port?

Impulse

If the momentum of an object changes, then either the mass or the velocity or both change. If the mass remains unchanged, as is most often the case, then the velocity changes and acceleration occurs. What produces acceleration? We know the answer is *force.* The greater the force acting on an object, the greater its change in velocity and, hence, the greater its change in momentum.

Timing is especially important when changing momentum.
Insights

But something else is important in changing momentum: time—how long a time the force acts. If you apply a brief force to a stalled automobile, you produce a change in its momentum. Apply the same force over an extended period of time, and you produce a greater change in the automobile's momentum. A force sustained for a long time produces more change in momentum than does the same force applied briefly. So, both force and time interval are important in changing momentum.

The quantity *force* × *time interval* is called **impulse.** In shorthand notation,

$$\text{Impulse} = Ft$$

FIGURE 6.3
When you push with the same force for twice the time, you impart twice the impulse and produce twice the change in momentum.

CHECK YOURSELF

1. Which has more momentum, a 1-ton car moving at 100 km/h or a 2-ton truck moving at 50 km/h?
2. Does a moving object have impulse?
3. Does a moving object have momentum?
4. For the same force, which cannon imparts a greater impulse to a cannonball—a long cannon or a short one?

Impulse Changes Momentum

The greater the impulse exerted on something, the greater will be the change in momentum. The exact relationship is

$$\text{Impulse} = \text{change in momentum}$$

We can express all terms in this relationship in shorthand notation and introduce the delta symbol Δ (a letter in the Greek alphabet used to denote "change in" or "difference in"):[1]

$$Ft = \Delta(mv)$$

The impulse–momentum relationship helps us to analyze many examples in which forces act and motion changes. Sometimes the impulse can be considered to be the cause of a change of momentum. Sometimes a change of momentum can be considered to be the cause of an impulse. It doesn't matter which way you think about it. The important thing is that impulse and change of momentum are always linked. Here we will consider some ordinary examples in which impulse is related to (1) increasing momentum, (2) decreasing momentum over a long time, and (3) decreasing momentum over a short time.

CHECK YOUR ANSWERS

1. Both have the same momentum (1 ton × 100 km/h = 2 ton × 50 km/h).
2. No, impulse is not something an object *has,* like momentum. Impulse is what an object can *provide* or what it can *experience* when it interacts with some other object. An object cannot possess impulse just as it cannot possess force.
3. Yes, but, like velocity, in a relative sense—that is, with respect to a frame of reference, usually Earth's surface. The momentum possessed by a moving object with respect to a stationary point on Earth may be quite different from the momentum it possesses with respect to another moving object.
4. The long cannon will impart a greater impulse because the force acts over a longer time. (A greater impulse produces a greater change in momentum, so a long cannon will impart more speed to a cannonball than a short cannon.)

[1]This relationship is derived by rearranging Newton's second law to make the time factor more evident. If we equate the formula for acceleration, $a = F/m$, with what acceleration actually is, $a = \Delta v/\Delta t$, we get $F/m = \Delta v/\Delta t$. From this we derive $F\Delta t = \Delta(mv)$. Calling Δt simply t, the time interval, $Ft = \Delta(mv)$.

FIGURE 6.4
The force of impact on a golf ball varies throughout the duration of impact.

Case 1: Increasing Momentum

To increase the momentum of an object, it makes sense to apply the greatest force possible for as long as possible. A golfer teeing off and a baseball player trying for a home run do both of these things when they swing as hard as possible and follow through with their swings. Following through extends the time of contact.

The forces involved in impulses usually vary from instant to instant. For example, a golf club that strikes a ball exerts zero force on the ball until it comes in contact; then the force increases rapidly as the ball is distorted (Figure 6.4). The force then diminishes as the ball comes up to speed and returns to its original shape. So, when we speak of such forces in this chapter, we mean the *average* force.

Case 2: Decreasing Momentum

If you were in a car that was out of control and you had to choose between hitting a concrete wall or a haystack, you wouldn't have to call on your knowledge of physics to make up your mind. Common sense tells you to choose the haystack. But, knowing the physics helps you to understand *why* hitting a soft object is entirely different than hitting a hard one. In the case of hitting either the wall or the haystack and coming to a stop, it takes the *same* impulse to decrease your momentum to zero. The same impulse does not mean the same amount of force or the same amount of time; rather it means the same *product* of force and time. By hitting the haystack instead of the wall, you extend the *time during which your momentum is brought to zero.* A longer time interval reduces the force and decreases the resulting deceleration. For example, if the time interval is extended 100 times, the force is reduced to a hundredth. Whenever we wish the force to be small, we extend the time of contact. Hence, the padded dashboards and airbags in motor vehicles.

FIGURE 6.5
If the change in momentum occurs over a long time, the hitting force is small.

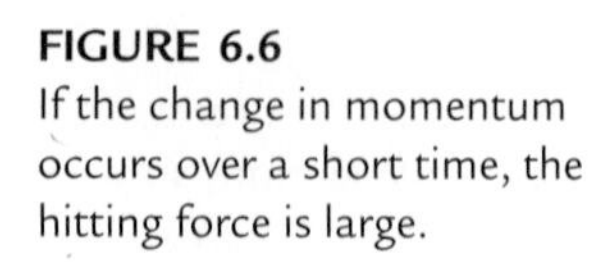

FIGURE 6.6
If the change in momentum occurs over a short time, the hitting force is large.

FIGURE 6.7
A large change in momentum over a long time requires a safely small average force.

When jumping from an elevated position down to the ground, what happens if you keep your legs straight and stiff? Ouch! Instead, you bend your knees when your feet make contact with the ground. By doing so you extend the time during which your momentum decreases by 10 to 20 times that of a stiff-legged, abrupt landing. The resulting force on your bones is reduced by 10 to 20 times. A wrestler thrown to the floor tries to extend his time of impact with the mat by relaxing his muscles and spreading the impact into a series of smaller ones as his foot, knee, hip, ribs, and shoulder successively hit the mat. Of course, falling on a mat is preferable to falling on a solid floor because the mat also increases the time during which the force acts.

The safety net used by circus acrobats is a good example of how to achieve the impulse needed for a safe landing. The safety net reduces the force experienced by a fallen acrobat by substantially increasing the time interval during which the force acts.

If you're about to catch a fast baseball with your bare hand, you extend your hand forward so you'll have plenty of room to let your hand move backward after you make contact with the ball. You extend the time of impact and thereby reduce the force of impact. Similarly, a boxer rides or rolls with the punch to reduce the force of impact (Figure 6.8).

Case 3: Decreasing Momentum over a Short Time

Changing Momentum—Follow-through
Decreasing Momentum over a Short Time

When boxing, if you move into a punch instead of away, you're in trouble. Likewise, if you catch a high-speed baseball while your hand moves toward the ball instead of away upon contact. Or, when your car is out of control, if you drive it into a concrete wall instead of a haystack, you're really in trouble. In these cases of short impact times, the impact forces are large. Remember that,

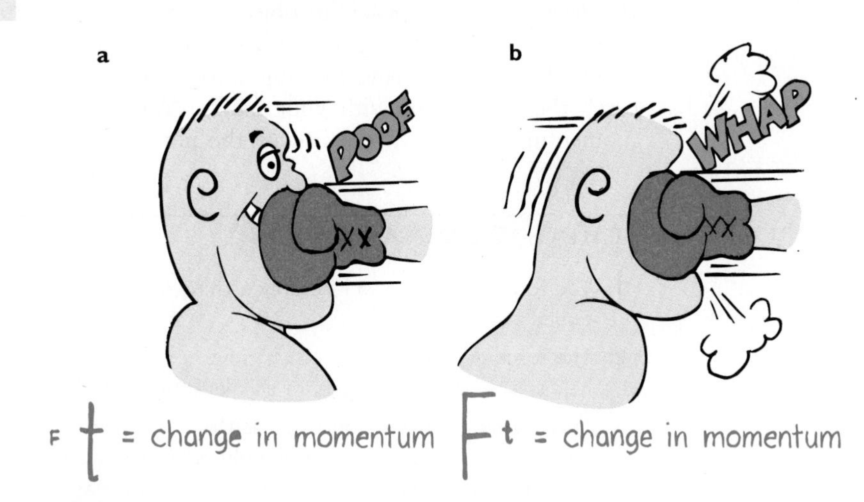

FIGURE 6.8
In both cases, the impulse provided by the boxer's jaw reduces the momentum of the punch. (a) When the boxer moves away (rides with the punch), he extends the time and diminishes the force. (b) If the boxer moves into the glove, the time is reduced and he must withstand a greater force.

FIGURE 6.9
Cassy imparts a large impulse to the bricks in a short time and produces a considerable force.

for an object brought to rest, the impulse is the same, no matter how it is stopped. But, if the time is short, the force will be large.

The idea of short time of contact explains how a karate expert can split a stack of bricks with the blow of her bare hand (Figure 6.9). She brings her arm and hand swiftly against the bricks with considerable momentum. This momentum is quickly reduced when she delivers an impulse to the bricks. The impulse is the force of her hand against the bricks multiplied by the time during which her hand makes contact with the bricks. By swift execution, she makes the time of contact very brief and correspondingly makes the force of impact huge. If her hand is made to bounce upon impact, the force is even greater.

CHECK YOURSELF

1. If the boxer in Figure 6.8 is able to increase the duration of impact three times as long by riding with the punch, by how much will the force of impact be reduced?
2. If the boxer instead moves *into* the punch such as to decrease the duration of impact by half, by how much will the force of impact be increased?
3. A boxer being hit with a punch contrives to extend time for best results, whereas a karate expert delivers a force in a short time for best results. Isn't there a contradiction here?
4. When does impulse equal momentum?

Bouncing

If a flowerpot falls from a shelf onto your head, you may be in trouble. If it bounces from your head, you may be in more serious trouble. Why? Because impulses are greater when an object bounces. The impulse required to bring an object to a stop and then to "throw it back again" is greater than the impulse required merely to bring the object to a stop. Suppose, for example, that you catch the falling pot with your hands. You provide an impulse to reduce its momentum to zero. If you throw the pot upward again, you have to provide

CHECK YOUR ANSWERS

1. The force of impact will be only a third of what it would have been if he hadn't pulled back.
2. The force of impact will be two times greater than it would have been if he had held his head still. Impacts of this kind account for many knockouts.
3. There is no contradiction because the best results for each are quite different. The best result for the boxer is reduced force, accomplished by maximizing time, and the best result for the karate expert is increased force delivered in minimum time.
4. Generally, impulse equals a *change* in momentum. If the initial momentum of an object is zero when the impulse is applied, then impulse = final momentum. And, if an object is brought to rest, impulse = initial momentum.

CONSERVATION LAWS

A conservation law specifies that certain quantities in a system remain precisely constant, regardless of what changes may occur within the system. It is a law of constancy during change. In this chapter, we see that momentum is unchanged during collisions. We say that momentum is conserved. In the next chapter, we'll learn that energy is conserved as it transforms—the amount of energy in light, for example, transforms completely to thermal energy when the light is absorbed. We'll see, in Chapter 8, that angular momentum is conserved—whatever the rotational motion of a planetary system, its angular momentum remains unchanged so long as it is free of outside influences. In Chapter 22, we'll learn that electric charge is conserved, which means that it can neither be created nor destroyed. When we study nuclear physics, we'll see that these and other conservation laws rule in the submicroscopic world. Conservation laws are a source of deep insights into the simple regularity of nature and are often considered the most fundamental of physical laws. Can you think of things in your own life that remain constant as other things change?

FIGURE 6.10
The Pelton wheel. The curved blades cause water to bounce and make a U-turn, which produces a greater impulse to turn the wheel.

additional impulse. This increased amount of impulse is the same that your head supplies if the flowerpot bounces from it.

The photo opener to this chapter shows physics instructor Howie Brand swinging a dart against a wooden block. When the dart has a nail at its nose, the dart comes to a halt as it sticks to the block. The block remains upright. When the nail is removed and the nose of the dart is half of a solid rubber ball, the dart bounces upon contact with the block. The block topples over. The force against the block is greater when bouncing occurs.

The fact that impulses are greater when bouncing occurs was used with great success during the California Gold Rush. The waterwheels used in gold-mining operations were not very effective. A man named Lester A. Pelton recognized a problem with the flat paddles on the waterwheels. He designed a curved paddle that caused the incoming water to make a U-turn upon impact with the paddle. Because the water "bounced," the impulse exerted on the waterwheel was increased. Pelton patented his idea, and he probably made more money from his invention, the Pelton wheel, than any of the gold miners earned. Physics can indeed richen your life in more ways than one.

CHECK YOURSELF

1. In reference to Figure 6.9, how does the force that Cassy exerts on the bricks compare with the force exerted on her hand?
2. How will the impulse resulting from the impact differ if her hand bounces back upon striking the bricks?

CHECK YOUR ANSWERS

1. In accord with Newton's third law, the forces will be equal. Only the resilience of the human hand and the training she has undergone to toughen her hand allow this feat to be performed without broken bones.
2. The impulse will be greater if her hand bounces from the bricks upon impact. If the time of impact is not correspondingly increased, a greater force is then exerted on the bricks (and her hand!).

Conservation of Momentum

From Newton's second law, you know that, to accelerate an object, a net force must be applied to it. This chapter states much the same thing, but in different language. If you wish to change the momentum of an object, exert an impulse on it.

Only an impulse external to a system can change the momentum of the system. Internal forces and impulses won't work. For example, the molecular forces within a baseball have no effect on the momentum of the baseball, just as a push against the dashboard of a car you're sitting in does not affect the momentum of the car. Molecular forces within the baseball and a push on the dashboard are internal forces. They come in balanced pairs that cancel to zero within the object. To change the momentum of the ball or the car, an external push or pull is required. If no external force is present, then no external impulse is present, and no change in momentum is possible.

As another example, consider the cannon being fired in Figure 6.11. The force on the cannonball inside the cannon barrel is equal and opposite to the force causing the cannon to recoil. Since these forces act for the same time, the impulses are also equal and opposite. Recall Newton's third law about action and reaction forces. It applies to impulses, too. These impulses are internal to the system comprising the cannon and the cannonball, so they don't change the momentum of the cannon–cannonball system. Before the firing, the system is at rest and the momentum is zero. After the firing, the net momentum, or total momentum, is *still* zero. Net momentum is neither gained nor lost.

Momentum is conserved for all collisions, elastic and inelastic (whenever external forces don't interfere).

Insights

Momentum, like the quantities velocity and force, has both direction and magnitude. It is a *vector quantity*. Like velocity and force, momentum can be cancelled. So, although the cannonball in the preceding example gains momentum when fired and the recoiling cannon gains momentum in the opposite direction, there is no gain in the cannon–cannonball *system*. The momenta (plural form of momentum) of the cannonball and the cannon are equal in magnitude and opposite in direction.[2] Therefore, these momenta cancel to zero for the

FIGURE 6.11
Interactive Figure
The momentum before firing is zero. After firing, the net momentum is still zero, because the momentum of the cannon is equal and opposite to the momentum of the cannonball.

[2]Here we neglect the momentum of ejected gases from the exploding gunpowder, which can be considerable. Firing a gun with blanks at close range is a definite no-no because of the considerable momentum of ejecting gases. More than one person has been killed by close-range firing of blanks. In 1998, a minister in Jacksonville, Florida, dramatizing his sermon before several hundred parishioners, including his family, shot himself in the head with a blank round from a .357-caliber Magnum. Although no slug emerged from the gun, exhaust gases did—enough to be lethal. So, strictly speaking, the momentum of the bullet (if any) + the momentum of the exhaust gases is equal to the opposite momentum of the recoiling gun.

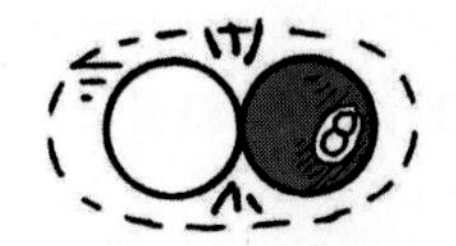

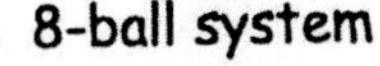

FIGURE 6.12
A cue ball hits an eight ball head-on. Consider this event in three systems: (a) An external force acts on the eight-ball system, and its momentum increases. (b) An external force acts on the cue-ball system, and its momentum decreases. (c) No external force acts on the cue-ball + eight-ball system, and momentum is conserved (simply transferred from one part of the system to the other).

system as a whole. Since no net external force acts on the system, there is no net impulse on the system and no net change in momentum. You can see that, *if no net force or net impulse acts on a system, the momentum of that system cannot change.*

When momentum, or any quantity in physics, does not change, we say it is *conserved.* The idea that momentum is conserved when no external force acts is elevated to a central law of mechanics, called the **law of conservation of momentum,** which states

> **In the absence of an external force, the momentum of a system remains unchanged.**

In any system wherein all forces are internal—as, for example, cars colliding, atomic nuclei undergoing radioactive decay, or stars exploding—the net momentum of the system before and after the event is the same.

CHECK YOURSELF

1. Newton's second law states that, if no net force is exerted on a system, no acceleration occurs. Does it follow that no change in momentum occurs?
2. Newton's third law states that the force a cannon exerts on a cannonball is equal and opposite to the force the cannonball exerts on the cannon. Does it follow that the *impulse* the cannon exerts on the cannonball is equal and opposite to the *impulse* the cannonball exerts on the cannon?

CHECK YOUR ANSWERS

1. Yes, because no acceleration means that no change occurs in velocity or in momentum (mass × velocity). Another line of reasoning is simply that no net force means there is no net impulse and thus no change in momentum.
2. Yes, because the interaction between both occurs during the same *time* interval. Since time is equal and the forces are equal and opposite, the impulses, *Ft*, are also equal and opposite. Impulse is a vector quantity and can be cancelled.

Collisions

Momentum is conserved in collisions—that is, the net momentum of a system of colliding objects is unchanged before, during, and after the collision. This is because the forces that act during the collision are internal forces—forces acting and reacting within the system itself. There is only a redistribution or sharing of whatever momentum exists before the collision.

In any collision, we can say

Net momentum before collision = net momentum after collision.

This is true no matter how the objects might be moving before they collide.

When a moving billiard ball makes a head-on collision with another billiard ball at rest, the moving ball comes to rest and the other ball moves with the speed of the colliding ball. We call this an **elastic collision;** ideally, the colliding objects rebound without lasting deformation or the generation of heat (Figure 6.13). But momentum is conserved even when the colliding objects become entangled during the collision. This is an **inelastic collision,** characterized by deformation, or the generation of heat, or both. In a perfectly inelastic collision, both objects stick together. Consider, for example, the case of a freight car moving along a track and colliding with another freight car at rest (Figure 6.14). If the freight cars are of equal mass and are coupled by the collision, can we predict the velocity of the coupled cars after impact?

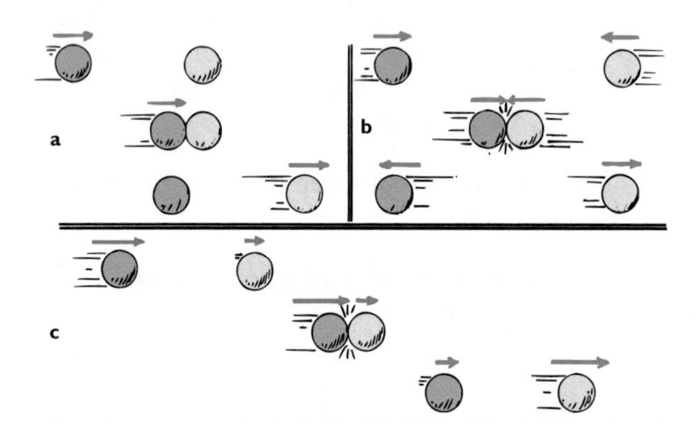

FIGURE 6.13
Interactive Figure
Elastic collisions of equally massive balls. (a) A green ball strikes a yellow ball at rest. (b) A head-on collision. (c) A collision of balls moving in the same direction. In each case, momentum is transferred from one ball to the other.

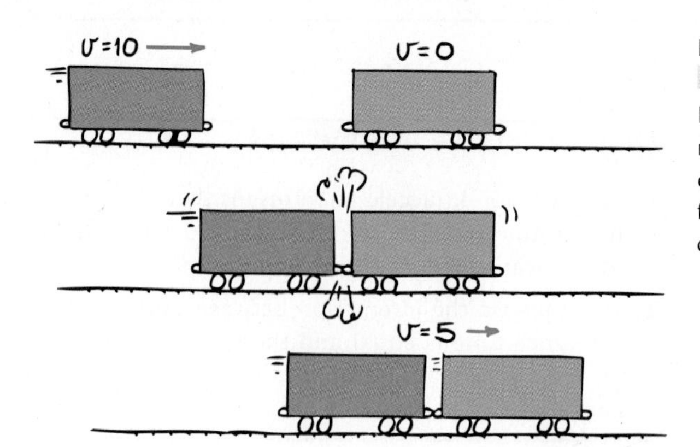

FIGURE 6.14
Interactive Figure
Inelastic collision. The momentum of the freight car on the left is shared with the freight car on the right after collision.

FIGURE 6.15
Interactive Figure
Inelastic collisions. The net momentum of the trucks before and after collision is the same.

Suppose the single car is moving at 10 meters per second, and we consider the mass of each car to be m. Then, from the conservation of momentum,

$$(\text{net } mv)_{\text{before}} = (\text{net } mv)_{\text{after}}$$
$$(m \times 10)_{\text{before}} = (2m \times V)_{\text{after}}$$

By simple algebra, $V = 5$ m/s. This makes sense because, since twice as much mass is moving after the collision, the velocity must be half as much as the velocity before collision. Both sides of the equation are then equal.

Note the inelastic collisions shown in Figure 6.15. If A and B are moving with equal momenta in opposite directions (A and B colliding head-on), then one of these is considered to be negative, and the momenta add algebraically to zero. After collision, the coupled wreck remains at the point of impact, with zero momentum.

If, on the other hand, A and B are moving in the same direction (A catching up with B), the net momentum is simply the addition of their individual momenta.

If A, however, moves east with, say, 10 more units of momentum than B moving west (not shown in the figure), after collision, the coupled wreck moves east with 10 units of momentum. The wreck will finally come to a rest, of course, because of the external force of friction by the ground. The time of impact is short, however, and the impact force of the collision is so much greater than the external friction force that momentum immediately before and after the collision is, for practical purposes, conserved. The net momentum just before the trucks collide (10 units) is equal to the combined momentum of the crumpled trucks just after impact (10 units). The same principle applies to gently docking spacecraft, where friction is entirely absent. Their net momentum just before docking is preserved as their net momentum just after docking.

FIGURE 6.16
Will Maynez demonstrates his air track. Blasts of air from tiny holes provide a friction-free surface for the carts to glide upon.

CHECK YOURSELF

Consider the air track in Figure 6.16. Suppose a gliding cart with a mass of 0.5 kg bumps into, and sticks to, a stationary cart that has a mass of 1.5 kg. If the speed of the gliding cart before impact is v_{before}, how fast will the coupled carts glide after collision?

FIGURE 6.17
Two fish make up a system, which has the same momentum just before lunch and just after lunch.

For a numerical example of momentum conservation, consider a fish that swims toward and swallows a smaller fish at rest (Figure 6.17). If the larger fish has a mass of 5 kg and swims 1 m/s toward a 1-kg fish, what is the velocity of the larger fish immediately after lunch? Neglect the effects of water resistance.

$$\text{Net momentum before lunch} = \text{net momentum after lunch}$$
$$(5 \text{ kg})(1 \text{ m/s}) + (1 \text{ kg})(0 \text{ m/s}) = (5 \text{ kg} + 1 \text{ kg})\,v$$
$$5 \text{ kg·m/s} = (6 \text{ kg})\,v$$
$$v = 5/6 \text{ m/s}$$

Here we see that the small fish has no momentum before lunch because its velocity is zero. After lunch, the combined mass of both fishes moves at velocity v, which, by simple algebra, is seen to be 5/6 m/s. This velocity is in the same direction as that of the larger fish.

Suppose the small fish in this example is not at rest, but swims toward the left at a velocity of 4 m/s. It swims in a direction opposite that of the larger

CHECK YOUR ANSWER

According to momentum conservation, the momentum of the 0.5-kg cart before the collision = momentum of both carts stuck together afterwards.

$$0.5v_{\text{before}} = (0.5 + 1.5)v_{\text{after}}$$
$$v_{\text{after}} = \left(\tfrac{0.5}{2.0}\right)v_{\text{before}} = \frac{v_{\text{before}}}{4}$$

This makes sense, because four times as much mass will be moving after the collision, so the coupled carts will glide more slowly. The same momentum means four times the mass glides 1/4 as fast.

fish—a negative direction, if the direction of the larger fish is considered positive. In this case,

$$\text{Net momentum before lunch} = \text{net momentum after lunch}$$
$$(5\text{ kg})(1\text{ m/s}) + (1\text{ kg})(-4\text{ m/s}) = (5\text{ kg} + 1\text{ kg})\,v$$
$$(5\text{ kg}\cdot\text{m/s}) - (4\text{ kg}\cdot\text{m/s}) = (6\text{ kg})\,v$$
$$1\text{ kg}\cdot\text{m/s} = 6\text{ kg}\,v$$
$$v = 1/6\text{ m/s}$$

Note that the negative momentum of the smaller fish before lunch effectively slows the larger fish after lunch. If the smaller fish were swimming twice as fast, then

$$\text{Net momentum before lunch} = \text{net momentum after lunch}$$
$$(5\text{ kg})(1\text{ m/s}) + (1\text{ kg})(-8\text{ m/s}) = (5\text{ kg} + 1\text{ kg})\,v$$
$$(5\text{ kg}\cdot\text{m/s}) - (8\text{ kg}\cdot\text{m/s}) = (6\text{ kg})\,v$$
$$-3\text{ kg}\cdot\text{m/s} = 6\text{ kg}\,v$$
$$v = -1/2\text{ m/s}$$

Here we see the final velocity is $-1/2$ m/s. What is the significance of the minus sign? It means that the final velocity is *opposite* to the initial velocity of the larger fish. After lunch, the two-fish system moves toward the left. We leave as a chapter-end problem finding the initial velocity of the smaller fish to halt the larger fish in its tracks.

More Complicated Collisions

The net momentum remains unchanged in any collision, regardless of the angle between the paths of the colliding objects. Expressing the net momentum when different directions are involved can be achieved with the parallelogram rule of vector addition. We will not treat such complicated cases in great detail here but will show some simple examples to convey the concept.

In Figure 6.18, we see a collision between two cars traveling at right angles to each other. Car A has a momentum directed due east, and car B's momentum

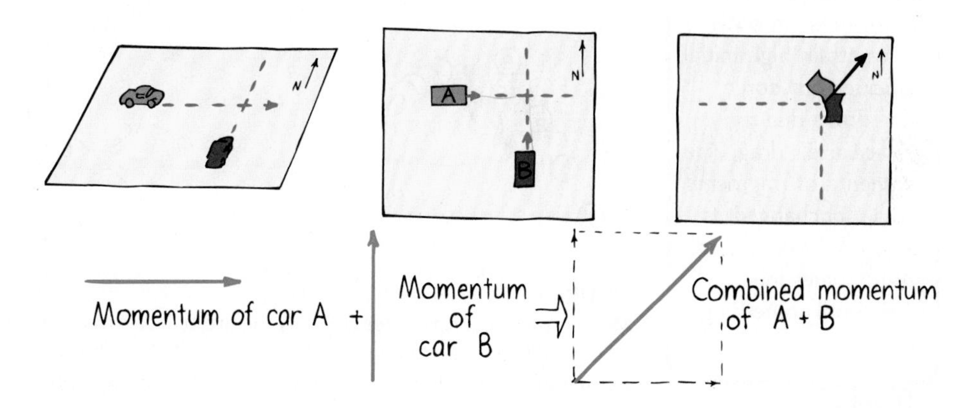

FIGURE 6.18
Interactive Figure
Momentum is a vector quantity.

is directed due north. If their individual momenta are equal in magnitude, then their combined momentum is in a northeastly direction. This is the direction the coupled cars will travel after collision. We see that, just as the diagonal of a square is not equal to the sum of two of the sides, the magnitude of the resulting momentum will not simply equal the arithmetic sum of the two momenta before collision. Recall the relationship between the diagonal of a square and the length of one of its sides, Figure 5.23 in Chapter 5—the diagonal is $\sqrt{2}$ times the length of the side of a square. So, in this example, the magnitude of the resultant momentum will be equal to $\sqrt{2}$ times the momentum of either vehicle.

Figure 6.19 shows a falling Fourth-of-July firecracker exploding into two pieces. The momenta of the fragments combine by vector addition to equal the original momentum of the falling firecracker. Figure 6.20*b* extends this idea to the microscopic realm, where the tracks of subatomic particles are revealed in a liquid hydrogen bubble chamber.

Whatever the nature of a collision or however complicated it is, the total momentum before, during, and after remains unchanged. This extremely useful law enables us to learn much from collisions without knowing any details about the forces that act in the collision. We will see, in the next chapter, that energy as well as momentum is conserved. By applying momentum and energy conservation to the collisions of subatomic particles as observed in various detection

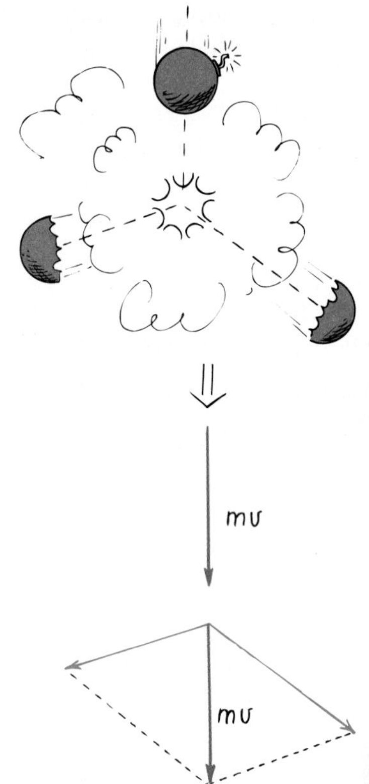

FIGURE 6.19
After the firecracker bursts, the momenta of its fragments add up (by vector addition) to the original momentum.

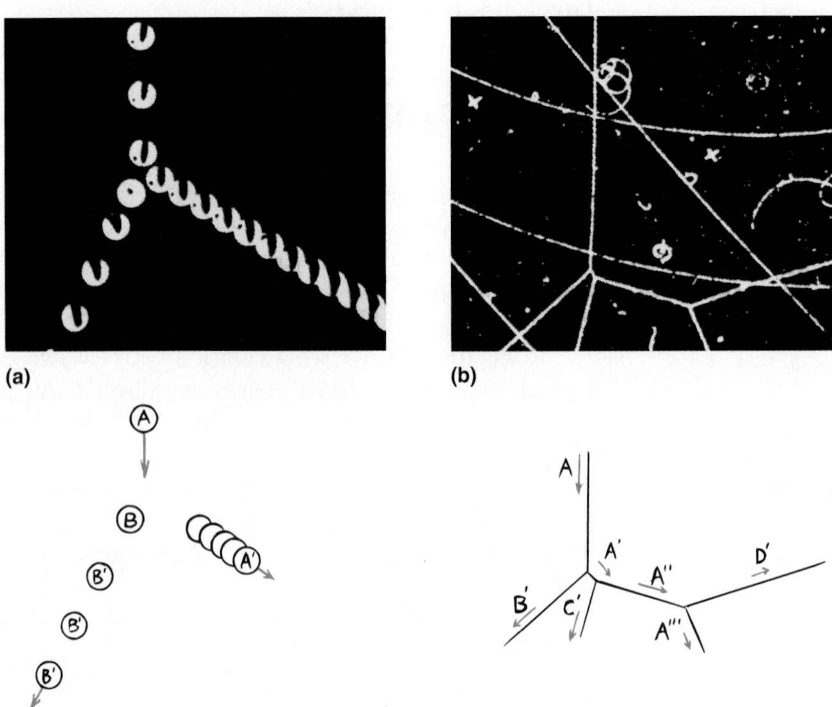

FIGURE 6.20
Momentum is conserved for colliding billiard balls and for colliding nuclear particles in a liquid hydrogen bubble chamber. In (*a*), billiard ball A strikes billiard ball B, which was initially at rest. In (*b*), proton A collides successively with protons B, C, and D. The moving protons leave tracks of tiny bubbles.

Unlike explosions on Earth, those in outer space attain high initial velocities but don't slow down due to gravity or air drag. The momenta of fragments remain unchanged as they travel outward in straight lines indefinitely until something is hit.

Insights

chambers, we can compute the masses of these tiny particles. We obtain this information by measuring momenta and energy before and after collisions. Remarkably, this achievement is possible without any exact knowledge of the forces that act.

Conservation of momentum and conservation of energy (which we will cover in the next chapter) are the two most powerful tools of mechanics. Applying them yields detailed information that ranges from facts about the interactions of subatomic particles to the structure and motion of entire galaxies.

Summary of Terms

Momentum The product of the mass of an object and its velocity.

Impulse The product of the force acting on an object and the time during which it acts.

Relationship of impulse and momentum Impulse is equal to the change in the momentum of the object that the impulse acts upon. In symbol notation,

$$Ft = \Delta mv$$

Law of conservation of momentum In the absence of an external force, the momentum of a system remains unchanged. Hence, the momentum before an event involving only internal forces is equal to the momentum after the event:

$$mv_{(\text{before event})} = mv_{(\text{after event})}$$

Elastic collision A collision in which colliding objects rebound without lasting deformation or the generation of heat.

Inelastic collision A collision in which the colliding objects become distorted, generate heat, and possibly stick together.

Review Questions

Momentum

1. Which has a greater momentum, a heavy truck at rest or a moving skateboard?

Impulse

2. How does impulse differ from force?
3. What are the two ways to increase impulse?
4. For the same force, why does a long cannon impart more speed to a cannonball than a small cannon?

Impulse Changes Momentum

5. Is the impulse–momentum relationship related to Newton's second law?
6. To impart the greatest momentum to an object, should you exert the largest force possible, extend that force for as long a time as possible, or both? Explain.
7. When you are in the way of a moving object and an impact force is your fate, are you better off decreasing its momentum over a short time or over a long time? Explain.
8. Why is it a good idea to have your hand extended forward when you are getting ready to catch a fast-moving baseball with your bare hand?
9. Why would it be a poor idea to have the back of your hand up against the outfield wall when you catch a long fly ball?
10. In karate, why is a force that is applied for a short time more advantageous?
11. In boxing, why is it advantageous to roll with the punch?

Bouncing

12. Which undergoes the greatest change in momentum: (1) a baseball that is caught, (2) a baseball that is thrown, or (3) a baseball that is caught and then thrown back, if all of the baseballs have the same speed just before being caught and just after being thrown?
13. In the preceding question, in which case is the greatest impulse required?

Conservation of Momentum

14. Can you produce a net impulse on an automobile by sitting inside and pushing on the dashboard? Can the internal forces within a soccer ball produce an impulse on the soccer ball that will change its momentum?
15. Is it correct to say that, if no net impulse is exerted on a system, then no change in the momentum of the system will occur?
16. What does it mean to say that momentum (or any quantity) is *conserved*?
17. When a cannonball is fired, momentum is conserved for the *system* of cannon + cannonball. Would momentum be conserved for the system if momentum were not a vector quantity? Explain.

Review #1, 3, 5, 8, 11, 17, 21, 25, 27, 31, 41, 47, 49

Collisions

18. Distinguish between an *elastic collision* and an *inelastic collision*. For which type of collision is momentum conserved?

19. Railroad car A rolls at a certain speed and makes a perfectly elastic collision with car B of the same mass. After the collision, car A is observed to be at rest. How does the speed of car B compare with the initial speed of car A?

20. If the equally massive cars of the previous question stick together after colliding inelastically, how does their speed after the collision compare with the initial speed of car A?

More Complicated Collisions

21. Suppose a ball of putty moving horizontally with 1 kg·m/s of momentum collides and sticks to an identical ball of putty moving vertically with 1 kg·m/s of momentum. Why is their combined momentum not simply the arithmetic sum, 2 kg·m/s?

22. In the preceding question, what is the total momentum of the balls of putty before and after the collision?

Project

When you get a bit ahead in your studies, cut classes some afternoon and visit your local pool or billiards parlor and bone up on momentum conservation. Note that, no matter how complicated the collision of balls, the momentum along the line of action of the cue ball before impact is the same as the combined momentum of all the balls along this direction after impact and that the components of momenta perpendicular to this line of action cancels to zero after impact, the same value as before impact in this direction. You'll see both the vector nature of momentum and its conservation more clearly when rotational skidding, "English," is not imparted to the cue ball. When English is imparted by striking the cue ball off center, rotational momentum, which is also conserved, somewhat complicates analysis. But, regardless of how the cue ball is struck, in the absence of external forces, both linear and rotational momentum are always conserved. Both pool and billiards offer a first-rate exhibition of momentum conservation in action.

One-Step Calculations

Momentum = *mv*

1. What is the momentum of an 8-kg bowling ball rolling at 2 m/s?

2. What is the momentum of a 50-kg carton that slides at 4 m/s across an icy surface?

Impulse = *Ft*

3. What impulse occurs when an average force of 10 N is exerted on a cart for 2.5 s?

4. What impulse occurs when the same force of 10 N acts on the cart for twice the time?

Impulse = change in momentum; $Ft = \Delta mv$

5. What is the impulse on an 8-kg ball rolling at 2 m/s when it bumps into a pillow and stops?

6. How much impulse stops a 50-kg carton sliding at 4 m/s when it meets a rough surface?

Conservation of momentum: $mv_{before} = mv_{after}$

7. A 2-kg blob of putty moving at 3 m/s slams into a 2-kg blob of putty at rest. Calculate the speed of the two stuck-together blobs of putty immediately after colliding.

8. Calculate the speed of the two blobs if the one at rest is 4 kg.

Exercises

1. To bring a supertanker to a stop, its engines are typically cut off about 25 km from port. Why is it so difficult to stop or turn a supertanker?

2. In terms of impulse and momentum, why do padded dashboards make automobiles safer?

3. In terms of impulse and momentum, why do air bags in cars reduce the chances of injury in accidents?

4. Why do gymnasts use floor mats that are very thick?

5. In terms of impulse and momentum, why are nylon ropes, which stretch considerably under tension, favored by mountain climbers?

6. Why is it a serious folly for a bungee jumper to use a steel cable rather than an elastic cord?

7. When jumping from a significant height, why is it advantageous to land with your knees bent?

8. A person can survive a feet-first impact at a speed of about 12 m/s (27 mi/h) on concrete; 15 m/s (34 mi/h) on soil; and 34 m/s (76 mi/h) on water. Why the different values for different surfaces?

9. When catching a foul ball at a baseball game, why is it important to extend your bare hands upward so they can move downward as the ball is being caught?

10. Automobiles were previously manufactured to be as rigid as possible, whereas today's autos are designed to crumple upon impact. Why?

11. In terms of impulse and momentum, why is it important that helicopter blades deflect air downward?

12. A lunar vehicle is tested on Earth at a speed of 10 km/h. When it travels as fast on the Moon, is its momentum more, less, or the same?

13. It is generally much more difficult to stop a heavy truck than a skateboard when they move at the same speed. State a case in which the moving skateboard could require more stopping force. (Consider relative times.)

14. If you throw a raw egg against a wall, you'll break it; but, if you throw it with the same speed into a sagging sheet, it won't break. Explain, using concepts from this chapter.

15. Why is it difficult for a firefighter to hold a hose that ejects large amounts of water at a high speed?

16. Would you care to fire a gun that has a bullet ten times as massive as the gun? Explain.

17. Why are the impulses that colliding objects exert on each other equal and opposite?

18. If a ball is projected upward from the ground with 10 kg·m/s of momentum, what is the momentum of recoil of Earth? Why do we not feel this?

19. When an apple falls from a tree and strikes the ground without bouncing, what becomes of its momentum?

20. Why is a punch more forceful with a bare fist than with a boxing glove?

21. Why do 6-ounce boxing gloves hit harder than 16-ounce gloves?

22. A boxer can punch a heavy bag for more than an hour without tiring, but will tire quickly when boxing with an opponent for a few minutes. Why? (Hint: When the boxer's fist is aimed at the bag, what supplies the impulse to stop the punches? When the boxer's fist is aimed at the opponent, what or who supplies the impulse to stop the punches that don't connect?)

23. Railroad cars are loosely coupled so that there is a noticeable time delay from the time the first car is moved until the last cars are moved from rest by the locomotive. Discuss the advisability of this loose coupling and slack between cars from the point of view of impulse and momentum.

24. If only an external force can change the velocity of a body, how can the internal force of the brakes bring a car to rest?

25. You are at the front of a floating canoe near a dock. You jump, expecting to land on the dock easily. Instead you land in the water. Explain.

26. Explain how a swarm of flying insects can have a net momentum of zero.

27. A fully dressed person is at rest in the middle of a pond on perfectly frictionless ice and must get to shore. How can this be accomplished?

28. If you throw a ball horizontally while standing on roller skates, you roll backward with a momentum that matches that of the ball. Will you roll backward if you go through the motions of throwing the ball, but instead hold on to it? Explain.

29. The examples of the two previous exercises can be explained in terms of momentum conservation and in terms of Newton's third law. Assuming you've answered them in terms of momentum conservation, answer them also in terms of Newton's third law (or vice versa, if you answered already in terms of Newton's third law).

30. In Chapter 5, rocket propulsion was explained in terms of Newton's third law. That is, the force that propels a rocket is from the exhaust gases pushing against the rocket, the reaction to the force the rocket exerts on the exhaust gases. Explain rocket propulsion in terms of momentum conservation.

31. Explain how the conservation of momentum is a consequence of Newton's third law.

32. Go back to Exercise 22 in Chapter 5 and answer it in terms of momentum conservation.

33. Your friend says that the law of momentum conservation is violated when a ball rolls down a hill and gains momentum. What do you say?

34. If you place a box on an inclined plane, it gains momentum as it slides down. What is responsible for this change in momentum?

35. What is meant by a system, and how is it related to the conservation of momentum?

36. If you toss a ball upward, is the momentum of the moving ball conserved? Is the momentum of the system consisting of ball + Earth conserved? Explain your answers.

37. The momentum of an apple falling to the ground is not conserved because the external force of gravity acts on it. But momentum is conserved in a larger system. Explain.

38. Drop a stone from the top of a high cliff. Identify the system wherein the net momentum is zero as the stone falls.

39. A car hurtles off a cliff and crashes on the canyon floor below. Identify the system wherein the net momentum is zero during the crash.

40. Bronco dives from a hovering helicopter and finds his momentum increasing. Does this violate the conservation of momentum? Explain.

41. An ice sailcraft is stalled on a frozen lake on a windless day. The skipper sets up a fan as shown. If all the wind bounces backward from the sail, will the craft be set in motion? If so, in what direction?

42. Will your answer to the preceding exercise be different if the air is brought to a halt by the sail without bouncing?

43. Discuss the advisability of simply removing the sail in the preceding exercises.

44. Which exerts the greater impulse on a steel plate—machine gun bullets that bounce from the plate, or the same bullets squashing and sticking to the plate?

45. As you toss a ball upward, is there a change in the normal force on your feet? Is there a change when you catch the ball? (Think of doing this while standing on a weighing scale.)

46. When you are traveling in your car at highway speed, the momentum of a bug is suddenly changed as it splatters onto your windshield. Compared to the change in momentum of the bug, by how much does the momentum of your car change?

47. If a tennis ball and a bowling ball collide in midair, does each undergo the same amount of momentum change? Defend your answer.

48. If a Mack truck and a Ford Escort have a head-on collision, which vehicle will experience the greater force of impact? The greater impulse? The greater change in momentum? The greater deceleration?

49. Would a head-on collision between two cars be more damaging to the occupants if the cars stuck together or if the cars rebounded upon impact?

50. A 0.5-kg cart on an air track moves 1.0 m/s to the right, heading toward a 0.8-kg cart moving to the left at 1.2 m/s. What is the direction of the two-cart system's momentum?

51. When vertically falling sand lands in a horizontally moving cart, the cart slows. Ignore any friction between the cart and the tracks. Give two reasons for this, one in terms of a horizontal force acting on the cart and one in terms of momentum conservation.

52. In a movie, the hero jumps straight down from a bridge onto a small boat that continues to move with no change in velocity. What physics is being violated here?

53. Suppose that there are three astronauts outside a spaceship and that they decide to play catch. All the astronauts weigh the same on Earth and are equally strong. The first astronaut throws the second one toward the third one and the game begins. Describe the motion of the astronauts as the game proceeds. How long will the game last?

54. To throw a ball, do you exert an impulse on it? Do you exert an impulse to catch it at the same speed? About how much impulse do you exert, in comparison, if you catch it and immediately throw it back again? (Imagine yourself on a skateboard.)

55. In reference to Figure 6.9, how will the impulse at impact differ if Cassy's hand bounces back upon striking the bricks? In any case, how does the force exerted on the bricks compare to the force exerted on her hand?

56. Light consists of tiny "corpuscles" called *photons* that possess momentum. This can be demonstrated with a radiometer, shown in the sketch below. Metal vanes painted black on one side and white on the other are free to rotate around the point of a needle mounted in a vacuum. When photons are incident on the black surface, they are absorbed; when photons are incident upon the white surface, they are reflected. Upon which surface is the impulse of incident light greater, and which way will the vanes rotate? (They rotate in the opposite direction in the more common radiometers in which air is present in the glass chamber; your instructor may tell you why.)

57. A deuteron is a nuclear particle of unique mass made up of one proton and one neutron. Suppose that a deuteron is accelerated up to a certain very high speed in a cyclotron and directed into an observation chamber, where it collides with and sticks to a target particle that is initially at rest and then is observed to move at exactly half the speed of the incident deuteron. Why do the observers state that the target particle is itself a deuteron?

58. When a stationary uranium nucleus undergoes fission, it breaks into two unequal chunks that fly apart. What can you conclude about the momenta of the chunks? What can you conclude about the speed of the chunks?

59. A billiard ball will stop short when it collides head-on with a ball at rest. The ball cannot stop short, however, if the collision is not exactly head-on—that is, if the second ball moves at an angle to the path of the first. Do you know why? (Hint: consider momentum before and after the collision along the initial direction of the first ball and also in a direction perpendicular to this initial direction.)

60. You have a friend who says that, after a golf ball collides with a bowling ball at rest, although the speed gained by the bowling ball is very small, its momentum exceeds the initial momentum of the golf ball. Your friend further asserts this is related to the “negative” momentum of the golf ball after collision. Another friend says this is hogwash—that momentum conservation would be violated. Which friend do you agree with?

Problems

1. Solve for the force of friction acting on a 50-kg carton sliding at 4 m/s if it stops in 3 s.

2. (a) Solve for the amount of force exerted on an 8-kg ball rolling at 2 m/s when it bumps into a pillow and stops in 0.5 s. (b) How much force does the pillow exert on the ball?

3. A car crashes into a wall at 25 m/s and is brought to rest in 0.1 s. Calculate the average force exerted on a 75-kg test dummy by the seat belt.

4. A 1000-kg car accidentally drops from a crane and crashes at 30 m/s to the ground below and comes to an abrupt halt. Consider questions a and b. Which question can be answered using the given information and which one cannot be answered? Explain. (a) What impulse acts on the car when it crashes? (b) What is the force of impact on the car?

5. At a ball game, consider a baseball of mass $m =$ 0.15 kg that falls directly downward at a speed $v = 40$ m/s into the hands of a fan. What impulse Ft must be supplied to bring the ball to rest? If the ball is stopped in 0.03 s, what is the average force of the ball on the catcher's hand?

6. Judy (mass 40.0 kg), standing on slippery ice, catches her leaping dog, Atti (mass 15 kg), moving horizontally at 3.0 m/s. What is the speed of Judy and her dog after the catch?

7. A railroad diesel engine weighs four times as much as a freight car. If the diesel engine coasts at 5 km/h into a freight car that is initially at rest, how fast do the two coast after they couple together?

8. A 5-kg fish swimming 1 m/s swallows an absent-minded 1-kg fish swimming toward it at a velocity that brings both fish to a halt immediately after lunch. What is the velocity v of the smaller fish before lunch?

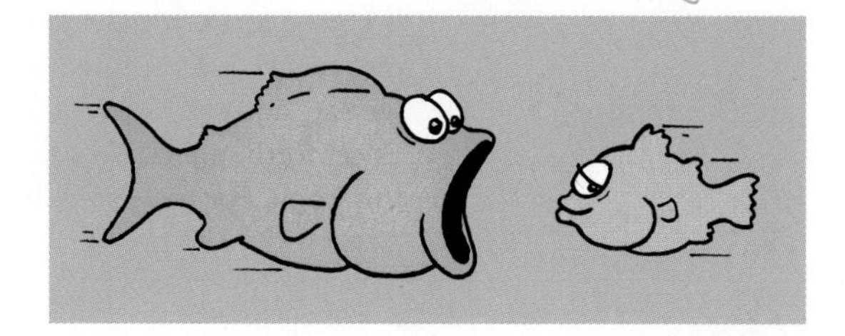

9. Comic-strip hero Superman meets an asteroid in outer space and hurls it at 800 m/s, as fast as a bullet. The asteroid is a thousand times more massive than Superman. In the strip, Superman is seen at rest after the throw. Taking physics into account, what would be his recoil velocity?

10. Two automobiles, each of mass 1000 kg, are moving at the same speed, 20 m/s, when they collide and stick together. In what direction and at what speed does the wreckage move (a) if one car was driving north and one south? (b) if one car was driving north and one east (as shown in Figure 6.18)?

CHAPTER 7

Energy

Annette Rappleyea uses a ballistic pendulum to calculate the speed of a fired ball.

Perhaps the concept most central to all of science is energy. The combination of energy and matter makes up the universe: Matter is substance, and energy is the mover of substance. The idea of matter is easy to grasp. Matter is stuff that we can see, smell, and feel. It has mass, and it occupies space. Energy, on the other hand, is abstract. We cannot see, smell, or feel most forms of energy. Surprisingly, the idea of energy was unknown to Isaac Newton, and its existence was still being debated in the 1850s. Although energy is familiar to us, it is difficult to define, because it is not only a "thing" but both a thing and a process—similar to both a noun and a verb. Persons, places, and things have energy, but we usually observe energy only when it is being transferred or being transformed. It comes to us in the form of electromagnetic waves from the Sun, and we feel it as thermal energy; it is captured by plants and binds molecules of matter together; it is in the foods we eat, and we receive it by digestion. Even matter itself is condensed, bottled-up energy, as set forth in Einstein's famous formula, $E = mc^2$, which we'll return to in the last part of this book. For now, we'll begin our study of energy by considering a related concept: *work*.

Work

The word *work*, in common usage, means physical or mental exertion. Don't confuse the physics definition of work with the everyday notion of work.

Insights

In the previous chapter, we saw that changes in an object's motion depend both on force and on how long the force acts. "How long" meant time. We called the quantity "force × time" *impulse*. But "how long" does not always mean time. It can mean distance also. When we consider the concept of force × *distance*, we are talking about an entirely different concept—**work.**

When we lift a load against Earth's gravity, work is done. The heavier the load or the higher we lift the load, the more work is done. Two things enter the picture whenever work is done: (1) application of a force, and (2) the movement of something by that force. For the simplest case, where the force is constant and the motion is in a straight line in the direction of the force,[1]

[1]More generally, work is the product of only the component of force that acts in the direction of motion and the distance moved. For example, if a force acts at an angle to the motion, the component of force parallel to motion is multiplied by the distance moved. When a force acts at right angles to the direction of motion, with no force component in the direction of motion, no work is done. A common example is a satellite in a circular orbit; the force of gravity is at right angles to its circular path and no work is done on the satellite. Hence, it orbits with no change in speed.

FIGURE 7.1
Work is done in lifting the barbell.

FIGURE 7.2
He may expend energy when he pushes on the wall, but, if the wall doesn't move, no work is done on the wall.

we define the work done on an object by an applied force as the product of the force and the distance through which the object is moved. In shorter form:

$$\text{Work} = \text{force} \times \text{distance}$$
$$W = Fd$$

If we lift two loads one story up, we do twice as much work as in lifting one load the same distance, because the *force* needed to lift twice the weight is twice as much. Similarly, if we lift a load two stories instead of one story, we do twice as much work because the *distance* is twice as great.

We see that the definition of work involves both a force and a distance. A weightlifter who holds a barbell weighing 1000 newtons overhead does no work on the barbell. He may get really tired holding the barbell, but, if it is not moved by the force he exerts, he does no work *on the barbell.* Work may be done on the muscles by stretching and contracting, which is force times distance on a biological scale, but this work is not done on the barbell. Lifting the barbell, however, is a different story. When the weightlifter raises the barbell from the floor, he does work on it.

Work generally falls into two categories. One of these is the work done against another force. When an archer stretches her bowstring, she is doing work against the elastic forces of the bow. Similarly, when the ram of a pile driver is raised, work is required to raise the ram against the force of gravity. When you do push-ups, you do work against your own weight. You do work on something when you force it to move against the influence of an opposing force—often friction.

The other category of work is work done to change the speed of an object. This kind of work is done in bringing an automobile up to speed or in slowing it down. In both categories, work involves a transfer of energy.

The unit of measurement for work combines a unit of force (N) with a unit of distance (m); the unit of work is the newton-meter (N·m), also called the *joule* (J), which rhymes with *cool.* One joule of work is done when a force of 1 newton is exerted over a distance of 1 meter, as in lifting an apple over your head. For larger values, we speak of kilojoules (kJ, thousands of joules), or megajoules (MJ, millions of joules). The weightlifter in Figure 7.1 does work in kilojoules. To stop a loaded truck going at 100 km/h takes megajoules of work.

Power

The definition of work says nothing about how long it takes to do the work. The same amount of work is done when carrying a load up a flight of stairs, whether we walk up or run up. So why are we more tired after running upstairs in a few seconds than after walking upstairs in a few minutes? To understand this difference, we need to talk about a measure of how fast the work is done—*power*. **Power** is equal to the amount of work done per the time it takes to do it:

$$\text{Power} = \frac{\text{work done}}{\text{time interval}}$$

A high-power engine does work rapidly. An automobile engine that delivers twice the power of another automobile engine does not necessarily produce

FIGURE 7.3
The three main engines of a space shuttle can develop 33,000 MW of power when fuel is burned at the enormous rate of 3400 kg/s. This is like emptying an average-size swimming pool in 20 s.

twice as much work or make a car go twice as fast as the less powerful engine. Twice the power means the engine can do twice the work in the same time or do the same amount of work in half the time. A more powerful engine can get an automobile up to a given speed in less time than a less powerful engine can.

Here's another way to look at power: A liter (L) of fuel can do a certain amount of work, but the power produced when we burn it can be any amount, depending on how *fast* it is burned. It can operate a lawnmower for a half hour or a jet engine for a half second.

The unit of power is the joule per second (J/s), also known as the watt (in honor of James Watt, the eighteenth-century developer of the steam engine). One watt (W) of power is expended when 1 joule of work is done in 1 second. One kilowatt (kW) equals 1000 watts. One megawatt (MW) equals 1 million watts. In the United States, we customarily rate engines in units of horsepower and electricity in kilowatts, but either may be used. In the metric system of units, automobiles are rated in kilowatts. (One horsepower is the same as three-fourths of a kilowatt, so an engine rated at 134 horsepower is a 100-kW engine.)

CHECK YOURSELF

If a forklift is replaced with a new forklift that has twice the power, how much greater a load can it lift in the same amount of time? If it lifts the same load, how much faster can it operate?

Mechanical Energy

When work is done by an archer in drawing a bowstring, the bent bow acquires the ability to do work on the arrow. When work is done to raise the heavy ram of a pile driver, the ram acquires the ability to do work on the object it hits when it falls. When work is done to wind a spring mechanism, the spring acquires the ability to do work on various gears to run a clock, ring a bell, or sound an alarm.

In each case, something has been acquired that enables the object to do work. It may be in the form of a compression of atoms in the material of an object, a physical separation of attracting bodies, or a rearrangement of electric charges in the molecules of a substance. This "something" that enables an object to do work is **energy**.[2] Like work, energy is measured in joules. It appears in many forms that will be discussed in the following chapters. For now, we will focus on the two most common forms of **mechanical energy**—the energy due to

CHECK YOUR ANSWER

The forklift that delivers twice the power will lift twice the load in the same time or the same load in half the time. Either way, the owner of the new forklift is happy.

[2]Strictly speaking, that which enables an object to do work is its available energy, for not all the energy in an object can be transformed to work.

the position of something or the movement of something. Mechanical energy can be in the form of potential energy, kinetic energy, or the sum of the two.

Potential Energy

An object may store energy by virtue of its position. The energy that is stored and held in readiness is called **potential energy** (PE) because in the stored state it has the potential for doing work. A stretched or compressed spring, for example, has the potential for doing work. When a bow is drawn, energy is stored in the bow. The bow can do work on the arrow. A stretched rubber band has potential energy because of the relative position of its parts. If the rubber band is part of a slingshot, it is capable of doing work.

The chemical energy in fuels is also potential energy. It is actually energy of position at the submicroscopic level. This energy is available when the positions of electric charges within and between molecules are altered—that is, when a chemical change occurs. Any substance that can do work through chemical action possesses potential energy. Potential energy is found in fossil fuels, electric batteries, and the foods we consume.

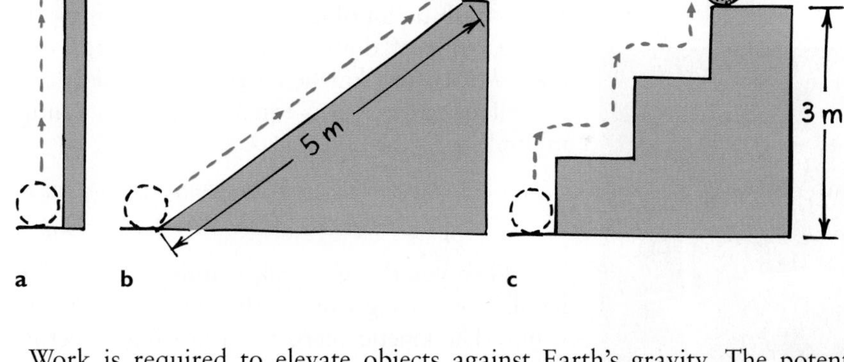

FIGURE 7.4
The potential energy of the 10-N ball is the same (30 J) in all three cases because the work done in elevating it 3 m is the same whether it is (a) lifted with 10 N of force, (b) pushed with 6 N of force up the 5-m incline, or (c) lifted with 10 N up each 1-m stair. No work is done in moving it horizontally (neglecting friction).

Work is required to elevate objects against Earth's gravity. The potential energy due to elevated positions is called *gravitational potential energy*. Water in an elevated reservoir and the raised ram of a pile driver both have gravitational potential energy. Whenever work is done, energy changes.

The amount of gravitational potential energy possessed by an elevated object is equal to the work done against gravity in lifting it. The work done equals the force required to move it upward times the vertical distance it is moved (remember $W = Fd$). The upward force required while moving at constant velocity is equal to the weight, mg, of the object, so the work done in lifting it through a height h is the product mgh.

$$\text{gravitational potential energy} = \text{weight} \times \text{height}$$

$$\text{PE} = mgh$$

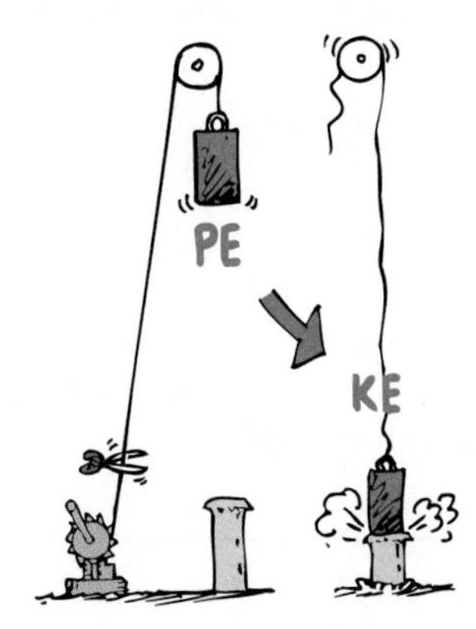

FIGURE 7.5
The potential energy of the elevated ram of the pile driver is converted to kinetic energy when it is released.

Note that the height is the distance above some chosen reference level, such as the ground or the floor of a building. The gravitational potential energy, mgh, is relative to that level and depends only on mg and h. We can see, in Figure 7.4, that the potential energy of the elevated ball does not depend on the path taken to get it there.

FIGURE 7.6
Both do the same work in elevating the block.

Potential energy, gravitational or otherwise, has significance only when it *changes*—when it does work or transforms to energy of some other form. For example, if the ball in Figure 7.4 falls from its elevated position and does 20 joules of work when it lands, then it has lost 20 joules of potential energy. How much *total* potential energy the ball had when it was elevated is relative to some reference level, and it isn't important. What's important is the amount of potential energy that is converted to some other form. Only *changes* in potential energy are meaningful. One of the kinds of energy into which potential energy can change is energy of motion, or *kinetic energy*.

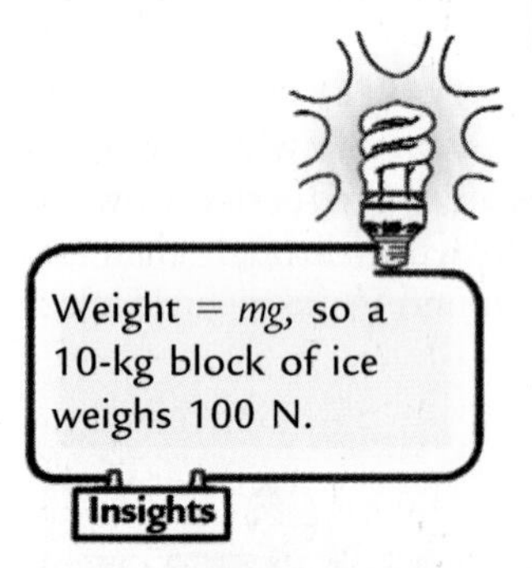

Weight = *mg*, so a 10-kg block of ice weighs 100 N.
Insights

CHECK YOURSELF

1. How much work is done in lifting the 100-N block of ice a vertical distance of 2 m, as shown in Figure 7.6?
2. How much work is done in pushing the same block of ice up the 4-m-long ramp? (The force needed is only 50 N, which is the reason ramps are used).
3. What is the increase in the block's gravitational potential energy in each case?

Kinetic Energy

If you push on an object, you can set it in motion. If an object is moving, then it is capable of doing work. It has energy of motion. We say it has *kinetic energy* (KE). The **kinetic energy** of an object depends on the mass of the object as well as its speed. It is equal to the mass multiplied by the square of the speed, multiplied by the constant $\frac{1}{2}$.

$$\text{Kinetic energy} = \tfrac{1}{2}\,\text{mass} \times \text{speed}^2$$
$$\text{KE} = \tfrac{1}{2}\,mv^2$$

When you throw a ball, you do work on it to give it speed as it leaves your hand. The moving ball can then hit something and push it, doing work on what it hits. The kinetic energy of a moving object is equal to the work required to bring it from rest to that speed, or the work the object can do while being brought to rest:

$$\text{Net force} \times \text{distance} = \text{kinetic energy}$$

or, in equation notation,

$$Fd = \tfrac{1}{2}\,mv^2$$

FIGURE 7.7
The potential energy of Tenny's drawn bow equals the work (average force × distance) that she did in drawing the arrow into position. When the arrow is released, most of the potential energy of the drawn bow will become the kinetic energy of the arrow.

Note that the speed is squared, so, if the speed of an object is doubled, its kinetic energy is quadrupled ($2^2 = 4$). Consequently, it takes four times the work to double the speed. Whenever work is done, energy changes.

CHECK YOUR ANSWERS

1. $W = Fd = 100\text{ N} \times 2\text{ m} = 200\text{ J}$.
2. $W = Fd = 50\text{ N} \times 4\text{ m} = 200\text{ J}$.
3. Either way increases the block's potential energy by 200 J. The ramp simply makes this work easier to perform.

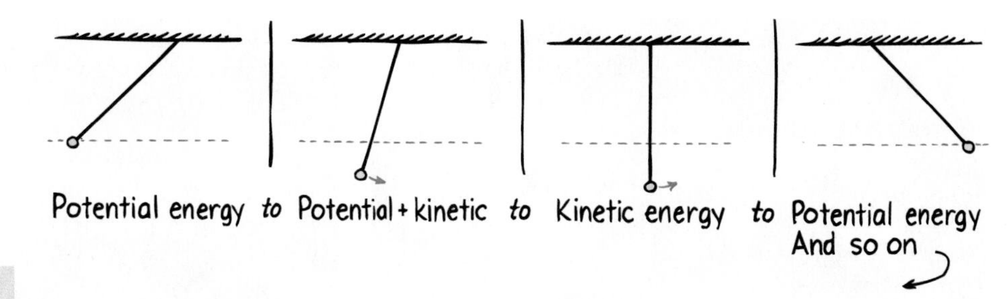

FIGURE 7.8
Energy transitions in a pendulum. PE is relative to the lowest point of the pendulum, when it is vertical.

Bowling Ball and Conservation of Energy

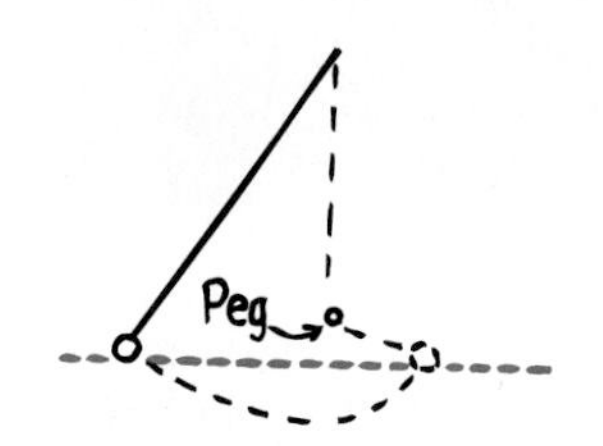

FIGURE 7.9
Interactive Figure
The pendulum bob will swing to its original height whether or not the peg is present.

FIGURE 7.10
The downhill "fall" of the roller coaster results in its roaring speed in the dip, and this kinetic energy sends it up the steep track to the next summit.

Work-Energy Theorem

When a car speeds up, its gain in kinetic energy comes from the work done on it. Or, when a moving car slows, work is done to reduce its kinetic energy. We can say[3]

$$\text{Work} = \Delta\text{KE}$$

Work equals *change* in kinetic energy. This is the **work-energy theorem.** The work in this equation is the *net* work—that is, the work based on the net force. If, for instance, you push on an object and friction also acts on the object, the change of kinetic energy is equal to the work done by the net force, which is your push minus friction. In this case, only part of the total work that you do changes the object's kinetic energy. The rest is soaked up by friction, which goes into heat. If the force of friction is equal and opposite to your push, the net force on the object is zero and no net work is done. Then there is zero change in the object's kinetic energy.

The work-energy theorem applies to decreasing speed as well. When you slam on the brakes of a car, causing it to skid, the road does work on the car. This work is the friction force multiplied by the distance over which the friction force acts.

Interestingly, the maximum friction that the brakes can supply is nearly the same whether the car moves slowly or quickly. In a panic stop with antilock brakes, the only way for the brakes to do more work is to act over a longer distance. A car moving at twice the speed of another takes four times ($2^2 = 4$) as much work to stop. Since the frictional force is nearly the same for both cars, the faster one takes four times as much distance to stop. The same rule applies to older-model brakes that can lock the wheels. The force of friction on a skidding tire is also nearly independent of speed. So, as accident investigators are well aware, an automobile going 100 kilometers per hour, with four times the kinetic energy that it would have at 50 kilometers per hour, skids four times as far with its wheels locked as it would from a speed of 50 kilometers per hour. Kinetic energy depends on speed *squared*.

Automobile brakes convert kinetic energy to heat. Professional drivers are familiar with another way to slow a vehicle—shift to low gear and allow the engine to do the braking. Today's hybrid cars do the same and divert braking energy to electrical storage batteries, where it is used to complement the energy produced by gasoline combustion. (Chapter 25 treats how they do this.) Hooray for hybrid cars!

---

[3]This can be derived as follows: If we multiply both sides of $F = ma$ (Newton's second law) by d, we get $Fd = mad$. Recall from Chapter 3 that, for constant acceleration, $d = \frac{1}{2}at^2$, so we can say $Fd = ma(\frac{1}{2}at^2) = \frac{1}{2}maat^2 = \frac{1}{2}m(at)^2$; and substituting $\Delta v = at$, we get $Fd = \Delta\frac{1}{2}mv^2$. That is, Work = ΔKE.

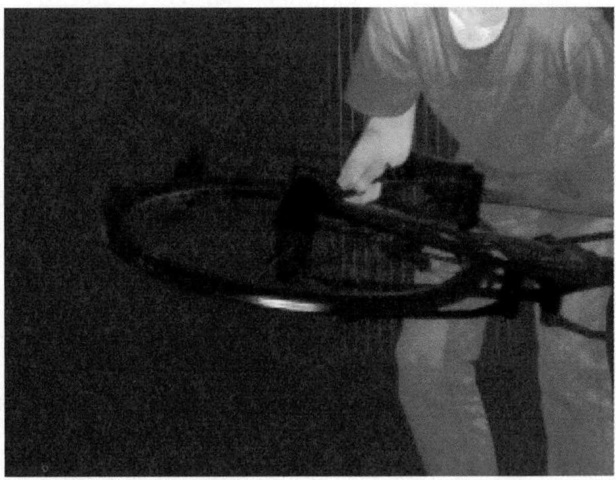

FIGURE 7.11
Due to friction, energy is transferred both into the floor and into the tire when the bicycle skids to a stop. An infrared camera reveals the heated tire track (the red streak on the floor, left) and the warmth of the tire (right). (Courtesy of Michael Vollmer.)

The work-energy theorem applies to more than changes in kinetic energy. Work can change the potential energy of a mechanical device, the heat energy in a thermal system, or the electrical energy in an electrical device. Work is not a form of energy, but a way of transferring energy from one place to another or one form to another.[4]

Kinetic energy and potential energy are two among many forms of energy, and they underlie other forms of energy, such as chemical energy, nuclear energy, sound, and light. Kinetic energy of random molecular motion is related to temperature; potential energies of electric charges account for voltage; and kinetic and potential energies of vibrating air define sound intensity. Even light energy originates from the motion of electrons within atoms. Every form of energy can be transformed into every other form.

CHECK YOURSELF

1. When you are driving at 90 km/h, how much more distance do you need to stop than if you were driving at 30 km/h?
2. Can an object have energy?
3. Can an object have work?

CHECK YOUR ANSWERS

1. Nine times farther. The car has nine times as much kinetic energy when it travels three times as fast: $\frac{1}{2}m(3v)^2 = \frac{1}{2}m9v^2 = 9(\frac{1}{2}mv^2)$. The friction force will ordinarily be the same in either case; therefore, nine times as much work requires nine times as much distance.
2. Yes, but in a relative sense. For example, an elevated object may possess PE relative to the ground below, but none relative to a point at the same elevation. Similarly, the KE that an object has is relative to a frame of reference, usually the Earth's surface. (We will see that material objects have *energy of being*, $E = mc^2$, the congealed energy that makes up their mass.) Read on!
3. No, unlike momentum or energy, work is not something that an object *has*. Work is something that an object *does* to some other object. An object can *do* work only if it has energy.

Conservation of Energy

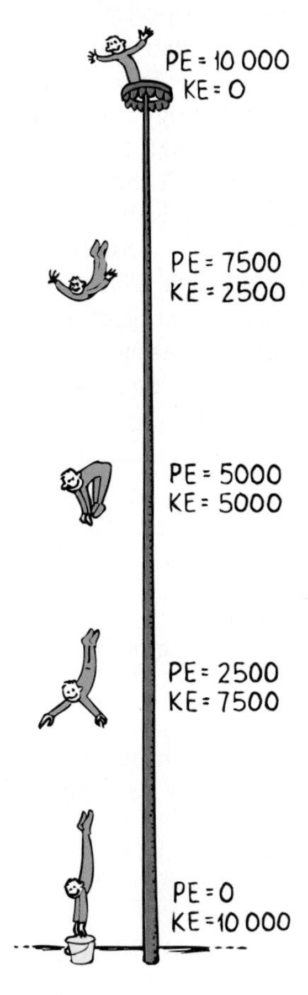

FIGURE 7.12
Interactive Figure
A circus diver at the top of a pole has a potential energy of 10,000 J. As he dives, his potential energy converts to kinetic energy. Note that, at successive positions one-fourth, one-half, three-fourths, and all the way down, the total energy is constant. (Adapted from K. F. Kuhn and J. S. Faughn, *Physics in Your World*. Philadelphia: Saunders, 1980.)

More important than knowing *what energy is* is understanding how it behaves—*how it transforms*. We can better understand the processes and changes that occur in nature if we analyze them in terms of energy *changes*—transformations from one form into another, or of transfers from one location to another. Energy is nature's way of keeping score.

Consider the changes in energy in the operation of the pile driver back in Figure 7.5. Work done to raise the ram, giving it potential energy, becomes kinetic energy when the ram is released. This energy transfers to the piling below. The distance the piling penetrates into the ground multiplied by the average force of impact is almost equal to the initial potential energy of the ram. We say *almost* because some energy goes into heating the ground and ram during penetration. Taking heat energy into account, we find energy transforms without net loss or net gain. Quite remarkable!

The study of various forms of energy and their transformations from one form into another has led to one of the greatest generalizations in physics—the law of **conservation of energy:**

> **Energy cannot be created or destroyed; it may be transformed from one form into another, but the total amount of energy never changes.**

When we consider any system in its entirety, whether it be as simple as a swinging pendulum or as complex as an exploding supernova, there is one quantity that isn't created or destroyed: energy. It may change form or it may simply be transferred from one place to another, but, conventional wisdom tells us, the total energy score stays the same. This energy score takes into account the fact that the atoms that make up matter are themselves concentrated bundles of energy. When the nuclei (cores) of atoms rearrange themselves, enormous amounts of energy can be released. The Sun shines because some of this nuclear energy is transformed into radiant energy.

Enormous compression due to gravity and extremely high temperatures in the deep interior of the Sun fuse the nuclei of hydrogen atoms together to form helium nuclei. This is *thermonuclear fusion,* a process that releases radiant energy, a small part of which reaches Earth. Part of the energy reaching Earth falls on plants (and on other photosynthetic organisms), and part of this, in turn, is later stored in the form of coal. Another part supports life in the food chain that begins with plants (and other photosynthesizers) and part of this energy later is stored in oil. Part of the energy from the Sun goes into the evaporation of water from the ocean, and part of this returns to Earth in rain that may be trapped behind a dam. By virtue of its elevated position, the water behind a dam has energy that may be used to power a generating plant below, where it will be transformed to electric energy. The energy travels through wires to homes, where it is used for lighting, heating, cooking, and operating electrical gadgets. How wonderful that energy transforms from one form to another!

[4]The work-energy theorem can be further stated as Work = $\Delta E + Q$, where Q is the energy transfer due to a temperature difference.

ENERGY AND TECHNOLOGY

Try to imagine life before energy was something that humans controlled. Imagine home life without electric lights, refrigerators, heating and cooling systems, the telephone, and radio and TV—not to mention the family automobile. We may romanticize a better life without these, but only if we overlook the hours of daily toil devoted to doing laundry, cooking, and heating our homes. We'd also have to overlook how difficult it was getting a doctor in times of emergency before the advent of the telephone—when a doctor had little more in his bag than laxatives, aspirins, and sugar pills—and when infant death rates were staggering.

We have become so accustomed to the benefits of technology that we are only faintly aware of our dependence on dams, power plants, mass transportation, electrification, modern medicine, and modern agricultural science for our very existence. When we dig into a good meal, we give little thought to the technology that went into growing, harvesting, and delivering the food on our table. When we turn on a light, we give little thought to the centrally controlled power grid that links the widely separated power stations by long-distance transmission lines. These lines serve as the productive life force of industry, transportation, and the electrification of our society. Anyone who thinks of science and technology as "inhuman" fails to grasp the ways in which they make our lives more human.

CHALLENGE QUESTIONS

1. Does an automobile consume more fuel when its air conditioner is turned on? When its lights are on? When its radio is on while it is sitting in the parking lot?
2. Rows of wind-powered generators are used in various windy locations to generate electric power. Does the power generated affect the speed of the wind? That is, would locations behind the "windmills" have more wind if the windmills weren't there?

Machines

A **machine** is a device for multiplying forces or simply changing the direction of forces. The principle underlying every machine is the **conservation of energy** concept. Consider one of the simplest machines, the **lever** (Figure 7.13). At the same time that we do work on one end of the lever, the other end does work on the load. We see that the direction of force is changed: if we push down, the load is lifted up. If the little work done by friction forces is small enough to neglect, the work input will be equal to the work output.

$$\text{Work input} = \text{work output}$$

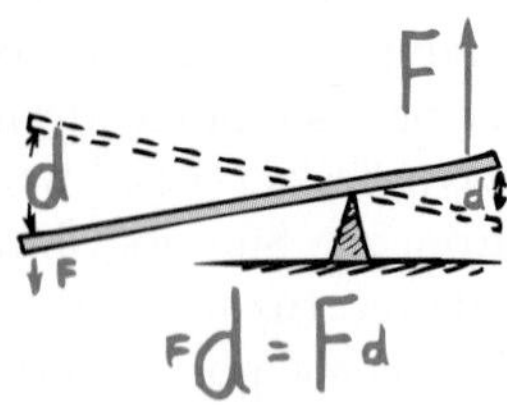

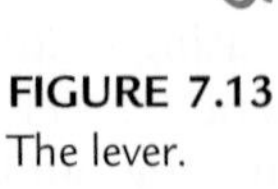

FIGURE 7.13
The lever.

CHECK YOUR ANSWERS

1. The answer to all three questions is *yes,* for the energy they consume ultimately comes from the fuel. Even the energy taken from the battery must be given back to the battery by the alternator, which is turned by the engine, which runs from the energy of the fuel. There's no free lunch!
2. Windmills generate power by taking KE from the wind, so the wind is slowed by interaction with the windmill blades. So, yes, it would be windier behind the windmills if they weren't there.

Machines: Pulleys

Since work equals force times distance, input force × input distance = output force × output distance.

$$(\text{Force} \times \text{distance})_{\text{input}} = (\text{force} \times \text{distance})_{\text{output}}$$

The point of support on which a lever rotates is called a *fulcrum*. When the fulcrum of a lever is relatively close to the load, then a small input force will produce a large output force. This is because the input force is exerted through a large distance and the load is moved through a correspondingly short distance. So a lever can be a force multiplier. But no machine can multiply work or multiply energy. That's a conservation-of-energy no-no!

The principle of the lever was understood by Archimedes, a famous Greek scientist in the third century BC. He said, "Give me where to stand, and I will move the Earth."

Today, a child can use the principle of the lever to jack up the front end of an automobile. By exerting a small force through a large distance, she can provide a large force that acts through a small distance. Consider the ideal example illustrated in Figure 7.14. Every time she pushes the jack handle down 25 centimeters, the car rises only a hundredth as far but with 100 times the force.

Another simple machine is a pulley. Can you see that it is a lever "in disguise"? When used as in Figure 7.15, it changes only the direction of the force; but, when used as in Figure 7.16, the output force is doubled. Force is increased and distance moved is decreased. As with any machine, forces can change while work input and work output are unchanged.

A block and tackle is a system of pulleys that multiplies force more than a single pulley can do. With the ideal pulley system shown in Figure 7.17, the man pulls 7 meters of rope with a force of 50 newtons and lifts a load of 500 newtons through a vertical distance of 0.7 meter. The energy the man expends in pulling the rope is numerically equal to the increased potential energy of the 500-newton block. Energy is transferred from the man to the load.

Any machine that multiplies force does so at the expense of distance. Likewise, any machine that multiplies distance, such as your forearm and elbow, does so at the expense of force. No machine or device can put out more energy than is put into it. No machine can create energy; it can only transfer energy or transform it from one form to another.

FIGURE 7.14
Applied force × applied distance = output force × output distance.

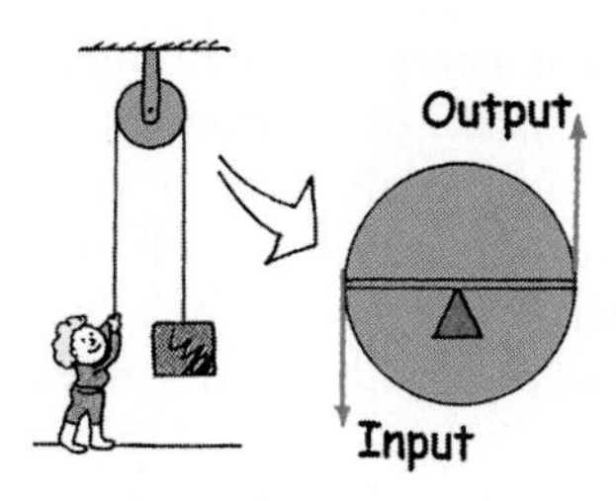

FIGURE 7.15
Interactive Figure
This pulley acts like a lever. It changes only the direction of the input force.

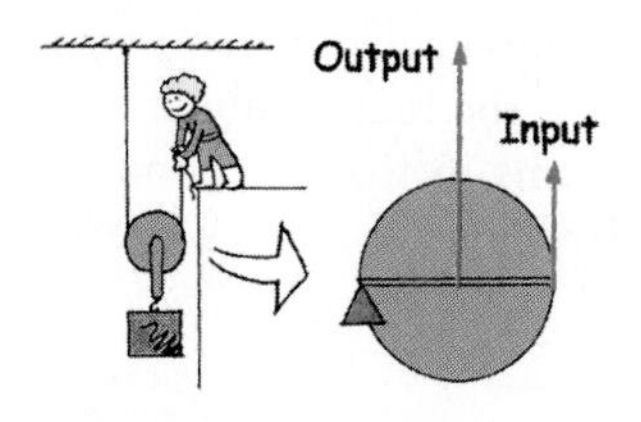

FIGURE 7.16
Interactive Figure
In this arrangement, a load can be lifted with half the input force.

A machine can multiply force, but never *energy*—no way!
Insights

FIGURE 7.17
Interactive Figure
Applied force × applied distance = output force × output distance.

Efficiency

The three previous examples were of *ideal machines*; 100% of the work input appeared as work output. An ideal machine would operate at 100% efficiency. In practice, this doesn't happen, and we can never expect it to happen. In any transformation, some energy is dissipated to molecular kinetic energy—thermal energy. This makes the machine and its surroundings warmer.

Even a lever rocks about its fulcrum and converts a small fraction of the input energy into thermal energy. We may do 100 joules of work and get out 98 joules of work. The lever is then 98% efficient, and we degrade only 2 joules of work input into thermal energy. If the girl back in Figure 7.14 puts in 100 joules of work and increases the potential energy of the car by 60 joules, the jack is 60% efficient; 40 joules of her input work has been applied against friction, making its appearance as thermal energy.

In a pulley system, a considerable fraction of input energy typically goes into thermal energy. If we do 100 joules of work, the forces of friction acting through the distances through which the pulleys turn and rub about their axles may dissipate 60 joules of energy as thermal energy. In that case, the work output is only 40 joules and the pulley system has an efficiency of 40%. The lower the efficiency of a machine, the greater the percentage of energy that is degraded to thermal energy.

Inefficiency exists whenever energy in the world around us is transformed from one form to another. **Efficiency** can be expressed by the ratio

$$\text{Efficiency} = \frac{\text{useful energy output}}{\text{total energy input}}$$

An automobile engine is a machine that transforms chemical energy stored in fuel into mechanical energy. The bonds between the molecules in the petroleum fuel break up when the fuel burns. Carbon atoms in the fuel combine with oxygen in the air to form carbon dioxide, hydrogen atoms in the fuel combine with oxygen to form water, and energy is released. How nice if all this energy

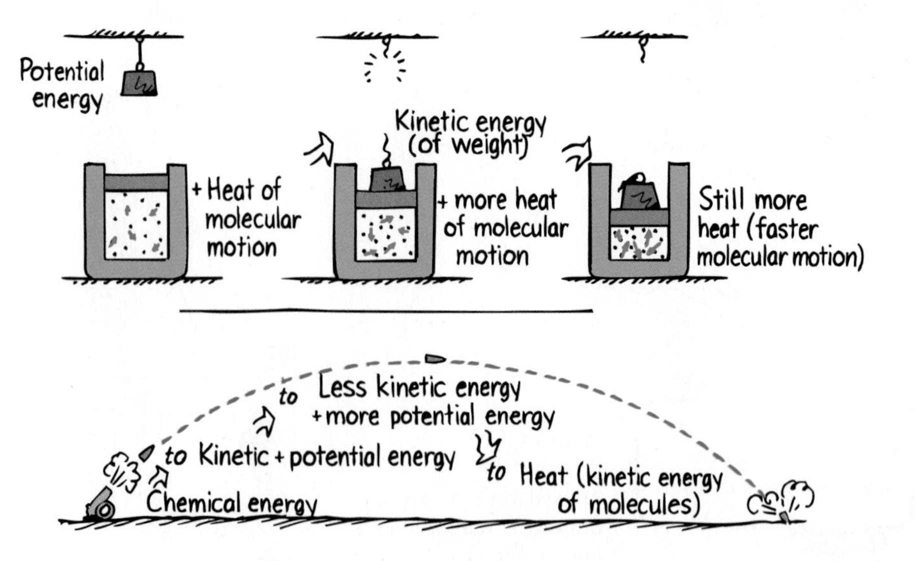

FIGURE 7.18
Energy transitions. The graveyard of mechanical energy is thermal energy.

could be converted into useful mechanical energy—that is to say, how nice it would be if we could have an engine that is 100% efficient. This is impossible, however, because much of the energy is transformed into thermal energy, a little of which may be used to warm passengers in the winter but most of which is wasted. Some goes out in the hot exhaust gases, and some is dissipated to the air through the cooling system or directly from hot engine parts.[5]

Conservation of Energy: Numerical Example

CHALLENGE QUESTION

Consider an imaginary miracle car that has a 100% efficient engine and burns fuel that has an energy content of 40 megajoules per liter. If the air drag and overall frictional forces on the car traveling at highway speed is 500 N, how far could the car travel per liter of fuel at this speed?

Look at the inefficiency that accompanies transformations of energy in this way: In any transformation, there is a dilution of available *useful energy*. The amount of usable energy decreases with each transformation until there is nothing left but thermal energy at ordinary temperature. When we study thermodynamics, we'll see that thermal energy is useless for doing work unless it can be transformed to a lower temperature. Once it reaches the lowest practical temperature, that of our environment, it cannot be used. The environment around us is the graveyard of useful energy.

Comparison of Kinetic Energy and Momentum[6]

Understanding the distinction between momentum and kinetic energy is high-level physics.

Insights

Kinetic energy and momentum are both properties of motion. But they are different. Momentum, like velocity, is a vector quantity. Energy, on the other hand, like mass, is a scalar quantity. When two objects move toward each other, their momenta may partially or fully cancel. Their total momentum is less than the momentum of either one alone. But their kinetic energies cannot

CHECK YOUR ANSWER

From the definition work = force × distance, simple rearrangement gives distance = work/force. If all 40 million J of energy in 1 L were used to do the work of overcoming the air drag and frictional forces, the distance would be:

$$\text{distance} = \frac{\text{work}}{\text{force}} = \frac{40{,}000{,}000 \text{ J/L}}{500 \text{ N}} = 80{,}000 \text{ m/L} = 80 \text{ km/L}$$

(This is about 190 mpg.) The important point here is that, even with a hypothetically perfect engine, there is an upper limit of fuel economy dictated by the conservation of energy.

[5]When you study thermodynamics in Chapter 18, you'll learn that an internal combustion engine *must* transform some of its fuel energy into thermal energy. A fuel cell, on the other hand, which could power future automobiles, doesn't have this limitation. Watch for fuel-cell–powered automobiles in the future!

[6]This section may be skipped in a light treatment of mechanics.

FIGURE 7.19
Compared with the metal bullet with the same momentum, the rubber bullet is more effective in knocking the block over because it bounces upon impact. The rubber bullet undergoes the greater change in momentum and thereby imparts the greater impulse, or wallop, to the block. Which bullet does more damage?

cancel. Since kinetic energies are always positive (or zero), the total kinetic energy of two moving objects is greater than the kinetic energy of either one alone.

For example, the momenta of two cars just before a head-on collision may add up to exactly zero, and the combined wreck after collision will have the same zero value for the momentum. But the kinetic energies add, and this energy is still there after the collision, although it's in different forms—mainly in the form of thermal energy. Or the momenta of two firecrackers approaching each other may cancel, but, when they explode, there is no way their energies can cancel. Energy comes in many forms; momentum has only one form. The vector quantity momentum is different from the scalar quantity kinetic energy.

Another difference is the velocity dependence of the two. Whereas momentum is proportional to velocity (mv), kinetic energy is proportional to the square of velocity ($\frac{1}{2}mv^2$). An object that moves with twice the velocity of another object of the same mass has twice the momentum, but it has four times the kinetic energy. It can provide twice the impulse to whatever it encounters, but it can do four times as much work.

Suppose you're carrying a football and are about to be slammed into and tackled by an approaching player whose oncoming momentum is equal but opposite to your own. The combined momentum of you and the approaching player before impact is zero, but, on impact, you both stop short in your tracks. The wallop or jarring exerted on each of you is the same. This is true whether you

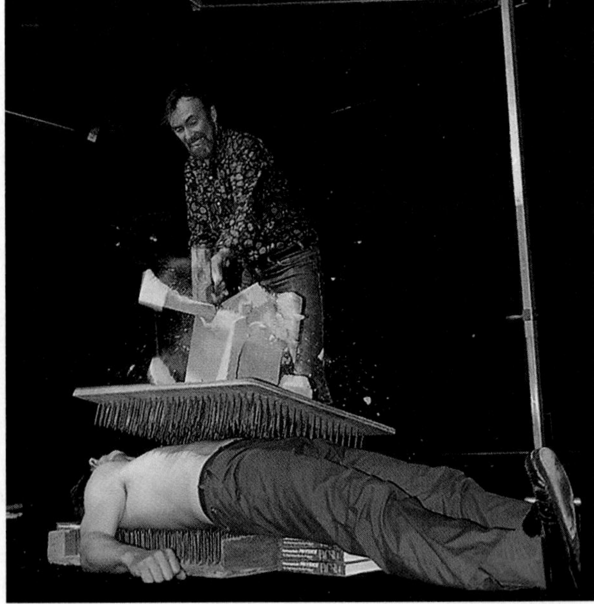

FIGURE 7.20
The author puts kinetic energy and momentum into the hammer, which strikes the block that rests on lab-manual author Paul Robinson, who is bravely sandwiched between beds of nails. Paul is not harmed. Why? Except for the flying cement fragments, every bit of the momentum of the hammer at impact is imparted to Paul, and subsequently to the table and the Earth that supports him. But the momentum only provides the wallop; the energy does the damage. Most of the kinetic energy never gets to him, for it goes into smashing the block apart and into thermal energy. What energy remains is distributed over the more than 200 nails that make contact with his body. The driving force per nail is not enough to puncture the skin.

are tackled by a slow-moving heavy player or a fast-moving light player. If the product of his mass and his velocity matches yours, you're stopped short. Stopping power is one thing, but what about damage? Experienced football players know that it hurts more to be stopped by a fast-moving light player than by a slow-moving heavy player. Why? Because a light player moving with the same momentum has more kinetic energy. If he has the same momentum as a heavy player but is only half as massive, he has twice the velocity. Twice the velocity and half the mass give the lighter player twice the kinetic energy of the heavier player.[7] He does twice the work on you, tends to deform you twice as much, and generally does twice the damage to you. Watch out for the fast-moving little guys!

Energy for Life

Your body is a machine—an extraordinarily wonderful machine. It is made up of smaller machines—living cells. Like any machine, a living cell needs a source of energy. In animals—including you—cells feed on various hydrocarbon compounds that release energy when they react with oxygen. Like gasoline burned in an automobile engine, there is more potential energy in the food molecules than there is in the reaction products after food metabolism. The energy difference is what sustains life.

Energy is nature's way of keeping score.
Insights

We see inefficiency at work in the food chain. Larger creatures feed on smaller creatures, who, in turn, eat smaller creatures, and so on down the line to land plants and ocean plankton that are nourished by the Sun. Advancing each step up the food chain involves inefficiency. In the African bush, 10 kilograms of grass may produce 1 kilogram of gazelle. However, it will require 10 kilograms of gazelle to sustain 1 kilogram of lion. We see that each energy transformation along the food chain contributes to overall inefficiency. Interestingly enough, some of the largest creatures on the planet, the elephant and the blue whale, consume lower down on the food chain. Humans also are considering such tiny organisms as krill and yeast as efficient sources of nourishment.

Sources of Energy

Except for nuclear and geothermal power, the source of practically all our energy is the Sun. Sunlight evaporates water, which later falls as rain; rainwater flows into rivers and turns water wheels or modern generator turbines and then returns to the sea. On a longer time scale, the energy of sunlight produces wood and, on a still longer time scale, petroleum, coal, and natural gas. These materials are the result of photosynthesis, a biological process that incorporates the Sun's radiant energy into plant tissue. The world's present supply of fossil energy is being exhausted in the wink of an eye (a few hundred years) relative to the time required to produce it (millions of years). Whereas petroleum and coal fueled the industries of the twentieth century, their roles will diminish in this present century. So we look to alternative sources of energy.

[7]Note that $\frac{1}{2}(m/2)(2v)^2 = mv^2$, which is twice the value $\frac{1}{2}mv^2$ of the heavier player of mass m and speed v.

FIGURE 7.21
Dry-rock geothermal power. (a) A hole is sunk several kilometers into dry granite. (b) Water is pumped into the hole at high pressure and fractures surrounding rock to form a cavity with increased surface area. (c) A second hole is sunk to intercept the cavity. (d) Water is circulated down one hole and through the cavity, where it is superheated, before rising through the second hole. After driving a turbine, it is recirculated into the hot cavity again, making a closed cycle.

Energy in sunlight is nicely captured by way of photovoltaic cells to generate electricity. They're found in solar-powered calculators and, more recently, in flexible solar shingles on the rooftops of buildings. Solar-energy technology offers a promising future because photovoltaic cells can generate large quantities of electricity for nations with sufficient sunlight and sufficient land area. To meet all the energy needs of the United States through photovoltaic sources, an area as large as Massachusetts might be needed. But it is not expected photovoltaic sources alone will supply the electricity we will need in the future.

Even the wind, caused by unequal warming of Earth's surface, is a form of solar power. The energy of wind can be used to turn generator turbines within specially equipped windmills. Because wind power can't be turned on and off at will, it is now only a supplement to fossil and nuclear fuels for large-scale power production. Harnessing the wind is most practical when the energy it produces is stored for future use, such as in the form of hydrogen.

Sooner or later, all the sunlight that falls on Earth will be radiated back into space. Energy in any ecosystem is always in transit—you can rent it, but you can't own it.

Insights

The most concentrated form of usable energy is in uranium—a nuclear fuel that could provide large quantities of energy for many decades. Advanced fission technology that involves breeder reactors and the use of thorium could extend that time line to many hundreds of years (Chapter 34). Nuclear power plants do not require much land area and are dependent on location only to the extent that they need cooling water. Present-day plants use nuclear fission, but it is likely that nuclear fusion will predominate in the future. Controlled nuclear fusion remains an intriguing energy alternative of vast magnitude. At the present time, public concern about anything nuclear prevents the growth of nuclear power in the United States. It is interesting to note that Earth's interior is kept hot because of a form of nuclear power, radioactive decay, which has been with us since time zero.

A by-product of radioactive decay in Earth's interior is geothermal energy—heat that can be tapped beneath the Earth's surface. Geothermal energy is commonly found in areas of volcanic activity, such as Iceland, New Zealand, Japan, and Hawaii, where heated water near Earth's surface is tapped to provide steam for running turbogenerators. In locations where heat from volcanic activity is near the ground surface in the absence of groundwater, another method holds promise for producing electricity: dry-rock geothermal power, where cavities are made in deep, dry, hot rock into which water is introduced. When the water turns to steam, it is piped to a turbine at the surface. Then it is returned to the cavity for reheating.

Inventors take heed: When introducing a new idea, first be sure it is in context with what is presently known. For example, it should be consistent with the conservation of energy.

Insights

Geothermal power, like solar, wind, and water power, is friendly to the environment. Other methods for obtaining energy have serious environmental consequences. Although nuclear power isn't a polluter of the atmosphere, it remains controversial because of the nuclear wastes that are generated. The combustion of fossil fuels, on the other hand, leads to increased atmospheric concentrations of carbon dioxide, sulfur dioxide, and other pollutants, and to excess warming of the atmosphere.

Current attention is acknowledging hydrogen-powered vehicles, part of a potential future hydrogen economy. It must be emphasized that hydrogen is *not* an energy source. Like electricity, hydrogen is a carrier or storehouse of energy that requires an energy source. Only part of the work required to separate hydrogen from water or hydrocarbons is the energy available for use. Again, for emphasis, hydrogen is *not* a source of energy.

As the world population increases, so does our need for energy—especially since per capita demand is also growing. With the rules of physics to guide them, technologists are presently researching newer and cleaner ways to develop energy sources. But they race to keep ahead of a growing world population and greater demand in the developing world. Unfortunately, so long as controlling population is politically and religiously incorrect, human misery becomes the check to unrestrained population growth. H.G. Wells once wrote (in *The Outline of History*), "Human history becomes more and more a race between education and catastrophe."

Summary of Terms

Work The product of the force and the distance moved by the force:

$$W = Fd$$

(More generally, work is the component of force in the direction of motion times the distance moved.)

Power The time rate of work:

$$\text{Power} = \frac{\text{work}}{\text{time}}$$

(More generally, power is the rate at which energy is expended.)

Energy The property of a system that enables it to do work.

Mechanical energy Energy due to the position of something or the movement of something.

Potential energy The energy that something possesses because of its position.

Kinetic energy Energy of motion, quantified by the relationship

$$\text{Kinetic energy} = \tfrac{1}{2}mv^2.$$

Work-energy theorem The work done on an object equals the change in kinetic energy of the object.

$$\text{Work} = \Delta\text{KE}$$

(Work can also transfer other forms of energy to a system.)

Conservation of energy Energy cannot be created or destroyed; it may be transformed from one form into another, but the total amount of energy never changes.

Machine A device, such as a lever or pulley, that increases (or decreases) a force or simply changes the direction of a force.

Conservation of energy for machines The work output of any machine cannot exceed the work input. In an ideal machine, where no energy is transformed into thermal energy, $\text{work}_{\text{input}} = \text{work}_{\text{output}}$; $(Fd)_{\text{input}} = (Fd)_{\text{output}}$.

Lever Simple machine consisting of a rigid rod pivoted at a fixed point called the fulcrum.

Efficiency The percentage of the work put into a machine that is converted into useful work output. (More generally, useful energy output divided by total energy input.)

Suggested Reading

Bodanis, David. $E = mc^2$: *A Biography of the World's Most Famous Equation*. New York: Berkley Publishing Group, 2002. A delightful and engaging history of our understanding of energy.

Review Questions

1. When is energy most evident?

Work

2. A force sets an object in motion. When the force is multiplied by the time of its application, we call the quantity *impulse,* which changes the *momentum* of that object. What do we call the quantity *force* × *distance*?
3. Cite an example in which a force is exerted on an object without doing work on the object.
4. Which requires more work—lifting a 50-kg sack a vertical distance of 2 m or lifting a 25-kg sack a vertical distance of 4 m?

Power

5. If both sacks in the preceding question are lifted their respective distances in the same time, how does the power required for each compare? How about for the case in which the lighter sack is moved its distance in half the time?

Mechanical Energy

6. Exactly what is it that enables an object to do work?

Potential Energy

7. A car is raised a certain distance in a service-station lift and therefore has potential energy relative to the floor. If it were raised twice as high, how much potential energy would it have?
8. Two cars are raised to the same elevation on service-station lifts. If one car is twice as massive as the other, how do their potential energies compare?
9. When is the potential energy of something significant?

Kinetic Energy

10. A moving car has kinetic energy. If it speeds up until it is going four times as fast, how much kinetic energy does it have in comparison?

Work-Energy Theorem

11. Compared with some original speed, how much work must the brakes of a car supply to stop a car that is moving four times as fast? How will the stopping distance compare?
12. If you push a crate horizontally with 100 N across a 10-m factory floor, and friction between the crate and the floor is a steady 70 N, how much kinetic energy is gained by the crate?
13. How does speed affect the friction between a road and a skidding tire?

Conservation of Energy

14. What will be the kinetic energy of a pile driver ram when it undergoes a 10 kJ decrease in potential energy?
15. An apple hanging from a limb has potential energy because of its height. If it falls, what becomes of this energy just before it hits the ground? When it hits the ground?
16. What is the source of energy in sunshine?

Machines

17. Can a machine multiply input force? Input distance? Input energy? (If your three answers are the same, seek help, for the last question is especially important.)
18. If a machine multiplies force by a factor of four, what other quantity is diminished, and by how much?
19. A force of 50 N is applied to the end of a lever, which is moved a certain distance. If the other end of the lever moves one-third as far, how much force can it exert?

Efficiency

20. What is the efficiency of a machine that miraculously converts all the input energy to useful output energy?
21. How does the useful work output of a machine and the total energy input relate to its efficiency?
22. What happens to the percentage of useful energy as it is transformed from one form to another?

Comparison of Kinetic Energy and Momentum

23. What does it mean to say that momentum is a vector quantity and that energy is a scalar quantity?
24. Can momenta cancel? Can energies cancel?
25. If a moving object doubles its speed, how much more momentum does it have? How much more energy?
26. If a moving object doubles its speed, how much more impulse does it provide to whatever it bumps into (how much more wallop)? How much more work does it do as it is stopped (how much more damage)?

Energy for Life

27. In what sense are our bodies machines?

Sources of Energy

28. What is the ultimate source of energies for the burning of fossil fuels, dams, and windmills?
29. What is the ultimate source of geothermal energy?
30. Can we correctly say that a new source of energy is hydrogen? Why or why not?

One-Step Calculations

Work = force × distance: $W = Fd$

1. Calculate the work done when a force of 1 N moves a book 2 m.
2. Calculate the work done when a 20-N force pushes a cart 3.5 m.
3. Calculate the work done in lifting a 500-N barbell 2.2 m above the floor. (What is the potential energy of the barbell when it is lifted to this height?)

Power = work/time: $P = W/t$

4. Calculate the watts of power expended when a force of 1 N moves a book 2 m in a time interval of 1 s?
5. Calculate the power expended when a 20-N force pushes a cart 3.5 m in a time of 0.5 s.
6. Calculate the power expended when a 500-N barbell is lifted 2.2 m in 2 s.

Gravitational potential energy = weight × height: PE = mgh

7. How many joules of potential energy does a 1-kg book gain when it is elevated 4 m? When it is elevated 8 m?
8. Calculate the increase in potential energy when a 20-kg block of ice is lifted a vertical distance of 2 m.
9. Calculate the change in potential energy of 8 million kg of water dropping 50 m over Niagara Falls.

Kinetic energy = 1/2 mass × speed²: KE = $1/2\, mv^2$

10. Calculate the number of joules of kinetic energy a 1-kg book has when tossed across the room at a speed of 2 m/s.
11. Calculate the kinetic energy of a 3-kg toy cart that moves at 4 m/s.
12. Calculate the kinetic energy of the same cart moving at twice the speed.

Work-energy theorem: Work = ΔKE

13. How much work is required to increase the kinetic energy of a car by 5000 J?
14. What change in kinetic energy does an airplane experience on takeoff if it is moved a distance of 500 m by a sustained net force of 5000 N?

Exercises

1. Why is it easier to stop a lightly loaded truck than a heavier one that has equal speed?
2. Which requires more work to stop—a light truck or a heavy truck moving with the same momentum?
3. How much work do you do on a 25-kg backpack when you walk a horizontal distance of 100 m?
4. If your friend pushes a lawnmower four times as far as you do while exerting only half the force, which one of you does more work? How much more?
5. Why does one get tired when pushing against a stationary wall when no work is done on the wall?
6. Which requires more work: stretching a strong spring a certain distance or stretching a weak spring the same distance? Defend your answer.
7. Two people who weigh the same climb a flight of stairs. The first person climbs the stairs in 30 s, while the second person climbs them in 40 s. Which person does more work? Which uses more power?
8. Is more work required to bring a fully loaded truck up to a given speed, than the same truck lightly loaded? Defend your answer.
9. In determining the potential energy of Tenny's drawn bow (Figure 7.7), would it be an underestimate or an overestimate to multiply the force with which she holds the arrow in its drawn position by the distance she pulled it? Why do we say the work done is the *average* force × distance?
10. A cart gains energy as it rolls down a hill. What is the force that does the work? (Don't just say "gravity.")
11. When a rifle with a longer barrel is fired, the force of expanding gases acts on the bullet for a longer distance. What effect does this have on the velocity of the emerging bullet? (Do you see why long-range cannons have such long barrels?)
12. Your friend says that the kinetic energy of an object depends on the reference frame of the observer. Explain why you agree or disagree.
13. You and a flight attendant toss a ball back and forth in an airplane in flight. Does the KE of the ball depend on the speed of the airplane? Carefully explain.
14. You watch your friend take off in a jet plane, and you comment on the kinetic energy she has acquired. But she says she has no such increase in kinetic energy. Who is correct?
15. When a jumbo jet lands, there is a decrease in both its kinetic and potential energy. Where does this energy go?
16. A baseball and a golf ball have the same momentum. Which has the greater kinetic energy?
17. You have a choice of catching a baseball or a bowling ball, both with the same KE. Which is safer?
18. Explain how the sport of pole vaulting dramatically changed when flexible fiberglass poles replaced stiff wooden poles.
19. At what point in its motion is the KE of a pendulum bob at a maximum? At what point is its PE at a

maximum? When its KE is at half its maximum value, how much PE does it have?

20. A physics instructor demonstrates energy conservation by releasing a heavy pendulum bob, as shown in the sketch, allowing it to swing to and fro. What would happen if, in his exuberance, he gave the bob a slight shove as it left his nose? Explain.

21. Does the International Space Station have gravitational PE? KE? Explain.

22. What does the work-energy theorem say about the speed of a satellite in circular orbit?

23. A moving hammer hits a nail and drives it into a wall. If the hammer hits the nail with twice the speed, how much deeper will the nail be driven? If it hits with three times the speed?

24. Why does the force of gravity do no work on (a) a bowling ball rolling along a bowling alley, and (b) a satellite in circular orbit about the Earth?

25. Why does the force of gravity do work on a car that rolls down a hill, but no work when it rolls along a level part of the road?

26. Does the string that supports a pendulum bob do work on the bob as its swings to and fro? Does the force of gravity do any work on the bob?

27. A crate is pulled across a horizontal floor by a rope. At the same time, the crate pulls back on the rope, in accord with Newton's third law. Does the work done on the crate by the rope then equal zero? Explain.

28. On a playground slide, a child has potential energy that decreases by 1000 J while her kinetic energy increases by 900 J. What other form of energy is involved, and how much?

29. Someone wanting to sell you a SuperBall claims that it will bounce to a height greater than the height from which it is dropped. Can this be?

30. Why can't a SuperBall released from rest reach its original height when it bounces from a rigid floor?

31. Consider a ball thrown straight up in the air. At what position is its kinetic energy at a maximum? Where is its gravitational potential energy at a maximum?

32. Discuss the design of the roller coaster shown in the sketch in terms of the conservation of energy.

33. Suppose that you and two classmates are discussing the design of a roller coaster. One classmate says that each summit must be lower than the previous one. Your other classmate says this is nonsense, for as long as the first one is the highest, it doesn't matter what height the others are. What do you say?

34. Consider the identical balls released from rest on Tracks A and B, as shown. When they reach the right ends of the tracks, which will have the greater speed? Why is this question easier to answer than the similar one (Exercise 40) in Chapter 3?

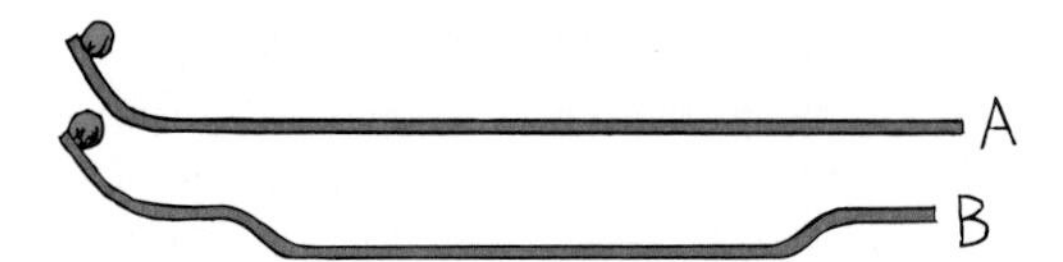

35. Suppose an object is set sliding, with a speed less than escape velocity, on an infinite frictionless plane in contact with the surface of the Earth, as shown. Describe its motion. (Will it slide forever at a constant velocity? Will it slide to a stop? In what way will its energy changes be similar to that of a pendulum?)

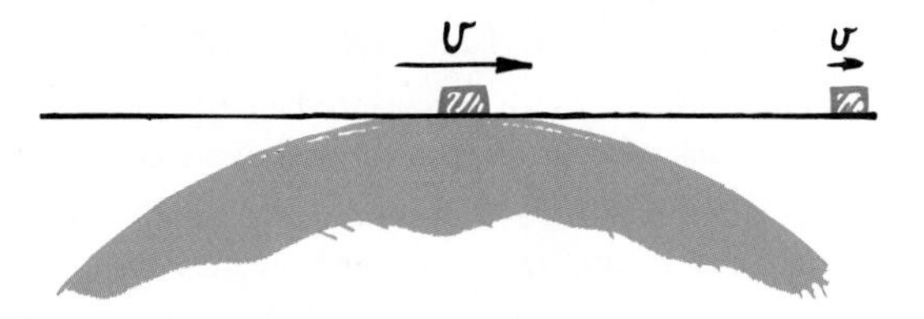

36. If a golf ball and a Ping-Pong ball both move with the same kinetic energy, can you say which has the greater speed? Explain in terms of the definition of KE. Similarly, in a gaseous mixture of massive molecules and light molecules with the same average KE, can you say which have the greater speed?

37. Does a car burn more gasoline when its lights are turned on? Does the overall consumption of gasoline depend on whether or not the engine is running while the lights are on? Defend your answer.

38. Running a car's air conditioner usually increases fuel consumption. But, at certain speeds, a car with its windows open and with the air conditioner turned off can consume more fuel. Explain.

39. When the girl in Figure 7.14 jacks up a car, how can applying so little force produce sufficient work to raise the car?

40. Why bother using a machine if it cannot multiply work input to achieve greater work output?

41. You tell your friend that no machine can possibly put out more energy than is put into it, and your friend states that a nuclear reactor puts out more energy than is put into it. What do you say?

42. What famous equation describes the relationship between mass and energy?

43. This may seem like an easy question for a physics type to answer: With what force does a rock that weighs 10 N strike the ground if dropped from a rest position 10 m high? In fact, the question cannot be answered unless you know more. Why?

44. Your friend is confused about ideas discussed in Chapter 4 that seem to contradict ideas discussed in this chapter. For example, in Chapter 4, we learned that the net force is zero for a car traveling along a level road at constant velocity, and, in this chapter, we learned that work is done in such a case. Your friend asks, "How can work be done when the net force equals zero?" Explain.

45. In the absence of air resistance, a ball thrown vertically upward with a certain initial KE will return to its original level with the same KE. When air resistance is a factor affecting the ball, will it return to its original level with the same, less, or more KE? Does your answer contradict the law of energy conservation?

46. You're on a rooftop and you throw one ball downward to the ground below and another upward. The second ball, after rising, falls and also strikes the ground below. If air resistance can be neglected, and if your downward and upward initial speeds are the same, how will the speeds of the balls compare upon striking the ground? (Use the idea of energy conservation to arrive at your answer.)

47. Going uphill, the gasoline engine in a gasoline–electric hybrid car provides 75 horsepower while the total power propelling the car is 90 horsepower. Burning gasoline provides the 75 horsepower. What provides the other 15 horsepower?

48. When a driver applies brakes to keep a car going downhill at constant speed and constant kinetic energy, the potential energy of the car decreases. Where does this energy go? Where does most of it go with a hybrid vehicle?

49. Does the KE of a car change more when it goes from 10 to 20 km/h or when it goes from 20 to 30 km/h?

50. Can something have energy without having momentum? Explain. Can something have momentum without having energy? Defend your answer.

51. When the mass of a moving object is doubled with no change in speed, by what factor is its momentum changed? By what factor is its kinetic energy changed?

52. When the velocity of an object is doubled, by what factor is its momentum changed? By what factor is its kinetic energy changed?

53. Which, if either, has greater momentum: a 1-kg ball moving at 2 m/s or a 2-kg ball moving at 1 m/s? Which has greater kinetic energy?

54. A car has the same kinetic energy when traveling north as when it turns around and travels south. Is the momentum of the car the same in both cases?

55. If an object's KE is zero, what is its momentum?

56. If your momentum is zero, is your kinetic energy necessarily zero also?

57. If two objects have equal kinetic energies, do they necessarily have the same momentum? Defend your answer.

58. Two lumps of clay with equal and opposite momenta have a head-on collision and come to rest. Is momentum conserved? Is kinetic energy conserved? Why are your answers the same or different?

59. You may choose between two head-on collisions with kids on skateboards. One is with a light kid moving rather fast, and the other is with a twice-as-heavy kid moving half as fast. Considering only the issues of mass and speed, which collision do you prefer?

60. Scissors for cutting paper have long blades and short handles, whereas metal-cutting shears have long handles and short blades. Bolt cutters have very long handles and very short blades. Why is this so?

61. Discuss the fate of the physics instructor sandwiched between the beds of nails (Figure 7.20) if the block were less massive and unbreakable and the beds contained fewer nails.

62. Consider the swinging-balls apparatus. If two balls are lifted and released, momentum is conserved as two balls pop out the other side with the same speed as the released balls at impact. But momentum would also be conserved if one ball popped out at twice the speed. Can you explain why this never happens? (And can you explain why this exercise is in Chapter 7 rather than in Chapter 6?)

63. An inefficient machine is said to "waste energy." Does this mean that energy is actually lost? Explain.

64. If an automobile were to have a 100% efficient engine, transferring all of the fuel's energy to work, would the engine be warm to your touch? Would its exhaust heat the surrounding air? Would it make any noise? Would it vibrate? Would any of its fuel go unused?

65. To combat wasteful habits, we often speak of "conserving energy," by which we mean turning off lights and hot water when they are not being used, and keeping thermostats at a moderate level. In this chapter, we also speak of "energy conservation." Distinguish between these two usages.

66. When an electric company can't meet its customers' demand for electricity on a hot summer day, should the problem be called an "energy crisis" or a "power crisis"? Explain.

67. Your friend says that one way to improve air quality in a city is to have traffic lights synchronized so that motorists can travel long distances at constant speed. What physics principle supports this claim?

68. The energy we require to live comes from the chemically stored potential energy in food, which is transformed into other energy forms during the metabolism process. What happens to a person whose combined work and heat output is less than the energy consumed? What happens when the person's work and heat output is greater than the energy consumed? Can an undernourished person perform extra work without extra food? Defend your answers.

69. Once used, can energy be regenerated? Is your answer consistent with the common term "renewable energy"?

70. What do international peace, cooperation, and security have to do with addressing the world's energy needs?

Problems

1. The second floor of a house is 6 m above street level. How much work is required to lift a 300-kg refrigerator to the second-floor level?

2. Belly-flop Bernie dives from atop a tall flagpole into a swimming pool below. His potential energy at the top is 10,000 J. What is his kinetic energy when his potential energy reduces to 1000 J?

3. Which produces the greater change in kinetic energy: exerting a 10-N force for a distance of 5 m, or exerting a 20-N force over a distance of 2 m? (Assume that all of the work goes into KE.)

4. This question is typical on some driver's-license exams: A car moving at 50 km/h skids 15 m with locked brakes. How far will the car skid with locked brakes at 150 km/h?

5. A lever is used to lift a heavy load. When a 50-N force pushes one end of the lever down 1.2 m, the load rises 0.2 m. Calculate the weight of the load.

6. In raising a 5000-N piano with a pulley system, the workers note that, for every 2 m of rope pulled down, the piano rises 0.2 m. Ideally, how much force is required to lift the piano?

7. In the hydraulic machine shown, it is observed that, when the small piston is pushed down 10 cm, the large piston is raised 1 cm. If the small piston is pushed down with a force of 100 N, what is the most force that the large piston could exert?

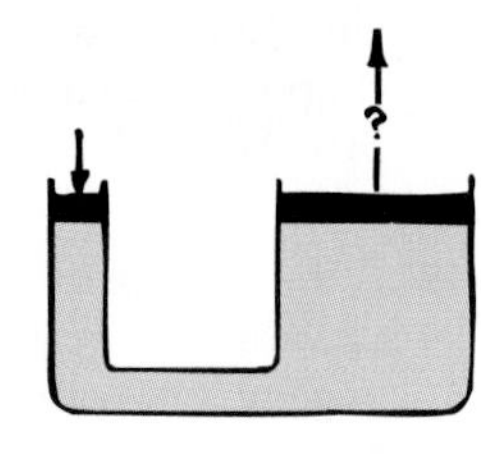

8. A 60-kg skydiver moving at terminal speed falls 50 m in 1 s. What power is the skydiver expending on the air?

9. Consider the inelastic collision between the two freight cars in the previous chapter (Figure 6.14). The momentum before and after the collision is the same. The KE, however, is less after the collision than before the collision. How much less, and what becomes of this energy?

10. Using the definitions of momentum and kinetic energy, $p = mv$ and $\text{KE} = (\frac{1}{2})mv^2$, show, by algebraic manipulation, that you can write $\text{KE} = p^2/2m$. This equation tells us that, if two objects have the same momentum, the one of less mass has the greater kinetic energy (see Exercise 2).

11. How many kilometers per liter will a car obtain if its engine is 25% efficient and it encounters an average retarding force of 500 N at highway speed? Assume that the energy content of gasoline is 40 MJ/liter.

12. The power we derive from metabolism can do work and can generate heat. (a) What is the mechanical efficiency of a relatively inactive person who expends 100 W of power to produce about 1 W of power in the form of work, while generating about 99 W of heat? (b) What is the mechanical efficiency of a cyclist who, in a burst of effort, produces 100 W of mechanical power from 1000 W of metabolic power?

CHAPTER 8

Rotational Motion

Mary Beth Monroe demonstrates a "torque feeler" before she passes it around for her students to try.

Rotational Motion

Which moves faster on a merry-go-round, a horse near the outside rail or one near the inside rail? Or do both have the same speed? Ask different people this question, and you'll get different answers. That's because it's easy to get linear speed confused with rotational speed.

Circular Motion

When an object turns about an internal axis, the motion is a *rotation,* or spin. A merry-go-round or a turntable rotates about a central internal axis. When an object turns about an external axis, the motion is a *revolution.* Earth makes one revolution about the Sun each year, while it rotates about its polar axis once per day.

Insights

Rotational Speed

Linear speed, which we simply called *speed* in previous chapters, is the distance traveled per unit of time. A point on the outside edge of a merry-go-round or turntable travels a greater distance in one complete rotation than a point nearer the center. Traveling a greater distance in the same time means a greater speed. Linear speed is greater on the outer edge of a rotating object than it is closer to the axis. The linear speed of something moving along a circular path can be called **tangential speed** because the direction of motion is tangent to the circumference of the circle. For circular motion, we can use the terms linear speed and tangential speed interchangeably. Units of linear or tangential speed are usually m/s or km/h.

Rotational speed (sometimes called *angular speed*) involves the number of rotations or revolutions per unit of time. All parts of the rigid merry-go-round and turntable turn about the axis of rotation in the same amount of time. Thus, all parts share the same rate of rotation, or the same *number of rotations or revolutions per unit of time.* It is common to express rotational rates in revolutions per minute (RPM).[1] For example, most phonograph turntables, which were common several years ago, rotate at $33\frac{1}{3}$ RPM. A ladybug sitting anywhere on the surface of the turntable revolves at $33\frac{1}{3}$ RPM.

Tangential speed and rotational speed are related. Have you ever ridden on a big, round, rotating platform in an amusement park? The faster it turns, the faster your tangential speed. This makes sense; the greater the RPMs, the faster your speed in meters per second. We say that tangential speed is *directly proportional* to rotational speed at any fixed distance from the axis of rotation.

[1]Physics types usually describe rotational speed, ω, in terms of the number of "radians" turned in a unit of time. There are a little more than 6 radians in a full rotation (2π radians, to be exact). When a direction is assigned to rotational speed, we call it *rotational velocity* (often called *angular velocity*). Rotational velocity is a vector whose magnitude is the rotational speed. By convention, the rotational velocity vector lies along the axis of rotation.

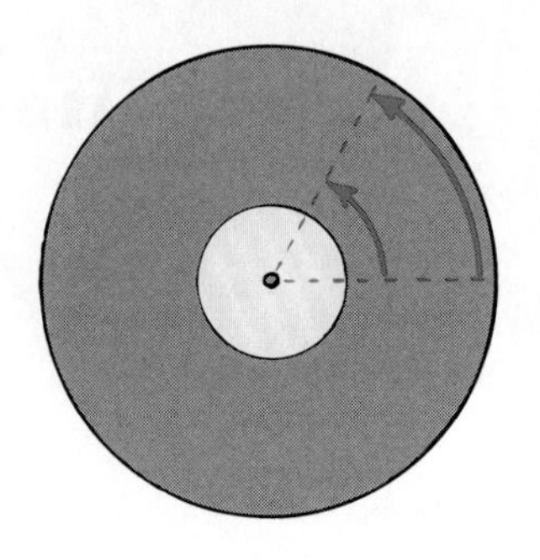

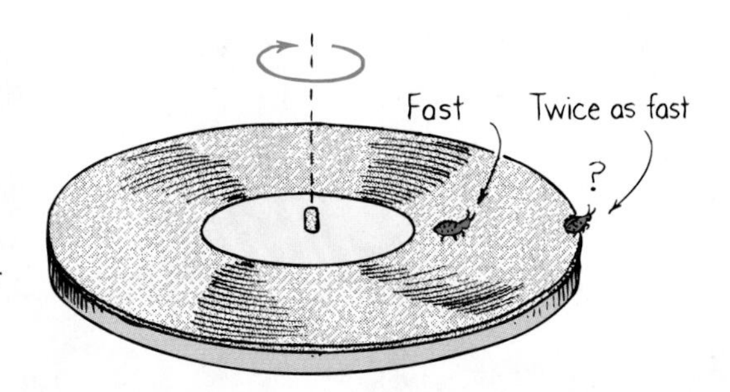

FIGURE 8.2
Interactive Figure
The entire turntable rotates at the same rotational speed, but ladybugs at different distances from the center travel at different tangential speeds. A ladybug twice as far from the center moves twice as fast.

FIGURE 8.1
Interactive Figure
When the turntable rotates, a ladybug farther from the center travels a longer path in the same time and has a greater tangential speed.

Tangential speed, unlike rotational speed, depends on radial distance (the distance from the axis). At the very center of the rotating platform, you have no speed at all; you merely rotate. But, as you approach the edge of the platform, you find yourself moving faster and faster. Tangential speed is directly proportional to distance from the axis for any given rotational speed.

So we see that tangential speed is proportional to both radial distance and rotational speed.[2]

$$\text{Tangential speed} \sim \text{radial distance} \times \text{rotational speed.}$$

In symbol form,

$$v \sim r\omega$$

where v is tangential speed and ω (Greek letter omega) is rotational speed. You move faster if the rate of rotation increases (bigger ω). You also move faster if you move farther from the axis (bigger r). Move out twice as far from the rotational axis at the center and you move twice as fast. Move out three times as far and you have three times as much tangential speed. If you find yourself in any kind of rotating system, your tangential speed depends on how far you are from the axis of rotation.

FIGURE 8.3
Interactive Figure
The tangential speed of each person is proportional to the rotational speed of the platform multiplied by the distance from the central axis.

CHECK YOURSELF

1. Imagine yourself on a rotating platform like the one shown in Figure 8.3. If you sit halfway between the rotational axis and the outer edge and have a rotational speed of 20 RPM and a tangential speed of 2 m/s, what will be the rotational and tangential speeds of your friend who sits at the outer edge?
2. Trains ride on a pair of tracks. For straight-line motion, both tracks are the same length. Not so for tracks along a curve. Which track is longer, the one on the outside of the curve or the one on the inside?

CHECK YOUR ANSWERS

1. Since the rotating platform is rigid, all parts have the same rotational speed, so your friend also rotates at 20 RPM. Tangential speed is a different story: Since she is twice as far from the axis of rotation, she moves twice as fast—4 m/s.
2. Similar to Figure 8.1, the outer track is longer—just as the circumference of a circle of greater radius is longer.

When a row of people, locked arm in arm at the skating rink, makes a turn, the motion of "tail-end Charlie" is evidence of a greater tangential speed.

[2]If you take a follow-up physics course, you will learn that, when the proper units are used for tangential speed v, rotational speed ω, and radial distance r, the direct proportion of v to both r and ω becomes the exact equation $v = r\omega$. So the tangential speed will be directly proportional to r when all parts of a system simultaneously have the same ω, as for a wheel, disk, or rigid wand. (The direct proportionality of v to r is not valid for the planets, because the planets have different rotational speeds.)

WHEELS ON RAILROAD TRAINS

Why does a moving railroad train stay on the tracks? Most people assume that the wheel flanges keep the wheels from rolling off. But, if you look at these flanges, you'll likely note they are rusty. They seldom touch the track, except when they follow slots that switch the train from one set of tracks to another. So how do the wheels of a train stay on the tracks? They stay on the track because their rims are slightly tapered.

If you roll a tapered cup across a surface, it makes a curved path (Figure 8.4). The wider part of the cup has a greater radius, rolls a greater distance per revolution, and therefore has a greater tangential speed than the narrower end. If you fasten a pair of cups together at their wide ends (simply taping them together) and roll the pair along a pair of parallel tracks (Figure 8.5), the cups will remain on the track and center themselves whenever they roll off center. This occurs because, when the pair rolls to the left of center, say, the wider part of the left cup rides on the left track while the narrow part of the right cup rides on the right track. This steers the pair toward the center. If it "overshoots" toward the right, the process repeats, this time toward the left, as the wheels tend to center themselves. Likewise for a railroad train, where passengers feel the train swaying as these corrective actions occur.

This tapered shape is essential on the curves of railroad tracks. On any curve, the distance along the outer part is longer than the distance along the inner part (as we saw in Figure 8.1). So, whenever a vehicle follows a curve, its outer wheels travel faster than its inner wheels. For an automobile, this is no problem, because the wheels are freewheeling and roll independently of each other. For a train, however, like the pair of fastened cups, pairs of wheels are firmly connected so that they rotate together. Opposite wheels have the same RPM at any time. But, due to the slightly tapered rim of the wheel, its speed along the track depends on whether it rides on the narrow part of the rim or the wide part. On the wide part, it travels faster. So, when a train rounds a curve, wheels on the outer track ride on the wider part of the tapered rims, while opposite wheels ride on their narrow parts. In this way, the wheels have different tangential speeds for the same rotational speed. This is $v \sim r\omega$ in action! Can you see that, if the wheels were not tapered, scraping would occur and the wheels would squeal when a train rounds a curve?

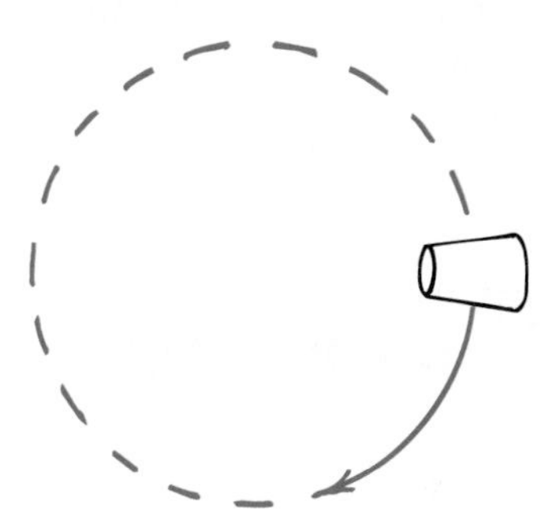

FIGURE 8.4
Because the wide part of the cup rolls faster than the narrow part, the cup rolls in a curve.

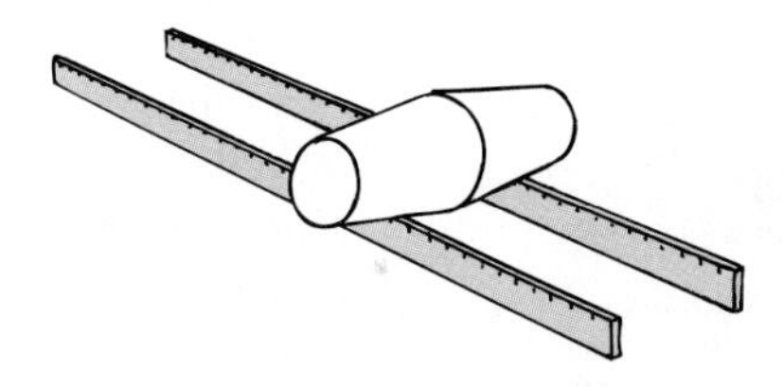

FIGURE 8.5
A pair of fastened cups stay on the tracks as they roll because, when they roll off center, the different tangential speeds due to the taper cause them to self-correct toward the center of the track.

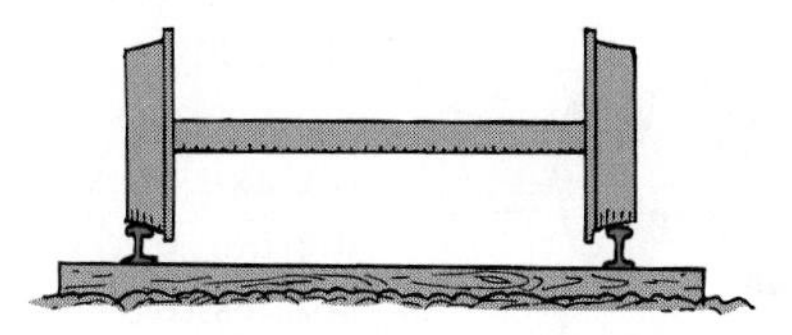

FIGURE 8.6
Wheels of a railroad train are slightly tapered (shown exaggerated here).

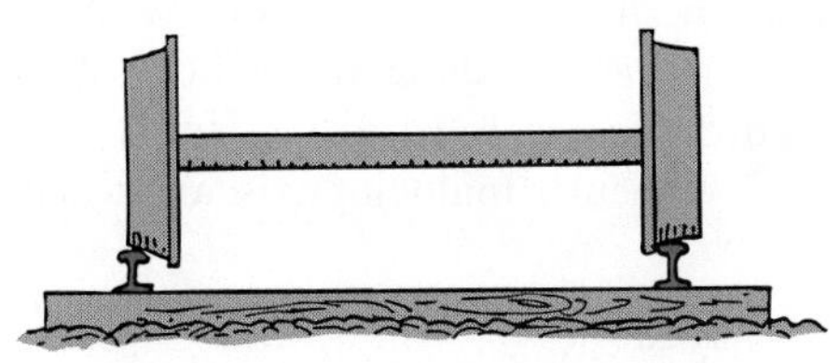

Narrow part of left wheel goes slower, so wheels curve to left

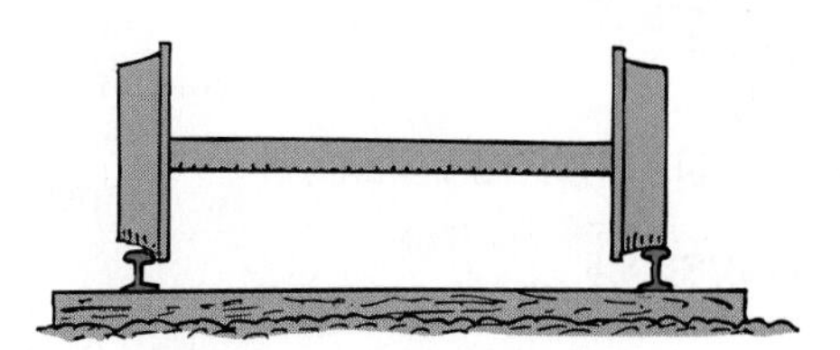

Wide part of left wheel goes faster, so wheels curve to right

FIGURE 8.7
(Left) Along a track that curves to the left, the right wheel rides on its wide part and goes faster while the left wheel rides on its narrow part and goes slower. (Right) The opposite is true when the track curves to the right.

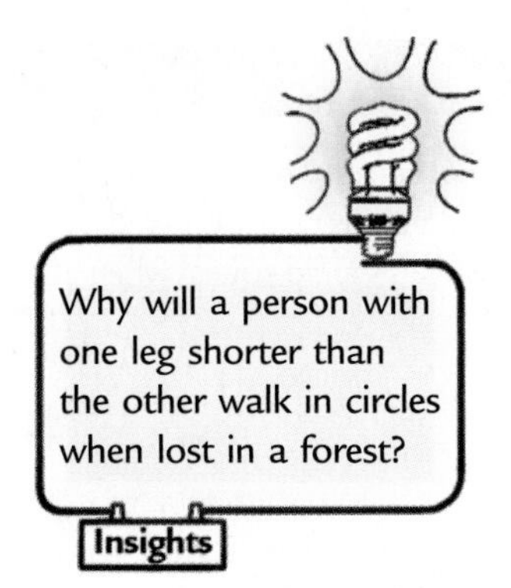

Why will a person with one leg shorter than the other walk in circles when lost in a forest?
Insights

When tangential speed undergoes change, we speak of a *tangential acceleration*. Any change in tangential speed indicates an acceleration in the direction of tangential motion. For example, a person on a rotating platform that speeds up or slows down undergoes a tangential acceleration. We'll soon see that anything moving in a curved path undergoes another kind of acceleration—one directed to the center of curvature. This is *centripetal acceleration,* which we'll return to later in this chapter.

Rotational Inertia

Just as an object at rest tends to stay at rest and an object in motion tends to remain moving in a straight line, *an object rotating about an axis tends to remain rotating about the same axis unless interfered with by some external influence.* (We shall see shortly that this external influence is properly called a *torque.*) The property of an object to resist changes in its rotational state of motion is called **rotational inertia.**[3] Bodies that are rotating tend to remain rotating, while nonrotating bodies tend to remain nonrotating. In the absence of outside influences, a rotating top keeps rotating, while a top at rest remains at rest.

Rotational Inertia Using Weighted Pipes
Rotational Inertia Using a Hammer
Rotational Inertia with a Weighted Rod

Like inertia for linear motion, rotational inertia depends on mass. The thick stone disk that rotates beneath a potter's wheel is very massive, and, once it is spinning, it tends to remain spinning. But, unlike linear motion, rotational inertia depends on the distribution of the mass about the axis of rotation. The greater the distance between an object's mass concentration and the axis, the greater the rotational inertia. This is evident in industrial flywheels that are constructed so that most of their mass is concentrated far from the axis, along the rim. Once rotating, they have a greater tendency to remain rotating. When at rest, they are more difficult to get rotating.

The greater the rotational inertia of an object, the greater the difficulty in changing its rotational state. This fact is employed by a circus tightrope walker who carries a long pole to aid balance. Much of the mass of the pole is far from the axis of rotation, its midpoint. The pole, therefore, has considerable rotational inertia. If the tightrope walker starts to topple over, a tight grip on the pole rotates the pole. But the rotational inertia of the pole resists, giving the tightrope walker time to readjust his or her balance. The longer the pole, the better. And better still if massive objects are attached to the ends. But a tightrope walker with no pole can at least extend his or her arms full length to increase the body's rotational inertia.

The rotational inertia of the pole, or of any object, depends on the axis about which it rotates.[4] Compare the different rotations of a pencil. Consider three axes—one, about its central core parallel to the length of the pencil, where the lead is; two, about the perpendicular midpoint axis; and three, about an axis perpendicular

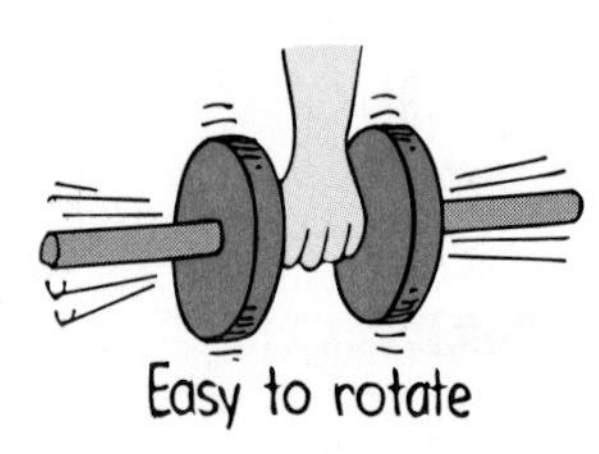

FIGURE 8.8
Rotational inertia depends on the distribution of mass relative to the axis of rotation.

[3]Often called *moment of inertia.*

[4]When the mass of an object is concentrated at the radius r from the axis of rotation (as for a simple pendulum bob or a thin ring), rotational inertia I is equal to the mass m multiplied by the square of the radial distance. For this special case, $I = mr^2$.

to one end. Rotational inertia is very small about the first position, and it's easy to rotate the pencil between your fingertips because most all the mass is very close to the axis. About the second axis, like that used by the tightrope walker in the illustration above, rotational inertia is greater. About the third axis, at the end of the pencil so that it swings like a pendulum, rotational inertia is greater still.

A long baseball bat held near its end has more rotational inertia than a short bat. Once it is swinging, it has a greater tendency to keep swinging, but it is harder to bring it up to speed. A short bat, with less rotational inertia, is easier to swing—which explains why baseball players sometimes "choke up" on a bat by grasping it closer to the more massive end. Similarly, when you run with your legs bent, you reduce their rotational inertia so you can rotate them back and forth more quickly. A long-legged person tends to walk with slower strides than a person with short legs. The different strides of creatures with different leg lengths are especially evident in animals. Giraffes, horses, and ostriches run with a slower gait than dachshunds, mice, and bugs.

An idea put forward in past years for boosting efficiency in railroad travel was massive disks—rotors—beneath the flooring of railroad cars. When brakes were applied to slow the cars, rather than converting braking energy to heat via friction, the energy would be diverted to revving the rotors. Their rotational kinetic energy was then to be used in operating the train. The massiveness of the rotors turned out to make the scheme impractical. But the idea hasn't been lost. Today's hybrid automobiles do much the same thing—not mechanically, but electrically. Braking energy is diverted to electric batteries, which are then used for operating the automobile.

Insights

FIGURE 8.9
The tendency of the pole to resist rotation aids the acrobat.

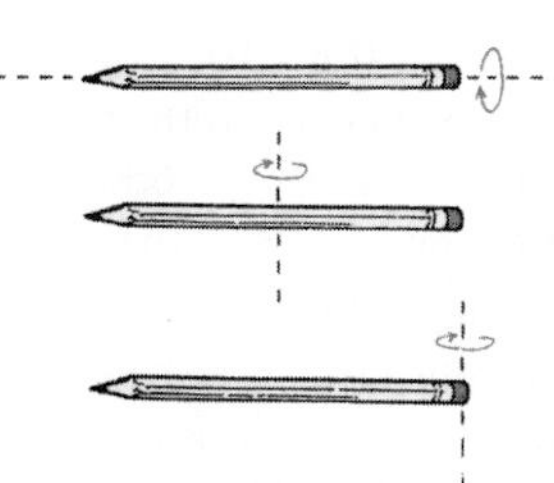

FIGURE 8.10
The pencil has different rotational inertias about different rotational axes.

FIGURE 8.11
Short legs have less rotational inertia than long legs. An animal with short legs has a quicker stride than one with long legs, just as a baseball batter can swing a short bat more quickly than a long one.

FIGURE 8.12
You bend your legs when you run to reduce rotational inertia.

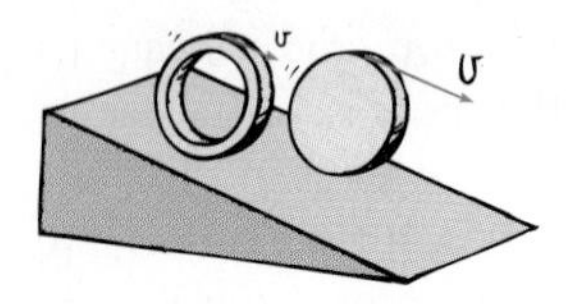

FIGURE 8.13
A solid cylinder rolls down an incline faster than a hoop, whether or not they have the same mass or outer diameter. A hoop has greater rotational inertia relative to its mass than a cylinder does.

Because of rotational inertia, a solid cylinder starting from rest will roll down an incline faster than a hoop. Both rotate about a central axis, and the shape that has most of its mass far from its axis is the hoop. So, for its weight, a hoop has more rotational inertia and is harder to start rolling. Any solid cylinder will outroll any hoop on the same incline. This doesn't seem plausible at first, but remember that any two objects, regardless of mass, will fall together when dropped. They will also slide together when released on an inclined plane. When rotation is introduced, the object with the larger rotational inertia *relative to its own mass* has the greater resistance to a change in its motion. Hence, any solid cylinder will roll down any incline with more acceleration than any hollow cylinder, regardless of mass or radius. A hollow cylinder has more "laziness per mass" than a solid cylinder. Try it and see!

Figure 8.14 compares rotational inertias for various shapes and axes. It is not important for you to learn these values, but you can see how they vary with the shape and axis.

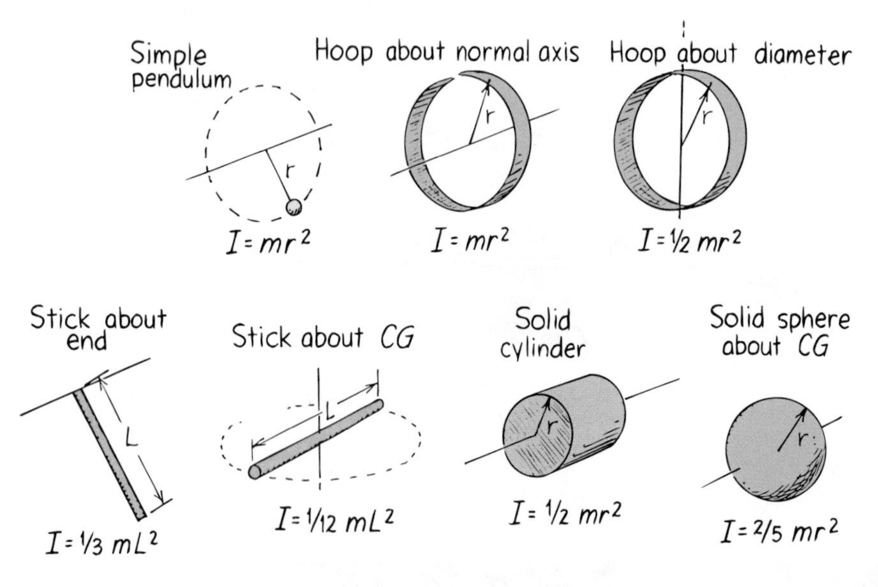

FIGURE 8.14
Rotational inertias of various objects, each of mass m, about indicated axes.

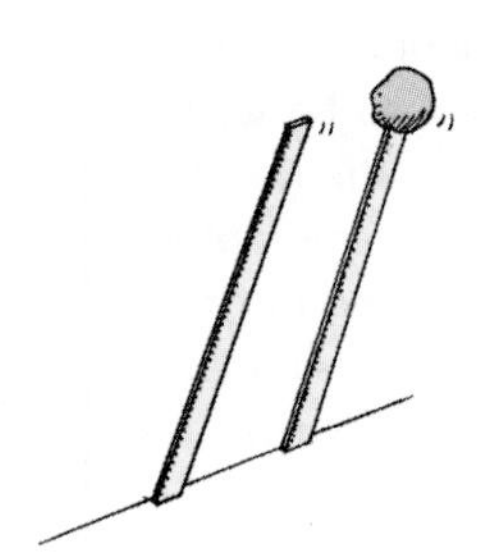

FIGURE 8.15
Which stick has the greater rotational inertia about its bottom end? When released, which will rotate to the floor first?

CHECK YOURSELF

1. Consider balancing a hammer upright on the tip of your finger. If the head of the hammer is heavy and the handle is long, would it be easier to balance with the end of the handle on your fingertip so that the head is at the top, or the other way around with the head at your fingertip and the end of the handle at the top?
2. Consider a pair of metersticks standing nearly upright against a wall. If you release them, they'll rotate to the floor in the same time. But what if one has a massive hunk of clay stuck to its top end (Figure 8.15)? Will it rotate to the floor in a longer or shorter time?
3. Just for fun, and since we're discussing round things, why are manhole covers circular in shape?

Torque

Difference Between Torque and Weight
Why a Ball Rolls Down a Hill

If and when all clocks go digital, will clockwise and counterclockwise have meaning?

Insights

Hold the end of a meterstick horizontally with your hand. Dangle a weight from it near your hand and you can feel the stick twist. Now slide the weight farther from your hand and you can feel that the twist is greater. But the weight is the same. The force acting on your hand is the same. What's different is the *torque*.

A torque (rhymes with *dork*) is the rotational counterpart of force. Force tends to change the motion of things; torque tends to twist or change the state of rotation of things. If you want to make a stationary object move, apply force. If you want to make a stationary object rotate, apply torque.

Just as rotational inertia differs from regular inertia, torque differs from force. Both rotational inertia and torque involve distance from the axis of rotation. In the case of torque, this distance, which provides leverage, is called the *lever arm*. It is the shortest distance between the applied force and the rotational axis. We define **torque** as the product of this lever arm and the force that tends to produce rotation:

$$\text{Torque} = \text{lever arm} \times \text{force}$$

Torques are intuitively familiar to youngsters playing on a seesaw. Kids can balance a seesaw even when their weights are unequal. Weight alone doesn't

FIGURE 8.17
Since ancient times, mass has been measured by balancing torques.

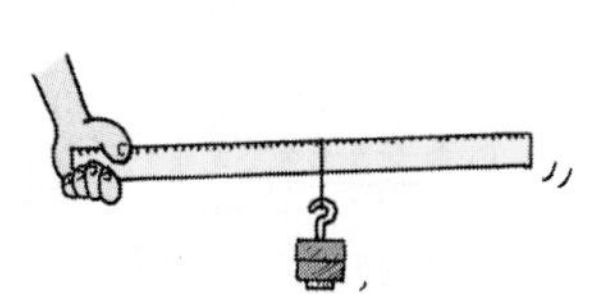

FIGURE 8.16
Move the weight farther from your hand and feel the difference between force and torque.

CHECK YOUR ANSWERS

1. Stand the hammer with the handle at your fingertip and the head at the top. Why? Because it will have more rotational inertia this way and be more resistant to a rotational change. Those acrobats you see on stage who balance a long pole have an easier task when their friends are at the top of the pole. A pole empty at the top has less rotational inertia and is more difficult to balance!
2. Try it and see! (If you don't have clay, fashion something equivalent.)
3. Not so fast on this one. Give it some thought if you haven't come up with an answer. *Then* look to the end of the chapter for an answer.

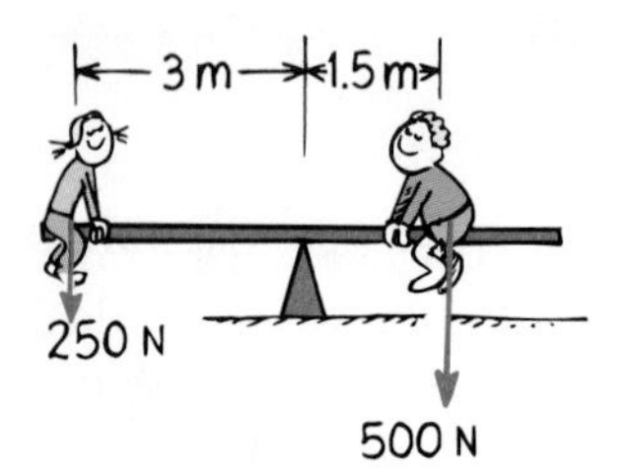

FIGURE 8.18
Interactive Figure
No rotation is produced when the torques balance each other.

produce rotation—torque does also—and children soon learn that the distance they sit from the pivot point is every bit as important as their weight. The torque produced by the boy on the right in Figure 8.18 tends to produce clockwise rotation, while torque produced by the girl on the left tends to produce counterclockwise rotation. If the torques are equal, making the net torque zero, no rotation is produced.

Recall the equilibrium rule in Chapter 2—that the sum of the forces acting on a body or any system must equal zero for mechanical equilibrium. That is, $\Sigma F = 0$. We now see an additional condition. The *net torque* on a body or on a system must also be zero for mechanical equilibrium ($\Sigma \mathrm{T} = 0$, where T stands for torque). Anything in mechanical equilibrium doesn't accelerate—neither linearly nor rotationally.

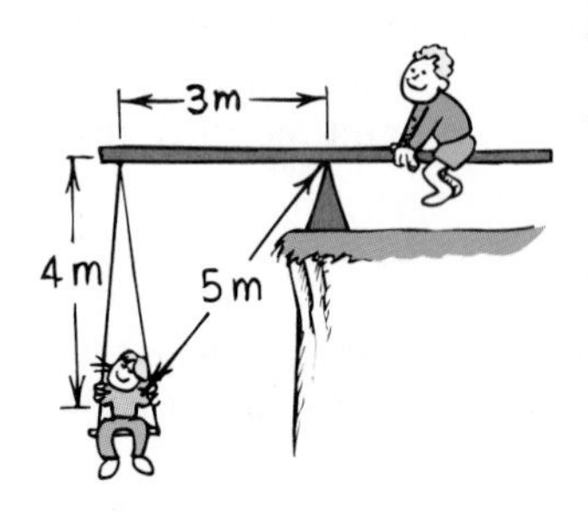

FIGURE 8.19
The lever arm is still 3 m.

Suppose that the seesaw is arranged so that the half-as-heavy girl is suspended from a 4-meter rope hanging from her end of the seesaw (Figure 8.19). She is now 5 meters from the fulcrum, and the seesaw is still balanced. We see that the lever-arm distance is still 3 meters and not 5 meters. The lever arm about any axis of rotation is the perpendicular distance from the axis to the line along which the force acts. This will always be the shortest distance between the axis of rotation and the line along which the force acts.

This is why the stubborn bolt shown in Figure 8.20 is more likely to turn when the applied force is perpendicular to the handle, rather than at an oblique angle as shown in the first figure. In the first figure, the lever arm is shown by the dashed line and is less than the length of the wrench handle. In the second figure, the lever arm is equal to the length of the wrench handle. In the third figure, the lever arm is extended with a piece of pipe to provide more leverage and a greater torque.

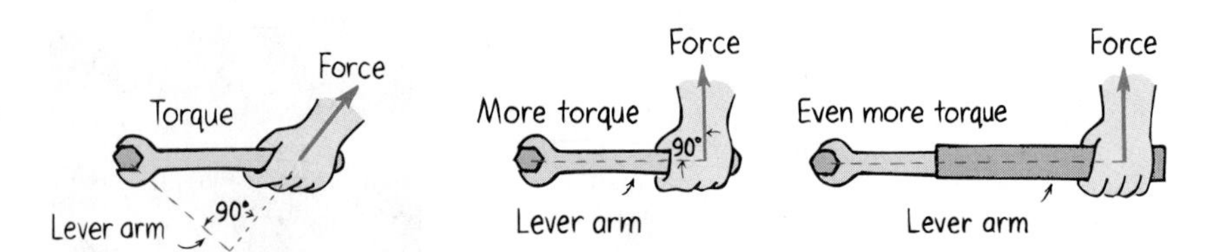

FIGURE 8.20
Interactive Figure
Although the magnitudes of the force are the same in each case, the torques are different.

CHECK YOURSELF

1. If a pipe effectively extends a wrench handle to three times its length, by how much will the torque increase for the same applied force?
2. Consider the balanced seesaw in Figure 8.18. Suppose the girl on the left suddenly gains 50 N, such as by being handed a bag of apples. Where should she sit in order to balance, assuming the heavier boy does not move?

CHECK YOUR ANSWERS

1. Three times more leverage for the same force produces three times more torque. (Caution: This method of increasing torque sometimes results in shearing off the bolt!)
2. She should sit $\frac{1}{2}$ m closer to the center. Then her lever arm is 2.5 m. This checks: $300 \text{ N} \times 2.5 \text{ m} = 500 \text{ N} \times 1.5 \text{ m}$.

Center of Mass and Center of Gravity

Toss a baseball into the air, and it will follow a smooth parabolic trajectory. Toss a baseball bat spinning into the air, and its path is not smooth; its motion is wobbly, and it seems to wobble all over the place. But, in fact, it wobbles about a very special place, a point called the **center of mass (CM).**

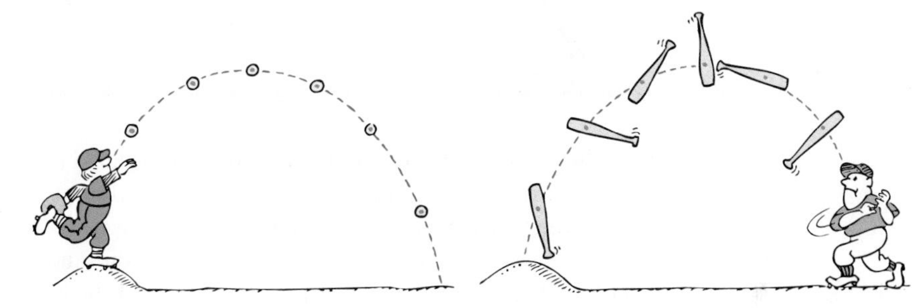

FIGURE 8.21
The center of mass of the baseball and that of the bat follow parabolic trajectories.

For a given body, the center of mass is the average position of all the mass that makes up the object. For example, a symmetrical object, such as a ball, has its center of mass at its geometrical center; by contrast, an irregularly shaped body, such as a baseball bat, has more of its mass toward one end. The center of mass of a baseball bat, therefore, is toward the thicker end. A solid cone has its center of mass exactly one-fourth of the way up from its base.

Center of gravity (CG) is a term popularly used to express center of mass. The center of gravity is simply the average position of weight distribution. Since weight and mass are proportional, center of gravity and center of mass refer to the same point of an object.[5] The physicist prefers to use the term *center of mass,* for an object has a center of mass whether or not it is under the influence of gravity. However, we shall use either term to express this concept, and we shall favor the term *center of gravity* when weight is part of the picture.

The multiple-flash photograph (Figure 8.23) shows a top view of a wrench sliding across a smooth horizontal surface. Note that its center of mass, indicated by the white dot, follows a straight-line path, while other parts of the wrench wobble as they move across the surface. Since there is no external force acting

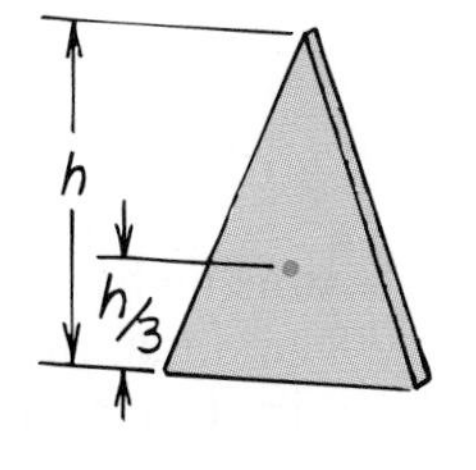

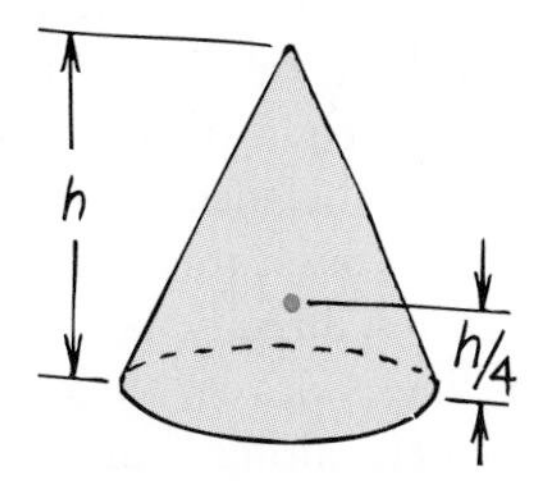

FIGURE 8.22
The center of mass for each object is shown by the red dot.

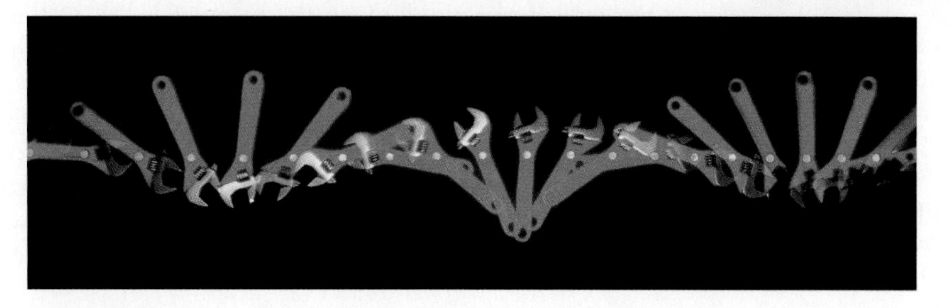

FIGURE 8.23
The center of mass of the spinning wrench follows a straight-line path.

[5]For almost all objects on and near Earth's surface, these terms are interchangeable. There can be a small difference between center of gravity and center of mass when an object is large enough for gravity to vary from one part to another. For example, the center of gravity of the Empire State Building is about 1 millimeter below its center of mass. This is due to the lower stories being pulled a little more strongly by Earth's gravity than the upper stories. For everyday objects (including tall buildings), we can use the terms *center of gravity* and *center of mass* interchangeably.

FIGURE 8.24
The center of mass of the cannonball and its fragments moves along the same path before and after the explosion.

on the wrench, its center of mass moves equal distances in equal time intervals. The motion of the spinning wrench is the combination of the straight-line motion of its center of mass and the rotational motion about its center of mass.

If the wrench were instead tossed into the air, no matter how it rotates, its center of mass (or center of gravity) would follow a smooth parabolic arc. The same is true for an exploding cannonball (Figure 8.24). The internal forces that act in the explosion do not change the center of gravity of the projectile. Interestingly enough, if air resistance can be neglected, the center of gravity of the dispersed fragments as they fly though the air will be in the same location as the center of gravity would have been if the explosion hadn't occurred.

CHECK YOURSELF

1. Where is the CG of a donut?
2. Can an object have more than one CG?

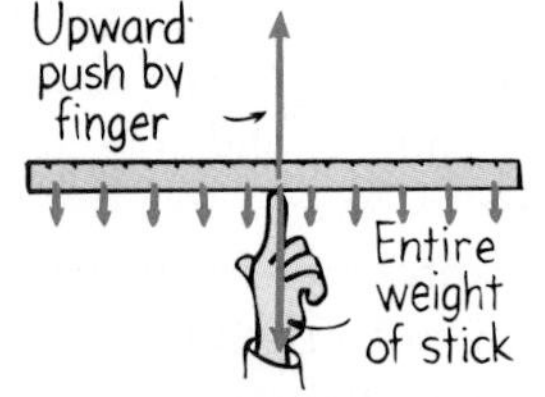

FIGURE 8.25
The weight of the entire stick behaves as if it were concentrated at the stick's center.

Locating the Center of Gravity

The center of gravity of a uniform object, such as a meterstick, is at its midpoint, for the stick acts as if its entire weight were concentrated there. Supporting that single point supports the entire stick. Balancing an object provides a simple method of locating its center of gravity. In Figure 8.25, the many small arrows represent the pull of gravity all along the meterstick. All of these can be combined into a resultant force acting through the center of gravity. The entire weight of the stick may be thought of as acting at this single point. Hence, we can balance the stick by applying a single upward force in a direction that passes through this point.

The center of gravity of any freely suspended object lies directly beneath (or at) the point of suspension (Figure 8.26). If a vertical line is drawn through the point of suspension, the center of gravity lies somewhere along that line. To determine exactly where it lies along the line, we have only to suspend the

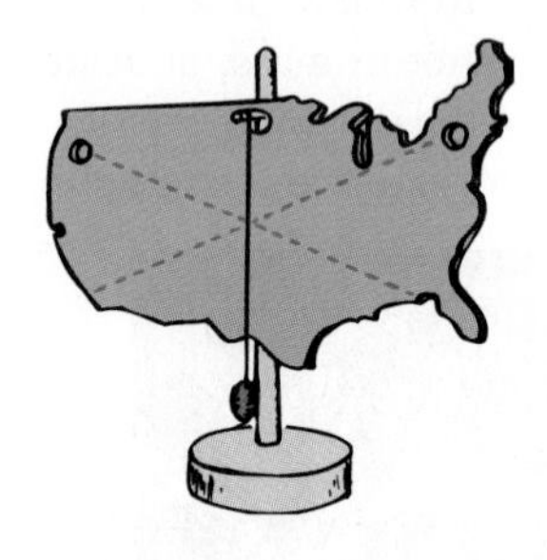

FIGURE 8.26
Finding the center of gravity for an irregularly shaped object.

CHECK YOUR ANSWERS

1. In the center of the hole!
2. No. A rigid object has one CG. If it is nonrigid, such as a piece of clay or putty, and is distorted into different shapes, then its CG may change as its shape changes. Even then, it has one CG for any given shape.

Locating the Center of Gravity

FIGURE 8.27
The athlete executes a "Fosbury flop" to clear the bar while his center of gravity passes beneath the bar.

object from some other point and draw a second vertical line through that point of suspension. The center of gravity lies where the two lines intersect.

The center of mass of an object may be a point where no mass exists. For example, the center of mass of a ring or a hollow sphere is at the geometrical center where no matter exists. Similarly, the center of mass of a boomerang is outside the physical structure, not within the material of the boomerang (Figure 8.28).

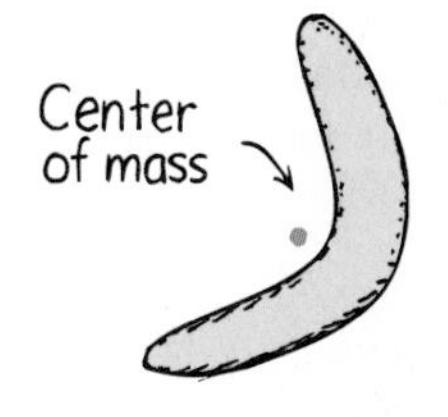

FIGURE 8.28
The center of mass can be outside the mass of a body.

CHECK YOURSELF

1. Where is the center of mass of Earth's atmosphere?
2. A uniform meterstick supported at the 25-cm mark balances when a 1-kg rock is suspended at the 0-cm end. What is the mass of the meterstick?

CHECK YOUR ANSWERS

1. Like a giant basketball, Earth's atmosphere is a spherical shell with its center of mass at Earth's center.
2. The mass of the meterstick is 1 kg. Why? The system is in equilibrium, so any torques must be balanced: The torque produced by the weight of the rock is balanced by the equal but oppositely directed torque produced by the weight of the stick applied at its CG, the 50-cm mark. The support force at the 25-cm mark is applied midway between the rock and the stick's CG, so the lever arms about the support point are equal (25 cm). This means that the weights (and hence the masses) of the rock and stick must also be equal. (Note that we don't have to go through the laborious task of considering the fractional parts of the stick's weight on either side of the fulcrum, for the CG of the whole stick really is at one point—the 50-cm mark!) Interestingly enough, the CG of the *rock + stick system* is at the 25-cm mark—directly above the fulcrum.

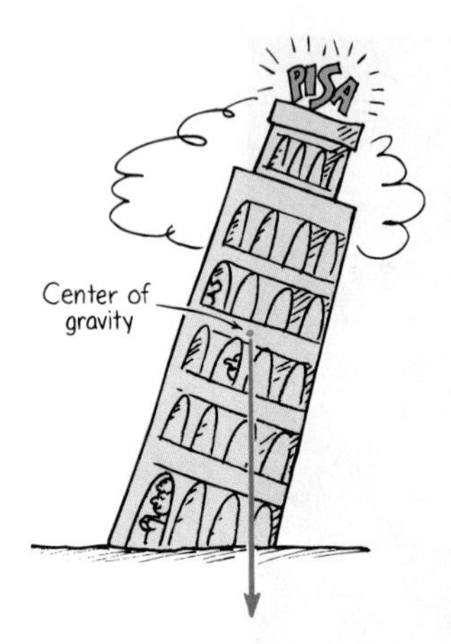

FIGURE 8.29
The center of gravity of the Leaning Tower of Pisa lies above its base of support, so the tower is in stable equilibrium.

Stability

The location of the center of gravity is important for stability (Figure 8.29). If we draw a line straight down from the center of gravity of an object of any shape and it falls inside the base of the object, it is in stable **equilibrium;** it will balance. If it falls outside the base, it is unstable. Why doesn't the famous Leaning Tower of Pisa topple over? As we can see in Figure 8.29, a line from the center of gravity of the tower to the level of the ground falls inside its base, so the Leaning Tower has stood for several centuries. If the tower leaned far enough so that the center of gravity extended beyond the base, an unbalanced torque would topple the tower.

To reduce the likelihood of tipping, it is usually advisable to design objects with a wide base and low center of gravity. The wider the base, the higher the center of gravity must be raised before the object will tip over.

When you stand erect (or lie flat), your center of gravity is within your body. Why is the center of gravity lower in an average woman than it is in an average man of the same height? Is your center of gravity always at the same point in your body? Is it always inside your body? What happens to it when you bend over?

If you are fairly flexible, you can bend over and touch your toes without bending your knees. Ordinarily, when you bend over and touch your toes, you extend your lower extremity as shown in the left half of Figure 8.31, so that your center of gravity is above a base of support, your feet. If you attempt to do this when standing against a wall, however, you cannot counterbalance yourself, and your center of gravity soon protrudes beyond your feet, as shown in the right half of Figure 8.31.

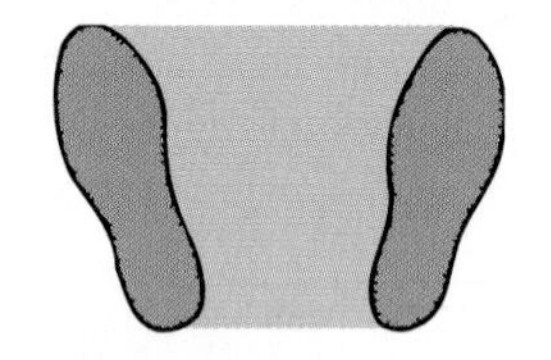

FIGURE 8.30
When you stand, your center of gravity is somewhere above the base area bounded by your feet. Why do you keep your legs far apart when you have to stand in the aisle of a bumpy-riding bus?

FIGURE 8.31
You can lean over and touch your toes without falling over only if your center of gravity is above the area bounded by your feet.

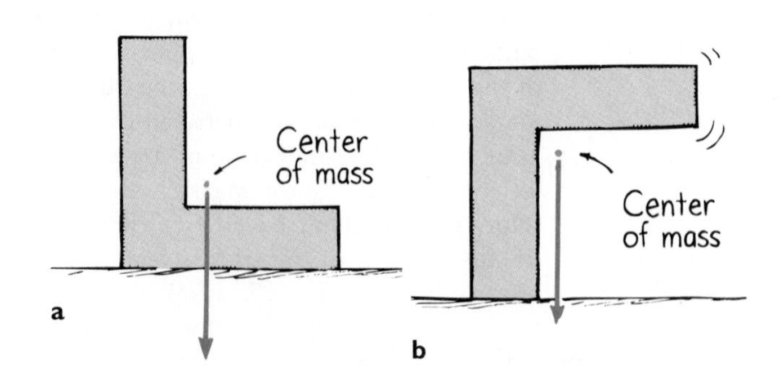

FIGURE 8.32
The center of mass of the L-shaped object is located where no mass exists. In (a), the center of mass is above the base of support, so the object is stable. In (b), it is not above the base of support, so the object is unstable and will topple over.

FIGURE 8.33
Where is Alexei's center of gravity relative to his hands?

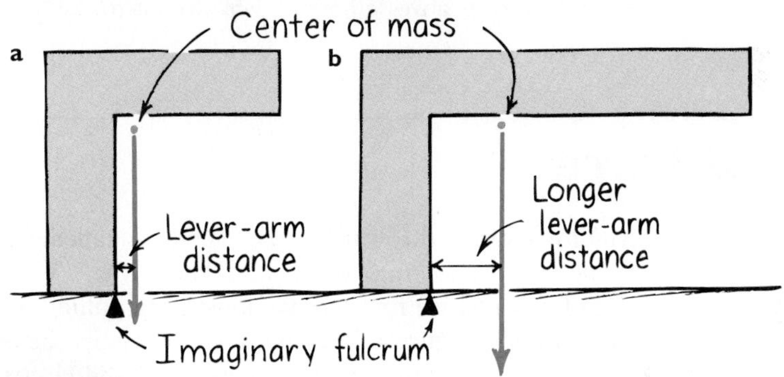

FIGURE 8.34
The greater torque acts on the figure in (b) for two reasons. What are they?

You rotate because of an unbalanced torque. This is evident in the two L-shaped objects shown in Figure 8.34. Both are unstable and will topple unless fastened to the level surface. It is easy to see that, even if both shapes have the same weight, the one on the right is more unstable. This is because of the greater lever arm and, hence, a greater torque.

Try balancing the pole end of a broom upright on the palm of your hand. The support base is quite small and relatively far beneath the center of gravity, so it's difficult to maintain balance for very long. After some practice, you can do it if you learn to make slight movements of your hand to respond exactly to variations in balance. You learn to avoid underresponding or overresponding to the slightest variations in balance. The intriguing Segway Human Transporter (Figure 8.35) does much the same. Variations in balance are quickly sensed by gyroscopes, and an internal high-speed computer regulates a motor to keep the vehicle upright. The computer regulates corrective adjustments of the wheel speed in a way quite similar to the way in which your brain coordinates your adjustive action when balancing a long pole on the palm of your hand. Both feats are truly amazing.

FIGURE 8.35
Gyroscopes and computer-assisted motors in the Segway vehicle make continual adjustments to keep the combined CG of Lillian and the vehicle above the wheel base.

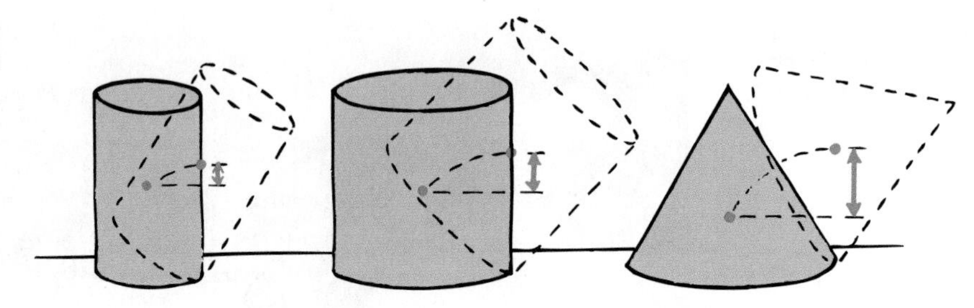

FIGURE 8.36
Stability is determined by the vertical distance that the center of gravity is raised in tipping. An object with a wide base and a low center of gravity is more stable.

CHECK YOURSELF

1. Why is it dangerous to slide open the top drawers of a fully loaded file cabinet that is not secured to the floor?
2. When a car drives off a cliff, why does it rotate forward as it falls?

Centripetal Force

Any force directed toward a fixed center is called a **centripetal force.** *Centripetal* means "center seeking" or "toward the center." When we whirl a tin can on the end of a string, we find that we must keep pulling on the string—exerting a centripetal force (Figure 8.37). The string transmits the centripetal force, which pulls the can into a circular path. Gravitational and electrical forces can produce centripetal forces. The Moon, for example, is held in an almost circular orbit by gravitational force directed toward the center of the Earth. The orbiting electrons in atoms experience an electrical force toward the central nuclei. Anything that moves in a circular path does so because it's acted upon by a centripetal force.

FIGURE 8.37
The force exerted on the whirling can is toward the center.

Centripetal Force

Centripetal force depends on the mass m, tangential speed v, and radius of curvature r of the circularly moving object. In lab, you'll likely use the exact relationship

$$F = mv^2/r.$$

Notice that speed is squared, so twice the speed needs four times the force. The inverse relationship with the radius of curvature tells us that half the radial distance requires twice the force.

FIGURE 8.38
The centripetal force (adhesion of mud on the spinning tire) is not great enough to hold the mud on the tire, so it flies off in straight-line directions.

CHECK YOUR ANSWERS

1. The filing cabinet is in danger of tipping because the CG may extend beyond the support base. If it does, then torque due to gravity tips the cabinet.
2. When all wheels are on the ground, the car's CG is above a support base. But, when the car drives off a cliff, the front wheels are first to leave the ground and the support base shrinks to the line between the rear wheels. So the car's CG extends beyond the support base and it rotates, as would the Leaning Tower of Pisa if its CG extended beyond its support base.

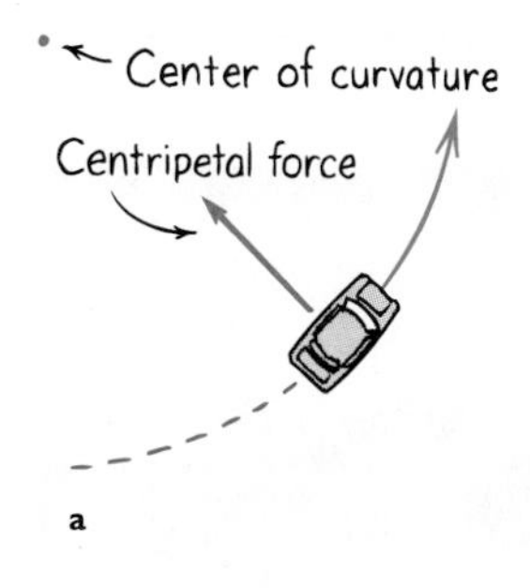

Centripetal force is not a basic force of nature; it is simply the name given to any force, whether string tension, gravitational, electrical, or whatever, that is directed toward a fixed center. If the motion is circular and executed at constant speed, this force is at right angles to the path of the moving object.

When an automobile rounds a corner, the friction between the tires and the road provides the centripetal force that holds the car in a curved path (Figure 8.39). If this friction is insufficient (due to oil or gravel on the pavement, for example), the tires slide sideways and the car fails to make the curve; the car tends to skid tangentially off the road.

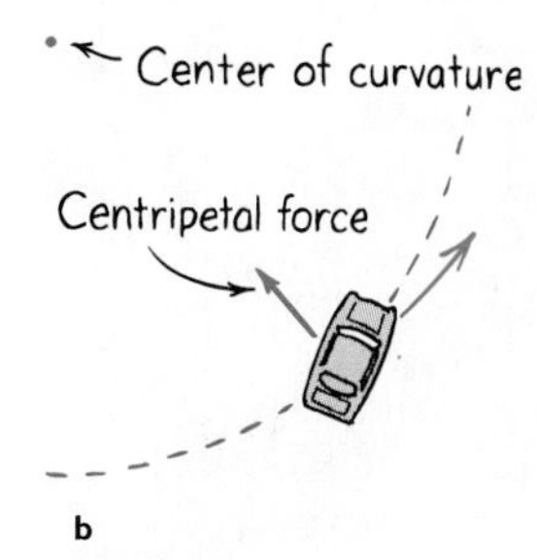

FIGURE 8.39
(a) When a car goes around a curve, there must be a force pushing the car toward the center of the curve. (b) A car skids on a curve when the centripetal force (friction of road on tires) is not great enough.

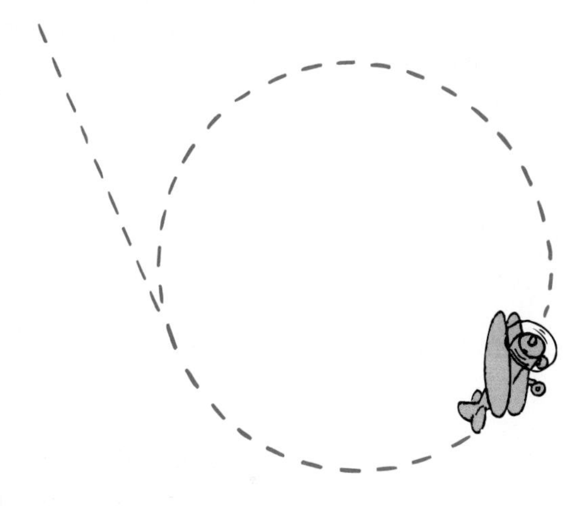

FIGURE 8.40
Large centripetal forces on the wings of an aircraft enable it to fly in circular loops. The acceleration away from the straight-line path the aircraft would follow in the absence of centripetal force is often several times greater than the acceleration due to gravity, *g*. For example, if the centripetal acceleration is 49 m/s^2 (five times as great as 9.8 m/s^2), we say that the aircraft is undergoing 5 *g*'s. At the bottom of the loop, the seat presses against the pilot with an *additional* force five times greater than his weight, so the seat pushes up on him with a total force six times his weight. Typical fighter aircraft are designed to withstand accelerations up to 8 or 9 *g*'s. The pilot as well as the aircraft must withstand the centripetal acceleration. Pilots of fighter aircraft wear pressure suits to prevent blood from flowing away from the head toward the legs, which could cause a blackout.

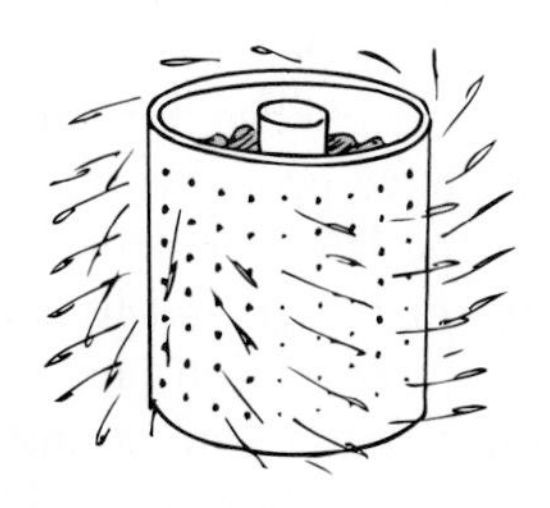

FIGURE 8.41
The clothes are forced into a circular path, but the water is not.

Centripetal force plays the main role in the operation of a centrifuge. A familiar example is the spinning tub in an automatic washing machine (Figure 8.41). In its spin cycle, the tub rotates at high speed and produces a centripetal force on the wet clothes, which are forced into a circular path by the inner wall of the tub. The tub exerts great force on the clothes, but the holes in the tub prevent the exertion of the same force on the water in the clothes. The water escapes tangentially out the holes. Strictly speaking, the clothes are forced away from the water; the water is not forced away from the clothes. Think about that.

Centrifugal Force

Although centripetal force is center directed, an occupant inside a rotating system seems to experience an outward force. This apparent outward force is called **centrifugal force.** *Centrifugal* means "center-fleeing" or "away from the center."

PRACTICING PHYSICS: WATER-BUCKET SWING

Half fill a bucket of water and swing it in a vertical circle, as Marshall Ellenstein demonstrates. The bucket and water accelerate toward the center of your swing. If you swing the bucket fast enough, the water won't fall out at the top. Interestingly, although it won't fall *out,* it still falls. The trick is to swing the bucket fast enough so that the bucket falls as fast as the water inside it falls. Can you see that, because the bucket is revolving, the water moves tangentially—and stays in the bucket? In Chapter 10, we'll learn that an orbiting space shuttle similarly falls while in orbit. The trick is to impart sufficient tangential velocity to the shuttle so that it falls around the curved Earth rather than into it.

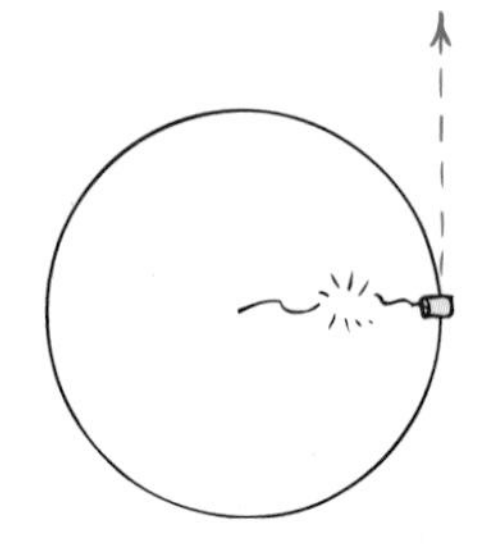

FIGURE 8.42
When the string breaks, the whirling can moves in a straight line, tangent to—not outward from—the center of its circular path.

In the case of the whirling can, it is a common misconception to believe that a centrifugal force pulls outward on the can. If the string holding the whirling can breaks (Figure 8.42), the can doesn't move radially outward, but goes off in a tangent straight-line path—because *no* force acts on it. We illustrate this further with another example.

Suppose you are a passenger in a car that suddenly stops short. You pitch forward toward the dashboard. When this occurs, you don't say that something forced you forward. In accord with the law of inertia, you pitched forward because of the absence of a force, which seat belts would have provided. Similarly, if you are in a car that rounds a sharp corner to the left, you tend to pitch outward to the right—not because of some outward or centrifugal force, but because there is no centripetal force holding you in circular motion (again, the purpose of seat belts). The idea that a centrifugal force bangs you against the car door is a misconception. (Sure, you push out against the door, but only because the door pushes in on you—Newton's third law.)

Likewise, when you swing a tin can in a circular path, no force pulls the can outward—the only force on the can is the string pulling it inward. There is no outward force on the can. Now suppose there is a ladybug inside the whirling

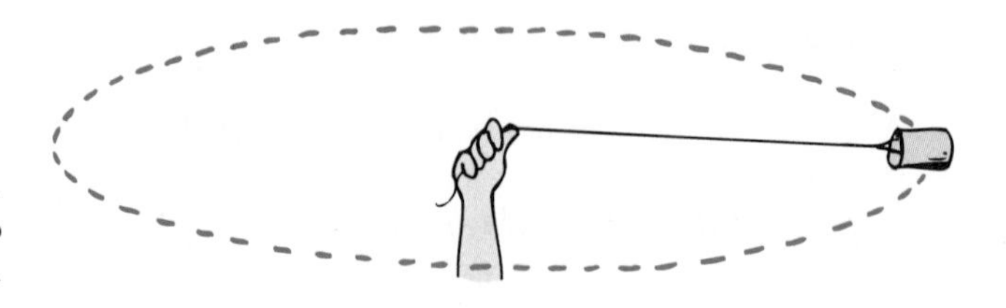

FIGURE 8.43
The only force that is exerted on the whirling can (neglecting gravity) is directed *toward* the center of circular motion. This is a centripetal force. No *outward* force acts on the can.

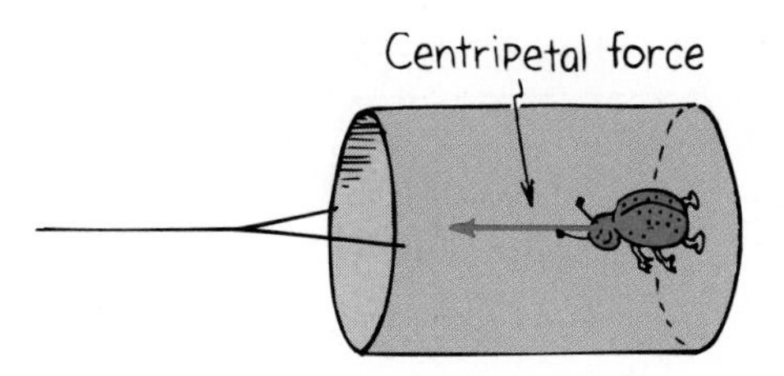

FIGURE 8.44
The can provides the centripetal force necessary to hold the ladybug in a circular path.

can (Figure 8.44). The can presses against the bug's feet and provides the centripetal force that holds it in a circular path. Neglecting gravity, the only force exerted on the ladybug is the can pressing on its feet. From our outside stationary frame of reference, we see no centrifugal force exerted on the ladybug, just as no centrifugal force banged us against the car door. The centrifugal force effect is caused not by a real force, but by inertia—the tendency of the moving object to follow a straight-line path. But try telling that to the ladybug!

Centrifugal Force in a Rotating Reference Frame

If we stand at rest and watch somebody whirling a can overhead in a horizontal circle, we see that the force on the can is centripetal, just as it is on a ladybug inside the can. The bottom of the can exerts a force on the ladybug's feet. Neglecting gravity, no other force acts on the ladybug. But, as viewed from inside the frame of reference of the revolving can, things appear very different.[6]

In the rotating frame of the ladybug, in addition to the force of the can on the ladybug's feet, there is an apparent centrifugal force that is exerted on the ladybug. Centrifugal force *in a rotating reference frame* is a force in its own right, as real as the pull of gravity. However, there is a fundamental difference. Gravitational force is an interaction between one mass and another. The gravity we experience is our interaction with Earth. But for centrifugal force in the rotating frame, no such agent exists—there is no interaction counterpart. Centrifugal force *feels* like gravity, but with no agent pulling. Nothing produces it; it is a result of rotation. For this reason, physicists call it an "inertial" force (or sometimes a *fictitious* force)—an *apparent* force—and not a real force like gravity, electromagnetic forces, and nuclear forces. Nevertheless, to observers who are in a rotating system, centrifugal force feels like, and is interpreted to be, a very real force. Just as gravity is ever present at Earth's surface, centrifugal force is ever present in a rotating system.

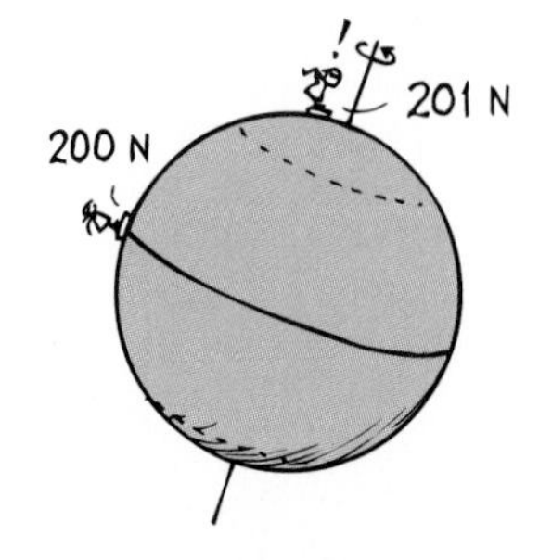

FIGURE 8.45
In the frame of reference of the spinning Earth, we experience a centrifugal force that slightly decreases our weight. Like an outside horse on a merry-go-round, we have the greatest tangential speed farthest from Earth's axis, at the equator. Centrifugal force is therefore maximum for us when we are at the equator and zero for us at the poles, where we have no tangential speed. So, strictly speaking, if you want to lose weight, walk toward the equator!

[6]A frame of reference wherein a free body exhibits no acceleration is called an *inertial* frame of reference. Newton's laws are seen to hold exactly in an inertial frame. A rotating frame of reference, in contrast, is an accelerating frame of reference. Newton's laws are not valid in an accelerating frame of reference.

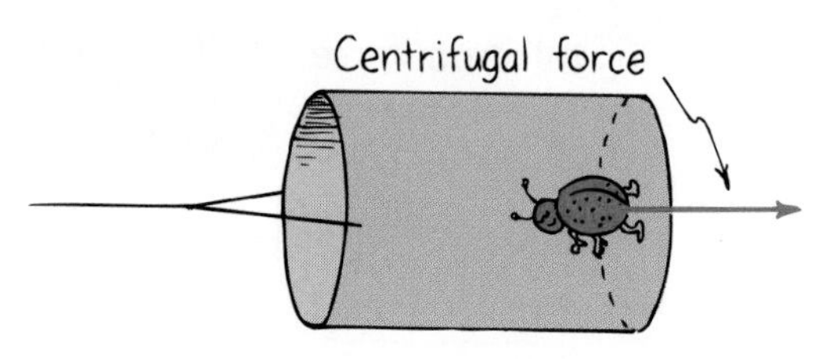

FIGURE 8.46
From the reference frame of the ladybug inside the whirling can, it is being held to the bottom of the can by a force that is directed away from the center of circular motion. The ladybug calls this outward force a *centrifugal* force, which is as real to it as gravity.

CHECK YOURSELF

A heavy iron ball is attached by a spring to the rotating platform, as shown in the sketch. Two observers, one in the rotating frame and one on the ground at rest, observe its motion. Which observer sees the ball being pulled outward, stretching the spring? Which sees the spring pulling the ball into circular motion?

Simulated Gravity

A rotating habitat need not be a huge wheel. Gravity could be simulated in a pair of circling pods connected by a long cable.

Insights

Consider a colony of ladybugs living inside a bicycle tire—a balloon tire with plenty of room inside. If we toss the bicycle wheel through the air or drop it from an airplane high in the sky, the ladybugs will be in a weightless condition. They will float freely while the wheel is in free fall. Now spin the wheel. The ladybugs will feel themselves pressed to the outer part of the tire's inner surface. If the wheel is spun at just the right speed, the ladybugs will experience *simulated gravity* that will feel like the gravity they are accustomed to. Gravity is simulated by centrifugal force. The "down" direction to the ladybugs will be what we would call radially outward, away from the center of the wheel.

Simulated Gravity

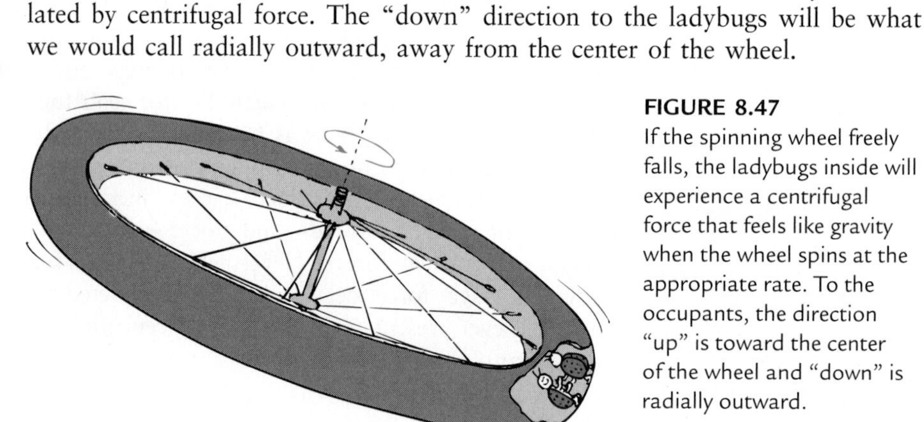

FIGURE 8.47
If the spinning wheel freely falls, the ladybugs inside will experience a centrifugal force that feels like gravity when the wheel spins at the appropriate rate. To the occupants, the direction "up" is toward the center of the wheel and "down" is radially outward.

CHECK YOUR ANSWER

The observer in the reference frame of the rotating platform states that a centrifugal force pulls radially outward on the ball, which stretches the spring. The observer in the rest frame states that a centripetal force supplied by the stretched spring pulls the ball into circular motion. Only the observer in the rest frame can identify an action–reaction pair of forces; where action is spring on ball, reaction is ball on spring. The rotating observer can't identify a reaction counterpart to the centrifugal force because there isn't any!

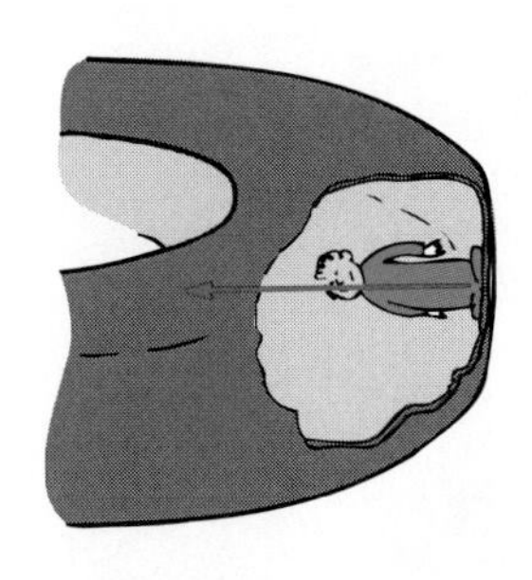

FIGURE 8.48
The interaction between the man and the floor, as seen from a stationary frame of reference outside the rotating system. The floor presses against the man (action) and the man presses back on the floor (reaction). The only force exerted on the man is by the floor. It is directed toward the center and is a centripetal force.

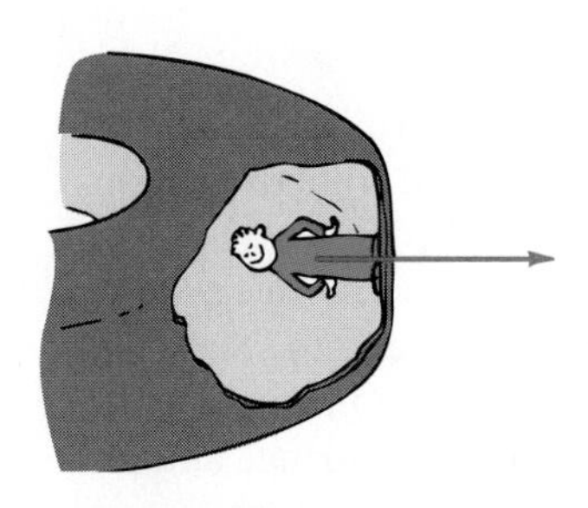

FIGURE 8.49
As seen from inside the rotating system, in addition to the man–floor interaction, there is a centrifugal force exerted on the man at his center of mass. It seems as real as gravity. Yet, unlike gravity, it has no reaction counterpart—there is nothing out there that he can pull back on. Centrifugal force is not part of an interaction, but it is a consequence of rotation. It is therefore called an apparent, or fictitious, force.

Humans today live on the outer surface of this spherical planet and are held here by gravity. The planet has been the cradle of humankind. But we will not stay in the cradle forever. We are becoming a spacefaring people. Many people, in coming years, will likely live in huge, lazily rotating habitats in space and will be held to the inner surfaces by centrifugal force. The rotating habitats will provide a simulated gravity so that the human body can function normally.

Occupants of today's space vehicles feel weightless because they experience no *support force*. They're not pressed against a supporting floor by gravity, nor do they experience a centrifugal force due to spinning. Over extended periods, this can cause loss of muscle strength and detrimental changes in the body, such as loss of calcium from the bones. But future space travelers need not be subject to weightlessness. Their space habitats will probably spin, like the ladybug's spinning wheel, effectively supplying a support force and nicely simulating gravity. Structures of small diameter would have to rotate at high rates to provide a simulated gravitational acceleration of 1 g. Sensitive and delicate organs in our inner ears sense rotation. Although there appears to be no difficulty at a single revolution per minute (RPM) or so, many people find it difficult to adjust to rates greater than 2 or 3 RPM (although some easily adapt to 10 or so RPM). To simulate normal Earth gravity at 1 RPM requires a large structure—one about 2 kilometers in diameter. This is an immense structure, compared with today's space vehicles. Finances have dictated the size of the first inhabited structures. Russia's early Mir space station housed a few people for months at a time for fourteen years. The International Space Station can support a larger crew, but, like Mir, it doesn't rotate. Therefore, its crew members have to adjust to living in a weightless environment. Rotating habitats may follow later.

Centrifugal acceleration is directly proportional to the radial distance from the hub, so a variety of g states is possible. If the structure rotates so that inhabitants on the inside of the outer edge experience 1 g, then halfway to the axis they would experience 0.5 g. At the axis itself they would experience weightlessness (zero g). The variety of fractions of g possible from the rim of a rotating space habitat holds promise for a most different and (at this writing) as yet unexperienced environment. In this still very hypothetical structure, we would be able to perform ballet at 0.5 g; diving and acrobatics at 0.2 g and lower-g states; and three-dimensional soccer (and new sports not yet conceived) in very low g states.

CHECK YOURSELF

If Earth were to spin faster about its axis, your weight would be less. If you were in a rotating space habitat that increased its spin rate, you'd "weigh" more. Explain why greater spin rates produce opposite effects in these cases.

CHECK YOUR ANSWER

You're on the *outside* of the spinning Earth, but you'd be on the *inside* of a spinning space habitat. A greater spin rate on the outside of the Earth tends to throw you *off* a weighing scale, causing it to show a decrease in weight, but *against* a weighing scale *inside* the space habitat to show an increase in weight.

FIGURE 8.50
This artist's rendering shows the interior of a fanciful space colony that would be occupied by a few thousand people.

Angular Momentum

Things that rotate, whether a colony in space, a cylinder rolling down an incline, or an acrobat doing a somersault, remain rotating until something stops them. A rotating object has an "inertia of rotation." Recall, from Chapter 6, that all moving objects have "inertia of motion" or *momentum*—the product of mass and velocity. This kind of momentum is **linear momentum.** Similarly, the "inertia of rotation" of rotating objects is called **angular momentum.**

A planet orbiting the Sun, a rock whirling at the end of a string, and the tiny electrons whirling about atomic nuclei all have angular momentum.

Angular momentum is defined as the product of rotational inertia and rotational velocity.

$$\text{Angular momentum} = \text{rotational inertia} \times \text{rotational velocity}$$

It is the counterpart of linear momentum:

$$\text{Linear momentum} = \text{mass} \times \text{velocity}$$

Like linear momentum, angular momentum is a vector quantity and has direction as well as magnitude. In this book, we won't treat the vector nature of angular momentum (or even of torque, which also is a vector), except to acknowledge the remarkable action of the gyroscope. The rotating bicycle wheel in Figure 8.51 shows what happens when a torque caused by Earth's gravity acts to change the direction of its angular momentum (which is along the wheel's axle). The pull of gravity that normally acts to topple the wheel over and change its rotational axis causes it instead to *precess* about a vertical axis. You must do this yourself while standing on a turntable to fully believe it. You probably won't fully understand it unless you do follow-up study sometime in the future.

FIGURE 8.51
Angular momentum keeps the wheel axle nearly horizontal when a torque supplied by Earth's gravity acts on it. Instead of causing the wheel to topple, the torque causes the wheel's axle to turn slowly around the circle of students. This is called *precession*.

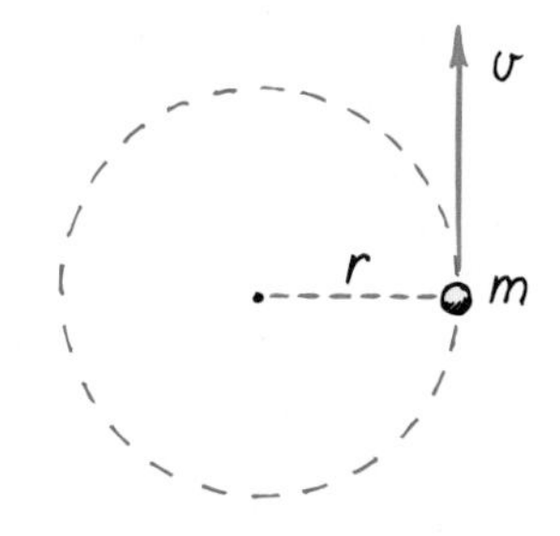

FIGURE 8.52
A small object of mass *m* whirling in a circular path of radius *r* with a speed *v* has angular momentum *mvr*.

For the case of an object that is small compared with the radial distance to its axis of rotation, such as a tin can swinging from a long string or a planet orbiting in a circle around the Sun, the angular momentum can be expressed as the magnitude of its linear momentum, mv, multiplied by the radial distance, r (Figure 8.52). In shorthand notation,

$$\text{Angular momentum} = mvr$$

Just as an external net force is required to change the linear momentum of an object, an external net torque is required to change the angular momentum of an object. We can state a rotational version of Newton's first law (the law of inertia):

> **An object or system of objects will maintain its angular momentum unless acted upon by an external net torque.**

Our solar system has angular momentum contributed by the Sun, the spinning and orbiting planets, and myriad other smaller bodies. The angular momentum of the solar system today will be its angular momentum for eons to come. Only an external torque from outside the solar system can change it. In the absence of such a torque, we say the angular momentum of the solar system is conserved.

Conservation of Angular Momentum

Just as the linear momentum of any system is conserved if no net force acts on the system, angular momentum is conserved if no net torque acts. The **law of conservation of angular momentum** states:

> **If no external net torque acts on a rotating system, the angular momentum of that system remains constant.**

This means that, with no external torque, the product of rotational inertia and rotational velocity at one time will be the same as at any other time.

FIGURE 8.53
Interactive Figure
Conservation of angular momentum. When the man pulls his arms and the whirling weights inward, he decreases his rotational inertia I, and his rotational speed ω correspondingly increases.

Conservation of Angular Momentum Using a Rotating Platform

Why do short acrobats have an advantage in tumbling or in other end-over-end rotational motions?

Insights

An interesting example illustrating the conservation of angular momentum is shown in Figure 8.53. The man stands on a low-friction turntable with weights extended. His rotational inertia, I, with the help of the extended weights, is relatively large in this position. As he slowly turns, his angular momentum is the product of his rotational inertia and rotational velocity—ω. When he pulls the weights inward, the rotational inertia of his body and the weights is considerably reduced. What is the result? His rotational speed increases! This example is best appreciated by the turning person who feels changes in rotational speed that seem to be mysterious. But it's straightforward physics! This procedure is used by a figure skater who starts to whirl with her arms and perhaps a leg extended and then draws her arms and leg in to obtain a greater rotational speed. Whenever a rotating body contracts, its rotational speed increases.

Similarly, when a gymnast is spinning freely in the absence of unbalanced torques on his or her body, angular momentum does not change. However, rotational speed can be changed by simply making variations in rotational inertia. This is done by moving some part of the body toward or away from the axis of rotation.

If a cat is held upside down and dropped, it is able to execute a twist and to land upright, even if it has no initial angular momentum. Zero-angular-momentum twists and turns are performed by turning one part of the body against the other. While falling, the cat rearranges its limbs and tail several times to change its rotational inertia repeatedly until it lands feet downward. During this maneuver the total angular momentum remains zero (Figure 8.55). When it is over, the cat is not turning. This maneuver rotates the body through an angle, but it does not create continuing rotation. To do so would violate angular momentum conservation.

Humans can perform similar twists without difficulty, though not as fast as a cat. Astronauts have learned to make zero-angular-momentum rotations as they orient their bodies in preferred directions when floating freely in space.

The law of conservation of angular momentum is seen in the motions of planets and the shapes of galaxies. It is fascinating to note that the conservation of angular momentum tells us that the Moon is getting farther away from the Earth. This is because the Earth's daily rotation is slowly decreasing due to the friction of ocean waters on the ocean bottom, just as an automobile's wheels

FIGURE 8.54
Rotational speed is controlled by variations in the body's rotational inertia as angular momentum is conserved during a forward somersault.

FIGURE 8.55
Time-lapse photo of a falling cat.

slow down when brakes are applied. This decrease in Earth's angular momentum is accompanied by an equal increase in the angular momentum of the Moon in its orbital motion about Earth, which results in the Moon's increasing distance from Earth and decreasing speed. This increase of distance amounts to one-quarter of a centimeter per rotation. Have you noticed that the Moon is getting farther away from us lately? Well, it is; each time we see another full Moon, it is one-quarter of a centimeter farther away!

Oh yes—before we end this chapter, we'll give an answer to Check Yourself question 3 back on page 136. Manhole covers are circular because a circular cover is the only shape that can't fall into the hole. A square cover, for example, can be tilted vertically and turned so it can drop diagonally into the hole. Likewise for every other shape. If you're working in a manhole and some fresh kids are horsing around above, you'll be glad the cover is round!

Summary of Terms

Tangential speed The linear speed tangent to a curved path, such as in circular motion.

Rotational speed The number of rotations or revolutions per unit of time; often measured in rotations or revolutions per second or per minute. (Scientists usually measure it in radians per second.)

Rotational inertia That property of an object that measures its resistance to any change in its state of rotation: if at rest, the body tends to remain at rest; if rotating, it tends to remain rotating and will continue to do so unless acted upon by a external net torque.

Torque The product of force and lever-arm distance, which tends to produce rotation.

$$\text{Torque} = \text{lever arm} \times \text{force}$$

Center of mass (CM) The average position of the mass of an object. The CM moves as if all the external forces acted at this point.

Center of gravity (CG) The average position of weight or the single point associated with an object where the force of gravity can be considered to act.

Equilibrium The state of an object in which it is not acted upon by a net force or a net torque.

Centripetal force A force directed toward a fixed point, usually the cause of circular motion: $F = mv^2/r$.

Centrifugal force An outward force apparent in a rotating frame of reference. It is apparent (fictitious) in the sense that it is not part of an interaction but is a result of rotation—with no reaction-force counterpart.

Angular momentum The product of a body's rotational inertia and rotational velocity about a particular axis. For an object that is small compared with the radial distance, it can be expressed as the product of mass, speed, and the radial distance of rotation.

Conservation of angular momentum When no external torque acts on an object or a system of objects, no change of angular momentum can occur. Hence, the angular momentum before an event involving only internal torques or no torques is equal to the angular momentum after the event.

Suggested Reading

Brancazio, P. J. *Sport Science.* New York: Simon & Schuster, 1984.

Clarke, A. C. *Rendezvous with Rama.* New York: Harcourt Brace Jovanovich, 1973. This is the first science-fiction novel to seriously consider habitation inside a spinning space facility.

Review Questions

Circular Motion

1. What is meant by tangential speed?
2. Distinguish between tangential speed and rotational speed.
3. What is the relationship between tangential speed and distance from the center of the rotational axis? Give an example.
4. A tapered cup rolled on a flat surface makes a circular path. What does this tell you about the tangential speed of the rim of the wide end of the cup compared with that of the rim of the narrow end?
5. How does the tapered rim of a wheel on a railroad train allow one part of the rim to have a greater tangential speed than another part when it is rolling on a track?

Rotational Inertia

6. What is rotational inertia, and is it similar to inertia as studied in previous chapters?
7. Inertia depends on mass; rotational inertia depends on mass and something else. What?
8. Does the rotational inertia of a particular object differ for different axes of rotation? Can one object have more than one rotational inertia?
9. Consider three axes of rotation for a pencil: along the lead; at right angles to the lead at the middle; at right angles to the lead at one end. Rate the rotational inertias about each axis from small to large.
10. Which is easier to get swinging, a baseball bat held at the end or one held closer to the massive end (choked up)?
11. Why does bending your legs when running enable you to swing your legs to and fro more rapidly?
12. Which will have the greater acceleration rolling down an incline, a hoop or a solid disk?

Torque

13. What does a torque tend to do to an object?
14. What is meant by the "lever arm" of a torque?
15. How do clockwise and counterclockwise torques compare when a system is balanced?

Center of Mass and Center of Gravity

16. If you toss a stick into the air, it appears to wobble all over the place. Specifically, what place?
17. Where is the center of mass of a baseball? Where is its center of gravity? Where are these centers for a baseball bat?

Locating the Center of Gravity

18. If you hang at rest by your hands from a vertical rope, where is your center of gravity with respect to the rope?
19. Where is the center of mass of a soccer ball?

Stability

20. What is the relationship between center of gravity and support base for an object in stable equilibrium?
21. Why doesn't the Leaning Tower of Pisa topple over?
22. In terms of center of gravity, support base, and torque, why can you not stand with heels and back to a wall and then bend over to touch your toes and return to your stand-up position?

Centripetal Force

23. When you whirl a can at the end of a string in a circular path, what is the direction of the force you exert on the can?
24. Is it an inward force or an outward force that is exerted on the clothes during the spin cycle of an automatic washing machine?

Centrifugal Force

25. If the string that holds a whirling can in its circular path breaks, what kind of force causes it to move in a straight-line path—centripetal, centrifugal, or no force? What law of physics supports your answer?

26. If you are in a car that rounds a curve, and you are not wearing a seat belt, and you slide across your seat and slam against a car door, what kind of force is responsible—centripetal, centrifugal, or no force? Support your answer.

Centrifugal Force in a Rotating Reference Frame

27. Why is centrifugal force in a rotating frame called a "fictitious force"?

Simulated Gravity

28. How can gravity be simulated in an orbiting space station?

Angular Momentum

29. Distinguish between linear momentum and angular momentum.

30. What is the law of inertia for rotating systems in terms of angular momentum?

Conservation of Angular Momentum

31. What does it mean to say that angular momentum is conserved?

32. If a skater who is spinning pulls her arms in so as to reduce her rotational inertia by half, by how much will her angular momentum increase? By how much will her rate of spin increase? (Why are your answers different?)

Projects

1. Write a letter to Grandpa and tell him how you're learning to distinguish between closely related concepts, using the examples of force and torque. Tell him how the two are similar, and how they're different. Suggest where he can find "hands-on" things in his home that can illustrate the difference between the two. Also cite an example that shows how the net force on an object can be zero while the net torque isn't, as well as an example showing vice versa. (Now make your grandpa's day and *send* the letter to him!)

2. Fasten a pair of foam cups together at their wide ends and roll them along a pair of metersticks that simulate railroad tracks. Note how they self-correct whenever their path departs from the center. Question: If you taped the cups together at their narrow ends, so they tapered oppositely, would the pair of cups self-correct or self-destruct when rolling slightly off center?

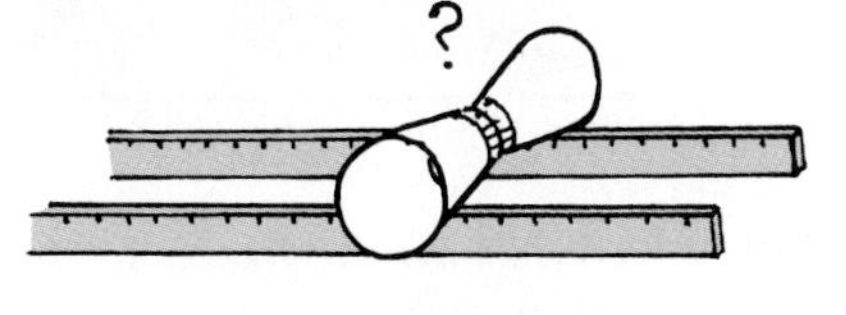

3. Fasten a fork, spoon, and wooden match together as shown. The combination will balance nicely—on the edge of a glass, for example. This happens because the center of gravity actually "hangs" below the point of support.

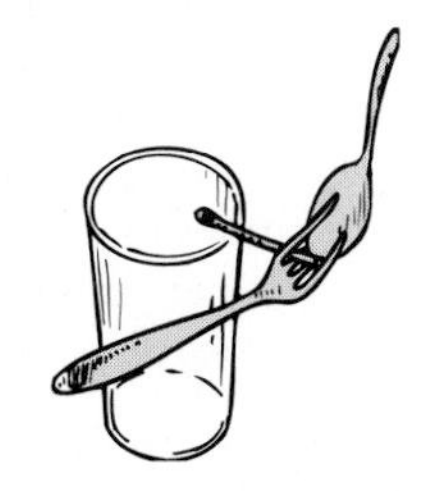

4. Stand with your heels and back against a wall and try to bend over and touch your toes. You'll find that you have to stand away from the wall to do so without toppling over. Compare the minimum distance of your heels from the wall with that of a friend of the opposite sex. Try the activity shown here, lifting a chair with your head against the wall. Why are females better than males in doing this?

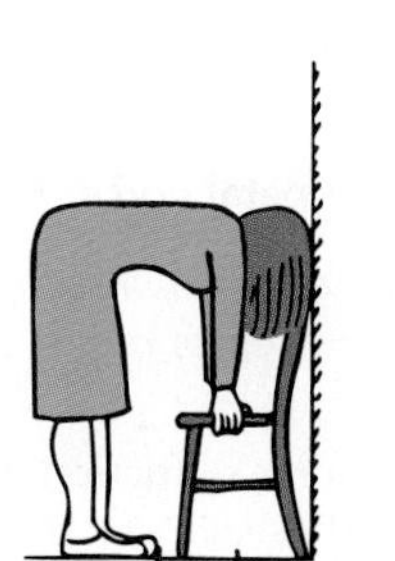

5. First ask a friend to stand facing a wall with her toes touching the wall, then ask her to stand on the balls of her feet without toppling backward. Your friend won't be able to do it. Now you explain why it can't be done.

6. Rest a meterstick on two extended forefingers as shown. Slowly bring your fingers together. At what part of the stick do your fingers meet? Can you explain why this always happens, no matter where you start your fingers?

7. Swing a pail of water around rapidly in a circle at arm's length and the water will not spill. Why?
8. Place the hook of a wire coat hanger over your finger. Carefully balance a coin on the straight wire on the bottom directly under the hook. You may have to flatten the wire with a hammer or fashion a tiny platform with tape. With a surprisingly small amount of practice you can swing the hanger and balanced coin back and forth and then in a circle. Centripetal force holds the coin in place.

One-Step Calculations

Torque = Lever Arm × Force

1. Calculate the torque produced by a 50-N perpendicular force at the end of a 0.2-m-long wrench.
2. Calculate the torque produced by the same 50-N force when a pipe extends the length of the wrench to 0.5 m.

Centripetal Force: $F = mv^2/r$

3. Calculate the tension in a string that whirls a 2-kg toy in a horizontal circle of radius 2.5 m when it moves at 3 m/s.
4. Calculate the force of friction that keeps a 75-kg person sitting on the edge of a horizontal rotating platform when the person sits 2 m from the center of the platform and has a tangential speed of 3 m/s.

Angular Momentum = mvr (also = $I\omega$)

5. Calculate the angular momentum of the person in the previous problem.
6. If the person's speed doubles and all else remains the same, what will be the person's angular momentum?

Exercises

1. When rewinding an audio or videotape, one reel winds fastest at the end. Which reel is this, and why does it speed up?
2. A large wheel is coupled to a wheel with half the diameter, as shown. How does the rotational speed of the smaller wheel compare with that of the larger wheel? How do the tangential speeds at the rims compare (assuming the belt doesn't slip)?

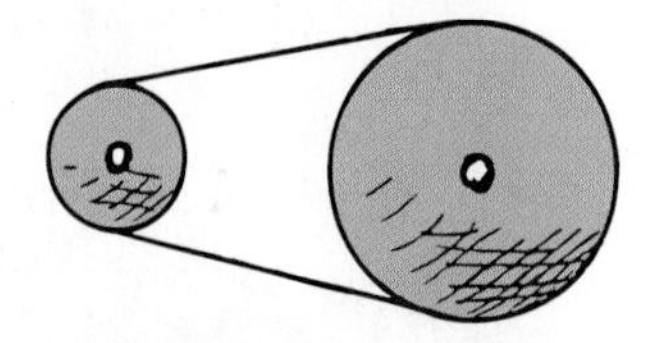

3. An automobile speedometer is configured to read speed proportional to the rotational speed of its wheels. If larger wheels, such as those of snow tires, are used, will the speedometer reading be high or low—or no different?
4. Dan and Sue cycle at the same speed. The tires on Dan's bike are larger in diameter than those on Sue's bike. Which wheels, if either, have the greater rotational speed?
5. The wheels of railroad trains are tapered, a feature especially important on curves. How, if at all, does the amount of taper relate to the curving of the tracks?
6. Use the equation $v = r\omega$ to explain why the end of a fly swatter moves much faster than your wrist when swatting a fly.
7. Flamingos are frequently seen standing on one leg with the other lifted. What can you say about the bird's center of mass with respect to the foot on which it stands?
8. In this chapter, we learn that an object may *not* be in mechanical equilibrium even when $\Sigma F = 0$. Explain.
9. The front wheels located far out in front of the racing vehicle help to keep the vehicle from nosing upward when it accelerates. What physics concepts play a role here?

10. When a car drives off a cliff, why does it rotate forward as it falls? (Consider the torque it experiences as it rolls off the cliff.)

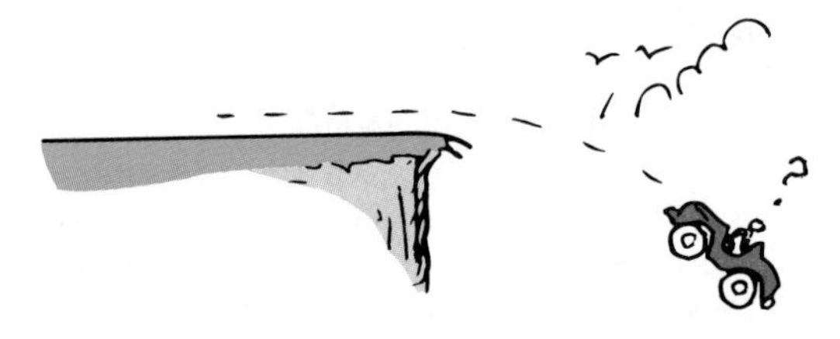

11. Why does a car nose up when accelerating and nose down when braking?

12. Which will have the greater acceleration rolling down an incline—a bowling ball or a volleyball? Defend your answer.

13. A softball and a basketball start from rest and roll down an incline. Which ball reaches the bottom first? Defend your answer.

14. How would a ramp help you to distinguish between two identical-looking spheres of the same weight, one solid and the other hollow?

15. Which will roll down an incline faster—a can of water or a can of ice?

16. Why are lightweight tires preferred over lightweight frames in bicycle racing?

17. A youngster who has entered a soap-box derby (in which 4-wheel unpowered vehicles roll from rest down a hill) asks if large massive wheels or lightweight ones should be used. Also, should the wheels have spokes or be solid? What advice do you offer?

18. Is the net torque changed when a partner on a seesaw stands or hangs from her end instead of sitting? (Does the weight or the lever arm change?)

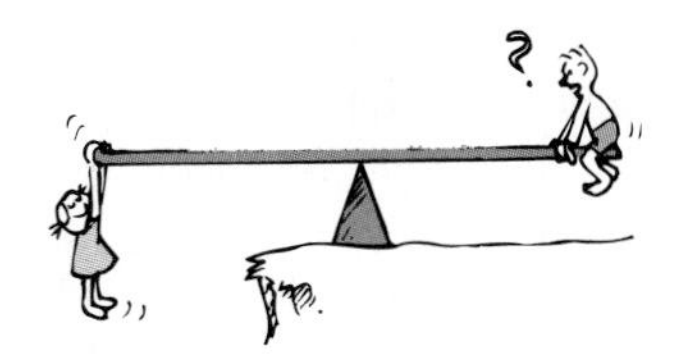

19. When you pedal a bicycle, maximum torque is produced when the pedal sprocket arms are in the horizontal position, and no torque is produced when they are in the vertical position. Explain.

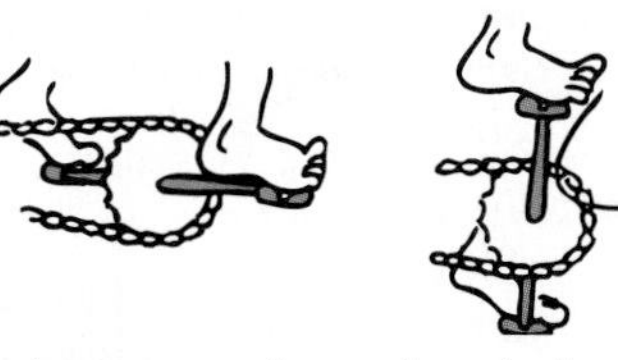

20. Can a force produce a torque when there is no lever arm?

21. When the line of action of a force intersects the center of mass of an object, can that force produce a torque about the object's center of mass?

22. The spool is pulled in three ways, as shown below. There is sufficient friction for rotation. In what direction will the spool roll in each case?

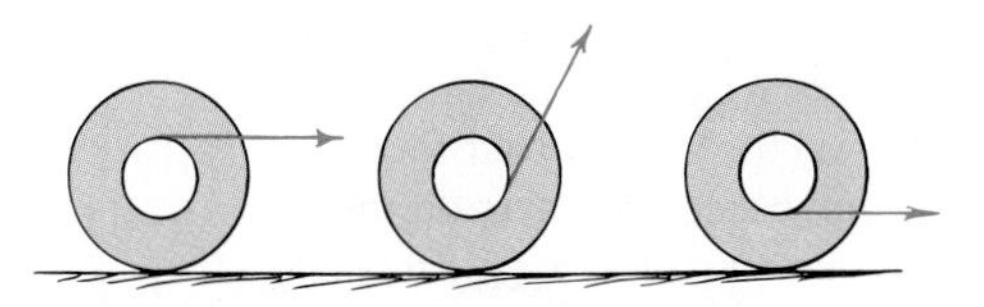

23. When a bowling ball leaves your hand, it may not spin. But farther along the alley, it does spin. What produces the spinning?

24. Why does sitting closest to the center of a vehicle provide the most comfortable ride in a bus traveling on a bumpy road, in a ship in a choppy sea, or in an airplane in turbulent air?

25. Which is more difficult—doing sit-ups with your knees bent, or with your legs straight out? Why?

26. Explain why a long pole is more beneficial to a tightrope walker if the pole droops.

27. Why is the wobbly motion of a single star an indication that the star has one or more planets orbiting around it?

28. Why must you bend forward when carrying a heavy load on your back?

29. Why is it easier to carry the same amount of water in two buckets, one in each hand, than in a single bucket?

30. Nobody at the playground wants to play with the obnoxious boy, so he fashions a seesaw as shown so he can play by himself. Explain how this is done.

31. Using the ideas of torque and center of gravity, explain why a ball rolls down a hill.

32. How can the three bricks be stacked so that the top brick has maximum horizontal displacement from the bottom brick? For example, stacking them like the dotted lines suggest would be unstable and the bricks would topple. (Hint: Start with the top brick and work down. At every interface the CG of the bricks above must not extend beyond the end of the supporting brick.)

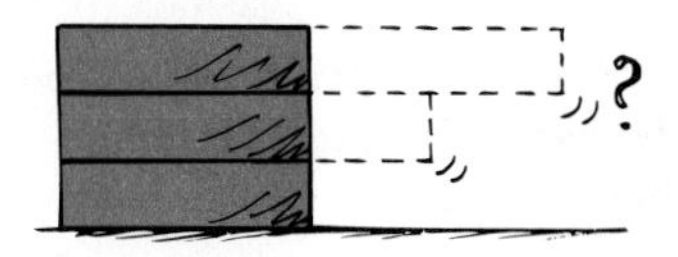

33. Where is the center of mass of the Earth's atmosphere?

34. Why is it important to secure file cabinets to the floor, especially for cabinets with heavy loads in top drawers?

35. Describe the comparative stabilities of the three objects shown in Figure 8.36, page 144, in terms of work and potential energy.

36. The centers of gravity of the three trucks parked on a hill are shown by the X's. Which truck(s) will tip over?

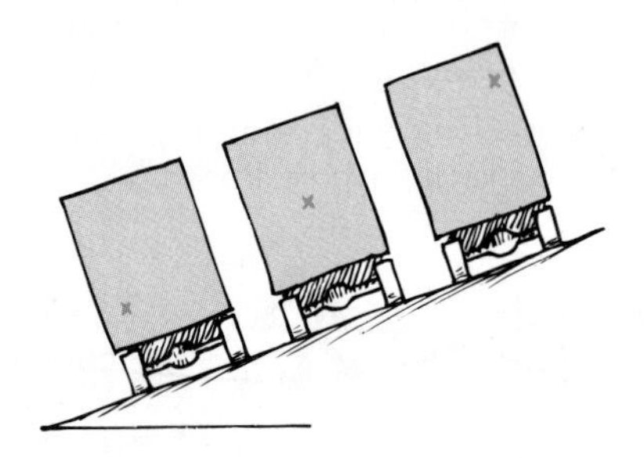

37. A long track balanced like a seesaw supports a golf ball and a more massive billiard ball with a compressed spring between the two. When the spring is released, the balls move away from each other. Does the track tip clockwise, tip counterclockwise, or remain in balance as the balls roll outward? What principles do you use for your explanation?

38. When a long-range cannonball is fired toward the equator from a northern (or southern) latitude, it lands west of its "intended" longitude. Why? (Hint: Consider a flea jumping from part way out to the outer edge of a rotating turntable.)

39. A racing car on a flat circular track needs friction between the tires and the track to maintain its circular motion. How much more friction is required for twice the speed?

40. Can an object move along a curved path if no force acts on it?

41. When you are in the front passenger seat of a car turning to the left, you may find yourself pressed against the right-side door. Why do you press against the door? Why does the door press on you? Does your explanation involve a centrifugal force, or Newton's laws?

42. Friction is needed for a car rounding a curve. But, if the road is banked, friction may not be required at all. What, then, supplies the needed centripetal force?

43. As a car speeds up when rounding a curve, does centripetal acceleration also increase? Use an equation to defend your answer.

44. Explain why a centripetal force does *not* do work on a circularly moving object.

45. Under what conditions could a car remain on a banked track covered with slippery ice?

46. The occupant inside a rotating space habitat of the future feels that she is being pulled by artificial gravity against the outer wall of the habitat (which becomes the "floor"). Explain what is going on in terms of Newton's laws and centripetal force.

47. The sketch shows a coin at the edge of a turntable. The weight of the coin is shown by the vector **W**. Two other forces act on the coin—the normal force and a force of friction that prevents it from sliding off the edge. Draw in force vectors for both of these.

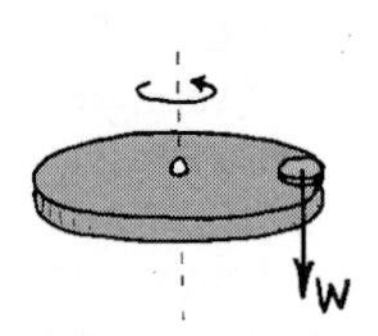

48. The sketch shows a conical pendulum. The bob swings in a circular path. The tension **T** and weight **W** are shown by vectors. Draw a parallelogram with these vectors and show that their resultant lies in the plane of the circle. (See the parallelogram rule in Chapter 5.) What is the name of this resultant force?

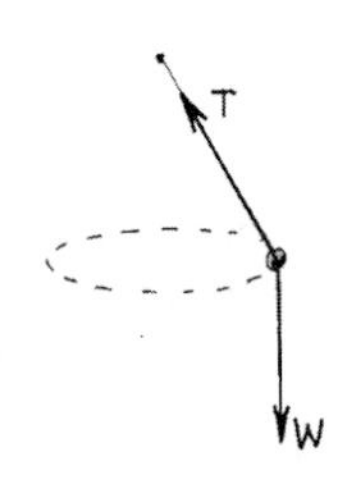

49. A motorcyclist is able to ride on the vertical wall of a bowl-shaped track as shown. Friction of the wall on the tires is shown by the vertical red vector. (a) How does the magnitude of this vertical vector compare with the weight of the motorcycle + rider? (b) Does the horizontal red vector represent the normal force acting on the bike and rider, the centripetal force, both, or neither? Defend your answer.

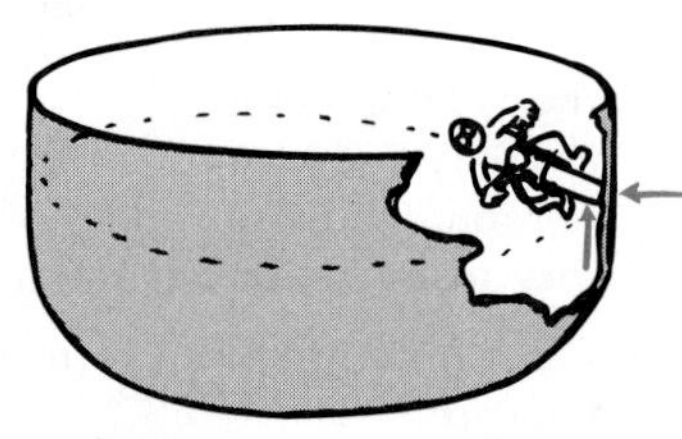

50. Consider a ball rolling around in a circular path on the inner surface of a cone. The weight of the ball is shown by the vector **W**. Without friction, only one other force acts on the ball—a normal force. (a) Draw in the normal vector (how long depends on b). (b) Using the parallelogram rule, show that the resultant of the two vectors is along the radial direction of the ball's circular path. (Yes, the normal is appreciably larger than the weight!)

51. You sit at the middle of a large turntable at an amusement park as it is set spinning and then allowed to spin freely. When you crawl toward the edge of the turntable, does the rate of the rotation increase, decrease, or remain unchanged? What physics principle supports your answer?

52. A sizable quantity of soil is washed down the Mississippi River and deposited in the Gulf of Mexico each year. What effect does this tend to have on the length of a day? (Hint: Relate this to Figure 8.53, page 152.)

53. Strictly speaking, as more and more skyscrapers are built on the surface of the Earth, does the day tend to become longer or shorter? And, strictly speaking, does the falling of autumn leaves tend to lengthen or shorten the 24-hour day? What physics principle supports your answers?

54. If all of Earth's inhabitants moved to the equator, how would this affect Earth's rotational inertia? How would it affect the length of a day?

55. If the world's populations moved to the North Pole and the South Pole, would the 24-hour day become longer, shorter, or stay the same?

56. If the polar ice caps of the Earth were to melt, the oceans would be deeper by about 30 m. What effect would this have on the Earth's rotation?

57. A toy train is initially at rest on a track fastened to a bicycle wheel, which is free to rotate. How does the wheel respond when the train moves clockwise? When the train backs up? Does the angular momentum of the wheel–train system change during these maneuvers? How would the resulting motions be affected if the train were much more massive than the track? Or vice versa?

58. Why does a typical small helicopter with a single main rotor have a second small rotor on its tail? Describe the consequence if the small rotor fails in flight.

59. We believe that our galaxy was formed from a huge cloud of gas. The original cloud was far larger than the present size of the galaxy, was more or less spherical, and was rotating very much more slowly than the galaxy is now. In this sketch, we see the original cloud and the galaxy as it is now (seen edgewise). Explain how the law of gravitation and the conservation of angular momentum contribute to the galaxy's present shape and why it rotates faster now than when it was a larger, spherical cloud.

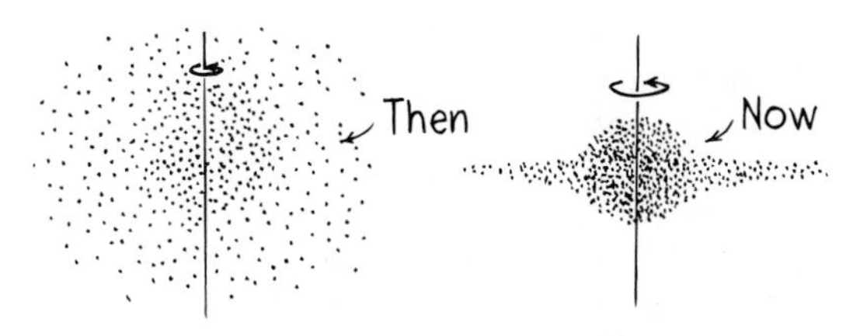

60. Earth is not spherical but bulges at the equator. Jupiter bulges more. What is the cause of these bulges?

Problems

1. Consider a bicycle that has wheels with a circumference of 2 m. What is the linear speed of the bicycle when the wheels rotate at 1 revolution per second?

2. What is the tangential speed of a passenger on a Ferris wheel that has a radius of 10 m and rotates once in 30 s?

3. Neglect the weight of the meterstick and consider only the two weights hanging from its ends. One is a 1-kg weight and the other is a 3-kg weight, as shown. Where is the center of mass of this system (the point of balance)? How does your answer relate to torque?

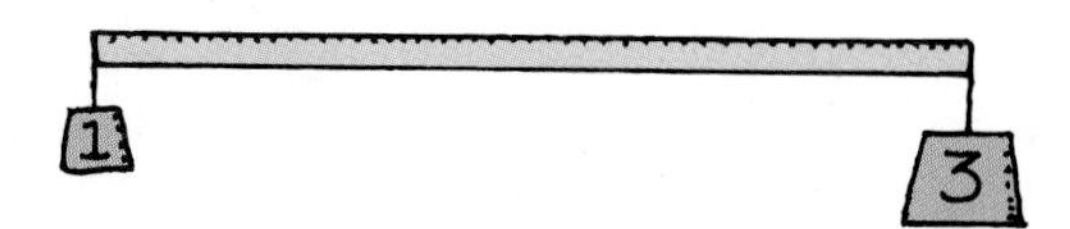

4. A 10,000-N vehicle is stalled one-quarter of the way across the bridge. Calculate the additional reaction

forces that are supplied at the supports on both ends of the bridge.

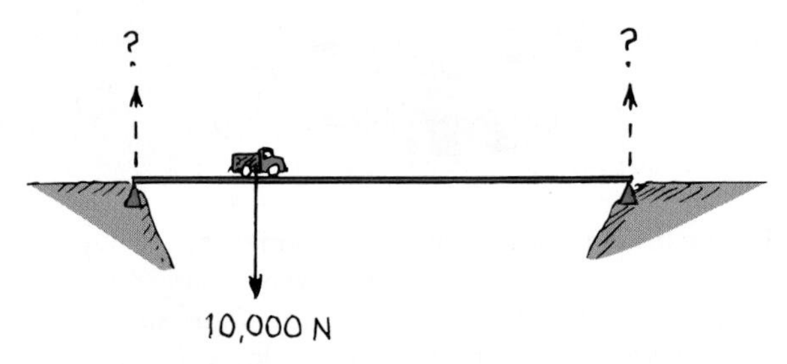

5. The rock has a mass of 1 kg. What is the mass of the measuring stick if it is balanced by a support force at the one-quarter mark?

6. To tighten a bolt, you push with a force of 80 N at the end of a wrench handle that is 0.25 m from the axis of the bolt. (a) What torque are you exerting? (b) If you move your hand inward to be only 0.10 m from the bolt, what force do you have to exert to achieve the same torque? (c) Do your answers depend on the direction of your push relative to the direction of the wrench handle?

7. Consider a too-small space habitat that consists of a rotating cylinder of radius 4 m. If a man standing inside is 2 m tall and his feet are at 1 g, what is the g force at the elevation of his head? (Do you see why projections call for large habitats?)

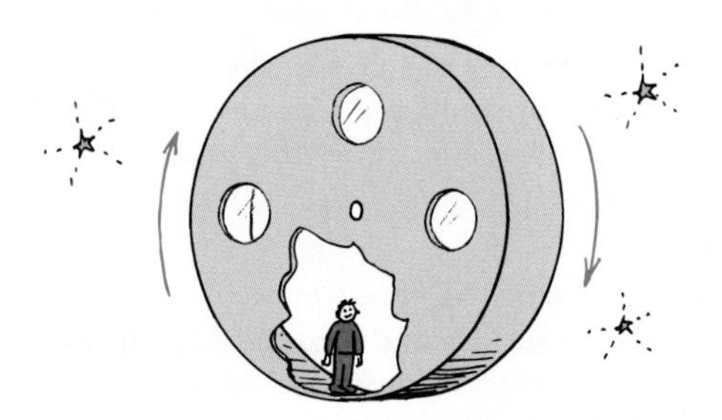

8. If the variation in g between one's head and feet is to be less than 1/100 g, then, compared to one's height, what should be the minimum radius of the space habitat?

9. If a trapeze artist rotates once each second while sailing through the air and contracts to reduce her rotational inertia to one-third of what it was, how many rotations per second will result?

10. How much greater is the angular momentum of the Earth orbiting about the Sun than the Moon orbiting about the Earth? (Set up a ratio of angular momenta using data on the inside back cover.)

In explaining spring and neap ocean tides, Praful Shah uses a model of the Sun, the Moon, and the Earth.

CHAPTER 9

Gravity

Newton did not discover gravity. That discovery dates back to earlier times, when Earth dwellers experienced the consequences of tripping and falling. What Newton discovered was that gravity is universal—that it is not unique to Earth, as others of his time assumed.

From the time of Aristotle, the circular motion of heavenly bodies was regarded as natural. The ancients believed that the stars, the planets, and the Moon move in divine circles, free from any impelling forces. As far as the ancients were concerned, this circular motion required no explanation. Isaac Newton, however, recognized that a force of some kind must act on the planets, whose orbits, he knew, were ellipses; otherwise, their paths would be straight lines. Others of his time, influenced by Aristotle, supposed that any force on a planet would be directed along its path. Newton, however, reasoned that the force on each planet would be directed toward a fixed central point—toward the Sun. This, the force of gravity, was the same force that pulls an apple off a tree. Newton's stroke of intuition, that the force between the Earth and an apple is the same force that pulls moons and planets and everything else in our universe, was a revolutionary break with the prevailing notion that there were two sets of natural laws: one for earthly events, and another, altogether different, for motion in the heavens. This union of terrestrial laws and cosmic laws is called the Newtonian synthesis.

The Universal Law of Gravity

According to popular legend, Newton was sitting under an apple tree when the idea struck him that gravity extends beyond Earth. Perhaps he looked up through tree branches toward the origin of the falling apple and noticed the Moon. In any event, Newton had the insight to see that the force between the Earth and a falling apple is the same force that pulls the Moon in an orbital path around the Earth, a path similar to a planet's path around the Sun.

To test this hypothesis, Newton compared the fall of an apple with the "fall" of the Moon. He realized that the Moon falls in the sense that *it falls away from the straight line it would follow if there were no forces acting on it.* Because of its tangential velocity, it "falls around" the round Earth (more about this in the next chapter). By simple geometry, the Moon's distance of fall per second could be compared with the distance that an apple or anything that far away would fall in one second. Newton's calculations didn't check. Disappointed, but recognizing that brute fact must always win over a beautiful hypothesis, he placed his papers in a drawer, where they remained for nearly 20 years. During this period, he founded and developed the field of geometric optics, for which he first became famous.

FIGURE 9.1
Could the gravitational pull on the apple reach to the Moon?

Newton's interest in mechanics was rekindled with the advent of a spectacular comet in 1680 and another two years later. He returned to the Moon problem at the prodding of his astronomer friend, Edmund Halley, for whom the second comet was later named. He made corrections in the experimental data used in his earlier method and obtained excellent results. Only then did he publish what is one of the most far-reaching generalizations of the human mind: the **law of universal gravitation.**[1]

Everything pulls on everything else in a beautifully simple way that involves only mass and distance. According to Newton, every body attracts every other body with a force that, for any two bodies, is directly proportional to the product of their masses and inversely proportional to the square of the distance separating them.

This statement can be expressed as

$$\text{Force} \sim \frac{\text{mass}_1 \times \text{mass}_2}{\text{distance}^2}$$

or symbolically as

$$F \sim \frac{m_1 m_2}{d^2}$$

where m_1 and m_2 are the masses of the bodies and d is the distance between their centers. Thus, the greater the masses m_1 and m_2, the greater the force of attraction between them, in direct proportion to the masses.[2] The greater the distance of separation d, the weaker the force of attraction, in inverse proportion to the square of the distance between their centers of mass.

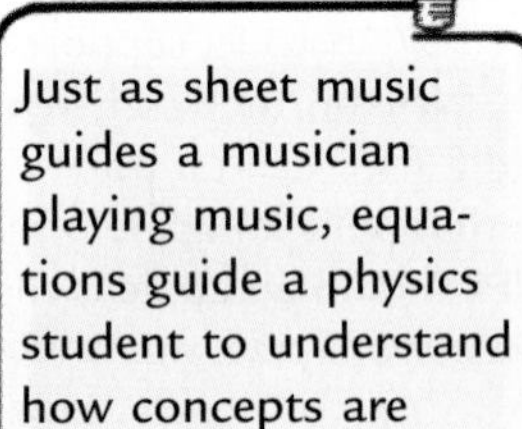

Just as sheet music guides a musician playing music, equations guide a physics student to understand how concepts are connected.

Insights

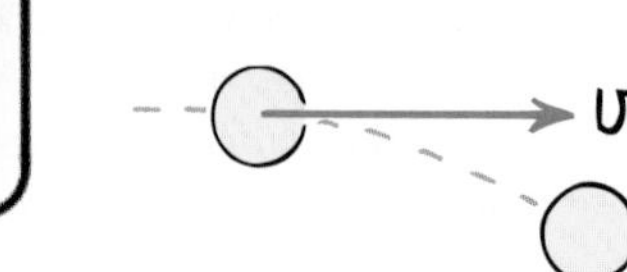

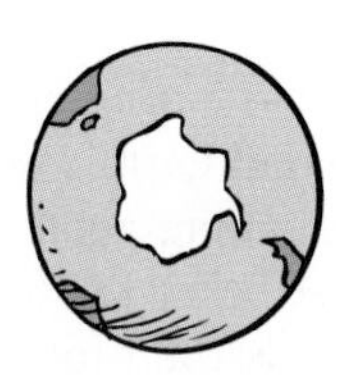

FIGURE 9.2
The tangential velocity of the Moon about the Earth allows it to fall around the Earth rather than directly into it. If this tangential velocity were reduced to zero, what would be the fate of the Moon?

[1]This is a dramatic example of the painstaking effort and cross-checking that go into the formulation of a scientific theory. Contrast Newton's approach with the failure to "do one's homework," the hasty judgments, and the absence of cross-checking that so often characterize the pronouncements of people advocating less-than-scientific theories.

[2]Note the different role of mass here. Thus far, we have treated mass as a measure of inertia, which is called *inertial mass*. Now we see mass as a measure of gravitational force, which in this context is called *gravitational mass*. It is experimentally established that the two are equal, and, as a matter of principle, the equivalence of inertial and gravitational mass is the foundation of Einstein's general theory of relativity.

CHECK YOURSELF

1. In Figure 9.2, we see that the Moon falls around Earth rather than straight into it. If the Moon's tangential velocity were zero, how would it move?
2. According to the equation for gravitational force, what happens to the force between two bodies if the mass of one of the bodies is doubled? If both masses are doubled?
3. Gravitational force acts on all bodies in proportion to their masses. Why, then, doesn't a heavy body fall faster than a light body?

The Universal Gravitational Constant, *G*

The proportionality form of the universal law of gravitation can be expressed as an exact equation when the constant of proportionality G is introduced. G is called the *universal gravitational constant*. Then the equation is

$$F = G\frac{m_1 m_2}{d^2}$$

Just as π relates circumference and diameter for circles, G relates gravitational force with mass and distance.

Insights

In words, the force of gravity between two objects is found by multiplying their masses, dividing by the square of the distance between their centers, and then multiplying this result by the constant G. The magnitude of G is identical to the magnitude of the force between a pair of 1-kg masses that are 1 meter apart: 0.0000000000667 newton. This small magnitude indicates an extremely weak force. In standard units and in scientific notation,[3]

$$G = 6.67 \times 10^{-11}\text{N}\cdot\text{m}^2/\text{kg}^2$$

An English physicist, Henry Cavendish, first measured G long after the time of Newton in the eighteenth century. He accomplished this by measuring the tiny force between lead masses with an extremely sensitive torsion balance. A simpler method was later developed by Philipp von Jolly, who attached a spherical flask of mercury to one arm of a sensitive balance (Figure 9.4). After the balance

FIGURE 9.3
As the rocket gets farther from the Earth, gravitational strength between the rocket and the Earth decreases.

CHECK YOUR ANSWERS

1. If the Moon's tangential velocity were zero, it would fall straight down and crash into Earth!
2. When one mass is doubled, the force between it and the other one doubles. If both masses double, the force is four times as much.
3. The answer goes back to Chapter 4. Recall Figure 4.12, in which heavy and light bricks fall with the same acceleration because both have the same ratio of weight to mass. Newton's second law ($a = F/m$) reminds us that greater force acting on greater mass does not result in greater acceleration.

[3]The numerical value of G depends entirely on the units of measurement we choose for mass, distance, and time. The international system of choice is: for mass, the kilogram; for distance, the meter; and for time, the second. Scientific notation is discussed in Appendix A at the end of this book.

Interestingly, Newton could calculate the product of G and Earth's mass, but not either one alone. Calculating G alone was first done by Henry Cavendish.

Because of the relative weakness of gravity, G is the least accurately known of all the fundamental constants of physics. Even so, it is now (2005) known to an accuracy of five significant figures.

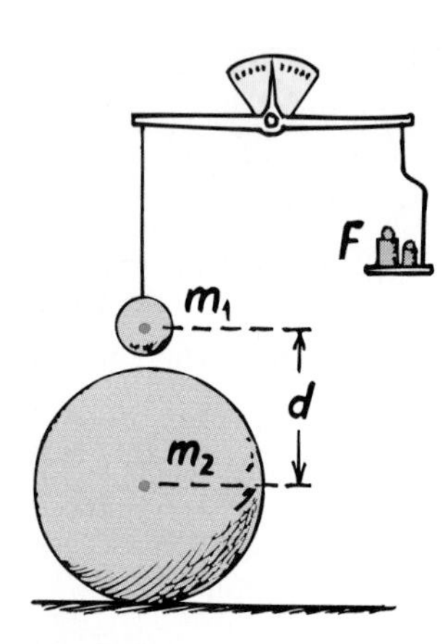

FIGURE 9.4
Jolly's method of measuring G. Balls of mass m_1 and m_2 attract each other with a force F equal to the weights needed to restore balance.

was put in equilibrium, a 6-ton lead sphere was rolled beneath the mercury flask. The gravitational force between the two masses was measured by the weight needed on the opposite end of the balance to restore equilibrium. All the quantities, m_1, m_2, F, and d were known, from which the constant G was calculated:

$$G = \frac{F}{\left(\frac{m_1 m_2}{d^2}\right)} = 6.67 \times 10^{-11}\,\frac{\text{N}}{\text{kg}^2/\text{m}^2} = 6.67 \times 10^{-11}\,\text{N}\cdot\text{m}^2/\text{kg}^2$$

The value of G indicates that the force of gravity is a very weak force. It is the weakest of the presently known four fundamental forces. (The other three are the electromagnetic force and two kinds of nuclear forces.) We sense gravitation only when masses like that of the Earth are involved. If you stand on a large ship, the force of attraction between you and the ship is too weak for ordinary measurement. The force of attraction between you and the Earth, however, can be measured. It is your weight.

Your weight depends not only on your mass but also on your distance from the center of the Earth. At the top of a mountain, your mass is the same as it is anywhere else, but your weight is slightly less than it is at ground level. That's because your distance from Earth's center is greater.

Von Jolly's Method of Measuring the Attraction Between Two Masses

Once the value of G was known, the mass of the Earth was easily calculated. The force that Earth exerts on a mass of 1 kilogram at its surface is 9.8 newtons. The distance between the 1-kilogram mass and the center of the Earth is Earth's radius, 6.4×10^6 meters. Therefore, from $F = G(m_1 m_2/d^2)$, where m_1 is the mass of the Earth,

$$9.8\text{ N} = 6.67 \times 10^{-11}\,\text{N}\cdot\text{m}^2/\text{kg}^2\,\frac{1\text{ kg} \times m_1}{(6.4 \times 10^6\text{m})^2}$$

from which the mass of the Earth is calculated to be $m_1 = 6 \times 10^{24}$ kilograms.

You can never change only one thing! Every equation reminds us of this—you can't change a term on one side without affecting the other side.

Insights

In the eighteenth century, when G was first measured, people all over the world were excited about it. That's because newspapers everywhere announced the discovery as one that measured the mass of the planet Earth. How exciting that Newton's formula gives the mass of the entire planet, with all its oceans, mountains, and inner parts yet to be discovered. G and the mass of Earth were measured when a great portion of Earth's surface was still undiscovered.

CHECK YOURSELF

If there is an attractive force between all objects, why do we not feel ourselves gravitating toward massive buildings in our vicinity?

CHECK YOUR ANSWER

Gravity certainly does pull us to massive buildings and everything else in the universe. Physicist Paul A. M. Dirac, recipient of the 1933 Nobel Prize, put it this way: "Pick a flower on Earth and you move the farthest star!" How much we are influenced by buildings or how much interaction there is between flowers and stars is another story. The forces between buildings and us are relatively small because their masses are small compared with the mass of the Earth. The forces due to the stars are extremely tiny because of their great distances from us. These tiny forces escape our notice when they are overwhelmed by the overpowering attraction to the Earth.

Gravity and Distance: The Inverse-Square Law

We can better understand how gravity is diluted with distance by considering how paint from a paint gun spreads with increasing distance (Figure 9.5). Suppose we position a paint gun at the center of a sphere with a radius of 1 meter, and a burst of paint spray travels 1 meter to produce a square patch of paint that is 1 millimeter thick. How thick would the patch be if the experiment were done in a sphere with twice the radius? If the same amount of paint travels in straight lines for 2 meters, it will spread to a patch twice as tall and twice as wide. The paint would then be spread over an area four times as big, and its thickness would be only 1/4 millimeter.

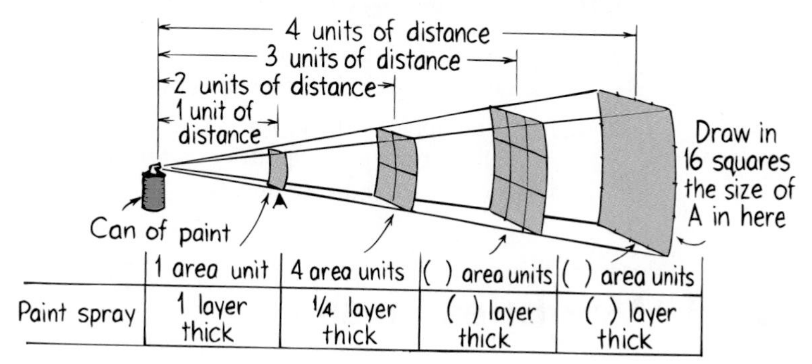

FIGURE 9.5
The inverse-square law. Paint spray travels radially away from the nozzle of the can in straight lines. Like gravity, the "strength" of the spray obeys the inverse-square law.

Can you see from the figure that, for a sphere of radius 3 meters, the thickness of the paint patch would be only 1/9 millimeter? Can you see the thickness of the paint decreases as the square of the distance increases? This is known as the **inverse-square law.** The inverse-square law holds for gravity and for all phenomena wherein the effect from a localized source spreads uniformly throughout the surrounding space: the electric field about an isolated electron, light from a match, radiation from a piece of uranium, and sound from a cricket.

It is important to emphasize that the distance term *d* in Newton's equation for gravity is the distance between the centers of masses of the objects. Note in Figure 9.6 that the apple that normally weighs 1 newton at Earth's surface

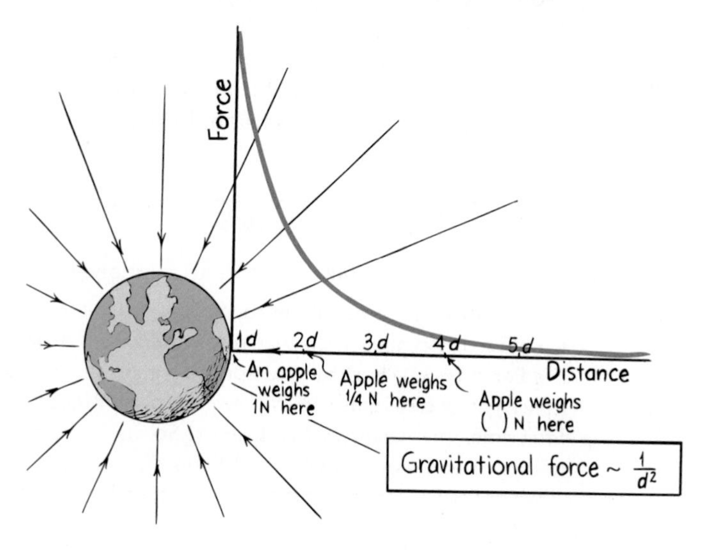

FIGURE 9.6
Interactive Figure
If an apple weighs 1 N at the Earth's surface, it would weigh only 1/4 N twice as far from the center of the Earth. At three times the distance, it would weigh only 1/9 N. Gravitational force versus distance is plotted in color. What would the apple weigh at four times the distance? Five times?

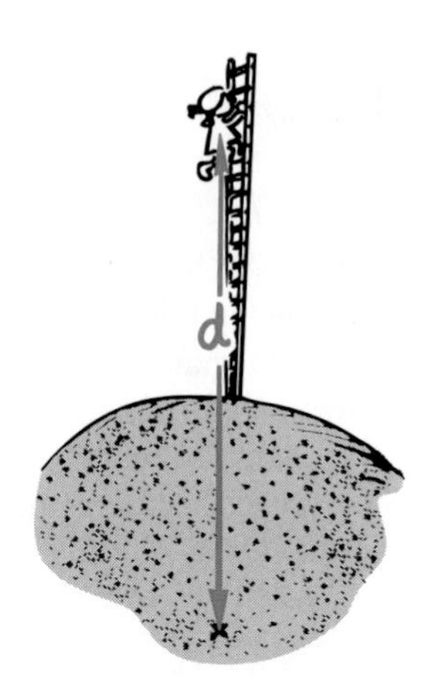

FIGURE 9.7
According to Newton's equation, her weight (not her mass) decreases as she increases her distance from Earth's center.

weighs only 1/4 as much when it is twice the distance from Earth's center. The greater the distance from Earth's center, the less the weight of an object. A child who weighs 300 newtons at sea level will weigh only 299 newtons atop Mt. Everest. For greater distances, force is less. For very great distances, Earth's gravitational force approaches zero. The force *approaches* zero, but it never gets there. Even if you were transported to the far reaches of the universe, the gravitational influence of home would still be with you. It may be overwhelmed by the gravitational influences of nearer and/or more massive bodies, but it is there. The gravitational influence of every material object, however small or however far, is exerted through all of space.

CHECK YOURSELF

1. By how much does the gravitational force between two objects decrease when the distance between their centers is doubled? Tripled? Increased tenfold?
2. Consider an apple at the top of a tree that is pulled by Earth's gravity with a force of 1 N. If the tree were twice as tall, would the force of gravity be only 1/4 as strong? Defend your answer.

Weight and Weightlessness

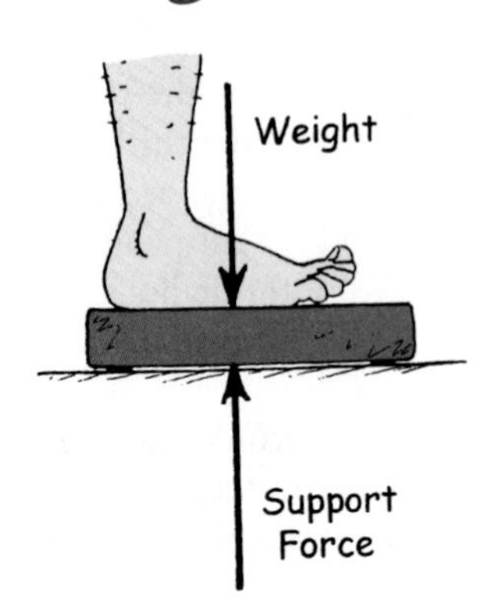

FIGURE 9.8
When you step on a weighing scale, two forces act on it; a downward force of gravity (your ordinary weight, *mg*, if there is no acceleration) and an upward support force. These equal and opposite forces squeeze a springlike device inside the scale that is calibrated to show weight.

When you step on a bathroom scale, you effectively compress a spring inside. When the pointer stops, the elastic force of the deformed spring balances the gravitational attraction between you and the Earth—nothing moves as you and the scale are in static equilibrium. The pointer is calibrated to show your **weight.** If you stand on a bathroom scale in a moving elevator, you'll find variations in your weight. If the elevator accelerates upward, the springs inside the bathroom scale are more compressed and your weight reading is greater. If the elevator accelerates downward, the springs inside the scale are less compressed and your weight reading is less. If the elevator cable breaks and the elevator falls freely, the reading on the scale goes to zero. According to the scale's reading, you would be **weightless.** Would you really be weightless? We can answer this question only if we agree on what we mean by *weight*.

In Chapters 2 and 4, we treated the weight of an object as the force due to gravity upon it. When in equilibrium on a firm surface, weight is evidenced by a support force, or, when in suspension, by a supporting rope tension. In either case,

CHECK YOUR ANSWERS

1. It decreases to one-fourth, one-ninth, and one-hundredth the original value.
2. No, because an apple at the top of the twice-as-tall apple tree is not twice as far from Earth's center. The taller tree would need a height equal to the radius of the Earth (6,370 km) for the apple's weight at its top to reduce to 1/4 N. Before its weight decreases by 1%, an apple or any object must be raised 32 km—nearly four times the height of Mt. Everest. So, as a practical matter, we disregard the effects of everyday changes in elevation.

FIGURE 9.9
Interactive Figure

Your weight equals the force with which you press against the supporting floor. If the floor accelerates up or down, your weight varies (even though the gravitational force *mg* that acts on you remains the same).

with no acceleration, weight equals *mg*. Then, when we discussed rotating environments in Chapter 8, we learned that a support force can occur without regard to gravity. So a broader definition of the weight of something is the force it exerts against a supporting floor or a weighing scale. According to this definition, you are as heavy as you feel; so, in an elevator that accelerates downward, the supporting force of the floor is less and you weigh less. If the elevator is in free fall, your weight is zero (Figure 9.9). Even in this weightless condition, however, there is still a gravitational force acting on you, causing your downward acceleration. But gravity now is not felt as weight because there is no support force.

Astronauts in orbit are without a support force and are in a sustained state of weightlessness. Astronauts sometimes experience "space sickness" until they become accustomed to a state of sustained weightlessness. Astronauts in orbit are in a state of continual free fall.

The International Space Station in Figure 9.11 provides a weightless environment. The station facility and astronauts all accelerate equally toward Earth,

FIGURE 9.10
Both are weightless.

FIGURE 9.11
The inhabitants in this laboratory and docking facility continually experience weightlessness. They are in free fall around the Earth. Does a force of gravity act on them?

Astronauts inside an orbiting space vehicle have no weight, even though the force of gravity between them and the Earth is only slightly less than at ground level.

Insights

at somewhat less than 1 g because of their altitude. This acceleration is not sensed at all; with respect to the station, the astronauts experience zero g. Over extended periods of time, this causes loss of muscle strength and other detrimental changes in the body. Future space travelers, however, need not be subjected to weightlessness. As mentioned in the previous chapter, lazily rotating giant wheels or pods at the end of a tether will likely take the place of today's nonrotating space habitats. Rotation effectively supplies a support force and nicely provides weight.

CHECK YOURSELF

In what sense is drifting in space far away from all celestial bodies like stepping down off a stepladder?

Ocean Tides

Seafaring people have always known that there is a connection between the ocean tides and the Moon, but no one could offer a satisfactory theory to explain the two high tides per day. Newton showed that the ocean tides are caused by *differences* in the gravitational pull between the Moon and the Earth on opposite sides of the Earth. Gravitational force between the Moon and the Earth is stronger on the side of the Earth nearer to the Moon, and it is weaker on the side of the Earth that is farther from the Moon. This is simply because the gravitational force is weaker with increased distance.

To understand why the difference in gravitational pulls by the Moon on opposite sides of the Earth produces tides, pretend you have a big spherical ball of Jell-O. If you were to exert the same force on every part of the ball, it would remain spherical as it accelerated. But, if you were to pull harder on one side than the other, there would be a difference in accelerations and the ball would become elongated (Figure 9.13). That is what happens to this big ball we're living on. The side closer to the Moon is pulled with a greater force and has a

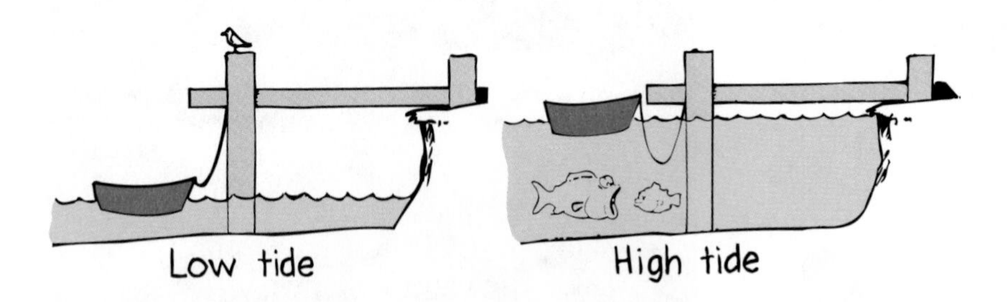

FIGURE 9.12 Ocean tides.

CHECK YOUR ANSWER

In both cases, you'd experience weightlessness. Drifting in deep space, you would remain weightless because no discernable force acts on you. Stepping from a stepladder, you would be only momentarily weightless because of a momentary lapse of support force.

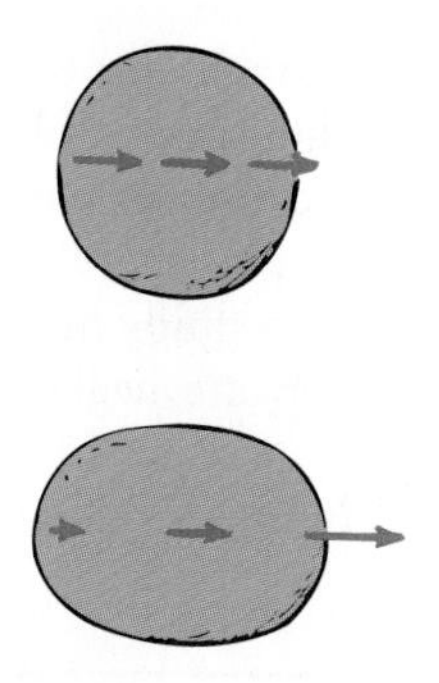

FIGURE 9.13
A ball of Jell-O remains spherical when all parts of it are pulled equally in the same direction. When one side is pulled more than the other, however, its shape is elongated.

greater acceleration toward the Moon than the far side—thus, the Earth is somewhat football shaped. But does the Earth accelerate toward the Moon? Yes, it must, because a force acts on it, and, where there is a net force, there is acceleration. It is a *centripetal* acceleration, for Earth circles the center of mass of the Earth–Moon system (a point within the Earth about three-quarters of the way from the center to the surface). Both the Earth and the Moon undergo centripetal acceleration as they circle each other about the Earth–Moon center of mass. This makes both the Earth and the Moon slightly elongated. The elongation of the Earth is mainly in its oceans, which bulge equally on opposite sides.

On a world average, the ocean bulges are nearly 1 meter above the average surface level of the ocean. Earth spins once per day, so a fixed point on Earth passes beneath both of these bulges each day. This produces two sets of ocean tides per day. Any part of the Earth that passes beneath one of the bulges has a high tide. When Earth has made a quarter turn 6 hours later, the water level at the same part of the ocean is nearly 1 meter below the average sea level. This is low tide. The water that "isn't there" is under the bulges that make up the high tides. A second high tidal bulge is experienced when Earth makes another quarter turn. So we have two high tides and two low tides daily. It turns out that, while the Earth spins, the Moon moves in its orbit and appears at the same position in our sky every 24 hours and 50 minutes, so the two-high-tide cycle is actually at 24-hour-and-50-minute intervals. That is why tides do not occur at the same time every day.

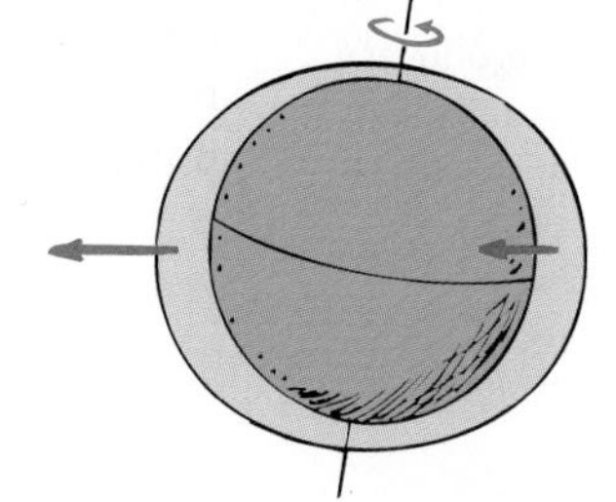

FIGURE 9.14
Two tidal bulges remain relatively fixed with respect to the Moon while the Earth spins daily beneath them.

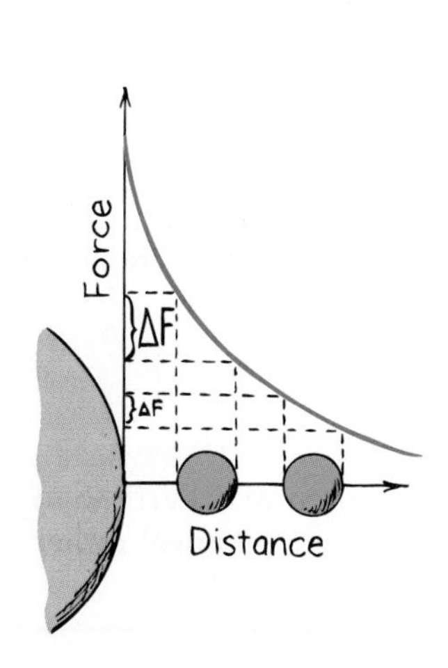

FIGURE 9.15
A plot of gravitational force versus distance (not to scale). The greater the distance from the Sun, the smaller the force F, which varies as $1/d^2$, and the smaller the difference in gravitational pulls on opposite sides of a planet, ΔF, which varies as $1/d^3$, and, hence, the smaller the tides.

The Sun also contributes to ocean tides, although it is less than half as effective as the Moon in raising tides—even though its pull on the Earth is 180 times greater than the pull of the Moon. Why doesn't the Sun cause tides 180 times greater than lunar tides? The answer has to do with a key word: *difference.* Because of the great distance of the Sun, the difference in its gravitational pull on opposite sides of the Earth is very small (Figure 9.15). The percentage difference in the Sun's pulls across the Earth is only about 0.017%, compared with 6.7% across the Earth by the Moon. It is only because the pull of the Sun is 180 times stronger than the Moon's that the Sun tides are almost half as high (180×0.017 percent $= 3\%$, nearly half of 6.7%).

Newton deduced that the difference in pulls decreases as the *cube* of the distance between the centers of the bodies—twice as far away produces 1/8 the tide; three times as far, only 1/27 the tide, and so on. Only relatively close distances result in appreciable tides, and so the nearby Moon produces larger tides than the enormously more massive but more distant Sun. The amount of tide

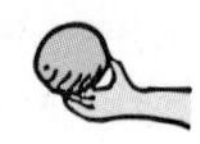

also depends on the size of the body experiencing tides. Although the Moon produces a considerable tide in the Earth's oceans, which are thousands of kilometers apart, it produces scarcely any tide in a lake. That's because no part of the lake is significantly closer to the Moon than any other part, so there is no significant *difference* in Moon pulls on the lake. Similarly for the fluids in your body. Any tides in the fluids of your body caused by the Moon are negligible. You're not tall enough for tides. What microtides the Moon may produce in your body are only about one two-hundredth the tides produced by a one-kilogram melon held one meter above your head (Figure 9.16).

FIGURE 9.16
The tidal force difference due to a 1 kg body 1 m over the head of an average height person is about 60 trillionths (6×10^{-11}) N/kg. For an overhead Moon, it is about 0.3 trillionth (3×10^{-13}) N/kg. So holding a melon over your head produces about 200 times as much tidal effect in your body as the Moon does.

CHECK YOURSELF

We know that both the Moon and the Sun produce our ocean tides. And we know the Moon plays the greater role because it is closer. Does its closeness mean that it pulls on the Earth's oceans with more gravitational force than the Sun?

When the Sun, Earth, and Moon are aligned, the tides due to the Sun and the Moon coincide. Then we have higher-than-average high tides and lower-than-average low tides. These are called **spring tides** (Figure 9.17). (Spring tides have nothing to do with the spring season.) You can tell when the Sun, Earth, and Moon are aligned by the full Moon or by the new Moon. When the Moon is full, the Earth is between the Sun and Moon. (If all three are *exactly* in line, then we have a lunar eclipse, for the full Moon passes into Earth's shadow.) A new Moon occurs when the Moon is between the Sun and the Earth, when the nonilluminated hemisphere of the Moon faces the Earth. (When this alignment is perfect, the Moon blocks the Sun and we have a solar eclipse.) Spring tides occur at the times of a new or full Moon.

FIGURE 9.17
When the attractions of the Sun and the Moon are lined up with each other, spring tides occur.

All spring tides are not equally high because Earth–Moon and Earth–Sun distances vary; the orbital paths of the Earth and the Moon are elliptical rather

CHECK YOUR ANSWER

No, the Sun's pull is much stronger. Gravitational pull weakens as the distance squared. But the *difference* in pulls across the Earth's oceans weakens as the distance cubed. When the distance to the Sun is squared, gravitation from the Sun is still stronger than gravitation from the closer Moon because of the Sun's enormous mass. But, when the distance to the Sun is cubed, as is the case for tidal forces, the Sun's influence is less than the Moon's. Differences in distance are the key to tidal forces. If the Moon were closer to Earth, the tides on both the Earth and the Moon would increase as the cube of the closer distance. Being too close could catastrophically tear the Moon into pieces—the likely cause of the planetary rings of Saturn and other planets.

FIGURE 9.18
When the attractions of the Sun and the Moon are about 90° apart (at the time of a half Moon), neap tides occur.

than circular. The Moon's distance from Earth varies by about 10% and its effect in raising tides varies by about 30%. Highest spring tides occur when the Moon and Sun are closest to Earth.

When the Moon is halfway between a new Moon and a full Moon, in either direction (Figure 9.18), the tides due to the Sun and the Moon partly cancel each other. Then, the high tides are lower than average and the low tides are not as low as average low tides. These are called **neap tides.**

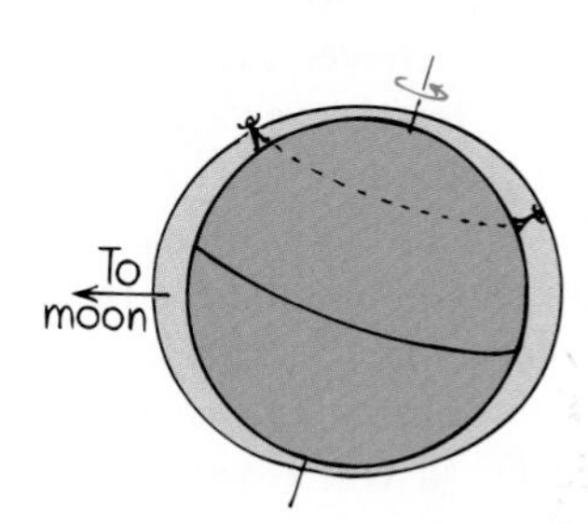

FIGURE 9.19
The inequality of the two high tides per day. Because of the Earth's tilt, a person may find the tide nearest the Moon much lower (or higher) than the tide half a day later. Inequalities of tides vary with the positions of the Moon and the Sun.

Another factor that affects the tides is the tilt of the Earth's axis (Figure 9.19). Even though the opposite tidal bulges are equal, Earth's tilt causes the two daily high tides experienced in most parts of the ocean to be unequal most of the time.

Our treatment of tides is quite simplified here, for tides are actually more complicated. Interfering land masses and friction with the ocean floor, for example, complicate tidal motions. In many places, the tides break up into smaller "basins of circulation," where a tidal bulge travels like a circulating wave that moves around in a small basin of water that is tilted. For this reason, the high tide may be hours away from an overhead Moon. In midocean, the variation in water level—the range of the tide—is usually about a meter. This range varies in different parts of the world; it is greatest in some Alaskan fjords and is most notable in the basin of the Bay of Fundy, between New Brunswick and Nova Scotia in eastern Canada, where tidal differences sometimes exceed 15 meters. This is largely due to the ocean floor, which funnels shoreward in a V-shape. The tide often comes in faster than a person can run. Don't dig clams near the water's edge at low tide in the Bay of Fundy!

Tides in the Earth and Atmosphere

Earth is not a rigid solid but, for the most part, is molten liquid covered by a thin, solid, and pliable crust. As a result, the Moon–Sun tidal forces produce Earth tides as well as ocean tides. Twice each day, the solid surface of Earth rises and falls as much as one-quarter meter! As a result, earthquakes and volcanic eruptions have a slightly higher probability of occurring when Earth is experiencing an Earth spring tide—that is, near a full or new Moon.

In addition to ocean tides, the Moon and Sun make atmospheric tides—highest then lowest during a full Moon. Does this explain why some of your friends are weird when the Moon is full?

We live at the bottom of an ocean of air that also experiences tides. Being at the bottom of the atmosphere, we don't notice these tides (just as a fish in deep water doesn't notice the ocean tides). In the upper part of the atmosphere is the ionosphere, so named because it contains many ions—electrically charged atoms that are the result of ultraviolet light and intense cosmic ray bombardment. Tidal effects in the ionosphere produce electric currents that alter the magnetic field that surrounds the Earth. These are magnetic tides. They, in turn, regulate the degree to which cosmic rays penetrate into the lower atmosphere. The cosmic-ray penetration is evident in subtle changes in the behaviors of living things. The highs and lows of magnetic tides are greatest when the atmosphere

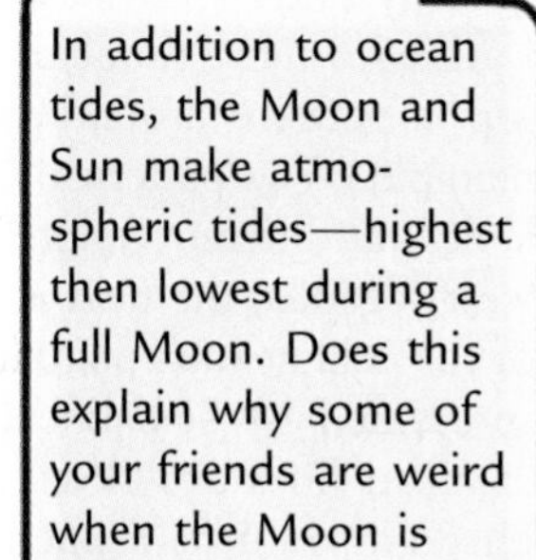

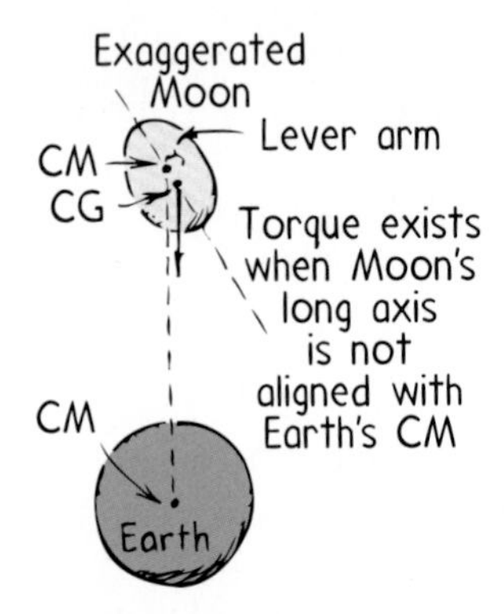

FIGURE 9.20
Earth's pull on the Moon at its center of gravity produces a torque about the Moon's center of mass, which tends to rotate the long axis of the Moon into alignment with Earth's gravitational field (like a compass needle that aligns with a magnetic field). That's why only one side of the Moon faces Earth.

is having its spring tides—again, near the full and new Moon. Have you noticed that some of your friends seem a bit weird at the time of a full Moon?

Tides on the Moon

There are two tidal bulges on the Moon for the same reason there are two tidal bulges on Earth—the near and far sides of each body are pulled differently. So the Moon is pulled slightly away from a spherical shape into a football shape, with its long axis pointing toward the Earth. But unlike Earth's tides, the tidal bulges remain in fixed locations, with no "daily" rising and falling of Moon tides. Since the Moon takes 27.3 days to make a single revolution about its own axis (and also about the Earth–Moon axis), the same lunar hemisphere faces the Earth all the time. This is because the elongated Moon's center of gravity is slightly displaced from its center of mass, so, whenever the Moon's long axis is not lined up toward the Earth (Figure 9.20), Earth exerts a small torque on the Moon. This tends to twist the Moon toward aligning with Earth's gravitational field, like the torque that aligns a compass needle with a magnetic field. So we see there is a reason why the Moon always shows us its same face.

Interestingly enough, this "tidal lock" is also working on the Earth. Our days are getting longer at the rate of 2 milliseconds per century. In a few billion years, our day will be as long as a month, and the Earth will always show the same face to the Moon. How about that!

Gravitational Fields

Earth and the Moon pull on each other. This is *action at a distance,* because Earth and Moon interact with each other even though they are not in contact. We can look at this in a different way: We can regard the Moon as interacting with the *gravitational field* of the Earth. The properties of the space surrounding any massive body can be considered to be altered in such a way that another massive body in this region experiences a force. This alteration of space is a **gravitational field.** It is common to think of rockets and distant space probes being influenced by the gravitational field at their locations in space, rather than by Earth and other planets or stars. The field concept plays an in-between role in our thinking about the forces between different masses.

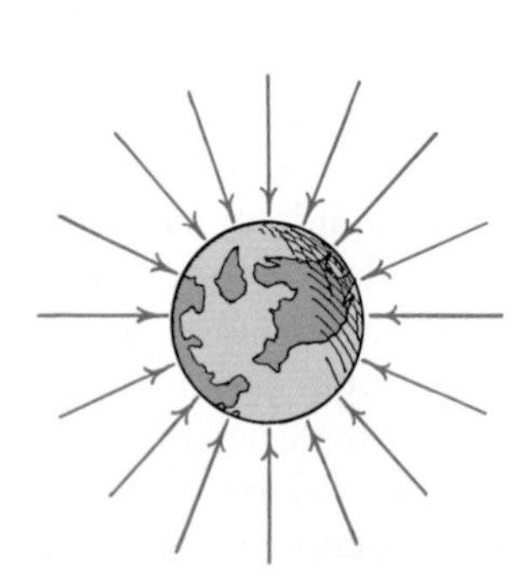

FIGURE 9.21
Field lines represent the gravitational field about the Earth. Where the field lines are closer together, the field is stronger. Farther away, where the field lines are farther apart, the field is weaker.

A gravitational field is an example of a *force field,* for any body with mass experiences a force in the field space. Another force field, perhaps more familiar, is a magnetic field. Have you ever seen iron filings lined up in patterns around a magnet? (Look ahead to Figure 24.2 on page 460, for example.) The pattern of the filings shows the strength and direction of the magnetic field at different points in the space around the magnet. Where the filings are closest together, the field is strongest. The direction of the filings shows the direction of the field at each point.

The pattern of Earth's gravitational field can be represented by field lines (Figure 9.21). Like the iron filings around a magnet, the field lines are closer together where the gravitational field is stronger. At each point on a field line, the direction of the field is along the line. Arrows show the field direction. A particle, astronaut, spaceship, or any body in the vicinity of Earth will be accelerated in the direction of the field line at that location.

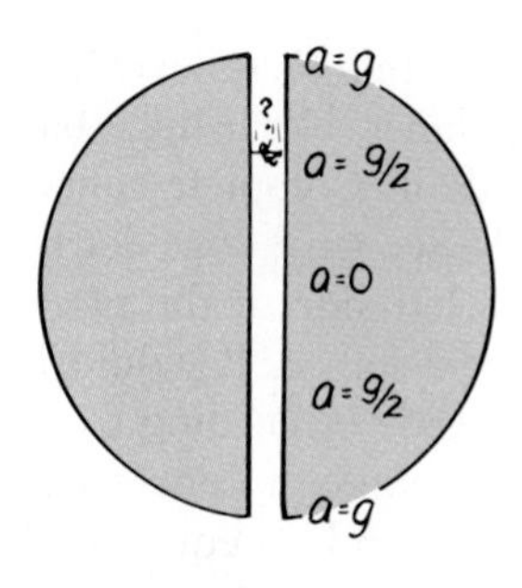

FIGURE 9.22
As you fall faster and faster in a hole bored completely through the Earth, your acceleration decreases because the part of Earth's mass beneath you gets smaller and smaller. Less mass means less attraction until, at the center, the net force is zero and acceleration is zero. Momentum carries you past the center and against a growing acceleration to the opposite end of the tunnel, where acceleration is again *g*, directed back toward the center.

The strength of Earth's gravitational field, like the strength of its force on objects, follows the inverse-square law. It is strongest near the Earth's surface and weakens with increasing distance from the Earth.[4]

The gravitational field at Earth's surface varies slightly from location to location. Above large subterranean lead deposits, for example, the field is slightly stronger than average. Above large caverns, perhaps filled with natural gas, the field is slightly weaker. To predict what lies beneath Earth's surface, geologists and prospectors of oil and minerals make precise measurements of Earth's gravitational field.

Gravitational Field Inside a Planet[5]

The gravitational field of the Earth exists inside Earth as well as outside. Imagine a hole drilled completely through the Earth from the North Pole to the South Pole. Forget about impracticalities, such as the high-temperature molten interior, and consider the motion you would undergo if you fell into such a hole. If you started at the North Pole end, you'd fall and gain speed all the way down to the center then lose speed all the way "up" to the South Pole. Without air drag, the one-way trip would take nearly 45 minutes. If you failed to grab the edge of the hole when you reached the South Pole, you'd fall back toward the center, and return to the North Pole in the same time.

Your acceleration, a, will be progressively less as you continue toward the center of the Earth. Why? Because, as you fall toward Earth's center, there is less mass pulling you toward the center. When you are at the center of the Earth, the pull down is balanced by the pull up, so the net force on you as you whiz with maximum speed past Earth's center is zero. Your acceleration is zero at the center—$a = 0$. The gravitational field of Earth at its center is zero![6]

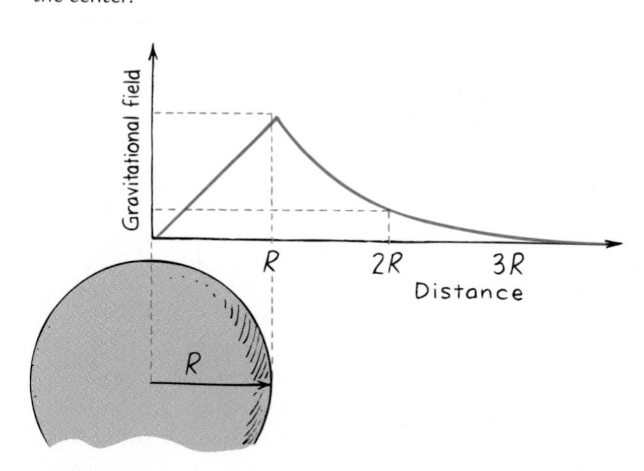

FIGURE 9.23
The gravitational field intensity inside a planet of uniform density is directly proportional to the radial distance from its center and is maximum at its surface. Outside, it is inversely proportional to the square of the distance from its center.

The composition of Earth varies, being most dense at its core and least dense at the surface. Inside a hypothetical planet of uniform density, however, the field inside increases linearly—that is, at a steady rate—from zero at its center to g at the surface. We won't go into why this is so, but perhaps your instructor will provide the explanation. In any event, a plot of the gravitational field intensity inside and outside a solid planet of uniform density is shown in Figure 9.23.

Imagine a spherical cavern at the center of a planet. The cavern would be gravity-free because of the cancellation of gravitational forces in every

[4]The strength of the gravitational field g at any point is equal to the force F per unit of mass placed there. So $g = Fm$, and its units are newtons per kilogram (N/kg). The field g also equals the free-fall acceleration of gravity. The units N/kg and m/s^2 are equivalent.

[5]This section may be skipped for a brief treatment of gravitational fields.

[6]Interestingly enough, during the first few kilometers beneath the Earth's surface, you'd actually gain acceleration because the density of the compact center is much greater than the density of the surface material. So gravity would be slightly stronger during the first part of a fall. Farther in, gravitation would decrease and would diminish to zero at the Earth's center.

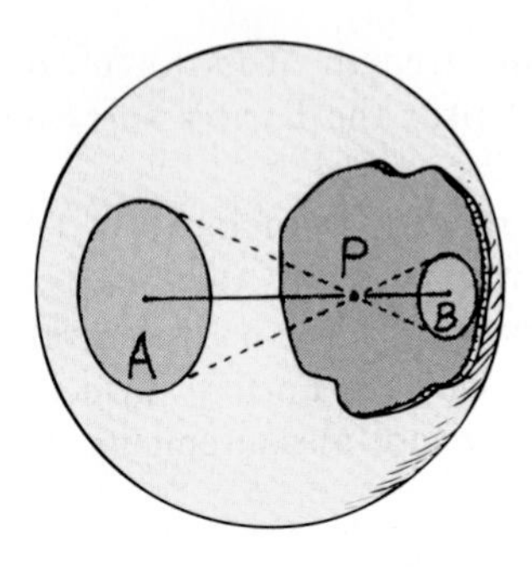

FIGURE 9.24
The gravitational field anywhere inside a spherical shell of uniform thickness and composition is zero, because the field components from all the particles of mass in the shell cancel one another. A mass at point P, for example, is attracted just as much to the larger but farther region A as it is to the smaller but closer region B.

direction. Amazingly, the size of the cavern doesn't change this fact—even if it constitutes most of the volume of the planet! A hollow planet, like a huge basketball, would have no gravitational field anywhere inside it. Complete cancellation of gravitational forces occurs everywhere inside. To see why, consider the particle P in Figure 9.24, which is twice as far from the left side of the planet as it is from the right side. If gravity depended only on distance, P would be attracted only 1/4 as much to the left side as to the right side (according to the inverse-square law). But gravity also depends on mass. Imagine a cone reaching to the left from P to encompass region A in the figure, and an equal-angle cone reaching to the right encompassing region B. Region A has four times the area and therefore four times the mass of region B. Since 1/4 of 4 is equal to 1, P is attracted to the farther but more massive region A with just as much force as it is to the closer but less massive region B. Cancellation occurs. More thought will show that cancellation will occur anywhere inside a planetary shell having uniform density and thickness. A gravitational field would exist at and beyond its outer surface and would behave as if all the mass of the planet were concentrated at its center; but everywhere inside the hollow part, the gravitational field is zero. Anyone inside would be weightless.

Gravitational Field Inside a Hollow Planet

The Weight of an Object Inside a Hollow Planet but Not at Its Center

CHECK YOURSELF

1. Suppose you stepped into a hole bored clear through the center of the Earth and made no attempt to grab the edges at either end. Neglecting air drag, what kind of motion would you experience?
2. Halfway to the center of the Earth, would the force of gravity on you be less than at the surface of the Earth?

Although gravity can be cancelled inside a body or between bodies, it cannot be shielded in the same way that electric forces can. In Chapter 22, we will see that electric forces can repel as well as attract, which makes shielding possible. Since gravitation only attracts, a similar kind of shielding cannot occur.

CHECK YOUR ANSWERS

1. You would oscillate back and forth. If Earth were an ideal sphere of uniform density and there were no air drag, your oscillation would be what is called *simple harmonic motion*. Each round trip would take nearly 90 minutes. Interestingly, we will see in the next chapter that an Earth satellite in close orbit about the Earth also takes the same 90 minutes to make a complete round trip. (This is no coincidence: If you study physics further, you'll learn that "back and forth" simple harmonic motion is simply the vertical component of uniform circular motion—interesting stuff.)
2. Gravitational force on you would be less, because there is less mass of Earth below you, which pulls you with less force. If the Earth were a uniform sphere of uniform density, gravitational force halfway to the center would be exactly half that at the surface. But since Earth's core is so dense (about seven times the density of surface rock), gravitational force halfway down would be somewhat more than half. Exactly how much depends on how Earth's density varies with depth, which is information that is unknown today.

Eclipses provide convincing evidence for this. The Moon is in the gravitational field of both the Sun and the Earth. During a lunar eclipse, Earth is directly between the Moon and the Sun, and any shielding of the Sun's field by Earth would result in a deviation of the Moon's orbit. Even a very slight shielding effect would accumulate over a period of years and show itself in the timing of subsequent eclipses. But there have been no such discrepancies; past and future eclipses are calculated to a high degree of accuracy using only the simple law of gravitation. No shielding effect in gravitation has ever been found.

Einstein's Theory of Gravitation

FIGURE 9.25
Warped space-time. Space-time near a star is curved in four dimensions in a way similar to the two-dimensional surface of a waterbed when a heavy ball rests on it.

In the early part of the twentieth century, a model for gravity quite unlike Newton's was presented by Einstein in his general theory of relativity. Einstein perceived a gravitational field as a geometrical warping of four-dimensional space and time; he realized that bodies put dents in space and time somewhat like a massive ball placed in the middle of a large waterbed dents the two-dimensional surface (Figure 9.25). The more massive the ball, the greater the dent or warp. If we roll a marble across the top of the bed but well away from the ball, the marble will roll in a straight-line path. But if we roll the marble near the ball, it will curve as it rolls across the indented surface of the waterbed. If the curve closes on itself, the marble will orbit the ball in an oval or circular path. If you put on your Newtonian glasses, so that you see the ball and marble but not the bed, you might conclude that the marble curves because it is attracted to the ball. If you put on your Einsteinian glasses, so that you see the marble and the indented waterbed but not the "distant" ball, you would likely conclude that the marble curves because the surface on which it moves is curved—in two dimensions for the waterbed and in four dimensions for space and time.[7] In Chapter 36, we will treat Einstein's theory of gravitation in more detail.

Black Holes

Suppose you were indestructible and could travel in a spaceship to the surface of a star. Your weight on the star would depend both on your mass and the star's mass and on the distance between the star's center and your belly button. If the star were to burn out and collapse to half its radius with no change in its mass, your weight at its surface, determined by the inverse-square law, would be four times as much (Figure 9.26). If the star were to collapse to a tenth of its radius, your weight at its surface would be 100 times as much. If the star kept shrinking, the gravitational field at the surface would become stronger. It would be more and more difficult for a starship to leave. The velocity required to escape, the *escape velocity,* would increase. If a star such as our Sun collapsed to a radius of less than 3 kilometers, the escape velocity from its surface would exceed the speed of light, and nothing—not even light—could escape! The Sun would be invisible. It would be a **black hole.**

[7]Don't be discouraged if you cannot visualize four-dimensional space-time. Einstein himself often told his friends, "Don't try. I can't do it either." Perhaps we are not too different from the great thinkers around Galileo who couldn't visualize a moving Earth!

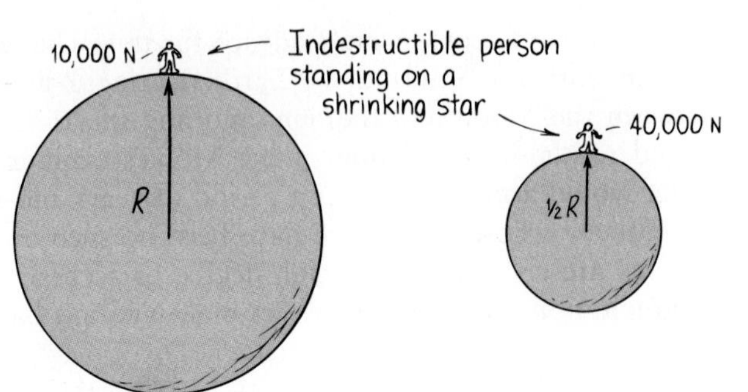

FIGURE 9.26
If a star collapses to half its radius and there is no change in its mass, gravitation at its surface is multiplied by 4.

Contrary to stories about black holes, they're nonaggressive and don't reach out and swallow innocents at a distance. Their gravitational fields are no stronger than the original fields about the stars before collapse—except at distances smaller than the original star radius. Except when they are too close, black holes shouldn't worry future astronauts.

Insights

The Sun, in fact, has too little mass to experience such a collapse, but, when some stars with greater mass—now estimated to be at least 1.5 solar masses or more—reach the end of their nuclear resources, they undergo collapse; and, unless rotation is high enough, the collapse continues until the stars reach infinite densities. Gravitation near these shrunken stars is so enormous that light cannot escape from their vicinity. They have crushed themselves out of visible existence. The results are black holes, which are completely invisible.

A black hole is no more massive than the star from which it collapsed, so the gravitational field in regions at and greater than the original star's radius is no different after the star's collapse than before. But, at closer distances near the vicinity of a black hole, the gravitational field can be enormous—a surrounding warp into which anything that passes too close—light, dust, or a spaceship—is drawn. Astronauts could enter the fringes of this warp and, if they were in a powerful spaceship, they could still escape. After a certain distance, however, they could not, and they would disappear from the observable universe. Any object falling into a black hole would be torn to pieces. No feature of the object would survive except its mass, its angular momentum (if any), and its electric charge (if any).

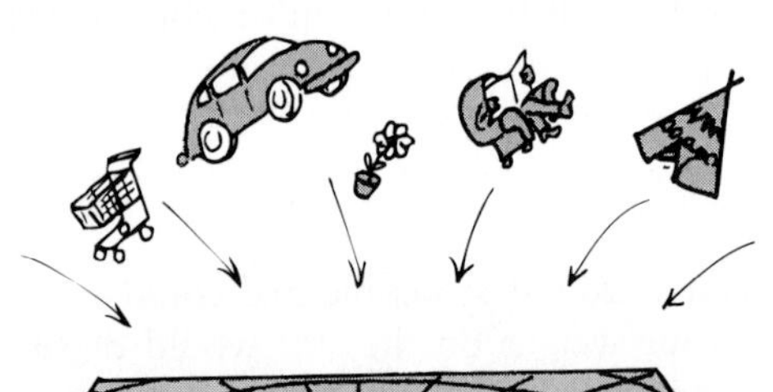

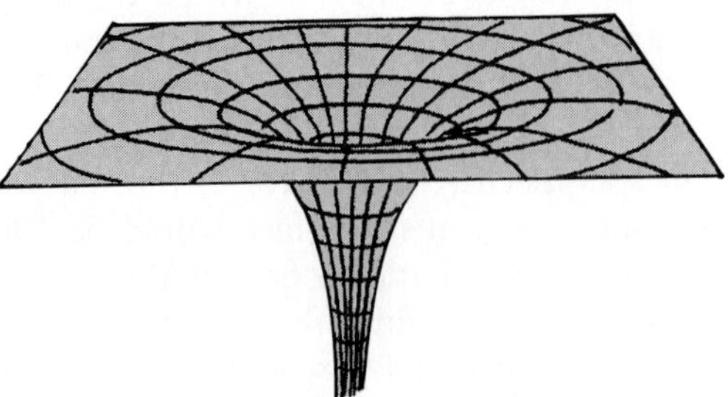

FIGURE 9.27
Anything that falls into a black hole is crushed out of existence. Only mass, angular momentum, and electric charge are retained by the black hole.

FIGURE 9.28
A speculative wormhole may be the portal to another part of our universe or even to another universe.

A theoretical entity with some similarity to a black hole is the "wormhole" (Figure 9.28). Like a black hole, a wormhole is an enormous distortion of space and time. But instead of collapsing toward an infinitely dense point, the wormhole opens out again in some other part of the universe—or even, conceivably, in some other universe! Whereas the existence of black holes has been confirmed, the wormhole remains an exceedingly speculative notion. Some science buffs imagine that the wormhole opens the possibility of time travel.[8]

How can a black hole be detected if there is literally no way to "see" it? It makes itself felt by its gravitational influence on nearby matter and on neighboring stars. There is now good evidence that some binary star systems consist of a luminous star and an invisible companion with black-hole-like properties orbiting around each other. Even stronger evidence points to more massive black holes at the centers of many galaxies. In a young galaxy, observed as a "quasar," the central black hole sucks in matter that emits great quantities of radiation as it plunges to oblivion. In an older galaxy, stars are observed circling in a powerful gravitational field around an apparently empty center. These galactic black holes have masses ranging from millions to more than one billion times the mass of our Sun. The center of our own galaxy, although not so easy to see as the centers of some other galaxies, almost surely hosts a black hole. Discoveries are coming faster than textbooks can report. Check your astronomy Web site for the latest update.

Universal Gravitation

A sphere has the smallest surface area for any volume of matter.

Insights

We all know that the Earth is round. But why is the Earth round? It is round because of gravitation. Everything attracts everything else, and so the Earth has attracted itself together as far as it can! Any "corners" of the Earth have been pulled in; as a result, every part of the surface is equidistant from the center of gravity. This makes it a sphere. Therefore, we see, from the law of gravitation, that the Sun, the Moon, and the Earth are spherical because they have to be (although rotational effects make them slightly ellipsoidal).

If everything pulls on everything else, then the planets must pull on each other. The force that controls Jupiter, for example, is not just the force from the Sun; there are also the pulls from the other planets. Their effect is small in comparison with the pull of the much more massive Sun, but it still shows. When Saturn is near Jupiter, its pull disturbs the otherwise smooth path traced by Jupiter. Both planets "wobble" about their expected orbits. The interplanetary forces causing

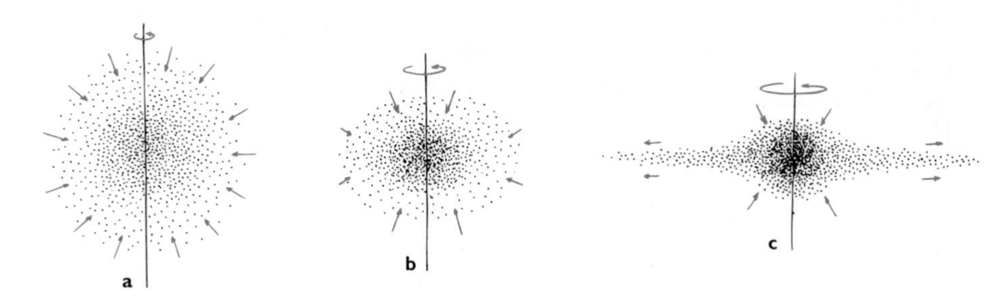

FIGURE 9.29
Formation of the solar system. A slightly rotating ball of interstellar gas (a) contracts due to mutual gravitation and (b) conserves angular momentum by speeding up. The increased momentum of individual particles and clusters of particles causes them (c) to sweep in wider paths about the rotational axis, producing an overall disk shape. The greater surface area of the disk promotes cooling and condensation of matter in swirling eddies—the birthplace of the planets.

[8]Stephen Hawking, a pioneering expert on black holes, was one of the first to speculate about the existence of wormholes. But, in 2003, to the dismay of many science buffs, he announced his belief that they cannot exist.

Discovery of Neptune

A widespread assumption is that when the Earth was first no longer considered the center of the universe, its place and humankind were demoted and no longer considered special. On the contrary, writings of the time suggest most Europeans viewed humans as filthy and sinful because of Earth's lowly position—farthest from heaven, with hell at its center. Human elevation didn't occur until the Sun, viewed positively, took a center position. We became special by showing we're not so special.

Insights

this wobbling are called *perturbations*. By the 1840s, studies of the most recently discovered planet at the time, Uranus, showed that the deviations of its orbit could not be explained by perturbations from all other known planets. Either the law of gravitation was failing at this great distance from the Sun or an unknown eighth planet was perturbing the orbit of Uranus. An Englishman and a Frenchman, J. C. Adams and Urbain Leverrier, each assumed Newton's law to be valid, and they independently calculated where an eighth planet should be. At about the same time, Adams sent a letter to the Greenwich Observatory in England and Leverrier sent a letter to the Berlin Observatory in Germany, both suggesting that a certain area of the sky be searched for a new planet. The request by Adams was delayed by misunderstandings at Greenwich, but Leverrier's request was heeded immediately. The planet Neptune was discovered that very night!

Subsequent tracking of the orbits of both Uranus and Neptune led to the prediction and discovery of Pluto in 1930 at the Lowell Observatory in Arizona. Whatever you may have learned in your early schooling, many astronomers now regard Pluto as an asteroid and not a true planet. Many asteroids of nearly planetary size have recently been, and continue to be, discovered beyond Neptune (Quaoar, for example, is an asteroid that has a moon of its own). Astronomers are faced with the problem of whether to classify the increasing list of Pluto's neighbors as planets or to reclassify Pluto. Rather than Pluto being the pipsqueak of planets, many now consider it to be the king among asteroids. Whatever its status, the object we call Pluto takes 248 years to make a single revolution about the Sun, so no one will see it in its discovered position again until the year 2178.

Recent evidence suggests that the universe is expanding and accelerating outward, pushed by an antigravity *dark energy* that makes up some 73% of the universe. Another 23% is composed of the yet-to-be-discovered particles of exotic *dark matter*. Ordinary matter, the stuff of stars, cabbages, and kings, makes up only about 4%. The concepts of dark energy and dark matter are late twentieth- and twenty-first-century confirmations. The present view of the universe has progressed appreciably beyond the universe as Newton perceived it.

Yet few theories have affected science and civilization as much as Newton's theory of gravity. The successes of Newton's ideas ushered in the Age of Enlightenment. Newton had demonstrated that, by observation and reason, people could uncover the workings of the physical universe. How profound that all the moons and planets and stars and galaxies have such a beautifully simple rule to govern them, namely,

$$F = G\frac{m_1 m_2}{d^2}$$

The formulation of this simple rule is one of the major reasons for the success in science that followed, for it provided hope that other phenomena of the world might also be described by equally simple and universal laws.

This hope nurtured the thinking of many scientists, artists, writers, and philosophers of the 1700s. One of these was the English philosopher John Locke, who argued that observation and reason, as demonstrated by Newton, should be our best judge and guide in all things. Locke urged that all of nature and even society should be searched to discover any "natural laws" that might exist. Using Newtonian physics as a model of reason, Locke and his followers modeled a system of government that found adherents in the thirteen British colonies across the Atlantic. These ideas culminated in the Declaration of Independence and the Constitution of the United States of America.

Summary of Terms

Law of universal gravitation Every body in the universe attracts every other body with a force that, for two bodies, is directly proportional to the product of their masses and inversely proportional to the square of the distance separating them:

$$F = G\frac{m_1 m_2}{d^2}$$

Inverse-square law A law relating the intensity of an effect to the inverse square of the distance from the cause:

$$\text{Intensity} \sim \frac{1}{\text{distance}^2}$$

Gravity follows an inverse-square law, as do the effects of electric, magnetic, light, sound, and radiation phenomena.

Weight The force that an object exerts on a supporting surface (or, if suspended, on a supporting string), which is often, but not always, due to the force of gravity.

Weightless Being without a support force, as in free fall.

Spring tides High or low tides that occur when the Sun, the Earth, and the Moon are all lined up so that the tides due to the Sun and the Moon coincide, making the high tides higher than average and the low tides lower than average.

Neap tides Tides that occur when the Moon is midway between new and full, in either direction. Tides due to the Sun and the Moon partly cancel, making the high tides lower than average and the low tides higher than average.

Gravitational field The influence that a massive body extends into the space around itself, producing a force on another massive body. It is measured in newtons per kilogram (N/kg).

Black hole A concentration of mass resulting from gravitational collapse, near which gravity is so intense that not even light can escape.

Suggested Reading

Cole, K.C. *The Hole in the Universe: How Scientists Peered over the Edge of Emptiness and Found Everything.* New York: Harcourt, 2001.

Einstein, A., and L. Infeld. *The Evolution of Physics.* New York: Simon & Schuster, 1938.

Gamow, G. *Gravity.* Science Study Series. Garden City, N.Y.: Doubleday (Anchor), 1962.

Review Questions

1. What did Newton discover about gravity?
2. What is the Newtonian synthesis?

The Universal Law of Gravity

3. In what sense does the Moon "fall"?
4. State Newton's law of universal gravitation in words. Then do the same with one equation.

The Universal Gravitational Constant, G

5. What is the magnitude of gravitational force between two 1-kilogram bodies that are 1 meter apart?
6. What is the magnitude of the gravitational force between the Earth and a 1-kilogram body?
7. What do we call the gravitational force between the Earth and your body?
8. When *G* was first measured by Henry Cavendish, newspapers of the time hailed his experiment as the "weighing the Earth experiment." Why?

Gravity and Distance: The Inverse-Square Law

9. How does the force of gravity between two bodies change when the distance between them is doubled?
10. How does the thickness of paint sprayed on a surface change when the sprayer is held twice as far away?
11. How does the brightness of light change when a point source of light is brought twice as far away?
12. Where do you weigh more—at the bottom of Death Valley or atop one of the peaks of the Sierra Nevada? Why?

Weight and Weightlessness

13. Would the springs inside a bathroom scale be more compressed or less compressed if you weighed yourself in an elevator that accelerated upward? Downward?
14. Would the springs inside a bathroom scale be more compressed or less compressed if you weighed yourself in an elevator that moved upward at *constant velocity*? Downward at *constant velocity*?
15. When is your weight equal to mg?
16. Give an example of when your weight is more than mg. Give an example of when it's zero.

Ocean Tides

17. Do tides depend more on the strength of gravitational pull or on the *difference* in strengths? Explain.
18. Why do both the Sun and the Moon exert a greater gravitational force on one side of the Earth than the other?
19. (Fill in the blank.) Gravitational force (units N) depends on the inverse square of distance. But tidal force, the difference in gravitational forces per unit mass (units N/kg), depends on the inverse _______ of distance.
20. Distinguish between *spring tides* and *neap tides.*

Tides in the Earth and Atmosphere

21. Do tides occur in the molten interior of the Earth for the same reason that tides occur in the oceans?

22. Why are all tides greatest at the time of a full Moon or new Moon?

Tides on the Moon

23. Does the fact that one side of the Moon always faces the Earth mean that the Moon rotates about its axis (like a top) or that it doesn't rotate about its axis?

24. Why is there a torque about the Moon's center of mass when the Moon's long axis is not aligned with Earth's gravitational field?

25. Is there a torque about the Moon's center of mass when the Moon's long axis is aligned with Earth's gravitational field? Explain how this compares with a magnetic compass.

Gravitational Fields

26. What is a gravitational field, and how can its presence be detected?

Gravitational Field Inside a Planet

27. What is the magnitude of the gravitational field at the Earth's center?

28. For a planet of uniform density, how would the magnitude of the gravitational field halfway to the center compare with the field at the surface?

29. What would be the magnitude of the gravitational field anywhere inside a hollow, spherical planet?

Einstein's Theory of Gravitation

30. Newton viewed the curving of the path of a planet as being caused by a force acting upon the planet. How did Einstein view the curving of the path of a planet?

Black Holes

31. If Earth shrank with no change in its mass, what would happen to your weight at the surface?

32. What happens to the strength of the gravitational field at the surface of a star that shrinks?

33. Why is a black hole invisible?

Universal Gravitation

34. What was the cause of perturbations discovered in the orbit of the planet Uranus? What greater discovery did this lead to?

35. What percentage of the universe is presently speculated to be composed of dark matter and dark energy?

Projects

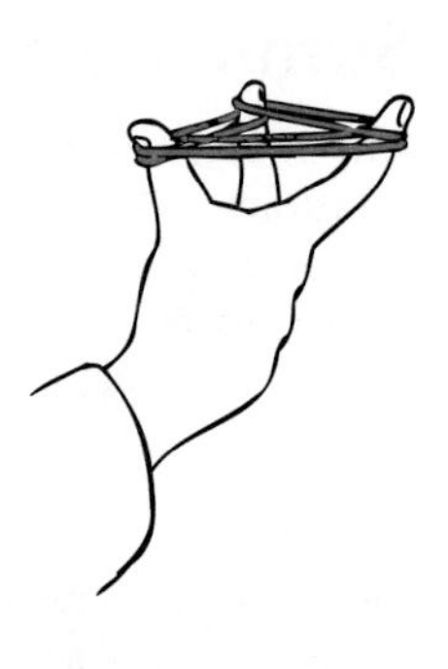

1. Hold up your thumb and first two fingers and make a V sign. Place a strong rubber band across your thumb and first finger. This represents the force of gravity between the Sun and the Earth. Place a medium-strength rubber band across your thumb and second finger to represent the force of gravity between the Sun and the Moon. Then place a weak rubber band across your first two fingers to represent the force of gravity between the Moon and the Earth. Note how all fingers pull on each other. Likewise for the gravitational pulls among the Sun, the Earth, and the Moon.

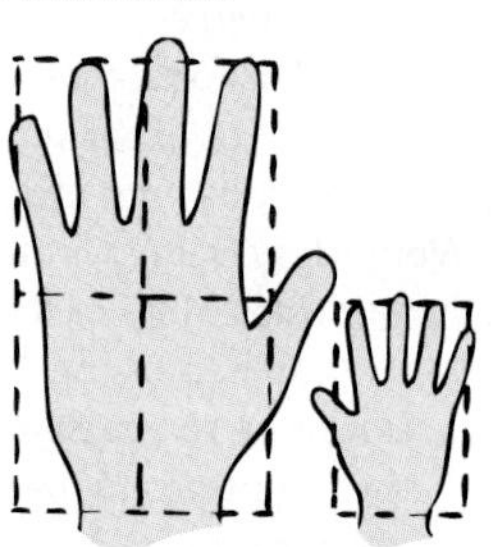

2. Hold your hands outstretched in front of you, one twice as far from your eyes as the other, and make a casual judgment as to which hand looks bigger. Most people see them to be about the same size, while many see the nearer hand as slightly bigger. Almost no one, upon casual inspection, sees the nearer hand as four times as big; but, by the inverse-square law, the nearer hand should appear to be twice as tall and twice as wide and therefore seem to occupy four times as much of your visual field as the farther hand. Your belief that your hands are the same size is so strong that you likely overrule this information. Now, if you overlap your hands slightly and view them with one eye closed, you'll see the nearer hand as clearly bigger. This raises an interesting question: What other illusions do you have that are not so easily checked?

3. Repeat the eyeballing experiment, only this time use two dollar bills—one regular, and the other folded along its middle lengthwise, and again width-wise, so it has 1/4 the area. Now hold the two in front of your eyes. Where do you hold the folded one so that it looks the same size as the unfolded one? Nice?

One-Step Calculations

$$F = G\frac{m_1 m_2}{d^2}$$

1. Calculate the force of gravity on a 1-kg mass at the Earth's surface. The mass of the Earth is 6×10^{24} kg, and its radius is 6.4×10^{6} m.

2. Calculate the force of gravity on the same 1-kg mass if it were 6.4×10^6 m above the Earth's surface (that is, if it were two Earth radii from the Earth's center).
3. Calculate the force of gravity between Earth (mass $= 6.0 \times 10^{24}$ kg) and the Moon (mass $= 7.4 \times 10^{22}$ kg). The average Earth–Moon distance is 3.8×10^8 m.
4. Calculate the force of gravity between Earth and the Sun (the Sun's mass $= 2.0 \times 10^{30}$ kg; average Earth–Sun distance $= 1.5 \times 10^{11}$ m).
5. Calculate the force of gravity between a newborn baby (mass = 3 kg) and the planet Mars (mass $= 6.4 \times 10^{23}$ kg), when Mars is at its closest to the Earth (distance $= 5.6 \times 10^{10}$ m).
6. Calculate the force of gravity between a newborn baby of mass 3 kg and the obstetrician of mass 100 kg, who is 0.5 m from the baby. Which exerts more gravitational force on the baby, Mars or the obstetrician? By how much?

Exercises

1. Comment on whether or not the following label on a consumer product should be cause for concern. *CAUTION: The mass of this product pulls on every other mass in the universe, with an attracting force that is proportional to the product of the masses and inversely proportional to the square of the distance between them.*
2. Gravitational force acts on all bodies in proportion to their masses. Why, then, doesn't a heavy body fall faster than a light body?
3. What would be the path of the Moon if somehow all gravitational forces on it vanished to zero?
4. Is the force of gravity stronger on a piece of iron than on a piece of wood if both have the same mass? Defend your answer.
5. Is the force of gravity stronger on a crumpled piece of paper than on an identical piece of paper that has not been crumpled? Defend your answer.
6. What is the relationship between force and distance in an inverse-square law?
7. Why could Newton not determine the magnitude of G with his equation?
8. A friend says that, since Earth's gravity is so much stronger than the Moon's gravity, rocks on the Moon could be dropped to the Earth. What is wrong with this assumption?
9. Another friend says that the Moon's gravity would prevent rocks dropping from the Moon to the Earth, but that, if the Moon's gravity somehow vanished to zero, then rocks on the Moon would fall to Earth. What is wrong with this assumption?
10. A friend says that astronauts in orbit are weightless because they're beyond the pull of Earth's gravity. Correct your friend's ignorance.
11. Somewhere between the Earth and the Moon, gravity from these two bodies on a space pod would cancel. Is this location nearer the Earth or the Moon?
12. An apple falls because of the gravitational attraction to Earth. How does the gravitational attraction of the Earth to the apple compare? (Does force change when you interchange m_1 and m_2 in the equation for gravity—m_2m_1 instead of m_1m_2?)
13. Larry weighs 300 N at the surface of the Earth. What is the weight of the Earth in the gravitational field of Larry?
14. The Earth and the Moon are attracted to each other by gravitational force. Does the more massive Earth attract the less massive Moon with a force that is greater, smaller, or the same as the force with which the Moon attracts the Earth? (With an elastic band stretched between your thumb and forefinger, which is pulled more strongly by the band, your thumb or your forefinger?)
15. If the Moon pulls the Earth as strongly as the Earth pulls the Moon, why doesn't the Earth rotate around the Moon, or why don't both rotate around a point midway between them?
16. Is the acceleration due to gravity more or less atop Mt. Everest than at sea level? Defend your answer.
17. An astronaut lands on a planet that has the same mass as Earth but twice the diameter. How does the astronaut's weight differ from that on Earth?
18. An astronaut lands on a planet that has twice the mass as Earth and twice the diameter. How does the astronaut's weight differ from that on Earth?
19. If the Earth somehow expanded to a larger radius, with no change in mass, how would your weight be affected? How would it be affected if the Earth instead shrunk? (*Hint:* Let the equation for gravitational force guide your thinking.)
20. The intensity of light from a central source varies inversely as the square of the distance. If you lived on a planet only half as far from the Sun as our Earth, how would the light intensity compare with that on Earth? How about a planet ten times farther away than the Earth?
21. A small light source located 1 m in front of a 1-m^2 opening illuminates a wall behind. If the wall is 1 m behind the opening (2 m from the light source), the illuminated area covers 4 m^2. How many square

meters will be illuminated if the wall is 3 m from the light source? 5 m? 10 m?

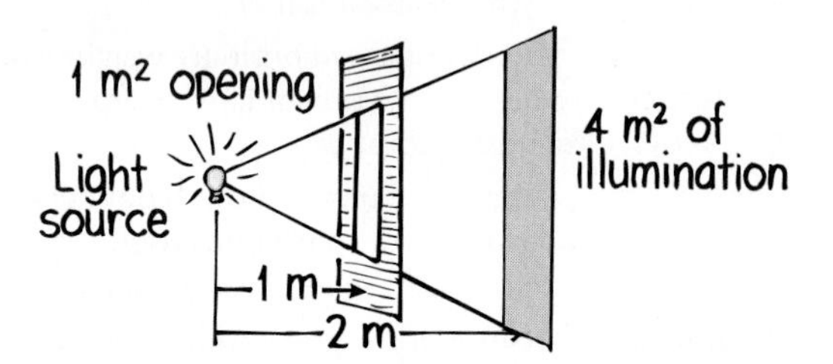

22. The planet Jupiter is more than 300 times as massive as Earth, so it might seem that a body on the surface of Jupiter would weigh 300 times as much as on Earth. But it so happens that a body would scarcely weigh three times as much on the surface of Jupiter as it would on the surface of the Earth. Can you think of an explanation for why this is so? (*Hint:* Let the terms in the equation for gravitational force guide your thinking.)

23. Why do the passengers in high-altitude jet planes feel the sensation of weight while passengers in an orbiting space vehicle, such as a space shuttle, do not?

24. Why does a person in free fall experience weightlessness, while a person falling at terminal velocity does not?

25. If you were in a car that drove off the edge of a cliff, why would you be momentarily weightless? Would gravity still be acting on you?

26. Is gravitational force acting on a person who falls off a cliff, or on an astronaut inside an orbiting space shuttle?

27. What two forces act on you while you are in a moving elevator? When are these forces of equal magnitude and when are they not?

28. If you were in a freely falling elevator and you dropped a pencil, it would hover in front of you. Is there a force of gravity acting on the pencil? Defend your answer.

29. Why does a bungee jumper feel weightless during the jump?

30. Since your weight when standing on the Earth is the gravitational attraction between you and the Earth, would your weight be greater if the Earth gained mass? If the Sun gained mass? Why are your answers the same or different?

31. Your friend says that the primary reason astronauts in orbit feel weightless is that they are beyond the main pull of Earth's gravity. Why do you agree or disagree?

32. Explain why the following reasoning is wrong. "The Sun attracts all bodies on the Earth. At midnight, when the Sun is directly below, it pulls on you in the same direction as the Earth pulls on you; at noon, when the Sun is directly overhead, it pulls on you in a direction opposite to the Earth's pull on you. Therefore, you should be somewhat heavier at midnight and somewhat lighter at noon."

33. When will the gravitational force between you and the Sun be greater—today at noon, or tomorrow at midnight? Defend your answer.

34. If the mass of the Earth increased, your weight would correspondingly increase. But, if the mass of the Sun increased, your weight would not be affected at all. Why?

35. Most people today know that the ocean tides are caused principally by the gravitational influence of the Moon, and most people therefore think that the gravitational pull of the Moon on the Earth is greater than the gravitational pull of the Sun on the Earth. What do you think?

36. If somebody tugged hard on your shirt sleeve, it would likely tear. But if all parts of your shirt were tugged equally, no tearing would occur. How does this relate to tidal forces?

37. Would ocean tides exist if the gravitational pull of the Moon (and the Sun) were somehow equal on all parts of the world? Explain.

38. Why aren't high ocean tides exactly 12 hours apart?

39. With respect to spring and neap ocean tides, when are the tides lowest? That is, when is it best for digging clams?

40. Whenever the ocean tide is unusually high, will the following low tide be unusually low? Defend your answer in terms of "conservation of water." (If you slosh water in a tub so that it is extra deep at one end, will the other end be extra shallow?)

41. The Mediterranean Sea has very little sediment churned up and suspended in its waters, mainly because of the absence of any substantial ocean tides. Why do you suppose the Mediterranean Sea has practically no tides? Similarly, are there tides in the Black Sea? In the Great Salt Lake? Your county reservoir? A glass of water? Explain.

42. The human body is composed mostly of water. Why does the Moon overhead cause appreciably less tidal effect in the fluid compartment of your body than a 1-kg melon held over your head?

43. If the Moon didn't exist, would the Earth still have ocean tides? If so, how often?

44. What would be the effect on the Earth's tides if the diameter of the Earth were very much larger than it is? If the Earth were as it presently is, but the Moon were very much larger and had the same mass?

45. Does the strongest tidal force on our bodies come from the Earth, the Moon, or the Sun?

46. Exactly why do tides occur in the Earth's crust and in the Earth's atmosphere?

47. The value of g at the Earth's surface is about 9.8 m/s^2. What is the value of g at a distance of twice the Earth's radius?

48. If the Earth were of uniform density (same mass/volume throughout), what would the value of g be inside the Earth at half its radius?

49. If the Earth were of uniform density, would your weight increase or decrease at the bottom of a deep mine shaft? Defend your answer.

50. It so happens that an actual *increase* in weight is found even in the deepest mine shafts. What does this tell us about how the Earth's density changes with depth?

51. Which requires more fuel—a rocket going from the Earth to the Moon or a rocket coming from the Moon to the Earth? Why?

52. If you could somehow tunnel inside a star, would your weight increase or decrease? If, instead, you somehow stood on the surface of a shrinking star, would your weight increase or decrease? Why are your answers different?

53. If our Sun shrank in size to become a black hole, show from the gravitational force equation that the Earth's orbit would not be affected.

54. If the Earth were hollow but still had the same mass and same radius, would your weight in your present location be more, less, or the same as it is now? Explain.

55. Some people dismiss the validity of scientific theories by saying they are "only" theories. The law of universal gravitation is a theory. Does this mean that scientists still doubt its validity? Explain.

56. We say, on page 174, that gravity cannot be shielded and, on the same page, that gravitational components cancel to zero inside a uniform shell. Why are these two statements not contradictory?

57. Strictly speaking, you weigh a tiny bit less when you are in the lobby of a massive skyscraper than you do at home. Why is this so?

58. A person falling into a black hole would probably be killed by tidal forces before being swallowed up by the hole. Explain why this would be so.

59. A spaceship near a shrinking star sees it collapse to a black hole. Is there a change in the gravitational field at the spaceship's location? Defend your answer.

60. Make up two multiple-choice questions—one that would check a classmate's understanding of the inverse-square law and another that would check a distinction between weight and weightlessness.

Problems

1. Find the change in the force of gravity between two planets when the distance between them is decreased by a factor of five.

2. The value of g at the Earth's surface is about 9.8 N/kg. What is the value of g at a distance from the Earth's center that is four times the Earth's radius?

3. Show, by algebraic reasoning, that your gravitational acceleration toward an object of mass M a distance d away is $a = GM/d^2$ and therefore doesn't depend on your mass.

4. The mass of a certain neutron star is 3.0×10^{30} kg (1.5 solar masses) and its radius is 8,000 m (8 km). What is the acceleration of gravity at the surface of this condensed, burned-out star?

5. Many people mistakenly believe that the astronauts that orbit the Earth are "above gravity." Calculate g for space-shuttle territory, 200 kilometers above the Earth's surface. Earth's mass is 6×10^{24} kg, and its radius is 6.38×10^6 m (6380 km). Your answer is what percentage of 9.8 m/s^2?

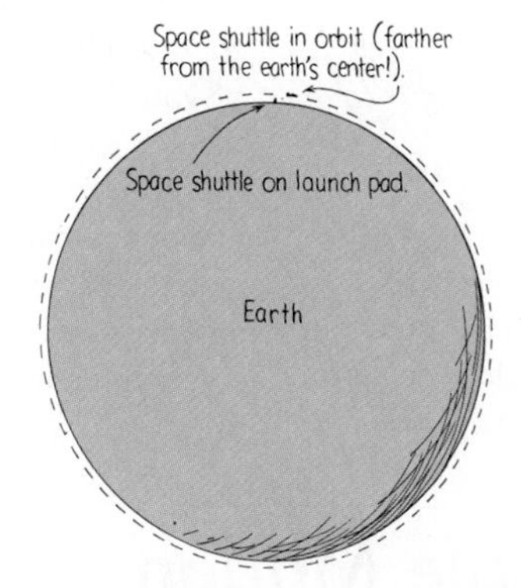

6. The difference in force per mass (N/kg) across a body, tidal force T_F, is approximated by $T_F = 4GMR/d^3$, where G is the gravitational constant, M the mass of the body causing tides, R the radius of the body having tides, and d the center-to-center distance between the bodies. Make two calculations: (a) T_F that the Moon exerts on you, and (b) T_F that a 1-kg melon 1 m above your head exerts on you. For simplicity, let R for you be 1 m (pretend you're 2 m tall). Let d to the Moon be 3.8×10^8 m and to the melon be 2 m. Mass M of the Moon is 7.3×10^{22} kg. (c) Compare your calculations. Which is larger, and by how much? Then share this information with any friends who say that the tidal forces of planets on people influence their lives.

CHAPTER 10

Projectile and Satellite Motion

Chuck Stone asks his class to predict the projectile's range.

Projectile Motion

THE Physics Place
Orbital Motion and Kepler's Laws

From the top of Mauna Kea, a dormant volcano in Hawaii (or from any high vantage point where the distant ocean horizon is sharp and clear), you can see the curvature of the Earth. You have to eyeball the line where ocean and sky meet against a long straightedge in front of your eyes. Otherwise, you can't be sure if your eyes are playing tricks on you. If you line up your sight so that the bottom of the middle of the straightedge just touches the juncture between sky and ocean, you'll note a space between sky and the ocean at the ends. You're seeing Earth's curvature. Now toss a rock horizontally toward the horizon. It quickly falls several meters to the ground below in front of you. It curves as it falls. You'll note that, the faster you throw the rock, the wider the curve. Then you wonder how fast Superman would have to throw the rock to clear the horizon ahead, and how fast he'd have to throw it so that its curved path would match the curve of the Earth. If he could do that, and if air drag were somehow eliminated, the rock would follow a curved path completely around the Earth and become an Earth satellite. A satellite is, after all, no more than a projectile moving fast enough to continually clear the horizon as it falls.

FIGURE 10.1
If Superman threw a rock fast enough, it would orbit Earth, if there were no air drag.

Projectile Motion

Without gravity, you could toss a rock at an angle skyward and it would follow a straight-line path. Because of gravity, however, the path curves. A tossed rock, a cannonball, or any object that is projected by some means and continues in motion by its own inertia is called a **projectile.** To the cannoneers of earlier centuries, the curved paths of projectiles seemed very complex. Today these paths are surprisingly simple when we look at the horizontal and vertical components of velocity separately.

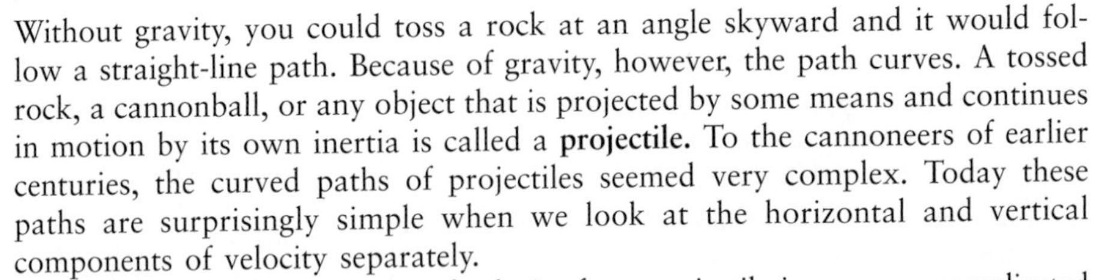

The horizontal component of velocity for a projectile is no more complicated than the horizontal velocity of a bowling ball rolling freely on the lane of a bowling alley. If the retarding effect of friction can be ignored, there is no horizontal force on the ball and its velocity is constant. It rolls of its own inertia and covers equal distances in equal intervals of time (Figure 10.2, top). The horizontal component of a projectile's motion is just like the bowling ball's motion along the lane.

Projectile Motion
More Projectile Motion

The vertical component of motion for a projectile following a curved path is just like the motion described in Chapter 3 for a freely falling object. The

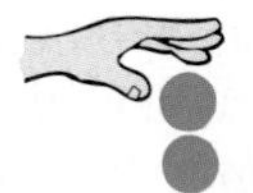

FIGURE 10.2
(Top) Roll a ball along a level surface, and its velocity is constant because no component of gravitational force acts horizontally. (Left) Drop it, and it accelerates downward and covers a greater vertical distance each second.

vertical component is exactly the same as for an object falling freely straight down, as shown at the left in Figure 10.2. The faster the object falls, the greater the distance covered in each successive second. Or, if the object is projected upward, the vertical distances of travel decrease with time on the way up.

The curved path of a projectile is a combination of horizontal and vertical motion. The horizontal component of velocity for a projectile is completely independent of the vertical component of velocity when air drag is small enough to ignore. Then the constant horizontal-velocity component is not affected by the vertical force of gravity. Each component is independent of the other. Their combined effects produce the trajectories of projectiles.

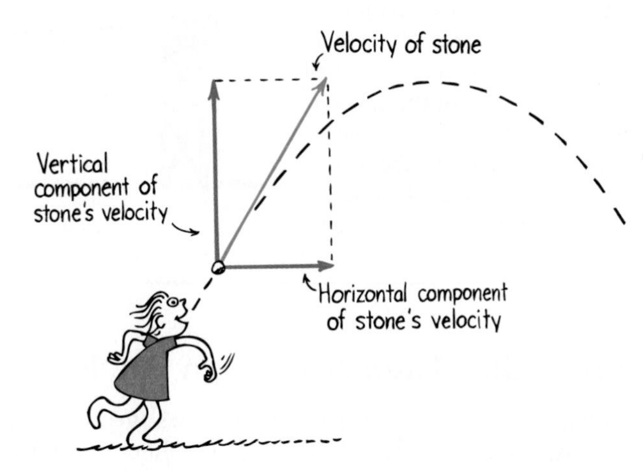

FIGURE 10.3
Vertical and horizontal components of a stone's velocity.

Projectiles Launched Horizontally

Projectile motion is nicely analyzed in Figure 10.4, which shows a simulated multiple-flash exposure of a ball rolling off the edge of a table. Investigate it carefully, for there's a lot of good physics there. On the left we notice equally timed

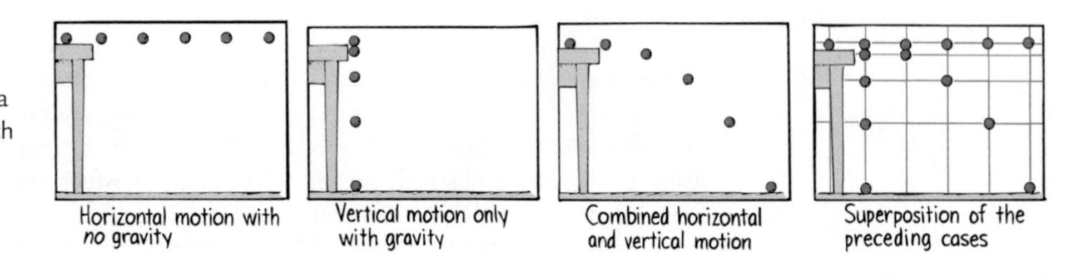

FIGURE 10.4
Interactive Figure
Simulated photographs of a moving ball illuminated with a strobe light.

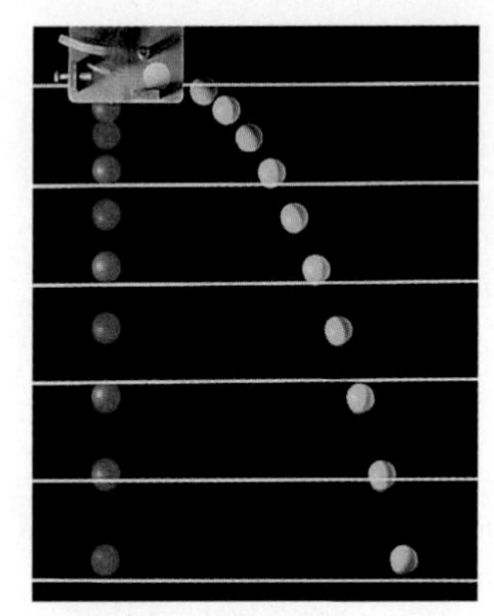

FIGURE 10.5
Interactive Figure
A strobe-light photograph of two golf balls released simultaneously from a mechanism that allows one ball to drop freely while the other is projected horizontally.

sequential positions of the ball without the effect of gravity. Only the effect of the ball's horizontal component of motion is shown. Next we see vertical motion without a horizontal component. The curved path in the third view is best analyzed by considering the horizontal and vertical components of motion separately. There are two important things to notice. The first is that the ball's horizontal component of velocity doesn't change as the falling ball moves forward. The ball travels the same horizontal distance in equal times between each flash. That's because there is no component of gravitational force acting horizontally. Gravity acts only *downward,* so the only acceleration of the ball is *downward.* The second thing to notice is that the vertical positions become farther apart with time. The vertical distances traveled are the same as if the ball were simply dropped. Note the curvature of the ball's path is the combination of horizontal motion, which remains constant, and vertical motion, which undergoes acceleration due to gravity.

The trajectory of a projectile that accelerates only in the vertical direction while moving at a constant horizontal velocity is a **parabola.** When air drag is small enough to neglect, as it is for a heavy object without great speed, the trajectory is parabolic.

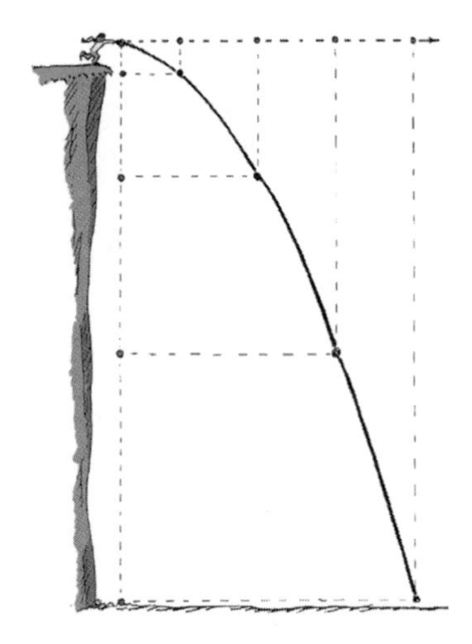

FIGURE 10.6
The vertical dashed line is the path of a stone dropped from rest. The horizontal dashed line would be its path if there were no gravity. The curved solid line shows the resulting trajectory that combines horizontal and vertical motion.

CHECK YOURSELF

At the instant a cannon fires a cannonball horizontally over a level range, another cannonball held at the side of the cannon is released and drops to the ground. Which ball, the one fired downrange or the one dropped from rest, strikes the ground first?

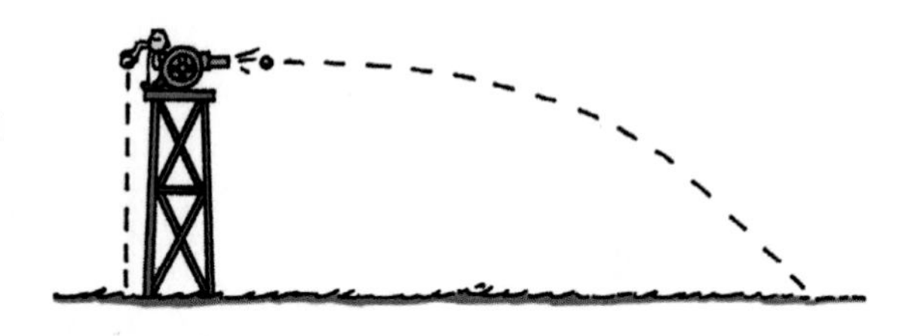

Projectiles Launched at an Angle

In Figure 10.7, we see the paths of stones thrown at an angle upward (left) and downward (right). The dashed straight lines show the ideal trajectories of the stones if there were no gravity. Notice that the vertical distance beneath the idealized straight-line paths is the same for equal times. This vertical distance is independent of what's happening horizontally.

CHECK YOUR ANSWER

Both cannonballs hit the ground at the same time, for both fall *the same vertical distance.* Can you see that the physics is the same as the physics of Figures 10.4 through 10.6? We can reason this another way by asking which one would hit the ground first if the cannon were pointed at an *upward* angle. Then the dropped cannonball would hit first, while the fired ball is still airborne. Now consider the cannon pointing *downward.* In this case, the fired ball hits first. So projected upward, the dropped one hits first; downward, the fired one hits first. Is there some angle at which there is a dead heat, where both hit at the same time? Can you see that this occurs when the cannon is horizontal?

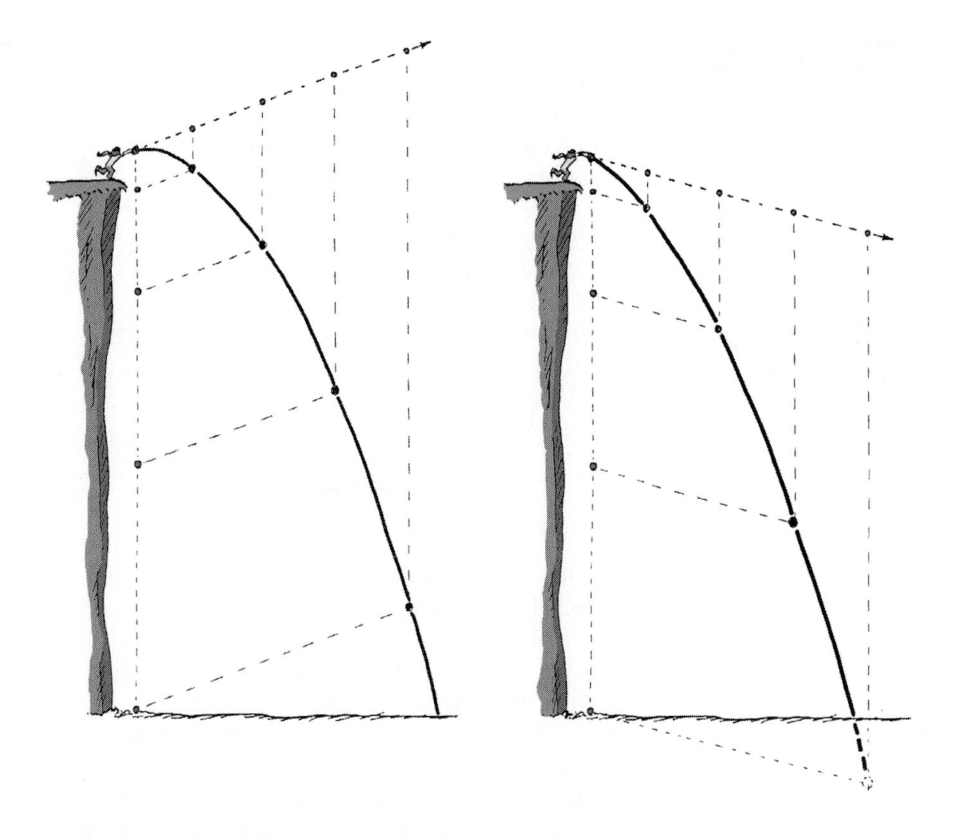

FIGURE 10.7
Whether launched at an angle upward or downward, the vertical distance of fall beneath the idealized straight-line path is the same for equal times.

Figure 10.8 shows specific vertical distances for a cannonball shot at an upward angle. If there were no gravity the cannonball would follow the straight-line path shown by the dashed line. But there is gravity, so this doesn't occur. What really happens is that the cannonball continuously falls beneath the imaginary line until it finally strikes the ground. Note that the vertical distance it falls beneath any point on the dashed line is the same vertical distance it would

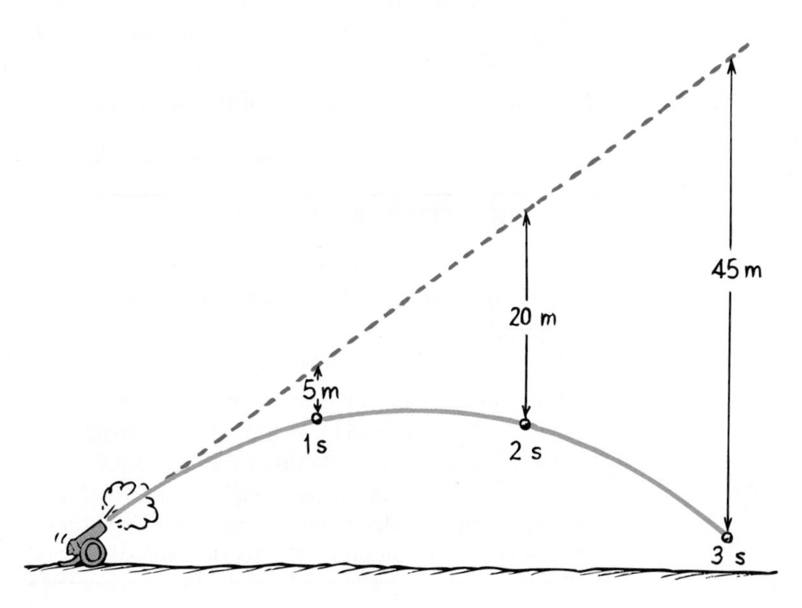

FIGURE 10.8
With no gravity, the projectile would follow a straight-line path (dashed line). But, because of gravity, the projectile falls beneath this line the same vertical distance it would fall if it were released from rest. Compare the distances fallen with those given in Table 3.3 in Chapter 3. (With $g = 9.8\ \text{m/s}^2$, these distances are more precisely 4.9 m, 19.6 m, and 44.1 m.)

PRACTICING PHYSICS

Hands-On Dangling Beads

Make your own model of projectile paths. Divide a ruler or a stick into five equal spaces. At position 1, hang a bead from a string that is 1 cm long, as shown. At position 2, hang a bead from a string that is 4 cm long. At position 3, do the same with a 9-cm length of string. At position 4, use 16 cm of string, and for position 5, use 25 cm of string. If you hold the stick horizontally, you will have a version of Figure 10.6. Hold it at a slight upward angle to show a version of Figure 10.7, left. Hold it at a downward angle to show a version of Figure 10.7, right.

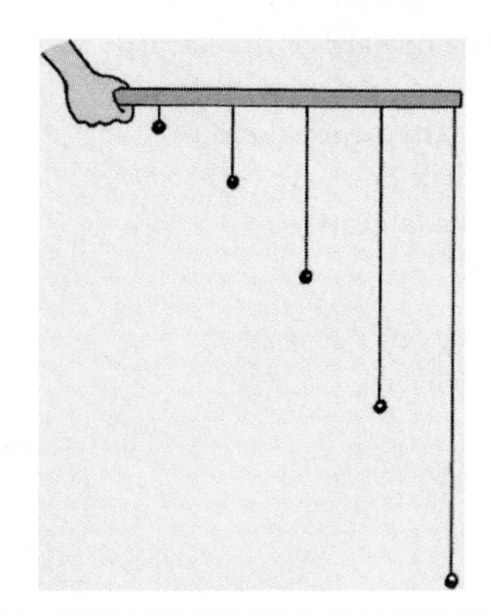

fall if it were dropped from rest and had been falling for the same amount of time. This distance, as introduced in Chapter 3, is given by $d = \frac{1}{2}gt^2$, where t is the elapsed time.

We can put it another way: Shoot a projectile skyward at some angle and pretend there is no gravity. After so many seconds t, it should be at a certain point along a straight-line path. But, because of gravity, it isn't. Where is it? The answer is that it's directly below this point. How far below? The answer in meters is $5t^2$ (or, more precisely, $4.9t^2$). How about that!

Note another thing from Figure 10.8 and previous figures. The ball moves equal horizontal distances in equal time intervals. That's because no acceleration takes place horizontally. The only acceleration is vertical, in the direction of Earth's gravity.

CHECK YOURSELF

1. Suppose the cannonball in Figure 10.8 were fired faster. How many meters below the dashed line would it be at the end of the 5 s?
2. If the horizontal component of the cannonball's velocity is 20 m/s, how far downrange will the cannonball be in 5 s?

CHECK YOUR ANSWERS

1. The vertical distance beneath the dashed line at the end of 5 s is 125 m [$d = 5t^2 = 5(5)^2 = 5(25) = 125$ m]. Interestingly enough, this distance doesn't depend on the angle of the cannon. If air drag is neglected, any projectile will fall $5t^2$ meters below where it would have reached if there were no gravity.
2. With no air drag, the cannonball will travel a horizontal distance of 100 m [$d = \bar{v}t = (20 \text{ m/s})(5 \text{ s}) = 100$ m]. Note that, since gravity acts only vertically and there is no acceleration in the horizontal direction, the cannonball travels equal horizontal distances in equal times. This distance is simply its horizontal component of velocity multiplied by the time (and not $5t^2$, which applies only to vertical motion under the acceleration of gravity).

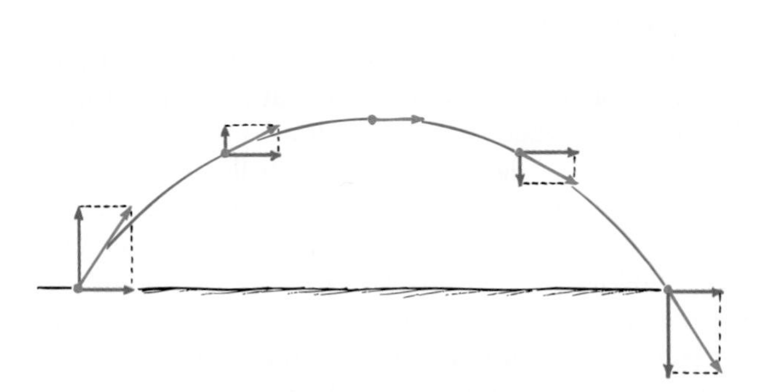

FIGURE 10.9 Interactive Figure

The velocity of a projectile at various points along its trajectory. Note that the vertical component changes and that the horizontal component is the same everywhere.

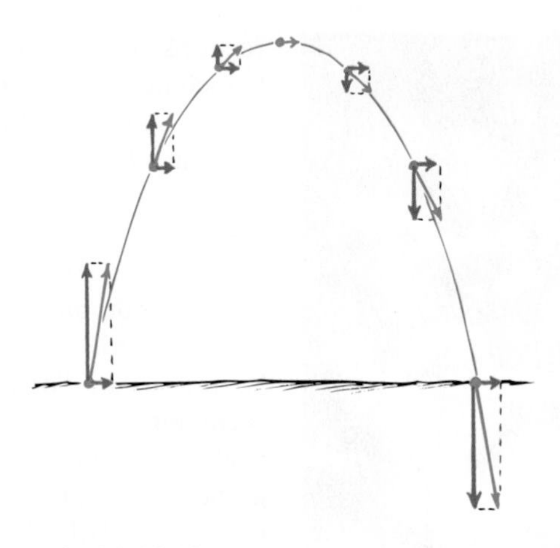

FIGURE 10.10

Trajectory for a steeper projection angle.

In Figure 10.9, we see vectors representing both horizontal and vertical components of velocity for a projectile following a parabolic trajectory. Notice that the horizontal component everywhere along the trajectory is the same, and only the vertical component changes. Note also that the actual velocity is represented by the vector that forms the diagonal of the rectangle formed by the vector components. At the top of the trajectory, the vertical component is zero, so the velocity at the zenith is only the horizontal component of velocity. Everywhere else along the trajectory, the magnitude of velocity is greater (just as the diagonal of a rectangle is greater than either of its sides).

Figure 10.10 shows the trajectory traced by a projectile launched with the same speed at a steeper angle. Notice the initial velocity vector has a greater vertical component than when the launch angle is smaller. This greater component results in a trajectory that reaches a greater height. But the horizontal component is less, and the range is less.

Figure 10.11 shows the paths of several projectiles, all with the same initial speed but different launching angles. The figure neglects the effects of air drag, so the trajectories are all parabolas. Notice that these projectiles reach different *altitudes,* or heights above the ground. They also have different *horizontal ranges,* or distances traveled horizontally. The remarkable thing to note from Figure 10.11 is that the same range is obtained from two different launching angles when the angles add up to 90°! An object thrown into the air at an angle of 60°, for example, will have the same range as if it were thrown at the same speed at an angle of 30°. For the smaller angle, of course, the object

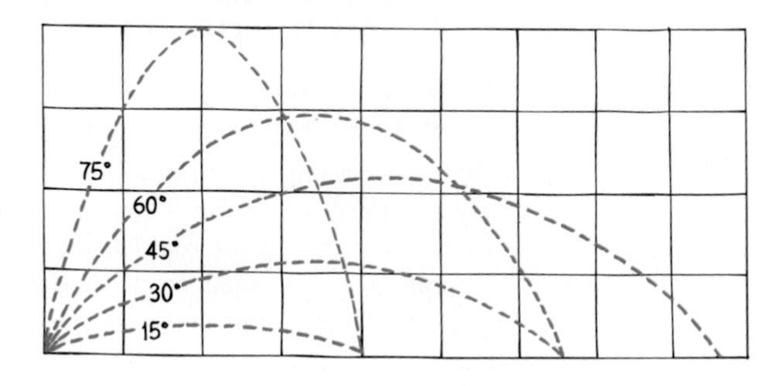

FIGURE 10.11

Interactive Figure

Ranges of a projectile shot at the same speed at different projection angles.

FIGURE 10.12
Maximum range is attained when a ball is batted at an angle of nearly 45°.

remains in the air for a shorter time. The greatest range occurs when the launching angle is 45°—and when air drag is negligible.

Without the effects of air, the maximum range for a baseball would occur when it is batted 45° above the horizontal. Because of air drag and lift due to spinning of the ball (Chapter 14), the best range occurs at batting angles noticeably less than 45°. Air drag and spin are more significant for golf balls, where angles less than 38° or so result in maximum range. For heavy projectiles like javelins and the shot, air has less effect on range. A javelin, being heavy and presenting a very small cross section to the air, follows an almost perfect parabola when thrown. So does a shot. For such projectiles, maximum range for equal launch speeds would occur for a launch angle of about 45° (slightly less because the launching height is above ground level). Aha, but launching speeds are *not* equal for such a projectile thrown at different angles. In throwing a javelin or putting a shot, a significant part of the launching *force* goes into combating gravity—the steeper the angle, the less speed it has when leaving the thrower's hand. So gravity plays a role before and after launching. You can test this yourself: Throw a heavy boulder horizontally, then vertically—you'll find the horizontal throw to be considerably faster than the vertical throw. So maximum range for heavy projectiles thrown by humans is attained for angles of less than 45°—and not because of air drag.

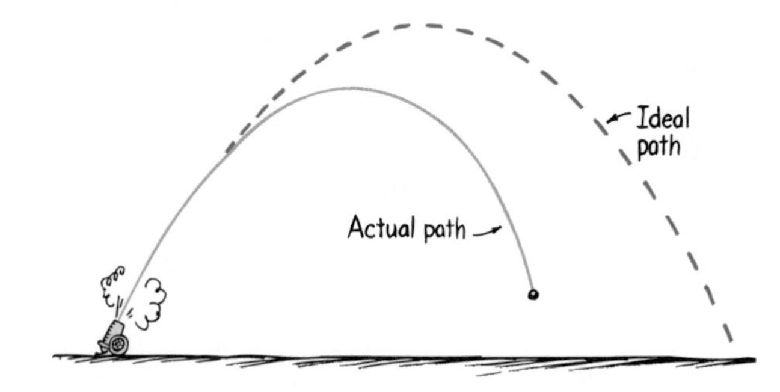

FIGURE 10.13
Interactive Figure
In the presence of air drag, the trajectory of a high-speed projectile falls short of the idealized parabolic path.

CHECK YOURSELF

1. A baseball is batted at an angle into the air. Once airborne, and neglecting air drag, what is the ball's acceleration vertically? Horizontally?
2. At what part of its trajectory does the baseball have minimum speed?
3. Consider a batted baseball following a parabolic path on a day when the Sun is directly overhead. How does the speed of the ball's shadow across the field compare with the ball's horizontal component of velocity?

CHECK YOUR ANSWERS

1. Vertical acceleration is g because the force of gravity is vertical. Horizontal acceleration is zero because no horizontal force acts on the ball.
2. A ball's minimum speed occurs at the top of its trajectory. If it is launched vertically, its speed at the top is zero. If launched at an angle, the vertical component of velocity is zero at the top, leaving only the horizontal component. So the speed at the top is equal to the horizontal component of the ball's velocity at any point. Doesn't this make sense?
3. They are the same!

HANG TIME REVISITED

In Chapter 3, we stated that airborne time during a jump is independent of horizontal speed. Now we see why this is so—horizontal and vertical components of motion are independent of each other. The rules of projectile motion apply to jumping. Once one's feet are off the ground, only the force of gravity acts on the jumper (neglecting air drag). Hang time depends only on the vertical component of lift-off velocity. It so happens, however, that the action of running can make a difference. When running, the lift-off force during jumping can be somewhat increased by pounding of the feet against the ground (and the ground pounding against the feet in action–reaction fashion), so hang time for a running jump can often exceed hang time for a standing jump. But, once the runner's feet are off the ground, only the vertical component of lift-off velocity determines hang time.

When air drag is small enough to be negligible, a projectile will rise to its maximum height in the same time it takes to fall from that height to its initial level (Figure 10.14). This is because its deceleration by gravity while going up is the same as its acceleration by gravity while coming down. The speed it loses while going up is therefore the same as the speed gained while coming down. So the projectile arrives at its initial level with the same speed it had when it was initially projected.

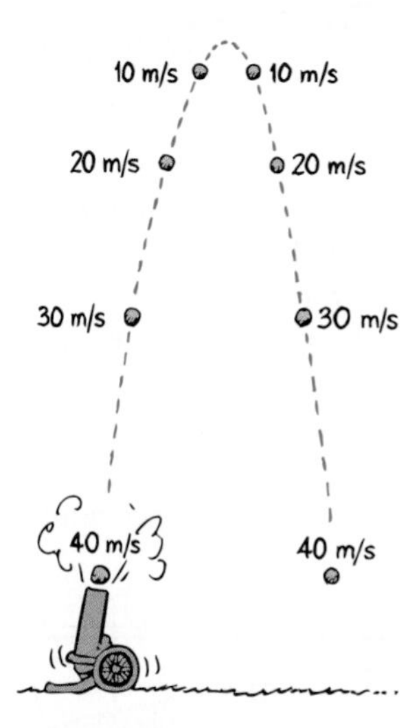

FIGURE 10.14
Without air drag, speed lost while going up equals speed gained while coming down: Time going up equals time coming down.

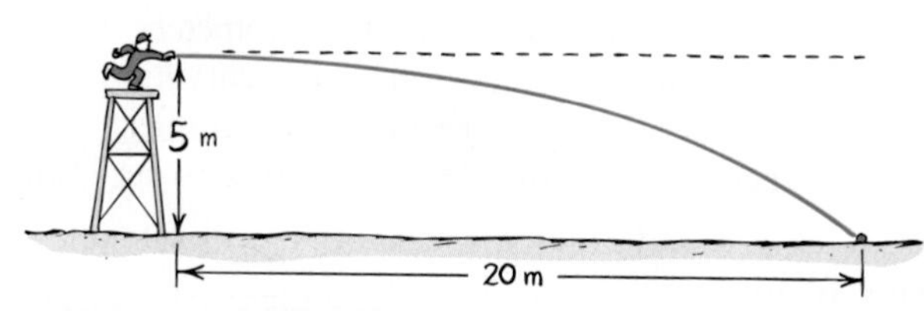

FIGURE 10.15
How fast is the ball thrown?

CHECK YOURSELF

The boy on the tower throws a ball 20 m downrange, as shown in Figure 10.15. What is his pitching speed?

Baseball games normally take place on level ground. For the short-range projectile motion on the playing field, Earth can be considered to be flat because the flight of the baseball is not affected by the Earth's curvature. For very long-range projectiles, however, the curvature of the Earth's surface must be taken into account. We'll now see that, if an object is projected fast enough, it will fall all the way around the Earth and become an Earth satellite.

CHECK YOUR ANSWER

The ball is thrown horizontally, so the pitching speed is horizontal distance divided by time. A horizontal distance of 20 m is given, but the time is not stated. However, knowing the vertical drop is 5 m, you remember that a 5-m drop takes 1 s! From the equation for constant speed (which applies to horizontal motion), $v = d/t = (20\text{ m})/(1\text{ s}) = 20\text{ m/s}$. It is interesting to note that the equation for constant speed, $v = d/t$, guides our thinking about the crucial factor in this problem—the *time*.

Fast-Moving Projectiles—Satellites

Consider the baseball pitcher on the tower in Figure 10.15. If gravity did not act on the ball, the ball would follow a straight-line path shown by the dashed line. But gravity does act, so the ball falls below this straight-line path. In fact, as just discussed, 1 second after the ball leaves the pitcher's hand it will have fallen a vertical distance of 5 meters below the dashed line—whatever the pitching speed. It is important to understand this, for it is the crux of satellite motion.

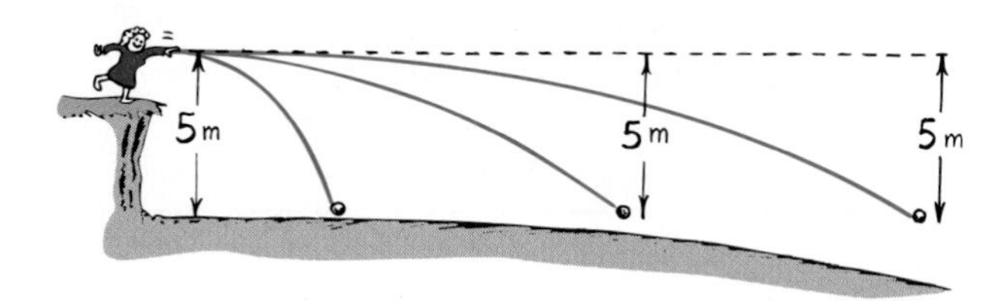

FIGURE 10.16
If you throw a stone at any speed, one second later it will have fallen 5 m below where it would have been without gravity.

An Earth **satellite** is simply a projectile that falls *around* Earth rather than *into* it. The speed of the satellite must be great enough to ensure that its falling distance matches the Earth's curvature. A geometrical fact about the curvature of the Earth is that its surface drops a vertical distance of 5 meters for every 8000 meters tangent to the surface (Figure 10.17). If a baseball could be thrown fast enough to travel a horizontal distance of 8 kilometers during the one second it takes to fall 5 meters, then it would follow the curvature of the Earth. This is a speed of 8 kilometers per second. If this doesn't seem fast, convert it to kilometers per hour and you get an impressive 29,000 kilometers per hour (or 18,000 miles per hour)!

Earth's curvature, dropping 5 m for each 8-km tangent, means that, if you were floating in a calm ocean, you'd be able to see only the top of a 5-m mast on a ship 8 km away.

Insights

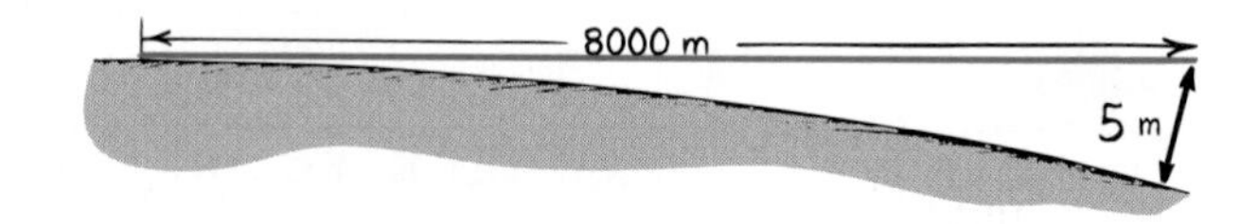

FIGURE 10.17
Earth's curvature—not to scale!

At this speed, atmospheric friction would burn the baseball—or even a piece of iron—to a crisp. This is the fate of bits of rock and other meteorites that enter the Earth's atmosphere and burn up, appearing as "falling stars." That is why satellites, such as the space shuttles, are launched to altitudes of 150 kilometers or more—to be above almost all of the atmosphere and to be nearly free of air drag. A common misconception is that satellites orbiting at high altitudes are free from gravity. Nothing could be further from the truth. The force of gravity on a satellite 200 kilometers above the Earth's surface is nearly as strong as it is at the surface. The high altitude is to position the satellite beyond the Earth's atmosphere, where air drag is almost totally absent, but not beyond Earth's gravity.

Satellite motion was understood by Isaac Newton, who reasoned that the Moon was simply a projectile circling Earth under the attraction of gravity. This concept is illustrated in a drawing by Newton (Figure 10.19). He compared the motion of the Moon to a cannonball fired from the top of a high mountain. He imagined that the mountaintop was above Earth's atmosphere, so that air drag would not impede the motion of the cannonball. If fired with a low horizontal speed, a cannonball would follow a curved path and soon hit the Earth

FIGURE 10.18
If the speed of the stone and the curvature of its trajectory are great enough, the stone may become a satellite.

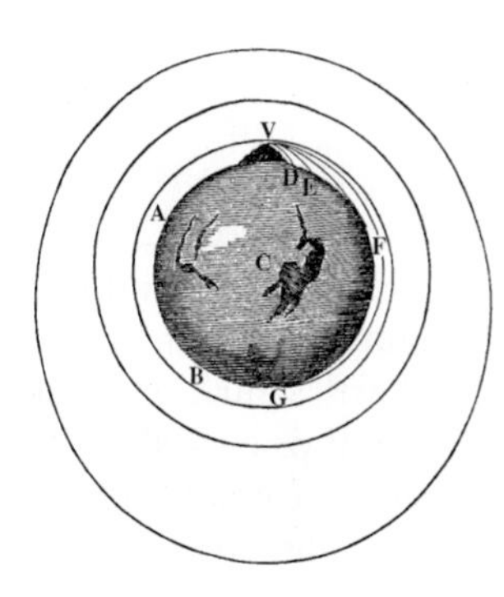

FIGURE 10.19
"The greater the velocity . . . with which (a stone) is projected, the farther it goes before it falls to the Earth. We may therefore suppose the velocity to be so increased, that it would describe an arc of 1, 2, 5, 10, 100, 1000 miles before it arrived at the Earth, till at last, exceeding the limits of the Earth, it should pass into space without touching." —Isaac Newton, *System of the World*.

below. If it were fired faster, its path would be less curved and it would hit the Earth farther away. If the cannonball were fired fast enough, Newton reasoned, the curved path would become a circle and the cannonball would circle the Earth indefinitely. It would be in orbit.

FIGURE 10.20
A space shuttle is a projectile in a constant state of free fall. Because of its tangential velocity, it falls around the Earth rather than vertically into it.

Both cannonball and Moon have tangential velocity (parallel to the Earth's surface) sufficient to ensure motion *around* the Earth rather than *into* it. If there is no resistance to reduce its speed, the Moon or any Earth satellite "falls" around and around the Earth indefinitely. Similarly, the planets continuously fall around the Sun in closed paths. Why don't the planets crash into the Sun? They don't because of their tangential velocities. What would happen if their tangential velocities were reduced to zero? The answer is simple enough: Their falls would be straight toward the Sun, and they would indeed crash into it. Any objects in the solar system without sufficient tangential velocities have long ago crashed into the Sun. What remains is the harmony we observe.

CHECK YOURSELF

One of the beauties of physics is that there are usually different ways to view and explain a given phenomenon. Is the following explanation valid? Satellites remain in orbit instead of falling to Earth because they are beyond the main pull of Earth's gravity.

CHECK YOUR ANSWER

No, no, a thousand times no! If any moving object were beyond the pull of gravity, it would move in a straight line and would not curve around the Earth. Satellites remain in orbit because they *are* being pulled by gravity, not because they are beyond it. For the altitudes of most Earth satellites, Earth's gravitational field is only a few percent weaker than it is at the Earth's surface.

Circular Satellite Orbits

FIGURE 10.21
Interactive Figure
Fired fast enough, the cannonball will go into orbit.

An 8-kilometers-per-second cannonball fired horizontally from Newton's mountain would follow the Earth's curvature and glide in a circular path around the Earth again and again (provided the cannoneer and the cannon got out of the way). Fired at a slower speed, the cannonball would strike Earth's surface; fired at a faster speed, it would overshoot a circular orbit, as we will discuss shortly. Newton calculated the speed for circular orbit, and because such a cannon-muzzle velocity was clearly impossible, he did not foresee the possibility of humans launching satellites (and likely didn't consider multistage rockets).

Note that, in circular orbit, the speed of a satellite is not changed by gravity: Only the direction changes. We can understand this by comparing a satellite in circular orbit with a bowling ball rolling along a bowling lane. Why doesn't the gravity that acts on the bowling ball change its speed? The answer is that gravity pulls straight downward with no component of force acting forward or backward.

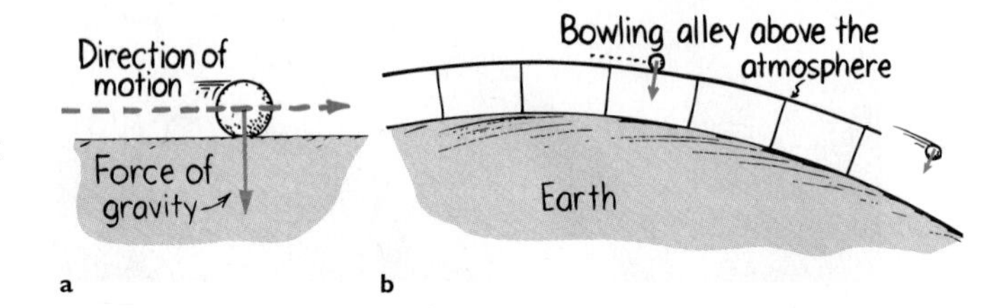

FIGURE 10.22
(a) The force of gravity on the bowling ball is at 90° to its direction of motion, so it has no component of force to pull it forward or backward, and the ball rolls at constant speed. (b) The same is true even if the bowling alley is larger and remains "level" with the curvature of the Earth.

Consider a bowling lane that completely surrounds the Earth, elevated high enough to be above the atmosphere and air drag. The bowling ball will roll at constant speed along the lane. If a part of the lane were cut away, the ball would roll off its edge and would hit the ground below. A faster ball encountering the gap would hit the ground farther along the gap. Is there a speed at which the ball will clear the gap (like a motorcyclist who drives off a ramp and clears a gap to meet a ramp on the other side)? The answer is yes: 8 kilometers per second will be enough to clear that gap—and any gap—even a 360° gap. The ball would be in circular orbit.

Circular Orbits

Note that a satellite in circular orbit is always moving in a direction perpendicular to the force of gravity that acts upon it. The satellite does not move in the direction of the force, which would increase its speed, nor does it move in a direction against the force, which would decrease its speed. Instead, the satellite moves at right angles to the gravitational force that acts upon it. No change in speed occurs—only a change in direction. So we see why a satellite in circular orbit sails parallel to the surface of the Earth at constant speed—a very special form of free fall.

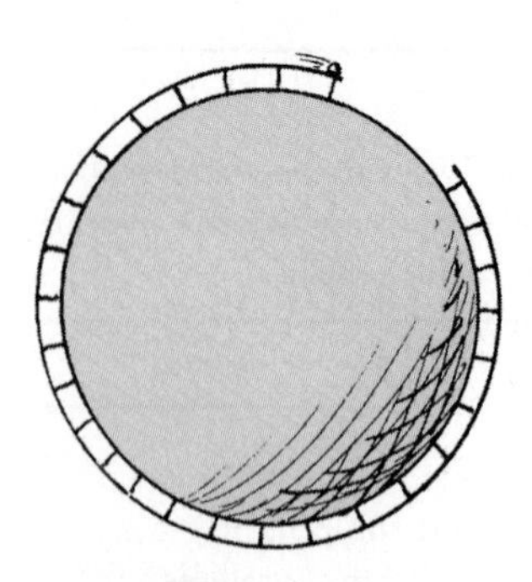

FIGURE 10.23
What speed will allow the ball to clear the gap?

For a satellite close to the Earth, the period (the time for a complete orbit about the Earth) is about 90 minutes. For higher altitudes, the orbital speed is less, distance is more, and the period is longer. For example, communication satellites located in orbit 5.5 Earth radii above the surface of the Earth have a period of 24 hours. This period matches the period of daily Earth rotation. For an orbit around the equator, these satellites remain always above the same point on the ground. The Moon is even farther away and has a period of 27.3 days.

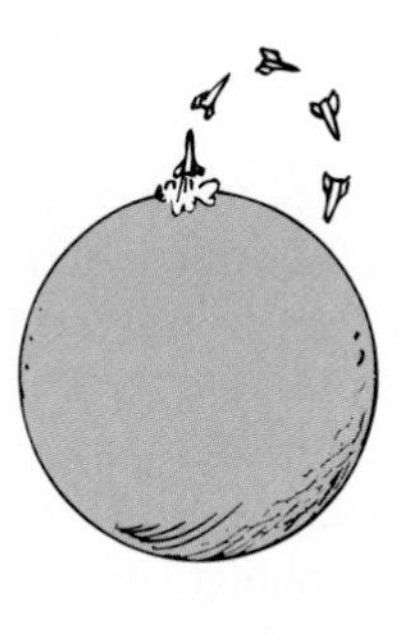

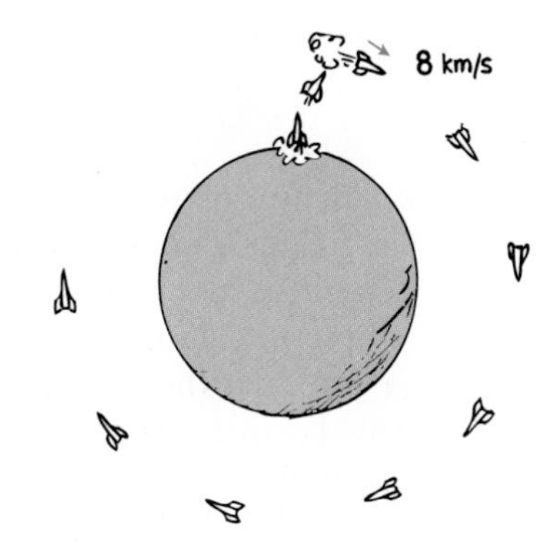

FIGURE 10.24
The initial thrust of the rocket pushes it up above the atmosphere. Another thrust to a tangential speed of at least 8 km/s is required if the rocket is to fall around rather than into the Earth.

The higher the orbit of a satellite, the less its speed, the longer its path, and the longer its period.[1]

Putting a payload into Earth orbit requires control over the speed and direction of the rocket that carries it above the atmosphere. A rocket initially fired vertically is intentionally tipped from the vertical course. Then, once above the drag of the atmosphere, it is aimed horizontally, whereupon the payload is given a final thrust to orbital speed. We see this in Figure 10.24, where, for the sake of simplicity, the payload is the entire single-stage rocket. With the proper tangential velocity, it falls around the Earth, rather than into it, and becomes an Earth satellite.

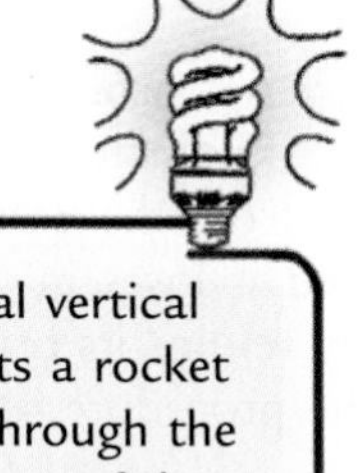

The initial vertical climb gets a rocket quickly through the denser part of the atmosphere. Eventually, the rocket must acquire enough tangential speed to remain in orbit without thrust, so it must tilt until its path is parallel to the Earth's surface.

Insights

CHECK YOURSELF

1. True or false: The space shuttle orbits at altitudes in excess of 150 kilometers to be above both gravity and the Earth's atmosphere.
2. Satellites in close circular orbit fall about 5 meters during each second of orbit. Why doesn't this distance accumulate and send satellites crashing into Earth's surface?

CHECK YOUR ANSWERS

1. False. What satellites are above is the atmosphere and air drag—*not* gravity! It's important to note that Earth's gravity extends throughout the universe in accord with the inverse-square law.
2. In each second, the satellite falls about 5 m below the straight-line tangent it would have followed if there were no gravity. Earth's surface also curves 5 m beneath a straight-line 8-km tangent. The process of falling with the curvature of the Earth continues from tangent line to tangent line, so the curved path of the satellite and the curve of the Earth's surface "match" all the way around the Earth. Satellites do, in fact, crash to the Earth's surface from time to time when they encounter air drag in the upper atmosphere that decreases their orbital speed.

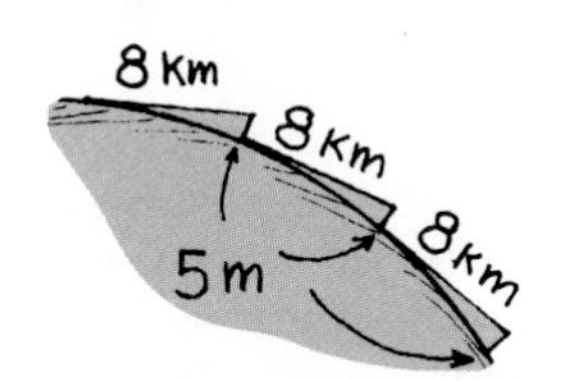

[1]The speed of a satellite in circular orbit is given by $v = \sqrt{GM/d}$ and the period of satellite motion is given by $T = 2\pi\sqrt{d^3/GM}$, where G is the universal gravitational constant (see previous Chapter 9), M is the mass of the Earth (or whatever body the satellite orbits), and d is the distance of the satellite from the center of the Earth or other parent body.

Elliptical Orbits

If a projectile just above the drag of the atmosphere is given a horizontal speed somewhat greater than 8 kilometers per second, it will overshoot a circular path and trace an oval path called an **ellipse.**

An ellipse is a specific curve: the closed path taken by a point that moves in such a way that the sum of its distances from two fixed points (called *foci*) is constant. For a satellite orbiting a planet, one focus is at the center of the planet; the other focus could be internal or external to the planet. An ellipse can be easily constructed by using a pair of tacks (one at each focus), a loop of string, and a pencil (Figure 10.25). The closer the foci are to each other, the closer the ellipse is to a circle. When both foci are together, the ellipse *is* a circle. So we can see that a circle is a special case of an ellipse.

Whereas the speed of a satellite is constant in a circular orbit, speed varies in an elliptical orbit. For an initial speed greater than 8 kilometers per second, the satellite overshoots a circular path and moves away from the Earth, against the force of gravity. It therefore loses speed. The speed it loses in receding is regained as it falls back toward the Earth, and it finally rejoins its original path with the same speed it had initially (Figure 10.27). The procedure repeats over and over, and an ellipse is traced each cycle.

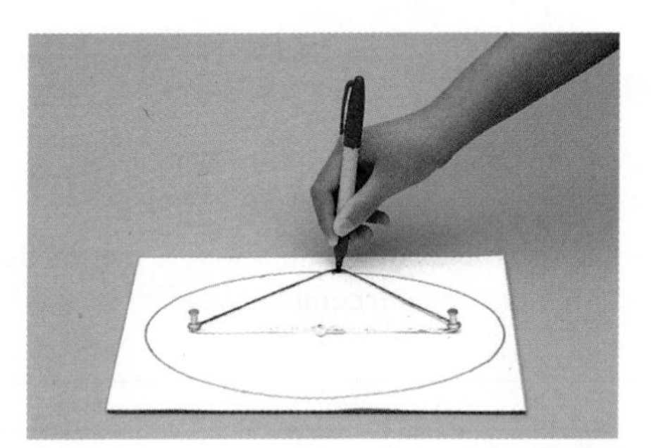

FIGURE 10.25 Interactive Figure
A simple method for constructing an ellipse.

Interestingly enough, the parabolic path of a projectile, such as a tossed baseball or a cannonball, is actually a tiny segment of a skinny ellipse that extends within and just beyond the center of the Earth (Figure 10.28a). In Figure 10.28b, we see several paths of cannonballs fired from Newton's mountain. All these ellipses have the center of the Earth as one focus. As muzzle

FIGURE 10.26
The shadows cast by the ball are all ellipses, one for each lamp in the room. The point at which the ball makes contact with the table is the common focus of all three ellipses.

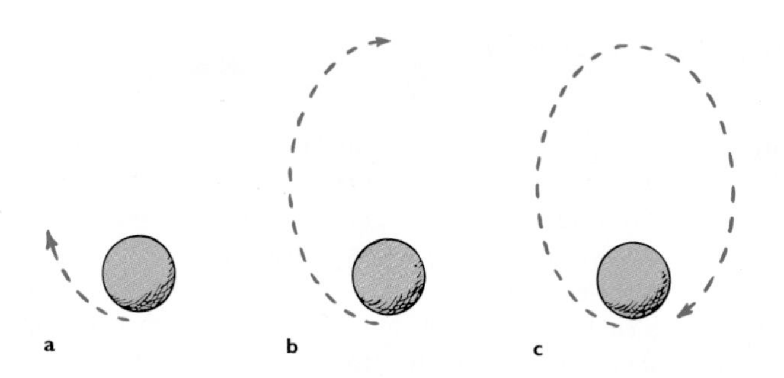

FIGURE 10.27
Elliptical orbit. An Earth satellite that has a speed somewhat greater than 8 km/s overshoots a circular orbit (a) and travels away from the Earth. Gravitation slows it to a point where it no longer moves farther from the Earth (b). It falls toward the Earth, gaining the speed it lost in receding (c), and follows the same path as before in a repetitious cycle.

velocity is increased, the ellipses are less eccentric (more nearly circular); and, when muzzle velocity reaches 8 kilometers per second, the ellipse rounds into a circle and does not intercept the Earth's surface. The cannonball coasts in circular orbit. At greater muzzle velocities, orbiting cannonballs trace the familiar external ellipses.

CHECK YOURSELF

The orbital path of a satellite is shown in the sketch. In which of the marked positions A through D does the satellite have the greatest speed? The lowest speed?

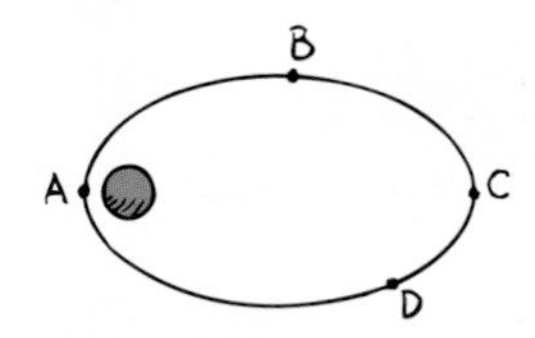

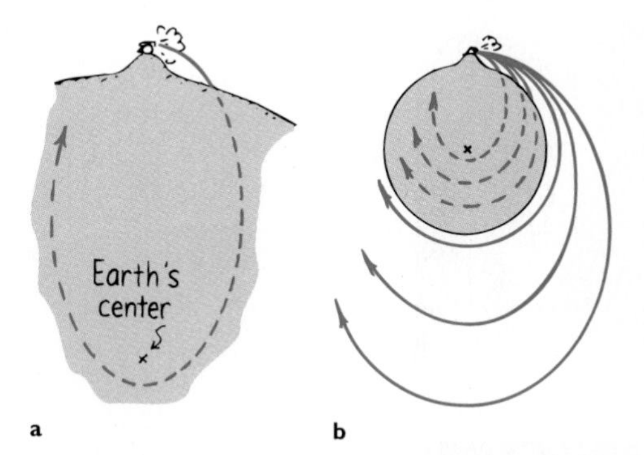

FIGURE 10.28
(a) The parabolic path of the cannonball is part of an ellipse that extends within the Earth. Earth's center is the far focus. (b) All paths of the cannonball are ellipses. For less than orbital speeds, the center of the Earth is the far focus; for a circular orbit, both foci are Earth's center; for greater speeds, the near focus is Earth's center.

CHECK YOUR ANSWER

The satellite has its greatest speed as it whips around A and has its lowest speed at position C. After passing C, it gains speed as it falls back to A to repeat its cycle.

WORLD MONITORING BY SATELLITE

Satellites are useful in monitoring our planet. Figure A shows the path traced in one period by a satellite in circular orbit launched in a northeasterly direction from Cape Canaveral, Florida. The path is curved only because the map is flat. Note that the path crosses the equator twice in one period, for the path describes a circle whose plane passes through the Earth's center. Note also that the path does not terminate where it begins. This is because Earth rotates beneath the satellite while it orbits. During the 90-minute period, Earth turns 22.6°, so, when the satellite makes a complete orbit, it begins its new sweep many kilometers to the west (about 2500 km at the equator). This is quite advantageous for Earth-monitoring satellites. Figure B shows the area monitored over 10 days by successive sweeps for a typical satellite. A dramatic but typical example of such monitoring is the three-year worldwide watch of the distribution of ocean phytoplankton (Figure C). Such extensive information would have been impossible to acquire before the advent of satellites.

FIGURE A
The path of a typical satellite launched in a northeasterly direction from Cape Canaveral. Because the Earth rotates while the satellite orbits, each sweep passes overhead some 2100 km farther west at the latitude of Cape Canaveral.

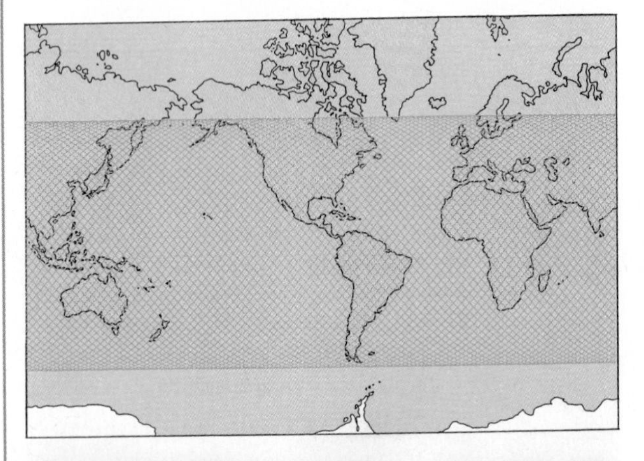

FIGURE B
Typical sweep pattern for a satellite over the period of a week.

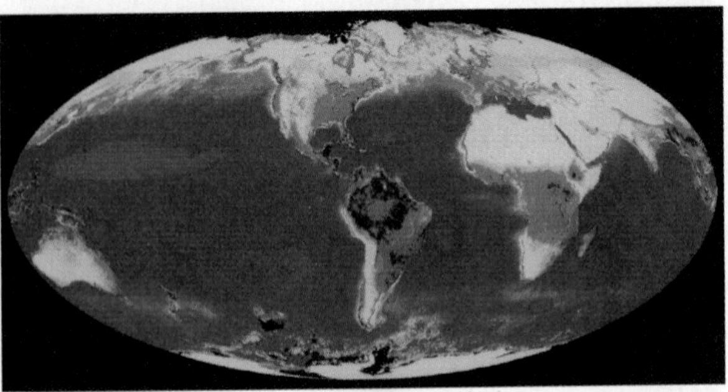

FIGURE C
Phytoplankton production in the Earth's oceans over a 3-year period. Magenta and yellow show the highest concentrations, while blue shows moderately high concentrations.

Kepler's Laws of Planetary Motion

Tycho Brahe (1546–1601)

Johannes Kepler (1571–1630)

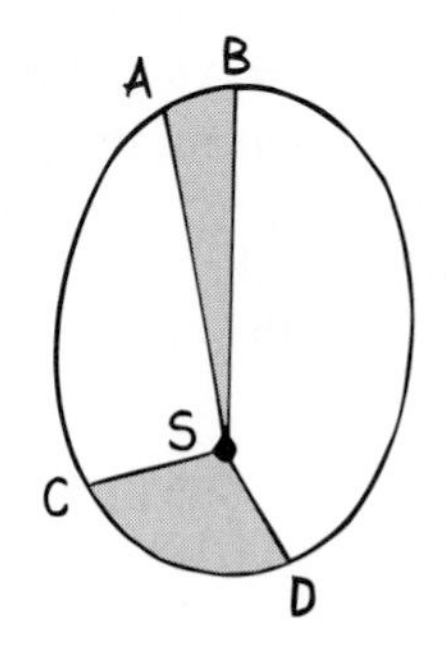

FIGURE 10.29
Equal areas are swept out in equal intervals of time.

Newton's law of gravitation was preceded by three important discoveries about planetary motion by the German astronomer Johannes Kepler, who started as a junior assistant to the famed Danish astronomer Tycho Brahe. Brahe directed the world's first great observatory, in Denmark, just before the advent of the telescope. Using huge, brass, protractor-like instruments called *quadrants,* Brahe measured the positions of planets over twenty years so accurately that his measurements are still valid today. Brahe entrusted his data to Kepler. After Brahe's death, Kepler converted Brahe's measurements to values that would be obtained by a stationary observer outside the solar system. Kepler's expectation that the planets would move in perfect circles around the Sun was shattered after years of effort. He discovered the paths to be ellipses. This is Kepler's first law of planetary motion:

The path of each planet around the Sun is an ellipse with the Sun at one focus.

Kepler also found that the planets do not revolve around the Sun at a uniform speed but move faster when they are nearer the Sun and slower when they are farther from the Sun. They do this in such a way that an imaginary line or spoke joining the Sun and the planet sweeps out equal areas of space in equal times. The triangular-shaped area swept out during a month when a planet is orbiting far from the Sun (triangle ASB in Figure 10.29) is equal to the triangular area swept out during a month when the planet is orbiting closer to the Sun (triangle CSD in Figure 10.29). This is Kepler's second law:

The line from the Sun to any planet sweeps out equal areas of space in equal time intervals.

Kepler was the first to coin the word *satellite*. He had no clear idea as to *why* the planets moved as he discovered. He lacked a conceptual model. Kepler didn't see that a satellite is simply a projectile under the influence of a gravitational force directed toward the body that the satellite orbits. You know that, if you toss a rock upward, it goes slower the higher it rises because it's moving against gravity. And you know that, when it returns, it's moving with gravity and its speed increases. Kepler didn't see that a satellite behaves in the same way. Going away from the Sun, it slows. Going toward the Sun, it speeds up. A satellite, whether a planet orbiting the Sun or one of today's satellites orbiting the Earth, moves slower against the gravitational field and faster with the field. Kepler didn't see this simplicity and instead fabricated complex systems of geometrical figures to find sense in his discoveries. These systems proved to be futile.

After ten years of searching by trial and error for a connection between the time it takes a planet to orbit the Sun and its distance from the Sun, Kepler discovered a third law. From Brahe's data, Kepler found that the square of any planet's period (T) is directly proportional to the cube of its average orbital radius (r). Law three is:

The square of the orbital period of a planet is directly proportional to the cube of the average distance of the planet from the Sun ($T^2 \sim r^3$ for all planets).

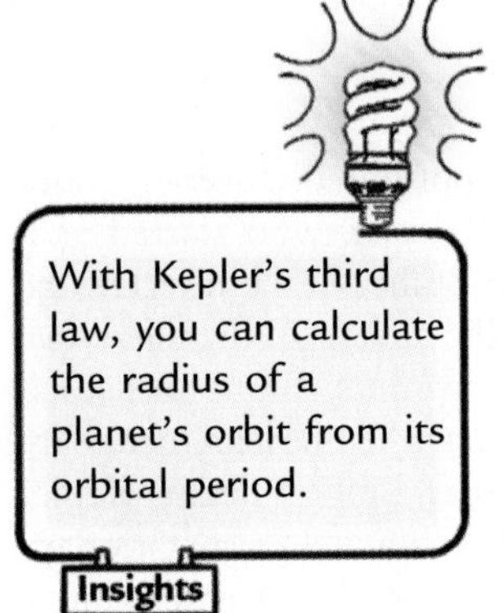

With Kepler's third law, you can calculate the radius of a planet's orbit from its orbital period.
Insights

This means that the ratio T^2/r^3 is the same for all planets. So, if a planet's period is known, its average orbital radial distance is easily calculated (or vice versa).

It is interesting to note that Kepler was familiar with Galileo's ideas about inertia and accelerated motion, but he failed to apply them to his own work. Like Aristotle, he thought that the force on a moving body would be in the same direction as the body's motion. Kepler never appreciated the concept of inertia. Galileo, on the other hand, never appreciated Kepler's work and held to his conviction that the planets move in circles.[2] Further understanding of planetary motion required someone who could integrate the findings of these two great scientists.[3] The rest is history, for this task fell to Isaac Newton.

Energy Conservation and Satellite Motion

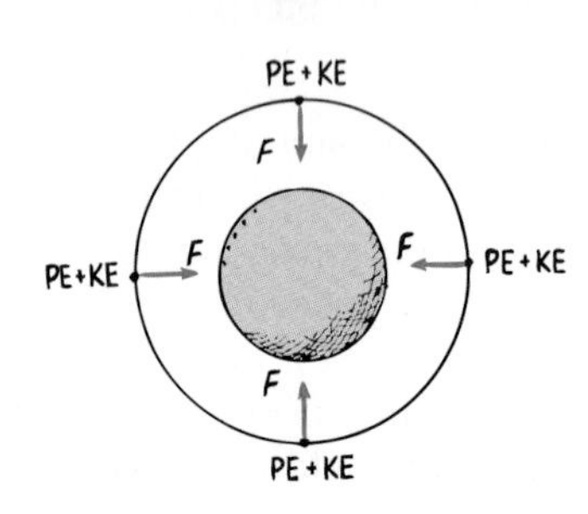

FIGURE 10.30
The force of gravity on the satellite is always toward the center of the body it orbits. For a satellite in circular orbit, no component of force acts along the direction of motion. The speed and thus the KE do not change.

Recall, from Chapter 7, that an object in motion possesses kinetic energy (KE) due to its motion. An object above Earth's surface possesses potential energy (PE) by virtue of its position. Everywhere in its orbit, a satellite has both KE and PE. The sum of the KE and PE is a constant all through the orbit. The simplest case occurs for a satellite in circular orbit.

In a circular orbit, the distance between the satellite and the center of the attracting body does not change, which means that the PE of the satellite is the same everywhere in its orbit. Then, by the conservation of energy, the KE must also be constant. So a satellite in circular orbit coasts at an unchanging PE, KE, and speed (Figure 10.30).

In an elliptical orbit, the situation is different. Both speed and distance vary. PE is greatest when the satellite is farthest away (at the *apogee*) and least when the satellite is closest (at the *perigee*). Note that the KE will be least when the PE is most, and the KE will be most when the PE is least. At every point in the orbit, the sum of KE and PE is the same (Figure 10.31).

At all points along the elliptical orbit, except at the apogee and the perigee, there is a component of gravitational force parallel to the direction of motion of the satellite. This component of force changes the speed of the satellite. Or we can say that (this component of force) × (distance moved) = ΔKE. Either way, when the satellite gains altitude and moves against this component, its speed and KE decrease. The decrease continues to the apogee. Once past the apogee, the satellite moves in the same direction as the component, and the speed and KE increase. The increase continues until the satellite whips past the perigee and repeats the cycle.

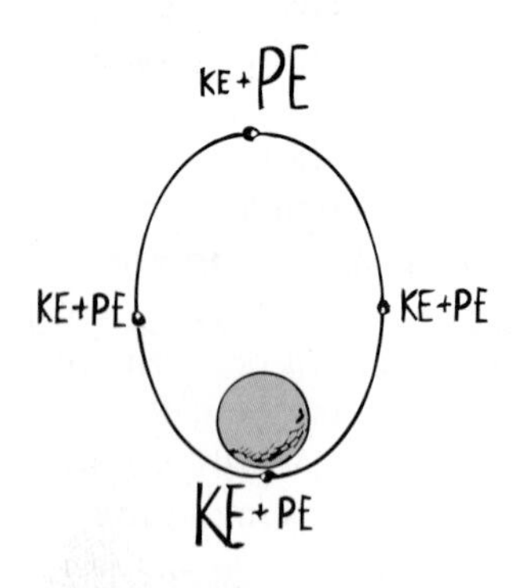

FIGURE 10.31
The sum of KE and PE for a satellite is a constant at all points along its orbit.

[2]It is not easy to look at familiar things through the new insights of others. We tend to see only what we have learned to see or wish to see. Galileo reported that many of his colleagues were unable or refused to see the moons of Jupiter when they peered skeptically through his telescopes. Galileo's telescopes were a boon to astronomy, but more important than a new instrument with which to see things better was a new way of understanding what is seen. Is this still true today?

[3]Perhaps your instructor will show that Kepler's third law results when Newton's inverse-square formula for gravitational force is equated to centripetal force, and how T^2/r^3 equals a constant that depends only on G and M, the mass of the body about which orbiting occurs. Intriguing stuff!

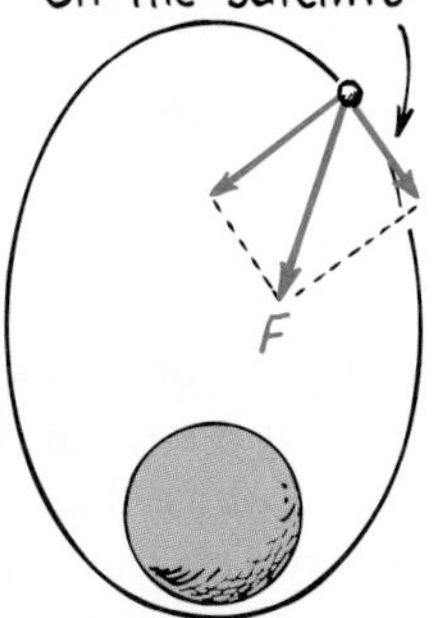

FIGURE 10.32
In elliptical orbit, a component of force exists along the direction of the satellite's motion. This component changes the speed and, thus, the KE. (The perpendicular component changes only the direction.)

CHECK YOURSELF

1. The orbital path of a satellite is shown in the sketch. In which marked positions A through D does the satellite have the greatest KE? The greatest PE? The greatest total energy?
2. Why does the force of gravity change the speed of a satellite when it is in an elliptical orbit but not when it is in a circular orbit?

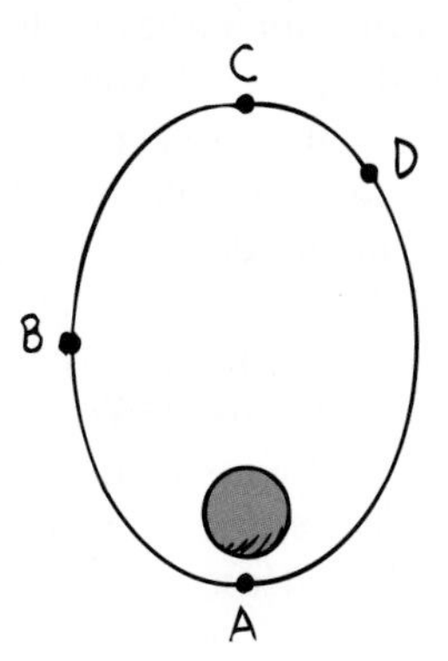

Escape Speed

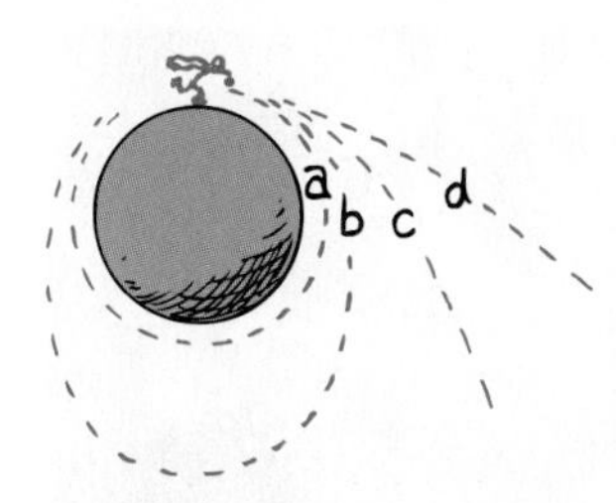

FIGURE 10.33
Interactive Figure
If Superman tosses a ball 8 km/s horizontally from the top of a mountain high enough to be just above air drag (a), then about 90 minutes later he can turn around and catch it (neglecting the Earth's rotation). Tossed slightly faster (b), it will take an elliptical orbit and return in a slightly longer time. Tossed at more than 11.2 km/s (c), it will escape the Earth. Tossed at more than 42.5 km/s (d), it will escape the solar system.

We know that a cannonball fired horizontally at 8 kilometers per second from Newton's mountain would find itself in orbit. But what would happen if the cannonball were instead fired at the same speed *vertically*? It would rise to some maximum height, reverse direction, and then fall back to Earth. Then the old saying "What goes up must come down" would hold true, just as surely as a stone tossed skyward will be returned by gravity (unless, as we shall see, its speed is great enough).

In today's space-faring age, it is more accurate to say, "What goes up *may* come down," for there is a critical starting speed that permits a projectile to outrun gravity and to escape the Earth. This critical speed is called the **escape speed** or, if direction is involved, the *escape velocity*. From the surface of the

CHECK YOUR ANSWERS

1. KE is maximum at the perigee A; PE is maximum at the apogee C; the total energy is the same everywhere in the orbit.
2. In circular orbit, the gravitational force is always perpendicular to the orbital path. With no component of gravitational force along the path, only the direction of motion changes—not the speed. In elliptical orbit, however, the satellite moves in directions that are not perpendicular to the force of gravity. Then components of force do exist along the path, which change the speed of the satellite. A component of force along (parallel to) the direction the satellite moves does work to change its KE.

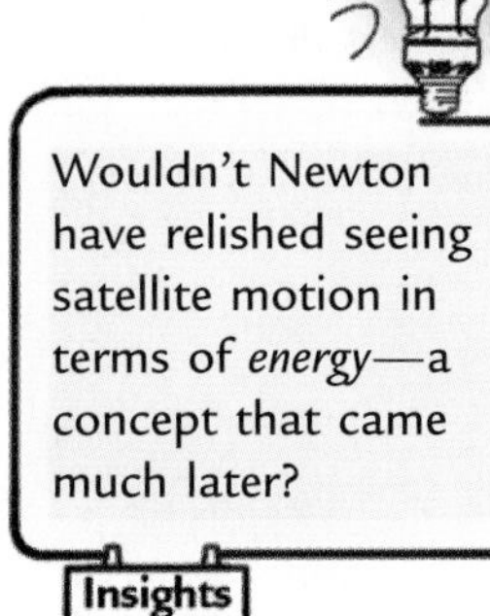

Wouldn't Newton have relished seeing satellite motion in terms of *energy*—a concept that came much later?

Insights

Earth, escape speed is 11.2 kilometers per second. If you launch a projectile at any speed greater than that, it will leave the Earth, traveling slower and slower, never stopping due to the Earth's gravity.[4] We can understand the magnitude of this speed from an energy point of view.

How much work would be required to lift a payload against the force of the Earth's gravity to a distance very, very far ("infinitely far") away? We might think that the change of PE would be infinite because the distance is infinite. But gravity diminishes with distance by the inverse-square law. The force of gravity on the payload would be strong only near the Earth. Most of the work done in launching a rocket occurs within 10,000 km or so of the Earth. It turns out that the change of PE of a 1-kilogram body moved from the surface of the Earth to an infinite distance is 62 million joules (62 MJ). So, to put a payload infinitely far from Earth's surface requires at least 62 million joules of energy per kilogram of load. We won't go through the calculation here, but 62 million joules per kilogram corresponds to a speed of 11.2 kilometers per second, whatever the total mass involved. This is the escape speed from the surface of the Earth.[5]

If we give a payload any more energy than 62 million joules per kilogram at the surface of the Earth or, equivalently, any more speed than 11.2 kilometers per second, then, neglecting air drag, the payload will escape from the Earth, never to return. As it continues outward, its PE increases and its KE decreases. Its speed becomes less and less, though it is never reduced to zero. The payload outruns the gravity of the Earth. It escapes.

The escape speeds from various bodies in the solar system are shown in Table 10.1. Note that the escape speed from the surface of the Sun is

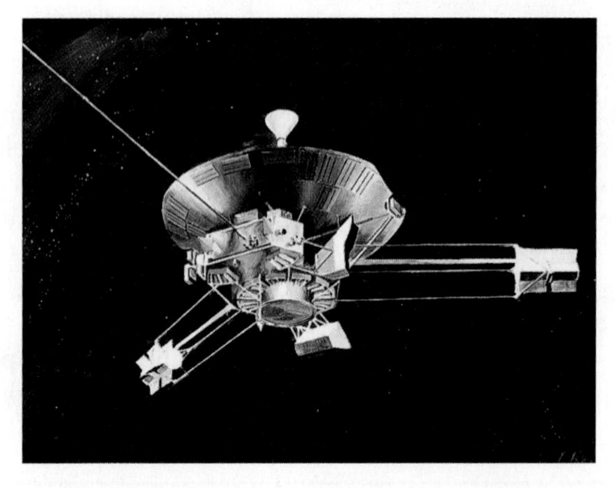

FIGURE 10.34
Pioneer 10, launched from Earth in 1972, passed the outermost planet in 1984 and is now wandering in our galaxy.

FIGURE 10.35
The European–U.S. spacecraft *Cassini* beams close-up images of Saturn and its giant moon Titan to Earth. It also measures surface temperatures, magnetic fields, and the size, speed, and trajectories of tiny surrounding space particles.

[4]Escape speed from any planet or any body is given by $v = \sqrt{2GM/d}$, where G is the universal gravitational constant, M is the mass of the attracting body, and d is the distance from its center. (At the surface of the body, d would simply be the radius of the body.) For a bit more mathematical insight, compare this formula with the one for orbital speed in Footnote 1 a few pages back.

[5]Interestingly enough, this might well be called the *maximum falling speed*. Any object, however far from the Earth, released from rest and allowed to fall to Earth only under the influence of Earth's gravity, would not exceed 11.2 km/s. (With air friction, it would be less.)

If you dropped a candy bar to Earth from a distance as far away as Pluto, wouldn't its speed of impact be about 11.2 km/s?

Insights

TABLE 10.1
Escape Speeds at the Surface of Bodies in the Solar System

Astronomical Body	*Mass (Earth masses)*	*Radius (Earth radii)*	*Escape Speed (km/s)*
Sun	333,000	109	620
Sun (at a distance of the Earth's orbit)		23,500	42.2
Jupiter	318	11	60.2
Saturn	95.2	9.2	36.0
Neptune	17.3	3.47	24.9
Uranus	14.5	3.7	22.3
Earth	1.00	1.00	11.2
Venus	0.82	0.95	10.4
Mars	0.11	0.53	5.0
Mercury	0.055	0.38	4.3
Moon	0.0123	0.27	2.4

Just as planets fall around the Sun, stars fall around the centers of galaxies. Those with insufficient tangential speeds are pulled into, and are gobbled up by, the galactic nucleus—usually a black hole.

Insights

620 kilometers per second. Even at a 150,000,000-kilometer distance from the Sun (Earth's distance), the escape speed to break free of the Sun's influence is 42.5 kilometers per second—considerably more than the escape speed from the Earth. An object projected from Earth at a speed greater than 11.2 kilometers per second but less than 42.5 kilometers per second will escape the Earth but not the Sun. Rather than recede forever, it will take up an orbit around the Sun.

The first probe to escape the solar system, *Pioneer 10,* was launched from Earth in 1972 with a speed of only 15 kilometers per second. The escape was accomplished by directing the probe into the path of oncoming Jupiter. It was whipped about by Jupiter's great gravitational field, picking up speed in the process—similar to the increase in the speed of a baseball encountering an oncoming bat. Its speed of departure from Jupiter was increased enough to exceed the escape speed from the Sun at the distance of Jupiter. *Pioneer 10* passed the orbit of Pluto in 1984. Unless it collides with another body, it will wander indefinitely through interstellar space. Like a note inside a bottle cast into the sea, *Pioneer 10* contains information about the Earth that might be of interest to extraterrestrials, in hopes that it will one day "wash up" and be found on some distant "seashore."

It is important to point out that the escape speed of a body is the initial speed given by a brief thrust, after which there is no force to assist motion. One could escape Earth at *any* sustained speed more than zero, given enough time. For example, suppose a rocket is launched to a destination such as the Moon. If the rocket engines burn out when still close to the Earth, the rocket needs a minimum speed of 11.2 kilometers per second. But if the rocket engines can be sustained for long periods of time, the rocket could reach the Moon without ever attaining 11.2 kilometers per second.

It is interesting to note that the accuracy with which an unmanned rocket reaches its destination is not accomplished by staying on a preplanned path or by getting back on that path if the rocket strays off course. No attempt is made to return the rocket to its original path. Instead, the control center in effect

asks, "Where is it now and what is its velocity? What is the best way to reach its destination, given its present situation?" With the aid of high-speed computers, the answers to these questions are used in finding a new path. Corrective thrusters direct the rocket to this new path. This process is repeated over and over again all the way to the goal.[6]

The mind that encompasses the universe is as marvelous as the universe that encompasses the mind.

Insights

[6]Is there a lesson to be learned here? Suppose you find that you are off course. You may, like the rocket, find it more fruitful to follow a course that leads to your goal as best plotted from your present position and circumstances, rather than try to get back on the course you plotted from a previous position and under, perhaps, different circumstances.

Summary of Terms

Projectile Any object that moves through the air or through space under the influence of gravity.

Parabola The curved path followed by a projectile under the influence of gravity only.

Satellite A projectile or small celestial body that orbits a larger celestial body.

Ellipse The oval path followed by a satellite. The sum of the distances from any point on the path to two points called foci is a constant. When the foci are together at one point, the ellipse is a circle. As the foci get farther apart, the ellipse becomes more "eccentric."

Kepler's laws Law 1: The path of each planet around the Sun is an ellipse with the Sun at one focus. Law 2: The line from the Sun to any planet sweeps out equal areas of space in equal time intervals. Law 3: The square of the orbital period of a planet is directly proportional to the cube of the average distance of the planet from the Sun ($T^2 \sim r^3$ for all planets).

Escape speed The speed that a projectile, space probe, or similar object must reach to escape the gravitational influence of the Earth or of another celestial body to which it is attracted.

Suggested Web Site

For information on space-faring projections, visit the Web site of the National Space Society (NSS) at www.nss.org.

Review Questions

1. Why must a horizontally moving projectile have a large speed to become an Earth satellite?

Projectile Motion

2. What exactly is a projectile?

Projectiles Launched Horizontally

3. Why does the vertical component of velocity for a projectile change with time, whereas the horizontal component of velocity doesn't?

Projectiles Launched at an Angle

4. A stone is thrown upward at an angle. What happens to the horizontal component of its velocity as it rises? As it falls?
5. A stone is thrown upward at an angle. What happens to the vertical component of its velocity as it rises? As it falls?
6. A projectile falls beneath the straight-line path it would follow if there were no gravity. How many meters does it fall below this line if it has been traveling for 1 s? For 2 s?
7. Do your answers to the previous question depend on the angle at which the projectile is launched?
8. A projectile is launched upward at an angle of 75° from the horizontal and strikes the ground a certain distance downrange. For what other angle of launch at the same speed would this projectile land just as far away?
9. A projectile is launched vertically at 100 m/s. If air resistance can be neglected, at what speed will it return to its initial level?

Fast-Moving Projectiles—Satellites

10. How can a projectile "fall around the Earth"?
11. Why will a projectile that moves horizontally at 8 km/s follow a curve that matches the curvature of the Earth?
12. Why is it important that the projectile in the previous question be above Earth's atmosphere?

Circular Satellite Orbits

13. Why doesn't the force of gravity change the speed of a satellite in circular orbit?

14. How much time is taken for a complete revolution of a satellite in close orbit about the Earth?

15. For orbits of greater altitude, is the period longer or shorter?

Elliptical Orbits

16. Why does the force of gravity change the speed of a satellite in an elliptical orbit?

17. At what part of an elliptical orbit does a satellite have the greatest speed? The slowest speed?

Kepler's Laws of Planetary Motion

18. Who gathered the data that showed that the planets travel in elliptical paths around the Sun? Who discovered this fact? Who explained this fact?

19. What did Kepler discover about the speed of planets and their distance from the Sun? Was this discovery aided by thinking of satellites as projectiles moving under the influence of the Sun?

20. In Kepler's thinking, what was the direction of force on a planet? In Newton's thinking, what was the direction of force?

Energy Conservation and Satellite Motion

21. Why is kinetic energy a constant for a satellite in a circular orbit but not for a satellite in an elliptical orbit?

22. With respect to the apogee and perigee of an elliptical orbit, where is the gravitational potential greatest? Where is it the least?

23. Is the sum of kinetic and potential energies a constant for satellites in circular orbits, in elliptical orbits, in or both?

Escape Speed

24. What is the minimum speed for orbiting the Earth in close orbit? The maximum speed? What happens above this speed?

25. We talk of 11.2 km/s as the escape speed from the Earth. Is it possible to escape from the Earth at half this speed? At one-quarter this speed? If so, how?

Exercises

1. In synchronized diving, divers remain in the air for the same time. Is this possible if they have different weights? Defend your answer.

2. Suppose you roll a ball off a tabletop. Will the time to hit the floor depend on the speed of the ball? (Will a fast ball take a longer time to hit the floor?) Defend your answer.

3. Suppose you roll a ball off a tabletop. Compared with a slow roll, will a faster-moving ball hit the floor with a higher *speed*? Defend your answer.

4. If you toss a ball vertically upward in a uniformly moving train, it returns to its starting place. Will it do the same if the train is accelerating? Explain.

5. A heavy crate accidentally falls from a high-flying airplane just as it flies directly above a shiny red Porsche smartly parked in a car lot. Relative to the Porsche, where will the crate crash?

6. Suppose you drop an object from an airplane traveling at constant velocity, and further suppose that air resistance doesn't affect the falling object. What will be its falling path as observed by someone at rest on the ground, not directly below but off to the side where a clear view can be seen? What will be the falling path as observed by you looking downward from the airplane? Where will the object strike the ground, relative to you in the airplane? Where will it strike in the more realistic case in which air resistance does affect the fall?

7. Fragments of fireworks beautifully illuminate the night sky. (a) What specific path is traced by each fragment? (b) What paths would the fragments trace in a gravity-free region?

8. In the absence of air drag, why does the horizontal component of a projectile's motion not change, while the vertical component does?
9. At what point in its trajectory does a batted baseball have its minimum speed? If air drag can be neglected, how does this compare with the horizontal component of its velocity at other points?
10. A friend claims that bullets fired by some high-powered rifles travel for many meters in a straight-line path before they start to fall. Another friend disputes this claim and states that all bullets from any rifle drop beneath a straight-line path a vertical distance given by $\frac{1}{2}gt^2$ and that the curved path is apparent for low velocities and less apparent for high velocities. Now it's your turn: Will all bullets drop the same vertical distance in equal times? Explain.
11. For maximum range, a football should be punted at about 45° to the horizontal—somewhat less due to air drag. But punts are often kicked at angles greater than 45°. Can you think of a reason why?
12. Two golfers each hit a ball at the same speed, but one at 60° with the horizontal and the other at 30°. Which ball goes farther? Which hits the ground first? (Ignore air resistance.)
13. When a rifle is being fired at a distant target, why isn't the barrel lined up so that it points exactly at the target?
14. A park ranger shoots a monkey hanging from a branch of a tree with a tranquilizing dart. The ranger aims directly at the monkey, not realizing that the dart will follow a parabolic path and thus will fall below the monkey. The monkey, however, sees the dart leave the gun and lets go of the branch to avoid being hit. Will the monkey be hit anyway? Does the velocity of the dart affect your answer, assuming that it is great enough to travel the horizontal distance to the tree before hitting the ground? Defend your answer.

15. A projectile is fired straight upward at 141 m/s. How fast is it moving at the instant it reaches the top of its trajectory? Suppose instead that it were fired upward at 45°. What would be its speed at the top of its trajectory?
16. When you jump upward, your hang time is the time your feet are off the ground. Does hang time depend on your vertical component of velocity when you jump, your horizontal component of velocity, or both? Defend your answer.
17. The hang time of a basketball player who jumps a vertical distance of 2 feet (0.6 m) is about 2/3 second. What will be the hang time if the player reaches the same height while jumping 4 feet (1.2 m) horizontally?
18. Since the Moon is gravitationally attracted to the Earth, why doesn't it simply crash into the Earth?
19. When the space shuttle coasts in a circular orbit at constant speed about the Earth, is it accelerating? If so, in what direction? If not, why not?
20. Which planets have a greater period than one Earth year, those closer to the Sun than to the Earth or those farther from the Sun than from the Earth?
21. Does the speed of a falling object depend on its mass? Does the speed of a satellite in orbit depend on its mass? Defend your answers.
22. On which does the speed of a circling satellite *not* depend: the mass of the satellite, the mass of the Earth, or the distance of the satellite from the Earth.
23. A circularly moving object requires a centripetal force. What supplies this force for satellites that orbit the Earth?
24. Mars has about 1/9 the mass of Earth. If Mars were somehow positioned into the same orbit as Earth's, how would its time to circle the Sun compare with Earth's? (Longer, shorter, or the same?)
25. If you have ever watched the launching of an Earth satellite, you may have noticed that the rocket starts vertically upward, then departs from a vertical course and continues its climb at an angle. Why does it start vertically? Why does it not continue vertically?
26. If a cannonball is fired from a tall mountain, gravity changes its speed all along its trajectory. But if it is fired fast enough to go into circular orbit, gravity does not change its speed at all. Explain.
27. A satellite can orbit at 5 km above the Moon, but not at 5 km above the Earth. Why?
28. In 2000–2001, NASA's Near Earth Asteroid Rendezvous (NEAR) spacecraft orbited around the 20-mile-long asteroid Eros. Was the orbital speed of this spacecraft greater or less than 8 km/s? Why?
29. Would the speed of a satellite in close circular orbit about Jupiter be greater than, equal to, or less than 8 km/s?

30. Why are satellites normally sent into orbit by firing them in an easterly direction, the direction in which the Earth spins?

31. When a satellite in circular orbit slows, perhaps due to the firing of a "retro rocket," it then gains more speed than it had initially. Why?

32. Of all the United States, why is Hawaii the most efficient launching site for nonpolar satellites? (*Hint*: Look at the spinning Earth from above either pole and compare it to a spinning turntable.)

33. Earth is closer to the Sun in December than in June. In which of these two months is Earth moving faster around the Sun?

34. Two planets are never seen at midnight. Which two, and why?

35. Why does a satellite burn up when it descends into the atmosphere when it doesn't burn up when it ascends through the atmosphere?

36. Neglecting air drag, could a satellite be put into orbit in a circular tunnel beneath the Earth's surface? Discuss.

37. In the sketch on the left, a ball gains KE when rolling down a hill because work is done by the component of weight (F) that acts in the direction of motion. Sketch in the similar component of gravitational force that does work to change the KE of the satellite on the right.

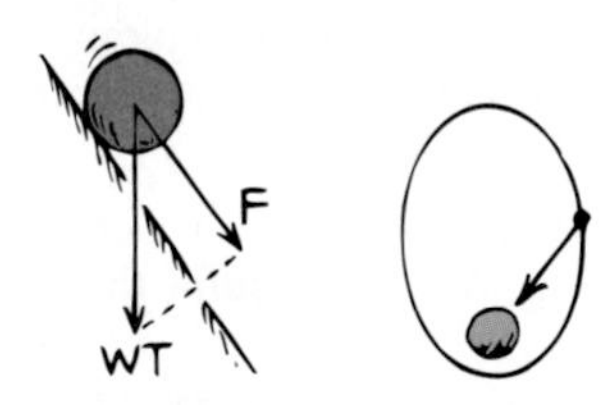

38. Why is work done by the force of gravitation on a satellite when it moves from one part of an elliptical orbit to another, but not when it moves from one part of a circular orbit to another?

39. What is the shape of the orbit when the velocity of the satellite is everywhere perpendicular to the force of gravity?

40. If a space shuttle circled the Earth at a distance equal to the Earth–Moon distance, how long would it take for it to make a complete orbit? In other words, what would be its period?

41. Can a satellite coast in a stable orbit in a plane that doesn't intersect the Earth's center? Defend your answer.

42. Can a satellite maintain an orbit in the plane of the Arctic Circle? Why or why not?

43. A communications satellite with a 24-hour period hovers over a fixed point on the Earth. Why is it placed in orbit only in the plane of the Earth's equator? (*Hint*: Think of the satellite's orbit as a ring around the Earth.)

44. A "geosynchronous" Earth satellite can remain directly overhead in Singapore, but not in San Francisco. Why?

45. When an Earth satellite is placed into a higher orbit, what happens to its period?

46. If a flight mechanic drops a wrench from a high-flying jumbo jet, it crashes to Earth. If an astronaut on the orbiting space shuttle drops a wrench, does it crash to Earth also? Defend your answer.

47. How could an astronaut in a space shuttle "drop" an object vertically to Earth?

48. A high-orbiting spaceship travels at 7 km/s with respect to the Earth. Suppose it projects a capsule rearward at 7 km/s with respect to the ship. Describe the path of the capsule with respect to the Earth.

49. A satellite in circular orbit about the Moon fires a small probe in a direction opposite to the velocity of the satellite. If the speed of the probe relative to the satellite is the same as the satellite's speed relative to the Moon, describe the motion of the probe. If the probe's relative speed is twice the speed of the satellite, why would it pose a danger to the satellite?

50. The orbital velocity of the Earth about the Sun is 30 km/s. If the Earth were suddenly stopped in its tracks, it would simply fall radially into the Sun. Devise a plan whereby a rocket loaded with radioactive wastes could be fired into the Sun for permanent disposal. How fast and in what direction with respect to the Earth's orbit should the rocket be fired?

51. If you stopped an Earth satellite dead in its tracks, it would simply crash into the Earth. Why, then, don't the communications satellites that "hover motionless" above the same spot on Earth crash into the Earth?

52. In an accidental explosion, a satellite breaks in half while in circular orbit about the Earth. One half is brought momentarily to rest. What is the fate of the half brought to rest? What happens to the other half?

53. A giant rotating wheel in space provides artificial gravity for its occupants, as discussed in Chapter 8. Instead of a full wheel, discuss the idea of a pair of capsules joined by a tether line and rotating about each other. Can such an arrangement provide artificial gravity for the occupants?

54. What advantage is there to launching space vehicles from high-flying aircraft instead of from the ground?

55. Escape speed from the surface of the Earth is 11.2 km/s, but a space vehicle could escape from the Earth at half this speed or less. Explain.

56. What is the maximum possible speed of impact upon the surface of the Earth for a faraway body initially at rest that falls to the Earth by virtue of the Earth's gravity only?

57. If Pluto were somehow stopped short in its orbit, it would fall into, rather than around, the Sun. How fast would it be moving when it hit the Sun?

58. At what point in its elliptical orbit about the Sun is the acceleration of the Earth toward the Sun at a maximum? At what point is it at a minimum? Defend your answers.

59. At which of the indicated positions does the satellite in elliptical orbit experience the greatest gravitational force? Have the greatest speed? The greatest velocity? The greatest momentum? The greatest kinetic energy? The greatest gravitational potential energy? The greatest total energy? The greatest angular momentum? The greatest acceleration?

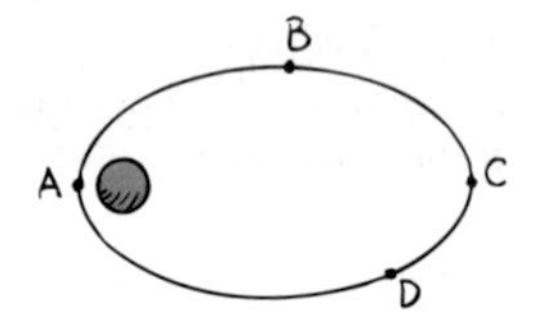

60. A rocket coasts in an elliptical orbit around the Earth. To attain the greatest amount of KE for escape using a given amount of fuel, should it fire its engines at the apogee or at the perigee? (*Hint*: Let the formula $Fd = \Delta \text{KE}$ be your guide to thinking. Suppose the thrust F is brief and of the same duration in either case. Then consider the distance d the rocket would travel during this brief burst at the apogee and at the perigee.)

Problems

1. A ball is thrown horizontally from a cliff at a speed of 10 m/s. What is its speed one second later?

2. An airplane is flying horizontally with speed 1000 km/h (280 m/s) when an engine falls off. Neglecting air resistance, assume it takes 30 s for the engine to hit the ground. (a) How high is the airplane? (b) How far horizontally does the engine travel while it falls? (c) If the airplane somehow continues to fly as if nothing had happened, where is the engine relative to the airplane at the moment the engine hits the ground?

3. A cannonball shot with an initial velocity of 141 m/s at an angle of 45° follows a parabolic path and hits a balloon at the top of its trajectory. Neglecting air resistance, how fast is the cannonball going when it hits the balloon?

4. Students in a lab measure the speed of a steel ball launched horizontally from a tabletop to be 4.0 m/s. If the tabletop is 1.5 m above the floor, where should they place a 20-cm-tall tin coffee can to catch the ball when it lands?

5. John and Tracy look from their 80-m high-rise balcony to a swimming pool below—not exactly below, but rather 20 m out from the bottom of their building. They wonder how fast they would have to jump horizontally to succeed in reaching the pool. What is the answer?

6. Ignoring air drag, what is the maximum speed possible for a horizontally moving tennis ball as it clears the net 1.0 m high and strikes within the court's border, 12.0 m distant?

7. Calculate a person's hang time who moves horizontally 3 m during a 1.25-m high jump. What is the hang time when moving 6 m horizontally during this jump?

8. Calculate the speed in m/s at which Earth revolves about the Sun. You may assume the orbit is nearly circular.

9. The Moon is about 3.8×10^5 km from the Earth. Find its average orbital speed about the Earth.

10. A certain satellite has a kinetic energy of 8 billion joules at perigee (the point at which it is closest to the Earth) and 5 billion joules at apogee (the point at which it is farthest from the Earth). As the satellite travels from apogee to perigee, how much work does the gravitational force do on it? Does its potential energy increase or decrease during this time, and by how much?

PART TWO

Properties of Matter

Like everyone, I'm made of atoms. They're so small and numerous that I inhale billions of trillions with each breath of air. I exhale some of them right away, but other atoms stay for awhile and become part of me, which I may exhale later. With each breath you take, some of my atoms are in it and become part of you (and likewise, yours become part of me). There are more atoms in a breath of air than the total number of humans since time zero, so in each breath you inhale, you recycle atoms that once were a part of every person who lived. Hey, in this sense, we're all one!

CHAPTER 11

The Atomic Nature of Matter

Extraordinary twentieth-century physicist Richard Feynman contributed enormously to our understanding of atoms and physics in general.

Imagine that you inhabit the world of Alice in Wonderland when she shrank in size. Pretend that you're standing on a chair and then you step off it and fall in slow motion to the floor—and, as you fall, you continually shrink. As you fall toward the wooden floor, you brace yourself for impact; and, as you get nearer and nearer, becoming smaller and smaller, you notice that the surface of the floor is not as smooth as it first looked. Great crevices appear that are the microscopic irregularities that are found in all wood. In falling into one of these canyon-sized crevices, you again brace yourself for impact, only to find that the bottom of the canyon consists of many other crevices. Falling farther while growing even smaller, you notice the solid walls throb and pucker. The throbbing surfaces consist of hazy blobs, mostly spherical, some egg-shaped, some larger than others, and all oozing into each other, making up long chains of complicated structures. Falling still farther, you brace for impact as you approach one of these cloudy spheres closer and closer, smaller and smaller, and—Wow! You have entered a new universe. You fall into a sea of emptiness, occupied by occasional specks whirling past at unbelievably high speeds. You are in an **atom,** as empty of matter as the solar system. The solid floor you have fallen into is, except for specks of matter here and there, mostly empty space. If you continue falling, you might fall many meters through "solid" matter before making a direct hit with a subatomic speck.

All matter, however solid it appears, is made up of tiny building blocks, which themselves are mostly empty space. These are atoms—which can combine to form molecules, which, in turn flock together to form the matter that we see around us.

The Atomic Hypothesis

The idea that matter is composed of atoms goes back to the Greeks in the fifth century BC. Investigators of nature back then wondered whether matter was continuous or not. We can break a rock into pebbles, and the pebbles into fine gravel. The gravel can be broken into fine sand, which then can be pulverized into powder. Perhaps it seemed to the fifth-century Greeks that there was a smallest bit of rock, an "atom," that could not be divided any further.

Aristotle, the most famous of the early Greek philosophers, didn't agree with the idea of atoms. In the fourth century BC, he taught that all matter is

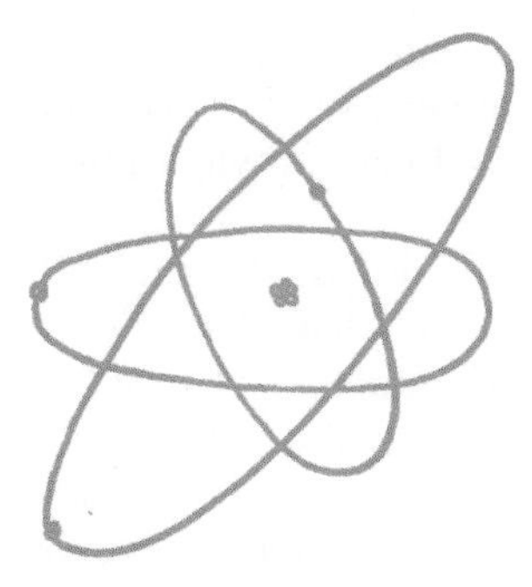

FIGURE 11.1
An early model of the atom, with a central nucleus and orbiting electrons, much like a solar system with orbiting planets.

composed of various combinations of four elements—earth, air, fire, and water. This view seemed reasonable because, in the world around us, matter is seen in only four forms; solids (earth), gases (air), liquids (water), and the state of flames (fire). The Greeks viewed fire as the element of change, since fire was observed to produce changes on substances that burned. Aristotle's ideas about the nature of matter lasted for more than 2000 years.

The atomic idea was revived in the early 1800s by an English meteorologist and school teacher, John Dalton. He successfully explained the nature of chemical reactions by proposing that all matter is made of atoms. He and others of the time, however, had no direct evidence for their existence. Then, in 1827, a Scottish botanist named Robert Brown noticed something very unusual in his microscope. He was studying grains of pollen suspended in water, and he saw that the grains were continually moving and jumping about. At first he thought the grains were some kind of moving life forms, but later he found that dust particles and grains of soot suspended in water moved in the same way. This perpetual jiggling of particles—now called **Brownian motion**—results from collisions between visible particles and invisible atoms. The atoms are invisible because they're so small. Although he couldn't see the atoms, Brown could see the effect they had on particles he *could* see. It's like a super-giant beach ball being bounced around by a crowd of people at a football game. From a high-flying airplane, you wouldn't see the people because they are small relative to the enormous ball, which you would be able to see. The pollen grains that Brown observed moved because they were constantly being jostled by the atoms (actually, by the atomic combinations referred to as molecules) that made up the water surrounding them.

Evidence for Atoms

All this was explained in 1905 by Albert Einstein, the same year that he announced the theory of special relativity. Until Einstein's explanation—which made it possible to find the masses of atoms—many prominent physicists remained skeptical about the existence of atoms. So we see that the reality of the atom was not firmly established until the early twentieth century.

We can't "see" atoms because they're too small. We can't see the farthest star either. There's much that we can't see. But that doesn't prevent investigation of such things or even collecting indirect evidence.

Insights

In 1963, the importance of atoms was emphasized by the American physicist Richard Feynman, who stated that, if some cataclysm were to destroy all scientific knowledge and only one sentence could be passed on to the next generation of creatures, the statement with the most information in the least words would be: "*All things are made of atoms—little particles that move around in perpetual motion, attracting each other when they are a little distance apart, but repelling upon being squeezed into one another.*" All matter—shoes, ships, sealing wax, cabbages, and kings—any material we can think of—is made of atoms. This is the atomic hypothesis, which now serves as a central foundation of all of science.

Characteristics of Atoms

Atoms, the building blocks of matter, are *incredibly tiny*. An atom is as many times smaller than you as an average star is larger than you. A nice way to say this is that we stand between the atoms and the stars. Or another way of

How long would it take you to count to one million? If each count takes one second, counting nonstop to a million would take 11.6 days. To count to a billion (10^9) would take 31.7 years. To count to a trillion (10^{12}) would take 31,700 years. Counting to 10^{22} would take about ten thousand times the age of the universe!

Insights

stating the smallness of atoms is that the diameter of an atom is to the diameter of an apple as the diameter of an apple is to the diameter of the Earth. So, to imagine an apple full of atoms, think of the Earth solid-packed with apples. Both have about the same number.

Atoms are numerous. There are about 100,000,000,000,000,000,000,000 atoms in a gram of water (a thimbleful). In scientific notation, that's 10^{23} atoms. The number 10^{23} is an enormous number, more than the number of drops of water in all the lakes and rivers of the world. So there are more atoms in a thimbleful of water than there are drops of water in the world's lakes and rivers. In the atmosphere, there are about 10^{22} atoms in a liter of air. Interestingly, the volume of the atmosphere contains about 10^{22} liters of air. That's an incredibly large number of atoms, and the same incredibly large number of liters of atmosphere. Atoms are so small and so numerous that there are about as many atoms in the air in your lungs at any moment as there are breathfuls of air in the Earth's atmosphere.

Atoms get around. Atoms are perpetually moving. They migrate from one location to another. In solids, the rate of migration is low; in liquids, it is greater; and in gases, migration is greatest. Drops of food coloring in a glass of water, for example, soon spread to color the entire glass of water. The same would be true of a cupful of food coloring thrown into an ocean: It would spread around and later be found in every part of the world's oceans.

Water dilution is a main reason for salmon being able to return to their birthplaces. Atoms and molecules from soil and vegetation in a lake or a stream make that water unique. Likewise with salmon spawning habitats. Once hatched, young salmon remain in local streams for two years before beginning their voyage to the ocean, where they remain for an average of four years. What also goes into the ocean, of course, is water from the region in which they grew up. The original water composition is diluted as it travels to the ocean. In the ocean, it is further diluted—but never to zero. When the time comes to return to their original habitat, salmon follow their noses. They swim in a direction where concentrations of familiar water become greater. In time, they'll encounter the source of that water. Humans can discern different bottled waters, and salmon have enormously more ability to sense the difference in waters, as bloodhounds have a similar sensitivity to air composition.

Smell discrimination for salmon has been measured in parts per *trillion*—quite incredible. Pacific salmon use other navigation devices in the open ocean, including sensitivity to polarized light, ocean currents, magnetic fields, and temperature and salinity gradients.

Insights

Atoms and molecules in the atmosphere spread around much more than they do in the ocean. Atoms and molecules in air zip around at speeds up to ten times the speed of sound. They spread rapidly, so oxygen that surrounds you today may have been halfway across the country a few days ago. Taking Figure 11.2 further, your exhaled breaths of air quite soon mix with other atoms in the atmosphere. After the few years it takes for your breath to mix uniformly in the atmosphere, anyone, anywhere on Earth, who inhales a breath of air will take in, on the average, one of the atoms in that exhaled breath of yours. But you exhale many, many breaths, so other people breathe in many, many atoms that were once in your lungs—that were once a part of you; and, of course, vice versa. Believe it or not, with each breath you take in, you breathe atoms that were once a part of everyone who ever lived! Considering that the atoms we exhale were part of our bodies (the nose of a dog

Atoms Are Recyclable

FIGURE 11.2
There are as many atoms in a normal breath of air as there are breathfuls of air in the atmosphere of the Earth.

has no trouble discerning this), it can be truly said that we are literally breathing one another.

Atoms are ageless. Many atoms in your body are nearly as old as the universe itself. When you breathe, for example, only some of the atoms that you inhale are exhaled in your next breath. The remaining atoms are taken into your body to become part of you, and they later leave your body by various means. You don't "own" the atoms that make up your body; you borrow them. We all share from the same atom pool, as atoms forever migrate around, within, and among us. So some of the atoms in the nose you scratch today could have been part of your neighbor's ear yesterday!

Most people know that we are all made of the same kinds of atoms. But what most people don't know is that we are made of the *same* atoms—atoms that cycle from person to person as we breathe and as our perspiration is vaporized. We recycle atoms on a grand scale.

So the origin of the lightest atoms goes back to the origin of the universe, and most heavier atoms are older than the Sun and the Earth. There are atoms in your body that have existed since the first moments of time, recycling throughout the universe among innumerable forms, both nonliving and living. You're the present caretaker of the atoms in your body. There will be many who will follow you.

CHECK YOURSELF

1. Which are older, the atoms in the body of an elderly person or those in the body of a baby?
2. World population grows each year. Does this mean that the mass of the Earth increases each year?
3. Are there really atoms that were once a part of Albert Einstein incorporated in the brains of all the members of your family?

CHECK YOUR ANSWERS

1. The age of the atoms in both is the same. Most of the atoms were manufactured in stars that exploded before the solar system came into existence.
2. The greater number of people increases the mass of the Earth by zero. The atoms that make up our bodies are the same atoms that were here before we were born—we are but dust, and unto dust we shall return. Human cells are merely rearrangements of material already present. The atoms that make up a baby forming in its mother's womb must be supplied by the food the mother eats. And those atoms originated in stars—some of them in far-away galaxies. (Interestingly, the mass of the Earth *does* increase by the incidence of roughly 40,000 tons of interplanetary dust each year, but not by the birth and survival of more people.)
3. Quite so, and of Oprah Winfrey too. However, these atoms are combined differently than they were previously. If you experience one of those days when you feel like you'll never amount to anything, take comfort in the thought that many of the atoms that now constitute your body will live forever in the bodies of all the people on Earth who are yet to be.

Life is not measured by the number of breaths we take, but by the moments that take our breath away. —*George Carlin*

Atomic Imagery

FIGURE 11.3
Information about the ship is revealed by passing waves because the distance between wave crests is small compared with the size of the ship. The passing waves reveal nothing about the chain.

Atoms are too small to be seen with visible light. You could connect an array of optical microscopes atop one another and never "see" an atom because light is made up of waves, and atoms are smaller than the wavelengths of visible light. The size of a particle visible under the highest magnification must be larger than the wavelengths of visible light. This is better understood by an analogy with water waves. A ship is much larger than the water waves that roll on by it. As Figure 11.3 shows, water waves can reveal features of the ship. The waves *diffract* as they pass the ship. But diffraction is nil for waves that pass by the anchor chain, revealing little or nothing about it. Similarly, waves of visible light are too coarse compared with the size of an atom to show details of the size and shape of atoms. Atoms are incredibly small.

Yet here in Figure 11.4 we see a picture of atoms—the historic 1970 image of chains of individual thorium atoms. The picture is not a photograph but an electron micrograph—it was not made with light but with a thin electron beam in a scanning electron microscope (SEM) developed by Albert Crewe at the University of Chicago's Enrico Fermi Institute. An electron beam, such as the one that sprays a picture on an early television screen, is a stream of particles that have wave properties. The wavelength of an electron beam is smaller than the wavelengths of visible light, and atoms are larger than the tiny wavelengths of an electron beam. Crewe's electron micrograph is the first high-resolution image of individual atoms.

In the mid-1980s, researchers developed a new kind of microscope—the scanning tunneling microscope (STM). It employs a sharp tip that is scanned over a surface at a distance of a few atomic diameters in a point-by-point and line-by-line fashion. At each point, a tiny electric current, called a tunneling current, is measured between the tip and the surface. Variations in the current reveal the surface topology. The image of Figure 11.5 beautifully shows the position of a ring of atoms. The ripples shown in the ring of atoms reveal the wave nature of matter. This image, among many others, underscores the delightful interplay of art and science.

Because we can't see inside an atom, we construct models. A model is an abstraction that helps us to visualize what we can't see, and, importantly, it enables us to make predictions about unseen portions of the natural world.

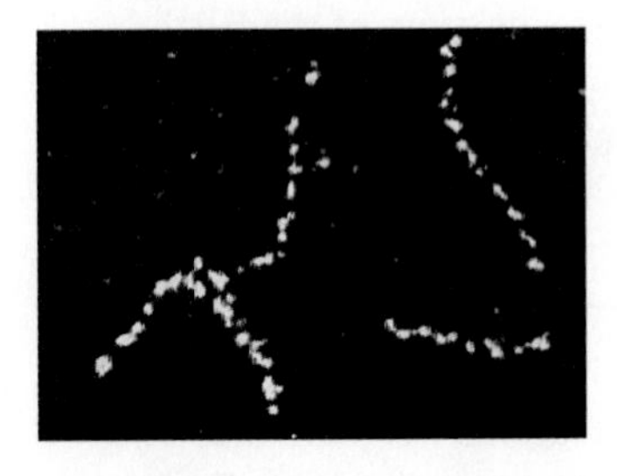

FIGURE 11.4
The strings of dots are chains of thorium atoms imaged with a scanning electron microscope. This historic photograph of individual atoms was taken in 1970 by researchers at the University of Chicago's Enrico Fermi Institute.

FIGURE 11.5
An image of 48 iron atoms positioned into a circular ring that "corrals" electrons on a copper crystal surface; taken with a scanning tunneling microscope at the IBM Almaden Laboratory in San Jose, California.

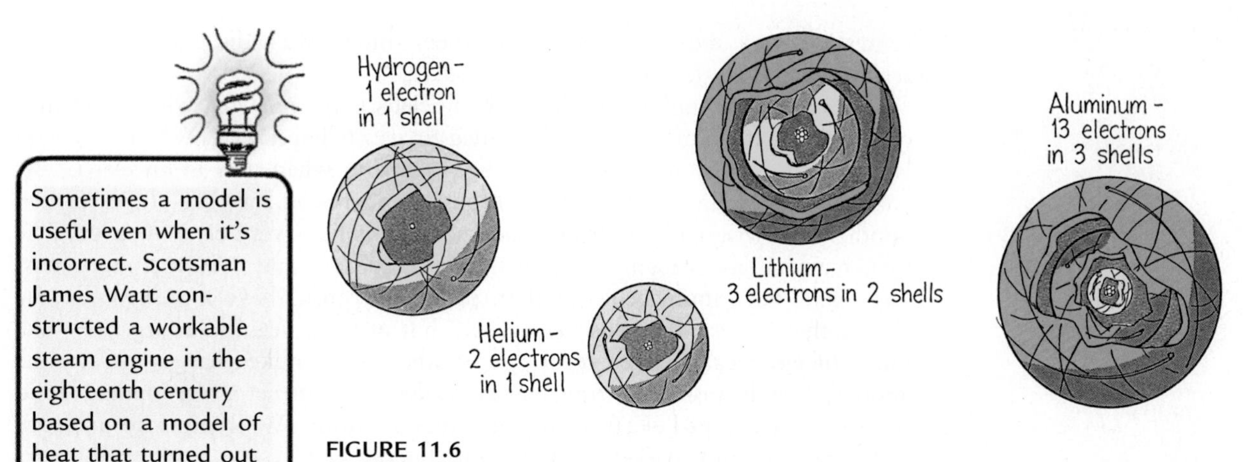

Sometimes a model is useful even when it's incorrect. Scotsman James Watt constructed a workable steam engine in the eighteenth century based on a model of heat that turned out to be quite incorrect.

Insights

FIGURE 11.6
The classical model of the atom consists of a tiny nucleus surrounded by electrons that orbit within spherical shells. As the charges of nuclei increase, electrons are pulled closer, and the shells become smaller.

The model of the atom most familiar to the general public is akin to that of the solar system. As with the solar system, most of an atom's volume is empty space. At the center is a tiny and very dense nucleus in which most of the mass is concentrated. Surrounding the nucleus are "shells" of orbiting electrons. These are the same electrically charged electrons that constitute the electric current in your calculator. Although electrons electrically repel other electrons, they are electrically attracted to the nucleus, which has a net positive charge. As the size and charge of the nuclei increase, electrons are pulled closer, and the shells become smaller. Interestingly, the uranium atom, with its 92 electrons, is not appreciably larger in diameter than the lightest atom, hydrogen. This model was first proposed in the early twentieth century, and it reflects a rather simplified understanding of the atom. It was soon discovered, for example, that electrons don't orbit the atom's center like planets orbit the Sun. Like most early models, however, the planetary atomic model served as a useful stepping stone to further understanding and more accurate models. Any atomic model, no matter how refined, is nothing more than a symbolic representation of the atom and not a physical picture of the actual atom.

Atomic Structure

Nearly all the mass of an atom is concentrated in the **atomic nucleus,** which occupies only a few quadrillionths of its volume. The nucleus, therefore, is extremely dense. If bare atomic nuclei could be packed against each other into a lump 1 centimeter in diameter (about the size of a large pea), the lump would weigh 133,000,000 tons! Huge electrical forces of repulsion prevent such close packing of atomic nuclei because each nucleus is electrically charged and repels all other nuclei. Only under special circumstances are the nuclei of two or more atoms squashed into contact. When this happens, a violent kind of nuclear reaction may occur. Such a reaction, a *thermonuclear fusion reaction,* occurs in the centers of stars and ultimately makes them shine. (We'll discuss these nuclear reactions in Chapter 34.)

The principal building block of the nucleus is the *nucleon,* which is in turn composed of fundamental particles called *quarks* (Chapter 32). When a nucleon is in an electrically neutral state, it is a *neutron;* when it is in an electrically charged state, it is a *proton.* All protons are identical; they are copies of one another. Likewise with neutrons: Each neutron is like every other neutron. The lighter nuclei have roughly equal numbers of protons and neutrons; more massive nuclei have more neutrons than protons. Protons have a positive electric charge that repels other positive charges but attracts negative charges. So like kinds of electrical charges repel one another, and unlike charges attract one another. It is the positive protons in the nucleus that attract a surrounding cloud of negatively charged electrons to constitute an atom. (We'll return to nucleons in Chapter 33 and to electric charge in Chapter 22.)

As we'll explore in Chapter 22, electrons make up the flow of electricity in electrical circuits. They are exceedingly light, almost 2000 times lighter than nucleons, and thus they contribute very little mass to the atom. An electron in one atom is identical to any electron in or out of any other atom. Electrons repel other electrons, but a multitude of electrons can be held together within an atom because of their attraction to the positively charged nucleus.

The number of protons in the nucleus is electrically balanced by an equal number of electrons whirling about the nucleus. The atom itself is electrically neutral, so it normally doesn't attract or repel other atoms. But when atoms are close together, the negative electrons on one atom may at times be closer to the positive nucleus of another atom, which results in a net attraction between the atoms. That is why some atoms combine to form molecules.

The fact that electrons repel other electrons has interesting consequences. When the atoms of your hand push against the atoms of a wall, for example, electrical repulsions prevent your hand from passing through the wall. These same electrical repulsions prevent us from falling through the solid floor. They also allow us the sense of touch. Interestingly, when you touch someone, your atoms don't meet the atoms of the one you touch. Instead, the atoms get close enough so that you sense electrical repulsion forces. There is still a tiny, though imperceptible, gap of space between you and the person you are touching.

CHECK YOURSELF

Why do electrically neutral atoms repel one another when they are close?

CHECK YOUR ANSWER

In an electrically neutral atom, the number of positive protons is balanced by the same number of negative electrons. Electrons, however, reside on the outer surface of the atom, meaning that the atomic surface is negatively charged. Two or more atoms, therefore, can't get together without an electrical repulsion between their outer surfaces. Hence, we can't walk through walls. Sometimes, however, electrons are able to jump from one atom to the next. This occurs during a chemical reaction, in which atoms are able to link together to form larger structures, such as molecules.

The Elements

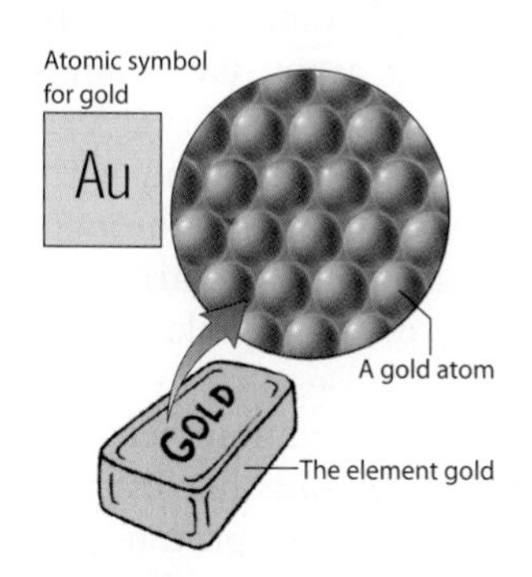

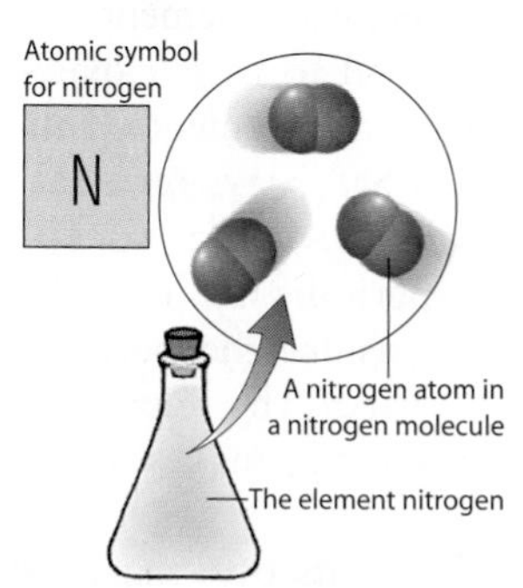

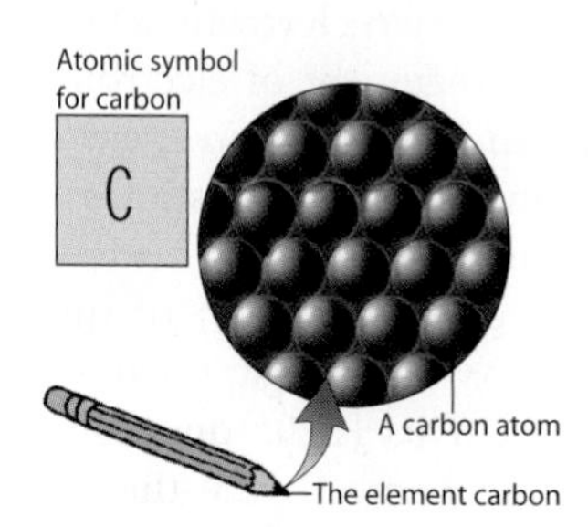

FIGURE 11.7
Any element consists only of one kind of atom. Gold consists only of gold atoms, a flask of gaseous nitrogen consists only of nitrogen atoms, and the carbon of a graphite pencil is composed only of carbon atoms.

When a substance is composed of atoms of the same kind, we call that substance an **element.** A pure 24-carat gold ring, for example, is composed only of gold atoms. A gold ring with a lower carat value is composed of gold and other elements, such as nickel. The silvery liquid in a barometer or thermometer is the element mercury. The entire liquid consists of only mercury atoms. Of course if a substance contains only a single kind of atom, it can correctly be called an element. An atom of a particular element is the smallest sample of that element. Although *atom* and *element* are often used interchangeably, *element* is preferred when referring to macroscopic quantities. For example, we speak of isolating a mercury *atom* from a flask of the *element* mercury.

The lightest element of all is hydrogen. In the universe at large, it is the most abundant element—more than 90% of the atoms in the known universe are hydrogen atoms. Helium, the second-lightest element, provides most of the remaining atoms in the universe. Heavier atoms in our surroundings were manufactured by the fusion of light elements in the hot, high-pressure cauldrons deep within the interiors of stars. The heaviest elements form when huge stars implode and then explode—supernovas. Nearly all the elements on Earth are remnants of stars that exploded long before the solar system came into existence.

Just as dots of only three colors of light combine to form almost every perceivable color on a television screen, only about 100 distinct elements form all the materials we know about. To date, more than 112 elements are known. Of these, about 90 occur in nature. The others are produced in the laboratory with high-energy atomic accelerators and nuclear reactors. These laboratory-produced elements are too unstable (radioactive) to occur naturally in appreciable quantities.

From a pantry containing less than 100 elements, we have the atoms that constitute almost every simple, complex, living, or nonliving substance in the known universe. More than 99% of the material on Earth is formed from only about a dozen of the elements. The other elements are relatively rare. Living things are composed primarily of five elements: oxygen (O), carbon (C), hydrogen (H), nitrogen (N), and calcium (Ca). The letters in parentheses represent the chemical symbols for these elements.

FIGURE 11.8
Both you and Leslie are made of stardust—in the sense that the carbon, oxygen, nitrogen, and other atoms that make up your body originated in the deep interiors of ancient stars that have long since exploded.

The Periodic Table of the Elements

Interestingly, of every 200 atoms in our bodies, 126 are hydrogen, 51 are oxygen, and just 19 are carbon.

Insights

Elements are classified by the number of protons their atoms contain, which is their **atomic number.** Hydrogen, containing one proton per atom, has atomic number 1; helium, containing two protons per atom, has atomic number 2; and so on in sequence to the heaviest naturally occurring element, uranium, with atomic number 92. The numbers continue beyond atomic number 92 through the artificially produced transuranic (beyond uranium) elements. The arrangement of elements by their atomic numbers makes up **the periodic table of the elements** (Figure 11.9).

The periodic table is a chart that lists atoms by their atomic number and also by their electrical arrangements. Like the rows of a calendar that list the days of the week, each element, from left to right, has one more proton and electron than the preceding element. Reading down the table, each element has one more shell than the one above. The inner shells are filled to their capacities, and the outer shell may or may not be, depending on the element. Only the elements at the far right of the table, like the column of Saturdays on a calendar, have their outer shells filled to capacity. These are the *noble gases*—helium, neon, argon, krypton, xenon, and radon. The periodic table is the chemist's road map—and much more. Most scientists consider the periodic table to be the most elegant organizational chart ever devised. The enormous human effort and ingenuity that went into finding its regularities makes a fascinating atomic detective story.[1]

Elements may have up to seven shells, and each shell has its own capacity for electrons. The first and innermost shell has a capacity for two electrons while the second shell has a capacity for eight electrons. The arrangement of electrons in the shells dictates such properties as melting and freezing temperatures, electrical conductivity, and the taste, texture, appearance, and color of substances. The arrangements of electrons quite literally give life and color to the world.

Models of the atom evolve with new findings. The classical model of the atom has given way to a model that views the electron as a standing wave—altogether different from the idea of an orbiting particle. This is the quantum mechanical model, introduced in the 1920s. **Quantum mechanics** is the theory of the small-scale world that includes predicted wave properties of matter. It deals with "lumps" occurring at the subatomic level—lumps of matter or lumps of such things as energy and angular momentum. (More about the quantum in Chapters 31 and 32).

Isotopes

Whereas the number of protons in a nucleus exactly matches the number of electrons around the nucleus in a neutral atom, the number of protons in the nucleus need not match the number of neutrons there. For example, all hydrogen

[1]A clearly written discussion of the periodic table by my nephew, John Suchocki, is in Chapter 5 of his *Conceptual Chemistry,* 2nd ed. (Benjamin Cummings, 2004). Also see Chapter 14 of *Conceptual Physical Science,* 3rd ed., by Hewitt, Suchocki, and Hewitt (Addison-Wesley, 2004). Interesting material!

Period	1	2	3	4	5	6	7	8	9	10	11	12	13	14	15	16	17	18
1	1 H Hydrogen 1.0079																	2 He Helium 4.003
2	3 Li Lithium 6.941	4 Be Beryllium 9.012											5 B Boron 10.811	6 C Carbon 12.011	7 N Nitrogen 14.007	8 O Oxygen 15.999	9 F Fluorine 18.998	10 Ne Neon 20.180
3	11 Na Sodium 22.990	12 Mg Magnesium 24.305											13 Al Aluminum 26.982	14 Si Silicon 28.086	15 P Phosphorus 30.974	16 S Sulfur 32.066	17 Cl Chlorine 35.453	18 Ar Argon 39.948
4	19 K Potassium 39.098	20 Ca Calcium 40.078	21 Sc Scadium 44.956	22 Ti Titanium 47.88	23 V Vanadium 50.942	24 Cr Chromium 51.996	25 Mn Manganese 54.938	26 Fe Iron 55.845	27 Co Cobalt 58.933	28 Ni Nickel 58.69	29 Cu Copper 63.546	30 Zn Zinc 65.39	31 Ga Gallium 69.723	32 Ge Germanium 72.61	33 As Arsenic 74.922	34 Se Selenium 78.96	35 Br Bromine 79.904	36 Kr Krypton 83.8
5	37 Rb Rubidium 85.468	38 Sr Strontium 87.62	39 Y Yttrium 88.906	40 Zr Zirconium 91.224	41 Nb Niobium 92.906	42 Mo Molybdenum 95.94	43 Tc Technetium (98)	44 Ru Ruthenium 101.07	45 Rh Rhodium 102.906	46 Pd Palladium 106.42	47 Ag Silver 107.868	48 Cd Cadmium 112.411	49 In Indium 114.82	50 Sn Tin 118.71	51 Sb Antimony 121.76	52 Te Tellurium 127.60	53 I Iodine 126.905	54 Xe Xenon 131.29
6	55 Cs Cesium 132.905	56 Ba Barium 137.327	57 La Lanthanum 138.906	72 Hf Hafnium 178.49	73 Ta Tantalum 180.948	74 W Tungsten 183.84	75 Re Rhenium 186.207	76 Os Osmium 190.23	77 Ir Iridium 192.22	78 Pt Platinum 195.08	79 Au Gold 196.967	80 Hg Mercury 200.59	81 Tl Thallium 204.383	82 Pb Lead 207.2	83 Bi Bismuth 208.980	84 Po Polonium (209)	85 At Astatine (210)	86 Rn Radon (222)
7	87 Fr Francium (223)	88 Ra Radium 226.025	89 Ac Actinium 227.028	104 Rf Rutherfordium (261)	105 Db Dubnium (262)	106 Sg Seaborgium (266)	107 Bh Bohrium (264)	108 Hs Hassium (269)	109 Mt Meitnerium (268)	110 Ds Darmstadtium (271)	111 Ro Roentgenium (272)	112 Uub (285)		114 Uuq (289)		116 Uuh (292)		

Lanthanides	58 Ce Cerium 140.115	59 Pr Praseodymium 140.908	60 Nd Neodymium 144.24	61 Pm Promethium (145)	62 Sm Samarium 150.36	63 Eu Europium 151.964	64 Gd Gadolinium 157.25	65 Tb Terbium 158.925	66 Dy Dyspsrosium 162.5	67 Ho Holmium 164.93	68 Er 68 Erbium 167.26	69 Tm Thulium 168.934	70 Yb Ytterbium 173.04	71 Lu Lutetium 174.967
Actinides	90 Th Thorium 232.038	91 Pa Protactinium 231.036	92 U Uranium 238.029	93 Np Neptunium 237.05	94 Pu Plutonium (244)	95 Am Americium (243)	96 Cm Curium (247)	97 Bk Berkelium (247)	98 Cf Californium (251)	99 Es Einsteinium (252)	100 Fm Fermium (257)	101 Md Mendelevium (258)	102 No Nobelium (259)	103 Lr Lawrencium (262)

Key: Metal; Metalloid; Nonmetal

FIGURE 11.9
The periodic table of the elements. The number above the chemical symbol is the *atomic number,* and the number below is the *atomic mass* averaged by isotopic abundance in the Earth's surface and expressed in atomic mass units (amu). Atomic masses for radioactive elements shown in parentheses are the whole number nearest the most stable isotope of that element.

Don't confuse an isotope with an *ion,* which is an atom that is electrically charged owing to an excess or a deficiency of electrons.

Insights

nuclei have a single proton but most have no neutrons. A small percentage contain one neutron and an even smaller percentage contain two neutrons. Similarly, most iron nuclei with 26 protons contain 30 neutrons, while a small percentage contain 29 neutrons. Atoms of the same element that contain different numbers of neutrons are **isotopes** of the element. The various isotopes of an element all have the same number of electrons, and so, for the most part, they behave identically. The hydrogen atoms in H_2O, for example, may or may not contain a neutron. The oxygen doesn't "know the difference." But if significant amounts of the hydrogen atoms have neutrons, then the H_2O is slightly heavier, and it's appropriately called "heavy water."

We identify isotopes by their *mass number,* which is the total number of protons and neutrons (in other words, the number of nucleons) in the nucleus. A hydrogen isotope with one proton and no neutrons, for example, has a mass number of 1, and is referred to as hydrogen-1. Likewise, an iron atom with 26 protons and 30 neutrons has a mass number of 56 and is referred to as iron-56. An iron atom with 26 protons and only 29 neutrons would be called iron-55.

In order for two chemicals to bond, they must first collide in the proper orientation. Second, they must have sufficient kinetic energies to initiate the breaking of chemical bonds so that new ones can form.

Insights

The total mass of an atom is called its *atomic mass*. This is the sum of the masses of all the atom's components (electrons, protons, and neutrons). Because electrons are so much less massive than protons and neutrons, their contribution to atomic mass is negligible. Atoms are so small that expressing their masses in units of grams or kilograms is not practical. Instead, scientists use a specially defined unit of mass known as the **atomic mass unit** or **amu.** A nucleon has a mass of about 1 amu. An atom with 12 nucleons, such as carbon-12, therefore, has a mass of about 12 amu. The periodic table lists atomic masses in units of amu.

Most elements have a variety of isotopes. The atomic mass for each element listed in the periodic table is the weighted average of the masses of these isotopes based on the occurrence of each isotope on Earth. For example, carbon with six protons and six neutrons has an atomic mass of 12.000 amu. About one percent of all carbon atoms, however, contain seven neutrons. The heavier isotope raises the average atomic mass of carbon from 12.000 amu to 12.011 amu.

CHECK YOURSELF

1. Which contributes more to an atom's mass, electrons or protons? Which contributes more to an atom's volume (its size)?
2. Which is represented by a whole number—the mass number or the atomic mass?
3. Do two isotopes of iron have the same *atomic number*? The same *atomic mass number*?

CHECK YOUR ANSWERS

1. Protons contribute more to an atom's mass; electrons contribute more to its size.
2. The mass number is always given as a whole number, such as hydrogen-1 or carbon-12. Atomic mass, by contrast, is the average mass of the various isotopes of an element and is thus represented by a fractional number.
3. The two isotopes of iron have the same atomic number 26, because they each have 26 protons in the nucleus. They have different atomic mass numbers if they have different numbers of neutrons in the nucleus.

Compounds and Mixtures

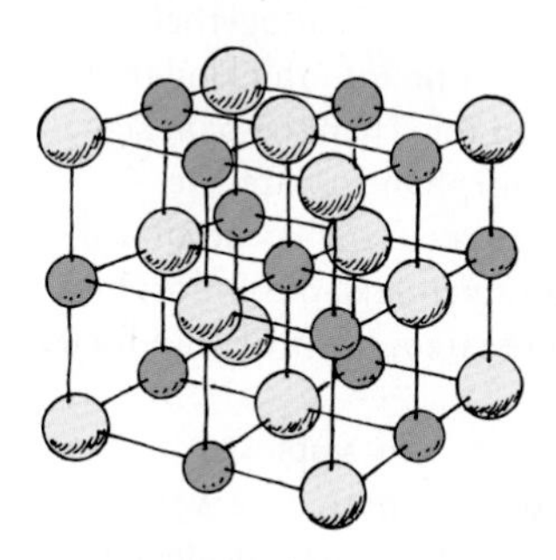

FIGURE 11.10
Table salt (NaCl) is a crystalline compound that is not made of molecules. The sodium and chlorine atoms are arranged in a repeating pattern in which each atom is surrounded by six atoms of the other kind.

A pure chemical material that consists of more than one kind of atom is called a **compound.** Examples of simple compounds include water, ammonia, and methane. A compound is uniquely different from the elements from which it is made, and it can only be separated into its constituent elements by chemical means. Sodium, for example, is a metal that reacts violently with water. Chlorine is a poisonous yellow-green gas. Yet the compound of these two elements is the harmless white crystal (NaCl) that you sprinkle on your potatoes. Consider also that, at ordinary temperatures, the elements hydrogen and oxygen are both gases. When combined, they form the compound water (H_2O), a liquid—quite different.

Not all substances react with one another chemically when they are brought close together. A substance that is mixed together without chemically bonding is called a **mixture.** Sand combined with salt is a mixture. Hydrogen and oxygen gas form a mixture until ignited, whereupon they form the compound water. A common mixture that we all depend on is nitrogen and oxygen together with a little argon and small amounts of carbon dioxide and other gases. It is the air that we breathe.

CHECK YOURSELF

Is common table salt an element, a compound, or a mixture?

Molecules

A **molecule** consists of two or more atoms held together by the sharing of electrons. (We say such atoms are *covalently bonded.*) A molecule may be as simple as the two-atom combination of oxygen (O_2) or the two-atom combination of nitrogen (N_2), which are the elements that constitute most of the air we

CHECK YOUR ANSWER

Salt is not an element. If it were, you'd see it listed in the periodic table. Pure table salt is a compound of the elements sodium and chlorine, represented in Figure 11.10. Notice that the sodium atoms (green) and the chlorine atoms (yellow) are arranged in a three-dimensional repeating pattern—a crystal. Each sodium atom is surrounded by six chlorine atoms, and each chlorine atom is surrounded by six sodiums. Interestingly, there are no separate sodium–chlorine groups that can be labeled molecules.[2]

[2]In a strict sense, common table salt *is* a mixture—often with small amounts of potassium iodide and sugar. The iodine in the potassium iodide has virtually wiped out a common affliction of earlier times, a swelling of the thyroid gland known as endemic goiter. Tiny amounts of sugar prevent oxidation of the salt, which otherwise would turn yellow.

THE PLACEBO EFFECT*

People have always sought healers for help with physical pain and fear. As treatment, traditional healers often administer herbs, or chant, or wave their hands over a patient's body. And improvement, more often than not, actually occurs! This is the *placebo effect*. A placebo may be a healing practice or a substance (pill) containing elements or molecules that have no medical value. But, remarkably, the placebo effect does have a biological basis. It so happens that, when you are fearful or in pain, your brain response is *not* to mobilize your body's healing mechanism—it instead prepares your body for some external threat. It's an evolutionary adaptation that assigns highest priority to preventing additional injury. Stress hormones released into the bloodstream increase respiration, blood pressure, and heart rate—changes that usually *impede* recovery. The brain prepares your body for action; recovery can wait.

That's why the first objective of a good healer or physician is to relieve stress. Most of us begin feeling better even before leaving the healer's (or doctor's) office. Prior to 1940, most medicine was based on the placebo effect, when about the only medicines doctors had in their bags were laxatives, aspirin, and sugar pills. In about half the cases, a sugar pill is as effective in stopping pain as an aspirin. Here's why: Pain is a signal to the brain that something is wrong and needs attention. The signal is induced at the site of inflammation by prostaglandins released by white blood cells. Aspirin blocks the production of prostaglandins and therefore relieves the pain. The mechanism for pain relief by a placebo is altogether different. The placebo fools the brain into thinking that whatever is wrong is being cared for. Then the pain signal is lowered by the release of endorphins, opiate-like proteins found naturally in the brain. So, instead of blocking the *production* of prostaglandins, the endorphins block their *effect*. With pain alleviated, the body can focus on healing.

The placebo effect has always been employed (and still is!) by healers and others who claim to have wondrous cures that lie outside modern medicine. These healers benefit from the public's tendency to believe that, if *B* follows *A*, then *B* is *caused* by *A*. The cure could be due to the healer, but it could also merely be the body repairing itself. Although the placebo effect can certainly influence the perception of pain, it has not been shown to influence the body's ability to fight infection or repair injury.

Is the placebo effect at work for those who believe that better health is bequeathed to those who wear crystals, magnets, or certain metal bracelets? If so, is there any harm in thinking so—even if there is no scientific evidence for it? Harboring positive beliefs is usually quite harmless—but not always. For serious problems requiring modern medical treatment, reliance on these aids can be disastrous if used to the exclusion of modern medical treatment. The placebo effect has real limitations.

*Adapted from *Voodoo Science: The Road from Foolishness to Fraud*, by Robert L. Park, Oxford University Press, New York, 2000.

breathe. Two atoms of hydrogen combine with a single atom of oxygen to form a water molecule (H_2O). Changing a molecule by one atom can make a big difference. In chlorophyll, for example, there is a ring of hydrogen, carbon, and oxygen atoms that surrounds a single magnesium atom. Substitute an iron atom for the magnesium atom and it becomes a ring most similar to that found in hemoglobin (an oxygen-carrying protein in our blood). So one atom can make the difference between a molecule that a plant can use and one that a person can use.

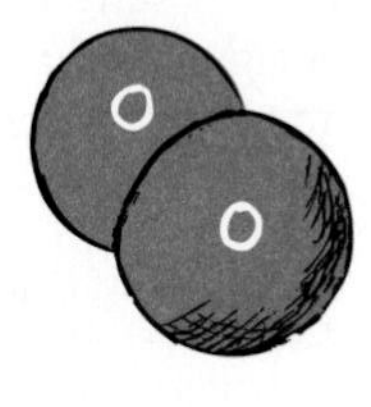

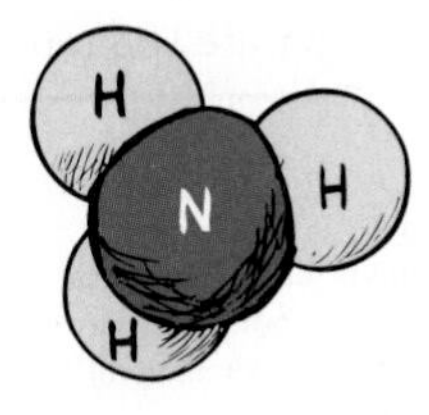

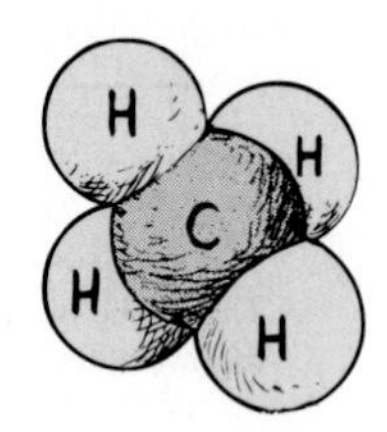

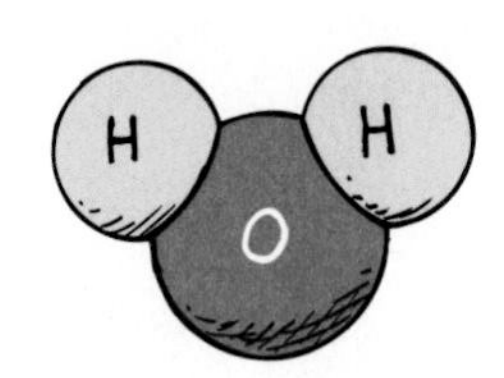

FIGURE 11.11
Models of simple molecules. The atoms in a molecule are not just mixed together but are joined in a well-defined way.

CHECK YOURSELF

How many atomic nuclei are in a single oxygen atom? In a single oxygen molecule?

Energy is required to pull molecules apart. We can understand this by considering a pair of magnets stuck together. Just as some "muscle energy" is required to pull the magnets apart, the breaking apart of molecules requires energy. During photosynthesis, plants use the energy of sunlight to break apart the bonds within atmospheric carbon dioxide and water. The major product of photosynthesis is carbohydrate molecules, which retain this solar energy until the plant is oxidized, either slowly by rotting or quickly by burning. Then the energy that came from the sunlight is released back into the environment. So the slow warmth of decaying compost or the quick warmth of a campfire is really the warmth of stored sunlight!

More things can burn besides those that contain carbon and hydrogen. Iron "burns" (oxidizes) too. That's what rusting is—the slow combination of oxygen atoms with iron atoms, releasing energy. When the rusting of iron is speeded up, it makes nice hand-warmer packs for skiers and winter hikers. Any process in which atoms rearrange to form different molecules is called a *chemical reaction*.

Our sense of smell is sensitive to exceedingly small quantities of molecules. Our olfactory organs easily detect small concentrations of such noxious gases as hydrogen sulfide (the stuff that smells like rotten eggs), ammonia, and ether. The smell of perfume is the result of molecules that rapidly evaporate and diffuse haphazardly in the air until some of them get close enough to your nose to be inhaled. They are just a few of the billions of jostling molecules that, in their aimless wanderings, happen to wind up in the nose. You can get an idea of the speed of molecular diffusion in the air when you are in your bedroom and smell food very soon after the oven door has been opened in the kitchen.

Antimatter

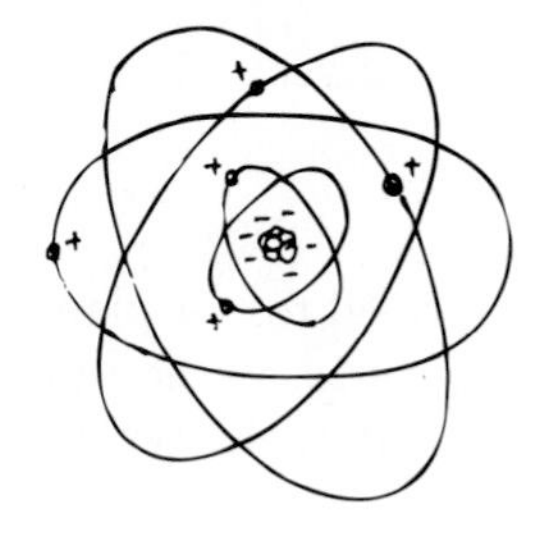

FIGURE 11.12
An atom of antimatter has a negatively charged nucleus surrounded by positrons.

Whereas matter is composed of atoms with positively charged nuclei and negatively charged electrons, **antimatter** is composed of atoms with negative nuclei and positive electrons, or *positrons*.

Positrons were first discovered in 1932, in cosmic rays bombarding the Earth's atmosphere. Today, antiparticles of all types are regularly produced in laboratories using large nuclear accelerators. A positron has the same mass as an electron and the same magnitude of charge but the opposite sign. Antiprotons have the same mass as protons but are negatively charged. The first complete artificial anti-atom, a positron orbiting an antiproton, was constructed in 1995. Every charged particle has an antiparticle of the same mass and opposite charge.

CHECK YOUR ANSWER

There is one nucleus in an oxygen atom (O), and two in the combination of two oxygen atoms—an oxygen molecule (O_2).

Neutral particles (such as the neutron) also have antiparticles, which are alike in mass and in some other properties but opposite in certain other properties. For every particle there is an antiparticle. There are even antiquarks.

Gravitational force does not distinguish between matter and antimatter—each attracts the other. Also, there is no way to indicate whether something is made of matter or antimatter by the light it emits. Only through much subtler, hard-to-measure nuclear effects could we determine whether a distant galaxy is made of matter or antimatter. But, if an antistar were to meet a star, it would be a different story. They would mutually annihilate each other, with most of the matter converting to radiant energy (this is what happened to the anti-atom created in 1995, which rapidly annihilated in a puff of energy). This process, more so than any other known, results in the maximum energy output per gram of substance—$E = mc^2$, with a 100% mass conversion.[3] (Nuclear fission and fusion, by contrast, convert less than 1% of the matter involved.)

There cannot be both matter and antimatter in our immediate environment, at least not in appreciable amounts or for appreciable times. That's because something made of antimatter would be completely transformed to radiant energy as soon as it touched matter, consuming an equal amount of normal matter in the process. If the Moon were made of antimatter, for example, a flash of energetic radiation would result as soon as one of our spaceships touched it. Both the spaceship and an equal amount of the antimatter Moon would disappear in a burst of radiant energy. We know the Moon is not antimatter because this didn't happen during the Moon missions. (Actually, astronauts weren't in this kind of danger, for previous evidence showed that the Moon is made of ordinary matter.) But what about other galaxies? There is strong reason to believe that in the part of the universe we know (the "observable universe"), galaxies are made only of normal matter—apart from the occasional transitory antiparticle. But what of the universe beyond? Or other universes? We don't know.

CHECK YOURSELF

If a one-gram body of antimatter meets a ten-gram body of matter, what mass survives?

Dark Matter

We know that the elements in the periodic table are not confined to our planet. From studies of radiation coming from other parts of the universe, we find that stars and other objects "out there" are composed of the same particles we have

CHECK YOUR ANSWER

Nine grams of matter survive (the other two grams are transformed into radiant energy).

[3]Some physicists speculate that, right after the Big Bang, the early universe had billions of times more particles than it has now, and that a near total extinction of matter and antimatter caused by their mutual anihilation left only the relatively small amount of matter now present in the universe.

on Earth. Stars emit light that produces the same "atomic spectra" (Chapter 30) as the elements in the periodic table. How wonderful to find that the laws that govern matter on Earth extend throughout the observable universe. Yet there remains one troubling detail. In the closing years of the twentieth century, astrophysicists found there is a lot more mass out there than we can directly see.

Astrophysicists talk of the **dark matter**—matter we can't see that tugs on stars and galaxies that we *can* see. Gravitational forces within galaxies are measured to be far greater than visible matter can account for. Only in this twenty-first century has it been confirmed that some 23% of matter in the universe is composed of the unseen dark matter. Whatever dark matter is, most or all of it is likely to be "exotic" matter—very different from the elements that make up the periodic table, and different from any extension of the present list of elements. Much of the rest of the universe is *dark energy* (briefly mentioned in Chapter 7), which pushes outward on the expanding universe. Both dark matter and dark energy make up some 90% of the universe. Dark matter and dark energy seem to be different stuff. At this writing, neither has been identified. Speculations abound about dark matter and dark energy, but we don't know what they are.

Finding the nature of the dark matter and the nature of the vacuum energy are high-priority quests in these times. What we will have learned by midcentury will likely dwarf all that we have ever known.

Insights

Richard Feynman often used to shake his head and say he didn't know anything. When he and other top physicists say they don't know anything, they mean that what they *do* know is closer to nothing than to what they *can* know. Scientists know enough to realize that they have a relatively small handle on an enormous universe still full of mysteries. From a looking-backward point of view, today's scientists know enormously more than their forebears a century ago, and scientists then knew much more than *their* forebears. But, from our present vantage point, looking forward, there is so much yet to be learned. Physicist John A. Wheeler, Feynman's graduate-school advisor, sees the next level of physics going beyond *how* to *why*—to meaning. We have scarcely scratched the surface.

Summary of Terms

Atom The smallest particle of an element that has all of the element's chemical properties.

Brownian motion The haphazard movement of tiny particles suspended in a gas or liquid resulting from their bombardment by the fast-moving atoms or molecules of the gas or liquid.

Atomic nucleus The core of an atom, consisting of two basic subatomic particles—protons and neutrons.

Element A pure substance consisting of only one kind of atom.

Atomic number The number that designates the identity of an element, which is the number of protons in the nucleus of an atom; in a neutral atom, the atomic number is also the number of electrons in the atom.

The periodic table of the elements A chart that lists the elements in horizontal rows by their atomic number and in vertical columns by their similar electron arrangements and chemical properties. (See Figure 11.9.)

Quantum mechanics The theory of the small-scale world that includes predicated wave properties of matter.

Isotopes Different forms of an atom that contains the same number of protons but different numbers of neutrons.

Atomic mass unit (amu) The standard unit of atomic mass, which is equal to one-twelfth the mass of the common atom of carbon, arbitrarily given the value of exactly 12. One amu has a mass of 1.661×10^{-24} grams.

Compound A material in which atoms of different elements are chemically bonded to one another.

Mixture A substance whose components are mixed together without combining chemically.

Molecule A group of atoms held together by a sharing of electrons. Atoms combine to form molecules.

Antimatter A "complementary" form of matter composed of antiparticles having the same mass as particles of ordinary matter but being opposite in charge.

Dark matter Unseen and unidentified matter that is evident by its gravitational pull on stars in the galaxies. Along with *dark energy,* dark matter constitutes perhaps 90% of the stuff of the universe.

Suggested Reading

Feynman, R. P., R. B. Leighton, and M. Sands. *The Feynman Lectures on Physics,* vol. I, chap. 1. Reading, Mass.: Addison-Wesley, 1963.

Rigden, John S. *Hydrogen: The Essential Element.* Cambridge, Mass.: Harvard University Press, 2002.

Suchocki, J. *Conceptual Chemistry,* 2nd ed., chap. 5. San Francisco: Benjamin Cummings, 2004. Contains an excellent treatment of the periodic table.

Review Questions

The Atomic Hypothesis

1. What causes dust particles and tiny grains of soot to move with Brownian motion?
2. Who first explained Brownian motion and made a convincing case for the existence of atoms?
3. According to Richard Feynman, when do atoms attract each other, and when do they repel?

Atom Characteristics

4. How does the approximate number of atoms in the air in your lungs compare with the number of breaths of air in the atmosphere of the Earth?
5. Are most of the atoms around us younger or older than the Sun?

Atomic Imagery

6. Why can atoms not be seen with a powerful optical microscope?
7. Why can atoms be seen with an electron beam?
8. What is the purpose of a model in science?

Atomic Structure

9. How does the mass of an atomic nucleus compare with the mass of an atom as a whole?
10. What is a nucleon?
11. How does the mass and electric charge of a proton compare with the mass and charge of an electron?
12. Since atoms are mostly empty space, why don't we fall through a floor we stand on?

The Elements

13. What is the lightest of the elements?
14. What is the most abundant element in the known universe?
15. How were elements heavier than hydrogen formed?
16. Where did the heaviest elements originate?
17. What are the five most common elements in living things?

Periodic Table of the Elements

18. What does the atomic number of an element tell you about the element?
19. What is characteristic of the columns in the periodic table?

Isotopes

20. What are isotopes?
21. Distinguish between *mass number* and *atomic mass.*

Compounds and Mixtures

22. What is a compound? Give three examples.
23. What is a mixture? Give three examples.

Molecules

24. How does a molecule differ from an atom?
25. Compared with the energy it takes to separate oxygen and hydrogen from water, how much energy is given off when they recombine? (What general physics principle is illustrated here?)

Antimatter

26. How do matter and antimatter differ?
27. What occurs when a particle of matter and a particle of antimatter meet?

Dark Matter

28. What is the evidence that dark matter exists?

Project

A candle will burn only if oxygen is present. Will a candle burn twice as long in an inverted liter jar as it will in an inverted half-liter jar? Try it and see.

Exercises

1. How many types of atoms can you expect to find in a pure sample of any element?
2. How many individual atoms are in a water molecule?
3. When a container of gas is heated, what happens to the average speed of its molecules?
4. The average speed of a perfume-vapor molecule at room temperature may be about 300 m/s, but you'll find the speed at which the scent travels across the room is much less. Why?

5. A cat strolls across your backyard. An hour later, a dog with his nose to the ground follows the trail of the cat. Explain this occurrence from a molecular point of view.

6. If no molecules in a body could escape, would the body have any odor?

7. Where were the atoms that make up a newborn infant "manufactured"?

8. Which of the following is not an element: hydrogen, carbon, oxygen, water?

9. Your friend says that what makes one element distinct from another is the number of electrons about the atomic nucleus. Do you agree wholeheartedly, partially, or not at all? Explain.

10. Can two different elements contain the same total number of protons? If so, give an example.

11. What is the cause of the Brownian motion of dust particles? Why aren't larger objects, such as baseballs, similarly affected?

12. Why is Brownian motion apparent only for microscopic particles?

13. Why don't equal masses of golf balls and Ping-Pong balls contain the same number of balls?

14. Why don't equal masses of carbon atoms and oxygen atoms contain the same number of particles?

15. Which contains more atoms: 1 kg of lead or 1 kg of aluminum?

16. Which of the following are pure elements: H_2, H_2O, He, Na, NaCl, H_2SO_4, U?

17. How many atoms are in a molecule of ethanol, C_2H_6O?

18. The atomic masses of two isotopes of cobalt are 59 and 60. (a) What is the number of protons and neutrons in each? (b) What is the number of orbiting electrons in each when the isotopes are electrically neutral?

19. A particular atom contains 29 electrons, 34 neutrons, and 29 protons. What is the atomic number of this element, and what is its name?

20. Gasoline contains only hydrogen and carbon atoms, yet nitrogen oxide and nitrogen dioxide are produced when gasoline burns. What is the source of the nitrogen and oxygen atoms?

21. A tree is composed mainly of carbon. Where does this carbon come from?

22. How do the number of protons in an atomic nucleus dictate the chemical properties of the element?

23. Which would be the more valued result: taking one proton from each nucleus in a sample of gold or adding one proton to each gold nucleus? Explain.

24. If two protons and two neutrons are removed from the nucleus of an oxygen atom, what nucleus remains?

25. What element results if you add a pair of protons to the nucleus of mercury? (See the periodic table.)

26. What element results if two protons and two neutrons are ejected from a radium nucleus?

27. How does an ion differ from an atom?

28. To become a negative ion, does an atom lose or gain an electron?

29. To become a positive ion, does an atom lose or gain an electron?

30. Fish don't live very long in water that has been boiled and brought back to room temperature. Provide an explanation for this fact.

31. You could swallow a capsule of germanium without ill effects. But, if a proton were added to each of the germanium nuclei, you would not want to swallow the capsule. Why? (Consult the periodic table.)

32. An ozone molecule and an oxygen molecule are both pure oxygen. How are they different?

33. If you eat metallic sodium or inhale chlorine gas, you stand a strong chance of dying. When these two elements combine, however, you can safely sprinkle the resulting compound on your popcorn for better taste. What is going on?

34. What results when water is chemically decomposed?

35. Helium is an inert gas, meaning that it doesn't readily combine with other elements. What five other elements would you also expect to be inert gases? (See the periodic table.)

36. What element results if one of the neutrons in a nitrogen nucleus is converted by radioactive decay into a proton?

37. Which of the following elements would you predict to have properties most like those of silicon (Si): aluminum (Al), phosphorus (P), or germanium (Ge)? (Consult the periodic table.)

38. Carbon, with a half-full outer shell of electrons—four in a shell that can hold eight—readily shares its electrons with other atoms and forms a vast number of molecules, many of which are the organic molecules that form the bulk of living matter. Looking at the periodic table, what other element do you think might play a role like carbon in life forms on some other planet?

39. Which contributes more to an atom's mass—electrons or protons? Which contributes more to an atom's size?

40. A hydrogen atom and a carbon atom move at the same speed. Which has the greater kinetic energy?

41. In a gaseous mixture of hydrogen and oxygen gas, both with the same average kinetic energy, which molecules move faster on average?

42. The atoms that constitute your body are mostly empty space, and structures such as the chair you're sitting on are composed of atoms that are also mostly empty space. So why don't you fall through the chair?

43. When 50 cubic centimeters (cm^3) of alcohol are mixed with 50 cm^3 of water, the volume of the mixture is only 98 cm^3. Can you offer an explanation for this?

44. (a) From an atomic point of view, why must you heat a solid to melt it? (b) If you have a solid and a liquid at room temperature, what conclusion can you draw about the relative strengths of their interatomic forces?

45. In what sense can you truthfully say that you are a part of every person in history? In what sense can you say that you will tangibly contribute to every person on Earth who will follow?

46. What are the chances that at least one of the atoms exhaled by your very first breath will be in your next breath?

47. Hydrogen and oxygen always react in a 1:8 ratio by mass to form water. Early investigators thought this meant that oxygen was eight times more massive than hydrogen. What chemical formula did these investigators assume for water?

48. When antimatter meets matter, what is produced, and how much percentage wise?

49. Somebody told your friend that, if an antimatter alien ever set foot upon the Earth, the whole world would explode into pure radiant energy. Your friend looks to you for verification or refutation of this claim. What do you say?

50. Make up a multiple-choice question that will test your classmates on the distinction between any two terms in the Summary of Terms list.

Problems

1. How many grams of oxygen are there in 18 g of water?

2. How many grams of hydrogen are there in 16 g of methane gas? (The chemical formula for methane is CH_4.)

3. Gas A is composed of diatomic molecules (two atoms to a molecule) of a pure element. Gas B is composed of monatomic molecules (one atom to a "molecule") of another pure element. Gas A has three times the mass of an equal volume of gas B at the same temperature and pressure. How do the atomic masses of elements A and B compare?

4. A teaspoon of an organic oil dropped on the surface of a quiet pond spreads out to cover almost an acre. The oil film has a thickness equal to the size of a molecule. In the lab, when you drop 0.001 milliliter (10^{-9} m^3) of the organic oil on the still surface of water, you find that it spreads to cover an area of 1.0 m^2. If the layer is one molecule thick, what is the size of a single molecule?

The problems that follow require some knowledge of exponents.

5. The diameter of an atom is about 10^{-10} m. (a) How many atoms make a line a millionth of a meter (10^{-6} m) long? (b) How many atoms cover a square a millionth of a meter on a side? (c) How many atoms fill a cube a millionth of a meter on a side? (d) If a dollar were attached to each atom, what could you buy with your line of atoms? With your square of atoms? With your cube of atoms?

6. There are approximately 10^{23} H_2O molecules in a thimbleful of water and 10^{46} H_2O molecules in the Earth's oceans. Suppose that Columbus threw a thimbleful of water into the ocean and that those water molecules have by now mixed uniformly with all the water molecules in the oceans. Can you show that, if you dip a sample thimbleful of water from anywhere in the ocean, you'll probably scoop up at least one of the molecules that was in Columbus's thimble? (*Hint:* The ratio of the number of molecules in a thimble to the number of molecules in the ocean will equal the ratio of the number of molecules in question to the number of molecules the thimble can hold.)

7. There are approximately 10^{22} molecules in a single medium-sized breath of air and approximately 10^{44} molecules in the atmosphere of the whole world. The number 10^{22} squared is equal to 10^{44}. So how many breaths of air are there in the world's atmosphere? How does this number compare with the number of molecules in a single breath? If all the molecules from Julius Caesar's last dying breath are now thoroughly mixed in the atmosphere, how many of these, on the average, do we inhale with each single breath?

8. Assume that the present world population of about 6×10^9 people is about 1/20 the number of people who ever lived on Earth. How does the number of people who ever lived compare to the number of air molecules in a single breath?

CHAPTER 12

Solids

Former American Association of Physics Teachers president John Hubisz displays an enlarged image of Eric Müller's famous micrograph.

Humans have been using solid materials for many thousands of years. The terms Stone Age, Bronze Age, and Iron Age tell us the importance of solid materials in the development of civilization. Wood and clay were also important to early peoples, and gems were put to use for art and adornment. The numbers and uses of materials multiplied over the centuries, yet there was little progress in understanding the nature of solids. This understanding had to await discoveries about atoms that occurred in the twentieth century. Armed with knowledge of the atom, today's chemists, metallurgists, and materials scientists invent new materials daily. Solid-state physicists explore semiconductors and other solids and tailor them to meet the demands of our information age.

Müller's Micrograph

A striking example showing the structure of solid matter is shown above. The picture is a micrograph produced back in 1958 by Erwin Müller. Dr. Müller used an extremely fine platinum needle with a hemisphere-shaped tip that was 40-millionths of a centimeter in diameter. The needle was enclosed in a tube of rarefied helium and subjected to a very high positive voltage (25,000 volts). This voltage created such an intense electric force that any helium atoms "settling" on the atoms of the needle tip were stripped of electrons to become *ions*. Positively charged helium ions streamed away from the platinum needle tip in a direction almost perpendicular to its surface at every point. They then struck a fluorescent screen, producing this picture of the needle tip, which magnifies the spacing of the atoms by approximately 750,000 times. Clearly, the platinum is crystalline, with the atoms arranged like oranges in a grocer's display. Although the photograph is not of the atoms themselves, it shows the positions of the atoms and reveals the microarchitecture of one of the solids that make up our world.

Crystal Structure

Metals, salts, and most minerals—the materials of the Earth—are made up of crystals. People have known about such crystals as salt and quartz for centuries, but it wasn't until the twentieth century that crystals were interpreted as regular arrays of atoms. X rays were used in 1912 to confirm that each crystal is a three-dimensional orderly arrangement—a crystalline latticework of atoms. The atoms in a crystal were measured to be very close together, about the same distance apart as the wavelength of X rays. The German physicist Max von Laue discovered that a beam of X rays directed upon a crystal is diffracted, or separated, into a characteristic pattern (Figure 12.1). X-ray *diffraction patterns* on photographic film show crystals to be neat mosaics of atoms sited on regular lattices, like three-dimensional chessboards or children's jungle gyms. Such metals as iron, copper, and gold have relatively simple crystal structures. Tin and cobalt are a little more complex. All metals contain a jumble of many crystals, each almost perfect, each with the same regular lattice but at some inclination to the crystal nearby. These metal crystals can be seen when a metal surface is *etched,* or cleaned, with acid. You can see the crystal structures on the surface of galvanized iron exposed to the weather, or on brass doorknobs etched by the perspiration of many hands.

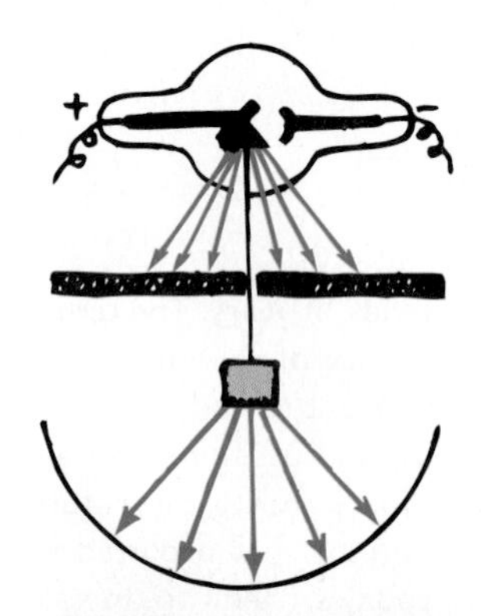

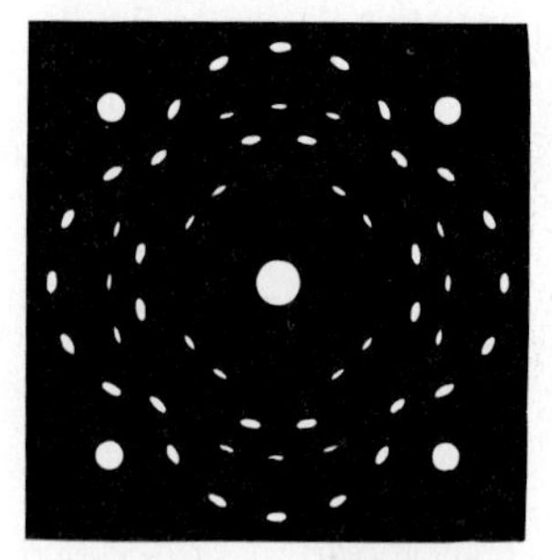

FIGURE 12.1
X-ray determination of crystal structure. The photograph of salt is a product of X-ray diffraction. Rays from the X-ray tube are blocked by a lead screen except for a narrow beam that hits the crystal of sodium chloride (common table salt). The radiation that penetrates the crystal and reaches the photographic film makes the pattern shown. The white spot in the center is due to the main unscattered beam of X rays. The size and arrangement of the other spots result from the latticework structure of sodium and chlorine ions in the crystal. A crystal of sodium chloride always produces this same design. Every crystalline structure has its own unique X-ray diffraction picture.

Von Laue's photographs of the X-ray diffraction patterns fascinated English scientists William Henry Bragg and his son William Lawrence Bragg. They developed a mathematical formula that showed just how X rays should scatter from the various regularly spaced atomic layers in a crystal. With this formula and an analysis of the pattern of spots in a diffraction pattern, they could determine distances between the atoms in a crystal. X-ray diffraction is a vital tool today in the biological and physical sciences.

Diffraction patterns of DNA (similar to the one of salt shown here), made by Maurice Wilkins and Rosalind Franklin in 1953, provided the data from which James D. Watson and Francis Crick discovered the double helix of DNA.

Insights

FIGURE 12.2
The crystalline structure of diamond is illustrated with sticks to represent the covalent bonds responsible for its extreme hardness.

CRYSTAL POWER

The regularly repeating internal structure of atoms in crystals gives them aesthetic properties that have long made them attractive in jewelry. Crystals also have properties that are very important to the electronics and optical industries, and they are used in just about every type of modern technology. In times past, crystals were valued for their alleged healing powers. This belief continues today, particularly among occultists and New Age healers. Crystals are said to channel "good energy" and ward off "bad energy." They carry "vibrations" that resonate with healing "frequencies" that help maintain a beneficial body balance. When properly arranged, crystals are said to provide protection against harmful electromagnetic forces emitted by power lines, cellular phones, computer monitors, microwave ovens, and other people. We are told that crystals are "medically proven" to heal and to protect, and that such claims are "based on Nobel Prize–winning physics."

Crystals *do* give off energy—as every other object does. We'll learn in Chapter 16 that all things radiate energy—and also that all things absorb energy. If a crystal or any substance radiates more energy than it receives, its temperature drops. Atoms in crystals *do* vibrate and *do* resonate with matching frequencies of external vibrations—just as molecules in all gases and in all liquids do. If purveyors of crystal power are talking about some kind of energy special to crystals, or to life, no scientific evidence supports this (the discoverer of such a special kind of energy would quickly become world famous). Of course, evidence for a new kind of energy, such as the dark energy discussed in the previous chapter, could one day be found, but this is not what is claimed by purveyors of crystal power, who assert that modern scientific evidence supports their claims.

The evidence for crystal power is not experimental; rather, it is confined to *testimonials.* As advertising illustrates, people are generally more persuaded by testimonials than by stated facts. Testimonials by people who are convinced of the personal benefits of crystals are common. Being convinced by scientific evidence is one thing; being convinced by wishful thinking, communal reinforcement, or by a placebo effect is quite another. None of the claims for the special powers of crystals have ever been backed up by scientific evidence.

Claims aside, wearing crystal pendants seems to give some people a good *feeling,* and even a feeling of protection. These feelings, and the aesthetic qualities of crystals, are their virtues. Some people feel that crystals bring good luck, just as carrying a rabbit's foot in your pocket is supposed to do. The difference between crystal power and rabbit's feet, however, is that the benefits of crystals are couched in scientific terms, whereas claims for the benefits of carrying a rabbit's foot are not. Hence, the purveyors of crystal power are into full-fledged pseudoscience.

Noncrystalline solids are said to be *amorphous*. In the amorphous state, atoms and molecules in a solid are distributed randomly. Rubber, glass, and plastic are among the materials that lack an orderly, repetitive arrangement of their basic particles. In many amorphous solids, the particles have some freedom to wander. This is evident in the elasticity of rubber and the tendency of glass to flow when subjected to stress over long periods of time.

Whether atoms are in a crystalline state or in an amorphous state, each atom or ion vibrates about its own position. Atoms are tied together by electrical bonding forces. We'll not discuss **atomic bonding** now, except to mention

FIGURE 12.3
The cubic crystalline structure of common salt as seen through a microscope. The cubic shape is a consequence of the cubic arrangement of sodium and chloride ions.

the four principal types of bonding in solids: ionic, covalent, metallic, and Van der Waals', the last being the weakest. Some properties of a solid are determined by the types of bonds it has. More information can be found about these bonds in almost any chemistry text.

Density

FIGURE 12.4
When the volume of the bread is reduced, its density increases.

Is iron heavier than wood? The question is ambiguous, for it depends on the amounts of iron and wood. A large log is clearly heavier than an iron nail. A better question to ask is if iron is *denser* than wood, in which case the answer is *yes,* iron is denser than wood. The masses of the atoms and the spacing between them determine the **density** of materials. We think of density as the "lightness" or "heaviness" of materials of the same size. It's a measure of the compactness of matter, of how much mass occupies a given space; it is the amount of mass per unit volume:

$$\text{Density} = \frac{\text{mass}}{\text{volume}}$$

The densities of a few materials are shown in Table 12.1. Density is usually expressed with metric units, generally kilograms per cubic meter, kilograms per liter, or grams per cubic centimeter. Water, for example has a mass density of 1000 kg/m^3, or, equivalently, 1 g/cm^3. So the mass of a cubic meter of fresh water is 1000 kilograms, and, equivalently, the mass of a cubic centimeter (about the size of a sugar cube) of fresh water is 1 g.

Density may be expressed in terms of weight rather than mass. This is weight density, which is defined as weight per unit volume:

$$\text{Weight density} = \frac{\text{weight}}{\text{volume}}$$

Weight density is measured in N/m^3. Because a 1-kg body has a weight of 9.8 N, weight density is numerically 9.8 × mass density. For example, the weight

TABLE 12.1
Densities of Common Substances (kg/m^3)

(For density in g/cm^3, divide by 1000)

Solids	Density
Iridium	22,650
Osmium	22,610
Platinum	21,090
Gold	19,300
Uranium	19,050
Lead	11,340
Silver	10,490
Copper	8,920
Brass	8,600
Iron	7,874
Tin	7,310
Aluminum	2,700
Concrete	2,300
Ice	919
Liquids	
Mercury	13,600
Glycerin	1,260
Seawater	1,025
Water at 4°C	1,000
Ethyl alcohol	785
Gasoline	680

FIGURE 12.5
Both Stephanie and the tree are composed mainly of hydrogen, oxygen, and carbon. Food ingestion supplies these to Stephanie, whereas the tree gets most of its oxygen and carbon from the air. In this sense, a tree can be thought of as "solid air."

density of water is 9800 N/kg^3. In the British system, 1 cubic foot (ft^3) of fresh water (almost 7.5 gallons) weighs 62.4 pounds. Thus, in the British system, fresh water has a weight density of 62.4 lb/ft^3.

Iridium, a hard, brittle, silvery-white metal in the platinum family, is the densest substance on Earth. Although the individual iridium atom is less massive than individual atoms of platinum, gold, lead, or uranium, the close spacing of iridium atoms in the crystalline form contributes to its greater density. More iridium atoms fit into a cubic centimeter than other more massive but more widely spaced atoms. Hence iridium has a whopping density of 22,650 kg/m^3.

A cubic meter is a sizable volume and contains a million cubic centimeters, so there are a million grams of water in a cubic meter (or, equivalently, a thousand kilograms of water in a cubic meter). Hence, 1 g/cm^3 = 1000 kg/m^3.

Insights

CHECK YOURSELF

1. *Here's an easy one:* When water freezes, it expands. What does this say about the density of ice compared with the density of water?
2. *Here's a slightly tricky one*: Which weighs more, a liter of ice or a liter of water?
3. Which has the greater density—100 kg of lead or 1000 kg of aluminum?
4. What is the density of 1000 kg of water?
5. What is the volume of 1000 kg of water?

Elasticity

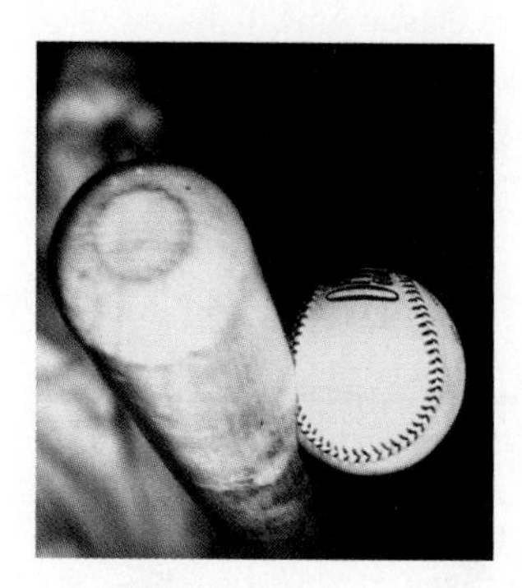

FIGURE 12.6
A baseball is elastic.

When an object is subjected to external forces, it undergoes changes in size, or in shape, or in both. The changes depend on the arrangement and bonding of the atoms in the material. A spring, for example, can be stretched or compressed by external forces.

A weight hanging on a spring stretches the spring. Additional weight stretches it further. If the weights are removed, the spring returns to its original length. We say that the spring is *elastic*. When a batter hits a baseball, the baseball temporarily changes its shape. An archer, about to shoot an arrow, first bends the bow, which springs back to its original shape when the arrow is released. The spring, the baseball, and the bow are examples of elastic objects. **Elasticity** is the property of a material wherein it changes shape when a deforming

CHECK YOUR ANSWERS

1. Ice is less dense than water (because it has more volume for the same mass), which is why ice floats on water.
2. Don't say that they weigh the same! A liter of water weighs more. If it is frozen, then its volume will be more than a liter. If you shave that part off so that the chunk of ice is the same size as the original liter of water, it will certainly weigh less.
3. Density is a *ratio* of mass and volume (or weight and volume), and this ratio is greater for any amount of lead than for any amount of aluminum—see Table 12.1.
4. The density of *any* amount of water is 1000 kg/m^3 (or 1 g/cm^3).
5. The volume of 1000 kg of water is 1 m^3.

force acts upon it, and returns to its original shape when the force is removed. Not all materials return to their original shape when a deforming force is applied and then removed. Materials that do not resume their original shape after being deformed are said to be *inelastic*. Clay, putty, and dough are inelastic materials. Lead is also inelastic, since it's easy to distort lead permanently.

When a weight is hung on a spring, a force (gravity) acts upon it. The stretch is directly proportional to the applied force (Figure 12.7). Similarly, when you climb into bed, the compression of bedsprings is directly proportional to the weight applied. This relationship, noted in the mid-seventeenth century by the British physicist Robert Hooke, a contemporary of Isaac Newton, is called **Hooke's law.** The amount of stretch or compression (change in length), Δx, is directly proportional to the applied force, F. In shorthand notation,

$$F \sim \Delta x$$

If an elastic material is stretched or compressed beyond a certain amount, it will not return to its original state and will remain distorted. The distance beyond which permanent distortion occurs is called the *elastic limit*. Hooke's law holds only as long as the force does not stretch or compress the material beyond its elastic limit.

Robert Hooke, one of England's greatest scientists, was the first to propose a wave theory of light and the first to describe the cell (for which he became known as the father of microscopy). As an artist and surveyor, he helped Christopher Wren rebuild London after the great fire in 1766. As a physicist, he collaborated with Robert Boyle and other physicists of his time and was chosen to head the Royal Society. Upon his death, Isaac Newton became president of the Royal Society, and he jealously destroyed everything he could of Hooke's work. No paintings or likenesses of Hooke survive today.

Insights

CHALLENGE YOURSELF

1. A 2-kg load is hung from the end of a spring. The spring then stretches a distance of 10 cm. If, instead, a 4-kg load is hung from the same spring, how much will the spring stretch? What if a 6-kg load were hung from the same spring? (Assume that none of these loads stretches the spring beyond its elastic limit.)
2. If a force of 10 N stretches a certain spring 4 cm, how much stretch will occur for an applied force of 15 N?

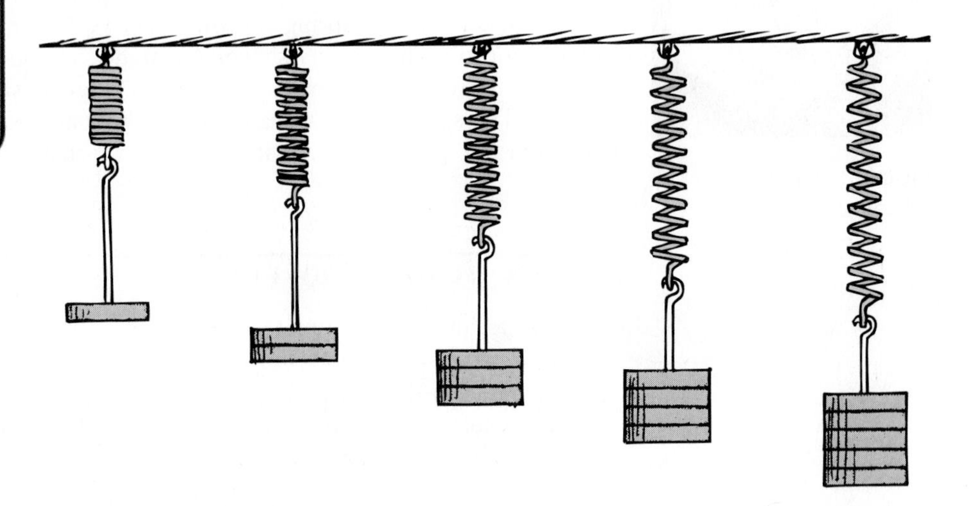

FIGURE 12.7
The stretch of the spring is directly proportional to the applied force. If the weight is doubled, the spring stretches twice as much.

Tension and Compression

When something is pulled on (stretched), it is said to be in *tension*. When it is pushed in (squashed), it is in *compression*. If you bend a ruler (or any stick), the part bent on the outside of the curve is in tension. The inner curved part, which is pushed in, is in compression. Compression causes things to get shorter and wider, whereas tension causes them to get longer and thinner. This is not obvious for most rigid materials, however, because the shortening or lengthening is very small.

Steel is an excellent elastic material, because it can withstand large forces and then return to its original size and shape. Because of its strength and elastic properties, it is used to make not only springs but construction girders as well. Vertical girders of steel used in the construction of tall buildings undergo only slight compression. A typical 25-meter-long vertical girder (column) used in high-rise construction is compressed about a millimeter when it carries a 10-ton load. This can add up. A 70- to 80-story building can compress the huge steel columns at its base by some 2.5 centimeters (a full inch) when construction is finished.

More deformation occurs when girders are used horizontally, where their tendency is to sag under heavy loads. When a horizontal beam is supported at one or both ends, it is under both tension and compression from its weight and the load it supports. Consider the horizontal beam supported at one end (known as a cantilever beam) in Figure 12.8. It sags because of its own weight and because of the load it carries at its end. A little thought will show that the top part of the beam tends to be stretched. Atoms tend to be pulled apart. The top

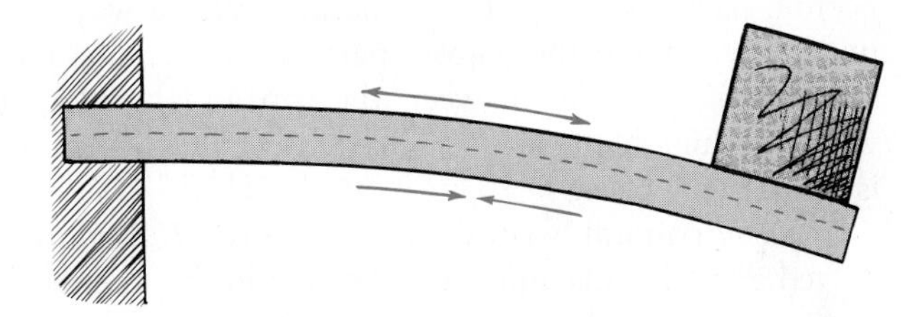

FIGURE 12.8
The top part of the beam is stretched and the bottom part is compressed. What happens in the middle portion, between top and bottom?

CHECK YOUR ANSWERS

1. A 4-kg load has twice the weight of a 2-kg load. In accord with Hooke's law, $F \sim \Delta x$, two times the applied force results in two times the stretch, so the spring should stretch 20 cm. The weight of the 6-kg load makes the spring stretch three times as far, 30 cm. (If the elastic limit were exceeded, the amount of stretch could not be predicted with the information given.)
2. The spring will stretch 6 cm. By ratio and proportion, (10 N)/(4 cm) = (15 N)/(6 cm), which is read: 10 newtons is to 4 centimeters as 15 newtons is to 6 centimeters. If taking a lab, you'll learn that the ratio of force to stretch is called the *spring constant* k (in this case k = 2.5 N/cm), and Hooke's law is expressed as the equation $F = k\Delta x$.

PRACTICING PHYSICS

If you nail four sticks together to form a rectangle, they can be deformed into a parallelogram without too much effort. But, if you nail three sticks together to form a triangle, the shape cannot be changed without actually breaking the sticks or dislodging the nails. The triangle is the strongest of all the geometrical figures, which is why you see triangular shapes in bridges. Go ahead and nail three sticks together and see, and then look at the triangles used in strengthening structures of many kinds.

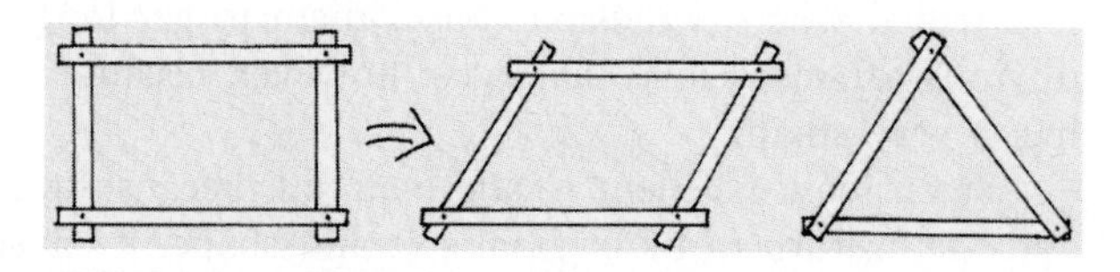

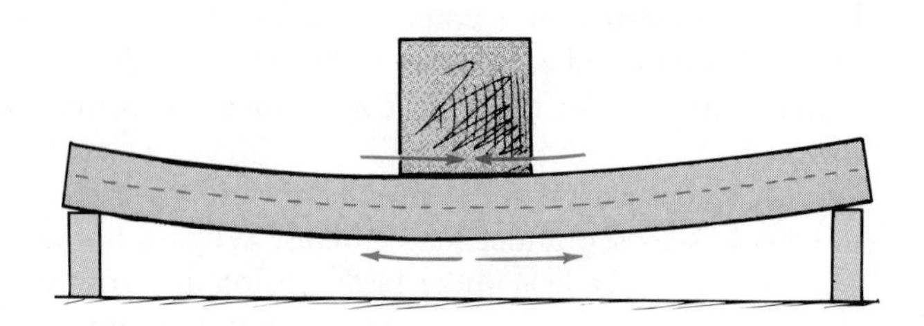

FIGURE 12.9
The top part of the beam is compressed and the bottom part is stretched. Where is the neutral layer (the part that is not under stress due to tension or compression)?

part is slightly longer and is under tension. More thought shows that the bottom part of the beam is under compression. Atoms there are squeezed together. The bottom part is slightly shorter because of the way it is bent. So the top part is under tension and the bottom part of the beam is under compression. Can you see that, somewhere between the top and bottom, there will be a region in which nothing happens, in which there is neither tension nor compression? This is the *neutral layer.*

The horizontal beam shown in Figure 12.9, known as a simple beam, is supported at both ends and carries a load in the middle. Here there is compression in the top of the beam and tension in the bottom part. Again, there is a neutral layer along the middle portion of the thickness of the beam all along its length.

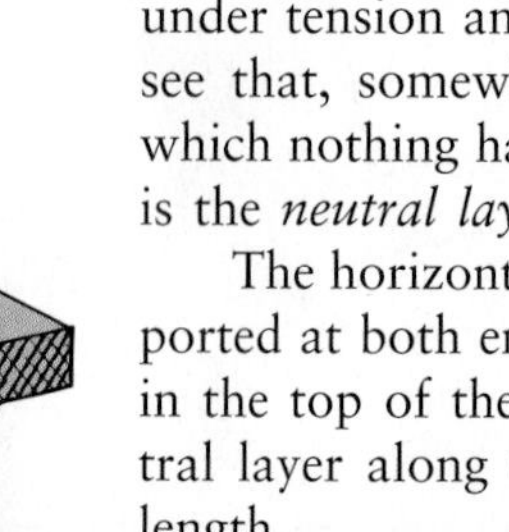

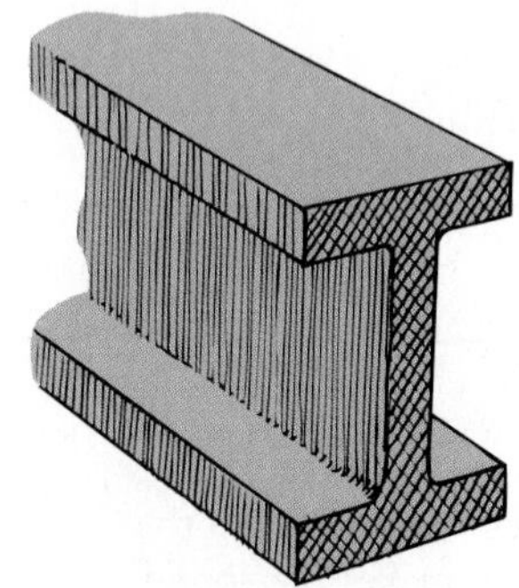

FIGURE 12.10
An I-beam is like a solid bar with some of the steel scooped from its middle where it is needed least. The beam is therefore lighter, but it has nearly the same strength.

With the neutral layer in mind, we can see why the cross section of steel girders has the form of the letter I (Figure 12.10). Most of the material in these I-beams is concentrated in the top and bottom *flanges*. When the beam is used horizontally in construction, the stress is predominantly in the top and bottom flanges. One flange is squeezed while the other is stretched, the two flanges carrying virtually all stresses in the beam. Between the flanges is a relatively stress-free region, the *web,* that acts principally to hold the top and bottom flanges well apart. Because of the web, comparatively little material is needed. An I-beam is nearly as strong as a solid rectangular bar of the same overall dimensions, with considerably less weight. A large rectangular steel beam on a certain span might fail under its own weight, whereas an I-beam of the same depth could carry much added load.

FIGURE 12.11
The upper half of each horizontal branch is under tension because of the branch's weight, while the lower half is under compression. In what location is the wood neither stretched nor compressed?

CHECK YOURSELF

1. When you walk on floorboards that sag due to your weight, where is the neutral layer?
2. Suppose you drill a hole horizontally through a tree branch as shown. Where will the hole weaken the branch the least—through the upper portion, the middle portion, or the lower portion?

Arches

FIGURE 12.12
Common semicircular stone arches, which have stood for centuries.

Stone breaks more easily under tension than compression. The roofs of stone structures built by the Egyptians during the time of the pyramids were constructed with many horizontal stone slabs. Because of the weakness of these slabs under the tension forces produced by gravity, many vertical columns had to be erected to support the roofs. Likewise for the temples in ancient Greece. Then came arches—and fewer vertical columns.

Look at the tops of the windows in old brick buildings. Chances are the tops are arched. Likewise for the shapes of old stone bridges. When a load is placed on a properly arched structure, the compression strengthens rather than weakens the structure. The stones are pushed together more firmly and are held together by compressive forces. With just the right shape of the arch, the stones do not even need cement to hold them together. When the load being supported is uniform and extends horizontally, as with a bridge, the proper shape is a

CHECK YOUR ANSWERS

1. The neutral layer is midway between the top and bottom surfaces of the boards.
2. Drill the hole horizontally through the middle of the branch, through the neutral layer—a hole there will hardly affect the branch's strength because fibers there are neither stretched nor compressed. Wood fibers in the top part are being stretched, so a hole there may result in fibers pulling apart. Fibers in the lower part are compressed, so a hole there might crush under compression.

FIGURE 12.13
The horizontal slabs of stone of the roof cannot be too long because stone easily breaks under tension. That is why many vertical columns are needed to support the roof.

FIGURE 12.14
Both the curve of a sagging chain and the Gateway Arch in St. Louis are catenaries.

parabola, the same curve followed by a thrown ball. The cables of a suspension bridge provide an example of an "upside-down" parabolic arch. If, on the other hand, the arch is supporting only its own weight, the curve that gives it maximum strength is called a *catenary*. A catenary is the curve formed by a rope or chain hung between two points of support. Tension along every part of the rope or chain is parallel to the curve. So, when a free-standing arch takes the shape of an inverted catenary, compression within it is everywhere parallel to the arch, just as tension between adjacent links of a hanging chain is everywhere parallel to the chain. The Gateway Arch, which graces the waterfront of St. Louis, Missouri, is a catenary (Figure 12.14).

If you rotate an arch through a complete circle, you have a dome. The weight of the dome, like that of an arch, produces compression. Modern domes, such as the Astrodome in Houston, are three-dimensional catenaries, and they cover vast areas without the interruption of columns. There are shallow domes (the Jefferson Monument) and tall ones (the United States Capitol). And long before these, there were the igloos in the Arctic.

A catenary arch could even be made with slippery blocks of ice! Provided that the compressive forces between blocks are parallel to the arch and that temperature doesn't rise enough to cause melting, the arch will remain stable.

Insights

FIGURE 12.15
The weight of the dome produces compression, not tension, so no support columns are needed in the middle.

CHECK YOURSELF

Why is it easier for a chicken inside an eggshell to poke its way out than it is for a chicken on the outside to poke its way in?

Scaling[1]

Scaling

Have you ever noticed how strong an ant is for its size? An ant can carry the weight of several ants on its back, whereas a strong elephant would have great difficulty carrying even a single elephant. How strong would an ant be if it were scaled up to the size of an elephant? Would this "super ant" be several times stronger than an elephant? Surprisingly, the answer is *no*. Such an ant would be unable to lift its own weight off the ground. Its legs would be too thin for its greater weight, and they would likely break.

There is a reason for the thin legs of the ant and the thick legs of an elephant. As the size of a thing increases, it grows heavier much faster than it grows stronger. You can support a toothpick horizontally at its ends and notice no sag. But support an entire tree trunk of the same kind of wood horizontally at its ends and you'll notice an appreciable sag. Relative to its weight, the toothpick is much stronger than the tree. **Scaling** is the study of how the volume and shape (size) of any object affect the relationship of its weight, strength, and surface area.

Strength is related to the area of the cross section (which is two-dimensional and is measured in *square* centimeters), whereas *weight* relates to volume (which is three-dimensional and is measured in *cubic* centimeters). To understand this square–cube relationship, consider the simplest case of a solid cube of matter, 1 centimeter on a side—say, a sugar cube. Any 1-cubic-centimeter cube has a cross section of 1 square centimeter. That is, if we sliced through the cube parallel to one of its faces, the sliced area would be 1 square centimeter.

CHECK YOUR ANSWER

To poke its way into a shell, a chicken on the outside must contend with compression, which greatly resists shell breakage. But, when poking from the inside, only the weaker shell tension must be overcome. To see that the compression of the shell requires greater force, try crushing an egg along its long axis by squeezing it between your thumb and fingers. Surprised? Try it along its shorter diameter. Surprised? (Do this over a sink with some protection, such as gloves, for possible shell splinters.)

[1]Scaling was studied by Galileo, who differentiated the bone sizes of various creatures. Material in this section is based on two delightful and informative essays: "On Being the Right Size," by J. B. S. Haldane, and "On Magnitude," by Sir D'Arcy Wentworth Thompson, both in James R. Newman (ed.), *The World of Mathematics*, vol. II. New York: Simon & Schuster, 1956.

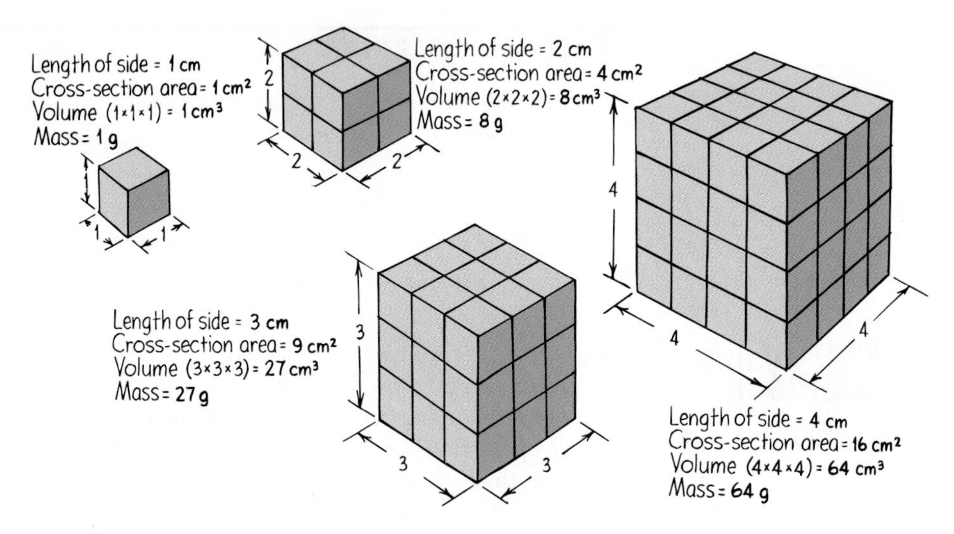

FIGURE 12.16 Interactive Figure

As the linear dimensions of an object change by some factor, the cross-sectional area changes as the square of this factor, and the volume (and hence the weight) changes as the cube of this factor. We see that, when the linear dimensions are doubled (factor = 2), the area grows by $2^2 = 4$, and the volume grows by $2^3 = 8$.

Compare this with a cube that has double the linear dimensions—a cube 2 centimeters on each side. As shown in the sketch, its cross-sectional area will be 2×2, or 4 square centimeters, and its volume will be $2 \times 2 \times 2$, or 8 cubic centimeters. Therefore, the cube will be four times as strong but eight times as heavy. Careful investigation of Figure 12.16 shows that, for increases of linear dimensions, the cross-sectional area and the total area grow as the square of the increase, whereas volume and weight grow as the cube of the increase.

Leonardo da Vinci was the first to report that the cross-sectional area of a tree trunk is equal to the combined surface areas produced by making a horizontal slice through all of the branches farther up on the tree.

Insights

Volume (and thus weight) increase much faster than the corresponding increase of cross-sectional area. Although the figure demonstrates the simple example of a cube, the principle applies to an object of any shape. Consider a football player who can do many pushups. Suppose he could somehow be scaled up to twice his size—that is, twice as tall and twice as broad, with his bones twice as thick and every linear dimension increased by a factor of two. Would he be twice as strong and be able to lift himself with twice the ease? The answer is *no*. Although his twice-as-thick arms would have four times the cross-sectional area and be four times as strong, he would be eight times as heavy. For comparable effort, he would be able to lift only half his weight. Relative to his new weight, he would be weaker than before.

We find in nature that large animals have disproportionately thick legs compared with those of small animals. This is because of the relationship between volume and area—the fact that volume (and weight) grow as the cube of the factor by which the linear dimension increases, while strength (and area) grow as the square of the increase factor. So we see that there is a reason for the thin legs of a deer or an antelope, as well as for the thick legs of a rhinoceros, a hippopotamus, or an elephant.

Surface Area vs. Volume

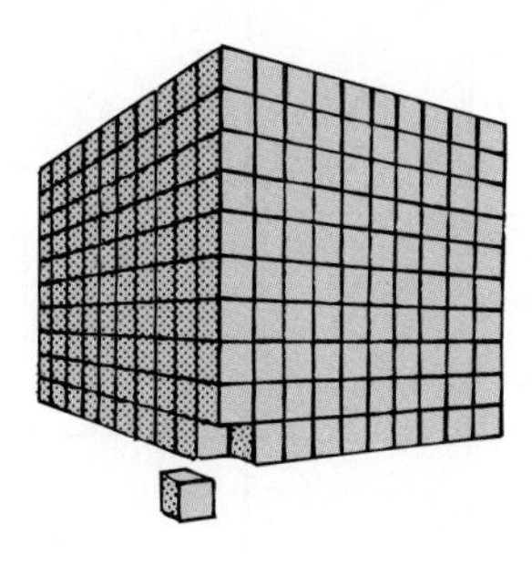

CHECK YOURSELF

1. Consider a 1-cubic-centimeter cube scaled up to a cube 10 centimeters long on each edge.
   - **a.** What would be the volume of the scaled-up cube?
   - **b.** What would be its cross-sectional surface area?
   - **c.** What would be its total surface area?
2. If you were somehow scaled up to twice your size while retaining your present proportions, would you be stronger or weaker? Explain your reasoning.

So the great strengths attributed to King Kong and other fictional giants cannot be taken seriously. The fact that the consequences of scaling are conveniently omitted is one of the differences between science and science fiction.

Important also is the comparison of total surface area to volume (Figure 12.17). Total surface area, just like cross-sectional area, grows in proportion to the square of an object's linear size, whereas volume grows in proportion to the cube of the linear size. So, as an object grows, its surface area and volume grow at different rates, with the result that the ratio of surface area to volume *decreases.* In other words, both the surface area and the volume of a growing object increase, but the growth of surface area *relative to* the growth of volume decreases. Not many people really understand this concept. The following examples may be helpful.

A chef knows that more potato peelings result from peeling 5 kilograms of small potatoes than from peeling 5 kilograms of large potatoes. Smaller objects have more surface area per kilogram. Thinner french fries cook faster in oil than fatter french fries. Flat hamburgers cook faster than meatballs of the same mass. Crushed ice will cool a drink much faster than a single ice cube of the same mass, because crushed ice presents more surface area to the beverage. Steel wool rusts away at the sink while steel knives rust more slowly. Rusting is a

CHECK YOUR ANSWERS

1. **a.** The volume of the scaled-up cube would be (length of side)^3 = (10 cm)^3, or 1000 cm^3.
   - **b.** Its cross-sectional surface area would be (length of side)^2 = (10 cm)^2, or 100 cm^2.
   - **c.** Its total surface area = 6 sides × area of a side = 600 cm^2.
2. Your scaled-up self would be four times as strong, because the cross-sectional area of your twice-as-thick bones and muscles increase by four. You could lift a load four times as heavy. But your weight would be eight times greater than before, so you would not be stronger relative to your greater weight. Having four times the strength carrying eight times the weight gives you a strength-to-weight ratio of only half its former value. So, if you can barely lift your own weight now, when scaled up you could lift only half your new weight. Your strength would increase, but your strength-to-weight ratio would decrease. Stay as you are!

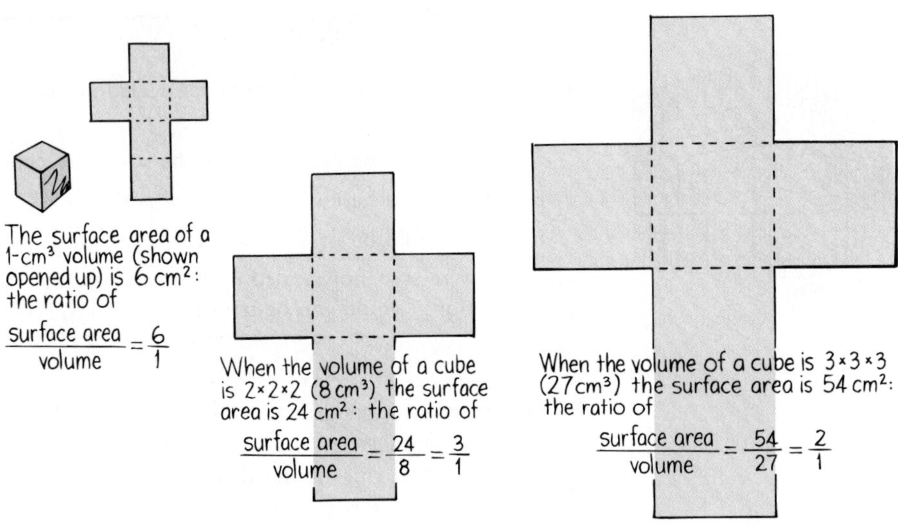

The surface area of a 1-cm³ volume (shown opened up) is 6 cm²: the ratio of
surface area / volume = 6/1
When the volume of a cube is 2×2×2 (8 cm³) the surface area is 24 cm²: the ratio of
surface area / volume = 24/8 = 3/1
When the volume of a cube is 3×3×3 (27cm³) the surface area is 54 cm²: the ratio of
surface area / volume = 54/27 = 2/1

FIGURE 12.17
As the size of an object increases, there is a greater increase in its volume than in its surface area; as a result, the ratio of surface area to volume decreases.

FIGURE 12.18
The long tail of the monkey not only helps the monkey to maintain its balance but also effectively radiates excess heat.

surface phenomenon. Iron rusts when exposed to air, but it rusts much faster, and is soon eaten away, if it is in the form of small strands or filings. Chunks of coal burn, while coal dust explodes when ignited. These are all consequences of the fact that volume and area are not in direct proportion to each other.

The large ears of African elephants are nature's way of compensating for the small ratio of surface area to volume for these large creatures. The large ears are not for better hearing but primarily for better cooling. The rate at which a creature dissipates heat is proportional to its surface area. If an African elephant didn't have large ears, it would not have enough surface area to cool its huge mass. The large ears greatly increase overall surface area, which facilitates cooling in hot climates.

FIGURE 12.19
The African elephant has less surface area relative to its weight than many other animals. It compensates for this with its large ears, which significantly increase its radiating surface area and promote cooling of the body.

Large raindrops fall faster than small raindrops, and large fish swim faster than small fish.

Insights

At the microscopic level, living cells must contend with the fact that the growth of volume is faster than the growth of surface area. Cells obtain nourishment by diffusion through their surfaces. As cells grow, their surface area increases, but not fast enough to keep up with their increasing volume. For example, if the surface area increases by four, the corresponding volume increases by eight. Eight times as much mass must be sustained by only four times as much nourishment. At some size, the surface area isn't large enough to allow sufficient nutrients to pass into the cell, placing a limit on how large cells can grow. So cells divide, and there is life as we know it. That's nice.

Not so nice is the fate of large creatures when they fall. The statement "The bigger they are, the harder they fall" holds true, and it is a consequence of the small ratio of surface area to weight. Resistance of the air to movement through it is proportional to the surface area of the moving object. If you fall out of a tree, even in the presence of air resistance, your rate of fall is very nearly 1 *g*. You don't have enough surface area relative to your weight to slow down to a safe speed—unless you wear a parachute. Small creatures, on the other hand, need no parachute. They have plenty of surface area relative to their small weights. An insect can fall from the top of a tree to the ground below without harm. The ratio of surface area to weight is in the insect's favor, for the insect is, in effect, its own parachute.

Raindrops can get as large as 8 mm in diameter. Drops about 1 mm in diameter take the shape of a sphere; at 2 mm, they flatten like a hamburger bun; and at 5 mm, air drag distorts them to parachute shapes. When circulation tosses raindrops back up into the air, often repeatedly, they may freeze and form hail. Hailstones have been recorded that are as large as 178 mm in diameter.

Insights

The dissimilar consequences of a fall are just one illustration of the different relationships that large organisms and small organisms have with the physical environment. The force of gravity is tiny for insects compared with cohesion forces (stickiness) between their feet and the surfaces they walk on. That's why a fly can crawl up a wall or along the ceiling completely ignoring gravity. Humans and elephants can't do that. The lives of small creatures are ruled not by gravity but by such forces as surface tension, cohesion, and capillarity, which we'll treat in the next chapter.

It is interesting to note that the heartbeat rate of a mammal is related to the size of the mammal. The heart of a tiny shrew beats about twenty times faster than the heart of an elephant. In general, small mammals live fast and die young; larger animals live at a leisurely pace and live longer. Don't feel bad about a pet hamster who doesn't live as long as a dog. All warm-blooded animals have about the same life span—not in terms of years, but in the average number of heartbeats (about 800 million). Humans are the exception: we live two to three times longer than other mammals of our size.

Researchers are finding that, when something shrinks enough—whether it's an electronic circuit, a motor, a film of lubricant, or an individual metal or ceramic crystal—it stops acting like a miniature version of its larger self and starts behaving in new and different ways. Palladium metal, for example, which is normally composed of grains about 1000 nanometers in size, is found to be five times as strong when formed from 5-nanometer grains.[2] Scaling is enormously important as more devices are being miniaturized.

[2] A nanometer is one billionth of a meter, so 1000 nanometers is one millionth of a meter, or one thousandth of a millimeter. Small indeed!

Summary of Terms

Atomic bonding The linking together of atoms to form larger structures, including solids.

Density The mass of a substance per unit volume:

$$\text{Density} = \frac{\text{mass}}{\text{volume}}$$

Weight density is weight per unit volume:

$$\text{Weight density} = \frac{\text{weight}}{\text{volume}}$$

Elasticity The property of a material wherein it changes shape when a deforming force acts upon it and returns to its original shape when the force is removed.

Hooke's law The amount of stretch or compression of an elastic material is directly proportional to the applied force: $F \sim \Delta x$. When the spring constant k is introduced, $F = k\Delta x$.

Scaling The study of how size affects the relationships among weight, strength, and surface.

Suggested Reading

For more about the relationships among the size, area, and volume of objects, read these essays: "On Being the Right Size," by J. B. S. Haldane, and "On Magnitude," by Sir D'Arcy Wentworth Thompson. Both are in J. R. Newman, ed., *The World of Mathematics*, Vol. II. New York: Simon & Schuster, 1956.

Bryson, Bill. *A Short History of Nearly Everything*. New York: Broadway Books, 2003.

Review Questions

Müller's Micrograph

1. What does the Müller micrograph indicate about the pattern of platinum atoms at the tip of a platinum needle?

Crystal Structure

2. How does the arrangement of atoms in a crystalline substance differ from that in a noncrystalline substance?
3. What evidence can you cite for the microscopic crystal nature of certain solids? For macroscopic crystal nature?

Density

4. What happens to the volume of a loaf of bread that is squeezed? The mass? The density?
5. Which is denser, something having a density of 1000 kg/m^3 or something having a density of 1 g/cm^3? Defend your answer.
6. Osmium is not the heaviest atom found in nature. What, then, accounts for osmium being the densest *substance* on Earth?
7. How does mass density differ from weight density?

Elasticity

8. Why do we say that a spring is elastic?
9. Why do we say a blob of putty is inelastic?
10. What is Hooke's law?
11. Does Hooke's law apply to elastic materials or to inelastic materials?
12. What is meant by the elastic limit for a particular object?
13. If the weight of a 1-kg body stretches a spring by 2 cm, how much will the spring be stretched when it supports a 3-kg load? (Assume the spring does not reach its elastic limit.)

Tension and Compression

14. Distinguish between *tension* and *compression*.
15. What and where is the neutral layer in a beam that supports a load?
16. Why are the cross sections of metal beams in the shape of the letter I instead of solid rectangles?

Arches

17. Why were so many vertical columns needed to support the roofs of stone buildings in ancient Egypt and Greece?
18. Is it *tension* or *compression* that strengthens an arch that supports a load?
19. Why is cement not needed between the stone blocks of an arch that has the shape of an inverted catenary?
20. Why are no vertical columns needed to support the middle of domed stadiums, such as the Houston Astrodome?

Scaling

21. Does the strength of a person's arm usually depend on the length of the arm or on its cross-sectional area?
22. What is the volume of a sugar cube that measures 1 cm on each side? What is the cross-sectional area of the cube?

23. If the linear dimensions of an object are doubled, by how much does the surface area increase? By how much does the volume increase?

24. As the volume of an object is increased, its surface area also increases. During this increase, does the *ratio* of square meters to cubic meters increase or does it decrease?

25. Which has more skin, an elephant or a mouse? Which has more skin *per unit of body weight,* an elephant or a mouse?

26. Why can small creatures fall considerable distances without harm, while people need parachutes to do the same?

Projects

1. If you live in a region where snow falls, collect some snowflakes on black cloth and examine them with a magnifying glass. You'll see that all the many shapes are hexagonal crystalline structures, among the most beautiful sights in nature.

2. Simulate atomic close packing with a couple of dozen or so pennies. Arrange them in a square pattern so each penny inside the perimeter makes contact with four others. Then arrange them hexagonally, so contact with six others occurs. Compare the areas occupied by the same number of pennies close packed both ways.

3. Are you slightly longer while lying down than you are tall when standing up? Make measurements and see.

4. Hold an egg vertically and dangle a small chain beside it. Can you see that the chain follows the contour of the egg—shallow sag for the more rounded end, and deeper sag for the more pointed end? Nature has not overlooked the catenary!

Exercises

1. You take 1000 milligrams of a vitamin. Your friend takes 1 gram of the same vitamin. Who takes more?

2. Your friend says that the primary difference between a solid and a liquid is the kind of atoms in the material. Do you agree or disagree, and why?

3. In what sense can it be said that a tree is solidified air?

4. Silicon is the chief ingredient of both glass and semiconductor devices, yet the physical properties of glass are different from those of semiconductor devices. Explain.

5. What evidence can you cite to support the claim that crystals are composed of atoms that are arranged in specific patterns?

6. What happens to the density of air in a common rubber balloon when it is heated?

7. Is iron necessarily heavier than cork? Explain.

8. How does the density of a 100-kg iron block compare with the density of an iron filing?

9. What happens to the density of water when it freezes to become ice?

10. In a deep dive, a whale is appreciably compressed by the pressure of the surrounding water. What happens to the whale's density during the dive?

11. The uranium atom is the heaviest and most massive atom among the naturally occurring elements. Why, then, isn't a solid bar of uranium the densest metal?

12. Which has more volume, a kilogram of gold or a kilogram of aluminum?

13. Which has more mass, a liter of ice or a liter of water?

14. How would you test the notion that a steel ball is more elastic than a rubber ball?

15. Why does the suspended spring stretch more at the top than at the bottom?

16. If the spring in the sketch above were supporting a heavy weight, how would the sketch be changed?

17. A thick rope is stronger than a thin rope of the same material. Is a long rope stronger than a short rope?

18. When you bend a meterstick, one side is under tension and the other is under compression. Which side is which?

19. Tension and compression occur in a partially supported horizontal beam when it sags due to gravity or when it supports a load. Make a simple sketch to show a means of supporting the beam so that tension occurs at the top part and compression at the bottom. Sketch another case in which compression is at the top and tension occurs at the bottom.

20. Suppose you're constructing a balcony that extends beyond the main frame of your house. In a concrete overhanging slab, should steel reinforcing rods be embedded in the top, middle, or bottom of the slab?

21. Can a horizontal I-beam support a greater load when the web is horizontal or when the web is vertical? Explain.

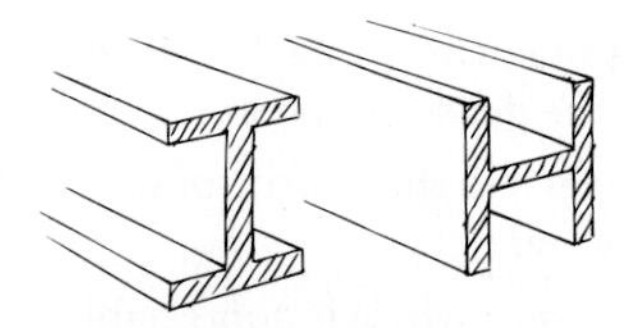

22. The sketches are of top views of a dam to hold back a lake. Which of the two designs is preferable? Why?

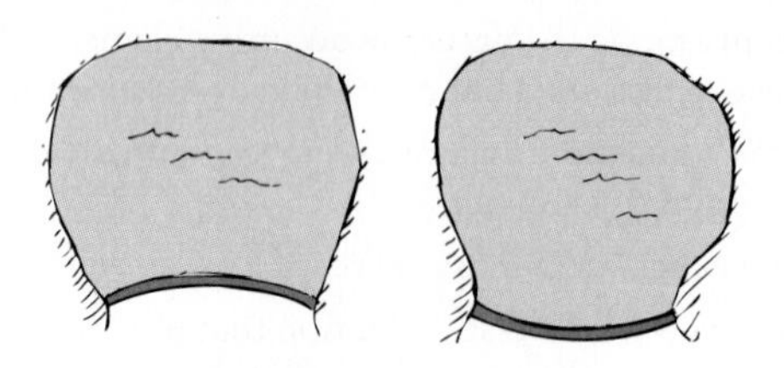

23. Consider a very large wooden barrel for storing wine. Should the "flat" ends be concave (bending inward) or convex (bending outward)? Why?

24. Why do you suppose that girders are so often arranged to form triangles in the construction of bridges and other structures? (Compare the stability of three sticks nailed together to form a triangle with four sticks nailed to form a rectangle, or any number of sticks to form multilegged geometric figures. Try it and see!)

25. Consider two bridges that are exact replicas of each other except that every dimension of the larger is exactly twice that of the other—that is, twice as long, structural elements twice as thick, etc. Which bridge is more likely to collapse under its own weight?

26. Only with great difficulty can you crush an egg when squeezing it along its long axis, but it breaks easily if you squeeze it sideways. Why?

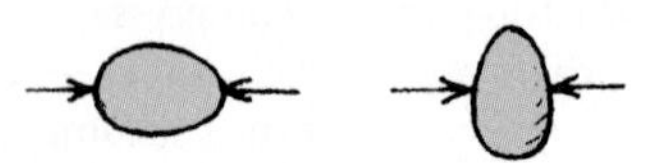

27. Archie designs an arch to serve as an outdoor sculpture in a park. The arch is to be of a certain width and of a certain height. To achieve the size and shape for the strongest possible arch, he suspends a chain from two equally elevated supports as far apart as the arch is wide and allows the chain to hang as low as the arch is high. He then designs the arch to have exactly the inverted shape of the hanging chain. Explain why.

28. The photo shows a semicircular arch of stone. Note that it must be held together with steel rods to prevent outward movement. If the shape of the arch were not a semicircle but the shape used by Archie in the previous exercise, would the steel rods be necessary? Explain.

29. A candymaker making taffy apples decides to use 100 kg of large apples rather than 100 kg of small apples. Will the candymaker need to make more or less taffy to cover the apples?

30. Why is it easier to start a fire with kindling rather than with large sticks and logs of the same kind of wood?

31. Why does a chunk of coal burn when ignited, whereas coal dust explodes?

32. Why does a two-story house that is roughly a cube suffer less heat loss than a rambling one-story house of the same volume?

33. Why is heating more efficient in large apartment buildings than in single-family dwellings?

34. Some environmentally conscious people build their homes in the shape of a dome. Why is less heat lost in a dome-shaped dwelling?

35. Why does crushed ice melt faster than the same mass of ice cubes?

36. Why do some animals curl up into a ball when they are cold?

37. Why is rust a greater problem for thin iron rods than for thick iron piles?

38. Why do thin french fries cook faster than thick fries?

39. If you are grilling hamburgers and getting impatient, why is it a good idea to flatten the burgers to make them wider and thinner?

40. If you use a batch of cake batter for cupcakes and bake them for the time suggested for baking a cake, what will be the result?

41. Why are mittens warmer than gloves on a cold day? And which parts of the body are most susceptible to frostbite? Why?

42. What is the advantage to a gymnast of being short in stature?

43. How does scaling relate to the fact that the heartbeat of large creatures is generally slower than the heartbeat of smaller creatures?

44. Nourishment is obtained from food through the inner surface area of the intestines. Why is it that a small organism, such as a worm, has a simple and relatively straight intestinal tract, while a large organism, such as a human being, has a complex and extensively folded intestinal tract?

45. The human lungs have a volume of only about 4 L, yet an internal surface area of nearly 100 m^2. Why is this important, and how is this possible?

46. What does the concept of scaling have to do with the fact that living cells in a whale are about the same size as those in a mouse?

47. Which fall faster, large or small raindrops?

48. Who has more need for drink in a dry desert climate, a child or an adult?

49. Why doesn't a hummingbird soar like an eagle and an eagle flap its wings like a hummingbird?

50. Can you relate the idea of scaling to the governance of small versus large groups of citizens? Explain.

Problems

1. Find the density of a 5-kg solid cylinder that is 10 cm tall and has a radius of 3 cm.

2. What is the weight of a cubic meter of cork? Could you lift it? (For the density of cork, use 400 kg/m^3.)

3. A certain spring stretches 4 cm when a load of 20 N is suspended from it. How much will the spring stretch if 45 N is suspended from it (and it doesn't reach its elastic limit)?

4. A certain spring stretches 4 cm when a load of 10 N is suspended from it. How much will the spring stretch if an identical spring also supports the load as shown in *a* and *b*? (Neglect the weights of the springs.)

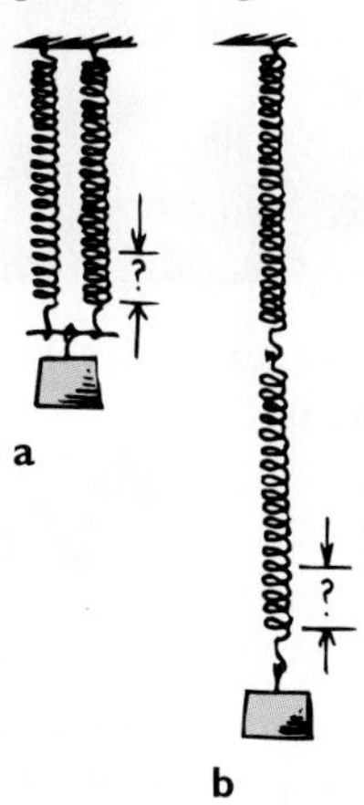

5. If a certain spring stretches 4 cm when a load of 10 N is suspended from it, how much will the spring stretch if it is cut in half and 10 N is suspended from it?

6. If the linear dimensions of a storage tank are reduced to half their former values, by how much does the overall surface area of the tank decrease? By how much does its volume decrease?

7. A cube 2 cm on a side is cut into cubes 1 cm on a side. (a) How many cubes result? (b) What was the surface area of the original cube and what is the total surface area of the eight smaller cubes? What is the ratio of surface areas? (c) What are the surface-to-volume ratios of the original cube and of all the smaller cubes combined?

8. Larger people at the beach need more suntan lotion than smaller people. Relative to a smaller person, how much lotion will a twice-as-heavy person use?

9. Consider eight 1-cm^3 sugar cubes stacked two by two to form a single bigger cube. What will be the volume of the combined cube? How does its surface area compare with the total surface area of the eight separate cubes?

10. Consider eight little spheres of mercury, each with a diameter of 1 mm. When they coalesce to form a single sphere, how big will it be? How does its surface area compare with the total surface area of the previous eight little spheres?

CHAPTER 13

Liquids

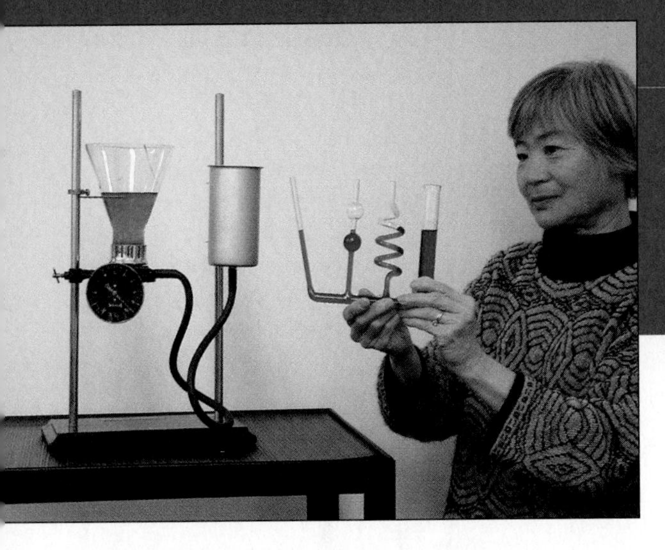

Tsing Bardin demonstrates the depth dependence of water pressure.

Molecules that make up a liquid are not confined to fixed positions, as they are in solids, but can flow from position to position by sliding over one another. While a solid maintains a particular shape, a liquid takes the shape of its container. Molecules of a liquid are close together and greatly resist compressive forces. Liquids, like solids, are difficult to compress. Gases, as we shall see in the next chapter, are easily compressed. Both liquids and gases can flow, so both are called *fluids*.

Pressure

A liquid contained in a vessel exerts forces against the walls of the vessel. To discuss the interaction between the liquid and the walls, it is convenient to introduce the concept of **pressure.** Pressure is a force divided by the area over which the force is exerted.[1]

$$\text{Pressure} = \frac{\text{force}}{\text{area}}$$

FIGURE 13.1
Although the weight of both blocks is the same, the upright block exerts greater pressure against the table.

As an illustration of the distinction between pressure and force, consider the two blocks in Figure 13.1. The blocks are identical, but one stands on its end and the other stands on its side. Both blocks are of equal weight and therefore exert the same force on the surface (if you were to put them on a bathroom scale, each would register the same), but the upright block exerts a greater *pressure* against the surface. If the block were tipped up so that its contact with the table were on a single corner, the pressure would be greater still.

[1]Pressure may be measured in any unit of force divided by any unit of area. The standard international (SI) unit of pressure, the newton per square meter, is called the *pascal* (Pa) after the seventeenth-century theologian and scientist Blaise Pascal. A pressure of 1 Pa is very small and approximately equals the pressure exerted by a dollar bill resting flat on a table. Science types more often use kilopascals (1 kPa = 1000 Pa).

Pressure in a Liquid

FIGURE 13.2
The tower does more than store water. The depth of water above ground level ensures substantial and reliable water pressure to the many homes it serves.

When you swim under water, you can feel the water pressure acting against your eardrums. The deeper you swim, the greater the pressure. The pressure you feel is due to the weight of water above you. As you swim deeper, there is more water above you and therefore greater pressure. The pressure a liquid exerts depends on its depth.

Liquid pressure also depends on the density of the liquid. If you were submerged in a liquid more dense than water, the pressure would be correspondingly greater. The pressure due to a liquid is precisely equal to the product of weight density and depth:[2]

$$\text{Liquid pressure} = \text{weight density} \times \text{depth}$$

Simply put, the pressure a liquid exerts against the sides and bottom of a container depends on the density and the depth of the liquid. If we neglect atmospheric pressure, at twice the depth, the liquid pressure against the bottom is twice as great; at three times the depth, the liquid pressure is threefold; and so on. Or, if the liquid is two or three times as dense, the liquid pressure is correspondingly two or three times as great for any given depth. Liquids are practically incompressible—that is, their volume can hardly be changed by pressure (water volume decreases by only 50 millionths of its original volume for each atmosphere increase in pressure). So, except for

Earth is the only planet in the solar system with a mostly liquid surface–its oceans. If the Earth were a little closer to the Sun, the oceans would turn to vapor. If the Earth were a little farther away, most of its surface, not just the polar regions, would be solid ice. It's nice that the Earth is located where it is.

FIGURE 13.3
The dependence of liquid pressure on depth is not a problem for the giraffe because of its large heart and its intricate system of valves and elastic, absorbent blood vessels in the brain. Without these structures, it would faint when suddenly raising its head, and it would be subject to brain hemorrhaging when lowering it.

[2]This is derived from the definitions of pressure and weight density. Consider an area at the bottom of a vessel of liquid. The weight of the column of liquid directly above this area produces pressure. From the definition

$$\text{Weight density} = \frac{\text{weight}}{\text{volume}}$$

we can express this weight of liquid as

$$\text{Weight} = \text{weight density} \times \text{volume}$$

where the volume of the column is simply the area multiplied by the depth. Then we get

$$\text{Pressure} = \frac{\text{force}}{\text{area}} = \frac{\text{weight}}{\text{area}} = \frac{\text{weight density} \times \text{volume}}{\text{area}}$$

$$= \frac{\text{weight density} \times (\cancel{\text{area}} \times \text{depth})}{\cancel{\text{area}}} = \text{weight density} \times \text{depth}.$$

For total pressure, we add to this equation the pressure due to the atmosphere on the surface of the liquid.

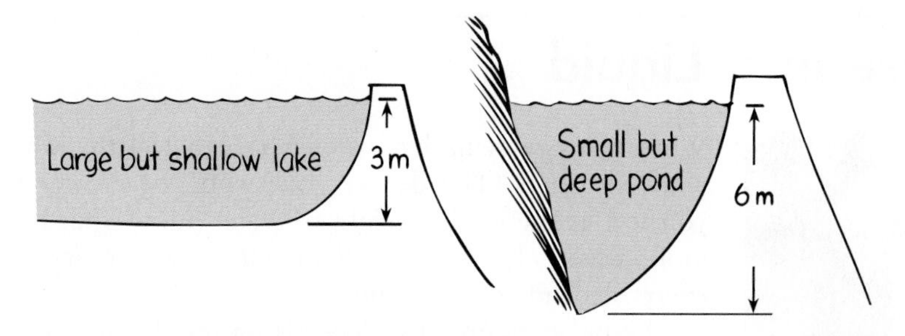

FIGURE 13.4
The average water pressure acting against the dam depends on the average depth of the water and not on the volume of water held back. The large, shallow lake exerts only one-half the average pressure that the small, deep pond exerts.

small changes produced by temperature, the density of a particular liquid is practically the same at all depths.[3]

If you press your hand against a surface, and somebody else presses against your hand in the same direction, then the pressure against the surface is greater than if you pressed alone. Likewise with the atmospheric pressure that presses on the surface of a liquid. The total pressure of a liquid, then, is weight density × depth *plus* the pressure of the atmosphere. When this distinction is important, we will use the term *total pressure*. Otherwise, our discussions of liquid pressure refer to pressure without regard to the normally ever-present atmospheric pressure. (You will learn more about atmospheric pressure in the next chapter.)

It is important to recognize that the pressure does not depend on the *amount* of liquid present. Volume is not the key—depth is. The average water pressure acting against a dam depends on the average *depth* of the water and not on the volume of water held back. For example, the large, shallow lake in Figure 13.4 exerts only half the average pressure that the small, deep pond exerts.

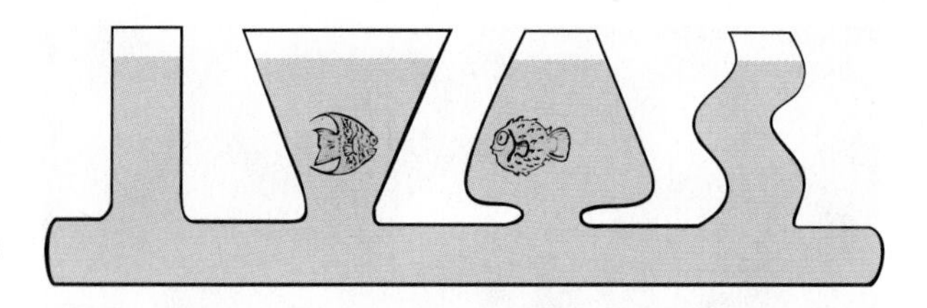

FIGURE 13.5
Liquid pressure is the same for any given depth below the surface, regardless of the shape of the containing vessel. Liquid pressure = weight density × depth (plus the air pressure at the top).

You'll feel the same pressure whether you dunk your head a meter beneath the surface of the water in a small pool or to the same depth in the middle of a large lake. The same is true for a fish. Refer to the connecting vases in Figure 13.5. If we hold a goldfish by its tail and dunk its head a couple of centimeters under the surface, the water pressure on the fish's head will be the same in any of the vases. If we release the fish and it swims a few centimeters deeper, the pressure on the fish will increase with depth and be the same no matter which vase the fish is in. If the fish swims to the bottom, the pressure will be greater, but it makes no difference what vase it is in. All vases are filled to equal depths, so the water pressure is the same at the bottom of each vase, regardless of its shape or volume. If water pressure at the bottom of a vase were greater than water pressure at the bottom of a neighboring narrower vase, the greater pressure would force water

[3]The density of freshwater is 1000 kg/m^3. Since the weight (*mg*) of 1000 kg is 1000 × 9.8 = 9800 N, the weight density of water is 9800 newtons per cubic meter (9800 N/m^3). The water pressure beneath the surface of a lake is simply equal to this density multiplied by the depth in meters. For example, water pressure is 9800 N/m^2 at a depth of 1m and 98,000 N/m^2 at a depth of 10 m. In SI units, pressure is measured in pascals, so this would be 9800 Pa and 98,000 Pa, respectively; or, in kilopascals, 9.8 kPa and 98 kPa, respectively. For the *total pressure* in these cases, add the pressure of the atmosphere, 101.3 kPa.

FIGURE 13.6
Roman aqueducts assured that water flowed slightly downhill from reservoir to city.

Some ancient pipe systems installed in Rome at the time the aqueducts were being built indicate that not all Romans then believed that water couldn't flow uphill.

Insights

sideways and then up the narrower vase to a higher level until the pressures at the bottom were equalized. But this doesn't happen. Pressure is depth dependent, not volume dependent, so we see that there is a reason why water seeks its own level.

The fact that water seeks its own level can be demonstrated by filling a garden hose with water and holding the two ends at the same height. The water levels will be equal. If one end is raised higher than the other, water will flow out of the lower end, even if it has to flow "uphill" part of the way. This fact was not fully understood by some of the early Romans, who built elaborate aqueducts with tall arches and roundabout routes to ensure that water would always flow slightly downward every place along its route from the reservoir to the city. If pipes were laid in the ground and followed the natural contour of the land, in some places the water would have to flow uphill, and the Romans were skeptical of this. Careful experimentation was not yet the mode; so, with plentiful slave labor, the Romans built unnecessarily elaborate aqueducts.

FIGURE 13.7
The forces of a liquid pressing against a surface add up to a net force that is perpendicular to the surface.

An experimentally determined fact about liquid pressure is that it is exerted equally in all directions. For example, if we are submerged in water, no matter which way we tilt our heads we feel the same amount of water pressure on our ears. Because a liquid can flow, the pressure isn't only downward. We know pressure acts sideways when we see water spurting sideways from a leak in the side of an upright can. We know pressure also acts upward when we try to push a beach ball beneath the surface of the water. The bottom of a boat is certainly pushed upward by water pressure.

When liquid presses against a surface, there is a net *force* that is *perpendicular* to the surface. Although pressure doesn't have a specific direction, force does. Consider the triangular block in Figure 13.7. Focus your attention only at the three points midway along each surface. Water is forced against each point from many directions, only a few of which are indicated. Components of the forces that are not perpendicular to the surface cancel each other out, leaving only a net perpendicular force at each point.

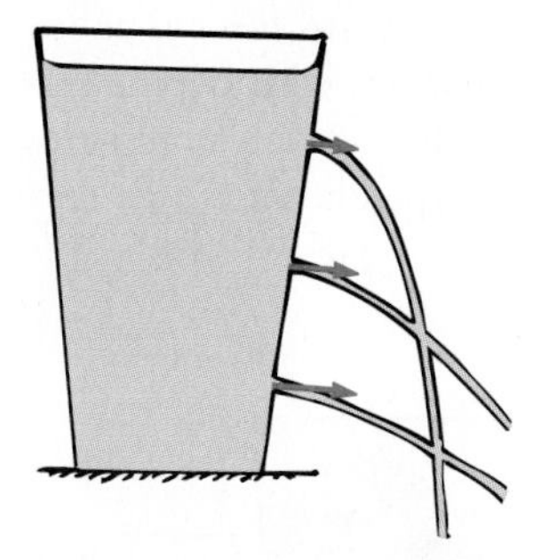

FIGURE 13.8
The force vectors act perpendicular to the inner container surface and increase with increasing depth.

That's why water spurting from a hole in a bucket initially exits the bucket in a direction at right angles to the surface of the bucket in which the hole is located. Then it curves downward due to gravity. The force exerted by a fluid on a smooth surface is always at right angles to the surface.[4]

[4]The speed of liquid out of the hole is $\sqrt{2gh}$, where h is the depth below the free surface. Interestingly enough, this is the same speed the water (or anything else) would have if freely falling the same vertical distance h.

WATER DOWSING

The practice of water dowsing, which goes back to ancient times in Europe and Africa, was carried across the Atlantic to America by some of the earliest settlers. Dowsing refers to the practice of using a forked stick, or a rod, or some similar device to locate underground water, minerals, or hidden treasure. In the classic method of dowsing, each hand grasps a fork, with palms upward. The pointed end of the stick points skyward at an angle of about 45°. The dowser walks back and forth over the area to be tested, and, when passing over a source of water (or whatever else is being sought), the stick is supposed to rotate downward. Some dowsers have reported an attraction so great that blisters formed on their hands. Some claim special powers that enable them to "see" through soil and rock, and some are mediums who go into trances when conditions are especially favorable. Although most dowsing is done at the actual site, some dowsers claim to be able to locate water simply by passing the stick over a map.

Since drilling a well is an expensive process, the dowser's fee is usually seen as reasonable. The practice is widespread, with thousands of dowsers active in the United States. This is because dowsing works. The dowser can hardly miss—not because of special powers, but because groundwater is within 100 meters of the surface at almost every spot on Earth.

If you drill a hole into the ground, you'll find that the wetness of the soil varies with depth. Near the surface, pores and open spaces are filled mostly with air. Deeper, the pores are saturated with water. The upper boundary of this water-saturated zone is called the *water table*. It usually rises and falls with the contour of the surface topography. Wherever you see a natural lake or pond, you're seeing a place where the water table extends above the surface of the land.

The depth, quantity, and quality of water below the water table are studied by hydrologists, who rely on a variety of technological techniques—water dowsing *not* being one of them. Findings of the U.S. Geological Survey conclude that water dowsing falls into the category of pseudoscience. As mentioned in Chapter 1, the real test of a water dowser would be finding a location in which water can't be found.

Buoyancy

Anyone who has ever lifted a heavy submerged object out of water is familiar with *buoyancy*, the apparent loss of weight experienced by objects submerged in a liquid. For example, lifting a large boulder off the bottom of a riverbed is a relatively easy task as long as the boulder is below the surface. When it is lifted above the surface, however, the force required to lift it is increased considerably. This is because, when the boulder is submerged, the water exerts an upward force on it that is exactly opposite to the direction of gravity's pull. This upward force is called the **buoyant force**, and it is a consequence of pressure increasing with depth. Figure 13.9 shows why the buoyant force acts upward. Forces due to water pressures are exerted everywhere against the boulder in a direction perpendicular to its surface, as shown by the vectors. Force vectors against the sides at equal depths cancel one another, so there is no horizontal buoyant force. Force vectors in the vertical direction, however, don't cancel. Pressure is greater against the bottom of the boulder because the bottom is deeper. So upward forces against the bottom are greater than downward forces against the top, producing a net force upward—the buoyant force.

FIGURE 13.9
Interactive Figure
The greater pressure against the bottom of a submerged object produces an upward buoyant force.

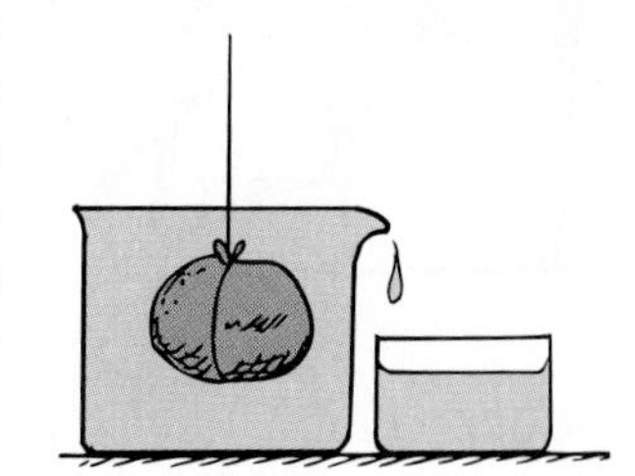

FIGURE 13.10
When a stone is submerged, it displaces water that has a volume equal to the volume of the stone.

Understanding buoyancy requires understanding the expression "volume of water displaced." If a stone is placed in a container that is brimful of water, some water will overflow (Figure 13.10). Water is *displaced* by the stone. A little thought will tell us that the

FIGURE 13.11
The increase in water level is the same as if you poured in a volume of water equal to the stone's volume.

volume of the stone—that is, the amount of space it takes up—is equal to the *volume of the water displaced.* If you place any object in a container partly filled with water, the level of the surface rises (Figure 13.11). By how much? By exactly the same amount as if a volume of water were poured in that equals the volume of the submerged object. This is a good method for determining the volume of irregularly shaped objects: *A completely submerged object always displaces a volume of liquid equal to its own volume.*

CHECK YOURSELF

A recipe calls for a specific amount of butter. How does the displacement method relate to the use of a kitchen measuring cup?

Archimedes' Principle

If you stick your foot in water, it's immersed. If you jump in and sink and immersion is total, you're submerged.

The relationship between buoyancy and displaced liquid was first discovered in the third century BC by the Greek scientist Archimedes. It is stated as follows:

An immersed object is buoyed up by a force equal to the weight of the fluid it displaces.

This relationship is called **Archimedes' principle.** It is true of liquids and gases, both of which are fluids. If an immersed object displaces 1 kilogram of fluid, the buoyant force acting on it is equal to the weight of 1 kilogram.[5] By *immersed,* we mean either *completely* or *partially submerged.* If we immerse a sealed 1-liter container halfway into the water, it will displace a half-liter of water and be buoyed up by a force equal to the weight of a half-liter of water—no matter what is in the container. If we immerse it completely (submerge it), it will be buoyed up by a force equivalent to the weight of a full liter of water (1 kilogram of mass). If the container is fully submerged and doesn't compress, the buoyant force will equal the weight of 1 kilogram of water at *any* depth. This is because, at any depth, the container can displace no greater volume of water than its own volume. And the weight of this displaced water (not the weight of the submerged object!) is equal to the buoyant force.

If a 30-kilogram object displaces 20 kilograms of fluid upon immersion, its apparent weight will be equal to the weight of 10 kilograms (98 newtons). Note that, in Figure 13.13, the 3-kilogram block has an apparent weight equal to the weight of 1 kilogram when submerged. The apparent weight of a submerged object is its usual weight in air minus the buoyant force.

FIGURE 13.12
A liter of water occupies a volume of 1000 cm^3, has a mass of 1 kg, and weighs 9.8 N. Its density may therefore be expressed as 1 kg/L and its weight density as 9.8 N/L. (Seawater is slightly denser, about 10.0 N/L).

CHECK YOUR ANSWER

Put some water in the cup before you add the butter. Note the water-level reading on the side of the cup. Then add the butter and watch the water level rise. Because butter floats, poke it beneath the surface. When you subtract the lower-level reading from the higher-level reading, you know the volume of the butter.

[5]In lab, you may find it convenient to express buoyant force in kilograms, even though a kilogram is a unit of mass and not a unit of force. So, strictly speaking, the buoyant force is the *weight* of 1 kg, which is 9.8 N. Or we could as well say that the buoyant force is 1 *kilogram weight,* not simply 1 kg.

FIGURE 13.13
Objects weigh more in air than in water. When submerged, this 3-N block appears to weigh only 1 N. The "missing" weight is equal to the weight of water displaced, 2 N, which equals the buoyant force.

CHECK YOURSELF

1. Does Archimedes' principle tell us that, if an immersed object displaces liquid weighing 10 N, the buoyant force on the object is 10 N?
2. A 1-liter container completely filled with lead has a mass of 11.3 kg and is submerged in water. What is the buoyant force acting on it?
3. A boulder is thrown into a deep lake. As it sinks deeper and deeper into the water, does the buoyant force upon it increase or decrease?
4. Since buoyant force is the net force that a fluid exerts on a body and, as we learned in Chapter 4, that net force produces acceleration, why doesn't a submerged body accelerate?

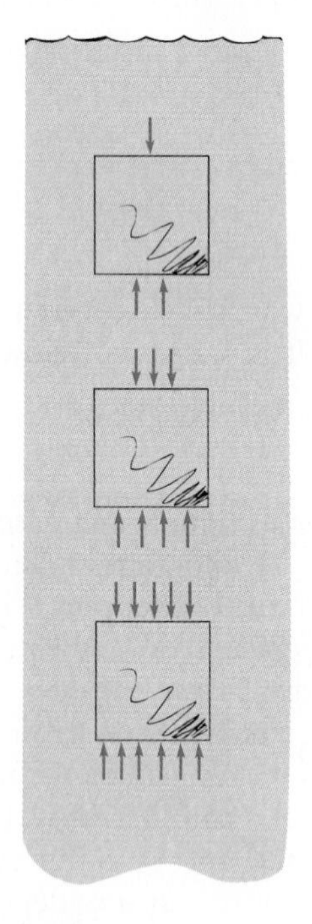

FIGURE 13.14
Interactive Figure
The difference in the upward and downward force acting on the submerged block is the same at any depth.

Perhaps your instructor will summarize Archimedes' principle by way of a numerical example to show that the difference between the upward-acting and the downward-acting forces due to the similar pressure differences on a submerged cube is numerically identical to the weight of fluid displaced. It makes no difference how deep the cube is placed because, although the pressures are greater with increasing depths, the *difference* between the pressure up against the bottom of the cube and the pressure down against the top of the cube is the same at any depth (Figure 13.14). Whatever the shape of the submerged body, the buoyant force is equal to the weight of fluid displaced.

CHECK YOUR ANSWERS

1. Yes. Looking at it another way, the immersed object pushes 10 N of fluid aside. The displaced fluid reacts by pushing back on the immersed object with 10 N.
2. The buoyant force is equal to the weight of the liter of water displaced. One L of water has a mass of 1 kg, and a weight of 9.8 N. So the buoyant force on it is 9.8 N. (The 11.3 kg of the lead is irrelevant; 1 L of anything submerged in water will displace 1 L and be buoyed upward with a force of 9.8 N, the weight of 1 kg.)
3. Buoyant force remains unchanged as the boulder sinks, because the boulder displaces the same volume of water at any depth.
4. A submerged body does accelerate if the buoyant force is not balanced by other forces acting on it—the force of gravity and fluid resistance. The net force on a submerged body is the result of the net force the fluid exerts (buoyant force), the weight of the body, and, if the submerged body is moving, the force of fluid friction.

What Makes an Object Sink or Float?

It's important to remember that the buoyant force acting on a submerged object depends on the *volume* of the object. Small objects displace small amounts of water and are acted on by small buoyant forces. Large objects displace large amounts of water and are acted on by larger buoyant forces. It is the *volume* of the submerged object—not its *weight*—that determines the buoyant force. The buoyant force is equal to the weight of the *volume of fluid* displaced. (Misunderstanding this idea is the root of much confusion that people have about buoyancy.)

The weight of an object does play a role, however, in floating. Whether an object will sink or float in a liquid depends on how the buoyant force *compares with the object's weight*. This, in turn, depends on the object's density. Consider these three simple rules:

1. **An object more dense than the fluid in which it is immersed will sink.**
2. **An object less dense than the fluid in which it is immersed will float.**
3. **An object having a density equal to the density of the fluid in which it is immersed will neither sink nor float.**

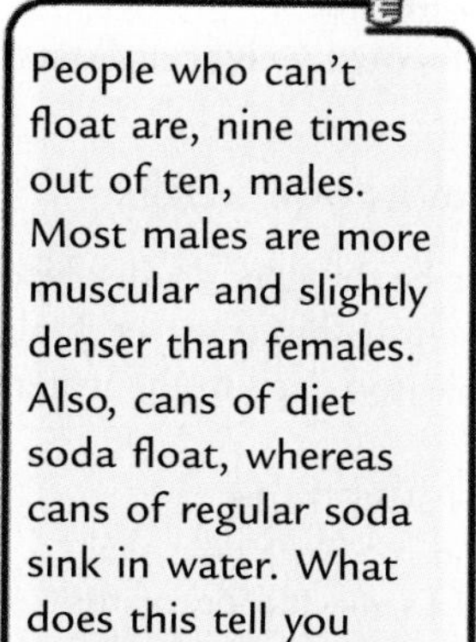
People who can't float are, nine times out of ten, males. Most males are more muscular and slightly denser than females. Also, cans of diet soda float, whereas cans of regular soda sink in water. What does this tell you about their relative densities?

Insights

Rule 1 seems reasonable enough, for objects denser than water sink to the bottom, regardless of the water's depth. Scuba divers near the bottoms of deep bodies of water may sometimes encounter a waterlogged piece of wood hovering above the ocean floor (with a density equal to that of water at that depth), but never do they encounter hovering rocks!

From Rules 1 and 2, what can you say about people who, try as they may, cannot float? They're simply too dense! To float more easily, you must reduce your density. The formula *weight density = weight/volume* says you must either reduce your weight or increase your volume. Wearing a life jacket increases volume while correspondingly adding very little to your weight. It reduces your overall density.

Rule 3 applies to fish, which neither sink nor float. A fish normally has the same density as water. A fish can regulate its density by expanding and contracting an air sac in its body that changes its volume. The fish can move upward by increasing its volume (which decreases its density) and downward by contracting its volume (which increases its density).

For a submarine, weight, not volume, is varied to achieve the desired density. Water is taken into or blown out of its ballast tanks. Similarly, the overall density of a crocodile increases when it swallows stones. From 4 to 5 kilograms of stones have been found in the stomachs of large crocodiles. Because of this increased density, the crocodile swims lower in the water, thus exposing itself less to its prey. (Figure 13.15).

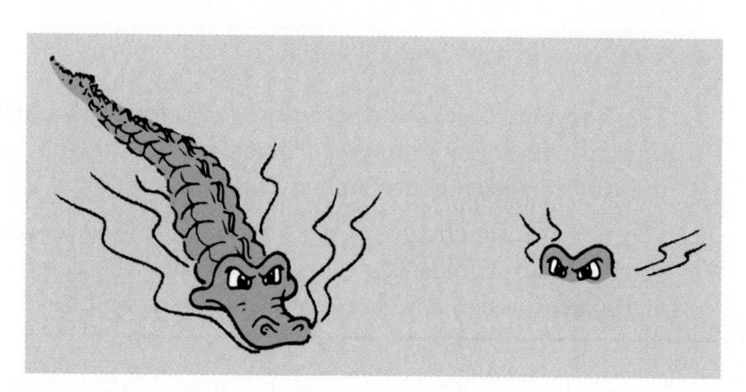

FIGURE 13.15
(*left*) A crocodile coming toward you in the water. (*right*) A stoned crocodile coming toward you in the water.

CHECK YOURSELF

1. Two solid blocks of identical size are submerged in water. One block is lead and the other is aluminum. Upon which is the buoyant force greater?
2. If a fish makes itself denser, it will sink; if it makes itself less dense, it will rise. In terms of buoyant force, why is this so?

Flotation

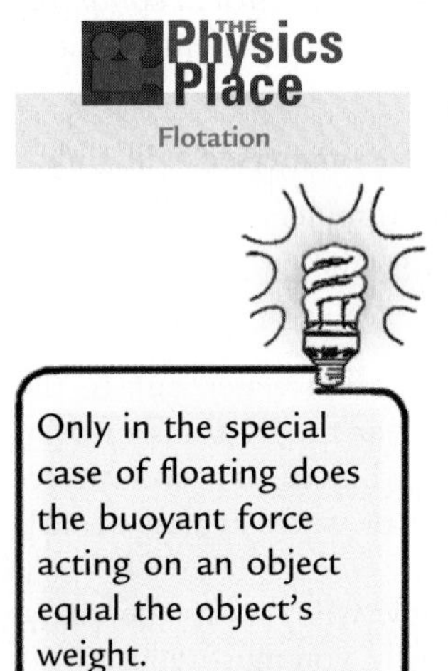

The Physics Place
Flotation
Only in the special case of floating does the buoyant force acting on an object equal the object's weight.
Insights

Primitive peoples made their boats of wood. Could they have conceived of an iron ship? We don't know. The idea of floating iron might have seemed strange. Today it is easy for us to understand how a ship made of iron can float.

Consider a 1-ton block of solid iron. Because iron is nearly eight times denser than water, it displaces only 1/8 ton of water when submerged, which is not enough to keep it afloat. Suppose we reshape the same iron block into a bowl (Figure 13.16). It still weighs 1 ton. But when we put it in water, it displaces a greater volume of water than when it was a block. The deeper the iron bowl is immersed, the more water it displaces and the greater the buoyant force acting on it. When the buoyant force equals 1 ton, it will sink no farther.

When any boat displaces a weight of water equal to its own weight, it floats. This is sometimes called the **principle of flotation:**

A floating object displaces a weight of fluid equal to its own weight.

Every ship, every submarine, and every dirigible must be designed to displace a weight of fluid equal to its own weight. Thus, a 10,000-ton ship must be built wide enough to displace 10,000 tons of water before it sinks too deep in the water.

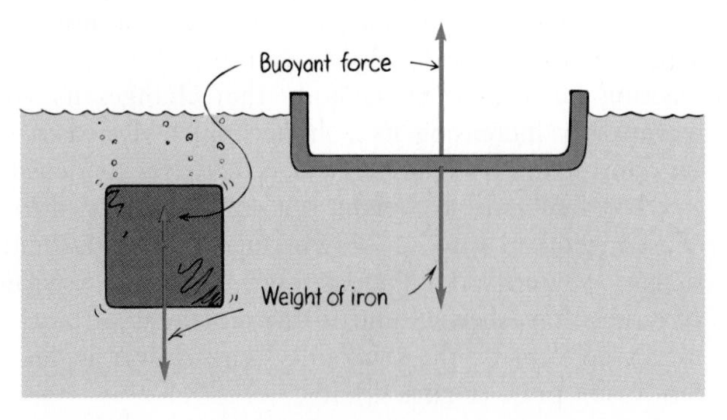

FIGURE 13.16
An iron block sinks, while the same quantity of iron shaped like a bowl floats.

CHECK YOUR ANSWERS

1. The buoyant force is the same on each, for both blocks displace the same volume of water. For submerged objects, only the volume of water displaced, not the object's weight, determines the buoyant force.
2. When the fish increases its density by decreasing its volume, it displaces less water, so the buoyant force decreases. When the fish decreases its density by expanding, it displaces a greater volume of water, and the buoyant force increases.

FIGURE 13.17
The weight of a floating object equals the weight of the water displaced by the submerged part.

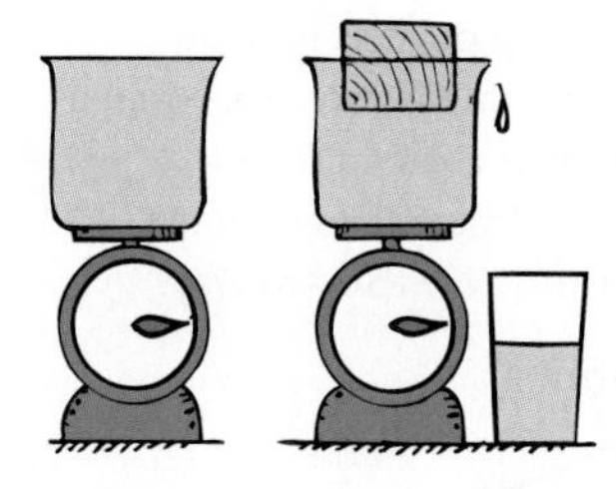

FIGURE 13.18
A floating object displaces a weight of fluid equal to its own weight.

The same holds true for vessels in air. A dirigible that weighs 100 tons displaces at least 100 tons of air. If it displaces more, it rises; if it displaces less, it falls. If it displaces exactly its weight, it hovers at constant altitude.

For a given volume of displaced fluid, a denser fluid exerts a greater buoyant force than a less dense fluid. A ship, therefore, floats higher in salt water than in freshwater because salt water is slightly denser. Similarly, a solid chunk of iron will float in mercury, even though it will sink in water.

CHECK YOURSELF

1. Why is it easier for you to float in salt water than in freshwater?
2. On a boat ride, the skipper gives you a life preserver filled with lead pellets. When he sees the skeptical look on your face, he says that you'll experience a greater buoyant force if you fall overboard than your friends who wear Styrofoam-filled life preservers. Is he being truthful?

FIGURE 13.19
The same ship empty and loaded. How does the weight of its load compare with the weight of extra water displaced?

CHECK YOUR ANSWERS

1. It's easier because a lesser amount of your body is immersed when displacing your weight—you don't "sink" as far. You'd float even higher in mercury (density 13.6 g/cm^3), and you'd sink completely in alcohol (density 0.8 g/cm^3).
2. He's truthful. But what he doesn't tell you is that you'll drown! Your life preserver will submerge and displace more water than those of your friends who float at the surface. Although the buoyant force on you will be greater, your increased weight is greater still! Whether you float or sink depends on whether or not the buoyant force equals your weight.

FLOATING MOUNTAINS

The tip of a floating iceberg above the ocean's surface is approximately 10% of the whole iceberg. That's because ice is 0.9 times the density of water, so 90% of it submerges in water. Similarly, a mountain floats on the Earth's semiliquid mantle with only its tip showing. That's because Earth's continental crust is about 0.85 times the density of the mantle it floats upon; thus, about 85% of a mountain extends beneath the Earth's surface. So, like floating icebergs, mountains are appreciably deeper than they are high.

There is an interesting gravitational sidelight to this: Recall, from Chapter 9, that the gravitational field at the Earth's surface varies slightly with varying densities of underlying rock (which is valuable information to geologists and oil prospectors), and that gravitation is less at the top of a mountain because of the greater distance to Earth's center. Combining these ideas, we see that, because the bottom of a mountain extends deep into the Earth's mantle, there is increased distance between a mountaintop and the denser mantle. This increased "gap" further reduces gravitation at the top of a mountain.

Another interesting fact about mountains: If you could shave off the top of an iceberg, the iceberg would be lighter and would be buoyed up to nearly its original height before being shaved. Similarly, when mountains erode, they are lighter, and they are pushed up from below to float to nearly their original heights. So, when a kilometer of mountain erodes away, some 85% of a kilometer of mountain thrusts up from below. That's why it takes so long for mountains to weather away.

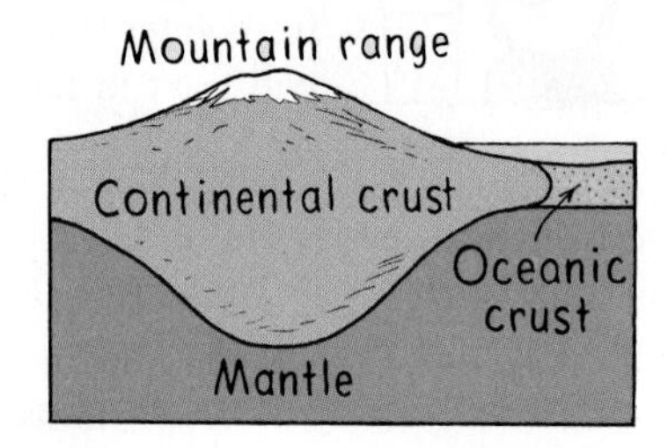

FIGURE 13.20
The continental crust is deeper beneath mountains.

CHECK YOURSELF

A river barge loaded with gravel approaches a low bridge that it cannot quite pass under. Should gravel be removed *from* or added *to* the barge?

Pascal's Principle

One of the most important facts about fluid pressure is that a change in pressure at one part of the fluid will be transmitted undiminished to other parts. For example, if the pressure of city water is increased at the pumping station by 10 units of pressure, the pressure everywhere in the pipes of the connected system will be increased by 10 units of pressure (provided that the water is at rest). This rule is called **Pascal's principle:**

> **A change in pressure at any point in an enclosed fluid at rest is transmitted undiminished to all points in the fluid.**

Pascal's principle was discovered in the seventeenth century by Blaise Pascal (who was an invalid at the age of 18 and remained so until his death at the age of 39), for whom the SI unit of pressure, the pascal ($1 \text{ Pa} = 1 \text{ N/m}^2$), was named.

DO YOU HAVE AN ANSWER?

Ho, ho, ho! Do you think ol' Hewitt is going to give *all* the answers to Check Yourself questions? Good teaching is asking good questions, not providing all answers. You're on your own with this one!

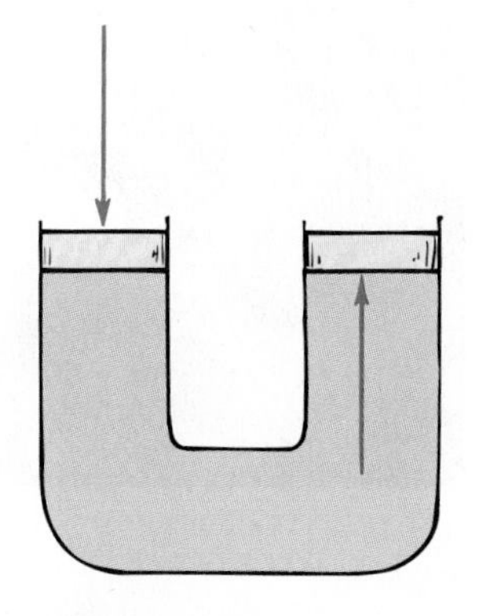

FIGURE 13.21
The force exerted on the left piston increases the pressure in the liquid and is transmitted to the right piston.

Fill a U-tube with water and place pistons at each end, as shown in Figure 13.21. Pressure exerted against the left piston will be transmitted throughout the liquid and against the bottom of the right piston. (The pistons are simply "plugs" that can slide freely but snugly inside the tube.) The pressure that the left piston exerts against the water will be exactly equal to the pressure the water exerts against the right piston. This is nothing to write home about. But suppose you make the tube on the right side wider and use a piston of larger area: Then the result will be impressive. In Figure 13.22 the piston on the right has 50 times the area of the piston on the left (let's say that the left piston has a cross-sectional area of 100 square centimeters and that the right piston has a cross-sectional area of 5000 square centimeters). Suppose a 10-kg load is placed on the left piston. Then an additional pressure (nearly 1 N/cm^2) due to the weight of the load is transmitted throughout the liquid and up against the larger piston. Here is where the difference between force and pressure comes in. The additional pressure is exerted against every square centimeter of the larger piston. Since there is 50 times the area, 50 times as much force is exerted on the larger piston. Thus, the larger piston will support a 500-kg load—fifty times the load on the smaller piston!

This *is* something to write home about, for we can multiply forces using such a device. One newton input produces 50 newtons output. By further increasing the area of the larger piston (or reducing the area of the smaller piston), we can multiply force, in principle, by any amount. Pascal's principle underlies the operation of the hydraulic press.

The hydraulic press does not violate energy conservation, because a decrease in distance moved compensates for the increase in force. When the small piston in Figure 13.22 is moved downward 10 centimeters, the large piston will be raised only one-fiftieth of this, or 0.2 centimeter. The input force multiplied by the distance moved by the smaller piston is equal to the output force multiplied by the distance moved by the larger piston; this is one more example of a simple machine operating on the same principle as a mechanical lever.

Seventeenth-century theologian and scientist Blaise Pascal is remembered scientifically for hydraulics, which changed the technological landscape more than he imagined. He is remembered theologically for his many assertions, one of which relates to centuries of human landscape: "Men never do evil so cheerfully and completely as when they do so from religious conviction."

Insights

Pascal's principle applies to all fluids, whether gases or liquids. A typical application of Pascal's principle for gases and liquids is the automobile lift seen in many service stations (Figure 13.23). Increased air pressure produced by an air compressor is transmitted through the air to the surface of oil in an underground reservoir. The oil, in turn, transmits the pressure to a piston, which lifts the automobile. The relatively low pressure that exerts the lifting force against the piston is about the same as the air pressure in automobile tires.

Hydraulics is employed by modern devices ranging from very small to enormous. Note the hydraulic pistons in almost all construction machines where heavy loads are involved (Figure 13.24)

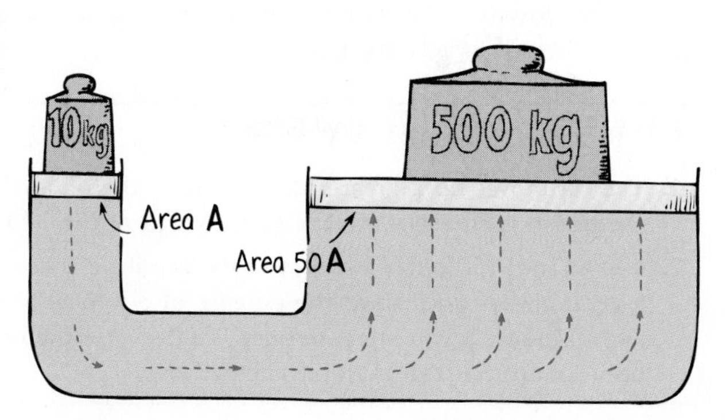

FIGURE 13.22
A 10-kg load on the left piston will support 500 kg on the right piston.

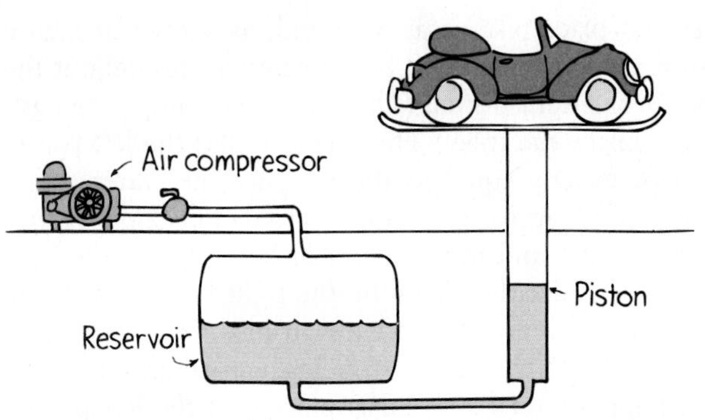

FIGURE 13.23
Pascal's principle in a service station.

FIGURE 13.24
Pascal's principle at work in the hydraulic devices on this incredible machine. We can only wonder whether Pascal envisioned the extent to which his principle would allow huge loads to be so easily lifted.

CHECK YOURSELF

1. As the automobile in Figure 13.23 is being lifted, how does the change in oil level in the reservoir compare with the distance the automobile moves?
2. If a friend commented that a hydraulic device is a common way of multiplying energy, what would you say?

Surface Tension

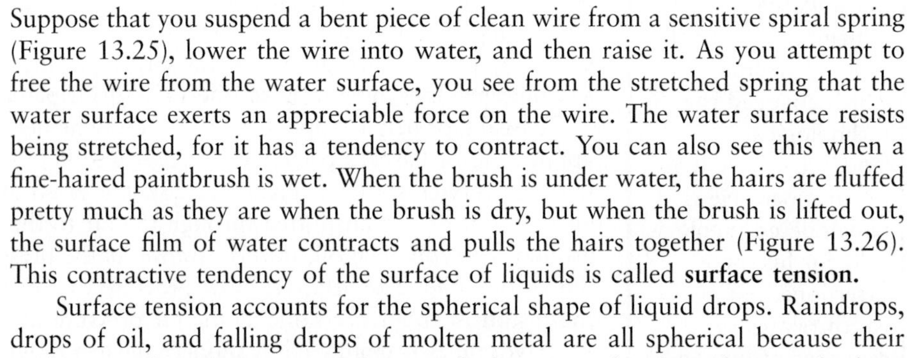

Suppose that you suspend a bent piece of clean wire from a sensitive spiral spring (Figure 13.25), lower the wire into water, and then raise it. As you attempt to free the wire from the water surface, you see from the stretched spring that the water surface exerts an appreciable force on the wire. The water surface resists being stretched, for it has a tendency to contract. You can also see this when a fine-haired paintbrush is wet. When the brush is under water, the hairs are fluffed pretty much as they are when the brush is dry, but when the brush is lifted out, the surface film of water contracts and pulls the hairs together (Figure 13.26). This contractive tendency of the surface of liquids is called **surface tension.**

Surface tension accounts for the spherical shape of liquid drops. Raindrops, drops of oil, and falling drops of molten metal are all spherical because their surfaces tend to contract and force each drop into the shape having the least surface area. This is a sphere, the geometrical figure that has the least surface area for a given volume. For this reason, the mist and dewdrops on spider webs or on the downy leaves of plants are nearly spherical blobs. (The larger they are, the more that gravity flattens them.)

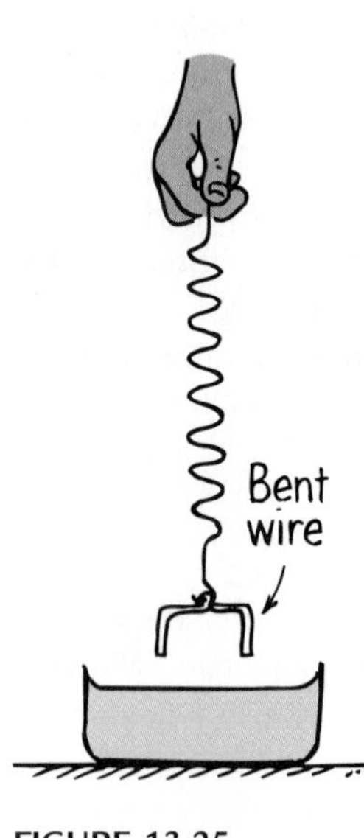
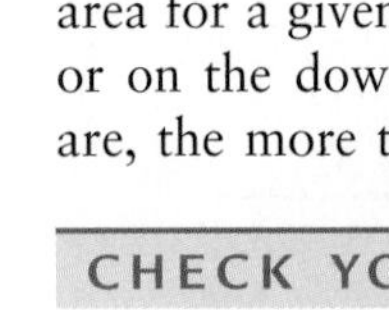

FIGURE 13.25
When the bent wire is lowered into the water and then raised, the spring will stretch because of surface tension.

CHECK YOUR ANSWERS

1. The car moves up a greater distance than the oil level drops, since the area of the piston is smaller than the surface area of the oil in the reservoir.
2. No, no, no! Although a hydraulic device, like a mechanical lever, can multiply *force*, it always does so at the expense of distance. Energy is the product of force and distance. If you increase one, you decrease the other. No device has ever been found that can multiply energy!

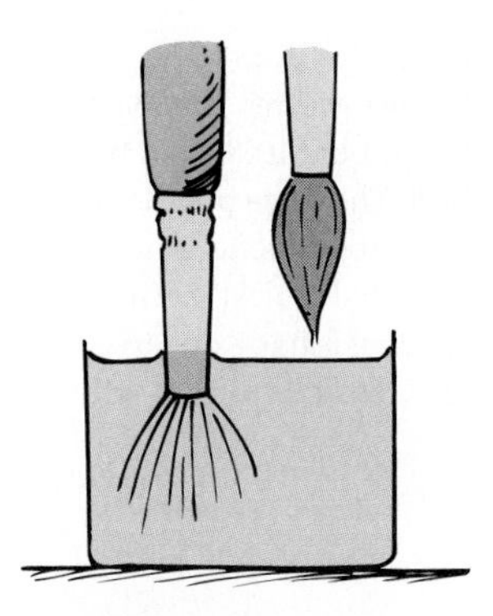

FIGURE 13.26
When the brush is taken out of the water, the hairs are held together by surface tension.

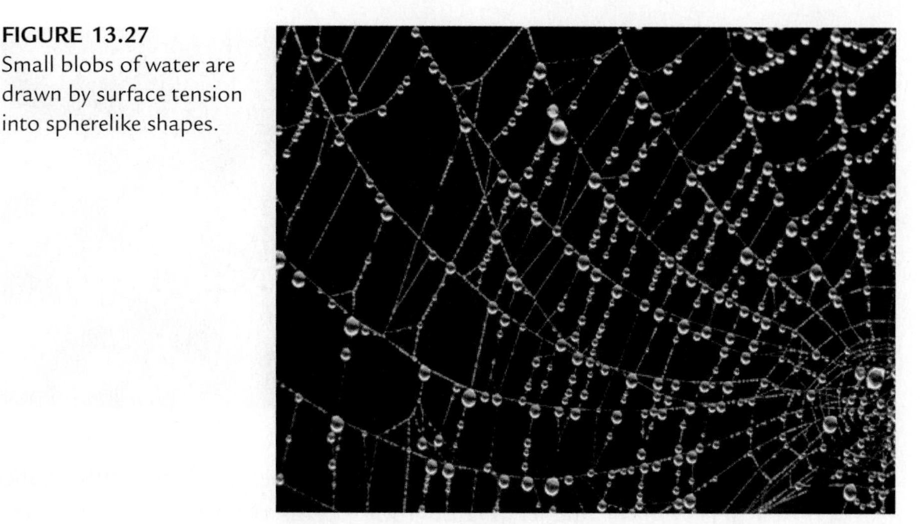

FIGURE 13.27
Small blobs of water are drawn by surface tension into spherelike shapes.

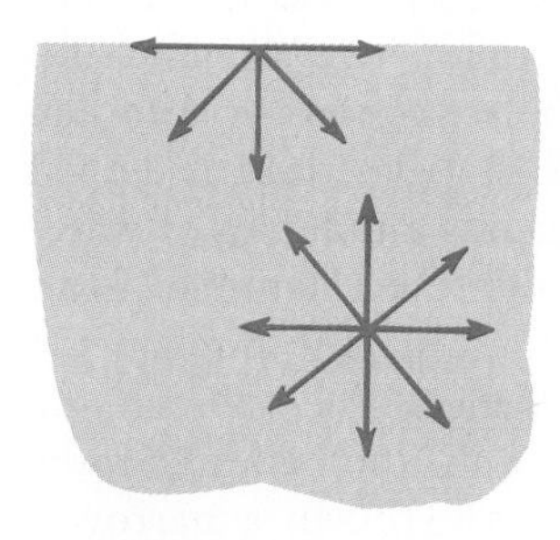

FIGURE 13.28
A molecule at the surface is pulled only sideways and downward by neighboring molecules. A molecule beneath the surface is pulled equally in all directions.

Surface tension is caused by molecular attractions. Beneath the surface, each molecule is attracted in every direction by neighboring molecules, resulting in no tendency to be pulled in any specific direction. A molecule on the surface of a liquid, however, is pulled only by neighbors on each side and downward from below; there is no pull upward (Figure 13.28). These molecular attractions thus tend to pull the molecule from the surface into the liquid, and this tendency minimizes the surface area. The surface behaves as if it were tightened into an elastic film. This is evident when dry steel needles or razor blades seem to float on water. They don't float in the usual sense, but they are supported by the surface molecules opposing an increase in surface area. The water surface sags like a piece of plastic wrap, which allows certain insects, such as water striders, to run across the surface of a pond.

The surface tension of water is greater than that of other common liquids, and pure water has a stronger surface tension than soapy water. We can see this when a little soap film on the surface of water is effectively pulled out over the entire surface. This minimizes the surface area of the water. The same thing happens for oil or grease floating on water. Oil has less surface tension than cold water, and it is drawn out into a film covering the whole surface. But hot water has less surface tension than cold water because the faster-moving molecules are not bonded as tightly. This allows the grease or oil in hot soups to float in little bubbles on the surface of the soup. When the soup cools and the surface tension of the water increases, the grease or oil is dragged out over the surface of the soup. The soup becomes "greasy." Hot soup tastes different from cold soup primarily because the surface tension of water in the soup changes with temperature.

Capillarity

When the end of a thoroughly clean glass tube with a small inside diameter is dipped into water, the water wets the inside of the tube and rises in it. In a tube with a bore of about $\frac{1}{2}$ millimeter in diameter, for example, the water rises

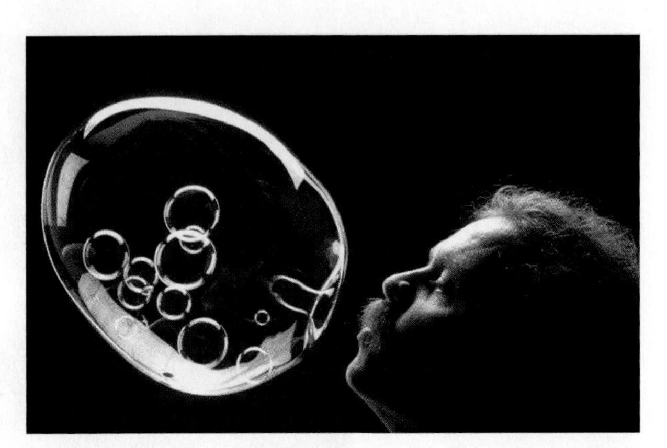

FIGURE 13.29
Bubble Master Tom Noddy blows bubbles within bubbles. The large bubble is elongated due to blowing, but it will quickly settle to a spherical shape due to surface tension.

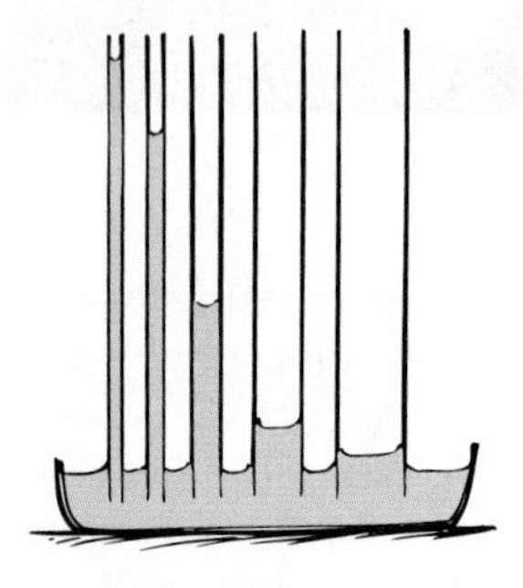

FIGURE 13.30
Capillary tubes.

slightly higher than 5 centimeters. With a still smaller bore, the water rises much higher (Figure 13.30). This rise of a liquid in a fine, hollow tube or in a narrow space is **capillarity.**

When thinking of capillarity, think of molecules as sticky balls. Water molecules stick to glass more than to each other. The attraction between unlike substances such as water and glass is called *adhesion.* The attraction between like substances, molecular stickiness, is called *cohesion.* When a glass tube is dipped into water, the adhesion between the glass and water causes a thin film of water to be drawn up over the inner and outer surfaces of the tube (Figure 13.31a). Surface tension causes this film to contract (Figure 13.31b). The film on the outer surface contracts enough to make a rounded edge. The film on the inner surface contracts more and raises water with it until the adhesive force is balanced by the weight of the water lifted (Figure 13.31c). In a narrower tube, the weight of the water in the tube is small and the water is lifted higher than it would be if the tube were wider.

If a paintbrush is dipped part way into water, the water will rise up into the narrow spaces between the bristles by capillary action. If your hair is long, let it hang into the sink or bathtub, and water will seep up to your scalp in the same way. This is how oil soaks upward in a lamp wick and water soaks into a bath towel when one end hangs in water. Dip one end of a lump of sugar in coffee, and the entire lump is quickly wet. Capillary action is essential for plant growth. It brings water to the roots of plants and carries sap and nourishment to high branches of trees. Just about everywhere we look, we can see capillary action at work. That's nice.

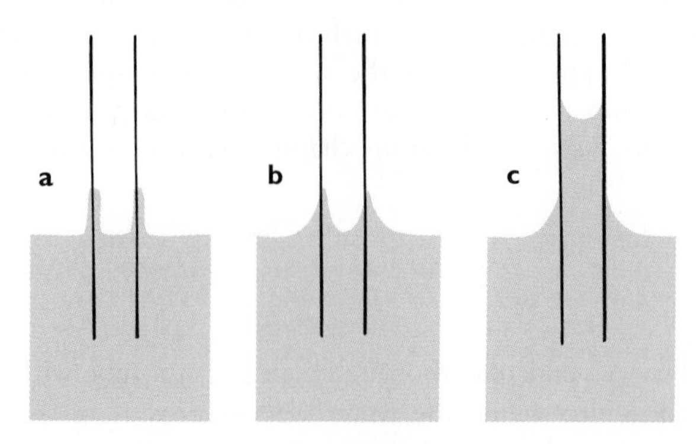

FIGURE 13.31
Hypothetical stages of capillary action, as seen in a cross-sectional view of a capillary tube.

But from the point of view of an insect, capillarity is not so nice. Recall from the previous chapter that, because of an insect's relatively large surface area, it falls slowly in air. Gravity poses almost no risk at all—but not so with capillarity. Being in the grip of water may be fatal to an insect—unless it is equipped for water like a water strider.

Summary of Terms

Pressure The ratio of force to the area over which that force is distributed:

$$\text{Pressure} = \frac{\text{force}}{\text{area}}$$

$$\text{Liquid pressure} = \text{weight density} \times \text{depth}$$

Buoyant force The net upward force that a fluid exerts on an immersed object.

Archimedes' principle An immersed body is buoyed up by a force equal to the weight of the fluid it displaces.

Principle of flotation A floating object displaces a weight of fluid equal to its own weight.

Pascal's principle The pressure applied to a motionless fluid confined in a container is transmitted undiminished throughout the fluid.

Surface tension The tendency of the surface of a liquid to contract in area and thus to behave like a stretched elastic membrane.

Capillarity The rise of a liquid in a fine, hollow tube or in a narrow space.

Suggested Reading

Rogers, E. *Physics for the Inquiring Mind.* Princeton, N.J.: Princeton University Press, 1960. Chapter 6 of this textbook, an oldie but goodie (and a wonderful influence on this author's writing), treats surface tension in fascinating detail.

Review Questions

1. Give two examples of a fluid.

Pressure

2. Distinguish between *force* and *pressure*.

Pressure in a Liquid

3. What is the relationship between liquid pressure and the depth of a liquid? Between liquid pressure and density?
4. If you swim beneath the surface in salt water, will the pressure be greater than in freshwater at the same depth? Why or why not?
5. How does water pressure one meter below the surface of a small pond compare with water pressure one meter below the surface of a huge lake?
6. If you punch a hole in a container filled with water, in what direction does the water initially flow outward from the container?

Buoyancy

7. Why does buoyant force act upward on an object submerged in water?
8. Why is there no horizontal buoyant force on a submerged object?
9. How does the volume of a completely submerged object compare with the volume of water displaced?

Archimedes' Principle

10. How does the buoyant force on a submerged object compare with the weight of water displaced?
11. Distinguish between a *submerged* body and an *immersed* body.
12. What is the mass of 1 L of water? What is its weight in newtons?
13. If a 1-L container is immersed halfway into water, what is the volume of water displaced? What is the buoyant force on the container?

What Makes an Object Sink or Float?

14. Is the buoyant force on a submerged object equal to the weight of the object itself or equal to the weight of the fluid displaced by the object?
15. There is a condition in which the buoyant force on an object does equal the weight of the object. What is this condition?
16. Does the buoyant force on a submerged object depend on the volume of the object or the weight of the object?
17. Fill in the blanks: An object denser than water will ______ in water. An object less dense than water will ______ in water. An object with the same density of water will ____________ in water.
18. How is the density of a fish controlled? How is the density of a submarine controlled?

Flotation

19. It was emphasized earlier that buoyant force does not equal an object's weight but does equal the weight of displaced water. Now we say buoyant force equals the object's weight. Isn't this a grand contradiction? Explain.
20. What weight of water is displaced by a 100-ton ship? What is the buoyant force that acts on a floating 100-ton ship?

Pascal's Principle

21. What happens to the pressure in all parts of a confined fluid if the pressure in one part is increased?
22. If the pressure in a hydraulic press is increased by an additional 10 N/cm^2, how much extra load will the output piston support if its cross-sectional area is 50 cm^2?

Surface Tension

23. What geometrical shape has the least surface area for a given volume?
24. What is the cause of surface tension?

Capillarity

25. Distinguish between *adhesive* and *cohesive* forces.
26. What determines how high water will climb in a capillary tube?

Projects

1. Place an egg in a pan of tap water. Then dissolve salt in the water until the egg floats. How does the density of an egg compare to that of tap water? To that of salt water?
2. If you punch a couple of holes in the bottom of a water-filled container, water will spurt out because of water pressure. Now drop the container, and, as it freely falls, note that the water no longer spurts out! If your friends don't understand this, could you figure it out and then explain it to them?
3. Float a water-soaked Ping-Pong ball in a can of water held more than a meter above a rigid floor. Then drop the can. Careful inspection will show the ball pulled beneath the surface as both the ball and the can drop. (What does this say about surface tension?). More dramatically, when the can makes impact with the floor, what happens to the ball, and why? Try it and you'll be astonished! (*Caution*: Unless you're wearing safety goggles, keep your head away from above the can when it makes impact.)
4. Soap greatly weakens the cohesive forces between water molecules. You can see this by putting some oil in a bottle of water and shaking it so that the oil and water mix. Notice that the oil and water quickly separate as soon as you stop shaking the bottle. Now add some soap to the mixture. Shake the bottle again and you will see that the soap makes a fine film around each little oil bead and that a longer time is required for the oil to gather after you stop shaking the bottle.

   This is how soap works in cleaning. It breaks the surface tension around each particle of dirt so that the water can reach the particles and surround them. The dirt is carried away in rinsing. Soap is a good cleaner only in the presence of water.

One-Step Calculations

Pressure = weight density × depth

(Neglect the pressure due to the atmosphere in the calculations below.)

1. Calculate the water pressure at the bottom of the 100-m-high water tower shown in Figure 13.2.
2. Calculate the water pressure at the base of a dam when the depth of water behind the dam is 100 m.
3. The top floor of a building is 50 m above the basement. Calculate how much greater the water pressure is in the basement compared with the pressure at the top floor.
4. Water pressure at the bottom of a 1-m-tall closed barrel is 98 kPa. What is the pressure at the bottom of the barrel when a 5-m pipe filled with water is inserted into the top of the barrel?

Exercises

1. What common liquid covers more than two-thirds of our planet, makes up 60% of our bodies, and sustains our lives and lifestyles in countless ways?
2. Which is more likely to hurt—being stepped on by a 200-lb man wearing loafers or being stepped on by a 100-lb woman wearing high heels?
3. Which do you suppose exerts more pressure on the ground—an elephant or a lady standing on spike heels? (Which will be more likely to make dents in a linoleum floor?) Approximate a rough calculation for each.
4. Stand on a bathroom scale and read your weight. When you lift one foot up so that you're standing on one foot, does the reading change? Does a scale read force or pressure?
5. Why are persons who are confined to bed less likely to develop bedsores on their bodies if they use a waterbed rather than an ordinary mattress?
6. You know that a sharp knife cuts better than a dull knife. Do you know why this is so? Defend your answer.

7. If water faucets upstairs and downstairs are turned fully on, will more water per second flow out of the upstairs faucets or the downstairs faucets?

8. The photo shows physics instructor Marshall Ellenstein walking barefoot on broken glass bottles in his class. What physics concept is Marshall demonstrating, and why is he careful that the broken pieces are small and numerous? (The Band-Aids on his feet are for humor!)

9. Why does your body get more rest when you're lying down than it does when you're sitting? And why is blood pressure measured in the upper arm, at the elevation of your heart? Is blood pressure in your legs greater?

10. When standing, blood pressure in your legs is greater than in your upper body. Would this be true for an astronaut in orbit? Defend your answer.

11. How does water pressure 1 meter beneath the surface of a lake compare with water pressure 1 meter beneath the surface of a swimming pool?

12. Which teapot holds more liquid?

13. The sketch shows a reservoir that supplies water to a farm. It is made of wood and is reinforced with metal hoops. (a) Why is it elevated? (b) Why are the hoops closer together near the bottom part of the tank?

14. A block of aluminum with a volume of 10 cm^3 is placed in a beaker of water filled to the brim. Water overflows. The same is done in another beaker with a 10-cm^3 block of lead. Does the lead displace more, less, or the same amount of water?

15. A block of aluminum with a mass of 1 kg is placed in a beaker of water filled to the brim. Water overflows. The same is done in another beaker with a 1-kg block of lead. Does the lead displace more, less, or the same amount of water?

16. A block of aluminum with a weight of 10 N is placed in a beaker of water filled to the brim. Water overflows. The same is done in another beaker with a 10-N block of lead. Does the lead displace more, less, or the same amount of water? (Why are your answers to this exercise and to Exercise 15 different from your answer to Exercise 14?)

17. In 1960, the U.S. Navy's bathyscaphe *Trieste* (a submersible) descended to a depth of nearly 11 kilometers in the Marianas Trench near the Philippines in the Pacific Ocean. Instead of a large viewing window, it was a small circular window 15 centimeters in diameter. What is your explanation for so small a window?

18. There is a story about Pascal in which it is said that he climbed a ladder and poured a small container of water into a tall, thin, vertical pipe inserted into a wooden barrel full of water below. The barrel burst when the water in the pipe reached about 12 m. This was all the more intriguing because the weight of added water in the tube was very small. What two physical principles was Pascal demonstrating?

19. There is a legend of a Dutch boy who bravely held back the whole North Sea by plugging a hole in a dike with his finger. Is this possible and reasonable? (See also Problem 4.)

20. If you've wondered about the flushing of toilets on the upper floors of city skyscrapers, how do you suppose the plumbing is designed so that there is not an enormous impact of sewage arriving at the basement level? (Check your speculations with someone who is knowledgeable about architecture.)

21. Why does water "seek its own level"?

22. Suppose that you wish to lay a level foundation for a home on hilly and bushy terrain. How can you use a garden hose filled with water to determine equal elevations for distant points?

23. When you are bathing on a stony beach, why do the stones hurt your feet less when you're standing in deep water?

24. If liquid pressure were the same at all depths, would there be a buoyant force on an object submerged in the liquid? Explain.

25. A can of diet soda floats in water, whereas a can of regular soda sinks. Explain this phenomenon first in terms of density, then in terms of weight versus buoyant force.

26. Why will a block of iron float in mercury but sink in water?

27. The mountains of the Himalayas are slightly less dense than the mantle material upon which they "float." Do you suppose that, like floating icebergs, they are deeper than they are high?

28. Why is a high mountain composed mostly of lead an impossibility on the planet Earth?

29. How much force is needed to push a nearly weightless but rigid 1-L carton beneath a surface of water?

30. Why will a volleyball held beneath the surface of water have more buoyant force than if it is floating?

31. Why does an inflated beach ball pushed beneath the surface of water swiftly shoot above the water surface when released?

32. Why is it inaccurate to say that heavy objects sink and that light objects float? Give exaggerated examples to support your answer.

33. Why is the buoyant force on a submerged submarine appreciably greater than the buoyant force on it while it is floating?

34. A piece of iron placed on a block of wood makes it float lower in the water. If the iron were instead suspended beneath the wood, would it float as low, lower, or higher? Defend your answer.

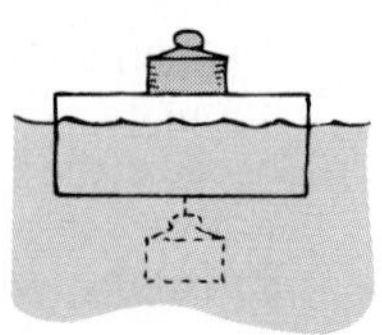

35. Compared with an empty ship, would a ship loaded with a cargo of Styrofoam sink deeper into the water or rise in the water? Defend your answer.

36. If a submarine starts to sink, will it continue to sink to the bottom if no changes are made? Explain.

37. A barge filled with scrap iron is in a canal lock. If the iron is thrown overboard, does the water level at the side of the lock rise, fall, or remain unchanged? Explain.

38. Would the water level in a canal lock go up or down if a battleship in the lock sank?

39. Will a rock gain or lose buoyant force as it sinks deeper in water? Or will the buoyant force remain the same at greater depths? Defend your answer.

40. Will a swimmer gain or lose buoyant force as she swims deeper in the water? Or will her buoyant force remain the same at greater depths? Defend your answer, and contrast it with your answer to Exercise 39.

41. A balloon is weighted so that it is barely able to float in water. If it is pushed beneath the surface, will it return to the surface, stay at the depth to which it is pushed, or sink? Explain. (*Hint*: Does the balloon's density change?)

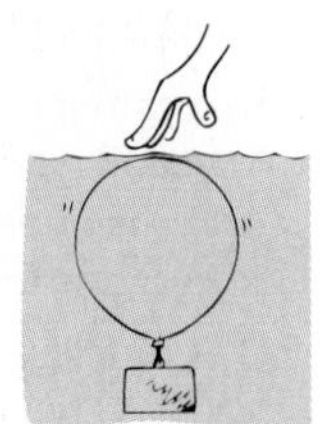

42. The density of a rock doesn't change when it is submerged in water, but your density changes when you are submerged. Explain.

43. In answering the question of why bodies float higher in salt water than in freshwater, your friend replies that the reason is that salt water is denser than freshwater. (Does your friend often answer questions by reciting only factual statements that relate to the answers but don't provide any concrete reasons?) How would you answer the same question?

44. A ship sailing from the ocean into a freshwater harbor sinks slightly deeper into the water. Does the buoyant force on the ship change? If so, does it increase or decrease?

45. Suppose that you are given the choice between two life preservers that are identical in size, the first a light one filled with Styrofoam and the second a very heavy one filled with gravel. If you submerge these life preservers in the water, upon which will the buoyant force be greater? Upon which will the buoyant force be ineffective? Why are your answers different?

46. The weight of the human brain is about 15 N. The buoyant force supplied by fluid around the brain is about 14.5 N. Does this mean that the weight of fluid surrounding the brain is at least 14.5 N? Defend your answer.

47. The relative densities of water, ice, and alcohol are 1.0, 0.9, and 0.8, respectively. Do ice cubes float higher or lower in a mixed alcoholic drink? What comment can you make about a cocktail in which the ice cubes lie submerged at the bottom of the glass?

48. When an ice cube in a glass of water melts, does the water level in the glass rise, fall, or remain unchanged? Does your answer change if the ice cube has many air bubbles? How about if the ice cube contains many grains of heavy sand?

49. When the wooden block is placed in the beaker, what happens to the scale reading? Answer the same question for an iron block.

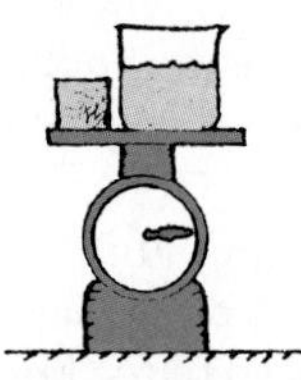

50. A half-filled bucket of water is on a spring scale. Will the reading of the scale increase or remain the same if a fish is placed in the bucket? (Will your answer be different if the bucket is initially filled to the brim?)

51. The weight of the container of water, as shown in *a*, is equal to the weight of the stand and the suspended solid iron ball. When the suspended ball is lowered into the water, as shown in *b*, the balance is upset. Will the additional weight needed on the right side to restore balance be greater than, equal to, or less than the weight of the solid iron ball?

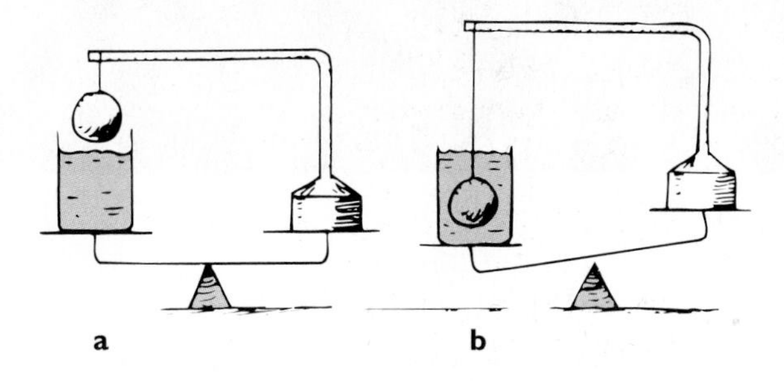

52. If the gravitational field of the Earth were to increase, would a fish float to the surface, sink, or stay at the same depth?

53. What would you experience when swimming in water in an orbiting space habitat where simulated gravity is $\frac{1}{2}g$? Would you float in the water as you do on Earth?

54. We say that the shape of a liquid is that of its container. But, with no container and no gravity, what is the natural shape of a blob of water? Why?

55. If you release a Ping-Pong ball beneath the surface of water, it will rise to the surface. Would it do the same if it were inside a big blob of water floating weightless in an orbiting spacecraft?

56. So you're on a run of bad luck, and you slip quietly into a small, quiet pool as hungry crocodiles lurking at the bottom are relying on Pascal's principle to help them to detect a tender morsel. What does Pascal's principle have to do with their delight at your arrival?

57. In the hydraulic arrangement shown, the larger piston has an area that is fifty times that of the smaller piston. The strong man hopes to exert enough force on the large piston to raise the 10 kg that rest on the small piston. Do you think he will be successful? Defend your answer.

58. In the hydraulic arrangement shown in Figure 13.22, the multiplication of force is equal to the ratio of the areas of the large and small pistons. Some people are surprised to learn that the area of the liquid surface in the reservoir of the arrangement shown in Figure 13.23 is immaterial. What is your explanation to clear up this confusion?

59. Why will hot water leak more readily than cold water through small leaks in a car radiator?

60. On the surface of a pond, it is common to see water striders, insects that can "walk" on the surface of water without sinking. What physics concept explains their ability?

Problems

1. The depth of water behind the Hoover Dam in Nevada is 220 m. What is the water pressure at the base of this dam? (Neglect the pressure due to the atmosphere.)

2. A 6-kg piece of metal displaces 1 liter of water when submerged. What is its density?

3. A rectangular barge, 5 m long and 2 m wide, floats in freshwater. (*a*) Find how much deeper it floats when its load is a 400-kg horse. (*b*) If the barge can only be pushed 15 cm deeper into the water before water overflows to sink it, how many 400-kg horses can it carry?

4. A dike in Holland springs a leak through a hole of area 1 cm^2 at a depth of 2 m below the water surface. How much force must a boy apply to the hole with his thumb to stop the leak? Could he do it?

5. A merchant in Kathmandu sells you a solid-gold 1-kg statue for a very reasonable price. When you return home, you wonder whether or not you got a bargain, so you lower the statue into a container of water and measure the volume of displaced water. What volume will verify that it's pure gold?

6. When a 2.0-kg object is suspended in water, it "masses" 1.5 kg. What is the density of the object?

7. An ice cube measures 10 cm on a side and floats in water. One cm extends above water level. If you shaved off the 1-cm part, how many cm of the remaining ice would extend above water level?

8. A swimmer wears a heavy belt to make her average density exactly equal to the density of water. Her mass, including the belt, is 60 kg. (*a*) What is the swimmer's weight in newtons? (*b*) What is the swimmer's volume in m^3? (*c*) At a depth of 2 m below the surface of a pond, what buoyant force acts on the swimmer? What net force acts on her?

9. A vacationer floats lazily in the ocean with 90% of his body below the surface. The density of the ocean water is 1,025 kg/m^3. What is the vacationer's average density?

10. In the hydraulic pistons shown in the sketch, the small piston has a diameter of 2 cm. The larger piston has a diameter of 6 cm. How much more force can the larger piston exert compared with the force applied to the smaller piston?

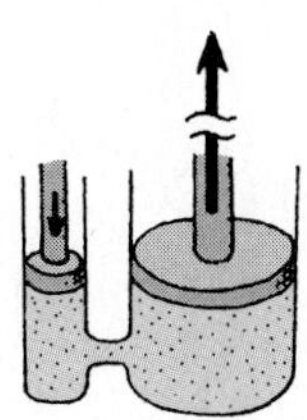

CHAPTER 14

Gases and Plasmas

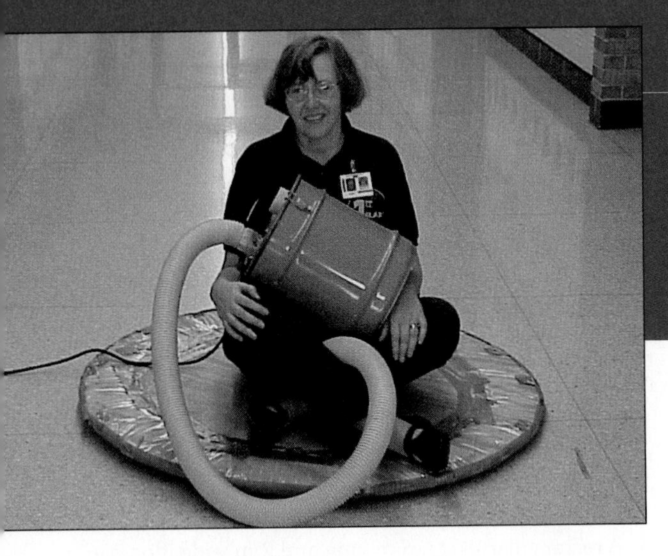

Ann Brandon fascinates her students when she rides on a cushion of air blown through a hole in the middle of this jumbo air puck.

Gases as well as liquids flow; hence, both are called *fluids*. The primary difference between a gas and a liquid is the distance between molecules. In a gas, the molecules are far apart and free from the cohesive forces that dominate their motions when in the liquid and solid phases. Their motions are less restricted. A gas expands indefinitely and fills all space available to it. Only when the quantity of gas is very large, such as in the Earth's atmosphere or in a star, do gravitational forces limit the size, or determine the shape, of the mass of a gas.

The Atmosphere

The thickness of our atmosphere is determined by two competing factors: the kinetic energy of its molecules, which tends to spread the molecules apart; and gravity, which tends to hold them near the Earth. If Earth gravity were somehow shut off, atmospheric molecules would dissipate and disappear. Or if gravity acted but the molecules moved too slowly to form a gas (as might occur on a remote, cold planet), our "atmosphere" would be a liquid or solid layer, just so much more matter lying on the ground. There would be nothing to breathe: again, no atmosphere.

But our atmosphere is a happy compromise between energetic molecules that tend to fly away and gravity that holds them back. Without the heat of the Sun, air molecules would lie on Earth's surface the way settled popcorn lies at the bottom of a popcorn machine. But, if heat is added to the popcorn and to the atmospheric gases, both will bumble their way up to higher altitudes. Pieces of popcorn in a popper attain speeds of a few kilometers per hour and reach altitudes up to a meter or two; molecules in the air move at speeds of about 1600 kilometers per hour and bumble up to many kilometers in altitude. Fortunately, there is an energizing Sun, there is gravity, and Earth has an atmosphere.

The exact height of the atmosphere has no real meaning, for the air gets thinner and thinner the higher one goes. Eventually, it thins out to emptiness in interplanetary space. Even in the vacuous regions of interplanetary space, however, there is a gas density of about one molecule per cubic centimeter. This is primarily hydrogen, the most plentiful element in the universe. About 50% of the atmosphere is below an altitude of 5.6 kilometers (18,000 ft), 75% is below 11 kilometers (36,000 ft), 90% is below 18 kilometers (60,000 ft), and

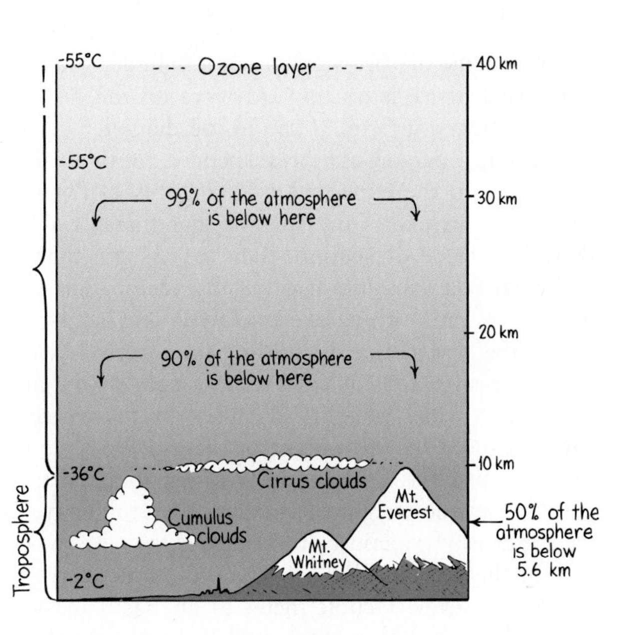

FIGURE 14.1
The atmosphere. Air is more compressed at sea level than at higher altitudes. Like feathers in a huge pile, what's at the bottom is more squashed than what's nearer the top.

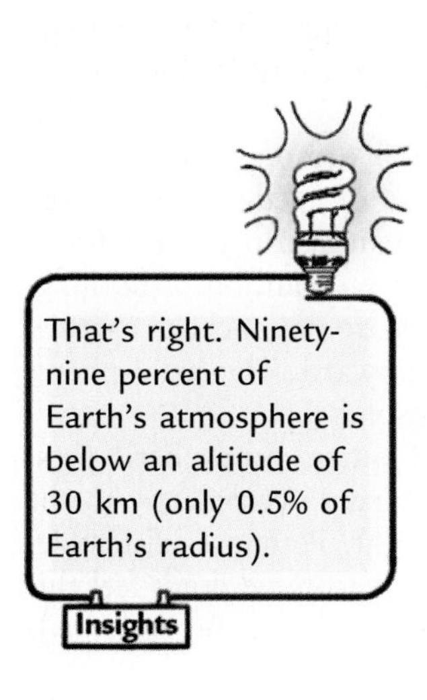

That's right. Ninety-nine percent of Earth's atmosphere is below an altitude of 30 km (only 0.5% of Earth's radius).
Insights

99% is below about 30 kilometers (100,000 ft) (Figure 14.1). A detailed description of the atmosphere can be found in any good Web site.

Atmospheric Pressure

We live at the bottom of an ocean of air. The atmosphere, much like water in a lake, exerts pressure. One of the most celebrated experiments demonstrating the pressure of the atmosphere was conducted in 1654 by Otto von Guericke, burgermeister of Magdeburg and inventor of the vacuum pump. Von Guericke placed together two copper hemispheres about $\frac{1}{2}$ meter in diameter to form a sphere, as shown in Figure 14.2. He fashioned an airtight joint with an oil-soaked leather gasket. When he evacuated the sphere with his vacuum pump, two teams of eight horses each were unable to pull the hemispheres apart.

Interestingly, von Guericke's demonstration preceded knowledge of Newton's third law. The forces on the hemispheres would have been the same if he used only one team of horses and tied the other end of the rope to a tree!

Insights

FIGURE 14.2
The famous "Magdeburg hemispheres" experiment of 1654, demonstrating atmospheric pressure. Two teams of horses couldn't pull the evacuated hemispheres apart. Were the hemispheres sucked together or pushed together? By what?

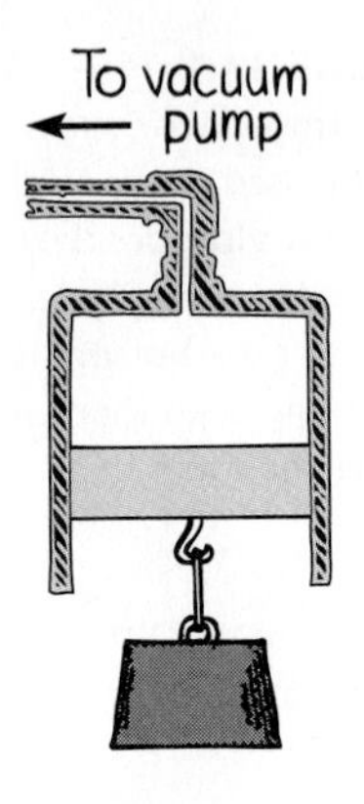

FIGURE 14.3
Is the piston that supports the load pulled up or pushed up?

TABLE 14.1
Densities of Various Gases

Gas	*Density* (kg/m^3)*
Dry air	
0°C	1.29
10°C	1.25
20°C	1.21
30°C	1.16
Hydrogen	0.090
Helium	0.178
Nitrogen	1.25
Oxygen	1.43

*At sea-level atmospheric pressure and at 0°C (unless otherwise specified)

When the air pressure inside a cylinder like the one shown in Figure 14.3 is reduced, there is an upward force on the piston. This force is large enough to lift a heavy weight. If the inside diameter of the cylinder is 10 centimeters or greater, a person can be suspended by this force.

What do the experiments of Figures 14.2 and 14.3 demonstrate? Do they show that air exerts pressure or that there is a "force of suction"? If we say there is a force of suction, then we assume that a vacuum can exert a force. But what is a vacuum? It is an absence of matter; it is a condition of nothingness. How can nothing exert a force? The hemispheres are not sucked together, nor is the piston that is holding the weight sucked upward. The hemispheres and the piston are pushed by the weight of the atmosphere.

Just as water pressure is caused by the weight of water, **atmospheric pressure** is caused by the weight of the air. We have adapted so completely to the invisible air that we sometimes forget that it has weight. Perhaps a fish "forgets" about the weight of water in the same way. The reason we don't feel this weight crushing against our bodies is that the pressure inside our bodies equals that of the surrounding air. There is no net force for us to sense.

At sea level, 1 cubic meter of air has a mass of about 1.25 kilograms. So the air in your kid sister's small bedroom weighs about as much as she does! The density of air decreases with altitude. At 10 kilometers, for example, 1 cubic meter of air has a mass of about 0.4 kilograms. To compensate for this, airplanes are pressurized; the additional air needed to fully pressurize a modern jumbo jet, for example, is more than 1000 kilograms. Air is heavy if you have enough of it. If your kid sister doesn't believe that air has weight, you can show her why she falsely perceives the air to be weight free. If you hand her a plastic bag of water, she'll tell you that it has weight. But, if you hand her the same bag of water while she's submerged in a swimming pool, she won't feel its weight. That's because she and the bag are surrounded by water. Likewise with the air that is all around us.

FIGURE 14.4
You don't notice the weight of a bag of water while you're submerged in water. Similarly, you don't notice the weight of air while you are submerged in an "ocean" of air.

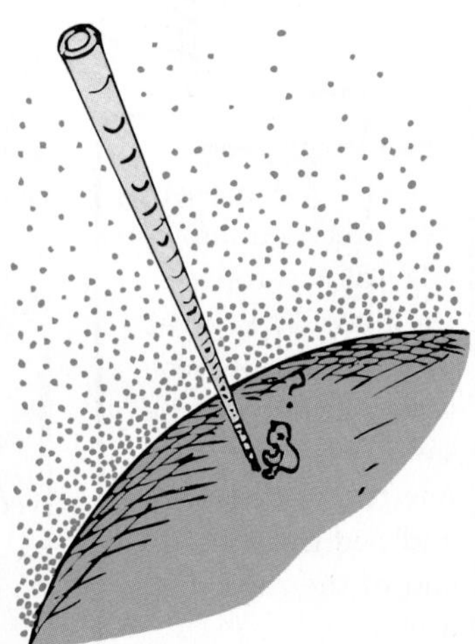

FIGURE 14.5
The mass of air that would occupy a bamboo pole that extends 30 km up—to the "top" of the atmosphere—is about 1 kg. This air weighs about 10 N.

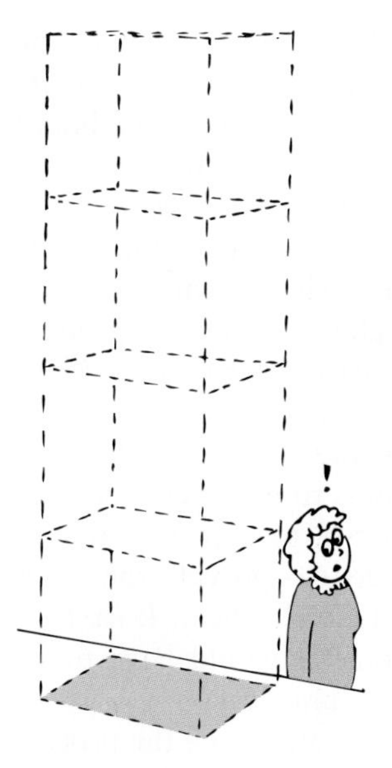

FIGURE 14.6
The weight of air bearing down on a 1-square-meter surface at sea level is about 100,000 N. In other words, atmospheric pressure is about 10^5 N/m^2, or about 100 kPa.

Consider the mass of air in an upright 30-kilometer-tall bamboo pole that has an inside cross-sectional area of 1 square centimeter. If the density of air inside the pole matches the density of air outside, the mass of enclosed air would be about 1 kilogram. The weight of this much air is about 10 newtons. So air pressure at the bottom of the bamboo pole would be about 10 newtons per square centimeter (10 N/cm^2). Of course, the same is true without the bamboo pole. There are 10,000 square centimeters in 1 square meter, so a column of air 1 square meter in cross section that extends up through the atmosphere has a mass of about 10,000 kilograms. The weight of this air is about 100,000 newtons (10^5 N). This weight produces a pressure of 100,000 newtons per square meter—or, equivalently, 100,000 pascals, or 100 kilopascals. To be more exact, the average atmospheric pressure at sea level is 101.3 kilopascals (101.3 kPa).[1]

The pressure of the atmosphere is not uniform. Besides altitude variations, there are variations in atmospheric pressure at any one locality due to moving weather fronts and storms. Measurement of changing air pressure is important to meteorologists when predicting weather.

CHECK YOURSELF

1. About how many kilograms of air occupy a classroom that has a 200-m^2 floor area and a 4-m-high ceiling? (Assume a chilly temperature of 10°C.)
2. Why doesn't the pressure of the atmosphere break windows?

Barometer

A common instrument used for measuring the pressure of the atmosphere is called a **barometer**. A simple mercury barometer is illustrated in Figure 14.7. A glass tube, longer than 76 centimeters and closed at one end, is filled with mercury and tipped upside down in a dish of mercury. The mercury in the tube runs out of the submerged open bottom until the level in the tube is 76 centimeters above the level in the dish. The empty space trapped above, except for some mercury vapor, is a vacuum. The vertical height of the mercury column remains constant even when the tube is tilted, unless the top of the tube is less than 76 centimeters above the level in the dish—in which case the mercury completely fills the tube.

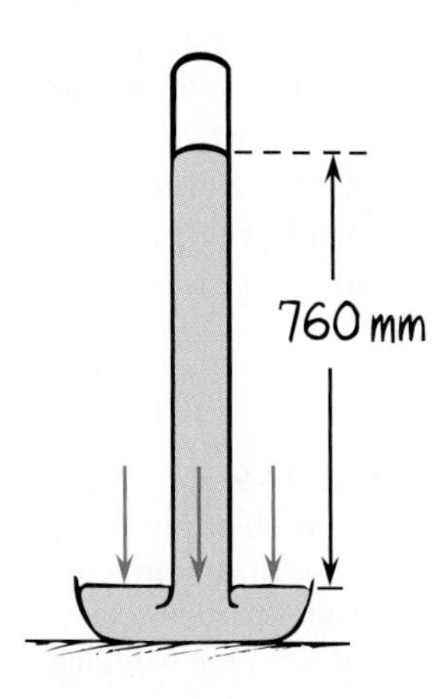

FIGURE 14.7
A simple mercury barometer.

CHECK YOUR ANSWERS

1. The mass of air is 1000 kg. The volume of air is 200 m^2 × 4 m = 800 m^3; each cubic meter of air has a mass of about 1.25 kg, so 800 m^3 × 1.25 kg/m^3 = 1000 kg.
2. Atmospheric pressure is exerted on both sides of a window, so no net force is exerted on the window. If, for some reason, the pressure is reduced or increased on one side only, as when a tornado passes by, then watch out! Reduced outside air pressure created by a tornado can cause a building to explode.

[1]The pascal (1 N/m^2) is the SI unit of measurement. The average pressure at sea level (101.3 kPa) is often called 1 atmosphere. In British units, the average atmospheric pressure at sea level is 14.7 lb/in^2.

FIGURE 14.8
Strictly speaking, these two do not suck the soda up the straws. They instead reduce pressure in the straws and allow the weight of the atmosphere to press the liquid up into the straws. Could they drink a soda this way on the Moon?

Why does mercury behave in this way? The explanation is similar to the reason why a simple seesaw will balance when the weights of people at its two ends are equal. The barometer "balances" when the weight of the liquid in the tube exerts the same pressure as the atmosphere outside. Whatever the width of the tube, a 76-centimeter column of mercury weighs the same as the air that would fill a super-tall 30-kilometer tube of the same width. If the atmospheric pressure increases, then the atmosphere pushes down harder on the mercury and the column of mercury is pushed higher than 76 centimeters. The mercury is literally pushed up into the tube of a barometer by the weight of the atmosphere.

Could water be used to make a barometer? The answer is *yes,* but the glass tube would have to be much longer—13.6 times as long, to be exact. You may recognize this number as the density of mercury relative to that of water. A volume of water 13.6 times that of mercury is needed to provide the same weight as the mercury in the tube. So the tube would have to be at least 13.6 times taller than the mercury column. A water barometer would have to be 13.6×0.76 meter, or 10.3 meters high—too tall to be practical.

What happens in a barometer is similar to what happens during the process of drinking through a straw. By sucking, you reduce the air pressure in the straw that is placed in a drink. The weight of the atmosphere on the drink pushes liquid up into the reduced-pressure region inside the straw. Strictly speaking, the liquid is not sucked up; it is pushed up by the pressure of the atmosphere. If the atmosphere is prevented from pushing on the surface of the drink, as in the party-trick bottle with the straw passing through an airtight cork stopper, one can suck and suck and get no drink.

If you understand these ideas, you can understand why there is a 10.3-meter limit on the height that water can be lifted with vacuum pumps. The old-fashioned farm-type pump, like the one shown in Figure 14.9, operates by producing a partial vacuum in a pipe that extends down into the water below. The weight of the atmosphere on the surface of the water simply pushes the water up into the region of reduced pressure inside the pipe. Can you see that, even with a perfect vacuum, the maximum height to which water can be lifted is 10.3 meters?

When the pump handle is raised, air in the pipe is "thinned" as it expands to fill a larger volume. Atmospheric pressure on the well surface pushes water up into the pipe, causing water to overflow at the spout.

Insights

FIGURE 14.9
The atmosphere pushes water from below up into a pipe that is partially evacuated of air by the pumping action.

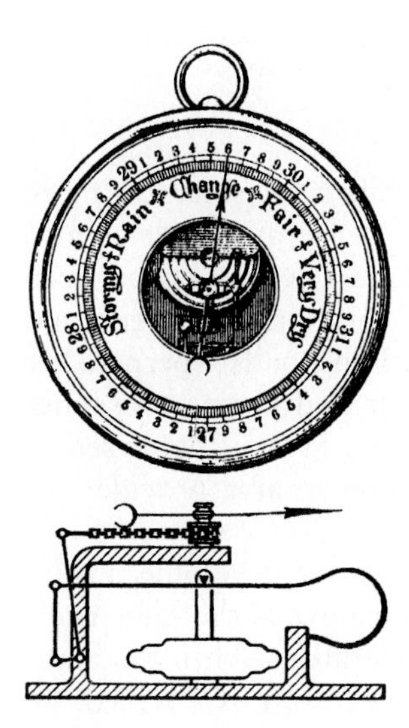

FIGURE 14.10
An aneroid barometer (top) and its cross section (bottom).

CHECK YOURSELF

What is the maximum height to which water can be sucked up through a straw?

A small portable instrument that measures atmospheric pressure is the aneroid barometer. The classic model shown in Figure 14.10 uses a metal box that is partially exhausted of air and has a slightly flexible lid that bends in or out with changes in atmospheric pressure. The motion of the lid is indicated on a scale by a mechanical spring-and-lever system. Since the atmospheric pressure decreases with increasing altitude, a barometer can be used to determine elevation. An aneroid barometer calibrated for altitude is called an *altimeter* (altitude meter). Some altimeters are sensitive enough to indicate a change in elevation of less than a meter.

The pressure (or vacuum) inside a television picture tube is about one-ten-thousandth pascal (10^{-4} Pa). At an altitude of about 500 kilometers, which is artificial satellite territory, gas pressure is about one-ten-thousandth of this (10^{-8} Pa). This is a pretty good vacuum by earthbound standards. Still greater vacuums exist in the wakes of satellites orbiting at this distance, and they can reach 10^{-13} Pa. This is called a "hard vacuum." Technologists requiring hard vacuums look to the prospects of laboratories orbiting in space.

Vacuums on Earth are produced by pumps, which work by virtue of a gas tending to fill its container. If a space with less pressure is provided, gas will flow from the region of higher pressure to the one of lower pressure. A vacuum pump simply provides a region of lower pressure into which fast-moving gas molecules randomly move. The air pressure is repeatedly lowered by piston and valve action (Figure 14.11). The best vacuums attainable with mechanical pumps are about one pascal. Better vacuums, down to 10^{-8} Pa, are attainable with vapor-diffusion or vapor-jet pumps. Sublimation pumps can reach 10^{-12} Pa. Greater vacuums are very difficult to attain.

Is atmospheric pressure actually different over a few centimeters' difference in altitude? The fact that it is is demonstrated with any helium-filled balloon that rises in air. Atmospheric pressure up against the bottom surface of the balloon is greater than atmospheric pressure down against the top.

Insights

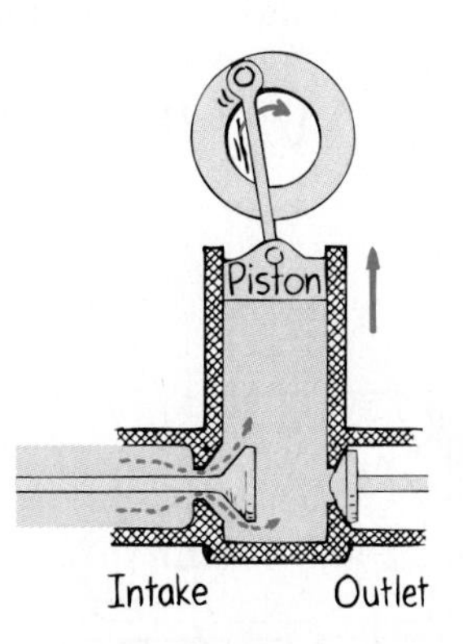

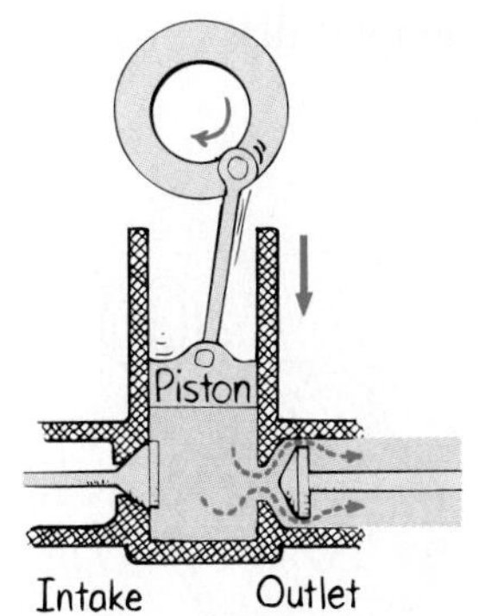

FIGURE 14.11
A mechanical vacuum pump. When the piston is lifted, the intake valve opens and air moves in to fill the empty space. When the piston is moved downward, the outlet valve opens and the air is pushed out. What changes would you make to convert this pump into an air compressor?

CHECK YOUR ANSWER

At sea level, however strong your lungs may be, or whatever device you use to make a vacuum in the straw, the water cannot be pushed up by the atmosphere higher than 10.3 m.

Boyle's Law

The air pressure inside the inflated tires of an automobile is considerably more than the atmospheric pressure outside. The density of the air inside is also more than the density of the air outside. To understand the relation between *pressure* and *density*, think of the molecules of air (primarily nitrogen and oxygen) inside the tire, which behave like tiny Ping-Pong balls—perpetually moving helter skelter and banging against one another and against the inner walls. Their impacts produce a jittery force that appears to our coarse senses as a steady push. This pushing force averaged over a unit of area provides the pressure of the enclosed air.

A tire pressure gauge at a service station doesn't measure absolute air pressure. A flat tire registers zero pressure on the gauge, but a pressure of about one atmosphere exists there. Gauges read "gauge" pressure—pressure greater than atmospheric pressure.

Insights

Suppose that there are twice as many molecules in the same volume (Figure 14.12). Then the air density is doubled. If the molecules move at the same average speed—or, equivalently, if they have the same temperature—then the number of collisions will be doubled. This means that the pressure is doubled. So pressure is proportional to density.

We can also double the air density by compressing air to half its volume. Consider the cylinder with the movable piston in Figure 14.13. If the piston is pushed downward so that the volume is half the original volume, the density of molecules will double and the pressure will correspondingly double. Decrease the volume to a third of its original value, and the pressure increases by three, and so forth (provided that the temperature remains the same).

Notice in these examples involving the piston that *pressure and volume are inversely proportional;* if you double one, for example, you halve the other:[2]

$$P_1V_1 = P_2V_2$$

Here P_1 and V_1 represent the original pressure and volume, respectively, and P_2 and V_2 represent the second pressure and volume. Or, put more graphically,

$$_{P}V = P_{V}$$

FIGURE 14.12
When the density of gas in the tire is increased, pressure is increased.

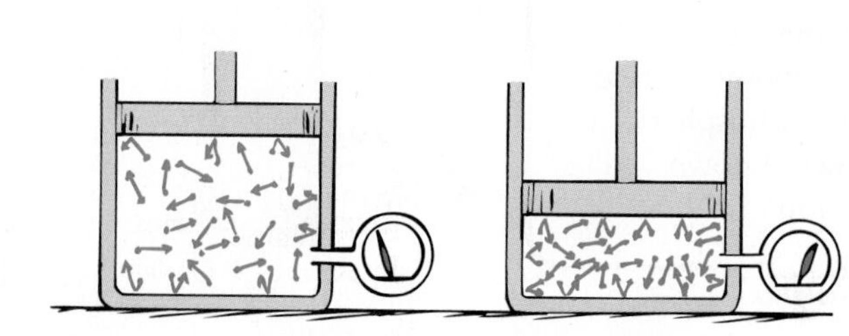

FIGURE 14.13
When the volume of gas is decreased, density and therefore pressure are increased.

[2]A general law that takes temperature changes into account is $P_1V_1/T_1 = P_2V_2/T_2$, where T_1 and T_2 represent the first and second *absolute* temperatures, measured in the SI unit called the *kelvin* (Chapters 15 and 18).

Workers in underwater construction work in an environment of compressed air. The air pressure in their underwater chambers is at least as much as the combined pressure of water and atmosphere outside.

Insights

In general, we can say that the product of pressure and volume for a given mass of gas is a constant as long as the temperature doesn't change. This relationship is called **Boyle's law,** after physicist Robert Boyle, who, with the help of fellow physicist Robert Hooke, made this discovery in the seventeenth century.

Boyle's law applies to ideal gases. An ideal gas is one in which the disturbing effects of the forces between molecules and the finite size of the individual molecules can be neglected. Air and other gases under normal pressures approach ideal-gas conditions.

CHECK YOURSELF

1. A piston in an airtight pump is withdrawn so that the volume of the air chamber is increased three times. What is the change in pressure?
2. A scuba diver 10.3 m deep breathes compressed air. If she were to hold her breath while returning to the surface, by how much would the volume of her lungs tend to increase?

Buoyancy of Air

A crab lives at the bottom of its ocean of water and looks upward at jellyfish floating above it. Similarly, we live at the bottom of our ocean of air and look upward at balloons drifting above us. A balloon is suspended in air, and a jellyfish is suspended in water for the same reason: Each is buoyed upward by a displaced weight of fluid equal to its own weight. In one case, the displaced fluid is air; and in the other case, it is water. As discussed in the previous chapter, objects in water are buoyed upward because the pressure acting up against the bottom of the object exceeds the pressure acting down against the top. Likewise, air pressure acting up against an object in air is greater than the pressure above pushing down. The buoyancy, in both cases, is numerically equal to the weight of fluid displaced. **Archimedes' principle** holds for air just as it does for water:

> **An object surrounded by air is buoyed up by a force equal to the weight of the air displaced.**

FIGURE 14.14
All bodies are buoyed up by a force equal to the weight of the air they displace. Why, then, don't all objects float like this balloon?

CHECK YOUR ANSWERS

1. The pressure in the piston chamber is reduced to one-third. This is the principle that underlies a mechanical vacuum pump.
2. Atmospheric pressure can support a column of water 10.3 m high, so the pressure in water due to its weight alone equals atmospheric pressure at a depth of 10.3 m. Taking the pressure of the atmosphere at the water's surface into account, the total pressure at this depth is twice atmospheric pressure. Unfortunately for the scuba diver, her lungs tend to inflate to twice their normal size if she holds her breath while rising to the surface. A first lesson in scuba diving is not to hold your breath when ascending. To do so can be fatal. (*Scuba* is an acronym for Self-Contained Underwater Breathing Apparatus.)

We know that a cubic meter of air at ordinary atmospheric pressure and room temperature has a mass of about 1.2 kilograms, so its weight is about 12 newtons. Therefore, any 1-cubic-meter object in air is buoyed up with a force of 12 newtons. If the mass of the 1-cubic-meter object is greater than 1.2 kilograms (so that its weight is greater than 12 newtons), it falls to the ground when released. If an object of this size has a mass less than 1.2 kilograms, it rises in the air. Any object that has a mass that is less than the mass of an equal volume of air will rise in air. Another way to say this is that any object less dense than air will rise in air. Gas-filled balloons that rise in air are less dense than air.

Greatest buoyancy would be achieved if a balloon were evacuated, but this isn't practical. The weight of a structure needed to keep an evacuated balloon from collapsing would more than offset the advantage of the extra buoyancy. So balloons are filled with gas less dense than ordinary air, which keeps the balloon from collapsing while keeping it light. In sport balloons, the gas is simply heated air. In balloons intended to reach very high altitudes or to stay up for a long time, helium is usually used. Its density is small enough so that the combined weight of helium, balloon, and whatever the cargo happens to be is less than the weight of the air it displaces.[3] Low-density gas is used in a balloon for the same reason that cork or Styrofoam is used in a swimmer's life preserver. The cork or Styrofoam possesses no strange tendency to be drawn toward the surface of water, and the gas possesses no strange tendency to rise. Both are buoyed upward like anything else. They are simply light enough for the buoyancy to be significant.

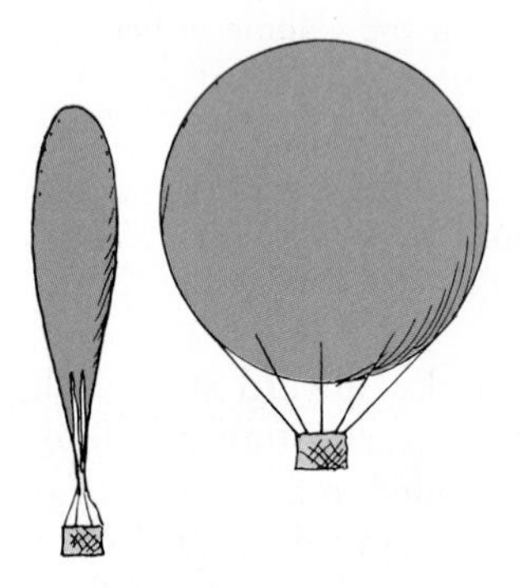

FIGURE 14.15
(Left) At ground level, the balloon is partially inflated. (Right) The same balloon is fully inflated at high altitudes where surrounding pressure is less.

Unlike water, the atmosphere has no definable surface. There is no "top." Furthermore, unlike water, the atmosphere becomes less dense with altitude. Whereas cork will float to the surface of water, a released helium-filled balloon does not rise to any atmospheric surface. How high will a balloon rise? We can state the answer in at least three ways. (1) A balloon will rise only so long as it displaces a weight of air greater than its own weight. Air becomes less dense with altitude, so, when the weight of displaced air equals the total weight of the balloon, upward acceleration of the balloon ceases. (2) We can also say that, when the buoyant force on the balloon equals its weight, the balloon will cease to rise. (3) Equivalently, when the average density of the balloon (including its load) equals the density of the surrounding air, the balloon will cease rising. Helium-filled toy balloons usually break when released in the air because, as the balloon rises to regions of less pressure, the helium in the balloon expands, increasing the volume and stretching the rubber until it ruptures.

CHECK YOURSELF

1. Is there a buoyant force acting on you? If there is, why are you not buoyed up by this force?
2. (This one calls for your best thinking!) How does buoyancy change as a helium-filled balloon ascends?

Large blimps are designed so that, when loaded, they will slowly rise in air; that is, their total weight is a little less than the weight of air displaced. When in motion, the blimp may be raised or lowered by means of horizontal "elevators."

[3]Hydrogen is the least dense gas, but, because it is highly flammable, it is seldom used.

Bernoulli's Principle

FIGURE 14.16
Because the flow is continuous, water speeds up when it flows through the narrow and/or shallow part of the brook.

Thus far, we have treated pressure only as it applies to stationary fluids. Now we shall consider fluids in motion—*fluid dynamics*.

Motion produces an additional influence on a fluid. Consider a continuous flow of water through a pipe. Because water doesn't "bunch up," the amount of water that flows past any given section of the pipe is the same as the amount that flows past any other section of the same pipe. This is true whether the pipe widens or narrows. As a consequence of continuous flow, the water will slow down in the wide parts and speed up in the narrow parts. You can observe this when you put your finger over the outlet of a garden hose.

Daniel Bernoulli, a Swiss scientist of the eighteenth century, advanced the theory of water flowing through pipes. He found that the pressure at the walls of the pipes decreases when the speed of the water increases. Bernoulli found this to be a principle both of liquids and of gases. **Bernoulli's principle,** in its simplest form, states:

> **When the speed of a fluid increases, internal pressure in the fluid decreases.**

Bernoulli's principle is consistent with the conservation of energy. For a steady flow of an ideal fluid free of internal friction, there are three kinds of energy: kinetic energy due to motion, work associated with pressure forces, and gravitational potential energy due to elevation. In a steady fluid flow where no energy is added or taken away, whatever work is done by one part of the fluid on another part makes its appearance as kinetic and potential energy. Then the sum of the three energy terms remains constant.[4] If the elevation of the flowing fluid does not change, then an increase in speed simply means a decrease in pressure, and vice versa.

Because the volume of water flowing through a pipe of different cross-sectional areas A remains constant, speed of flow v is high where the area is small, and the speed is low where the area is large. This is stated in the equation of continuity:

$$A_1v_1 = A_2v_2$$

The product A_1v_1 at point 1 equals the product A_2v_2 at point 2.

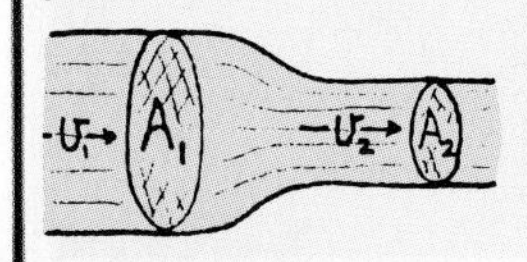

Insights

CHECK YOUR ANSWERS

1. There is a buoyant force acting on you, and you *are* buoyed upward by it. You don't notice it only because your weight is so much greater.
2. If the balloon is free to expand as it rises, the increase in volume is counteracted by a decrease in the density of higher-altitude air. So, interestingly, the greater volume of displaced air doesn't weigh more, and buoyancy stays the same. If a balloon is not free to expand, buoyancy will decrease as a balloon rises because of the less-dense displaced air. Usually, balloons expand as they rise initially, and, if they don't finally rupture, the stretching of their fabric reaches a maximum, and they settle where their buoyancy matches their weight.

[4]In mathematical form: $\frac{1}{2}mv^2 + mgy + pV =$ constant (along a streamline); where m is the mass of some small volume V, v is its speed, g is the acceleration due to gravity, y is its elevation, and p is its internal pressure. If mass m is expressed in terms of density ρ, where $\rho = m/V$, and each term is divided by V, Bernoulli's equation takes the form: $\frac{1}{2}\rho v^2 + \rho gy + p =$ constant. Then all three terms have units of pressure. If y does not change, an increase in v means a decrease in p, and vice versa. Note that, when v is zero, Bernoulli's equation reduces to $\Delta p = -\rho g\Delta y$ (weight density $\times$ depth).

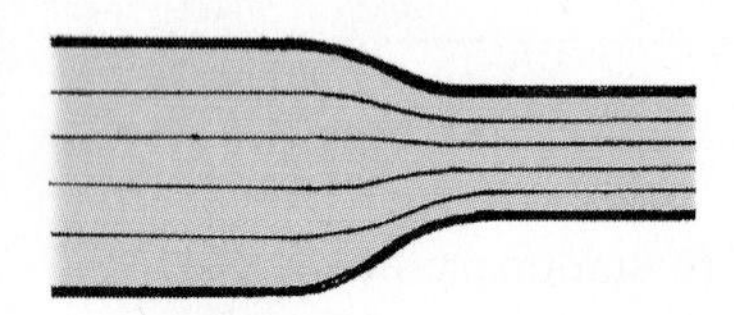

FIGURE 14.17
Water speeds up when it flows into the narrower pipe. The close-together streamlines indicate increased speed and decreased internal pressure.

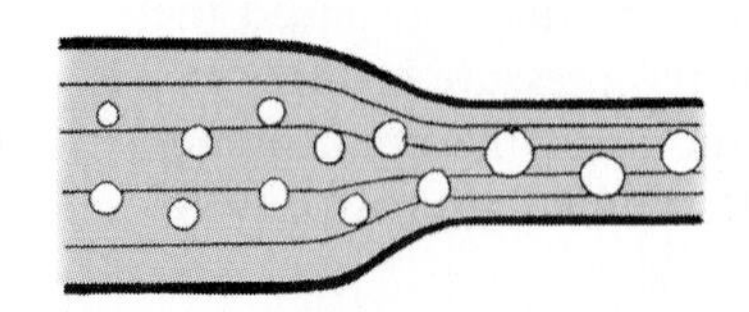

FIGURE 14.18
Internal pressure is greater in slower-moving water in the wide part of the pipe, as evidenced by the more-squeezed air bubbles. The bubbles are bigger in the narrow part because internal pressure there is less.

The decrease of fluid pressure with increasing speed may at first seem surprising, particularly if we fail to distinguish between the pressure *in* the fluid and the pressure exerted *by* the fluid on something that interferes with its flow. The pressure within the fast-moving water in a fire hose is relatively low, whereas the pressure that the water can exert on anything in its path to slow it down may be huge. We distinguish between the internal pressure in a fluid and the pressure a fluid may exert on anything that changes its momentum.

In steady flow, one small bit of fluid follows along the same path as a bit of fluid in front of it. The motion of a fluid in steady flow follows *streamlines,* which are represented by thin lines in Figure 14.17 and later figures. Streamlines are the smooth paths, or trajectories, of the bits of fluid. The lines are closer together in the narrower regions, where the flow is faster and pressure is less.

Pressure differences are nicely evident when a liquid contains air bubbles. The volume of an air bubble depends on the pressure of the surrounding liquid. Where the liquid gains speed, pressure is lowered and bubbles are bigger. As Figure 14.18 indicates, bubbles are squeezed smaller in slower, higher-pressure liquids.

Bernoulli's principle holds primarily for steady flow. If the flow speed is too great, the flow may become turbulent and follow a changing, curling path known as an eddy. In that case, Bernoulli's principle does not hold.

Applications of Bernoulli's Principle

Hold a sheet of paper in front of your mouth, as shown in Figure 14.19. When you blow across the top surface, the paper rises. That's because the internal pressure of moving air across the curved top of the paper is less than the atmospheric pressure beneath it.

If you ever ride in a convertible car with the canvas top up, notice that the top puffs upward as the car moves: Bernoulli again. The pressure outside, where air speeds up getting over the top, is less than the static atmospheric

FIGURE 14.19
The paper rises when Tim blows across its top surface.

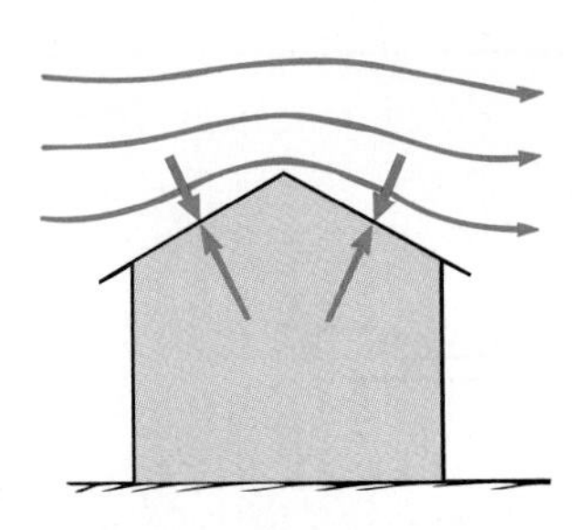

FIGURE 14.20
Air pressure above the roof is less than air pressure beneath the roof.

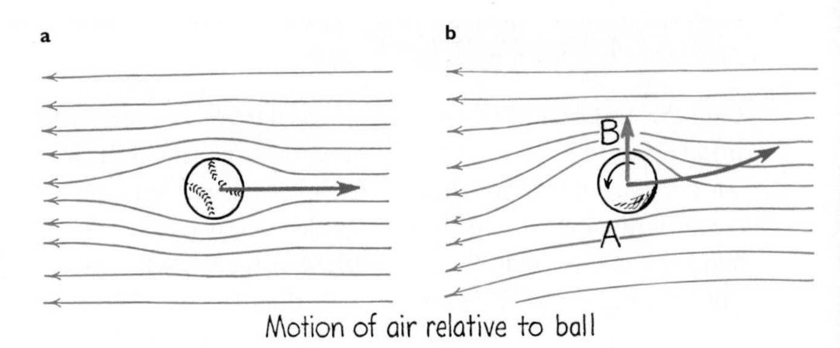

FIGURE 14.21
(a) The streamlines are the same on either side of a nonspinning baseball. (b) A spinning ball produces a crowding of streamlines. The resulting "lift" (red arrow) causes the ball to curve, as shown by the blue arrow.

pressure on the inside. The difference in pressures on the canvas fabric causes it to bulge upward.

Consider wind blowing across a peaked roof. The wind speeds up as it flows over the top, as the crowding of streamlines in Figure 14.20 indicates. Pressure along the streamlines is reduced where they are closer together. Unless the building is well vented, greater pressure inside and beneath the roof can push it off. Even a small pressure difference over a large roof area can produce a large upward "lifting" force.

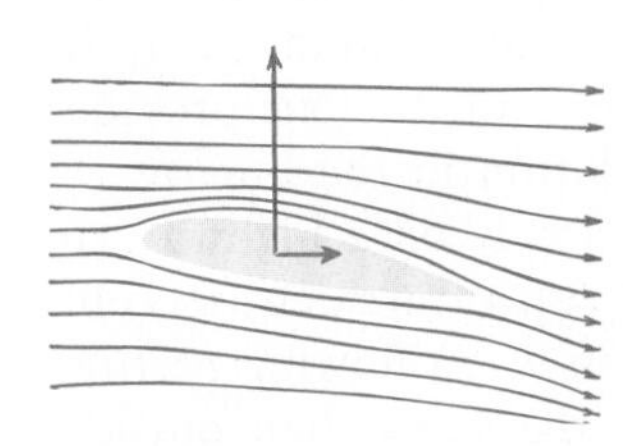

FIGURE 14.22
The vertical vector represents the net upward force (lift) that results from more air pressure below the wing than above the wing. The horizontal vector represents air drag.

If we think of the blown-off roof as an airplane wing, we can better understand the lifting force that supports a heavy airplane. In both cases, a greater pressure below pushes the roof or wing into a region of reduced pressure above. Wings come in a variety of designs. What they all have in common is that the air is made to flow faster over the top surface than across the bottom surface. This is mainly accomplished by the tilt of the wing, called the *angle of attack*. Air flows faster across the top for much the same reason that air flows faster in a narrowed pipe or in any constricted region. Most often, but not always, different speeds of airflow over and beneath a wing are enhanced by a difference in surface curvatures—more curvature on the top than on the bottom (camber). When the difference in pressures produces a net upward force, we have lift.[5] When lift equals weight, horizontal flight is possible. The lift is greater for higher speeds and larger wing areas. Hence, low-speed gliders have very large wings relative to the size of the fuselage. The wings of faster-moving aircraft are relatively small.

FIGURE 14.23
Why does the liquid in the reservoir go up the tube?

Bernoulli's principle also plays a role in the curved path of spinning balls. When a moving baseball, tennis ball, or any kind of ball spins, unequal air pressures are produced on opposite sides of the ball. Note, in Figure 14.21b, that the streamlines are closer at B than at A for the direction of spin shown. Air pressure is greater at A, and the ball curves as indicated. Curving may be increased by threads or fuzz, which help to drag a thin layer of air with the ball and to produce further crowding of streamlines on one side.

A familiar sprayer, such as a perfume atomizer, utilizes Bernoulli's principle. When you squeeze the bulb, air rushes across the open end of a tube

[5]Pressure differences are only one way to understand wing lift. Another way uses Newton's third law. The wing forces air downward (action) and the air forces the wing upward (reaction). Air is deflected downward by the wing tilt, the angle of attack—even when flying upside down! When riding in a car, place your hand out the window and pretend it's a wing. Tip it up slightly so air is forced downward. Up goes your hand! Air lift provides a nice example to remind us that there is often more than one way to explain the behavior of nature.

PRACTICING PHYSICS

Fold the end of a filing card down so that you make a little bridge. Stand it on a table and blow through the arch as shown. No matter how hard you blow, you will not succeed in blowing the card off the table (unless you blow against the side of it). Try this with your friends who have not taken physics. Then explain it to them!

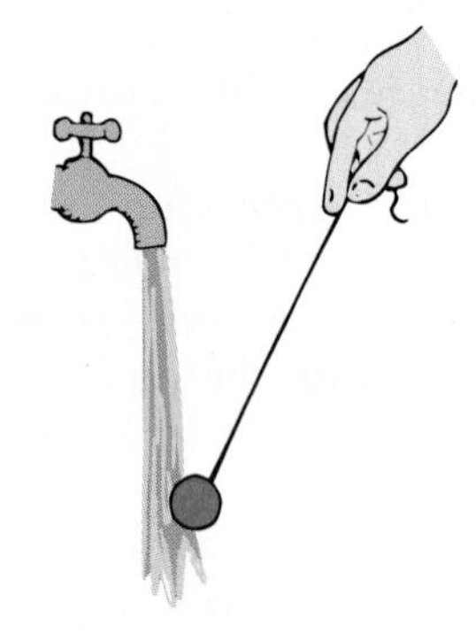

FIGURE 14.24
Pressure is greater in the stationary fluid (air) than in the moving fluid (water stream). The atmosphere pushes the ball into the region of reduced pressure.

inserted into the perfume. This reduces pressure in the tube, whereupon atmospheric pressure on the liquid below pushes it up into the tube, where it is carried away by the stream of air.

Bernoulli's principle plays an important role for animals living in underground burrows. Entrances to their burrows are usually mound shaped, producing variations in wind speed across different entrances. This provides necessary pressure differences of air to enable circulation in the burrow.

Bernoulli's principle explains why trucks passing closely on a highway are drawn to each other, and why passing ships run the risk of a sideways collision. Water flowing between the ships travels faster than water flowing past the outer sides. Streamlines are closer together between the ships than outside. Hence, water pressure acting against the hulls is reduced between the ships. Unless the ships are steered to compensate for this, the greater pressure against the outer sides of the ships forces them together. Figure 14.25 shows a demonstration of this, which you can do in your kitchen sink.

Bernoulli's principle can play a role when your bathroom shower curtain swings toward you in the shower when water is turned on full blast. Air near the water stream flows into the lower-pressure stream and is swept downward with the falling water. Air pressure inside the curtain is thus reduced, and the atmospheric pressure outside pushes the curtain inward (providing an escape route for the downward-swept air). Although convection produced by temperature differences likely plays a bigger role, the next time you're taking a shower and the curtain swings in against your legs, think of Daniel Bernoulli!

CHECK YOURSELF

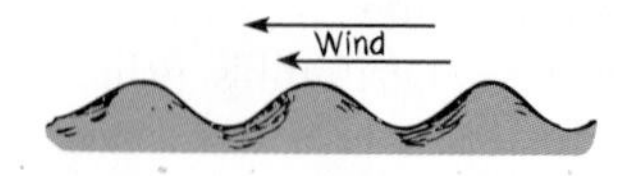

On a windy day, waves in a lake or the ocean are higher than their average height. How does Bernoulli's principle contribute to the increased height?

CHECK YOUR ANSWER

The troughs of the waves are partially shielded from the wind, so air travels faster over the crests. Pressure there is therefore lower than down below in the troughs. The greater pressure in the troughs pushes water into the even higher crests.

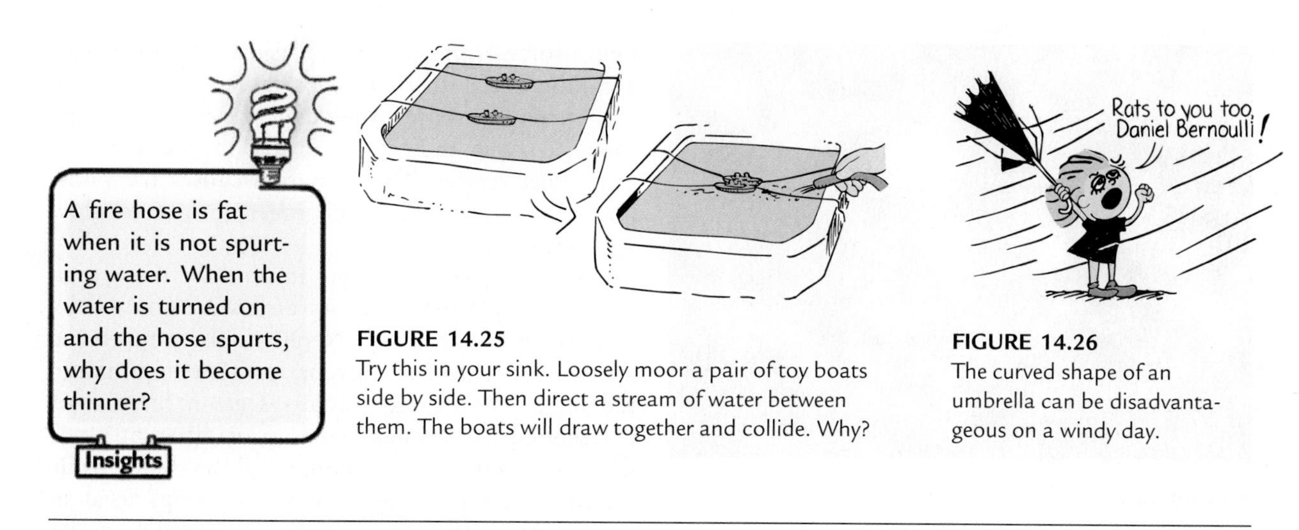

A fire hose is fat when it is not spurting water. When the water is turned on and the hose spurts, why does it become thinner?

Insights

FIGURE 14.25
Try this in your sink. Loosely moor a pair of toy boats side by side. Then direct a stream of water between them. The boats will draw together and collide. Why?

FIGURE 14.26
The curved shape of an umbrella can be disadvantageous on a windy day.

Plasma

In addition to solids, liquids, and gases, there is a fourth phase of matter—**plasma** (not to be confused with the clear liquid part of blood, also called plasma). It is the least common phase in our everyday environment, but it is the most prevalent phase of matter in the universe as a whole. The Sun and other stars are largely plasma.

A plasma is an electrified gas. The atoms that make it up are *ionized*, stripped of one or more electrons, with a corresponding number of free electrons. Recall that a neutral atom has as many positive protons inside the nucleus as it has negative electrons outside the nucleus. When one or more of these electrons is stripped from the atom, the atom has more positive charge than negative charge and becomes a *positive ion*. (Under some conditions, it may have extra electrons, in which case it is a *negative ion*.) Although the electrons and ions are themselves electrically charged, the plasma as a whole is electrically neutral because there are still equal numbers of positive and negative charges, just as there are in an ordinary gas. Nevertheless, a plasma and a gas have very different properties. The plasma readily conducts electric current, it absorbs certain kinds of radiation that pass unhindered through a gas, and it can be shaped, molded, and moved by electric and magnetic fields.

Our Sun is a ball of hot plasma. Plasmas on Earth are created in laboratories by heating gases to very high temperatures, making them so hot that electrons are "boiled" off the atoms. Plasmas may also be created at lower temperatures by bombarding atoms with high-energy particles or radiation.

Plasma in the Everyday World

If you're reading this by light emitted by a fluorescent lamp, you don't have to look far to see plasma in action. Within the glowing tube of the lamp is plasma that contains argon and mercury ions (as well as many neutral atoms of these elements). When you turn the lamp on, a high voltage between electrodes at each end of the tube causes electrons to flow. These electrons

FIGURE 14.27
Streets are illuminated at night by glowing plasmas.

ionize some atoms, forming plasma, which provides a conducting path that keeps the current flowing. The current activates some mercury atoms, causing them to emit radiation, mostly in the invisible ultraviolet region. This radiation causes the phosphor coating on the tube's inner surface to glow with visible light.

Similarly, the neon gas in an advertising sign becomes a plasma when its atoms are ionized by electron bombardment. Neon atoms, after being activated by electric current, emit predominantly red light. The different colors seen in these signs correspond to plasmas made up of different kinds of atoms. Argon, for example, glows blue, while helium glows pink. Sodium vapor lamps used in street lighting emit yellow light stimulated by glowing plasmas (Figure 14.27).

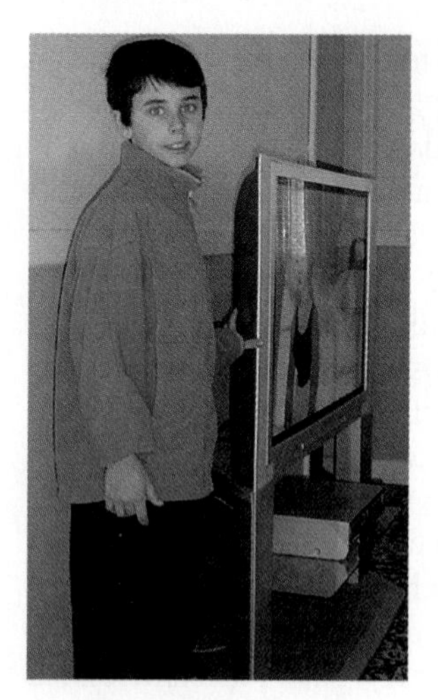

FIGURE 14.28
Plasma in action—a television set with a flat panel display. Hundreds of thousands of tiny pixels are lit up red, green, and/or blue by glowing plasmas. By combining these colors in different proportions, the television can produce the entire color spectrum.

A recent plasma innovation is the flat plasma TV screen. The screen is made up of many thousands of pixels, each of which is composed of three separate subpixel cells. One cell has a phosphor that fluoresces red, another has a phosphor that fluoresces green, and the other blue. The pixels are sandwiched between a network of electrodes that are charged thousands of times in a small fraction of a second, producing electric currents that flow through gases in the cells. As in a fluorescent lamp, the gases convert to glowing plasmas that release ultraviolet light that stimulates the phosphors. The combination of cell colors makes up the pixel color. The image on the screen is the blend of pixel colors activated by the TV control signal.

The aurora borealis and the aurora australis (called the northern and southern lights, respectively) are glowing plasmas in the upper atmosphere. Layers of low-temperature plasma encircle the whole Earth. Occasionally, showers of electrons from outer space and radiation belts enter "magnetic windows" near Earth's poles, crashing into the layers of plasma and producing light.

These layers of plasma, which extend upward some 80 kilometers, make up the ionosphere, and they act as mirrors to low-frequency radio waves. Higher-frequency radio and TV waves pass through the ionosphere. This is why you can pick up radio stations from long distances on your lower-frequency AM radio, but you have to be in the "line of sight" of broadcasting or relay antennas to pick up higher-frequency FM and TV signals. Have you ever noticed that, at night, you can sometimes receive very distant stations on your AM radio? This is because plasma layers settle closer together in the absence of the energizing sunlight and consequently are better reflectors of radio waves.

Plasma Power

A higher-temperature plasma is the exhaust of a jet engine. It is a weakly ionized plasma; but, when small amounts of potassium salts or cesium metal are added, it becomes a very good conductor; and, when it is directed into a magnet, electricity is generated! This is MHD power, the **m**agneto**h**ydro**d**ynamic interaction between a plasma and a magnetic field. (We will treat the mechanics of generating electricity in this way in Chapter 25.) Low-pollution MHD power is in operation at a few places in the world already. Looking forward, we can expect to see more plasma power with MHD.

An even more promising achievement will be plasma power of a different kind—the controlled fusion of atomic nuclei. We will treat the physics of fusion in Chapter 34. The benefits of controlled fusion may be far reaching. Fusion power may not only make electrical energy abundant, but it may also provide the energy and means to recycle and even synthesize elements.

Humankind has come a long way with the mastery of the first three phases of matter. Our mastery of the fourth phase may bring us much farther.

Summary of Terms

Atmospheric pressure The pressure exerted against bodies immersed in the atmosphere. It results from the weight of air pressing down from above. At sea level, atmospheric pressure is about 101 kPa.

Barometer Any device that measures atmospheric pressure.

Boyle's law The product of pressure and volume is a constant for a given mass of confined gas, as long as temperature remains unchanged:

$$P_1V_1 = P_2V_2$$

Archimedes' principle (for air) An object in the air is buoyed up with a force equal to the weight of displaced air.

Bernoulli's principle When the speed of a fluid increases, internal pressure in the fluid decreases.

Plasma An electrified gas containing ions and free electrons. Most of the matter in the universe is in the plasma phase.

Review Questions

The Atmosphere

1. What is the energy source for the motion of gas in the atmosphere? What prevents atmospheric gases from flying off into space?
2. How high would you have to go in the atmosphere for half of the mass of air to be below you?

Atmospheric Pressure

3. What is the cause of atmospheric pressure?
4. What is the mass of a cubic meter of air at room temperature (20°C)?
5. What is the approximate mass of a column of air 1 cm^2 in area that extends from sea level to the upper atmosphere? What is the weight of this amount of air?
6. What is the pressure at the bottom of the column of air referred to in the previous question?

Barometer

7. How does the pressure at the bottom of a 76-cm column of mercury in a barometer compare with air pressure at the bottom of the atmosphere?
8. How does the weight of mercury in a barometer compare with the weight of an equal cross section of air from sea level to the top of the atmosphere?
9. Why would a water barometer have to be 13.6 times taller than a mercury barometer?
10. When you drink liquid through a straw, is it more accurate to say the liquid is pushed up the straw rather than sucked up the straw? What exactly does the pushing? Defend your answer.
11. Why will a vacuum pump not operate for a well that is more than 10.3 m deep?
12. Why is it that an aneroid barometer is able to measure altitude as well as atmospheric pressure?

Boyle's Law

13. By how much does the density of air increase when it is compressed to half its volume?
14. What happens to the air pressure inside a balloon when it is squeezed to half its volume at constant temperature?
15. What is an ideal gas?

Buoyancy of Air

16. A balloon that weighs 1 N is suspended in air, drifting neither up nor down. (a) How much buoyant force acts on it? (b) What happens if the buoyant force decreases? (c) If it increases?
17. Does the air exert buoyant force on all objects in air or only on objects such as balloons that are very light for their size?
18. What usually happens to a toy helium-filled balloon that rises high into the atmosphere?

Bernoulli's Principle

19. What are streamlines? Is pressure greater or less in regions where streamlines are crowded?

20. What happens to the internal pressure in a fluid flowing in a horizontal pipe when its speed increases?

21. Does Bernoulli's principle refer to changes in internal pressure of a fluid or to pressures the fluid may exert on objects?

Applications of Bernoulli's Principle

22. How does Bernoulli's principle apply to the flight of airplanes?

23. Why does a spinning ball curve in its flight?

24. Why do ships passing each other in open seas run a risk of sideways collisions?

Plasma

25. How does a plasma differ from a gas?

Plasma in the Everyday World

26. Cite at least three examples of plasma in your daily environment.

27. Why is AM radio reception better at night?

Plasma Power

28. What can be produced when a plasma beam is directed into a magnet?

Projects

1. Find the pressure exerted by the tires of your car on the road and compare it with the air pressure in the tires. For this project, you need the weight of your car, which you can get from a manual or a dealer. You divide the weight by four to get the approximate weight held up by one tire. You can closely approximate the area of contact of a tire with the road by tracing the edges of tire contact on a sheet of paper marked with one-inch squares beneath the tire. After you calculate the pressure of the tire against the road, compare it with the air pressure in the tire. Are they nearly equal? If not, which is greater?

2. Try this in the bathtub or when you're washing dishes. Lower a drinking glass, mouth downward, over a small floating object (which makes the inside water level visible). What do you observe? How deep will the glass have to be pushed in order to compress the enclosed air to half its volume? (You won't be able to get that much air compression in your bathtub unless it's 10.3 m deep!)

3. You ordinarily pour water from a full glass into an empty glass simply by placing the full glass above the empty glass and tipping. Have you ever poured air from one glass into another? The procedure is similar. Lower two glasses in water, mouths downward. Let one fill with water by tilting its mouth upward. Then hold the water-filled glass mouth downward above the air-filled glass. Slowly tilt the lower glass and let the air escape, filling the upper glass. You will be pouring air from one glass into another!

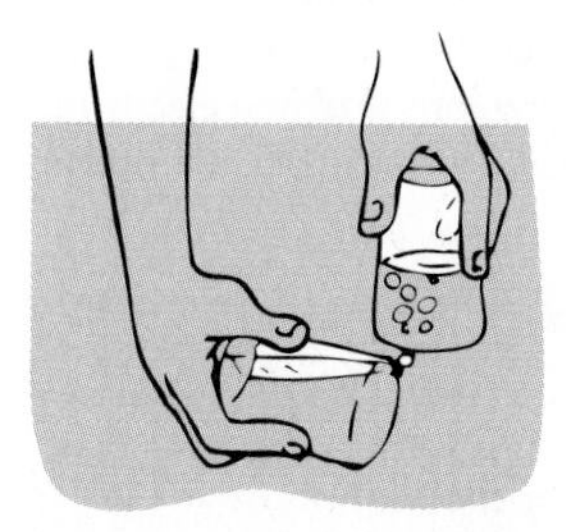

Pouring Air from One Glass to Another

4. Hold a glass under water, letting it fill with water. Then turn it upside down and raise it, but with its mouth beneath the surface. Why does the water not run out? How tall would a glass have to be before water began to run out? (If you could find such a glass, you might need to cut holes in your ceiling and roof to make room for it!)

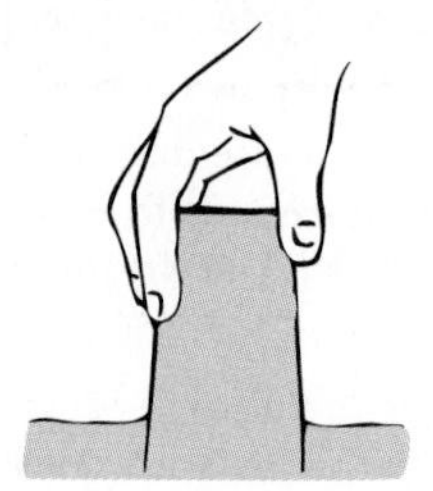

5. Place a card over the open top of a glass filled to the brim with water and invert it. Why does the card stay in place? Try it sideways.

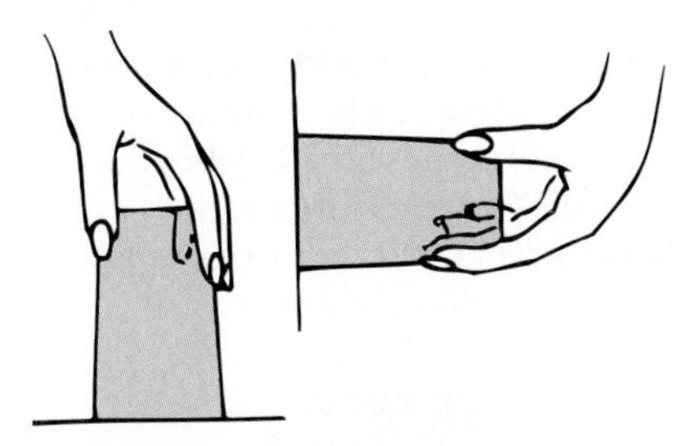

6. Invert a water-filled pop bottle or a small-necked jar. Notice that the water doesn't simply fall out but gurgles out of the container. Air pressure won't let it get out until some air has pushed its way up inside the bottle to occupy the space above the liquid. How would an inverted, water-filled bottle empty on the Moon?
7. Do as Professor Dan Johnson does below. Pour about half a cup of water into a 5-or-so-liter metal can with a screw top. Place the can *open* on a stove and heat it until the water boils and steam comes out of the opening. Quickly remove the can and screw the cap on tightly. Allow the can to stand. Steam inside condenses, which can be hastened by cooling the can with a dousing of cold water. What happens to the vapor pressure inside? (Don't do this with a can you expect to use again.)

8. Heat a small amount of water to boiling in an aluminum soda-pop can and invert it quickly into a dish of cooler water. Surprisingly dramatic!
9. Make a small hole near the bottom of an open tin can. Fill the can with water, which will proceed to spurt from the hole. If you cover the top of the can firmly with the palm of your hand, the flow stops. Explain.

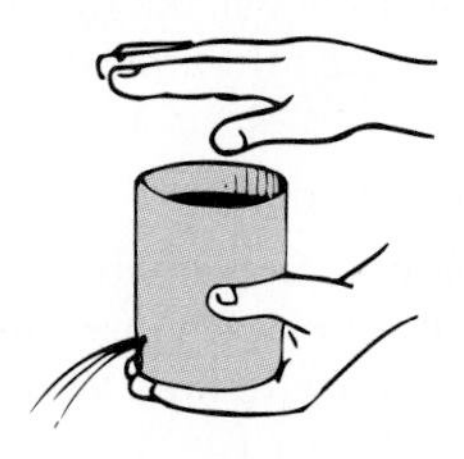

10. Lower a narrow glass tube or drinking straw into water and place your finger over the top of the tube. Lift the tube from the water and then lift your finger from the top of the tube. What happens? (You'll do this often if you enroll in a chemistry lab.)

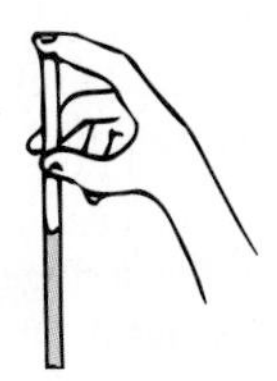

11. Push a pin through a small card and place it in the hole of a thread spool. Try to blow the card from the spool by blowing through the hole. Try it in all directions.

12. Hold a spoon in a stream of water as shown and feel the effect of the differences in pressure.

Exercises

1. It is said that a gas fills all the space available to it. Why, then, doesn't the atmosphere go off into space?
2. Why is there no atmosphere on the Moon?
3. Count the tires on a large tractor-trailer that is unloading food at your local supermarket, and you may be surprised to count 18 tires. Why so many tires? (*Hint:* Consider Project 1.)
4. The valve stem on a tire must exert a certain force on the air within to prevent any of that air from leaking out. If the diameter of the valve stem were doubled, by how much would the force exerted by the valve stem increase?
5. Why is the pressure in an automobile's tires slightly greater after the car has been driven several kilometers?
6. Why is a soft underinflated football at sea level much firmer when it is taken to a high elevation in the mountains?
7. What is the purpose of the ridges that prevent the funnel from fitting tightly in the mouth of a bottle?

8. How does the density of air in a deep mine compare with the air density at the Earth's surface?
9. When an air bubble rises in water, what happens to its mass, volume, and density?
10. Two teams of eight horses each were unable to pull the Magdeburg hemispheres apart (Figure 14.2). Why? Suppose two teams of nine horses each could pull them apart. Then would one team of nine horses succeed if the other team were replaced with a strong tree? Defend your answer.
11. When boarding an airplane, you bring a bag of chips (or any other item packaged in an airtight foil package) and, while you are in flight, you notice that the bag puffs up. Explain why this happens.
12. Why do you suppose that airplane windows are smaller than bus windows?
13. A half cup or so of water is poured into a 5-L can and is placed over a source of heat until most of the water has boiled away. Then the top of the can is screwed on tightly and the can is removed from the heat and allowed to cool. What happens to the can and why?
14. We can understand how pressure in water depends on depth by considering a stack of bricks. The pressure below the bottom brick is determined by the weight of the entire stack. Halfway up the stack, the pressure is half because the weight of the bricks above is half. To explain atmospheric pressure, we should consider compressible bricks, like those made of foam rubber. Why is this so?

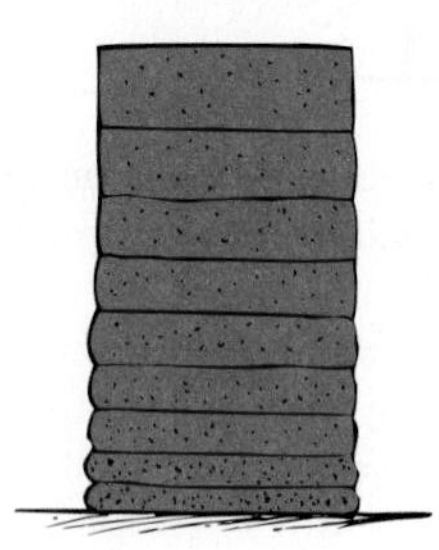

15. The "pump" in a vacuum cleaner is merely a high-speed fan. Would a vacuum cleaner pick up dust from a rug on the Moon? Explain.
16. Suppose that the pump shown in Figure 14.9 worked with a perfect vacuum. From how deep a well could water be pumped?
17. If a liquid only half as dense as mercury were used in a barometer, how high would its level be on a day of normal atmospheric pressure?
18. Why does the size of the cross-sectional area of a mercury barometer not affect the height of the enclosed mercury column?
19. From how deep a container could mercury be drawn with a siphon?
20. If you could somehow replace the mercury in a mercury barometer with a denser liquid, would the height of the liquid column be greater than or less than the mercury? Why?
21. Would it be slightly more difficult to draw soda through a straw at sea level or on top of a very high mountain? Explain.
22. The pressure exerted against the ground by an elephant's weight distributed evenly over its four feet is less than 1 atmosphere. Why, then, would you be crushed beneath the foot of an elephant, while you're unharmed by the pressure of the atmosphere?
23. Your friend says that the buoyant force of the atmosphere on an elephant is significantly greater than the buoyant force of the atmosphere on a small helium-filled balloon. What do you say?

24. Which will register the greater weight: an empty flattened balloon, or the same balloon filled with air? Defend your answer, then try it and see.

25. On a sensitive balance, weigh an empty, flat, thin plastic bag. Then weigh the bag filled with air. Will the readings differ? Explain.

26. Why is it so difficult to breathe when snorkeling at a depth of 1 m, and practically impossible at a 2-m depth? Why can't a diver simply breathe through a hose that extends to the surface?

27. How does the concept of buoyancy complicate the old question "Which weighs more, a pound of lead or a pound of feathers?"

28. Why does the weight of an object in air differ from its weight in a vacuum (remembering that weight is the force exerted against a supporting surface)? Cite an example in which this would be an important consideration.

29. A little girl sits in a car at a traffic light holding a helium-filled balloon. The windows are closed and the car is relatively airtight. When the light turns green and the car accelerates forward, her head pitches backward but the balloon pitches forward. Explain why.

30. Would a bottle of helium gas weigh more or less than an identical bottle filled with air at the same pressure? Than an identical bottle with the air pumped out?

31. When you replace helium in a balloon with less-dense hydrogen, does the buoyant force on the balloon change if the balloon remains the same size? Explain.

32. A steel tank filled with helium gas doesn't rise in air, but a balloon containing the same helium rises easily. Why?

33. If the number of gas atoms in a container is doubled, the pressure of the gas doubles (assuming constant temperature and volume). Explain this pressure increase in terms of molecular motion of the gas.

34. What, if anything, happens to the volume of gas in an atmospheric research-type balloon when it is heated?

35. What, if anything, happens to the pressure of the gas in a rubber balloon when the balloon is squeezed smaller?

36. What happens to the size of the air bubbles released by a diver as they rise?

37. You and Tim float a long string of closely spaced helium-filled balloons over his used-car lot. You secure both ends of the string to the ground several meters apart so that the balloons float over the lot in an arc. What is the name of this arc? (Why could this exercise have been included in Chapter 12?)

38. The gas pressure inside an inflated rubber balloon is always greater than the air pressure outside. Explain.

39. Two identical balloons of the same volume are pumped up with air to more than atmospheric pressure and suspended on the ends of a stick that is horizontally balanced. One of the balloons is then punctured. Is the balance of the stick upset? If so, which way does it tip?

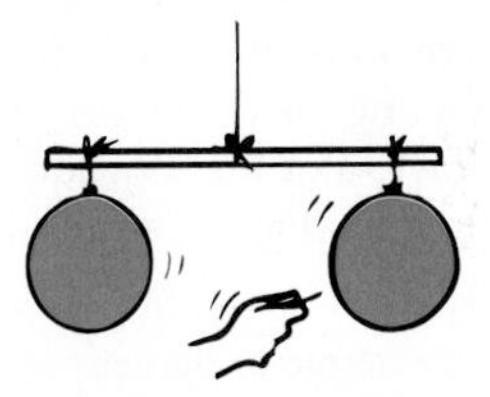

40. Two balloons that have the same weight and volume are filled with equal amounts of helium. One is rigid and the other is free to expand as the pressure outside decreases. When released, which will rise higher? Explain.

41. The force of the atmosphere at sea level against the outside of a 10-m^2 store window is about a million N. Why does this not shatter the window? Why might the window shatter in a strong wind blowing past the window?

42. Why does the fire in a fireplace burn more briskly on a windy day?

43. What happens to the pressure in water as it speeds up when it is ejected by the nozzle of a garden hose?

44. Why do airplanes normally take off facing the wind?

45. What provides the lift to keep a Frisbee in flight?

46. Imagine a huge space colony that consists of a rotating air-filled cylinder. How would the density of air at "ground level" compare to the air densities "above"?

47. Would a helium-filled balloon "rise" in the atmosphere of a rotating space habitat? Defend your answer.

48. When a steadily flowing gas flows from a larger-diameter pipe to a smaller-diameter pipe, what happens to (a) its speed, (b) its pressure, and (c) the spacing between its streamlines?

49. Compare the spacing of streamlines around a tossed baseball that doesn't spin in flight with the spacing of streamlines around one that does. Why does the spinning baseball veer from the course of a nonspinning one?

50. Why is it easier to throw a curve with a tennis ball than a baseball?

51. Why do airplanes extend wing flaps that increase the area of the wing during takeoffs and landings? Why are these flaps pulled in when the airplane has reached cruising speed?

52. How is an airplane able to fly upside down?

53. Why are runways longer for takeoffs and landings at high-altitude airports, such as those in Denver and Mexico City?

54. When a jet plane is cruising at high altitude, the flight attendants have more of a "hill" to climb as they walk forward along the aisle than when the plane is cruising at a lower altitude. Why does the pilot have to fly with a greater angle of attack at a high altitude than at a low altitude?

55. What physics principle underlies these three observations? When passing an oncoming truck on the highway, your car tends to sway toward the truck. The canvas roof of a convertible automobile bulges upward when the car is traveling at high speeds. The windows of older trains sometimes break when a high-speed train passes by on the next track.

56. How will two dangling vertical sheets of paper move when you blow between them? Try it and see.

57. A steady wind blows over the waves of an ocean. Why does the wind increase the peaks and troughs of the waves?

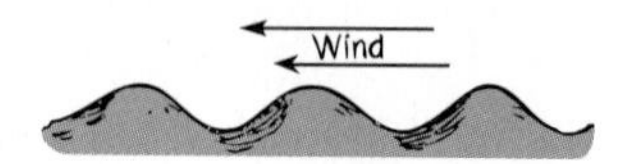

58. Wharves are made with pilings that permit the free passage of water. Why would a solid-walled wharf be disadvantageous to ships attempting to pull alongside?

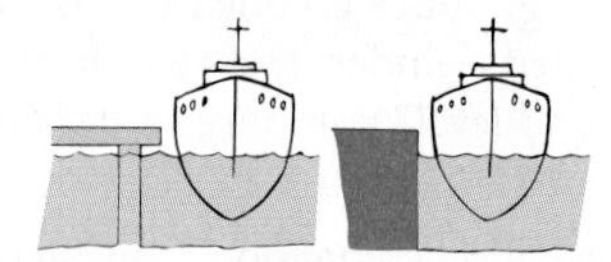

59. Is lower pressure the result of fast-moving air, or is fast-moving air the result of lower pressure? Give one example supporting each point of view. (In physics, when two things are related—such as force and acceleration or speed and pressure—it is usually arbitrary which one we call cause and which one we call effect.)

60. Why is the reception for far-away radio stations clearer at nighttime on your AM radio?

Problems

1. What change in pressure occurs in a party balloon that is squeezed to one-third its volume with no change in temperature?

2. Air in a cylinder is compressed to one-tenth its original volume with no change in temperature. What happens to its pressure?

3. In the previous problem, if a valve is opened to let out enough air to bring the pressure back down to its original value, what percentage of the molecules escape?

4. Estimate the buoyant force that air exerts on you. (To do this, you can estimate your volume by knowing your weight and by assuming that your weight density is a bit less than that of water.)

5. Nitrogen and oxygen in their liquid states have densities only 0.8 and 0.9 that of water. Atmospheric pressure is due primarily to the weight of nitrogen and oxygen gas in the air. If the atmosphere were somehow liquefied, would its depth be greater or less than 10.3 m?

6. A mountain-climber friend with a mass of 80 kg ponders the idea of attaching a helium-filled balloon to himself to effectively reduce his weight by 25% when he climbs. He wonders what the approximate size of such a balloon would be. Hearing of your physics skills, he asks you. What answer can you provide, showing your calculations?

7. On a perfect fall day, you are hovering at low altitude in a hot-air balloon, accelerated neither upward nor downward. The total weight of the balloon, including its load and the hot air in it, is 20,000 N. (a) What is the weight of the displaced air? (b) What is the volume of the displaced air?

8. How much lift is exerted on the wings of an airplane that have a total surface area of 100 m^2 when the difference in air pressure below and above the wings is 4% of atmospheric pressure?

PART THREE

Heat

Although the temperature of these sparks exceeds 2000°C, the heat they impart when striking my skin is very small—which illustrates that *temperature* and *heat* are different concepts. Learning to distinguish between closely related concepts is the challenge and essence of *Conceptual Physics.*

CHAPTER 15

Temperature, Heat, and Expansion

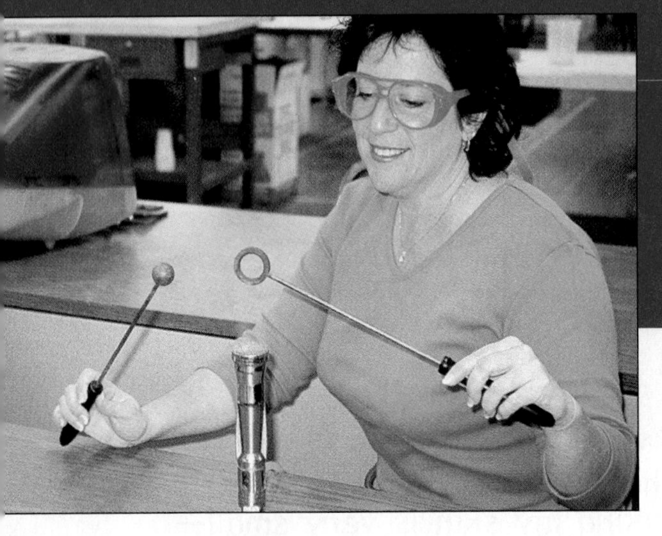

Ellyn Daugherty asks her class to predict whether the hole in the ring expands or contracts when heated.

All matter—solid, liquid, and gas—is composed of continuously jiggling atoms or molecules. Because of this random motion, the atoms and molecules in matter have kinetic energy. The average kinetic energy of the individual particles produces in an effect we can sense—warmth. Whenever something becomes warmer, the kinetic energy of its atoms or molecules increases. If you strike a solid penny with a hammer, it becomes warmer because the hammer's blow causes the atoms in the metal to jostle faster. If you put a flame to a liquid, it too becomes warmer. If you rapidly compress air in a tire pump, the air inside becomes warmer. When a solid, liquid, or gas gets warmer, its atoms or molecules move faster. They have more kinetic energy.

Temperature

FIGURE 15.1
Can we trust our sense of hot and cold? Will both fingers feel the same temperature when placed in the warm water?

The quantity that indicates how warm or cold an object is with respect to some standard is called **temperature.** The first "thermal meter" for measuring temperature, the *thermometer,* was invented by Galileo in 1602 (the word *thermal* is from the Greek term for "heat"). The familiar mercury-in-glass thermometer came into widespread use some seventy years later. (You can expect to see mercury thermometers phased out in coming years because of the danger of mercury poisoning.) We express the temperature of some quantity of matter by a number that corresponds to its degree of hotness or coldness on some chosen scale.

Nearly all materials expand when their temperature is raised and contract when their temperature is lowered. Most thermometers measure temperature by means of the expansion or contraction of a liquid, usually mercury or colored alcohol, in a glass tube with a scale.

On the most widely used temperature scale, the international scale, the number 0 is assigned to the temperature at which water freezes and the number 100 to the temperature at which water boils (at standard atmospheric pressure). The space between is divided into 100 equal parts called *degrees;* hence, a thermometer so calibrated has been called a *centigrade thermometer* (from *centi,* "hundredth," and *gradus,* "step"). However, it is now called a *Celsius thermometer* in honor of the man who first suggested the scale, the Swedish astronomer Anders Celsius (1701–1744).

FIGURE 15.2
A testament to Fahrenheit outside his home (now in Gdansk, Poland).

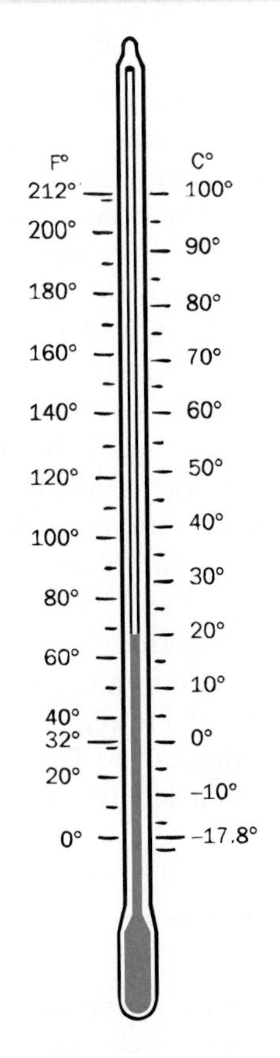

FIGURE 15.3
Fahrenheit and Celsius scales on a thermometer.

Another temperature scale is popular in the United States. On this scale, the number 32 is assigned to the temperature at which water freezes, and the number 212 is assigned to the temperature at which water boils. Such a scale makes up a Fahrenheit thermometer, named after its illustrious originator, the German physicist Gabriel Daniel Fahrenheit (1686–1736). The Fahrenheit scale will become obsolete if and when the United States goes metric.[1]

Still another temperature scale, favored by scientists, is the Kelvin scale, named after the Scottish physicist William Thomson, 1st Baron Kelvin (1824–1907). This scale is calibrated not in terms of the freezing and boiling points of water but in terms of energy itself. The number 0 is assigned to the lowest possible temperature—**absolute zero,** at which a substance has absolutely no kinetic energy to give up.[2] Absolute zero corresponds to −273°C on the Celsius scale. Units on the Kelvin scale have the same size increments as degrees on the Celsius scale, so the temperature of melting ice is 273 K. There are no negative numbers on the Kelvin scale. We won't treat this scale further until we study thermodynamics in Chapter 18.

Arithmetic formulas used for converting from Fahrenheit to Celsius, and vice versa, are popular in classroom exams. Such arithmetic exercises are not really physics, and the probability that you'll have the occasion to do this task elsewhere is small, so we will not be concerned with them here. Besides, this conversion can be very closely approximated by simply reading the corresponding temperature from the side-by-side scales in Figure 15.3.

Temperature is related to the random motion of atoms and molecules in a substance. (For brevity, hereafter in this chapter, we'll simply say *molecules* to mean *atoms and molecules.*) More specifically, temperature is proportional to the average "translational" kinetic energy of molecular motion (motion that carries the molecule from one place to another). Molecules may also rotate or vibrate, with associated rotational or vibrational kinetic energy—but these motions are not translational and don't directly affect temperature.

The effect of translational kinetic energy versus rotational and vibrational kinetic energy is dramatically demonstrated by a microwave oven. The microwaves that bombard your food cause certain molecules in the food, mainly water molecules, to flip to and fro and to oscillate with considerable rotational kinetic energy. But oscillating molecules don't cook food. What does raise the temperature and cook the food is the translational kinetic energy imparted to neighboring molecules that are bounced off the oscillating water molecules. (To picture this, imagine a bunch of marbles set flying in all directions after encountering the spinning blades of a fan. Also, see page 424.) If neighboring molecules didn't interact with the oscillating water molecules, the temperature of the food would be no different than before the microwave oven was turned on.

[1]The conversion to Celsius will put the United States in step with the rest of the world, where the Celsius scale is the standard. Americans are slow to convert. Changing any long-established custom is difficult, and the Fahrenheit scale does have some advantages in everyday use. For example, its degrees are smaller (1°F = 5/9°C), which gives greater accuracy when reporting the weather in whole-number temperature readings. Then, too, people somehow attribute a special significance to numbers increasing by an extra digit. Thus, when the temperature of a hot day is reported to reach 100°F, the idea of heat is conveyed more dramatically than by stating that the temperature is 38°C. Like so much of the British system of measure, the Fahrenheit scale is geared to human beings.

[2]Even at absolute zero, a substance has what is called "zero-point energy," which is unavailable energy that cannot be transferred to a different substance. Helium, for example, has enough motion at absolute zero to avoid freezing. The explanation for this involves quantum theory.

CHECK YOURSELF

True or false? Temperature is a measure of the total kinetic energy in a substance.

Interestingly enough, what a thermometer really displays is its *own* temperature. When a thermometer is in thermal contact with something whose temperature we wish to know, energy will flow between the two until their temperatures are equal and thermal equilibrium is established. If we know the temperature of the thermometer, we then know the temperature of the something. A thermometer should be small enough that it doesn't appreciably alter the temperature of the something being measured. If you are measuring the air temperature of a room, then your thermometer is small enough. But, if you are measuring the temperature of a drop of water, contact between the drop and the thermometer may change the drop's temperature—a classic case of the measuring process changing the thing that is being measured.

Heat

FIGURE 15.4
There is more molecular kinetic energy in the container filled with warm water than in the small cupful of higher-temperature water.

If you touch a hot stove, energy enters your hand because the stove is warmer than your hand. When you touch a piece of ice, however, energy transfers from your hand into the colder ice. The direction of spontaneous energy transfer is always from a warmer object to a neighboring cooler object. The energy transferred from one object to another because of a temperature difference between them is called **heat.**

It is important to point out that matter does not *contain* heat. Matter contains molecular kinetic energy and possibly potential energy, *not heat.* Heat is *energy in transit* from a body of higher temperature to one of lower temperature. Once transferred, the energy ceases to be heat. (As an analogy, work is also energy in transit. A body does not *contain* work. It *does* work or has work done on it.) In previous chapters, we called the energy resulting from heat flow *thermal energy* to make clear its link to heat and temperature. In this chapter, we will use the term that scientists prefer, *internal energy.*

Internal energy is the grand total of all energies inside a substance. In addition to the translational kinetic energy of jostling molecules in a substance, there is energy in other forms. There is rotational kinetic energy of molecules and kinetic energy due to internal movements of atoms within molecules. There is also potential energy due to the forces between molecules. So a substance does not contain heat—it contains internal energy.

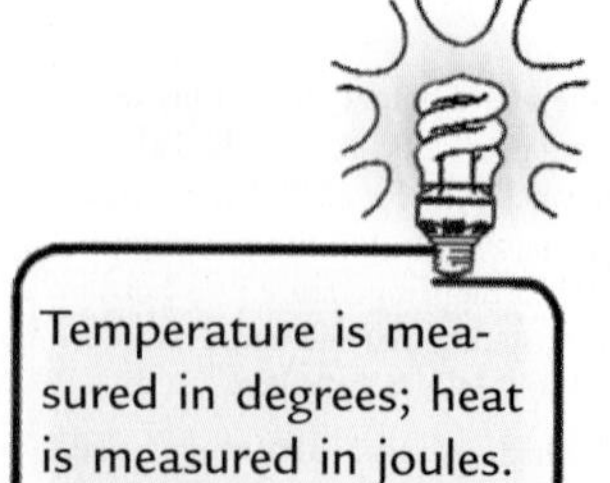

Temperature is measured in degrees; heat is measured in joules.
Insights

CHECK YOUR ANSWER

False. Temperature is a measure of the *average* (not *total*!) translational kinetic energy of molecules in a substance. For example, there is twice as much total molecular kinetic energy in 2 L of boiling water as in 1 L—but the temperatures of the two volumes of water are the same because the *average* translational kinetic energy per molecule is the same in each.

Just as dark is the absence of light, cold is the absence of thermal energy.
Insights

When a substance absorbs or gives off heat, the internal energy of the substance increases or decreases. In some cases, as when ice is melting, the added heat does not increase molecular kinetic energy but goes instead into other forms of energy. The substance undergoes a change of phase, which we will cover in detail in Chapter 17.

For two things in thermal contact, heat flow is from the higher-temperature substance to the lower-temperature substance. This is not necessarily a flow from a substance with more internal energy to a substance with less internal energy. There is more internal energy in a bowl of warm water than there is in a red-hot thumbtack; yet, if the tack is immersed in the water, heat flows from the hot tack to the warm water and not the other way around. Heat never flows of itself from a lower-temperature substance into a higher-temperature substance.

How much heat flows depends not only on the temperature difference between substances but on the amount of material as well. For example, a barrelful of hot water will transfer more heat to a cooler substance than a cupful of water will. There is more internal energy in the larger volume of water.

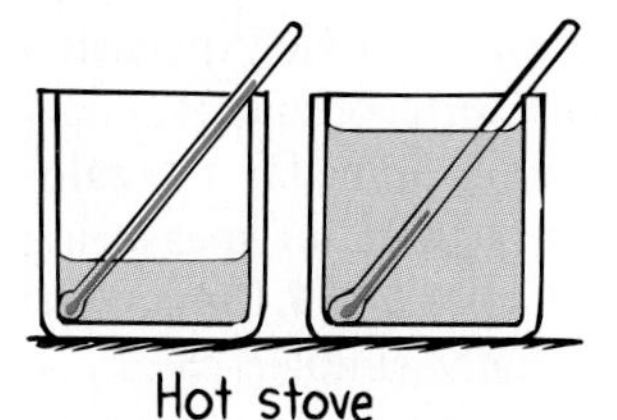

FIGURE 15.5
Although the same quantity of heat is added to both containers, the temperature increases more in the container with the smaller amount of water.

CHECK YOURSELF

1. Suppose you apply a flame to 1 L of water for a certain time and its temperature rises by 2°C. If you apply the same flame for the same time to 2 L of water, by how much will its temperature rise?
2. If a fast marble hits a random scatter of slow marbles, does the fast marble usually speed up or slow down? Which lose(s) kinetic energy and which gain(s) kinetic energy—the initially fast-moving marble or the initially slow ones? How do these questions relate to the direction of heat flow?

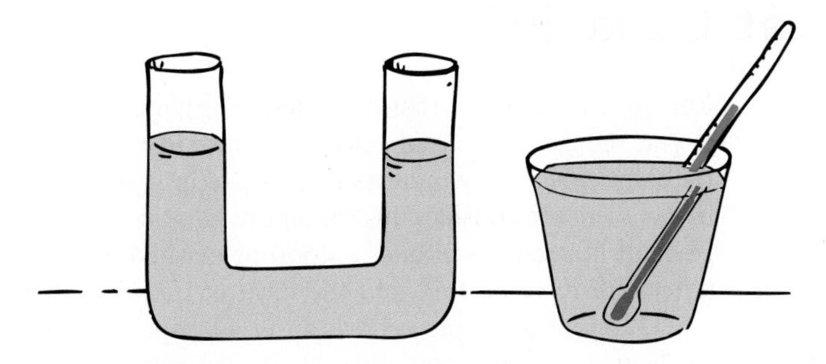

FIGURE 15.6
Just as water in the two arms of the U-tube seeks a common level (where the pressures at any given depth are the same), the thermometer and its immediate surroundings reach a common temperature (at which the average molecular KE is the same for both).

CHECK YOUR ANSWERS

1. Its temperature will rise by only 1°C, because there are twice as many molecules in 2 L of water, and each molecule receives only half as much energy on the average.
2. A fast-moving marble slows when it hits slower-moving marbles. It transfers some of its kinetic energy to the slower ones. Likewise with the flow of heat. Molecules with more kinetic energy that are in contact with molecules having less kinetic energy transfer some of their excess energy to the less energetic ones. The direction of energy transfer is from hot to cold. For both the marbles and the molecules, however, the *total* energy before and after contact is the same.

FIGURE 15.7
To the weight watcher, the peanut contains 10 Calories; to the physicist, it releases 10,000 calories (or 41,840 joules) of energy when burned or consumed.

Measuring Heat

So heat is the flow of energy from one thing to another due to a temperature difference. Since heat is a form of energy, it is measured in joules. In the United States, a more common unit of heat is the *calorie*. The calorie is defined as the amount of heat required to change the temperature of 1 gram of water by 1 Celsius degree.[3]

The energy ratings of foods and fuels are determined by burning them and measuring the energy released. (Your body "burns" food at a slow rate.) The heat unit used to label foods is actually the kilocalorie, which is 1000 calories (the heat required to raise the temperature of 1 kilogram of water by 1°C). To distinguish this unit from the smaller calorie, the food unit is sometimes called a *Calorie* (written with a capital C). It is important to remember that the calorie and Calorie are units of energy. These names are historical carryovers from the early idea that heat is an invisible fluid called *caloric*. This view persisted into the nineteenth century. We now know that heat is a form of energy, not a separate substance, so it doesn't need its own separate unit. Someday the calorie may give way to the joule, an SI unit, as the common unit for measuring heat. (The relationship between calories and joules is that 1 calorie = 4.184 joules.) In this book, we'll learn about heat with the conceptually simpler calorie—but, in the lab, you may use the joule equivalent, where an input of 4.184 joules raises the temperature of 1 gram of water by 1°C.

CHECK YOURSELF

An iron thumbtack and a big iron bolt are removed from a hot oven. Both are red-hot and have the same temperature. When dropped into identical containers of water of equal temperature, which one raises the water temperature more?

Specific Heat Capacity

FIGURE 15.8
The filling of hot apple pie may be too hot to eat even though the crust is not.

You've likely noticed that some foods remain hotter much longer than others do. If you remove a piece of toast from a toaster and pour hot soup into a bowl at the same time, a few minutes later the soup is still pleasantly warm, while the toast has cooled off considerably. Similarly, if you wait a short while before eating a piece of hot roast beef and a scoop of mashed potatoes, both initially at the same temperature, you'll find that the meat has cooled off more than the potatoes.

Different substances have different capacities for storing internal energy. If we heat a pot of water on a stove, we might find that it requires 15 minutes to raise it from room temperature to its boiling temperature. But, if we put an

CHECK YOUR ANSWER

The larger piece of iron has more internal energy to impart to the water and warms it more than the thumbtack. Although they have the same initial temperature (the same *average* kinetic energy per molecule), the more massive bolt has more molecules and therefore more *total* energy—internal energy. This example underscores the difference between temperature and internal energy.

[3]A less common unit of heat is the British thermal unit (BTU). The BTU is defined as the amount of heat required to change the temperature of 1 lb of water by 1 Fahrenheit degree. One BTU equals 1054 J.

equal mass of iron on the same flame, we would find that it would rise through the same temperature range in only about 2 minutes. For silver, the time would be less than a minute.

Different materials require different quantities of heat to raise the temperature of a given mass of the material by a specified number of degrees. Different materials absorb energy in different ways. The energy may increase the jiggling motion of molecules, which raises the temperature; or it may increase the amount of internal vibration or rotation within the molecules and go into potential energy, which does not raise the temperature. Generally, a combination of both occurs.

If you add 1 calorie of heat to 1 gram of water, you'll raise its temperature by 1°C.

Insights

Whereas 1 gram of water requires 1 calorie of energy to raise its temperature 1 Celsius degree, it takes only about one-eighth as much energy to raise the temperature of a gram of iron by the same amount. Water absorbs more heat per gram than iron for the same change in temperature. We say water has a higher **specific heat capacity** (sometimes simply called *specific heat*).[4]

The specific heat capacity of any substance is defined as the quantity of heat required to change the temperature of a unit mass of the substance by 1 degree.

We can think of specific heat capacity as thermal inertia. Recall that inertia is a term used in mechanics to signify the resistance of an object to a change in its state of motion. Specific heat capacity is a sort of thermal inertia since it signifies the resistance of a substance to a change in its temperature.

CHECK YOURSELF

Which has a higher specific heat capacity, water or sand?

The High Specific Heat Capacity of Water

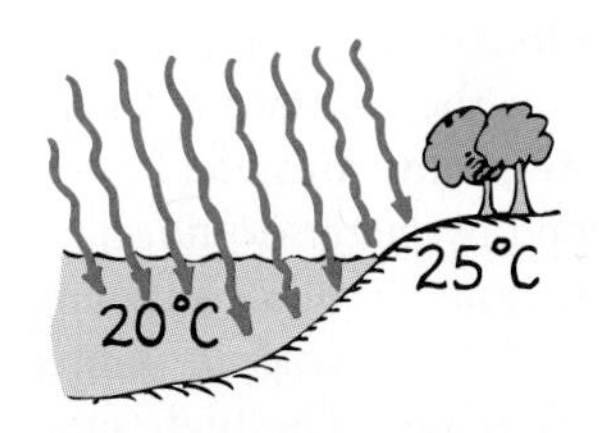

FIGURE 15.9
Because water has a high specific heat capacity and is transparent, it takes more energy to warm the water than to warm the land. Solar energy incident on the land is concentrated at the surface, but that hitting the water extends beneath the surface and so is "diluted."

Water has a much higher capacity for storing energy than all but a few uncommon materials. A relatively small amount of water absorbs a large quantity of heat for a correspondingly small temperature rise. Because of this, water is a very useful cooling agent and is used in the cooling systems of automobiles and other engines. If a liquid of lower specific heat capacity were used in cooling systems, its temperature would rise higher for a comparable absorption of heat.

Water also takes a long time to cool, a fact that explains why, in previous times, hot-water bottles were employed on cold winter nights. (Electric blankets have, for the most part, replaced them.) This tendency on the part of water to resist changes in temperature improves the climate in many locations. The

CHECK YOUR ANSWER

Water has a higher specific heat capacity. Water has more thermal inertia and takes a longer time to warm in the hot sunlight and a longer time to cool on a cold night. Sand has a low heat capacity, as evidenced by how quickly the surface warms in the morning sunlight and how quickly it cools at night. (Walking or running barefoot across scorching sand in the daytime is a much different experience from doing the same in the evening).

[4]If we know the specific heat capacity c, the formula for the quantity of heat Q involved when a mass m of a substance undergoes a change in temperature ΔT is $Q = cm\Delta T$. Or, heat transferred = specific heat capacity × mass × temperature change.

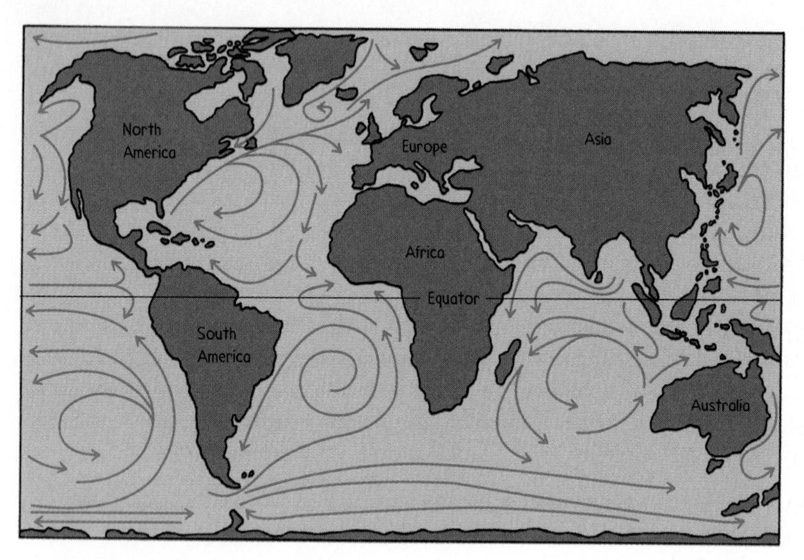

FIGURE 15.10
Many ocean currents, shown in blue, distribute heat from the warmer equatorial regions to the colder polar regions.

next time you are looking at a world globe, notice the high latitude of Europe. If water did not have a high specific heat capacity, the countries of Europe would be as cold as the northeastern regions of Canada, for both Europe and Canada receive about the same amount of sunlight per square kilometer. The Atlantic current known as the Gulf Stream carries warm water northeast from the Caribbean. It retains much of its internal energy long enough to reach the North Atlantic off the coast of Europe, where it then cools. The energy released, about 1 calorie per degree for each gram of water that cools, transfers to the air, where it is carried by the westerly winds over the European continent.

Water is king when it comes to specific heat capacity.

Insights

A similar effect occurs in the United States. The winds in the latitudes of North America are westerly. On the West Coast, air moves from the Pacific Ocean to the land. Because of water's high specific heat capacity, an ocean does not vary much in temperature from summer to winter. The water is warmer than the air in the winter and cooler than the air in the summer. In winter, the water warms the air that moves over it, and the air warms the coastal regions of North America. In summer, the water cools the air, and the coastal regions are cooled. On the East Coast, air moves from the land to the Atlantic Ocean. Land, with a lower specific heat capacity, gets hot in the summer but cools rapidly in the winter. As a result of water's high specific heat capacity and the wind directions, the West Coast city of San Francisco is warmer in the winter and cooler in the summer than the East Coast city of Washington, D.C., which is at about the same latitude.

Islands and peninsulas that are more or less surrounded by water do not have the same extremes of temperatures that are observed in the interior of a continent. When air is hot in summer months, water cools it. When air is cold in winter months, water warms it. Water moderates temperature extremes. The high summer and low winter temperatures common in Manitoba and the Dakotas, for example, are largely due to the absence of large bodies of water. Europeans, islanders, and people living near ocean air currents should be glad that water has such a high specific heat capacity. San Franciscans are!

FIGURE 15.11
The temperature of the sparks is very high, about 2000°C. That's a lot of energy per molecule of spark. Because of the few molecules per spark, however, internal energy is safely small. Temperature is one thing; transfer of energy is another.

CHECK YOURSELF

Why does a piece of watermelon stay cool for a longer time than sandwiches do when both are removed from a picnic cooler on a hot day?

CHECK YOUR ANSWER

Water in the melon has more "thermal inertia" than sandwich ingredients, and it resists changes in temperature much more. This thermal inertia is specific heat capacity.

Thermal Expansion

FIGURE 15.12
One end of the bridge is fixed, while the end shown rides on rockers to allow for thermal expansion.

When the temperature of a substance is increased, its molecules or atoms jiggle faster and move farther apart, on the average. The result is an expansion of the substance. With few exceptions, all forms of matter—solids, liquids, gases, and plasmas—generally expand when they are heated and contract when they are cooled.

In most cases involving solids, these changes in volume are not very noticeable, but careful observation usually detects them. Telephone wires become longer and sag more on a hot summer day than they do on a cold winter day. A metal lid on a glass jar can often be loosened by heating just the lid under hot water. If one part of a piece of glass is heated or cooled more rapidly than adjacent parts, the resulting expansion or contraction may break the glass, especially if the glass is thick. Pyrex heat-resistant glassware is an exception because it's specially formulated to expand very little with increasing temperature (about a third as much as ordinary glass).

The expansion of substances must be accommodated in structures and devices of all kinds. A dentist uses filling material that has the same rate of expansion as teeth have. The aluminum pistons of some automobile engines are just enough smaller in diameter than the steel cylinders to allow for the much greater expansion rate of aluminum. A civil engineer uses reinforcing steel with the same expansion rate as concrete. Long steel bridges commonly have one end fixed while the other rests on rockers (Figure 15.12). The Golden Gate Bridge in San Francisco contracts more than a meter in cold weather. The roadway itself is segmented with tongue-and-groove–type gaps called *expansion joints* (Figure 15.13). Similarly, concrete roadways and sidewalks are intersected by gaps, sometimes filled with tar, so that the concrete can expand freely in summer and contract in winter.

In times past, railroad tracks were laid in 39-foot segments connected by joint bars, with gaps left for thermal expansion. In summer months, the tracks expanded and the gaps were narrow. In winter, the gaps widened, which was responsible for a more pronounced clickity-clack when the trains rolled along the tracks. We don't hear clickity-clacks these days because someone got the bright idea to eliminate the gaps by welding the tracks together. Doesn't expansion in the summer heat cause the welded tracks to buckle, as shown in Figure 15.14? Not if the tracks are laid and welded on the hottest summer days! Track shrinkage on cold winter days stretches the tracks, which doesn't cause buckling. Stretched tracks are okay.

FIGURE 15.13
This gap in the roadway of a bridge is called an expansion joint; it allows the bridge to expand and contract. (Was this photo taken on a warm or a cold day?)

Different substances expand at different rates. When two strips of different metals—say, one of brass and the other of iron—are welded or riveted together, the greater expansion of one metal results in the bending shown in Figure 15.15. Such a compound thin bar is called a *bimetallic strip*. When the strip is heated, one side of the double strip becomes longer than the other, causing the strip to bend into a curve. On the other hand, when the strip is cooled, it tends to bend in the opposite direction, because the metal that expands more also shrinks more. The movement of the strip may be used to turn a pointer, regulate a valve, or close a switch.

FIGURE 15.14
Thermal expansion. The extreme heat of a July day in Asbury Park, New Jersey, caused the buckling of these railroad tracks. (*Wide World Photos*)

A practical application of this is the thermostat (Figure 15.16). The back-and-forth bending of the bimetallic coil opens and closes an electric circuit. When the room becomes too cold, the coil bends toward the brass side, and, in so doing, it activates an electrical switch that turns on the heat. When the room becomes too warm, the coil bends toward the iron side, which activates an electrical contact that turns off the heating unit. Refrigerators are equipped with thermostats to prevent them from becoming either too warm or too cold. Bimetallic strips are used in oven thermometers, electric toasters, automatic chokes on carburetors, and various other devices.

Liquids expand appreciably with increases in temperature. In most cases the expansion of liquids is greater than the expansion of solids. The gasoline overflowing a car's tank on a hot day is evidence for this. If the tank and contents expanded at the same rate, they would expand together and no overflow would occur. Similarly, if the expansion of the glass of a thermometer were as great as the expansion of the mercury, the mercury would not rise with increasing temperature. The reason the mercury in a thermometer rises with increasing temperature is that the expansion of liquid mercury is greater than the expansion of glass.

How a Thermostat Works

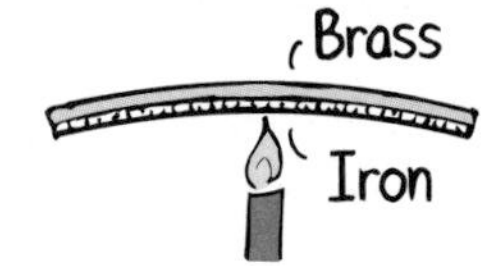

FIGURE 15.15
A bimetallic strip. Brass expands more when heated than iron does and contracts more when cooled. Because of this behavior, the strip bends as shown.

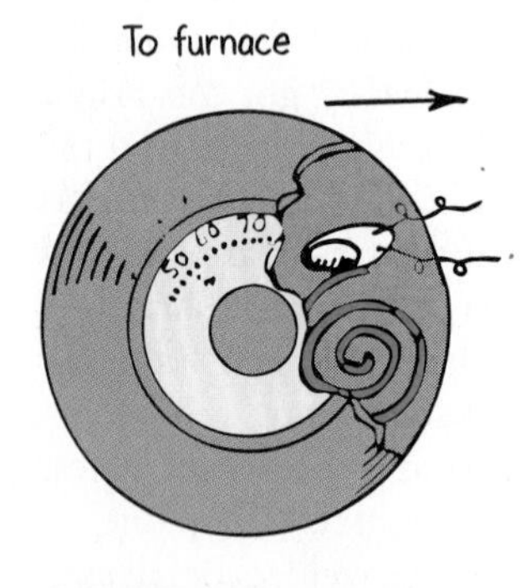

FIGURE 15.16
A thermostat. When the bimetallic coil expands, the drop of liquid mercury rolls away from the electrical contacts and breaks the electrical circuit. When the coil contracts, the mercury rolls against the contacts and completes the circuit.

FIGURE 15.17
Place a dented Ping-Pong ball in boiling water, and you'll remove the dent. Why?

CHECK YOURSELF

Why is it advisable to allow telephone lines to sag when stringing them between poles in summer?

Expansion of Water

Water, like most other substances, expands when heated. But, interestingly, it *doesn't* expand in the temperature range between 0°C and 4°C. Something quite fascinating happens in this range. Ice has a crystalline structure, with open-structured crystals. Water molecules in this open structure occupy a greater volume than they do in the liquid phase (Figure 15.18). This means that ice is less dense than water.

When ice melts, not all the open-structured crystals collapse. Some microscopic crystals remain in the ice-water mixture, making up a microscopic slush that slightly "bloats" the water, increasing its volume slightly (Figure 15.20). This results in ice water being less dense than slightly warmer water. As the temperature of water at 0°C is increased, more of the remaining ice crystals collapse. The melting of these crystals further decreases the volume of the water. The water undergoes two processes at the same time—contraction and expansion. Volume tends to decrease as ice crystal collapse, while volume tends to increase due to greater molecular motion. The collapsing effect dominates until the temperature reaches 4°C. After that, expansion overrides contraction, because most of the microscopic ice crystals have melted by then (Figure 15.21).

When ice water freezes to become solid ice, its volume increases tremendously, and its density is much lower. That's why ice floats on water. Like most

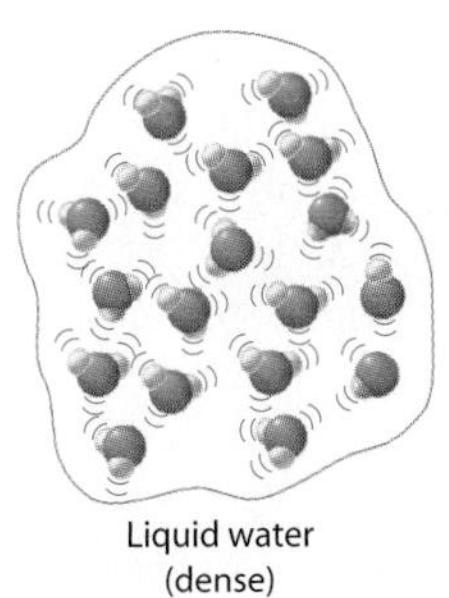

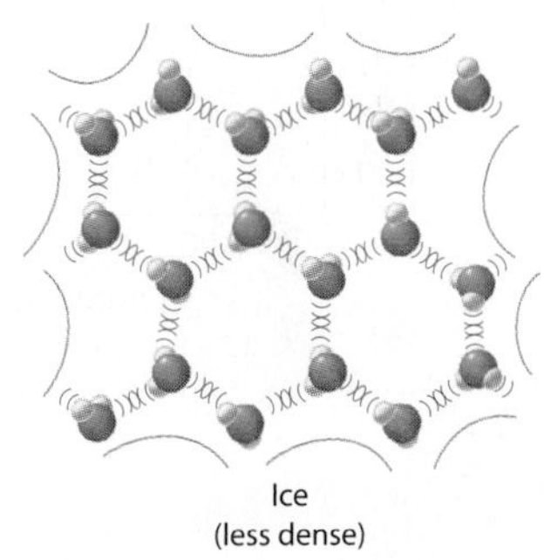

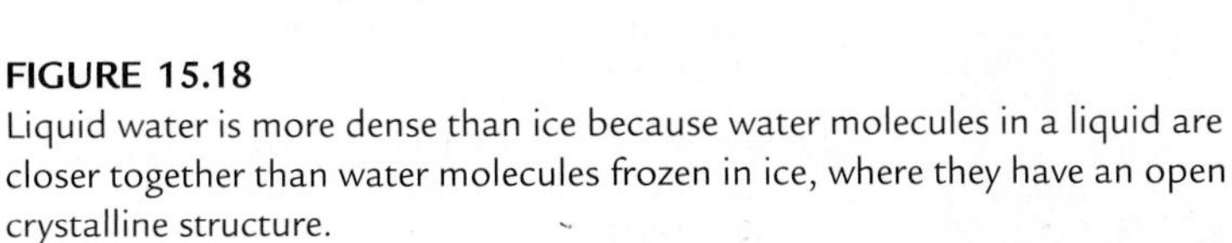

FIGURE 15.18
Liquid water is more dense than ice because water molecules in a liquid are closer together than water molecules frozen in ice, where they have an open crystalline structure.

FIGURE 15.19
The six-sided structure of a snowflake is a result of the six-sided ice crystals that make it up.

CHECK YOUR ANSWER

Telephone lines are longer in summer, when they are warmer, and shorter in winter, when they are cooler. They therefore sag more on hot summer days than in winter. If the telephone lines are not strung with enough sag in summer, they might contract too much and snap during the winter.

other substances, solid ice contracts with further cooling. This behavior of water is very important in nature. If water were most dense at 0°C, it would settle to the bottom of a pond or lake. Water at 0°C, however, is less dense and "floats" at the surface. That's why ice forms at the surface.

So a pond freezes from the surface downward. In a cold winter, the ice will be thicker than in a milder winter. Water at the bottom of an ice-covered pond is 4°C, which is relatively warm for the organisms that live there. Interestingly, very deep bodies of water are not ice covered even in the coldest of winters. This is because all the water must be cooled to 4°C before lower temperatures can be reached. For deep water, the winter is not long enough to reduce an entire pond to 4°C. Any 4°C water lies at the bottom. Because of water's high specific heat and poor ability to conduct heat, the bottom of a deep body of water in a cold region remains at a constant 4°C year-round. Fish should be glad that this is so.

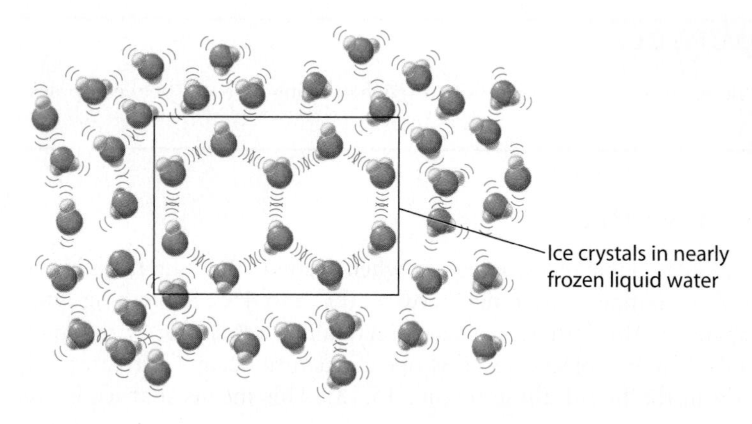

FIGURE 15.20
Close to 0°C, liquid water contains crystals of ice. The open structure of these 3-D cagelike crystals increases the volume of the water slightly.

FIGURE 15.21
Between 0°C and 4°C, the volume of liquid water decreases as temperature increases. Above 4°C, water behaves the way other substances do: Its volume increases as its temperature increases. The volumes shown here are for a 1-gram sample.

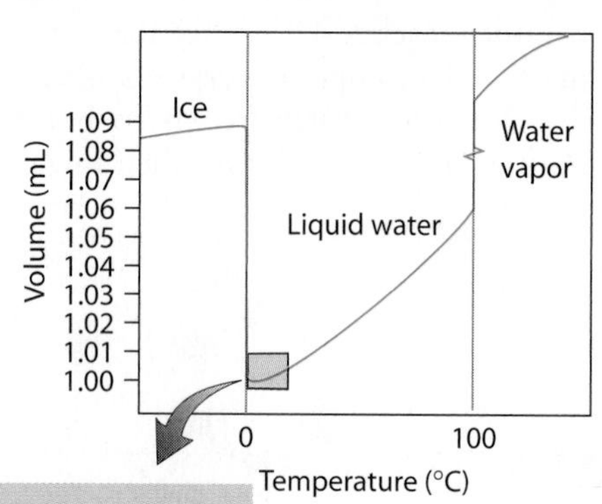

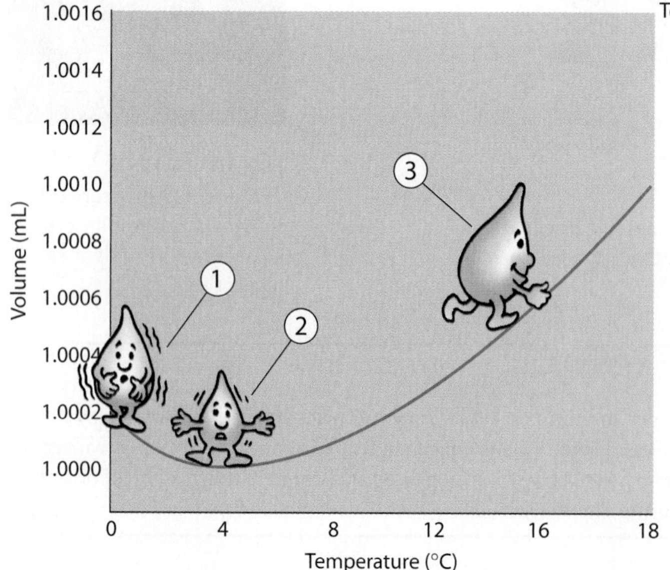

(1) Liquid water below 4°C is bloated with ice crystals.

(2) Upon warming, the crystals collapse, resulting in a smaller volume for the liquid water.

(3) Above 4°C, liquid water expands as it is heated because of greater molecular motion.

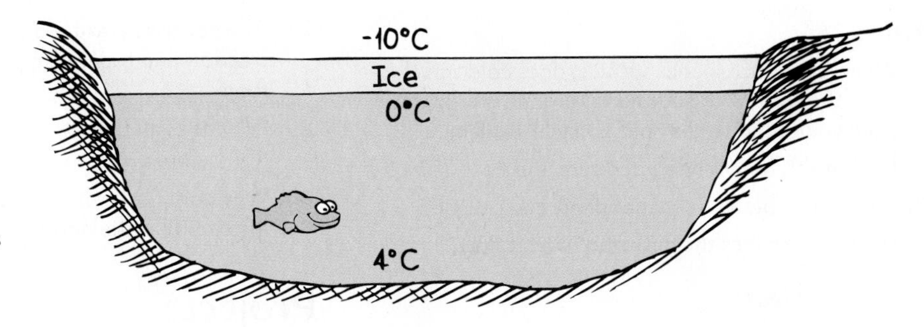

FIGURE 15.22
As water is cooled at the surface, it sinks until the temperature of the entire lake is 4°C. Only then can the surface water cool to 0°C without sinking. Once ice has formed, temperatures lower than 4°C can extend down into the lake.

CHECK YOURSELF

1. What was the precise temperature at the bottom of Lake Michigan, where the water is deep and the winters long, on New Year's Eve in 1901?
2. What's inside the open spaces of the water crystals shown in Figure 15.18? Is it air, water vapor, or nothing?

CHECK YOUR ANSWERS

1. 4°C, because the temperature at the bottom of any body of water containing any 4°C water has a bottom temperature of 4°C, for the same reason that rocks are at the bottom. Rocks are more dense than water, and 4°C water is more dense than water at any other temperature. So both rocks and 4°C water sink to the bottom. Water is also a poor heat conductor, so, if the body of water is deep and in a region of long winters and short summers, the water at the bottom likely remains a constant 4°C year-round.
2. There's nothing at all in the open spaces. It's empty space—a void. If there were air or vapor in the open spaces, the illustration should show molecules there—oxygen and nitrogen for air and H_2O for water vapor.

Summary of Terms

Temperature A measure of the average translational kinetic energy per molecule in a substance, measured in degrees Celsius or Fahrenheit or in kelvins (K).

Absolute zero The lowest possible temperature that a substance may have—the temperature at which molecules of the substance have their minimum kinetic energy.

Heat The energy that flows from a substance of higher temperature to a substance of lower temperature, commonly measured in calories or joules.

Internal energy The total of all molecular energies, kinetic plus potential, that are internal to a substance.

Specific heat capacity The quantity of heat per unit mass required to raise the temperature of a substance by 1 Celsius degree.

Review Questions

1. Why does a penny become warmer when it is struck by a hammer?

Temperature

2. What are the temperatures for freezing water on the Celsius and Fahrenheit scales? For boiling water?
3. What are the temperatures for freezing water and boiling water on the Kelvin temperature scale?
4. What is meant by "translational" kinetic energy?
5. Which affects temperature—translational kinetic energy, rotational kinetic energy, vibrational kinetic energy, or all of these?
6. What is meant by the statement that a thermometer measures its own temperature?

Heat

7. When you touch a cold surface, does cold travel from the surface to your hand or does energy travel from your hand to the cold surface? Explain.
8. Distinguish between temperature and heat.
9. Distinguish between heat and internal energy.
10. What determines the direction of heat flow?

Measuring Heat

11. How is the energy value of foods determined?
12. Distinguish between a calorie and a Calorie.
13. Distinguish between a calorie and a joule.

Specific Heat Capacity

14. Which warms up faster when heat is applied—iron or silver?
15. Does a substance that heats up quickly have a high or a low specific heat capacity?
16. Does a substance that cools off quickly have a high or a low specific heat capacity?
17. How does the specific heat of water compare with the specific heats of other common materials?
18. Northeastern Canada and much of Europe receive about the same amount of sunlight per unit area. Why then is Europe generally warmer in the winter?
19. According to the law of conservation of energy, if ocean water cools, something else should warm. What is it that warms?
20. Why is the temperature fairly constant for land masses surrounded by large bodies of water?

Thermal Expansion

21. Why do substances expand when temperature is increased?
22. Why does a bimetallic strip bend with changes in temperature?
23. Which generally expands more for an equal increase in temperature—solids or liquids?

Expansion of Water

24. When the temperature of ice-cold water is increased slightly, does it undergo a net expansion or a net contraction?
25. What is the reason for ice being less dense than water?
26. Does "microscopic slush" in water tend to make it more dense or less dense?
27. What happens to the amount of "microscopic slush" in cold water when its temperature is increased?
28. At what temperature do the combined effects of contraction and expansion produce the smallest volume for water?
29. Why does all the water in a lake have to be cooled to 4°C before surface water can be cooled below 4°C?
30. Why does ice form at the surface of a body of water instead of at the bottom?

Project

Write a letter to Grandpa saying how you're learning to see the connections in nature. Also give him examples of how you're learning to distinguish between closely related ideas. Use temperature and heat as examples.

Exercises

1. In a meeting room, there are chairs, a table, and people. Which of these things has a temperature (a) lower than, (b) greater than, or (c) equal to the temperature of the air?
2. Which is greater—an increase in temperature of 1 Celsius degree or an increase of 1 Fahrenheit degree?
3. In a glass of water at room temperature, do all the molecules have the same speed?
4. Why wouldn't you expect all the molecules in a gas to have the same speed?
5. Why can't you establish whether you are running a high temperature by touching your own forehead?
6. Which has more kinetic energy—the molecules in a gram of ice or the molecules in a gram of steam? Defend your answer.
7. Which has the greater amount of internal energy—an iceberg or a cup of hot coffee? Defend your answer.
8. When a mercury thermometer is heated, the mercury expands and rises in the thin tube of glass. What does this indicate about the relative rates of expansion for mercury and glass? What would happen if their expansion rates were the same?
9. Which is the largest unit of heat transfer—Calorie, calorie, or joule?
10. If you drop a hot rock into a pail of water, the temperature of the rock and the water will change until both are equal. The rock will cool and the water will warm. Does this hold true if the hot rock is dropped into the Atlantic Ocean? Explain.
11. Consider two glasses, one filled with water and the other half full, with the water in the two glasses being at the same temperature. In which glass are the water molecules moving faster? In which is there greater

internal energy? In which will more heat be required to increase the temperature by 1°C?

12. Would you expect the temperature of water at the bottom of Niagara Falls to be slightly higher than the temperature at the top of the falls? Why?

13. Thermometers in a physics lab often use gas rather than mercury. Whereas changes in volume indicate temperature in a mercury thermometer, what changes in a gas do you think indicate temperature in a gas thermometer?

14. Why does the pressure of gas enclosed in a rigid container increase as the temperature increases?

15. Adding the same amount of heat to two different objects does not necessarily produce the same increase in temperature. Why not?

16. A certain quantity of heat is supplied to both a kilogram of water and to a kilogram of iron. Which undergoes the greater change in temperature? Defend your answer.

17. Which has the greater specific heat—an object that cools quickly, or an object of the same mass that cools more slowly?

18. If the specific heat of water were less, would a nice hot bath be a longer or a shorter experience?

19. In addition to the random motions of molecules from place to place that are associated with temperature, some molecules can absorb large amounts of energy that go into vibrations and rotations of the molecule itself. Would you expect materials composed of such molecules to have a high or a low specific heat? Explain.

20. What role does specific heat capacity play in a watermelon staying cool after removal from a cooler on a hot day?

21. Ethyl alcohol has about one-half the specific heat capacity of water. If equal masses of each at the same temperature are supplied with equal quantities of heat, which will undergo the greater change in temperature?

22. When a 1-kg metal pan containing 1 kg of cold water is removed from the refrigerator and set on a table, which absorbs more heat from the room—the pan or the water?

23. Bermuda is about as far north of the equator as North Carolina, but, unlike North Carolina, it has a subtropical climate year-round. Why is this so?

24. Iceland, so named to discourage conquest by expanding empires, is not at all ice covered like Greenland and parts of Siberia, even though it is not far from the Arctic Circle. The average winter temperature of Iceland is considerably higher than it is in regions at the same latitude in eastern Greenland and central Siberia. Why is this so?

25. Why does the presence of large bodies of water tend to moderate the climate of nearby land—to make it warmer in cold weather and cooler in hot weather?

26. If the winds at the latitude of San Francisco and Washington, D.C., were from the east rather than from the west, why might San Francisco be able to grow only cherry trees and Washington, D.C., both cherry trees and palm trees?

27. In times past, on a cold winter night, it was common to bring a hot object to bed with you. Which would keep you warmer through the cold night—a 10-kg iron brick or a 10-kg jug of hot water at the same temperature? Explain.

28. Desert sand is very hot in the day and very cool at night. What does this indicate about its specific heat?

29. Cite an exception to the claim that all substances expand when heated.

30. Would a bimetallic strip function if the two different metals have the same rates of expansion? Is it important that they expand at different rates? Explain.

31. Steel plates are commonly attached to each other with rivets, which are slipped into holes in the plates and rounded over with hammers. The hotness of the rivets makes them easier to round over, but their hotness has another important advantage in providing a tight fit. What is it?

32. An old method for breaking boulders was to put them in a hot fire and then to douse them with cold water. Why would this fracture the boulders?

33. After you have driven a car for some distance, why does the air pressure in the tires increase?

34. Structural groaning noises are sometimes heard in the attic of old buildings on cold nights. Give an explanation in terms of thermal expansion.

35. An old remedy for a pair of nested drinking glasses that stick together is to run water at different temperatures into the inner glass and over the surface of the outer glass. Which water should be hot, and which cold?

36. Why is it important that glass mirrors used in astronomical observatories be composed of glass with a low "coefficient of expansion"?

37. In terms of thermal expansion, why is it important that a key and its lock be made of the same or similar materials?

38. Any architect will tell you that chimneys are never used as a weight-bearing part of a wall. Why?

39. Looking at the expansion joint in the photo of Figure 15.13, would you say it was taken on a warm day or a cold day? Why?

40. Would you or the gas company gain by having gas warmed before it passed through your gas meter?

41. After filling your gas tank to the top and parking your car where is exposed to the hot Sun, why does the gasoline overflow?

42. A metal ball is just able to pass through a metal ring. When the ball is heated, however, it will not pass through the ring. What would happen if the ring, rather than the ball, were heated? Does the size of the hole increase, stay the same, or decrease?

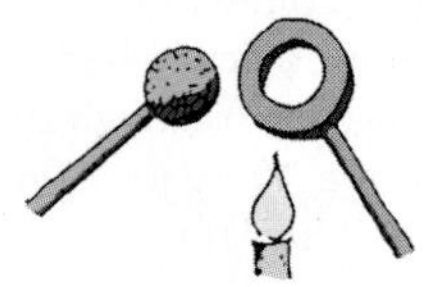

43. Consider a pair of brass balls of the same diameter, one hollow and the other solid. Both are heated with equal increases in temperature. Compare the diameters of the heated balls.

44. After a machinist very quickly slips a hot, snugly fitting iron ring over a very cold brass cylinder, there is no way that the two can be separated intact. Can you explain why this is so?

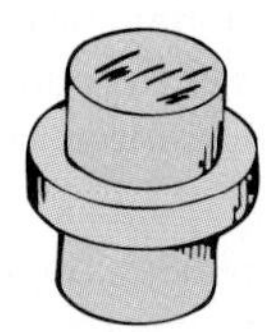

45. Suppose that you cut a small gap in a metal ring. If you were to heat the ring, would the gap become wider or narrower?

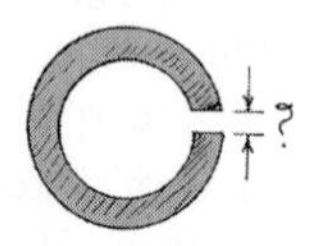

46. When a mercury thermometer is warmed, the mercury level momentarily goes down before it rises. Can you give an explanation for this?

47. Why do long steam pipes often have one or more relatively large U-shaped sections of pipe?

48. Why are incandescent bulbs typically made of very thin glass?

49. One of the reasons the first lightbulbs were expensive was that the electrical lead wires into the bulb were made of platinum, which expands at about the same rate as glass when heated. Why is it important that the metal leads and the glass have the same coefficient of expansion?

50. After you measure the dimensions of a plot of land with a steel tape on a hot day, you return and remeasure the same plot on a cold day. On which day do you determine the larger area for the land?

51. What was the precise temperature at the bottom of Lake Superior at 12:01 a.m. on October 31, 1894?

52. Suppose that water is used in a thermometer instead of mercury. If the temperature is at 4°C and then changes, why can't the thermometer indicate whether the temperature is rising or falling?

53. A piece of solid iron sinks in a container of molten iron. A piece of solid aluminum sinks in a container of molten aluminum. Why does a piece of solid water (ice) not sink in a container of "molten" (liquid) water? Explain, using molecular terms.

54. How does the combined volume of the billions and billions of hexagonal open spaces in the structures of ice crystals in a piece of ice compare with the portion of ice that floats above the water line?

55. How would the shape of the curve in Figure 15.21 differ if density rather than volume were plotted against temperature? Make a rough sketch.

56. What happens to the volume of water as it is cooled from 3°C to 1°C?

57. State whether water at the following temperatures will expand or contract when warmed a little: 0°C; 4°C; 6°C.

58. Why is it important to protect water pipes so that they don't freeze?

59. If cooling occurred at the bottom of a pond instead of at the surface, would the pond freeze from the bottom up? Explain.

60. If water had a lower specific heat, would ponds be more likely to freeze or less likely to freeze?

Problems

Quantity of heat, Q, is equal to the specific heat capacity of the substance c multiplied by its mass m and the temperature change ΔT; that is, $Q = cm\Delta T$.

1. What would be the final temperature of a mixture of 50 g of 20°C water and 50 g of 40°C water?
2. If you wish to warm 100 kg of water by 20°C for your bath, how much heat is required? (Give your answer in calories and joules.)
3. The specific heat capacity of copper is 0.092 calories per gram per degree Celsius. How much heat is required to raise the temperature of a 10-g piece of copper from 0°C to 100°C? How does this compare with the heat needed to raise the temperature of the same mass of water through the same temperature difference?
4. What would be the final temperature when 100 g of 25°C water is mixed with 75 g of 40°C water? (*Hint:* Equate the heat gained by the cool water with the heat lost by the warm water.)
5. What will be the final temperature of 100 g of 20°C water when 100 g of 40°C iron nails are submerged in it? (The specific heat of iron is 0.12 cal/g C°. Here you should equate the heat gained by the water to the heat lost by the nails.)

To solve the following problems, you need to know the average coefficient of linear expansion, α, which differs for different materials. We define α to be the change in length per unit length—or the fractional change in length—for a temperature change of 1 C°. That is, $\alpha = \Delta L/L$ per C°. For aluminum, $\alpha = 24 \times 10^{-6}$/C°, and for steel, $\alpha = 11 \times 10^{-6}$/C°.

The change in length ΔL of a material is given by $\Delta L = L\alpha\Delta T$.

6. Suppose that a bar 1 m long expands 0.5 cm when heated. By how much will a bar 100 m long of the same material expand when similarly heated?
7. Suppose that the 1.3-km main span of steel for the Golden Gate Bridge had no expansion joints. How much longer would it be for an increase in temperature of 15°C?
8. A 10.00-m long steel wire supports a pendulum bob at its end. How many millimeters longer is the wire when the temperature increases by 20.0°C?
9. Two equal-length strips of aluminum and steel are heated. Which expands more? How much more—that is, by what factor is one expansion greater than the other?
10. Consider a 40,000-km steel pipe that forms a ring to fit snugly all around the circumference of the Earth. Suppose people along its length breathe on it so as to raise its temperature 1 Celsius degree. The pipe gets longer. It also is no longer snug. How high does it stand above ground level? (To simplify, consider only the expansion of its radial distance from the center of the Earth, and apply the geometry formula that relates circumference C and radius r, $C = 2\pi r$. The result is surprising!)

CHAPTER 16

Heat Transfer

John Suchocki demonstrates the low conductivity of red-hot coals with his bare feet.

The Secret to Walking on Hot Coals

The spontaneous transfer of heat is always from warmer objects to cooler objects. If several objects near one another have different temperatures, then those that are warm become cooler and those that are cool become warmer, until all have a common temperature. This equalization of temperatures is brought about in three ways: by *conduction,* by *convection,* and by *radiation.*

Conduction

Hold one end of an iron nail in a flame. Before long, the end you're holding will quickly become too hot to hold. Heat enters the metal nail at the end that is kept in the flame and is transmitted along the entire length of the nail. The transmission of heat in this manner is called **conduction.** The fire causes the atoms at the heated end of the nail to move more rapidly. These atoms vibrate against neighboring atoms, which, in turn, do the same. More important, free electrons that can drift through the metal are made to jostle and transfer energy by colliding with atoms and other free electrons within the nail.

How well a solid object conducts heat depends on the bonding within its atomic or molecular structure. Solids built of atoms that have one or more "loose" outer electrons conduct heat (and electricity) well. Metals have the "loosest" outer electrons, which are free to carry energy by collisions throughout the metal. They are excellent conductors of heat and electricity for this reason. Silver is the best, copper is next, and, among the common metals, aluminum and then iron are next in order. Wool, wood, straw, paper, cork, and Styrofoam, on the other hand, are poor conductors of heat. The outer electrons in the atoms of these materials are firmly attached. Poor conductors are called *insulators.*

FIGURE 16.1
When you touch a nail stuck in ice, does cold flow from the nail to your hand, or does energy flow from your hand to the nail?

Because wood is a good insulator, it is used for handles of cookware. Even when it is hot, you can grasp the wooden handle of a pot with your bare hand and quickly remove the pot from a hot stove without harm. Grasping an iron handle of the same temperature would surely burn your hand. Wood is a good insulator even when it's red hot, which is why firewalking professor John Suchocki can walk barefoot on red-hot wooden coals without burning his feet (chapter-opener photo, above). (CAUTION: Don't try this on your own;

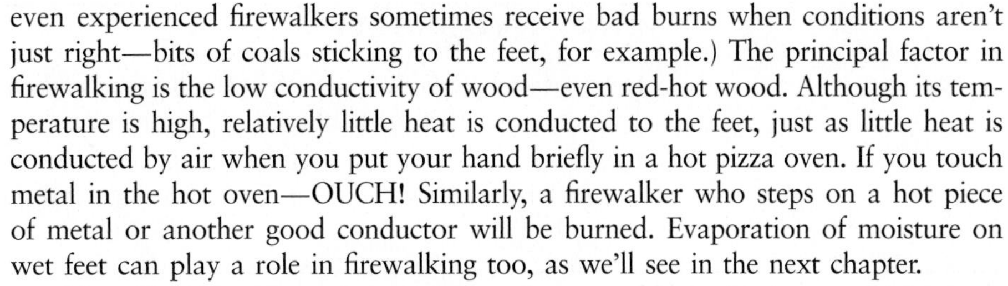

The feeling of warmth or cold for different materials involves rates of heat transfer, not necessarily temperatures.

Insights

even experienced firewalkers sometimes receive bad burns when conditions aren't just right—bits of coals sticking to the feet, for example.) The principal factor in firewalking is the low conductivity of wood—even red-hot wood. Although its temperature is high, relatively little heat is conducted to the feet, just as little heat is conducted by air when you put your hand briefly in a hot pizza oven. If you touch metal in the hot oven—OUCH! Similarly, a firewalker who steps on a hot piece of metal or another good conductor will be burned. Evaporation of moisture on wet feet can play a role in firewalking too, as we'll see in the next chapter.

Most liquids and gases are poor conductors of heat. Air is a very poor conductor, which is why your hand isn't harmed if placed briefly in a hot pizza oven. The good insulating properties of such things as wool, fur, and feathers are largely due to the air spaces they contain. Other porous substances are likewise good insulators because of their many small air spaces. Be glad that air is a poor conductor; if it weren't, you'd feel quite chilly on a 20°C (68°F) day!

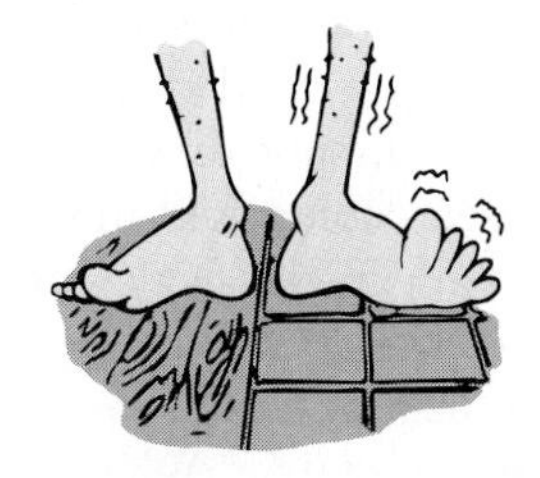

FIGURE 16.2
The tile floor feels colder than the wooden floor, even though both floor materials are at the same temperature. This is because tile is a better conductor of heat than wood, and so heat is more readily conducted out of the foot touching the tile.

Snow is a poor conductor (a good insulator)—about the same as dry wood. Hence, a blanket of snow can literally keep the ground warm in winter. Snowflakes are formed of crystals, which collect into feathery masses, imprisoning air and thereby interfering with the escape of heat from Earth's surface. Traditional Arctic winter dwellings are shielded from the cold by their snow covering. Animals in the forest find shelter from the cold in snowbanks and in holes in the snow. The snow doesn't provide warmth; it simply slows down the loss of the heat that the animals generate.

Heat is transmitted from a higher to a lower temperature. We often hear people say that they wish to keep the cold out of their homes. A better way to put this is to say that they want to prevent the heat from escaping. There is no "cold" that flows into a warm home (unless a cold wind blows into it). If the home becomes colder, it is because heat flows out. Homes are insulated with rock wool or spun glass to prevent heat from escaping rather than to prevent cold from entering.

It is important to note that no insulator can totally prevent heat from getting through it. An insulator just reduces the rate at which heat penetrates. In winter, even the best-insulated warm homes will gradually cool. Insulation slows heat transfer.

FIGURE 16.3
Snow patterns on the roof of a house show areas of conduction and insulation. Bare parts show where heat from inside has leaked through the roof and melted the snow.

CHECK YOURSELF

1. In desert regions that are hot in the daytime and cold at nighttime, the walls of houses are often made of mud. Why is it important that the mud walls be thick?
2. Why is it that you can place your hand briefly inside a hot pizza oven without harm, but you are burned if you touch the metal sides of the oven?

Convection

Convection ovens are simply ovens with a fan inside, which speeds up cooking by the circulation of warmed air.

Insights

Liquids and gases transmit heat mainly by **convection,** which is heat transfer due to the actual motion of the fluid itself. Unlike conduction (in which heat is transferred by successive collisions of electrons and atoms), convection involves the motion of "blobs" of matter—the overall motion of a fluid. Convection can occur in all fluids, whether liquids or gases. Whether we heat water in a pan or heat air in a room, the process is the same (Figure 16.4). As the fluid is heated from below, the molecules at the bottom begin moving faster, spreading apart more, becoming less dense, and are buoyed upward. Denser, cooler fluid moves in to take the place of the now-warmed fluid at the bottom. In this way, convection currents keep the fluid stirred up as it heats—warmer fluid moving away from the heat source and cooler fluid moving toward the heat source.

Convection currents occur in the atmosphere, affecting the weather. When air is warmed, it expands. In expanding, it becomes less dense than the surrounding air. Like a balloon, it is buoyed upward. When the rising air reaches an altitude at which its density matches that of the surrounding air, it no longer rises. This is evident when we see smoke from a fire rise and then level off as it cools to match the density of the surrounding air. Rising warm air expands as it rises because less atmospheric pressure squeezes it at higher altitudes. As the air expands, it cools. (Do the following experiment right now. With your mouth open, blow on your hand. Your breath is warm. Now repeat, but this time pucker your lips to make a small hole so your breath expands as it leaves your mouth. Note that your breath is appreciably cooler! Expanding air cools.) This is the opposite of what occurs when air is compressed. If you've ever compressed air with a tire pump, you probably noticed that both air and pump become quite hot.

We can understand the cooling of expanding air by thinking of molecules of air as tiny Ping-Pong balls bouncing against one another. A ball picks up speed when it is hit by another that approaches with a greater speed. But when

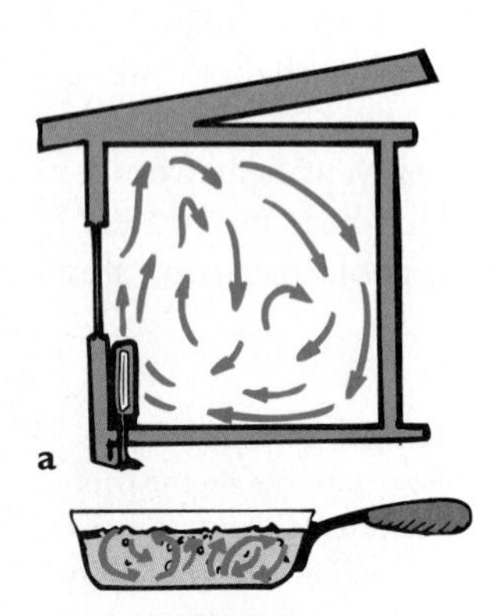

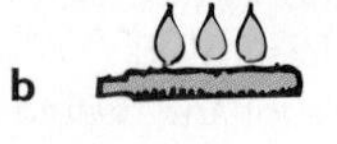

FIGURE 16.4
(a) Convection currents in air. (b) Convection currents in liquid.

CHECK YOUR ANSWERS

1. A wall of appropriate thickness keeps the house warm at night by slowing the flow of heat from inside to outside, and it keeps the house cool in the daytime by slowing the flow of heat from outside to inside. Such a wall has "thermal inertia."
2. When your hand is in the air of the hot oven, you're not harmed because air is a poor conductor—heat doesn't travel well between the hot air and your hand. Air also has a low specific heat capacity, so the total amount of thermal energy in the air that can be transferred to your skin is small. Touching the hot metal sides of the oven is another story, for metal is an excellent conductor and has a much larger specific heat capacity, so considerable heat flows into your hand.

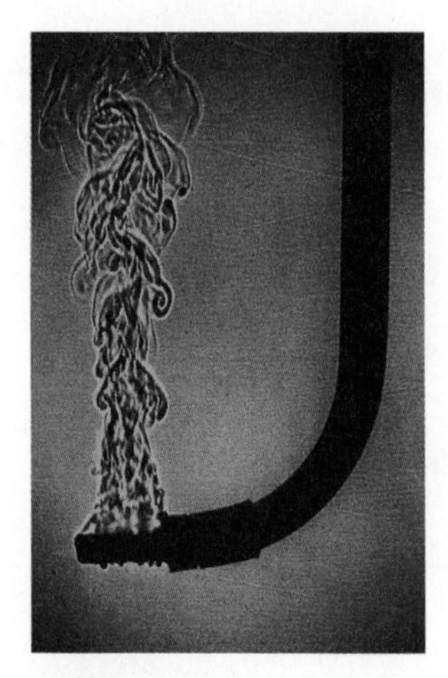

FIGURE 16.5
A heater at the tip of a J-tube submerged in water produces convection currents, which are revealed as shadows (caused by deflections of light in water of different temperatures).

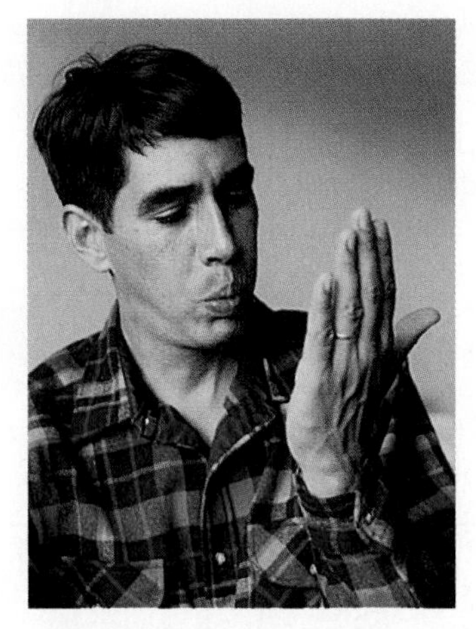

FIGURE 16.6
Blow warm air onto your hand from your wide-open mouth. Now reduce the opening between your lips so that the air expands as you blow. Do you notice a difference in air temperature?

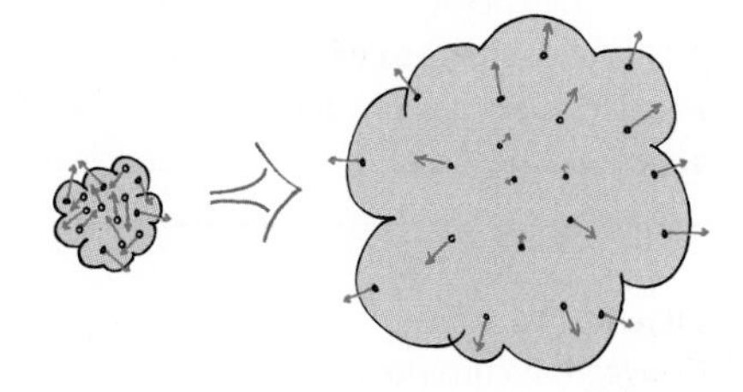

FIGURE 16.7
Molecules in a region of expanding air collide more often with receding molecules than with approaching ones. Their rebound speeds therefore tend to decrease and, as a result, the expanding air cools.

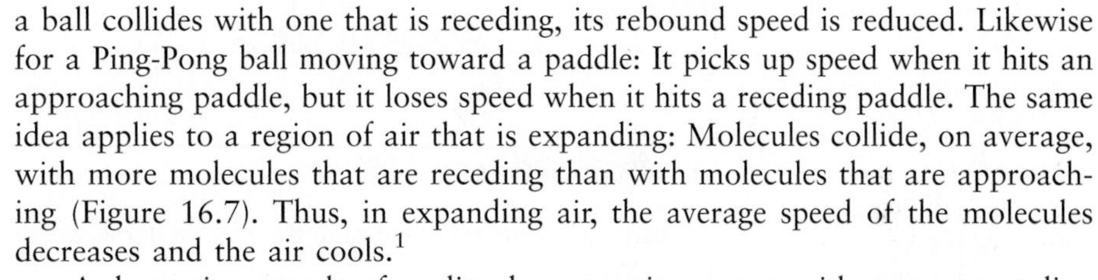

a ball collides with one that is receding, its rebound speed is reduced. Likewise for a Ping-Pong ball moving toward a paddle: It picks up speed when it hits an approaching paddle, but it loses speed when it hits a receding paddle. The same idea applies to a region of air that is expanding: Molecules collide, on average, with more molecules that are receding than with molecules that are approaching (Figure 16.7). Thus, in expanding air, the average speed of the molecules decreases and the air cools.[1]

FIGURE 16.8
The hot steam expands from the pressure cooker and is cool to Millie's touch.

A dramatic example of cooling by expansion occurs with steam expanding through the nozzle of a pressure cooker (Figure 16.8). The cooling effect of both expansion and rapid mixing with cooler air allows you to hold your hand comfortably in the jet of condensed vapor. (Caution: If you try this, be sure to place your hand high above the nozzle at first and then lower it to a comfortable distance. If you put your hand at the nozzle where no steam appears, watch out! Steam is invisible near the nozzle before it has sufficiently expanded and cooled. The cloud of "steam" that you see is actually condensed water vapor, which is much cooler.)

Convection currents stirring the atmosphere result in winds. Some parts of Earth's surface absorb heat from the Sun more readily than others, and, as a result, the air near the surface is heated unevenly and convection currents form. This is evident at the seashore. In the daytime, the shore warms more easily than the water; air over the shore is pushed up (we say it rises) by cooler air that comes in from above the water to take its place. The result is a sea breeze. At night, the process reverses, because the shore cools off more quickly than the water, and then the warmer air is over the sea (Figure 16.9). If you build a fire on the beach, you'll notice that the smoke sweeps landward during the day and seaward at night.

[1]Where does the energy go in this case? We will see, in Chapter 18, that it goes into work done on the surrounding air as the expanding air pushes outward.

CONVECTION POWER TOWER

Imagine, in a hot desert, a huge greenhouse—a circular, glass-roofed enclosure some several kilometers in diameter with a kilometer-high chimney in the middle. Such a huge greenhouse preheats the desert air, which flows to the center and rises in the chimney. In the chimney updraft are wind turbines, generating megawatts of clean power. Such power plants are similar to wind turbines, but they are more reliable because they produce their own wind. Watch for the advent of these twenty-first-century clean power sources.

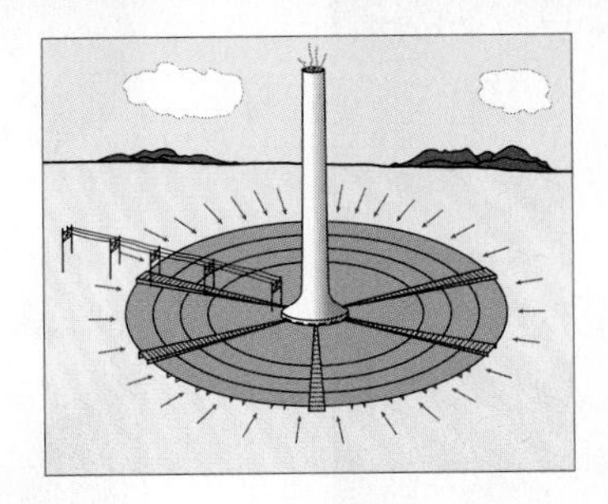

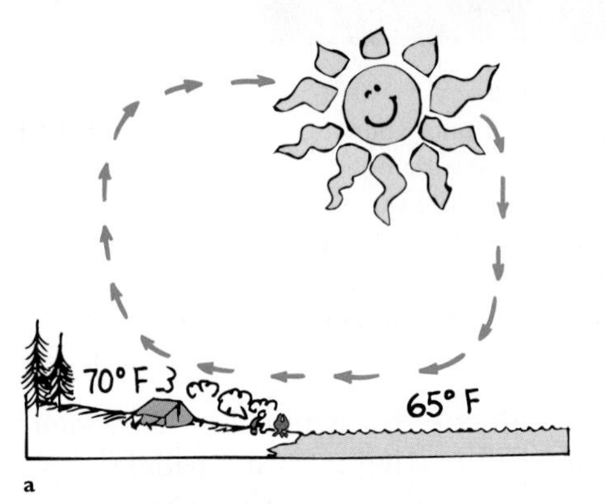

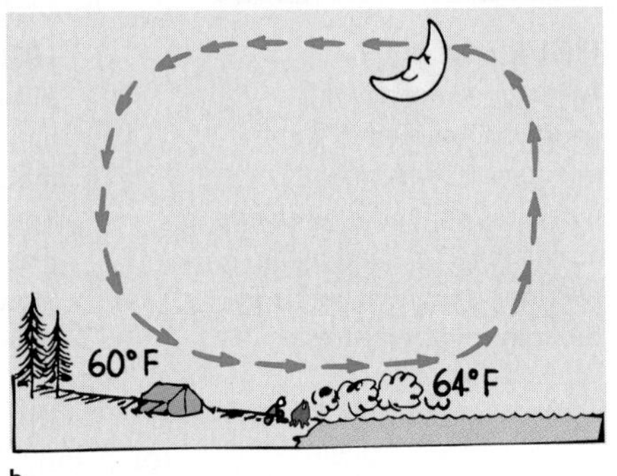

FIGURE 16.9
Convection currents produced by unequal heating of land and water. (a) During the day, warm air above the land rises, and cooler air over the water moves in to replace it. (b) At night, the direction of air flow is reversed because then the water is warmer than the land.

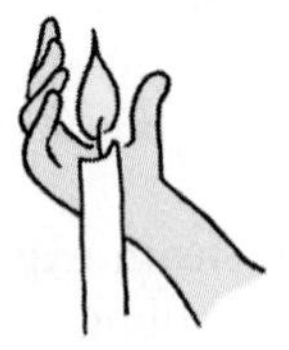

CHECK YOURSELF

You can hold your fingers beside the candle flame without harm, but not above the flame. Why?

Radiation

Energy from the Sun passes through space and then through the Earth's atmosphere and warms Earth's surface. This energy does not pass through the atmosphere by conduction, because air is a poor conductor. Nor does it pass through by convection, for convection begins only after the Earth is warmed. We also know that neither convection nor conduction is possible in the empty space between our atmosphere and the Sun. We can see that energy must be transmitted by some other means—by **radiation.**[2] The energy so radiated is called *radiant energy*.

CHECK YOUR ANSWER

Hot air travels upward by air convection. Since air is a poor conductor, very little heat travels sideways to your fingers.

[2]The radiation we are talking about here is electromagnetic radiation, including visible light. Don't confuse this with radioactivity, a process of the atomic nucleus that we'll discuss in Part 7.

PRACTICING PHYSICS

Hold the bottom end of a test tube full of cold water in your hand. Heat the top part in a flame until the water boils. The fact that you can still hold the bottom of the test tube shows that glass and water are poor conductors of heat and that convection does not move the hot water downward. This is even more dramatic when you wedge chunks of ice at the bottom with some steel wool; then the water above can be brought to a boil without melting the ice. Try it and see.

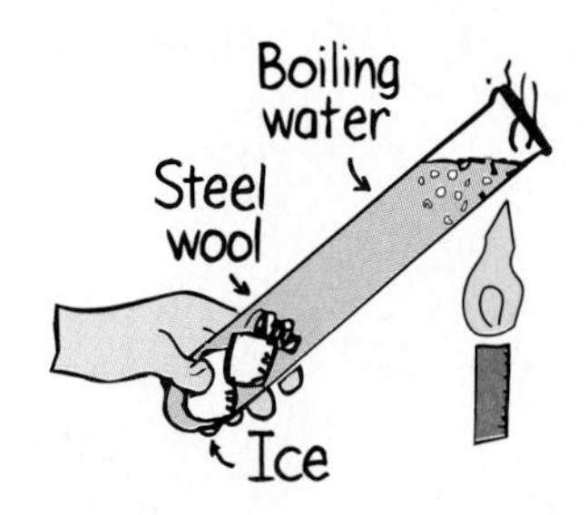

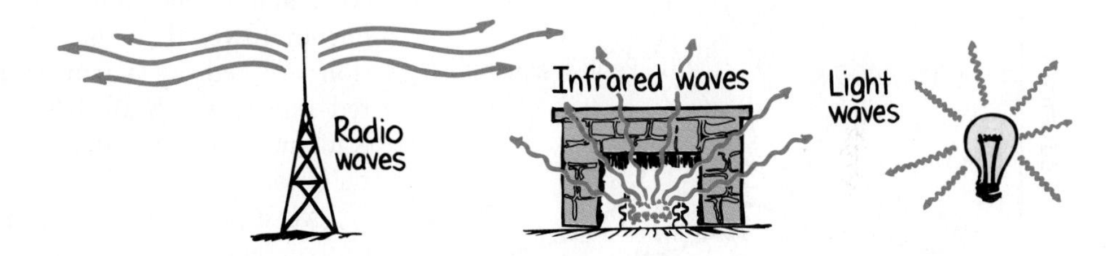

FIGURE 16.10
Types of radiant energy (electromagnetic waves).

Radiant energy is in the form of *electromagnetic waves*. It includes radio waves, microwaves, infrared radiation, visible light, ultraviolet radiation, X-rays, and gamma rays. These types of radiant energy are listed in order of wavelength, from longest to shortest. Infrared (below-the-red) radiation has longer wavelengths than visible light. The longest visible wavelengths are for red light, and the shortest are for violet light. Ultraviolet (beyond-the-violet) radiation has shorter wavelengths. (Wavelength is treated in more detail in Chapter 19, and electromagnetic waves are covered in more detail in Chapters 25 and 26.)

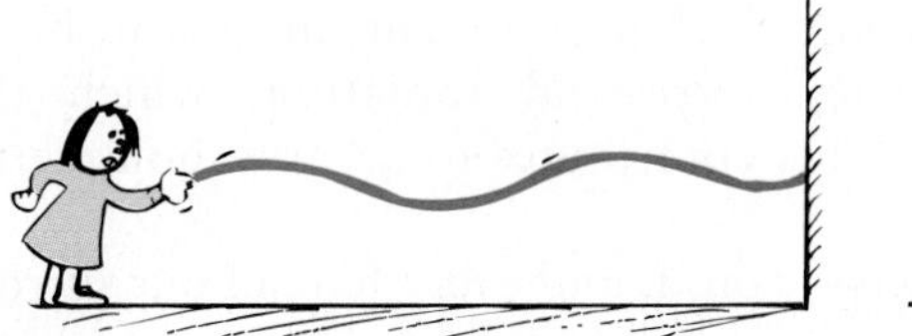
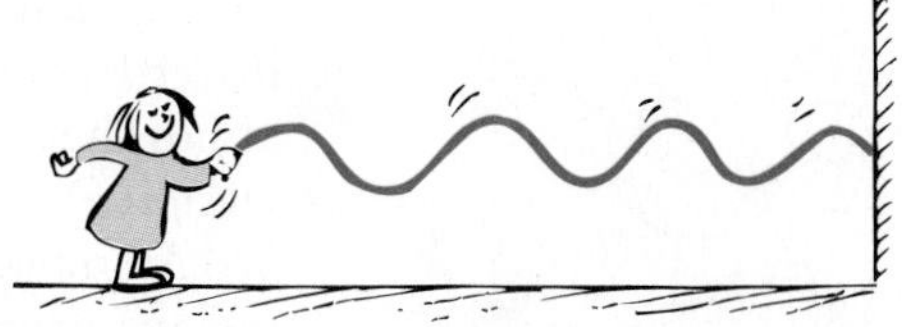

FIGURE 16.11
A wave of long wavelength is produced when the rope is shaken gently (at a low frequency). When it is shaken more vigorously (at a high frequency), a wave of shorter wavelength is produced.

The wavelength of radiation is related to the *frequency* of radiation. Frequency is the rate of vibration of a wave. The girl in Figure 16.11 shakes a rope at a low frequency (left), and a higher frequency (right). Note that the low-frequency shake produces a long, lazy wave, and the higher-frequency one produces shorter waves. Likewise with electromagnetic waves. We will see, in Chapter 26, that vibrating electrons emit electromagnetic waves. High-frequency vibrations produce short waves and low-frequency vibrations produce longer waves.

Emission of Radiant Energy

All substances at any temperature above absolute zero emit radiant energy. The peak frequency $\bar{f}$ of the radiant energy is directly proportional to the absolute (Kelvin) temperature T of the emitter (Figure 16.12):

$$\bar{f} \sim T$$

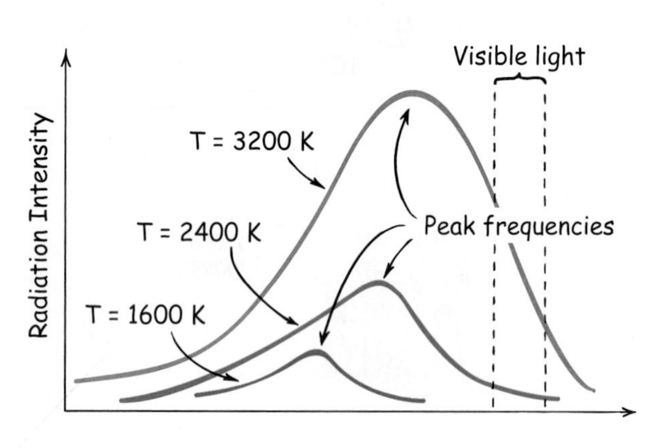

FIGURE 16.12
Interactive Figure
Radiation curves for different temperatures. The peak frequency of radiant energy is directly proportional to the absolute temperature of the emitter.

The surface of the Sun has a high temperature (by earthly standards) and therefore emits radiant energy at a high frequency—much of it in the visible portion of the electromagnetic spectrum. The surface of the Earth, by comparison, is relatively cool, and so the radiant energy it emits has a frequency lower than that of visible light. The radiation emitted by Earth is in the form of infrared waves, which are below our threshold of sight. Radiant energy emitted by Earth is called **terrestrial radiation.**

Most people know that the Sun glows and emits radiant energy, and many educated people know that the source of the Sun's radiant energy involves nuclear reactions in its deep interior. However, relatively few people know that Earth similarly glows (with terrestrial radiation) largely because of nuclear reactions. These nuclear reactions are the simple radioactive decay of uranium and other elements in the Earth's interior. A very different kind of nuclear reaction, thermonuclear fusion, energizes the Sun. (We'll treat radioactive decay in Chapter 33 and thermonuclear fusion in Chapter 34.)

Visit any deep mine and you'll find it's warm down there—year-round. This is ultimately due to radioactivity in the Earth's interior. Much of this heat conducts to the surface, where it is radiated as terrestrial radiation. So radiant energy is emitted by both the Sun and the Earth. The principal difference is that the Sun emits far more energy, and at a higher frequency. We'll learn shortly how the atmosphere is transparent to the high-frequency solar radiation, which passes right through it, but is opaque to much of the lower-frequency terrestrial radiation, which therefore remains in the atmosphere. This contributes to a "greenhouse effect" and, likely, to global warming.

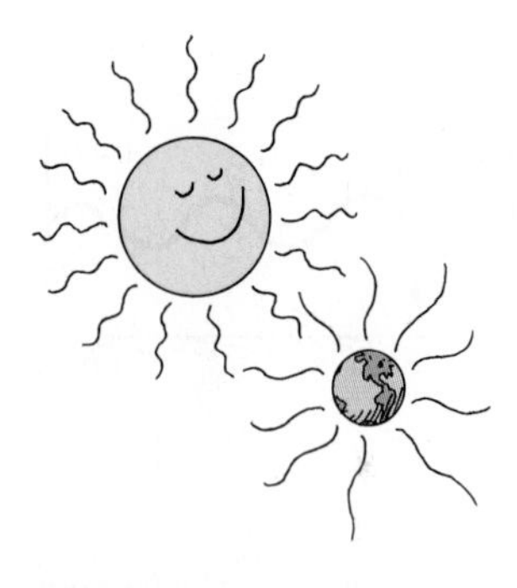

FIGURE 16.13
Both the Sun and the Earth emit the same kind of radiant energy. The Sun's glow is visible to the eye; Earth's glow is of longer waves and so is not visible to the eye.

All objects—you, I, and everything in our surroundings—continually emit radiant energy in a mixture of frequencies and corresponding wavelengths. High-temperature objects like the Sun emit high-frequency waves of short wavelengths, as well as lower-frequency waves in the long-wavelength end of the infrared region. Infrared waves absorbed by our skin produce the sensation of heat. Hence, infrared radiation is often called *heat radiation.* Common sources that give the sensation of heat are the burning embers in a fireplace, a lamp filament, and the Sun. All of these emit infrared radiation in addition to visible light. When this radiant energy encounters an object, it is partly reflected and partly absorbed. The part that is absorbed increases the internal energy of the object. If this object happens to be your skin, you feel the radiation as warmth.

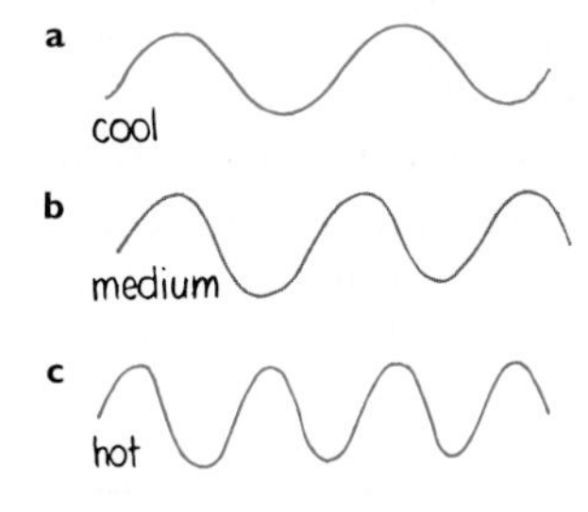

FIGURE 16.14
(a) A low-temperature (cool) source emits primarily low-frequency, long-wavelength waves. (b) A medium-temperature source emits primarily medium-frequency, medium-wavelength waves. (c) A high-temperature (hot) source emits primarily high-frequency, short-wavelength waves.

When an object is hot enough, some of the radiant energy emitted is in the range of visible light. The glow of flowing lava is a good example. At a temperature of about 500°C, lava (or anything else) emits the longest waves we can see—low-frequency red light. Where the temperatures are higher, we see a yellowish light—a blend of red and higher frequencies of light. At even higher temperatures, beginning at about 1200°C, the blends produce white light (more about this in Chapter 27): All the different waves to which the eye is sensitive are emitted, and we see the hot object as "white hot." The filament of an incandescent lamp is at least 1200°C when it emits white light, and more commonly reaches about 2500°C.

CHECK YOURSELF

Do any of the following *not* give off radiant energy? (a) The Sun. (b) Lava from a volcano. (c) Red-hot coals. (d) This book that you're reading.

Absorption of Radiant Energy

If everything is emitting energy, why doesn't everything finally run out of it? The answer is that everything is also absorbing energy. Good emitters of radiant energy are also good absorbers; poor emitters are poor absorbers. For example, a radio antenna constructed to be a good emitter of radio waves is also, by its very design, a good receiver (absorber) of them. A poorly designed transmitting antenna is also a poor receiver.

It's interesting to note that, if a good emitter were not also a good absorber, black objects would remain warmer than lighter-colored objects and the two would never reach a common temperature. Objects in thermal contact, given sufficient time, reach the same temperature. A blacktop pavement and dark automobile body may remain hotter than their surroundings on a hot day, but, at nightfall, these dark objects cool faster! Sooner or later, all objects come to thermal equilibrium. So a dark object that absorbs a lot of radiant energy must emit a lot as well.

FIGURE 16.15
When the containers are filled with hot (or cold) water, the blackened one cools (or warms) faster.

You can verify this with a pair of metal containers of the same size and shape, one having a white or mirrorlike surface and the other having a blackened surface (Figure 16.15). Fill the containers with hot water and place a thermometer in each. You will find that the black container cools faster. The blackened surface is a better emitter. Coffee or tea stays hot longer in a shiny pot than in a blackened one. The same experiment can be done in reverse. This time, fill each container with ice water and place the containers in front of a fireplace or outside on

Everything around you both radiates and absorbs energy continuously!

Insights

CHECK YOUR ANSWER

Let's hope you didn't answer (d), the book. Why? Because the book, like the other substances listed, has temperature—though not as much. According to the rule $\bar{f} \sim T$, it therefore emits radiation whose peak frequency $\bar{f}$ is quite low compared with the radiation frequencies emitted by the other substances. Everything with any temperature above absolute zero emits electromagnetic radiation. That's right—*everything*!

a sunny day—wherever there is a good source of radiant energy. You'll find that the black container warms up faster. An object that emits well also absorbs well.

Whether a surface plays the role of net emitter or net absorber depends on whether its temperature is above or below that of its surroundings. If it's hotter than its surroundings, the surface will be a net emitter and will cool. If it's colder than its surroundings, it will be a net absorber and become warmer. Every surface, hot or cold, both absorbs and emits radiant energy.

A hot pizza placed outside on a winter day is a net emitter. The same pizza placed in a hotter oven is a net absorber.

Insights

CHECK YOURSELF

1. If a good absorber of radiant energy were a poor emitter, how would its temperature compare with the temperature of its surroundings?
2. A farmer turns on the propane burner in his barn on a cold morning and heats the air to 20°C (68°F). Why does he still feel cold?

Reflection of Radiant Energy

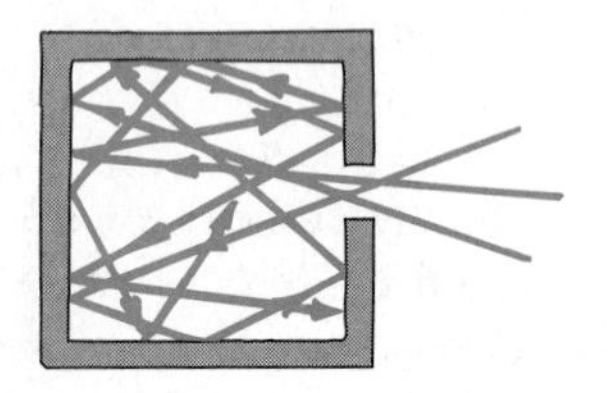

FIGURE 16.16
Radiation that enters the cavity has little chance of leaving because most of it is absorbed. For this reason, the opening to any cavity looks black to us.

Absorption and reflection are opposite processes. A good absorber of radiant energy reflects very little radiant energy, including visible light. Hence, a surface that reflects very little or no radiant energy looks dark. So a good absorber appears dark and a perfect absorber reflects no radiant energy and appears completely black. The pupil of the eye, for example, allows light to enter with no reflection, which is why it appears black. (An exception occurs in flash photography when pupils appear pink, which occurs when very bright light is reflected off the eye's inner pink surface and back through the pupil.)

Look at the open ends of pipes in a stack; the holes appear black. Look at open doorways or windows of distant houses in the daytime, and they, too, look black. Openings appear black because the light that enters them is reflected back and forth on the inside walls many times and is partly absorbed at each reflection. As a result, very little or none of the light remains to come back out of the opening and travel to your eyes (Figure 16.16).

Good reflectors, on the other hand, are poor absorbers. Clean snow is a good reflector and therefore does not melt rapidly in sunlight. If the snow is dirty, it absorbs radiant energy from the Sun and melts faster. Dropping black soot from an aircraft onto snow-covered mountains in late winter is a technique sometimes used to prevent flooding in the spring. The soot causes the snow to start melting earlier than it normally would. This spreads the runoff over a longer period of time. So, we get a steady flow of snow melt rather than a sudden runoff all at once when the outside temperatures start increasing.

Emission and absorption in the visible part of the spectrum are affected by color. But emission and absorption in the infrared part of the spectrum are more affected by surface texture. A dull finish emits/absorbs better in the infrared than a polished one, whatever its color.

Insights

CHECK YOUR ANSWERS

1. If a good absorber were not also a good emitter, there would be a net absorption of radiant energy and the temperature of the absorber would remain higher than the temperature of the surroundings. Things around us approach a common temperature only because good absorbers are, by their very nature, also good emitters.
2. The walls of the barn are still cold. The farmer radiates more energy to the walls than the walls radiate back, and he is chilled. (Inside your home or your classroom, you are comfortable only if the walls are warm, not just the air.)

CHALLENGE

Is it more efficient to paint the radiators in your home black or silver?

FIGURE 16.17
The hole in the box that Helen holds looks perfectly black and seems to indicate a black interior when, in fact, the interior has been painted a bright white.

Cooling at Night by Radiation

Bodies that radiate more energy than they receive become cooler. This happens at night when solar radiation is absent. An object left out in the open at night radiates energy into space and, because of the absence of any warmer bodies in the vicinity, receives very little energy from space in return. Thus, it gives out more energy than it receives and becomes cooler. But if the object is a good conductor of heat—such as metal, stone, or concrete—heat conducts to it from the ground, which somewhat stabilizes its temperature. On the other hand, materials such as wood, straw, and grass are poor conductors, and little heat is conducted into them from the ground. These insulating materials are net radiators and get *colder than the air*. It is common for frost to form on these kinds of materials even when the temperature of the air does not go down to freezing. Have you ever seen a frost-covered lawn or field on a chilly but above-freezing morning before the Sun is up? The next time you see this, notice that the frost forms only on the grass, straw, or other poor conductors, while none forms on the cement, stone, or other good conductors.

Hard-core gardeners will cover their favorite plants with a tarp when they expect a frost. The plants radiate just as before, but now they are receiving radiant energy from the tarp rather than from the dark night sky. Because the tarp radiates as an object at the temperature of the surroundings rather than at the temperature of the cold, dark sky, frost doesn't form on the plants' leaves. This is the same reason that plants on a covered porch won't have frost on them, whereas plants exposed to the open sky will.

Earth itself exchanges radiation with its surroundings. The Sun is a dominant part of Earth's surroundings during the day. The sunlit half of the Earth absorbs more radiant energy than it emits. At night, if the air is relatively transparent, Earth radiates more energy to deep space than it gets back. As the Bell Laboratories researchers Arno Penzias and Robert Wilson learned in 1965, outer space has a temperature—about 2.7 K (2.7 degrees above absolute zero). Space itself emits weak radiation characteristic of that low temperature.[3]

WHAT'S YOUR THINKING?

Most of the heat provided by a heating radiator is accomplished by convection, so color makes little difference (a better name for this type of heater would be a *convector*). For optimum efficiency, however, a silver-painted radiator will radiate less, become and remain hotter, and do a better job of heating the air. Go silver!

[3]Penzias and Wilson shared a Nobel Prize for this discovery, deemed to be a relic of the Big Bang. By studying this "cosmic background radiation," scientists are learning much about the early history of the universe and its present shape.

FIGURE 16.18
Patches of frost crystals betray the hidden entrances to mouse burrows. Each cluster of crystals is frozen mouse breath!

CHECK YOURSELF

1. Which is likely to be colder—a night when you can see the stars, or a night when you cannot?
2. In winter, why does the road surface on a bridge tend to be more icy than the road surfaces at either end of the bridge?

Newton's Law of Cooling

FIGURE 16.19
The long stem of a wine glass helps to prevent heat from the hand from warming the wine.

An object at a different temperature from its surroundings will ultimately come to a common temperature with its surroundings. A relatively hot object cools as it warms its surroundings; a cool object warms as it cools its surroundings. When considering how quickly (or slowly) something cools, we speak of its *rate* of cooling—how many degrees' change in temperature per unit of time.

The rate of cooling of an object depends on how much hotter the object is than its surroundings. The temperature change per minute of a hot apple pie will be more if the hot pie is put in a cold freezer than if it is placed on the kitchen table. When the pie cools in the freezer, the temperature difference between it and its surroundings is greater. On a cold day, a warm home will leak heat to the outside at a greater rate when there is a large difference between the inside and outside temperatures. Keeping the inside of your home at a high temperature on a cold day is more costly than keeping it at a lower temperature. If you keep the temperature difference small, the rate of cooling will be correspondingly low.

CHECK YOUR ANSWERS

1. It is colder on the starry night, when the Earth's surface radiates directly to frigid deep space. On a cloudy night, net radiation is less, because the clouds radiate energy back to the Earth's surface.
2. Energy radiated by roads on land is partly replenished by heat conducted from the warmer ground below the pavement. But there's an absence of thermal contact between the road surfaces of bridges and the ground, so they receive very little, if any, replenishing energy conducted from the ground. This is why road surfaces on bridges get colder than roads on land, which increases the chance of ice formation on bridges. Understanding heat transfer can make you a safer driver!

Interestingly, Newton's law of cooling is an empirical relationship, and not a fundamental law like Newton's laws of motion.

Insights

The rate of cooling of an object—whether by conduction, convection, or radiation—is approximately proportional to the temperature difference ΔT between the object and its surroundings.

$$\text{Rate of cooling} \sim \Delta T$$

This is known as **Newton's law of cooling.** (Guess who is credited with discovering this?)

The law holds also for warming. If an object is cooler than its surroundings, its rate of warming up is also proportional to ΔT.[4] Frozen food will warm up faster in a warm room than in a cold room.

The rate of cooling we experience on a cold day can be increased by the added convective effect of the wind. We speak of this in terms of *wind chill.* For example, a wind chill of −20°C means we are losing heat at the same rate as if the temperature were −20°C without wind.

CHECK YOURSELF

Since a hot cup of tea loses heat more rapidly than a lukewarm cup of tea, would it be correct to say that a hot cup of tea will cool to room temperature before a lukewarm cup of tea will?

The Greenhouse Effect

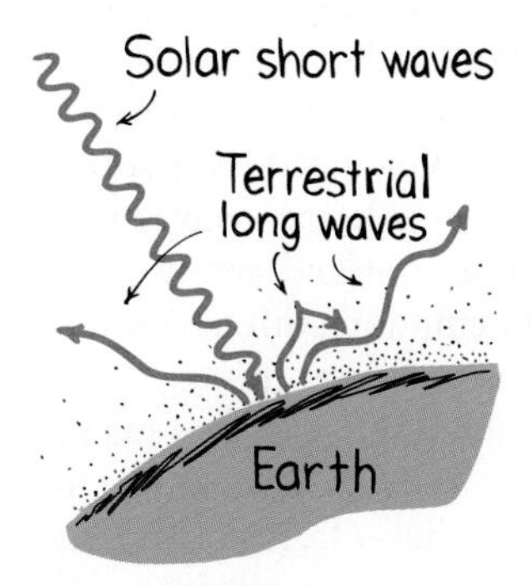

FIGURE 16.20
Interactive Figure
The hot Sun emits short waves, and the cool Earth emits long waves—terrestrial radiation. Water vapor, carbon dioxide, and other "greenhouse gases" in the atmosphere retain heat that would otherwise be radiated from Earth into space.

Earth and its atmosphere gain energy when they absorb radiant energy from the Sun. This warms the surface of the Earth, which, in turn, emits terrestrial radiation, much of which escapes to outer space. Absorption and emission continue at equal rates to produce an average equilibrium temperature. Over the last 500,000 years, the average temperature of the Earth has fluctuated between 19°C and 27°C, and is presently at the high point, 27°C. Earth's temperature increases when either the radiant energy coming in increases or there is a decrease in the escape of terrestrial radiation.

The **greenhouse effect** is the warming of the lower atmosphere, the effect of atmospheric gases on the balance of terrestrial and solar radiation. Because

CHECK YOUR ANSWER

No! Although the rate of cooling is greater for the hotter cup, it has farther to cool to reach thermal equilibrium. The extra time is equal to the time it takes to cool to the initial temperature of the lukewarm cup of tea. Cooling *rate* and cooling *time* are not the same thing.

[4] A warm object that contains a *source* of energy may remain warmer than its surroundings indefinitely. The heat it emits doesn't necessarily cool it, and Newton's law of cooling doesn't apply. Thus, an automobile engine that is running remains warmer than the automobile's body and the surrounding air. But, after the engine is shut off, it cools in accordance with Newton's law of cooling and gradually approaches the same temperature as its surroundings. Likewise, the Sun will remain hotter than its surroundings as long as its nuclear furnace is running—another five billion years or so.

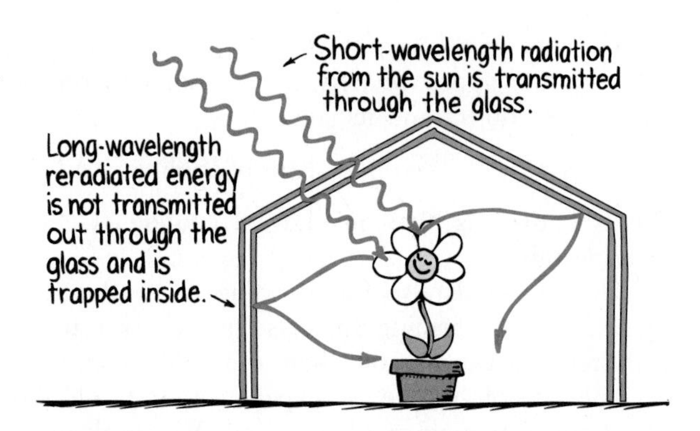

Short-wavelength radiation from the sun is transmitted through the glass.
Long-wavelength reradiated energy is not transmitted out through the glass and is trapped inside.

FIGURE 16.21
Glass is transparent to short-wavelength radiation but opaque to long-wavelength radiation. Reradiated energy from the plant is of long wavelength because the plant has a relatively low temperature.

Volcanoes put more particulate matter into the atmosphere, by far, than industries and all human activity.

Insights

of the Sun's high temperature, high-frequency waves make up solar radiation—ultraviolet, visible light, and short-wavelength infrared waves. The atmosphere is transparent to much of this radiation, especially the visible light, so solar energy reaches the Earth's surface and is absorbed. The Earth's surface, in turn, reradiates part of this energy. But since Earth's surface is relatively cool, it reradiates the energy at low frequencies—mainly long-wavelength infrared. Certain atmospheric gases (mainly water vapor and carbon dioxide) absorb and reemit much of this long-wave radiation back to Earth. So the long-wave radiation that doesn't escape Earth's atmosphere helps to keep Earth warm. This process is very nice, because Earth would be a frigid −18°C otherwise. Our present environmental concern is that excess carbon dioxide and other "greenhouse gases" (from fossil-fuel combustion and other industrial processes) will trap too much energy and make the Earth too warm.

The atmospheric greenhouse effect gets its name from the glass structures used by farmers and florists to "trap" solar energy. Glass is transparent to waves of visible light but opaque to ultraviolet and infrared waves. Glass acts as a sort of one-way valve. It allows visible light to enter, but it prevents longer waves from leaving. So short waves of sunlight enter through the glass roof and are absorbed by the soil and plants inside. The soil and plants, in turn, emit long infrared waves. This energy cannot penetrate the glass, and the greenhouse warms up.

Interestingly enough, in the farmers' or florists' greenhouse, heating is mainly due to the ability of glass to prevent convection currents from combining the cooler outside air with the warmer inside air. The greenhouse effect plays a bigger role in the warming of the Earth than in the warming of greenhouses.

CHECK YOURSELF

1. What eventually happens to the solar energy that falls on Earth?
2. What does it mean to say that the greenhouse effect is like a one-way valve?

CHECK YOUR ANSWERS

1. Sooner or later, it will be radiated back into space. Energy is always in transit—you can rent it, but you can't own it.
2. The transparent material—atmosphere for Earth and glass for the greenhouse—passes only incoming short waves and blocks outgoing long waves. As a result, radiant energy is trapped within the "greenhouse."

Solar Power

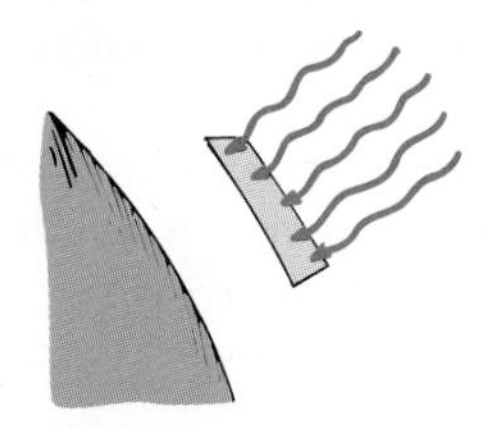

FIGURE 16.22
Over each square meter of area perpendicular to the Sun's rays at the top of the atmosphere, the Sun pours 1400 J of radiant energy each second. Hence the solar constant is 1.4 kJ/s/m^2, or 1.4 kW/m^2.

If you step from the shade into the sunshine, you're noticeably warmed. The warmth you feel isn't so much because the Sun is hot, for its surface temperature of 6000°C is no hotter than the flames of some welding torches. We are warmed principally because the Sun is so *big.*[5] As a result, it emits enormous amounts of energy, less than one part in a billion of which reaches Earth. Nonetheless, the amount of radiant energy received each second over each square meter at right angles to the Sun's rays at the top of the atmosphere is 1400 joules (1.4 kJ). This input of energy is called the **solar constant.** This is equivalent, in power units, to 1.4 kilowatts per square meter (1.4 kW/m^2). The amount of **solar power** that reaches the ground is attenuated by the atmosphere and reduced by nonperpendicular elevation angles of the Sun. Also, of course, it ceases at night. The solar power received in the United States, averaged over day and night, summer and winter, is about 13% of the solar constant (0.18 kW/m^2). This amount of power, falling on the roof area of a typical American house, is twice the power needed to comfortably heat and cool the house year-round. More and more homes are using solar energy for space heating and water heating. Also gaining in popularity are photovoltaic shingles used in roofing buildings.

Solar heating needs a distribution system to move solar energy from the collector to the storage or living space. When the distribution system requires external energy to operate fans or pumps, we have an active system. When the distribution is by natural means (conduction, convection, or radiation), we have a passive system. At the present time, passive systems are essentially problem-free and serve as an economical supplement to conventional heating—even in the northern parts of the United States and in Canada.

FIGURE 16.23
A simple but effective use of solar power.

On a larger scale, the problems of utilizing solar power to generate electricity are greater. First, there is the fact that no energy arrives at night. This calls for supplemental sources of energy or efficient energy-storage devices. Variations in weather, particularly in cloud cover, produce a variable energy supply from day to day and from season to season. Even in clear daylight hours, the Sun is high in the sky for only part of the day. At the time of this writing, solar-energy collecting and concentration systems, whether arrays of mirrors or photovoltaic cells, are not yet competitive in cost with electrical power generated by conventional power sources. Projections indicate that the story may be different later in the twenty-first century.

FIGURE 16.24
Higher-tech solar water heaters are covered with glass to provide a greenhouse effect, which further heats the water. Why are the collectors a dark color?

[5]To visualize how big the Sun is, realize that its diameter is more than three times the distance between the Earth and the Moon. So if the Earth and the Moon were inside the Sun, with Earth at the solar center, the Moon would be deep inside. The Sun is *really* big!

PRACTICING PHYSICS

Is it the distance from the Sun or the angle of the Sun's rays on Earth that accounts for frigid polar regions and tropical equatorial regions? You can see the answer for yourself if you hold a flashlight above a surface and note its brightness. When the light strikes perpendicularly, light energy is concentrated; but, when the surface is tipped, keeping the same distance, incident light is more spread out. Can you see that the same energy over a larger area is akin to the low temperatures of the Arctic and Antarctic regions of the Earth?

The sketch below is of the Earth with parallel rays of light coming from the Sun. Count the number of rays that strike Region A and equal-area Region B. Where is the energy per unit area less? How does this relate to climate?

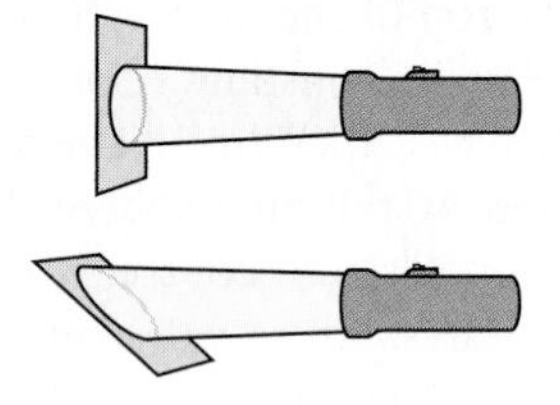

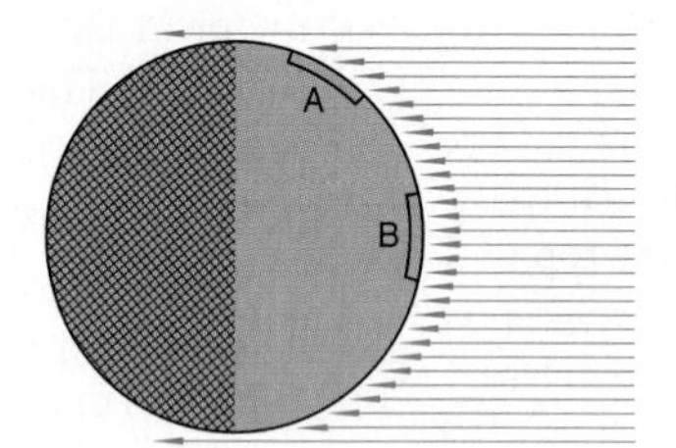

Controlling Heat Transfer

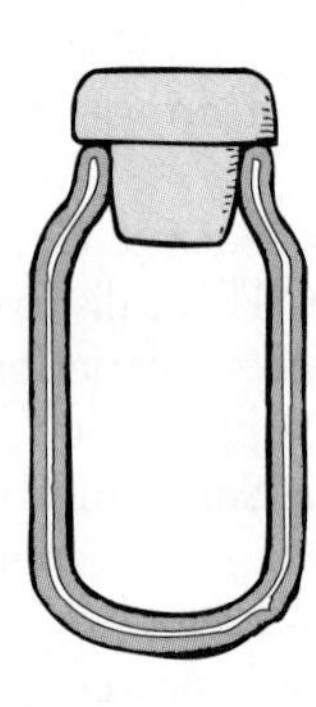

FIGURE 16.25
A Thermos bottle.

A nice way to review the methods of heat transfer is by considering a device that inhibits all three methods: the vacuum bottle. A vacuum bottle (which you may know by the trademark name Thermos) consists of a double-walled glass container with a vacuum between the walls. (There is usually an outer covering as well.) The glass surfaces that face each other are silvered. A close-fitting stopper made of cork or plastic seals the bottle. Any liquid in a vacuum bottle—hot or cold—will remain close to its original temperature for many hours.

1. Heat transfer by *conduction* through the vacuum is impossible. Some heat escapes by conduction through the glass and stopper, but this is a slow process, as glass, plastic, and cork are poor conductors.
2. The vacuum has no fluid to convect, so there is no heat loss through the walls by *convection.*
3. Heat loss by *radiation* is reduced by the silvered surfaces of the walls, which reflect heat waves back into the bottle.

Summary of Terms

Conduction The transfer of heat energy by molecular and electron collisions within a substance (especially a solid).

Convection The transfer of heat energy in a gas or liquid by means of currents in the heated fluid. The fluid moves, carrying energy with it.

Radiation The transfer of energy by means of electromagnetic waves.

Terrestrial radiation The radiation emitted by the Earth to outer space.

Newton's law of cooling The rate of loss of heat from an object is proportional to the temperature difference between the object and its surroundings.

Greenhouse effect Warming of the lower atmosphere by short-wavelength radiation from the Sun that penetrates the atmosphere, is absorbed by the Earth, and is reradiated at longer wavelengths that cannot easily escape Earth's atmosphere.

Solar constant 1400 J/m^2 received from the Sun each second at the top of Earth's atmosphere on an area perpendicular to the Sun's rays; expressed in terms of power, 1.4 kW/m^2.

Solar power Energy per unit time derived from the Sun.

Review Questions

1. What are the three common ways in which heat is transferred?

Conduction

2. What is the role of "loose" electrons in heat conductors?
3. If you touch the metal sides inside of a hot pizza oven with a bare hand, you're in trouble. But hold your hand briefly in the oven air and you're okay. What does this tell you about the relative conductivities of metal and air?
4. If you step quickly on red-hot coals with bare feet, you're probably okay. What is the explanation?
5. Why are such materials as wood, fur, feathers, and even snow good insulators?
6. Does a good insulator prevent heat from getting through it, or does it simply slow its passage?

Convection

7. How is heat transferred from one place to another by convection?
8. What happens to the volume of air as it rises? What happens to its temperature?
9. When an air molecule is hit by an approaching faster-moving molecule, does its rebound speed increase or decrease? How about when it hits a receding molecule?
10. How are the speeds of molecules of air affected when the air is compressed by the action of a tire pump?
11. How are the speeds of molecules of air affected when the air rapidly expands?
12. Why is Millie's hand not burned when she holds it above the escape valve of the pressure cooker (Figure 16.8)?
13. Why does the direction of coastal winds change from day to night?

Radiation

14. In what form does radiant energy travel?
15. Relatively speaking, do high-frequency waves have long wavelengths or short wavelengths?

Emission of Radiant Energy

16. How does the frequency of radiant energy relate to the absolute temperature of the radiating source?
17. What is *terrestrial radiation*?
18. Cite a primary difference between waves of solar radiation and waves of terrestrial radiation.
19. What is *heat radiation*?

Absorption of Radiant Energy

20. Since all objects emit energy to their surroundings, why don't the temperatures of all objects continuously decrease?
21. What determines whether an object is a net absorber or a net emitter of radiant energy at a given time?
22. Which will normally warm faster—a black pot of cold water or a silvered pot of cold water? Explain.

Reflection of Radiant Energy

23. Can an object be both a good absorber and a good reflector at the same time? Why, or why not?
24. Why does the pupil of the eye appear black?

Cooling at Night by Radiant Energy

25. What happens to the temperature of something that radiates energy without absorbing the same amount in return?
26. An object radiating energy at night is in contact with the relatively warm Earth. How does its conductivity affect whether or not it becomes appreciably colder than the air?

Newton's Law of Cooling

27. If you want a room-temperature can of beverage to cool quickly, should you put it in the freezer compartment or in the main part of your refrigerator? Or does it matter?
28. Which will undergo the greater rate of cooling—a red-hot poker in a warm oven or a red-hot poker in a cold room (or do both cool at the same rate)?
29. Does Newton's law of cooling apply to warming as well as to cooling?

The Greenhouse Effect

30. What would be the consequence of completely eliminating the greenhouse effect?
31. In what way does glass act like a one-way valve for a conventional greenhouse? Does the atmosphere play the same role?

Solar Power

32. What is the solar constant, and would it be greater in Maine or in Florida?

Controlling Heat Transfer

33. Cite three ways in which a Thermos bottle inhibits heat transfer.

Projects

1. Wrap a piece of paper around a thick metal bar and place it in a flame. Notice that the paper will not catch fire. Can you explain this in terms of the conductivity of the metal bar? (Paper generally will not ignite until its temperature reaches about 230°C.)

2. Turn a common incandescent lamp on and off quickly while holding your hand a few inches from the bulb. You feel its heat, but, when you touch the bulb, it isn't hot. Can you explain this in terms of radiant energy and the bulb's transparency?
3. Write a letter to Grandma and share your knowledge about why air temperature is cooler on clear nights and warmer on cloudy nights. With reasoned examples, convince her that all things are continually emitting energy—and absorbing energy.

Exercises

1. On a cold day, why does a metal doorknob feel colder than the wooden door?
2. What is the explanation for feather beds being warm?
3. Wrap a fur coat around a thermometer. Will the temperature rise?
4. If 70°F air feels warm and comfortable to us, why does 70°F water feel cool when we swim in it?
5. At what common temperature will a block of wood and a block of metal both feel neither hot nor cold to the touch?
6. If you hold one end of a metal nail against a piece of ice, the end in your hand soon becomes cold. Does cold flow from the ice to your hand? Explain.
7. What is the purpose of a layer of copper or aluminum on the bottom of stainless steel cookware?
8. In terms of physics, why do restaurants serve baked potatoes wrapped in aluminum foil?
9. Many tongues have been injured by licking a piece of metal on a very cold day. Why would no harm result if a clean piece of wood were licked on the same day?
10. Wood is a better insulator than glass, yet fiberglass is commonly used as an insulator in wooden buildings. Explain.
11. Visit a snow-covered cemetery and note that the snow does not slope upward against the gravestones. Instead, it forms depressions as shown. Can you think of a reason for this?

12. Why are mittens warmer than gloves on a cold day?
13. If you were caught in freezing weather with only your own body heat as a source, would you be warmer in an Arctic igloo or in a wooden shack? Defend your answer.
14. What physics is involved in explaining why you can safely hold your bare hand in a hot pizza oven for a few seconds, but, if you momentarily touch the metal inside, you'll burn yourself?
15. If you touch a piece of metal that is 200°C (392°F)—OUCH! But placing your bare hand in 200°C air for a short period of time, such as in a hot oven, does not result in pain. What is your explanation?
16. Wood conducts heat very poorly—it has a very low conductivity. Does wood still have a low conductivity if it is hot? Could you quickly and safely grasp the wooden handle of a pan from a hot oven with your bare hand? Although the pan handle is hot, is much heat conducted from it to your hand if grasped briefly? Why would it be a poor idea to do the same with an iron handle? Explain.
17. Does wood have a low conductivity if it is very hot—that is, in the stage of smoldering, red-hot coals? Could you safely walk across a bed of red-hot wooden coals with bare feet? Although the coals are hot, does much heat conduct from them to your feet if you step quickly? Could you do the same on red-hot iron coals? Explain. (Caution: Coals can stick to your feet, so—OUCH!—don't try it!)
18. Is it possible for heat to flow between two objects with the same internal energy? Can heat flow from an object with less internal energy to one with more internal energy? Defend your answers.
19. When two cups of hot chocolate, one at 55°C and the other at 60°C, are poured into a bowl, why will the temperature of the mixture be between 55°C and 60°C?
20. Why is it incorrect to say that, when a hot object warms a cold object, temperature flows between them?

21. Why is it incorrect to say that, when a hot object warms a cold one, the increase in temperature of the cold one is equal to the decrease in temperature of the hot one? When is this statement correct?

22. A friend says that the molecules in a mixture of gas in thermal equilibrium have the same average kinetic energy. Do you agree or disagree? Explain.

23. Your friend states that the average speed of all hydrogen and nitrogen molecules in a gas is the same. Do you agree or disagree, and why?

24. Why would you not expect all the molecules of air in your room to have the same average speed?

25. In a mixture of hydrogen and oxygen gases at the same temperature, which molecules move faster? Why?

26. One container is filled with argon gas and the other with krypton gas. If both gases have the same temperature, in which container are the atoms moving faster? Why?

27. Which atoms, on average, move slower in a mixture, U-238 or U-235? How would this affect diffusion through a porous membrane of otherwise identical gases made from these isotopes?

28. Solid uranium can be converted chemically to uranium fluoride, UF_6, which can be cooked up into a dense vapor that diffuses through a porous barrier. Which is likely to diffuse at a greater rate, a gas with isotopes U-235 or U-238?

29. Consider two equal-size rooms connected by an open door. One room is maintained at a higher temperature than the other one. Which room contains more air molecules?

30. Some ceiling fans are reversible so that they drive air down or pull it up. In which direction should the fan drive the air during winter? In which direction for summer?

31. In a still room, smoke from a candle will sometimes rise only so far, not reaching the ceiling. Explain why.

32. Why does helium, released into the atmosphere, eventually disappear into space?

33. Ice cubes float in a glass of iced tea. Why would cooling be less if the cubes were instead on the bottom of the drink?

34. The ratio of oxygen molecules to nitrogen molecules in the atmosphere decreases with increasing altitude. Why?

35. Release a single molecule in an evacuated region and it will fall as fast as, and no differently from, a baseball released in the same region. Explain.

36. Isn't it true that gravity tends to decrease upward movement of air and to increase downward movement? And isn't it true that air density is always less above any point in air than below, providing an upward "migration window"? Make a hypothesis about how these two opposite effects affect the air.

37. What does the high specific heat of water have to do with convection currents in the air at the seashore?

38. If we warm a volume of air, it expands. Does it then follow that, if we expand a volume of air, it warms? Explain.

39. How could you change the drawing in Figure 16.7 to make it illustrate the heating of air when it is compressed? Make a sketch of this case.

40. A snow-making machine used for ski areas blows a mixture of compressed air and water through a nozzle. The temperature of the mixture may initially be well above the freezing temperature of water, yet crystals of snow are formed as the mixture is ejected from the nozzle. Explain how this happens.

41. More radiant energy is emitted by a dull-black steam radiator than by a silver one. However, silver is the most efficient color for a steam radiator that heats a room. Why?

42. In which form of heat transfer is a medium not required?

43. Turn an incandescent lamp on and off quickly while you are standing near it. You can feel its heat, but you find when you touch the bulb that it is not hot. Explain why you felt heat from it when you first turned it on.

44. Why does a good *emitter* of heat radiation appear black at room temperature?

45. A number of bodies at different temperatures placed in a closed room share radiant energy and ultimately reach a common temperature. Would this thermal equilibrium be possible if good absorbers were poor emitters and poor absorbers were good emitters? Explain.

46. From the rules that a good absorber of radiation is a good radiator and a good reflector is a poor absorber, state a rule relating the reflecting and radiating properties of a surface.

47. Since energy is radiated by all objects, why can't we see them in the dark?

48. The heat of volcanoes and natural hot springs comes from trace amounts of radioactive minerals in common rock in the Earth's interior. Why isn't the same kind of rock at the Earth's surface warm to the touch?

49. Suppose that, at a restaurant, you are served coffee before you are ready to drink it. In order that it be hottest when you are ready for it, would you be wiser to add cream to it right away or when you are ready to drink it?

50. Even though metal is a good conductor, frost can be seen on parked cars in the early morning even when the air temperature is above freezing. Can you explain this?
51. When there is morning frost on the ground in an open park, why is frost unlikely to be found on the ground directly beneath park benches?
52. Why is whitewash sometimes applied to the glass of florists' greenhouses in the summer?
53. On a very cold sunny day, you wear a black coat and a transparent plastic coat. Which should be worn on the outside for maximum warmth?
54. If the composition of the upper atmosphere were changed so that it permitted a greater amount of terrestrial radiation to escape, what effect would this have on Earth's climate?
55. Is it important to convert temperatures to the Kelvin scale when we use Newton's law of cooling? Why or why not?
56. If you wish to save fuel and you're going to leave your warm house for a half hour or so on a very cold day, should you turn your thermostat down a few degrees, turn it off altogether, or let it remain at the room temperature you desire?
57. If you wish to save fuel and you're going to leave your cool house for a half hour or so on a very hot day, should you turn your air conditioning thermostat up a bit, turn it off altogether, or let it remain at the room temperature you desire?
58. Why is the insulation in an attic commonly thicker than the insulation in the walls of a house?
59. As more energy from fossil fuels and other nonrenewable fuels is consumed on Earth, the overall temperature of the Earth tends to rise. Regardless of the increase in energy, however, the temperature does not rise indefinitely. By what process is an indefinite rise prevented? Explain your answer.
60. Make up a multiple-choice question that would test a classmate's understanding of the distinction between conduction and convection. Make another in which the term *radiation* is the correct answer.

Problems

1. Will burns a 0.6-g peanut beneath 50 g of water, which increases in temperature from 22°C to 50°C.
   a. Assuming 40% efficiency, what is the food value, in calories, of the peanut?
   b. What is the food value in calories per gram?

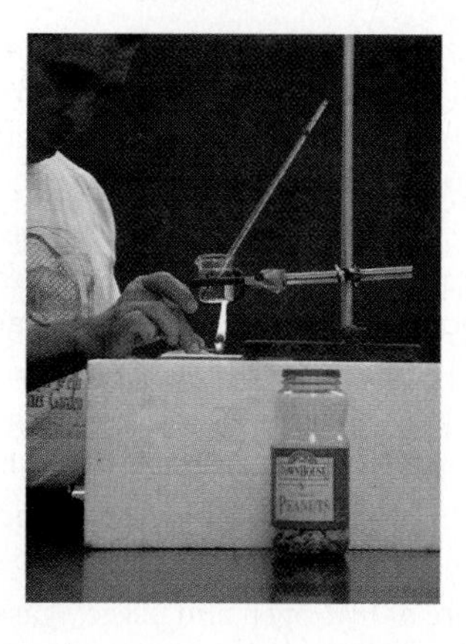

2. Radioactive decay of granite and other rocks in Earth's interior provides sufficient energy to keep the interior molten, to heat lava, and to provide warmth to natural hot springs. This is due to the average release of about 0.03 J per kilogram each year. How many years are required for a chunk of thermally insulated granite to increase in temperature by 500°C? (Assume that the specific heat capacity of granite is 800 J/kg·C°.)
3. Pounding a nail into wood makes the nail warmer. Suppose that a hammer exerts an average force of 500 N on a 6-cm steel nail when it drives into a piece of wood. The nail becomes hotter. If all the heat goes to the nail, calculate its increase in temperature. (Assume that the specific heat capacity of steel is 450 J/kg·C°.)
4. A container of hot water at 80°C cools to 79°C in 15 seconds when it is placed in a 20°C room. Use Newton's law of cooling to estimate the time it will take for the container to cool from 50°C to 49°C, and, still later, to cool from 40°C to 39°C.
5. In a 25°C room, hot coffee in a vacuum flask cools from 75°C to 50°C in eight hours. What will its temperature be after another eight hours?
6. At a certain location, the solar power per unit area reaching the Earth's surface is 200 W/m^2, averaged over a 24-hour day. If you live in a house whose average power requirement is 3 kW and you can convert solar power to electric power with 10% efficiency, how large a collector area will you need to meet all your household energy requirements from solar energy? Will it fit in your yard?

Dean Baird demonstrating Figure 17.15—regelation.

CHAPTER 17

Change of Phase

Matter around us exists in four common *phases* (or *states*). Ice, for example, is the *solid* phase of H_2O. Add energy, and you add motion to the rigid molecular structure, which breaks down to form H_2O in the *liquid* phase, water. Add more energy, and the liquid changes to the *gaseous* phase. Add still more energy, and the molecules break into ions and electrons, giving the *plasma* phase. The phase of matter depends on its temperature and the pressure that is exerted on it. Changes of phase almost always require a transfer of energy.

Evaporation

Water in an open container will eventually evaporate, or dry up. The liquid that disappears becomes water vapor in the air. **Evaporation** is a change of phase from liquid to gas that occurs at the surface of a liquid.

The temperature of any substance is related to the average kinetic energy of its particles. Molecules in liquid water have a wide variety of speeds, moving about in all directions and bumping against one another. At any moment, some move at very high speeds while others move hardly at all. In the next moment, the slowest may become the fastest due to molecular collisions. Some gain kinetic energy while others lose kinetic energy. Molecules at the surface that gain kinetic energy by being bumped from below may have enough energy to break free from the liquid. They can leave the surface and fly into the space above the liquid. In this way, they become molecules of vapor.

The increased kinetic energy of molecules bumped hard enough to break free from the liquid comes from molecules remaining in the liquid. This is "billiard-ball physics": When balls bump into one another and some gain kinetic energy, the others lose the same amount. Molecules about to be propelled out of the liquid are the gainers, while the losers of energy remain in the liquid. Thus the average kinetic energy of the molecules remaining in the liquid is lowered—evaporation is a cooling process. Interestingly, the fast molecules that break free from the surface are slowed as they fly away due to their attraction to the surface. So although the water is cooled by evaporation, the air above is not correspondingly warmed by the process.

FIGURE 17.1
When wet, the cloth covering on the sides of the canteen promotes cooling. The fastest-moving water molecules evaporate from the wet cloth, decreasing its temperature, cooling the metal, and, in turn, cooling the water within.

The canteen shown in Figure 17.1 keeps cool because of evaporation when the cloth that covers the sides is kept wet. As the faster-moving water molecules break free from the cloth, the temperature of the cloth decreases. The cool wet cloth, in turn, cools the metal canteen by conduction, which, in turn, cools the water inside. So energy is transferred from the water in the canteen to the air outside. In this way, the water is cooled appreciably below the air temperature outside.

The cooling effect of evaporation is strikingly evident when some rubbing alcohol is applied to your back. The alcohol evaporates very rapidly, cooling the surface of the body quickly. The more rapid the evaporation, the faster the cooling.

When our bodies overheat, our sweat glands produce perspiration. This is part of nature's thermostat, for the evaporation of perspiration cools us and helps us to maintain a stable body temperature. Many animals have very few sweat glands or none at all and must cool themselves by other means (Figures 17.2 and 17.3).

FIGURE 17.3
Pigs, having no sweat glands, wallow in the mud to cool themselves.

FIGURE 17.2
Hanz cools himself by panting. In this way, evaporation occurs in the mouth and within the bronchial tract.

CHECK YOURSELF

Would evaporation be a cooling process if molecules of every speed had an equal chance to escape from the surface of a liquid?

CHECK YOUR ANSWER

No. If molecules of all speeds escaped equally easily from the surface, the molecules left behind would have the same range of speeds as before any escaped, and there would be no change of the liquid's temperature. When only the faster molecules can break free, those left behind are slower, and the liquid becomes cooler.

The rate of evaporation is greater at higher temperatures because there is a greater proportion of molecules having sufficient kinetic energy to escape the liquid. Water evaporates at lower temperatures, too, but at a lower rate. A puddle of water, for example, may slowly evaporate to dryness on a cool day.

Even frozen water "evaporates." This form of evaporation, in which molecules are bumped directly from a solid to a gaseous phase, is called **sublimation.** Because water molecules are so tightly held in the solid phase, frozen water does not evaporate (sublime) as readily as liquid water. Sublimation, however, does account for the loss of significant portions of snow and ice, and it is especially high on sunny days in dry climates.

Condensation

FIGURE 17.4
Heat is given up by steam when it condenses inside the radiator.

FIGURE 17.5
If you're chilly outside the shower stall, step back inside and be warmed by the condensation of the excess water vapor.

The opposite of evaporation is **condensation**—the changing of a gas to a liquid. When gas molecules near the surface of a liquid are attracted to the liquid, they strike the surface with increased kinetic energy and become part of the liquid. In collisions with low-energy molecules in the liquid, excess kinetic energy is shared with the liquid, increasing the liquid temperature. Condensation is a warming process.

A dramatic example of the warming that results from condensation is the energy released by steam when it condenses—a painful experience if it condenses on you. That's why a steam burn is much more damaging than a burn from boiling water of the same temperature; the steam releases considerable energy when it condenses to a liquid and wets the skin. This energy release by condensation is utilized in steam-heating systems.

Steam is water vapor at a high temperature, usually 100°C or more. Cooler water vapor also releases energy when it condenses. In taking a shower, for example, you're warmed by condensation of vapor in the shower region—even vapor from a cold shower—if you remain in the moist shower area. You quickly sense the difference if you step outside. Away from the moisture, net evaporation takes place quickly and you feel chilly. But, if you remain in the shower stall, even with the water turned off, the warming effect of condensation counteracts the cooling effect of evaporation. If as much moisture condenses as evaporates, you feel no change in body temperature. If condensation exceeds evaporation, you are warmed. If evaporation exceeds condensation, you are cooled. So now you know why you can dry yourself with a towel much more comfortably if you remain in the shower stall. To dry yourself thoroughly, you can finish the job in a less moist area.

Spend a July afternoon in dry Tucson or Phoenix where evaporation is appreciably greater than condensation. The result of this pronounced evaporation is a much cooler feeling than you would experience in a same-temperature July afternoon in New York City or New Orleans. In these humid locations, condensation noticeably counteracts evaporation, and you feel the warming effect as vapor in the air condenses on your skin. You are literally being bombarded by the impact of H_2O molecules in the air slamming into you. Put more mildly, you are warmed by the condensation of vapor in the air upon your skin.

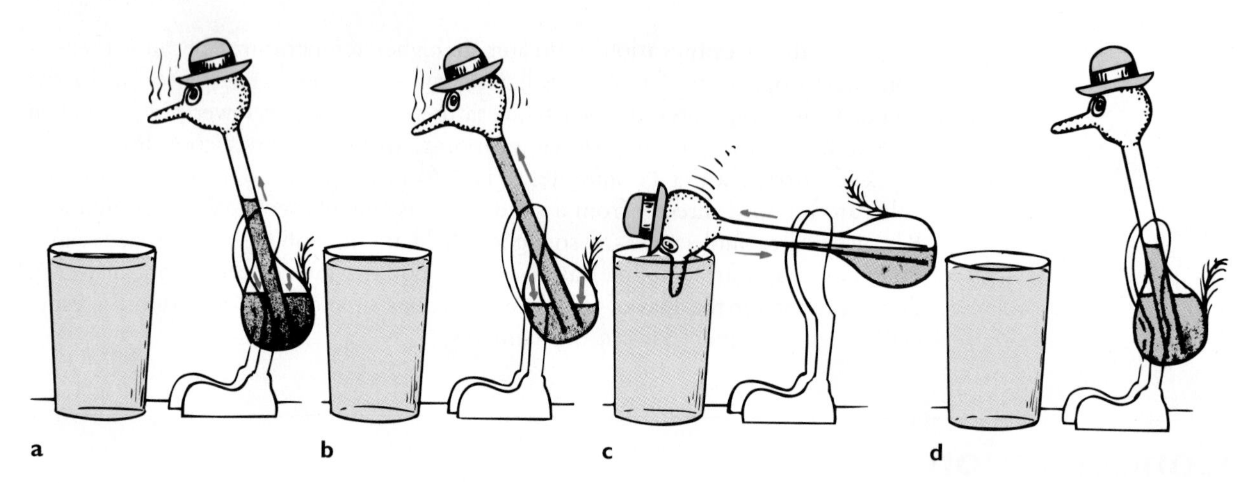

FIGURE 17.6
The toy drinking bird operates by the evaporation of ether inside its body and by the evaporation of water from the outer surface of its head. The lower body contains liquid ether, which evaporates rapidly at room temperature. As it (a) vaporizes, it (b) increases pressure (inside arrows), which pushes ether up the tube. Ether in the upper part does not vaporize because the head is cooled by the evaporation of water from the outer felt-covered beak and head. When the weight of ether in the head is sufficient, the bird (c) pivots forward, permitting the ether to flow back to the body. Each pivot wets the felt surface of the beak and head, and the cycle is repeated.

CHECK YOURSELF

If the water level in a dish of water remains unchanged from one day to the next, can you conclude that no evaporation or condensation has occurred?

Condensation in the Atmosphere

Condensation Is a Warming Process

There is always some water vapor in the air. A measure of the amount of this water vapor is called *humidity* (mass of water per volume of air). Weather reports often use the term *relative humidity*—the ratio of the amount of water vapor currently in the air at a given temperature to the largest amount of water vapor the air can contain at that temperature.[1]

Air that contains as much vapor as it can is saturated. Saturation occurs when the air temperature drops and water-vapor molecules in the air begin condensing. Water molecules tend to stick together. Because of their normally high average

CHECK YOUR ANSWER

Not at all, for there is much activity taking place at the molecular level. Both evaporation and condensation occur continuously. The fact that the water level remains constant simply indicates equal rates of evaporation and condensation, not that nothing's happening. As many molecules leave the surface by evaporation as return by condensation, so no *net* evaporation or condensation takes place. The two processes cancel each other.

[1]Relative humidity is a good indicator of comfort. For most people, conditions are ideal when the temperature is about 20°C and the relative humidity is between 50% and 60%. When the relative humidity is higher, moist air feels "muggy" as condensation counteracts the evaporation of perspiration.

FIGURE 17.7
Condensation of water vapor.

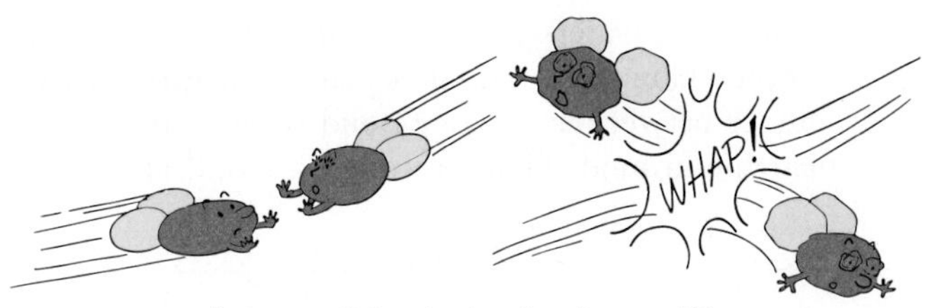
WHAP!
Fast-moving H_2O molecules rebound upon collision

Slow-moving H_2O molecules coalesce upon collision

speeds in air, however, most water molecules do not stick together when they collide. Instead, these fast-moving molecules bounce back when they collide and therefore remain in the gaseous phase. Some water molecules move more slowly than average, though, and these slow ones are more likely to stick to one another upon collision (Figure 17.7). (This can be understood by thinking of a fly making a grazing contact with sticky flypaper. At a high speed, it has enough momentum and energy to rebound from the flypaper into the air without sticking, but, at a low speed, it is more likely to get stuck.) So slow-moving water molecules are the ones most likely to condense and to form droplets of water in saturated air. Because lower air temperatures are characterized by slower-moving molecules, saturation and condensation are more likely to occur in cool air than in warm air. Warm air can contain more water vapor than cold air.

Clouds are normally denser than air. So why don't clouds fall from the sky? The answer is, clouds *do* fall! A stable cloud falls as fast as the upwelling of air beneath it, so it remains stationary.

Insights

CHECK YOURSELF

Why does dew form on the surface of a cold soft-drink can?

Fog and Clouds

Warm air rises. As it rises, it expands. As it expands, it chills. As the air chills, water-vapor molecules are slowed. Lower-speed molecular collisions result in water molecules sticking together. If there are larger and slower-moving particles or ions present, water vapor condenses upon these particles, and, with sufficient buildup, we have a cloud. If these particles are not present, we can stimulate cloud formation by "seeding" the air with appropriate particles or ions.

Warm breezes blow over the ocean and become moist. When the moist air moves from warmer to cooler waters or from warm water to cool land, it chills.

FIGURE 17.8
Why is it common for clouds to form where there are updrafts of warm moist air?

CHECK YOUR ANSWER

Water vapor in the air is chilled when it makes contact with the cold can. What is the fate of chilled water molecules? They are slower when they collide, and they stick. This is condensation, which is why the surface of a cold can is wet.

As it chills, water-vapor molecules begin sticking rather than bouncing off one another. Condensation takes place near ground level, and we have fog. The difference between fog and a cloud is basically altitude. Fog is a cloud that forms near the ground. Flying through a cloud is much like driving through fog.

Boiling

It is common to say that we boil water, meaning that we add heat to it. Actually, the boiling process cools the water.

Insights

Under the right conditions, evaporation can take place beneath the surface of a liquid, forming bubbles of vapor that are buoyed to the surface, where they escape. This change of phase *throughout* a liquid rather than only at the surface is called **boiling.** Bubbles in the liquid can form only when the pressure of the vapor within the bubbles is great enough to resist the pressure of the surrounding liquid. Unless the vapor pressure is great enough, the surrounding pressure will collapse any bubbles that tend to form. At temperatures below the boiling point, the vapor pressure in bubbles is not great enough, so bubbles do not form until the boiling point is reached. At this temperature—100°C for water at atmospheric pressure—molecules are energetic enough to exert a vapor pressure as great as the pressure of the surrounding water (which is due mainly to atmospheric pressure).

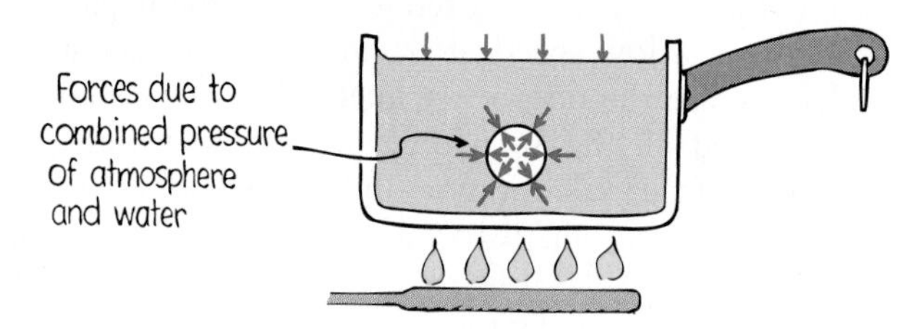

FIGURE 17.9
The motion of water-vapor molecules in the bubble of steam (much enlarged) creates a gas pressure (called the *vapor pressure*) that counteracts the atmospheric and water pressure against the bubble.

If pressure is increased, the molecules in the vapor must move faster to exert enough pressure to keep the bubble from collapsing. Extra pressure can be provided either by going deeper below the surface of the liquid (as in geysers, discussed below) or by increasing the air pressure above the liquid's surface—which is how a pressure cooker works. A pressure cooker has a tight-fitting lid that does not allow vapor to escape until it reaches a certain pressure greater than normal air pressure. As the evaporating vapor builds up inside the sealed pressure cooker, pressure on the surface of the liquid increases, which at first prevents boiling. Bubbles that would normally form are crushed. Continued heating increases the temperature beyond 100°C. Boiling does not occur until the vapor pressure within the bubbles overcomes the increased pressure on the water. The boiling point is raised. Conversely, lowered pressure (as at high altitudes) decreases the boiling point of the liquid. So we see that boiling depends not only on temperature but on pressure as well.

FIGURE 17.10
The tight lid of a pressure cooker holds pressurized water vapor above the water's surface, and this inhibits boiling. In this way, the boiling temperature of the water is increased to above 100°C.

At high altitudes, water boils at a lower temperature. For example, in Denver, Colorado, the Mile-High City, water boils at 95°C instead of at the 100°C boiling temperature characteristic of sea level. If you try to cook food in boiling water of a lower temperature, you must wait a longer time for proper cooking. A 3-minute boiled egg in Denver is yucky. If the temperature of the boiling water

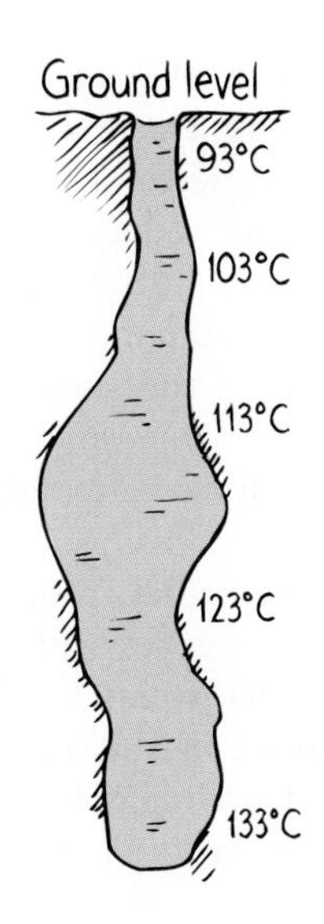

FIGURE 17.11
A geyser of the Old Faithful type.

is too low, food will not cook at all. It is important to note that it is the high temperature of the water that cooks the food, not the boiling process itself.

Geysers

A geyser is a periodically erupting pressure cooker. It consists of a long, narrow, vertical hole into which underground streams seep (Figure 17.11). The column of water is heated by volcanic heat below to temperatures exceeding 100°C. This can happen because the relatively deep vertical column of water exerts pressure on the deeper water, thereby increasing the boiling point. The narrowness of the shaft shuts off convection currents, which allows the deeper portions to become considerably hotter than the water surface. Water at the surface is less than 100°C, but the water temperature below, where it is being heated, is more than 100°C, high enough to permit boiling before water at the top reaches the boiling point. Boiling therefore begins near the bottom, where the rising bubbles push out the column of water above, and the eruption starts. As the water gushes out, the pressure on the remaining water is reduced. It then boils rapidly and erupts with great force.

THE Physics Place
Boiling Is a Cooling Process

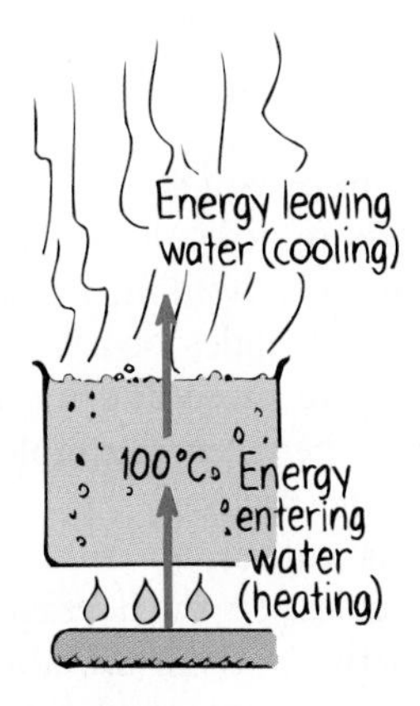

FIGURE 17.12
Heating warms the water, and boiling cools it.

Boiling Is a Cooling Process

Evaporation is a cooling process. So is boiling. At first thought, this may seem surprising—perhaps because we usually associate boiling with heating. But heating water is one thing; boiling is another. When 100°C water at atmospheric pressure is boiling, its temperature remains constant. That means it cools as fast as it warms. By what mechanism? By boiling. If cooling didn't take place, continued input of energy to a pot of boiling water would result in a continued increase in temperature. The reason a pressure cooker reaches higher temperatures is because it prevents normal boiling, which, in effect, prevents cooling.

CHECK YOURSELF

Since boiling is a cooling process, would it be a good idea to cool your hot and sticky hands by dipping them into boiling water?

Mountaineering pioneers in the nineteenth century, without altimeters, used the boiling point of water to determine their altitudes.

Insights

Boiling and Freezing at the Same Time

We usually boil water by the application of heat. But we can boil water by the reduction of pressure. We can dramatically show the cooling effect of evaporation

CHECK YOUR ANSWER

No, no, no! When we say boiling is a cooling process, we mean that the water (not your hands!) is being cooled relative to the higher temperature it would attain otherwise. Because of cooling, it remains at 100°C instead of getting hotter. A dip in 100°C water would be most uncomfortable for your hands!

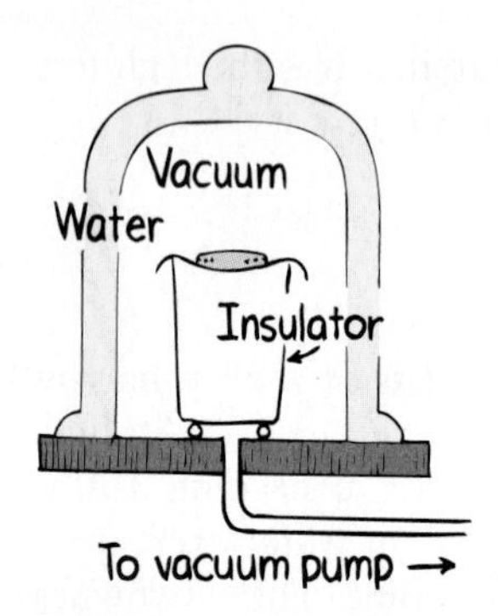

FIGURE 17.13
Apparatus to demonstrate that water will freeze and boil at the same time in a vacuum. A gram or two of water is placed in a dish that is insulated from the base by a polystyrene cup.

and boiling when room-temperature water is placed in a vacuum jar (Figure 17.13). If the pressure in the jar is slowly reduced by a vacuum pump, the water will start to boil. The boiling process removes heat from the water left in the dish, which cools to a lower temperature. As the pressure is further reduced, more and more of the slower-moving molecules boil away. Continued boiling results in a lowering of temperature until the freezing point of approximately 0°C is reached. Continued cooling by boiling causes ice to form over the surface of the bubbling water. Boiling and freezing are taking place at the same time! This must be witnessed to be appreciated. Frozen bubbles of boiling water are a remarkable sight.[2]

If you spray some drops of coffee into a vacuum chamber, they, too, will boil until they freeze. Even after they are frozen, the water molecules will continue to evaporate into the vacuum until little crystals of coffee solids are left. This is how freeze-dried coffee is made. The low temperature of this process tends to keep the chemical structure of coffee solids from changing. When hot water is added, much of the original flavor of the coffee is restored. Boiling really is a cooling process!

Melting and Freezing

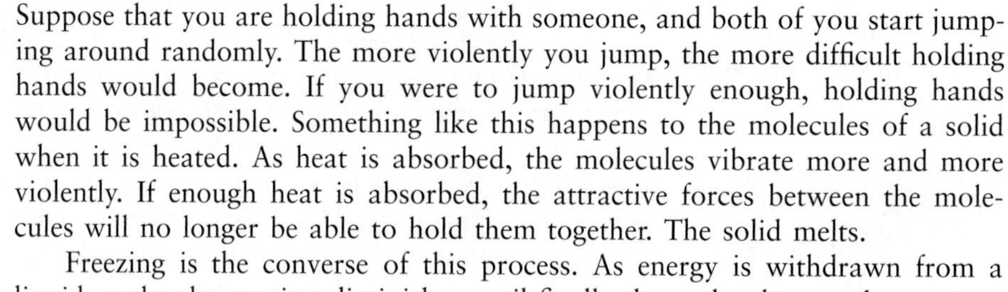

Suppose that you are holding hands with someone, and both of you start jumping around randomly. The more violently you jump, the more difficult holding hands would become. If you were to jump violently enough, holding hands would be impossible. Something like this happens to the molecules of a solid when it is heated. As heat is absorbed, the molecules vibrate more and more violently. If enough heat is absorbed, the attractive forces between the molecules will no longer be able to hold them together. The solid melts.

Freezing is the converse of this process. As energy is withdrawn from a liquid, molecular motion diminishes until finally the molecules, on the average, are moving slowly enough so that the attractive forces between them are able to cause cohesion. The molecules then vibrate about fixed positions and form a solid.

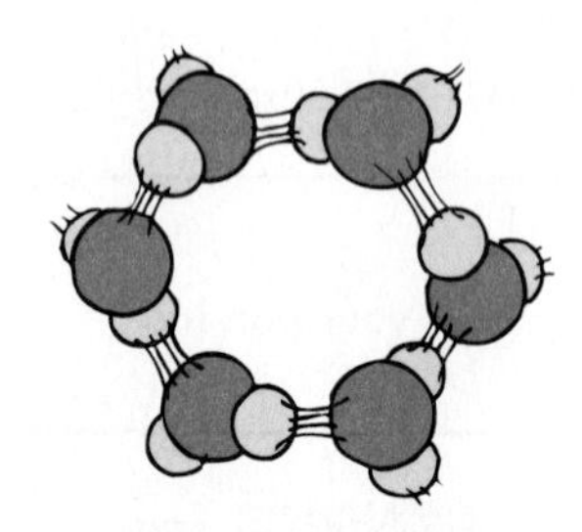

FIGURE 17.14
The open structure of pure ice crystals that normally fuse at 0°C. When other kinds of molecules or ions are introduced, crystal formation is interrupted, and the freezing temperature is lowered.

At atmospheric pressure, water freezes at 0°C—unless such substances as sugar or salt are dissolved in it. Then the freezing point is lower. In the case of salt, chlorine ions grab electrons from the hydrogen atoms in H_2O and impede crystal formation. The result of this interference by "foreign" ions is that slower motion is required for the formation of the six-sided ice-crystal structures. As ice crystals form, the interference is intensified because the proportion of "foreign" molecules or ions among nonfused water molecules increases. Connections become more and more difficult. Only when the water molecules move slowly enough for attractive forces to play an unusually large part in the process can freezing be completed. The ice first formed is almost always pure H_2O.

[2]"Water Freezer" is my favorite exhibit at the Exploratorium in San Francisco. Room-temperature water is placed in a vacuum chamber where it rapidly boils and turns to ice. The exhibit is featured in the opening photograph of the following chapter.

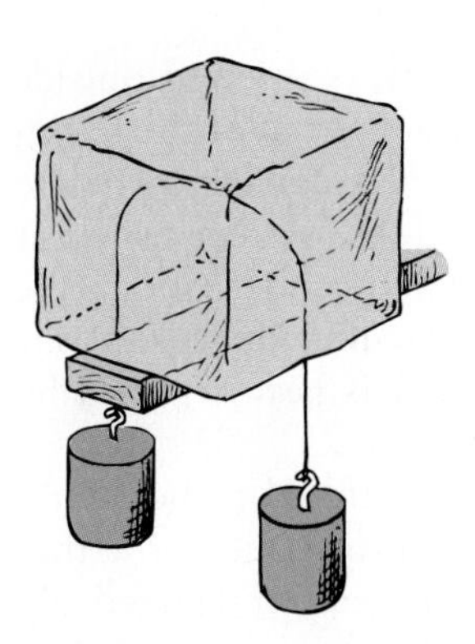

FIGURE 17.15
Regelation. The wire gradually passes through the ice without cutting it in half.

Regelation

Because H_2O molecules form open structures in the solid phase (Figure 17.14), the application of pressure can cause ice to melt. The crystals are simply crushed to the liquid phase. (The temperature of the melting point is only slightly lowered, by 0.007°C for each additional atmosphere of added pressure.) When the pressure is removed, molecules crystallize and refreezing occurs. This phenomenon of melting under pressure and freezing again when the pressure is reduced is called **regelation.** It is one of the properties of water that distinguishes it from other materials.

Regelation is nicely illustrated in Figure 17.15 (and in the opening photo of this chapter). A fine copper wire with weights attached to its ends is hung over a block of ice.[3] The wire will slowly cut its way through the ice, but its track will be left full of ice. So the wire and the weights will fall to the floor, leaving the ice a solid block.

The making of snowballs is a good example of regelation. When we compress the snow with our hands, we cause a slight melting of the ice crystals; when pressure is removed, refreezing occurs and binds the snow together. Making snowballs is difficult in very cold weather because the pressure we can apply is not enough to melt the snow.

Energy and Changes of Phase

If we continually add heat to a solid or a liquid, the solid or liquid will eventually change phase. A solid will liquefy, and a liquid will vaporize. Energy input is required for both liquefaction of a solid and vaporization of a liquid. Conversely, energy must be extracted from a substance to change its phase in the direction from gas to liquid to solid (Figure 17.16).

The cooling cycle of a refrigerator nicely employs the concepts shown in Figure 17.16. A refrigerator is a **heat pump** that "pumps" heat from a cold environment to a warm one. This is accomplished by a liquid of low boiling point, the refrigerant, which is pumped into the cooling unit, where it turns into a gas.[4] When changing phase from liquid to gas, heat is drawn from the

FIGURE 17.16
Interactive Figure
Energy changes with change of phase.

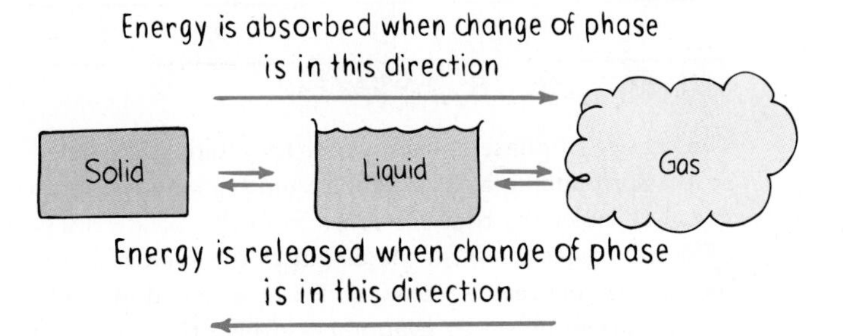
Energy is absorbed when change of phase
is in this direction
Solid
Liquid
Gas
Energy is released when change of phase
is in this direction

[3]Changes of phase are occurring as ice melts and the water refreezes. We will see that energy is needed for these changes. When the water immediately above the wire refreezes, it gives up energy. How much? Sufficient to melt an equal amount of ice immediately under the wire. This energy must be conducted through the thickness of the wire. Hence, this demonstration requires that the wire be an excellent conductor of heat. String, for example, won't do.

[4]Present-day research is directed to making thermoelectric devices in which electrons take the place of a fluid. Electric currents undergo expansion (cooling) and compression (heating) when moving between materials having different electron configurations. Watch for motorless refrigerators in the future!

A refrigerator is a "heat pump." It transfers heat out of a cold environment and into a warm environment. When the process is reversed, the heat pump is an air conditioner. In both cases, external energy operates the device.

Insights

interior where the food is. The gas, with its added energy, is directed outside to the condensation coils, located at the back side or at the bottom of the refrigerator. At the coils, heat is released into the air as the gas condenses to the liquid phase again. A pump forces the refrigerant through the system, where it is made to undergo cyclic phase changes of vaporization and condensation. The next time you're near a refrigerator, place your hand near the condenser coils in the rear, or at the bottom, and you'll feel the air that has been warmed by the energy extracted from inside.

Heat pumps of various designs are increasingly being utilized to heat (and cool) homes. What these heat pumps have in common with one another is that they function like a standard refrigerator. Whereas a refrigerator unavoidably warms a room by taking heat from food inside and depositing it at its condensation coils, heat pumps warm a room *deliberately*. Instead of extracting heat from food, they can extract heat from water that is pumped in from nearby underground pipes.[5] Water underground is relatively warm. Subsoil temperatures depend on latitude. In the Midwest and the Central Plains, subsoil temperature below a meter deep is about 13°C (55°F) year-round—warmer than the air in wintertime. Underground pipes outside the home carry 13°C water to a heat pump inside the home. Heat is extracted from the water (just as from food in a refrigerator) by the vaporization of a common refrigerant. The vaporized refrigerant is then pumped to condensation coils, where it condenses and gives off heat to warm the home. The cooled water is returned to the ground outside, where it warms back up to ground temperature and repeats the cycle.

In summer, the process can be reversed, turning the heat pump into a cooler. An air conditioner is a heat pump in reverse. Employing the same principles, it simply pumps heat energy from inside the home to the outside. That's why outside air temperatures are elevated in a crowded city where air conditioners are operating.

So we see that a solid must absorb energy to melt, and a liquid must absorb energy to vaporize. Conversely, a gas must release energy to liquefy, and a liquid must release energy to solidify.

CHECK YOURSELF

When H_2O in the vapor phase condenses, is the surrounding air warmed or cooled?

CHECK YOUR ANSWER

The change of phase is from vapor to liquid, which releases energy (Figure 17.16), so the surrounding air is warmed. Another way to see this is via Figure 17.7, where the H_2O molecules that condense from the air are the slower-moving ones. Removal of slower molecules from the air raises the average kinetic energy of remaining molecules—hence the warming. This goes hand in hand with water cooling when faster-moving molecules evaporate—where the ones remaining in the liquid phase have a lowered average kinetic energy.

[5]Depending on the amount of heat needed, about 200 to 500 meters of piping is normally placed outside the home in trenches about 1.0 to 1.8 meters beneath the ground surface. The configuration of the piping may be horizontal loops or deeper-reaching vertical U-shapes.

Heat of vaporization is either the energy required to separate molecules from the liquid phase or the energy released when gas condenses to the liquid phase.

Insights

Heat of fusion is either the energy needed to separate molecules from the solid phase or the energy released when bonds form in a liquid that change it to the solid phase.

Insights

Let's look, in particular, at the changes of phase that occur in H_2O. To make the numbers simple, consider a 1-gram piece of ice at a temperature of −50°C in a closed container that is placed on a stove to heat. A thermometer in the container reveals a slow increase in temperature up to 0°C. Then an amazing thing happens. The temperature remains at 0°C even though heat input continues. Rather than getting warmer, the ice begins to melt. In order for the whole gram of ice to melt, 80 calories (335 joules) of energy is absorbed by the ice, not even raising its temperature a fraction of a degree. Only when all the ice melts will each additional calorie (4.18 joules) absorbed by the water increase its temperature by 1°C until the boiling temperature, 100°C, is reached. Again, as energy is added, the temperature remains constant while more and more of the gram of water is boiled away and becomes steam. The water must absorb 540 calories (2255 joules) of heat energy to vaporize the whole gram. Finally, when all the water has become steam at 100°C, the temperature begins to rise once more. It will continue to rise as long as energy is added. This process is graphed in Figure 17.17.

The 540 calories (2255 joules) required to vaporize a gram of water is a large amount of energy—much more than is required to transform a gram of ice at absolute zero to boiling water at 100°C. Although the molecules in steam and boiling water at 100°C have the same average kinetic energy, steam has more potential energy because the molecules are relatively free of each other and are not bound together as in the liquid phase. Steam contains a vast amount of energy that can be released during condensation.

So we see that the energies required to melt ice (80 calories, or 335 joules, per gram) and to boil water (540 calories, or 2255 joules, per gram) are the same as the amounts released when the phase changes are in the opposite direction. The processes are reversible.

The amount of energy required to change a unit mass of any substance from solid to liquid (and vice versa) is called the **latent heat of fusion** for the substance. (The word *latent* reminds us that this is thermal energy hidden from the thermometer.) For water, we have seen that this is 80 calories per gram (335 joules per gram). The amount of energy required to change any substance

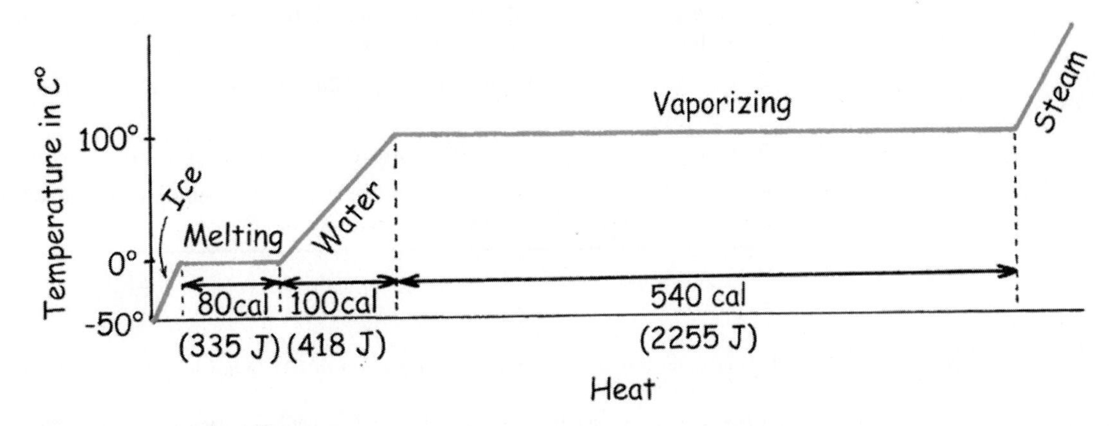

FIGURE 17.17
A graph showing the energy involved in the heating and the changes of phase of 1 g of H_2O.

from liquid to gas (and vice versa) is called the **latent heat of vaporization** for the substance. For water, we have seen this is a whopping 540 calories per gram (2255 joules per gram).[6] It so happens that these relatively high values are due to the strong forces between water molecules—hydrogen bonds.

Water's heat of vaporization is huge. The energy needed to vaporize a quantity of boiling water is nearly seven times the energy needed to melt the same amount of ice.

Insights

CHECK YOURSELF

1. How much energy is transferred when 1 gram of steam at 100°C condenses to water at 100°C?
2. How much energy is transferred when 1 gram of boiling water at 100°C cools to ice water at 0°C?
3. How much energy is transferred when 1 gram of ice water at 0°C freezes to ice at 0°C?
4. How much energy is transferred when 1 gram of steam at 100°C turns to ice at 0°C?

FIGURE 17.18
On a cold day, hot water freezes faster than warm water because of the energy that leaves the hot water during rapid evaporation.

The large value of 540 calories per gram for the latent heat of vaporization of water explains why, under some conditions, hot water will freeze faster than warm water.[7] This phenomenon is evident when a thin layer of water is distributed over a large area—like when you wash a car with hot water on a cold winter day or flood a skating rink with hot water, which melts, smooths out the rough spots, and refreezes quickly. The rate of cooling by rapid evaporation is very high because each evaporating gram of water draws at least 540 calories from the water left behind. This is an enormous amount of energy compared with the 1 calorie per Celsius degree that is drawn from each gram of water that cools by thermal conduction. Evaporation is truly a cooling process.

CHECK YOUR ANSWERS

1. One gram of steam at 100°C transfers 540 calories of energy when it condenses to become water at the same temperature.
2. One gram of boiling water transfers 100 calories when it cools 100°C to become ice water.
3. One gram of ice water at 0°C transfers 80 calories to become ice at 0°C.
4. One gram of steam at 100°C transfers to the surroundings a grand total of the above values, 720 calories, to become ice at 0°C.

---

[6]In SI units, the heat of vaporization of water is expressed 2.255 megajoules per kilogram (MJ/kg), and the heat of fusion of water is 0.335 MJ/kg.

[7]Boiling-hot water will not freeze before cold water does, but it will freeze before moderately hot water does. For example, boiling-hot water will freeze before water warmer than about 60°C, but not before water cooler than 60°C. Try it and see.

PRACTICING PHYSICS

Fill in the number of calories or joules at each step in changing the phase of 1 gram of 0°C ice to 100°C steam.

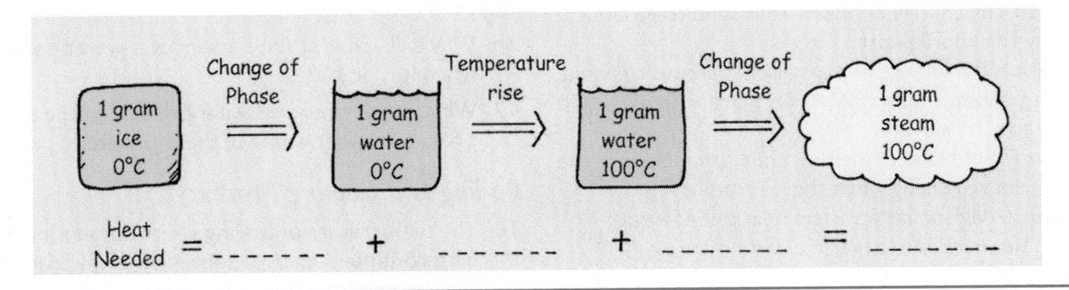

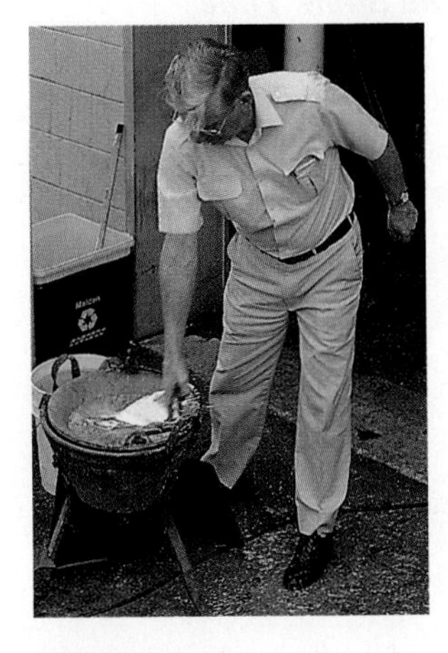

FIGURE 17.19
Paul Ryan tests the hotness of molten lead by dragging his wetted finger through it.

CHECK YOURSELF

Suppose that 4 grams of boiling water is poured onto a cold surface and that 1 gram rapidly evaporates. If evaporation takes 540 calories from the remaining 3 grams of water and no other heat transfer takes place, what will be the temperature of the remaining 3 grams?

You dare not touch your dry finger to a hot skillet on a hot stove, but you can certainly do so without harm if you first wet your finger and touch the skillet very briefly. You can even touch it a few times in succession as long as your finger remains wet. This is because energy that ordinarily would go into burning your finger goes instead into changing the phase of the moisture on your finger. The energy converts the moisture to a vapor, which then provides an insulating layer between your finger and the skillet. Similarly, you are able to judge the hotness of a hot clothes iron if you touch it very briefly with a wet finger.

Former supervisor Paul Ryan of the Department of Public Works in Malden, Massachusetts, years ago used molten lead to seal pipes in certain plumbing operations. He startled onlookers by dragging his finger through molten lead to judge its hotness (Figure 17.19). He was sure that the lead was very hot and that his finger was thoroughly wet before doing this. (Do not try this on your own, for if the lead is not hot enough it will stick to your finger and you could be seriously burned!) Similarly, firewalkers walking barefoot on red-hot coals often prefer wet feet (others such as Dave Willey in Figure 17.20 prefer dry feet, claiming hot coals more readily stick to wet feet—OUCH!). The low conductivity of wooden coals, however (as discussed in the previous chapter), is the principal reason the feet of barefoot firewalkers are not burned.

FIGURE 17.20
Professor Dave Willey walks barefoot across red-hot wooden coals without harm.

CHECK YOUR ANSWER

The remaining 3 grams will turn to 0°C. 540 calories from 3 grams means each gram gives up 180 calories. 100 calories from a gram of boiling water reduces its temperature to 0°C, and removal of 80 more calories turns it to ice. This is why hot water so quickly turns to ice in a freezing-cold environment. (In practice, because of other heat transfer, more than 1 gram of the original 4 grams would need to evaporate to freeze the rest.)

Summary of Terms

Evaporation The change of phase from liquid to gaseous.

Sublimation The change of phase from solid to gaseous, skipping the liquid phase.

Condensation The change of phase from gaseous to liquid.

Boiling Rapid evaporation that takes place within a liquid as well as at its surface.

Regelation The process of melting under pressure and the subsequent refreezing when the pressure is removed.

Heat pump A device that transfers heat out of a cool environment and into a warm environment.

Latent heat of fusion The amount of energy required to change a unit mass of a substance from solid to liquid (and vice versa).

Latent heat of vaporization The amount of energy required to change a unit mass of a substance from liquid to gas (and vice versa).

Review Questions

1. What are the four common phases of matter?

Evaporation

2. Do the molecules in a liquid all have about the same speed, or do they have a wide variety of speeds?
3. What is evaporation, and why is it a cooling process? Exactly what is it that cools?
4. Why does warmer water evaporate more readily than cold water?
5. What is sublimation?

Condensation

6. Distinguish between condensation and evaporation.
7. Why is a steam burn more damaging than a burn from boiling water of the same temperature?
8. Why do you feel uncomfortably warm on a hot and humid day?

Condensation in the Atmosphere

9. Distinguish between humidity and relative humidity.
10. Why does water vapor in the air condense when the air is chilled?

Fog and Clouds

11. Why does warm moist air form clouds when it rises?
12. What is the basic difference between a cloud and fog?

Boiling

13. Distinguish between evaporation and boiling.
14. Does increased atmospheric pressure increase or decrease the boiling point of water? Why is this so?
15. Is it the boiling of water or the higher temperature of water that cooks food faster in a pressure cooker?

Geysers

16. Why will water at the bottom of a geyser not boil when it is at 100°C?
17. What happens to the water pressure at the bottom of a geyser when some of the water above gushes out?

Boiling Is a Cooling Process

18. The temperature of boiling water doesn't increase with continued energy input. Why is this evidence that boiling is a cooling process?

Boiling and Freezing at the Same Time

19. When will water boil at a temperature of less than 100°C?
20. What evidence can you cite for the claim that water can boil at a temperature of 0°C?

Melting and Freezing

21. Why does increasing the temperature of a solid make it melt?
22. Why does decreasing the temperature of a liquid make it freeze?
23. Why does freezing of water not occur at 0°C when foreign ions are present?

Regelation

24. What happens to the hexagonal open structure of ice when sufficient pressure is applied to it?
25. Why does a wire not simply cut a block of ice in two when it passes through the ice?

Energy and Changes of Phase

26. Does a liquid give off energy or absorb energy when it turns into a gas?
27. Does a liquid give off energy or absorb energy when it turns into a solid?
28. Is heat discharged at the back of a refrigerator and by a heat pump given off by vaporization of the refrigerating fluid or by condensation?
29. How many calories are needed to change the temperature of 1 g of water by 1°C? To melt 1 g of ice at 0°C? To vaporize 1 g of boiling water at 100°C?
30. Cite two reasons why firewalkers don't burn their wetted feet when walking barefoot on red-hot coals.

Projects

1. Place a Pyrex funnel mouth-down in a saucepan full of water so that the straight tube of the funnel sticks

out above the water. Rest a part of the funnel on a nail or a coin so that water can get under it. Place the pan on a stove and watch the water as it begins to boil. Where do the bubbles form first? Why? As the bubbles rise, they expand rapidly and push water ahead of them. The funnel confines the water, which is forced up the tube and driven out at the top. Now do you know how a geyser and a coffee percolator work?

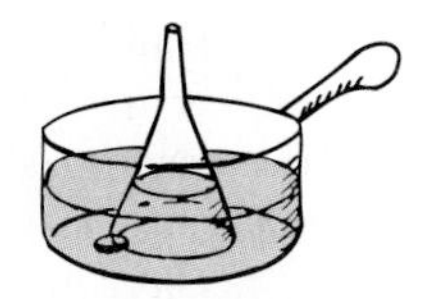

2. Watch the spout of a teakettle of boiling water. Notice that you cannot see the steam that issues from the spout. The cloud that you see farther away from the spout is not steam but condensed water droplets. Now hold the flame of a candle in the cloud of condensed steam. Can you explain your observations?
3. You can make rain in your kitchen. Put a cup of water in a Pyrex saucepan or a Silex coffeemaker and heat it slowly over a low flame. When the water is warm, place a saucer filled with ice cubes on top of the container. As the water below is heated, droplets form at the bottom of the cold saucer and combine until they are large enough to fall, producing a steady "rainfall" as the water below is gently heated. How does this resemble, and how does it differ from, the way in which natural rain is formed?
4. Measure the temperature of boiling water and the temperature of a boiling solution of salt and water. How do they compare?
5. Do as Dean Baird demonstrates on page 325, or as the sketch shows, and suspend a heavy weight by copper wire over an ice cube. In a matter of minutes, the wire will be pulled through the ice. The ice will melt beneath the wire and refreeze above it, leaving a visible path if the ice is clear.

6. Place a tray of boiling-hot water and a tray of water from a hot-water tap in a freezer. The trays should be filled to about the same depth. See which water will freeze first.
7. If you suspend an open-topped container of water in a pot of boiling water, water in the inner container will reach 100°C but will not boil. Can you explain why this is so?
8. Write a letter to Grandma and tell her why bringing water to a boil when she's making tea is actually a process that *cools* the water. Explain how she could convince her teatime friends of this intriguing concept.

Exercises

1. Alcohol evaporates more quickly than water at the same temperature. Which produces more cooling—alcohol or the same amount of water on your skin?
2. You can determine wind direction by wetting your finger and holding it up in the air. Explain.
3. When you step out of a swimming pool on a hot, dry day in the Southwest, you feel quite chilly. Why?
4. Why is sweating an efficient mechanism for cooling off on a hot day?
5. Why does blowing over hot soup cool the soup?
6. Can you give two reasons why pouring a cup of hot coffee into a saucer results in faster cooling?
7. A covered glass of water sits for days with no drop in water level. Strictly speaking, can you say that nothing has happened, that no evaporation or condensation has taken place? Explain.
8. How might water be desalinized by freezing?
9. Pretend that all the molecules in a liquid have the same speed, not random speeds. Would evaporation of this liquid cause the remaining liquid to be cooled? Explain.
10. Where does the energy come from that keeps the dunking bird in Figure 17.6 operating?
11. Does a common electric fan cool the air in a room? If not, then why is it used in an overly warm room?
12. An inventor claims to have developed a new perfume that lasts a long time because it doesn't evaporate. Comment on this claim.
13. Porous canvas bags filled with water are used by travelers in hot weather. When the bags are slung on the outside of a fast-moving car, the water inside is cooled considerably. Explain.
14. Why will wrapping a bottle in a wet cloth at a picnic often produce a cooler bottle than placing the bottle in a bucket of cold water?
15. The human body can maintain its customary temperature of 37°C on a day when the temperature is above 40°C. How is this done?
16. Double-pane windows have nitrogen gas or very dry air between the panes. Why is ordinary air a poor idea?
17. Why are icebergs often surrounded by fog?
18. How does Figure 17.7 help explain the moisture that forms on the inside of car windows when you're parking with your date on a cool night?
19. You know that the windows in your warm house get wet on a cold day. But can moisture form on the windows if the interior of your house is cold on a hot day? How is this different?

20. On freezing days, frost often forms on windows. Why is there usually more frost on the bottom portions of the windows?

21. Why do clouds often form above mountain peaks? (*Hint:* Consider the updrafts.)

22. Why will clouds tend to form above either a flat or a mountainous island in the middle of the ocean? (*Hint:* Compare the specific heat of the land with that of the water and the subsequent convection currents in the air.)

23. A great amount of water vapor changes phase to become water in the clouds that form a thunderstorm. Does this release thermal energy or absorb it?

24. When can you add heat to something without raising its temperature?

25. When can you withdraw heat from something without lowering its temperature?

26. Why does the temperature of boiling water remain the same as long as the heating and boiling continue?

27. Why do vapor bubbles in a pot of boiling water get larger as they rise in the water?

28. Why does the boiling temperature of water increase when the water is under increased pressure?

29. Why does the boiling temperature of water decrease when the water is under reduced pressure, such as when it is at a higher altitude?

30. Place a jar of water on a small stand within a saucepan of water so that the bottom of the jar is held above the bottom of the pan. When the pan is placed on a stove, the water in the pan will boil, but not the water in the jar. Why?

31. Hydrothermal vents are openings in the ocean floor that discharge very hot water. Water emerging at nearly 280°C from one such vent off the Oregon coast, some 2400 m beneath the surface, is not boiling. Provide an explanation.

32. Why should you not grasp a hot skillet with a wet dishcloth?

33. Water will boil spontaneously in a vacuum—on the surface of the Moon, for example. Could you cook an egg in this boiling water? Explain.

34. Our inventor friend proposes a design for cookware that will allow boiling to occur at a temperature less than 100°C so that food can be cooked with less energy consumption. Comment on this idea.

35. If water that boils due to reduced pressure is not hot, then is ice formed by the reduced pressure not cold? Explain.

36. How can water be brought to a boil without heating it?

37. Your instructor hands you a closed flask partly filled with room-temperature water. When you hold it, the heat transfer between your bare hands and the flask causes the water to boil. Quite impressive! How is this accomplished?

38. When you boil potatoes, will your cooking time be reduced with vigorously boiling water instead of gently boiling water? (Directions for cooking spaghetti call for vigorously boiling water—not to lessen cooking time but to prevent something else. If you don't know what it is, ask a cook.)

39. Why does placing a lid over a pot of water on a stove shorten the time for the water to come to a boil, whereas, after the water is boiling, the use of a lid only slightly shortens the cooking time?

40. In the power plant of a nuclear submarine, the temperature of the water in the reactor is above 100°C. How is this possible?

41. Explain why the eruptions of many geysers repeat with notable regularity.

42. Why does the water in a car radiator sometimes boil explosively when the radiator cap is removed?

43. Can ice be colder than 0°C? What is the temperature of an ice–water mixture?

44. Why is very cold ice "sticky"?

45. Would regelation occur if ice crystals did not have an open structure? Explain.

46. People who live where snowfall is common will tell you that air temperatures are higher when it's snowing than when it's clear. Some misinterpret this by stating that snowfall can't occur on very cold days. Explain this misinterpretation.

47. A piece of metal and an equal mass of wood are both removed from a hot oven at equal temperatures and are dropped onto blocks of ice. The metal has a lower specific heat capacity than the wood. Which will melt more ice before cooling to 0°C?

48. How does melting ice change the temperature of the surrounding air?

49. Why does dew form by condensation on a cold soft-drink can but not on the same can at room temperature?

50. What accounts for the bulged ends of a can of soda that has been frozen?

51. Why is half-frozen fruit punch always sweeter than completely melted fruit punch?

52. Air-conditioning units contain no water whatever, yet it is common to see water dripping from them when they're running on a hot day. Explain.

53. Is it condensation or vaporization that occurs on the cold outside coils of an operating air conditioner?

54. Some old-timers found that, when they wrapped newspaper around the ice in their iceboxes, melting was inhibited. Discuss the advisability of this practice.

55. When ice in a pond melts, what effect does this have on the temperature of the nearby air?

56. Why is it that a tub of water placed in a farmer's canning cellar helps prevent canned food from freezing in cold winters?

57. Why will spraying fruit trees with water before a frost help to protect the fruit from freezing?

58. There are several theories to explain the beginning of an ice age. Consider the following one: If the temperature of the Earth increases, evaporation of the oceans increases, precipitation increases, and snowfall increases. This results in more snow left in cooler places at the end of each summer, with successively increasing depths each winter that become hard packed from the bottom up to become ice. All the while, the ice reflects more solar radiation that would otherwise be absorbed. This, in turn, cools the Earth further and allows for more freezing. Can you continue this sequence and see how the process reverses itself and how the ice age withdraws?

59. Low-temperature geothermal power utilizes the small temperature difference between locations above and below ground, which can be sufficient for changing the phase of a refrigerant. Can devices that heat homes in this manner in winter cool homes in summer?

60. Why does an overheated dog pant?

Problems

1. The quantity of heat Q that changes the temperature ΔT of a mass m of a substance is given by $Q = cm\Delta T$, where c is the specific heat capacity of the substance. For example, for H_2O, $c = 1$ cal/g·C°. And for a change of phase, the quantity of heat Q that changes the phase of a mass m is $Q = mL$, where L is the heat of fusion or heat of vaporization of the substance. For example, for H_2O, the heat of fusion is 80 cal/g (or 80 kcal/kg) and the heat of vaporization is 540 cal/g (or 540 kcal/kg). Use these relationships to determine the number of calories to change (a) 1 kg of 0°C ice to 0°C ice water, (b) 1 kg of 0°C ice water to 1 kg of 100°C boiling water, (c) 1 kg of 100°C boiling water to 1 kg of 100°C steam, and (d) 1 kg of 0°C ice to 1 kg of 100°C steam.

2. The specific heat capacity of ice is about 0.5 cal/g·C°. Supposing that it remains at that value all the way to absolute zero, calculate the number of calories it would take to change a 1-g ice cube at absolute zero (−273°C) to 1 g of boiling water. How does this number of calories compare to the number of calories required to change the same gram of 100°C boiling water to 100°C steam?

3. Find the mass of 0°C ice that 10 g of 100°C steam will completely melt.

4. If 50 g of hot water at 80°C is poured into a cavity in a very large block of ice at 0°C, what will be the final temperature of the water in the cavity? How much ice must melt in order to cool the hot water down to this temperature?

5. A 50-g chunk of 80°C iron is dropped into a cavity in a very large block of ice at 0°C. How many grams of ice will melt? (The specific heat capacity of iron is 0.11 cal/g·C°.)

6. Calculate the height that a block of ice at 0°C must be dropped to completely melt upon impact. Assume no air resistance and that all the energy goes into melting the ice. [*Hint*: Equate the joules of gravitational potential energy to the product of the mass of ice and its heat of fusion (in SI units, 335,000 J/kg). Do you see why the answer doesn't depend on mass?]

7. A 10-kg iron ball is dropped onto a pavement from a height of 100 m. If half of the heat generated goes into warming the ball, find the temperature increase of the ball. (In SI units, the specific heat capacity of iron is 450 J/kg·C°.) Why is the answer the same for a ball of any mass?

8. The heat of vaporization of ethyl alcohol is about 200 cal/g. If 2 kg of this fluid were allowed to vaporize in a refrigerator, how many grams of ice would be formed from 0°C water?

Remember, review questions provide you with a self check of whether or not you grasp the central ideas of the chapter. The exercises and problems are extra "pushups" for you to try after you have at least a fair understanding of the chapter and can handle the review questions.

CHAPTER 18

Thermodynamics

Ron Hipschman at the Exploratorium removes a piece of ice from the Water Freezer exhibit. When room-temperature water was placed in the chamber and a vacuum was drawn, rapid evaporation cooled the water to ice.

The study of heat and its transformation to mechanical energy is called **thermodynamics** (which stems from Greek words meaning "movement of heat"). The science of thermodynamics was developed in the early nineteenth century, before the atomic and molecular theory of matter was understood. Because the early workers in thermodynamics had only vague notions of atoms, and knew nothing about electrons and other microscopic particles, the models they used invoked macroscopic notions—such as mechanical work, pressure, and temperature—and their roles in energy transformations. The foundation of thermodynamics is the conservation of energy and the fact that heat flows spontaneously from hot to cold and not the other way around. Thermodynamics provides the basic theory of heat engines, from steam turbines to nuclear reactors, and the basic theory of refrigerators and heat pumps. We begin our study of thermodynamics with a look at one of its early concepts—a lower limit to temperature.

Absolute Zero

In principle, there is no upper limit to temperature. As thermal motion increases, a solid object first melts and then vaporizes; as the temperature is further increased, molecules break up into atoms, and atoms lose some or all of their electrons, thereby forming a cloud of electrically charged particles—a plasma. This situation exists in stars, where the temperature is many millions of degrees Celsius.

In contrast, there is a definite limit at the other end of the temperature scale. Gases expand when heated and contract when cooled. Nineteenth-century experiments found that all gases, regardless of their initial pressures or volumes, change by 1/273 of their volume at 0°C for each degree Celsius change in temperature, provided the pressure is held constant. So, if a gas at 0°C were cooled down by 273°C, it would contract, according to this rule, by 273/273 of its volume and be reduced to zero volume. Clearly, we cannot have a substance with zero volume. Scientists also found that the pressure of any gas in any container of *fixed* volume changes by 1/273 of its pressure at 0°C for each degree Celsius change in temperature. So a gas in a container of fixed volume cooled to 273°C below zero would have no pressure whatsoever. In practice, every gas

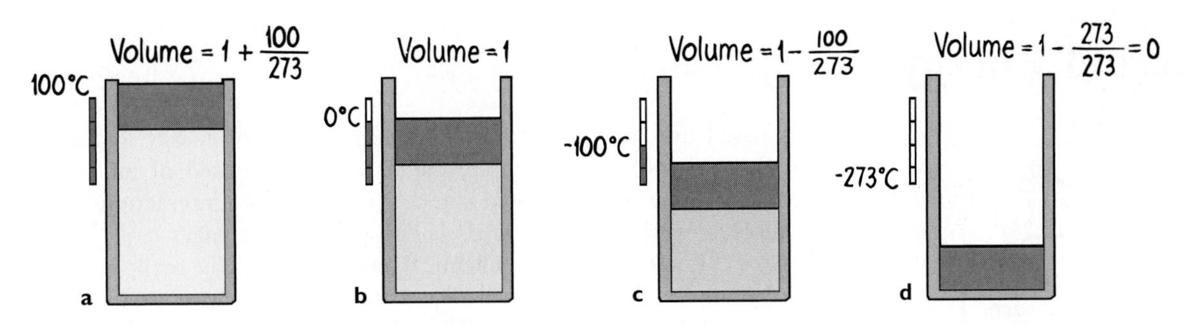

FIGURE 18.1
The gray piston in the vessel goes down as the volume of gas (blue) shrinks. The volume of gas changes by 1/273 its volume at 0°C with each 1°C change in temperature when the pressure is held constant. (a) At 100°C, the volume is 100/273 greater than it is at (b), when its temperature is 0°C. (c) When the temperature is reduced to −100°C, the volume is reduced by 100/273. (d) At −273°C, the volume of the gas would be reduced by 273/273 and so would be zero.

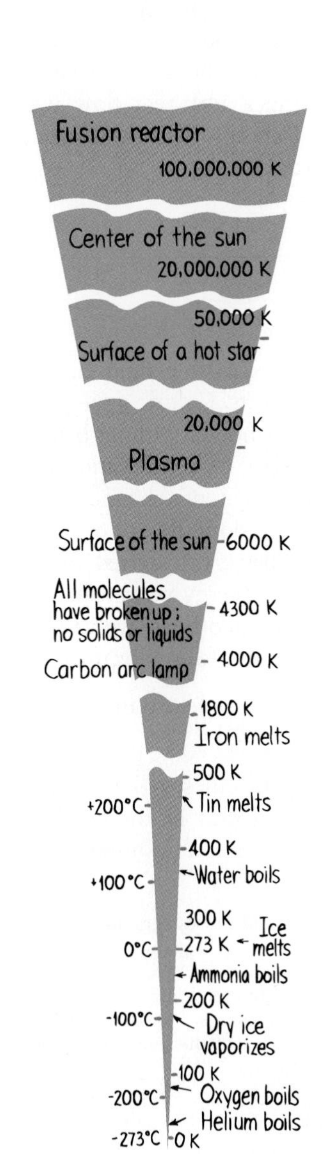

FIGURE 18.2
Some absolute temperatures.

liquefies before it gets this cold. Nevertheless, these decreases by 1/273 increments suggested the idea of a lowest temperature: −273°C. So there is a limit to coldness. When atoms and molecules lose all available kinetic energy, they reach the **absolute zero** of temperature. At absolute zero, as briefly discussed in Chapter 15, no more energy can be extracted from a substance and no further lowering of its temperature is possible. This limiting temperature is actually 273.15° below zero on the Celsius scale (and 459.7° below zero on the Fahrenheit scale).

The absolute temperature scale is called the Kelvin scale, after the nineteenth-century Scottish physicist William Thomson, 1st Baron Kelvin, who coined the word *thermodynamics* and was the first to suggest this thermodynamic temperature scale. Absolute zero is 0 K (short for "0 kelvin," rather than "0 degrees kelvin"). There are no negative numbers on the Kelvin scale. Degrees on the Kelvin scale are calibrated with the same-sized divisions as on the Celsius scale. Thus, the melting point of ice is 273.15 K, and the boiling point of water is 373.15 K.

CHECK YOURSELF

1. Which is larger, a Celsius degree or a kelvin?
2. A flask of helium gas has a temperature of 0°C. If a second, identical flask containing an equal mass of helium is twice as hot (has twice the internal energy), what is its temperature in degrees Celsius?

CHECK YOUR ANSWERS

1. Neither. They are equal.
2. A container of helium twice as hot has twice the absolute temperature, or two times 273 K. This would be 546 K, or 273°C. (Simply subtract 273 from the Kelvin temperature to convert to Celsius degrees. Can you see why?)

Internal Energy

Absolute zero isn't the coldest you can reach. It's the coldest you can hope to approach. (Researchers have been within a billionth of a degree of it.)

Insights

As briefly discussed in Chapter 15, there is a vast amount of energy locked in all materials. In this book, for example, the paper is composed of molecules that are in constant motion. They have kinetic energy. Due to interactions with neighboring molecules, they also have potential energy. The pages can be easily burned, so we know they store chemical energy, which is really electric potential energy at the molecular level. We know there are vast amounts of energy associated with atomic nuclei. Then there is the "energy of being," described by the celebrated equation $E = mc^2$ (mass energy). Energy within a substance is found in these and other forms, which, when taken together, are called **internal energy.**[1] Although the internal energy in even the simplest substance can be quite complex, in our study of heat changes and heat flow we will be concerned only with the *changes* in the internal energy of a substance. Changes in temperature indicate these changes in internal energy.

First Law of Thermodynamics

Looking for a diet plan? Consume fewer calories than you burn. This is the only diet plan firmly based on the first law of thermodynamics.

Insights

Some two hundred years ago, heat was thought to be an invisible fluid called *caloric,* which flowed like water from hot objects to cold objects. Caloric appeared to be conserved—that is, it seemed to flow from one place to another without being created or destroyed. This idea was the forerunner of the law of conservation of energy. By the middle of the nineteenth century, it became apparent that the flow of heat was nothing more than the flow of energy itself. The caloric theory of heat was gradually abandoned.[2] Today we view heat as energy being transferred from one place to another, usually by molecular collisions. Heat is energy in transit.

When the law of energy conservation is enlarged to include heat, we call it the **first law of thermodynamics.**[3] We state it generally in the following form:

When heat flows to or from a system, the system gains or loses an amount of energy equal to the amount of heat transferred.

By *system,* we mean a well-defined group of atoms, molecules, particles, or objects. The system may be the steam in a steam engine or it may be the Earth's entire atmosphere. It can even be the body of a living creature. The important

[1]If this book were poised at the edge of a table and ready to fall, it would possess gravitational potential energy; if it were tossed into the air, it would possess kinetic energy. But these are not examples of internal energy, because they involve more than just the elements of which the book is composed. They include gravitational interactions with the Earth and motion with respect to the Earth. If we wish to include these forms, we must do so in terms of a larger "system"—a system enlarged to include both the book and the Earth. They are not part of the internal energy of the book itself.

[2]Popular ideas, when proved wrong, are seldom suddenly discarded. People tend to identify with the ideas that characterize their time; hence, it is often the young who are more prone to discover and accept new ideas and to push the human adventure forward.

[3]There is also a zeroth law of thermodynamics (bearing this charming name because it was formulated *after* the first and second laws were formulated), which states that two systems, each in thermal equilibrium with a third system, are in equilibrium with each other. A third law states that no system can have its absolute temperature reduced to zero.

point is that we must be able to define what is contained *within* the system and what is *outside* it. If we add heat to the steam in a steam engine, to the Earth's atmosphere, or to the body of a living creature, we are adding energy to the system. The system can "use" this heat to increase its own internal energy or to do work on its surroundings. So adding heat does one or both of two things: (1) it increases the internal energy of the system, if it remains in the system, or (2) it does work on things external to the system, if it leaves the system. More specifically, the first law states:

Heat added to a system = increase in internal energy + external work done by the system.

The first law is an overall principle that is not concerned with the inner workings of the system itself. Whatever the details of molecular behavior in the system, the added heat energy serves only two functions: It increases the internal energy of the system or it enables the system to do external work (or both). Our ability to describe and predict the behavior of systems that may be too complicated to analyze in terms of atomic and molecular processes is one of the beauties of thermodynamics. Thermodynamics bridges the microscopic and macroscopic worlds.

Put an airtight can of air on a hot stove and heat it up. Let's define the air within the can as the "system." Because the can is of fixed volume, the air can't do any work on it (work involves movement by a force). All the heat that goes into the can increases the internal energy of the enclosed air, so its temperature rises. If the can is fitted with a movable piston, then the heated air can do work as it expands and pushes the piston outward. Can you see that, as it does work, the temperature of the enclosed air must be less than if no work were done on the piston? If heat is added to a system that does no external work, then the amount of heat added equals the internal energy increase of the system. If the system does external work, then the increase in internal energy is correspondingly less.

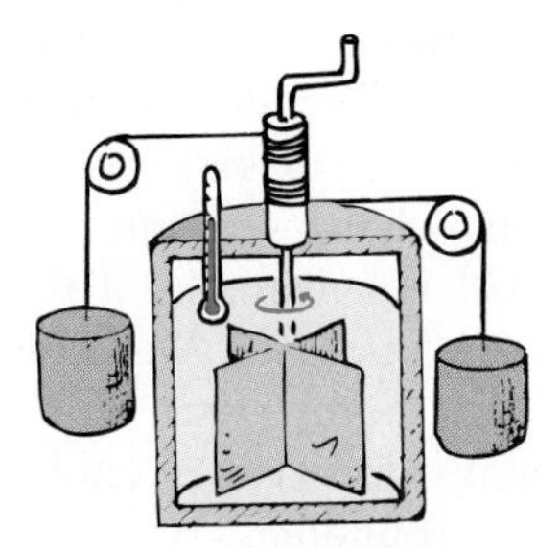

FIGURE 18.3
Paddle-wheel apparatus used to compare heat with mechanical energy. As the weights fall, they give up potential energy (mechanical), which is converted to heat that warms the water. This equivalence of mechanical and heat energy was first demonstrated by James Joule, for whom the unit of energy was named.

Consider a given amount of energy supplied to a steam engine. The amount supplied will be evident in the increase in the internal energy of the steam and in the mechanical work done. The sum of the increase in internal energy and the work done will equal the energy input. In no way can energy output exceed energy input. The first law of thermodynamics is simply the thermal version of the law of conservation of energy.

CHECK YOURSELF

1. If 100 J of heat is added to a system that does no external work, by how much is the internal energy of that system raised?
2. If 100 J of heat is added to a system that does 40 J of external work, by how much is the internal energy of that system raised?

Adding heat to a system so the system can do mechanical work is only one application of the first law of thermodynamics. If, instead of adding heat, we

CHECK YOUR ANSWERS

1. 100 J.
2. 60 J. We see from the first law that 100 J = 60 J + 40 J.

do mechanical work on the system, the first law tells us what we can expect: an increase in internal energy. Rub your palms together and they become warmer. Or rub two dry sticks together and they'll certainly become hotter. Or pump the handle of a bicycle pump and the pump becomes hot. Why? Because we are primarily doing mechanical work on the system and raising its internal energy. If the process happens so quickly that very little heat is conducted out of the system, then most of the work input goes into increasing the internal energy, and the temperature rises.

Adiabatic Processes

Compressing or expanding a gas while no heat enters or leaves the system is said to be an **adiabatic process** (from the Greek for "impassable"). Adiabatic conditions can be achieved by thermally insulating a system from its surroundings (with Styrofoam, for example) or by performing the process so rapidly that heat has no time to enter or leave. In an adiabatic process, therefore, because no heat enters or leaves the system, the "heat added" part of the first law of thermodynamics must be zero. Then, under adiabatic conditions, changes in internal energy are equal to the work done on or by the system.[4] For example, if we do work on a system by compressing it, its internal energy increases: We raise its temperature. We notice this by the warmth of a bicycle pump when air is compressed. If work is done by the system, its internal energy decreases: It cools. When a gas adiabatically expands, it does work on its surroundings, releasing internal energy as it becomes cooler. Expanding air cools.

You can demonstrate the cooling of air as it expands by repeating the personal experiment of blowing air on your hand, discussed in Chapter 16. Blow air on your hand—first with your mouth wide open, then with puckered lips so the air expands (Figure 16.6, Chapter 16). Your breath is noticeably cooler when the air expands!

FIGURE 18.4
If you do work on the pump by pressing down on the piston, you compress the air inside. What happens to the temperature of the enclosed air? What happens to its temperature if it expands and pushes the piston outward?

Meteorology and the First Law

Thermodynamics is useful to meteorologists when analyzing weather. Meteorologists express the first law of thermodynamics in the following form:

Air temperature rises as heat is added or as pressure is increased.

The temperature of the air may be changed by adding or subtracting heat, by changing the pressure of the air (which involves work), or by both. Heat is added

[4] Δ Heat = Δ internal energy + work. When there is no heat transfer, Δ Heat = 0. So
$0 = \Delta$ internal energy + work
Then we can say
$-$Work = Δ internal energy

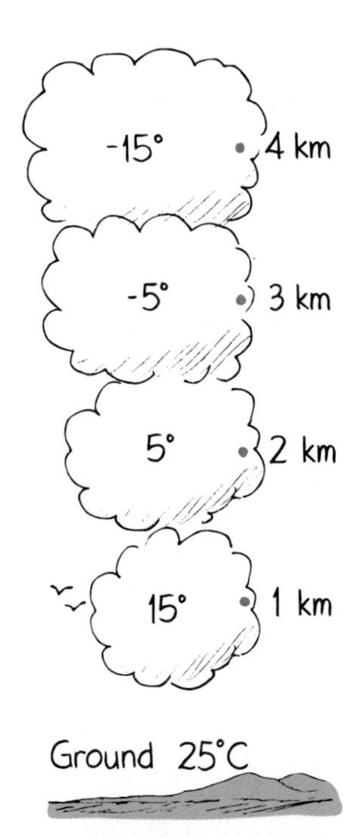

FIGURE 18.5
The temperature of a parcel of dry air that expands adiabatically decreases by about 10°C for each kilometer of elevation.

by solar radiation, by long-wave Earth radiation, by moisture condensation, or by contact with the warm ground. An increase in air temperature results. The atmosphere may lose heat by radiation to space, by evaporation of rain falling through dry air, or by contact with cold surfaces. The result is a drop in air temperature.

There are some atmospheric processes in which the amount of heat added or subtracted is very small—small enough so that the process is nearly adiabatic. Then we have the adiabatic form of the first law:

Air temperature rises (or falls) as pressure increases (or decreases).

Adiabatic processes in the atmosphere are characteristic of parts of the air, called *parcels,* which have dimensions on the order of tens of meters to kilometers. These parcels are large enough that outside air doesn't appreciably mix with the air inside them during the minutes to hours that they exist. They behave as if they are enclosed in giant, tissue-light garment bags. As a parcel flows up the side of a mountain, its pressure lessens, allowing it to expand and cool. The reduced pressure results in reduced temperature.[5] Measurements show that the temperature of a parcel of dry air will decrease by 10°C for a decrease in pressure that corresponds to an increase in altitude of 1 kilometer. So dry air cools 10°C for each kilometer it rises (Figure 18.5). Air flowing over tall mountains or rising in thunderstorms or cyclones may change elevation by several kilometers. Thus, if a parcel of dry air at ground level with a comfortable temperature of 25°C were lifted to 6 kilometers, the temperature would be a frigid −35°C. On the other hand, if air at a typical temperature of −20°C at an altitude of 6 kilometers descends to the ground, its temperature would be a whopping 40°C. A dramatic example of this adiabatic warming is the *chinook*—a wind that blows down from the Rocky Mountains across the Great Plains. Cold air moving down the slopes of the mountains is compressed into a smaller volume and is appreciably warmed (Figure 18.6). The effect of expansion or compression on gases is quite impressive.[6]

FIGURE 18.6
Chinooks, which are warm, dry winds, occur when high-altitude air descends and is adiabatically warmed.

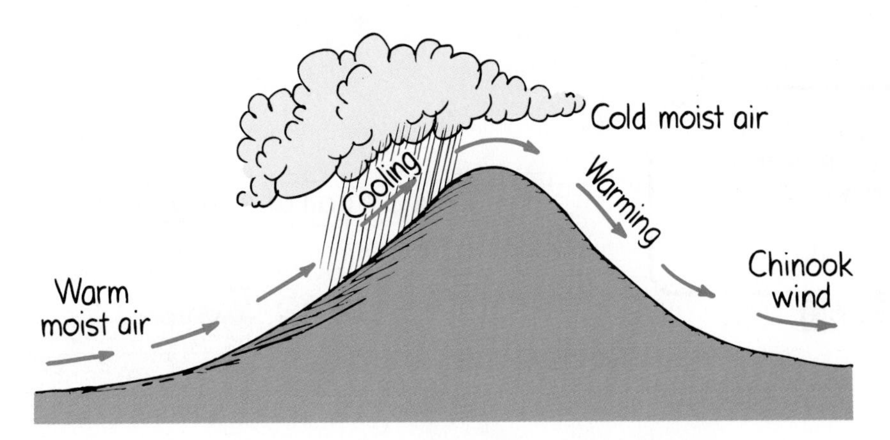

[5]Recall that, in Chapter 16, we treated the cooling of expanding air at the microscopic level by considering the behavior of colliding molecules. With thermodynamics, we consider only the macroscopic measurements of temperature and pressure to come up with the same results. It's nice to analyze things from more than one point of view.

[6]Interestingly enough, when you're flying at high altitudes where outside air temperature is typically −35°C, you're quite comfortable in your warm cabin—but not because of heaters. The process of compressing outside air to a cabin pressure that approximates atmospheric pressure at sea level would normally heat the air to a roasting 55°C (131°F). So air conditioners must be used to extract heat from the pressurized air.

FIGURE 18.7
A thunderhead is the result of the rapid adiabatic cooling of a rising mass of moist air. It derives energy from condensation of water vapor.

A rising parcel cools as it expands. But the surrounding air is cooler at increased elevations also. The parcel will continue to rise as long as it is warmer (and therefore less dense) than the surrounding air. If it gets cooler (denser) than its surroundings, it will sink. Under some conditions, large parcels of cold air sink and remain at a low level, with the result that the air above is warmer. When the upper regions of the atmosphere are warmer than the lower regions, we have a **temperature inversion.** If any rising warm air is denser than this upper layer of warm air, it will rise no farther. It is common to see evidence of this over a cold lake where visible gases and particles, such as smoke, spread out in a flat layer above the lake rather than rise and dissipate higher in the atmosphere (Figure 18.8). Temperature inversions trap smog and other thermal pollutants. The smog of Los Angeles is trapped by such an inversion, caused by low-level cold air from the ocean over which is a layer of hot air that has moved over the mountains from the hotter Mojave Desert. The mountains aid in holding the trapped air (Figure 18.9). The mountains on the edge of Denver play a similar role in trapping smog beneath a temperature inversion.[7]

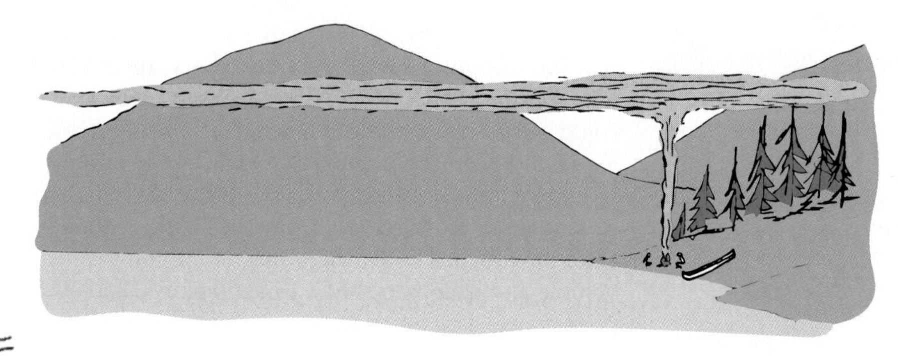

FIGURE 18.8
The layer of campfire smoke over the lake indicates a temperature inversion. The air above the smoke is warmer than the smoke, and the air below is cooler.

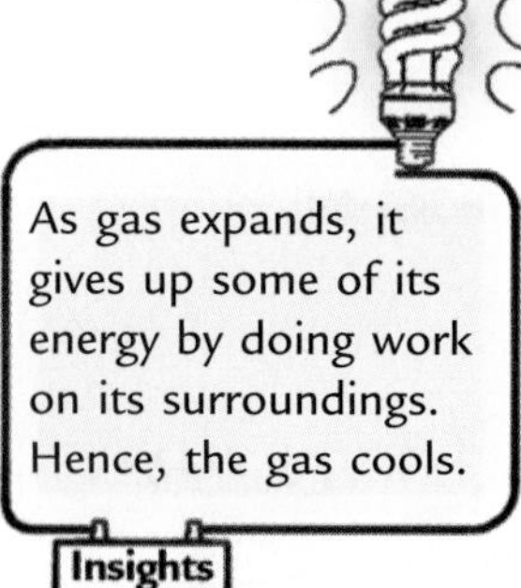

As gas expands, it gives up some of its energy by doing work on its surroundings. Hence, the gas cools.
Insights

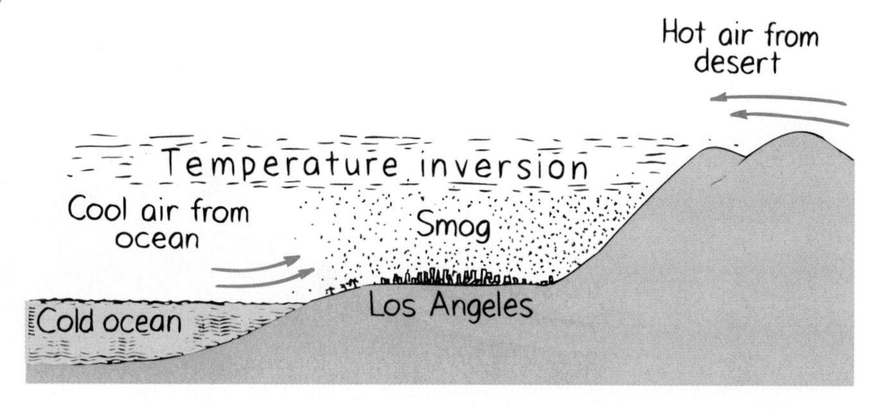

FIGURE 18.9
Smog in Los Angeles is trapped by the mountains and a temperature inversion caused by warm air from the Mojave Desert overlying cool air from the Pacific Ocean.

[7]Strictly speaking, meteorologists call any temperature profile that thwarts upward convection a temperature inversion—including instances in which the upper regions of air are cooler but not cool enough to allow continued upward convection.

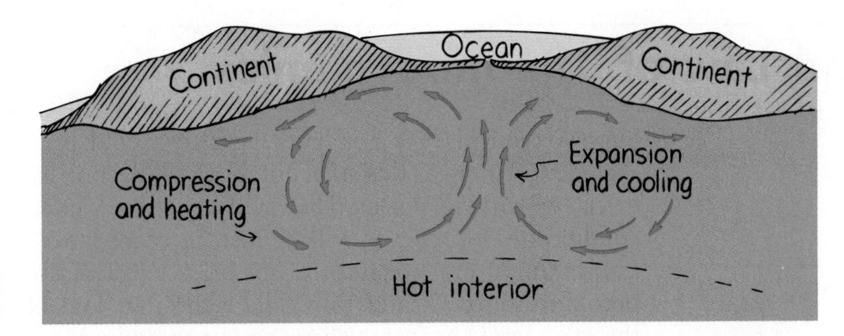

FIGURE 18.10
Interactive Figure

Do convection currents in the Earth's mantle drive the continents as they drift across the global surface? Do rising parcels of molten material cool faster or slower than the surrounding material? Do sinking parcels heat to temperatures above or below those of the surroundings? The answers to these questions are not known at this writing.

Adiabatic parcels are not restricted to the atmosphere, and changes in them do not necessarily happen quickly. Some deep ocean currents take thousands of years for circulation. The water masses are so huge and conductivities are so low that no appreciable quantities of heat are transferred to or from these parcels over these long periods of time. They are warmed or cooled adiabatically by changes in pressure. Changes in adiabatic ocean convection, as evidenced by the recurring El Niño, have a great effect on Earth's climate. Ocean convection is influenced by the temperature of the ocean floor, which, in turn, is influenced by convection currents in the molten material that lies beneath the Earth's crust (Figure 18.10). Knowledge of the behavior of molten material in Earth's mantle is difficult to acquire. Once a parcel of hot liquid material deep within the mantle begins to rise, will it continue to rise to Earth's crust? Or will its rate of adiabatic cooling find it cooler and denser than its surroundings, at which point it will sink? Is convection self-perpetuating? Geophysical types are currently pondering these questions.

CHECK YOURSELF

1. If a parcel of dry air initially at 0°C expands adiabatically while flowing upward alongside a mountain a vertical distance of 1 km, what will its temperature be? What will its temperature be when it has risen 5 km?
2. What happens to the air temperature in a valley when cold air blowing across the mountaintops descends into the valley?
3. Imagine a giant dry-cleaner's garment bag full of air at a temperature of −10°C floating like a balloon with a string hanging from it 6 km above the ground. If you were able to yank it suddenly to the ground, what would its approximate temperature be?

CHECK YOUR ANSWERS

1. At 1 km elevation, its temperature will be −10°C; at 5 km, −50°C.
2. The air is adiabatically compressed and the temperature in the valley is increased. In this way, residents of some valley towns in the Rocky Mountains, such as Salida, Colorado, experience "banana-belt" weather in midwinter.
3. If it is pulled down so quickly that heat conduction is negligible, it would be adiabatically compressed by the atmosphere and its temperature would rise to a piping-hot 50°C (122°F), just as air is heated by compression in a bicycle pump.

Second Law of Thermodynamics

Suppose you place a hot brick next to a cold brick in a thermally insulated region. You know that the hot brick will cool as it gives heat to the cold brick, which will warm. They will arrive at a common temperature: thermal equilibrium. No energy will be lost, in accordance with the first law of thermodynamics. But pretend the hot brick extracts heat from the cold brick and becomes hotter. Would this violate the first law of thermodynamics? Not if the cold brick becomes correspondingly colder so that the combined energy of both bricks remains the same. If this were to happen, it would not violate the first law. But it would violate the **second law of thermodynamics.** The second law identifies the direction of energy transformation in natural processes. The second law of thermodynamics can be stated in many ways, but, most simply, it is this:

Heat of itself never flows from a cold object to a hot object.

In winter, heat flows from inside a warm, heated home to the cold air outside. In summer, heat flows from the hot air outside into the cooler interior. The direction of spontaneous heat flow is from hot to cold. Heat can be made to flow the other way, but only by doing work on the system or by adding energy from another source—as occurs with heat pumps and air conditioners, both of which cause heat to flow from cooler to warmer places.

The huge amount of internal energy in the ocean cannot be used to light a single flashlight bulb without external effort. Energy will not of itself flow from the lower-temperature ocean to the higher-temperature bulb filament. Without external effort, the direction of heat flow is *from* hot *to* cold.

The body temperature of camels can rise several degrees above normal without causing heatstroke. Their excess heat is dissipated when air temperature drops at night.

Insights

Heat Engines

It is easy to change work completely into heat—simply rub your hands together briskly. The heat that is created then adds to the internal energy of your hands, making them warmer. Or push a crate at constant speed along a floor. All the work you do in overcoming friction is completely converted to heat, which warms the crate and the floor. But the reverse process, changing heat completely into work, can never occur. The best that can be done is the conversion of some heat to mechanical work. The first heat engine to do this was the steam engine, invented three centuries ago.

A **heat engine** is any device that changes internal energy into mechanical work. The basic idea behind a heat engine, whether a steam engine, internal combustion engine, or jet engine, is that mechanical work can be obtained only when heat flows from a high temperature to a low temperature. In every heat engine, only some of the heat can be transformed into work.

In considering heat engines, we talk about *reservoirs*. Heat flows out of a high-temperature reservoir and into a low-temperature one. Every heat engine (1) gains heat from a reservoir of higher temperature, increasing the engine's internal energy; (2) converts some of this energy into mechanical

An important source of water for a camel is not its hump but its oversized nose, which allows it to withdraw water from its own exhaled breath. Its inner nostrils are structured to recapture effectively most of the moisture contained in the warm water–saturated air moving out of its lungs.

Insights

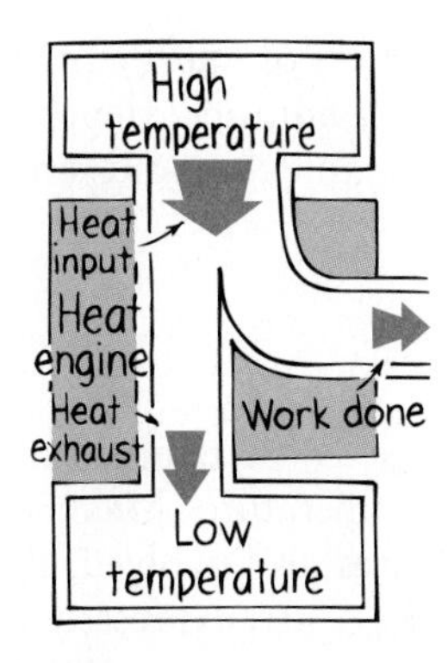

FIGURE 18.11
When heat in a heat engine flows from the high-temperature reservoir to the low-temperature sink, part of the heat can be turned into work. (If work is put into a heat engine, the flow of heat may be from the low-temperature sink to the high-temperature reservoir, as in a refrigerator or air conditioner.)

work; and (3) expels the remaining energy as heat to some lower-temperature reservoir, usually called a *sink* (Figure 18.11). In a gasoline engine, for example, (1) products of burning fuel in the combustion chamber provide the high-temperature reservoir, (2) hot gases do mechanical work on the piston, and (3) heat is expelled to the environment via the cooling system and exhaust (Figure 18.12).

The second law tells us that no heat engine can convert all the heat supplied into mechanical energy. Only *some* of the heat can be transformed into work, with the remainder expelled in the process. Applied to heat engines, the second law may be stated:

When work is done by a heat engine operating between two temperatures, T_{hot} and T_{cold}, only some of the input heat at T_{hot} can be converted to work, and the rest is expelled at T_{cold}.

Every heat engine discards some heat, which may be desirable or undesirable. Hot air expelled in a laundry on a cold winter day may be quite desirable, while the same hot air on a hot summer day is something else. When expelled heat is undesirable, we call it *thermal pollution.*

Before scientists understood the second law, many people thought that a very low-friction heat engine could convert nearly all the input heat energy to useful work. But not so. In 1824, the French engineer Nicolas Léonard Sadi Carnot[8] analyzed the functioning of a heat engine and made a

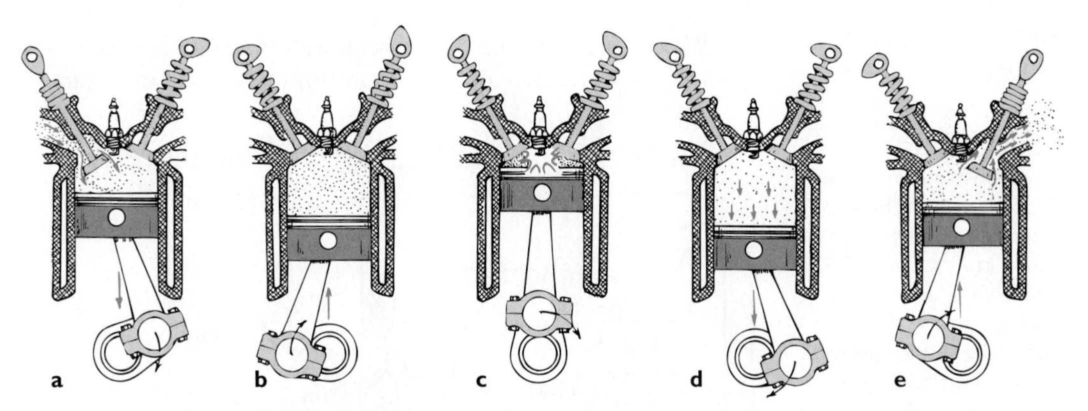

FIGURE 18.12 Interactive Figure
A four-cycle internal-combustion engine. (a) A fuel–air mixture from the carburetor fills the cylinder as the piston moves down. (b) The piston moves upward and compresses the mixture—adiabatically, because no appreciable heat is transferred in or out. (c) The spark plug fires, ignites the mixture, and raises it to a high temperature. (d) Adiabatic expansion pushes the piston downward, the power stroke. (e) The burned gases are pushed out the exhaust pipe. Then the intake valve opens and the cycle repeats. These stages can be put differently: (a) suck, (b) squeeze, (c) bang, (d) push, and (e) blow.

[8]Carnot was the son of Lazare Nicolas Marguerite Carnot (pronounced car-no), who created the fourteen armies after the revolution that defended France against all of Europe. After his defeat at Waterloo, Napoleon said to Lazare, "Monsieur Carnot, I came to know you too late." A few years after producing his famous equation, Nicolas Léonard Sadi Carnot died tragically at the age of 36 during a cholera epidemic that swept through Paris.

fundamental discovery. He showed that the greatest fraction of energy input that can be converted to useful work, even under ideal conditions, depends on the temperature difference between the hot reservoir and the cold sink. His equation is

$$\text{Ideal efficiency} = \frac{T_{\text{hot}} - T_{\text{cold}}}{T_{\text{hot}}}$$

where T_{hot} is the temperature of the hot reservoir and T_{cold} is the temperature of the cold sink.[9] Ideal efficiency depends only on the temperature difference between input and exhaust. Whenever ratios of temperatures are involved, the absolute temperature scale must be used. So T_{hot} and T_{cold} are expressed in kelvins. For example, when the hot reservoir in a steam turbine is 400 K (127°C) and the sink is 300 K (27°C), the ideal efficiency is

$$\frac{400 - 300}{400} = \frac{1}{4}$$

This means that, even under ideal conditions, only 25% of the heat provided by the steam can be converted into work, while the remaining 75% is expelled as waste. This is why steam is superheated to high temperatures in steam engines and power plants. The higher the steam temperature driving a motor or turbogenerator, the higher the possible efficiency of power production. [Increasing operating temperature in the example to 600 K yields an efficiency (600 − 300)/600 = 1/2, which is twice the efficiency at 400 K.]

We can see the role of temperature difference between heat reservoir and sink in the operation of the steam-turbine engine in Figure 18.13. The hot

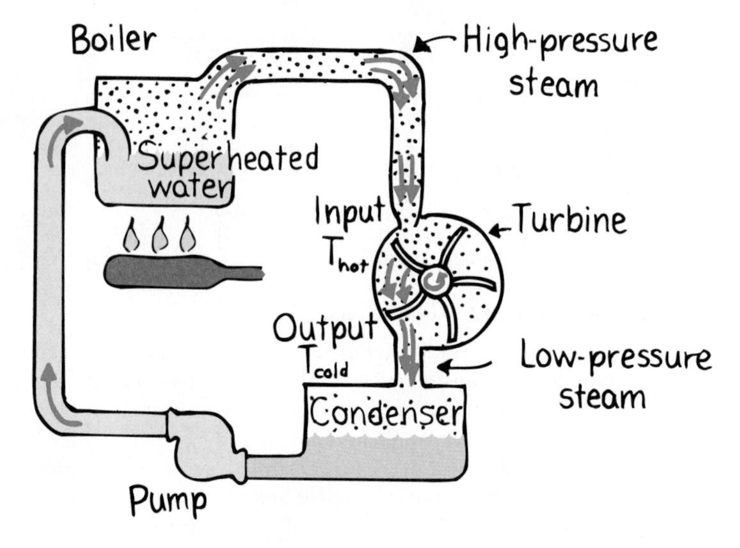

FIGURE 18.13
A simplified steam turbine. The turbine turns because pressure exerted by high-temperature steam on the front side of the turbine blades is greater than that exerted by low-temperature steam on the back side of the blades. Without a pressure difference, the turbine would not rotate and deliver energy to an external load (an electric generator, for example). The presence of steam pressure on the back side of the blades, even in the absence of friction, prevents the turbine from being a perfectly efficient engine.

[9]Efficiency = work output/heat input.
From energy conservation, heat input = work output + heat that flows out at low temperature (See Figure 18.11). So work output = heat input − heat output.
So efficiency = (heat input − heat output)/(heat input).
In the ideal case, it can be shown that the ratio (heat out)/(heat in) = $T_{\text{cold}}/T_{\text{hot}}$. Then we can say Ideal efficiency = $(T_{\text{hot}} - T_{\text{cold}})/T_{\text{hot}}$.

THERMODYNAMICS DRAMATIZED!

Put a small amount of water in an aluminum soda-pop can and heat it on a stove until steam issues from the opening. When this occurs, air has been driven out and replaced by steam. Then, with a pair of tongs, quickly invert the can into a pan of water. Crunch! The can is crushed by atmospheric pressure! Why? When the molecules of steam encounter water from the pan, condensation occurs, leaving a very low pressure in the can, whereupon the surrounding atmospheric pressure crunches the can. Here we see, dramatically, how pressure is reduced by condensation. Can you now better understand the role of condensation in the turbine of Figure 18.13?

FIGURE 18.14

reservoir is steam from the boiler and the cold sink is the exhaust region in the condenser. The hot steam exerts pressure and does work on the blades when it pushes on their front sides. This is nice. But what if the same steam pressure were also exerted on the *back side* of the blades? This would be countereffective and not so nice. It is vital that pressure on the back side of the blades be reduced. How to do this? The same way that pressure inside the can of steam is reduced in Figure 18.14. If you condense the steam, the pressure on the back sides is greatly reduced. We know that, with confined steam, temperature and pressure go hand in hand: Increase temperature and you increase pressure; decrease temperature and you decrease pressure. So the pressure difference necessary for the operation of a heat engine is directly related to the temperature difference between source and sink. The greater the temperature difference, the greater the efficiency.[10]

Carnot's equation states the upper limit of efficiency for all heat engines, whether in an automobile, a nuclear-powered ship, or a jet aircraft. In practice,

> Sharks rely on a gel under their skin that detects extremely small changes in ocean temperature—less than a thousandth of a degree Celsius. This capability likely assists in detecting subtle temperature boundaries where their prey are found.
>
> **Insights**

[10]Physicist Victor Weisskopf tells the story of an engineer who is explaining the operation of a steam engine to a peasant. The engineer explains in detail the engine's steam cycle, whereupon the peasant asks, "Yes, I understand all that, but where's the horse?" It's difficult to abandon our way of looking at the world when a newer method comes along to replace established ways.

friction is always present in all engines, and efficiency is always less than ideal.[11] So, whereas friction is solely responsible for the inefficiencies of many devices, the overriding concept in the case of heat engines is the second law of thermodynamics: Only some of the heat input can be converted to work—even without friction.

CHECK YOURSELF

1. What would be the ideal efficiency of an engine if its hot reservoir and exhaust were the same temperature—say, 400 K?
2. What would be the ideal efficiency of a machine having a hot reservoir at 400 K and a cold reservoir somehow maintained at absolute zero?

Order Tends to Disorder

FIGURE 18.15
The Transamerica Pyramid and some other buildings are heated by electric lighting, which is why the lights are on most of the time.

The first law of thermodynamics states that energy can neither be created nor destroyed. It speaks of the *quantity* of energy. The second law qualifies this by adding that the form energy takes in transformations "deteriorates" to less useful forms. It speaks of the *quality* of energy, as energy becomes more diffuse and ultimately degenerates into waste. Another way to say this is that organized energy (concentrated, and therefore usable, high-quality energy) degenerates into disorganized energy (nonusable, low-quality energy). Once water flows over a waterfall, it loses its potential for useful work. Similarly for gasoline, where organized energy degrades as it burns in a car engine. Useful energy degenerates to nonuseful forms and is unavailable for doing the same work again, such as driving another car engine. Heat, diffused into the environment as thermal energy, is the graveyard of useful energy.

The quality of energy is lowered with each transformation, as energy of an organized form tends to degrade into disorganized forms. With this broader perspective, the second law can be stated another way:

In natural processes, high-quality energy tends to transform into lower-quality energy—order tends toward disorder.

Biological systems are enormously complex and, while living, never reach thermal equilibrium.

Insights

CHECK YOUR ANSWERS

1. Zero efficiency: $(400-400)/400 = 0$. So no work output is possible for any heat engine unless a temperature difference exists between the reservoir and the sink.
2. $(400-0)/400 = 1$. Only in this idealized case is an ideal efficiency of 100% possible.

[11]The ideal efficiency of an automobile internal-combustion engine is more than 50%, but, in practice, the actual efficiency is about 25%. Engines of higher operating temperatures (compared to sink temperatures) would be more efficient, but the melting point of engine materials limits the upper temperatures at which they can operate. Higher efficiencies await engines made with new materials with higher melting points. Watch for ceramic engines!

FIGURE 18.16
If you push a heavy crate across a rough floor, all your work goes into heating the floor and the crate. Work against friction produces heat, which cannot do any work on the crate. Ordered energy is transformed into disordered energy.

FIGURE 18.17
Molecules of perfume readily go from the bottle (a more ordered state) to the air (a less ordered state), and not vice versa.

Consider a system consisting of a stack of pennies on a table, all heads up. Somebody walks by and accidentally bumps the table and the pennies topple to the floor, certainly not landing all heads up. Order becomes disorder. Molecules of gas all moving in harmony make up an orderly state—and also an unlikely state. On the other hand, molecules of gas moving in haphazard directions with a broad range of speeds make up a disorderly, chaotic (and more likely) state. If you remove the lid of a perfume bottle, molecules escape into the room and make up a more disorderly state. Relative order becomes disorder. You would not expect the reverse to happen by itself; that is, you would not expect the perfume molecules to spontaneously order themselves back into the bottle and thereby return to the more ordered containment.

CHECK YOURSELF

In your bedroom are probably some 10^{27} air molecules. If they were all to congregate on the opposite side of the room, you could suffocate. But this is unlikely. Is such a spontaneous congregation of molecules less likely, more likely, or of the same probability if there are fewer molecules in the room?

Processes in which disorder returns to order without any external help don't occur in nature. Interestingly, time is given a direction via this thermodynamic rule. Time's arrow always points from order to disorder.[12]

Disordered energy can be changed to ordered energy only with organizational effort or work input. For example, water in a refrigerator freezes and becomes more ordered because work is put into the refrigeration cycle; gas can be ordered into a small region if a compressor supplied with outside energy does work. Processes in which the net effect is an increase in order always require an external input of energy. But, for such processes, there is always an increase of disorder somewhere else to more than offset the increase of order.

CHECK YOUR ANSWER

Fewer molecules mean a greater probability of their spontaneously congregating on the opposite side of your room. Exaggeration makes this credible: If there is only one molecule in the room, there's a 50% chance of it being in the other half of the room. If you have two molecules, the chances of both being on one side at the same time is 25%. If there are three molecules, the chance of your being breathless is one-eighth (12.5%). The greater the number of molecules, the greater the chances of there being nearly equal numbers of molecules on both sides of the room.

[12]Reversible systems look sensible when a movie film of them is run backward. Remember those old movies where a train comes to a stop inches away from a heroine tied to the tracks? How was this done without mishap? Simple. The train began at rest, inches away from the heroine, and went *backward*, gaining speed. When the film was reversed, the train was seen to move *toward* the heroine. Watch closely for the telltale smoke that *enters* the smokestack!

Entropy

FIGURE 18.18
Entropy.

The idea of lowering the "quality" of energy is embodied in the idea of **entropy,** a measure of the *amount of disorder* in a system.[13] More entropy means more degradation of energy. Since energy tends to degrade and to disperse with time, the total amount of entropy in any system tends to increase with time. Whenever a physical system is allowed to distribute its energy freely, it always does so in a manner such that entropy increases while the energy of the system remaining available for doing work decreases.

The net entropy in the universe is continually increasing (continually running "downhill"). We say *net* because there are some regions in which energy is actually being organized and concentrated. This occurs in living organisms, which survive by concentrating and organizing energy from food sources. All living organisms, from bacteria to trees to human beings, extract energy from their surroundings and use it to increase their own organization. In living organisms, entropy decreases. But the order in life forms is maintained by increasing entropy elsewhere; resulting in a net increase in entropy. Energy must be transformed into the living system to support life. When it isn't, the organism soon dies and tends toward disorder.[14]

The first law of thermodynamics is a universal law of nature to which no exceptions have been observed. The second law, however, is a probabilistic statement. Given enough time, even the most improbable states may occur; entropy may sometimes decrease. For example, the haphazard motions of air molecules could momentarily become harmonious in a corner of the room, just as a

FIGURE 18.19
Why is the motto of this contractor—"Increasing entropy is our business"—so appropriate?

Old riddle: "How do you unscramble an egg?" Answer: "Feed it to a chicken." But even then you won't get your original egg back. Making eggs takes energy and increases entropy.

Insights

[13]Entropy can be expressed mathematically. The increase in entropy ΔS of a thermodynamic system is equal to the amount of heat added to the system ΔQ divided by the temperature T at which the heat is added: $\Delta S = \Delta Q/T$

[14]Interestingly enough, the American writer Ralph Waldo Emerson, who lived during the time the second law of thermodynamics was the new science topic of the day, philosophically speculated that not everything becomes more disordered with time and cited the example of human thought. Ideas about the nature of things grow increasingly refined and better organized as they pass through the minds of succeeding generations. Human thought is evolving toward more order.

barrelful of pennies spilled on the floor could all come up heads. These situations are possible, but they are not probable. The second law tells us the most probable course of events, not the only possible one.

The laws of thermodynamics are often stated this way: You can't win (because you can't get any more energy out of a system than you put into it), you can't break even (because you can't get as much useful energy out as you put in), and you can't get out of the game (entropy in the universe is always increasing).

Summary of Terms

Thermodynamics The study of heat and its transformation to different forms of energy.

Absolute zero The lowest possible temperature that a substance may have; the temperature at which particles of a substance have their minimum kinetic energy.

Internal energy The total energy (kinetic plus potential) of the submicroscopic particles that make up a substance. *Changes* in internal energy are of principal concern in thermodynamics.

First law of thermodynamics A restatement of the law of energy conservation, applied to systems in which energy is transferred by heat and/or work. The heat added to a system equals its increase in internal energy plus the external work it does on its environment.

Adiabatic process A process, often of fast expansion or compression, wherein no heat enters or leaves a system.

Temperature inversion A condition in which upward convection of air ceases, often because an upper region of the atmosphere is warmer than the region below it.

Second law of thermodynamics Thermal energy never spontaneously flows from a cold object to a hot object. Also, no machine can be completely efficient in converting heat to work; some of the heat supplied to the machine at high temperature is dissipated as waste heat at lower temperature. And finally, all systems tend to become more and more disordered as time goes by.

Heat engine A device that uses heat as input and supplies mechanical work as output, or that uses work as input and moves heat "uphill" from a cooler to a warmer place.

Entropy A measure of the disorder of a system. Whenever energy freely transforms from one form to another, the direction of transformation is toward a state of greater disorder and therefore toward one of greater entropy.

Review Questions

1. From where does the word *thermodynamics* stem, and what does it mean?
2. Is the study of thermodynamics concerned primarily with microscopic processes or with macroscopic ones?

Absolute Zero

3. By how much does the volume of gas at 0°C contract for each decrease in temperature of 1 Celsius degree when the pressure is held constant?
4. By how much does the pressure of gas at 0°C decrease for each decrease in temperature of 1 Celsius degree when the volume is held constant?
5. If we assume the gas does not condense to a liquid, what volume is approached for a gas at 0°C cooled by 273 Celsius degrees?
6. What is the lowest possible temperature on the Celsius scale? On the Kelvin scale?

Internal Energy

7. What else besides molecular kinetic energy contributes to the internal energy of a substance?
8. Is the principal concern in the study of thermodynamics the *amount* of internal energy in a system or the *changes* in internal energy in a substance?

First Law of Thermodynamics

9. How does the law of the conservation of energy relate to the first law of thermodynamics?
10. What is the relationship between heat added to a system, change in its internal energy, and external work done by the system?
11. What happens to the internal energy of a system when mechanical work is done on it? What happens to its temperature?

Adiabatic Processes

12. What condition is necessary for a process to be adiabatic?
13. If work is done *on* a system, does the internal energy of the system increase or decrease? If work is done *by* a system, does the internal energy of the system increase or decrease?

Meteorology and the First Law

14. How do meteorologists express the first law of thermodynamics?
15. What is the adiabatic form of the first law?
16. What generally happens to the temperature of rising air? Of sinking air?
17. What is a temperature inversion?
18. Do adiabatic processes apply only to gases? Defend your answer.

Second Law of Thermodynamics

19. How does the second law of thermodynamics relate to the direction of heat flow?

Heat Engines

20. What three processes occur in every heat engine?
21. What exactly is thermal pollution?
22. How does the second law relate to heat engines?
23. Why is the condensation part of the cycle in a steam turbine so essential?

Order Tends to Disorder

24. Distinguish between high-quality energy and low-quality energy in terms of organized and disorganized energy. Give an example of each.
25. How can the second law be stated with regard to high-quality and lower-quality energy?
26. With respect to orderly and disorderly states, what do natural systems tend to do? Can a disorderly state ever transform to an orderly state? Explain.

Entropy

27. What is the physicist's term for *measure of amount of disorder*?
28. Distinguish between the first and second laws of thermodynamics in terms of whether or not exceptions occur.

Exercises

1. A friend said the temperature inside a certain oven is 500 and the temperature inside a certain star is 50,000. You're unsure whether your friend meant Celsius degrees or kelvins. How much difference does it make in each case?
2. The temperature of the Sun's interior is about 10^7 degrees. Does it matter whether this is degrees Celsius or kelvins? Explain.
3. When heat flows from a warm object in contact with a cool object, do both objects undergo the same amount of temperature change?
4. Heat always flows spontaneously from an object with a higher temperature to an object with a lower temperature. Is this the same thing as saying that heat always flows from an object with a greater internal energy to one with a lower internal energy? Explain.
5. Helium has the special property that its internal energy is directly proportional to its absolute temperature. Consider a flask of helium with a temperature of 10°C. If it is heated until it has twice the internal energy, what will its temperature be?
6. If you vigorously shake a can of liquid back and forth for more than a minute, will the temperature of the liquid increase? (Try it and see.)
7. When air is quickly compressed, why does its temperature increase?
8. Suppose you do 100 J of work in compressing a gas. If 80 J of heat escapes in the process, what is the change in internal energy of the gas?
9. When you pump a tire with a bicycle pump, the cylinder of the pump becomes hot. Give two reasons why this is so.
10. What happens to the gas pressure within a sealed gallon can when it is heated? When it is cooled? Why?
11. Is it possible to wholly convert a given amount of mechanical energy into thermal energy? Is it possible to wholly convert a given amount of thermal energy into mechanical energy? Cite examples to illustrate your answers.
12. Why do diesel engines require no spark plugs?
13. Everybody knows that warm air rises. So it might seem that the air temperature should be higher at the tops of mountains than down below. But the opposite is most often the case. Why?
14. What is the ultimate source of energy in coal, oil, and wood? Why do we call energy from wood renewable but energy from coal and oil nonrenewable?
15. What is the ultimate source of energy in a hydroelectric power plant?
16. The combined molecular kinetic energies of molecules in a cool lake are greater than the combined molecular kinetic energies of molecules in a cup of hot tea. Pretend you partially immerse the teacup in the lake and that the tea *absorbs* 10 calories from the water and becomes hotter, while the water that gives up 10 calories becomes cooler. Would this energy

transfer violate the first law of thermodynamics? The second law of thermodynamics? Defend your answers.

17. Why is *thermal pollution* a relative term?

18. Figure 18.14 shows the crushing of an inverted evacuated can placed in a pan of water. Would the water need to be cold? Would crushing occur if the water were hot but not boiling? Would it crush in boiling water? (Try it and see!)

19. Why is it advantageous to use steam that is as hot as possible in a steam-driven turbine?

20. Nearly all automobiles are powered by internal-combustion engines. A few are powered by *motors*—devices that convert electrical energy to mechanical energy. Which requires fuel for operation (or do both)? Defend your answer.

21. How does the ideal efficiency of an automobile relate to the temperature of the engine and the temperature of the environment in which it operates? Be specific.

22. Will the efficiency of a car engine increase, decrease, or remain the same if the muffler is removed? If it is driven on a very cold day? Defend your answers.

23. What happens to the efficiency of a heat engine when the temperature of the reservoir into which thermal energy is transferred is lowered?

24. To increase the efficiency of a heat engine, would it be preferable to produce the same temperature increment by increasing the temperature of the reservoir while maintaining the temperature of the sink constant, or to decrease the temperature of the sink while maintaining the temperature of the reservoir constant? Explain.

25. Under what conditions would a heat engine be 100% efficient?

26. Could you cool a kitchen by leaving the refrigerator door open and closing the kitchen door and windows? Explain.

27. Could you warm a kitchen by leaving the door of a hot oven open? Explain.

28. An electric fan not only doesn't decrease the temperature of air, but it actually increases air temperature. How, then, are you cooled by a fan on a hot day?

29. Why will a refrigerator containing a fixed amount of food consume more energy in a warm room than in a cold room?

30. A refrigerator moves heat from cold to warm. Why does this not violate the second law of thermodynamics?

31. What happens to the density of a quantity of gas when its temperature is decreased and its pressure is held constant?

32. If you squeeze an air-filled balloon and no heat escapes, what happens to the internal energy of the gas in the balloon?

33. In buildings that are being electrically heated, is it at all wasteful to turn all the lights on? Is turning all the lights on wasteful if the building is being cooled by air conditioning?

34. A wet bathing suit spontaneously chills itself (and its occupant). How can this happen without violating the second law of thermodynamics? (*Hint:* Is the bathing suit just transferring heat to its warmer surroundings, or is it doing more than that?)

35. Using both the first and the second laws of thermodynamics, defend the statement that 100% of the electrical energy that goes into lighting a lamp is converted to thermal energy.

36. Molecules in the combustion chamber of a rocket engine are in a high state of random motion. When the molecules are expelled through a nozzle in a more ordered state, will their temperature be more, less, or the same as their temperature in the chamber before being exhausted?

37. Is the total energy of the universe becoming more unavailable with time? Explain.

38. According to the second law of thermodynamics, is the universe moving to a more ordered state or to a more disordered state?

39. Comment on this statement: The second law of thermodynamics is one of the most fundamental laws of nature, yet it is not an exact law at all.

40. The ocean possesses enormous numbers of molecules, all with kinetic energy. Can this energy be extracted and used as a power source? Defend your answer.

41. Why do we say a substance in a liquid phase is more disordered than the same substance in a solid phase?

42. Water evaporates from a salt solution and leaves behind salt crystals that have a higher degree of molecular order than the more randomly moving molecules in the salt water. Has the entropy principle been violated? Why or why not?

43. Water put into the freezer compartment of your refrigerator goes to a state of less molecular disorder when it freezes. Is this an exception to the entropy principle? Explain.

44. As a chicken grows from an egg, it becomes more ordered with time. Does this violate the principle of entropy? Explain.

45. The United States Patent and Trademark Office rejects claims for perpetual motion machines (in which energy output is as great or greater than energy input) without even studying them. Why is this?

46. It is generally assumed that perpetual motion machines are impossible to construct. Is it inconsistent to say that molecules are in perpetual motion?

47. What is your response to a friend who says perpetual motion is an impossible state?

48. (a) If you spent ten minutes repeatedly shaking and throwing down a pair of coins, would you expect to see two heads come up at least once? (b) If you spent an hour shaking a handful of ten coins and throwing them down, would you expect to see all ten come up heads at least once? (c) If you stirred a box of 10,000 coins and dumped them repeatedly on the floor all day long, would you expect to see all 10,000 come up heads at least once?

49. In your bedroom are probably some 10^{27} air molecules. If they all happened to congregate on one side of the room, you could suffocate. But this is unlikely. Is such a circumstance less likely, more likely, or the same if there are many times fewer molecules in the room?

50. Make up two multiple-choice questions that would check a classmate's understanding of the distinction between heat and internal energy.

Problems

1. During a certain thermodynamic process, a sample of gas expands and cools, reducing its internal energy by 3000 J, while no heat is added or taken away. How much work is done during this process?

2. What is the ideal efficiency of an automobile engine in which fuel is heated to 2700 K and the outdoor air is at 270 K?

3. Calculate the Carnot efficiency of an OTEC power plant that operates on the temperature difference between deep 4°C water and 25°C surface water.

4. On a chilly 10°C day, your friend who likes cold weather says she wishes it were twice as cold. Taking this literally, what temperature would this be?

5. Imagine a giant dry-cleaner's bag full of air at a temperature of −35°C floating like a balloon with a string hanging from it 10 km above the ground. Estimate what its temperature would be if you were able to yank it suddenly back to the Earth's surface.

6. A power station with an efficiency of 0.4 generates 10^8 W of electric power and dissipates 1.5×10^8 W of thermal energy to the cooling water that flows through it. Knowing that the specific heat of water in SI units is 4,184 J/kg°C, calculate how many kg of water flows through the plant each second if the water is heated through 3 Celsius degrees.

7. A heat pump moves heat from a cooler to a warmer place, and is the heart of a refrigerator or an air conditioner; it is used to heat houses. The minimum work input needed to move energy "uphill" from T_{cold} to T_{hot} is

$$\text{Min work} = (\text{Energy moved}) \times (T_{hot} - T_{cold})/T_{cold}.$$

Calculate the minimum work need to move 1 J of energy (a) from inside a room at $T_{cold} = 295$ K to the outdoors at $T_{hot} = 308$ K, (b) from inside a lab freezer at $T_{cold} = 173$ K to a room at $T_{hot} = 293$ K, and (c) from a helium refrigerator whose inside temperature is $T_{cold} = 4$ K to a room with $T_{hot} = 300$ K. Comment on the differences.

8. Construct a table of all the possible combination of numbers that can come up when you throw two dice. Your friend says, "Yes, I know that seven is the most likely total number when two dice are thrown. But *why* seven?" Based on your table, answer your friend, and explain that, in thermodynamics, the situations that are likely to be observed are those that can be formed in the greatest number of ways.

PART FOUR

Sound

This CD is the pits—billions of them, inscribed in an array that is scanned by a laser beam at millions of pits per second. It's the sequence of pits detected as light and dark spots that forms a binary code that is converted into a continuous audio waveform. Digitized music! Whoever thought that something as complex as Beethoven's Fifth Symphony could be reduced to a series of ones and zeros? Sound physics!

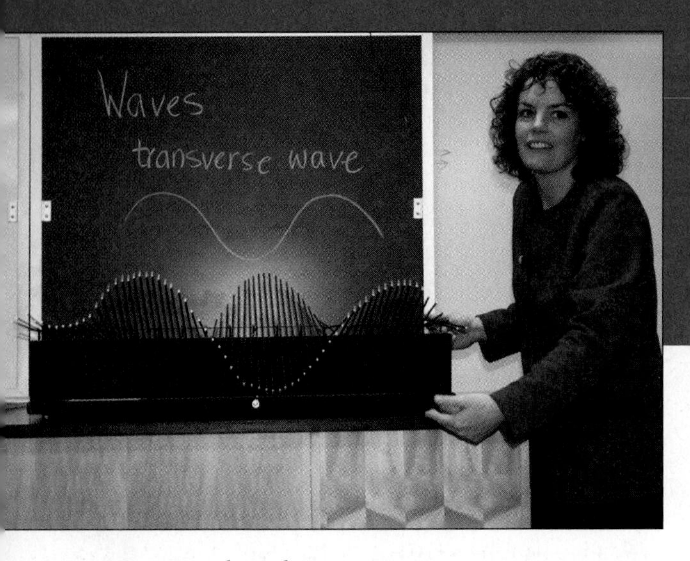

CHAPTER 19

Vibrations and Waves

Diane Riendeau shows how a vibration produces a wave.

In a general sense, anything that moves back and forth, to and fro, side to side, in and out, or up and down is vibrating. A *vibration* is a wiggle in time. A wiggle in both space and time is a *wave*. A wave extends from one place to another. Light and sound are both vibrations that propagate throughout space as waves, but two very different kinds of waves. Sound is the propagation of vibrations through a material medium—a solid, liquid, or gas. If there is no medium to vibrate, then no sound is possible. Sound cannot travel in a vacuum. But light can, for, as we shall learn in later chapters, light is a vibration of electric and magnetic fields—a vibration of pure energy. Although light can pass through many materials, it needs none. This is evident when it propagates through the vacuum between the Sun and the Earth. The source of all waves—sound, light, or whatever—is something that is vibrating. We shall begin our study of vibrations and waves by considering the motion of a simple pendulum.

Vibration of a Pendulum

If we suspend a stone at the end of a piece of string, we have a simple pendulum. Pendulums swing to and fro with such regularity that, for a long time, they were used to control the motion of most clocks. They can still be found in grandfather clocks and cuckoo clocks. Galileo discovered that the time a pendulum takes to swing to and fro through small distances depends only on the *length of the pendulum.*[1] The time of a to-and-fro swing, called the *period,* does not depend on the mass of the pendulum or on the size of the arc through which it swings.

A long pendulum has a longer period than a short pendulum; that is, it swings to and fro less frequently than a short pendulum. A grandfather clock's pendulum with a length of about 1 meter, for example, swings with a leisurely period of 2 seconds, while the much shorter pendulum of a cuckoo clock swings with a period that is less than a second.

In addition to length, the period of a pendulum depends on the acceleration of gravity. Oil and mineral prospectors use very sensitive pendulums to

[1]The exact equation for the period of a simple pendulum for small arcs is $T = 2\pi\sqrt{l/g}$, where T is the period, l is the length of the pendulum, and g is the acceleration of gravity.

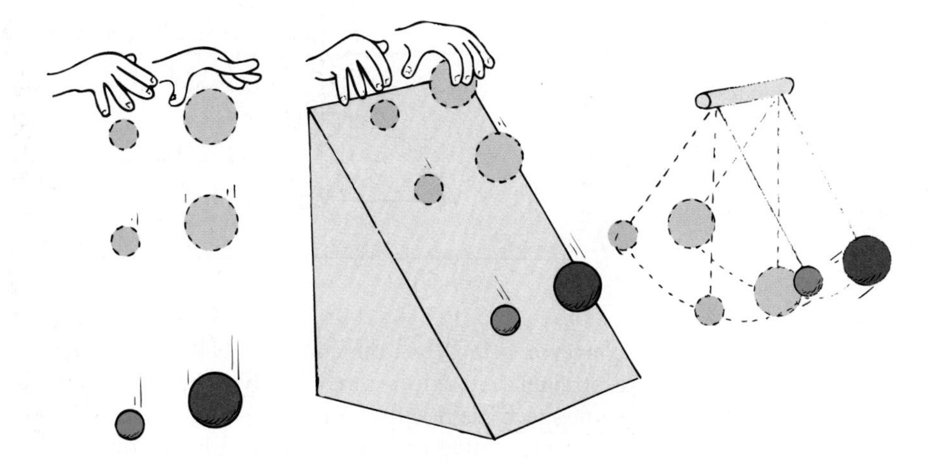

FIGURE 19.1
If you drop two balls of different mass, they accelerate at g. Let them slide without friction down the same incline, and they slide together at the same fraction of g. Tie them to strings of the same length so they are pendulums, and they swing to and fro in unison. In all cases, the motions are independent of mass.

detect slight differences in this acceleration, which is affected by the densities of underlying formations.

Wave Description

The to-and-fro vibratory motion (often called *oscillatory* motion) of a swinging pendulum in a small arc is called *simple harmonic motion*.[2] The pendulum bob filled with sand in Figure 19.2 exhibits simple harmonic motion above a

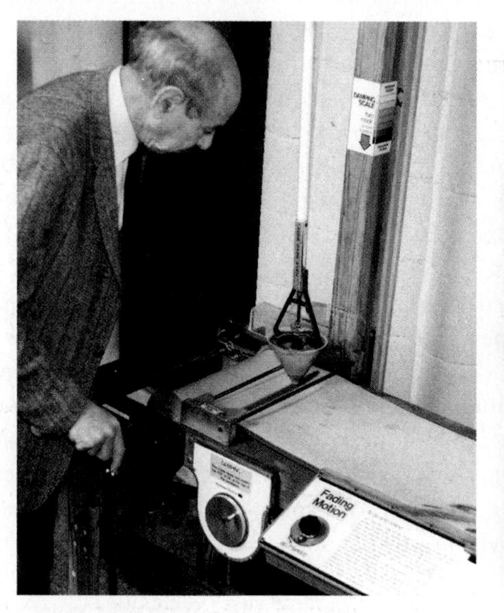

a

b

FIGURE 19.2
Frank Oppenheimer at the San Francisco Exploratorium demonstrates (a) a straight line traced by a swinging pendulum bob that leaks sand on the stationary conveyor belt. (b) When the conveyor belt is uniformly moving, a sine curve is traced.

[2]The condition for simple harmonic motion is that the restoring force is proportional to the displacement from equilibrium. This condition is met, at least approximately, for most vibrations. The component of weight that restores a displaced pendulum to its equilibrium position is directly proportional to the pendulum's displacement (for small angles)—likewise for a bob attached to a spring. Recall, from page 234 in Chapter 12, Hooke's law for a spring: $F = k\Delta x$, where the force to stretch (or compress) a spring is directly proportional to the distance stretched (or compressed).

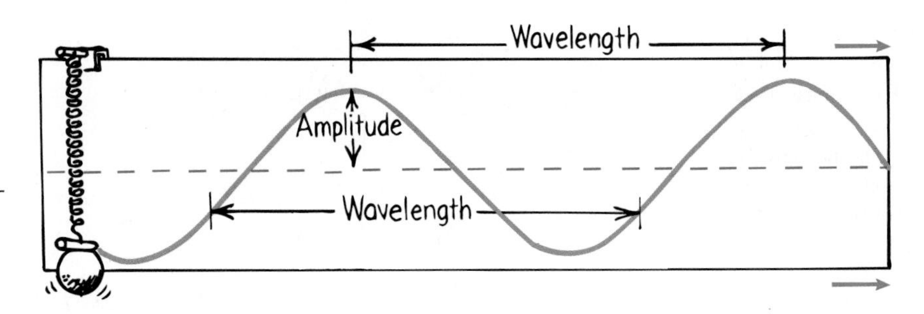

FIGURE 19.3
Interactive Figure
When the bob vibrates up and down, a marking pen traces out a sine curve on paper that is moved horizontally at constant speed.

conveyor belt. When the conveyor belt is not moving (left), the sand traces out a straight line. More interestingly, when the conveyor belt is moving at constant speed (right), the sand traces out a special curve known as a **sine curve.**

A sine curve can also be traced by a bob attached to a spring undergoing vertical simple harmonic motion (Figure 19.3). A sine curve is a pictorial representation of a wave. As for a water wave, the high points of a sine wave are called *crests* and the low points are called *troughs.* The straight dashed line represents the "home" position, or midpoint of the vibration. The term **amplitude** refers to the distance from the midpoint to the crest (or trough) of the wave. So the amplitude equals the maximum displacement from equilibrium.

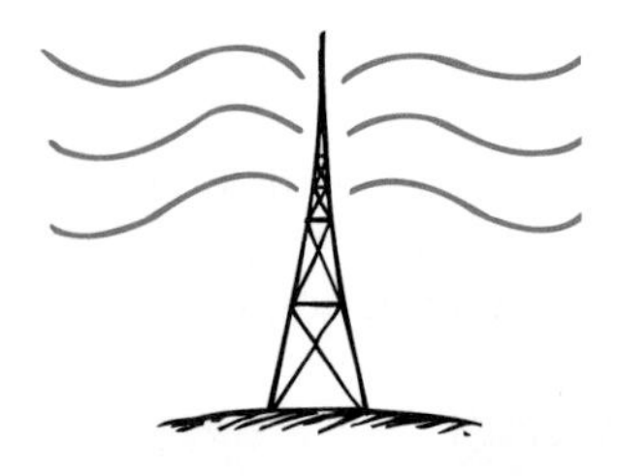

FIGURE 19.4
Electrons in the transmitting antenna vibrate 940,000 times each second and produce 940-kHz radio waves.

The **wavelength** of a wave is the distance from the top of one crest to the top of the next one. Or, equivalently, the wavelength is the distance between any successive identical parts of the wave. The wavelengths of waves at the beach are measured in meters, the wavelengths of ripples in a pond are measured in centimeters, and the wavelengths of light are measured in billionths of a meter (nanometers).

How frequently a vibration occurs is described by its **frequency.** The frequency of a vibrating pendulum, or of an object on a spring, specifies the number of to-and-fro vibrations it makes in a given time (usually one second). A complete to-and-fro oscillation is one vibration. If it occurs in one second, the frequency is one vibration per second. If two vibrations occur in one second, the frequency is two vibrations per second.

The unit of frequency is called the **hertz** (Hz), after Heinrich Hertz, who demonstrated radio waves in 1886. One vibration per second is 1 hertz; two vibrations per second is 2 hertz, and so on. Higher frequencies are measured in kilohertz (kHz, thousands of hertz), and still higher frequencies in megahertz (MHz, millions of hertz) or gigahertz (GHz, billions of hertz). AM radio waves are measured in kilohertz, while FM radio waves are measured in megahertz; radar and microwave ovens operate at gigahertz frequencies. A station at 960 kHz on the AM radio dial, for example, broadcasts radio waves that have a frequency of 960,000 vibrations per second. A station at 101.7 MHz on the FM dial broadcasts radio waves with a frequency of 101,700,000 hertz. These radio-wave frequencies are the frequencies at which electrons are forced to vibrate in the antenna of a radio station's transmitting tower. The source of all waves is something that vibrates. The frequency of the vibrating source and the frequency of the wave it produces are the same.

The frequency of a "classical" wave—such as a sound wave, water wave, or radio wave—matches the frequency of its vibrating source. (In the quantum world of atoms and photons, the rules are different.)

The **period** of a vibration or wave is the time for one complete vibration. If an object's frequency is known, its period can be calculated, and vice versa. Suppose, for example, that a pendulum makes two vibrations in one second. Its frequency is 2 Hz. The time needed to complete one vibration—that is, the period of vibration—is $\frac{1}{2}$ second. Or, if the vibration frequency is 3 Hz, then the period is $\frac{1}{3}$ second. The frequency and period are the inverse of each other:

$$\text{Frequency} = \frac{1}{\text{period}}$$

or, vice versa,

$$\text{Period} = \frac{1}{\text{frequency}}$$

CHECK YOURSELF

1. What is the frequency of a wave, given that its period is about 0.01667 second?
2. Gusts of wind make the Sears Building in Chicago sway back and forth at a vibration frequency of about 0.1 Hz. What is its period of vibration?

Wave Motion

Most information about our surroundings comes to us in some form of waves. It is through wave motion that sounds come to our ears, light to our eyes, and electromagnetic signals to our radios and television sets. Through *wave motion,* energy can be transferred from a source to a receiver without the transfer of matter between the two points.

Wave motion can be most easily understood by first considering the simple case of a horizontally stretched rope. If one end of such a rope is shaken up and down, a rhythmic disturbance travels along the rope. Each particle of the rope moves up and down, while, at the same time, the disturbance moves along the length of the rope. The medium, rope or whatever, returns to its initial condition after the disturbance has passed. What is propagated is the disturbance, not the medium itself.

Perhaps a more familiar example of wave motion is provided by a water wave. If a stone is dropped into a quiet pond, waves will travel outward in expanding circles, the centers of which are at the source of the disturbance. In this case, we might think that water is being transported with the waves, since water is splashed onto previously dry ground when the waves meet the shore. We should realize, however, that, barring obstacles, the water will flow back into the pond, and things will be much as they were in the beginning. The surface of the water will have been disturbed, but the water itself will have gone nowhere. A leaf on the surface will bob up and down as the waves pass, but will end up where it started. Again, the medium returns to its initial condition after the disturbance has passed—even in the extreme case of a tsunami.

CHECK YOUR ANSWERS

1. Frequency = 1/period = 1/0.01667 s = 60 Hz. (0.01667 = 1/60.) So the wave vibrates 60 times per second and has a period of 1/60 second.
2. The period is 1/frequency = 1/(0.1 Hz) = 1/(0.1 vibration/s) = 10 s. Each vibration therefore takes 10 s.

Let us consider another example of a wave to illustrate that what is transported from one place to another is a disturbance in a medium, not the medium itself. If you view a field of tall grass from an elevated position on a gusty day, you will see waves travel across the grass. The individual stems of grass do not leave their places; instead, they swing to and fro. Furthermore, if you stand in a narrow footpath, the grass that blows over the edge of the path, brushing against your legs, is very much like the water that doused the shore in our earlier example. While wave motion continues, the tall grass swings back and forth, vibrating between definite limits but going nowhere. When the wave motion stops, the grass returns to its initial position.

Wave Speed

The speed of periodic wave motion is related to the frequency and wavelength of the waves. We can understand this by considering the simple case of water waves (Figures 19.5 and 19.6). Imagine that we fix our eyes on a stationary point on the surface of water and observe the waves passing by this point. We can measure how much time passes between the arrival of one crest and the arrival of the next one (the period), and also observe the distance between crests (the wavelength). We know that speed is defined as distance divided by time. In this case, the distance is one wavelength and the time is one period, so **wave speed** = wavelength/period.

For example, if the wavelength is 10 meters and the time between crests at a point on the surface is 0.5 second, the wave moves 10 meters in 0.5 seconds and its speed is 10 meters divided by 0.5 seconds, or 20 meters per second.

Since period is equal to the inverse of frequency, the formula wave speed = wavelength/period can also be written

$$\text{Wave speed} = \text{wavelength} \times \text{frequency}$$

This relationship holds true for all kinds of waves, whether they are water waves, sound waves, or light waves.

The speed v of a wave can be expressed by the equation $v = f\lambda$, where f is the wave frequency and λ (the Greek letter lambda) is the wavelength of the wave.

Insights

FIGURE 19.5
Water waves.

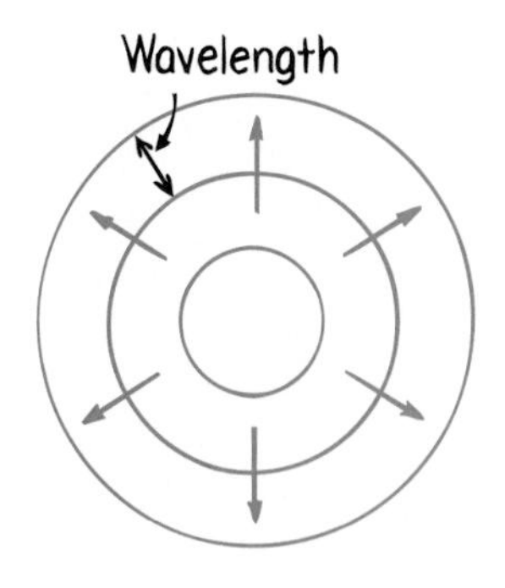

FIGURE 19.6
A top view of water waves.

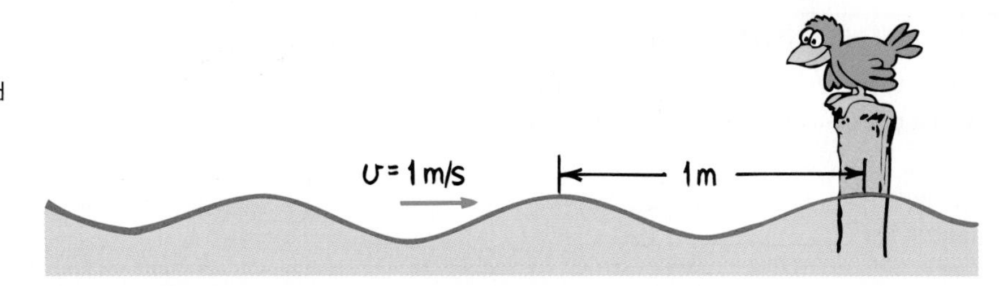

FIGURE 19.7
Interactive Figure
If the wavelength is 1 m, and one wavelength per second passes the pole, then the speed of the wave is 1 m/s.

CHECK YOURSELF

1. If a train of freight cars, each 10 m long, rolls by you at the rate of three cars each second, what is the speed of the train?
2. If a water wave oscillates up and down three times each second and the distance between wave crests is 2 m, what is its frequency? What is its wavelength? What is its wave speed?

Transverse Waves

Fasten one end of a rope to a wall and hold the free end in your hand. If you suddenly twitch the free end up and then down, a pulse will travel along the rope and back (Figure 19.8). In this case, the motion of the rope (up and down arrows) is at right angles to the direction of wave speed. The right-angled, or sideways, motion is called *transverse motion.* Now shake the rope with a regular, continuing up-and-down motion, and the series of pulses will produce a wave. Since the motion of the medium (the rope, in this case) is transverse to the direction the wave travels, this type of wave is called a **transverse wave.**

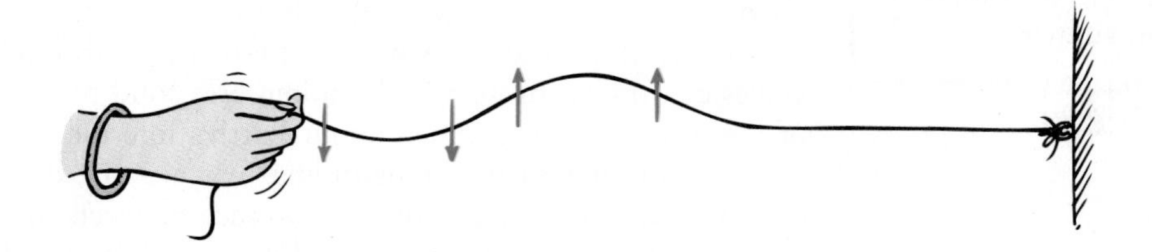

FIGURE 19.8
A transverse wave.

CHECK YOUR ANSWERS

1. 30 m/s. We can see this in two ways. (a) According to the speed definition from Chapter 2, $v = d/t = (3 \times 10\text{ m})/1\text{ s} = 30\text{ m/s}$, since 30 m of train passes you in 1 s. (b) If we compare our train to wave motion, where wavelength corresponds to 10 m and frequency is 3 Hz, then speed = wavelength × frequency = 10 m × 3 Hz = 10 m × 3/s = 30 m/s.
2. The frequency of the wave is 3 Hz, its wavelength is 2 m, and its wave speed = wavelength × frequency = 2 m × 3/s = 6 m/s. It is customary to express this as the equation $v = \lambda f$ where v is wave speed, λ (the Greek letter lambda) is wavelength, and f is wave frequency.

PRACTICING PHYSICS

Here we have a sine curve that represents a transverse wave. With a ruler, measure the wavelength and amplitude of the wave.

Wavelength = ____________

Amplitude = ____________

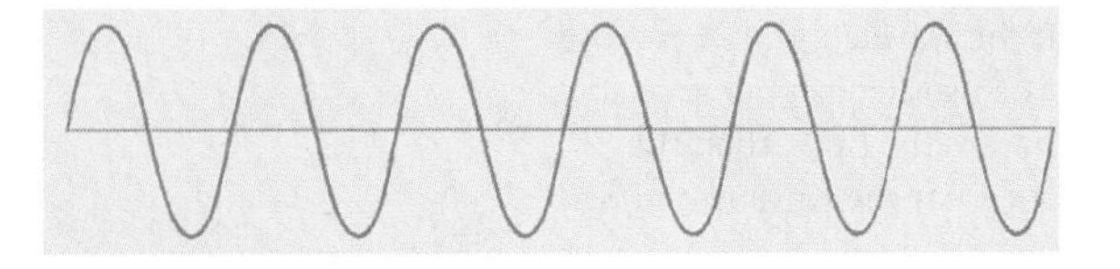

Waves in the stretched strings of musical instruments and on the surfaces of liquids are transverse. We will see later that electromagnetic waves, which make up radio waves and light, are also transverse.

Longitudinal Waves

Sound requires a medium. It can't travel in a vacuum because there's nothing to compress and stretch.

Insights

Not all waves are transverse. Sometimes parts that make up a medium move to and fro in the same direction in which the wave travels. Motion is *along* the direction of the wave rather than at right angles to it. This produces a **longitudinal wave.**

Both a transverse wave and a longitudinal wave can be demonstrated with a spring or a Slinky stretched out on the floor, as shown in Figure 19.9. A transverse wave is demonstrated by shaking the end of a Slinky from side to side. A longitudinal wave is demonstrated by rapidly pulling and pushing the end of the Slinky toward and away from you. In this case, we see that the medium vibrates parallel to the direction of energy transfer. Part of the Slinky is compressed, and a wave of *compression* travels along the spring. In between successive compressions is a stretched region, called a *rarefaction*. Both compressions and rarefactions travel in the same direction along the Slinky. Sound waves are longitudinal waves.

Waves that have been generated by earthquakes and that travel in the ground are of two main types: longitudinal P waves and transverse S waves. (Geology students often remember P waves as "push–pull" waves, and S waves as "side-to-side" waves.) S waves cannot travel through liquid matter, while P waves can travel through both molten and solid parts of Earth's interior. Study of these waves reveals much about Earth's interior.

The wavelength of a longitudinal wave is the distance between successive compressions or, equivalently, the distance between successive rarefactions. The most common example of longitudinal waves is sound in air. Elements of air vibrate to and fro about some equilibrium position as the waves move by. We will treat sound waves in detail in the next chapter.

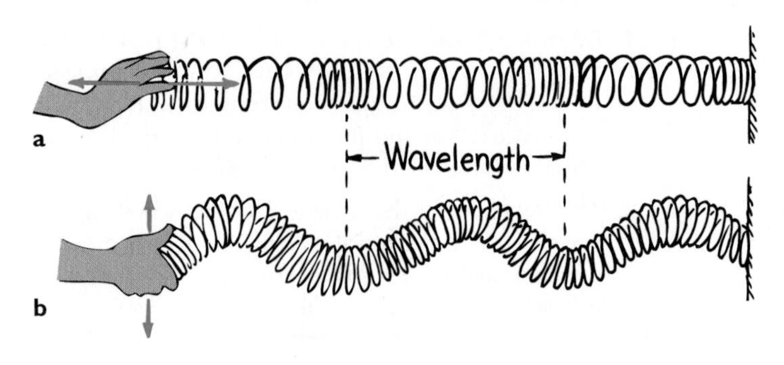

FIGURE 19.9
Interactive Figure
Both waves transfer energy from left to right. (a) When the end of the Slinky is pushed and pulled rapidly along its length, a longitudinal wave is produced. (b) When it's shaken from side to side, a transverse wave is produced.

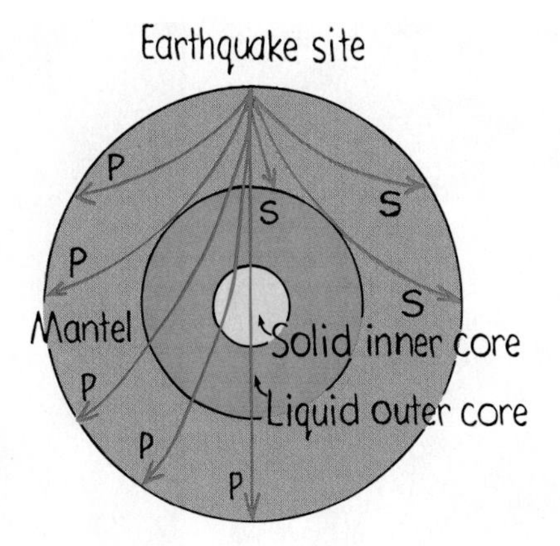

FIGURE 19.10
Waves generated by an earthquake. P waves are longitudinal and travel through both molten and solid materials. S waves are transverse and travel only through solid materials. Reflections and refractions of the waves provide information about the Earth's interior.

Interference

Whereas a material object such as a rock will not share its space with another rock, more than one vibration or wave can exist at the same time in the same space. If we drop two rocks in water, the waves produced by each can overlap and form an **interference pattern.** Within the pattern, wave effects may be increased, decreased, or neutralized.

When more than one wave occupies the same space at the same time, the displacements add at every point. This is the *superposition principle.* So, when the crest of one wave overlaps the crest of another, their individual effects add together to produce a wave of increased amplitude. This is called *constructive interference* (Figure 19.11). When the crest of one wave overlaps the trough of another, their individual effects are reduced. The high part of one wave simply fills in the low part of another. This is called *destructive interference.*

A bee flaps its wings 600 times each second—honey power!
Insights

Wave interference is easiest to see in water. In Figure 19.12 we see the interference pattern made when two vibrating objects touch the surface of water. We can see the regions where a crest of one wave overlaps the trough of another to produce regions of zero amplitude. At points along these regions, the waves arrive out of step. We say they are *out of phase* with each other.

Interference is characteristic of all wave motion, whether the waves are water waves, sound waves, or light waves. We will treat the interference of sound in the next chapter and the interference of light in Chapter 29.

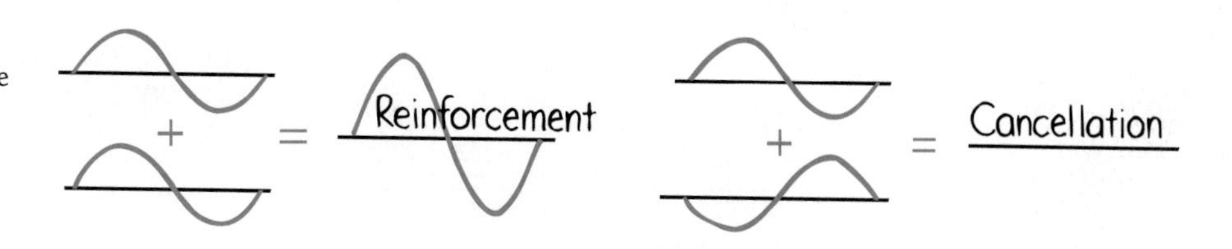

FIGURE 19.11
Constructive and destructive interference in a transverse wave.

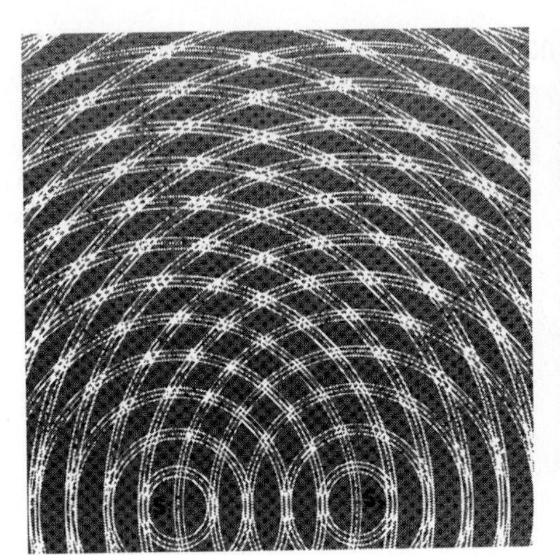

FIGURE 19.12
Two sets of overlapping water waves produce an interference pattern. The left image is an idealized drawing of the expanding waves from the two sources. The right image is a photograph of an actual interference pattern.

Standing Waves

If we tie a rope to a wall and shake the free end up and down, we produce a train of waves in the rope. The wall is too rigid to shake, so the waves are reflected back along the rope. By shaking the rope just right, we can cause the incident and reflected waves to form a **standing wave,** where parts of the rope, called the *nodes,* are stationary. Nodes are the regions of minimal or zero displacement, with minimal or zero energy. *Antinodes* (not labeled in Figure 19.13), on the other hand, are the regions of maximum displacement and maximum energy. You can hold your fingers just over and under the nodes and the rope doesn't touch them. Other parts of the rope, especially the antinodes, would make contact with your fingers. Antinodes occur halfway between nodes.

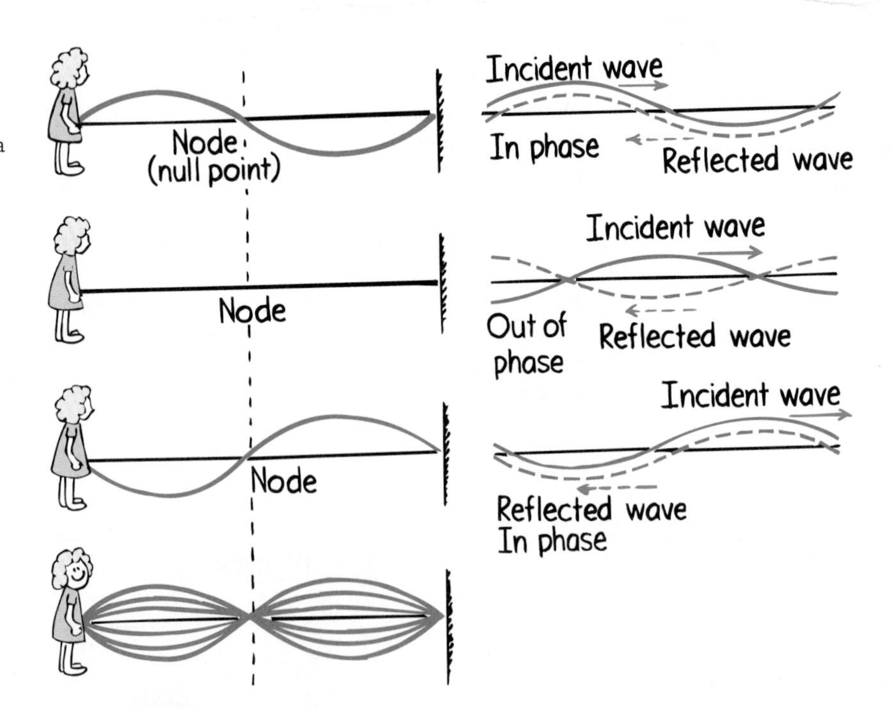

FIGURE 19.13
Interactive Figure
The incident and reflected waves interfere to produce a standing wave.

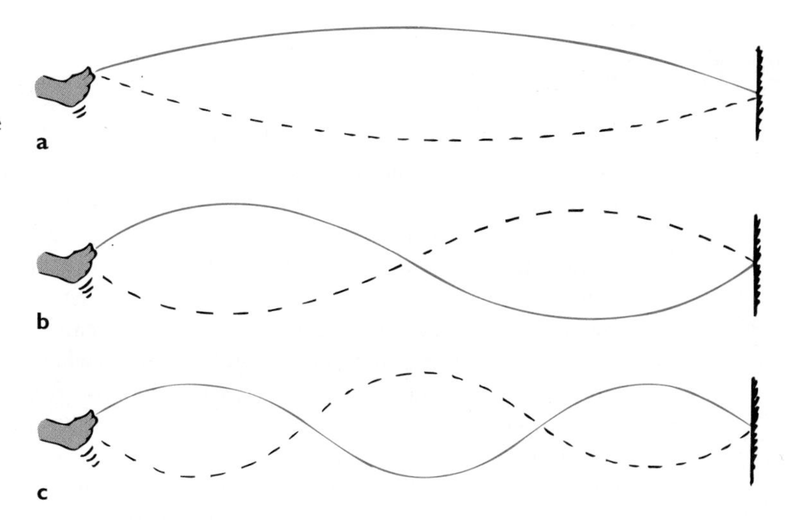

FIGURE 19.14
Interactive Figure

(a) Shake the rope until you set up a standing wave of one segment ($\frac{1}{2}$ wavelength). (b) Shake with twice the frequency and produce a wave with two segments (1 wavelength). (c) Shake with three times the frequency and produce three segments ($1\frac{1}{2}$ wavelengths).

Standing waves are the result of interference (and, as we will see in the next chapter, *resonance*). When two sets of waves of equal amplitude and wavelength pass through each other in opposite directions, the waves are steadily in and out of phase with each other. This occurs for a wave that reflects upon itself. Stable regions of constructive and destructive interference are produced.

It is easy to make standing waves yourself. Tie a rope—or better, a rubber tube—between two firm supports. Shake the tube from side to side with your hand near one of the supports. If you shake the tube with the right frequency, you will set up a standing wave, as shown in Figure 19.14a. Shake the tube with twice the frequency, and a standing wave of half the previous wavelength, having two loops, will result. (The distance between successive nodes is a half wavelength; two loops constitute a full wavelength.) If you triple the frequency, a standing wave with one-third the original wavelength, having three loops, will result; and so forth.

Standing waves are set up in the strings of musical instruments when plucked, bowed, or struck. They are set up in the air in an organ pipe, a trumpet, or a clarinet, and the air of a soda-pop bottle when air is blown over the top. Standing waves can be set up in a tub of water or in a cup of coffee by sloshing the tub or cup back and forth with the right frequency. Standing waves can be produced with either transverse or longitudinal vibrations.

CHECK YOURSELF

1. Is it possible for one wave to cancel another wave so that no amplitude remains?
2. Suppose you set up a standing wave of three segments, as shown in Figure 19.14c. If you shake with twice as much frequency, how many wave segments will occur in your new standing wave? How many wavelengths?

CHECK YOUR ANSWERS

1. Yes. This is called destructive interference. In a standing wave in a rope, for example, parts of the rope have no amplitude—the nodes.
2. If you impart twice the frequency to the rope, you'll produce a standing wave with twice as many segments. You'll have six segments. Since a full wavelength has two segments, you'll have three complete wavelengths in your standing wave.

Doppler Effect

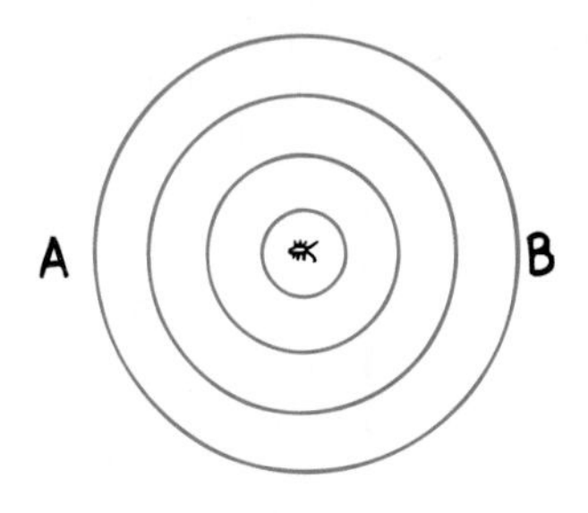

FIGURE 19.15
Top view of water waves formed by a stationary bug jiggling in still water.

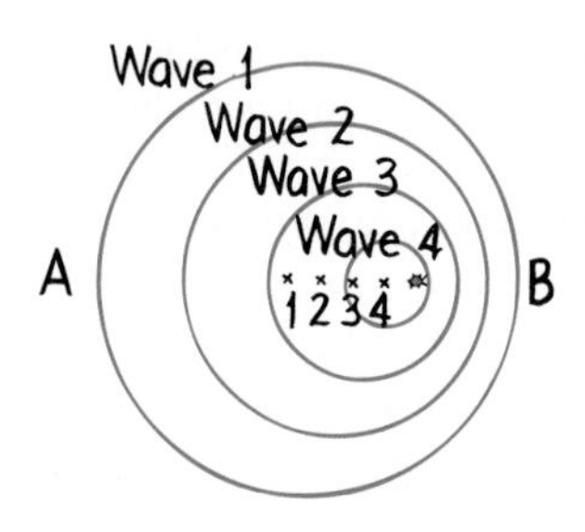

FIGURE 19.16
Interactive Figure
Water waves formed by a bug swimming in still water toward point B.

Doppler Effect

A pattern of water waves produced by a bug jiggling its legs and bobbing up and down in the middle of a quiet puddle is shown in Figure 19.15. The bug is not going anywhere but is merely treading water in a fixed position. The waves it makes are concentric circles, because wave speed is the same in all directions. If the bug bobs in the water at a constant frequency, the distance between wave crests (the wavelength) is the same in all directions. Waves encounter point A as frequently as they encounter point B. This means that the frequency of wave motion is the same at points A and B, or anywhere in the vicinity of the bug. This wave frequency is the same as the bobbing frequency of the bug.

Suppose the jiggling bug moves across the water at a speed less than the wave speed. In effect, the bug chases part of the waves it has produced. The wave pattern is distorted and is no longer made of concentric circles (Figure 19.16). The center of the outer wave was made when the bug was at the center of that circle. The center of the next smaller wave was made when the bug was at the center of that circle, and so forth. The centers of the circular waves move in the direction of the swimming bug. Although the bug maintains the same bobbing frequency as before, an observer at B would see the waves coming more often. The observer would measure a higher frequency. This is because each successive wave has a shorter distance to travel and therefore arrives at B more frequently than if the bug weren't moving toward B. An observer at A, on the other hand, would measure a *lower* frequency because of the longer time between wave-crest arrivals. This is because, in order to reach A, each crest has to travel farther than the one ahead of it due to the bug's motion. This change in frequency due to the motion of the source (or receiver) is called the **Doppler effect** (after the Austrian scientist Christian Doppler, 1803–1853).

Water waves spread over the flat surface of the water. Sound and light waves, on the other hand, travel in three-dimensional space in all directions like an expanding balloon. Just as circular waves are closer together in front of the swimming bug, spherical sound or light waves ahead of a moving source are closer together and reach a receiver more frequently.

The Doppler effect is evident when you hear the changing pitch of an ambulance siren as it passes you. When the vehicle approaches, the pitch is higher than normal (like a higher note on a musical scale). This is because the crests of the sound waves encounter your ear more frequently. And, when the vehicle passes and moves away, you hear a drop in pitch because the crests of the waves are hitting your ear less frequently.

The Doppler effect also occurs for light. When a light source approaches, there is an increase in its measured frequency; and, when it recedes, there is a decrease in its frequency. An increase in frequency is called a *blue shift*, because the increase is toward the high-frequency (or blue) end of the color spectrum. A decrease in frequency is called a *red shift*, referring to a shift toward the lower-frequency (or red) end of the color spectrum. Distant galaxies, for example, show a red shift in the light they emit. A measurement of this shift permits a calculation of their speeds of recession. A rapidly spinning star shows red-shifted light from the side turning away from us and blue-shifted light from the side turning toward us. This enables astronomers to calculate the star's spin rate.

Be clear about the distinction between *frequency* and *speed*. How frequently a wave vibrates is altogether different from how fast it moves from one location to another.

FIGURE 19.17
Interactive Figure
The pitch (frequency) of sound increases when a source moves toward you, and it decreases when the source moves away.

CHECK YOURSELF

While you're at rest, a sound source moves toward you. Do you measure an increase or a decrease in the speed of its sound wave?

Bow Waves

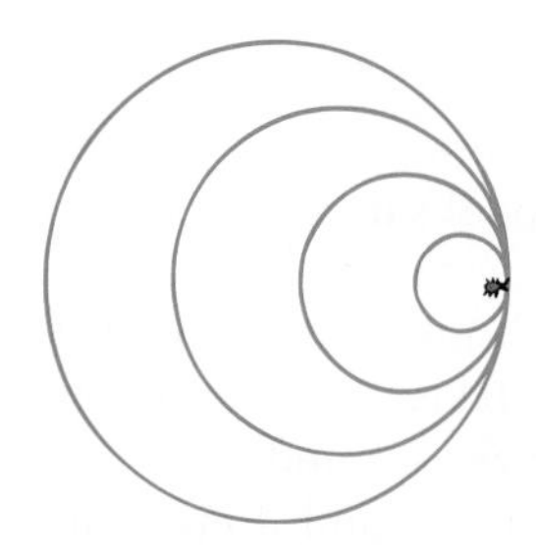

FIGURE 19.18
Wave pattern created by a bug swimming at wave speed.

When the speed of a source is as great as the speed of the waves it produces, something interesting occurs. The waves pile up in front of the source. Consider the bug in our previous example when it swims as fast as the wave speed. Can you see that the bug will keep up with the waves it produces? Instead of the waves moving ahead of the bug, they superimpose and hump up on one another directly in front of the bug (Figure 19.18). The bug moves right along with the leading edge of the waves it is producing.

A similar thing happens when an aircraft travels at the speed of sound. In the early days of jet aircraft, it was believed that this pile-up of sound waves in front of the airplane imposed a "sound barrier" and that, in order to go faster than the speed of sound, the plane would have to "break the sound barrier." What actually happens is that the overlapping wave crests disrupt the flow of air over the wings, making it more difficult to control the craft. But the barrier is not real. Just as a boat can easily travel faster than the waves it produces, an aircraft with sufficient power easily travels faster than the speed of sound. Then we say that it is *supersonic*. A supersonic airplane flies into smooth, undisturbed air because no sound wave can propagate out in front of it. Similarly, a bug swimming faster than the speed of water waves finds itself always entering into water with a smooth, unrippled surface.

When the bug swims faster than wave speed, ideally it produces a wave pattern, as shown in Figure 19.19. It outruns the waves it produces. The waves overlap at the edges and produce a V shape, called a **bow wave,** which appears to be dragging behind the bug. The familiar bow wave generated by a speedboat knifing through the water is not a typical oscillatory wave. It is a disturbance produced by the overlapping of many circular waves.

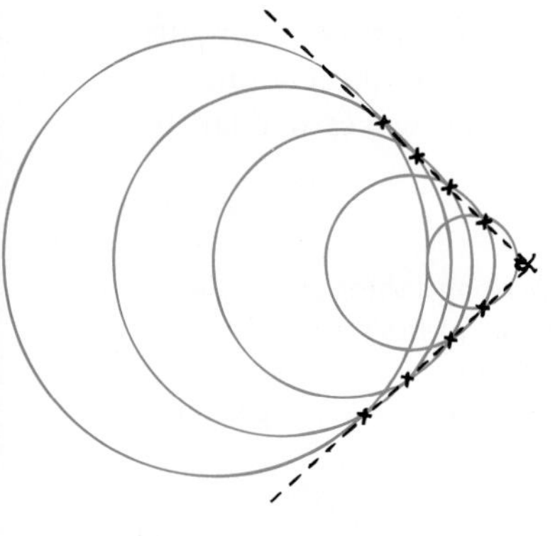

FIGURE 19.19
A bow wave, the pattern produced by a bug swimming faster than wave speed. The points at which adjacent waves overlap (x) produce the V shape.

CHECK YOUR ANSWER

Neither! It is the *frequency* of a wave that undergoes a change where there is motion of the source, not the *wave speed*. Be clear about the distinction between frequency and speed. How frequently a wave vibrates is altogether different from how fast the disturbance moves from one place to another.

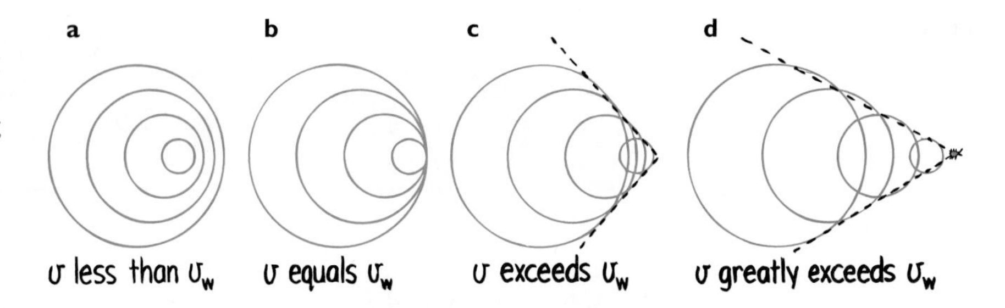

FIGURE 19.20
Patterns produced by a bug swimming at successively greater speeds. Overlapping at the edges occurs only when the bug swims faster than wave speed.

Some wave patterns made by sources moving at various speeds are shown in Figure 19.20. Note that, after the speed of the source exceeds wave speed, increased speed of the source produces a narrower V shape.[3]

Shock Waves

FIGURE 19.21
This aircraft is producing a cloud of water vapor that has just condensed out of the rapidly expanding air in the rarefied region behind the wall of compressed air.

A speedboat knifing through the water generates a two-dimensional bow wave. A supersonic aircraft similarly generates a three-dimensional **shock wave.** Just as a bow wave is produced by overlapping circles that form a V, a shock wave is produced by overlapping spheres that form a cone. And, just as the bow wave of a speedboat spreads until it reaches the shore of a lake, the conical wave generated by a supersonic craft spreads until it reaches the ground.

The bow wave of a speedboat that passes by can splash and douse you if you are at the water's edge. In a sense, you can say that you are hit by a "water boom." In the same way, when the conical shell of compressed air that sweeps behind a supersonic aircraft reaches listeners on the ground below, the sharp crack they hear is described as a **sonic boom.**

We don't hear a sonic boom from slower-than-sound, or subsonic, aircraft because the sound waves reaching our ears are perceived as one continuous tone. Only when the craft moves faster than sound do the waves overlap to

FIGURE 19.22
Shock wave of a bullet piercing a sheet of Plexiglass. Light deflecting as the bullet passes through the compressed air makes the shock visible. Look carefully and see the second shock wave originating at the tail of the bullet.

[3]Bow waves generated by boats in water are more complex than is indicated here. Our idealized treatment serves as an analogy for the production of the less complex shock waves in air.

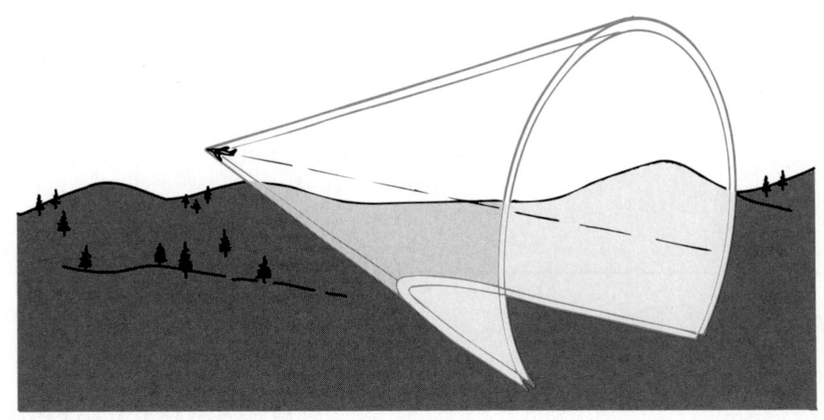

FIGURE 19.23
A shock wave.

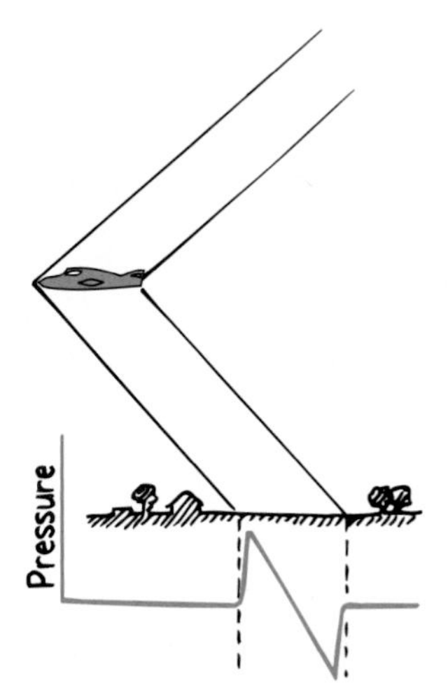

FIGURE 19.24
The shock wave is actually made up of two cones—a high-pressure cone with the apex at the bow of the aircraft and a low-pressure cone with the apex at the tail. A graph of the air pressure at ground level between the cones takes the shape of the letter N.

reach the listener in a single burst. The sudden increase in pressure is much the same in effect as the sudden expansion of air produced by an explosion. Both processes direct a burst of high-pressure air to the listener. The ear is hard pressed to distinguish between the high pressure from an explosion and the high pressure from many overlapping waves.

A water skier is familiar with the fact that next to the high crest of the V-shaped bow wave is a V-shaped depression. The same is true of a shock wave, which usually consists of two cones: a high-pressure cone generated at the bow of the supersonic aircraft and a low-pressure cone that follows at the tail of the craft.[4] The edges of these cones are visible in the photograph of the supersonic bullet in Figure 19.22. Between these two cones, the air pressure rises sharply to above atmospheric pressure, then falls below atmospheric pressure before sharply returning to normal beyond the inner tail cone (Figure 19.24). This overpressure suddenly followed by underpressure intensifies the sonic boom.

A common misconception is that sonic booms are produced when an aircraft flies through the "sound barrier"—that is, just as the aircraft surpasses the speed of sound. This is the same as saying that a boat produces a bow wave when it first overtakes its own waves. This is not so. The fact is that a shock wave and its resulting sonic boom are swept continuously behind and below an aircraft traveling faster than sound, just as a bow wave is swept continuously behind a speedboat. In Figure 19.25, listener B is in the process of hearing a sonic boom. Listener C has already heard it, and listener A will hear it shortly. The aircraft that generated this shock wave may have broken through the sound barrier many minutes ago!

It is not necessary that the moving source be "noisy" to produce a shock wave. Once an object is moving faster than the speed of sound, it will *make* sound. A supersonic bullet passing overhead produces a crack, which is a small

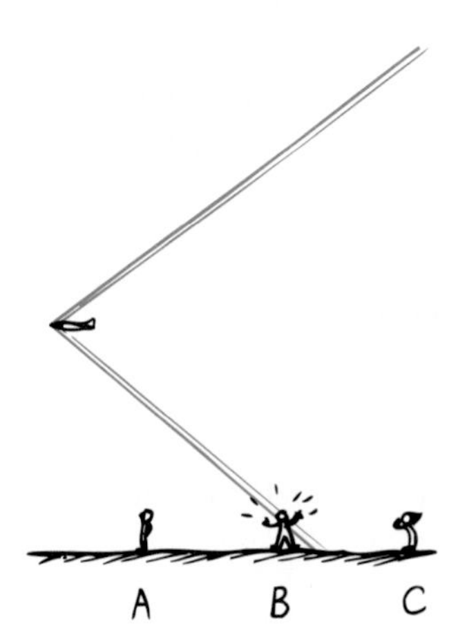

FIGURE 19.25
The shock wave has not yet reached listener A, but it is now reaching listener B and has already reached listener C.

[4]Shock waves are often more complex and involve multiple cones.

sonic boom. If the bullet were larger and disturbed more air in its path, the crack would be more boomlike. When a lion tamer cracks a circus whip, the cracking sound is actually a sonic boom produced by the tip of the whip when it travels faster than the speed of sound. Both the bullet and the whip are not in themselves sound sources, but, when traveling at supersonic speeds, they produce their own sound as they generate shock waves.

Summary of Terms

Sine curve The waveform traced by simple harmonic motion, which can be made visible on a moving conveyor belt by a pendulum swinging at right angles above the moving belt.

Amplitude For a wave or vibration, the maximum displacement on either side of the equilibrium (midpoint) position.

Wavelength The distance between successive crests, troughs, or identical parts of a wave.

Frequency For a vibrating body or medium, the number of vibrations per unit time. For a wave, the number of crests that pass a particular point per unit time.

Hertz The SI unit of frequency. One hertz (symbol Hz) equals one vibration per second.

Period The time in which a vibration is completed. The period of a wave equals the period of the source, and is equal to 1/frequency.

Wave speed The speed with which waves pass a particular point:

$$\text{Wave speed} = \text{wavelength} \times \text{frequency}$$

Transverse wave A wave in which the medium vibrates in a direction perpendicular (transverse) to the direction in which the wave travels. Light waves and water waves are transverse.

Longitudinal wave A wave in which the medium vibrates in a direction parallel (longitudinal) to the direction in which the wave travels. Sound waves are longitudinal.

Interference pattern The pattern formed by superposition of different sets of waves that produces reinforcement in some places and cancellation in others.

Standing wave A stationary wave pattern formed in a medium when two sets of identical waves pass through the medium in opposite directions.

Doppler effect The shift in received frequency due to motion of a vibrating source toward or away from a receiver.

Bow wave The V-shaped disturbance created by an object moving across a liquid surface at a speed greater than the wave speed.

Shock wave The cone-shaped disturbance created by an object moving at supersonic speed through a fluid.

Sonic boom The loud sound resulting from the incidence of a shock wave.

Review Questions

1. What is a *wiggle in time* called? What is a *wiggle in space and time* called?
2. What is the source of all waves?

Vibration of a Pendulum

3. What is meant by the *period* of a pendulum?
4. Which has the longer period, a short or a long pendulum?

Wave Description

5. How is a sine curve related to a wave?
6. Distinguish between these different parts of a wave: period, amplitude, wavelength, and frequency.
7. How many vibrations per second are represented in a radio wave of 101.7 MHz?
8. How do *frequency* and *period* relate to each other?

Wave Motion

9. In one word, what is it that moves from source to receiver in wave motion?
10. Does the medium in which a wave travels move with the wave? Give examples to support your answer.

Wave Speed

11. What is the relationship among frequency, wavelength, and wave speed?

Transverse Waves

12. In what direction are the vibrations relative to the direction of wave travel in a transverse wave?

Longitudinal Waves

13. In what direction are the vibrations relative to the direction of wave travel in a longitudinal wave?

#1, 5, 7, 9, 11, 13, 17, 19, 23, 25, 35, 37, 39, 43
pg 379 #1–5

14. The wavelength of a transverse wave is the distance between successive crests (or troughs). What is the wavelength of a longitudinal wave?

Interference

15. What is meant by the *superposition principle*?
16. Distinguish between *constructive interference* and *destructive interference*.
17. What kinds of waves can show interference?

Standing Waves

18. What is a *node*? What is an *antinode*?
19. Are standing waves a property of transverse waves, of longitudinal waves, or of both?

Doppler Effect

20. In the Doppler effect, does frequency change? Does wavelength change? Does wave speed change?
21. Can the Doppler effect be observed with longitudinal waves, with transverse waves, or with both?
22. What is meant by a blue shift and a red shift for light?

Bow Waves

23. How fast must a bug swim to keep up with the waves it produces? How fast must it move to produce a bow wave?
24. How fast does a supersonic aircraft fly compared with the speed of sound?
25. How does the V shape of a bow wave depend on the speed of the source?

Shock Waves

26. A bow wave on the surface of water is two dimensional. How about a shock wave in air?
27. True or false: A sonic boom occurs only when an aircraft is breaking through the sound barrier. Defend your answer.
28. True or false: In order for an object to produce a sonic boom, it must be "noisy." Give two examples to support your answer.

Projects

1. Tie a rubber tube, a spring, or a rope to a fixed support and produce standing waves. See how many nodes you can produce.
2. Wet your finger and rub it slowly around the rim of a thin-rimmed, stemmed wine glass while you hold the base of the glass firmly to a tabletop with your other hand. The friction of your finger will excite standing waves in the glass, much like the wave made on the strings of a violin by the friction from a violin bow. Try it with a metal bowl.
3. Write a letter to Grandma and tell her how waves can cancel one another, as well as what some of the applications of this physical phenomenon are these days.

Exercises

1. Does the period of a pendulum depend on the mass of the bob? On the length of the string?
2. A heavy person and a light person swing to and fro on swings of the same length. Who has the longer period?
3. A grandfather pendulum clock keeps perfect time. Then it is brought to a summer home high in the mountains. Does it run faster, slower, or the same? Explain.
4. If a pendulum is shortened, does its frequency increase or decrease? What about its period?
5. You pick up an empty suitcase and let it swing to and fro at its natural frequency. If the case were filled with books, would the natural frequency be lower, greater, or the same as before?
6. Is the time required to swing to and fro (the period) on a playground swing longer or shorter when you stand rather than sit? Explain.
7. Why does it make sense that the mass of a bob in a simple pendulum doesn't affect the frequency of the pendulum?
8. What happens to the period of a wave when the frequency decreases?
9. What happens to the wavelength of a wave when the frequency decreases?
10. If the speed of a wave doubles while the frequency remains the same, what happens to the wavelength?
11. If the speed of a wave doubles while the wavelength remains the same, what happens to the frequency?
12. If you clamp one end of a hacksaw blade in a vice and twang the free end, it vibrates. Now repeat, but first place a wad of clay on the free end. How, if at all, will the frequency of vibration differ? Would it make a difference if the wad of clay were stuck to the middle? Explain. (Why could this question have been asked back in Chapter 8?)
13. The needle of a sewing machine moves up and down in simple harmonic motion. Its driving force comes from a rotating wheel that is powered by an electric motor. How do you suppose the period of the up-and-down needle compares with the period of the rotating wheel?

14. If you shake the end of a spring to produce a wave, how does the frequency of the wave compare with the frequency of your shaking hand? Does your answer depend on whether you're producing a transverse wave or a longitudinal wave? Defend your answer.
15. What kind of motion should you impart to the nozzle of a garden hose so that the resulting stream of water approximates a sine curve?
16. What kind of motion should you impart to a stretched coiled spring (or Slinky) to provide a transverse wave? To provide a longitudinal wave?
17. What kind of wave is each of the following? (a) An ocean wave rolling toward Waikiki Beach. (b) The sound of one whale calling another whale under water. (c) A pulse sent down a stretched rope by snapping one end of it.
18. If a gas tap is turned on for a few seconds, someone a couple of meters away will hear the gas escaping long before she smells it. What does this indicate about the speed of sound and the motion of molecules in the sound-carrying medium?
19. If we double the frequency of a vibrating object or the wave it produces, what happens to the period?
20. Do the two terms *wave speed* and *wave frequency* refer to the same thing? Defend your answer.
21. Red light has a longer wavelength than violet light. Which has the greater frequency?
22. Consider a wave traveling along a thick rope tied to a thin rope. Which of these three wave characteristics does *not* undergo change—speed, frequency, or wavelength?
23. What is the frequency of the second hand of a clock? The minute hand? The hour hand?
24. What is the source of wave motion?
25. If you dip your finger repeatedly into a puddle of water, it creates waves. What happens to the wavelength if you dip your finger more frequently?
26. How does the frequency of vibration of a small object floating in water compare with the number of waves passing it each second?
27. How far, in terms of wavelength, does a wave travel in one period?
28. How many nodes, not including the end points, are there in a standing wave that is two wavelengths long? Three wavelengths long?
29. A rock is dropped in water, and waves spread over the flat surface of the water. What becomes of the energy in these waves when they die out?
30. The wave patterns seen in Figure 19.6 are composed of circles. What does this tell you about the speed of waves moving in different directions?
31. Why is lightning seen before thunder is heard?
32. A banjo player plucks the middle of an open string. Where are the nodes of the standing wave in the string? What is the wavelength of the vibrating string?
33. Violinists sometimes bow a string to produce maximum vibration (antinodes) at one-quarter and three-quarters of the string length rather than at the middle of the string. Then the string vibrates with a wavelength equal to the string length rather than twice the string length. (See Figures 19.14a and b). What is the effect on frequency when this occurs?
34. A bat chirps as it flies toward a wall. Is the frequency of the echoed chirps it receives higher, lower, or the same as the emitted ones?
35. Why is there a Doppler effect when the source of sound is stationary and the listener is in motion? In which direction should the listener move to hear a higher frequency? A lower frequency?
36. A railroad locomotive is at rest with its whistle shrieking, then starts moving toward you. (a) Does the frequency that you hear increase, decrease, or stay the same? (b) How about the wavelength reaching your ear? (c) How about the speed of sound in the air between you and the locomotive?
37. When you blow your horn while driving toward a stationary listener, the listener hears an increase in frequency of the horn. Would the listener hear an increase in horn frequency if he or she were also in a car traveling at the same speed in the same direction as you are? Explain.
38. Is there an appreciable Doppler effect when the motion of the source is at right angles to a listener? Explain.
39. How does the Doppler effect aid police in detecting speeding motorists?
40. Astronomers find that light emitted by a particular element at one edge of the Sun has a slightly higher frequency than light from that element at the opposite edge. What do these measurements tell us about the Sun's motion?
41. Would it be correct to say that the Doppler effect is the apparent change in the speed of a wave due to motion of the source? (Why is this question a test of reading comprehension as well as a test of physics knowledge?)
42. How does the phenomenon of interference play a role in the production of bow waves or shock waves?

43. What can you say about the speed of a boat that makes a bow wave?

44. Does the conical angle of a shock wave open wider, narrow down, or remain constant as a supersonic aircraft increases its speed?

45. If the sound of an airplane does not come from the part of the sky where the plane is seen, does this imply that the airplane is traveling faster than the speed of sound? Explain.

46. Does a sonic boom occur at the moment when an aircraft exceeds the speed of sound? Explain.

47. Why is it that a subsonic aircraft, no matter how loud it may be, cannot produce a sonic boom?

48. Imagine a superfast fish that is able to swim faster than the speed of sound in water. Would such a fish produce a "sonic boom"?

49. Make up a multiple-choice question that would check a classmate's understanding of the distinction between a transverse wave and a longitudinal wave.

50. Make up two multiple-choice questions that would check a classmate's understanding of the terms that describe a wave.

Problems

1. What is the frequency, in hertz, that corresponds to each of the following periods? (a) 0.10 s, (b) 5 s, (c) 1/60 s.

2. What is the period, in seconds, that corresponds to each of the following frequencies? (a) 10 Hz, (b) 0.2 Hz, (c) 60 Hz.

3. A skipper on a boat notices wave crests passing his anchor chain every 5 s. He estimates the distance between wave crests to be 15 m. He also correctly estimates the speed of the waves. What is this speed?

4. A weight suspended from a spring is seen to bob up and down over a distance of 20 cm twice each second. What is its frequency? Its period? Its amplitude?

5. Radio waves travel at the speed of light— 300,000 km/s. What is the wavelength of radio waves received at 100.1 MHz on your FM radio dial?

6. A mosquito flaps its wings 600 vibrations per second, which produces the annoying 600-Hz buzz. How far does the sound travel between wing beats? In other words, find the wavelength of the mosquito's sound.

7. On a keyboard, you strike middle C, whose frequency is 256 Hz. (a) What is the period of one vibration of this tone? (b) As the sound leaves the instrument at a speed of 340 m/s, what is its wavelength in air?

8. Pretend you foolishly play your keyboard instrument under water, where the speed of sound is 1,500 m/s. (a) What would be the wavelength of the middle-C tone in water? (b) Explain *why* middle C (or any other tone) has a longer wavelength in water than in air.

9. The wavelength of the signal from TV Channel 6 is 3.42 m. Does Channel 6 broadcast on a frequency above or below the FM radio band, which is 88 to 108 MHz?

10. As shown in the drawing, the half-angle of the shock-wave cone generated by a supersonic transport is 45°. What is the speed of the plane relative to the speed of sound?

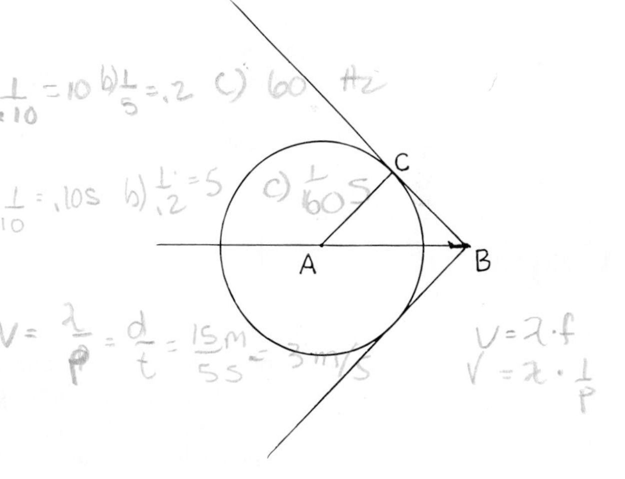

CHAPTER 20

Sound

Chris Chiaverina, former president of the American Association of Physics Teachers, and acoustical physicist Tom Rossing produce sounds on a hang, *a new hand-played steel instrument.*

If a tree fell in the middle of a deep forest hundreds of kilometers away from any living being, would there be a sound? Different people will answer this question in different ways. "No," some will say, "sound is subjective and requires a listener. If there is no listener, there will be no sound." "Yes," others will say, "a sound is not something in a listener's head. A sound is an objective thing." Discussions like this one often are beyond agreement because the participants fail to realize that they are arguing not about the nature of sound but about the definition of the word. Either side is correct, depending on the definition taken, but investigation can proceed only when a definition has been agreed upon. Physicists, such as the two shown above, usually take the objective position and define sound as a form of energy that exists whether or not it is heard, and they go on from there to investigate its nature.

Origin of Sound

Most sounds are waves produced by the vibrations of material objects. In a piano, a violin, and a guitar, the sound is produced by the vibrating strings; in a saxophone, by a vibrating reed; in a flute, by a fluttering column of air at the mouthpiece. Your voice results from the vibration of your vocal chords.

In each of these cases, the original vibration stimulates the vibration of something larger or more massive, such as the sounding board of a stringed instrument, the air column within a reed or wind instrument, or the air in the throat and mouth of a singer. This vibrating material then sends a disturbance through the surrounding medium, usually air, in the form of longitudinal waves. Under ordinary conditions, the frequency of the vibrating source and the frequency of the sound waves produced are the same.

We describe our subjective impression about the frequency of sound by the word *pitch*. Frequency corresponds to pitch: A high-pitched sound from a piccolo has a high frequency of vibration, while a low-pitched sound from a foghorn has a low frequency of vibration. The ear of a young person can normally hear pitches corresponding to the range of frequencies between about 20 and 20,000 hertz. As we grow older, the limits of this human hearing range shrink, especially at the high-frequency end. Sound waves with frequencies below 20 hertz are **infrasonic,** and those with frequencies above 20,000 hertz are called **ultrasonic.** We cannot hear infrasonic and ultrasonic sound waves.

Nature of Sound in Air

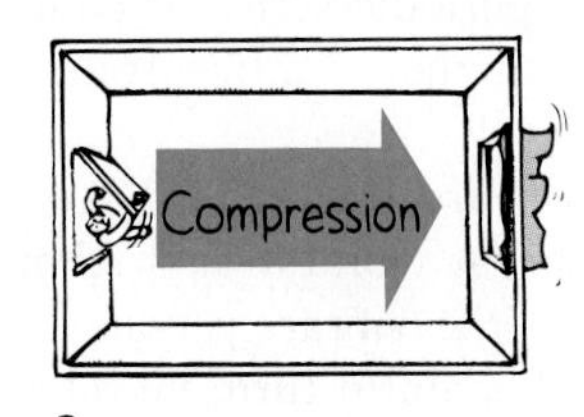

a

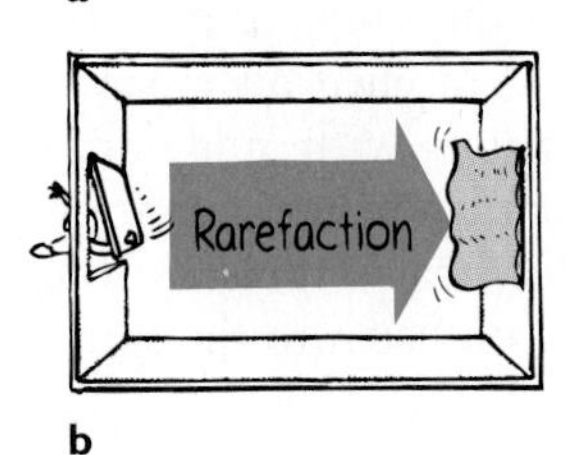

b

FIGURE 20.1
(a) When the door is opened, a compression travels across the room. (b) When the door is closed, a rarefaction travels across the room. (Adapted from A. V. Baez, *The New College Physics: A Spiral Approach*. San Francisco: W. H. Freeman and Company. Copyright © 1967.)

When we clap our hands, the sound produced is nonperiodic. It consists of a wave *pulse* that travels out in all directions. The pulse disturbs the air in the same way that a similar pulse would disturb a coiled spring or a Slinky. Each particle moves to and fro along the direction of the expanding wave.

For a clearer picture of this process, consider a long room, as shown in Figure 20.1a. At one end is an open window with a curtain over it. At the other end is a door. When we open the door, we can imagine the door pushing the molecules next to it away from their initial positions and into their neighbors. The neighboring molecules, in turn, push into their neighbors, and so on, like a compression traveling along a spring, until the curtain flaps out the window. A pulse of compressed air has moved from the door to the curtain. This pulse of compressed air is called a **compression.**

When we close the door (Figure 20.1b), the door pushes some air molecules out of the room. This produces an area of low pressure behind the door. Neighboring molecules then move into it, leaving a zone of lower pressure behind them. We say this zone of lower-pressure air is *rarefied.* Other molecules farther away from the door, in turn, move into these rarefied regions, and a disturbance again travels across the room. This is evidenced by the curtain, which flaps inward. This time the disturbance is a **rarefaction.**

As with all wave motion, it is not the medium itself that travels across the room but the energy-carrying pulse. In both cases, the pulse travels from the door to the curtain. We know this because, in both cases, the curtain moves after the door is opened or closed. If you continually swing the door open and closed in periodic fashion, you can set up a wave of periodic compressions and rarefactions that will cause the curtain to swing in and out of the window. On a much smaller but more rapid scale, this is what happens when a tuning fork is struck. The periodic vibrations of the tuning fork and the waves produced are considerably higher in frequency and lower in amplitude than

Atoms and molecules in a medium tremble as they transmit sound. Sound cannot travel in a vacuum because there's nothing to do the trembling.

Insights

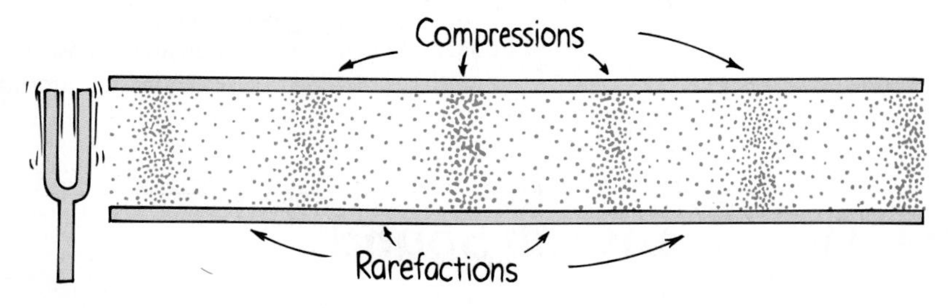

FIGURE 20.2
Compressions and rarefactions travel (both at the same speed in the same direction) from the tuning fork through the air in the tube.

FIGURE 20.3
A Ping-Pong paddle vibrating in the midst of Ping-Pong balls produces vibration of the balls.

those caused by the swinging door. You don't notice the effect of sound waves on the curtain, but you are well aware of them when they meet your sensitive eardrums.

Consider sound waves in the tube shown in Figure 20.2. For simplicity, only the waves that travel in the tube are depicted. When the prong of the tuning fork next to the tube moves toward the tube, a compression enters the tube. When the prong swings away in the opposite direction, a rarefaction follows the compression. It's like a Ping-Pong paddle moving to and fro in a room packed with Ping-Pong balls. As the source vibrates, a periodic series of compressions and rarefactions is produced. The frequency of the vibrating source and the frequency of the wave it produces are the same.

Pause to reflect on the physics of sound while you are listening to your radio sometime. The radio loudspeaker is a paper cone that vibrates in rhythm with an electrical signal. Air molecules next to the vibrating cone of the speaker are themselves set into vibration. This air, in turn, vibrates against neighboring particles, which, in turn, do the same, and so on. As a result, rhythmic patterns of compressed and rarefied air emanate from the loudspeaker, showering the whole room with undulating motions. The resulting vibrating air sets your eardrum into vibration, which, in turn, sends cascades of rhythmic electrical impulses along the cochlea nerve canal and into the brain. And you listen to the sound of music.

FIGURE 20.4
Waves of compressed and rarefied air, produced by the vibrating cone of the loudspeaker, make up the pleasing sound of music.

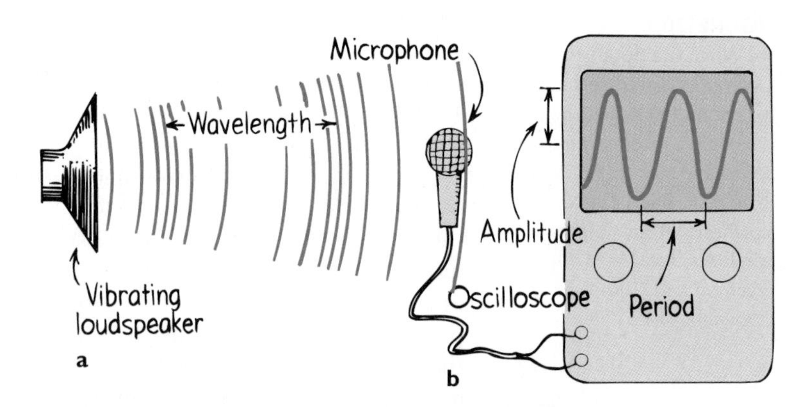

FIGURE 20.5
(a) The radio loudspeaker is a paper cone that vibrates in rhythm with an electric signal. The sound produced sets up similar vibrations in the microphone, which are displayed on an oscilloscope. (b) The shape of the waveform on the oscilloscope screen reveals information about the sound.

Media That Transmit Sound

Most sounds that we hear are transmitted through the air. However, any elastic substance—whether solid, liquid, gas, or plasma—can transmit sound. Elasticity is the ability of a material that has changed shape in response to an

PRACTICING PHYSICS

Suspend the wire grille from a refrigerator or an oven from a string, holding the ends of the string to your ears. Let a friend gently stroke the grille with pieces of broom straw and with other objects. The effect is best appreciated when you are in a relaxed condition with your eyes closed. Be sure to try this!

applied force to resume its initial shape once the distorting force is removed. Steel is an elastic substance. In contrast, putty is inelastic.[1] In elastic liquids and solids, the atoms are relatively close together, respond quickly to one another's motions, and transmit energy with little loss. Sound travels about four times faster in water than in air and about fifteen times faster in steel than in air.

Your two ears are so sensitive to the differences in sound reaching them that you can tell from what direction a sound is coming with almost pinpoint accuracy. With only one ear, you would have no idea (and, in an emergency, might not know which way to jump).

Insights

Relative to solids and liquids, air is a poor conductor of sound. You can hear the sound of a distant train more clearly if your ear is placed against the rail. Similarly, a watch placed on a table beyond hearing distance can be heard if you place your ear to the table. Or click some rocks together under water while your ear is submerged. You'll hear the clicking sound very clearly. If you've ever been swimming in the presence of motorized boats, you probably noticed that you can hear the boats' motors much more clearly under water than above water. Liquids and crystalline solids are generally excellent conductors of sound—much better than air. The speed of sound is generally greater in solids than in liquids, and greater in liquids than in gases. Sound won't travel in a vacuum, because the transmission of sound requires a medium. If there is nothing to compress and expand, there can be no sound.

Speed of Sound in Air

If we watch a person at a distance chopping wood, or see a far-away baseball player hit a ball, we can easily see that the blow takes place an appreciable time before its sound reaches our ears. Thunder is heard after a flash of lightning is seen. These common experiences show that sound requires a recognizable time to travel from one place to another. The speed of sound depends on wind conditions, temperature, and humidity. It does not depend on the loudness or the frequency of the sound; all sounds travel at the same speed. The speed of sound in dry air at 0°C is about 330 meters per second, nearly 1200 kilometers per hour (a little more than one millionth the speed of light). Water vapor in the air increases this speed slightly. Sound travels faster through warm air than cold air. This is to be expected because the faster-moving molecules in

[1]Elasticity is not "stretchiness," like a rubber band. Some very stiff materials are elastic—like steel.

warm air bump into one another more often and therefore can transmit a pulse in less time.[2] For each degree rise in temperature above 0°C, the speed of sound in air increases by 0.6 meter per second. So, in air at a normal room temperature of about 20°C, sound travels at about 340 meters per second.

CHECK YOURSELF

1. Do compressions and rarefactions in a sound wave travel in the same direction or in opposite directions from one another?
2. What is the approximate distance of a thunderstorm when you note a 3-s delay between the flash of lightning and the sound of thunder?

Reflection of Sound

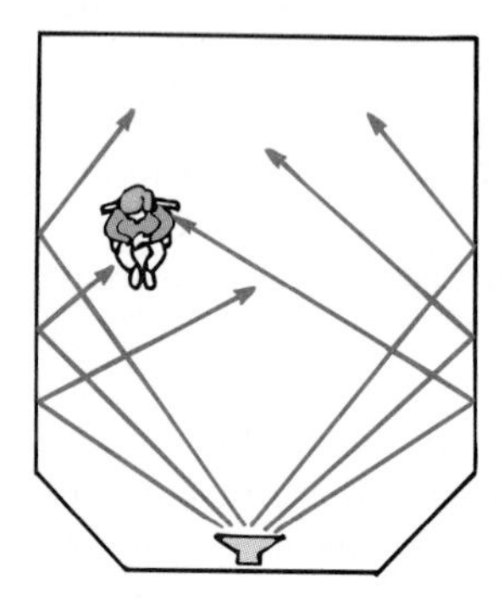

FIGURE 20.6
The angle of incident sound is equal to the angle of reflected sound.

We call the reflection of sound an *echo*. The fraction of energy carried by the reflected sound wave is large if the surface is rigid and smooth and less if the surface is soft and irregular. Sound energy not carried by the reflected sound wave is carried by the "transmitted" (absorbed) wave.

Sound reflects from a smooth surface the same way that light does—the angle of incidence is equal to the angle of reflection (Figure 20.6). Sometimes, when sound reflects from the walls, ceiling, and floor of a room, the reflecting surfaces are too reflective and the sound becomes garbled. This is due to multiple reflections called **reverberations.** On the other hand, if the reflective surfaces are too absorbent, the sound level will be low and the hall will sound dull and lifeless. Reflection of sound in a room makes it sound lively and full, as you have probably discovered while singing in the shower. In the design of an auditorium or concert hall, a balance must be achieved between reverberation and absorption. The study of sound properties is called *acoustics*.

It is often advantageous to place highly reflective surfaces behind the stage to direct sound out to an audience. Reflecting surfaces are suspended above the stage in some concert halls. The ones in Davies Hall in San Francisco are large shiny plastic surfaces that also reflect light (Figure 20.7). A listener can look up at these reflectors and see the reflected images of the members of the orchestra. The plastic reflectors are somewhat curved, which increases the field of view. Both sound and light obey the same law of reflection, so, if a reflector is oriented so that you can see a particular musical instrument, rest assured that you will hear it also. Sound from the instrument will follow the line of sight to the reflector and then to you.

One reason for the superior hearing of great grey owls is the feathers on their faces that approximate parabolic reflectors.

Insights

CHECK YOUR ANSWERS

1. They travel in the same direction.
2. Assuming that the speed of sound in air is about 340 m/s, in 3 s it will travel (340 m/s × 3 s) = 1020 m. There is no appreciable delay for the flash, so the storm is slightly more than 1 km away.

[2]The speed of sound in a gas is about 3/4 the average speed of the gas molecules.

FIGURE 20.7
The plastic plates above the orchestra reflect both light and sound. Adjusting them is quite simple: What you see is what you hear.

Refraction of Sound

Sound waves bend when parts of the wave fronts travel at different speeds. This happens in uneven winds or when sound is traveling through air of uneven temperatures. This bending of sound is called **refraction.** On a warm day, the air near the ground may be appreciably warmer than the rest of the air, so the speed of sound near the ground increases. Sound waves therefore tend to bend away from the ground, resulting in sound that does not seem to travel well. Different speeds of sound produce refraction.

We hear thunder when the lightning is reasonably close, but we often fail to hear the thunder for distant lightning because of refraction. The sound travels slower at higher altitudes and bends away from the ground. The opposite often occurs on a cold day or at night when the layer of air near the ground is colder than the air above. Then the speed of sound near the ground is reduced. The higher speed of the wave fronts above causes a bending of the sound toward the ground, resulting in sound that can be heard over considerably longer distances (Figure 20.8).

The direction in which sound travels, like the direction of a light ray, is always at right angles to its wave front.

Insights

The refraction of sound occurs under water, where the speed of sound varies with temperature. This poses a problem for surface vessels that bounce ultrasonic waves off the bottom of the ocean to chart the sea bottom's features. Refraction is a blessing to submarines that wish to escape detection. Because of thermal gradients and layers of water at different temperatures, the refraction of sound leaves gaps or "blind spots" in the water. This is where submarines hide. If it weren't for refraction, submarines would be easier to detect.

The multiple reflections and refractions of ultrasonic waves are used by physicians in a technique for harmlessly viewing inside the body without the

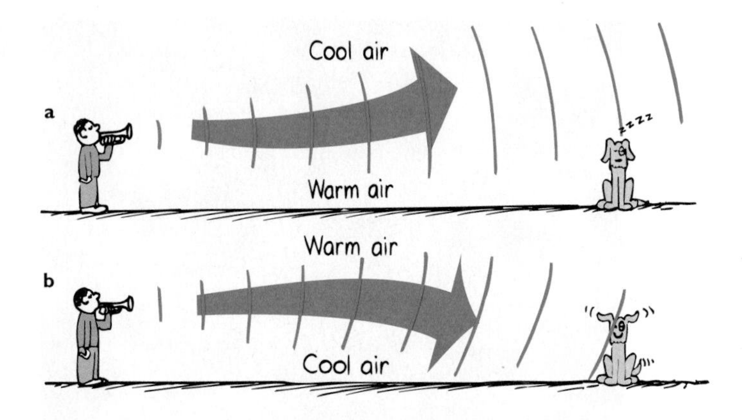

FIGURE 20.8
Sound waves are bent in air of uneven temperatures.

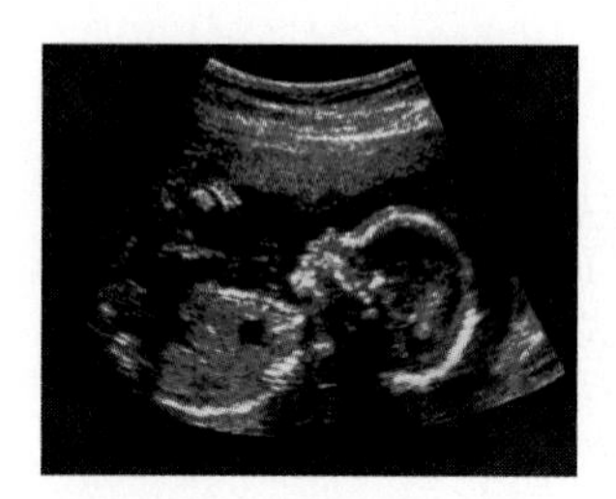

FIGURE 20.9
A 5-month-old fetus displayed on a viewing screen by ultrasound.

use of X rays. When high-frequency sound (ultrasound) enters the body, it is reflected more strongly from the outside of an organ than from its interior, and a picture of the outline of the organ is obtained. When ultrasound is incident upon a moving object, the reflected sound has a slightly different frequency. Using this Doppler effect, a physician can "see" the beating heart of a fetus as early as 11 weeks after conception (Figure 20.9).

The ultrasound echo technique may be relatively new to humans, but not to bats or dolphins. It is well known that bats emit ultrasonic squeaks and locate objects by their echoes. Dolphins do this and more.[3] The ultrasonic waves emitted by a dolphin enable it to "see" through the bodies of other animals and people. Skin, muscle, and fat are almost transparent to dolphins, so they "see" only a thin outline of the body—but the bones, teeth, and gas-filled cavities are clearly apparent. Physical evidence of cancers, tumors, heart attacks, and even emotional state can all be "seen" by the dolphin—as humans have only recently been able to accomplish with ultrasound.

FIGURE 20.10
A dolphin emits ultrahigh-frequency sound to locate and identify objects in its environment. Distance is sensed by the time delay between sending sound and receiving its echo, and direction is sensed by differences in time for the echo to reach its two ears. A dolphin's main diet is fish and, since hearing in fish is limited to fairly low frequencies, they are not alerted to the fact they are being hunted.

[3]The primary sense of the dolphin is acoustic, for vision is not a very useful sense in the often murky and dark depths of the ocean. Whereas sound is a passive sense for us, it is an active sense for the dolphin who sends out sounds and then perceives its surroundings on the basis of the echoes that come back. What's more interesting, the dolphin can reproduce the sonic signals that paint the mental image of its surroundings; thus the dolphin probably communicates its experience to other dolphins by communicating the full acoustic image of what is "seen," placing it directly in the minds of other dolphins. It needs no word or symbol for "fish," for example, but communicates an image of the real thing—perhaps with emphasis highlighted by selective filtering, as we similarly communicate a musical concert to others via various means of sound reproduction. Small wonder that the language of the dolphin is very unlike our own!

CHECK YOURSELF

An oceanic depth-sounding vessel surveys the ocean bottom with ultrasonic sound that travels 1530 m/s in seawater. How deep is the water if the time delay of the echo from the ocean floor is 2 s?

Energy in Sound Waves

Wave motion of all kinds possesses energy of varying degrees. Electromagnetic waves from the Sun, for instance, bring us the enormous quantities of energy that are necessary for life on Earth. By comparison, the energy in sound is extremely small. That's because producing sound requires only a small amount of energy. For example, 10,000,000 people talking at the same time would produce sound energy equal only to the energy needed to light a common flashlight. Hearing is possible only because of our remarkably sensitive ears. Only the most sensitive microphone can detect sound that is softer than what we can hear.

Sound energy dissipates to thermal energy while sound travels in air. For waves of higher frequency, the sound energy is transformed into internal energy more rapidly than for waves of lower frequencies. As a result, sound of low frequencies will travel farther through air than sound of higher frequencies. That's why the foghorns of ships are of a low frequency.

Forced Vibrations

If we strike an unmounted tuning fork, the sound from it may be rather faint. If we hold the same fork against a table after striking it, the sound is louder. This is because the table is forced to vibrate, and, with its larger surface, the table will set more air in motion. The table will be forced into vibration by a fork of any frequency. This is a case of **forced vibration.**

The mechanism in a music box is mounted on a sounding board. Without the sounding board, the sound the music-box mechanism produces is barely audible. Sounding boards are important in all stringed musical instruments.

Natural Frequency

When someone drops a wrench on a concrete floor, we are not likely to mistake its sound for that of a baseball bat hitting the floor. This is because the two objects vibrate differently when they are struck. Tap a wrench and the vibrations it makes are different from the vibrations of a baseball bat, or of anything else. Any object composed of an elastic material when disturbed will vibrate at its own special set of frequencies, which together form its special sound. We speak of an object's **natural frequency,** which depends on such factors

CHECK YOUR ANSWER

1530 m. (1 s down and 1 s up.)

as the elasticity and shape of the object. Bells and tuning forks, of course, vibrate at their own characteristic frequencies. And, interestingly enough, most things from planets to atoms and almost everything else in between have a springiness to them and vibrate at one or more natural frequencies.

Resonance

FIGURE 20.11
The natural frequency of the smaller bell is higher than that of the larger bell, and it rings at a higher pitch.

FIGURE 20.12
Interactive Figure
Pumping a swing in rhythm with its natural frequency produces a large amplitude.

When the frequency of forced vibrations on an object matches the object's natural frequency, a dramatic increase in amplitude occurs. This phenomenon is called **resonance.** Literally, *resonance* means "resounding," or "sounding again." Putty doesn't resonate because it isn't elastic, and a dropped handkerchief is too limp. In order for something to resonate, it needs a force to pull it back to its starting position and enough energy to keep it vibrating.

A common experience illustrating resonance occurs on a swing. When pumping a swing, we pump in rhythm with the natural frequency of the swing. More important than the force with which we pump is the timing. Even small pumps or small pushes from someone else, if delivered in rhythm with the frequency of the swinging motion, produce large amplitudes. A common classroom demonstration of resonance is illustrated with a pair of tuning forks adjusted to the same frequency and spaced a meter or so apart. When one of the forks is struck, it sets the other fork into vibration. This is a small-scale version of pushing a friend on a swing—it's the timing that's important. When a series of sound waves impinge on the fork, each compression gives the prong of the fork a tiny push. Since the frequency of these pushes corresponds to the natural frequency of the fork, the pushes will successively increase the amplitude of vibration. This is because the pushes occur at the right time and repeatedly occur in the same direction as the instantaneous motion of the fork. The motion of the second fork is often called a *sympathetic vibration.*

If the forks are not adjusted for matched frequencies, the timing of pushes is off, and resonance will not occur. When you tune your radio set, you are similarly adjusting the natural frequency of the electronics in the set to match one of the many surrounding signals. The set then resonates to one station at a time instead of playing all stations at once.

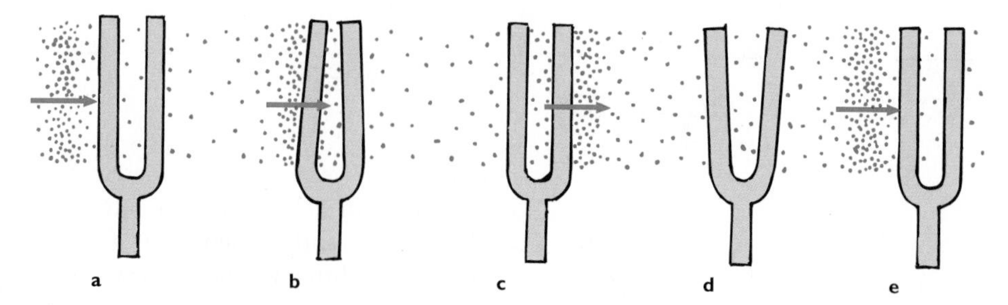

FIGURE 20.13
Stages of resonance. (Blue arrows indicate sound waves traveling to the right.) (a) The first compression meets the fork and gives it a tiny and momentary push. (b) The fork bends and then (c) returns to its initial position just at the time a rarefaction arrives and (d) overshoots in the opposite direction. Just when the fork returns to its initial position (e), the next compression arrives to repeat the cycle. Now it bends farther because it is moving.

FIGURE 20.14
In 1940, four months after being completed, the Tacoma Narrows Bridge in Washington State was destroyed by wind-generated resonance. A mild gale produced an irregular force in resonance with the natural frequency of the bridge, steadily increasing the amplitude of vibration until the bridge collapsed.

Resonance
Resonance and Bridges

Resonance is not restricted to wave motion. It occurs whenever successive impulses are applied to a vibrating object in rhythm with its natural frequency. Cavalry troops marching across a footbridge near Manchester, England, in 1831 inadvertently caused the bridge to collapse when they marched in rhythm with the bridge's natural frequency. Since then, it is customary to order troops to "break step" when crossing bridges in order to prevent resonance. A century later another major bridge disaster was caused by wind-generated resonance (Figure 20.14).

The effects of resonance are all about us. Resonance underscores not only the sound of music but the color of autumn leaves, the height of ocean tides, the operation of lasers, and a vast multitude of phenomena that add to the beauty of the world.

Interference

Sound waves, like any waves, can be made to exhibit **interference.** Recall that wave interference was discussed in the previous chapter. A comparison of interference for transverse waves and longitudinal waves is shown in Figure 20.15. In either case, when the crests of one wave overlap the crests of another wave, increased amplitude results. Or, when the crest of one wave overlaps the trough of another wave, decreased amplitude results. In the case of sound, the crest of a wave corresponds to a compression and the trough of a wave corresponds to a rarefaction. Interference occurs for all waves, both transverse and longitudinal.

An interesting case of sound interference is illustrated in Figure 20.16. If you are an equal distance from two sound speakers that emit identical tones of fixed frequency, the sound is louder because the effects of the two speakers add. The compressions and rarefactions of the tones arrive in step, or in *phase*. However, if you move to the side so that the paths from the speakers to you differ by a half wavelength, then the rarefactions from one speaker will be filled in by the compressions from the other speaker. This is destructive interference. It is just as

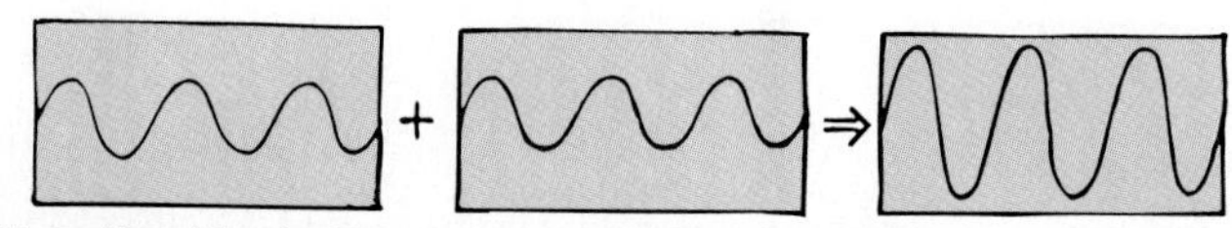

The superposition of two identical transverse waves in phase produces a wave of increased amplitude.

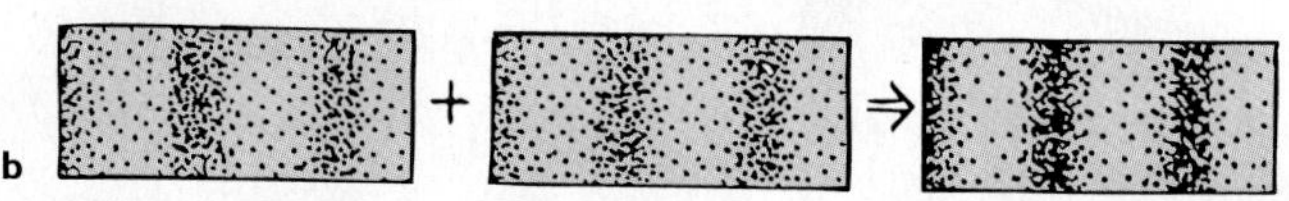

The superposition of two identical longitudinal waves in phase produces a wave of increased intensity.

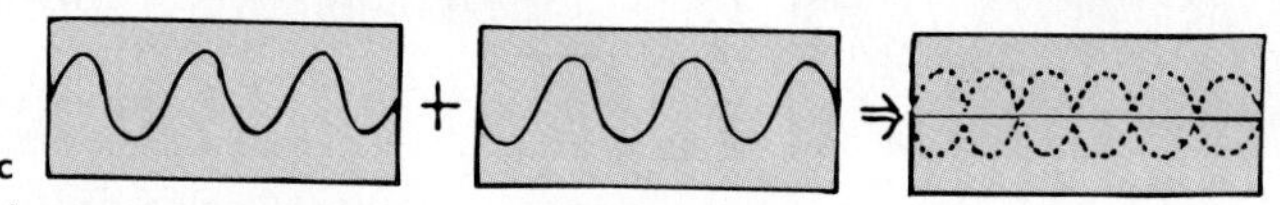

Two identical transverse waves that are out of phase destroy each other when they are superimposed.

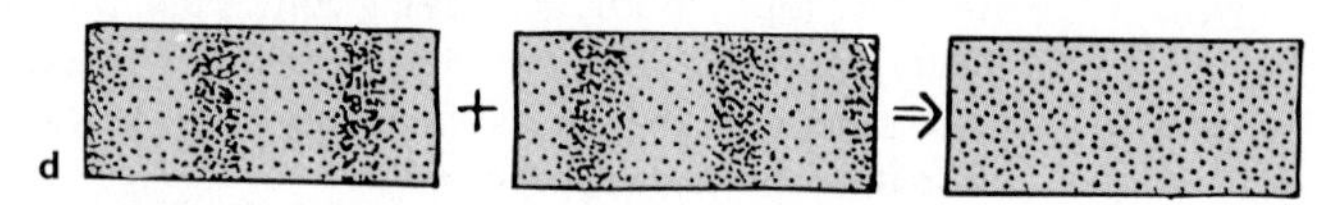

Two identical longitudinal waves that are out of phase destroy each other when they are superimposed.

FIGURE 20.15 Interactive Figure

Constructive (a,b) and destructive (c,d) wave interference in transverse and longitudinal waves.

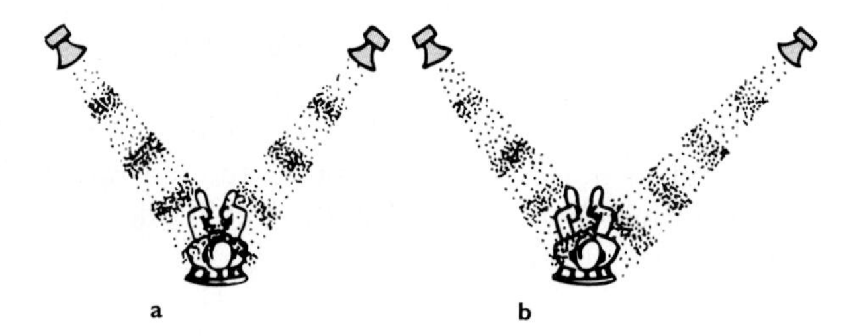

FIGURE 20.16
Interactive Figure

Interference of sound waves. (a) Waves arrive in phase and interfere constructively when the path lengths from the speakers are the same. (b) Waves arrive out of phase and interfere destructively when the path lengths differ by half a wavelength (or 3/2, 5/2, etc.).

if the crest of one water wave exactly filled in the trough of another water wave. If the region is devoid of any reflecting surfaces, little or no sound will be heard!

If the speakers emit a whole range of sounds with different frequencies, only some waves destructively interfere for a given difference in path lengths. So interference of this type is usually not a problem, because there is usually enough reflection of sound to fill in cancelled spots. Nevertheless, "dead spots" are sometimes evident in poorly designed theaters or music halls, where sound waves reflect off walls and interfere with nonreflected waves to produce zones of low amplitude. When you move your head a few centimeters in either direction, you may hear a noticeable difference.

Sound interference is dramatically illustrated when monaural sound is played by stereo speakers that are out of phase. Speakers are out of phase when the input wires to one speaker are interchanged (positive and negative wire inputs reversed). For a monaural signal, this means that, when one speaker is sending a compression of sound, the other is sending a rarefaction. The sound produced is not as full and not as loud as from speakers properly connected in phase because the longer waves are being cancelled by interference. Shorter

FIGURE 20.17
The positive and negative wire inputs to one of the stereo speakers have been interchanged, resulting in speakers that are out of phase. When the speakers are far apart, monaural sound is not as loud as from properly phased speakers. When they are brought face to face, very little sound is heard. Interference is nearly complete as the compressions of one speaker fill in the rarefactions of the other!

waves are cancelled as the speakers are brought closer together, and, when the pair of speakers are brought face to face against each other, very little sound is heard! Only the sound waves having the highest frequencies survive cancellation. You must try this to appreciate it.

Destructive sound interference is a useful property in *antinoise technology*. Such noisy devices as jackhammers are being equipped with microphones that send the sound of the device to electronic microchips, which create mirror-image wave patterns of the sound signals. For the jackhammer, this mirror-image sound signal is fed to earphones worn by the operator. Sound compressions (or rarefactions) from the hammer are cancelled by mirror-image rarefactions (or compressions) in the earphones. The combination of signals cancels the jackhammer noise. Noise-cancelling earphones are already common for pilots. The cabins of some airplanes are now quieted with antinoise technology.

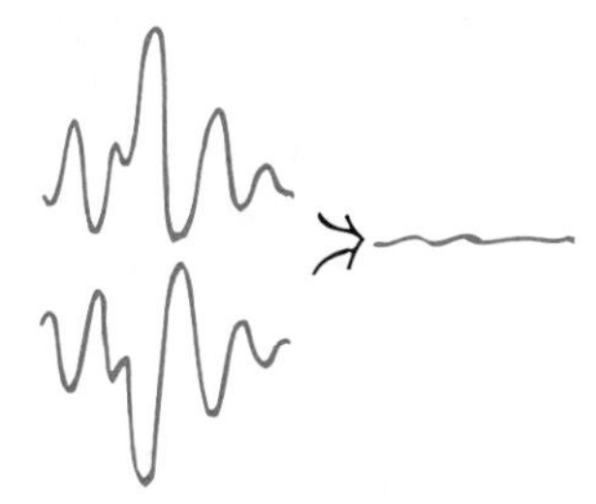

FIGURE 20.18
When a mirror image of a sound signal combines with the sound itself, the sound is cancelled.

FIGURE 20.19
Ken Ford tows gliders in quiet comfort when he wears his noise-canceling earphones.

Beats

Interference and Beats

When two tones of slightly different frequency are sounded together, a fluctuation in the loudness of the combined sounds is heard; the sound is loud, then faint, then loud, then faint, and so on. This periodic variation in the loudness of sound is called **beats** and is due to interference. Strike two slightly mismatched tuning forks and, because one fork vibrates at a different frequency than the other, the vibrations of the forks will be momentarily in step, then out of step, then in again, and so on. When the combined waves reach our ears in step—say, when a compression from one fork overlaps a compression from the other—the sound is a maximum. A moment later, when

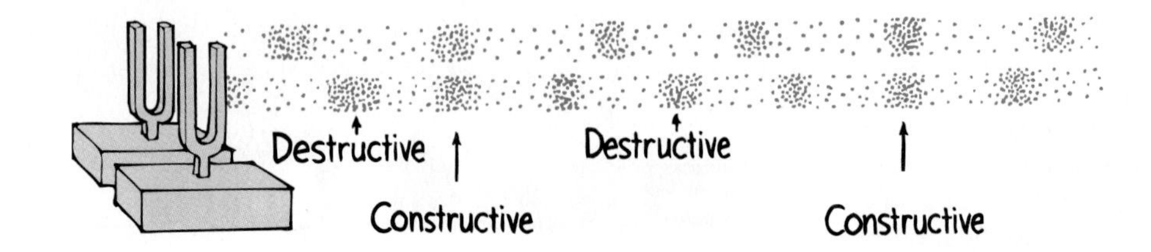

FIGURE 20.20
The interference of two sound sources of slightly different frequencies produces beats.

the forks are out of step, a compression from one fork is met with a rarefaction from the other, resulting in a minimum. The sound that reaches our ears throbs between maximum and minimum loudness and produces a tremolo effect.

Why does Hollywood persist in playing engine noises whenever a spacecraft in outer space passes by? Wouldn't seeing them float by silently be far more dramatic?

Insights

We can understand beats by considering the analogous case of a tall person and short person walking side by side with different strides. At some moment they will be in step, a little later out of step, then in step again, and so on. Imagine that the tall person with longer legs takes exactly seventy steps in 1 minute, and the shorter person takes seventy-two steps in the same time. The shorter person gains two steps per minute on the taller person. A little thought will show that they will both be momentarily in step twice each minute. In general, if two people with different strides walk together, the number of times they are in step in each unit of time is equal to the difference in the frequencies of the steps. This applies also to the pair of tuning forks. If one fork undergoes 264 vibrations each second and the other fork vibrates 262 times per second, they will be in step twice each second. A beat frequency of 2 hertz will be heard. The overall tone will correspond to the average frequency, 263 hertz.

CHECK YOURSELF

What is the beat frequency when a 262-Hz tuning fork and a 266-Hz tuning fork are sounded together? A 262-Hz fork and a 272-Hz fork?

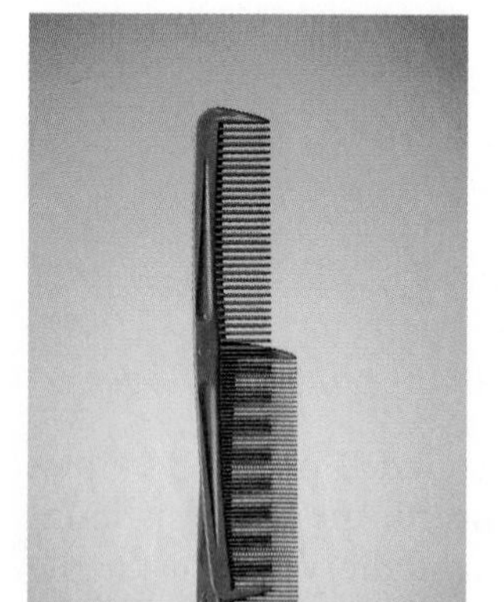

FIGURE 20.21
The unequal spacings of the combs produce a moiré pattern that is similar to beats.

If we overlap two combs with different teeth spacing, we'll see a moiré pattern that is related to beats (Figure 20.21). The number of beats per length will equal the difference in the number of teeth per length for the two combs.

Beats can occur with any kind of wave and can provide a practical way to compare frequencies. To tune a piano, for example, a piano tuner listens for beats produced between a standard frequency and the frequency of a particular string on the piano. When the frequencies are identical, the beats disappear. Beats can help you tune a variety of musical instruments. Simply listen for beats

CHECK YOUR ANSWERS

For the 262-Hz and 266-Hz forks, the ear will hear 264 Hz, which will beat at 4 Hz (266 − 262). For the 272-Hz and 262-Hz forks, 267 Hz will be heard, and some people will hear it throb ten times each second. Beat frequencies greater than 10 Hz are normally too rapid to be heard.

RADIO BROADCASTS

A radio receiver emits sound but, interestingly enough, it doesn't receive sound waves. A radio receiver, like a television set, receives *electromagnetic waves*—actually low-frequency light waves. These waves, which we will treat in detail in Part Six, are fundamentally different from sound waves—both in their completely different natures and their extremely high frequencies, which are way beyond the range of hearing.

Every radio station has an assigned frequency at which it broadcasts. The electromagnetic wave transmitted at this frequency is the *carrier wave*. The relatively low-frequency sound signal to be communicated is superimposed on the much higher-frequency carrier wave in two principal ways: by slight variations in amplitude that match the audio frequency or by slight variations in frequency. This impression of the sound wave on the higher-frequency radio wave is *modulation*. When the *amplitude* of the carrier wave is modulated, we call it AM, or *amplitude modulation*. AM stations broadcast in the range of 535 to 1605 kilohertz. When the *frequency* of the carrier wave is modulated, we call it FM, or *frequency modulation*. FM stations broadcast in the higher-frequency range of 88 to 108 megahertz. Amplitude modulation is similar to rapidly changing the brightness of a constant-color lightbulb. Frequency modulation is similar to rapidly changing the color of a constant-intensity lightbulb.

Turning the knob of a radio receiver to select a particular station is like adjusting movable masses on the prongs of a tuning fork to make it resonate to the sound produced by another fork. In choosing a radio station, you adjust the frequency of an electrical circuit inside the radio receiver to match and resonate with the frequency of the desired station. You sort out one carrier wave from many. Then the impressed sound signal is separated from the carrier wave, amplified, and fed to the loudspeaker. It's nice to hear only one station at a time!

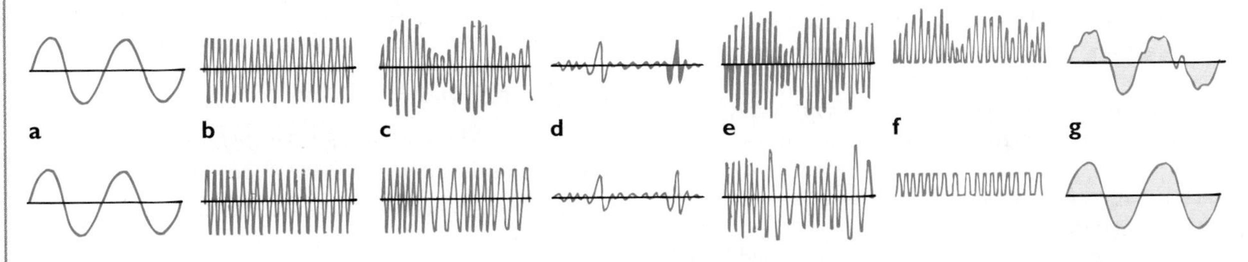

FIGURE 20.22
AM and FM radio signals. (a) Sound waves that enter a microphone. (b) Radio-frequency carrier wave produced by a transmitter without a sound signal. (c) Carrier wave modulated by signal. (d) Static interference. (e) Carrier wave and signal affected by static. (f) Radio receiver cuts out negative half of carrier wave. (g) Signal remaining is rough for AM because of static but is smooth for FM because the tips of the waveform are clipped without loss to the signal.

between the tone of your instrument and a standard tone produced by a piano or some other instrument.

Beats are utilized by dolphins in surveying the motions of things around them. When a dolphin sends out sound signals, beats may be produced when the echoes it receives interfere with the sound it sends. When there is no relative motion between the dolphin and the object returning the sound, the sending and receiving frequencies are the same and no beats occur. But when there is relative motion, the echo has a different frequency due to the Doppler effect, and beats are produced when the echo and emitted sound combine. The same principle is applied by the radar guns used by police officers. The beats between the signal that is sent and the one that is reflected are used to determine how fast the car that reflected the signal is moving.

CHECK YOURSELF

Is it correct to say that, in every case, without exception, any radio wave travels faster than any sound wave?

CHECK YOUR ANSWER

Yes, because any radio wave travels at the speed of light. A radio wave is an electromagnetic wave—in a very real sense, a low-frequency light wave (or we can say a light wave is a high-frequency radio wave!). A sound wave, on the other hand, is a mechanical disturbance propagated through a material medium by material particles that vibrate against one another. In air, the speed of sound is about 340 m/s, about one-millionth the speed of a radio wave. Sound travels faster in other media, but in no case at the speed of light. No sound wave can travel as fast as light.

Summary of Terms

Infrasonic Describes a sound that has a frequency too low to be heard by the normal human ear.

Ultrasonic Describes a sound that has a frequency too high to be heard by the normal human ear.

Compression Condensed region of the medium through which a longitudinal wave travels.

Rarefaction Rarefied region (of reduced pressure) of the medium through which a longitudinal wave travels.

Reverberation Persistence of sound, as in an echo, due to multiple reflections.

Refraction Bending of sound or any wave caused by a difference in wave speeds.

Forced vibration The setting up of vibrations in an object by a vibrating force.

Natural frequency A frequency at which an elastic object naturally tends to vibrate if it is disturbed and the disturbing force is removed.

Resonance The response of a body when a forcing frequency matches its natural frequency.

Interference A result of superposing different waves, often of the same wavelength. Constructive interference results from crest-to-crest reinforcement; destructive interference results from crest-to-trough cancellation.

Beats A series of alternate reinforcements and cancellations produced by the interference of two waves of slightly different frequencies, heard as a throbbing effect in sound waves.

Suggested Reading

Chiaverina, Chris, and Tom Rossing. *Light Science: Physics for the Visual Arts.* New York: Springer, 1999. Enjoyable reading from the two physicists having fun with sound on page 380.

Review Questions

1. How does a physicist usually define sound?

Origin of Sound

2. What is the relationship between *frequency* and *pitch*?
3. What is the average range of a young person's hearing?
4. Distinguish between *infrasonic* and *ultrasonic* sound waves.

Nature of Sound in Air

5. Distinguish between a *compression* and a *rarefaction*.
6. Cite evidence to support the fact that compressions and rarefactions travel in the same direction in a wave.

Media That Transmit Sound

7. Relative to solids and liquids, how does air rank as a conductor of sound?
8. Why will sound not travel in a vacuum?

Speed of Sound in Air

9. What factors does the speed of sound depend upon? What are some factors that it does *not* depend upon?
10. What is the speed of sound in dry air at 0°C?
11. Does sound travel faster in warm air than in cold air? Defend your answer.

Reflection of Sound

12. What is an *echo*?
13. What is a *reverberation*?

#1, 7, 13, 19, 21, 23, 25, 27, 35, 41, 44

Refraction of Sound

14. What is the cause of refraction?
15. Does sound tend to bend upward or downward when its speed is less near the ground?
16. Why does sound sometimes refract under water?

Energy in Sound Waves

17. Which is normally greater, the energy in ordinary sound or the energy in ordinary light?
18. What ultimately becomes of the energy of sound in the air?

Forced Vibrations

19. Why will a struck tuning fork sound louder when it is held against a table?

Natural Frequency

20. Give at least two factors that determine the natural frequency of an object.

Resonance

21. How do forced *vibrations* relate to *resonance*?
22. When you listen to a radio, why do you hear only one station at a time rather than hearing all stations at once?
23. How did wind-generated resonance affect the Tacoma Narrows Bridge in the state of Washington in 1940?

Interference

24. When is it possible for one wave to cancel another?
25. What kind of waves can exhibit interference?

Beats

26. What physical phenomenon underlies the production of beats?
27. What beat frequency will occur when a 370-Hz and a 374-Hz sound source are sounded together?
28. How does a radio wave differ from a sound wave?

Projects

1. In the bathtub, submerge your head and listen to the sound you make when clicking your fingernails together or tapping the tub beneath the water surface. Compare the sound by doing the same when both the source and your ears are above the water. At the risk of getting the floor wet, slide back and forth in the tub at different frequencies and see how the amplitude of the sloshing waves quickly builds up when you slide in rhythm with the waves. (The latter of these projects is most effective when you are alone in the tub.)

2. Stretch a piece of balloon rubber not too tightly over a radio loudspeaker. Glue a small, very lightweight piece of mirror, aluminum foil, or polished metal near one edge. Project a narrow beam of light on the mirror while your favorite music is playing and observe the beautiful patterns that are reflected on a screen or wall.

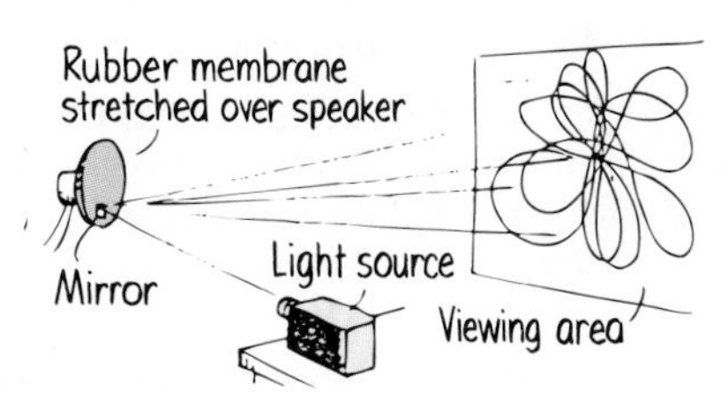

Exercises

1. Why do you not hear the sound of distant fireworks until after you see them?
2. If the Moon blew up, why would we not hear it?
3. Why would it be futile to attempt to detect sounds from other planets, even given the very best in audio detectors?
4. If you toss a stone in still water, concentric circles are formed. What form will waves have if a stone is tossed into smoothly flowing water?
5. Why do flying bees buzz?
6. A cat can hear sound frequencies up to 70,000 Hz. Bats send and receive ultrahigh-frequency squeaks up to 120,000 Hz. Which hears sound of shorter wavelengths, cats or bats?
7. What does it mean to say that a radio station is "at 101.1 on your FM dial"?
8. Sound from Source A has twice the frequency of sound from Source B. Compare the wavelengths of sound from the two sources.
9. Suppose a sound wave and an electromagnetic wave have the same frequency. Which has the longer wavelength?

10. What happens to the wavelength of sound as the frequency increases?
11. In the stands of a racetrack, you notice smoke from the starter's gun before you hear it fire. Explain.
12. In Olympic competition, a microphone detects the sound of the starter's gun and sends it electrically to speakers at every runner's starting block. Why?
13. When a sound wave moves past a point in air, are there changes in the density of air at this point? Explain.
14. At the instant that a high-pressure region is created just outside the prongs of a vibrating tuning fork, what is being created inside between the prongs?
15. Why is it so quiet after a snowfall?
16. If a bell is ringing inside a bell jar, we can no longer hear it when the air is pumped out, but we can still see it. What differences in the properties of sound and light does this indicate?
17. Why is the Moon described as a "silent planet"?
18. As you pour water into a glass, you repeatedly tap the glass with a spoon. As the tapped glass is being filled, does the pitch of the sound increase or decrease? (What should you do to answer this question?)
19. If the speed of sound were dependent on its frequency, would you enjoy a concert sitting in the second balcony? Explain.
20. If the frequency of sound is doubled, what change will occur in its speed? In its wavelength?
21. Why does sound travel faster in warm air?
22. Why does sound travel faster in moist air? (*Hint:* At the same temperature, water vapor molecules have the same average kinetic energy as the heavier nitrogen and oxygen molecules in the air. How, then, do the average speeds of H_2O molecules compare with those of N_2 and O_2 molecules?)
23. Would the refraction of sound be possible if the speed of sound were unaffected by wind, temperature, and other conditions? Defend your answer.
24. Why can the tremor of the ground from a distant explosion be felt before the sound of the explosion can be heard?
25. What kinds of wind conditions would make sound more easily heard at long distances? Less easily heard at long distances?
26. Ultrasonic waves have many applications in technology and medicine. One advantage is that large intensities can be used without danger to the ear. Cite another advantage of their short wavelength. (*Hint:* Why do microscopists use blue light rather than white light to see detail?)
27. Why is an echo weaker than the original sound?
28. If the distance from a bugle is tripled, by what factor does the sound intensity decrease? Assume that no reflections affect the sound.
29. What two physics mistakes occur in a science fiction movie that shows a distant explosion in outer space, where you see and hear the explosion at the same time?
30. A rule of thumb for estimating the distance in kilometers between an observer and a lightning stroke is to divide the number of seconds in the interval between the flash and the sound by 3. Is this rule correct? Defend your answer.
31. If a single disturbance at an unknown distance emits both transverse and longitudinal waves that travel with distinctly different speeds in the medium, such as in the ground during an earthquake, how can the distance to the disturbance be determined?
32. Why will marchers at the end of a long parade following a band be out of step with marchers near the front?
33. Why is it a sensible procedure for soldiers to break step when marching over a bridge?
34. Why is the sound of a harp soft in comparison with the sound of a piano?
35. Apartment dwellers will testify that bass notes are more distinctly heard from music played in nearby apartments. Why do you suppose lower-frequency sounds travel through walls, floors, and ceilings more easily?
36. If the handle of a tuning fork is held solidly against a table, the sound from the tuning fork becomes louder. Why? How will this affect the length of time the fork keeps vibrating? Explain.
37. The sitar, an Indian musical instrument, has a set of strings that vibrate and produce music, even though they are never plucked by the player. These "sympathetic strings" are identical to the plucked strings and are mounted below them. What is your explanation?
38. Why does a dance floor heave only when certain kinds of dance steps are being performed?
39. A pair of loudspeakers on two sides of a stage are emitting identical pure tones (tones of a fixed frequency and fixed wavelength in air). When you stand in the center aisle, equally distant from the two speakers, you hear the sound loud and clear. Why does the intensity of the sound diminish considerably when you step to one side? Suggestion: Use a diagram to make your point.

40. A special device can transmit out-of-phase sound from a noisy jackhammer to its operator using earphones. Over the noise of the jackhammer, the operator can easily hear your voice while you are unable to hear his. Explain.

41. When two out-of-phase speakers are brought together as shown in Figure 20.17, which waves are most cancelled, long waves or short waves? Why?

42. How can a certain note sung by a singer cause a crystal glass to shatter?

43. An object resonates when the frequency of a vibrating force either matches its natural frequency or is a submultiple of its natural frequency. Why will it not resonate to multiples of its natural frequency? (Think of pushing a child in a swing.)

44. Are beats the result of interference, or of the Doppler effect, or of both?

45. Can it correctly be said that beats of sound are much the same thing as the rhythmic "beat" of music? Defend your answer.

46. Two sound waves of the same frequency can interfere, but, in order to produce beats, the two sound waves must be of different frequencies. Why?

47. Walking beside you, your friend takes 50 strides per minute while you take 48 strides per minute. If you start in step, you'll soon be out of step. When will you be in step again?

48. Suppose a piano tuner hears 3 beats per second when listening to the combined sound from a tuning fork and the piano wire being tuned. After slightly tightening the string, 5 beats per second are heard. Should the string be loosened or tightened?

49. A piano tuner using a 264-Hz tuning fork hears 4 beats per second. What are two possible frequencies of vibration of the piano wire?

50. A human cannot hear sound at a frequency of 100 kHz, or sound at 102 kHz. But, if you walk into a room in which two sources are emitting sound waves, one at 100 kHz and the other at 102 kHz, you'll hear sound. Explain.

Problems

1. What is the wavelength of a 340-Hz tone in air? What is the wavelength of a 34,000-Hz ultrasonic wave in air?

2. For years, marine scientists were mystified by sound waves detected by underwater microphones in the Pacific Ocean. These so-called T-waves were among the purest sounds in nature. Eventually they traced the source to underwater volcanoes whose rising columns of bubbles resonated like organ pipes. What is the wavelength of a typical T-wave whose frequency is 7Hz? (The speed of sound in seawater is 1530 m/s.)

3. An oceanic depth-sounding vessel surveys the ocean bottom with ultrasonic waves that travel 1530 m/s in seawater. How deep is the water directly below the vessel if the time delay of the echo to the ocean floor and back is 6 s?

4. A bat flying in a cave emits a sound and receives its echo 0.1 s later. How far away is the cave wall?

5. You watch a distant person driving nails into a front porch at a regular rate of 1 stroke per second. You hear the sound of the blows exactly synchronized with the blows you see. And then you hear one more blow after you see the hammering stop. How far away is the nail-driving person?

6. Imagine a Rip van Winkle type who lives in the mountains. Just before going to sleep, he yells, "WAKE UP," and the sound echoes off the nearest mountain and returns 8 hours later. How far away is that mountain?

7. Two speakers are wired to emit identical sounds in unison. The wavelength in air of the sounds is 6 m. State whether the sounds interfere constructively or destructively when you are at a distance of (a) 12 m from both speakers, (b) 9 m from both speakers, and (c) 9 m from one speaker and 12 m from the other.

8. What is the frequency of the sound emitted by the speakers in the previous problem? Is this pitch low or high, relative to the range of human hearing?

9. A grunting porpoise emits sound at 57 Hz. What is the wavelength of this sound in water, where the speed of sound is 1,500 m/s?

10. What beat frequencies are possible with tuning forks of frequencies 256, 259, and 261 Hz?

Norm Whitlatch shows the first five harmonics, in blue, from a Fourier analysis of sound from a guitar pitch pipe, which is displayed on the upper green oscilloscope trace.

CHAPTER 21

Musical Sounds

Most of the sounds we hear are noises. The impact of a falling object, the slamming of a door, the roaring of a motorcycle, and most of the sounds from traffic in city streets are noises. Noise corresponds to an irregular vibration of the eardrum produced by some irregular vibration in our surroundings. The sound of music has a different character, having periodic tones—or musical "notes." Although noise doesn't have these characteristics, the line that separates music and noise is thin and subjective. To some contemporary composers, it is nonexistent.

Some people consider contemporary music and music from other cultures to be noise. Differentiating these types of music from noise becomes a problem of aesthetics. However, differentiating traditional music—that is, Western classical music and most types of popular music—from noise presents no problem. A person with total hearing loss could distinguish between these by using an oscilloscope. Recall, from Figure 20.5, that, when an electrical signal from a microphone is fed into an oscilloscope, patterns of air-pressure variations with time are nicely displayed that make it easy to distinguish between noise and traditional music (Figure 21.1).

Musicians usually speak of musical tones in terms of three principal characteristics: pitch, loudness, and quality.

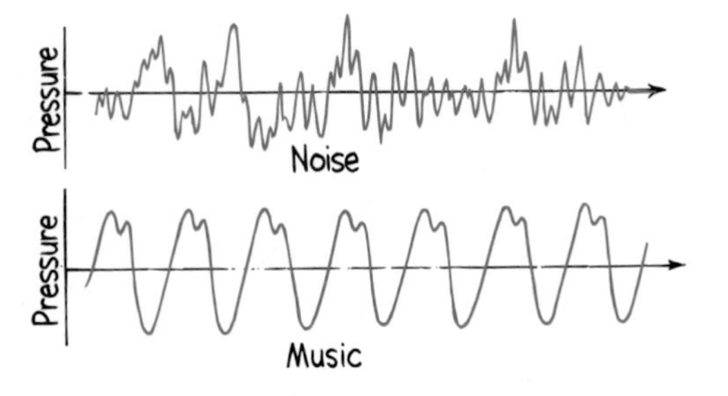

FIGURE 21.1
Graphical representations of noise and music.

Pitch

The **pitch** of a sound relates to its frequency. Most sounds are composites of frequencies, in which case the pitch corresponds to the lowest frequency component. Rapid vibrations of the sound source (high frequency) produce a high note, whereas slow vibrations (low frequency) produce a low note. We speak of the pitch of a sound in terms of its position on the musical scale. When concert A is struck on a piano, a hammer strikes two or three strings, each of which vibrates 440 times in 1 second. The pitch of concert A corresponds to 440 hertz.[1]

Different musical notes are obtained by changing the frequency of the vibrating sound source. This is usually done by altering the size, the tightness, or the mass of the vibrating object. A guitarist or violinist, for example, adjusts the tightness, or tension, of the strings of the instrument when tuning them. Then different notes can be played by altering the length of each string by "stopping" it with the fingers.

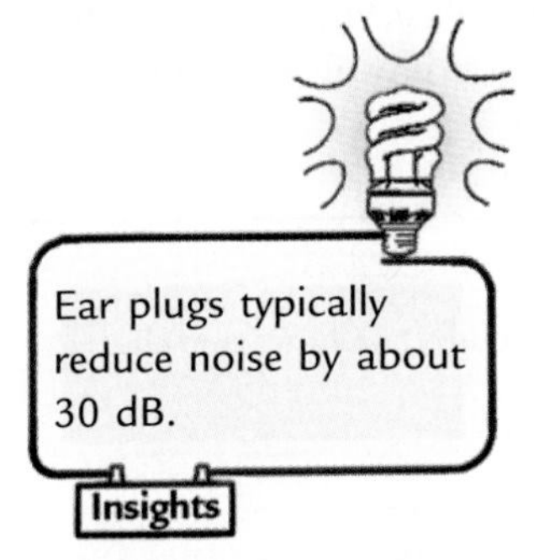

In wind instruments, the length of the vibrating air column can be altered (trombone and trumpet) or holes in the side of the tube can be opened and closed in various combinations (saxophone, clarinet, flute) to change the pitch of the note produced.

High-pitched sounds used in music are most often less than 4000 hertz, but the average human ear can hear sounds with frequencies up to 18,000 hertz. Some people and most dogs can hear tones of higher pitch than this. In general, the upper limit of hearing in people gets lower as they grow older. A high-pitched sound is often inaudible to an older person and yet may be clearly heard by a younger one. So, by the time you can really afford that high-fidelity music system, you may not be able to appreciate the difference.

Sound Intensity and Loudness

Ear plugs typically reduce noise by about 30 dB.

Insights

The **intensity** of sound depends on the amplitude of pressure variations within the sound wave. (And, as with all waves, intensity is directly proportional to the square of the wave amplitude.) Intensity is measured in units of watts/square meter. The human ear responds to intensities covering the enormous range from 10^{-12} W/m^2 (the threshold of hearing) to more than 1 W/m^2 (the threshold of pain). Because the range is so great, intensities are scaled by factors of ten, with the barely audible 10^{-12} W/m^2 as a reference intensity—called 0 *bel* (a unit named after Alexander Graham Bell). A sound ten times more intense has an intensity of 1 bel (10^{-11} W/m^2) or 10 *decibels.* Table 21.1 lists typical sounds and their intensities.

A sound of 10 decibels is 10 times as intense as 0 decibels, the threshold of hearing. Accordingly, 20 decibels is 100, or 10^2, times the intensity of the threshold of hearing, 30 decibels is 10^3 times the threshold of hearing, and 40 decibels is 10^4 times. So 60 decibels represents sound intensity a million times (10^6) greater than 0 decibels; 80 decibels represents sound 10^2 times as intense as 60 decibels.[2]

Physiological hearing damage begins at exposure to 85 decibels, the degree of damage depending on the length of exposure and on frequency characteristics. Damage from loud sounds can be temporary or permanent, depending on whether the organ of Corti, the receptor organ in the inner ear, is impaired or destroyed. A single burst of sound can produce vibrations in the organ intense enough to tear it apart. Less intense, but severe, noise can interfere with cellular processes in the organ and cause its eventual breakdown. Unfortunately, the cells of the organ do not regenerate.

[1]Interestingly enough, concert A varies from as low as 436 Hz to as high as 448 Hz in different symphony orchestras.

[2]The decibel scale is called a logarithmic scale. The decibel rating is proportional to the logarithm of the intensity.

TABLE 21.1
Common Sources and Sound Intensities

Source of Sound	*Intensity* (W/m^2)	*Sound Level* (dB)
Jet airplane 30 m away	10^2	140
Air-raid siren, nearby	1	120
Disco music, amplified	10^{-1}	115
Riveter	10^{-3}	100
Busy street traffic	10^{-5}	70
Conversation in home	10^{-6}	60
Quiet radio in home	10^{-8}	40
Whisper	10^{-10}	20
Rustle of leaves	10^{-11}	10
Threshold of hearing	10^{-12}	0

FIGURE 21.2
James displays a sound signal on an oscilloscope.

Sound intensity is a purely objective and physical attribute of a sound wave, and it can be measured by various acoustical instruments (and the oscilloscope in Figure 21.2). **Loudness,** on the other hand, is a physiological sensation. The ear senses some frequencies much better than others. A 3500-Hz sound at 80 decibels, for example, sounds about twice as loud to most people as a 125-Hz sound at 80 decibels; humans are more sensitive to the 3500-Hz range of frequencies. The loudest sounds we can tolerate have intensities a trillion times greater than the faintest sounds. The difference in perceived loudness, however, is much less than this amount.

CHECK YOURSELF

Is hearing permanently impaired when attending concerts, clubs, or functions that feature very loud music?

Quality

We have no trouble distinguishing between the tone from a piano and a tone of the same pitch from a clarinet. Each of these tones has a characteristic sound that differs in **quality,** or timbre. Most musical sounds are composed of a superposition of many tones differing in frequency. The various tones are called **partial tones,** or simply *partials*. The lowest frequency, called the **fundamental frequency,**

CHECK YOUR ANSWER

Yes, depending on how loud, how long, how near, and how often. Some music groups have emphasized loudness over quality. Tragically, as hearing becomes more and more impaired, members of the group (and their fans) require louder and louder sounds for stimulation. Hearing loss caused by sounds is particularly common in the frequency range of 2000–5000 Hz. Human hearing is normally most sensitive around 3000 Hz.

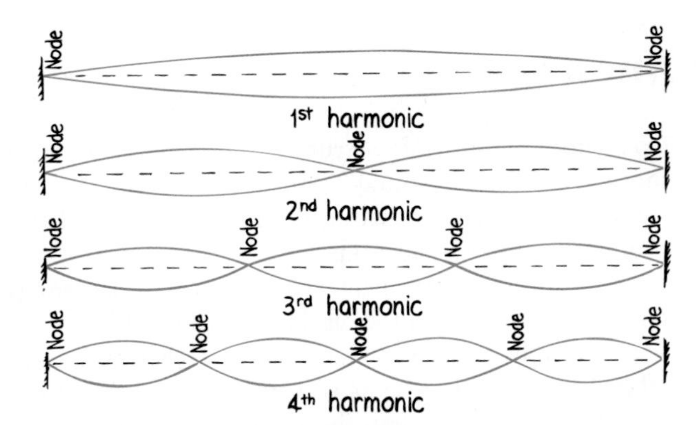

FIGURE 21.3
Modes of vibration of a guitar string.

determines the pitch of the note. A partial tone whose frequency is a whole number multiple of the fundamental frequency is called a **harmonic.** A tone that has twice the frequency of the fundamental is the second harmonic; a tone with three times the fundamental frequency is the third harmonic; and so on (Figure 21.3).[3] It is the variety of partial tones that gives a musical note its characteristic quality.

Thus, if we strike middle C on the piano, we produce a fundamental tone with a pitch of about 262 hertz and also a blending of partial tones of two, three, four, five, and so on, times the frequency of middle C. The number and relative loudness of the partial tones determine the quality of sound associated with the piano. Sound from practically every musical instrument consists of a fundamental and partials. Pure tones, those having only one frequency, can be produced electronically. Electronic synthesizers produce pure tones, and mixtures of these tones, to give a vast variety of musical sounds.

The quality of a tone is determined by the presence and relative intensity of the various partials. The sound produced by a certain tone from the piano and a clarinet of the same pitch have different qualities that the ear recognizes because their partials are different. A pair of tones of the same pitch with different qualities have either different partials or a difference in the relative intensity of the partials.

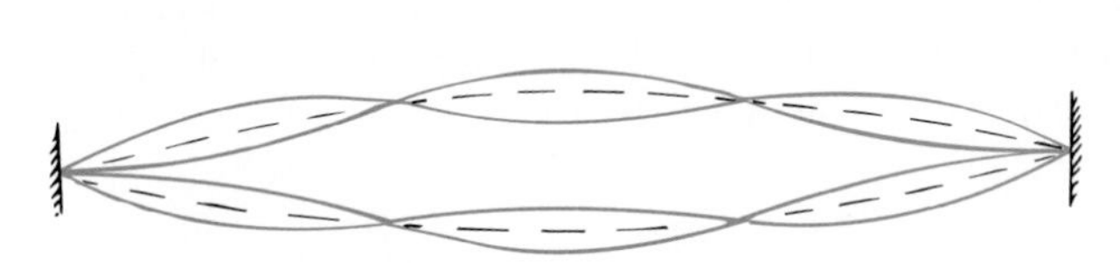

FIGURE 21.4
A composite vibration of the fundamental mode and the third harmonic.

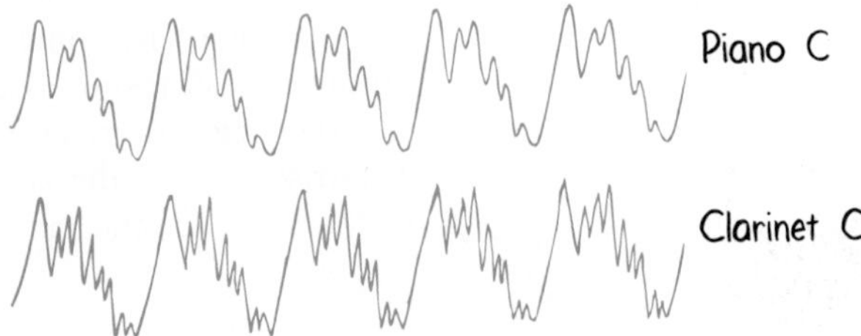

FIGURE 21.5
Sounds from the piano and clarinet differ in quality.

[3]In the terminology often used in music, the second harmonic is called the *first overtone,* the third harmonic is called the *second overtone,* and so on.

Not all partial tones present in a complex tone are integer multiples of the fundamental. Unlike the harmonics of woodwinds and brasses, stringed instruments, such as pianos, produce "stretched" partial tones that are nearly, but not quite, harmonics. This is an important factor in tuning pianos, and it occurs because the stiffness of the strings adds a little bit of restoring force to the tension.

Musical Instruments

Conventional musical instruments can be grouped into one of three classes: those in which the sound is produced by vibrating strings, those in which the sound is produced by vibrating air columns, and those in which the sound is produced by *percussion*—as with the vibrating of a two-dimensional surface.

In a stringed instrument, the vibration of the strings is transferred to a sounding board and then to the air, but with low efficiency. To compensate for this, we find relatively large string sections in orchestras. A smaller number of the high-efficiency wind instruments sufficiently balances a much larger number of violins.

In a wind instrument, the sound is a vibration of an air column in the instrument. There are various ways to set air columns into vibration. In brass instruments, such as trumpets, French horns, and trombones, vibrations of the player's lips interact with standing waves that are set up by acoustic energy reflected within the instrument by the flared bell. The lengths of the vibrating air columns are manipulated by pushing valves that add or subtract extra segments or by extending the length of the tube.[4] In woodwinds, such as clarinets, oboes, and saxophones, a stream of air produced by the musician sets a reed vibrating, whereas in fifes, flutes, and piccolos, the musician blows air against the edge of a hole to produce a fluttering stream that sets the air column into vibration.

In percussion instruments such as drums and cymbals, a two-dimensional membrane or elastic surface is struck to produce sound. The fundamental tone produced depends on the geometry, the elasticity, and, in some cases, the tension of the surface. Changes in pitch result from changing the tension in the vibrating surface; depressing the edge of a drum membrane with the hand is one way of accomplishing this. A variety of modes of vibration can be set up by striking the surface in different places. In the kettledrum, for example, the shape of the kettle changes the frequency of the drum. As in all musical sounds, the quality depends on the number and relative loudness of the partial tones.

Electronic musical instruments differ markedly from conventional musical instruments. Instead of strings that must be bowed, plucked, or struck, or reeds over which air must be blown, or diaphragms that must be tapped to produce sounds, some electronic instruments use electrons to generate the signals that produce musical sounds. Others start with sound from an acoustical instrument and then modify it. Electronic music demands of the composer and player an expertise beyond the knowledge of musicology. It brings a powerful new tool to the hands of the musician.

Fourier Analysis

Did you ever look closely at the grooves in an old phonograph record, the kind that provided music for Grandma and Granddad? Variations in the width of the grooves, seen in Figure 21.6, cause the phonograph needle (stylus) that rides in the groove to vibrate. These mechanical vibrations, in turn, are transformed into electrical vibrations to produce the sound you hear from a record. Isn't it

[4]A bugle has neither valves nor variable length. A bugler must be adept in creating various overtones to get various notes.

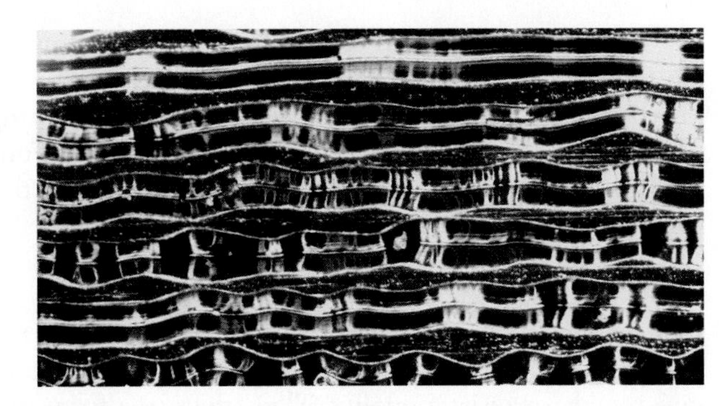

FIGURE 21.6
A microscopic view of the grooves in a phonograph record.

remarkable that all the distinct vibrations made by the various pieces of an orchestra are captured by the single-wave groove of the record? The sound of an oboe, when captured by the groove of phonograph records and displayed on an oscilloscope screen, looks like Figure 21.7*a*. This wave corresponds to the electrical signal produced by the vibrating needle. It also corresponds to the amplified signal that activates the loudspeaker of the sound system and to the amplitude of air vibrating against the eardrum. Figure 21.7*b* shows the wave appearance of a clarinet. When oboe and clarinet are sounded together, the principle of superposition is evident as their individual waves combine to produce the wave form shown in Figure 21.7*c*.

The shape of the wave in Figure 21.7*c* is the net result of shapes *a* and *b* superposing (interfering). If we know *a* and *b*, it is a simple thing to create *c*. But it is a far different problem to discern in *c* the shapes of *a* and *b* that make it up. Looking only at shape *c*, we cannot unscramble the oboe from the clarinet.

But if we play the record on the phonograph, our ears will at once know what instruments are being played, what notes they are playing, and what their relative loudness is. Our ears break the overall signal into its component parts automatically.

In 1822, the French mathematician Joseph Fourier discovered a mathematical regularity to the component parts of periodic wave motion. He found that even the most complex periodic wave motion can be disassembled into simple

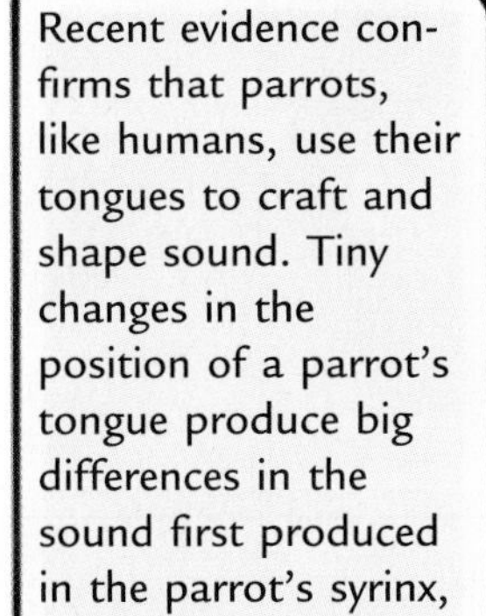

Recent evidence confirms that parrots, like humans, use their tongues to craft and shape sound. Tiny changes in the position of a parrot's tongue produce big differences in the sound first produced in the parrot's syrinx, a voice-box organ nestled between the trachea and the lungs.

Insights

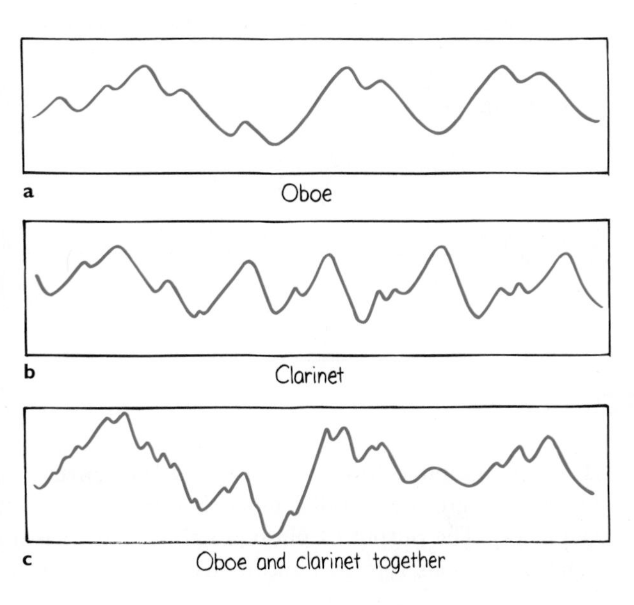

FIGURE 21.7
Wave forms of (a) an oboe, (b) a clarinet, and (c) the oboe and clarinet sounded together.

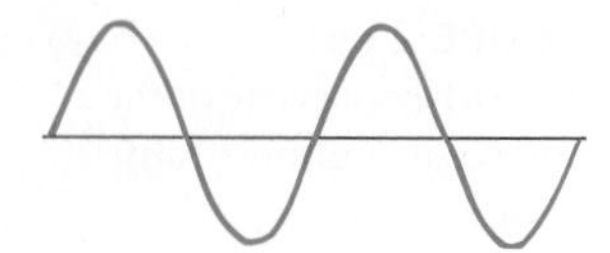

FIGURE 21.8
A sine wave.

sine waves that add together. A sine wave is the simplest of waves, having a single frequency (Figure 21.8). Fourier found that all periodic waves may be broken down into constituent sine waves of different amplitudes and frequencies. The mathematical operation for doing this is called **Fourier analysis.** We will not explain the mathematics here but simply point out that, by such analysis, one can find the pure sine waves that add to compose the tone of, say, a violin. When these pure tones are sounded together, as by striking a number of tuning forks or by selecting the proper keys on an electric organ, they combine to give the tone of the violin. The lowest-frequency sine wave is the fundamental and determines the pitch of the note. The higher-frequency sine waves are the partials that give the characteristic quality. Thus, the wave form of any musical sound is no more than a sum of simple sine waves.

FIGURE 21.9
Does each listener hear the same music?

Since the wave form of music is a multitude of various sine waves, to duplicate sound accurately by radio, record player, or tape recorder, we should be able to process as large a range of frequencies as possible. The notes of a piano keyboard range from 27 hertz to 4200 hertz, but, to duplicate the music of a piano composition accurately, the sound system must have a range of frequencies up to 20,000 hertz. The greater the range of the frequencies of an electrical sound system, the closer the musical output approximates the original sound, and hence the wide range of frequencies in a high-fidelity sound system.

Our ear performs a sort of Fourier analysis automatically. It sorts out the complex jumble of air pulsations that reach it and transforms them into pure tones composed of sine waves. We recombine various groupings of these pure tones when we listen. What combinations of tones we have learned to focus our attention on determines what we hear when we listen to a concert. We can direct our attention to the sounds of the various instruments and discern the faintest tones from the loudest; we can delight in the intricate interplay of instruments and still detect the extraneous noises of others around us. This is a most incredible feat.

Who better appreciates music—one knowledgeable about it, or the casual listener?
Insights

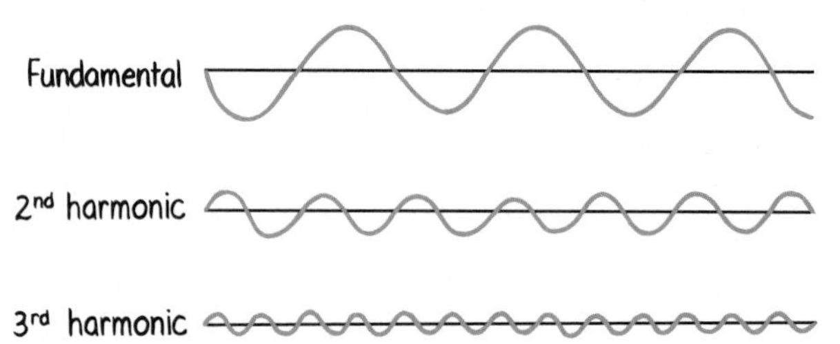

FIGURE 21.10
The fundamental and its harmonics combine to produce a composite wave.

Compact Discs

The phonograph records of old utilized a conventional stylus that vibrated in the squiggly groove of a disc more than twice the diameter of today's CDs (compact discs). The output of phonograph records was a signal like those shown in Figure 21.7. This type of continuous wave form is called an *analog* signal. The

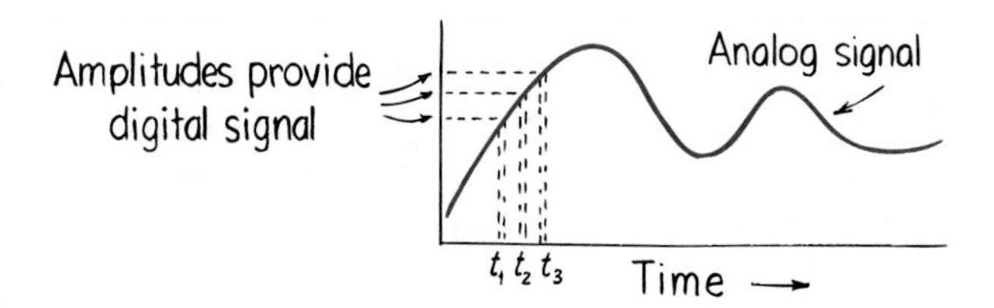

FIGURE 21.11
The amplitude of the analog wave form is measured at successive split seconds to provide digital information that is recorded in binary form on the reflective surface of the laser disc.

analog signal can be changed to a *digital* signal by measuring the numerical value of its amplitude during each split second (Figure 21.11). This numerical value can be expressed in a number system that is convenient for computers, called *binary*. In the binary code, any number can be expressed as a succession of ones and zeros; for example, the number 1 is 1, 2 is 10, 3 is 11, 4 is 100, 5 is 101, 17 is 10001, etc. So the shape of the analog waveform can be expressed as a series of "on" and "off" pulses that corresponds to a series of ones and zeros in binary code. That's where the laser compact disc, or CD, comes in.

CD digital players utilize a laser beam directed onto a plastic reflective disc having a series of microscopic pits about one-thirtieth the diameter of a strand of human hair (Figure 21.12). When the laser beam falls on a flat portion of the reflective surface, it is reflected directly into the player's optical system; this gives an "on" pulse. When the beam is incident upon a passing pit, very little of the laser beam returns to the optical sensor; this gives an "off" pulse. A flickering of "on" and "off" pulses generates the "one" and "zero" digits of the binary code.

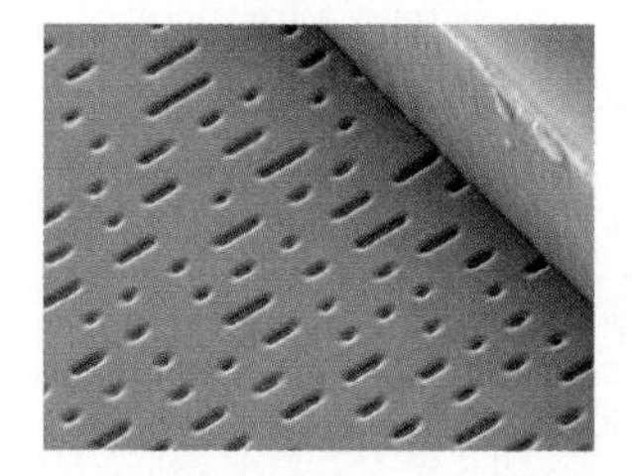

FIGURE 21.12
A microscopic view of the pits on a laser disc.

On a CD, the rate at which these tiny pits are sampled is 44,100 times per second. If you could gather all these pits together without overlapping they would be the size of the period at the end of this sentence. Billions of bits of information are encoded on the reflective surface, which is covered with a protective layer of clear plastic.

Same-size digital video discs (DVDs) have up to 6 times the information-carrying capacity of CDs, 4.7 GB versus 700 MB. DVD discs have smaller pits, which, in effect, makes the length of the spiral track more than twice as long as that of a CD. The smaller pits of the DVD are read with laser light of shorter wavelength and also by way of a more powerful focusing lens. Whereas the pits on a CD lie on a single reflecting surface, a DVD can have multiple layers. By precision focusing, laser light reads the pits in the desired layer.

Watch for the advent of the blue-laser DVD player. The shorter-wavelength blue light reads even more pits stored on a disc. This feature will result in the availability of high definition (HD) DVDs to present incredibly sharp pictures.

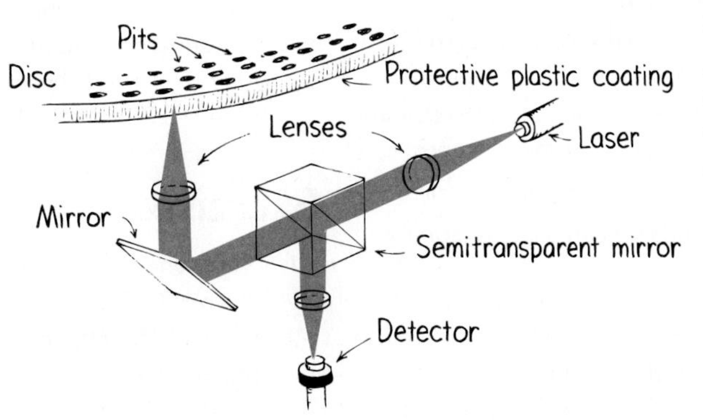

FIGURE 21.13
A tightly focused laser beam reads digital information represented by a series of pits on the laser disc.

Summary of Terms

Pitch The "highness" or "lowness" of a tone, as on a musical scale, which is principally governed by frequency. A high-frequency vibrating source produces a sound of high pitch; a low-frequency vibrating source produces a sound of low pitch.

Intensity The power per square meter carried by a sound wave, often measured in decibels.

Loudness The physiological sensation directly related to sound intensity or volume.

Quality The characteristic timbre of a musical sound, which is governed by the number and relative intensities of partial tones.

Partial tone Single-frequency component sound wave of a complex tone. When the frequency of a partial tone is an integer multiple of the lowest frequency, it is referred to as a *harmonic.*

Fundamental frequency The lowest frequency of vibration, or first harmonic, in a musical tone.

Harmonic A partial tone whose frequency is an integer multiple of the fundamental frequency. The second harmonic has twice the frequency of the fundamental, the third harmonic three times the frequency, and so on in sequence.

Fourier analysis A mathematical method that disassembles any periodic wave form into a combination of simple sine waves.

Review Questions

1. Distinguish between noise and music.
2. What are the three principal characteristics of musical tones?

Pitch

3. How does a high-pitch musical note compare with a low one in terms of frequency?
4. How does the highest pitch one can hear vary with age?

Sound Intensity and Loudness

5. What is a decibel, and how many decibels correspond to the lowest intensity sound we can hear?
6. Is the sound of 30 dB 30 times greater than the threshold of hearing, or 10^3 (a thousand) times greater?
7. Distinguish between sound intensity and loudness.
8. How do the loudest sounds we can tolerate compare with the faintest sounds?

Quality

9. What is it that determines the pitch of a note?
10. If the fundamental frequency of a note is 200 Hz, what is the frequency of the second harmonic? The third harmonic?
11. What exactly determines the musical quality of a note?
12. Why do the same notes plucked on a banjo and on a guitar have distinctly different sounds?

Musical Instruments

13. What are the three principal classes of musical instruments?
14. Why do orchestras generally have a greater number of stringed instruments than wind instruments?

Fourier Analysis

15. What did Fourier discover about complex periodic wave patterns?
16. A high-fidelity sound system may have a frequency range that extends up to or beyond 20,000 hertz. Of what use is this extended range?

Compact Discs

17. How was the sound signal captured on phonograph records of the twentieth century? How is the sound signal captured on a CD?

Projects

1. Test to see which ear has the better hearing by covering one ear and finding how far away your open ear can hear the ticking of a clock; repeat for the other ear. Notice also how the sensitivity of your hearing improves when you cup your ears with your hands.
2. Make the lowest-pitched sound you are capable of; then keep doubling the pitch to see how many octaves your voice can span. If you are a singer, what is your range?
3. On a sheet of graph paper, construct one full cycle (one period of the fundamental) of the composite wave of Figure 21.10 by superposing various vertical displacements of the fundamental and first two partial tones. Your instructor can show you how this is done. Then find the composite waves of partial tones of your own choosing.

Exercises

1. Your friend says that frequency is a quantitative measure of pitch. Do you agree or disagree?
2. As the pitch of sound increases, what happens to the frequency?

3. The yellow-green light emitted by street lights matches the yellow-green color to which the human eye is most sensitive. Consequently, a 100-watt street light emits light that is better seen at night. Similarly, the monitored sound intensities of television commercials are louder than the sound from regular programming, yet don't exceed the regulated intensities. At what frequencies do advertisers concentrate the commercial's sound?
4. The highest frequencies humans can hear is about 20,000 Hz. What is the wavelength of sound in air at this frequency? What is the wavelength for the lowest sounds we can hear, about 20 Hz?
5. Explain how you can lower the pitch of a note on a guitar by altering (a) the length of the string, (b) the tension of the string, or (c) the thickness or the mass of the string.
6. Why should guitars be played offstage before they are brought onstage for a concert? (*Hint*: Think thermally.)
7. If sound becomes louder, which wave characteristic is likely increasing—frequency, wavelength, amplitude, or speed?
8. Does the pitch of a note depend on sound frequency, loudness, quality, or on all of these?
9. A guitar and a flute are in tune with each other. Explain how a change in temperature could alter this situation.
10. When a guitar string is struck, a standing wave is produced that oscillates with a large sustained amplitude, pushing back and forth against the surrounding air to generate sound. How does the frequency of the resulting sound compare with the frequency of the standing wave in the string?
11. When a guitar player lessens the tension in a guitar string, will the string produce higher- or lower-pitched sound?
12. The strings on a harp are of different lengths and produce different notes. How are different notes produced on a guitar, where all strings have the same length?
13. If a vibrating string is made shorter (as by holding a finger on it), how does this affect the frequency of vibration and pitch?
14. A nylon guitar string vibrates in a standing-wave pattern, as shown below. What is the wavelength of the wave?

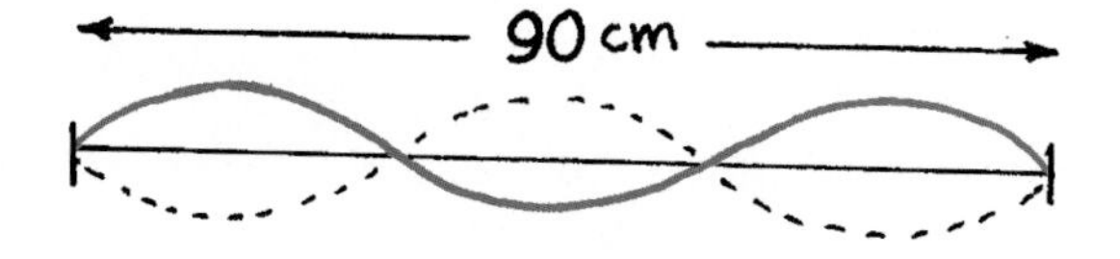

15. Why do tuning forks with long tines vibrate at a lower frequency than short-tined forks? (*Hint:* This question could have been asked back in Chapter 8.)
16. Why is the thickness greater for the bass strings of a guitar than for the treble strings?
17. Will the thicker or thinner of two guitar strings of the same tension and length vibrate at the higher frequency?
18. Why does a vibrating guitar string not sound as loud when it is mounted on a work bench as it does when mounted on a guitar?
19. Would a plucked guitar string vibrate for a longer time or a shorter time if the instrument had no sounding board? Why?
20. If you very lightly touch a guitar string at its midpoint, you can hear a tone that is one octave above the fundamental for that string. (An octave is a factor of two in frequency.) Explain.
21. If a guitar string vibrates in two segments, where can a tiny piece of folded paper be supported without flying off? How many pieces of folded paper could similarly be supported if the wave form were of three segments?
22. A violin string playing the note "A" oscillates at 440 Hz. What is the period of the string's oscillation?
23. The string of a cello playing the note C oscillates at 264 Hz. What is the period of the string's oscillation?
24. Why do the same notes on a trumpet and on a saxophone sound different when both are played with the same pitch and loudness?
25. The amplitude of a transverse wave in a stretched string is the maximum displacement of the string from its equilibrium position. What does the amplitude of a longitudinal sound wave in air correspond to?
26. Which of the two musical notes displayed one at a time on an oscilloscope screen has the higher pitch?

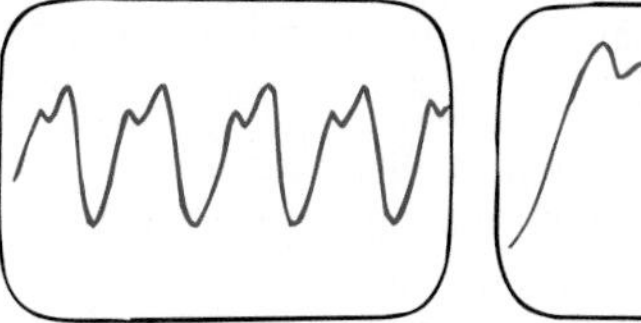
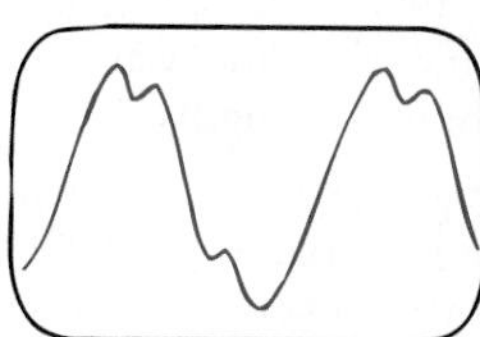

27. In the oscilloscopes shown above, which screen shows the louder sound (assuming detection by equivalent microphones)?
28. A loudspeaker produces a musical sound by means of the oscillation of a diaphragm. Does the loudness of sound produced depend on the frequency of oscillation, the amplitude of oscillation, the kinetic energy of oscillation, or all of these?

29. In a hi-fi speaker system, why is the woofer (low-frequency speaker) larger than the tweeter (high-frequency speaker)?

30. Which is a more objective measurement—sound intensity or loudness? Defend your answer.

31. One person has a threshold of hearing of 5 dB, and another of 10 dB. Which person has the more acute hearing?

32. How much more intense is sound at 40 dB than sound at zero dB?

33. How much more intense is sound at 110 dB than sound at 50 dB?

34. How is an electronic organ able to imitate the sounds made by various musical instruments?

35. After inhaling helium gas, a person talks with a high-pitched voice. One of the reasons for this is the higher speed of sound in helium than in air. Why does sound travel faster in helium?

36. Why does your voice sound fuller in the shower?

37. The frequency range for a telephone is between 500 Hz and 4000 Hz. Why does a telephone not do a very good job of transmitting music?

38. How many octaves does normal human hearing span? How many octaves are on a common piano keyboard? (If you're not sure, look and see.)

39. The note middle C on a piano has a fundamental frequency of 262 Hz. What is the frequency of the second harmonic of this note?

40. If the fundamental frequency of a guitar string is 220 Hz, what is the frequency of the second harmonic? Of the third harmonic?

41. If the fundamental frequency of a violin string is 440 Hz, what is the frequency of the second harmonic? Of the third harmonic?

42. How many nodes, not including the end points, are there in a standing wave that is three wavelengths long? How many are there in a standing wave that is four wavelengths long?

43. How can you tune the note A_3 on a piano to its proper frequency of 220 Hz with the aid of a tuning fork whose frequency is 440 Hz?

44. At an outdoor concert, the pitch of musical tones is *not* affected on a windy day. Explain.

45. A trumpet has keys and valves that permit the trumpeter to change the length of the vibrating air column and the position of the nodes. A bugle has no such keys and valves, yet it can sound various notes. How do you think the bugler achieves different notes?

46. The human ear is sometimes called a Fourier analyzer. What does this mean, and why is it an appropriate description?

47. The width of a laser beam is significant in reading CDs and DVDs. The thinner the beam, the closer the series of pits can be. Why will blue laser light allow closer pits than red laser light?

48. Do all the people in a group hear the same music when they listen attentively? (Do all see the same sight when looking at a painting? Do all taste the same flavor when sipping the same wine? Do all perceive the same aroma when smelling the same perfume? Do all feel the same texture when touching the same fabric? Do all come to the same conclusion when listening to a logical presentation of ideas?)

49. Why is it a safe prediction that you, presently reading this, will have a significantly greater loss of hearing in your later years than your grandparents experienced?

50. Make up a multiple-choice question that distinguishes between any of the terms listed in the Summary of Terms.

Problems

1. How much more intense than the threshold of hearing is a sound of 10 dB? 30 dB? 60 dB?

2. How much more intense is a sound of 40 dB than a sound of 30 dB?

3. A certain note has a frequency of 1000 Hz. What is the frequency of a note one octave above it? Two octaves above it? One octave below it? Two octaves below it?

4. Starting with a fundamental tone, how many harmonics are there between the first and second octaves? Between the second and third octaves? (Look at Figure 21.3 to get started.)

5. A cello string 0.75 m long has a 220-Hz fundamental frequency. Find the wave speed along the vibrating string.

PART FIVE

Electricity and Magnetism

How intriguing that this magnet outpulls the whole world when it lifts these nails. The pull between the nails and the Earth I call a **gravitational force,** and the pull between the nails and the magnet I call a **magnetic force.** I can name these forces, but I don't yet understand them. My learning begins by realizing there's a big difference in knowing the names of things and really understanding those things.

Electrostatics

Jim Stith, former president of the American Association of Physics Teachers, demonstrates a Wimshurst generator, which produces miniature lightning strokes.

Electricity is the name given to a wide range of electrical phenomena that, in one form or another, underlie just about everything around us. From lightning in the sky to the glowing of a lamp, from what holds atoms together as molecules to the nerve impulses that travel along your nervous system, electricity is all around us. The control of electricity is evident in many devices, from microwave ovens to computers. In this technological age, it is important to understand the basics of electricity and of how these basic ideas are used to sustain and enhance our current comfort, safety, and prosperity.

In this chapter, we will investigate electricity at rest, or **electrostatics.** Electrostatics involves electric charges, the forces between them, the aura that surrounds them, and their behavior in materials. In Chapter 23, we'll investigate the motion of electric charges, or *electric currents*. We'll also study the voltages that produce currents and how they can be controlled. In Chapter 24, we'll study the relationship of electric currents to magnetism, and, in Chapter 25, we'll learn how magnetism and electricity can be controlled to operate motors and other electrical devices and how electricity and magnetism connect to become light.

An understanding of electricity requires a step-by-step approach, for one concept is the foundation on which the next concept is based. So please put in extra attention in studying this material. It can be difficult, confusing, and frustrating if you're hasty; but, with careful effort, it can be comprehensible and rewarding. Onward!

Electrical Forces

What if there were a universal force that, like gravity, varies inversely as the square of the distance but is billions upon billions of times stronger? If there were such a force and if it were attractive like gravity, the universe would be pulled together into a tight ball, with all matter pulled as close together as it could get. But suppose this force were a repelling force, with every bit of matter repelling every other bit. What then? Then the universe would be an ever-expanding gaseous cloud. Suppose, however, that the universe consisted of two kinds of particles—positives and negatives, say. Suppose that positives repelled positives but attracted negatives and that negatives repelled negatives but attracted positives. In other words, like kinds repel and unlike kinds attract

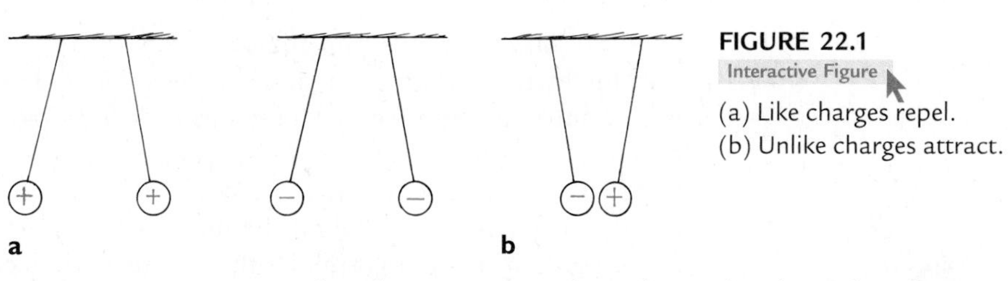

FIGURE 22.1
Interactive Figure
(a) Like charges repel.
(b) Unlike charges attract.

(Figure 22.1). Suppose that there were equal numbers of each so that this strong force were perfectly balanced! What would the universe be like? The answer is simple: It would be like the one we are living in. For there are such particles, and there is such a force. We call it the *electric force.*

Which charges are called positive and which are called negative is the result of a choice made by Benjamin Franklin. It could have been the other way around.

Insights

Clusters of positive and negative particles have been pulled together by the enormous attraction of the electrical force. By forming the compact and evenly mixed clusters of positives and negatives, the huge forces have balanced themselves out almost perfectly. These clusters are the atoms of matter. When two or more atoms join to form a molecule, the molecule also contains balanced positives and negatives. And when trillions of molecules combine to form a speck of matter, the electrical forces balance again. Between two pieces of ordinary matter, there is scarcely any electrical attraction or repulsion at all, because each piece contains equal numbers of positives and negatives. Between the Earth and the Moon, for example, there is no net electrical force. The much weaker gravitational force, which only attracts, remains as the predominant force between these bodies.

Electric Charges

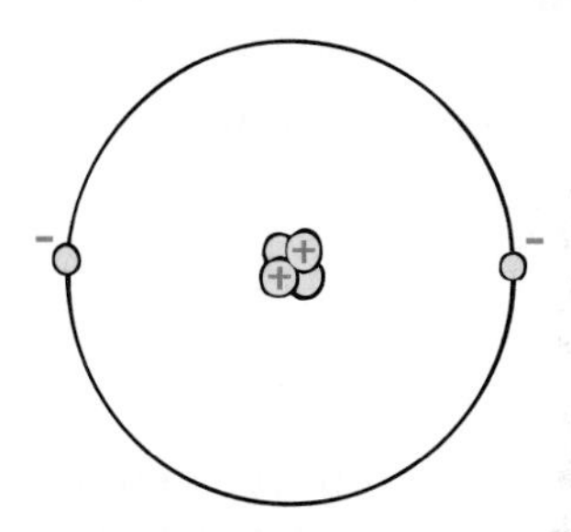

FIGURE 22.2
Interactive Figure
Model of a helium atom. The atomic nucleus is made up of two protons and two neutrons. The positively charged protons attract two negative electrons. What is the net charge of this atom?

The terms *positive* and *negative* refer to electric *charge,* the fundamental quantity that underlies all electrical phenomena. The positively charged particles in ordinary matter are protons, and the negatively charged particles are electrons. The attractive force between these particles causes them to lump together into incredibly small units—atoms. (Atoms also contain neutral particles called neutrons.) When two atoms get close together, the balance of attractive and repelling forces is not perfect, because electrons fly around within the volume of each atom. The atoms may then attract each other and form a molecule. In fact, all the chemical bonding forces that hold atoms together to form molecules are electrical in nature. Anyone planning to study chemistry should first know something about electrical attraction and repulsion and, before studying electrical phenomena, should know something about atoms. Here are some important facts about atoms:

1. Every atom is composed of a positively charged *nucleus* surrounded by negatively charged electrons.
2. The electrons of all atoms are identical. Each has the same quantity of negative charge and the same mass.
3. Protons and neutrons compose the nucleus. (The common form of the hydrogen atom, which has no neutron, is the only exception.) Protons are about 1800 times more massive than electrons, but they carry an amount of positive charge equal to the negative charge of electrons. Neutrons have slightly more mass than protons and have no net charge.
4. Atoms usually have as many electrons as protons, so the atom has zero *net* charge.

Why don't protons pull the oppositely charged electrons into the nucleus? You might think that the answer is the same as the reason why planets aren't pulled into the Sun by the gravitational force—because the electrons orbit around the nucleus. Unfortunately, this planetary explanation is invalid for electrons. When the nucleus was discovered in 1911, scientists knew that electrons couldn't orbit placidly around the nucleus in the way the Earth orbits the Sun. In only about a hundred-millionth of a second, according to classical physics, the electron would spiral into the nucleus, emitting electromagnetic radiation as it did so. So a new theory was needed, the theory called quantum mechanics. In describing electron motion, we still use old terminology, *orbit* and *orbital,* although the preferred word is *shell,* which suggests that the electrons are spread out over a spherical region. Today, the explanation for the atom's stability has to do with the wave nature of electrons. An electron behaves like a wave and requires a certain amount of space related to its wavelength. When we treat quantum mechanics in Chapter 32, we'll see that atomic size is determined by the minimum amount of "elbow room" that an electron requires.

Why don't the protons in the nucleus mutually repel and fly apart? What holds the nucleus together? The answer is that, in addition to electrical forces in the nucleus, there are even stronger nonelectrical nuclear forces that are able to hold the protons together in spite of the electrical repulsion. The neutrons play a role in putting more distance between the protons. We will treat the nuclear force further in Chapter 33.

CHECK YOURSELF

1. Beneath the complexities of electrical phenomena, there lies a fundamental rule from which nearly all other effects stem. What is this fundamental rule?
2. How does the charge of an electron differ from the charge of a proton?

Conservation of Charge

In a neutral atom, there are as many electrons as protons, so there is no net charge. The positive balances the negative exactly. If an electron is removed from an atom, then it is no longer neutral. The atom then has one more positive charge (proton) than negative charge (electron) and is said to be positively charged.[1] A charged atom is called an *ion.* A *positive ion* has a net positive charge. A *negative ion,* an atom with one or more extra electrons, is negatively charged.

CHECK YOUR ANSWERS

1. Like charges repel; opposite charges attract.
2. The charge of an electron is equal in magnitude, but opposite in sign, to the charge of a proton.

[1]Each proton has a charge $+e$, equal to $+1.6 \times 10^{-19}$ coulomb. Each electron has a charge $-e$, equal to -1.6×10^{-19} coulomb. Why such different particles have the same magnitude of charge is an unanswered question in physics. The equality of the magnitudes has been tested to high accuracy.

FIGURE 22.3
Electrons are transferred from the fur to the rod. The rod is then negatively charged. Is the fur charged? How much compared to the rod? Positively or negatively?

Material objects are made of atoms, which means they are composed of electrons and protons (and neutrons). Objects ordinarily have equal numbers of electrons and protons and are, therefore, electrically neutral. But, if there is a slight imbalance in the numbers, the object is electrically charged. An imbalance comes about when electrons are added to, or removed from, an object. Although electrons closest to the atomic nucleus, the innermost electrons, are bound very tightly to the oppositely charged atomic nucleus, the electrons farthest from the nucleus, the outermost electrons, are bound very loosely and can be easily dislodged. How much work is required to tear an electron away from an atom varies from one substance to another. The electrons are held more firmly in rubber and plastic than in your hair, for example. Thus, when a comb is passed through your hair, electrons transfer from the hair to the comb. The comb then has an excess of electrons and is said to be *negatively charged*. Your hair, in turn, has a deficiency of electrons and is said to be *positively charged*. To take another example, if you rub a glass or plastic rod with silk, the rod becomes positively charged. Silk has a greater affinity for electrons than does glass or plastic. Electrons are, therefore, rubbed off the rod and onto the silk.

So we see that an object having unequal numbers of electrons and protons is electrically charged. If it has more electrons than protons, it is negatively charged. If it has fewer electrons than protons, it is positively charged.

It is important to note that, when we charge something, no electrons are created or destroyed. Electrons are simply transferred from one material to another. Charge is *conserved*. In every event, whether it is a large-scale event or an event at the atomic and nuclear level, the principle of **conservation of charge** has always been found to apply. No case of the creation or destruction of electric charge has ever been found. The conservation of charge is a cornerstone in physics, ranking with the conservation of energy and the conservation of momentum.

Charge is like a baton in a relay race. It can be passed from one object to another but it cannot be lost.

Any object that is electrically charged has an excess or deficiency of some whole number of electrons—electrons cannot be divided into fractions of electrons. This means that the charge of the object is a whole-number multiple of the charge of an electron. It cannot have a charge equal to the charge of $1\frac{1}{2}$ or $1000\frac{1}{2}$ electrons, for example. Charge is "grainy," or made up of elementary units called *quanta*. We say that charge is *quantized*, with the smallest quantum of charge being that of the electron (or proton). No smaller units of charge have ever been observed.[2] All charged objects to date have a charge that is a whole-number multiple of the charge of a single electron or proton.

CHECK YOURSELF

If you scuff electrons onto your feet while walking across a rug, are you negatively or positively charged?

CHECK YOUR ANSWER

You have more electrons after you scuff your feet, so you are negatively charged (and the rug is positively charged).

[2]Within the atomic nucleus, however, elementary particles called *quarks* carry charges $\frac{1}{3}$ and $\frac{2}{3}$ the magnitude of the electron's charge. Each proton and each neutron is made up of three quarks. Since quarks always exist in such combinations and have never been found separated, the whole-number-multiple rule of electron charge holds for nuclear processes as well.

Coulomb's Law

The electrical force, like gravitational force, decreases inversely as the square of the distance between charged bodies. This relationship was discovered by Charles Coulomb in the eighteenth century and is called **Coulomb's law.** It states that, for two charged objects that are much smaller than the distance between them, the force between the two objects varies directly as the product of their charges and inversely as the square of the separation distance. (Review the inverse-square law in Figure 9.5 back on page 165.) The force acts along a straight line from one charged object to the other. Coulomb's law can be expressed as

$$F = k\frac{q_1 q_2}{d^2}$$

where d is the distance between the charged particles, q_1 represents the quantity of charge of one particle, q_2 represents the quantity of charge of the other particle, and k is the proportionality constant.

Coulomb's law is like Newton's law of gravity. But, unlike gravity, electric forces can be attractive or repulsive.

Insights

The unit of charge is the **coulomb,** abbreviated C. It turns out that a charge of 1C is the charge associated with 6.25 billion billion electrons. This might seem like a great number of electrons, but it only represents the amount of charge that passes through a common 100-watt lightbulb in a little over a second.

The proportionality constant k in Coulomb's law is similar to G in Newton's law of gravitation. Instead of being a very small number like G (6.67×10^{-11}), the electrical proportionality constant k is a very large number. It is approximately

$$k = 9{,}000{,}000{,}000 \ \mathrm{N \cdot m^2/C^2}$$

or, in scientific notation, $k = 9 \times 10^9 \ \mathrm{N \cdot m^2/C^2}$. The unit $\mathrm{N \cdot m^2/C^2}$ is not central to our interest here; it simply converts the right-hand side of the equation to the unit of force, the newton (N). What is important is the large magnitude of k. If, for example, a pair of like charged particles of 1 coulomb each were 1 meter apart, the force of repulsion between them would be 9 billion newtons.[3] That would be more than ten times the weight of a battleship! Obviously, such amounts of net charge do not exist in our everyday environment.

So Newton's law of gravitation for massive bodies is similar to Coulomb's law for electrically charged bodies.[4] Whereas the gravitational force of attraction

---

[3]In comparing the magnitudes of G and k, we must note that they depend on the units chosen for mass and electric charge, which could have been chosen differently. So our comparison only reminds us that electric forces are usually enormous compared with gravitational forces. Contrast the 9 billion newtons between the two unit charges 1 meter apart with the gravitational force of attraction between two unit masses (kilograms) 1 m apart: 6.67×10^{-11} N—an extremely small force. For the force to be 1 N, the masses at 1 m apart would have to be nearly 122,000 kg each! Gravitational forces between ordinary objects are much too small to be detected except in delicate experiments. But electrical forces between ordinary objects can be relatively huge. Even for highly charged objects, however, the imbalance of electrons to protons is normally less than one part per trillion.

[4]According to quantum theory, a force varies inversely as the square of the distance if it involves the exchange of particles with no mass. Exchange of massless photons is responsible for the electrical force, and exchange of massless gravitons accounts for the gravitational force. Some scientists have pursued an even deeper relationship between gravity and electricity. Albert Einstein spent the latter part of his life searching with little success for a "unified field theory." More recently, the electrical force has been unified with one of the two nuclear forces, the *weak force,* which plays a role in radioactive decay.

between such particles as an electron and a proton is extremely small, the electrical force between these particles is relatively enormous. Other than the big difference in strength, the most important difference between gravitational and electrical forces is that electrical forces may be either attractive or repulsive, whereas gravitational forces are only attractive.

CHECK YOURSELF

1. The proton that is the nucleus of the hydrogen atom attracts the electron that orbits it. Relative to this force, does the electron attract the proton with less force, with more force, or with the same amount of force?
2. If a proton at a particular distance from a charged particle is repelled with a given force, by how much will the force decrease when the proton is three times farther away from the particle? When it is five times farther away?
3. What is the sign of charge of the particle in this case?

Conductors and Insulators

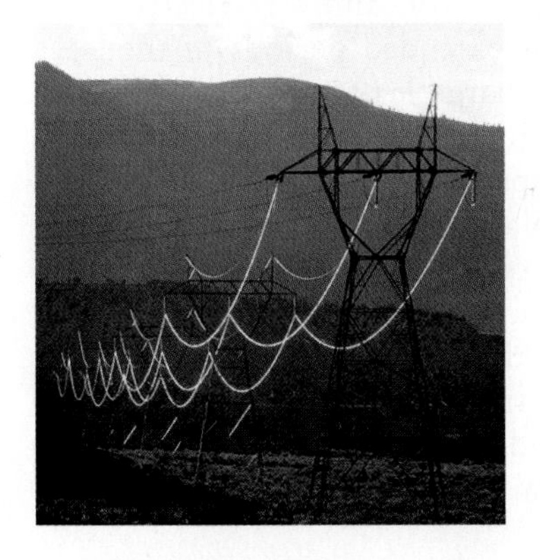

FIGURE 22.4
It is easier to establish an electric current through hundreds of kilometers of metal wire than through a few centimeters of insulating material.

It is easy to establish an electric current in metals because one (or more) of the electrons in the outer shell of the atoms in a metal is not anchored to the nuclei of particular atoms but is free to wander in the material. Such a material is called a good **conductor.** Any metal is a good conductor of electric current for the same reason it is a good heat conductor—the electrons in the outer atomic shell of its atoms are "loose."

The electrons in other materials—rubber and glass, for example—are tightly bound and belong to particular atoms. They are not free to wander about among other atoms in the material. Consequently, it isn't easy to make them flow. These materials are poor conductors of electric current for the same reason they are generally poor heat conductors. Such a material is called a good **insulator.**

All substances can be arranged in order of their ability to conduct electric charge. Those at the top of the list are conductors and those at the bottom are insulators. The ends of the list are very far apart. The conductivity of a metal, for example, can be more than a million trillion times greater than the conductivity of an insulator such as glass. In a common appliance cord, electrons flow through several meters of wire rather than flowing directly across from one wire to the other through a small fraction of a centimeter of vinyl or rubber insulation.

CHECK YOUR ANSWERS

1. The same amount of force, in accord with Newton's third law—basic mechanics! Recall that a force is an interaction between two things—in this case, between the proton and the electron. They pull on each other equally.
2. It decreases to 1/9 its original value; to 1/25.
3. Positive.

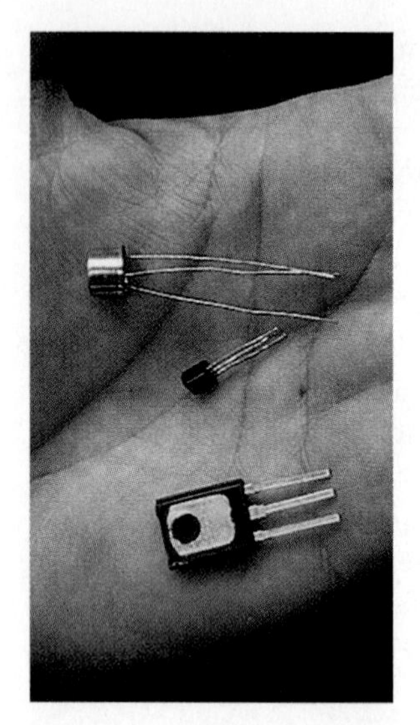

a

b

FIGURE 22.5
(a) Three transistors.
(b) Many transistors in an integrated circuit.

Semiconductors

Whether a substance is classified as a conductor or as an insulator depends on how tightly the atoms of the substance hold their electrons. A piece of copper is a good conductor, while a piece of wood is a good insulator. Some materials, such as germanium and silicon, however, are neither good conductors nor good insulators. These materials fall in the middle of the range of electrical resistivity, being fair insulators in their pure crystalline form and becoming excellent conductors when even one atom in 10 million is replaced with an impurity that adds or removes an electron from the crystal structure. A material that can be made to behave sometimes as an insulator and sometimes as a conductor is called a **semiconductor.** Thin layers of semiconducting materials sandwiched together make up *transistors,* which are used to control the flow of currents in circuits, to detect and amplify radio signals, and to produce oscillations in transmitters; they also act as digital switches. These tiny solids were the first electrical components in which materials with different electrical characteristics were not interconnected by wires but were physically joined in one structure. They require very little power, and they last indefinitely in normal use.

A semiconductor will also conduct when light of the proper color shines on it. A pure selenium plate is normally a good insulator, and any electric charge built up on its surface will remain there for extended periods in the dark. If the plate is exposed to light, however, the charge leaks away almost immediately. If a charged selenium plate is exposed to a pattern of light, such as the pattern of light and dark that makes up this page, the charge will leak away only from the areas exposed to light. If a black plastic powder were brushed across its surface, the powder would stick only to the charged areas where the plate had not been exposed to light. Now if a piece of paper with an electric charge on the back of it were put over the plate, the black plastic powder would be drawn to the paper to form the same pattern as, say, the one on this page. If the paper were then heated to melt the plastic and to fuse it to the paper, you might pay a nickel or a dime for it and call it a Xerox copy.

Superconductors

An ordinary conductor has only a small resistance to the flow of electric charge. An insulator has much greater resistance (we'll treat the topic of electric resistance in the following chapter). Remarkably, in certain materials at sufficiently low temperatures, electrical resistance disappears. The materials acquire zero resistance (infinite conductivity) to the flow of charge. Such a material is called a **superconductor.** Once electric current is established in a superconductor, the electrons flow indefinitely. With no electrical resistance, current passes through a superconductor without losing energy; no heat loss occurs when charges flow. Superconductivity in metals near absolute zero was discovered in 1911. In 1987, superconductivity at a "high" temperature (above 100 K) was discovered in a nonmetallic compound. At this writing, superconductivity at both "high" and low temperatures is being intensely investigated. Applications include long-distance transmission of power without loss, and high-speed, magnetically levitated vehicles to replace traditional rail trains.

Charging

We charge things by transferring electrons from one place to another. We can do this by physical *contact,* as occurs when substances are rubbed together or simply touched. Or we can redistribute the charge on an object simply by putting a charged object near it—this is called *induction.*

Static electricity is a growing problem at gasoline pumps. Even the tiniest of sparks can ignite vapors coming from the gasoline and cause fires—frequently lethal ones. A good rule is to touch metal in order to discharge static charge from your body before you fuel. Also, don't use a cell phone when fueling.

Insights

Charging by Friction and Contact

We are all familiar with the electrical effects produced by friction. We can stroke a cat's fur and hear the crackle of sparks that are produced, or comb our clean, dry hair in front of a mirror in a dark room and see as well as hear the sparks. We can scuff our shoes across a rug and feel a tingle as we reach for the doorknob. Talk to old-timers and they'll tell you about the surprising shock that was typical of sliding across a plastic seat cover while parked in an automobile (Figure 22.6). In all these cases, electrons are transferred by friction when one material rubs against another.

Electrons can be transferred from one material to another by simply touching. For example, when a negatively charged rod is placed in contact with a neutral object, some electrons will move to the neutral object. This method of charging is simply called **charging by contact.** If the object is a good conductor, electrons will spread to all parts of its surface because the transferred electrons repel one another. If it is a poor conductor, it may be necessary to touch the rod at several places on the object in order to get a more-or-less uniform distribution of charge.

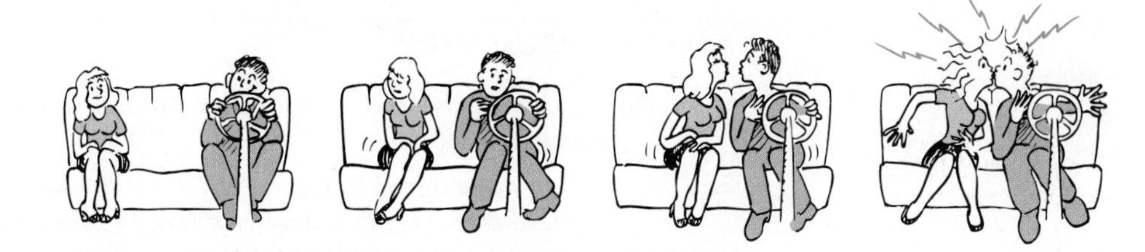

FIGURE 22.6
Charging by friction and then by contact.

Charging by Induction

If you bring a charged object *near* a conducting surface, you will cause electrons to move in the surface material, even though there is no physical contact. Consider the two insulated metal spheres, A and B, in Figure 22.7. (a) They

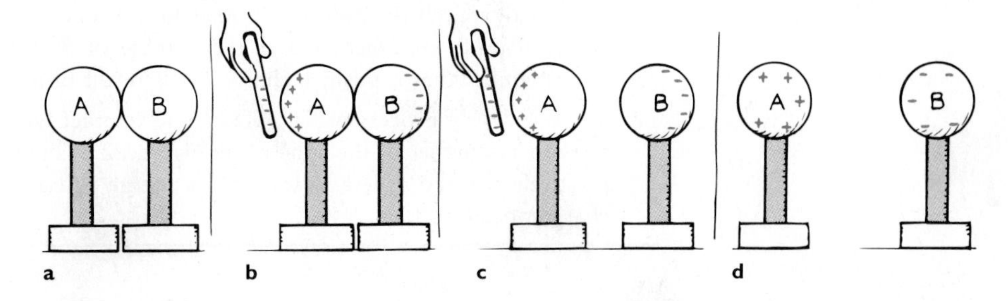

FIGURE 22.7
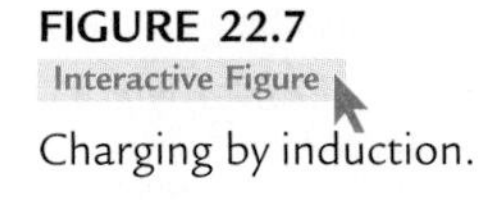

Charging by induction.

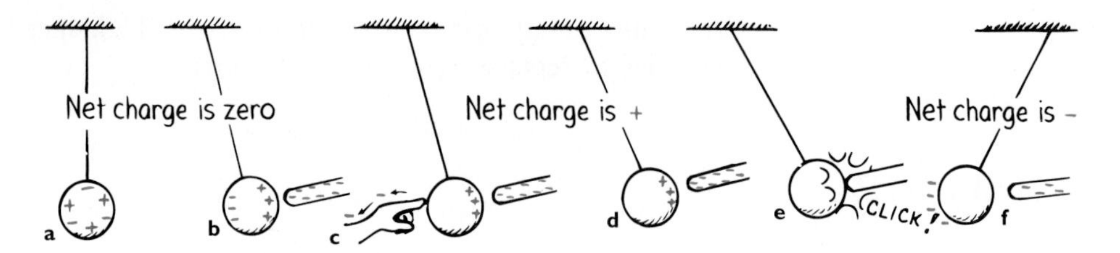

FIGURE 22.8 Interactive Figure

Stages of charge induction by grounding. (a) The net charge on the metal ball is zero. (b) Charge redistribution is induced on the ball by the presence of the charged rod. The net charge on the ball is still zero. (c) Touching the negative side of the ball removes electrons by contact. (d) This leaves the ball positively charged. (e) The ball is more strongly attracted to the negative rod, and, when it touches, charging by contact occurs. (f) The negative ball is repelled by the still somewhat negatively charged rod.

touch each other, so in effect they form a single uncharged conductor. (b) When a negatively charged rod is brought near A, electrons in the metal, being free to move, are repelled as far as possible until their mutual repulsion is big enough to balance the influence of the rod: The charge is redistributed. (c) If A and B are separated while the rod is still present, (d) each will be equal and oppositely charged. This is **charging by induction.** The charged rod has never touched them, and the rod retains the same charge it had initially.

We can similarly charge a single sphere by induction if we touch it when different parts of it are differently charged. Consider the metal sphere that hangs from a nonconducting string, as shown in Figure 22.8. When we touch the metal surface with a finger, we are providing a path for charge to flow to or from a very large reservoir for electric charge—the ground. We say that we are *grounding* the sphere, a process that may leave it with a net charge. We will return to this idea of grounding in the next chapter when we discuss electric currents.

CHECK YOURSELF

1. Would the charges induced on the spheres A and B of Figure 22.7 necessarily be exactly equal and opposite?
2. Why does the negative rod in Figure 22.7 have the same charge before and after the spheres are charged, but not when charging takes place, as in Figure 22.8?

CHECK YOUR ANSWERS

1. Yes. The charges must be equal and opposite on both spheres, because each single positive charge on sphere A is the result of a single electron being taken from A and moved to B. This is like taking bricks from the surface of a brick road and placing them all on the sidewalk. The number of bricks on the sidewalk will exactly match the number of holes in the road. Likewise, the number of extra electrons on B will exactly match the number of "holes" (positive charges) left in A. Remember that a positive charge is the result of an absent electron.
2. In the charging process of Figure 22.7, no contact was made between the negative rod and either of the spheres. In Figure 22.8, however, the rod touched the positively charged sphere. A transfer of charge by contact reduced the negative charge on the rod.

FIGURE 22.9
The negative charge at the bottom of the cloud induces a positive charge at the surface of the ground below.

FIGURE 22.10
The lightning rod is connected by heavy-duty wire so that it can conduct a very large current to the ground if it draws a lightning bolt to it. Most often, charge leaks off the pointed tip to prevent the occurrence of lightning.

Charging by induction occurs during thunderstorms. The negatively charged bottoms of clouds induce a positive charge on the surface of the Earth below. Benjamin Franklin was the first to demonstrate this when his famous kite-flying experiment proved that lightning is an electrical phenomenon.[5] Lightning is an electrical discharge between a cloud and the oppositely charged ground or between oppositely charged parts of clouds.

Franklin also found that charge flows readily to or from sharp metal points, and he fashioned the first lightning rod. If the rod is placed above a building and is connected to the ground, the point of the rod collects electrons from the air, preventing a large buildup of positive charge on the building by induction. This continual "leaking" of charge prevents a charge buildup that might otherwise lead to a sudden discharge between the cloud and the building. The primary purpose of the lightning rod, then, is to prevent a lightning discharge from occurring. If, for any reason, sufficient charge does not leak from the air to the rod and lightning strikes anyway, it may be attracted to the rod and take a direct path to the ground, thereby sparing the building. The overall purpose of the lightning rod is to prevent a fire caused by lightning.

Charge Polarization

Charging by induction is not restricted to conductors. When a charged rod is brought near an insulator, there are no free electrons that can migrate throughout the insulating material. Instead, there is a rearrangement of charges within

[5]Benjamin Franklin was careful to isolate himself from his apparatus and to keep out of the rain when he conducted this experiment, so he wasn't electrocuted as were others who attempted to duplicate his experiment. In addition to being a great statesman, Franklin was a first-rate scientist. He introduced the terms *positive* and *negative* as they relate to electricity, but nevertheless supported the one-fluid theory of electric charge, and contributed to our understanding of grounding and insulation. He also published a newspaper, formed the first fire-insurance company, and invented a safer, more efficient stove—a very busy man! Only a task as important as helping to form the United States' system of government prevented him from devoting even more of his time to his favorite activity—the scientific investigation of nature.

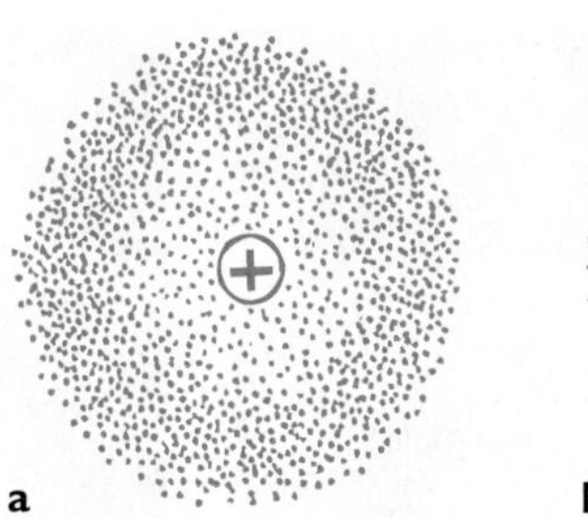

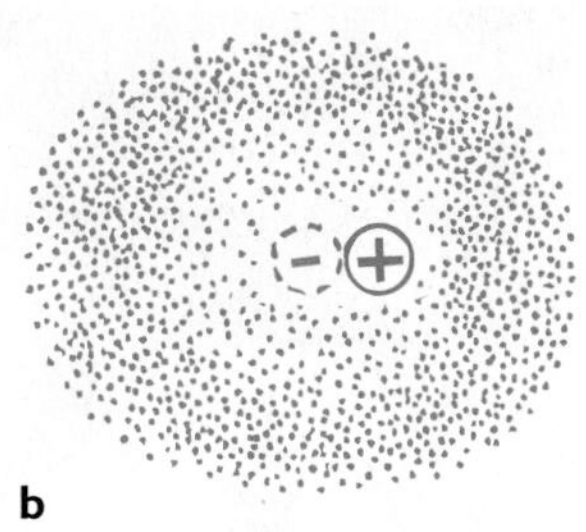

FIGURE 22.11
An electron buzzing around the atomic nucleus makes up an electron cloud. (a) The center of the negative cloud coincides with the center of the positive nucleus in an atom. (b) When an external negative charge is brought nearby to the right, as on a charged balloon, the electron cloud is distorted so that the centers of negative and positive charge no longer coincide. The atom is now electrically polarized.

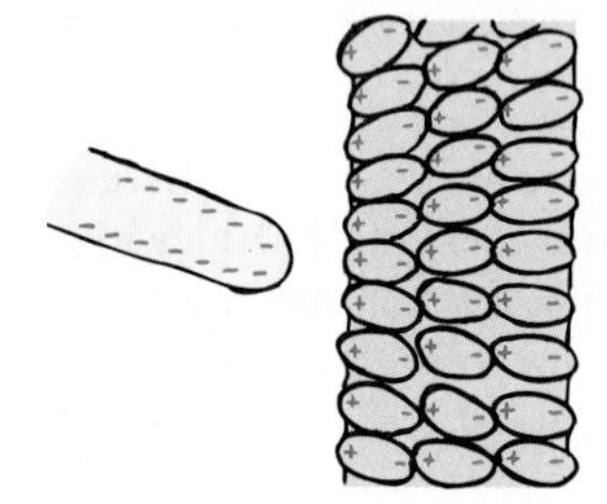

FIGURE 22.12
Interactive Figure
All the atoms or molecules near the surface become electrically polarized. Surface charges of equal magnitude and opposite sign are induced on opposite surfaces of the material.

the atoms and molecules themselves (Figure 22.11). Although atoms don't move from their relatively fixed positions, their "centers of charge" are moved. One side of the atom or molecule is induced into becoming more negative (or positive) than the opposite side. The atom or molecule is said to be **electrically polarized.** If the charged rod is negative, say, then the positive part of the atom or molecule is tugged in a direction toward the rod, and the negative side of the atom or molecule is pushed in a direction away from the rod. The positive and negative parts of the atoms and molecules become aligned. They are electrically polarized.

We can understand why electrically neutral bits of paper are attracted to a charged object—a comb passed through your hair, for example. When the charged comb is brought nearby, molecules in the paper are polarized. The sign of charge closest to the comb is opposite to the comb's charge. Charges of the same sign are slightly more distant. Closeness wins, and the bits of paper experience a net attraction. Sometimes they will cling to the comb and then suddenly fly off. This repulsion occurs because the paper bits acquire the same sign of charge as the charged comb when they come in contact.

Rub an inflated balloon on your hair, and it becomes charged. Place the balloon against the wall, and it sticks. This is because the charge on the balloon induces an opposite surface charge on the wall. Again, closeness wins, for the charge on the balloon is slightly closer to the opposite induced charge than to the charge of same sign (Figure 22.14). Many molecules—H_2O, for

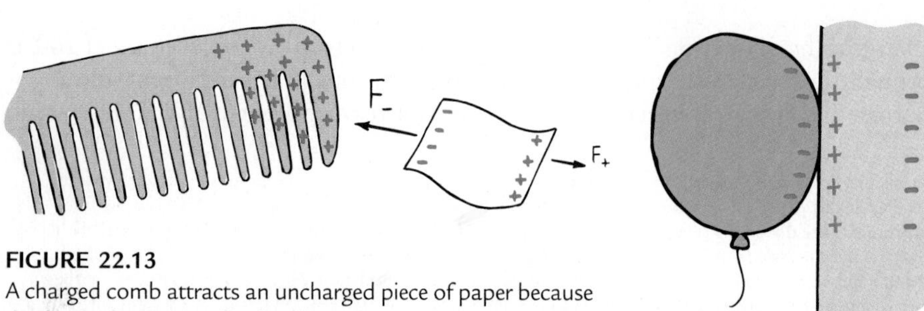

FIGURE 22.13
A charged comb attracts an uncharged piece of paper because the force of attraction for the closer charge is greater than the force of repulsion for the farther charge.

FIGURE 22.14
The negatively charged balloon polarizes atoms in the wooden wall and creates a positively charged surface, so the balloon sticks to the wall.

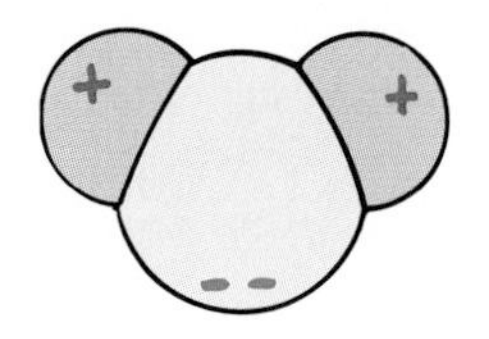

FIGURE 22.15
An H_2O molecule is an electric dipole.

example—are electrically polarized in their normal states. The distribution of electric charge is not perfectly even. There is a little more negative charge on one side of the molecule than the other (Figure 22.15). Such molecules are said to be *electric dipoles*.

CHECK YOURSELF

1. A negatively charged rod is brought close to some small pieces of neutral paper. The positive sides of molecules in the paper are attracted to the rod and the negative sides of the molecules are repelled. Since negative and positive sides are equal in number, why don't the attractive and repulsive forces cancel out?
2. In a humorous vein, if you rub a balloon on your hair and put your head to the wall, will it stick to the wall like the balloon would?

Electric Field

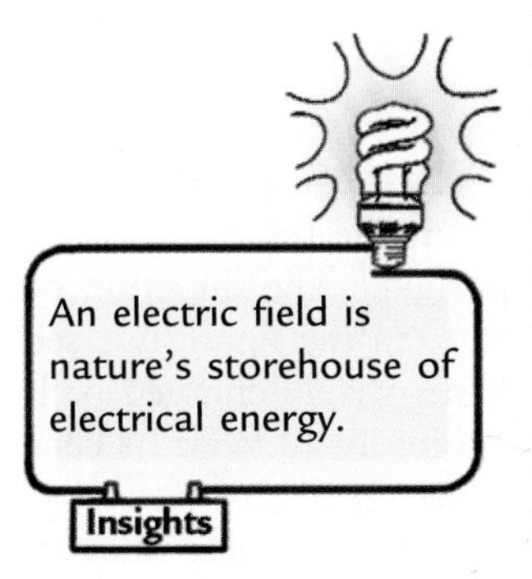

An electric field is nature's storehouse of electrical energy.
Insights

Electrical forces, like gravitational forces, act between things that are not in contact with each other. Both for electricity and for gravitation, a *force field* exists that influences charged and massive bodies, respectively. Recall, from Chapter 9, that the properties of space surrounding any massive body are altered such that another massive body introduced to this region will experience a force. The force is gravitational, and the altered space surrounding a massive body is its *gravitational field*. We can think of any other massive body as interacting with the field and not directly with the massive body producing it. For example, when an apple falls from a tree, we say it is interacting with the Earth, but we can also think of the apple as interacting with the gravitational field of the Earth. The field plays an intermediate role in the force between bodies. It is common to think of distant rockets and the like as interacting with gravitational fields rather than with the masses of the Earth and other bodies responsible for the fields. Just as the space around a planet (and around every other massive body) is filled with a gravitational field, the space around every electrically charged body is filled with an **electric field**—a kind of aura that extends through space.

An electric field has both magnitude (strength) and direction. The magnitude of the field at any point is simply the force per unit of charge. If a body

CHECK YOUR ANSWERS

1. The positive sides are simply closer to the rod. In accord with Coulomb's law, they therefore experience a greater electrical force than the negative sides, which are farther away. Hence we say that closeness wins. This greater force between positive and negative is attractive, so the neutral paper is attracted to the charged rod. Can you see that, if the rod were positive, attraction would still occur?
2. It would, if you were an airhead—that is, if the mass of your head were about that of the balloon, so that the force that would be produced would be evident.

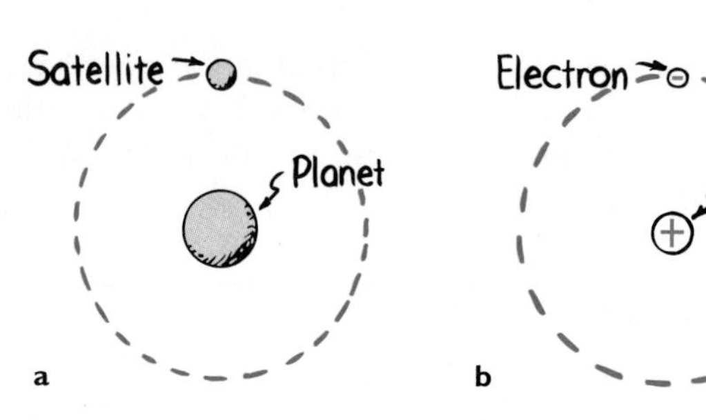

FIGURE 22.16
(a) A gravitational force holds the satellite in orbit about the planet, and (b) an electrical force holds the electron in orbit about the proton. In both cases, there is no contact between the bodies. We say that the orbiting bodies interact with the *force fields* of the planet and proton and are everywhere in contact with these fields. Thus, the force that one electrically charged body exerts on another can be described as the interaction between one body and the field set up by the other.

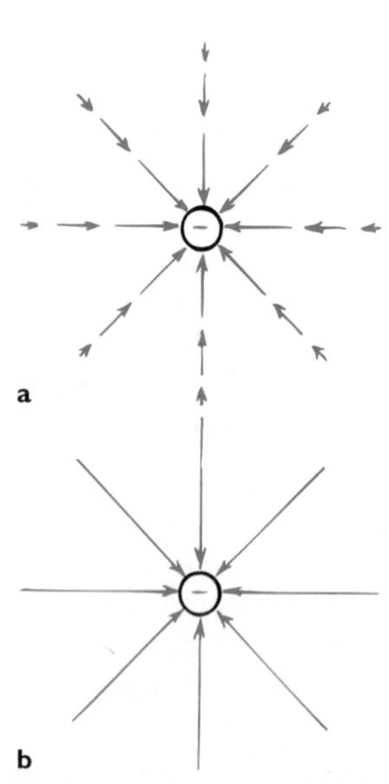

FIGURE 22.17
Interactive Figure
Electric-field representations about a negative charge.
(a) A vector representation.
(b) A lines-of-force representation.

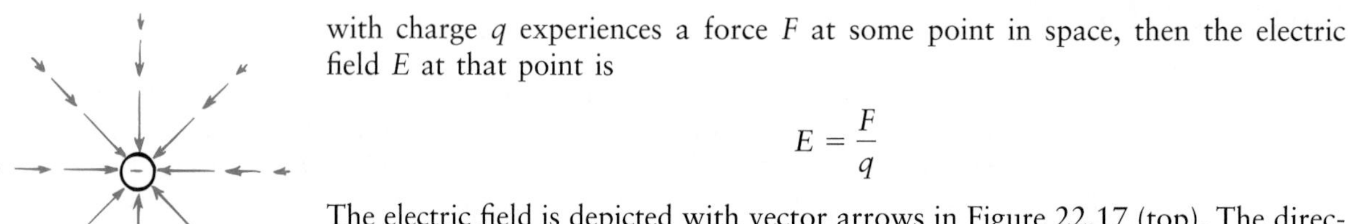

with charge q experiences a force F at some point in space, then the electric field E at that point is

$$E = \frac{F}{q}$$

The electric field is depicted with vector arrows in Figure 22.17 (top). The direction of the field is shown by the vectors and is defined to be the direction in which a small positive test charge at rest would be pushed.[6] The direction of the force and that of the field at any point are the same. In the figure, we see that all the vectors therefore point to the center of the negatively charged ball. If the ball were positively charged, the vectors would point away from its center because a positive test charge in the vicinity would be repelled.

A more useful way to describe an electric field is with electrical lines of force, as shown in Figure 22.17 (bottom). The lines of force shown in the figure represent a small number of the infinitely numerous possible lines that indicate the direction of the field. The figure is a two-dimensional representation of three dimensions. Where the lines are farther apart, the field is weaker. For an isolated charge, the lines extend to infinity; for two or more opposite charges, we represent the lines as emanating from a positive charge and terminating on a negative charge. Some electric field configurations are shown in Figure 22.18, and photographs of field patterns are shown in Figure 22.19. The photographs show bits of thread that are suspended in an oil bath surrounding charged conductors. The ends of the thread bits are charged by induction and tend to line up end-to-end with the field lines, like iron filings in a magnetic field.

The electric-field concept helps us understand not only the forces between isolated stationary charged bodies but also what happens when charges move. When charges move, their motion is communicated to neighboring charged bodies in the form of a field disturbance. The disturbance emanates from the

[6]The test charge is small so that it does not appreciably influence the sources of the field being measured. Recall, from our study of heat, the similar need for a thermometer of small mass when measuring the temperature of bodies of larger masses.

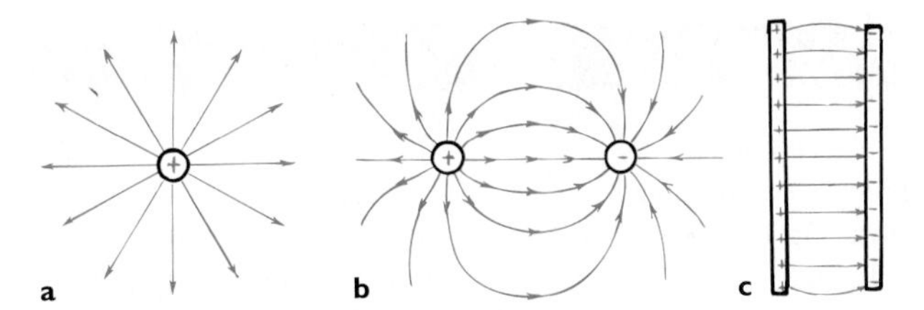

FIGURE 22.18 Interactive Figure

Some electric field configurations. (a) Lines of force emanating from a single positively charged particle. (b) Lines of force for a pair of equal but oppositely charged particles. Note that the lines emanate from the positive particle and terminate on the negative particle. (c) Uniform lines of force between two oppositely charged parallel plates.

accelerating charged body at the speed of light. We will learn that the electric field is a storehouse of energy and that energy can be transported over long distances in an electric field. Energy that is traveling in an electric field may be directed through, and guided by, metal wires, or it may be teamed up with a magnetic field to move through empty space. We will return to this idea in the next chapter and later when we learn about electromagnetic radiation.

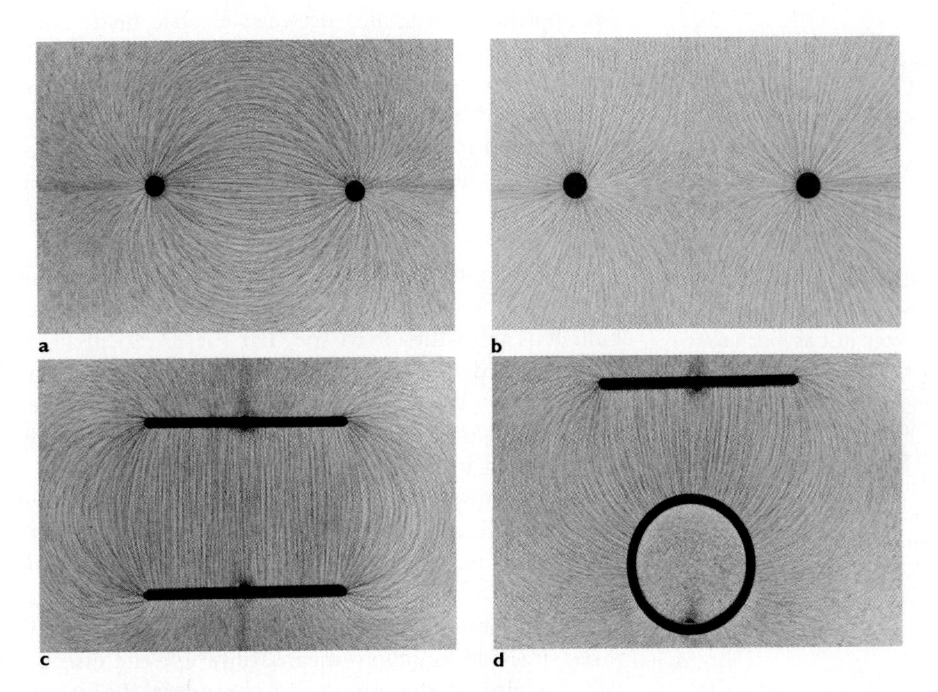

FIGURE 22.19

The electric field due to a pair of charged conductors is shown by bits of thread suspended in an oil bath surrounding the conductors. Note that the threads line up end to end along the direction of the electric field. (a) Equal and oppositely charged conductors (like those in Figure 22.18b). (b) Equal and identically charged conductors. (c) Oppositely charged plates. (d) Oppositely charged cylinder and plate.

MICROWAVE OVENS

Imagine an enclosure with Ping-Pong balls among a few batons, all at rest. Now imagine the batons suddenly flipping back and forth like semirotating propellers, striking neighboring Ping-Pong balls. The balls are energized, moving in all directions. A microwave oven operates similarly. The batons are water molecules (or other polar molecules) made to flip back and forth in rhythm with the microwaves in the enclosure. The Ping-Pong balls are nonpolar molecules that make up the bulk of the food being cooked.

Each H_2O molecule is an electric dipole that aligns with an electric field, like a compass needle aligns with a magnetic field. When the electric field is made to oscillate, the H_2O molecules oscillate also. The H_2O molecules move quite energetically when the oscillation frequency matches their natural frequency—at resonance. Food is cooked by a sort of "kinetic friction" as flip-flopping H_2O molecules (or other polar molecules) impart thermal motion to surrounding molecules. The metal enclosure reflects microwaves to and fro throughout the oven for rapid cooking.

Dry paper, foam plates, or other materials recommended for use in microwave ovens contain no water or any other polar molecules, so microwaves pass through them with no effect. Likewise for ice, in which the H_2O molecules are locked in position and can't rotate to and fro.

A note of caution is due when boiling water in a microwave oven. Water can sometimes heat faster than bubbles can form, and the water then heats beyond its boiling point—it becomes superheated. If the water is bumped or jarred just enough to cause the bubbles to form rapidly, they'll violently expel the hot water from its container. More than one person has had boiling water blast into his or her face.

Electric Shielding

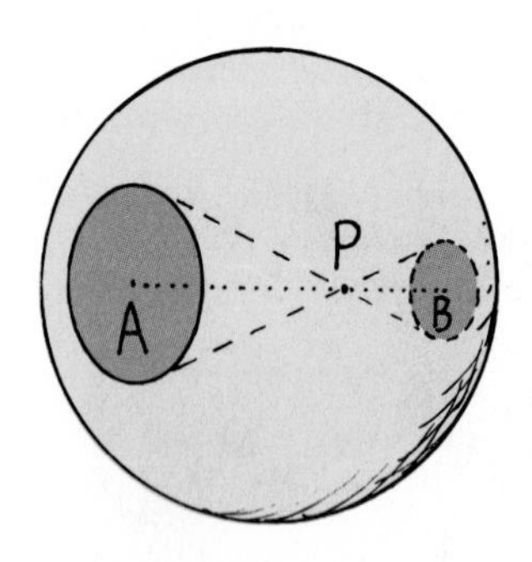

FIGURE 22.20
The test charge at P is attracted just as much to the greater amount of charge at farther region A as it is to the smaller amount of charge at closer region B. The net force on the test charge is zero—there, or anywhere inside the conductor. The electric field everywhere inside is also zero.

An important difference between electric fields and gravitational fields is that electric fields can be shielded by various materials, while gravitational fields cannot. The amount of shielding depends on the material used for shielding. For example, air makes the electric field between two charged objects slightly weaker than it would be in a vacuum, while oil placed between the objects can diminish the field to nearly a hundredth of its original strength. Metal can completely shield an electric field. When no current is flowing, the electric field inside metal is zero, regardless of the field strength outside.

Consider, for example, electrons on a spherical metal ball. Because of mutual repulsion, the electrons will spread out uniformly over the outer surface of the ball. It is not difficult to see that the electrical force exerted on a sample test charge placed at the exact center of the ball is zero, because opposing forces balance in every direction. Interestingly enough, complete cancellation occurs anywhere inside a conducting sphere. Understanding why this is true requires more thought and involves the inverse-square law and a bit of geometry. Consider the test charge at point P in Figure 22.20. The test charge is twice as far from the left side of the charged sphere as it is from the right side. If the electrical force between the test charge and the charges depended only on distance, then the test charge would be attracted only 1/4 as much to the left side as to the right side. (Remember the inverse-square law: Twice as far away means only 1/4 the effect, three times as far away means only 1/9 the effect, and so on.) But the force depends also on the amount of charge. In the figure, the cones extending from point P to areas A and B have the same apex angle, but one has twice the altitude of the other. This means that area A at the base of the longer cone is four times area B at the base of the shorter cone, which is true for any apex angle. Since 1/4 of 4 is equal to 1, a test charge at P is attracted equally to each side. Cancellation occurs. A similar argument applies if the cones emanating from point P are oriented in any direction. Complete cancellation occurs at all points

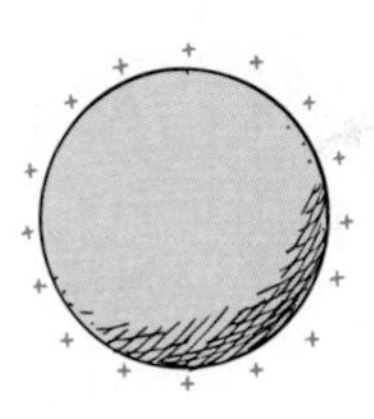

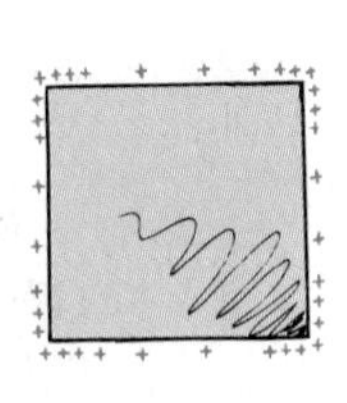

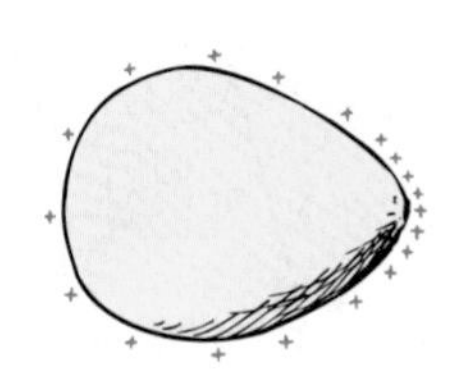

FIGURE 22.21
Electric charge distributes itself on the surface of all conductors in such a way that the electric field inside the conductors is zero.

FIGURE 22.22
Electrons from the lightning bolt mutually repel to the outer metal surface. Although the electric field they set up may be great *outside* the car, the net electric field *inside* the car is zero.

within the sphere. (Recall this argument back in Figure 9.4 in Chapter 9 for the cancellation of gravity inside a hollow planet. The metal ball behaves the same way, whether hollow or not, because all of its charge gathers on its outer surface.)

If the conductor is not spherical, then the charge distribution will not be uniform. The charge distribution over conductors of various shapes is shown in Figure 22.21. Most of the charge on a conducting cube, for example, is mutually repelled toward the corners. The remarkable thing is this: The exact charge distribution over the surface of a conductor is such that the electric field everywhere inside the conductor is zero. Look at it this way: If there were an electric field inside a conductor, then free electrons inside the conductor would be set in motion. How far would they move? Until equilibrium is established—which is to say, when the positions of all the electrons produce a zero field inside the conductor.

We cannot shield ourselves from gravity, because gravity only attracts. There are no repelling parts of gravity to offset attracting parts. Shielding electric fields, however, is quite simple. Surround yourself, or whatever you wish to shield, with a conducting surface. Put this surface in an electric field of any field strength. The free charges in the conducting surface will arrange themselves on the surface of the conductor in such a way that all field contributions inside cancel one another. That's why certain electronic components are encased in metal boxes and why certain cables have a metal covering—to shield them from outside electrical activity.

CHECK YOURSELF

Small bits of aligned thread vividly show the electric fields in the four photos of Figure 22.19. But the threads are not aligned inside the cylinder in Figure 22.19d. Why?

Electric Potential

When we studied energy in Chapter 7, we learned that an object has gravitational potential energy because of its location in a gravitational field. Similarly, a charged object has potential energy by virtue of its location in an electric field. Just as work is required to lift a massive object against the gravitational field of the Earth, work is required to push a charged particle against the electric field of a charged body. This work changes the electric potential energy of the charged

CHECK YOUR ANSWER

The electric field is shielded inside the cylinder, shown as a circle in the two-dimensional photograph. Hence, the threads are without alignment. The electric field inside any conductor is zero—so long as no electric charge is flowing.

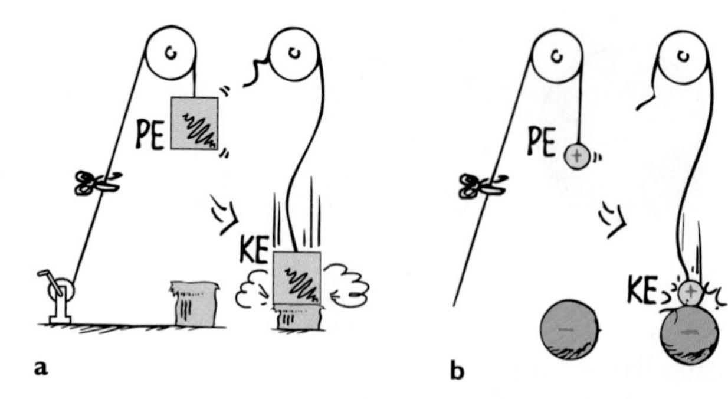

FIGURE 22.23
(a) The PE (gravitational potential energy) of a mass held in a gravitational field, when released, transforms to KE (kinetic energy). (b) The PE of a charged particle held in an electric field, when released, becomes KE. How does the KE acquired by each compare with the decrease in PE?

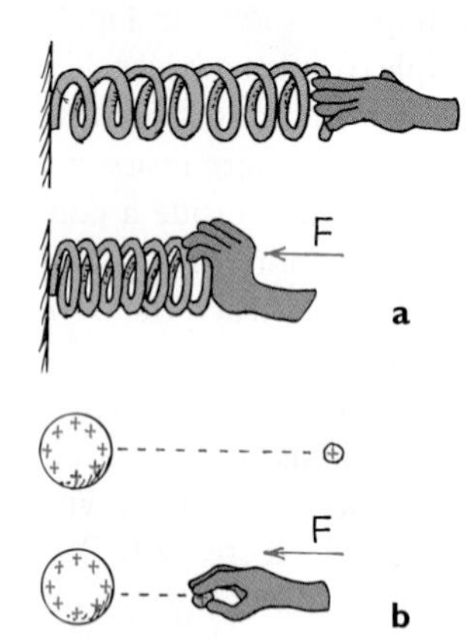

FIGURE 22.24
(a) The spring has more mechanical PE when compressed. (b) The charged particle similarly has more electrical PE when pushed closer to the charged sphere. In both cases, the increased PE is the result of work input.

particle.[7] Consider the particle with the small positive charge located at some distance from a positively charged sphere in Figure 22.24b. If you push the particle closer to the sphere, you will expend energy to overcome electrical repulsion; that is, you will do work in pushing the charged particle against the electric field of the sphere. This work done in moving the particle to its new location increases its energy. We call the energy the particle possesses by virtue of its location **electric potential energy.** If the particle is released, it accelerates in a direction away from the sphere, and its electric potential energy changes to kinetic energy.

If we instead push a particle with twice the charge, we do twice as much work pushing it, so the doubly charged particle in the same location has twice as much electric potential energy as before. A particle with three times the charge has three times as much potential energy; ten times the charge, ten times the potential energy; and so on. Rather than dealing with the potential energy of a charged body, it is convenient, when working with charged particles in an electric field, to consider the electric potential energy *per unit of charge*. We simply divide the amount of potential energy in any case by the amount of charge. For example, a particle with ten times as much charge as another in the same location will have ten times as much potential energy—but having ten times as much charge means that the energy per charge is the same. The concept of potential energy per unit charge is called **electric potential;** that is,

$$\text{electric potential} = \frac{\text{electric potential energy}}{\text{charge}}$$

The unit of measurement for electric potential is the volt, so electric potential is often called *voltage*. A potential of 1 volt (V) equals 1 joule (J) of energy per 1 coulomb (C) of charge.

$$1 \text{ volt} = 1 \frac{\text{joule}}{\text{coulomb}}$$

Thus, a 1.5-volt battery gives 1.5 joules of energy to every 1 coulomb of charge passing through the battery. Both the names *electric potential* and *voltage* are common, so either may be used. In this book, the names will be used interchangeably.

The significance of electric potential (voltage) is that a definite value for it can be assigned to a location. We can speak about the electric potentials at different locations in an electric field whether or not charges occupy those locations (once the location of zero voltage has been specified). Likewise with voltages at

[7]This work is positive if it increases the electric potential energy of the charged particle and negative if it decreases it.

FIGURE 22.25
Of the two charged bodies near the charged dome, the one carrying the greater charge has the higher electrical PE in the field of the dome. But the *electric potential* of each is the same—likewise for any amount of charge in the same location. Why?

In a nutshell: *Electric potential* and *potential* mean the same thing—electrical potential energy per unit charge—in units of volts. On the other hand, *potential difference* is the same as *voltage*—the *difference* in electrical potential between two points—also in units of volts.

Insights

various locations in an electric circuit. In the next chapter, you will see that the location of the positive terminal of a 12-volt battery is maintained at a voltage 12 volts higher than the location of the negative terminal. When a conducting medium connects these terminals with differing voltages, charges in the medium will move between them.

CHECK YOURSELF

1. If there were twice as many coulombs in the test charge near the charged sphere in Figure 22.24, would the electric potential energy of the test charge with respect to the charged sphere be the same or would it be twice as great? Would the electric potential of the test charge be the same or would it be twice as great?
2. What does it mean to say that your car has a 12-volt battery?

If you rub a balloon on your hair, the balloon becomes negatively charged—perhaps to several thousand volts! That would be several thousand joules of energy, if the charge were 1 coulomb. However, 1 coulomb is a very large amount of charge. The charge on a balloon rubbed on hair is more typically much less than a millionth of a coulomb. Therefore, the amount of energy associated with the charged balloon is very, very small. A high voltage means a lot of energy only if a lot of charge is involved. There is an important difference between electric potential energy and electric potential.

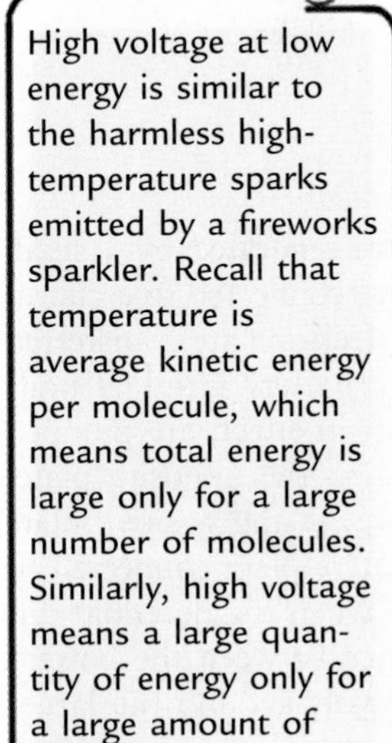

High voltage at low energy is similar to the harmless high-temperature sparks emitted by a fireworks sparkler. Recall that temperature is average kinetic energy per molecule, which means total energy is large only for a large number of molecules. Similarly, high voltage means a large quantity of energy only for a large amount of charge.

Insights

CHECK YOUR ANSWERS

1. Twice as many coulombs would find the test charge with twice as much electric potential energy. (This is because it would have taken twice as much work to put the charge at that location.) But the electric potential would be the same. This is because the electric potential is total electric potential energy divided by total charge. For example, ten times the energy divided by ten times the charge will give the same value as two times the energy divided by two times the charge. Electric potential is not the same thing as electric potential energy. Be sure you understand this before you study further.
2. It means that one of the battery terminals is 12 V higher in potential than the other one. In the next chapter, you will see that it also means that, when a circuit is connected between these terminals, each coulomb of charge in the resulting current will be given 12 J of energy as it passes through the battery.

FIGURE 22.26
Although the electric potential (voltage) of the charged balloon is high, the electric potential energy is low because of the small amount of charge. Therefore, very little energy transfers when the balloon is discharged.

Electric Energy Storage

Electric energy can be stored in a common device called a **capacitor,** which is found in nearly all electronic circuits. A capacitor is used as a storehouse for energy. Capacitors store the energy in common photoflash units. The rapid release of this energy is evident in the short duration of the flash. Similarly, but on a grander scale, enormous amounts of energy are stored in banks of capacitors that power giant lasers in national laboratories.

Watch for the advent of capacitors for energy storage in hybrid automobiles.

Insights

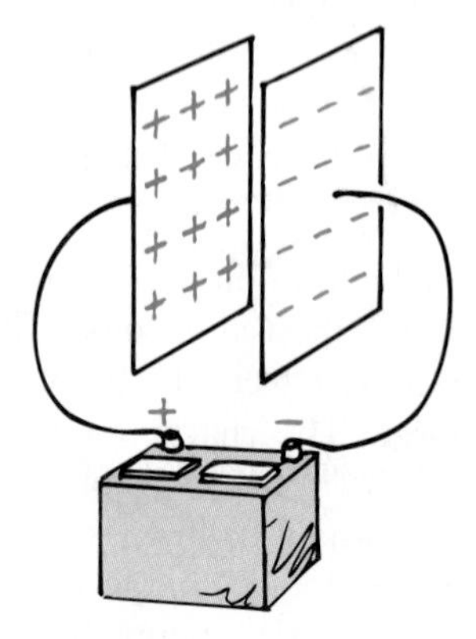

FIGURE 22.27
A capacitor consisting of two closely spaced parallel metal plates. When connected to a battery, the plates acquire equal and opposite charges. The voltage between the plates then matches the electric potential difference between the battery terminals.

FIGURE 22.28
Mona El Tawil-Nassar adjusts demonstration capacitor plates.

The simplest capacitor is a pair of conducting plates separated by a small distance, but not touching each other. When the plates are connected to a charging device, such as the battery shown in Figure 22.27, electrons are transferred from one plate to the other. This occurs as the positive battery terminal pulls electrons from the plate connected to it. These electrons, in effect, are pumped through the battery and through the negative terminal to the opposite plate. The capacitor plates then have equal and opposite charges—the positive plate connected to the positive battery terminal, and the negative plate connected to the negative terminal. The charging process is complete when the potential difference between the plates equals the potential difference between the battery terminals—the battery voltage. The greater the battery voltage, and the larger and closer the plates, the greater the charge that can be stored. In practice, the plates may be thin metallic foils separated by a thin sheet of paper. This "paper sandwich" is then rolled up to save space and inserted into a cylinder. Such a practical capacitor is shown with others in Figure 22.29. (We will consider the role of capacitors in circuits in the next chapter.)

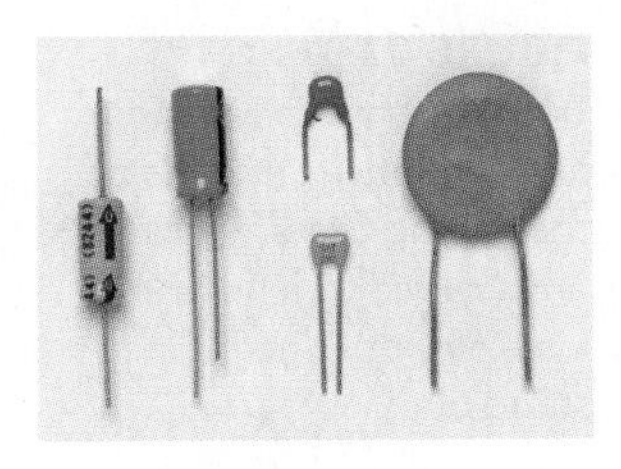

FIGURE 22.29
Practical capacitors.

A charged capacitor is discharged when a conducting path is provided between the plates. Discharging a capacitor can be a shocking experience if you happen to be the conducting path. The energy transfer that occurs can be fatal where high voltages are present, such as the power supply in a TV set—even after the set has been turned off. This is the main reason for the warning signs on such devices.

CHECK YOURSELF

What is the net charge of a charged capacitor?

The energy stored in a capacitor comes from the work required to charge it. The energy is stored in the electric field between its plates. Between parallel plates, the electric field is uniform, as indicated in Figures 22.18c and Figure 22.19c. So the energy stored in a capacitor is the energy of its electric field. In Chapter 25, we will see how energy from the Sun is radiated in the form of electric and magnetic fields. The fact that energy is contained in electric fields is truly far-reaching.

Van de Graaff Generator

A common laboratory device for building up high voltages is the *Van de Graaff generator.* This is one of the lightning machines that mad scientists used in old science-fiction movies. A simple model of the Van de Graaff generator is shown in Figure 22.30. A large, hollow metal sphere is supported by a cylindrical

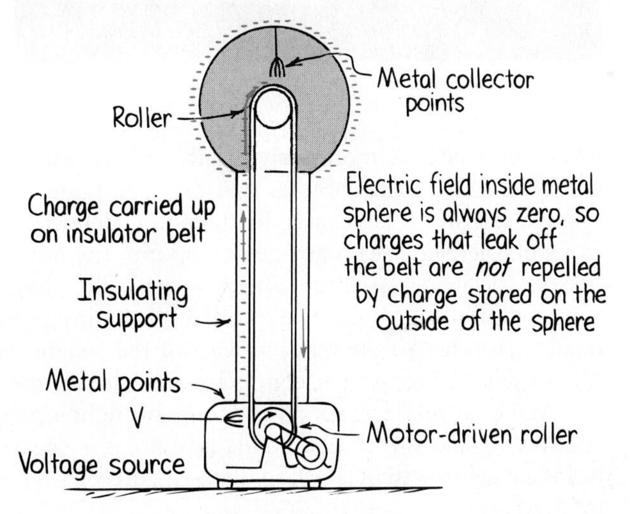

FIGURE 22.30
A simple model of a Van de Graaff generator.

CHECK YOUR ANSWER

The net charge of a charged capacitor is zero because the charges on its two plates are equal in number and opposite in sign. Even when the capacitor is discharged—say, by providing a path for charge flow between the oppositely charged plates—the net charge of the capacitor remains zero, for then each plate has zero charge.

FIGURE 22.31
Both the physics enthusiast and the spherical dome of the Van de Graaff generator are charged to a high voltage. Why does her hair stand out?

insulating stand. A motor-driven rubber belt inside the support stand passes a comblike set of metal needles that are maintained at a large negative potential relative to ground. Discharge by the points deposits a continuous supply of electrons on the belt, which are carried up into the hollow conducting sphere. Since the electric field inside the sphere is zero, the charge leaks onto metal points (tiny lightning rods) and is deposited on the inside of the sphere. The electrons repel each other to the outer surface of the sphere. Static charge always lies on the outside surface of any conductor. This leaves the inside uncharged and able to receive more electrons as they are brought up by the belt. The process is continuous and the charge builds up until the negative potential on the sphere is much greater than at the voltage source at the bottom—on the order of millions of volts.

A sphere with a radius of 1 meter can be raised to a potential of 3 million volts before electrical discharge occurs through the air. The voltage can be further increased by increasing the radius of the sphere or by placing the entire system in a container filled with high-pressure gas. Van de Graaff generators can produce voltages as high as 20 million volts. These voltages are used to accelerate charged particles that can be used as projectiles for penetrating the nuclei of atoms. Touching one can be a hair-raising experience.

Summary of Terms

Electricity General term for electrical phenomena, much like gravity has to do with gravitational phenomena, or sociology with social phenomena.

Electrostatics The study of electric charge at rest (not *in motion,* as in electric currents).

Conservation of charge Electric charge is neither created nor destroyed. The total charge before an interaction equals the total charge after.

Coulomb's law The relationship between electrical force, charge, and distance:

$$F = k\frac{q_1q_2}{d^2}$$

If the charges are alike in sign, the force is repulsive; if the charges are unlike, the force is attractive.

Coulomb The SI unit of electrical charge. One coulomb (symbol C) is equal to the total charge of 6.25×10^{18} electrons.

Conductor Any material having free charged particles that easily flow through it when an electric force acts on them.

Insulator A material without free charged particles and through which charge does not easily flow.

Semiconductor A device composed of material not only with properties that fall between a conductor and an insulator but with resistance that changes abruptly when other conditions change, such as temperature, voltage, and electric or magnetic fields.

Superconductor A material that is a perfect conductor with zero resistance to the flow of electric charge.

Charging by contact Transfer of electric charge between objects by rubbing or simple touching.

Charging by induction Redistribution of electric charges in and on objects caused by the electrical influence of a charged object close by but not in contact.

Electrically polarized Term applied to an atom or molecule in which the charges are aligned so that one side has a slight excess of positive charge and the other side a slight excess of negative charge.

Electric field Defined as force per unit charge, it can be considered to be an "aura" surrounding charged objects and is a storehouse of electric energy. About a charged point, the field decreases with distance according to the inverse-square law, like a gravitational field. Between oppositely charged parallel plates, the electric field is uniform.

Electric potential energy The energy a charged object possesses by virtue of its location in an electric field.

Electric potential The electric potential energy per unit of charge, measured in volts, and often called *voltage:*

Voltage = electric potential energy/amount of charge

Capacitor An electrical device—in its simplest form, a pair of parallel conducting plates separated by a small distance—that stores electric charge and energy.

Suggested Reading

Brands, H. W. *The First American: The Life and Times of Benjamin Franklin*. New York: Doubleday, 2000.

Bodanis, David. *Electric Universe—The Shocking True Story of Electricity.* New York: Crown, 2005.

Review Questions

Electrical Forces

1. Why does the gravitational force between Earth and Moon predominate over electrical forces?

Electric Charges

2. What part of an atom is *positively* charged and what part is *negatively* charged?
3. How does the charge of one electron compare to that of another electron? How does it compare with the charge of a proton?
4. How do the numbers of protons in the atomic nucleus normally compare to the number of electrons that orbit the nucleus?
5. What is normally the net charge of an atom?

Conservation of Charge

6. What is a positive ion? A negative ion?
7. What is meant by saying charge is *conserved?*
8. What is meant by saying charge is *quantized?*
9. What particle has exactly one quantum unit of charge?

Coulomb's Law

10. How does one *coulomb* of charge compare with the charge of a *single* electron?
11. How is Coulomb's law similar to Newton's law of gravitation? How is it different?

Conductors and Insulators

12. Why are metals good conductors both of heat and of electricity?
13. Why are materials such as glass and rubber good insulators?

Semiconductors

14. How does a *semiconductor* differ from a *conductor* or an *insulator?*

#3, 5, 7, 9, 11, 15, 21, 26, 27, 30, 31, 33, 36, 37, 41, 43, 45, 55, 57

15. What is a transistor composed of, and what are some of its functions?

Superconductors

16. How does the flow of current differ in a superconductor compared with the flow in ordinary conductors?

Charging

17. What happens to electrons in any charging process?

Charging by Friction and Contact

18. Cite an example of something charged by friction.
19. Cite an example of something charged by contact.

Charging by Induction

20. Give an example of something charged by induction.
21. What is the purpose of a lightning rod?

Charge Polarization

22. How does an electrically *polarized* object differ from an electrically *charged* object?
23. What is an electric dipole?

Electric Field

24. Give two examples of common force fields.
25. How is the magnitude of an electric field defined?
26. How is the direction of an electric field defined?

Electric Shielding

27. Why is there no electric field at the center of a charged spherical conductor?
28. Does an electric field exist within a charged spherical conductor at points other than its center?
29. When charges mutually repel and distribute themselves on the surface of conductors, what effect occurs inside the conductor?

Electric Potential

30. How much energy is given to each coulomb of charge that flows through a 1.5-volt battery?
31. A balloon may easily be charged to several thousand volts. Does that mean it has several thousand joules of energy? Explain.

Electric Energy Storage

32. How does the charge on one plate of a capacitor compare with that on the opposite plate?
33. Where is the energy stored in a capacitor?

Van de Graaff Generator

34. What is the magnitude of the electric field inside the dome of a charged van de Graaff generator?

Projects

1. Demonstrate charging by friction and discharging from points with a friend who stands at the far end of a carpeted room. Scuff your way across the rug until your noses are close together. This can be a delightfully tingling experience, depending on how dry the air is and how pointed your noses are.
2. Briskly rub a comb through your hair or on a woolen garment and bring it near a small but smooth stream of running water. Is the stream of water deflected?

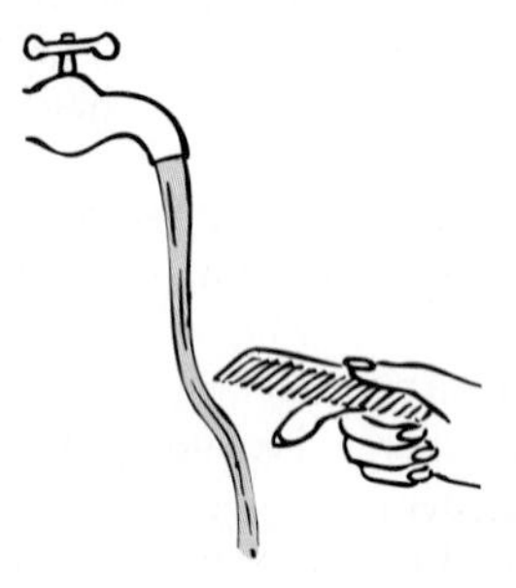

3. Write a letter to Grandpa and tell him why he'd be safe in a lightning storm if he's inside an automobile.

Exercises

1. We do not feel the gravitational forces between ourselves and the objects around us because these forces are extremely small. Electrical forces, in comparison, are extremely huge. Since we and the objects around us are composed of charged particles, why don't we usually feel electrical forces?
2. At the atomic level, what is meant by saying something is electrically charged?
3. Why is charge usually transferred by electrons rather than by protons?
4. Why are objects with vast amounts of electrons normally not electrically charged?
5. Why do clothes often cling together after tumbling in a clothes dryer?
6. Why will dust be attracted to a CD wiped with a dry cloth?
7. When you remove your wool suit from the dry cleaner's garment bag, the bag becomes positively charged. Explain how this occurs.

8. Plastic wrap becomes electrically charged when pulled from its container. As a result, it is attracted to objects such as food containers. Does the wrap stick better to plastic containers or to metal containers?

9. When combing your hair, you scuff electrons from your hair onto the comb. Is your hair then positively or negatively charged? How about the comb?

10. At some automobile toll booths, a thin metal wire protrudes from the road, making contact with cars before they reach the toll collector. What is the purpose of this wire?

11. Why are the tires for trucks carrying gasoline and other flammable fluids manufactured to be electrically conducting?

12. An electroscope is a simple device consisting of a metal ball that is attached by a conductor to two thin leaves of metal foil protected from air disturbances in a jar, as shown. When the ball is touched by a charged body, the leaves that normally hang straight down spread apart. Why? (Electroscopes are useful not only as charge detectors but also for measuring the quantity of charge: the more charge transferred to the ball, the more the leaves diverge.)

13. The leaves of a charged electroscope collapse in time. At higher altitudes, they collapse more rapidly. Why is this true? (*Hint:* The existence of cosmic rays was first indicated by this observation.)

14. Is it necessary for a charged body actually to touch the ball of the electroscope for the leaves to diverge? Defend your answer.

15. Strictly speaking, when an object acquires a positive charge by the transfer of electrons, what happens to its mass? What happens to its mass when it acquires a negative charge? (Think small!)

16. Strictly speaking, will a penny be slightly more massive if it has a negative charge or a positive charge? Explain.

17. A crystal of salt consists of electrons and positive ions. How does the net charge of the electrons compare with the net charge of the ions? Explain.

18. How can you charge an object negatively with only the help of a positively charged object?

19. It is relatively easy to strip the outer electrons from a heavy atom like that of uranium (which then becomes a uranium ion), but it is very difficult to remove the inner electrons. Why do you suppose this is so?

20. When one material is rubbed against another, electrons jump readily from one to the other but protons do not. Why is this? (Think in atomic terms.)

21. If electrons were positive and protons were negative, would Coulomb's law be written the same or differently?

22. What does the inverse-square law tell you about the relationship between force and distance?

23. The five thousand billion billion (5×10^{21}) freely moving electrons in a penny repel one another. Why don't they fly out of the penny?

24. How does the magnitude of electrical force between a pair of charged particles change when the particles are moved half as far apart? One third as far apart?

25. How does the magnitude of electric force compare between a pair of charged particles when they are brought to half their original distance of separation? To one-quarter their original distance? To four times their original distance? (What law guides your answers?)

26. When you double the distance between a pair of charged particles, what happens to the force between them? Does it depend on the sign of the charges? What law defends your answer?

27. When you double the charge on only one of a pair of particles, what effect does this have on the force between them? Does the effect depend on the sign of the charge?

28. When you double the charge on both particles in a pair, what effect does this have on the force between them? Does it depend on the sign of the charge?

29. The proportionality constant k in Coulomb's law is huge in ordinary units, whereas the proportionality constant G in Newton's law of gravitation is tiny. What does this indicate about the relative strengths of these two forces?

30. How do electrical field lines indicate the strength of an electric field?

31. How is the direction of an electric field indicated with electrical field lines?

32. Suppose that the strength of the electric field about an isolated point charge has a certain value at a distance of 1 m. How will the electric field strength compare at a distance of 2 m from the point charge? What law guides your answer?

33. How does a semiconductor differ from a conductor or an insulator?

34. In the phenomenon of superconductivity, what happens to electrical resistance at low temperatures?

35. Measurements show that there is an electric field surrounding the Earth. Its magnitude is about 100 N/C at the Earth's surface, and it points inward toward the Earth's center. From this information, can you state whether the Earth is negatively or positively charged?

36. Why are lightning rods normally taller than the buildings they protect?

37. Why are metal-spiked shoes not a good idea for golfers on a stormy day?

38. If you are caught outdoors in a thunderstorm, why should you not stand under a tree? Can you think of a reason why you should not stand with your legs far apart? Or why lying down can be dangerous? (*Hint:* Consider electric potential difference.)

39. If a large enough electric field is applied, even an insulator will conduct an electric current, as is evident in lightning discharges through the air. Explain how this happens, taking into account the opposite charges in an atom and how ionization occurs.

40. Why is a good conductor of electricity also a good conductor of heat?

41. If you rub an inflated balloon against your hair and place it against a door, by what mechanism does it stick? Explain.

42. How can a charged atom (an ion) attract a neutral atom?

43. When a car is moved into a painting chamber, a mist of paint is sprayed around its body. When the body is given a sudden electric charge and mist is attracted to it—presto—the car is quickly and uniformly painted. What does the phenomenon of polarization have to do with this?

44. If you place a free electron and a free proton in the same electric field, how will the forces acting on them compare? How will their accelerations compare? Their directions of travel?

45. Two pieces of plastic, a full ring and a half ring, have the same radius and charge density. Which electric field at the center has the greater magnitude? Defend your answer.

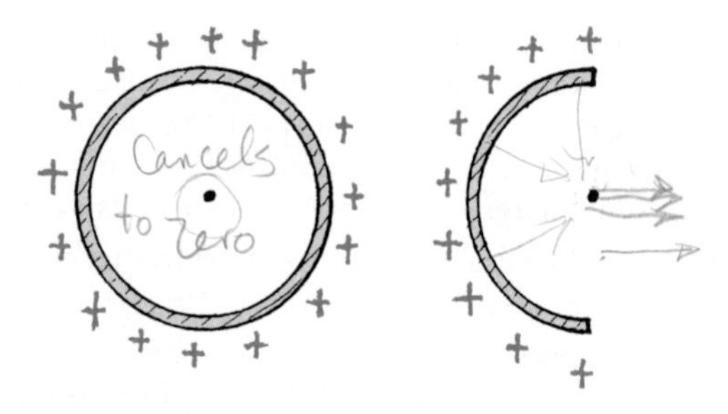

46. Why is the magnitude of the electric field zero midway between identical point charges?

47. Imagine a proton at rest a certain distance from a negatively charged plate. It is released and collides with the plate. Then imagine the similar case of an electron at rest the same distance away from a plate of equal and opposite charge. In which case will the moving particle have the greater speed when the collision occurs? Why?

48. A gravitational field vector points toward the Earth; an electric field vector points toward an electron. Why do electric field vectors point away from protons?

49. By what specific means do the bits of fine threads align in the electric fields shown in Figure 22.19?

50. Suppose that a metal file cabinet is charged. How will the charge concentration at the corners of the cabinet compare with the charge concentration on the flat parts of the cabinet?

51. If you were to expend 10 joules of work to push a 1-coulomb charge against an electric field, what would be its voltage with respect to its voltage in its starting position? When released, what will be its kinetic energy if it flies past its starting position?

52. You are not harmed by contact with a charged metal ball, even though its voltage may be very high. Is the reason similar to why you are not harmed by the greater-than-1000°C sparks from a Fourth-of-July sparkler? Defend your answer in terms of the energies that are involved.

53. What is the voltage at the location of a 0.0001-C charge that has an electric potential energy of 0.5 J (both measured relative to the same reference point)?

54. Why is it safe to remain inside a car during a lightning storm?

55. Why are the charges on opposing plates of a capacitor always of equal magnitude?

56. In order to store more energy in a parallel-plate capacitor whose plates differ by a fixed voltage, what change would you make in the plates?

57. Why is it dangerous to touch the terminals of a high-voltage capacitor even after the charging circuit is turned off?

58. An electron volt, eV, is a unit of energy. Which is larger, a GeV or an MeV?

59. Would you feel any electrical effects if you were inside the charged sphere of a Van de Graaff generator? Why or why not?

60. A friend says that the reason one's hair stands out while touching a charged Van de Graaff generator is simply that the hair strands become charged and are light enough so that the repulsion between strands is visible. Do you agree or disagree?

Problems

1. Two point charges are separated by 6 cm. The attractive force between them is 20 N. Find the force between them when they are separated by 12 cm. (Why can you solve this problem without knowing the magnitudes of the charges?)
2. If the charges attracting each other in the problem above have equal magnitude, what is the magnitude of each charge?
3. Two pellets, each with a charge of 1 microcoulomb (10^{-6} C), are located 3 cm (0.03 m) apart. What is the electric force between them? What would be the mass of an object that would experience this same force in the Earth's gravitational field?
4. Electronic types neglect the force of gravity on electrons. To see why, compute the force of Earth's gravity on an electron and compare it with the force exerted on the electron by an electric field of magnitude 10,000 V/m (a relatively small field). The mass and charge of an electron are given on the inside back cover.
5. Atomic physicists ignore the effect of gravity within an atom. To see why, calculate and compare the gravitational and electrical forces between an electron and a proton separated by 10^{-10} m. The needed charges and masses are given on the inside back cover of this book.
6. A droplet of ink in an industrial ink-jet printer carries a charge of 1.6×10^{-10} C and is deflected onto paper by a force of 3.2×10^{-4} N. Find the strength of the electric field to produce this force.
7. The potential difference between a storm cloud and the ground is 100 million volts. If a charge of 2 C flashes in a bolt from cloud to Earth, what is the change of potential energy of the charge?
8. An energy of 0.1 J is stored in the metal ball on top of a Van de Graaff generator. A spark carrying 1 microcoulomb (10^{-6} C) discharges the ball. What was the ball's potential relative to ground?
9. In 1909, Robert Millikan was the first to find the charge of an electron in his now-famous oil-drop experiment. In that experiment tiny oil drops were sprayed into a uniform electric field between a horizontal pair of oppositely charged plates. The drops were observed with a magnifying eyepiece, and the electric field was adjusted so that the upward force on some negatively charged oil drops was just sufficient to balance the downward force of gravity. That is, when suspended, upward force qE just equaled mg. Millikan accurately measured the charges on many oil drops and found the values to be whole-number multiples of 1.6×10^{-19}C—the charge of the electron. For this he won the Nobel prize. Questions: (a) If a drop of mass 1.1×10^{-14} kg remains stationary in an electric field of 1.68×10^{5} N/C, what is the charge of this drop? (b) How many extra electrons are on this particular oil drop (given the presently known charge of the electron)?

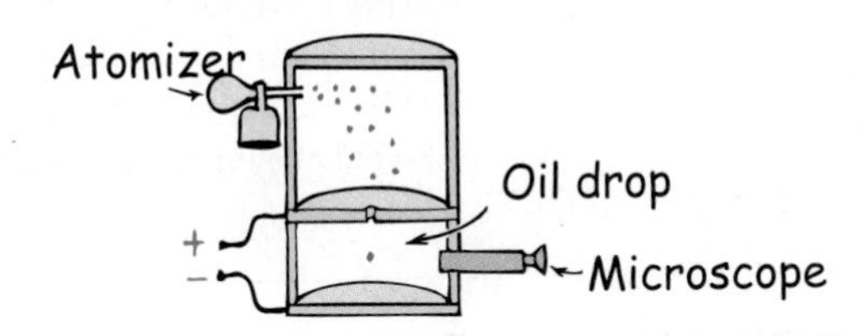

10. Find the voltage change when (a) an electric field does 12 J of work on a 0.0001-C charge; (b) the same electric field does 24 J of work on a 0.0002-C charge.

Electric Current

David Yee constructs a parallel circuit by fastening lamps to extended terminals of a common car battery.

In the previous chapter, we treated the concept of electric potential, which is measured in volts. Now we'll see that this voltage acts like an "electrical pressure" that can produce a flow of charge, or *current,* which is measured in amperes (or, simply, amps, and abbreviated A), and that the *resistance* that restrains this flow is measured in ohms (Ω). When the flow is in only one direction, we'll call it *direct current* (dc); when the flow is back and forth, we'll call it *alternating current* (ac). Electric current can deliver electric *power,* measured, just like mechanical power, in watts (W) or in thousands of watts, or kilowatts (kW). We see here many terms to be sorted out. This is easier to do when we have some understanding of the concepts these terms represent, which, in turn, are better understood if we know how they relate to one another. In this chapter, we'll discuss these terms and what they mean in detail. We begin with the flow of electric charge.

Flow of Charge

We often think of current flowing through a circuit, but don't say this around somebody who is picky about grammar, because the expression "current flows" is redundant. More properly, charge flows—which *is* current.

Insights

From our study of heat and temperature, recall that, when the ends of a conducting material are at different temperatures, heat energy flows from the higher temperature to the lower temperature. The flow ceases when both ends reach the same temperature. Similarly, when the ends of an electrical conductor are at different electric potentials—when there is a **potential difference**—charge flows from one end to the other.[1] The flow of charge persists for as long as there is a potential difference. Without a potential difference, no charge flows. Connect one end of a wire to the charged sphere of a Van de Graaff generator, for example, and the other end to the ground, and a surge of charge will flow through the wire. The flow will be brief, however, because the sphere will quickly reach a common potential with the ground.

To attain a sustained flow of charge in a conductor, some arrangement must be provided to maintain a difference in potential while charge flows from one end to the other. The situation is analogous to the flow of water from a higher

[1]When we say that charge flows, we mean that charged *particles* flow. Charge is a property of particular particles, most significantly electrons, protons, and ions. When the flow is of negative charge, electrons or negative ions constitute the flow. When the flow of charge is positive, protons or positive ions are flowing.

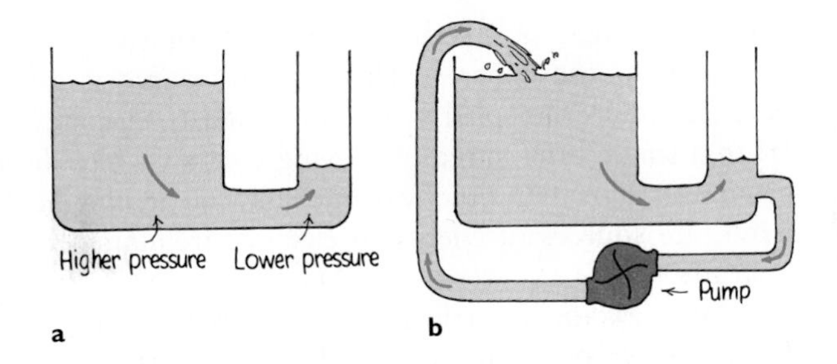

FIGURE 23.1
(a) Water flows from the reservoir of higher pressure to the reservoir of lower pressure. The flow ceases when the difference in pressure ceases. (b) Water continues to flow because a difference in pressure is maintained with the pump.

reservoir to a lower one (Figure 23.1a). Water will flow in a pipe that connects the reservoirs only as long as a difference in water level exists. The flow of water in the pipe, like the flow of charge in the wire that connects the Van de Graaff generator to the ground, will cease when the pressures at each end are equal (we imply this when we say that water seeks its own level). A continuous flow is possible if the difference in water levels—hence the difference in water pressures—is maintained with the use of a suitable pump (Figure 23.1b).

Electric Current

FIGURE 23.2
Each coulomb of charge made to flow in a circuit that connects the ends of this 1.5-V flashlight cell is energized with 1.5 J.

Just as water current is the flow of H_2O molecules, **electric current** is simply the flow of electric charge. In circuits of metal wires, electrons make up the flow of charge. This is because one or more electrons from each metal atom are free to move throughout the atomic lattice. These charge carriers are called *conduction electrons*. Protons, on the other hand, do not move because they are bound inside the nuclei of atoms that are more or less locked in fixed positions. In conducting fluids, however—such as in a car battery—positive ions typically compose the flow of electric charge.

The *rate* of electrical flow is measured in *amperes*. One ampere is a rate of flow equal to 1 coulomb of charge per second. (Recall that 1 coulomb, the standard unit of charge, is the electric charge of 6.25 billion billion electrons.) In a wire that carries 5 amperes, for example, 5 coulombs of charge pass any cross section in the wire each second. So that's a lot of electrons! In a wire that carries 10 amperes, twice as many electrons pass any cross section each second.

It is interesting to note that a current-carrying wire is not electrically charged. Under ordinary conditions, negative conduction electrons swarm through the atomic lattice made up of positively charged atomic nuclei. So there are as many electrons as protons in the wire. Whether a wire carries a current or not, the net charge of the wire is normally zero at every moment.

Voltage Sources

Charges flow only when they are "pushed" or "driven." A sustained current requires a suitable pumping device to provide a difference in electric potential—a voltage. An "electrical pump" is some sort of voltage source. If we charge

FIGURE 23.3
An unusual source of voltage. The electric potential between the head and tail of the electric eel (*Electrophorus electricus*) can be up to 600 V.

one metal sphere positively and another negatively, we can develop a large voltage between the spheres. This voltage source is not a good electrical pump because, when the spheres are connected by a conductor, the potentials equalize in a single brief surge of moving charges—like discharging a Van de Graaff generator. It is not practical. Generators or chemical batteries, on the other hand, are sources of energy in electric circuits and are capable of maintaining a steady flow.

Batteries and electric generators do work to pull negative charges away from positive ones. In chemical batteries, this work is usually, but not always, done by the chemical disintegration of zinc or lead in acid, and the energy stored in the chemical bonds is converted to electric potential energy.[2] Generators, such as alternators in automobiles, separate charge by electromagnetic induction, a process we will describe in Chapter 25. The work done by whatever means in separating the opposite charges is available at the terminals of the battery or generator. These different values of energy per charge create a difference in potential (voltage). This voltage provides the "electrical pressure" to move electrons through a circuit joined to these terminals.

The unit of electric potential difference (voltage) is the *volt*.[3] A common automobile battery will provide an electrical pressure of 12 volts to a circuit connected across its terminals. Then 12 joules of energy are supplied to each coulomb of charge that is made to flow in the circuit.

Store your batteries in a cool, dry place. If you put them in a refrigerator they'll last a bit longer.
Insights

There is often some confusion about charge flowing *through* a circuit and voltage placed, or impressed, *across* a circuit. We can distinguish between these ideas by considering a long pipe filled with water. Water will flow *through* the pipe if there is a difference in pressure *across* (or between) its ends. Water flows from the high-pressure end to the low-pressure end. Only the water flows, not the pressure. Similarly, electric charge flows because of the differences in electrical pressure (voltage). You say that charges flow *through* a circuit because of an applied voltage *across* the circuit. You don't say that voltage flows through a circuit. Voltage doesn't go anywhere, for it is the charges that move. Voltage produces current (if there is a complete circuit).

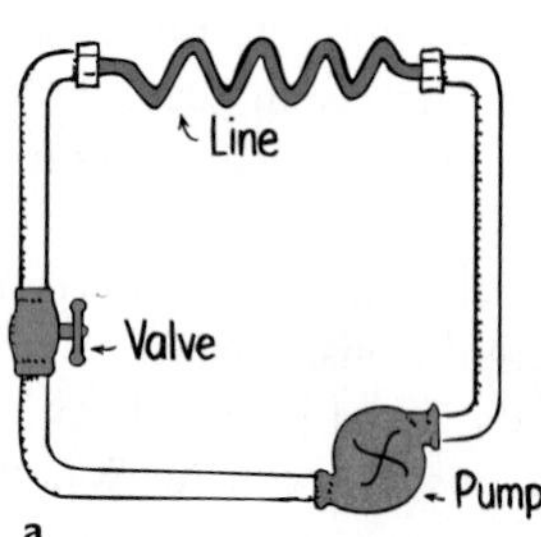

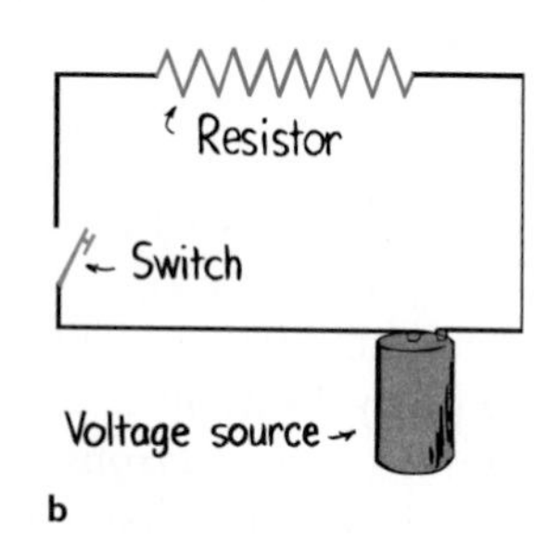

FIGURE 23.4
(a) In a hydraulic circuit, a narrow pipe (green) offers resistance to water flow.
(b) In an electric circuit, a lamp or other device (shown by the zigzag symbol for resistance) offers resistance to electron flow.

[2]The life of a battery depends on the length of time it shares its chemical energy with circuit devices. Like water pipes that become clogged with overuse and time, batteries build up resistance that further shortens their useful lives. An explanation of how batteries operate can be found in almost any chemistry textbook.

[3]The terminology of this area of physics can be confusing, so here's a quick summary of terms: *Electric potential* and *potential* mean the same thing—electrical potential energy per unit charge. Its units are volts. On the other hand, *potential difference* is the same thing as *voltage*—the difference in electrical potential between two points in a conducting path. Units of voltage are also volts.

Electrical Resistance

FIGURE 23.5
More water flows through a thick hose than through a thin one connected to a city's water system (same water pressure). Likewise for electric current in thick and thin wires connected across the same potential difference.

We know that a battery or generator of some kind is the prime mover and source of voltage in an electric circuit. How much current exists depends not only on the voltage but also on the **electrical resistance** the conductor offers to the flow of charge. This is similar to the rate of water flow in a pipe, which depends not only on the pressure difference between the ends of the pipe but also on the resistance offered by the pipe itself. A short pipe offers less resistance to water flow than a long pipe; the wider the pipe, the less the resistance. Likewise for the resistance of wires that carry current. The resistance of a wire depends both on the thickness and length of the wire and on its particular conductivity. Thick wires have less resistance than thin wires. Longer wires have more resistance than short wires. Copper wire has less resistance than steel wire of the same size. Electrical resistance also depends on temperature. The greater the jostling about of atoms within the conductor, the greater resistance the conductor offers to the flow of charge. For most conductors, increased temperature means increased resistance.[4] The resistance of some materials reaches zero at very low temperatures. These are the superconductors discussed briefly in the previous chapter.

Electrical resistance is measured in units called *ohms*. The Greek letter *omega*, Ω, is commonly used as the symbol for the ohm. This unit is named after Georg Simon Ohm, a German physicist who, in 1826, discovered a simple and very important relationship among voltage, current, and resistance.

Ohm's Law

The unit of electrical resistance is the ohm, Ω, like the song of old: "Ω, Ω on the Range."

Insights

The relationship among voltage, current, and resistance is summarized by a statement called **Ohm's law.** Ohm discovered that the current in a circuit is directly proportional to the voltage established across the circuit and is inversely proportional to the resistance of the circuit. In short,

$$\text{Current} = \frac{\text{voltage}}{\text{resistance}}$$

Or, in units form,

$$\text{Amperes} = \frac{\text{volts}}{\text{ohms}}$$

So, for a given circuit of constant resistance, current and voltage are proportional to each other.[5] This means that we'll get twice the current for twice the voltage. The greater the voltage, the greater the current. But, if the resistance of a circuit is doubled, the current will be half what it would be otherwise. The greater the resistance, the smaller the current. Ohm's law makes good sense.

[4]Carbon is an interesting exception. As temperature increases, more carbon atoms shake loose an electron. This increases the electric current. So the resistance of carbon lowers with increasing temperature. This and (primarily) its high melting point are why carbon is used in arc lamps.

[5]Many texts use V for voltage, I for current, and R for resistance, and express Ohm's law as $V = IR$. It then follows that $I = V/R$, or $R = V/I$, so, if any two variables are known, the third can be found. Units are abbreviated V for volts, A for amperes, and Ω for ohms.

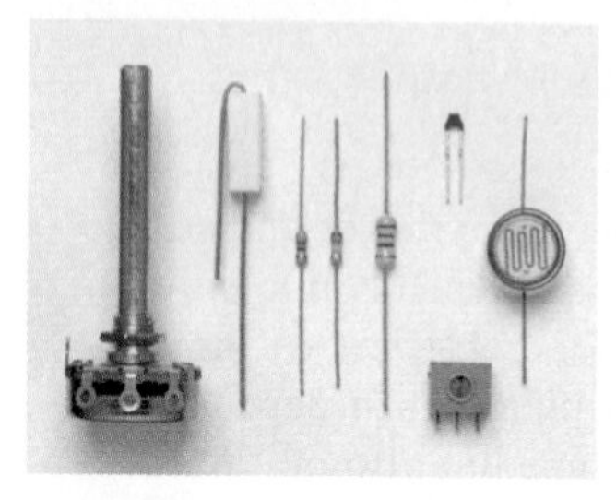

FIGURE 23.6
Resistors. The symbol of resistance in an electric circuit is —/\/\/\—.

Ohm's law tells us that a potential difference of 1 volt established across a circuit that has a resistance of 1 ohm will produce a current of 1 ampere. If a voltage of 12 volts is impressed across the same circuit, the current will be 12 amperes. The resistance of a typical lamp cord is much less than 1 ohm, while a typical lightbulb has a resistance of more than 100 ohms. An iron or an electric toaster has a resistance of 15–20 ohms. Remember that, for a given potential difference, less resistance means more current. In the interior of such electrical devices as computers and television receivers, current is regulated by circuit elements called *resistors,* whose resistance may be a few ohms or millions of ohms.

CHECK YOURSELF

1. How much current is drawn by a lamp that has a resistance of 60 Ω when a voltage of 12 V is impressed across it?
2. What is the resistance of an electric frying pan that draws 12 A when connected to a 120-V circuit?

Ohm's Law and Electric Shock

Current is a flow of charge, pressured into motion by voltage and hampered by resistance.

Handling Electric Wires

What causes electric shock in the human body—current or voltage? The damaging effects of shock are the result of current passing through the body. From Ohm's law, we can see that this current depends both on the voltage that is applied and on the electrical resistance of the human body. The resistance of one's body, which depends on its condition, ranges from about 100 ohms if it is soaked with salt water to about 500,000 ohms if the skin is very dry. If we touch the two electrodes of a battery with dry fingers, completing the circuit from one hand to the other, we can expect to offer a resistance of about 100,000 ohms. We usually cannot feel the current produced by 12 volts, and 24 volts just barely tingles. If our skin is moist, 24 volts can be quite uncomfortable. Table 23.1 describes the effects of different amounts of current on the human body.

TABLE 23.1
Effect of Electric Currents on the Body

Current (A)	*Effect*
0.001	Can be felt
0.005	Is painful
0.010	Causes involuntary muscle contractions (spasms)
0.015	Causes loss of muscle control
0.070	If through the heart, causes serious disruption; probably fatal if current lasts for more than 1 s

CHECK YOUR ANSWERS

1. 1/5 A. This is calculated from Ohm's law: 12 V/60 Ω = 0.2 A.
2. 10 Ω. Rearrange Ohm's law to read

$$\text{Resistance} = \text{voltage}/\text{current} = 120 \text{ V}/12 \text{ A} = 10\ \Omega.$$

CHECK YOURSELF

1. At a resistance of 100,000 Ω, what will be the current in your body if you touch the terminals of a 12-volt battery?
2. If your skin is very moist—so that your resistance is only 1000 Ω—and you touch the terminals of a 12-volt battery, how much current do you receive?

Many people are killed each year by current from common 120-volt electric circuits. If you touch a faulty 120-volt light fixture with your hand while you are standing on the ground, there is a 120-volt "electrical pressure" between your hand and the ground, so the current would probably not be enough to do serious harm. But, if you are standing barefoot in a wet bathtub connected through its plumbing to the ground, the resistance between you and the ground is very small. Your overall resistance is so low that the 120-volt potential difference may produce a harmful current in your body. Handling electrical devices while taking a bath is a definite no-no.

Drops of water that collect around the on–off switches of such devices as a hair dryer can conduct current to the user. Although distilled water is a good insulator, the ions in ordinary tap water greatly reduce the electrical resistance. These ions are contributed by dissolved materials, especially salts. There is usually a layer of salt left from perspiration on your skin, which, when your skin is wet, lowers your skin resistance to a few hundred ohms or less, depending on the distance over which the voltage acts.

FIGURE 23.7
The bird can stand harmlessly on one wire of high potential, but it had better not reach over and grab a neighboring wire! Why not?

Birds and High-Voltage Wires

An electric shock requires a *difference* in electric potential—a voltage difference—between one part of your body and another part. Most of the current will pass along the path of least electrical resistance connecting these two points. Suppose you fell from a bridge and managed to grab onto a high-voltage power line, halting your fall. So long as you touch nothing else of different potential, you will receive no shock at all. Even if the wire is thousands of volts above ground potential, and even if you hang by it with two hands, no appreciable charge will flow from one hand to the other. This is because there is no appreciable difference in electric potential between your hands. If, however, you reach over with one hand and grab onto a wire of different potential . . . zap! We have all seen birds perched on high-voltage wires. Every part of their bodies is at the same high potential as the wire, so they feel no ill effects.

Most electric plugs and sockets today are wired with three, instead of two, connectors. The principal two flat prongs on an electrical plug are for the

CHECK YOUR ANSWERS

1. $\dfrac{12\text{ V}}{100{,}000\ \Omega} = 0.00012\text{ A.}$
2. $\dfrac{12\text{ V}}{1000\ \Omega} = 0.012\text{ A.}$ Ouch!

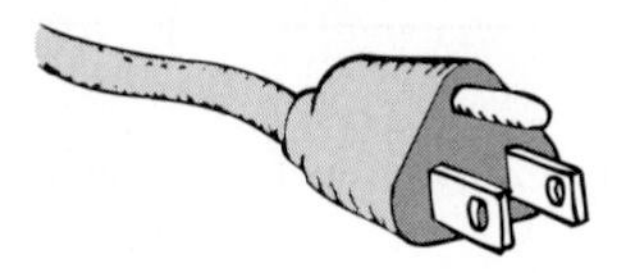

FIGURE 23.8
The round prong connects the body of the appliance directly to ground (Earth). Any charge that builds up on an appliance is therefore conducted to the ground, thereby preventing accidental shock.

current-carrying double wire, one part of which is "live" (energized) and the other neutral, while the round prong connects to a wire in the electrical system that is grounded—connected directly to the ground (Figure 23.8). The electric appliance at the other end of the plug is, therefore, connected to all three wires. If the live wire of the plugged appliance accidentally comes in contact with the metal surface of the appliance and you touch the appliance, you could receive a dangerous shock. This won't occur when the appliance casing is grounded via the ground wire, which assures that the appliance casing is always at zero ground potential.

Electric shock can overheat tissues in the body and disrupt normal nerve functions. It can upset the rhythmic electrical patterns that maintain proper heartbeat, and it can upset the nerve center that controls breathing. In rescuing shock victims, the first thing to do is to locate and turn off the power source. Then do CPR until help arrives. For heart-attack victims, on the other hand, a properly administered electric shock can sometimes be beneficial in getting the heartbeat started again.

CHECK YOURSELF

What causes electric shock—current or voltage?

Direct Current and Alternating Current

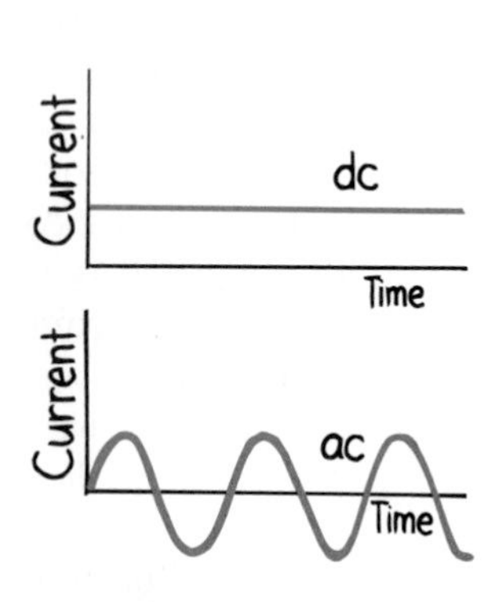

FIGURE 23.9
Time graphs of dc and ac.

Electric current may be dc or ac. By *dc,* we mean **direct current,** which refers to the flowing of charges in *one direction.* A battery produces direct current in a circuit because each terminal of a battery always has the same sign: the positive terminal is always positive, and the negative terminal is always negative. Electrons move from the repelling negative terminal toward the attracting positive terminal, always moving through the circuit in the same direction. Even if the current occurs in unsteady pulses, so long as electrons move in one direction only, it is dc.

Alternating current (ac) acts as the name implies. Electrons in the circuit are moved first in one direction and then in the opposite direction, alternating to and fro about relatively fixed positions. This is accomplished by alternating the polarity of voltage at the generator or other voltage source. Nearly all commercial ac circuits in North America involve voltages and currents that alternate back and forth at a frequency of 60 cycles per second. This is 60-hertz current.

Alternating Current

CHECK YOUR ANSWER

Electric shock *occurs* when current is produced in the body, which is *caused* by an impressed voltage. So the initial *cause* is the voltage, but it is the current that does the damage.

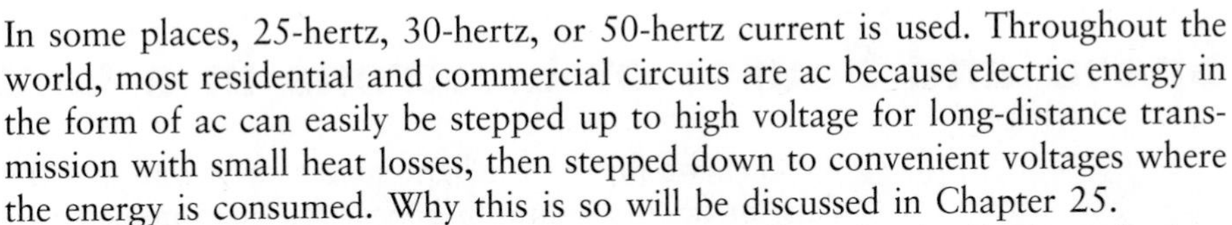

In some places, 25-hertz, 30-hertz, or 50-hertz current is used. Throughout the world, most residential and commercial circuits are ac because electric energy in the form of ac can easily be stepped up to high voltage for long-distance transmission with small heat losses, then stepped down to convenient voltages where the energy is consumed. Why this is so will be discussed in Chapter 25.

The voltage of ac in North America is normally 120 volts. In the early days of electricity, higher voltages burned out the filaments of electric lightbulbs. Tradition has it that 110 volts was first adopted as the standard because it made bulbs of the day glow as brightly as a gas lamp. So the hundreds of power plants built in the U.S. prior to 1900 produced electricity at 110 volts (or 115 or 120 volts). By the time electricity became popular in Europe, engineers had figured out how to manufacture lightbulbs that would not burn out so fast at higher voltages. Power transmission is more efficient at higher voltages, so Europe adopted 220 volts as their standard. The U.S. continued with 110 volts (today officially 120 volts) because so much 110-volt equipment was already installed. (Certain appliances, such as electric stoves and clothes dryers, use higher voltage.)

The primary use of electric current, whether dc or ac, is to transfer energy quietly, flexibly, and conveniently from one place to another.

In ac circuits, 120 volts is the "root-mean-square" average of the voltage. The actual voltage in a 120-volt ac circuit varies between +170 volts and −170 volts, delivering the same power to an iron or a toaster as a 120-volt dc circuit.

Insights

Converting ac to dc

Household current is ac. The current in a battery-operated device, such as a pocket calculator, is dc. You can operate these devices on ac instead of batteries with an ac–dc converter. In addition to a transformer to lower the voltage (Chapter 25), the converter uses a *diode,* a tiny electronic device that acts as a one-way valve to allow electron flow in one direction only (Figure 23.10). Since alternating current changes its direction each half cycle, current passes through a diode only half of each period. The output is a rough dc, and it is off half the time. To maintain continuous current while smoothing the bumps, a capacitor is used (Figure 23.12).

Recall, from the previous chapter, that a capacitor acts as a storage reservoir for charge. Just as it takes time to raise the water level in a reservoir when water is added, it takes time to add or remove electrons from the plates of a capacitor. A capacitor, therefore, produces a retarding effect on changes in current flow. It opposes changes in voltage and smooths the pulsed output.

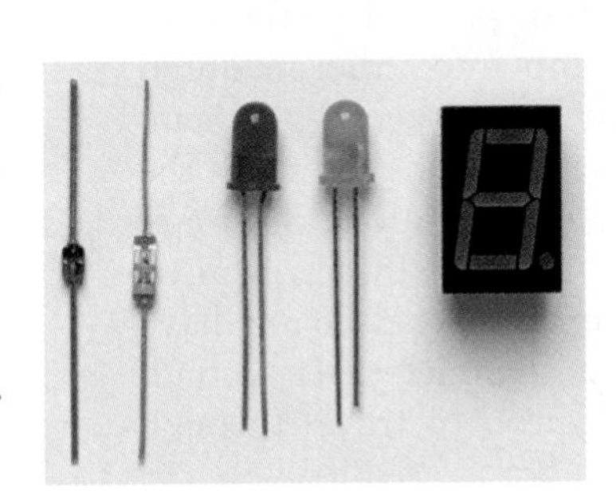

FIGURE 23.10
Diodes. As the symbol —▶|— suggests, current flows in the direction of the arrow but not in the reverse direction.

FIGURE 23.11
Water input to the reservoir may be in repeated spurts or pulses, but the output is a fairly smooth stream. Likewise with a capacitor.

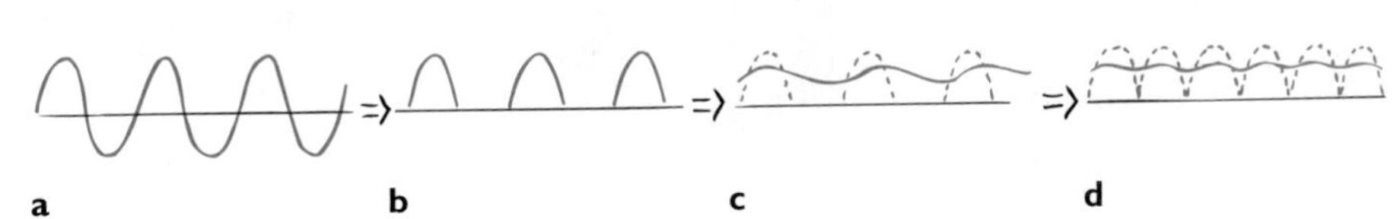

FIGURE 23.12
(a) When input to a diode is ac, (b) output is pulsating dc. (c) Slow charging and discharging of a capacitor provides continuous and smoother current. (d) In practice, a pair of diodes is used, so there are no gaps in current output. The pair of diodes reverses the polarity of alternate half-cycles instead of eliminating them.

Speed and Source of Electrons in a Circuit

When we flip on the light switch on a wall and the circuit is completed, either ac or dc, the lightbulb appears to glow immediately. When we make a noncellular telephone call, the electrical signal carrying our voice travels through connecting wires at seemingly infinite speed. This signal is transmitted through the conductors at nearly the speed of light.

It is *not* the electrons that move at this speed.[6] Although electrons inside metal at room temperature have an average speed of a few million kilometers per hour, they make up no current because they are moving in all possible directions. There is no net flow in any preferred direction. But, when a battery or generator is connected, an electric field is established inside the conductor. The electrons continue their random motions while simultaneously being nudged by this field. It is the electric field that can travel through a circuit at nearly the speed of light. The conducting wire acts as a guide or "pipe" for electric field lines (Figure 23.13). In the space outside the wire, the electric field has a pattern determined by the location of electric charges, including charges in the wire. Inside the wire, the electric field is directed along the wire.

After failing more than 6000 times before perfecting the first electric lightbulb, Edison stated that his trials were not failures, because he successfully discovered 6000 ways that don't work.

Insights

If the voltage source is dc, like the battery shown in Figure 23.13, the electric field lines are maintained in one direction in the conductor. Conduction electrons are accelerated by the field in a direction parallel to the field lines. Before they gain appreciable speed, they "bump into" the anchored metallic ions in their paths and transfer some of their kinetic energy to them. This is why current-carrying wires become hot. These collisions interrupt the motion of the electrons, so the speed at which they migrate along a wire is extremely low. This net flow of electrons is the *drift velocity*. In a typical dc circuit—the electrical system of an automobile, for example—electrons have a drift velocity that averages about a hundredth of a centimeter per second. At this rate, it would take about 3 hours for an electron to travel through

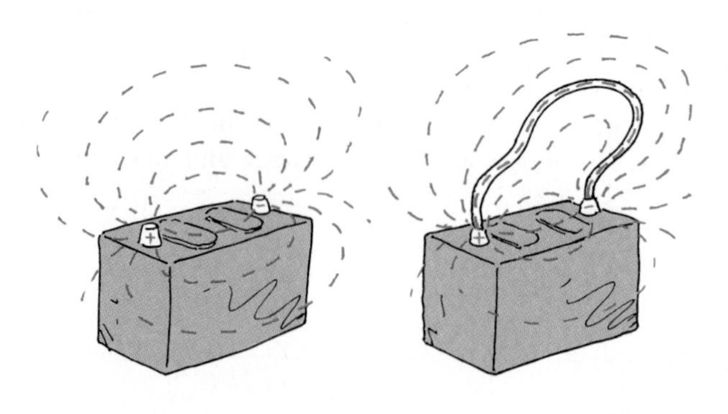

FIGURE 23.13
The electric field lines between the terminals of a battery are directed through a conductor that joins the terminals. A thick metal wire is shown here, but the path from one terminal to the other is usually an electric circuit. (You won't be shocked if you touch this connecting wire, but you might be burned because the wire would likely be very hot!)

[6]Much effort and expense are expended in building particle accelerators that accelerate electrons and protons to speeds near that of light. If electrons in a common circuit traveled that fast, one would only have to bend a wire at a sharp angle and electrons traveling through the wire would possess so much momentum that they would fail to make the turn and would fly off, providing a beam comparable to that produced by the accelerators!

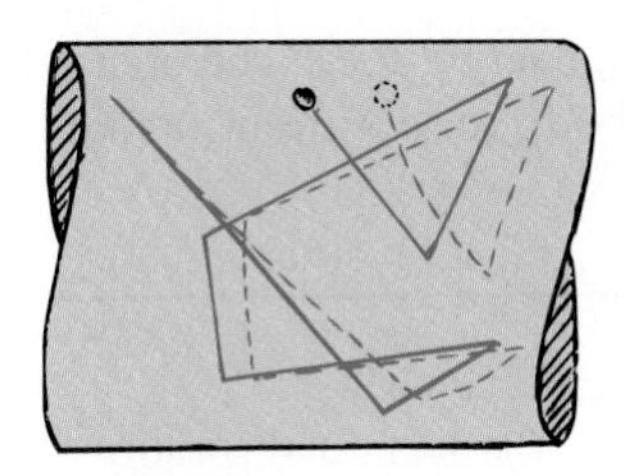

FIGURE 23.14
The solid lines suggest the random path of an electron jostling about in an atomic lattice at an average speed of about 1/200 the speed of light. The dashed lines suggest an exaggerated and idealized view of how this path would be altered if an electric field were applied. The electron would drift toward the right with a *drift velocity* much slower than a snail's pace.

1 meter of wire! Large currents are possible because of large numbers of flowing electrons. So although an electric signal travels at nearly the speed of light in a wire, the electrons that move in response to this signal actually travel slower than a snail's pace.

In an ac circuit, the conduction electrons don't progress along the wire at all. They oscillate rhythmically to and fro about relatively fixed positions. When you talk to your friend on a conventional telephone, it is the *pattern* of oscillating motion that is carried across town at nearly the speed of light. The electrons already in the wires vibrate to the rhythm of the traveling pattern.

A common misconception regarding electrical currents is that the current is propagated through the conducting wires by electrons bumping into one another—that an electrical pulse is transmitted in a manner similar to the way the pulse of a tipped domino is transferred along a row of closely spaced standing dominoes. This simply isn't true. The domino idea is a good model for the transmission of sound, but not for the transmission of electric energy. Electrons that are free to move in a conductor are accelerated by the electric field impressed upon them, but not because they bump into one another. True, they do bump into one another and other atoms, but this slows them down and offers resistance to their motion. Electrons throughout the entire closed path of a circuit all react simultaneously to the electric field.

Another common misconception about electricity is the source of electrons. In a hardware store, you can buy a water hose that is empty of water. But you can't buy a piece of wire, an "electron pipe," that is empty of electrons. The source of electrons in a circuit is the conducting circuit material itself. Some people think that the electrical outlets in the walls of their homes are a source of electrons. They incorrectly assume that electrons flow from the power utility through the power lines and into the wall outlets of their homes. This assumption is false. The outlets in homes are ac. Electrons make no net migration through a wire in an ac circuit.

When you plug a lamp into an outlet, *energy* flows from the outlet into the lamp, not electrons. Energy is carried by the pulsating electric field and causes vibratory motion of the electrons that already exist in the lamp filament. If a voltage of 120 volts is impressed on a lamp, then an average of 120 joules of energy is dissipated by each coulomb of charge that is made to vibrate. Most of this electrical energy appears as heat, while some transforms to light. Power utilities do not sell electrons. They sell *energy*. You supply the electrons.

So, when you are jolted by an electric shock, the electrons making up the current in your body originate in your body. Electrons do not emerge from the

Thomas Edison did much more than invent a functioning incandescent bulb in 1879. He solved the problems of building the dynamos, cable systems, and connections to light New York City. He made the phone work properly, and he gave us recorded music and the movies. He also invented a method of inventing: His New Jersey lab was the forerunner of the modern industrial research lab.

Insights

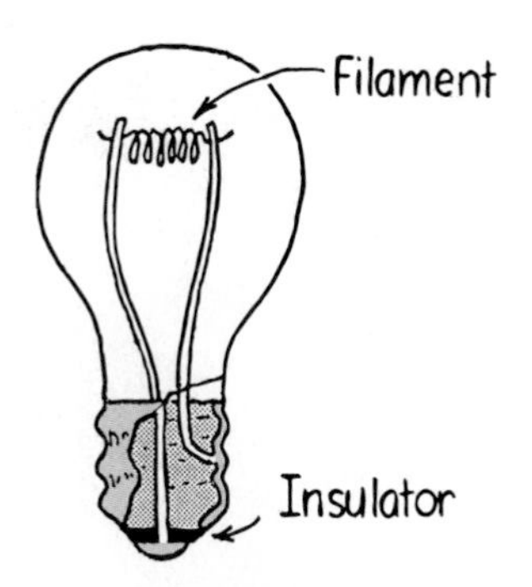

FIGURE 23.15
The conduction electrons that surge to and fro in the filament of the lamp do not originate in the voltage source. They are in the filament to begin with. The voltage source simply provides them with surges of energy.

Why is it correct to say that the energy from a car battery ultimately comes from fuel in the gas tank?

Insights

wire and pass through your body and into the ground; energy does. The energy simply causes free electrons already existing in your body to vibrate in unison. Small vibrations tingle; large vibrations can be fatal.

CHECK YOURSELF

1. Consider members of a marching band standing at rest. You can set them into motion in two ways: (1) Give the last person in line a shove that cascades to the first person in line. (2) Issue the command "forward, march." Which is analogous to the way electrons in a circuit move when a switch is closed, and which is analogous to the way sound travels?
2. In the lamp shown in Figure 23.15, why is light emitted by the filament and not by the connecting wires?

Electric Power

FIGURE 23.16
The power and voltage on the lightbulb read "100 W 120 V." How many amperes will flow through the bulb?

Unless it is in a superconductor, a charge moving in a circuit expends energy. This may result in heating the circuit or in turning a motor. The rate at which electric energy is converted into another form, such as mechanical energy, heat, or light, is called **electric power.** Electric power is equal to the product of current and voltage.[7]

$$\text{Power} = \text{current} \times \text{voltage}$$

If the voltage is expressed in volts and the current in amperes, then the power is expressed in watts. So, in units form,

$$\text{Watts} = \text{amperes} \times \text{volts}$$

If a lamp rated at 120 watts operates on a 120-volt line, you can see that it will draw a current of 1 ampere (120 watts = 1 ampere × 120 volts). A 60-watt lamp draws $\frac{1}{2}$ ampere on a 120-volt line. This relationship becomes a practical matter when you wish to know the cost of electrical energy, which is usually a few cents per kilowatt-hour, depending on locality. A kilowatt is 1000 watts, and a kilowatt-hour represents the amount of energy consumed in 1 hour at

CHECK YOUR ANSWERS

1. Issuing the command "forward, march" is analogous to the way electrons move when they sense the electric field that energizes the circuit when the switch is closed. One marcher lurching against the other is analogous to the way sound travels.
2. Energy is delivered to and dissipated at locations of circuit resistance. Nearly all the resistance in the lamp is in its filament. Hence, the filament glows with visible light.

[7]Recall, from Chapter 7, that Power = work/time; 1 Watt = 1 J/s. Note that the units for mechanical power and electrical power check (work and energy are both measured in joules):

$$\text{Power} = \frac{\cancel{\text{charge}}}{\text{time}} \times \frac{\text{energy}}{\cancel{\text{charge}}} = \frac{\text{energy}}{\text{time}}$$

FUEL CELLS

A battery is an energy-storage device. Once its stored chemical energy is converted to electrical energy, its energy is depleted. Then it must be discarded (if it is a disposable battery) or recharged with an opposite flow of electricity.

A *fuel cell,* on the other hand, converts the chemical energy of a fuel to electrical energy continuously and indefinitely, as long as fuel is supplied to it. In one version, hydrogen fuel and oxygen from the air react chemically to produce electrons and ions—and water. The ions flow internally within the cell in one direction; the electrons flow externally through an attached circuit in the other direction. Because this reaction directly converts chemical energy to electricity, it is more efficient than if the fuel were burned to produce heat, which, in turn, produces steam to turn turbines to generate electricity. The only "waste product" of such a fuel cell is pure water, suitable for drinking!

The space shuttle uses hydrogen fuel cells to meet its electrical needs. (Its hydrogen and oxygen are both brought on board in pressurized containers.) The cells also produce more than 100 gallons of drinking water for the astronauts during a typical week-long mission. Back on Earth, researchers are developing fuel cells for buses and automobiles. Experimental fuel-cell buses are already operating in several cities, such as Vancouver, British Columbia, and Chicago, Illinois. In the future, commercial buildings as well as individual homes may be outfitted with fuel cells as an alternative to receiving electricity from regional power stations.

So why aren't fuel cells widespread today? Currently, they are more expensive than other energy devices, such as gasoline or diesel engines—in fact, nearly 100 times more expensive per unit of energy produced. And there is the question of the availability of the choice fuel—hydrogen. Although hydrogen is the most plentiful element in the universe, and is plentiful in our immediate surroundings, it is locked away in water and hydrocarbon molecules. It isn't available in a free state. Energy is required to separate it from molecules in which it is tightly bonded. The energy needed to make hydrogen is presently supplied by conventional energy sources.

Hydrogen is, in effect, an energy-storage medium. Like electricity, it is created in one place and used in another. Fuel cells will be attractive in the future when their cost comes down and when the hydrogen needed to fuel them is generated by alternative energy sources, such as wind power.

Watch for a wide variety of contenders for producing hydrogen fuel in the projected hydrogen economy.

Insights

the rate of 1 kilowatt.[8] Therefore, in a locality in which electric energy costs 5 cents per kilowatt-hour, a 100-watt electric lightbulb can operate for 10 hours at a cost of 5 cents, or a half cent for each hour. A toaster or iron, which draws much more current and therefore much more energy, costs about ten times as much to operate.

CHECK YOURSELF

1. If a 120-V line to a socket is limited to 15 A by a safety fuse, will it operate a 1200-W hair dryer?
2. At 10¢/kWh, what does it cost to operate the 1200-W hair dryer for 1 h?

CHECK YOUR ANSWERS

1. Yes. From the expression watts = amperes × volts, we see that current = 1200 W/120 V = 10 A, so the hair dryer will operate when connected to the circuit. But two hair dryers on the same circuit will blow the fuse.
2. 12¢ (1200 W = 1.2 kW; 1.2 kW × 1h × 10¢/kWh = 12¢).

[8]Since Power = energy/time, simple rearrangement gives Energy = power × time; thus, energy can be expressed in the unit *kilowatt-hours* (kWh).

Electric Circuits

Electric Circuits

Any path along which electrons can flow is a *circuit.* For a continuous flow of electrons, there must be a complete circuit with no gaps. A gap is usually provided by an electric switch that can be opened or closed to either cut off or allow energy flow. Most circuits have more than one device that receives electric energy. These devices are commonly connected in a circuit in one of two ways, *series* or *parallel.* When connected in series, they form a single pathway for electron flow between the terminals of the battery, generator, or wall socket (which is simply an extension of these terminals). When connected in parallel, they form branches, each of which is a separate path for the flow of electrons. Both series and parallel connections have their own distinctive characteristics. We shall briefly treat circuits using these two types of connections.

Series Circuits

A simple **series circuit** is shown in Figure 23.17. Three lamps are connected in series with a battery. The same current exists almost immediately in all three lamps when the switch is closed. The greater the current in a lamp, the brighter it glows. Electrons do not "pile up" in any lamp but flow *through* each lamp—simultaneously. Some electrons move away from the negative terminal of the battery, some move toward the positive terminal, and some move through the filament of each lamp. Eventually, the electrons may move all the way around the circuit (the same amount of current passes through the battery). This is the only path of the electrons through the circuit. A break anywhere in the path results in an open circuit, and the flow of electrons ceases. Burning out one of the lamp filaments or simply opening the switch could cause such a break.

The circuit shown in Figure 23.17 illustrates the following important characteristics of series connections:

1. Electric current has but a single pathway through the circuit. This means that the current passing through the resistance of each electrical device along the pathway is the same.
2. This current is resisted by the resistance of the first device, the resistance of the second, and that of the third also, so the total resistance to current in the circuit is the sum of the individual resistances along the circuit path.

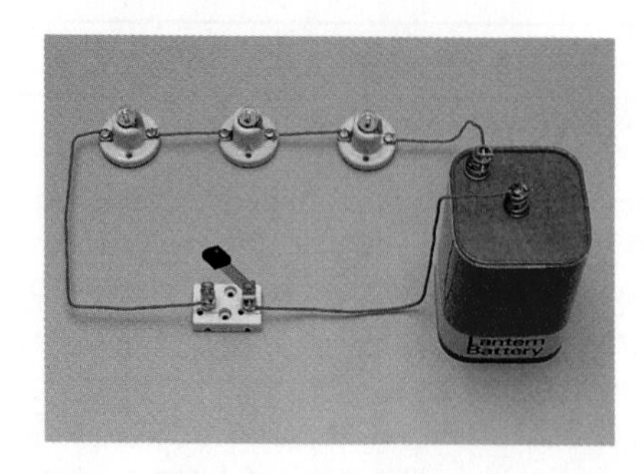

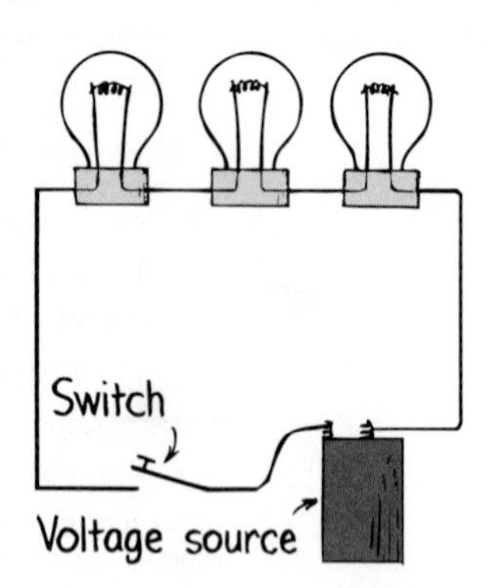

FIGURE 23.17
Interactive Figure
A simple series circuit. The 6-V battery provides 2 V across each lamp.

3. The current in the circuit is numerically equal to the voltage supplied by the source divided by the total resistance of the circuit. This is in accord with Ohm's law.
4. The total voltage impressed across a series circuit divides among the individual electrical devices in the circuit so that the sum of the "voltage drops" across the resistance of each individual device is equal to the total voltage supplied by the source. This characteristic follows from the fact that the amount of energy supplied to the total current is equal to the sum of energies supplied to each device.
5. The voltage drop across each device is proportional to its resistance—Ohm's law applies separately to each device. This follows from the fact that more energy is dissipated when a current passes through a large resistance than when the same current passes through a small resistance.

CHECK YOURSELF

1. What happens to current in other lamps if one lamp in a series circuit burns out?
2. What happens to the brightness of light from each lamp in a series circuit when more lamps are added to the circuit?

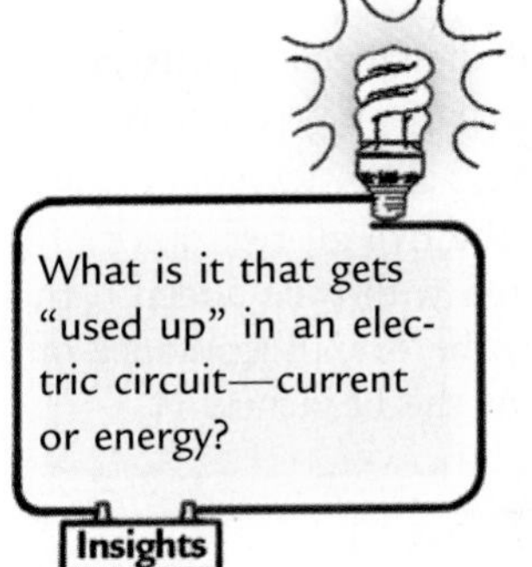

What is it that gets "used up" in an electric circuit—current or energy?
Insights

It is easy to see the main disadvantage of a series circuit: If one device fails, current in the whole circuit ceases. Some cheap Christmas tree lights are connected in series. When one bulb burns out, it's fun and games (or frustration) trying to locate which bulb to replace.

Most circuits are wired so that it is possible to operate several electrical devices, each independently of the other. In your home, for example, a lamp can be turned on or off without affecting the operation of other lamps or electrical devices. This is because these devices are connected not in series but in parallel with one another.

Parallel Circuits

A simple **parallel circuit** is shown in Figure 23.18. Three lamps are connected to the same two points A and B. Electrical devices connected to the same two points of an electrical circuit are said to be *connected in parallel.* The pathway for current from one terminal of the battery to the other is completed if only *one* lamp is lit. In this illustration, the circuit branches into three separate pathways

CHECK YOUR ANSWERS

1. If one of the lamp filaments burns out, the path connecting the terminals of the voltage source will break and current will cease. All lamps will go out.
2. The addition of more lamps in a series circuit results in a greater circuit resistance. This decreases the current in the circuit and therefore in each lamp, which causes dimming of the lamps. Since all voltages have to add up to the same total voltage, the voltage drop across each lamp will be less.

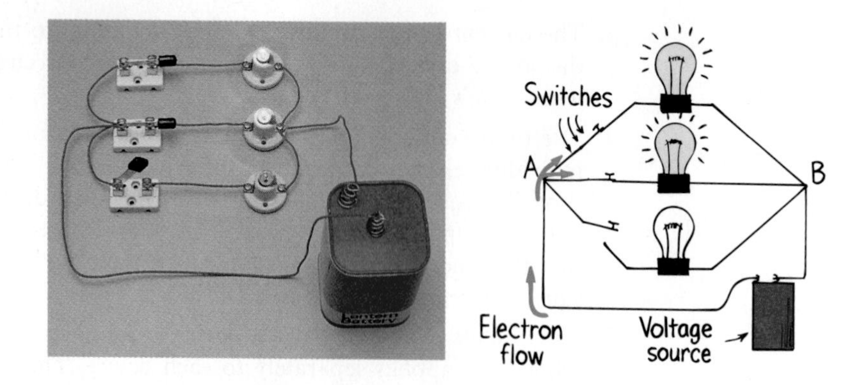

FIGURE 23.18
Interactive Figure
A simple parallel circuit. A 6-V battery provides 6 V across each lamp.

from A to B. A break in any one path does not interrupt the flow of charge in the other paths. Each device operates independently of the other devices.

The circuit shown in Figure 23.18 illustrates the following major characteristics of parallel connections:

1. Each device connects the same two points A and B of the circuit. The voltage is therefore the same across each device.
2. The total current in the circuit divides among the parallel branches. Since the voltage across each branch is the same, the amount of current in each branch is inversely proportional to the resistance of the branch—Ohm's law applies separately to each branch.
3. The total current in the circuit equals the sum of the currents in its parallel branches. This sum equals the current in the battery or other voltage source.
4. As the number of parallel branches is increased, the overall resistance of the circuit is *decreased.* Overall resistance is lowered with each added path between any two points of the circuit. This means the overall resistance of the circuit is less than the resistance of any one of the branches.

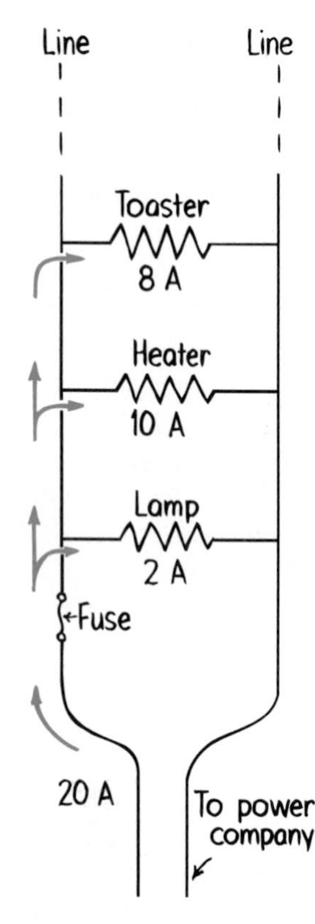

FIGURE 23.19
Circuit diagram for appliances connected to a household circuit.

CHECK YOURSELF

1. What happens to the current in other lamps if one of the lamps in a parallel circuit burns out?
2. What happens to the brightness of light from each lamp in a parallel circuit when more lamps are added in parallel?

Parallel Circuits and Overloading

Electricity is usually fed into a home by way of two wires called *lines.* These lines, which are very low in resistance, branch into parallel circuits connecting ceiling lights and wall outlets in each room. Lights and wall outlets are connected in parallel, so all are impressed with the same voltage, usually about 110–120 volts. As more devices are plugged in and turned on, more pathways for current result in lowering of the combined resistance of each circuit. Therefore, a greater amount of current occurs in the circuits. The sum of these currents equals the line current, which may be more than is safe. The circuit is then said to be *overloaded.*

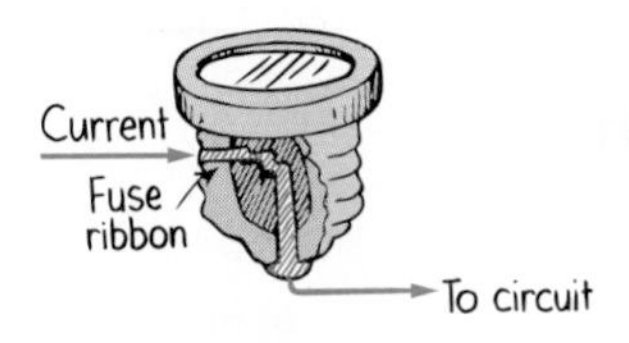

FIGURE 23.20
A safety fuse.

We can see how overloading occurs by considering the circuit in Figure 23.19. The supply line is connected in parallel to an electric toaster that draws 8 amperes, to an electric heater that draws 10 amperes, and to an electric lamp that draws 2 amperes. When only the toaster is operating and drawing 8 amperes, the total line current is 8 amperes. When the heater is also operating, the total line current increases to 18 amperes (8 amperes to the toaster and 10 amperes to the heater). If you turn on the lamp, the line current increases to 20 amperes. Connecting any more devices increases the current still more. Connecting too many devices into the same circuit results in overheating that may cause a fire.

Safety Fuses

FIGURE 23.21
Electrician Dave Hewitt with a safety fuse and a circuit breaker. He favors the old fuses, which he's found more reliable.

To prevent overloading in circuits, fuses are connected in series along the supply line. In this way, the entire line current must pass through the fuse. The fuse shown in Figure 23.20 is constructed with a wire ribbon that will heat up and melt at a given current. If the fuse is rated at 20 amperes, it will pass 20 amperes, but no more. A current above 20 amperes will melt the fuse, which "blows out" and breaks the circuit. Before a blown fuse is replaced, the cause of overloading should be determined and remedied. Often, insulation that separates the wires in a circuit erodes and allows the wires to touch. This greatly reduces the resistance in the circuit, effectively shortening the circuit path, and is called a *short circuit*.

In modern buildings, fuses have been largely replaced by circuit breakers, which use magnets or bimetallic strips to open a switch when the current is too great. Utility companies use circuit breakers to protect their lines all the way back to the generators.

CHECK YOUR ANSWERS

1. If one lamp burns out, the other lamps will be unaffected. This is because current in each branch, according to Ohm's law, is equal to voltage/resistance, and since neither voltage nor resistance is affected in the other branches, the current in those branches is unaffected. The total current in the overall circuit (the current through the battery), however, is decreased by an amount equal to the current drawn by the lamp in question before it burned out. But the current in any other single branch is unchanged.
2. The brightness of each lamp is unchanged as other lamps are introduced (or removed). Only the total resistance and total current in the total circuit changes, which is to say that the current in the battery changes. (There is resistance in a battery also, which we assume is negligible here.) As lamps are introduced, more paths are available between the battery terminals, which effectively decreases total circuit resistance. This decreased resistance is accompanied by an increased current, the same increase that feeds energy to the lamps as they are introduced. Although changes of resistance and current occur for the circuit as a whole, no changes occur in any individual branch in the circuit.

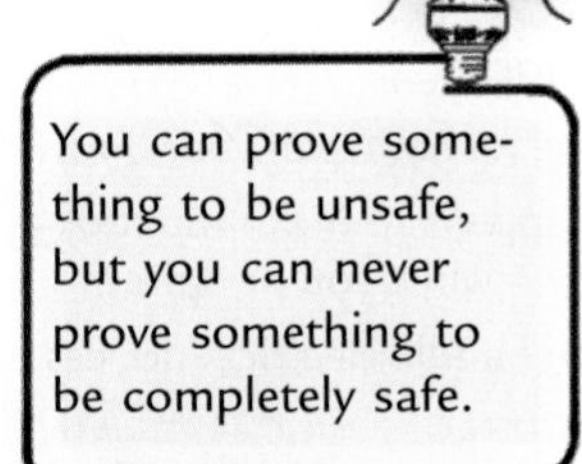

You can prove something to be unsafe, but you can never prove something to be completely safe.

Summary of Terms

Potential difference The difference in electric potential between two points, measured in volts. When two points of different electric potential are connected by a conductor, charge flows so long as a potential difference exists. (Synonymous with *voltage difference.*)

Electric current The flow of electric charge that transports energy from one place to another. Measured in amperes, where 1 A is the flow of 6.25×10^{18} electrons per second, or 1 coulomb per second.

Electrical resistance The property of a material that resists electric current. Measured in ohms (Ω).

Ohm's law The statement that the current in a circuit varies in direct proportion to the potential difference or voltage across the circuit and inversely with the circuit's resistance.

$$\text{Current} = \frac{\text{voltage}}{\text{resistance}}$$

A potential difference of 1 V across a resistance of 1 Ω produces a current of 1 A.

Direct current (dc) Electrically charged particles flowing in one direction only.

Alternating current (ac) Electrically charged particles that repeatedly reverse direction, vibrating about relatively fixed positions. In the United States, the vibrational rate is commonly 60 Hz.

Electric power The rate of energy transfer, or the rate of doing work; the amount of energy per unit time, which electrically can be measured by the product of current and voltage.

$$\text{Power} = \text{current} \times \text{voltage}$$

Electric power is measured in watts (or kilowatts), where $1\text{ A} \times 1\text{ V} = 1\text{ W}$.

Series circuit An electric circuit in which electrical devices are connected along a single wire such that the same electric current exists in all of them.

Parallel circuit An electric circuit in which electrical devices are connected in such a way that the same voltage acts across each one and any single one completes the circuit independently of all the others.

Suggested Reading

Bryson, Bill. *A Short History of Nearly Everything.* New York: Broadway Books, 2003. Much on scientific discoveries, many of which follow three stages: denying what is true; denying the importance of it; crediting the wrong person.

Review Questions

Flow of Charge

1. What condition is necessary for the flow of heat? What analogous condition is necessary for the flow of charge?
2. What condition is necessary for the sustained flow of water in a pipe? What analogous condition is necessary for the sustained flow of charge in a wire?

Electric Current

3. Why are *electrons,* rather than *protons,* the principal charge carriers in metal wires?
4. What exactly is an *ampere?*
5. Why is a current-carrying wire normally not electrically charged?

Voltage Sources

6. Name two kinds of practical "electric pumps."
7. How much energy is supplied to each coulomb of charge that flows through a 12-V battery?
8. Does charge flow *through* a circuit or *into* a circuit? Does voltage flow *through* a circuit, or is voltage established *across* a circuit?

Electrical resistance

9. Will water flow more easily through a wide pipe or a narrow pipe? Will current flow more easily through a thick wire or a thin wire?
10. Does heating a metal wire increase or decrease its electrical resistance?

Ohm's Law

11. If the voltage impressed across a circuit is held constant while the resistance doubles, what change occurs in the current?
12. If the resistance of a circuit remains constant while the voltage across the circuit decreases to half its former value, what change occurs in the current?

Ohm's Law and Electric Shock

13. How does wetness affect the resistance of your body?
14. For a given voltage, what happens to the amount of current that flows in your skin when you perspire?
15. Why is it a very poor idea to handle electrical devices while in the bathtub?
16. What is the function of the round third prong in a modern household electric plug?

Direct Current and Alternating Current

17. Distinguish between dc and ac.

18. Does a battery produce dc or ac? Does the generator at a power station produce dc or ac?

19. What does it mean to say that a certain current is 60 Hz?

Converting ac to dc

20. What property of a diode enables it to convert ac to pulsed dc?

21. A diode converts ac to pulsed dc. What electrical device smooths the pulsed dc to a smoother dc?

Speed and Source of Electrons in a Circuit

22. What is the error in saying that electrons in a common battery-driven circuit travel at about the speed of light?

23. Why does a wire that carries electric current become hot?

24. What is meant by *drift velocity*?

25. A tipped domino sends a pulse along a row of standing dominoes. Is this a good analogy for the way electric current, sound, or both travel?

26. What is the error in saying the source of electrons in a circuit is the battery or generator?

27. When you make your household electric payment at the end of the month, which of the following are you paying for: voltage, current, power, energy?

28. From where do the electrons originate that produce an electric shock when you touch a charged conductor?

Electric Power

29. What is the relationship among electric power, current, and voltage?

30. Which of these is a unit of power and which is a unit of energy—a watt, a kilowatt, a kilowatt-hour?

31. Distinguish between a *kilowatt* and a *kilowatt-hour*.

Electric Circuits

32. What is an *electric circuit*?

Series Circuits

33. In a circuit of two lamps in series, if the current through one lamp is 1 A, what is the current through the other lamp? Defend your answer.

34. If a voltage of 6 V is impressed across the above circuit and the voltage across the first lamp is 2 V, what is the voltage across the second lamp? Defend your answer.

35. What is a main shortcoming of a series circuit?

Parallel Circuits

36. In a circuit of two lamps in parallel, if there is a voltage of 6 V across one lamp, what is the voltage across the other lamp?

37. How does the sum of the currents through the branches of a simple parallel circuit compare with the current that flows through the voltage source?

38. As more lanes are added to toll booths, the resistance to vehicles passing through is reduced. How is this similar to what happens when more branches are added to a parallel circuit?

Parallel Circuits and Overloading

39. Are household circuits normally wired in series or in parallel, and when are they overloaded?

Safety Fuses

40. What is the function of fuses or circuit breakers in a circuit?

Projects

1. An electric cell is made by placing two plates made of different materials that have different affinities for electrons in a conducting solution. The voltage of a cell depends on the materials used and the solutions they are placed in, not on the size of the plates. (A battery is actually a series of cells.) You can make a simple 1.5-V cell by placing a strip of copper and a strip of zinc in a tumbler of salt water.

An easy cell to construct is the citrus cell. Stick a straightened paper clip and a piece of copper wire into a lemon. Hold the ends of the wire close together, but not touching, and place the ends on your tongue. The slight tingle you feel and the metallic taste you experience result from a slight electric current from the citrus cell through the wires when your moist tongue closes the circuit.

2. Examine the electric meter in your house. It is probably in the basement or on the outside of the house. You will see that, in addition to the clocklike dials in the meter, there is a circular aluminum disk that spins between the poles of magnets when electric current goes into the house. The more electric current, the faster the disk turns. The speed of the disk is directly proportional to the number of watts used; for example, it spins five times as fast for 500 W as for 100 W.

#1, 3, 5, 9, 13, 15, 21, 23, 29, 31, 33, 41, 43, 47, 57, 59

You can use the meter to determine how many watts an electrical device uses. First, see that all electrical devices in your home are disconnected (you may leave electric clocks connected, for the 2 watts they use will hardly be noticeable). The disk will be practically stationary. Then connect a 100-W bulb and note how many seconds it takes for the disk to make five complete revolutions. The black spot painted on the edge of the disk makes this easy. Disconnect the 100-W bulb and plug in a device of unknown wattage. Again, count the seconds for five revolutions. If it takes the same time, it's a 100-W device; if it takes twice the time, it's a 50-W device; half the time, a 200-W device; and so forth. In this way you can estimate the power consumption of devices fairly accurately.

3. Write a letter to Grandma and convince her that whatever electric shocks she may have received over the years have been due to the movement of electrons already in her body—not electrons coming from somewhere else.

One-Step Calculations

Ohm's Law: $I = V/R$

1. Calculate the current in a toaster that has a heating element of 15 Ω when connected to a 120-V outlet.
2. Calculate how much current warms your feet with electric socks that have a 90-ohm heating element powered by a 9-volt battery.
3. Calculate the current that moves through your fingers (resistance 1000 Ω) when you touch them to the terminals of a 6-volt battery.
4. Calculate the current in the 240-ohm filament of a bulb connected to a 120-V line.

Power $= IV$

5. Calculate the power of a device that carries 0.5 amperes when impressed with 120 volts.
6. Calculate the power of a hair dryer that operates on 120 volts and draws a current of 10 amperes.

Exercises

1. What two things can be done to increase the amount of flow in a water pipe? Similarly, what two things can be done to increase the current in an electrical circuit?
2. Consider a water pipe that branches into two smaller pipes. If the flow of water is 10 gallons per minute in the main pipe and 4 gallons per minute in one of the branches, how much water per minute flows in the other branch?
3. Consider a circuit with a main wire that branches into two other wires. If the current is 10 amperes in the main wire and 4 amperes in one of the branches, how much current is in the other branch?
4. One example of a water system is a garden hose that waters a garden. Another is the cooling system of an automobile. Which of these exhibits behavior more analogous with that of an electric circuit? Why?
5. What happens to the brightness of light emitted by a lightbulb when the current flowing through it increases?
6. Your friend says that a battery supplies the electrons in an electric circuit. Do you agree or disagree? Defend your answer.
7. Is a current-carrying wire electrically charged?
8. Your tutor tells you that an *ampere* and a *volt* really measure the same thing, and that the different terms only serve to make a simple concept seem confusing. Why should you consider getting a different tutor?
9. In which of the circuits below does a current exist to light the bulb?

10. Does more current flow out of a battery than into it? Does more current flow into a lightbulb than out of it? Explain.
11. Something gets "used up" in a battery that eventually dies and goes flat. One friend says that current is used up. Another friend says that energy is used up. Who, if either, do you agree with, and why?
12. Suppose you leave your car lights on while at a movie. When you return, your battery is too "weak" to start your car. A friend comes and gives you a jump start with his battery and battery cables. What physics is occurring here?
13. Your friend says that, when jump-starting a dead battery, you should connect your live battery in parallel with the dead battery, which, in effect, replaces the dead one. Do you agree?
14. An electron moving in a wire collides repeatedly with atoms and travels an average distance between collisions that is called the *mean free path*. If the mean free path is less in some metals, what can you say about

the resistance of these metals? For a given conductor, what can be done to lengthen the mean free path?

15. Why will the resistance of a wire be slightly different immediately after you have held it in your hand?

16. Why is the current in an incandescent bulb greater immediately after it is turned on than it is a few moments later?

17. A simple lie detector consists of an electric circuit, one part of which is part of your body—like from one finger to another. A sensitive meter shows the current that flows when a small voltage is applied. How does this technique indicate that a person is lying? (And when does this technique *not* tell when someone is lying?)

18. Only a small percentage of the electric energy fed into a common lightbulb is transformed into light. What happens to the remaining energy?

19. Why are thick wires rather than thin wires usually used to carry large currents?

20. Why does the filament of a lightbulb glow while the connecting wires do not?

21. Will a lamp with a thick filament draw more current or less current than a lamp with a thin filament?

22. A 1-mile-long copper wire has a resistance of 10 Ω. What will be its new resistance when it is shortened by (a) cutting it in half or (b) doubling it over and using it as "one" wire?

23. What is the effect on the current in a wire if both the voltage across it and its resistance are doubled? If both are halved?

24. Will the current in a lightbulb connected to a 220-V source be greater or less than the current in the same bulb when it is connected to a 110-V source?

25. Which will do less damage—plugging a 110-V appliance into a 220-V circuit or plugging a 220-V appliance into a 110-V circuit? Explain.

26. If a current of one- or two-tenths of an ampere were to flow into one of your hands and out the other, you would probably be electrocuted. But, if the same current were to flow into your hand and out the elbow above the same hand, you would survive even though the current might be large enough to burn your flesh. Explain.

27. Would you expect to find dc or ac in the filament of a lightbulb in your home? In the headlight of an automobile?

28. Are automobile headlights wired in parallel or in series? What is your evidence?

29. A car's headlights dissipate 40 W on low beam, and 50 W on high beam. Is there more or less resistance in the high-beam filament?

30. What unit is represented by (a) joule per coulomb, (b) coulomb per second, (c) watt·second?

31. To connect a pair of resistors so that their combined (equivalent) resistance will be greater than the resistance of either one, should you connect them in series or in parallel?

32. To connect a pair of resistors so that their combined (equivalent) resistance will be less than the resistance of either one, should you connect them in series or in parallel?

33. Between current and voltage, which remains the same for a 10-Ω and a 20-Ω resistor in series in a series circuit?

34. Between current and voltage, which remains the same for a 10-Ω and a 20-Ω resistor in parallel in a parallel circuit?

35. The damaging effects of electric shock result from the amount of current that flows in the body. Why, then, do we see signs that read "Danger—High Voltage" rather than "Danger—High Current"?

36. Comment on the warning sign shown in the sketch.

37. Is the following label on a household product cause for concern? "Caution: This product contains tiny, electrically charged particles moving at speeds in excess of 100,000,000 kilometers per hour."

38. Why are the wingspans of birds a consideration in determining the spacing between parallel wires in a power line?

39. Estimate the number of electrons that a power company delivers annually to the homes of a typical city of 50,000 people.

40. If electrons flow very slowly through a circuit, why does it not take a noticeably long time for a lamp to glow when you turn on a distant switch?

41. Why is the speed of an electric signal so much greater than the speed of sound?

42. If a glowing lightbulb is jarred and oxygen leaks inside, the bulb will momentarily brighten considerably before burning out. Putting excess current through a lightbulb will also burn it out. What physical change occurs when a lightbulb burns out?

43. Consider a pair of flashlight bulbs connected to a battery. Will they glow brighter if they are connected in series or in parallel? Will the battery run down faster if they are connected in series or in parallel?

44. What happens to the brightness of Bulb A when the switch is closed and Bulb B lights up?

45. If several bulbs are connected in series to a battery, they may feel warm to the touch but not visibly glow. What is your explanation?

46. In the circuit shown, how do the brightnesses of the identical lightbulbs compare? Which bulb draws the most current? What will happen if Bulb A is unscrewed? If Bulb C is unscrewed?

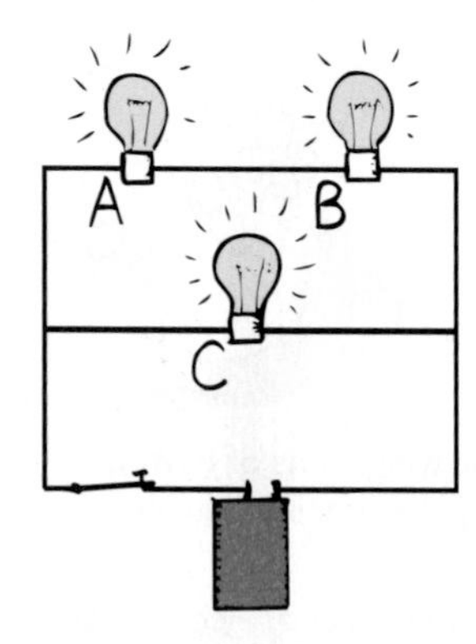

47. As more and more bulbs are connected in series to a flashlight battery, what happens to the brightness of each bulb? Assuming that heating inside the battery is negligible, what happens to the brightness of each bulb when more and more bulbs are connected in parallel?

48. What changes occur in the line current when more devices are introduced in a series circuit? In a parallel circuit? Why are your answers different?

49. Why is there no effect on other branches in a parallel circuit when one branch of the circuit is opened or closed?

50. Your friend says that the equivalent (combined) resistance of resistors connected in series is always more than the resistance of the largest resistor. Do you agree?

51. Your friend says that the equivalent (combined) resistance of resistors connected in parallel is always less than the resistance of the smallest resistor. Do you agree?

52. Your electronics friend needs a 20-ohm resistor but has only 40-ohm resistors. He tells you that he can combine them to produce a 20-ohm resistor. How?

53. Your electronics friend needs a 10-ohm resistor, but only has 40-ohm ones. How can he combine them to produce an equivalent resistance of 10 ohms?

54. When a pair of identical resistors are connected in series, which of the following is the same for both resistors—(a) voltage across each, (b) power dissipated in each, (c) current through each? Do any of your answers change if the resistors are different from each other?

55. When two identical resistors are connected in parallel, which of the following is the same for both resistors—(a) voltage across each, (b) power dissipated in each, (c) current through each? Do any of your answers change if the resistors are different from each other?

56. Because a battery has internal resistance, if the current it supplies increases, the voltage it supplies decreases. If too many bulbs are connected in parallel across a battery, will their brightness diminish? Explain.

57. Are these three circuits equivalent to one another? Why or why not?

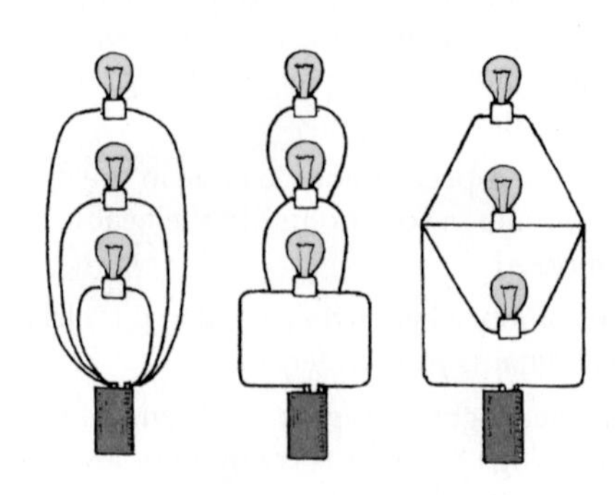

58. Figure 23.20 shows a fuse placed in household circuitry. In what other locations might a fuse be placed in this circuit to be useful, melting only if a problem arises?

59. Is the resistance of a 100-W bulb greater or less than the resistance of a 60-W bulb? Assuming the filaments in each bulb are of the same length and made of the same material, which bulb has the thicker filament?

60. If a 60-W bulb and a 100-W bulb are connected in series in a circuit, across which bulb will there be a greater voltage drop? How about if they are connected in parallel?

Problems

1. The wattage marked on a lightbulb is not an inherent property of the bulb but depends on the voltage to which it is connected, usually 110 or 120 V. How many amperes flow through a 60-W bulb connected in a 120-V circuit?

2. Rearrange the equation Current = voltage/resistance to express *resistance* in terms of current and voltage. Then solve the following: A certain device in a 120-V circuit has a current rating of 20 A. What is the resistance of the device (how many ohms)?

3. Using the formula Power = current × voltage, find the current drawn by a 1200-W hair dryer connected to 120 V. Then, using the method from the previous problem, find the resistance of the hair dryer.

4. The total charge that an automobile battery can supply without being recharged is given in terms of ampere-hours. A typical 12-V battery has a rating of 60 ampere-hours (60 A for 1 h, 30 A for 2 h, and so on). Suppose that you forget to turn the headlights off in your parked automobile. If each of the two headlights draws 3 A, how long will it be before your battery is "dead"?

5. How much does it cost to operate a 100-W lamp continuously for 1 week if the power utility rate is 15¢/kWh?

6. A 4-W night-light is plugged into a 120-V circuit and operates continuously for 1 year. Find the following: (a) the current it draws, (b) the resistance of its filament, (c) the energy consumed in a year, and (d) the cost of its operation for a year at the utility rate of 15¢/kWh.

7. An electric iron connected to a 110-V source draws 9 A of current. How much heat (in joules) does it generate in a minute?

8. How many coulombs of charge flow through the iron in the previous problem in one minute?

9. A certain lightbulb with a resistance of 95 ohms is labeled "150 W." Was this bulb designed for use in a 120-V circuit or a 240-V circuit?

10. In periods of peak demand, power companies lower their voltage. This saves them power (and saves you money!). To see the effect, consider a 1200-W toaster that draws 10 A when connected to 120 V. Suppose the voltage is lowered by 10% to 108 V. By how much does the current decrease? By how much does the power decrease? (*Caution*: The 1200-W label is valid only when 120 V is applied. When the voltage is lowered, it is the resistance of the toaster, not its power, that remains constant.)

Ken Ganezer shows the bluish-green glow of electrons circling the magnetic field lines inside a Thompson tube.

CHAPTER 24

Magnetism

Youngsters are fascinated with magnets, largely because they act at a distance. One can move a nail with a nearby magnet even when a piece of wood is in between. Likewise, a neurosurgeon can guide a pellet through brain tissue to inoperable tumors, pull a catheter into position, or implant electrodes while doing little harm to brain tissue. The use of magnets grows daily.

The term *magnetism* comes from the name Magnesia, a coastal district of ancient Thessaly, Greece, where certain stones were found by the Greeks more than 2000 years ago. These stones, called *lodestones,* had the unusual property of attracting pieces of iron. Magnets were first fashioned into compasses and used for navigation by the Chinese in the twelfth century.

In the sixteenth century, William Gilbert, Queen Elizabeth's physician, made artificial magnets by rubbing pieces of iron against lodestone, and he suggested that a compass always points north and south because the Earth has magnetic properties. Later, in 1750, John Michell, an English physicist and astronomer, found that magnetic poles obey the inverse-square law, and his results were confirmed by Charles Coulomb. The subjects of magnetism and electricity developed almost independently of each other until 1820, when a Danish physicist named Hans Christian Oersted discovered, in a classroom demonstration, that an electric current affects a magnetic compass.[1] He saw confirming evidence that magnetism was related to electricity. Shortly thereafter, the French physicist André Marie Ampere proposed that electric currents are the source of all magnetic phenomena.

In days gone by, Dick Tracy comic strips, in addition to predicting the advent of cell phones, featured the heading, "He who controls magnetism controls the universe."

Insights

Magnetic Forces

In Chapter 22, we discussed the forces that electrically charged particles exert on one another: The force between any two charged particles depends on the magnitude of the charge on each and their distance of separation, as specified in Coulomb's law. But Coulomb's law is not the whole story when the charged particles are moving with respect to each other. The force between electrically charged particles depends also, in a complicated way, on their motion. We find that,

[1]We can only speculate about how often such relationships become evident when they "aren't supposed to" and are dismissed as "something wrong with the apparatus." Oersted, however, had the insight—characteristic of a good scientist—to see that nature was revealing another of its secrets.

in addition to the force we call *electrical,* there is a force due to the motion of the charged particles that we call the **magnetic force.** The source of magnetic force is the motion of charged particles, usually electrons. Both electrical and magnetic forces are actually different aspects of the same phenomenon of electromagnetism.

Magnetic Poles

FIGURE 24.1
A horseshoe magnet.

The forces that magnets exert on one another are similar to electrical forces, for they can both attract and repel without touching, depending on which ends of the magnets are held near one another. Like electrical forces also, the strength of their interaction depends on the separation distance between the two magnets. Whereas electric charge is central to electrical forces, regions called *magnetic poles* give rise to magnetic forces.

If you suspend a bar magnet at its center by a piece of string, you'll have a compass. One end, called the *north-seeking pole,* points northward, and the opposite end, called the *south-seeking pole,* points southward. More simply, these are called the *north* and *south poles.* All magnets have both a north and a south pole (some have more than one of each). Refrigerator magnets, popular in recent years, have narrow strips of alternating north and south poles. These magnets are strong enough to hold sheets of paper against a refrigerator door, but they have a very short range because the north and south poles cancel. In a simple bar magnet, a single north pole and a single south pole are located at opposite ends. A common horseshoe magnet is simply a bar magnet that has been bent into a U shape. Its poles are also at its two ends (Figure 24.1).

When the north pole of one magnet is brought near the north pole of another magnet, they repel.[2] The same is true of a south pole near a south pole. If opposite poles are brought together, however, attraction occurs. We find that

Like poles repel each other; opposite poles attract.

This rule is similar to the rule for the forces between electric charges, where like charges repel one another and unlike charges attract. But there is a very important difference between magnetic poles and electric charges. Whereas electric charges can be isolated, magnetic poles cannot. Negatively charged electrons and positively charged protons are entities by themselves. A cluster of electrons need not be accompanied by a cluster of protons, and vice versa. But a north magnetic pole never exists without the presence of a south pole, and vice versa.

If you break a bar magnet in half, each half still behaves as a complete magnet. Break the pieces in half again, and you have four complete magnets. You can continue breaking the pieces in half and never isolate a single pole.[3] Even when your piece is one atom thick, there are two poles, which suggests that atoms themselves are magnets.

A magstripe on a credit card contains millions of tiny magnetic domains held together by a resin binder. Data are encoded in binary code, with zeros and ones distinguished by the frequency of domain reversals. It's quite amazing how quickly your name pops up when an airline reservationist swipes your card.

Insights

[2]The force of interaction between magnetic poles is given by $F \sim \frac{p_1 p_2}{d^2}$, where p_1 and p_2 represent magnetic pole strengths and d represents the separation distance between the poles. Note the similarity of this relationship to Coulomb's law.

[3]Theoretical physicists have speculated for more than 70 years about the possible existence of discrete magnetic "charges," called *magnetic monopoles.* These tiny particles would carry either a single north or a single south magnetic pole and would be the counterparts to the positive and negative charges in electricity. Various attempts have been made to find monopoles, but none has proved successful. All known magnets always have at least one north pole and one south pole.

CHECK YOURSELF

Does every magnet necessarily have a north and south pole?

Magnetic Fields

If you sprinkle some iron filings on a sheet of paper placed on a magnet, you'll see that the filings trace out an orderly pattern of lines that surround the magnet. The space around the magnet contains a **magnetic field.** The shape of the field is revealed by the filings, which align with the magnetic field lines that spread out from one pole and return to the other. It is interesting to compare the field patterns in Figures 24.2 and 24.4 with the electric field patterns in Figure 22.19 back in Chapter 22.

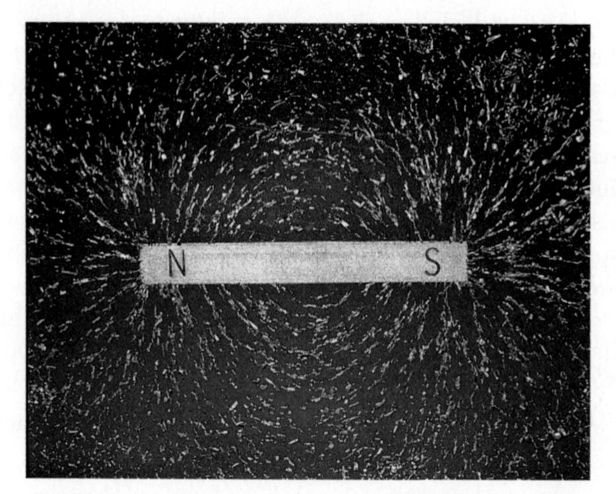

FIGURE 24.2
Interactive Figure

Top view of iron filings sprinkled around a magnet. The filings trace out a pattern of *magnetic field lines* in the space surrounding the magnet. Interestingly, the magnetic field lines continue inside the magnet (not revealed by the filings) and form closed loops.

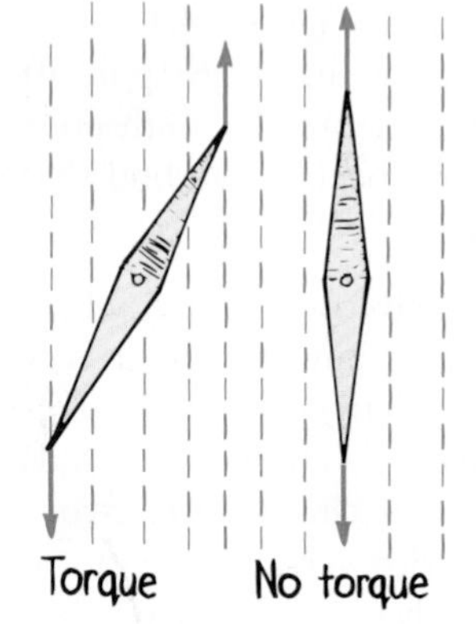

FIGURE 24.3
When the compass needle is not aligned with the magnetic field (left), the oppositely directed forces on the needle produce a pair of torques (called a *couple*) that twist the needle into alignment (right).

The direction of the field outside a magnet is from the north pole to the south pole. Where the lines are closer together, the field is stronger. The concentration of iron filings at the poles of the magnet in Figure 24.2 shows the magnetic field strength is greater there. If we place another magnet or a small compass anywhere in the field, its poles line up with the magnetic field.

Magnetism is very much related to electricity. Just as an electric charge is surrounded by an electric field, the same charge is also surrounded by a magnetic field if it is moving. This magnetic field is due to the "distortions" in the electric field caused by motion and was explained by Albert Einstein in 1905 in his special theory of relativity. We won't go into the details except to acknowledge that a magnetic field is a relativistic by-product of the electric

CHECK YOUR ANSWER

Yes, just as every coin has two sides, a "head" and a "tail." Some "trick" magnets have more than one pair of poles, but, nevertheless, poles always occur in pairs.

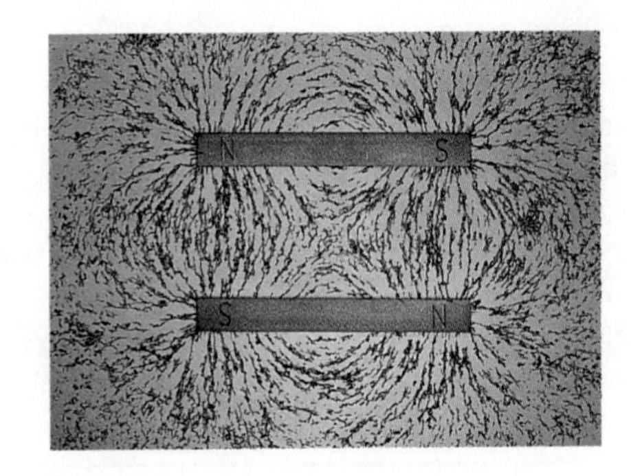

a

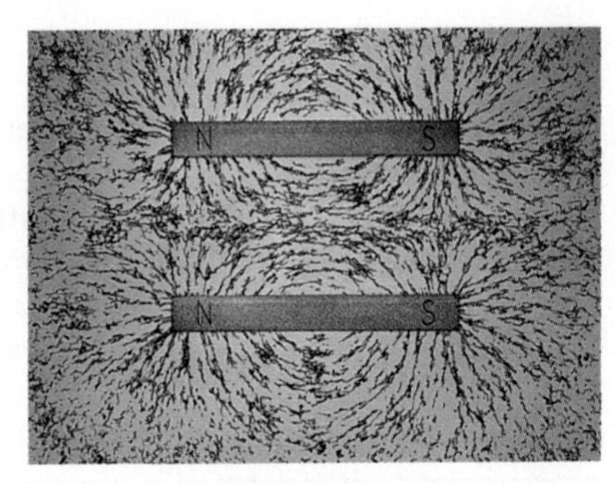

b

FIGURE 24.4
The magnetic field patterns for a pair of magnets. (a) Opposite poles are nearest each other, and (b) like poles are nearest each other.

FIGURE 24.5
Wai Tsan Lee shows iron nails that have become induced magnets.

field. Charged particles in motion have associated with them both an electric field and a magnetic field. A magnetic field is produced by the motion of electric charge.[4]

If the motion of electric charges produces magnetism, where is this motion in a common bar magnet? The answer is, in the electrons of the atoms that make up the magnet. These electrons are in constant motion. Two kinds of electron motion contribute to magnetism: electron spin and electron revolution. Electrons spin about their own axes like tops, and they also revolve about the atomic nucleus. In most common magnets, electron spin is the chief contributor to magnetism.

Every spinning electron is a tiny magnet. A pair of electrons spinning in the same direction makes up a stronger magnet. A pair of electrons spinning in opposite directions, however, work against each other. The magnetic fields cancel. This is why most substances are not magnets. In most atoms, the various fields cancel one another because the electrons spin in opposite directions. In such materials as iron, nickel, and cobalt, however, the fields do not cancel each other entirely. Each iron atom has four electrons whose spin magnetism is uncancelled. Thus, each iron atom is a tiny magnet. The same is true, to a lesser extent, for the atoms of nickel and cobalt. Most common magnets are made from alloys containing iron, nickel, and cobalt in various proportions.[5]

Magnetic Domains

The magnetic field of an individual iron atom is so strong that interactions among adjacent atoms cause large clusters of them to line up with one another. These clusters of aligned atoms are called **magnetic domains.** Each domain is made up of billions of aligned atoms. The domains are microscopic (Figure 24.6), and there are many of them in a crystal of iron. Like the alignment of iron atoms within domains, domains themselves can align with one another.

[4]Interestingly, since motion is relative, the magnetic field is relative. For example, when a charge moves by you, there is a definite magnetic field associated with the moving charge. But, if you move along with the charge so that there is no motion relative to you, you'll find no magnetic field associated with the charge. Magnetism is relativistic. In fact, it was Albert Einstein who first explained this when he published his first paper on special relativity, "On the Electrodynamics of Moving Bodies." (More on relativity in Chapters 35 and 36.)

[5]Electron spin contributes virtually all of the magnetic properties in magnets made from alloys containing iron, nickel, cobalt, and aluminum. In the rare earth metals, such as gadolinium, the orbital motion is more significant.

PRACTICING PHYSICS

Most iron objects around you are magnetized to some degree. A filing cabinet, a refrigerator, or even cans of food on your pantry shelf, have north and south poles induced by Earth's magnetic field. If you bring a magnetic compass near the tops of iron or steel objects in your home, you will find that the north pole of the compass needle points to the tops of these objects, and the south pole of the compass needle points to the bottoms. This shows that the objects are magnets, having a south pole on top and a north pole on the bottom. Turn cans of food that have been in a vertical position upside down and see how many days it takes for the poles to reverse themselves!

FIGURE 24.6
A microscopic view of magnetic domains in a crystal of iron. Each domain consists of billions of aligned iron atoms. The blue arrows pointing in different directions tell us that these domains are not aligned.

Not every piece of iron, however, is a magnet. This is because the domains in ordinary iron are not aligned. Consider a common iron nail: The domains in the nail are randomly oriented. Many of them are induced into alignment, however, when a magnet is brought nearby. (It is interesting to listen with an amplified stethoscope to the clickity-clack of domains undergoing alignment in a piece of iron when a strong magnet approaches.) The domains align themselves much as electrical charges in a piece of paper align themselves in the presence of a charged rod. When you remove the nail from the magnet, ordinary thermal motion causes most or all of the domains in the nail to return to a random arrangement. If the field of the permanent magnet is very strong, however, the nail may retain some permanent magnetism of its own after the two are separated.

Permanent magnets are made by simply placing pieces of iron or certain iron alloys in strong magnetic fields. Alloys of iron differ; soft iron is easier to magnetize than steel. It helps to tap the iron to nudge any stubborn domains into alignment. Another way of making a permanent magnet is to stroke a piece of iron with a magnet. The stroking motion aligns the domains in the iron. If a permanent magnet is dropped or heated, some of the domains are jostled out of alignment and the magnet becomes weaker.

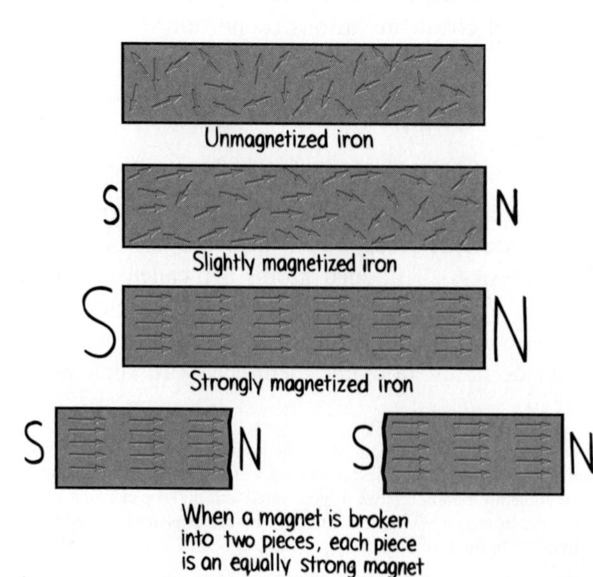

FIGURE 24.7
Interactive Figure
Pieces of iron in successive stages of magnetization. The arrows represent domains; the head is a north pole and the tail is a south pole. Poles of neighboring domains neutralize each other's effects, except at the two ends of a piece of iron.

MAGNETIC THERAPY*

Back in the eighteenth century, a celebrated "magnetizer" from Vienna, Franz Mesmer, brought his magnets to Paris and established himself as a healer in Parisian society. He healed patients by waving magnetic wands above their heads.

At that time, Benjamin Franklin, the world's leading authority on electricity, was visiting Paris as a U.S. representative. He suspected that Mesmer's patients did benefit from his ritual, but only because it kept them away from the bloodletting practices of other physicians. At the urging of the medical establishment, King Louis XVI appointed a royal commission to investigate Mesmer's claims. The commission included Franklin and Antoine Lavoisier, the founder of modern chemistry. The commissioners designed a series of tests in which some subjects thought they were receiving Mesmer's treatment when they weren't, while others received the treatment but were led to believe they had not. The results of these blind experiments established beyond any doubt that Mesmer's success was due solely to the power of suggestion. To this day, the report is a model of clarity and reason. Mesmer's reputation was destroyed, and he retired to Austria.

Now some two hundred years later, with increased knowledge of magnetism and physiology, hucksters of magnetism are attracting even larger followings. But there is no government commission of Franklins and Lavoisiers to challenge their claims. Instead, magnetic therapy is another of the untested and unregulated "alternative therapies" given official recognition by Congress in 1992.

Although testimonials about the benefits of magnets are many, there is no scientific evidence whatever for magnets boosting body energy or combating aches and pains. None. Yet millions of therapeutic magnets are sold in stores and catalogs. Consumers are buying magnetic bracelets, insoles, wrist and knee bands, back and neck braces, pillows, mattresses, lipstick, and even water. They are told that magnets have powerful effects on the body, mainly increasing blood flow to injured areas. The idea that blood is attracted by a magnet is bunk, for the type of iron that occurs in blood doesn't respond to a magnet. Furthermore, most therapeutic magnets are of the refrigerator type, with a very limited range. To get an idea of how quickly the field of these magnets drops off, see how many sheets of paper one of these magnets will hold on a refrigerator or any iron surface. The magnet will fall off after a few sheets of paper separate it from the iron surface. The field doesn't extend much more than one millimeter, and it wouldn't penetrate the skin, let alone into muscles. And even if it did, there is no scientific evidence that magnetism has any beneficial effects on the body at all. But, again, testimonials are another story.

Sometimes an outrageous claim has some truth to it. For example, the practice of bloodletting in previous centuries was, in fact, beneficial to a small percentage of men. These men suffered the genetic disease (*hemochromatosis,* excess iron in the blood—women being less afflicted partly due to menstruation). Although the number of men who benefited from bloodletting was small, testimonials of its success prompted the widespread practice that killed many.

No claim is so outrageous that testimonials can't be found to support it. Claims that the Earth is flat or claims for the existence of flying saucers are quite harmless and may amuse us. Magnetic therapy may likewise be harmless for many ailments, but not when it is used to treat a serious disorder in place of modern medicine. Pseudoscience may be promoted to intentionally deceive or it may be the result of flawed and wishful thinking. In either case, pseudoscience is very big business. The market is enormous for therapeutic magnets and other such fruits of unreason.

Scientists must keep open minds, must be prepared to accept new findings, and must be ready to be challenged by new evidence. But scientists also have a responsibility to inform the public when they are being deceived and, in effect, robbed by pseudoscientists whose claims are without substance.

*Adapted from *Voodoo Science: The Road from Foolishness to Fraud,* by Robert L. Park; Oxford University Press, 2000.

We each need a *knowledge filter* to tell the difference between what is true and what seems to be true. The best knowledge filter ever invented is science.

Insights

CHECK YOURSELF

How can a magnet attract a piece of iron that is not magnetized?

Electric Currents and Magnetic Fields

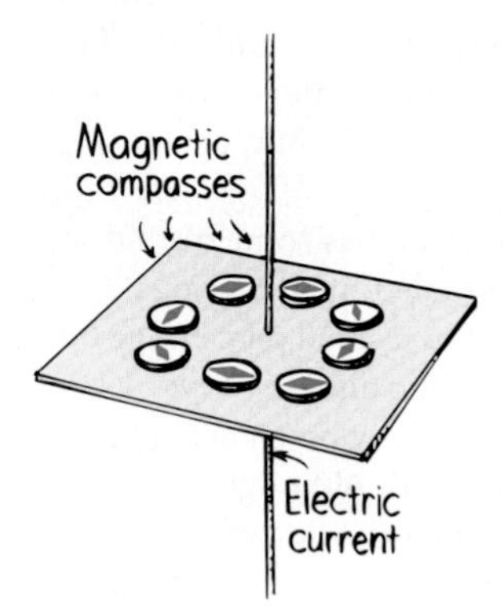

FIGURE 24.8
The compasses show the circular shape of the magnetic field surrounding the current-carrying wire.

Since a moving charge produces a magnetic field, it follows that a current of charges also produces a magnetic field. The magnetic field that surrounds a current-carrying conductor can be demonstrated by arranging an assortment of compasses around a wire (Figure 24.8) and passing a current through it. The compass needles line up with the magnetic field produced by the current and they show the field to be a pattern of concentric circles about the wire. When the current reverses direction, the compass needles turn around, showing that the direction of the magnetic field changes also. This is the effect that Oersted first demonstrated in the classroom.

If the wire is bent into a loop, the magnetic field lines become bunched up inside the loop (Figure 24.9). If the wire is bent into another loop, overlapping the first, the concentration of magnetic field lines inside the loops is doubled. It follows that the magnetic field intensity in this region is increased as the number of loops is increased. The magnetic field intensity is appreciable for a current-carrying coil of many loops.

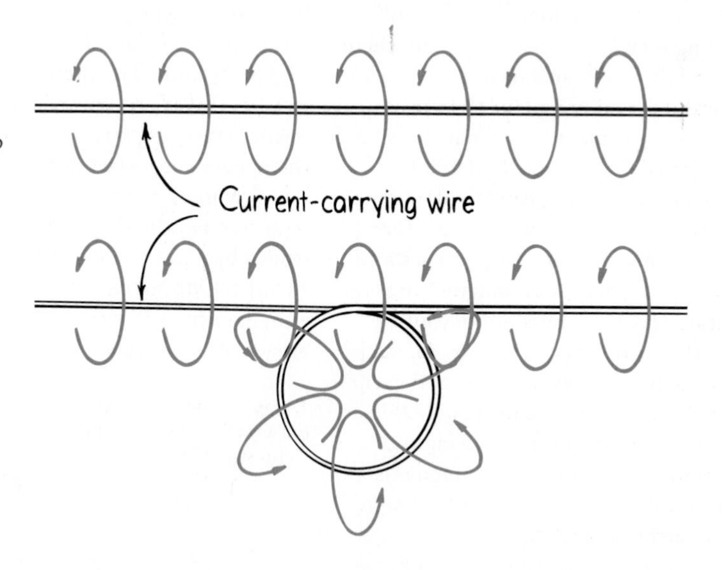

FIGURE 24.9
Magnetic field lines about a current-carrying wire crowd up when the wire is bent into a loop.

CHECK YOUR ANSWER

Domains in the unmagnetized piece of iron are induced into alignment by the magnetic field of the nearby magnet. See the similarity of this with Figure 22.13 back in Chapter 22. Like the pieces of paper that jump to the comb, pieces of iron will jump to a strong magnet when it is brought nearby. But, unlike the pieces of paper, they are not then repelled. Can you think of the reason why?

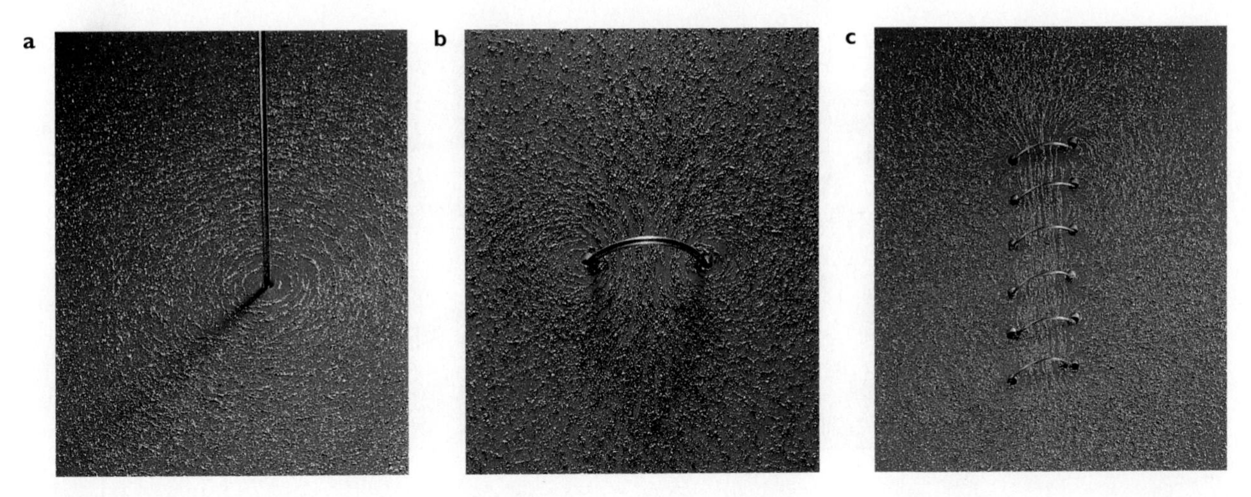

FIGURE 24.10
Iron filings sprinkled on paper reveal the magnetic field configurations about (a) a current-carrying wire, (b) a current-carrying loop, and (c) a current-carrying coil of loops.

Electromagnets

A current-carrying coil of wire is an **electromagnet.** The strength of an electromagnet is increased by simply increasing the current through the coil. Industrial magnets gain additional strength by having a piece of iron within the coil. Magnetic domains in the iron are induced into alignment, adding to the field. For extremely strong electromagnets, such as those used to control charged-particle beams in high-energy accelerators, iron is not used, because, beyond a certain point, all of its domains are aligned and it no longer adds to the field.

Electromagnets powerful enough to lift automobiles are a common sight in junkyards. The strength of these electromagnets is limited by heating of the current-carrying coils (due to electrical resistance) and saturation of magnetic-domain alignment in the core. The most powerful electromagnets, which do not have iron cores, use superconducting coils through which large electrical currents flow with ease.

FIGURE 24.11
A permanent magnet levitates above a superconductor because its magnetic field cannot penetrate the superconducting material.

Superconducting Electromagnets

Recall, from Chapter 22, that there is no electrical resistance in a superconductor to limit the flow of electric charge and, therefore, no heating, even if the current is enormous. Electromagnets that utilize superconducting coils produce extremely strong magnetic fields—and they do so very economically because there are no heat losses (although energy is used to keep the superconductors cold). At the Fermi National Accelerator Laboratory (Fermilab), near Chicago, superconducting electromagnets guide high-energy particles around an accelerator 4 miles in circumference. Superconducting magnets can also be found in magnetic resonance imaging (MRI) devices in hospitals.

Another application to watch for is magnetically levitated, or "maglev," transportation. Figure 24.12 shows the scale model of a maglev system developed in the United States. The vehicle, called a magplane, carries superconducting

FIGURE 24.12
A magnetically levitated vehicle—a *magplane.* Whereas conventional trains vibrate as they ride on rails at high speeds, magplanes can travel vibration-free at high speeds because they make no physical contact with the guideway they float above.

coils on its underside. Moving along an aluminum trough, these coils generate currents in the aluminum that act as mirror-image magnets and repel the magplane. It floats a few centimeters above the guideway, and its speed is limited only by air friction and passenger comfort.

A maglev train built by German engineers is currently operating at speeds up to 460 km/h between downtown Shanghai and its airport. It covers some 30 km in less than eight minutes. A high-speed line is projected that will connect Shanghai with 1380-km distant Beijing, reducing the customary 14-hour trip by half. Watch for the proliferation of this relatively new technology.

Magnetic Force on Moving Charged Particles

A charged particle at rest will not interact with a static magnetic field. But if the charged particle is moving in a magnetic field, the magnetic character of a charge in motion becomes evident. It experiences a deflecting force.[6] The force is greatest when the particle moves in a direction perpendicular to the magnetic field lines. At other angles, the force is less, and it becomes zero when the particles move parallel to the field lines. In any case, the direction of the force is always perpendicular to the magnetic field lines and to the velocity of the charged particle (Figure 24.13). So a moving charge is deflected when it crosses through a magnetic field, but, when it travels parallel to the field, no deflection occurs.

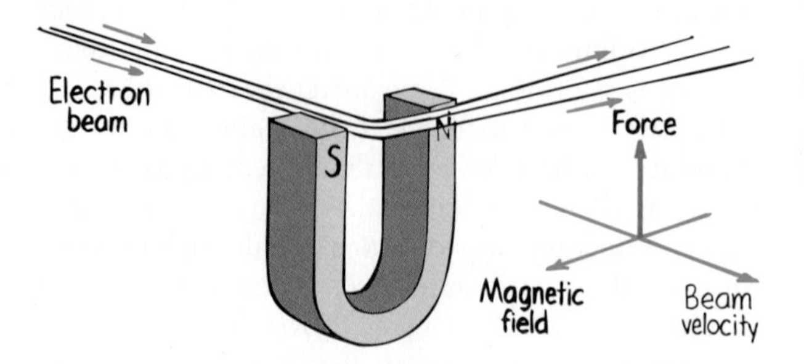

FIGURE 24.13
A beam of electrons is deflected by a magnetic field.

[6]When particles of electric charge q and velocity v move perpendicularly into a magnetic field of strength B, the force F on each particle is simply the product of the three variables: $F = qvB$. For nonperpendicular angles, v in this relationship must be the component of velocity perpendicular to B.

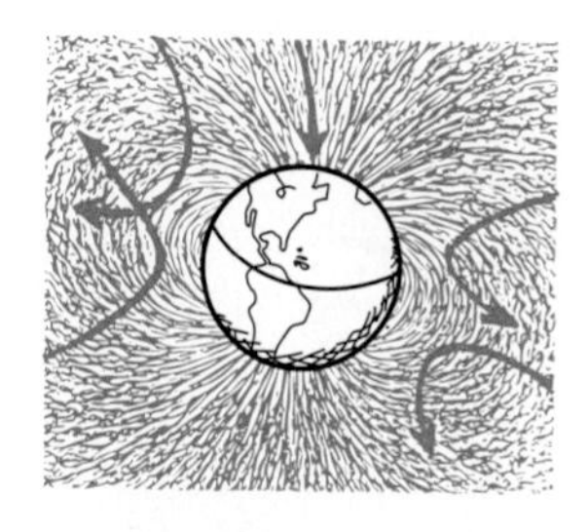

FIGURE 24.14
The magnetic field of the Earth deflects many charged particles that make up cosmic radiation.

This deflecting force is very different from the forces that occur in other interactions, such as the gravitational forces between masses, the electric forces between charges, and the magnetic forces between magnetic poles. The force that acts on a moving charged particle does not act along the line that joins the sources of interaction but, instead, acts perpendicularly to both the magnetic field and the electron beam.

We are fortunate that charged particles are deflected by magnetic fields. This fact is employed to guide electrons onto the inner surface of a TV tube, where they produce an image. More interestingly, charged particles from outer space are deflected by Earth's magnetic field. The intensity of harmful cosmic rays striking the Earth's surface would otherwise be much greater.

Magnetic Force on Current-Carrying Wires

THE Physics Place
Magnetic Forces on Current-Carrying Wires

Simple logic tells you that, if a charged particle moving through a magnetic field experiences a deflecting force, then a current of charged particles moving through a magnetic field experiences a deflecting force also. If the particles are trapped inside a wire when they respond to the deflecting force, the wire will also be pushed (Figure 24.15).

If we reverse the direction of the current, the deflecting force acts in the opposite direction. The force is strongest when the current is perpendicular to the magnetic field lines. The direction of force is not along the magnetic field lines or along the direction of current. The force is perpendicular to both field lines and current. It is a sideways force.

In the Problem Solving supplement, you'll learn the "simple" right-hand rule!
F
B
Insights

We see that, just as a current-carrying wire will deflect a magnet such as a compass needle (as discovered by Oersted in a classroom in 1820), a magnet will deflect a current-carrying wire. Discovering these complementary links between electricity and magnetism created much excitement, and people began harnessing the electromagnetic force for useful purposes almost immediately—with great sensitivity in electric meters and with great force in electric motors.

CHECK YOURSELF

What law of physics tells you that, if a current-carrying wire produces a force on a magnet, a magnet must produce a force on a current-carrying wire?

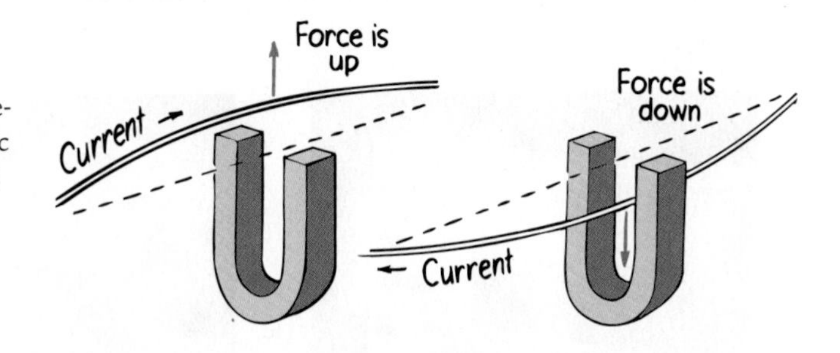

FIGURE 24.15
Interactive Figure
A current-carrying wire experiences a force in a magnetic field. (Can you see that this is a follow-up of what happens in Figure 24.13?)

CHECK YOUR ANSWER

Newton's third law, which applies to *all* forces in nature.

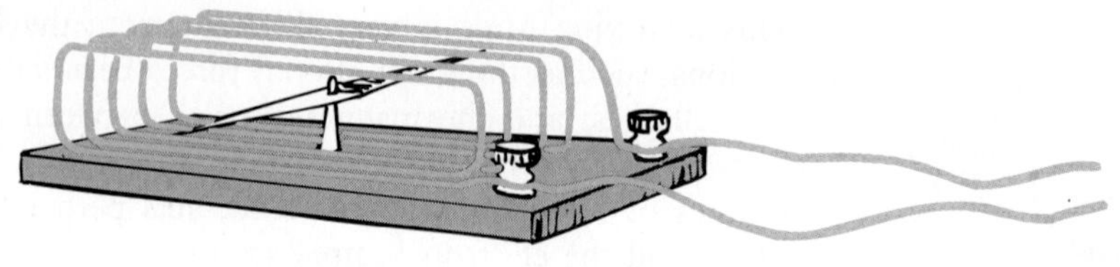

FIGURE 24.16
A very simple galvanometer.

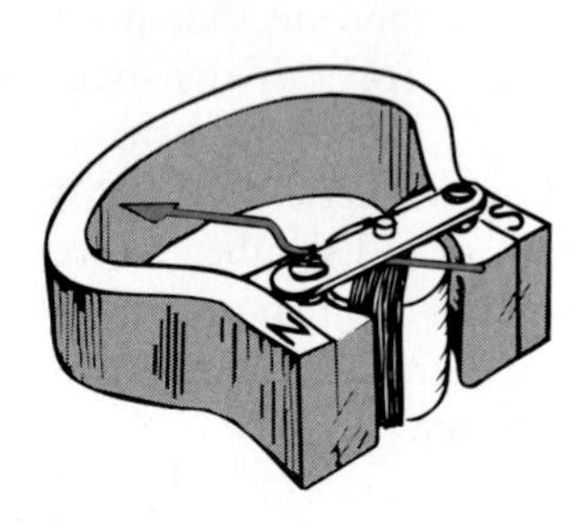

FIGURE 24.17
A common galvanometer design.

Electric Meters

The simplest meter to detect electric current is simply a magnet that is free to turn—a compass. The next most simple meter is a compass in a coil of wires (Figure 24.16). When an electric current passes through the coil, each loop produces its own effect on the needle, so a very small current can be detected. A sensitive current-indicating instrument is called a *galvanometer*.

A more common design is shown in Figure 24.17. It employs more loops of wire and is therefore more sensitive. The coil is mounted for movement, and the magnet is held stationary. The coil turns against a spring, so the greater the current in its windings, the greater its deflection. A galvanometer may be calibrated to measure current (amperes), in which case it is called an *ammeter*. Or it may be calibrated to measure electric potential (volts), in which case it is called a *voltmeter*.

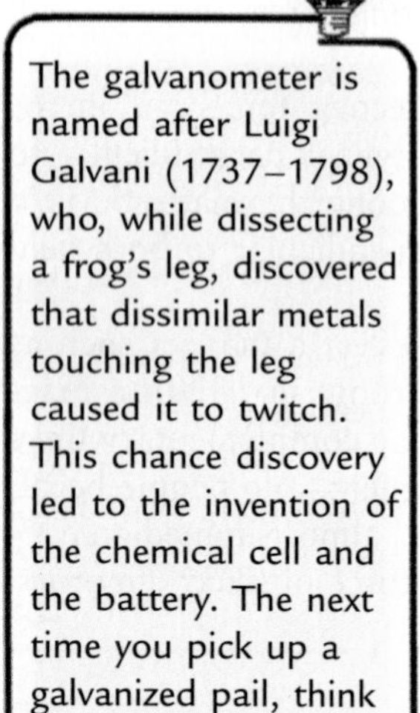

Electric Motors

If we modify the design of the galvanometer slightly, so deflection makes a complete rather than a partial rotation, we have an *electric motor*. The principal difference is that, in a motor, the current is made to change direction every time the coil makes a half rotation. After being forced to turn one-half rotation, the coil continues in motion just in time for the current to reverse, whereupon, instead of the coil reversing direction, it is forced to continue another half rotation in the same direction. This happens in cyclic fashion to produce continuous rotation, which has been harnessed to run clocks, operate gadgets, and lift heavy loads.

In Figure 24.19, we can see the principle of the electric motor in bare outline. A permanent magnet produces a magnetic field in a region in which a rectangular loop of wire is mounted to turn about the dashed axis shown.

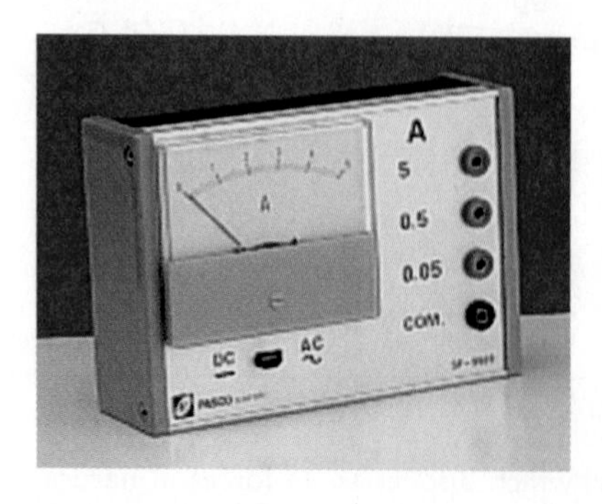

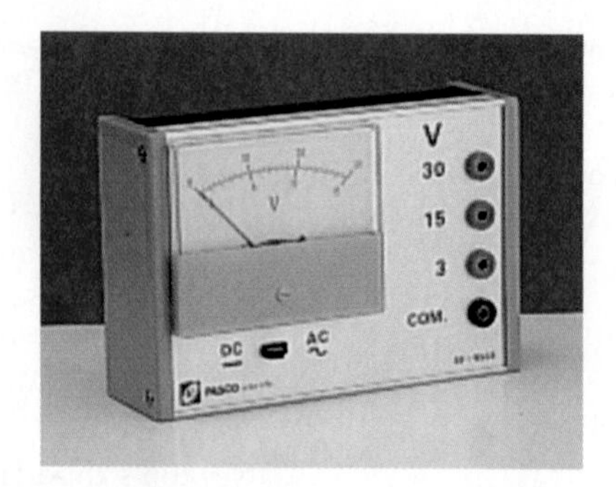

FIGURE 24.18
Both the ammeter and the voltmeter are basically galvanometers. (The electrical resistance of the instrument is made to be very low for the ammeter and very high for the voltmeter.)

A motor and a generator are actually the same device, with input and output reversed. The electrical device in a hybrid car operates both ways.

Insights

Any current in the loop has one direction in the upper side of the loop and the opposite direction in the lower side (because charges flowing into one end of the loop must flow out the other end). If the upper side of the loop is forced to the left by the magnetic field, the lower side is forced to the right, as if it were a galvanometer. But, unlike the situation in a galvanometer, the current in a motor is reversed during each half revolution by means of stationary contacts on the shaft. The parts of the wire that rotate and brush against these contacts are called *brushes*. In this way, the current in the loop alternates so that the forces on the upper and lower regions do not change directions as the loop rotates. The rotation is continuous as long as current is supplied.

We have described here only a very simple dc motor. Larger motors, dc or ac, are usually manufactured by replacing the permanent magnet by an electromagnet that is energized by the power source. Of course, more than a single loop is used. Many loops of wire are wound about an iron cylinder, called an *armature,* which then rotates when the wire carries current.

The advent of electric motors brought to an end much human and animal toil in many parts of the world. Electric motors have greatly changed the way people live.

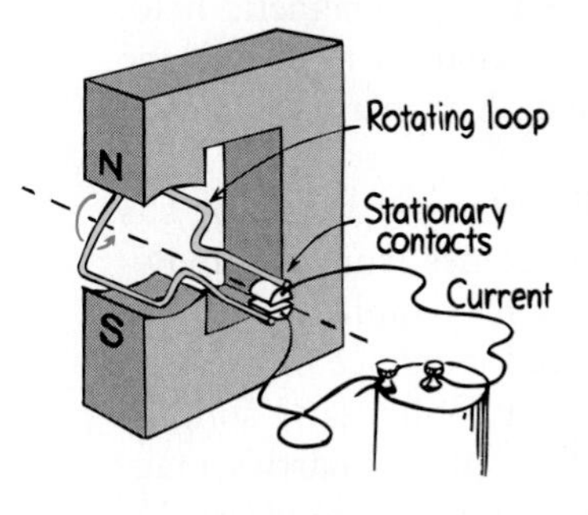

FIGURE 24.19
Interactive Figure
A simplified electric motor.

CHECK YOURSELF

What is the major similarity between a galvanometer and a simple electric motor? What is the major difference?

Earth's Magnetic Field

A suspended magnet or compass points northward because Earth itself is a huge magnet. The compass aligns with the magnetic field of the Earth. The magnetic poles of the Earth, however, do not coincide with the geographic poles—in fact, the magnetic and geographical poles are widely separated. The magnetic pole in the northern hemisphere, for example, is now located nearly 1800 kilometers from the geographic pole, somewhere in the Hudson Bay region of northern Canada. The other pole is located south of Australia (Figure 24.20). This means that compasses do not generally point to the true north. The discrepancy between the orientation of a compass and true north is known as the *magnetic declination.*

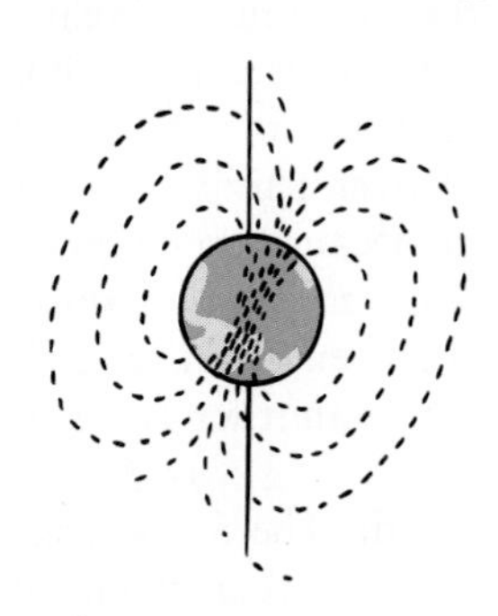

FIGURE 24.20
The Earth is a magnet.

CHECK YOUR ANSWERS

A galvanometer and a motor are similar in that they both employ coils positioned in a magnetic field. When a current passes through the coils, forces on the wires rotate the coils. The major difference is that the maximum rotation of the coil in a galvanometer is one-half turn, whereas, in a motor, the coil (wrapped on an armature) rotates through many complete turns. This is accomplished by alternating the current with each half turn of the armature.

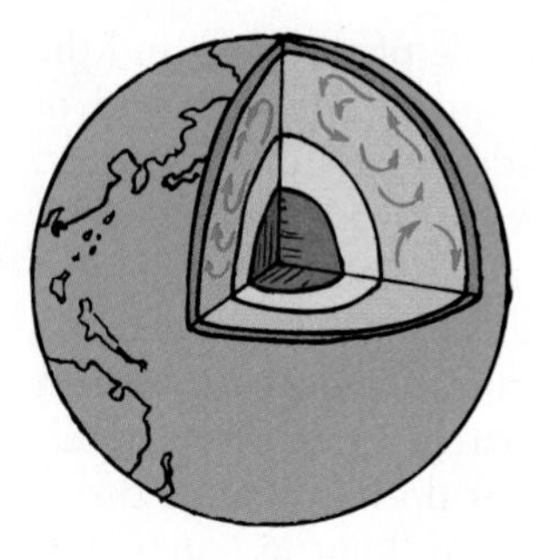

FIGURE 24.21
Convection currents in the molten parts of the Earth's interior may drive electric currents to produce the Earth's magnetic field.

Like tape from a tape recorder, history of the ocean's bottom is preserved in a magnetic record.
Insights

We do not know exactly why Earth itself is a magnet. The configuration of Earth's magnetic field is like that of a strong bar magnet placed near the center of Earth. But Earth is not a magnetized chunk of iron like a bar magnet. It is simply too hot for individual atoms to hold to a proper orientation. So the explanation must lie with electric currents deep in the interior. About 2000 kilometers below the outer rocky mantle (which itself is almost 3000 kilometers thick) lies the molten part that surrounds the solid center. Most Earth scientists think that moving charges looping around within the molten part of Earth create the magnetic field. Some Earth scientists speculate that the electric currents are the result of convection currents—from heat rising from the central core (Figure 24.21)—and that such convection currents combined with the rotational effects of the Earth produce Earth's magnetic field. Because of the Earth's great size, the speed of moving charges need only be about a millimeter per second to account for the field. A firmer explanation awaits more study.

Whatever the cause, the magnetic field of the Earth is not stable; it has wandered throughout geologic time. Evidence of this comes from the analysis of magnetic properties of rock strata. Iron atoms in a molten state are disoriented because of thermal motion, but a slight predominance of the iron atoms align with the magnetic field of the Earth. When cooling and solidification occurs, this predominance records the direction of the Earth's magnetic field in the resulting igneous rock. It's similar for sedimentary rocks, where magnetic domains in grains of iron that settle in sediments tend to align themselves with Earth's magnetic field and become locked into the rock that forms. The slight magnetism that results can be measured with sensitive instruments. As samples of rock are tested from different strata formed throughout geologic time, the magnetic field of the Earth for different periods can be charted. This evidence shows that there have been times when the magnetic field of the Earth has diminished to zero, followed by reversal of the poles. More than twenty reversals have taken place in the past 5 million years. The most recent occurred 700,000 years ago. Prior reversals happened 870,000 and 950,000 years ago. Studies of deep-sea sediments indicate that the field was virtually switched off for 10,000 to 20,000 years just over 1 million years ago. We cannot predict when the next reversal will occur because the reversal sequence is not regular. But there is a clue in recent measurements that show a decrease of more than 5% of the Earth's magnetic field strength in the last 100 years. If this change is maintained, we may well have another reversal within 2000 years.

The reversal of magnetic poles is not unique to the Earth. The Sun's magnetic field reverses regularly, with a period of 22 years. This 22-year magnetic cycle has been linked, through evidence in tree rings, to periods of drought on Earth. Interestingly enough, the long-known, 11-year sunspot cycle is just half the time during which the Sun gradually reverses its magnetic polarity.

Varying ion winds in Earth's atmosphere cause more rapid but much smaller fluctuations in Earth's magnetic field. Ions in this region are produced by the energetic interactions of solar ultraviolet rays and X-rays with atmospheric atoms. The motion of these ions produces a small but important part of Earth's magnetic field. Like the lower layers of air, the ionosphere is churned by winds. The variations in these winds are responsible for nearly all of the fast fluctuations in the Earth's magnetic field.

Cosmic Rays

FIGURE 24.22
The Van Allen radiation belts, shown here undistorted by the solar wind (cross section).

The universe is a shooting gallery of charged particles. They are called **cosmic rays** and consist of protons, alpha particles, and other atomic nuclei, as well as some high-energy electrons. The protons may be remnants of the Big Bang; the heavier nuclei probably boiled off from exploding stars. In any event, they travel through space at fantastic speeds and make up the cosmic radiation that is hazardous to astronauts. This radiation is intensified when the Sun is active and contributes added energetic particles. Cosmic rays are also a hazard to electronic instrumentation in space; impacts of heavily ionizing cosmic-ray nuclei can cause computer memory bits to "flip" or small microcircuits to fail. Fortunately for those of us on Earth's surface, most of these charged particles are deflected away by the magnetic field of the Earth. Some of them are trapped in the outer reaches of the Earth's magnetic field and make up the Van Allen radiation belts (Figure 24.22).

FIGURE 24.23
The aurora borealis lighting of the sky caused by charged particles in the Van Allen belts striking atmospheric molecules.

The Van Allen radiation belts consist of two doughnut-shaped rings about Earth, named after James A. Belts, who suggested their existence from data gathered by the U.S. satellite *Explorer I* in 1958.[7] The inner ring is centered about 3200 kilometers above the Earth's surface, and the outer ring, which is a larger and wider doughnut, is centered about 16,000 kilometers overhead. Astronauts orbit at safe distances well below these belts of radiation. Most of the charged particles—protons and electrons—trapped in the outer belt probably come from the Sun. Storms on the Sun hurl charged particles out in great fountains, many of which pass near the Earth and are trapped by its magnetic field. The trapped particles follow corkscrew paths around the magnetic field lines of the Earth and bounce between Earth's magnetic poles high above the atmosphere. Disturbances in Earth's field often allow the ions to dip into the atmosphere, causing it to glow like a fluorescent lamp. This is the beautiful *aurora borealis* (or northern lights); in the Southern Hemisphere, it is an *aurora australis*.

The particles trapped in the inner belt probably originated from Earth's atmosphere. This belt gained newer electrons from high-altitude hydrogen-bomb explosions in 1962.

In spite of Earth's protective magnetic field, many "secondary" cosmic rays reach Earth's surface.[8] These are particles created when "primary" cosmic rays—those coming from outer space—strike atomic nuclei high in the atmosphere. Cosmic-ray bombardment is greatest at the magnetic poles, because charged particles that hit the Earth there do not travel *across* the magnetic field lines but, rather, *along* the field lines and are not deflected. Cosmic-ray bombardment decreases away

[7]Humor aside, the name is actually James A. Van Allen (with his permission).

[8]Some biological scientists speculate that Earth's magnetic changes played a significant role in the evolution of life forms. One hypothesis is that, in the early phases of primitive life, Earth's magnetic field was strong enough to shield the delicate life forms from high-energy charged particles. But, during periods of zero strength, cosmic radiation and the spilling of the Van Allen belts increased the rate of mutation of more robust life forms—not unlike the mutations produced by X-rays in the famous heredity studies of fruit flies. Coincidences between the dates of increased life changes and the dates of the magnetic pole reversals in the last few million years lend support to this hypothesis.

FIGURE 24.24
In September 1997, a magnetometer on the spacecraft *Surveyor* detected a weak magnetic field about Mars that had one eight-hundredth the strength of the magnetic field at the Earth's surface. If the field was stronger in the past, we wonder if it played a role in shielding living material from the solar wind and cosmic rays.

from the poles, and it is smallest in equatorial regions. At middle latitudes, about five particles strike each square centimeter each minute at sea level; this number increases rapidly with altitude. So cosmic rays are penetrating your body as you are reading this—and even when you aren't reading this!

Biomagnetism

FIGURE 24.25
The pigeon may well be able to sense direction because of a built-in magnetic "compass" within its skull.

Certain bacteria biologically produce single-domain grains of magnetite (a compound equivalent to iron ore) that they string together to form internal compasses. They then use these compasses to detect the dip of Earth's magnetic field. Equipped with a sense of direction, the organisms are able to locate food supplies. Amazingly, those bacteria that live south of the equator build the same single-domain magnets as their counterparts that live north of the equator, but they align them in the opposite direction to coincide with the oppositely directed magnetic field in the Southern Hemisphere! Bacteria are not the only living organisms with built-in magnetic compasses: Pigeons have recently been found to have multiple-domain magnetite magnets within their skulls that are connected with a large number of nerves to the pigeon brain. Pigeons have a magnetic sense, and not only can they discern longitudinal directions along Earth's magnetic field, they can also detect latitude by the dip of Earth's field. Magnetic material has also been found in the abdomens of bees, whose behavior is affected by small magnetic fields. Some wasps, Monarch butterflies, sea turtles, and fish join the ranks of creatures with a magnetic sense. Magnetite crystals that resemble the crystals found in magnetic bacteria have been found in human brains. No one knows if these are linked to our sensations. Like the creatures mentioned above, we may share a common magnetic sense.

MRI: MAGNETIC RESONANCE IMAGING

Magnetic resonance image scanners provide high-resolution pictures of the tissues inside a body. Superconducting coils produce a strong magnetic field up to 60,000 times stronger than the intensity of the Earth's magnetic field, which is used to align the protons of hydrogen atoms in the body of the patient.

Like electrons, protons have a "spin" property, so they will align with a magnetic field. Unlike a compass needle that aligns with Earth's magnetic field, the proton's axis wobbles about the applied magnetic field. Wobbling protons are slammed with a burst of radio waves tuned to push the proton's spin axis sideways, perpendicular to the applied magnetic field. When the radio waves pass and the protons quickly return to their wobbling pattern, they emit faint electromagnetic signals whose frequencies depend slightly on the chemical environment in which the proton resides. The signals, which are detected by sensors, are then analyzed by a computer to reveal varying densities of hydrogen atoms in the body and their interactions with surrounding tissue. The images clearly distinguish fluid and bone, for example.

It is interesting to note that MRI was formerly called NMRI (nuclear magnetic resonance imaging), because hydrogen nuclei resonate with the applied fields. Because of public phobia of anything "nuclear," the devices are now called MRI scanners. Tell phobic friends that every atom in their bodies contains a nucleus!

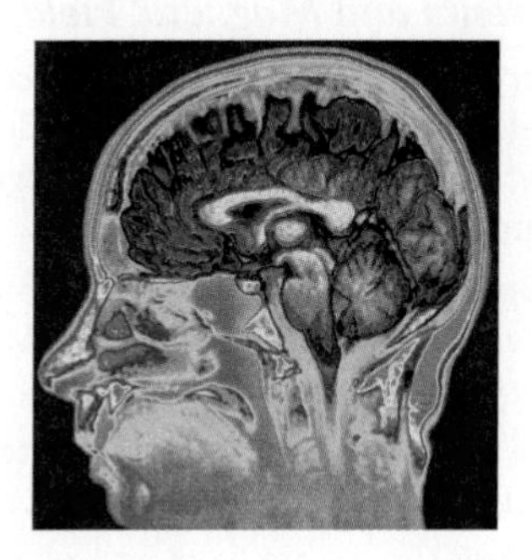

Summary of Terms

Magnetic force (1) Between magnets, it is the attraction of unlike magnetic poles for each other and the repulsion between like magnetic poles. (2) Between a magnetic field and a moving charged particle, it is a deflecting force due to the motion of the particle: The deflecting force is perpendicular to the velocity of the particle and perpendicular to the magnetic field lines. This force is greatest when the charged particle moves perpendicular to the field lines and is smallest (zero) when it moves parallel to the field lines.

Magnetic field The region of magnetic influence around a magnetic pole or a moving charged particle.

Magnetic domains Clustered regions of aligned magnetic atoms. When these regions themselves are aligned with one another, the substance containing them is a magnet.

Electromagnet A magnet whose field is produced by an electric current. It is usually in the form of a wire coil with a piece of iron inside the coil.

Cosmic rays Various high-speed particles that travel throughout the universe and originate in violent events in stars.

Review Questions

1. By whom, and in what setting, was the relationship between electricity and magnetism discovered?

Magnetic Forces

2. The force between electrically charged particles depends on the magnitude of charge, the distance of separation, and what else?
3. What is the source of magnetic force?

Magnetic Poles

4. Is the rule for the interaction between magnetic poles similar to the rule for the interaction between electrically charged particles?
5. In what way are *magnetic poles* very different from *electric charges?*

Magnetic Fields

6. How does magnetic field strength relate to the closeness of magnetic field lines about a bar magnet?
7. What produces a magnetic field?

8. What two kinds of rotational motion are exhibited by electrons in an atom?

Magnetic Domains

9. What is a magnetic domain?
10. At the micro level, what is the difference between an unmagnetized iron nail and a magnetized iron nail?
11. Why will dropping an iron magnet on a hard floor make it a weaker magnet?

Electric Currents and Magnetic Fields

12. In Chapter 22, we learned that the direction of the electric field about a point charge is radial to the charge. What is the direction of the magnetic field surrounding a current-carrying wire?
13. What happens to the direction of the magnetic field about an electric current when the direction of the current is reversed?
14. Why is the magnetic field strength greater inside a current-carrying loop of wire than about a straight section of wire?

Electromagnets

15. Why does a piece of iron in a current-carrying loop increase the magnetic field strength?

Superconducting Electromagnets

16. Why are the magnetic fields of superconducting magnets stronger than those of conventional magnets?

Magnetic Force on Moving Charged Particles

17. In what direction relative to a magnetic field does a charged particle move in order to experience maximum deflecting force? Minimum deflecting force?
18. What effect does Earth's magnetic field have on the intensity of cosmic rays striking the Earth's surface?

Magnetic Force on Current-Carrying Wires

19. Since a magnetic force acts on a moving charged particle, does it make sense that a magnetic force also acts on a current-carrying wire? Defend your answer.
20. What relative direction between a magnetic field and a current-carrying wire results in the greatest force?

Electric Meters

21. How does a galvanometer detect electric current?
22. What is a galvanometer called when it has been calibrated to read current? When it has been calibrated to read voltage?

Electric Motors

23. How often is current reversed in the loops of an electric motor?

Earth's Magnetic Field

24. What is meant by magnetic declination?
25. Why are there probably no permanently aligned magnetic domains in Earth's core?
26. What are *magnetic pole reversals,* and do they occur in the Sun as well as in the Earth?

Cosmic Rays

27. What is the cause of the aurora borealis (northern lights)?

Biomagnetism

28. Name at least six creatures that are known to harbor tiny magnets within their bodies.

Projects

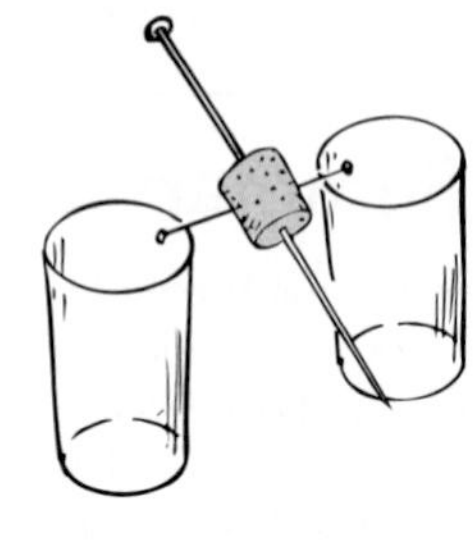

1. Find the direction and dip of the Earth's magnetic field lines in your locality. Magnetize a large steel needle or a straight piece of steel wire by stroking it a couple of dozen times with a strong magnet. Run the needle or wire through a cork in such a way that, when the cork floats, your thin magnet remains horizontal (parallel to the water's surface). Float the cork in a plastic or wooden container of water. The needle will point toward the magnetic pole. Then press a pair of unmagnetized common pins into the sides of the cork. Rest the pins on the rims of a pair of drinking glasses so that the needle or wire points toward the magnetic pole. It should dip in line with Earth's magnetic field.
2. An iron bar can be easily magnetized by aligning it with the magnetic field lines of the Earth and striking it lightly a few times with a hammer. This works best if the bar is tilted down to match the dip of Earth's field. The hammering jostles the domains so they are better able to fall into alignment with the Earth's field. The bar can be demagnetized by striking it when it is oriented in an east–west direction.

Exercises

1. Many dry cereals are fortified with iron, which is added to the cereal in the form of small iron particles. How might these particles be separated from the cereal?
2. In what sense are all magnets electromagnets?
3. All atoms have moving electric charges. Why, then, aren't all materials magnetic?
4. To make a compass, point an ordinary iron nail along the direction of the Earth's magnetic field (which, in the Northern Hemisphere, is angled downward as well as northward) and repeatedly strike it for a few seconds with a hammer or a rock. Then suspend it at its center of gravity by a string. Why does the act of striking magnetize the nail?
5. If you place a chunk of iron near the north pole of a magnet, attraction will occur. Why will attraction also occur if you place the same iron near the south pole of the magnet?
6. Do the poles of a horseshoe magnet attract each other? If you bend the magnet so that the poles get closer together, what happens to the force between the poles?
7. Why is it inadvisable to make a horseshoe magnet from a flexible material?
8. What kind of force field surrounds a stationary electric charge? What additional field surrounds it when it moves?
9. Your study buddy claims that an electron always experiences a force in an electric field, but not always in a magnetic field. Do you agree? Why or why not?
10. What is different about the magnetic poles of common refrigerator magnets and those of common bar magnets?
11. A friend tells you that a refrigerator door, beneath its layer of white-painted plastic, is made of aluminum. How could you check to see if this is true (without any scraping)?
12. Why will a magnet attract an ordinary nail or paper clip but not a wooden pencil?
13. Will either pole of a magnet attract a paper clip? Explain what is happening inside the attracted paper clip. (*Hint:* Consider Figure 24.13.)
14. Why aren't permanent magnets really permanent?
15. One way to make a compass is to stick a magnetized needle into a piece of cork and float it in a glass bowl full of water. The needle will align itself with the horizontal component of the Earth's magnetic field. Since the north pole of this compass is attracted northward, will the needle float toward the north side of the bowl? Defend your answer.
16. A "dip needle" is a small magnet mounted on a horizontal axis so that it can swivel up or down (like a compass turned on its side). Where on Earth will a dip needle point most nearly vertically? Where will it point most nearly horizontally?
17. In what direction would a compass needle point, if it were free to point in all directions, when located at the Earth's north magnetic pole in Canada?
18. What is the net magnetic force on a compass needle? By what mechanism does a compass needle align with a magnetic field?
19. Since the iron filings that align with the magnetic field of the bar magnet shown in Figure 24.2 are not solely little individual magnets, by what mechanism do they align themselves with the field of the magnet?
20. The north pole of a compass is attracted to the north magnetic pole of the Earth, yet like poles repel. Can you resolve this apparent dilemma?
21. We know that a compass points northward because Earth is a giant magnet. Will the northward-pointing needle point northward when the compass is brought to the Southern Hemisphere?
22. Your friend says that, when a compass is taken across the equator, it turns around and points in the opposite direction. Your other friend says that this is not true, that people in the Southern Hemisphere use the south magnetic pole of the compass to point toward the nearest pole. You're on; what do you say?
23. In what position can a current-carrying loop of wire be located in a magnetic field so that it doesn't tend to rotate?
24. Magnet A has twice the magnetic field strength of magnet B (at equal distance) and, at a certain distance, it pulls on magnet B with a force of 50 N. With how much force, then, does magnet B pull on magnet A?
25. In Figure 24.15 we see a magnet exerting a force on a current-carrying wire. Does a current-carrying wire exert a force on a magnet? Why or why not?
26. A strong magnet attracts a paper clip to itself with a certain force. Does the paper clip exert a force on the strong magnet? If not, why not? If so, does it exert as much force on the magnet as the magnet exerts on it? Defend your answers.
27. A current-carrying wire is in a north–south orientation. When a compass needle is placed below or above it, in what direction does the compass needle point?

28. A loudspeaker consists of a cone attached to a current-carrying coil located in a magnetic field. What is the relationship between vibrations in the current and vibrations of the cone?

29. Will a superconducting magnet use less electric power than a traditional copper-wire electromagnet or the same amount of power? Defend your answer.

30. When iron-hulled naval ships are built, the location of the shipyard and the orientation of the ship in the shipyard are recorded on a brass plaque permanently attached to the ship. Why?

31. Can an electron at rest in a magnetic field be set into motion by the magnetic field? What if it were at rest in an electric field?

32. A beam of electrons passes through a magnetic field without being deflected. What can you conclude about the orientation of the beam relative to the magnetic field? (Neglect any other fields.)

33. A cyclotron is a device for accelerating charged particles to high speed as they follow an expanding spiral path. The charged particles are subjected to both an electric field and a magnetic field. One of these fields increases the speed of the charged particles, and the other field causes them to follow a curved path. Which field performs which function?

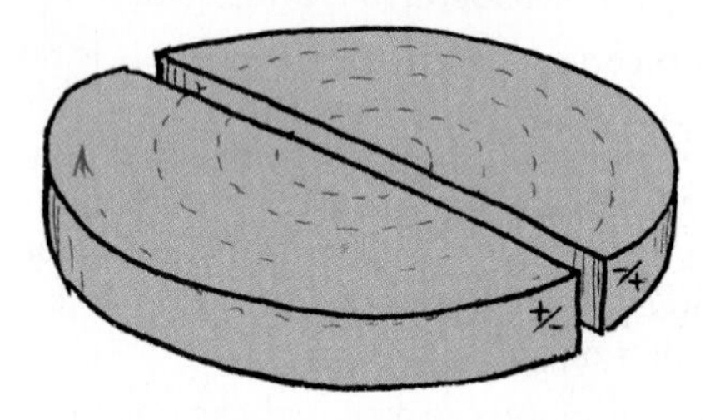

34. A proton moves in a circular path perpendicular to a constant magnetic field. If the field strength of the magnet is increased, does the diameter of the circular path increase, decrease, or remain the same?

35. A beam of high-energy protons emerges from a cyclotron. Do you suppose that there is a magnetic field associated with these particles? Why or why not?

36. A magnet can exert a force on a moving charged particle, but it cannot change the particle's kinetic energy. Why not?

37. A magnetic field can deflect a beam of electrons, but it cannot do work on the electrons to change their speed. Why?

38. Two charged particles are projected into a magnetic field that is perpendicular to their velocities. If the particles are deflected in opposite directions, what does this tell you about them?

39. Inside a laboratory room there is said to be either an electric field or a magnetic field, but not both. What experiments might be performed to establish what kind of field is in the room?

40. Why do astronauts keep to altitudes beneath the Van Allen radiation belts when doing space walks?

41. Residents of northern Canada are bombarded by more intense cosmic radiation than residents of Mexico. Why is this so?

42. What changes in cosmic-ray intensity at the Earth's surface would you expect during periods in which the Earth's magnetic field passed through a zero phase while undergoing pole reversals?

43. In a mass spectrometer (Figure 34.14), ions are directed into a magnetic field, where they curve and strike a detector. If a variety of singly ionized atoms travel at the same speed through the magnetic field, would you expect them all to be deflected by the same amount, or would different ions be bent different amounts? Defend your answer.

44. One way to shield a habitat in outer space from cosmic rays is with an absorbing blanket of some kind, which would function much like the atmosphere that protects the Earth. Speculate on a second way for shielding the habitat that would also be similar to the Earth's natural shielding.

45. If you had two bars of iron—one magnetized and one unmagnetized—and no other materials at hand, how could you determine which bar was the magnet?

46. Historically, replacing dirt roads with paved roads reduced friction on vehicles. Replacing paved roads with steel rails reduced friction further. What recent step reduces rail friction of vehicles? What friction remains after rail friction is eliminated?

47. Will a pair of parallel current-carrying wires exert forces on each other?

48. What is the magnetic effect of placing two wires with equal but oppositely directed currents close together or twisted about each other?

49. When a current is passed through a helically coiled spring, the spring contracts as if it's compressed. What's your explanation?

50. When preparing to undergo an MRI scan, why are patients advised to remove eyeglasses, watches, jewelry, and other metal objects?

Jean Curtis prompts a check-your-neighbor discussion to explain why the copper ring levitates about the iron core of the electromagnet.

CHAPTER 25

Electromagnetic Induction

In the early 1800s, the only current-producing devices were voltaic cells, which produced small currents by dissolving metals in acids. These were the forerunners of present-day batteries. In 1820, Oersted found that magnetism was produced by current-carrying wires. The question arose as to whether electricity could be produced from magnetism. The answer was provided in 1831 by two physicists, Michael Faraday in England and Joseph Henry in the United States—each working without knowledge of the other. Their discovery changed the world by making electricity commonplace—powering industries by day and lighting cities at night.

Electromagnetic Induction

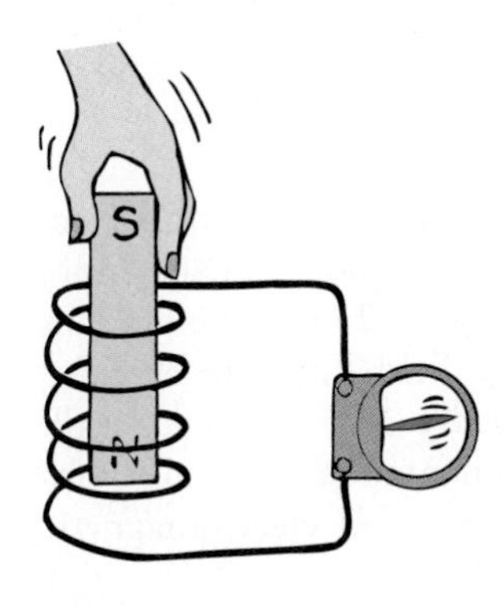

FIGURE 25.1
When the magnet is plunged into the coil, voltage is induced in the coil and charges in the coil are set in motion.

Faraday and Henry both discovered that electric current can be produced in a wire simply by moving a magnet in or out of a coiled part of the wire (Figure 25.1). No battery or other voltage source is needed—only the motion of a magnet in a wire loop. They found that voltage is caused, or *induced,* by the relative motion between a wire and a magnetic field. Voltage is induced whether the magnetic field of a magnet moves near a stationary conductor or the conductor moves in a stationary magnetic field (Figure 25.2). The results are the same for the same *relative* motion.

The greater the number of loops of wire that move in a magnetic field, the greater the induced voltage (Figure 25.3). Pushing a magnet into twice as many loops will induce twice as much voltage; pushing it into ten times as many loops will induce ten times as much voltage; and so on.[1] It may seem that we get something (energy) for nothing simply by increasing the number of loops in a coil of wire. But, assuming that the coil is connected to a resistor or other energy-dissipating device, we don't: We find that it is more difficult to push the magnet into a coil made up of a greater number of loops.

[1]Multiple loops of wire must be insulated, for bare wire loops touching each other make a short circuit. Interestingly, Joseph Henry's wife tearfully sacrificed part of the silk in her wedding gown to cover the wires of Henry's first electromagnets.

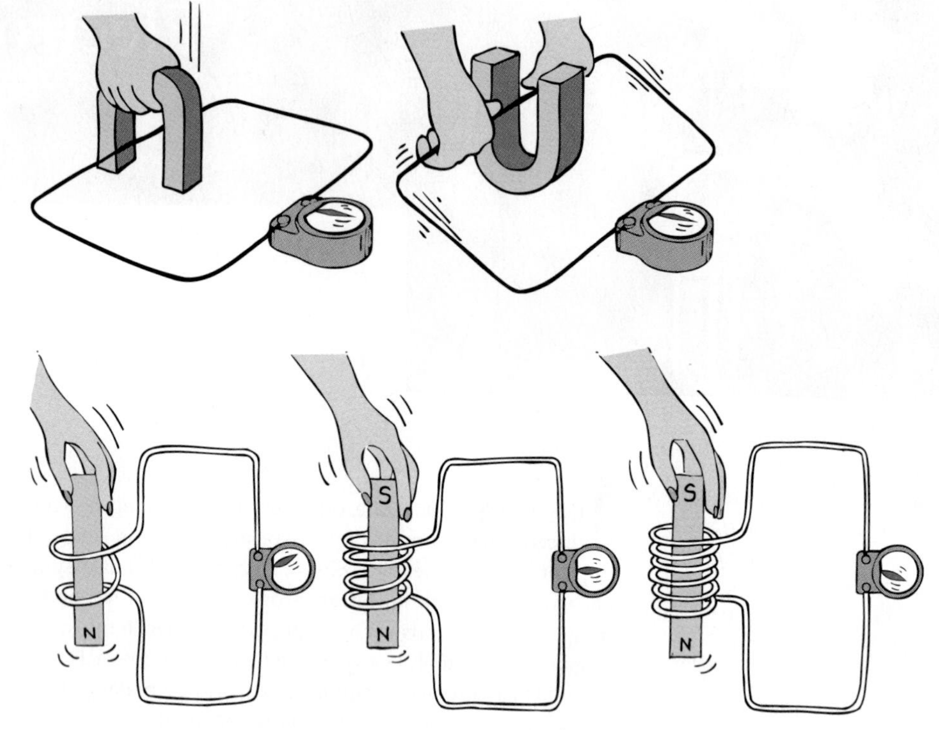

FIGURE 25.2
Voltage is induced in the wire loop either when the magnetic field moves past the wire or the wire moves through the magnetic field.

FIGURE 25.3 Interactive Figure
When a magnet is plunged into a coil of twice as many loops as another, twice as much voltage is induced. If the magnet is plunged into a coil with three times as many loops, then three times as much voltage is induced.

FIGURE 25.4
It is more difficult to push the magnet into a coil with more loops because the magnetic field of each current loop resists the motion of the magnet.

This is because the induced voltage makes a current, which makes an electromagnet, which repels the magnet in our hand. More loops means more voltage, which means we do more work to induce it (Figure 25.4). The amount of voltage induced depends on how fast the magnetic field lines are entering or leaving the coil. Very slow motion produces hardly any voltage at all. Quick motion induces a greater voltage. This phenomenon of inducing voltage by changing the magnetic field in a coil of wire is called **electromagnetic induction.**

Faraday's Law

Electromagnetic induction is summarized by **Faraday's law,** which states,

> **The induced voltage in a coil is proportional to the product of the number of loops and the rate at which the magnetic field changes within those loops.**

The amount of *current* produced by electromagnetic induction depends not only on the induced voltage but also on the resistance of the coil and the circuit to which it is connected.[2] For example, we can plunge a magnet in and out of a closed loop of rubber and in and out of a closed loop of copper. The voltage induced in each is the same, provided that the loops are the same size and the magnet moves with the same speed. But the current in each is quite different. Electrons in the rubber sense the same electric field as those in the copper, but their bonding to the fixed atoms prevents the movement of charge that so freely occurs in copper.

Changing a magnetic field in a closed loop induces voltage. If the loop is in an electrical conductor, then current is induced.

Insights

CHECK YOURSELF

1. What happens when a magnetically stored bit of information on a computer disk spins under a reading head that contains a small coil?
2. If you push a magnet into a coil connected to a resistor, as shown in Figure 25.4, you'll feel a resistance to your push. Why is this resistance greater in a coil with more loops?

Shake flashlights need no batteries. Shake the flashlight for 30 seconds or so and generate up to 5 minutes of bright illumination. Electromagnetic induction occurs as a built-in magnet slides to and fro between coils that charge a capacitor. When brightness diminishes, shake again. You provide the energy to charge the capacitor.

Insights

We have mentioned two ways in which voltage can be induced in a loop of wire: by moving the loop near a magnet or by moving a magnet near the loop. There is a third way, by changing a current in a nearby loop. All three cases possess the same essential ingredient—a changing magnetic field in the loop.

Electromagnetic induction is all around us. On the road, it triggers traffic lights when a car drives over—and changes the magnetic field in—coils of wire beneath the road surface. Hybrid cars utilize it to convert braking energy into electric energy in their batteries. We sense electromagnetic induction in the security systems of airports as we walk through upright coils, and, if we are carrying any significant quantities of iron, change the magnetic field of the coils and trigger an alarm. We use it with an ATM card when its magnetic strip is swiped through a scanner. We hear its effects every time we play a tape recorder. Electromagnetic induction is everywhere. As we shall see, at the end of this chapter and at the beginning of the following chapter, it even underlies the electromagnetic waves we call light.

CHECK YOUR ANSWERS

1. The changing magnetic field in the coil induces voltage. In this way, information stored magnetically on the disk is converted to electric signals.
2. Simply put, more work is required to provide more energy to be dissipated by more current in the resistor. You can also look at it this way: When you push a magnet into a coil, you cause the coil to become a magnet (an electromagnet). The more loops there are in the coil, the stronger the electromagnet that you produce and the stronger it pushes back against the magnet you are moving. (If the coil's electromagnet attracted your magnet instead of repelling it, energy would be created from nothing and the law of energy conservation would be violated. So the coil must repel your magnet.)

[2] Current also depends on the "inductance" of the coil. Inductance measures the tendency of a coil to resist a change in current because the magnetism produced by one part of the coil acts to oppose the change of current in other parts of the coil. In ac circuits, it is akin to resistance, depending on the frequency of the ac source and the number of loops in the coil. We'll not cover this topic here.

Generators and Alternating Current

FIGURE 25.5
Guitar pickups are tiny coils with magnets inside them. The magnets magnetize the steel strings. When the strings vibrate, voltage is induced in the coils and boosted by an amplifier, and sound is produced by a speaker.

When one end of a magnet is repeatedly plunged into and back out of a coil of wire, the direction of the induced voltage alternates. As the magnetic field strength inside the coil is increased (as the magnet enters the coil), the induced voltage in the coil is directed one way. When the magnetic field strength diminishes (as the magnet leaves the coil), the voltage is induced in the opposite direction. The frequency of the alternating voltage that is induced equals the frequency of the changing magnetic field within the loop.

It is more practical to induce voltage by moving a coil than it is by moving a magnet. This can be done by rotating the coil in a stationary magnetic field (Figure 25.6). This arrangement is called a **generator.** The construction of a generator is, in principle, identical to that of a motor. They look the same, but the roles of input and output are reversed. In a motor, electric energy is the input and mechanical energy is the output; in a generator, mechanical energy is the input and electric energy is the output. Both devices simply transform energy from one form to another.

It is interesting to compare the physics of a motor and a generator and to see that both operate under the same underlying principle: that moving electrons experience a force that is mutually perpendicular to both their velocity and the magnetic field they traverse (Figure 25.7). We will call the deflection of the wire the *motor effect,* and we will call what happens as a

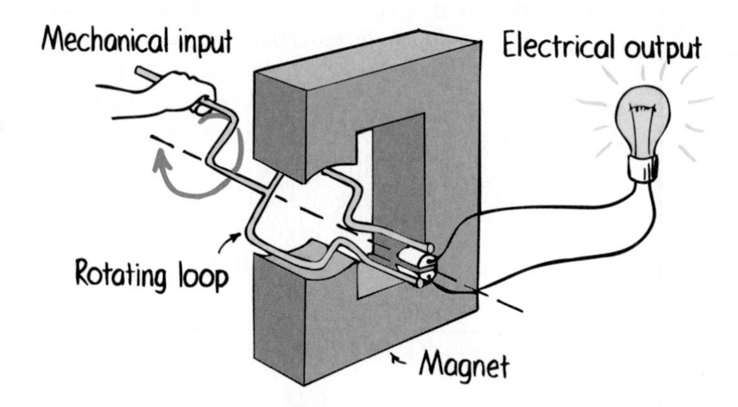

FIGURE 25.6
Interactive Figure
A simple generator. Voltage is induced in the loop when it is rotated in the magnetic field.

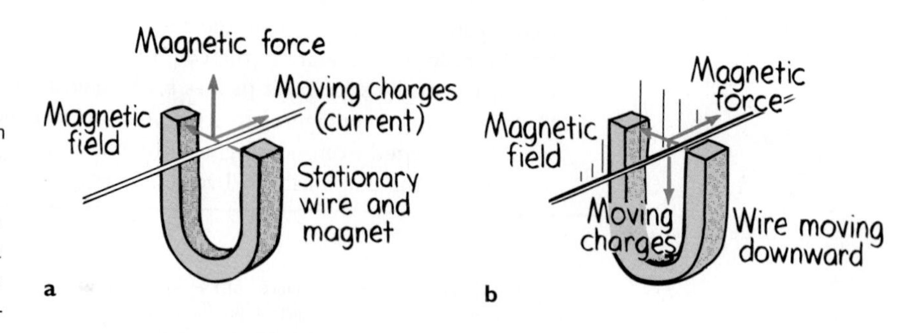

FIGURE 25.7
Interactive Figure
(a) Motor effect: When a current moves along the wire, there is a perpendicular upward force on the electrons. Since there is no conducting path upward, the wire is tugged upward along with the electrons. (b) Generator effect: When a wire with no initial current is moved downward, the electrons in the wire experience a deflecting force perpendicular to their motion. There is a conducting path in this direction that the electrons follow, thereby constituting a current.

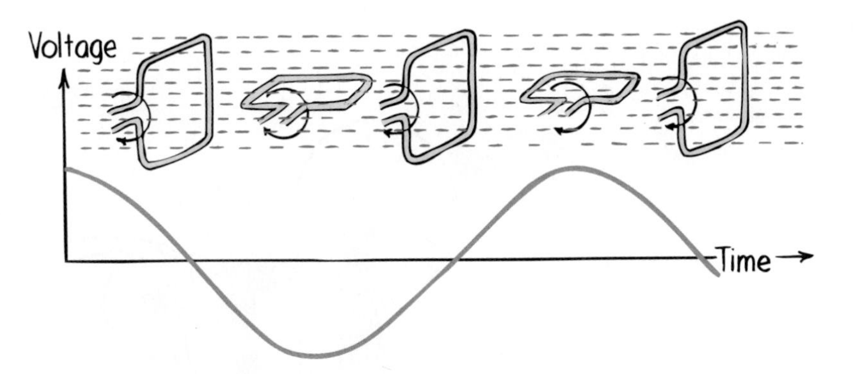

FIGURE 25.8
As the loop rotates, the induced voltage (and current) changes in magnitude and direction. One complete rotation of the loop produces one complete cycle in voltage (and in current).

When you step on the brakes in a hybrid car, the electric motor becomes a generator and charges a battery.

Insights

Application of E & M Induction

result of the law of induction the *generator effect*. These effects are summarized in (a) and (b) of the figure. Study them. Can you see that the two effects are related?

We can see the electromagnetic induction cycle in Figure 25.8. Note that, when the loop of wire is rotated in the magnetic field, there is a change in the number of magnetic field lines within the loop. When the plane of the loop is perpendicular to the field lines, the maximum number of lines is enclosed. As the loop rotates, it in effect chops the lines, so that fewer lines are enclosed. When the plane of the loop is parallel to the field lines, none are enclosed. Continued rotation increases and decreases the number of enclosed lines in cyclic fashion, with the greatest rate of change of field lines occurring when the number of enclosed field lines goes through zero. Hence, the induced voltage is greatest as the loop rotates through its parallel-to-the-lines orientation. Because the voltage induced by the generator alternates, the current produced is ac, an alternating current.[3] The alternating current in our homes is produced by generators standardized so that the current goes through 60 cycles of change each second—60 hertz.

Power Production

Nikola Tesla (1857–1943)

Fifty years after Michael Faraday and Joseph Henry discovered electromagnetic induction, Nikola Tesla and George Westinghouse put those findings to practical use and showed the world that electricity could be generated reliably and in sufficient quantities to light entire cities.

Turbogenerator Power

Tesla built generators much like those in operation today, but quite a bit more complicated than the simple model we have discussed. Tesla's generators had armatures—iron cores wrapped with bundles of copper wire—that were made to spin within strong magnetic fields by means of a turbine, which, in turn, was spun by the energy of steam or falling water. The rotating loops of wire in the

[3]With appropriate brushes and by other means, the ac in the loop(s) can be converted to dc to make a dc generator.

FIGURE 25.9
Steam drives the turbine, which is connected to the armature of the generator.

In making a great discovery, being at the right place at the right time is not enough—curiosity and hard work are also important.

Insights

armature cut through the magnetic field of the surrounding electromagnets, thereby inducing alternating voltage and current.

We can look at this process from an atomic point of view. When the wires in the spinning armature cut through the magnetic field, oppositely directed electromagnetic forces act on the negative and positive charges. Electrons respond to this force by momentarily swarming relatively freely in one direction throughout the crystalline copper lattice; the copper atoms, which are actually positive ions, are forced in the opposite direction. Because the ions are anchored in the lattice, however, they hardly move at all. Only the electrons move, sloshing back and forth in alternating fashion with each rotation of the armature. The energy of this electronic sloshing is tapped at the electrode terminals of the generator.

MHD Power

An interesting device similar to the turbogenerator is the MHD (**m**agneto**h**ydro**d**ynamic) generator, which eliminates the turbine and spinning armature altogether. Instead of making charges move in a magnetic field via a rotating armature, a plasma of electrons and positive ions expands through a nozzle and moves at supersonic speed through a magnetic field. Like the armature in a turbogenerator, the motion of charges through a magnetic field gives rise to a voltage and flow of current in accordance with Faraday's law of induction. Whereas in a conventional generator "brushes" carry the current to the external load circuit, in the MHD generator the same function is performed by conducting plates, or *electrodes* (Figure 25.10). Unlike the turbogenerator, the MHD generator can operate at any temperature to which the plasma can be heated, either by combustion or nuclear processes. The high temperature results in a high thermodynamic efficiency, which means more power for the same amount of fuel and less waste heat. Efficiency is further boosted when the "waste" heat is used to change water into steam and to run a conventional steam-turbine generator.

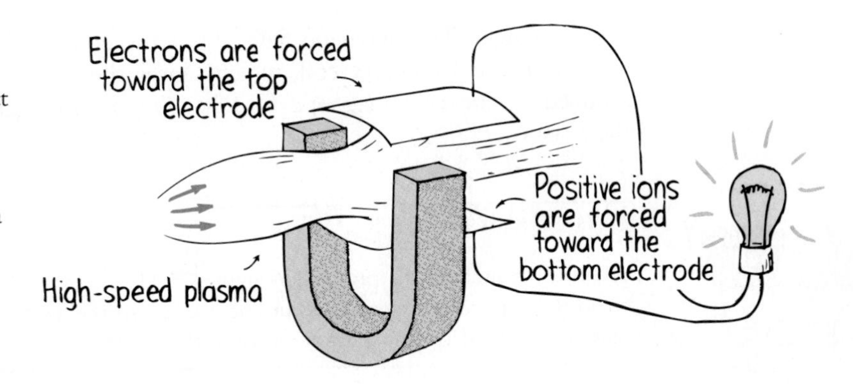
Electrons are forced
toward the top
electrode
Positive ions
are forced
toward the
bottom electrode
High-speed plasma

FIGURE 25.10
A simplified MHD generator. Oppositely directed forces act on the positive and negative particles in the high-speed plasma moving through the magnetic field. The result is a voltage difference between the two electrodes. Current then flows from one electrode to the other through an external circuit. There are no moving parts; only the plasma moves. In practice, superconducting electromagnets are used.

This substitution of a flowing plasma for rotating copper coils in a generator has become operational only since development of the technology to produce sufficiently high-temperature plasmas. Current plants use a high-temperature plasma formed by combustion of fossil fuels in air or oxygen.[4]

It's important to know that generators don't produce energy—they simply convert energy from some other form to electric energy. As we discussed in Chapter 3, energy from a source, whether fossil or nuclear fuel or wind or water, is converted to mechanical energy to drive the turbine. The attached generator converts most of this mechanical energy to electrical energy. Some people think that electricity is a primary source of energy. It is not. It is a carrier of energy that requires a source.

Transformers

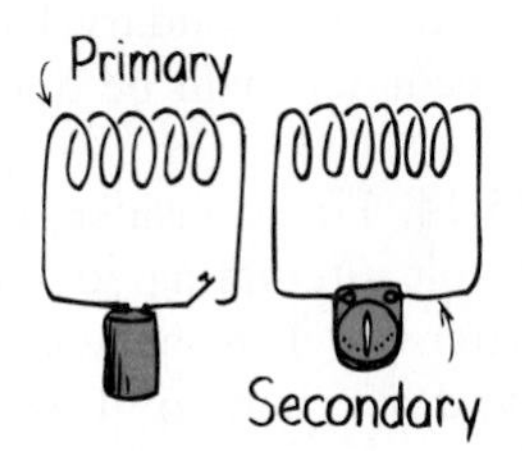
Primary
Secondary

FIGURE 25.11
Whenever the primary switch is opened or closed, voltage is induced in the secondary circuit.

Electric energy can certainly be carried along wires, and now we'll see how it can be carried across empty space. Energy can be transferred from one device to another with the simple arrangement shown in Figure 25.11. Note that one coil is connected to a battery and the other is connected to a galvanometer. It is customary to refer to the coil connected to the power source as the *primary* (input) and to the other as the *secondary* (output). As soon as the switch is closed in the primary and current passes through its coil, a current occurs in the secondary also—even though there is no material connection between the two coils. Only a brief surge of current occurs in the secondary, however. Then, when the primary switch is opened, a surge of current again registers in the secondary, but in the opposite direction.

This is the explanation: A magnetic field builds up around the primary when the current begins to flow through the coil. This means that the magnetic field is growing (that is, *changing*) about the primary. But, since the coils are near each other, this changing field extends into the secondary coil, thereby inducing a voltage in the secondary. This induced voltage is only temporary, for when the current and the magnetic field of the primary reach a steady state—that is, when the magnetic field is no longer changing—no further voltage is induced in the secondary. But, when the switch is turned off, the current in the primary

[4]Lower temperatures are sufficient when the electrically conducting fluid is liquid metal, usually lithium. A liquid-metal MHD power system is referred to as a LMMHD power system.

drops to zero. The magnetic field about the coil collapses, thereby inducing a voltage in the secondary coil, which senses the change. We see that voltage is induced whenever a magnetic field is *changing* through the coil, regardless of the reason.

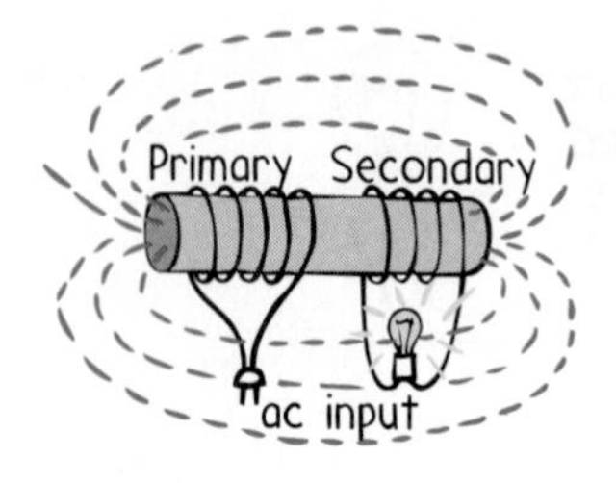

FIGURE 25.12
A simple transformer.

CHECK YOURSELF

When the switch of the primary in Figure 25.11 is opened or closed, the galvanometer in the secondary registers a current. But when the switch remains closed, no current is registered on the galvanometer of the secondary. Why?

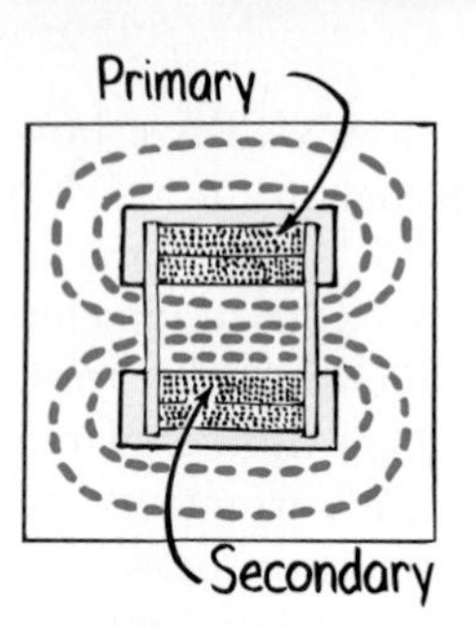

FIGURE 25.13
A practical and more efficient transformer. Both primary and secondary coils are wrapped on the inner part of the iron core (yellow), which guides alternating magnetic lines (green) produced by ac in the primary. The alternating field induces ac voltage in the secondary. Thus power at one voltage from the primary is transferred to the secondary at a different voltage.

If you place an iron core inside the primary and secondary coils of the arrangement of Figure 25.11, the magnetic field within the primary is intensified by the alignment of magnetic domains. The field is also concentrated in the core and extends into the secondary, which intercepts more of the field change. The galvanometer will show greater surges of current when the switch of the primary is opened or closed. Instead of opening and closing a switch to produce the change of magnetic field, suppose that alternating current is used to power the primary. Then the frequency of periodic changes in the magnetic field is equal to the frequency of the alternating current. Now we have a **transformer** (Figure 25.12). A more efficient arrangement is shown in Figure 25.13.

If the primary and secondary have equal numbers of wire loops (usually called *turns*), then the input and output alternating voltages will be equal. But if the secondary coil has more turns than the primary, the alternating voltage produced in the secondary coil will be greater than that produced in the primary. In this case, the voltage is said to be *stepped up*. If the secondary has twice as many turns as the primary, the voltage in the secondary will be double that of the primary.

We can see this with the arrangements in Figure 25.14. First consider the simple case of a single primary loop connected to a 1-volt alternating source and a single secondary loop connected to the ac voltmeter (a). The secondary intercepts the changing magnetic field of the primary, and a voltage of 1 volt is induced in the secondary. If another loop is wrapped around the core so the transformer has two secondaries (b), it intercepts the same magnetic field change. We see that 1 volt is induced in it also. There is no need to keep both secondaries separate, for we could join them (c) and still have a total induced voltage of 1 volt + 1 volt, or 2 volts. This is equivalent to saying that a voltage of 2 volts will be induced in a single secondary that has twice the number of loops as the primary. If the secondary is wound with three times as many loops, then three times as much voltage will be induced. Stepped-up voltage may light a neon sign or send power over a long distance.

CHECK YOUR ANSWER

When the switch remains in the closed position, there is a steady current in the primary and a steady magnetic field about the coil. This field extends to the secondary, but, unless there is a *change* in the field, electromagnetic induction does not occur.

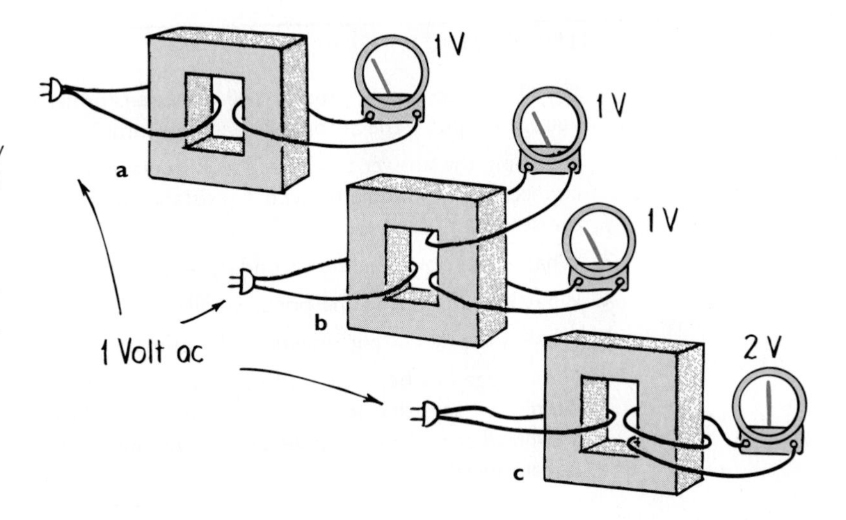

FIGURE 25.14
(a) The voltage of 1 V induced in the secondary equals the voltage of the primary. (b) A voltage of 1 V is also induced in the added secondary because it intercepts the same magnetic field change from the primary. (c) The voltages of 1 V each induced in the two one-turn secondaries are equivalent to a voltage of 2 V induced in a single two-turn secondary.

If the secondary has fewer turns than the primary, the alternating voltage produced in the secondary will be *lower* than that produced in the primary. The voltage is said to be *stepped down*. This stepped-down voltage may safely operate a toy electric train. If the secondary has half as many turns as the primary, then only half as much voltage is induced in the secondary. So electric energy can be fed into the primary at a given alternating voltage and taken from the secondary at a greater or lower alternating voltage, depending on the relative number of turns in the primary and secondary coil windings.

The relationship between primary and secondary voltages with respect to the relative number of turns is given by

$$\frac{\text{Primary voltage}}{\text{Number of primary turns}} = \frac{\text{secondary voltage}}{\text{number of secondary turns}}$$

It might seem that we get something for nothing with a transformer that steps up the voltage. Not so, for energy conservation always regulates what can happen. When voltage is stepped up, current in the secondary is less than in the primary. The transformer actually transfers energy from one coil to the other. Make no mistake about this point: It can in no way step up energy—a conservation of energy no-no. A transformer steps voltage up or down without a change in energy. The rate at which energy is transferred is called *power*. The power used in the secondary is supplied by the primary. The primary gives no more than the secondary uses, in accord with the law of conservation of energy. If the slight power losses due to heating of the core are neglected, then

$$\text{Power into primary} = \text{power out of secondary}$$

Electric power is equal to the product of voltage and current, so we can say

$$(\text{Voltage} \times \text{current})_{\text{primary}} = (\text{voltage} \times \text{current})_{\text{secondary}}$$

We see that, if the secondary has more voltage than the primary, it will have less current than the current in the primary. The ease with which voltages can be stepped up or down with a transformer is the principal reason that most electric power is ac rather than dc.

CHECK YOURSELF

1. If 100 V of ac is put across a 100-turn transformer primary, what will be the voltage output if the secondary has 200 turns?
2. Assuming the answer to the previous question is 200 V, and the secondary is connected to a floodlamp with a resistance of 50 Ω, what will be the ac current in the secondary circuit?
3. What is the power in the secondary coil?
4. What is the power in the primary coil?
5. What is the ac current drawn by the primary coil?
6. The voltage has been stepped up, and the current has been stepped down. Ohm's law says that increased voltage will produce increased current. Is there a contradiction here, or does Ohm's law not apply to circuits that have transformers?

Self-Induction

Current-carrying loops in a coil interact not only with loops of other coils but also with loops of the same coil. Each loop in a coil interacts with the magnetic field around the current in other loops of the same coil. This is *self-induction.* A self-induced voltage is produced. This voltage is always in a direction opposing the changing voltage that produces it and is commonly called the "back electromotive force," or simply "back emf."[5] We won't treat self-induction and back emfs here, except to acknowledge a common and dangerous effect. Suppose that a coil with a large number of turns is used as

CHECK YOUR ANSWERS

1. From 100 V/100 primary turns = (?) V/200 secondary turns, you can see that the secondary puts out 200 V.
2. From Ohm's law, 200 V/50 Ω = 4 A.
3. Power = 200 V × 4 A = 800 W.
4. By the law of conservation of energy, the power in the primary is the same, 800 W.
5. 800 W = 100 V × (?) A, so you see the primary draws 8 A. (Note that the voltage is stepped up from primary to secondary and that the current is correspondingly stepped down.)
6. Ohm's law still holds in the secondary circuit. The voltage induced across the secondary circuit, divided by the load (resistance) of the secondary circuit, equals the current in the secondary circuit. In the primary circuit, on the other hand, there is no conventional resistance. What "resists" the current in the primary is the transfer of energy to the secondary.

[5]The opposition of an induced effect to an inducing cause is called *Lenz's law;* it is a consequence of the conservation of energy.

FIGURE 25.15
When the switch is opened, the magnetic field of the coil collapses. This sudden change in the field can induce a huge voltage.

an electromagnet and is powered with a dc source, perhaps a small battery. Current in the coil is then accompanied by a strong magnetic field. When we disconnect the battery by opening a switch, we had better be prepared for a surprise. When the switch is opened, the current in the circuit falls rapidly to zero and the magnetic field in the coil undergoes a sudden decrease (Figure 25.15). What happens when a magnetic field suddenly changes in a coil, even if it is the same coil that produced it? The answer is that a voltage is induced. The rapidly collapsing magnetic field with its store of energy may induce an enormous voltage, large enough to develop a strong spark across the switch—or to you, if you are opening the switch! For this reason, electromagnets are connected to a circuit that absorbs excess charge and prevents the current from dropping too suddenly. This reduces the self-induced voltage. This is also, by the way, why you should disconnect appliances by turning off the switch, not by pulling out the plug. The circuitry in the switch may prevent a sudden change in current.

Power Transmission

Almost all electric energy sold today is in the form of ac, traditionally because of the ease with which it can be transformed from one voltage to another.[6] Large currents in wires produce heat and energy losses, so power is transmitted great distances at high voltages and correspondingly low currents (power = voltage × current). Power is generated at 25,000 V or less and is stepped up near the power station to as much as 750,000 V for long-distance transmission, then stepped down in stages at substations and distribution points to voltages needed in industrial applications (often 440 V or more) and for the home (240 and 120 V).

Energy, then, is transferred from one system of conducting wires to another by electromagnetic induction. It is but a short step further to find that the same principles account for eliminating wires and sending energy from a radio-transmitter antenna to a radio receiver many kilometers away. Extend these principles just a tiny step further to the transformation of energy of vibrating electrons in the Sun to life energy on Earth. The effects of electromagnetic induction are very far reaching.

FIGURE 25.16
Power transmission.

[6]Nowadays, power utilities can transform dc voltages using semiconductor technology. Keep an eye on the present advances in superconductor technology, and watch for resulting changes in power transmission.

Field Induction

Electromagnetic induction explains the induction of voltages and currents. Actually, the more basic *fields* are at the root of both voltages and currents. The modern view of electromagnetic induction states that electric and magnetic fields are induced. These, in turn, produce the voltages we have considered. So induction occurs whether or not a conducting wire or any material medium is present. In this more general sense, Faraday's law states:

An electric field is induced in any region of space in which a magnetic field is changing with time.

There is a second effect, an extension of Faraday's law. It is the same except that the roles of electric and magnetic fields are interchanged. It is one of nature's many symmetries. This effect was advanced by the British physicist James Clerk Maxwell in about 1860 and is known as **Maxwell's counterpart to Faraday's law:**

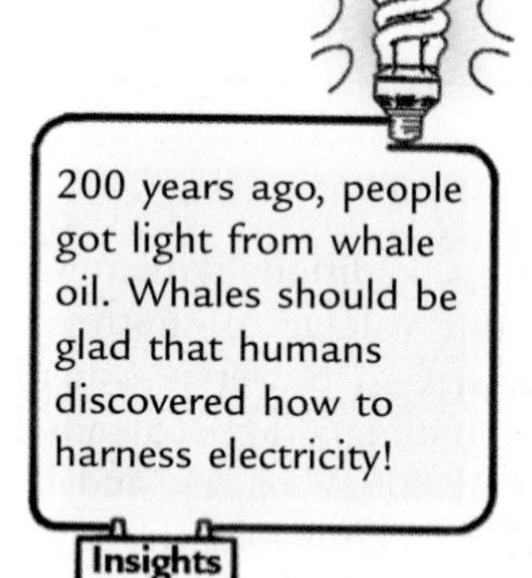

200 years ago, people got light from whale oil. Whales should be glad that humans discovered how to harness electricity!
Insights

A magnetic field is induced in any region of space in which an electric field is changing with time.

In each case, the strength of the induced field is proportional to the rate of change of the inducing field. Induced electric and magnetic fields are at right angles to each other.

Maxwell saw the link between electromagnetic waves and light.[7] If electric charges are set into vibration in the range of frequencies that match those of light, waves are produced that *are* light! Maxwell discovered that light is simply electromagnetic waves in the range of frequencies to which the eye is sensitive.

FIGURE 25.17
Sheron Snyder converts mechanical energy into electromagnetic energy, which, in turn, converts to light.

[7]On the eve of his discovery, Maxwell had a date with a young woman he was later to marry. While walking in a garden, his date remarked about the beauty and wonder of the stars. Maxwell asked how she would feel to know that she was walking with the only person in the world who knew what starlight really was. For it was true. At that time, James Clerk Maxwell was the only person in the world to know that light of any kind is energy carried in waves of electric and magnetic fields that continually regenerate each other.

ELECTROMAGNETIC FIELDS AND CANCER

Consider this recipe for gaining fame as an advocate for a better world: Write a best-selling book about a grave but hidden danger, become a hero in the public eye, and, to top it off, make a lot of money. The recipe is simple enough if you are willing to sacrifice objectivity and, perhaps, integrity: Just identify what can scare people, find a culprit who can fit the role of a bad guy, then seek and report testimonials (not studies!) that point an accusing finger at the culprit. When knowledgeable people oppose you, accuse them of concealing the truth and joining forces with the culprit. This recipe has been successful time and again.

In 1989, for example, such an alarmist published a series of sensational articles in a prominent magazine that aroused public fears about power lines and cancer. His claim was that people living near power lines were at great risk of getting cancer. Living near power lines, he stated, was the greatest health hazard facing the American public. He fanned flames that originated some ten years earlier when another alarmist reported increased leukemia in children who lived near power transformers in Denver. The leukemia-and-transformer scare was soon generalized to various cancers and power lines across the country. Not surprisingly, the magazine writer cited data that confirmed his accusations while ignoring data that didn't. This is like finding bullet holes in the side of a barn, painting a bullseye around them, then reporting the high correlation of bullets to the target. Yes, there are a lot of bullet holes (cancers) in the target area (near power lines), but there are a lot of bullet holes in other areas too! Not surprisingly, the writer found it more effective to relate frightening anecdotes of suffering and death from cancer than to report the results of published studies on the subject. He became a public hero, appeared on the most popular television shows, and published his series of magazine articles as a best-selling book with the morbid title *Currents of Death*. The writer was Paul Brodeur, who has since died.

Magnetic fields produced by electric power in most homes and workplaces are about 1% as strong as Earth's natural magnetic field. The overwhelming consensus among scientists was that no power-line hazard existed—which was seen by Brodeur as evidence that the scientific community was in league with the utilities and government in a massive cover-up. Studies accumulated. In 1994 a four-year study of 223,000 Canadian and French electrical workers showed no overall increase in cancer risk associated with occupational exposure to electromagnetic fields. An exhaustive review study in 1995 by the American Physical Society found no link between cancer and power lines.

Many spurious scientific claims are made by sincere people who truly believe their own rhetoric, who don't look deeply or critically into what they're talking about. Their ill-founded claims can confuse even educated listeners who are presented with an apparent glut of disparate scientific opinion. A pseudoscientific snowball, once started, can gain considerable momentum—and cost a lot of money. Whether Mr. Brodeur was a true believer or a charlatan we can't know for sure, but we do know the cost of his intemperate rhetoric: eighteen years of paranoia and billions of dollars needlessly spent. During all that time, there was not a single successful lawsuit based on health effects from electromagnetic fields.

The worry and fear generated by studies of the possible connection between electromagnetic fields and cancer produced no progress whatever in cancer prevention and gave peace of mind to no one. The dollars spent gave not one new piece of information on the cause or cure of cancer. Imagine the benefit if only a fraction of the money spent to counteract a nonexistent threat had been spent instead on pursuing valid research into the biologic causes of cancer.

In Perspective[8]

The ancient Greeks discovered that, when a piece of amber (a natural fossil resin) was rubbed, it picked up little pieces of papyrus. They found strange rocks in Magnesia (a region in Thessaly) that attracted iron. Probably because the air in that coastal district of Greece was relatively humid, they never noticed

[8]Adapted from R. P. Feynman, R. B. Leighton, and M. Sands, *The Feynman Lectures on Physics,* Vol. II, pp. 1-10 and 1-11. (Reading, Mass.: Addison-Wesley-Longman, 1964.) Richard P. Feynman, who was a Nobel laureate in physics and professor of physics at California Institute of Technology, is considered by many physicists to be among the most brilliant and inspirational physicists of his time—as well as the most colorful. He died in 1988.

(and therefore never studied) the static electric charge effects common in dry climates. Further development of our knowledge of electrical and magnetic phenomena did not occur until 400 years ago. The human world shrank as more was learned about electricity and magnetism. It became possible first to signal by telegraph over long distances, then to talk to another person many kilometers away through wires, then not only to talk but also to send pictures over many kilometers with no physical connections in between.

Energy, so vital to civilization, could be transmitted over hundreds of kilometers. The energy of elevated rivers was diverted into pipes that fed giant "waterwheels" connected to assemblages of twisted and interwoven copper wires that rotated about specially designed chunks of iron-revolving monsters called *generators*. Out of these, energy was pumped through copper rods as thick as your wrist and sent to huge coils wrapped around transformer cores, boosting it to high voltages for efficient long-distance transmission to cities. Then the transmission lines split into branches, then to more transformers, then more branching and spreading, until, finally, the energy of the river was spread throughout entire cities—turning motors, making heat, making light, working gadgetry. There was the miracle of hot lights from cold water hundreds of kilometers away—a miracle made possible by specially designed bits of copper and iron that turned because people had discovered the laws of electromagnetism.

These laws were discovered at about the time the American Civil War was being fought. From a long view of human history, there can be little doubt that events such as the American Civil War will pale into provincial insignificance in comparison with the more significant event of the nineteenth century: the discovery of the electromagnetic laws.

Summary of Terms

Electromagnetic induction The induction of voltage when a magnetic field changes with time. If the magnetic field within a closed loop changes in any way, a voltage is induced in the loop:

$$\text{Voltage induced} \sim \text{no. of loops} \times \frac{\Delta \text{ magnetic field}}{\Delta \text{ time}}$$

This is a statement of Faraday's law. The induction of voltage is actually the result of a more fundamental phenomenon: the induction of an electric *field*, as defined for the more general case below.

Faraday's law An electric field is created in any region of space in which a magnetic field is changing with time. The magnitude of the induced electric field is proportional to the rate at which the magnetic field changes. The direction of the induced field is at right angles to the changing magnetic field.

Generator An electromagnetic induction device that produces electric current by rotating a coil within a stationary magnetic field. A generator converts mechanical energy to electrical energy.

Transformer A device for transferring electric power from one coil of wire to another, by means of electromagnetic induction, for the purpose of transforming one value of voltage to another.

Maxwell's counterpart to Faraday's law A magnetic field is created in any region of space in which an electric field is changing with time. The magnitude of the induced magnetic field is proportional to the rate at which the electric field changes. The direction of the induced magnetic field is at right angles to the changing electric field.

Review Questions

Electromagnetic Induction

1. Exactly what did Michael Faraday and Joseph Henry discover?
2. What must change in order for electromagnetic induction to occur?

Faraday's Law

3. In addition to induced voltage, on what does the current produced by electromagnetic induction depend?
4. What are the three ways in which voltage can be induced in a wire?

Generators and Alternating Current

5. How does the frequency of induced voltage relate to how frequently a magnet is plunged in and out of a coil of wire?
6. What is the basic *similarity* between a generator and an electric motor? What is the basic *difference* between them?
7. Where in the rotation cycle of a simple generator is the induced voltage at a maximum?
8. Why does a generator produce alternating current?

Power Production

9. Who discovered electromagnetic induction, and who put it to practical use?

Turbogenerator Power

10. What is an armature?
11. What commonly supplies the energy input to a turbine?

MHD Power

12. What are the principal differences between an *MHD generator* and a *conventional generator*?
13. Does an MHD generator employ Faraday's law of induction?

Transformers

14. Electric energy can certainly be carried along wires, but can it be carried across empty space? If so, how?
15. Is electromagnetic induction a key feature of a transformer?
16. Why does a transformer require ac?
17. If a transformer is efficient enough, can it step up energy? Explain.
18. What name is given to the rate at which energy is transferred?
19. What is the principal advantage of ac over dc?

Self-Induction

20. When the magnetic field changes in a coil of wire, voltage in each loop of the coil is induced. Will voltage be induced in a loop if the source of the magnetic field is the coil itself?

Power Transmission

21. What is the purpose of transmitting power at high voltages over long distances?
22. Does the transmission of electric energy require electrical conductors between the source and receiver? Cite an example to defend your answer.

Field Induction

23. What is induced by the rapid alternation of a *magnetic field*?
24. What is induced by the rapid alternation of an *electric field*?

In Perspective

25. How can part of the energy of a cold river become the energy of a hot lamp hundreds of kilometers away?

Project

Write a letter to Grandma and tell her the answer to what has been a mystery for centuries—what light is. Tell her how light is related to electricity and magnetism.

Exercises

1. Why does the word *change* occur so frequently in this chapter?
2. A common pickup for an electric guitar consists of a coil of wire around a small permanent magnet, as described in Figure 25.5. Why will this type of pickup fail with nylon strings?
3. Why does an iron core increase the magnetic induction of a coil of wire?
4. Why are the armature and field windings of an electric motor usually wound on an iron core?
5. Why is a generator armature harder to rotate when it is connected to a circuit and supplying electric current?
6. Why does a motor also tend to act as a generator?
7. Will a cyclist coast farther if the lamp connected to the generator on his bicycle is turned off? Explain.
8. When an automobile moves over a wide, closed loop of wire embedded in a road surface, is the magnetic field of the Earth within the loop altered? Is a pulse of current produced? Can you cite a practical application for this at a traffic intersection?
9. At the security area of an airport, people walk through a large coil of wire and through a weak ac magnetic field. What is the result of a small piece of metal on a person that slightly alters the magnetic field in the coil?
10. A piece of plastic tape coated with iron oxide is magnetized more in some parts than in others. When the tape is moved past a small coil of wire, what happens in the coil? What is a practical application of this?
11. Joseph Henry's wife tearfully sacrificed part of her wedding gown for silk to cover the wires of Joseph's electromagnets. What was the purpose of the silk covering?

12. A certain simple earthquake detector consists of a little box firmly anchored to the Earth. Suspended inside the box is a massive magnet that is surrounded by stationary coils of wire fastened to the box. Explain how this device works, applying two important principles of physics—one studied in Chapter 2, and the other in this chapter.
13. How do the direction of the magnetic force and its effects differ between the motor effect and the generator effect, as shown in Figure 25.7?
14. When you turn the shaft of an electric motor by hand, what occurs in the interior coils of wire?
15. Your friend says that, if you crank the shaft of a dc motor manually, the motor becomes a dc generator. Do you agree or disagree?
16. Does the voltage output increase when a generator is made to spin faster? Defend your answer.
17. An electric saw running at normal speed draws a relatively small current. But, if a piece of wood being sawed jams and the motor shaft is prevented from turning, the current dramatically increases and the motor overheats. Why?
18. If you place a metal ring in a region in which a magnetic field is rapidly alternating, the ring may become hot to your touch. Why?
19. A magician places an aluminum ring on a table. Underneath is a hidden electromagnet. When the magician says "abracadabra" (and pushes a switch that starts current flowing through the coil under the table), the ring jumps into the air. Explain his "trick."
20. In the chapter-opening photograph, Jean Curtis asks her class why the copper ring levitates about the iron core of the electromagnet. What is the explanation, and does it involve ac or dc?
21. How could a lightbulb near, but not touching, an electromagnet be lit? Is ac or dc required? Defend your answer.
22. A length of wire is bent into a closed loop and a magnet is plunged into it, inducing a voltage and, consequently, a current in the wire. A second length of wire, twice as long, is bent into two loops of wire, and a magnet is similarly plunged into it. Twice the voltage is induced, but the current is the same as that produced in the single loop. Why?
23. Two separate but similar coils of wire are mounted close to each other, as shown below. The first coil is connected to a battery and has a direct current flowing through it. The second coil is connected to a galvanometer. 1. How does the galvanometer respond when the switch in the first circuit is closed? 2. After being closed how does the meter respond when the current is steady? 3. When the switch is opened?

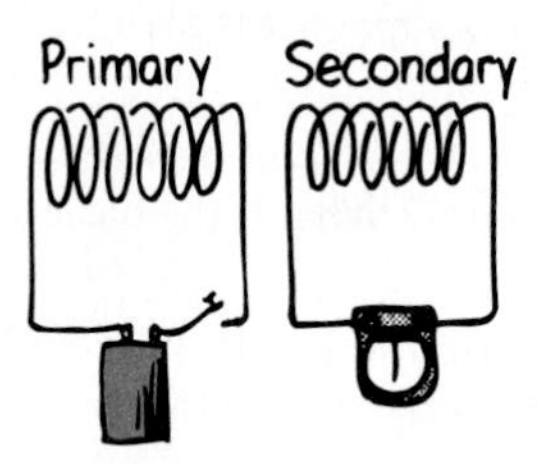

24. Why will more voltage be induced with the apparatus shown above if an iron core is inserted in the coils?
25. Why will a transformer not work if you are using dc?
26. How does the current in the secondary of a transformer compare with the current in the primary when the secondary voltage is twice the primary voltage?
27. In what sense can a transformer be considered to be an electrical lever? What does it multiply? What does it *not* multiply?
28. What is the principal difference between a step-up transformer and a step-down transformer?
29. Why can a hum usually be heard when a transformer is operating?
30. Why doesn't a transformer work with direct current? Why is ac required?
31. Why is it important that the core of a transformer pass through both coils?
32. Are the primary and secondary coils in a transformer physically linked, or is there space between the two? Explain.
33. In the circuit shown, how many volts are impressed across the lightbulb and how many amps flow through it?

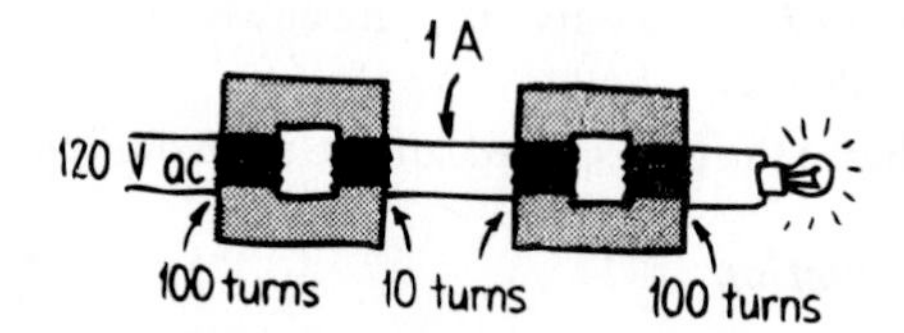

34. In the circuit shown, how many volts are impressed across the meter and how many amps flow through it?

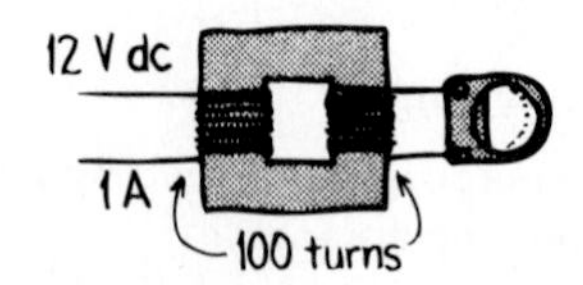

35. How would you answer the previous question if the input were 12 V ac?

36. Can an efficient transformer step up energy? Defend your answer.

37. Your friend says that, according to Ohm's law, high voltage produces high current. Then your friend asks, "So how can power be transmitted at high voltage and *low* current in a power line?" What is your illuminating response?

38. If a bar magnet is thrown into a coil of high-resistance wire, it will slow down. Why?

39. Your physics instructor drops a magnet through a long vertical copper pipe and it moves slowly compared with the drop of a nonmagnetized object. Provide an explanation.

40. This exercise is similar to the previous one. Why will a bar magnet fall slower and reach terminal velocity in a vertical copper or aluminum tube but not in a cardboard tube?

41. Although copper and aluminum are not magnetic, why is a sheet of either metal more difficult to pass between the pole pieces of a magnet than a sheet of cardboard?

42. A metal bar, pivoted at one end, oscillates freely in the absence of a magnetic field. But, when it oscillates between the poles of a magnet, its oscillations are quickly damped. Why? (Such magnetic damping is used in a number of practical devices.)

43. The metal wing of an airplane acts like a "wire" flying through the Earth's magnetic field. A voltage is induced between the wing tips, and a current flows along the wing, but only for a short time. Why does the current stop even though the airplane continues flying through the Earth's magnetic field?

44. What is wrong with this scheme? To generate electricity without fuel, arrange a motor to operate a generator that will produce electricity that is stepped up with transformers so that the generator can operate the motor and simultaneously furnish electricity for other uses.

45. With no magnets in the vicinity, why will current flow in a large coil of wire waved around in the air?

46. We know that the source of a sound wave is a vibrating object. What is the source of an electromagnetic wave?

47. What does an incident radio wave do to the electrons in a receiving antenna?

48. How do you suppose the frequency of an electromagnetic wave compares with the frequency of the electrons it sets into oscillation in a receiving antenna?

49. A friend says that changing electric and magnetic fields generate one another, and that this gives rise to visible light when the frequency of change matches the frequencies of light. Do you agree? Explain.

50. Would electromagnetic waves exist if changing magnetic fields could produce electric fields, but changing electric fields could not, in turn, produce magnetic fields? Explain.

Problems

1. The primary coil of a step-up transformer draws 100 W. Find the power provided by the secondary coil.

2. An ideal transformer has 50 turns in its primary and 250 turns in its secondary. 12 V ac is connected to the primary. Find: (a) volts ac available at the secondary; (b) current in a 10-ohm device connected to the secondary; and (c) power supplied to the primary.

3. A model electric train requires 6 V to operate. If the primary coil of its transformer has 240 windings, how many windings should the secondary have, if the primary is connected to a 120-V household circuit?

4. Neon signs require about 12,000 V for their operation. What should be the ratio of the number of loops in the secondary to the number of loops in the primary for a neon-sign transformer that operates from 120-V lines?

5. 100 kW (10^5 W) of power is delivered to the other side of a city by a pair of power lines between which the voltage is 12,000 V. (a) What current flows in the lines? (b) Each of the two lines has a resistance of 10 ohms. What is the voltage change *along* each line? (Think carefully. This voltage change is along each line, not between the lines.) (c) What power is expended as heat in both lines together (distinct from power delivered to customers)? Do you see why it is so important to step voltages up with transformers for long-distance transmission?

Remember, review questions provide you with a self check of whether or not you grasp the central ideas of the chapter. The exercises and problems are extra "pushups" for you to try after you have at least a fair understanding of the chapter and can handle the review questions.

PART SIX

Light

How nice that energetic photons in sunlight are stimulating vibrations of zillions of electrons in the molecular structure of this leaf. The more vigorous vibrations produce heat, while more subtle ones send out new photons, revealing the translucence and delicate structure of the leaf in intricate detail. And the radiating electrons don't vibe at any old frequency, by golly! They dance to an average rhythm of 6×10^{14} vibes per second, which is why the leaf is green!

CHAPTER 26

Properties of Light

Roy Unruh demonstrates the conversion of light energy to electric energy with model solar-powered vehicles.

Light is the only thing we can really see. But what *is* light? We know that, during the day, the primary source of light is the Sun and the secondary source is the brightness of the sky. Other common sources are flames, white-hot filaments in lightbulbs, and glowing gas in glass tubes. Light originates from the accelerated motion of electrons. It is an electromagnetic phenomenon and only a tiny part of a larger whole—a wide range of electromagnetic waves called the *electromagnetic spectrum.* We begin our study of light by investigating its electromagnetic properties. In the next chapter, we'll discuss its appearance—color. In Chapter 28, we'll learn how light behaves—how it reflects and refracts. Then we'll learn about its wave nature in Chapter 29 and its quantum nature in Chapters 30 and 31.

Light is the only thing we see. Sound is the only thing we hear.

Insights

Electromagnetic Waves

FIGURE 26.1
Shake an electrically charged object to and fro, and you produce an electromagnetic wave.

If you shake the end of a stick back and forth in still water, you will produce waves on the surface of the water. Similarly, if you shake an electrically charged rod to and fro in empty space, you will produce electromagnetic waves in space. This is because the moving charge is actually an electric current. What surrounds an electric current? The answer is, a magnetic field. What surrounds a changing electric current? The answer is, a changing magnetic field. Recall, from the previous chapter, that a changing magnetic field generates an electric field—electromagnetic induction. If the magnetic field is oscillating, the electric field that it generates will be oscillating, too. And what does an oscillating electric field do? In accordance with Maxwell's counterpart to Faraday's law of electromagnetic induction, it induces an oscillating magnetic field. The vibrating electric and magnetic fields regenerate each other to make up an **electromagnetic wave,** which emanates (moves outward) from the vibrating charge. There is only one speed, it turns out, for which the electric and magnetic fields remain in perfect balance, reinforcing each other as they carry energy through space. Let's see why this is so.

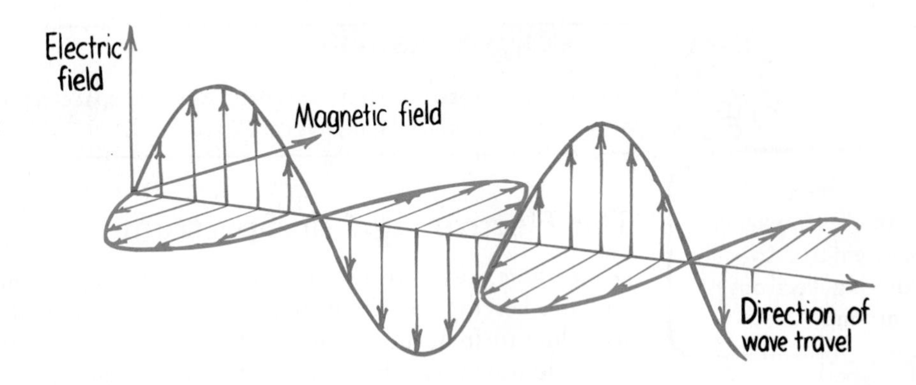

FIGURE 26.2
Interactive Figure
The electric and magnetic fields of an electromagnetic wave are perpendicular to each other and to the direction of motion of the wave.

Electromagnetic Wave Velocity

A spacecraft cruising through space may gain or lose speed, even if its engines are shut off, because gravity can accelerate it forward or backward. But an electromagnetic wave traveling through space never changes its speed. Not because gravity doesn't act on light, for it does. Gravity can change the frequency of light or deflect light, but it can't change the speed of light. What keeps light moving always at the same, unvarying speed in empty space? The answer has to do with electromagnetic induction and energy conservation.

If light were to slow down, its changing electric field would generate a weaker magnetic field, which, in turn, would generate a weaker electric field, and so on, until the wave dies out. No energy would be transported from one place to another. So light cannot travel slower than it does.

James Clerk Maxwell (1831–1879)

If light were to speed up, the changing electric field would generate a stronger magnetic field, which, in turn, would generate a stronger electric field, and so on, a crescendo of ever-increasing field strength and ever-increasing energy—clearly a no-no with respect to energy conservation. At only one speed does mutual induction continue indefinitely, carrying energy forward without loss or gain. From his equations of electromagnetic induction, James Clerk Maxwell calculated the value of this critical speed and found it to be 300,000 kilometers per second. In his calculation, he used only the constants in his equations determined by simple laboratory experiments with electric and magnetic fields. He didn't *use* the speed of light. He *found* the speed of light!

Maxwell quickly realized that he had discovered the solution to one of the greatest mysteries of the universe—the nature of light. He discovered that light is simply electromagnetic radiation within a particular frequency range, 4.3×10^{14} to 7×10^{14} vibrations per second. Such waves activate the "electrical antennae" in the retina of the eye. The lower-frequency waves appear red, and the higher-frequency waves appear violet.[1] Maxwell realized, at the same time, that electromagnetic radiation of *any* frequency propagates at the same speed as light.

[1]It is common to describe sound and radio by *frequency* and light by *wavelength*. In this book, however, we stay with the single concept of frequency in describing light.

Light is energy carried in an electromagnetic wave emitted by vibrating electrons in atoms.

Insights

CHECK YOURSELF

The unvarying speed of electromagnetic waves in space is a remarkable consequence of what central principle in physics?

The Electromagnetic Spectrum

In a vacuum, all electromagnetic waves move at the same speed and differ from one another in their frequency. The classification of electromagnetic waves according to frequency is the **electromagnetic spectrum** (Figure 26.3). Electromagnetic waves have been detected with a frequency as low as 0.01 hertz (Hz). Electromagnetic waves with frequencies of several thousand hertz (kHz) are classified as very low frequency radio waves. One million hertz (MHz) lies in the middle of the AM radio band. The very high frequency (VHF) television band of waves starts at about 50 MHz, and FM radio waves are between 88 and 108 MHz. Then come ultrahigh frequencies (UHF), followed by microwaves, beyond which are infrared waves, often called "heat waves." Further still is visible light, which makes up less than 1 millionth of 1% of the measured electromagnetic spectrum. The lowest frequency of light visible to our eyes appears red. The highest frequencies of visible light, which are nearly twice the frequency of red light, appear violet. Still higher frequencies are ultraviolet. These higher-frequency waves cause sunburns. Higher frequencies beyond ultraviolet extend into the X-ray and gamma-ray regions. There are no sharp boundaries between the regions, which actually overlap each other. The spectrum is separated into these arbitrary regions for classification.

In empty space, there is light but no sound. In air, light travels a million times faster than sound.

Insights

The concepts and relationships we treated earlier in our study of wave motion (Chapter 19) apply here. Recall that the frequency of a wave is the same as the frequency of the vibrating source. The same is true here: The frequency of an electromagnetic wave as it vibrates through space is identical to the frequency of

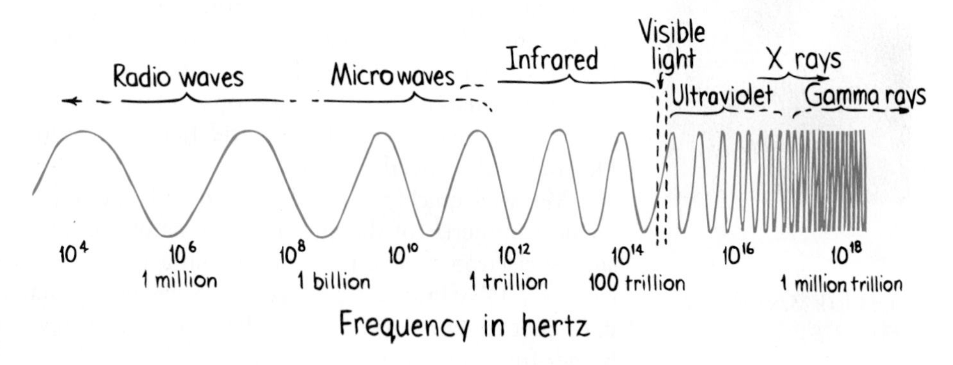

FIGURE 26.3
Interactive Figure
The electromagnetic spectrum is a continuous range of waves extending from radio waves to gamma rays. The descriptive names of the sections are merely a historical classification, for all waves are the same in nature, differing principally in frequency and wavelength; all travel at the same speed.

CHECK YOUR ANSWER

The underlying principle that makes light and all other electromagnetic waves travel at one fixed speed is the conservation of energy.

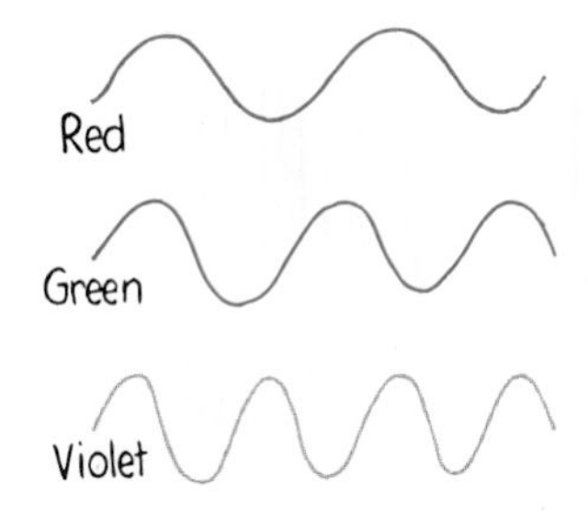

FIGURE 26.4
Interactive Figure
Relative wavelengths of red, green, and violet light. Violet light has nearly twice the frequency of red light, and half the wavelength.

the oscillating electric charge generating it.[2] Different frequencies correspond to different wavelengths—waves of low frequency have long wavelengths and waves of high frequencies have short wavelengths. For example, since the speed of the wave is 300,000 kilometers per second, an electric charge oscillating once per second (1 hertz) will produce a wave with a wavelength of 300,000 kilometers. This is because only one wavelength is generated in 1 second. If the frequency of oscillation were 10 hertz, then 10 wavelengths would be formed in 1 second, and the corresponding wavelength would be 30,000 kilometers. A frequency of 10,000 hertz would produce a wavelength of 30 kilometers. So the higher the frequency of the vibrating charge, the shorter the wavelength of radiant energy.[3]

We tend to think of space as empty, but only because we cannot see the montages of electromagnetic waves that permeate every part of our surroundings. We see some of these waves, of course, as light. These waves constitute only a microportion of the electromagnetic spectrum. We are unconscious of radio waves, which engulf us every moment. Free electrons in every piece of metal on the Earth's surface continually dance to the rhythms of these waves. They jiggle in unison with the electrons being driven up and down along radio- and television-transmitting antennae. A radio or television receiver is simply a device that sorts and amplifies these tiny currents. There is radiation everywhere. Our first impression of the universe is one of matter and void, but actually the universe is a dense sea of radiation in which occasional concentrates are suspended.

Before the advent of microwave ovens, there were infrared ones—simply called "ovens."

Insights

CHECK YOURSELF

Is it correct to say that a radio wave can be considered a low-frequency light wave? Can a radio wave also be considered to be a sound wave?

Transparent Materials

Light and Transparent Materials

Light is an energy-carrying electromagnetic wave that emanates from vibrating electrons in atoms. When light is transmitted through matter, some of the electrons in the matter are forced into vibration. In this way, vibrations in the

CHECK YOUR ANSWERS

Both a radio wave and a light wave are electromagnetic waves, which originate in the vibrations of electrons. Radio waves have lower frequencies than light waves, so a radio wave may be considered to be a low-frequency light wave (and a light wave, similarly, can be considered to be a high-frequency radio wave). But a sound wave is a mechanical vibration of matter and is not electromagnetic. A sound wave is fundamentally different from an electromagnetic wave. So a radio wave is definitely not a sound wave.

[2]This is a rule of classical physics, valid when charges are oscillating over dimensions that are large compared with the size of a single atom (for instance, in a radio antenna). Quantum physics permits exceptions. Radiation emitted by a single atom or molecule can differ in frequency from the frequency of the oscillating charge within the atom or molecule.

[3]The relationship is $c = f\lambda$, where c is the wave speed (constant), f is the frequency, and λ is the wavelength.

FIGURE 26.5
Just as a sound wave can force a sound receiver into vibration, a light wave can force electrons in materials into vibration.

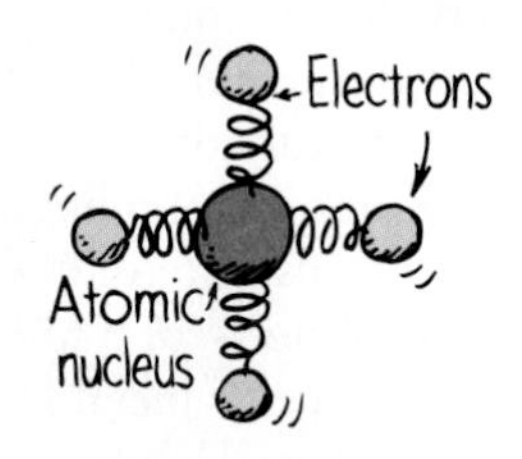

FIGURE 26.6
The electrons of atoms in glass have certain natural frequencies of vibration and can be modeled as particles connected to the atomic nucleus by springs.

emitter are transmitted to vibrations in the receiver. This is similar to the way sound is transmitted (Figure 26.5).

Thus the way a receiving material responds when light is incident upon it depends on the frequency of the light and on the natural frequency of the electrons in the material. Visible light vibrates at a very high frequency, some 100 trillion times per second (10^{14} hertz). If a charged object is to respond to these ultrafast vibrations, it must have very, very little inertia. Because the mass of electrons is so tiny, they can vibrate at this rate.

Such materials as glass and water allow light to pass through in straight lines. We say they are **transparent** to light. To understand how light travels through a transparent material, visualize the electrons in the atoms of transparent materials as if they were connected to the nucleus by springs (Figure 26.6).[4] When a light wave is incident upon them, the electrons are set into vibration.

Materials that are springy (elastic) respond more to vibrations at some frequencies than at others (Chapter 20). Bells ring at a particular frequency, tuning forks vibrate at a particular frequency, and so do the electrons of atoms and molecules. The natural vibration frequencies of an electron depend on how strongly it is attached to its atom or molecule. Different atoms and molecules have different "spring strengths." Electrons in the atoms of glass have a natural vibration frequency in the ultraviolet range. Therefore, when ultraviolet waves shine on glass, resonance occurs and the vibration of electrons builds up to large amplitudes, just as pushing someone at the resonant frequency on a swing builds to a large amplitude. The energy any glass atom receives is either reemitted or passed on to neighboring atoms by collisions. Resonating atoms in the glass can hold onto the energy of the ultraviolet light for quite a long time (about 100 millionths of a second). During this time, the atom makes about 1 million vibrations, and it collides with neighboring atoms and gives up its energy as heat. Thus, glass is not transparent to ultraviolet light.

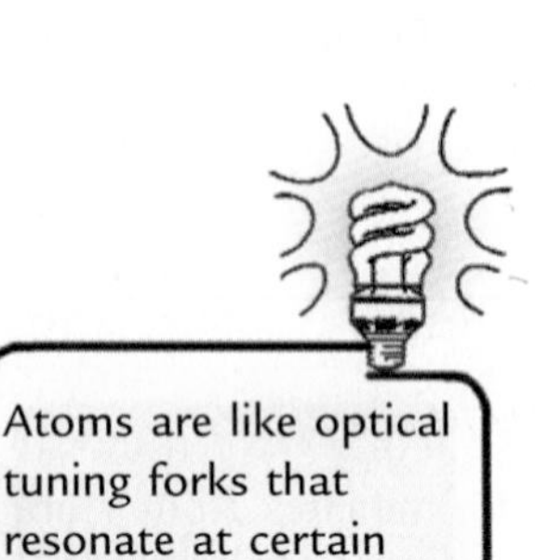

Atoms are like optical tuning forks that resonate at certain frequencies.
Insights

At lower wave frequencies, such as those of visible light, electrons in the glass atoms are forced into vibration, but at lower amplitudes. The atoms hold the energy for a shorter time, with less chance of collision with neighboring atoms, and with less energy transformed to heat. The energy of vibrating electrons is reemitted as light. Glass is transparent to all the frequencies of

[4]Electrons, of course, are not really connected by springs. Their "vibration" is actually orbital as they move around the nucleus, but the "spring model" helps us to understand the interaction of light with matter. Physicists devise such conceptual models to understand nature, particularly at the submicroscopic level. The worth of a model lies not in whether it is "true" but in whether it is useful. A good model not only is consistent with and explains observations, but also predicts what may happen. If predictions of the model are contrary to what happens, the model is usually either refined or abandoned. The simplified model that we present here—of an atom whose electrons vibrate as if on springs, with a time interval between absorbing energy and reemitting energy—is quite useful for understanding how light passes through transparent solids.

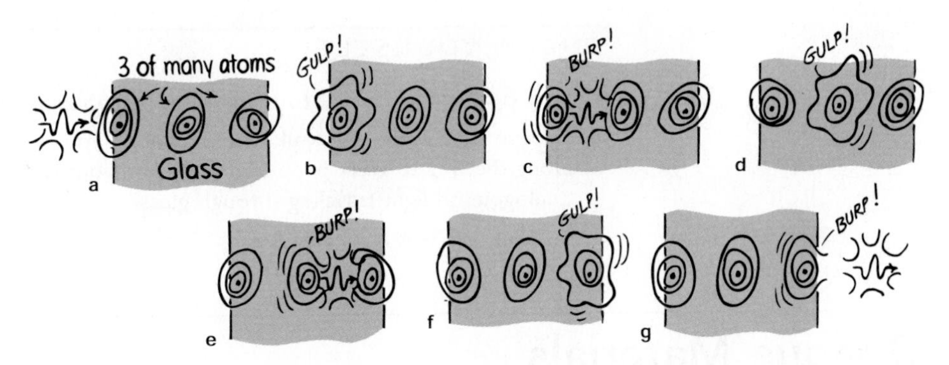

FIGURE 26.7
A wave of visible light incident upon a pane of glass sets up in atoms vibrations that produce a chain of absorptions and reemissions, which pass the light energy through the material and out the other side. Because of the time delay between absorptions and reemissions, the light travels through the glass more slowly than through empty space.

visible light. The frequency of the reemitted light that is passed from atom to atom is identical to the frequency of the light that produced the vibration in the first place. However, there is a slight time delay between absorption and reemission.

> Different materials have different molecular structures and therefore absorb or reflect light from various spectral ranges differently.
>
> **Insights**

It is this time delay that results in a lower average speed of light through a transparent material (Figure 26.7). Light travels at different average speeds through different materials. We say *average speeds* because the speed of light in a vacuum, whether in interstellar space or in the space between molecules in a piece of glass, is a constant 300,000 kilometers per second. We call this speed of light *c*.[5] The speed of light in the atmosphere is slightly less than in a vacuum, but it is usually rounded off as *c*. In water, light travels at 75% of its speed in a vacuum, or 0.75 *c*. In glass, light travels at about 0.67 *c*, depending on the type of glass. In a diamond, light travels at less than half its speed in a vacuum, only 0.41 *c*. When light emerges from these materials into the air, it travels at its original speed, *c*.

Infrared waves, with frequencies lower than those of visible light, vibrate not only the electrons, but entire atoms or molecules in the structure of the glass. This vibration increases the internal energy and temperature of the structure, which is why infrared waves are often called *heat waves*. So we see that glass is transparent to visible light, but not to ultraviolet and infrared light.

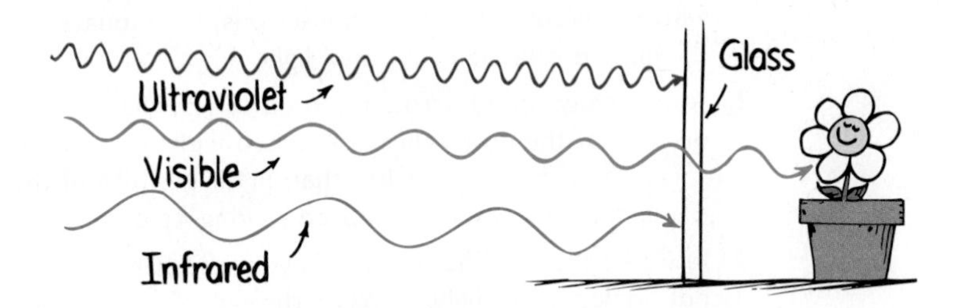

FIGURE 26.8
Glass blocks both infrared and ultraviolet, but it is transparent to visible light.

[5]The presently accepted value is 299,792 km/s, rounded to 300,000 km/s. (This corresponds to 186,000 mi/s.)

CHECK YOURSELF

1. Why is glass transparent to visible light but opaque to ultraviolet and infrared?
2. Pretend that, while you walk across a room, you make several momentary stops along the way to greet people who are "on your wavelength." How is this analogous to light traveling through glass?
3. In what way is it not analogous?

Opaque Materials

Most things around us are **opaque**—they absorb light without reemitting it. Books, desks, chairs, and people are opaque. Vibrations given by light to their atoms and molecules are turned into random kinetic energy—into internal energy. They become slightly warmer.

Metals are opaque. Because the outer electrons of atoms in metals are not bound to any particular atom, they are free to wander with very little restraint throughout the material (which is why metal conducts electricity and heat so well). When light shines on metal and sets these free electrons into vibration, their energy does not "spring" from atom to atom in the material but, instead, is reflected. That's why metals are shiny.

Earth's atmosphere is transparent to some ultraviolet light, to all visible light, and to some infrared light, but it is opaque to high-frequency ultraviolet light. The small amount of ultraviolet that does get through is responsible for sunburns. If it all got through, we would be fried to a crisp. Clouds are semitransparent to ultraviolet, which is why you can get a sunburn on a cloudy day. Dark skin absorbs ultraviolet before it can penetrate too far, whereas it travels deeper in fair skin. With mild and gradual exposure, fair skin develops a tan and increases

Longer-wavelength ultraviolet, called UV-A, is close to visible light and isn't harmful. Short-wavelength ultraviolet, called UV-C, would be harmful if it reached us, but is almost completely stopped by the atmosphere's ozone layer. It is the intermediate ultraviolet, UV-B, that can cause eye damage, sunburn, and skin cancer.

Insights

CHECK YOUR ANSWERS

1. Because the natural vibration frequency for electrons in glass is the same as the frequency of ultraviolet light, resonance occurs when ultraviolet waves shine on glass. The absorbed energy is passed on to other atoms as heat, not reemitted as light, making the glass opaque at ultraviolet frequencies. In the range of visible light, the forced vibrations of electrons in the glass are at smaller amplitudes—vibrations are more subtle, reemission of light (rather than the generation of heat) occurs, and the glass is transparent. Lower-frequency infrared light causes whole molecules, rather than electrons, to resonate; again, heat is generated and the glass is opaque to infrared light.
2. Your average speed across the room is less than it would be in an empty room because of the time delays associated with your momentary stops. Likewise, the speed of light in glass is less than in air because of the time delays caused by the light's interactions with atoms along its path.
3. In walking across the room, it is you who begin and complete the walk. This is not analogous to light traveling through glass because, according to our model for light passing through a transparent material, the light absorbed by the first electron that is made to vibrate is not the same light that is reemitted—even though the two, like identical twins, are indistinguishable.

FIGURE 26.9
Metals are shiny because light that shines on them forces free electrons into vibration, and these vibrating electrons then emit their "own" light waves as reflection.

protection against ultraviolet. Ultraviolet light is also damaging to the eyes—and to tarred roofs. Now you know why tarred roofs are covered with gravel.

Have you noticed that things look darker when they are wet than they do when they are dry? Light incident on a dry surface bounces directly to your eye, while light incident on a wet surface bounces around inside the transparent wet region before it reaches your eye. What happens with each bounce? Absorption! So more absorption of light occurs in a wet surface, and the surface looks darker.

Shadows

A thin beam of light is often called a *ray*. When we stand in the sunlight, some of the light is stopped while other rays continue in a straight-line path. We cast a **shadow**—a region where light rays cannot reach. If you are close to your own shadow, the outline of your shadow is sharp because the Sun is so far away. Either a large, far-away light source or a small, nearby light source will produce a sharp shadow. A large, nearby light source produces a somewhat blurry shadow (Figure 26.10). There is usually a dark part on the

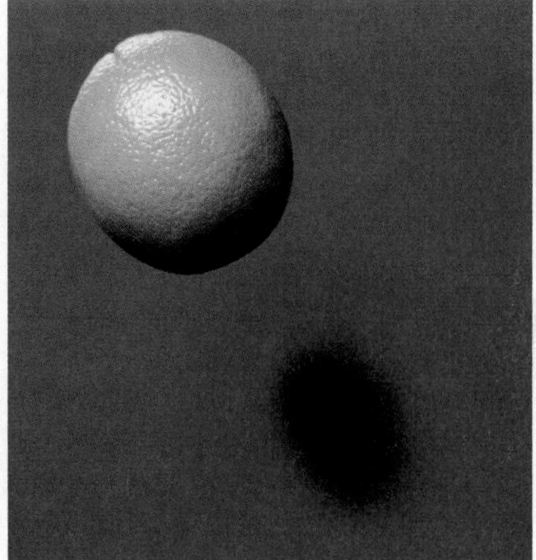

FIGURE 26.10
A small light source produces a sharper shadow than a larger source.

FIGURE 26.11
An object held close to a wall casts a sharp shadow because light coming from slightly different directions does not spread much behind the object. As the object is moved farther away from the wall, penumbras are formed and the umbra becomes smaller. When the object is farther away, the shadow is less distinct. When the object is very far away (not shown), no shadow is evident because all the penumbras mix together into a big blur.

inside and a lighter part around the edges of a shadow. A total shadow is called an **umbra** and a partial shadow is called a **penumbra.** A penumbra appears where some of the light is blocked but where other light fills it in (Figure 26.11). A penumbra also occurs where light from a broad source is only partially blocked.

Both the Earth and the Moon cast shadows when sunlight is incident upon them. When the path of either of these bodies crosses into the shadow cast by the other, an eclipse occurs (Figure 26.12). A dramatic example of the umbra and penumbra occurs when the shadow of the Moon falls on the Earth during a **solar eclipse.** Because of the large size of the Sun, the rays taper to provide an umbra and a surrounding penumbra (Figure 26.13). If you stand in the umbra part of the shadow, you experience darkness during the day—a total eclipse. If you stand in the penumbra, you experience a partial eclipse, for you see a crescent of the Sun.[6] In a **lunar eclipse,** the Moon passes into the shadow of the Earth.

CHECK YOURSELF

1. Which type of eclipse—a solar eclipse, a lunar eclipse, or both—is dangerous to view with unprotected eyes?
2. Why are lunar eclipses more commonly seen than solar eclipses?

[6]People are cautioned not to look at the Sun at the time of a solar eclipse because the brightness and the ultraviolet light of direct sunlight are damaging to the eyes. This good advice is often misunderstood by those who then think that sunlight is more damaging at this special time. But staring at the Sun when it is high in the sky is harmful whether or not an eclipse occurs. In fact, staring at the bare Sun is more harmful than when part of the Moon blocks it! The reason for special caution at the time of an eclipse is simply that more people are interested in looking at the Sun during this time.

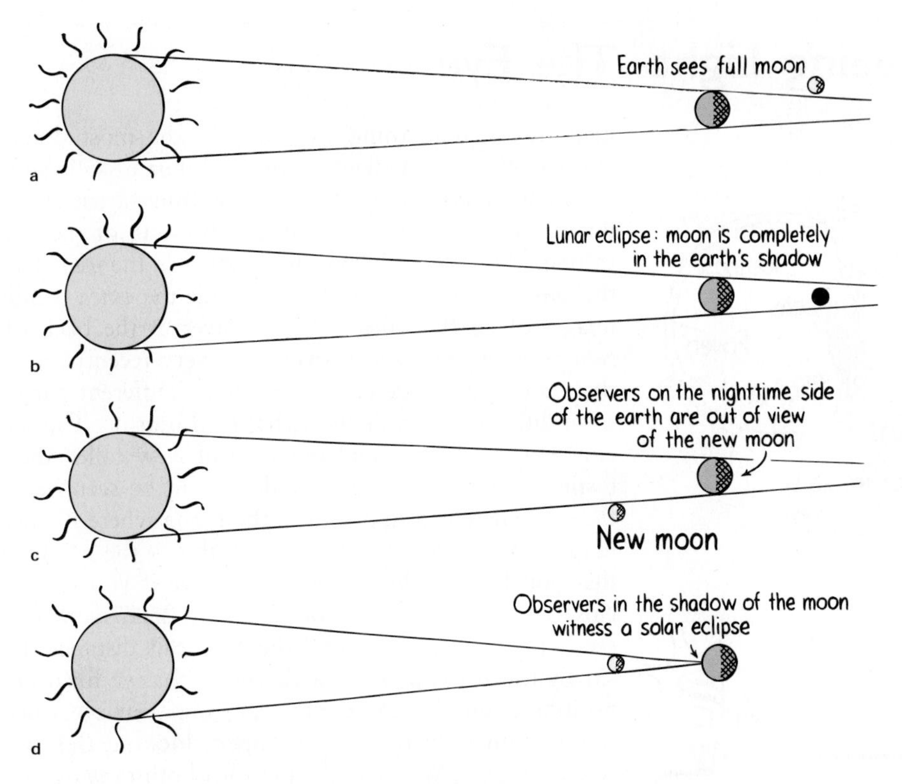

FIGURE 26.12 Interactive Figure
(a) A full Moon is seen when the Earth is between the Sun and the Moon. (b) When this alignment is perfect, the Moon is in Earth's shadow, and a lunar eclipse is produced. (c) A new Moon occurs when the Moon is between the Sun and Earth. (d) When this alignment is perfect, the Moon's shadow falls on part of the Earth to produce a solar eclipse.

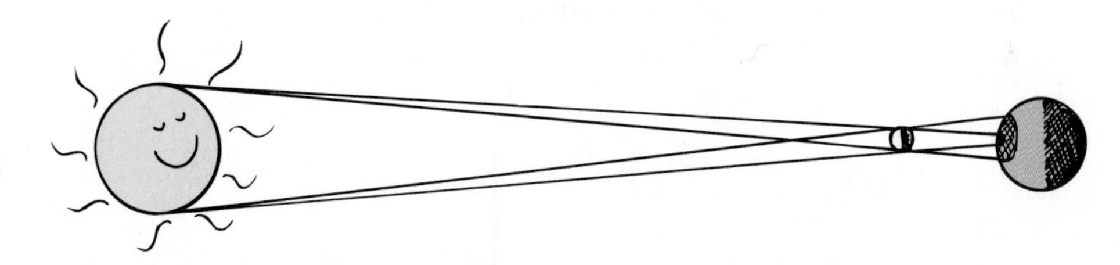

FIGURE 26.13
Interactive Figure
Details of a solar eclipse. A total eclipse is seen by observers in the umbra, and a partial eclipse is seen by observers in the penumbra. Most Earth observers see no eclipse at all.

CHECK YOUR ANSWERS

1. Only a solar eclipse is harmful when viewed directly because one views the Sun directly. During a lunar eclipse, one views a very dark Moon. It is not completely dark because Earth's atmosphere acts as a lens and bends some light into the shadow region. Interestingly enough, this is the light of red sunsets and sunrises all around the world, which is why the Moon appears a faint, deep red during a lunar eclipse.
2. Because the shadow of the relatively small Moon on the large Earth covers a very small part of the Earth's surface, only a relatively few people are in the shadow of the Moon in a solar eclipse. But the shadow of the Earth completely covers the Moon during a total lunar eclipse, so everybody who views the nighttime sky can see the shadow of the Earth on the Moon.

Seeing Light—The Eye

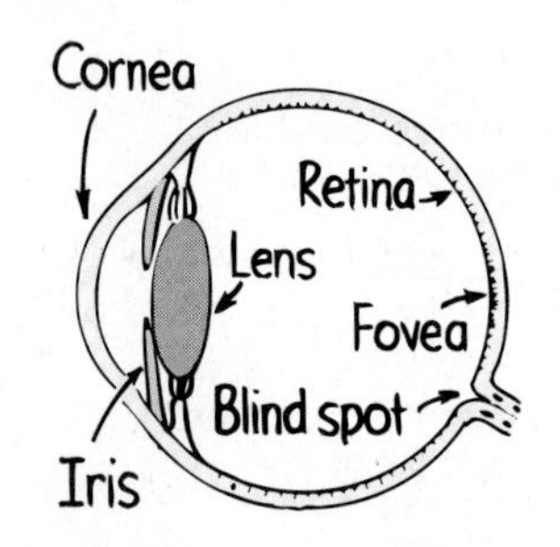

FIGURE 26.14
The human eye.

Light is the only thing we see with the most remarkable optical instrument known—the eye. A diagram of the human eye is shown in Figure 26.14.

Light enters the eye through the transparent cover called the *cornea,* which does about 70% of the necessary bending of the light before the light passes through the pupil (which is an aperture in the iris). The light then passes through the lens, which is used only to provide the extra bending power needed to focus images of nearby objects on the layer at the back of the eye. This layer—the *retina*—is extremely sensitive: Until very recently, it was more sensitive to light than any artificial detector ever made. Different parts of the retina receive light from different parts of the visual field outside. The retina is not uniform. There is a spot in the center of our field of view called the *fovea,* the region of most distinct vision. Much greater detail can be seen here than at the side parts of the eye. There is also a spot in the retina where the nerves carrying all the information exit along the optic nerve; this is the *blind spot.* You can demonstrate that you have a blind spot in each eye if you hold this book at arm's length, close your left eye, and look at Figure 26.15 with your right eye only. You can see both the round dot and the X at this distance. If you now move the book slowly toward your face, with your right eye fixed upon the dot, you'll reach a position about 20–25 centimeters from your eye where the X disappears. Now repeat with only the left eye open, looking this time at the X, and the dot will disappear. When you look with both eyes open, you are not aware of the blind spot, mainly because one eye "fills in" the part to which the other eye is blind. Amazingly, the brain fills in the "expected" view even with one eye open. Repeat the exercise of Figure 26.15 with small objects on various backgrounds. Note that, instead of seeing nothing, your brain gratuitously fills in the appropriate background. So you not only see what's there—you also see what's *not* there!

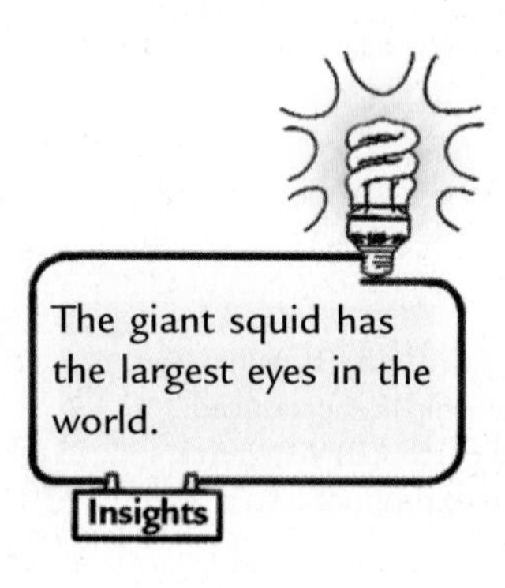

The giant squid has the largest eyes in the world.
Insights

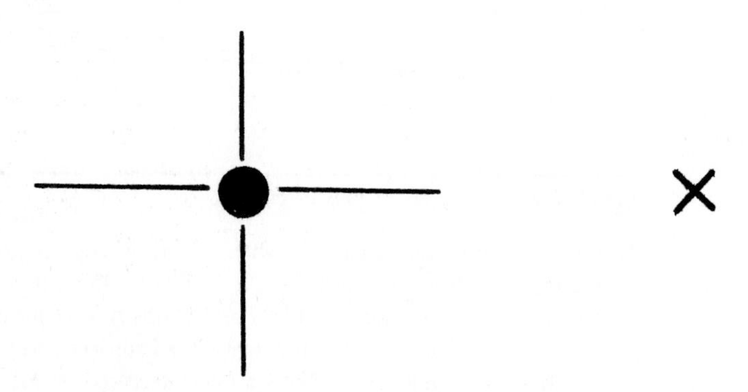

FIGURE 26.15
The blind-spot experiment. Close your left eye and look with your right eye at the round dot. Adjust your distance and find the blind spot that erases the X. Switch eyes and look at the X and the dot disappears. Does your brain fill in crossed lines where the dot was?

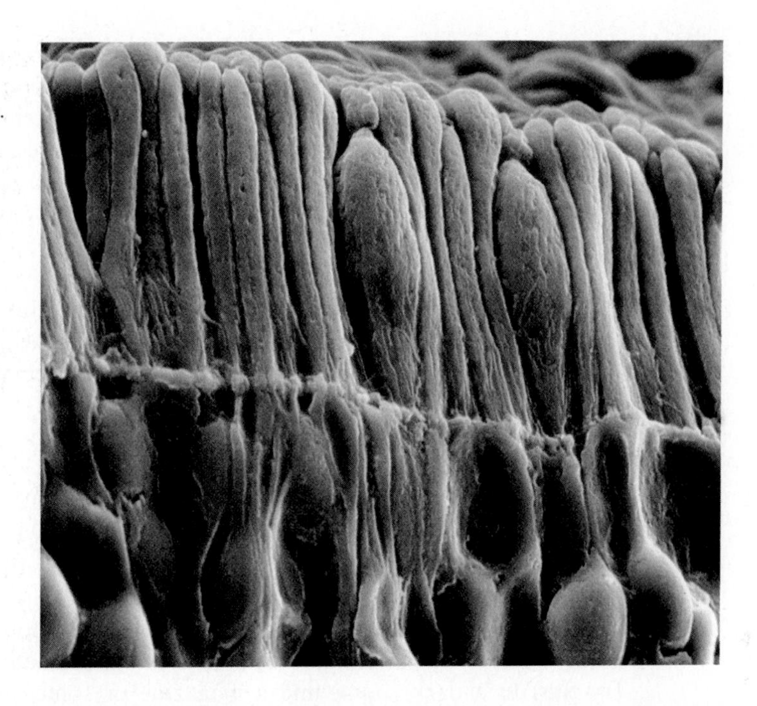

FIGURE 26.16
Magnified view of the rods and cones in the human eye.

The retina is composed of tiny antennae that resonate to the incoming light. There are two basic kinds of antennae, the rods and the cones (Figure 26.16). As the names imply, some of the antennae are rod-shaped and some cone-shaped. There are three types of cones: those that are stimulated by low-frequency light, those that are stimulated by light of intermediate frequencies, and those that are stimulated by light of higher frequencies. The rods predominate toward the periphery of the retina, while the three types of cones are denser toward the fovea. The cones are very dense in the fovea itself, and, since they are packed so tightly, they are much finer or narrower there than elsewhere in the retina. Color vision is possible because of the cones. Hence we see color most acutely by focusing an image on the fovea, where there are no rods. Primates and a species of ground squirrel are the only mammals that have the three types of cones and experience full color vision. The retinas of other mammals consist primarily of rods, which are sensitive only to lightness or darkness, like a black-and-white photograph or movie.

In the human eye, the number of cones decreases as we move away from the fovea. It's interesting that the color of an object disappears if it is viewed on the periphery of the visual field. This can be tested by having a friend enter your periphery of vision with some brightly colored objects. You will find that you can see the objects before you can see what color they are.

Another interesting fact is that the periphery of the retina is very sensitive to motion. Although our vision is poor from the corner of our eye, we are sensitive to anything moving there. We are "wired" to look for something jiggling to the side of our visual field, a feature that must have been important in our evolutionary development. So have your friend shake those brightly colored objects when she brings them into the periphery of your vision. If you can just barely see the objects when they shake, but not at all when they're stationary, then you won't be able to tell what color they are (Figure 26.17). Try it and see!

FIGURE 26.17
On the periphery of your vision, you can see an object and its color only if it is moving.

Another distinguishing feature of the rods and cones is the intensity of light to which they respond. The cones require more energy than the rods before they will "fire" an impulse through the nervous system. If the intensity of light is very low, the things we see have no color. We see low intensities with our rods. Dark-adapted vision is almost entirely due to the rods, while vision in bright light is due to the cones. Stars, for example, look white to us. Yet most stars are actually brightly colored. A time exposure of the stars with a camera reveals reds and red-oranges for the "cooler" stars and blues and blue-violets for the "hotter" stars. The starlight is too weak, however, to fire the color-perceiving cones in the retina. So we see the stars with our rods and perceive them as white or, at best, as only faintly colored. Females have a slightly lower threshold of firing for the cones, however, and can see a bit more color than males. So if she says she sees colored stars and he says she doesn't, she is probably right!

We find that the rods "see" better than the cones toward the blue end of the color spectrum. The cones can see a deep red where the rods see no light at all. Red light may as well be black as far as the rods are concerned. Thus, if you have two colored objects—say, one blue and one red—the blue one will appear much brighter than the red in dim light, although the red one might be much brighter than the blue one in bright light. The effect is quite interesting. Try this: In a dark room, find a magazine or something that has colors, and, before you know for sure what the colors are, judge the lighter and darker areas. Then carry the magazine into the light. You should see a remarkable shift between the brightest and dimmest colors.[7]

The rods and cones in the retina are not connected directly to the optic nerve, but, interestingly enough, are connected to many other cells that are joined to one another. While many of these cells are interconnected, only a few carry information to the optic nerve. Through these interconnections, a certain amount of information is combined from several visual receptors and "digested" in the retina. In this way, the light signal is "thought about" before it goes to the optic nerve and thence to the main body of the brain. So some brain functioning occurs in the eye itself. The eye does some of our "thinking" for us.

She loves you...

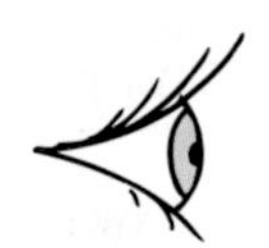
She loves you not?

FIGURE 26.18
The size of your pupils depends on your mood.

This thinking is betrayed by the iris, the colored part of the eye that expands and contracts and regulates the size of the pupil, admitting more or less light as the intensity of light changes. It so happens that the relative size of this enlargement or contraction is also related to our emotions. If we see, smell, taste, or hear something that is pleasing to us, our pupils automatically increase in size. If we see, smell, taste, or hear something repugnant to us, our pupils automatically contract. Many card players have betrayed the value of a hand by the size of their pupils! (The study of the size of the pupil as a function of attitudes is called *pupilometrics.*)

The brightest light that the human eye can perceive without damage is some 500 million times brighter than the dimmest light that can be perceived. Look at a nearby lightbulb. Then turn to look into a dimly lit closet. The difference in light intensity may be more than a million to one. Because of an effect called *lateral inhibition,* we don't perceive the actual differences in brightness. The

---

[7]This phenomenon is called the *Purkinje effect* after the Czech physiologist who discovered it.

FIGURE 26.19
Both rectangles are equally bright. Cover the boundary between them with your pencil and see.

brightest places in our visual field are prevented from outshining the rest, for whenever a receptor cell on our retina sends a strong brightness signal to our brain, it also signals neighboring cells to dim their responses. In this way, we even out our visual field, which allows us to discern detail in very bright areas and in dark areas as well. (Camera film is not so good at this. A photograph of a scene with strong differences of intensity may be overexposed in one area and underexposed in another.) Lateral inhibition exaggerates the difference in brightness at the edges of places in our visual field. Edges, by definition, separate one thing from another. So we accentuate differences. The gray rectangle on the left in Figure 26.19 appears darker than the gray rectangle on the right when the edge that separates them is in our view. But cover the edge with your pencil or your finger, and they look equally bright. That's because both rectangles *are* equally bright; each rectangle is shaded lighter to darker, moving from left to right. Our eye concentrates on the boundary where the dark edge of the left rectangle joins the light edge of the right rectangle, and our eye–brain system assumes that the rest of the rectangle is the same. We pay attention to the boundary and ignore the rest.

Questions to ponder: Is the way the eye selects edges and makes assumptions about what lies beyond similar to the way in which we sometimes make judgments about other cultures and other people? Don't we, in the same way, tend to exaggerate the differences on the surface while ignoring the similarities and subtle differences within?

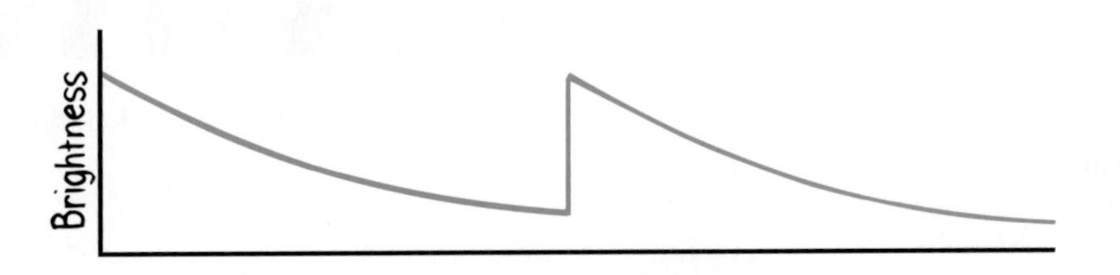

FIGURE 26.20
Graph of brightness levels for the rectangles in Figure 26.19.

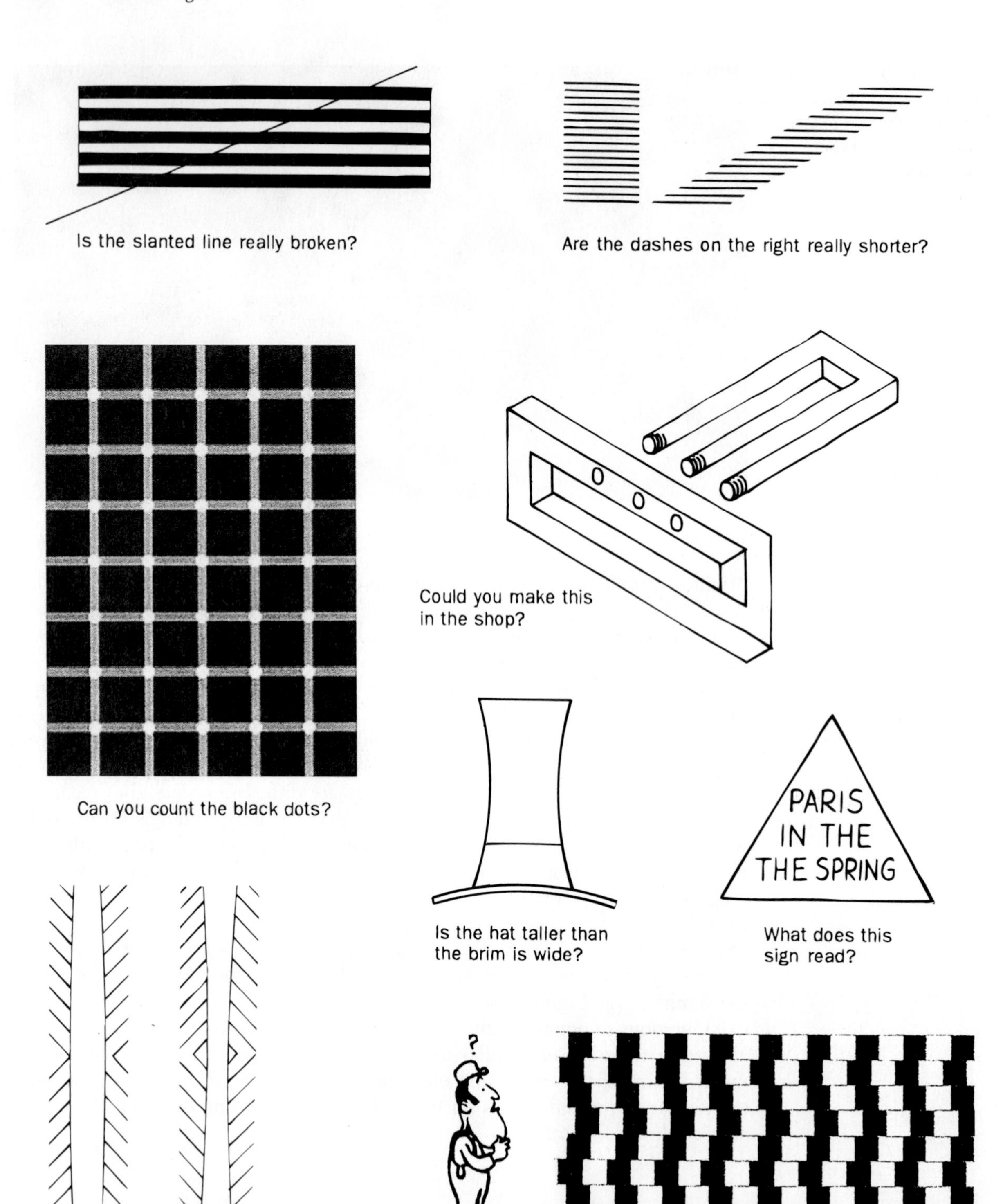

Is the slanted line really broken?
Are the dashes on the right really shorter?
Can you count the black dots?
Could you make this
in the shop?
Is the hat taller than
the brim is wide?
PARIS
IN THE
THE SPRING
What does this
sign read?
Are the vertical
lines parallel?
?
Are the rows of tiles really crooked?

FIGURE 26.21
Optical illusions.

Summary of Terms

Electromagnetic wave An energy-carrying wave emitted by a vibrating charge (often electrons) that is composed of oscillating electric and magnetic fields that regenerate one another.

Electromagnetic spectrum The range of electromagnetic waves extending in frequency from radio waves to gamma rays.

Transparent The term applied to materials through which light can pass in straight lines.

Opaque The term applied to materials that absorb light without reemission and thus through which light cannot pass.

Shadow A shaded region that appears where light rays are blocked by an object.

Umbra The darker part of a shadow where all the light is blocked.

Penumbra A partial shadow that appears where some but not all of the light is blocked.

Solar eclipse An event wherein the Moon blocks light from the Sun and the Moon's shadow falls on part of the Earth.

Lunar eclipse An event wherein the Moon passes into the shadow of the Earth.

Suggested Reading

Falk, D. S., D. R. Brill, and D. Stork. *Seeing the Light: Optics in Nature.* New York: Harper & Row, 1986.

For more on illusions, see www.michaelbach.de/ot

Review Questions

Electromagnetic Waves

1. What does a *changing magnetic field* induce?
2. What does a *changing electric field* induce?
3. What produces an electromagnetic wave?

Electromagnetic Wave Velocity

4. How is the fact that an electromagnetic wave in space never slows down consistent with the conservation of energy?
5. How is the fact that an electromagnetic wave in space never speeds up consistent with the conservation of energy?
6. What do electric and magnetic fields contain and transport?

The Electromagnetic Spectrum

7. What is the principal difference between a *radio wave* and *light*? Between *light* and an *X-ray*?
8. About how much of the measured electromagnetic spectrum does light occupy?
9. What is the color of visible light of the lowest frequencies? Of the highest frequencies?
10. How does the frequency of a radio wave compare to the frequency of the vibrating electrons that produce it?
11. How is the wavelength of light related to its frequency?
12. What is the wavelength of a wave that has a frequency of 1 Hz and travels at 300,000 km/s?
13. What do we mean when we say that outer space is not really empty?

Transparent Materials

14. The sound coming from one tuning fork can force another to vibrate. What is the analogous effect for light?
15. In what region of the electromagnetic spectrum is the resonant frequency of electrons in glass?
16. What is the fate of the energy in ultraviolet light that is incident upon glass?
17. What is the fate of the energy in visible light that is incident upon glass?
18. How does the frequency of reemitted light in a transparent material compare with the frequency of the light that stimulates its reemission?
19. How does the average speed of light in glass compare with its speed in a vacuum?
20. Why are infrared waves often called *heat waves*?

Opaque Materials

21. Why do opaque materials become warmer when light shines on them?
22. Why are metals shiny?
23. Why do wet objects normally look darker than the same objects when dry?

Shadows

24. Distinguish between an *umbra* and a *penumbra*.
25. Do the Earth and the Moon always cast shadows? What do we call the occurrence where one passes within the shadow of the other?

Seeing Light—The Eye

26. Distinguish between the *rods* and *cones* of the eye and between their functions.

Projects

1. Compare the size of the Moon on the horizon with its size higher in the sky. One way to do this is to hold at arm's length various objects that will just barely block out the Moon. Experiment until you find something just right, perhaps a thick pencil or a pen. You'll find that the object will be less than a centimeter, depending on the length of your arms. Is the Moon really bigger when it is near the horizon?
2. Which eye do you use more? To test which you favor, hold a finger up at arm's length. With both eyes open, look past it at a distant object. Now close your right eye. If your finger appears to jump to the right, then you use your right eye more. Check with friends who are both left-handed and right-handed. Is there a correlation between dominant eye and dominant hand?

Exercises

1. A friend says, in a profound tone, that light is the only thing we can see. Is your friend correct?
2. Your friend goes on to say that light is produced by the connection between electricity and magnetism. Is your friend correct?
3. What is the fundamental source of electromagnetic radiation?
4. Which have the longest wavelengths—light waves, X-rays, or radio waves?
5. Which has the shorter wavelengths, ultraviolet or infrared? Which has the higher frequencies?
6. How is it possible to take photographs in complete darkness?
7. What is it, exactly, that waves in a light wave?
8. We hear people talk of "ultraviolet light" and "infrared light." Why are these terms misleading? Why are we less likely to hear people talk of "radio light" and "X-ray light"?
9. Knowing that interplanetary space consists of a vacuum, what is your evidence that electromagnetic waves can travel through a vacuum?
10. What is the principal difference between a gamma ray and an infrared ray?
11. What is the speed of X-rays in a vacuum?
12. Which travels faster through a vacuum—an infrared ray or a gamma ray?
13. Your friend says that microwaves and ultraviolet light have different wavelengths but travel through space at the same speed. Do you agree or disagree?
14. Your friend says that any radio wave travels appreciably faster than any sound wave. Do you agree or disagree, and why?
15. Your friend says that outer space, instead of being empty, is chock full of electromagnetic waves. Do you agree or disagree?
16. Are the wavelengths of radio and television signals longer or shorter than waves detectable by the human eye?
17. Suppose a light wave and a sound wave have the same frequency. Which has the longer wavelength?
18. Which requires a physical medium in which to travel—light, sound, or both? Explain.
19. Do radio waves travel at the speed of sound, or at the speed of light, or somewhere in between?
20. When astronomers observe a supernova explosion in a distant galaxy, they see a sudden, simultaneous rise in visible light and other forms of electromagnetic radiation. Is this evidence to support the idea that the speed of light is independent of frequency? Explain.
21. What are the similarities and differences between radio waves and light?
22. A helium-neon laser emits light of wavelength 633 nanometers (nm). Light from an argon laser has a wavelength of 515 nm. Which laser emits the higher-frequency light?
23. Why would you expect the speed of light to be slightly less in the atmosphere than in a vacuum?
24. If you fire a bullet through a tree, it will slow down inside the tree and emerge at a speed that is less than the speed at which it entered. Does light, then, similarly slow down when it passes through glass and also emerge at a lower speed? Defend your answer.
25. Pretend that a person can walk only at a certain pace—no faster, no slower. If you time her uninterrupted walk across a room of known length, you can calculate her walking speed. If, however, she stops momentarily along the way to greet others in the room, the extra time spent in her brief interactions gives an *average* speed across the room that is less than her walking speed. How is this

similar to light passing through glass? In what way does it differ?

26. Is glass transparent or opaque to light of frequencies that match its own natural frequencies? Explain.

27. Short wavelengths of visible light interact more frequently with the atoms in glass than do longer wavelengths. Does this interaction time tend to speed up or to slow down the average speed of short-wavelength light in glass?

28. What determines whether a material is transparent or opaque?

29. You can get a sunburn on a cloudy day, but you can't get a sunburn even on a sunny day if you are behind glass. Explain.

30. Suppose that sunlight falls both on a pair of reading glasses and on a pair of dark sunglasses. Which pair of glasses would you expect to become warmer? Defend your answer.

31. Why does a high-flying airplane cast little or no shadow on the ground below, while a low-flying airplane casts a sharp shadow?

32. Only some of the people on the daytime side of the Earth can witness a solar eclipse when it occurs, whereas all the people on the nighttime side of the Earth can witness a lunar eclipse when it occurs. Why is this so?

33. Lunar eclipses are always eclipses of a full Moon. That is, the Moon is always seen full just before and after the Earth's shadow passes over it. Why is this? Why can we never have a lunar eclipse when the Moon is in its crescent or half-moon phase?

34. Do planets cast shadows? What is your evidence?

35. In 2004, the planet Venus passed between the Earth and the Sun. What kind of eclipse, if any, occurred?

36. What astronomical event would be seen by observers on the Moon at the time the Earth experiences a lunar eclipse? At the time the Earth experiences a solar eclipse?

37. Light from a location on which you concentrate your attention falls on your fovea, which contains only cones. If you wish to observe a weak source of light, like a faint star, why should you not look *directly* at the source?

38. Why do objects illuminated by moonlight lack color?

39. Why do we not see color at the periphery of our vision?

40. From your experimentation with Figure 26.15, is your blind spot located noseward from your fovea or to the outside of it?

41. Why should you be skeptical when your sweetheart holds you and looks at you with constricted pupils and says, "I love you"?

42. Can we infer that a person with large pupils is generally happier than a person with small pupils? If not, why not?

43. The intensity of light decreases as the inverse square of the distance from the source. Does this mean that light energy is lost? Explain.

44. Light from a camera flash weakens with distance in accord with the inverse-square law. Comment on an airline passenger who takes a flash photo of a city at nighttime from a high-flying plane.

45. Ships determine the ocean depth by bouncing sonar waves from the ocean bottom and measuring the round-trip time. How do some airplanes similarly determine their distance to the ground below?

46. The planet Jupiter is more than five times as far from the Sun as the planet Earth. How does the brightness of the Sun appear at this greater distance?

47. When you look at the night sky, some stars are brighter than others. Can you correctly say that the brightest stars emit more light? Defend your answer.

48. When you look at a distant galaxy through a telescope, how is it that you're looking backward in time?

49. When we look at the Sun, we are seeing it as it was 8 minutes ago. So we can only see the Sun "in the past." When you look at the back of your own hand, do you see it "now" or in "the past"?

50. "20/20 vision" is an arbitrary measure of vision—meaning that you can read what an average person can read at a distance of 20 feet in daylight. What is this distance in meters?

Problems

1. In about 1675, the Danish astronomer Olaus Roemer, measuring the times when one of Jupiter's moons appeared from behind Jupiter in its successive trips around that planet, and noticing the delays in these appearances as the Earth got farther from Jupiter, concluded that light took an extra 22 min to travel 300,000,000 km across the diameter of the Earth's orbit around the Sun. What approximate value for the speed of light did Roemer deduce? How much does it differ from our modern value? (Roemer's measurement, although not accurate by modern standards, was the first demonstration that light travels at a finite, not an infinite, speed.)

2. In one of Michelson's experiments, a beam from a revolving mirror traveled 15 km to a stationary mirror. How long a time interval elapsed before the beam returned to the revolving mirror?

3. The Sun is 1.50×10^{11} meters from the Earth. How long does it take for the Sun's light to reach the Earth? How long does it take light to cross the diameter of Earth's orbit? Compare this time with the time measured by Roemer in the seventeenth century (Problem 1).
4. How long does it take for a pulse of laser light to reach the Moon and to bounce back to the Earth?
5. The nearest star beyond the Sun is Alpha Centauri, 4.2×10^{16} meters away. If we were to receive a radio message from this star today, how long ago would it have been sent?
6. The wavelength of yellow sodium light in air is 589 nm. What is its frequency?
7. Blue-green light has a frequency of about 6×10^{14} Hz. Use the relationship $c = f\lambda$ to find the wavelength of this light in air. How does this wavelength compare with the size of an atom, which is about 10^{-10} m?
8. The wavelength of light changes as light goes from one medium to another, while the frequency remains the same. Is the wavelength longer or shorter in water than in air? Explain in terms of the equation speed = frequency $\times$ wavelength. A certain blue-green light has a wavelength of 600 nm (6×10^{-7} m) in air. What is its wavelength in water, where light travels at 75% of its speed in air? In Plexiglas, where light travels at 67% of its speed in air?
9. A certain radar installation that is used to track airplanes transmits electromagnetic radiation of wavelength 3 cm. (a) What is the frequency of this radiation, measured in billions of hertz (GHz)? (b) What is the time required for a pulse of radar waves to reach an airplane 5 km away and return?
10. A ball with the same diameter as a lightbulb is held halfway between the bulb and a wall, as shown in the sketch. Construct light rays (similar to those in Figure 26.13) and show that the diameter of the umbra on the wall is the same as the diameter of the ball and that the diameter of the penumbra is three times the diameter of the ball.

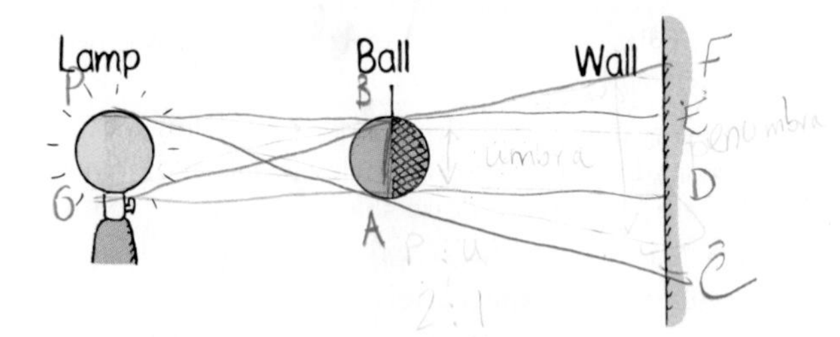

CHAPTER 27

Color

Lab-manual author Paul Robinson displays a variety of colors when he is illuminated by only red, green, and blue lamps.

Roses are red and violets are blue; colors intrigue artists and physics types too. To the physicist, the colors of objects are not in the substances of the objects themselves or even in the light they emit or reflect. Color is a physiological experience and is in the eye of the beholder. So when we say that light from a rose is red, in a stricter sense we mean that it *appears* red. Many organisms, including people with defective color vision, do not see the rose as red at all.

The colors we see depend on the frequency of the light we see. Lights of different frequencies are perceived as different colors; the lowest-frequency light we can detect appears to most people as the color red, and the highest frequency as violet. Between them range the infinite number of hues that make up the color spectrum of the rainbow. By convention, these hues are grouped into the seven colors of red, orange, yellow, green, blue, indigo, and violet. These colors together appear white. The white light from the Sun is a composite of all the visible frequencies.

Selective Reflection

FIGURE 27.1
The colors of things depend on the colors of the light that illuminates them.

Except for such light sources as lamps, lasers, and gas discharge tubes (which we will treat in Chapter 30), most of the objects around us reflect rather than emit light. They reflect only part of the light that is incident upon them, the part that gives them their color. A rose, for example, doesn't emit light; it reflects light (Figure 27.1). If we pass sunlight through a prism and then place a deep-red rose in various parts of the spectrum, the petals appear brown or black in all parts of the spectrum except in the red. In the red part of the spectrum, the petal appears red, but the green stem and leaves appear black. This shows that the red petals have the ability to reflect red light but not light of other colors; likewise, the green leaves have the ability to reflect green light but not light of other colors. When the rose is held in white light, the petals appear red and the leaves appear green because the petals reflect the red part of the white light and the leaves reflect the green part. To understand why objects reflect specific colors of light, we must turn our attention to the atom.

Light is reflected from objects in a manner similar to the way in which sound is "reflected" from a tuning fork when a nearby tuning fork sets it into vibration.

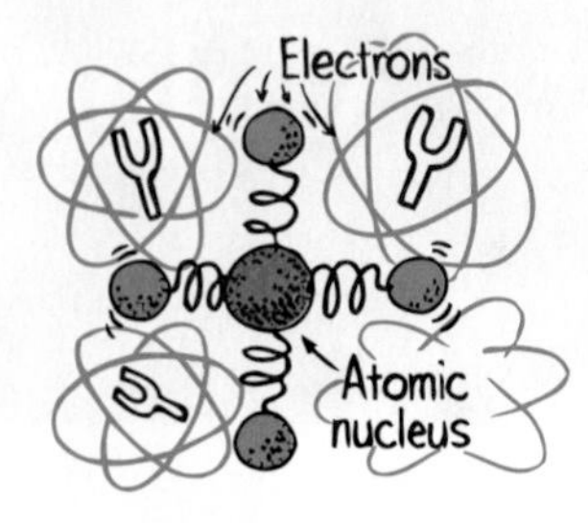

FIGURE 27.2
The outer electrons in an atom vibrate and resonate just as weights on springs would do. As a result, atoms and molecules behave somewhat like optical tuning forks.

One tuning fork can make another vibrate even when the frequencies are not matched, although at significantly reduced amplitudes. The same is true of atoms and molecules. The outer electrons that buzz about the atomic nucleus can be forced into vibration by the vibrating electric fields of electromagnetic waves.[1] Once vibrating, these electrons send out their own electromagnetic waves, just as vibrating acoustical tuning forks send out sound waves.

a

b

c

FIGURE 27.3
(a) Red ball seen under white light. The red color is due to the ball reflecting only the red part of the illuminating light. The rest of the light is absorbed by the surface. (b) Red ball seen under red light. (c) Red ball seen under green light. The ball appears black because the surface absorbs green light—there is no source of red light for it to reflect.

Different materials have different natural frequencies for absorbing and emitting radiation. In one material, electrons oscillate readily at certain frequencies; in another material, they oscillate readily at different frequencies. At the resonant frequencies at which the amplitudes of oscillation are large, light is absorbed; but, at frequencies below and above the resonant frequencies, light is reemitted. If the material is transparent, the reemitted light passes through it. If the material is opaque, the light passes back into the medium from which it came. This is reflection.

Usually, a material absorbs light of some frequencies and reflects the rest. If a material absorbs most of the visible light that is incident upon it but reflects red, for example, it appears red. That's why the petals of a red rose are red and the stem is green. The atoms of the petals absorb all visible light except red, which they reflect; the atoms of the stem absorb all light except green, which they reflect. An object that reflects light of all the visible frequencies, such as the white part of this page does, is the same color as the light that shines upon it. If a material absorbs all the light that shines upon it, it reflects none and is seen as black.

FIGURE 27.4
Most of the bunny's fur reflects light of all frequencies and appears white in sunlight. The bunny's dark fur absorbs all of the radiant energy in incident sunlight and therefore appears black.

Interestingly, the petals of most yellow flowers, such as daffodils, reflect red and green as well as yellow. Yellow daffodils reflect a broad band of frequencies. The reflected colors of most objects are not pure single-frequency colors but are composed of a spread of frequencies.

---

[1]The words *oscillation* and *vibration* both refer to periodic motion—motion that regularly repeats.

FIGURE 27.5
Color depends on the light source.

An object can reflect only those frequencies present in the illuminating light. The appearance of a colored object, therefore, depends on the kind of light that illuminates it. An incandescent lamp, for instance, emits more light in the lower than in the higher frequencies, enhancing any reds viewed in this light. In a fabric having only a little bit of red in it, the red is more apparent under an incandescent lamp than it is under a fluorescent lamp. Fluorescent lamps are richer in the higher frequencies, and so blues are enhanced under them. Usually we define an object's "true" color as the color it has in daylight. So, when you're shopping, the color of a garment you see in artificial light is not quite its true color (Figure 27.5).

Selective Transmission

The color of a transparent object depends on the color of the light it transmits. A red piece of glass appears red because it absorbs all the colors that compose white light, except red, which it *transmits*. Similarly, a blue piece of glass appears blue because it transmits primarily blue light and absorbs light of the other colors that illuminate it. The piece of glass contains dyes or *pigments*—fine particles that selectively absorb light of certain frequencies and selectively transmit others. From an atomic point of view, electrons in the pigment atoms selectively absorb illuminating light of certain frequencies. Light of other frequencies is reemitted from molecule to molecule in the glass. The energy of the absorbed light increases the kinetic energy of the molecules, and the glass is warmed. Ordinary window glass is colorless because it transmits light of all visible frequencies equally well.

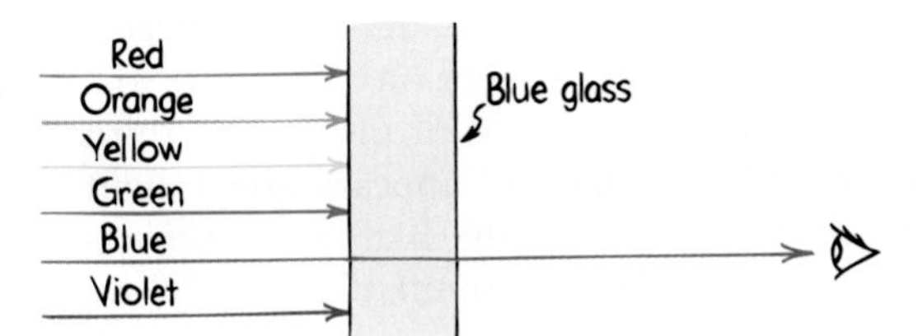

FIGURE 27.6
Only energy having the frequency of blue light is transmitted; energy of the other frequencies is absorbed and warms the glass.

CHECK YOURSELF

1. When red light shines on a red rose, why do the leaves become warmer than the petals?
2. When green light shines on a rose, why do the petals look black?
3. If you hold any small source of white light between you and a piece of red glass, you'll see two reflections from the glass: one from the front surface and one from the back surface. What color is each reflection?

CHECK YOUR ANSWERS

1. The leaves absorb rather than reflect the red light and so become warmer.
2. The petals absorb rather than reflect the green light. Because green is the only color illuminating the rose and because green contains no red to be reflected, the rose reflects no color and appears black.
3. The reflection from the front surface is white because the light doesn't go far enough into the colored glass to allow absorption of nonred light. Only red light reaches the back surface because the pigments in the glass absorb all the other colors, and so the back reflection is red.

Mixing Colored Light

All the colors added together produce white. The absence of all color is black.

Insights

The Physics Place
Yellow-Green Peak of Sunlight

The fact that white light from the Sun is a composite of all the visible frequencies is easily demonstrated by passing sunlight through a prism and observing the rainbow-colored spectrum. The intensity of light from the Sun varies with frequency, being most intense in the yellow-green part of the spectrum. It is interesting to note that our eyes have evolved to have maximum sensitivity in this range. That's why more fire engines these days are painted yellow-green, particularly at airports, where visibility is vital. Our sensitivity to yellow-green light also explains why we see better at night under the illumination of yellow sodium-vapor lamps than under common tungsten-filament lamps of the same brightness.

The graphical distribution of brightness versus frequency is called the *radiation curve* of sunlight (Figure 27.7). Most whites produced from reflected sunlight share this frequency distribution.

All the colors combined make white. Interestingly, the perception of white also results from the combination of only red, green, and blue light. We can understand this by dividing the solar radiation curve into three regions, as in Figure 27.8. Three types of cone-shaped receptors in our eyes perceive color. Light in the lowest third of the spectral distribution stimulates the cones sensitive to low frequencies and appears red; light in the middle third stimulates the cones sensitive to middle frequencies and appears green; light in the highest third stimulates the cones sensitive to the higher frequencies and appears blue. When all three types of cones are stimulated equally, we see white.

The Physics Place
Colored Shadows

Project red, green, and blue lights on a screen. Where they all overlap, white is produced. Where two of the three colors overlap, another color is produced (Figure 27.9). In the language of physicists, colored lights that overlap are said to *add* to each other. So we say that red, green, and blue light *add to produce white light,* and that any two of these colors of light add to produce another color. Various amounts of red, green, and blue, the colors to which each of our three types of cones are sensitive, produce any color in the spectrum. For this reason, red, green, and blue are called the **additive primary colors.** A close examination of the picture on most color television tubes will reveal that the

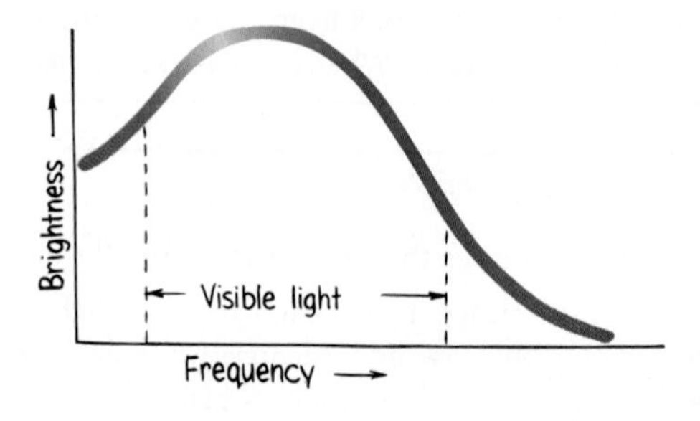

FIGURE 27.7
The radiation curve of sunlight is a graph of brightness versus frequency. Sunlight is brightest in the yellow-green region, in the middle of the visible range.

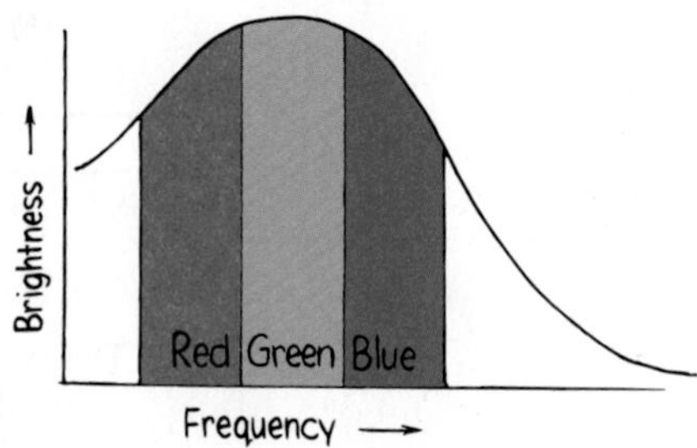

FIGURE 27.8
Radiation curve of sunlight divided into three regions, red, green, and blue. These are the additive primary colors.

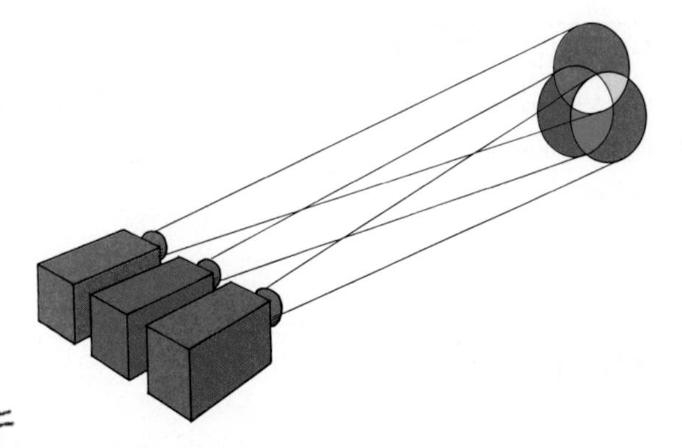

FIGURE 27.9
Interactive Figure
Color addition by the mixing of colored lights. When three projectors shine red, green, and blue light on a white screen, the overlapping parts produce different colors. White is produced where all three overlap.

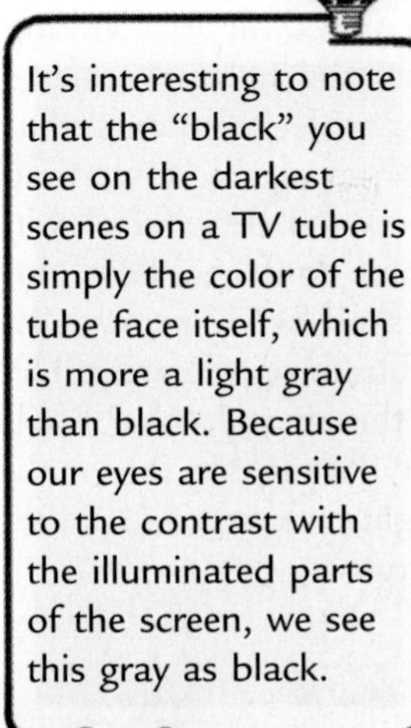

It's interesting to note that the "black" you see on the darkest scenes on a TV tube is simply the color of the tube face itself, which is more a light gray than black. Because our eyes are sensitive to the contrast with the illuminated parts of the screen, we see this gray as black.

Insights

picture is an assemblage of tiny spots, each less than a millimeter across. When the screen is lit, some of the spots are red, some green, some blue; the mixtures of these primary colors at a distance provide a complete range of colors, plus white.

Complementary Colors

Here's what happens when two of the three additive primary colors are combined:

Red + Blue = Magenta
Red + Green = Yellow
Blue + Green = Cyan

We say that magenta is the opposite of green; cyan is the opposite of red; and yellow is the opposite of blue. Now, when we add each of these colors to its opposite, we get white.

Magenta + Green = White (= Red + Blue + Green)
Yellow + Blue = White (= Red + Green + Blue)
Cyan + Red = White (= Blue + Green + Red)

When two colors are added together to produce white, they are called **complementary colors.** Every hue has some complementary color that when added to it will result in white.

The fact that a color and its complement combine to produce white light is nicely used in lighting stage performances. Blue and yellow lights shining on performers, for example, produce the effect of white light—except where one of the two colors is absent, as in the shadows. The shadow of one lamp, say the blue, is illuminated by the yellow lamp and appears yellow. Similarly, the shadow cast by the yellow lamp appears blue. This is a most interesting effect.

We see this effect in Figure 27.10, where red, green, and blue light shine on the golf ball. Note the shadows cast by the ball. The middle shadow is cast by the green spotlight and is not dark because it is illuminated by the red and blue lights, which make magenta. The shadow cast by the blue light appears yellow because it is illuminated by red and green light. Can you see why the shadow cast by the red light appears cyan?

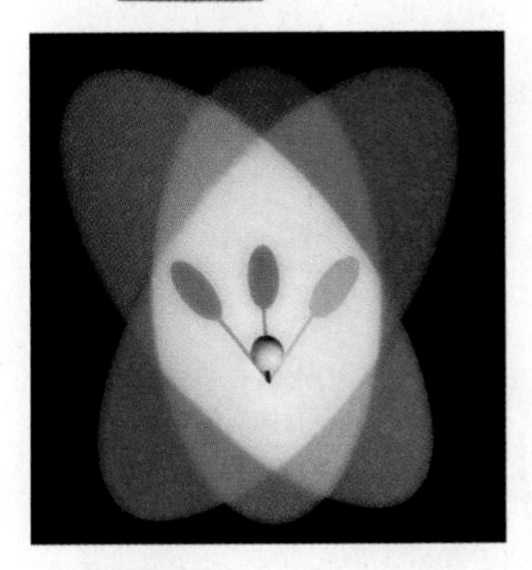

FIGURE 27.10
Interactive Figure
The white golf ball appears white when illuminated with red, green, and blue lights of equal intensities. Why are the shadows of the ball cyan, magenta, and yellow?

CHECK YOURSELF

1. From Figure 27.9, find the complements of cyan, of yellow, and of red.
2. Red + blue = ____.
3. White − red = ____.
4. White − blue = ____.

Mixing Colored Pigments

Every artist knows that if you mix red, green, and blue paint, the result will not be white but a muddy dark brown. Red and green paint certainly do not combine to form yellow, as is the rule for mixing colored lights. Mixing pigments in paints and dyes is entirely different from mixing lights. Pigments are tiny particles that absorb specific colors. For example, pigments that produce the color red absorb the complementary color cyan. So something painted red absorbs mostly cyan, which is why it reflects red. In effect, cyan has been *subtracted* from white light. Something painted blue absorbs yellow, and so reflects all the colors except yellow. Take yellow away from white and you've got blue. The colors magenta, cyan, and yellow are the **subtractive primaries.** The variety of colors you see in the colored photographs in this or any other book are the result of magenta, cyan, and yellow dots. Light illuminates the book, and light of some frequencies is subtracted from the light reflected. The rules of color subtraction differ from the rules of light addition.

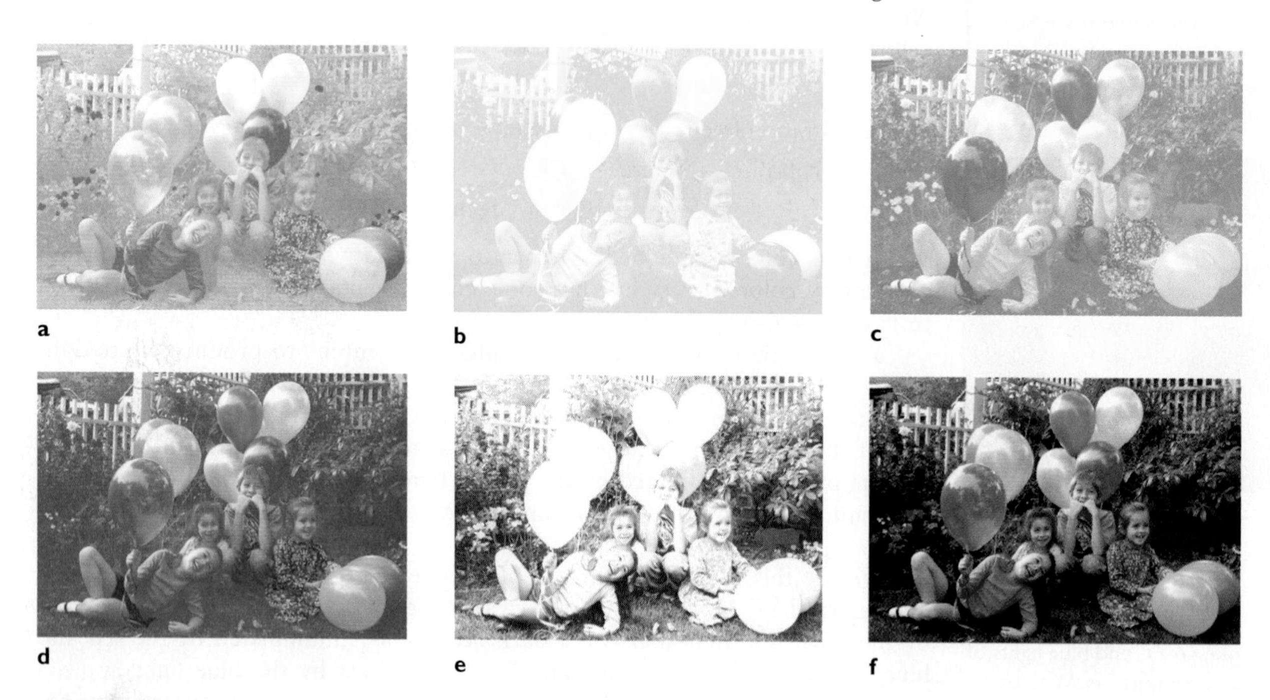

FIGURE 27.11
Only four colors of ink are used to print color illustrations and photographs—(a) magenta, (b) yellow, (c) cyan, and black. When magenta, yellow, and cyan are combined, they produce (d). Addition of black (e) produces the finished result (f).

FIGURE 27.12
Dyes or pigments, as in the three transparencies shown, absorb and effectively subtract light of some frequencies and transmit only part of the spectrum. The subtractive primary colors are yellow, magenta, and cyan. When white light passes through overlapping sheets of these colors, light of all frequencies is blocked (subtracted) and we have black. Where only yellow and cyan overlap, light of all frequencies except green is subtracted. Various proportions of yellow, cyan, and magenta dyes will produce nearly any color in the spectrum.

FIGURE 27.13
The rich colors of Sneezlee represent many frequencies of light. The photo, however, is a mixture of only yellow, magenta, cyan, and black.

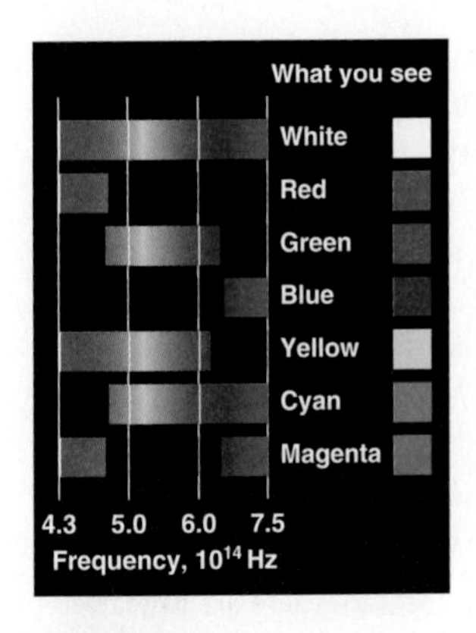

FIGURE 27.14
The approximate ranges of the frequencies we sense as the additive primary colors and the subtractive primary colors.

Color printing is an interesting application of color mixing. Three photographs (color separations) are taken of the illustration to be printed: one through a magenta filter, one through a yellow filter, and one through a cyan filter. Each of the three negatives has a different pattern of exposed areas that corresponds to the filter used and the color distribution in the original illustration. Light is shone through these negatives onto metal plates specially treated to hold printer's ink only in areas that have been exposed to light. The ink deposits are regulated on different parts of the plate by tiny dots. Inkjet printers deposit various combinations of magenta, cyan, yellow, and black inks. Examine the color in any of the figures in this or any book with a magnifying glass and see how the overlapping dots of these colors give the appearance of many colors. Or look at a billboard up close.

We see that all the rules of color addition and subtraction can be deduced from Figures 27.9, 27.10, and 27.12.

When we look at the colors on a soap bubble or soap film, we see cyan, magenta and yellow predominantly. What does this tell us? It tells us that some primary colors have been subtracted from the original white light! (How this happens will be discussed in Chapter 29.)

Why the Sky Is Blue

Why the Sky Is Blue and Why the Sunset Is Red

Not all colors are the result of the addition or subtraction of light. Some colors, like the blue of the sky, are the result of selective scattering. Consider the analogous case of sound: If a beam of a particular frequency of sound is directed to

CHECK YOUR ANSWERS

1. Red, blue, cyan. **2.** Magenta. **3.** Cyan. **4.** Yellow.

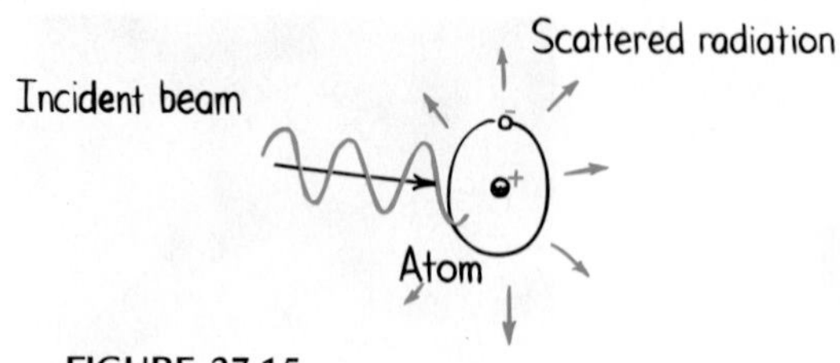

FIGURE 27.15
A beam of light falls on an atom and increases the vibrational motion of electrons in the atom. The vibrating electrons reemit the light in various directions. Light is scattered.

a tuning fork of similar frequency, the tuning fork is set into vibration and redirects the beam in multiple directions. The tuning fork *scatters* the sound. A similar process occurs with the scattering of light from atoms and particles that are far apart from one another, as they are in the atmosphere.[2]

Recall Figure 27.2, where we learned that atoms behave like tiny optical tuning forks and reemit light waves that shine on them. Molecules and larger collections of atoms do the same. The tinier the particle, the greater the amount of higher-frequency light it will reemit. This is similar to the way small bells ring with higher notes than larger bells. The nitrogen and oxygen molecules that make up most of the atmosphere are like tiny bells that "ring" with high frequencies when energized by sunlight. Like sound from the bells, the reemitted light is sent in all directions. When light is reemitted in all directions, we say the light is *scattered.*

Of the visible frequencies of sunlight, violet is scattered the most by nitrogen and oxygen in the atmosphere, followed by blue, green, yellow, orange, and red, in that order. Red is scattered only a tenth as much as violet. Although violet light is scattered more than blue, our eyes are not very sensitive to violet light. Therefore, the blue scattered light is what predominates in our vision, and we see a blue sky.

Isn't it true that knowing why the sky is blue and why sunsets are red *adds* to their beauty? Knowledge doesn't subtract.

Insights

The blue of the sky varies in different locations under different conditions. A principal factor is the water-vapor content of the atmosphere. On clear, dry days, the sky is a much deeper blue than on clear days with high humidity. In locations where the upper air is exceptionally dry, such as Italy and Greece, beautifully blue skies have inspired painters for centuries. Where the atmosphere contains a lot of particles of dust and other particles larger than oxygen and nitrogen molecules, light of the lower frequencies is also scattered strongly. This

FIGURE 27.16
In clean air, the scattering of high-frequency light provides a blue sky. When the air is full of particles larger than molecules, lower-frequency light is also scattered, which adds to the blue to give a whitish sky.

FIGURE 27.17
There are no blue pigments in the feathers of a blue jay. Instead, there are tiny alveolar cells in the barbs of its feathers that scatter light—mainly high-frequency light. So a blue jay is blue for the same reason the sky is blue—scattering.

[2]This type of scattering, which is called *Rayleigh scattering,* occurs whenever the scattering particles are much smaller than the wavelength of incident light and have resonances at frequencies higher than those of the scattered light. Scattering is much more complex than our simplified treatment here.

makes the sky less blue, and it takes on a whitish appearance. After a heavy rainstorm when the particles have been washed away, the sky becomes a deeper blue.

The grayish haze in the skies over large cities is the result of particles emitted by car and truck engines and by factories. Even when idling, a typical automobile engine emits more than 100 billion particles per second. Most particles are invisible, but they act as tiny centers to which other particles adhere. These are the primary scatterers of lower-frequency light. The largest of these particles absorb rather than scatter light, and a brownish haze is produced. Yuk!

Why Sunsets Are Red

Atmospheric soot heats Earth's atmosphere by absorbing light, while cooling local regions by blocking sunlight from reaching the ground. Soot particles in the air may trigger severe rains in one region and cause droughts and dust storms in another.

Insights

Light that isn't scattered is light that is transmitted. Because red, orange, and yellow light are the least scattered by the atmosphere, light of these lower frequencies is better transmitted through the air. Red, which is scattered the least—and, therefore, is transmitted the most—passes through more atmosphere than any other color. So the thicker the atmosphere through which a beam of sunlight travels, the more time there is to scatter all the higher-frequency components of the light. This means that the light that makes it through best is red. As Figure 27.18 shows, sunlight travels through more atmosphere at sunset, and that is why sunsets (and sunrises) are red.

At noon, sunlight travels through the least amount of atmosphere to reach the Earth's surface. Only a small amount of high-frequency light is scattered from the sunlight, enough to make the Sun look yellowish. As the day progresses and the Sun descends lower in the sky, the path through the atmosphere is longer, and more violet and blue are scattered from the sunlight. The removal of violet and blue leaves the transmitted light redder. The Sun becomes progressively redder, going from yellow to orange and finally to a red-orange at sunset. Sunsets and sunrises are unusually colorful following volcanic eruptions, because particles larger than atmospheric molecules are then more abundant in the air.[3]

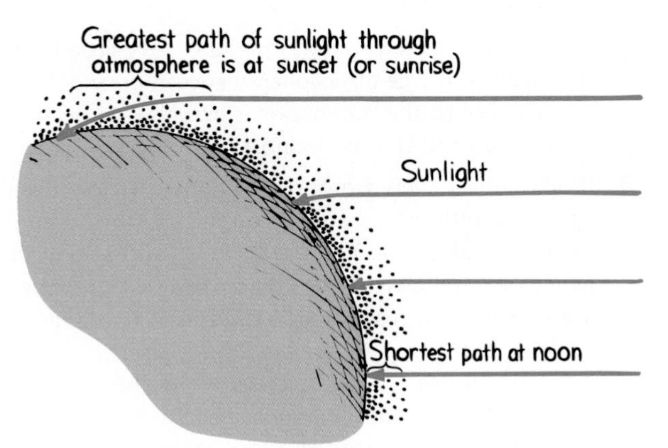
Greatest path of sunlight through atmosphere is at sunset (or sunrise)
Sunlight
Shortest path at noon

FIGURE 27.18
Interactive Figure
A sunbeam must travel through more of the atmosphere at sunset than at noon. As a result, more blue is scattered from the beam at sunset than at noon. By the time a beam of initially white light reaches the ground, only light of the lower frequencies survives to produce a red sunset.

[3]Sunsets and sunrises would be unusually colorful if particles larger than atmospheric molecules were more abundant in the air. This was the case all over the world for three years following the eruption, in 1883, of the volcano Krakatau (or Krakatoa) in what is now Indonesia, when micrometer-sized particles were spewed out in abundance and spread throughout the world's atmosphere. This occurred to a lesser extent following the 1991 eruption of Mount Pinatubo in the Philippines. Next?

PRACTICING PHYSICS

You can simulate a sunset with a fish tank full of water in which you've dropped a tiny bit of milk. A few drops will do. Then shine a flashlight beam through the water and you'll see that it looks bluish from the side. Milk particles are scattering the higher frequencies of light in the beam. Light emerging from the far end of the tank will have a reddish tinge. That's the light that wasn't scattered.

The colors of the sunset are consistent with our rules for color mixing. When blue is subtracted from white light, the complementary color that is left is yellow. When higher-frequency violet is subtracted, the resulting complementary color is orange. When medium-frequency green is subtracted, magenta is left. The combinations of resulting colors vary with atmospheric conditions, which change from day to day, giving us a variety of sunsets to enjoy.

CHECK YOURSELF

1. If molecules in the sky scattered low-frequency light more than high-frequency light, what color would the sky be? What color would sunsets be?
2. Distant dark mountains are bluish. What is the source of this blueness? (*Hint:* What is between us and the mountains we see?)
3. Distant snow-covered mountains reflect a lot of light and are bright. Very distant ones look yellowish. Why? (*Hint:* What happens to the reflected white light as it travels from the mountains to us?)

CHECK YOUR ANSWERS

1. If low-frequency light were scattered, the noontime sky would appear reddish-orange. At sunset, more reds would be scattered by the longer path of the sunlight, and the sunlight would be predominantly blue and violet. So sunsets would appear blue!
2. If we look at distant dark mountains, very little light from them reaches us, and the blueness of the atmosphere between us and them predominates. The blueness we attribute to the mountains is actually the blueness of the low-altitude "sky" between us and the mountains!
3. Bright snow-covered mountains appear yellow because the blue in the white light they reflect is scattered on its way to us. By the time the light reaches us, it is weak in the high frequencies and strong in the low frequencies—hence, it is yellowish. For greater distances, farther away than mountains are usually seen from, they would appear orange for the same reason a sunset appears orange.

   Why do we see the scattered blue when the background is dark but not when the background is bright? Because the scattered blue is faint. A faint color will show itself against a dark background but not against a bright background. For example, when we look from Earth's surface at the atmosphere against the darkness of space, the atmosphere is sky blue. But astronauts above who look down through the same atmosphere to the bright surface of the Earth do not see the same blueness.

Why Clouds Are White

FIGURE 27.19
A cloud is composed of water droplets of various sizes. The tiniest droplets scatter blue light, slightly larger ones scatter green light, and still larger ones scatter red light. The result is a white cloud.

Water droplets in a variety of sizes make up clouds. The different-size droplets produce a variety of scattered frequencies: The tiniest scatter more blue than other colors; slightly larger droplets scatter light of slightly higher frequencies, such as green; and still larger droplets scatter more red. The overall result is a white cloud. Electrons close to one another in a droplet vibrate together and in step, which results in a greater intensity of scattered light than from the same number of electrons vibrating separately. Hence, clouds are bright!

Larger assortments of droplets absorb much of the light incident upon them, and so the intensity of the scattered light is less. This contributes to the darkness of clouds composed of larger droplets. Further increase in the size of the droplets causes them to fall as raindrops, and we have rain.

The next time you find yourself admiring a crisp blue sky, or delighting in the shapes of bright clouds, or watching a beautiful sunset, think about all those ultratiny optical tuning forks vibrating away—you'll appreciate these everyday wonders of nature even more!

Why Water Is Greenish Blue

We often see a beautiful deep blue when we look at the surface of a lake or the ocean. But that isn't the color of water; it's the reflected color of the sky. The color of water itself, as you can see by looking at a piece of white material under water, is a pale greenish blue.

Although water is transparent to light of nearly all the visible frequencies, it strongly absorbs infrared waves. This is because water molecules resonate to the frequencies of infrared. The energy of the infrared waves is transformed into internal energy in the water, which is why sunlight warms water. Water molecules resonate somewhat in the visible red, which causes red light to be a little more strongly absorbed in water than blue light. Red light is reduced to one-quarter of its initial brightness by 15 meters of water. There is very little red light in the sunlight that penetrates below 30 meters of water. When red is removed from white light, what color remains? This question can be asked in another way: What is the complementary color of red? The complementary

FIGURE 27.20
Water is cyan because it absorbs red light. The froth in the waves is white because, like clouds, it is composed of a variety of tiny water droplets that scatter light of all the visible frequencies.

FIGURE 27.21
The extraordinary blue of Canadian Rocky Mountain lakes is produced by scattering from extremely fine particles of glacial silt suspended in the water.

color of red is cyan—a bluish-green color. In seawater, everything at these depths has a cyan color.

Many crabs and other sea creatures that appear black in deep water are found to be red when they are raised to the surface. At these depths, black and red look the same. Apparently the selection mechanism of evolution could not distinguish between black and red at such depths in the ocean.

Whereas the bluish-green color of water is produced by selective absorption of light, the intriguingly vivid blue of lakes in the Canadian Rocky Mountains is due to scattering.[4] The lakes are fed by runoff from melting glaciers that contain fine particles of silt, called rock flour, which remain suspended in the water. Light scatters from these tiny particles and gives the water its eerily vivid color (Figure 27.21). (Tourists who photograph these lakes are advised to inform their photo processors *not* to adjust the color to a "real" blue!)

Interestingly enough, the color we see is not in the world around us—the color is in our heads. The world is filled with a montage of vibrations—electromagnetic waves that stimulate the sensation of color when the vibrations interact with the cone-shaped receiving antennae in the retinas of our eyes. How nice that eye–brain interactions produce the beautiful colors we see.

[4]Scattering by small, widely spaced particles in the irises of blue eyes, rather than any pigments, accounts for their color. Absorption by pigments accounts for brown eyes.

Summary of Terms

Additive primary colors The three colors—red, blue, and green—that, when added in certain proportions, produce any other color in the visible-light part of the electromagnetic spectrum and can be mixed equally to produce white light.

Complementary colors Any two colors that, when added, produce white light.

Subtractive primary colors The three colors of absorbing pigments—magenta, yellow, and cyan—that, when mixed in certain proportions, reflect any other color in the visible-light part of the electromagnetic spectrum.

Suggested Reading

Murphy, Pat, and Paul Doherty. *The Color of Nature.* San Francisco: Chronicle Books, 1996.

Review Questions

1. What is the relationship between the frequency of light and its color?

Selective Reflection

2. What occurs when the outer electrons that buzz about the atomic nucleus encounter electromagnetic waves?
3. What happens to light when it falls upon a material that has a natural frequency equal to the frequency of the light?
4. What happens to light when it falls upon a material that has a natural frequency above or below the frequency of the light?

Selective Transmission

5. What color light is transmitted through a piece of red glass?
6. What is a *pigment*?
7. Which warms more quickly in sunlight—a colorless or a colored piece of glass? Why?

Mixing Colored Light

8. What is the evidence for the statement that white light is a composite of all the colors of the spectrum?
9. What is the color of the peak frequency of solar radiation?
10. To what color of light are our eyes most sensitive?
11. What is a *radiation curve*?
12. What frequency ranges of the radiation curve do red, green, and blue light occupy?
13. Why are red, green, and blue called the *additive primary colors*?

Complementary Colors

14. What is the resulting color of equal intensities of red light and cyan light combined?
15. Why are red and cyan called *complementary colors*?

Mixing Colored Pigments

16. When something is painted red, what color is most absorbed?
17. What are the *subtractive primary colors*?
18. If you look with a magnifying glass at pictures printed in full color in this or other books or magazines, you'll notice three colors of ink plus black. What are these colors?

Why the Sky Is Blue

19. Which interact more with high-pitched sounds—small bells or large bells?
20. Which interact more with high-frequency light—small particles or large particles?
21. Why is it incorrect to say the sky is blue because oxygen and nitrogen molecules are blue in color?
22. Why does the sky sometimes appear whitish?

Why Sunsets Are Red

23. Why does the Sun look reddish at sunrise and sunset but not at noon?
24. Why does the color of sunsets vary from day to day?

Why Clouds Are White

25. What is the evidence for a variety of droplet sizes in a cloud?
26. What is the effect on the color of a cloud when it contains an abundance of large droplets?

Why Water Is Greenish Blue

27. What part of the electromagnetic spectrum is most absorbed by water?
28. What part of the *visible* electromagnetic spectrum is most absorbed by water?
29. What color results when red is subtracted from white light?
30. Why does water appear cyan?

Projects

1. Stare at a piece of colored paper for 45 seconds or so. Then look at a plain white surface. The cones in your retina receptive to the color of the paper become fatigued, so you see an afterimage of the complementary color when you look at a white area. This is because the fatigued cones send a weaker signal to the brain. All the colors produce white, but all the colors minus one produce the complement to the missing color. Try it and see!
2. Cut a disk a few centimeters or so in diameter from a piece of cardboard; punch two holes a bit off-center, big enough to loop a piece of string as shown in the sketch. Twirl the disk as shown, so the string winds up like a rubber band on a model airplane. Then, if you tighten the string by pulling outward, the disk will spin. If half the disk is colored yellow and the other half blue, when it is spun the colors will be mixed and appear nearly white. (How close to white depends on the hues of the colors). Try this for other complementary colors.

3. Fashion a cardboard tube covered at each end with metal foil. Punch a hole in each end with a pencil, one about 3 or so millimeters in diameter and the other twice as big. Place your eye to the small hole and look through the tube at the colors of things against the black background of the tube. You'll see colors that look very different from how they appear against ordinary backgrounds.
4. Write a letter to Grandma and tell her what details you've learned that explain why the sky is blue, sunsets are red, and clouds are white. Discuss whether or not this information adds to or decreases your perception of the beauty of nature.

Exercises

1. What color of visible light has the longest wavelength? The shortest?
2. Why is red paint red?
3. In a dress shop with only fluorescent lighting, a customer insists on taking dresses into the daylight at the doorway to check their color. Is she being reasonable? Explain.

4. Why will the leaves of a red rose be warmed more than the petals when illuminated with red light? How does this relate to people in the hot desert wearing white clothes?
5. If the sunlight were somehow green instead of white, what color garment would be most advisable on an uncomfortably hot day? On a very cold day?
6. Why do we not list black and white as colors?
7. Why are the interiors of optical instruments intentionally black?
8. Fire engines used to be red. Yellow-green is now the preferred color. Why the change?
9. What is the color of common tennis balls, and why?
10. The radiation curve of the Sun (Figures 27.7 and 27.8) show that the brightest light from the Sun is yellow-green. Why, then, do we see the Sun as whitish instead of yellow-green?
11. What color does red cloth appear to be when illuminated by sunlight? By light from a neon sign? By cyan light?
12. Why does a white piece of paper appear white in white light, red in red light, blue in blue in blue light, and so on for every color?
13. A spotlight is coated so that it won't transmit yellow light from its white-hot filament. What color is the emerging beam of light?
14. How could you use the spotlights at a play to make the yellow clothes of the performers suddenly change to black?
15. Suppose that two flashlight beams are shone on a white screen, one through a pane of blue glass and the other through a pane of yellow glass. What color appears on the screen where the two beams overlap? Suppose, instead, that the two panes of glass are placed in the beam of a single flashlight. What then?
16. Does color television operate by color addition or by color subtraction? Defend your answer.
17. On a TV screen, red, green, and blue spots of fluorescent materials are illuminated at a variety of relative intensities to produce a full spectrum of colors. What dots are activated to produce yellow? Magenta? White?
18. What colors of ink do color ink-jet printers use to produce a full range of colors? Do the colors form by color addition or by color subtraction?
19. Physical-science author Suzanne Lyons is shown with son Tristan wearing red and daughter Simone wearing green. Note that the negative of the photo shows these colors differently. What is your explanation?

20. Your friend reasons that magenta and yellow paint mixed together will produce red because magenta is a combination of red and blue and yellow is a combination of red and green—and that the color in common is red. Do you agree or disagree, and why?
21. Streetlights that use high-pressure sodium vapor produce light that is mainly yellow with some red. Why are dark blue police cars not advisable in a community that uses these streetlights?
22. What color of light will be transmitted through overlapping cyan and magenta filters?
23. Look at your red, sunburned feet when they are under water. Why don't they look as red as when they are above water?
24. Why does the blood of injured deep-sea divers look greenish-black in underwater photographs taken with natural light, but red when flashbulbs are used?
25. By reference to Figure 27.9, complete the following equations:

    Yellow light + blue light = ________ light.
    Green light + ________ light = white light.
    Magenta + yellow + cyan = ________ light.

26. Check Figure 27.9 to see if the following three statements are accurate. Then fill in the last statement. (All colors are combined by the addition of light.)

    Red + green + blue = white.
    Red + green = yellow = white − blue.
    Red + blue = magenta = white − green.
    Green + blue = cyan = white − ________.

27. Your friend says that red and cyan light produce white light because cyan is green + blue, and so red + green + blue = white. Do you agree or disagree, and why?
28. In which of these cases will a ripe banana appear black—when illuminated with red, yellow, green, or blue light?
29. When white light is shone on red ink dried on a glass plate, the color that is transmitted is red. But the color that is reflected is not red. What is it?
30. Stare intently at an American flag. Then turn your view to a white area on a wall. What colors do you see in the image of the flag that appears on the wall?
31. Why can't we see stars in the daytime?
32. Why is the sky a darker blue when you are at high altitudes? (*Hint:* What color is the "sky" on the Moon?)
33. There is no atmosphere on the Moon to produce scattering of light. How does the daytime sky of the Moon appear when viewed from the Moon's surface?
34. Can stars be seen from the Moon's surface in the "daytime" when the Sun is shining?
35. What is the color of the setting Sun as seen on the Moon?
36. At the beach, you can get a sunburn while under the shade of an umbrella. What is your explanation?
37. Pilots sometimes wear glasses that transmit yellow light and absorb light of most other colors. Why does this help them see more clearly?
38. Does light travel faster through the lower atmosphere or through the upper atmosphere?
39. Why does smoke from a campfire look blue against trees near the ground but yellow against the sky?
40. Your friend says that the reason the distant dark mountains appear blue is because you're looking at the sky between you and the mountains. Do you agree or disagree?
41. Comment on the statement "Oh, that beautiful red sunset is just the leftover colors that weren't scattered on their way through the atmosphere."
42. If the sky on a certain planet in the solar system were normally orange, what color would sunsets be?
43. Volcanic emissions spew fine ashes in the air that scatter red light. What color does a full Moon appear to be through these ashes?
44. Tiny particles, like tiny bells, scatter high-frequency waves more than low-frequency waves. Large particles, like large bells, mostly scatter low-frequency waves. Intermediate-size particles and bells mostly scatter waves of intermediate frequencies. How does this relate to the whiteness of clouds?
45. Why is the foam of root beer white, while the beverage is dark brown?
46. Very big particles, like droplets of water, absorb more radiation than they scatter. How does this relate to the darkness of rain clouds?
47. How would the whiteness of snow appear if Earth's atmosphere were several times thicker?
48. The atmosphere of Jupiter is more than 1000 km thick. From the surface of Jupiter, would you expect to see a white Sun?
49. Red sunrises occur for the same reason as red sunsets. But sunsets are usually more colorful than sunrises—especially near cities. What is your explanation?
50. You're explaining to a youngster at the seashore why the water is cyan colored. The youngster points to the whitecaps of overturning waves and asks why they are white. What is your answer?

CHAPTER 28

Reflection and Refraction

Peter Hopkinson boosts class interest by this zany demonstration of standing astride a large mirror as he lifts his right leg while his unseen left leg provides support behind the mirror.

Most of the things we see around us do not emit their own light. They are visible because they reemit light reaching their surface from a primary source, such as the Sun or a lamp, or from a secondary source, such as the illuminated sky. When light falls on the surface of a material, it is either reemitted without change in frequency or is absorbed into the material and turned into heat.[1] We say light is *reflected* when it is returned into the medium from which it came—the process is **reflection.** When light crosses from one transparent material into another, we say it is *refracted*—the process is **refraction.** Usually some degree of reflection, refraction, and absorption occurs when light interacts with matter. In this chapter, we ignore the light absorbed and converted to heat energy and concentrate on the light that continues to be light after it meets a surface.

Reflection

When this page is illuminated with sunlight or lamplight, electrons in the atoms of the paper and ink vibrate more energetically in response to the oscillating electric fields of the illuminating light. The energized electrons reemit the light by which you see the page. When the page is illuminated by white light, the paper appears white, which reveals that the electrons reemit all the visible frequencies. Very little absorption occurs. The ink is a different story. Except for a bit of reflection, it absorbs all the visible frequencies and therefore appears black.

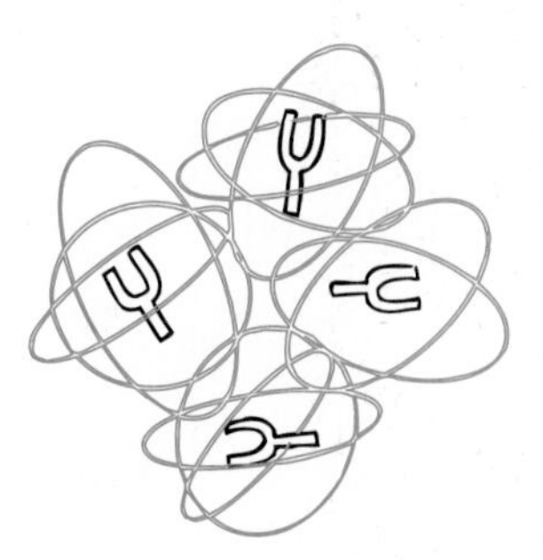

FIGURE 28.1
Light interacts with atoms as sound interacts with tuning forks.

[1]Another less common fate is absorption followed by reemission at lower frequencies—fluorescence (Chapter 30).

Principle of Least Time[2]

We all know that light ordinarily travels in straight lines. In going from one place to another, light will take the most efficient path and travel in a straight line. This is true if there is nothing to obstruct the passage of light between the places under consideration. If light is reflected from a mirror, the kink in the otherwise straight-line path is described by a simple formula. If light is refracted, as when it goes from air into water, still another formula describes the deviation of light from the straight-line path. Before thinking of light with these formulas, we will first consider an idea that underlies all the formulas that describe light paths. This idea, which was formulated by the French scientist Pierre Fermat in about 1650, is called **Fermat's principle of least time.** Fermat's idea was this: Out of all possible paths that light might travel to get from one point to another, it travels the path that requires the *shortest time.*

Law of Reflection

We can understand reflection by the principle of least time. Consider the following situation. In Figure 28.2, we see two points, A and B, and an ordinary plane mirror beneath. How can we get from A to B in the shortest time? The answer is simple enough—go straight from A to B! But, if we add the condition that the light must strike the mirror in going from A to B in the shortest time, the answer is not so easy. One way would be to go as quickly as possible to the mirror and then to B, as shown by the solid lines in Figure 28.3. This gives us a short path to the mirror but a very long path from the mirror to B. If we instead consider a point on the mirror a little to the right, we slightly increase the first distance, but we considerably decrease the second distance, and so the total path length shown by the dashed lines—and therefore the travel time—is less. How can we find the exact point on the mirror for which the time is shortest? We can find it very nicely by a geometric trick.

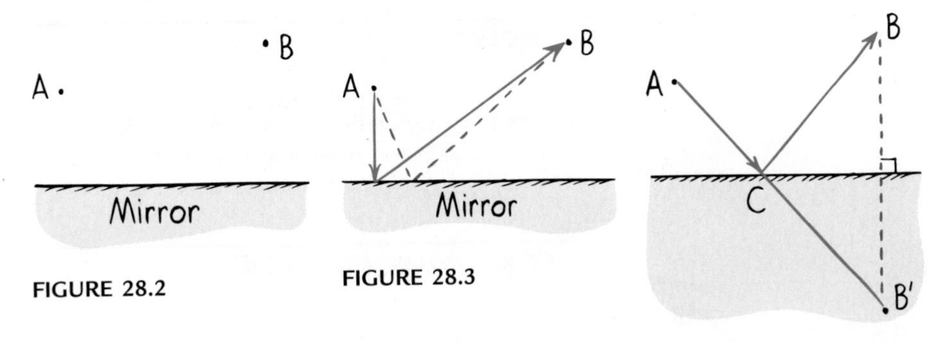

FIGURE 28.2

FIGURE 28.3

FIGURE 28.4

[2]This material and many of the examples of least time are adapted from R. P. Feynman, R. B. Leighton, and M. Sands, *The Feynman Lectures on Physics,* Vol. I, Chap. 26 (Reading, Mass.: Addison-Wesley, 1963).

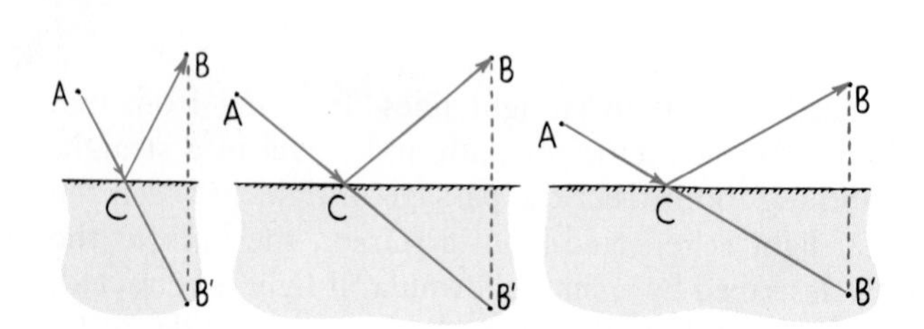

FIGURE 28.5
Reflection.

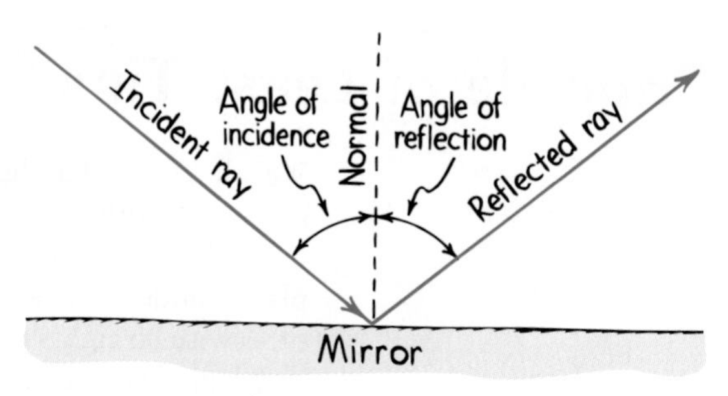

FIGURE 28.6 Interactive Figure
The law of reflection.

We construct, on the opposite side of the mirror, an artificial point, B′, which is the same distance "through" and below the mirror as the point B is above the mirror (Figure 28.4). The shortest distance between A and this artificial point B′ is simple enough to determine: It's a straight line. Now this straight line intersects the mirror at a point C, the precise point of reflection for the shortest path and hence the path of least time for the passage of light from A to B. Inspection will show that the distance from C to B equals the distance from C to B′. We see that the length of the path from A to B′ through C is equal to the length of the path from A to B bouncing off point C along the way.

Further inspection and a little geometrical reasoning will show that the angle of incident light from A to C is equal to the angle of reflection from C to B. This is the **law of reflection,** and it holds for all angles (Figure 28.5):

The angle of incidence equals the angle of reflection.

The law of reflection is illustrated with arrows representing light rays in Figure 28.6. Instead of measuring the angles of incident and reflected rays from the reflecting surface, it is customary to measure them from a line perpendicular to the plane of the reflecting surface. This imaginary line is called the *normal.* The incident ray, the normal, and the reflected ray all lie in the same plane.

EXERCISE

The construction of artificial points B′ in Figures 28.4 and 28.5 shows how light encounters point C in reflecting from A to B. By similar construction, show that light that originates at B and reflects to A also encounters the same point C.

CHECK YOUR ANSWER

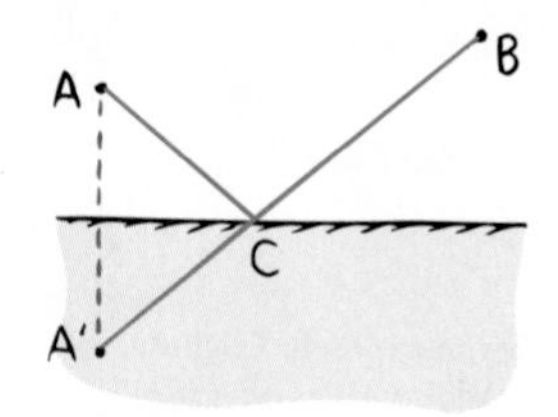

Construct an artificial point A′ as far below the mirror as A is above; then draw a straight line from B to A′ to find C, as shown at the left. Both constructions superimposed, at right, show that C is common to both. We see that light will follow the same path if it goes in the opposite direction. Whenever you see somebody else's eyes in a mirror, be assured that they can also see yours.

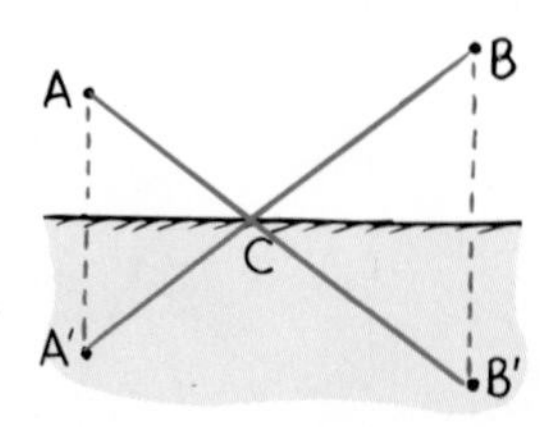

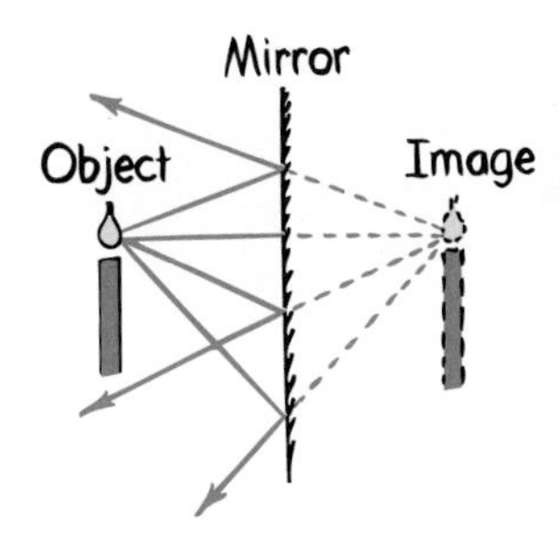

FIGURE 28.7
A virtual image is formed behind the mirror and is located at the position where the extended reflected rays (dashed lines) converge.

Plane Mirrors

Suppose a candle flame is placed in front of a plane mirror. Rays of light are sent from the flame in all directions. Figure 28.7 shows only four of the infinite number of rays leaving one of the infinite number of points on the candle. When these rays encounter the mirror, they are reflected at angles equal to their angles of incidence. The rays diverge from the flame and, on reflection, diverge from the mirror. These divergent rays appear to emanate from a particular point behind the mirror (where the dashed lines intersect). An observer sees an image of the flame at this point. The light rays do not actually come from this point, so the image is called a *virtual image*. The image is as far behind the mirror as the object is in front of the mirror, and image and object have the same size. When you view yourself in a mirror, for example, the size of your image is the same as the size your twin would appear if located as far behind the mirror as you are in front—as long as the mirror is flat (we call a flat mirror a *plane mirror*).

When the mirror is curved, the sizes and distances of object and image are no longer equal. We will not discuss curved mirrors in this text, except to say that the law of reflection still holds. A curved mirror behaves as a succession of flat mirrors, each at a slightly different angular orientation from the one next to it. At each point, the angle of incidence is equal to the angle of reflection (Figure 28.9). Note that, in a curved mirror, unlike in a plane mirror, the normals (shown by the dashed black lines to the left of the mirror) at different points on the surface are not parallel to one another.

Whether the mirror is plane or curved, the eye–brain system cannot ordinarily differentiate between an object and its reflected image. So the illusion that an object exists behind a mirror (or in some cases in front of a concave

Image Formation in a Mirror

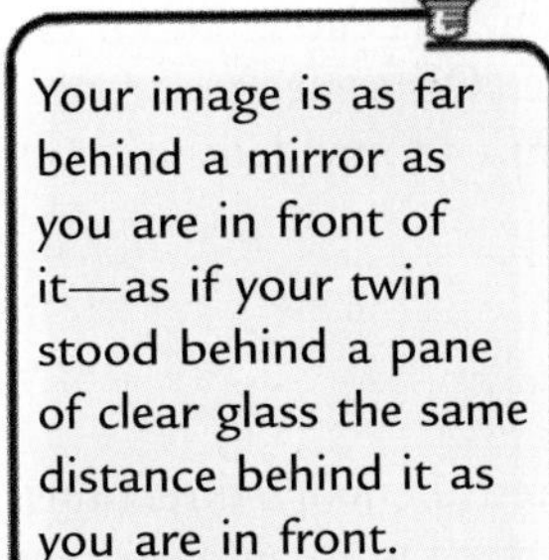

Your image is as far behind a mirror as you are in front of it—as if your twin stood behind a pane of clear glass the same distance behind it as you are in front.

Insights

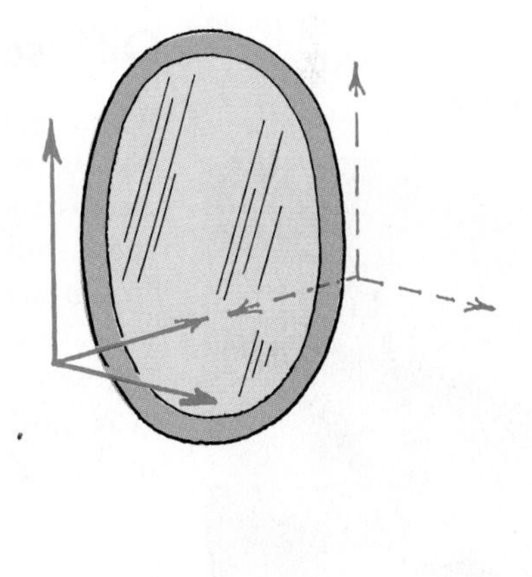

FIGURE 28.8
Marjorie's image is as far behind the mirror as she is in front. Note that she and her image have the same color of clothing—evidence that light doesn't change frequency upon reflection. Interestingly, her left–right axis is no more reversed than her up–down axis. The axis that *is* reversed, as shown to the right, is front–back. That's why it seems her left hand faces the right hand of her image.

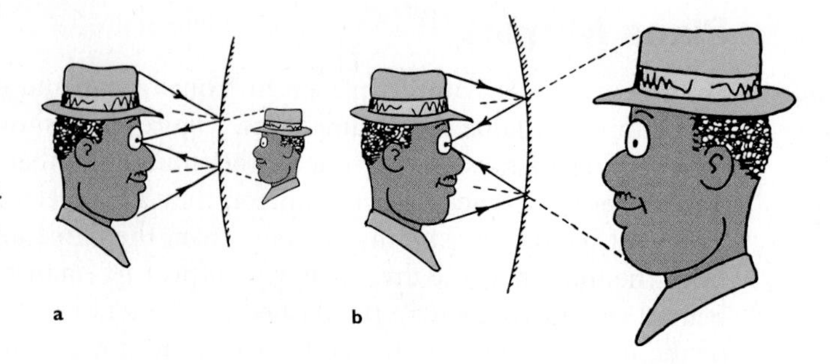

FIGURE 28.9
(a) The virtual image formed by a *convex* mirror (a mirror that curves outward) is smaller and closer to the mirror than the object. (b) When the object is close to a *concave* mirror (a mirror that curves inward like a "cave"), the virtual image is larger and farther away than the object. In either case, the law of reflection holds for each ray.

mirror) is merely due to the fact that the light from the object enters the eye in exactly the same manner, physically, as it would have entered if the object really were at the image location.

CHECK YOURSELF

1. What evidence can you cite to support the claim that the frequency of light does not change upon reflection?
2. If you wish to take a picture of your image while standing 5 m in front of a plane mirror, for what distance should you set your camera to provide the sharpest focus?

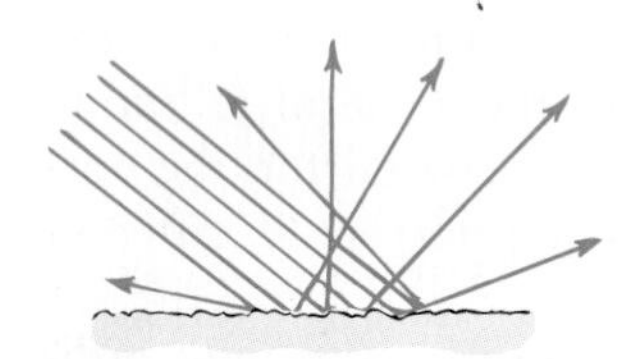

FIGURE 28.10
Diffuse reflection. Although each ray obeys the law of reflection, the many different surface angles that light rays encounter in striking a rough surface cause reflection in many directions.

Only part of the light that strikes a surface is reflected. On a surface of clear glass, for example, and for normal incidence (light perpendicular to the surface), only about 4% is reflected from each surface. On a clean and polished aluminum or silver surface, however, about 90% of incident light is reflected.

Diffuse Reflection

When light is incident on a rough surface, it is reflected in many directions. This is called **diffuse reflection** (Figure 28.10). If the surface is so smooth that the distances between successive elevations on the surface are less than about one-eighth the wavelength of the light, there is very little diffuse reflection, and the surface is said to be *polished.* A surface, therefore, may be polished for radiation of a long wavelength but not polished for light of a short wavelength. The wire-mesh "dish" shown in Figure 28.11 is very rough for light waves and so is hardly mirrorlike; but, for long-wavelength radio waves, it is "polished" and therefore an excellent reflector.

FIGURE 28.11
The open-mesh parabolic dish is a diffuse reflector for short-wavelength light but a polished reflector for long-wavelength radio waves.

CHECK YOUR ANSWERS

1. The color of an image is identical to the color of the object forming the image. When you look at yourself in a mirror, the color of your eyes doesn't change. The fact that the color is the same is evidence that the frequency of light doesn't change upon reflection.
2. Set your camera for 10 m; the situation is equivalent to standing 5 m in front of an open window and viewing your twin standing 5 m beyond the window.

FIGURE 28.12
A magnified view of the surface of ordinary paper.

Light reflecting from this page is diffuse. The page may be smooth to a radio wave, but to a light wave it is rough. Rays of light that strike this page encounter millions of tiny flat surfaces facing in all directions. The incident light therefore is reflected in all directions. This is a desirable circumstance. It enables us to see objects from any direction or position. You can see the road ahead of your car at night, for instance, because of diffuse reflection by the road surface. When the road is wet, there is less diffuse reflection, and it is harder to see. Most of our environment is seen by diffuse reflection.

An undesirable circumstance related to diffuse reflection is the ghost image that occurs on a TV set when the TV signal bounces off buildings and other obstructions. For antenna reception, this difference in path lengths for the direct signal and the reflected signal produces a slight time delay. The ghost image is normally displaced to the right, the direction of scanning in the TV tube, because the reflected signal arrives at the receiving antenna later than the direct signal. Multiple reflections may produce multiple ghosts.

Refraction

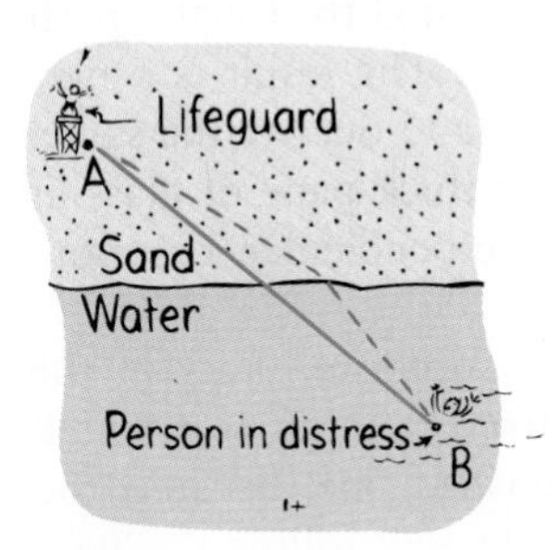

FIGURE 28.13
Refraction.

Recall, from Chapter 26, that the average speed of light is lower in glass and other transparent materials than through empty space. Light travels at different speeds in different materials.[3] It travels at 300,000 kilometers per second in a vacuum, at a slightly lower speed in air, and at about three-fourths that speed in water. In a diamond, light travels at about 40% of its speed in a vacuum. As mentioned at the opening of this chapter, when light bends in passing obliquely from one medium to another, we call the process **refraction.** It is a common observation that a ray of light bends and takes a longer path when it encounters glass or water at an oblique angle. But the longer path taken is nonetheless the path requiring the least time. A straight-line path would take a longer time. We can illustrate this with the following situation.

Imagine that you are a lifeguard at a beach and you spot a person in distress in the water. We show the relative positions of you, the shoreline, and the person in distress in Figure 28.13. You are at point A, and the person is at point B. You can run faster than you can swim. Should you travel in a straight line to get to B? A little thought will show that a straight-line path would not be the best choice because, if you instead spent a little bit more time traveling farther on land, you would save a lot more time in swimming a lesser distance in the water. The path of shortest time is shown by the dashed-line path, which clearly is not the path of the shortest distance. The amount of bending at the shoreline depends, of course, on how much faster you can run than swim. The situation is similar for a ray of light incident upon a body of water, as shown in Figure 28.14. The angle of incidence is larger than the

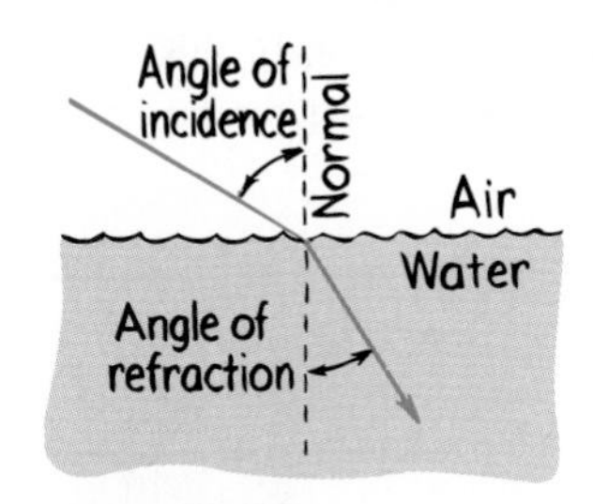

FIGURE 28.14
Interactive Figure
Refraction.

[3]Just how much the speed of light differs from its speed in a vacuum is given by the index of refraction, n, of the material:

$$n = \frac{\text{speed of light in vacuum}}{\text{speed of light in material}}$$

For example, the speed of light in a diamond is 124,000 km/s, and so the index of refraction for diamond is

$$n = \frac{300{,}000 \text{ km/s}}{124{,}000 \text{ km/s}} = 2.42$$

For a vacuum, $n = 1$.

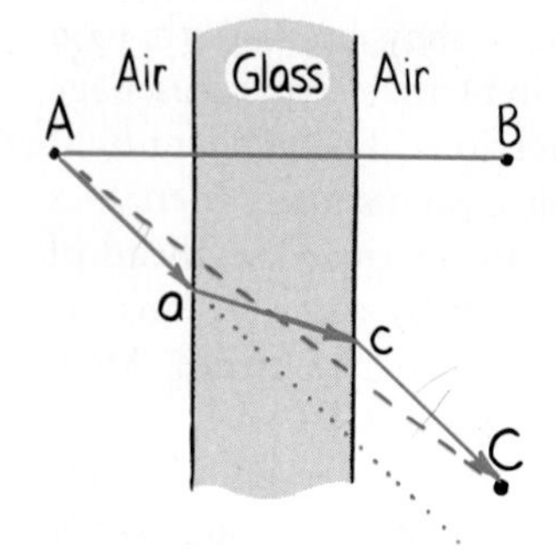

FIGURE 28.15
Refraction through glass. Although dashed line AC is the shortest path, light takes a slightly longer path through the air from A to a, then a shorter path through the glass to c, and then to C. The emerging light is displaced but parallel to the incident light.

angle of refraction by an amount that depends on the relative speeds of light in air and in water.

CHECK YOURSELF

Suppose the lifeguard in the preceding example were a seal instead of a human being. How would its path of least time from A to B differ?

Consider the pane of thick window glass in Figure 28.15. When light goes from point A through the glass to point B, it will go in a straight-line path. In this case, light encounters the glass perpendicularly, and we see that the shortest distance through both air and glass corresponds to the shortest time. But what about light that goes from point A to point C? Will it travel in the straight-line path shown by the dashed line? The answer is *no,* because if it did so it would be spending more time inside the glass, where light travels more slowly than in air. The light will instead take a less-inclined path through the glass. The time saved by taking the resulting shorter path through the glass more than compensates for the added time required to travel the slightly longer path through the air. The overall path is the path of least time. The result is a parallel displacement of the light beam, because the angles in and out are the same. You'll notice this displacement when you look through a thick pane of glass at an angle. The more your angle of viewing differs from perpendicular, the more pronounced the displacement.

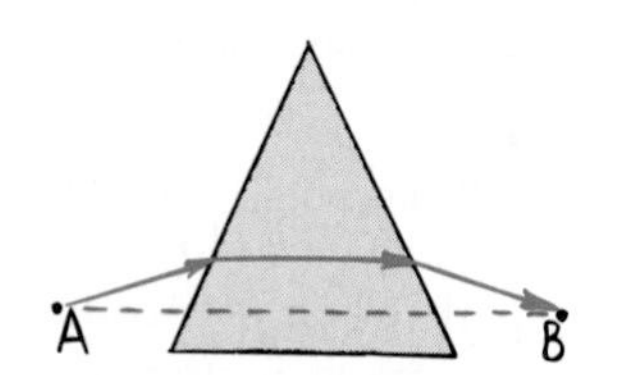

FIGURE 28.16
A prism.

Another example of interest is the prism, in which opposite faces of the glass are not parallel (Figure 28.16). Light that goes from point A to point B will not follow the straight-line path shown by the dashed line, because too much time would be spent in the glass. Instead, the light will follow the path shown by the solid line—a path that is a bit farther through the air—and pass through a thinner section of the glass to make its trip to point B. By this reasoning, one might think that the light should take a path closer to the upper vertex of the prism and seek the minimum thickness of glass. But if it did, the extra distance through the air would result in an overall longer time of travel. The path followed is the path of least time.

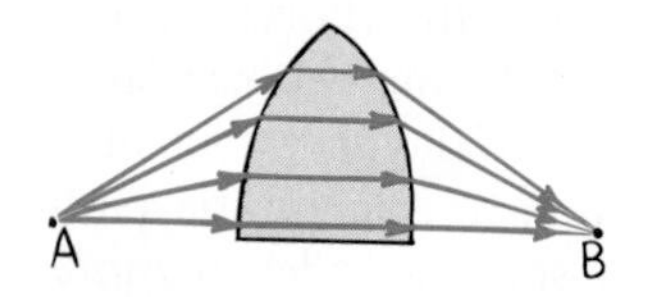

FIGURE 28.17
A curved prism.

It is interesting to note that a properly curved prism will provide many paths of equal time from a point A on one side to a point B on the opposite side (Figure 28.17). The curve decreases the thickness of the glass correctly to compensate for the extra distances light travels to points higher on the surface. For appropriate positions of A and B and for the appropriate curve on the surfaces of this modified prism, all light paths are of exactly equal time. In this case, all the light from A that is incident on the glass surface is focused on point B.

CHECK YOUR ANSWER

The seal can swim faster than it can run, and its path would bend as shown; likewise with light emerging from the bottom of a piece of glass into air.

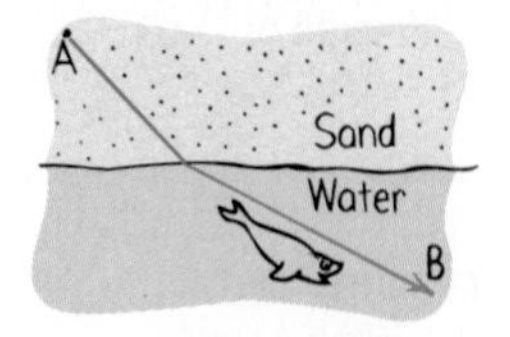

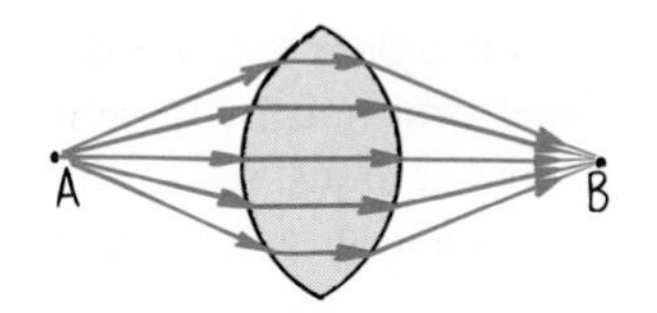

FIGURE 28.18
A converging lens.

We see that this shape is simply the upper half of a converging lens (Figure 28.18, and treated in more detail later in this chapter).

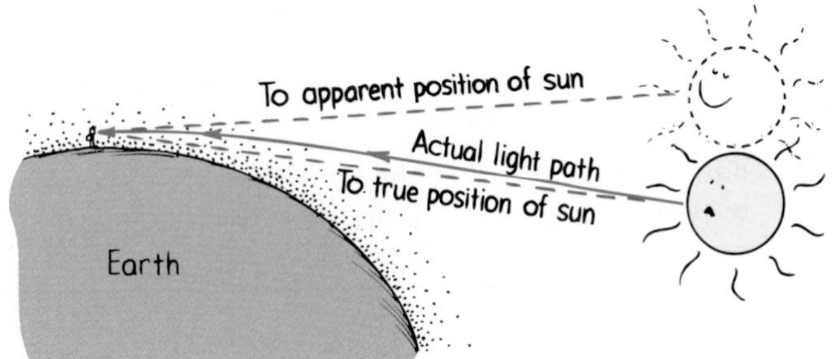

FIGURE 28.19
Because of atmospheric refraction, when the Sun is near the horizon, it appears to be higher in the sky.

FIGURE 28.20
The Sun's shape is distorted by differential refraction.

Whenever we watch a sunset, we see the Sun for several minutes after it has sunk below the horizon. Earth's atmosphere is thin at the top and dense at the bottom. Since light travels faster in thin air than it does in dense air, light from the Sun can reach us more quickly if, instead of traveling in a straight line, it avoids the denser air by taking a higher and longer path to penetrate the atmosphere at a steeper tilt (Figure 28.19). Since the density of the atmosphere changes gradually, the light path bends gradually to produce a curved path. Interestingly enough, this path of least time provides us with a slightly longer period of daylight each day. Furthermore, when the Sun (or Moon) is near the horizon, the rays from the lower edge are bent more than the rays from the upper edge. This produces a shortening of the vertical diameter, causing the Sun to appear elliptical (Figure 28.20).

Mirage

We are all familiar with the mirage we see while driving on a hot road. The sky appears to be reflected from water on the distant road, but, when we get there, the road is dry. Why is this so? The air is very hot just above the road surface and cooler above. Light travels faster through the thinner hot air than through the denser cool air above. So light, instead of coming to us from the sky in straight lines, also has least-time paths by which it curves down into the hotter region near the road for a while before reaching our eyes (Figure 28.21). A mirage is not, as many people mistakenly believe, a "trick of the mind." A mirage is formed by real light and can be photographed, as shown in Figure 28.22.

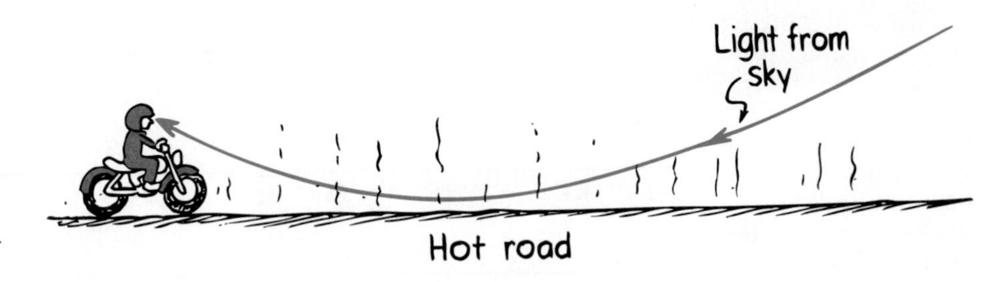

FIGURE 28.21
Light from the sky picks up speed in the air near the ground because that air is warmer and less dense than the air above. When the light grazes the surface and bends upward, the observer sees a mirage.

FIGURE 28.22
A mirage. The apparent wetness of the road is not reflection of the sky by water but, rather, refraction of sky light through the warmer and less-dense air near the road surface.

When we look at an object over a hot stove or over hot pavement, we see a wavy, shimmering effect. This is due to the various least-time paths of light as it passes through varying temperatures and therefore varying densities of air. The twinkling of stars results from similar phenomena in the sky, where light passes through unstable layers in the atmosphere.

CHECK YOURSELF

If the speed of light were the same in air of various temperatures and densities, would there still be slightly longer daytimes, twinkling stars at night, mirages, and slightly squashed Suns at sunset?

In the foregoing examples, how does light seemingly "know" what conditions exist and what compensations a least-time path requires? When approaching window glass at an angle, how does light know to travel a bit farther in air to save time in taking a less-inclined angle and therefore a shorter path through the pane of glass? In approaching a prism or a lens, how does light know to travel a greater distance in air to reach a thinner portion of the glass? How does light from the Sun know to travel above the atmosphere an extra distance before taking a shortcut through the denser air to save time? How does sky light above know that it can reach us in minimum time if it dips toward a hot road before tilting upward to our eyes? The principle of least time appears to be noncausal. It seems as if light has a mind of its own, that it can "sense" all the possible paths, calculate the times for each, and choose the one that requires the least time. Is this the case? As intriguing as all this may seem, there is a simpler explanation that doesn't assign foresight to light—that refraction is a consequence of light having different average speeds in different media.

Cause of Refraction

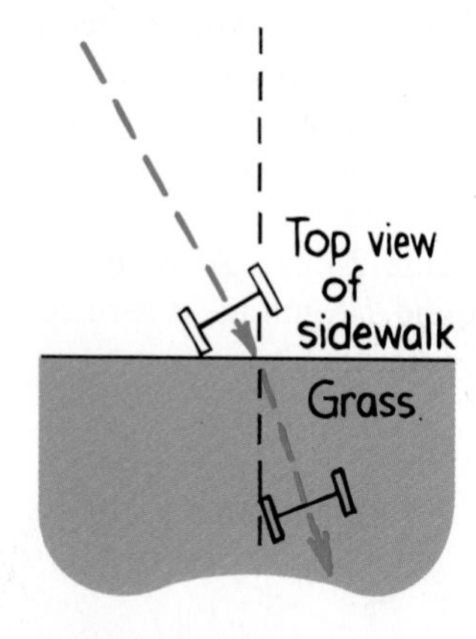

FIGURE 28.23
The direction of the rolling wheels changes when one wheel slows down before the other one does.

Refraction occurs when the average speed of light *changes* in going from one transparent medium to another. We can understand this by considering the action of a pair of toy cart wheels connected to an axle as the wheels roll gently downhill from a smooth sidewalk onto a grass lawn. If the wheels meet the grass at some angle (Figure 28.23), they will be deflected from their straight-line course. The direction of the rolling wheels is shown by the dashed line. Note that, on meeting the lawn, where the wheels roll more slowly owing to interaction with the grass, the left wheel slows down first. This is because it meets the grass while the right wheel is still on the smooth sidewalk. The faster-moving right wheel tends to pivot about the slower-moving left wheel because, during the same time interval, the right wheel travels farther than the left one. This action

CHECK YOUR ANSWER

No.

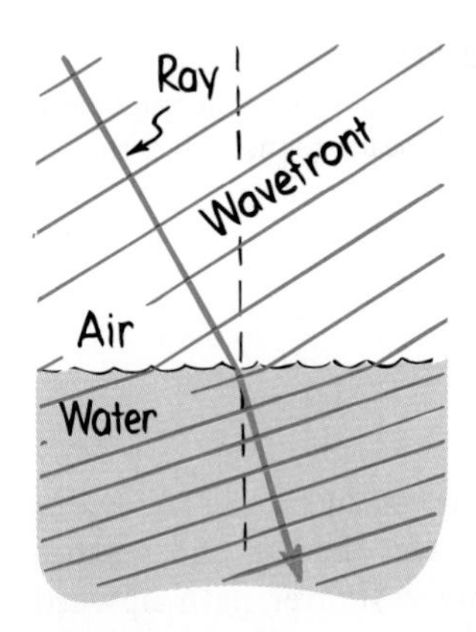

FIGURE 28.24
The direction of the light waves changes when one part of each wave slows down before the other part.

bends the direction of the rolling wheels toward the "normal," the lightly dashed line perpendicular to the grass–sidewalk border in Figure 28.23.

A light wave bends in a similar way, as shown in Figure 28.24. Note the direction of light, shown by the solid arrow (the light ray) and also note the *wave fronts* at right angles to the ray. (If the light source were close, the wave fronts would appear as segments of circles; but, if we assume the distant Sun is the source, the wave fronts form practically straight lines.) The wave fronts are everywhere perpendicular to the light rays. In the figure, the wave meets the water surface at an angle, so the left portion of the wave slows down in the water while the part still in the air travels at speed c. The ray or beam of light remains perpendicular to the wave front and bends at the surface, just as the wheels bend to change direction when they roll from the sidewalk onto the grass. In both cases, the bending is a consequence of a change in speed.[4]

The changeable speed of light provides a wave explanation for mirages. Sample wave fronts coming from the top of a tree on a hot day are shown in Figure 28.25. If the temperature of the air were uniform, the average speed of light would be the same in all parts of the air; light traveling toward the ground would meet the ground. But the air is warmer and less dense near the ground, and the wave fronts pick up speed as they travel downward, making them bend upward. So, when the observer looks downward, he sees the top of the tree—this is a mirage.

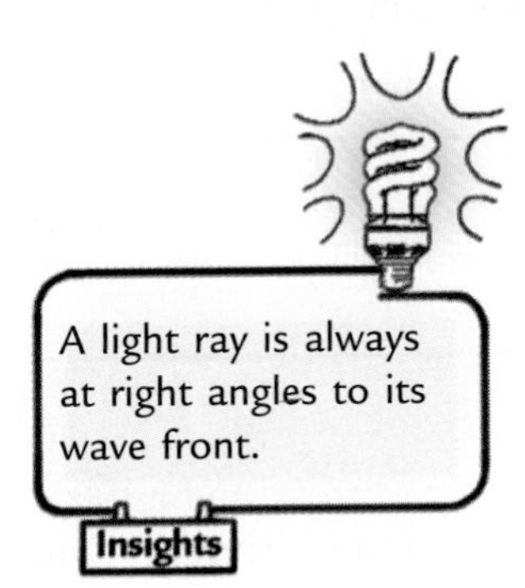

A light ray is always at right angles to its wave front.
Insights

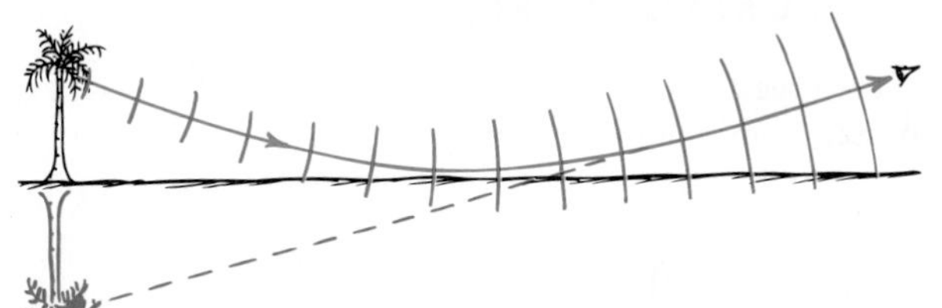

FIGURE 28.25
A wave explanation of a mirage. Wave fronts of light travel faster in the hot air near the ground and bend upward.

The refraction of light is responsible for many illusions; one of them is the apparent bending of a stick partly immersed in water. The submerged part seems closer to the surface than it really is. Likewise when you view a fish in the water. The fish appears nearer to the surface and closer (Figure 28.27). Because of refraction, submerged objects appear to be magnified. If we look straight down into water, an object submerged 4 meters beneath the surface will appear to be only 3 meters deep.

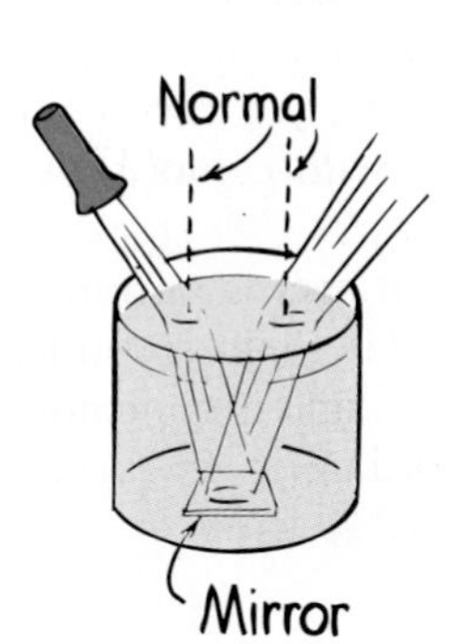

FIGURE 28.26
When light slows down in going from one medium to another, such as going from air to water, it refracts toward the normal. When it speeds up in traveling from one medium to another, such as going from water to air, it refracts away from the normal.

[4]The quantitative law of refraction, called *Snell's law,* is credited to W. Snell, a seventeenth-century Dutch astronomer and mathematician: $n_1 \sin \theta_1 = n_2 \sin \theta_2$, where n_1 and n_2 are the indices of refraction of the media on either side of the surface and θ_1 and θ_2 are the respective angles of incidence and refraction. If three of these values are known, the fourth can be calculated from this relationship.

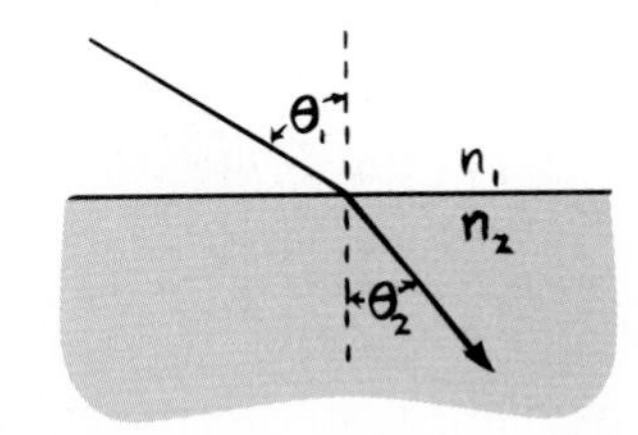

Model of Refraction

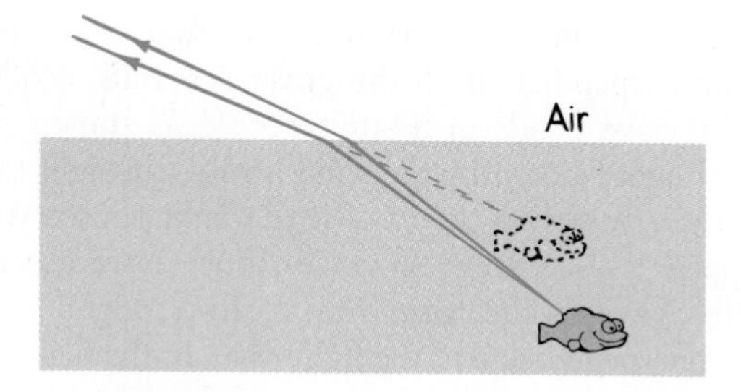

FIGURE 28.27
Because of refraction, a submerged object appears to be nearer to the surface than it actually is.

FIGURE 28.28
Because of refraction, the full root-beer mug appears to hold more root beer than it actually does.

We see that we can interpret the bending of light at the surface of the water in at least two ways. We can say that the light that leaves the fish and reaches the observer's eye does so in the least time by taking a shorter path upward toward the surface of the water and a correspondingly longer path through the air. In this view, least time dictates the path taken. Or we can say that the waves of light that happen to be directed upward at an angle toward the surface are bent off-kilter as they speed up when emerging into the air and that these waves reach the observer's eye. In this view, the change in speed from water to air dictates the path taken, and this path turns out to be a least-time path. Whichever view we choose, the results are the same.

CHECK YOURSELF

If the speed of light were the same in all media, would refraction still occur when light passes from one medium to another?

Dispersion

We know that the average speed of light is less than *c* in a transparent medium; how much less depends on the nature of the medium and on the frequency of light. The speed of light in a transparent medium depends on its frequency. Recall, from Chapter 26, that light whose frequencies match the natural or resonant frequencies of the electron oscillators in the atoms and molecules of the transparent medium is absorbed, and light with frequencies near the resonant frequencies interacts more often in the absorption/reemission sequence and therefore travels more slowly. Since the natural or resonant frequency of most transparent materials is in the ultraviolet part of the spectrum, higher-frequency light travels more slowly than lower-frequency light. Violet light travels about 1% slower in ordinary glass than red light. Light waves with colors between red and violet travel at their own intermediate speeds.

Because different frequencies of light travel at different speeds in transparent materials, they refract by different amounts. When white light is refracted twice, as in a prism, the separation of the different colors of light is quite noticeable. This separation of light into colors arranged according to frequency is called *dispersion* (Figure 28.29). It is what enabled Isaac Newton to form a spectrum when he held a glass prism in sunlight.

CHECK YOUR ANSWER

No.

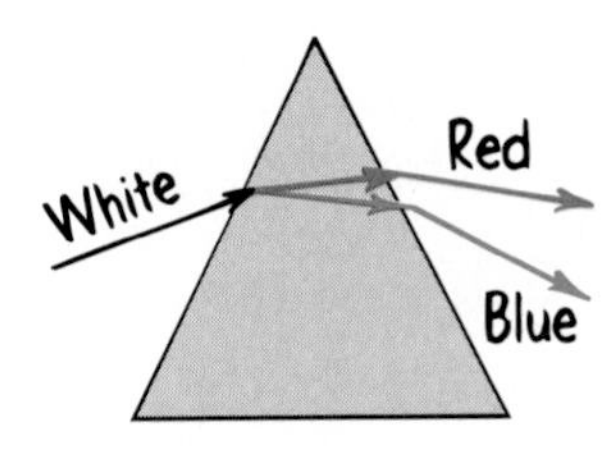

FIGURE 28.29
Dispersion by a prism makes the components of white light visible.

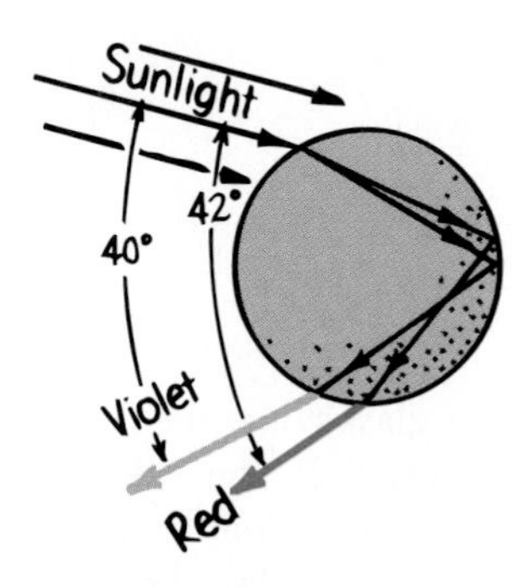

FIGURE 28.30
Dispersion of sunlight by a single raindrop.

Rainbows

A most spectacular illustration of dispersion is a rainbow. For a rainbow to be seen, the Sun must be shining in one part of the sky and water drops in a cloud or in falling rain must be present in the opposite part of the sky. When we turn our backs toward the Sun, we see the spectrum of colors in a bow. Seen from an airplane near midday, the bow forms a complete circle. All rainbows would be completely round if the ground were not in the way.

The beautiful colors of rainbows are dispersed from the sunlight by millions of tiny spherical water droplets that act like prisms. We can better understand this by considering an individual raindrop, as shown in Figure 28.30. Follow the ray of sunlight as it enters the drop near its top surface. Some of the light here is reflected (not shown), and the remainder is refracted into the water. At this first refraction, the light is dispersed into its spectrum colors, violet being deviated the most and red the least. Reaching the opposite side of the drop, each color is partly refracted out into the air (not shown) and partly reflected back into the water. Arriving at the lower surface of the drop, each color is again reflected (not shown) and refracted into the air. This second refraction is similar to that of a prism, where refraction at the second surface increases the dispersion already produced at the first surface.

Two refractions and a reflection can actually result in the angle between the incoming and outgoing rays being anything between zero and about 42° (0° corresponding to a full 180° reversal of the light). There is a strong concentration of light intensity, however, near the maximum angle of 42°. That is what is shown in Figure 28.30.

Although each drop disperses a full spectrum of colors, an observer is in a position to see the concentrated light of only a single color from any one drop (Figure 28.31). If violet light from a single drop reaches the eye of an observer, red light from the same drop is incident elsewhere toward the feet. To see red light, one must look to a drop higher in the sky. The color red will be seen when the angle between a beam of sunlight and the light sent back by a drop is 42°. The color violet is seen when the angle between the sunbeams and deflected light is 40°.

Why does the light dispersed by the raindrops form a bow? The answer to this involves a little geometric reasoning. First of all, a rainbow is not the flat two-dimensional arc it appears to be. It appears flat for the same reason a spherical burst of fireworks high in the sky appears as a disc—because of a lack of distance cues. The rainbow you see is actually a three-dimensional cone

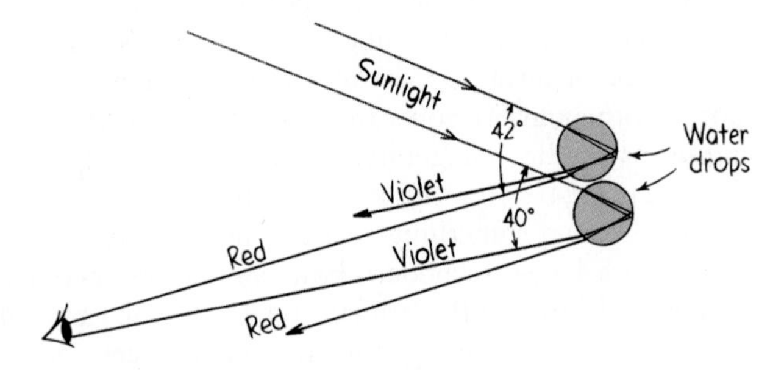

FIGURE 28.31
Sunlight incident on two sample raindrops, as shown, emerges from them as dispersed light. The observer sees the red light from the upper drop and the violet light from the lower drop. Millions of drops produce the whole spectrum of visible light.

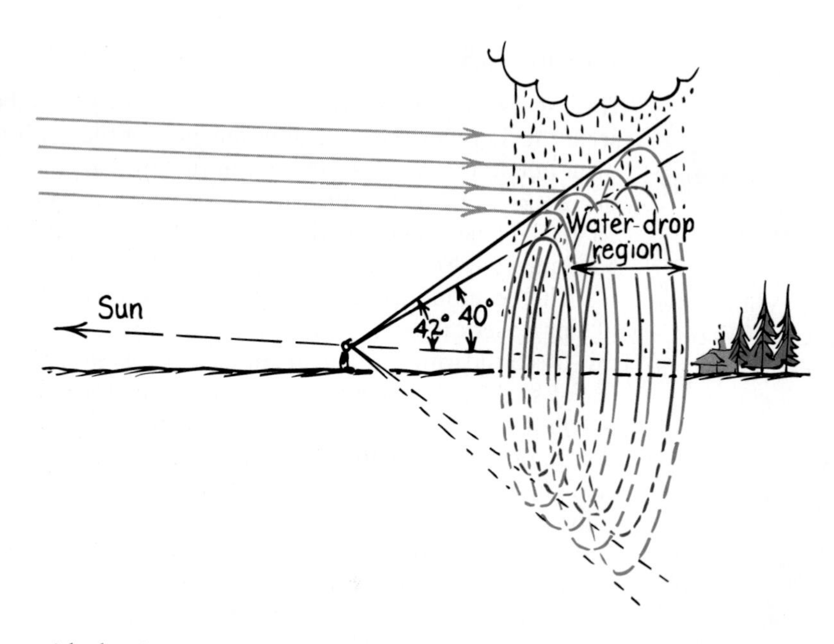

FIGURE 28.32
When your eye is located between the Sun (not shown off to the left) and a water-drop region, the rainbow you see is the edge of a three-dimensional cone that extends through the water-drop region. (Innumerable layers of drops form innumerable two-dimensional arcs like the four suggested here.)

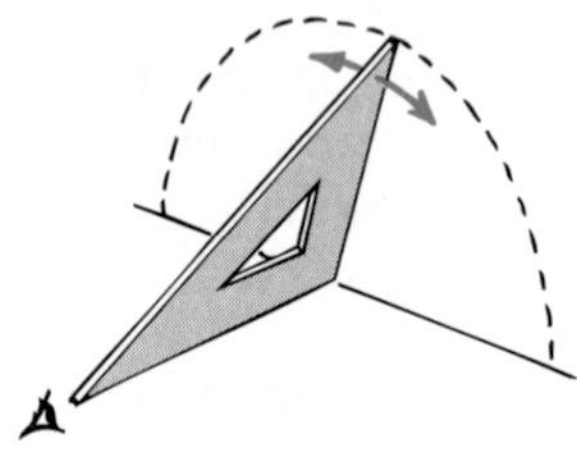

FIGURE 28.33
Only raindrops along the dashed line disperse red light to the observer at a 42° angle; hence, the light forms a bow.

with the tip (apex) at your eye (Figure 28.32). Consider a glass cone, the shape of those paper cones you sometimes see at drinking fountains. If you held the tip of such a glass cone against your eye, what would you see? You'd see the glass as a circle. Likewise with a rainbow. All the drops that disperse the rainbow's light toward *you* lie in the shape of a cone—a cone of different layers with drops that disperse red to your eye on the outside, orange beneath the red, yellow beneath the orange, and so on, all the way to violet on the inner conical surface. The thicker the region containing water drops, the thicker the conical edge you look through, and the more vivid the rainbow.

To see this further, consider only the deflection of red light. You see red when the angle between the incident rays of sunlight and dispersed rays is 42°. Of course, beams are dispersed 42° from drops all over the sky in all directions—up, down, and sideways. But the only red light *you* see is from drops that lie on a cone with a side-to-axis angle of 42°. Your eye is at the apex of this cone, as shown in Figure 28.33. To see violet, you look 40° from the conical axis (so the thickness of glass in the cone of the previous paragraph is tapered—very thin at the tip and thicker with increased distance from the tip).

Your cone of vision that intersects the cloud of drops that creates your rainbow is different from that of a person next to you. So when a friend says, "look at the pretty rainbow," you can reply, "Okay, move aside so I can see it too." Everybody sees his or her own personal rainbow.

Another fact about rainbows: A rainbow always faces you squarely, because of the lack of distance cues mentioned earlier. When you move, your rainbow moves with you. So you can never approach the side of a rainbow, or see it end-on as in the exaggerated view of Figure 28.32. You *can't* get to its end. Hence the expression "looking for the pot of gold at the end of the rainbow" means pursuing something you can never reach.

Often a larger, secondary bow with colors reversed can be seen arching at a greater angle around the primary bow. We won't treat this secondary bow except to say that it is formed by similar circumstances and is a result of double reflection within the raindrops (Figure 28.34). Because of this extra reflection (and extra refraction loss), the secondary bow is much dimmer, and it colors are reversed.

FIGURE 28.34
Two refractions and a reflection in water droplets produce light at all angles up to about 42°, with the intensity concentrated where we see the rainbow at 40° to 42°. No light emerges from the water droplet at angles greater than 42° unless it undergoes two or more reflections inside the drop. So the sky is brighter inside the rainbow than outside it. Notice the weak secondary rainbow to the right of the primary.

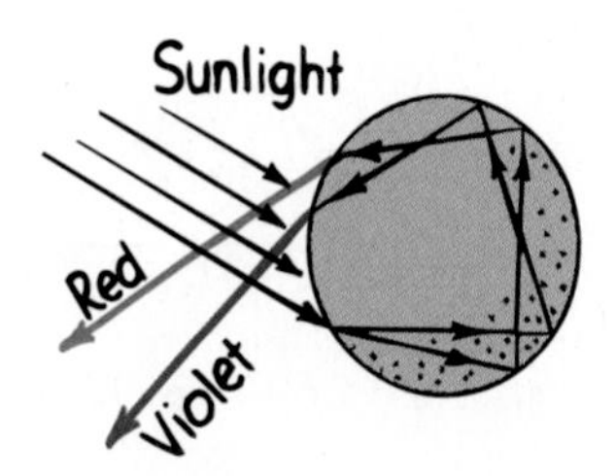

FIGURE 28.35
Double reflection in a drop produces a secondary bow.

CHECK YOURSELF

1. If you point to a wall with your arm extended to make about a 42° angle to the normal to the wall, then rotate your arm in a full circle while keeping the 42° angle to the wall normal, what shape does your arm describe? What shape on the wall does your finger sweep out?
2. If light traveled at the same speed in raindrops as it does in air, would we still have rainbows?

Total Internal Reflection

Some Saturday night when you're taking your bath, fill the tub extra deep and bring a waterproof flashlight into the tub with you. Switch off the bathroom light. Shine the submerged light straight up and then slowly tip it away from the surface. Note how the intensity of the emerging beam diminishes and how more light is reflected from the surface of the water to the bottom of the tub. At a certain angle, called the *critical angle,* you'll notice that the beam no longer emerges into the air above the surface. The intensity of the emerging beam reduces to zero where it tends to graze the surface. The **critical angle** is the minimum angle of incidence inside a medium at which a light ray is totally reflected. When the flashlight is tipped beyond the critical angle (48° from the normal for water), you'll notice that all the light is reflected back into the tub. This is **total internal reflection.** The light striking the air–water surface obeys the law of reflection: The angle of incidence is equal to the angle of reflection. The only light emerging from the surface of the water is that which is diffusely reflected from the bottom of the bathtub. This procedure is shown in Figure 28.37. The

CHECK YOUR ANSWERS

1. Your arm describes a cone, and your finger sweeps out a circle. Likewise with rainbows.
2. No.

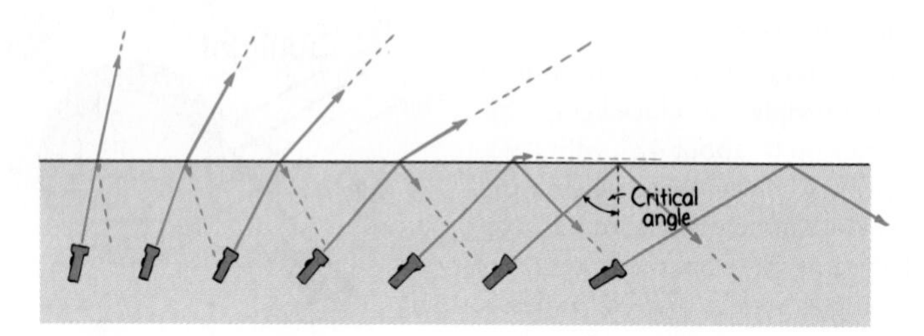

FIGURE 28.36 Interactive Figure

Light emitted in the water is partly refracted and partly reflected at the surface. The blue dashes show the direction of light and the length of the arrows indicates the proportions refracted and reflected. Beyond the critical angle, the beam is totally internally reflected.

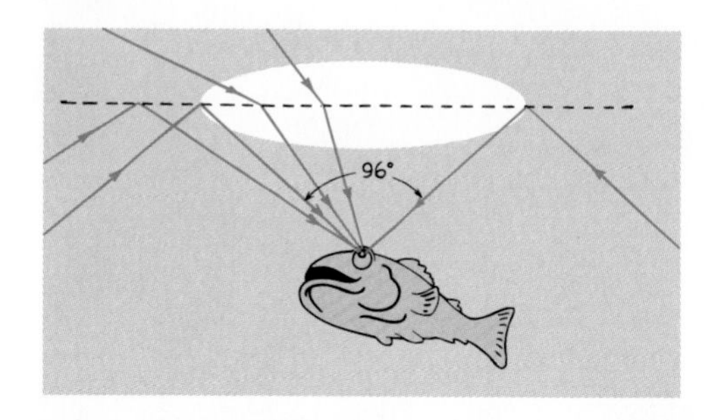

FIGURE 28.37

An observer underwater sees a circle of light at the still surface. Beyond a cone of 96° (twice the critical angle), an observer sees a reflection of the water interior or bottom.

Would you like to become rich? Be the first to invent a surface that will reflect 100% of the light incident upon it.

Insights

proportion of light refracted and light internally reflected is indicated by the relative lengths of the arrows.

Total internal reflection occurs in materials in which the speed of light is less than the speed of light outside. The speed of light is less in water than in air, so all light rays in water that reach the surface at more than an incident angle of 48° are reflected back into the water. So your pet goldfish in the bathtub looks up to see a reflected view of the sides and bottom of the bathtub. Directly above, it sees a compressed view of the outside world (Figure 28.37). The outside 180° view from horizon to opposite horizon is seen through an angle of 96°—twice the critical angle. A lens that similarly compresses a wide view, called a *fisheye lens,* is used for special-effect photography.

Total internal reflection occurs in glass surrounded by air, because the speed of light in glass is less than in air. The critical angle for glass is about 43°, depending on the type of glass. So light in the glass that is incident at angles greater than 43° to the surface is totally internally reflected. No light escapes beyond this angle; instead, all of it is reflected back into the glass—even if the outside surface is marred by dirt or dust. Hence the usefulness of glass prisms (Figure 28.38). A little light is lost by reflection before it enters the prism, but, once the light is inside, reflection from the 45° slanted face is total—100%. In contrast, silvered or aluminized mirrors reflect only about 90% of incident light. Hence the use of prisms instead of mirrors in many optical instruments.

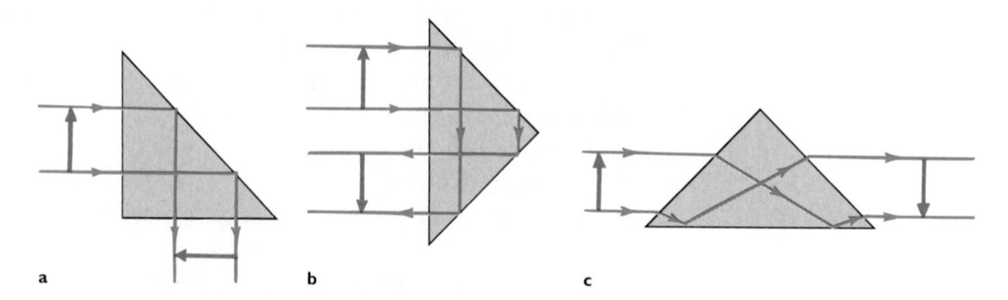

FIGURE 28.38

Total internal reflection in a prism. The prism changes the direction of the light beam (a) by 90°, (b) by 180°, and (c) not at all. Note that, in each case, the orientation of the image is different from the orientation of the object.

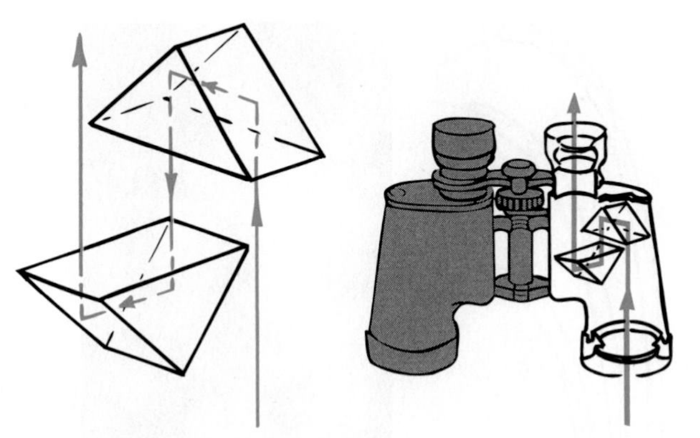

FIGURE 28.39
Total internal reflection in a pair of prisms, common in binoculars.

A pair of prisms, each reflecting light through 180°, is shown in Figure 28.39. Binoculars use pairs of prisms to lengthen the light path between lenses and thus eliminate the need for long barrels. So a compact set of binoculars is as effective as a longer telescope. Another advantage of prisms is that, whereas the image in a straight telescope is upside down, reflection by the prisms in binoculars reinverts the image, so things are seen right-side up.

> One of the many beauties of physics is the redness of a fully eclipsed Moon—resulting from the refraction of sunsets and sunrises that completely circle the Earth.
>
> **Insights**

The critical angle for a diamond is about 24.5°, smaller than for any other known substance. The critical angle varies slightly for different colors, because the speed of light varies slightly for different colors. Once light enters a diamond gemstone, most is incident on the sloped backsides at angles greater than 24.5° and is totally internally reflected (Figure 28.40). Because of the great slowdown in speed as light enters a diamond, refraction is pronounced, and, because of the frequency-dependence of the speed, there is great dispersion. Further dispersion occurs as the light exits through the many facets at its face. Hence we see unexpected flashes of a wide array of colors. Interestingly, when these flashes are narrow enough to be seen by only one eye at a time, the diamond "sparkles."

Total internal reflection also underlies the operation of optical fibers, or light pipes (Figure 28.41). An optical fiber "pipes" light from one place to another by a series of total internal reflections, much as a bullet ricochets down a steel pipe. Light rays bounce along the inner walls, following the twists and turns of the fiber. Optical fibers are used in decorative table lamps and to illuminate instrument displays on automobile dashboards from a single bulb. Dentists use them with flashlights to get light where they want it. Bundles of thin flexible glass or plastic fibers are used to see what is going on in inaccessible

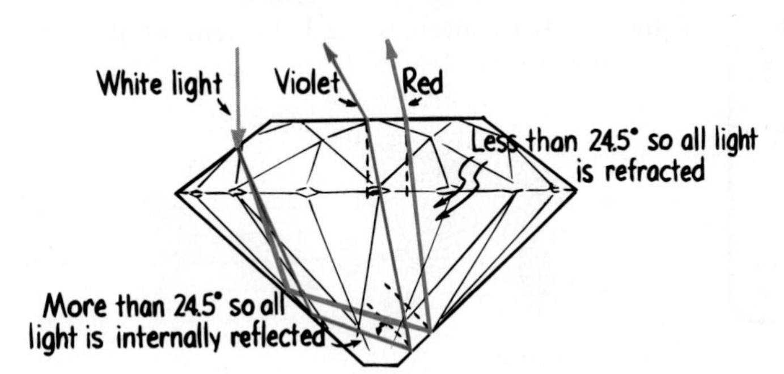
White light
Violet
Red
Less than 24.5° so all light is refracted
More than 24.5° so all light is internally reflected

FIGURE 28.40
Paths of light in a diamond. Rays that strike the inner surface at angles greater than the critical angle are internally reflected and exit via refraction at the top surface.

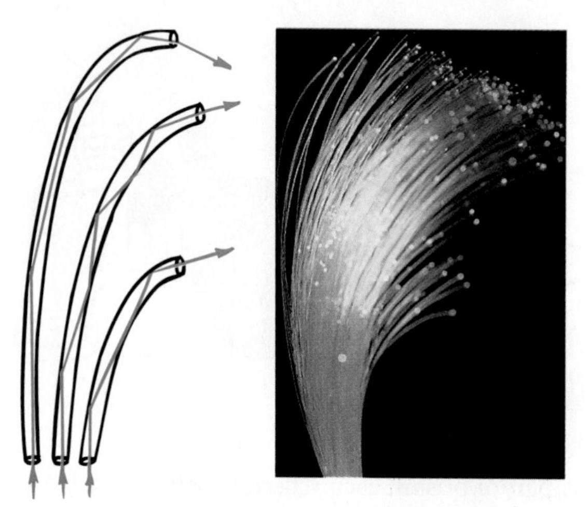

FIGURE 28.41
The light is "piped" from below by a succession of total internal reflections until it emerges at the top ends.

places, such as the interior of a motor or a patient's stomach. They can be made small enough to snake through blood vessels or through narrow canals in the body, such as the urethra. Light shines down some of the fibers to illuminate the scene and is reflected back along others.

Optical fibers are important in communications because they offer a practical alternative to copper wires and cables. In many places, thin glass fibers now replace thick, bulky, expensive copper cables to carry thousands of simultaneous telephone messages among the major switching centers. In many aircraft, control signals are fed from the pilot to the control surfaces by means of optical fibers. Signals are carried in the modulations of laser light. Unlike electricity, light is indifferent to temperature and fluctuations in surrounding magnetic fields, and so the signal is clearer. Also, it is much less likely to be tapped by eavesdroppers.

Lenses

Learning about lenses is a hands-on activity. Not fiddling with lenses while learning about them is like taking swimming lessons away from water.

Insights

A very practical case of refraction occurs in lenses. We can understand a lens by analyzing equal-time paths as we did earlier, or we can assume that it consists of a set of several matched prisms and blocks of glass arranged in the order shown in Figure 28.42. The prisms and blocks refract incoming parallel light rays so they converge to (or diverge from) a point. The arrangement shown in Figure 28.42*a* converges the light, and we call such a lens a **converging lens.** Note that it is thicker in the middle.

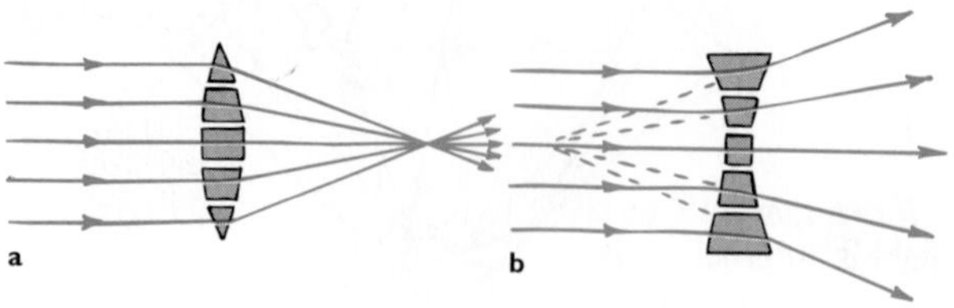

a b

FIGURE 28.42
A lens may be thought of as a set of blocks and prisms.
(a) A converging lens.
(b) A diverging lens.

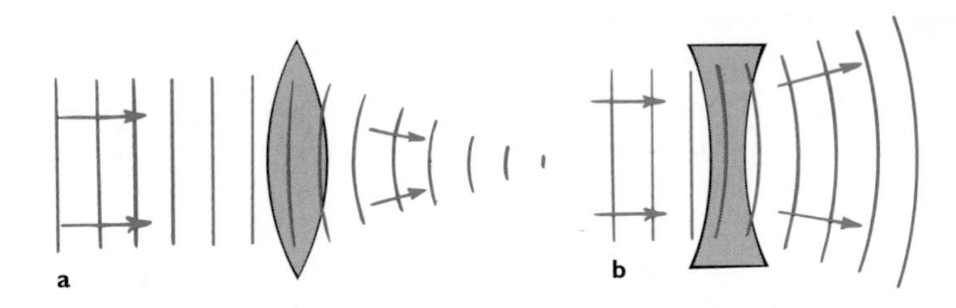

FIGURE 28.43
Wave fronts travel more slowly in glass than in air. (a) The waves are retarded more through the center of the lens, and convergence results. (b) The waves are retarded more at the edges, and divergence results.

The arrangement in *b* is different. The middle is thinner than the edges, and it diverges the light; such a lens is called a **diverging lens.** Note that the prisms diverge the incident rays in a way that makes them appear to come from a single point in front of the lens. In both lenses, the greatest deviation of rays occurs at the outermost prisms, for they have the greatest angle between the two refracting surfaces. No deviation occurs exactly in the middle, for in that region the glass faces are parallel to each other. Real lenses are not made of prisms, of course, as is indicated in Figure 28.42; they are made of a solid piece of glass with surfaces that are ground usually to a circular curve. In Figure 28.43, we see how smooth lenses refract waves.

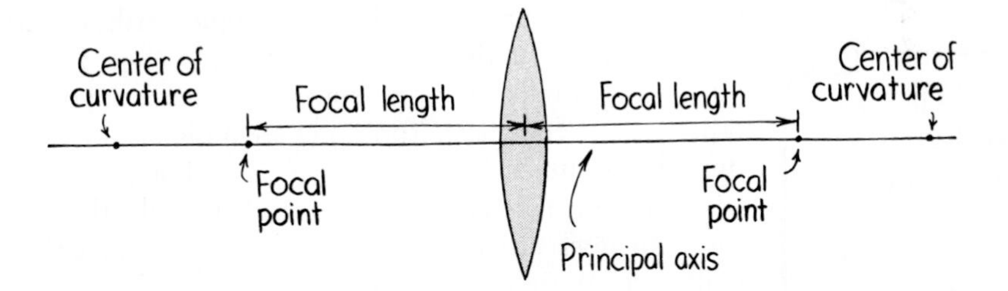

FIGURE 28.44
Key features of a converging lens.

Some key features in lens description are shown for a converging lens in Figure 28.44. The *principal axis* of a lens is the line joining the centers of curvatures of its surfaces. The *focal point* is the point to which a beam of parallel light, parallel to the principal axis, converges. Incident parallel beams that are not parallel to the principal axis focus at points above or below the focal point. All such possible points make up a *focal plane.* Because a lens has two surfaces, it has two focal points and two focal planes. When the lens of a camera is set for distant objects, the film is in the focal plane behind the lens in the camera. The *focal length* of the lens is the distance between the center of the lens and either focal point.

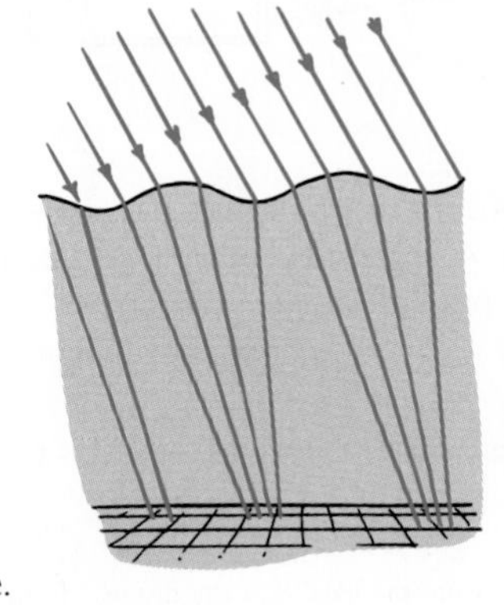

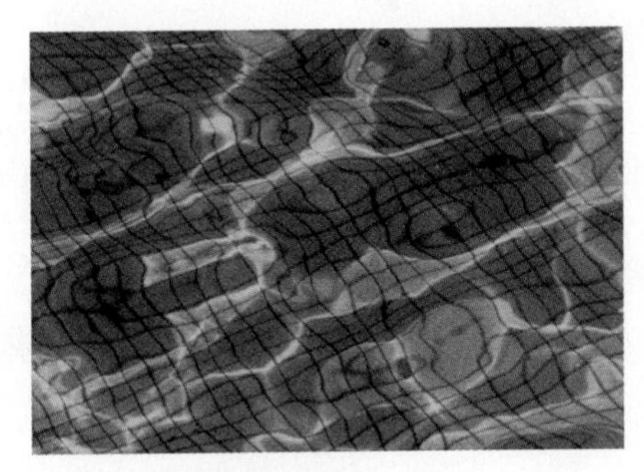

FIGURE 28.45
The moving patterns of bright and dark areas at the bottom of the pool result from the uneven surface of the water, which behaves like a blanket of undulating lenses. Just as we see the pool bottom shimmering, a fish looking upward at the Sun would see it shimmering too. Because of similar irregularities in the atmosphere, we see the stars twinkle.

PRACTICING PHYSICS

Make a pinhole camera. Cut out one end of a small cardboard box, and cover the end with semitransparent tracing or tissue paper. Make a clean-cut pinhole at the other end. (If the cardboard is thick, you can make the pinhole through a piece of tinfoil placed over a larger opening in the cardboard.) Aim the camera at a bright object in a darkened room, and you will see an upside-down image on the tracing paper. The tinier the pinhole, the dimmer and sharper the image. If, in a dark room, you replace the tracing paper with unexposed photographic film, cover the back so that it is light-tight and cover the pinhole with a removable flap. You're ready to take a picture. Exposure times differ depending principally on the kind of film and the amount of light. Try different exposure times, starting with about 3 seconds. Also try boxes of various lengths. The lens on a commercial camera is much bigger than the pinhole and therefore admits more light in less time—hence the name *snapshots*.

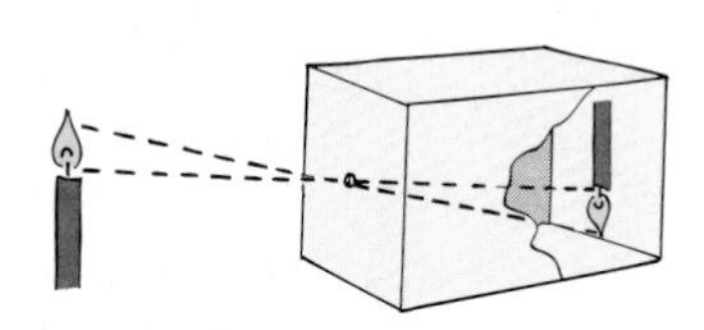

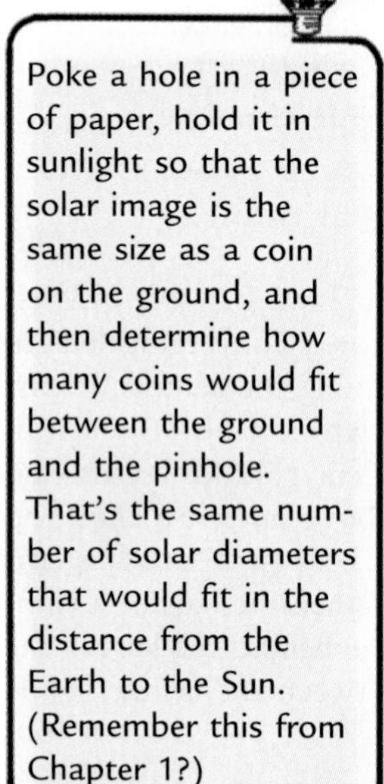

Poke a hole in a piece of paper, hold it in sunlight so that the solar image is the same size as a coin on the ground, and then determine how many coins would fit between the ground and the pinhole. That's the same number of solar diameters that would fit in the distance from the Earth to the Sun. (Remember this from Chapter 1?)
Insights

Image Formation by a Lens

At this moment, light is reflecting from your face onto this page. Light that reflects from your forehead, for example, strikes every part of the page. Likewise for the light that reflects from your chin. Every part of the page is illuminated with reflected light from your forehead, your nose, your chin, and every other part of your face. You don't see an image of your face on the page because there is too much overlapping of light. But put a barrier with a pinhole in it between your face and the page, and the light that reaches the page from your forehead does not overlap the light from your chin. Likewise for the rest of your face. Without this overlapping, an image of your face is formed on the page. It will be very dim, because very little light reflected from your face gets through the pinhole. To see the image, you'd have to shield the page from other light sources. The same is true of the vase and flowers in Figure 28.46*b*.[5]

The first cameras had no lenses and admitted light through a small pinhole. You can see why the image is upside down by the sample rays in Figure 28.46*b*

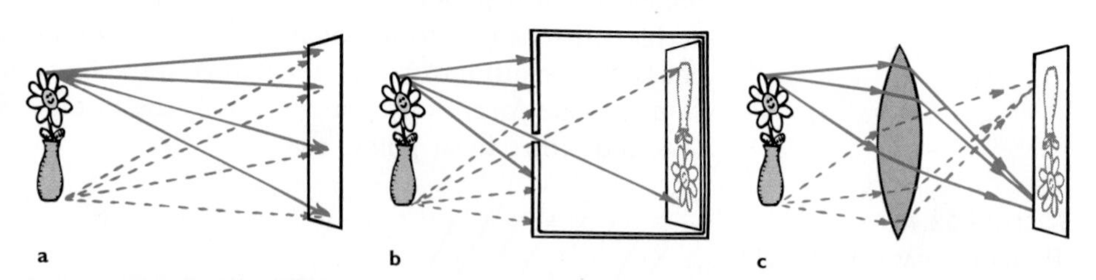

FIGURE 28.46 Interactive Figure

Image formation. (a) No image appears on the wall because rays from all parts of the object overlap all parts of the wall. (b) A single small opening in a barrier prevents overlapping rays from reaching the wall; a dim upside-down image is formed. (c) A lens converges the rays upon the wall without overlapping; more light makes a brighter image.

[5] A quantitative way of relating object distances with image distances is given by the thin-lens equation

$$\frac{1}{d_o} + \frac{1}{d_i} = \frac{1}{f} \quad \text{or} \quad d_i = \frac{d_o f}{d_o - f}$$

where d_o is the distance of the object from the lens, d_i is the distance from the lens to the image, and f is the focal length of the lens.

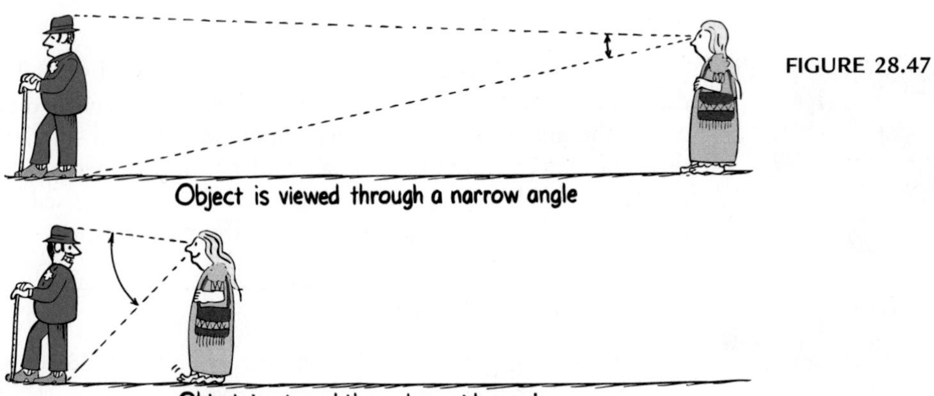

Object is viewed through a narrow angle
Object is viewed through a wide angle

FIGURE 28.47

and *c*. Long exposure times were required because of the small amount of light admitted by the pinhole. A somewhat larger hole would admit more light, but overlapping rays would produce a blurry image. Too large a hole would allow too much overlapping and no image would be discernible. That's where a converging lens comes in (Figure 28.46*c*). The lens converges light onto the screen without the unwanted overlapping of rays. Whereas the first pinhole cameras were useful only for still objects because of the long exposure time required, moving objects can be photographed with the lens camera because of the short exposure time—which, as previously mentioned, is why photographs taken with lens cameras came to be called *snapshots*.

Can you see why the image in Figure 28.46b is upside down? And is it true that, when your photographs are developed and printed, they're all upside down?

Insights

The simplest use of a converging lens is a magnifying glass. To understand how it works, think about how you examine objects near and far. With unaided vision, a far-away object is seen through a relatively narrow angle of view and a close object is seen through a wider angle of view (Figure 28.47). To see the details of a small object, you want to get as close to it as possible for the widest-angle view. But your eye can't focus when too close. That's where the magnifying glass comes in. When you are close to the object, the magnifying glass gives you a clear image that would be blurry otherwise.

When we use a magnifying glass, we hold it close to the object we wish to examine. This is because a converging lens provides an enlarged, right-side-up image only when the object is inside the focal point (Figure 28.48). If a screen is placed at the image distance, no image appears on it because no light is directed to the image position. The rays that reach our eye, however, behave *as if* they came from the image position; we call the result a **virtual image.**

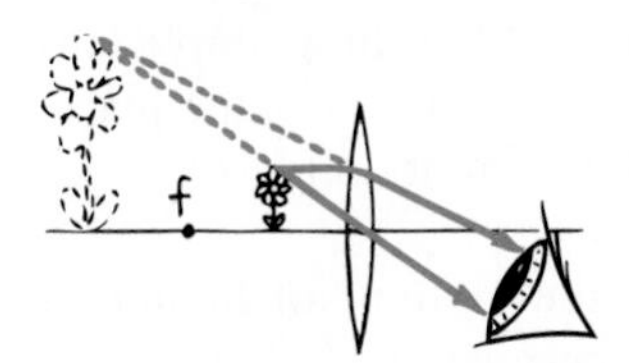

FIGURE 28.48
When an object is near a converging lens (inside its focal point *f*), the lens acts as a magnifying glass to produce a virtual image. The image appears larger and farther from the lens than the object.

When the object is far away enough to be outside the focal point of a converging lens, a **real image** is formed instead of a virtual image. Figure 28.49 shows a case in which a converging lens forms a real image on a screen. A real image is upside down. A similar arrangement is used for projecting slides and motion pictures on a screen and for projecting a real image on the film of a camera. Real images with a single lens are always upside down.

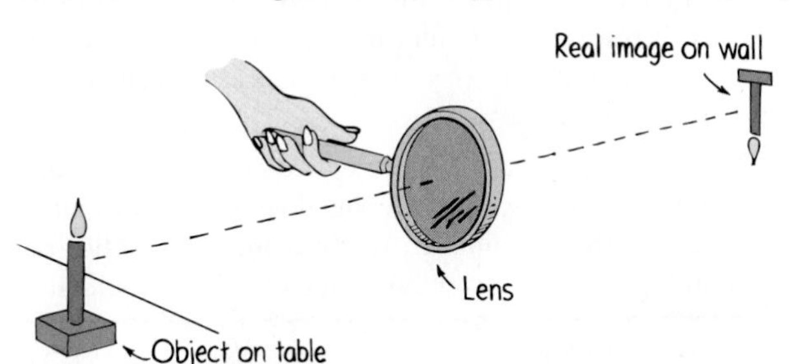

FIGURE 28.49
When an object is far from a converging lens (beyond its focal point), a real upside-down image is formed.

FIGURE 28.50
A diverging lens forms a virtual, right-side-up image of Jamie and his cat.

A diverging lens used alone produces a reduced virtual image. It makes no difference how far or how near the object is. When a diverging lens is used alone, the image is always virtual, right-side up, and smaller than the object. A diverging lens is often used as a "finder" on a camera. When you look at the object to be photographed through such a lens, you see a virtual image that approximates the same proportions as the photograph.

CHECK YOURSELF

1. Why is the greater part of the photograph in Figure 28.50 out of focus?
2. When the lens of a slide projector is half covered, what happens to the image on the screen?

Lens Defects

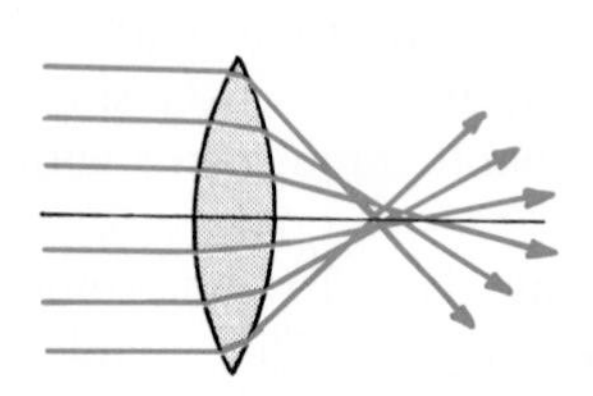
FIGURE 28.51
Spherical aberration.

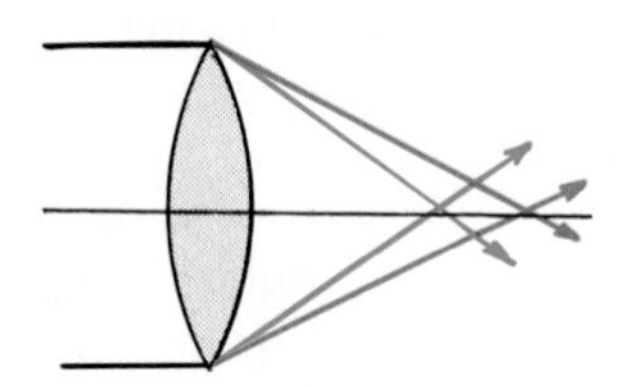
FIGURE 28.52
Chromatic aberration.

No lens provides a perfect image. A distortion in an image is called an **aberration.** By combining lenses in certain ways, aberrations can be minimized. For this reason, most optical instruments use compound lenses, each consisting of several simple lenses, instead of single lenses.

Spherical aberration results from light that passes through the edges of a lens focusing at a slightly different place from where light passing near the center of the lens focuses (Figure 28.51). This can be remedied by covering the edges of a lens, as with a diaphragm in a camera. Spherical aberration is corrected in good optical instruments by a combination of lenses.

Chromatic aberration is the result of light of different colors having different speeds and hence different refractions in the lens (Figure 28.52). In a simple lens (as in a prism), different colors of light do not come to focus in the same place. *Achromatic lenses,* which combine simple lenses of different kinds of glass, correct this defect.

The pupil of the eye changes in size to regulate the amount of light that enters. Vision is sharpest when the pupil is smallest because light then passes through only the central part of the eye's lens, where spherical and chromatic aberrations are minimal. Also, the eye then acts more like a pinhole camera, so

CHECK YOUR ANSWERS

1. Both Jamie and his cat and the virtual image of Jamie and his cat are "objects" for the lens of the camera that took this photograph. Because the objects are at different distances from the lens, their respective images are at different distances with respect to the film in the camera. So only one can be brought into focus. The same is true of your eyes. You cannot focus on near and far objects at the same time.
2. The image for a half-covered lens is dimmer because it is formed with half as much light. Importantly, this does not mean half the *image* is formed. The full image, though dimmer, is still there. (More rays, rather than the few we've chosen in Figure 28.46 to show image location, would show this.)

minimum focusing is required for a sharp image. You see better in bright light because in such light your pupils are smaller.

Astigmatism of the eye is a defect that results when the cornea is curved more in one direction than the other, somewhat like the side of a barrel. Because of this defect, the eye does not form sharp images. The remedy is eyeglasses with cylindrical lenses that have more curvature in one direction than in another.

If you wear glasses and have ever misplaced them, or if you find it difficult to read small print, try squinting, or, even better, try holding a pinhole (in a piece of paper or whatever) in front of your eye, close to the page. You'll see the print clearly, and, because you're close to it, it is magnified. Try it and see!

Insights

CHECK YOURSELF

1. If light traveled at the same speed in glass and in air, would glass lenses alter the direction of light rays?
2. Why is there chromatic aberration in light that passes through a lens but none in light that reflects from a mirror?

An option for those with poor sight today is wearing eyeglasses. The advent of eyeglasses probably occurred in China and in Italy in the late 1200s. (Curiously enough, the telescope wasn't invented until some 300 years later. If, in the meantime, anybody viewed objects through a pair of lenses separated along their axes, such as lenses fixed at the ends of a tube, there is no record of it.) In recent times, a present-day alternative to wearing eyeglasses is contact lenses. Now there is an alternative both to eyeglasses and to contact lenses for people with poor eyesight. Laser technology now allows eye surgeons to shave the cornea of the eye to a proper shape for normal vision. In tomorrow's world, the wearing of eyeglasses and contact lenses—at least for younger people—may be a thing of the past. We really do live in a rapidly changing world. And that can be nice.

CHECK YOUR ANSWERS

1. No.
2. Different frequencies travel at different speeds in a transparent medium and therefore refract at different angles, which produces chromatic aberration. The angles at which light reflects, however, has nothing to do with its frequency. One color reflects the same as any other color. In telescopes, therefore, mirrors are preferable to lenses because there is no chromatic aberration.

Summary of Terms

Reflection The return of light rays from a surface.

Refraction The bending of an oblique ray of light when it passes from one transparent medium to another.

Fermat's principle of least time Light takes the path that requires the least time when it goes from one place to another.

Law of reflection The angle of reflection equals the angle of incidence.

Diffuse reflection Reflection in irregular directions from an irregular surface.

Critical angle The minimum angle of incidence inside a medium at which a light ray is totally reflected.

Total internal reflection The total reflection of light traveling within a denser medium when it strikes the boundary with a less dense medium at an angle greater than the critical angle.

Converging lens A lens that is thicker in the middle than at the edges and that refracts parallel rays to a focus.

Diverging lens A lens that is thinner in the middle than at the edges, causing parallel rays to diverge as if from a point.

Virtual image An image formed by light rays that do not converge at the location of the image.

Real image An image formed by light rays that converge at the location of the image. A real image, unlike a virtual image, can be displayed on a screen.

Aberration Distortion in an image produced by a lens, which to some degree is present in all optical systems.

Suggested Reading

Bohren, Craig F. *Clouds in a Glass of Beer*. New York: Wiley, 1987.

Bohren, Craig F. *What Light Through Yonder Window Breaks?* New York: Wiley, 1991.

Greenler, R. *Rainbows, Halos, and Glories*. New York: Cambridge University Press, 1980.

Greenler, R. *Chasing the Rainbow: Recurrences in the Life of a Scientist*. New York: Cambridge University Press, 2000.

Review Questions

1. Distinguish between *reflection* and *refraction*.

Reflection

2. How does incident light that falls on an object affect the motion of electrons in the atoms of the object?
3. What do the electrons affected by illumination do when they are made to vibrate with greater energy?

Principle of Least Time

4. What is Fermat's principle of least time?

Law of Reflection

5. Cite the law of reflection.

Plane Mirrors

6. Relative to the distance of an object in front of a plane mirror, how far behind the mirror is the image?
7. What fraction of the light shining straight at a piece of clear glass is reflected from the first surface?

Diffuse Reflection

8. Can a surface be polished for some waves and not for others?

Refraction

9. Light bends when it passes obliquely from one medium to another, taking a somewhat longer path from one point to another. What does this longer path have to do with the time of travel of light?
10. How does the angle at which a ray of light strikes a pane of window glass compare with the angle at which it passes out the other side?
11. How does the angle at which a ray of light strikes a prism compare with the angle at which it passes out the other side?
12. In which medium does light travel faster, in thin air or in dense air? What does this difference in speed have to do with the length of a day?

Mirage

13. Is a mirage the result of reflection or refraction?

Cause of Refraction

14. When the wheel of a cart rolls from a smooth sidewalk onto a plot of grass, the interaction of the wheel with blades of grass slows the wheel. What slows light when it passes from air into glass or water?
15. What is the relationship between refraction and changes in the speed of light in a material?
16. Does the refraction of light make a swimming pool seem deeper or shallower?

Dispersion

17. What happens to light of a certain frequency when it is incident upon a material whose natural frequency matches the frequency of the light?
18. Which travels more slowly in glass, red light or violet light?

Rainbows

19. Does a single raindrop illuminated by sunlight deflect light of a single color or does it disperse a spectrum of colors?
20. Does a viewer see a single color or a spectrum of colors coming from a single faraway drop?
21. Why is a secondary rainbow dimmer than a primary bow?

Total Internal Reflection

22. What is meant by *critical angle*?
23. At what angle inside glass is light totally internally reflected? At what angle inside a diamond is light totally internally reflected?
24. Light normally travels in straight lines, but it "bends" in an optical fiber. Explain.

Lenses

25. Distinguish between a *converging lens* and a *diverging lens*.
26. What is the *focal length* of a lens?

#1, 3, 7, 9, 11, 15, 17, 19, 21, 25, 29, 43, 47, 49, 51, 55, 57

Image Formation by a Lens

27. Distinguish between a *virtual image* and a *real image*.

28. What kind of lens can be used to produce a real image? A virtual image?

Lens Defects

29. Why is vision sharpest when the pupils of the eye are very small?

30. What is astigmatism, and how can it be corrected?

Projects

1. Write a letter to Grandma and convince her that, in order to see her full-length image in a mirror, the mirror need be only half her height. Discuss also the intriguing role of distance in a mirror being half size. Perhaps rough sketches to accompany your explanations will help.

2. You can produce a spectrum by placing a tray of water in bright sunlight. Lean a pocket mirror against the inside edge of the tray and adjust it until a spectrum appears on the wall or ceiling. Aha! You've produced a spectrum without a prism.

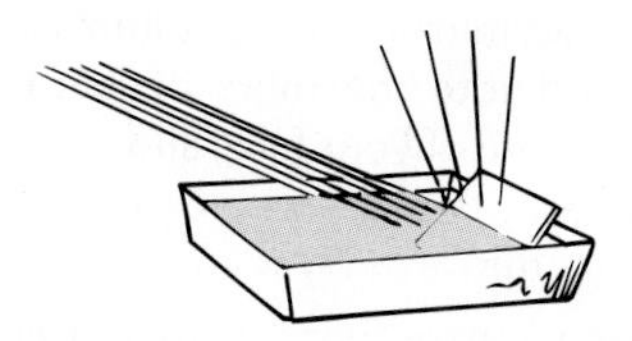

3. Set up two pocket mirrors at right angles and place a coin between them. You'll see four coins. Change the angle of the mirrors and see how many images of the coin you can see. With the mirrors at right angles, look at your face. Then wink. What do you see? You now see yourself as others see you. Hold a printed page up to the double mirrors and contrast its appearance with the reflection of a single mirror.

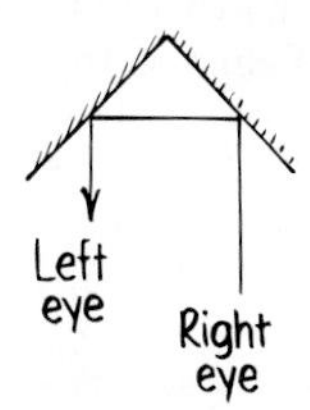

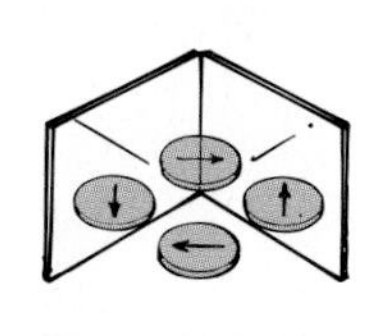

4. Look at yourself in a pair of mirrors at right angles to each other. You see yourself as others see you. Rotate the mirrors, still at right angles to each other. Does your image rotate also? Now place the mirrors 60° apart so you again see your face. Again rotate the mirrors and see if your image rotates also. Amazing?

5. Determine the magnifying power of a lens by focusing on the lines of a ruled piece of paper. Count the spaces between the lines that fit into one magnified space, and you have the magnifying power of the lens. You can do the same with binoculars and a distant brick wall. Hold the binoculars so that only one eye looks at the bricks through the eyepiece while the other eye looks directly at the bricks. The number of bricks seen with the unaided eye that will fit into one magnified brick gives the magnification of the instrument.

6. Look at the reflections of overhead lights from the two surfaces of eyeglasses, and you will see two fascinatingly different images. Why are they different?

Exercises

1. Fermat's principle is of least time rather than of least distance. Would least distance apply as well for reflection? For refraction? Why are your answers different?

2. This chapter opened with a photo of a physics instructor seeming to hover above the table. He isn't. Explain how he creates this illusion.

3. Her eye at point P looks into the mirror. Which of the numbered cards can she see reflected in the mirror?

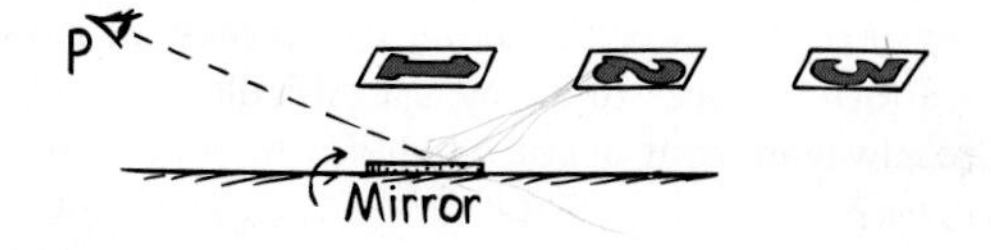

4. Cowboy Joe wishes to shoot his assailant by ricocheting a bullet off a mirrored metal plate. To do so, should he simply aim at the mirrored image of his assailant? Explain.
5. Trucks often have signs on their back ends that say; "If you can't see my mirrors, I can't see you." Explain the physics here.
6. Why is the lettering on the front of some vehicles "backward"?

AMBULANCE

7. When you look at yourself in the mirror and wave your right hand, your beautiful image waves the left hand. Then why don't the feet of your image wiggle when you shake your head?
8. Car mirrors are uncoated on the front surface and silvered on the back surface. When the mirror is properly adjusted, light from behind reflects from the silvered surface into the driver's eyes. Good. But this is not so good at nighttime with the glare of headlights behind. This problem is solved by the wedge shape of the mirror (see sketch). When the mirror is tilted slightly upward to the "nighttime" position, glare is directed upward toward the ceiling, away from the driver's eyes. Yet the driver can still see cars behind in the mirror. Explain.

9. To reduce glare of the surroundings, the windows of some department stores, rather than being vertical, slant inward at the bottom. How does this reduce glare?
10. A person in a dark room looking through a window can clearly see a person outside in the daylight, whereas the person outside cannot see the person inside. Explain.
11. What is the advantage of having matte (nonglossy) pages in this book rather than pages with a glossier surface?
12. Which kind of road surface is easier to see when driving at night—a pebbled, uneven surface, or a mirror-smooth surface? And why is it difficult to see the roadway in front of you when driving on a rainy night?
13. How many mirrors were involved in creating the photo below, taken by physics teacher Fred Myers of himself and his daughter McKenzie?

14. What must be the minimum length of a plane mirror in order for you to see a full image of yourself?
15. What effect does your distance from the plane mirror have in your answer to the preceding exercise? (Try it and see!)
16. Hold a pocket mirror almost at arm's length from your face and note how much of your face you can see. To see more of your face, should you hold the mirror closer or farther away, or would you have to have a larger mirror? (Try it and see!)
17. On a steamy mirror, wipe away just enough to see your full face. How tall will the wiped area be compared with the vertical dimension of your face?
18. The diagram shows a person and her twin at equal distances on opposite sides of a thin wall. Suppose a window is to be cut in the wall so each twin can see a complete view of the other. Show the size and location of the smallest window that can be cut in the wall to do the job. (*Hint:* Draw rays from the top of each twin's head to the other twin's eyes. Do the same from the feet of each to the eyes of the other.)

19. You can tell whether a person is nearsighted or farsighted by looking at how their face appears through their glasses. When a person's eyes seem magnified, is the person nearsighted or farsighted?

20. We see the bird and its reflection. Why do we not see the bird's feet in the reflection?

21. Why does reflected light from the Sun or Moon appear as a column in the body of water as shown? How would the reflected light appear if the water surface were perfectly smooth?

22. What is wrong with the cartoon of the man looking at himself in the mirror? (Have a friend face a mirror as shown, and you'll see.)

23. A pair of toy cart wheels are rolled obliquely from a smooth surface onto two plots of grass, a rectangular plot and a triangular plot as shown. The ground is on a slight incline so that, after slowing down in the grass, the wheels will speed up again when emerging on the smooth surface. Finish each sketch by showing some positions of the wheels inside the plots and on the other sides, thereby indicating the direction of travel.

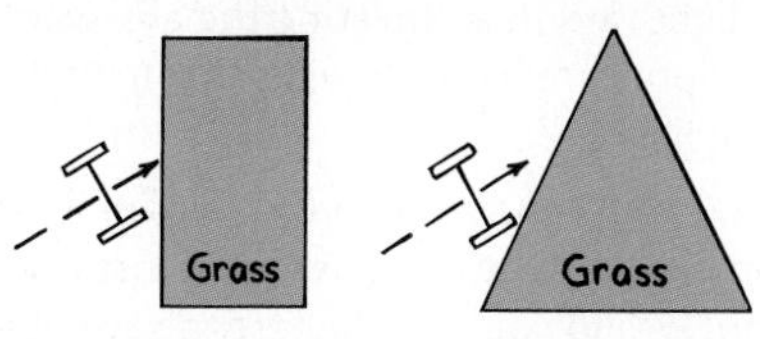

24. Your friend says that the wavelength of waves is less in water than in air, and cites Figure 28.24 as evidence. Do you agree or disagree, and why?

25. A pulse of red light and a pulse of blue light enter a glass block normal to its surface at the same time. Strictly speaking, after passing through the block, which pulse exits first?

26. During a lunar eclipse, the Moon is not completely dark but is usually a deep red in color. Explain this in terms of the refraction of all the sunsets and sunrises around the world.

27. If you place a glass test tube in water, you can see the tube. If you place it in clear soybean oil, you may not be able to see it. What does this indicate about the speed of light in the oil and in the glass?

28. A beam of light bends as shown in *a,* while the edges of the immersed square bend as shown in *b*. Do these pictures contradict each other? Explain.

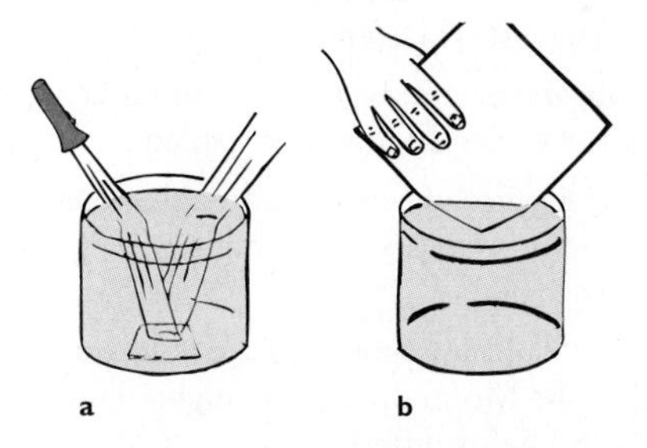

29. If, while standing on a bank, you wish to spear a fish beneath the water surface in front of you, should you aim above, below, or directly at the observed fish to make a direct hit? If, instead, you zap the fish with a laser, should you aim above, below, or directly at the observed fish? Defend your answers.

30. If the fish in the previous exercise were small and blue, and your laser light were red, what corrections should you make? Explain.

31. When a fish in a pond looks upward at an angle of 45°, does it see the sky above the water's surface or a reflection from the water–air boundary of the bottom of the pond? Defend your answer.

32. Upward rays of light in water toward the water–air boundary at angles greater than 48° to the normal are totally reflected. No rays beyond 48° refract outside. How about the reverse? Is there an angle at which light rays in air meeting the air–water boundary will totally reflect? Or will some light be refracted at all angles?

33. If you were to send a beam of laser light to a space station above the atmosphere and just above the horizon, would you aim the laser above, below, or at the visible space station? Defend your answer.

34. What exactly are you seeing when you observe a "water-on-the-road" mirage?

35. What accounts for the large shadows cast by the ends of the thin legs of the water strider?

36. When you stand with your back to the Sun, you see a rainbow as a circular arc. Could you move off to one side and then see the rainbow as the segment of an ellipse rather than the segment of a circle (as Figure 28.32 suggests)? Defend your answer.

37. Two observers standing apart from one another do not see the "same" rainbow. Explain.

38. A rainbow viewed from an airplane may form a complete circle. Where will the shadow of the airplane appear? Explain.

39. How is a rainbow similar to the halo sometimes seen around the Moon on a frosty night? How do rainbows and halos differ?

40. Transparent plastic swimming-pool covers called *solar heat sheets* have thousands of small air-filled bubbles that resemble lenses. The bubbles in these sheets are advertised to focus heat from the Sun into the water, thereby raising its temperature. Do you think these bubbles direct more solar energy into the water? Defend your answer.

41. Would the average intensity of sunlight measured by a light meter at the bottom of the pool in Figure 28.45 be different if the water were still?

42. When your eye is submerged in water, is the bending of light rays from water to your eyes more, less, or the same as in air?

43. Why will goggles allow a swimmer under water to focus more clearly on what he or she is looking at?

44. If a fish wore goggles above the water surface, why would vision be better for the fish if the goggles were filled with water? Explain.

45. Does a diamond under water sparkle more or less than in air? Defend your answer.

46. Cover the top half of a camera lens. What effect does this have on the pictures taken?

47. What will happen to the image projected onto a screen by a lens when you cover half the lens? (Try it and see!)

48. How could a converging lens be made for sound waves? (Such a lens is a feature of San Francisco's Exploratorium.)

49. Would refracting telescopes and microscopes magnify if light had the same speed in glass as in air? Defend your answer.

50. There is less difference between the speed of light in glass and the speed of light in water than there is between the speed of light in glass and the speed of light in air. Does this mean that a magnifying glass will magnify more or magnify less when it is used under water rather than in air?

51. Waves don't overlap in the image of a pinhole camera. Does this feature contribute to sharpness or to a blurry image?

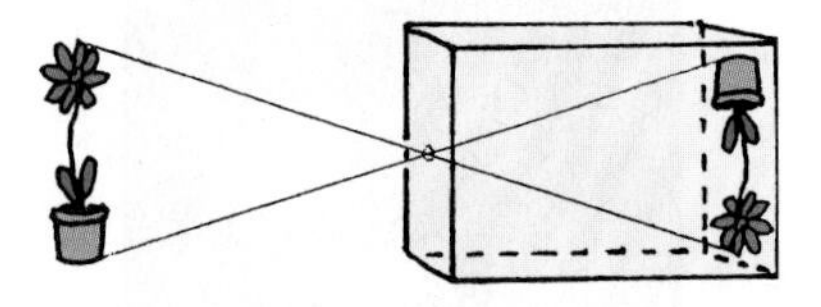

52. Why does the sharpness of the image in a pinhole camera not depend on the position of the viewing screen?

53. Whereas pinholes provide sharp images, lenses with large apertures are advantageous for spy cameras in high-flying aircraft. Why?

54. Why can't you take a photograph of your image in a plane mirror and focus the camera on both your image and on the mirror frame?

55. In terms of focal length, how far behind the camera lens is the film located when very distant objects are being photographed?

56. Why do you put slides into a slide projector upside down?

57. The image produced by a converging lens is upside down. Our eyes have converging lenses. Does this mean the images we see are upside down on our retinas? Explain.

58. The images produced by a converging camera lens are upside down. Does this mean the photographs taken with cameras are upside down?

59. Maps of the Moon are upside down. Why?

60. Why do older people who do not wear glasses read books farther away from their eyes than younger people do?

Problems

1. Show with a simple diagram that, when a mirror with a fixed beam incident upon it is rotated through a certain angle, the reflected beam is rotated through an angle twice as large. (This doubling of displacement makes irregularities in ordinary window glass more evident.)

2. A butterfly at eye level is 20 cm in front of a plane mirror. You are behind the butterfly, 50 cm from the mirror. What is the distance between your eye and the image of the butterfly in the mirror?

3. If you take a photograph of your image in a plane mirror, how many meters away should you set your focus if you are 2 m in front of the mirror?

4. Suppose you walk toward a mirror at 2 m/s. How fast do you and your image approach each other? (The answer is *not* 2 m/s.)

5. When light strikes glass perpendicularly, about 4% is reflected at each surface. How much light is transmitted through a pane of window glass?

6. No glass is perfectly transparent. Mainly because of reflections, about 92% of light passes through an average sheet of clear windowpane. The 8% loss is not noticed through a single sheet, but, through several sheets, the loss is apparent. How much light is transmitted by a "double-glazed" window (one with two sheets of glass)?

7. The diameter of the Sun makes an angle of 0.53° from Earth. How many minutes does it take the Sun to move one solar diameter in an overhead sky? (Remember that it takes 24 hours, or 1440 minutes, for the Sun to move through 360°.) How does your answer compare with the time it takes the Sun to disappear, once its lower edge meets the horizon at sunset? (Does refraction affect your answer?)

8. Consider two ways that light might hypothetically get from its starting point S to its final point F by way of a mirror: by reflecting from point A or by reflecting from point B. Since light travels at a fixed speed in air, the path of least time will also be the path of least distance. Show by calculation that the path SBF is shorter than the path SAF. How does this result tend to support the principle of least time?

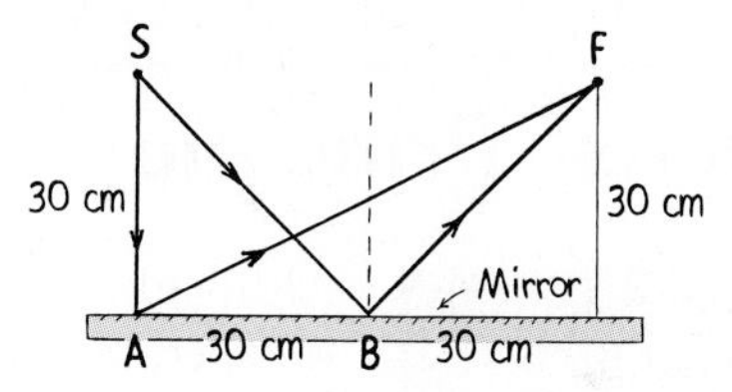

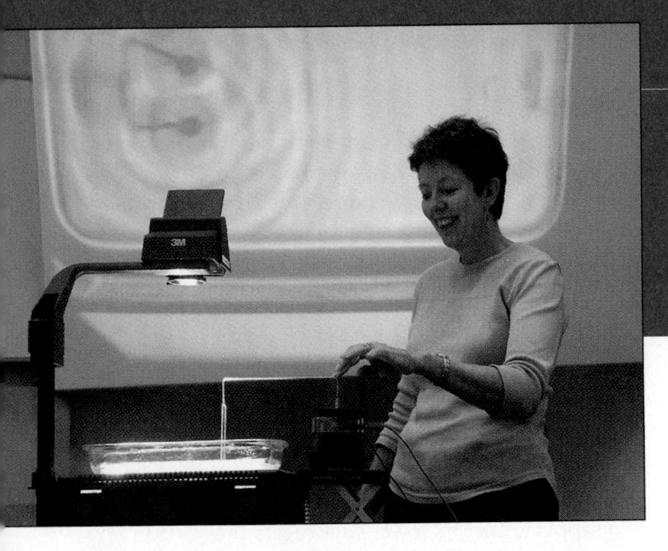

Jennie McKelvie showing that a ripple tank works just fine in New Zealand.

CHAPTER 29

Light Waves

Throw a rock in a quiet pool, and waves appear along the surface of the water. Strike a tuning fork, and waves of sound spread in all directions. Light a match, and waves of light similarly expand in all directions–at the enormously high speed of 300,000 kilometers per second. In this chapter, we'll study the wave nature of light. In the next chapter, we'll see that light also has a particle nature. Here we'll investigate some of the wave properties of light: diffraction, interference, and polarization.

Huygens' Principle

FIGURE 29.1
Water waves.

Although Galileo is credited as being the first to design a pendulum device to operate a system of toothed wheels, it was a Dutchman, Christian Huygens, who made the first pendulum clock. Huygens is most remembered, however, for his idea about waves.[1] The wave crests shown in Figure 29.1 form concentric circles—called *wave fronts*. Huygens proposed that the wave fronts of light waves spreading out from a point source can be regarded as the overlapped crests of tiny secondary waves (Figure 29.2)—that wave fronts are made up of tinier wave fronts. This idea is called **Huygens' principle.**

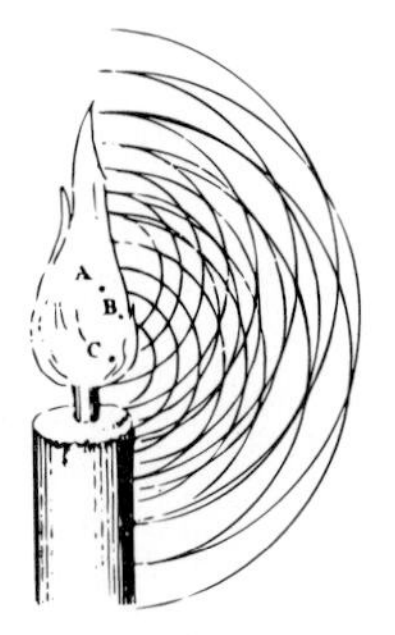

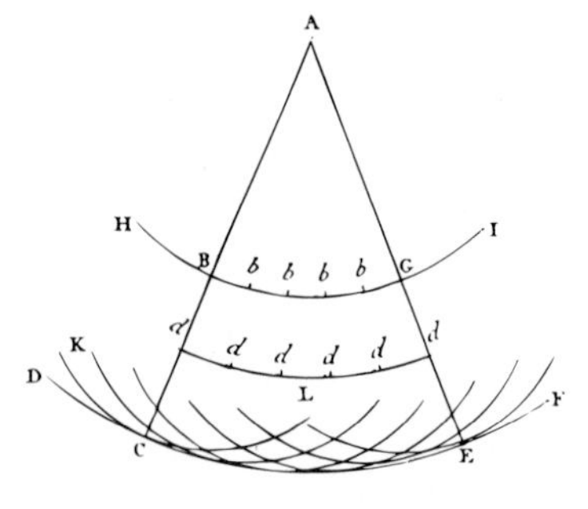

FIGURE 29.2
These drawings are from Huygens' book *Treatise on Light.* Light from A expands in wave fronts, every point of which behaves as if it were a new source of waves. Secondary wavelets starting at *b,b,b,b* form a new wave front (*d,d,d,d*); secondary wavelets starting at *d,d,d,d* form still another new wave front (DCEF).

[1]In 1665, 13 years before Huygens made his hypothesis about wave fronts, the English physicist Robert Hooke first proposed a wave theory of light.

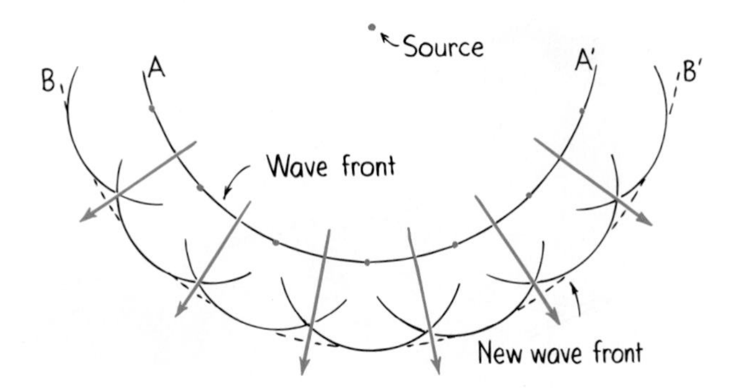

FIGURE 29.3
Interactive Figure
Huygens' principle applied to a spherical wave front.

Consider the spherical wave front in Figure 29.3. We can see that, if all points along the wave front AA′ are sources of new wavelets, a short time later the new overlapping wavelets will form a new surface, BB′, which can be regarded as the envelope of all the wavelets. In the figure we show only a few of the infinite number of wavelets from a few secondary point sources along AA′ that combine to produce the smooth envelope BB′. As the wave spreads, a segment appears less curved. Very far from the original source, the waves nearly form a plane—as do waves from the Sun, for example. A Huygens wavelet construction for plane wave fronts is shown in Figure 29.4. We see the laws of reflection and refraction illustrated via Huygens' principle in Figure 29.5.

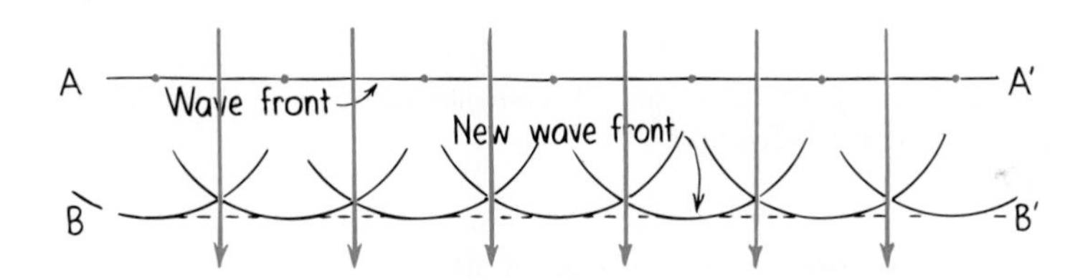

FIGURE 29.4
Interactive Figure
Huygens' principle applied to a plane wave front.

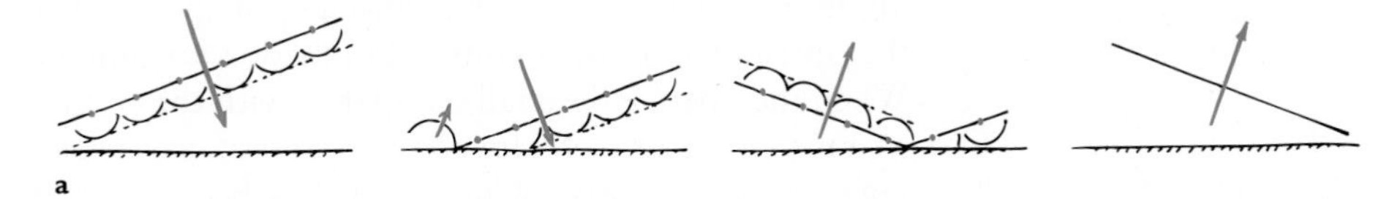

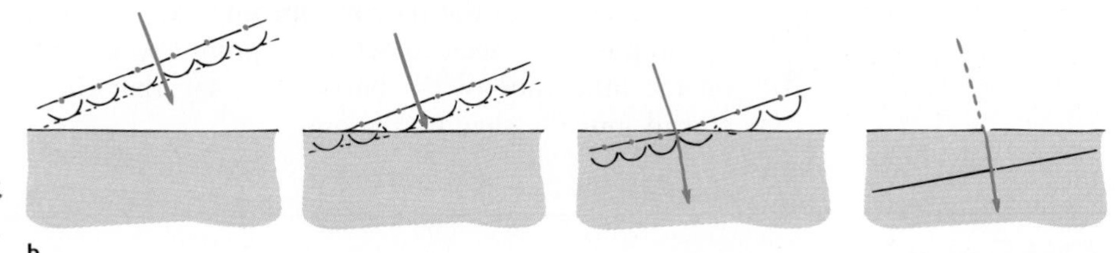

FIGURE 29.5
Huygens' principle applied to (a) reflection and (b) refraction.

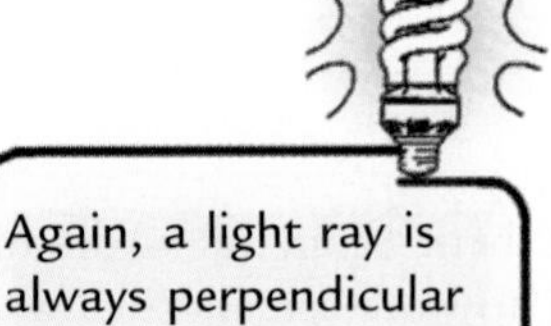

Again, a light ray is always perpendicular to a wave front.

Insights

Plane waves can be generated in water by successively dipping a horizontally held straightedge, such as a meterstick, into the surface (Figure 29.6). The photographs in Figure 29.7 are top views of a ripple tank in which plane waves are incident upon openings of various sizes (the straightedge is not shown). In *a*, where the opening is wide, we see the plane waves continue through the opening

FIGURE 29.6
The oscillating meterstick makes plane waves in the tank of water. Water oscillating in the opening acts as a source of waves that fan out on the other side of the barrier. Water diffracts through the opening.

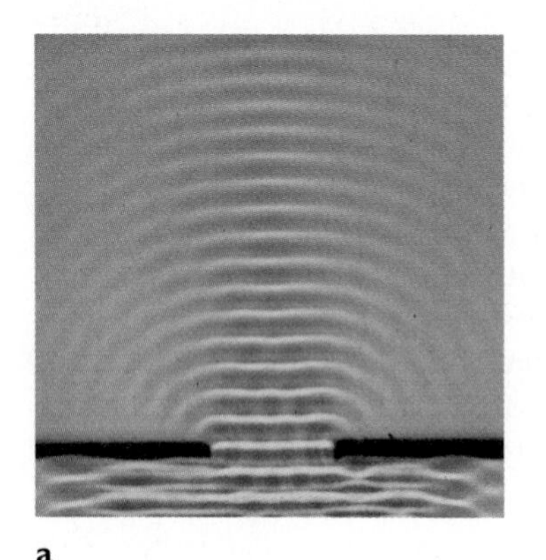

a

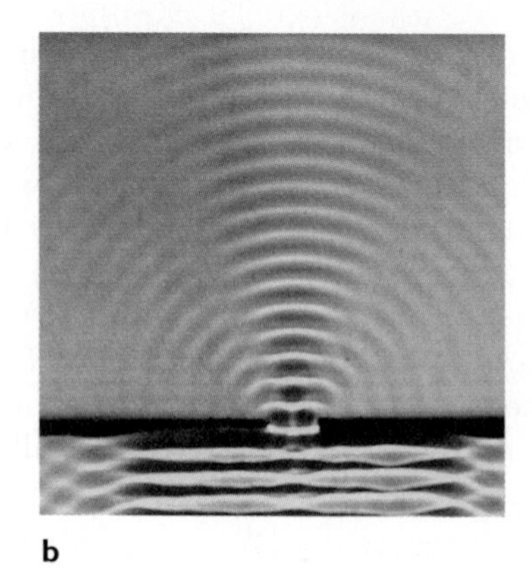

b

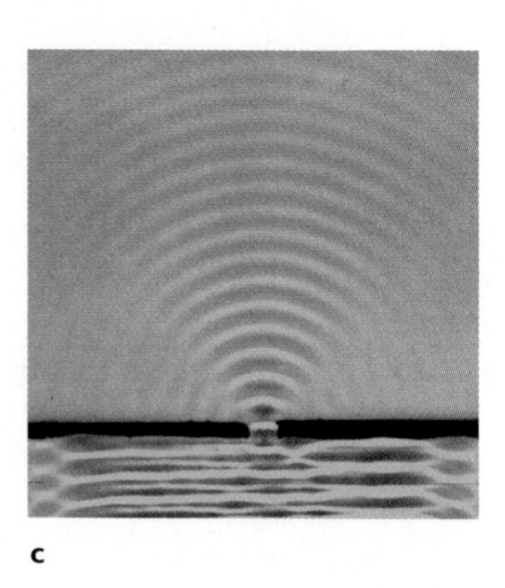

c

FIGURE 29.7
Plane waves passing through openings of various sizes. The smaller the opening, the greater the bending of the waves at the edges—in other words, the greater the diffraction.

without change—except at the corners, where the waves are bent into the shadow region, as predicted by Huygens' principle. As the width of the opening is narrowed, as in *b,* less and less of the incident wave is transmitted, and the spreading of waves into the shadow region becomes more pronounced. When the opening is small compared with the wavelength of the incident wave, as in *c,* the truth of Huygens' idea that every part of a wave front can be regarded as a source of new wavelets becomes quite apparent. As the waves are incident upon the narrow opening, the water sloshing up and down in the opening is easily seen to act as a "point" source of the new waves that fan out on the other side of the barrier. We say that the waves are *diffracted* as they spread into the shadow region.

Diffraction

In the previous chapter, we saw that light can be bent from its ordinary straight-line path by reflection and by refraction, and now we see another way in which light bends. Any bending of light by means other than reflection and refraction is called **diffraction.** The diffraction of plane water waves shown in Figure 29.7 occurs for all kinds of waves, including light waves.

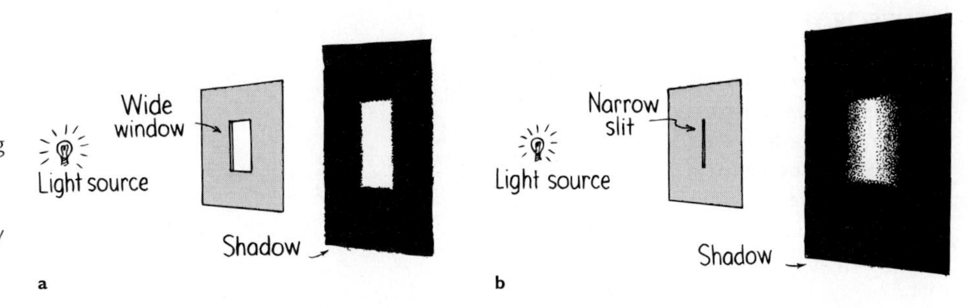

FIGURE 29.8
Interactive Figure
(a) Light casts a sharp shadow with some fuzziness at its edges when the opening is large compared with the wavelength of the light.
(b) When the opening is very narrow, diffraction is more apparent and the shadow is fuzzier.

When light passes through an opening that is large compared with the wavelength of light, it casts a shadow such as the one shown in Figure 29.8*a*. We see a rather sharp boundary between the light and dark area of the shadow. But, if we pass light through a thin razor slit in a piece of opaque cardboard, we see that the light diffracts (Figure 29.8*b*). The sharp boundary between the light and dark area disappears, and the light spreads out like a fan to produce a bright area that fades into darkness without sharp edges. The light is diffracted.

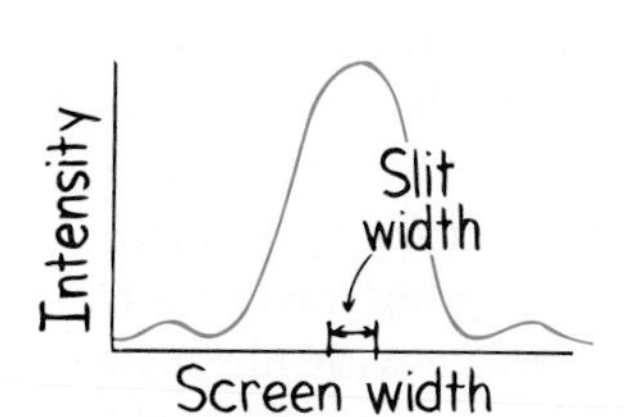

FIGURE 29.9
Graphic interpretation of diffracted light through a single thin slit.

A graph of the intensity distribution for diffracted light through a single thin slit appears in Figure 29.9. Because of diffraction, there is a gradual increase in light intensity rather than an abrupt change from dark to light. A photodetector sweeping across the screen would sense a gradual change from no light to maximum light. (Actually, there are slight fringes of intensity to either side of the main pattern; we will see shortly that these are evidence of interference that is more pronounced with a double slit or multiple slits.)

Diffraction is not confined to narrow slits or to openings in general but can be seen for all shadows. On close examination, even the sharpest shadow is blurred slightly at the edge. When the light is of a single color (monochromatic), diffraction can produce *diffraction fringes* at the edge of the shadow, as in Figure 29.10. In white light, the fringes merge together to create a fuzzy blur at the edge of a shadow.

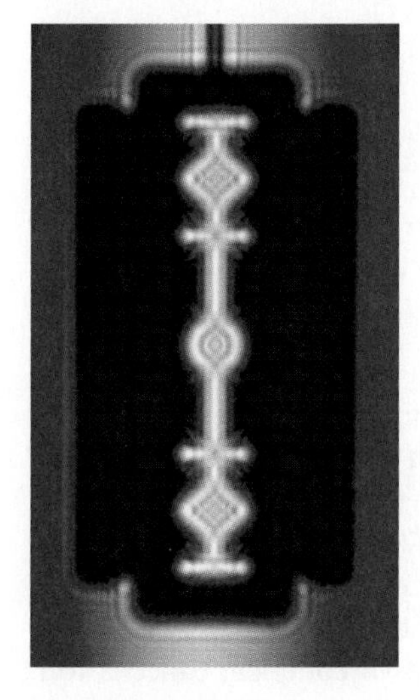

FIGURE 29.10
Diffraction fringes are evident in the shadows of monochromatic (single-frequency) laser light. These fringes would be filled in by multitudes of other fringes if the source were white light.

The amount of diffraction depends on the wavelength of the wave compared with the size of the obstruction that casts the shadow. Longer waves diffract more. They're better at filling in shadows, which is why the sounds of foghorns are low-frequency long waves—to fill in any "blind spots." Likewise for radio waves of the standard AM broadcast band, which are very long compared with the size of most objects in their path. The wavelengths of AM radio waves range from 180 to 550 meters, and the waves readily bend around buildings and other objects that might otherwise obstruct them. A long-wavelength radio wave doesn't "see" a relatively small building in its path, but a short-wavelength radio wave does. The radio waves of the FM band range from 2.8 to 3.4 meters and don't bend very well around buildings. This is one of the reasons that FM reception is often poor in localities where AM comes in loud and clear. In the case of radio reception, we don't wish to "see" objects in the path of radio waves, so diffraction is not a bad thing.

Diffraction is not so nice for viewing very small objects with a microscope. If the size of an object is about the same as the wavelength of light, diffraction blurs the image. If the object is smaller than the wavelength of light, no structure can be seen. The entire image is lost due to diffraction. No amount of magnification or perfection of microscope design can defeat this fundamental diffraction limit.

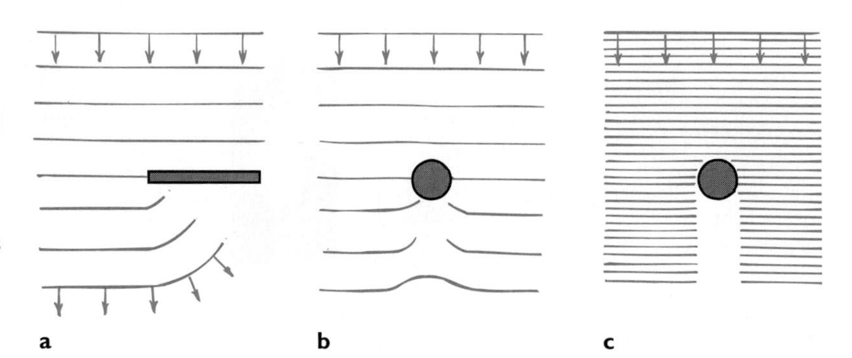

FIGURE 29.11
(a) Waves tend to spread into the shadow region. (b) When the wavelength is about the size of the object, the shadow is soon filled in. (c) When the wavelength is short relative to the object's size, a sharper shadow is cast.

To minimize this problem, microscopists illuminate tiny objects with electron beams rather than with light. Relative to light waves, electron beams have extremely short wavelengths. *Electron microscopes* take advantage of the fact that all matter has wave properties: A beam of electrons has a wavelength shorter than those of visible light. In an electron microscope, electric and magnetic fields, rather than optical lenses, are used to focus and magnify images.

The fact that smaller details can be seen better with shorter wavelengths is neatly employed by the dolphin in scanning its environment with ultrasound. The echoes of long-wavelength sound give the dolphin an overall image of objects in its surroundings. To examine more detail, the dolphin emits sound of shorter wavelengths. The dolphin has always done naturally what physicians are now able to do with an ultrasonic imaging devices.

CHECK YOURSELF

Why does a microscopist use blue light rather than white light to illuminate objects being viewed?

Interference

Spectacular illustrations of diffraction are shown in Figure 29.12. Physicist Chuck Manka made these by placing pieces of photographic film in the shadow of a screw illuminated with laser light. The fringes appear in both figures. These fringes are produced by **interference,** which we first discussed in Chapter 19. Constructive and destructive interference is reviewed in Figure 29.13. We see that the adding, or *superposition,* of a pair of identical waves in phase with each other produces a wave of the same frequency but twice the amplitude. If the waves are exactly one-half wavelength out of phase, their superposition results in complete cancellation. If they are out of phase by other amounts, partial cancellation occurs.

CHECK YOUR ANSWER

There is less diffraction with blue light, and so the microscopist sees more detail (just as a dolphin beautifully investigates fine detail in its environment by the echoes of ultra-short wavelengths of sound).

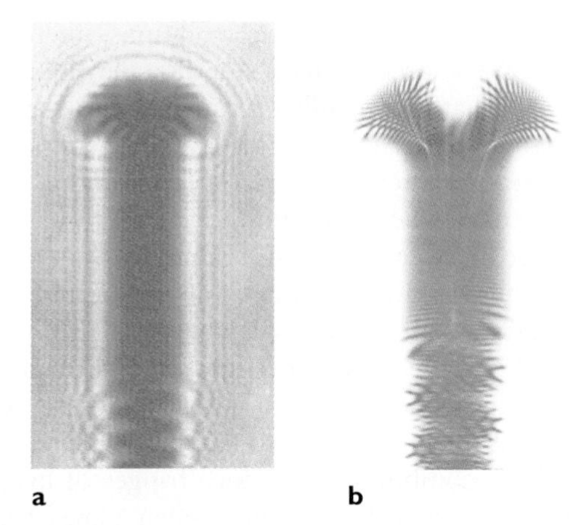

FIGURE 29.12
(a) The shadow of a screw in laser light shows fringes produced by destructive interference of the diffracted light. (b) A longer exposure shows fringes within the shadow produced by constructive and destructive interference.

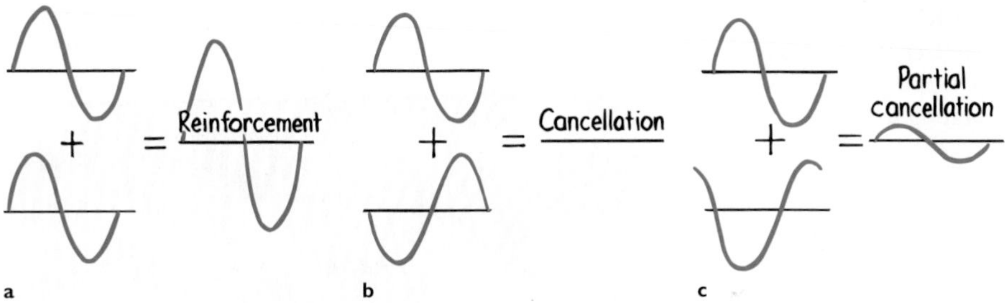

FIGURE 29.13
Wave interference.

+ = Reinforcement
+ = Cancellation
+ = Partial cancellation

a b c

The interference of water waves is a common sight, as shown in Figure 29.14. In some places, crests overlap crests; in other places, crests overlap troughs of other waves.

FIGURE 29.14
Interference of water waves.

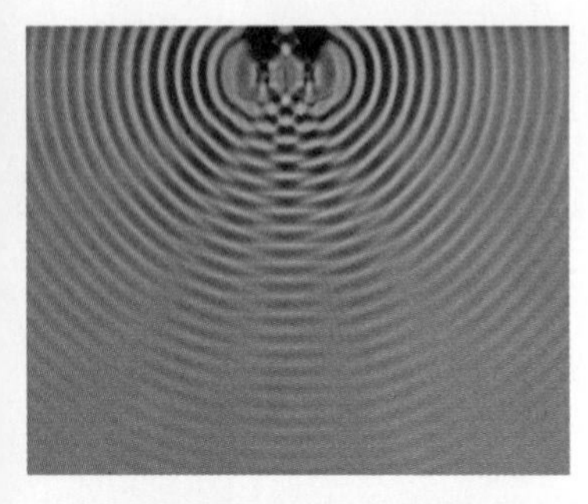

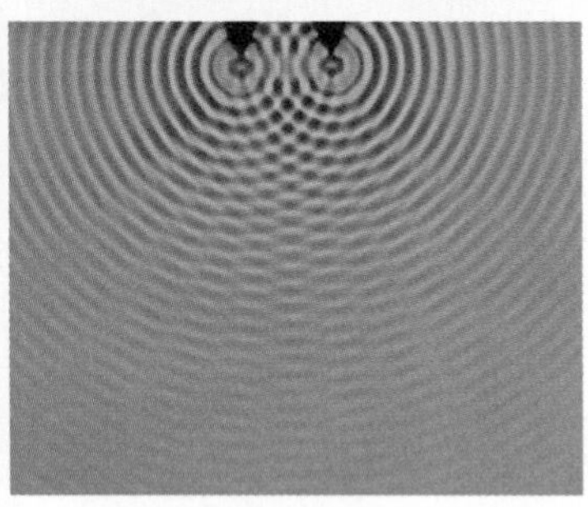

FIGURE 29.15
Interference patterns of overlapping waves from two vibrating sources.

Under more carefully controlled conditions, interesting patterns are produced by a pair of wave sources placed side by side (Figure 29.15). Drops of water are allowed to fall at a controlled frequency into shallow tanks of water (ripple tanks) while their patterns are photographed from above. Note that areas of constructive and destructive interference extend as far as the right-side edges of the ripple tanks, where the number of these regions and their size depend on the distance between the wave sources and on the wavelength (or frequency) of the waves. Interference is not restricted to easily seen water waves but is a property of all waves.

In 1801, the wave nature of light was convincingly demonstrated when the British physicist and physician Thomas Young performed his now famous interference experiment.[2] Young found that light directed through two closely spaced pinholes recombines to produce fringes of brightness and darkness on a screen behind. The bright fringes form when a crest from the light wave through one hole and a crest from the light wave through the other hole arrive at the screen at the same time. The dark fringes form when a crest from one wave and a trough from the other arrive at the same time. Figure 29.16 shows Young's drawing of the pattern of superimposed waves from the two sources. When his experiment is done with two closely spaced slits instead of pinholes, the fringe patterns are straight lines (Figure 29.18).

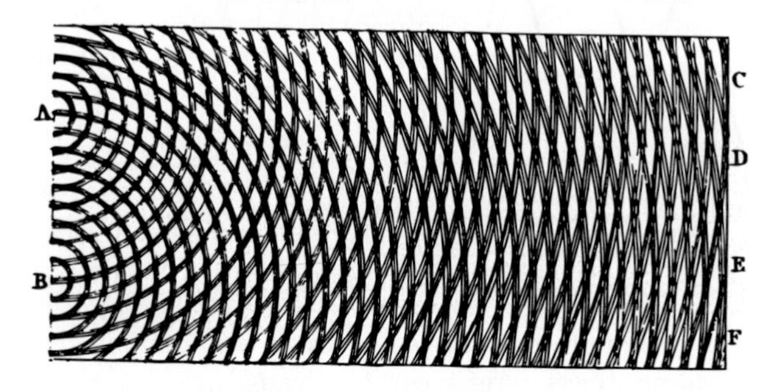

FIGURE 29.16
Thomas Young's original drawing of a two-source interference pattern. The dark circles represent wave crests; the white spaces between the crests represent troughs. Constructive interference occurs where crests overlap crests or troughs overlap troughs. Letters C, D, E, and F mark regions of destructive interference.

FIGURE 29.17
Bright fringes occur when waves from both slits arrive in phase; dark areas result from the overlapping of waves that are out of phase.

[2]Thomas Young read fluently at the age of 2; by 4, he had read the Bible twice; by 14, he knew eight languages. In his adult life, he was a physician and scientist, contributing to an understanding of fluids, work and energy, and the elastic properties of materials. He was the first person to make progress in deciphering Egyptian hieroglyphics. No doubt about it—Thomas Young was a bright guy!

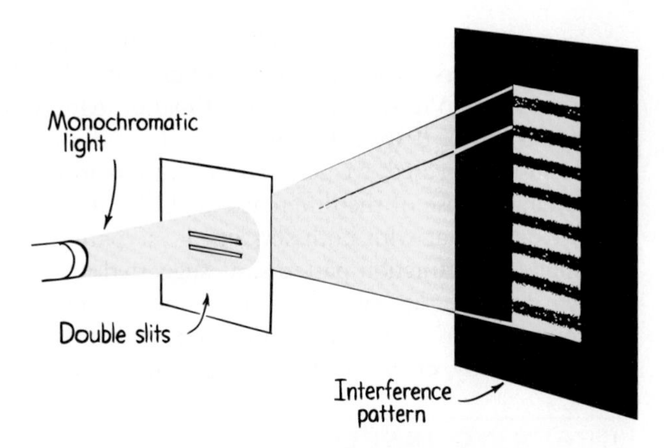

FIGURE 29.18
Interactive Figure
When monochromatic light passes through two closely spaced slits, a striped interference pattern is produced.

We see in Figure 29.19 how the series of bright and dark lines results from the different path lengths from the slits to the screen.[3] For the central bright fringe, the paths from the two slits are the same length and so the waves arrive in phase and reinforce each other. The dark fringes on either side of the central fringe result from one path being longer (or shorter) by one-half wavelength, so that the waves arrive half a wavelength out of phase. The other sets of dark fringes occur where the paths differ by odd multiples of one-half wavelength: 3/2, 5/2, and so on.

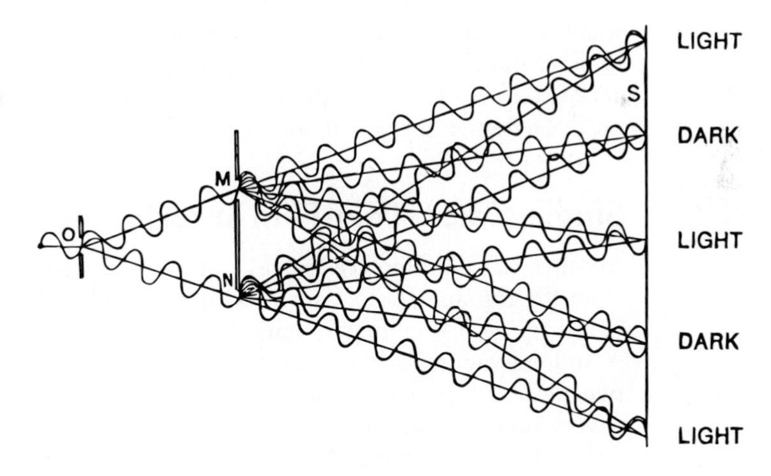

FIGURE 29.19
Interactive Figure
Light from O passes through slits M and N and produces an interference pattern on the screen S.

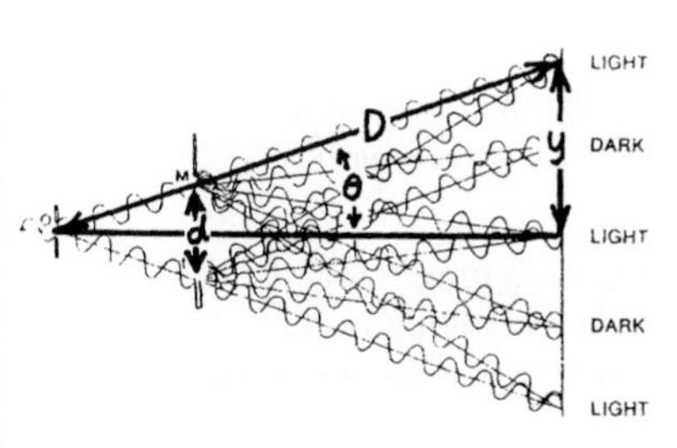

[3]In lab, you may determine the wavelength of light using measurements based on Figure 29.19. The equation for the first off-center interference maximum from two or more slits is

$$\lambda = d \sin \theta$$

where λ is the wavelength of light being diffracted, d is the distance between adjacent slits, and θ is the angle between lines to the central fringe of light and the first off-center constructive-interference fringe. From the diagram, $\sin \theta$ is the ratio of distance y to distance D, where y is the distance on the screen between the central fringe of light and the first constructive-interference fringe on either side. D is the distance from the fringe to the slits (which, in practice, is much greater than shown here).

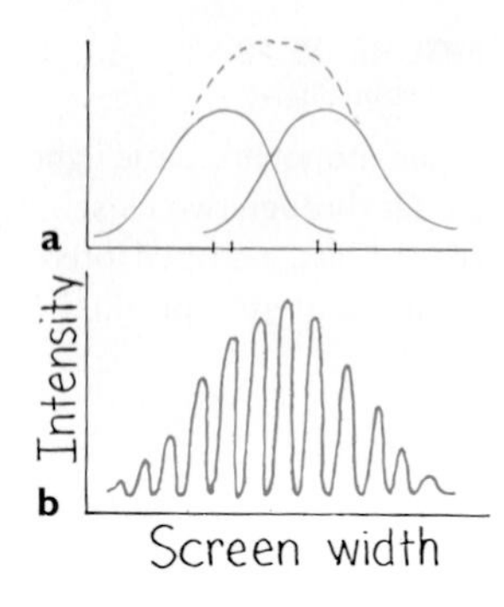

FIGURE 29.20
The light that diffracts through each of the double slits does not form a superposition of intensities as suggested in (a). The intensity pattern, because of interference, is as shown in (b).

In performing this double-slit experiment, suppose we cover one of the slits so that light passes through only a single slit. Then light will fan out and illuminate the screen to form a simple diffraction pattern, as discussed earlier (Figure 29.8*b* and 29.9). If we cover the other slit and allow light to pass only through the slit just uncovered, we get the same illumination on the screen, only displaced somewhat because of the difference in slit location. If we didn't know better, we might expect, that with both slits open, the pattern would simply be the sum of the single-slit diffraction patterns, as suggested in Figure 29.20*a*. But this doesn't happen. Instead, the pattern formed is one of alternating light and dark bands, as shown in *b*. We have an interference pattern. Interference of light waves does not, by the way, create or destroy light energy; it merely redistributes it.

CHECK YOURSELF

1. If the double slits were illuminated with monochromatic (single-frequency) red light, would the fringes be more widely or more closely spaced than if they were illuminated with monochromatic blue light?
2. Why is it important that monochromatic light be used?

Interference patterns are not limited to single and double slits. A multitude of closely spaced slits makes up a *diffraction grating*. These devices, like prisms, disperse white light into colors. Whereas a prism separates the colors of light by refraction, a diffraction grating separates colors by interference. These are used in devices called *spectrometers*, which we will discuss in the next chapter, and more commonly in such items as costume jewelry and automobile bumper stickers. These materials are ruled with tiny grooves that diffract light into a brilliant spectrum of colors. This is also seen in the colors dispersed by the feathers of some birds, and in the beautiful colors dispersed by the microscopic pits on the reflective surface of compact discs.

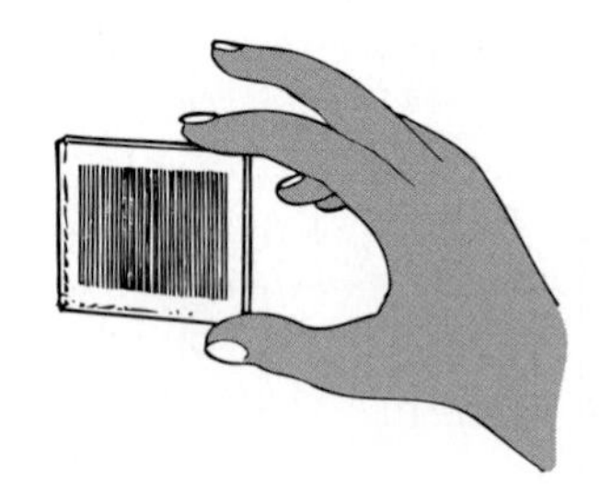

FIGURE 29.21
Because of the interference it causes, a diffraction grating disperses light into colors. It may be used in place of a prism in a spectrometer.

Single-Color Thin-Film Interference

Another way interference fringes can be produced is by the reflection of light from the top and bottom surfaces of a thin film. A simple demonstration can be set up with a monochromatic light source and a couple of pieces of glass. A sodium-vapor lamp provides a good source of monochromatic light. The two pieces of glass are placed one atop the other, as shown in Figure 29.22. A very thin piece of paper is placed between the plates at one edge. This leaves a very thin wedge-shaped film of air between the plates. If the eye is in a position

CHECK YOUR ANSWERS

1. More widely spaced. Can you see in Figure 29.19 that a slightly longer—and therefore a slightly more displaced—path from entrance slit to screen would result for the longer waves of red light?
2. If light of various wavelengths were diffracted by the slits, dark fringes for one wavelength would be filled in with bright fringes for another, resulting in no distinct fringe pattern. If you haven't seen this, be sure to ask your instructor to demonstrate it.

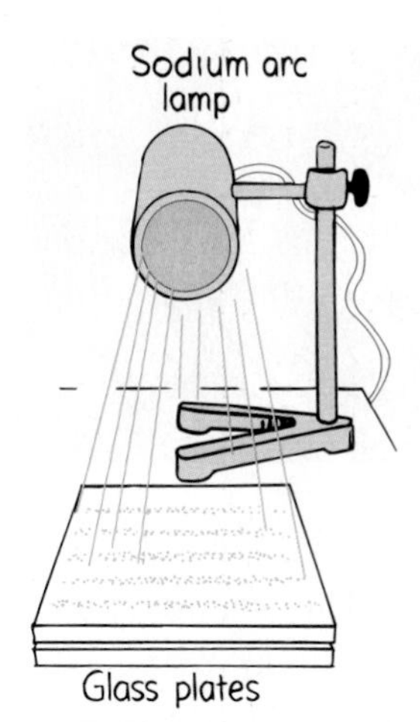

FIGURE 29.22
Interference fringes produced when monochromatic light is reflected from two plates of glass with an air wedge between them.

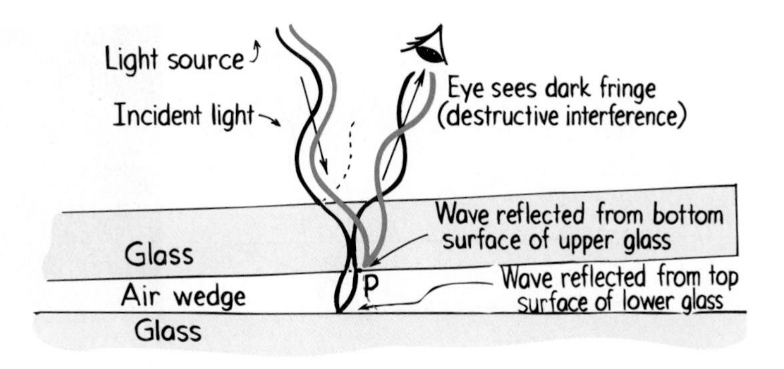

FIGURE 29.23
Reflection from the upper and lower surfaces of a "thin film of air."

to see the reflected image of the lamp, the image will not be continuous but will be made up of dark and bright bands.

The cause of these bands is the interference between the waves reflected from the glass on the top and bottom surfaces of the air wedge, as shown in the exaggerated view in Figure 29.23. The light reflecting from point P comes to the eye by two different paths. In one of these paths, the light is reflected from the top of the air wedge; in the other path, it is reflected from the lower side. If the eye is focused on point P, both rays reach the same place on the retina of the eye. But these rays have traveled different distances and may meet in phase or out of phase, depending on the thickness of the air wedge—that is, on how much farther one ray has traveled than the other. When we examine the entire surface of the glass, we see alternate dark and bright regions—the dark portions, where the air thickness is just right to produce destructive interference, and the bright portions, where the air wedge is just the proper amount thinner or thicker to result in the reinforcement of light. So the dark and bright bands are caused by the interference of light waves reflected from the two sides of the thin film.[4]

If the surfaces of the glass plates are perfectly flat, the bands are uniform. But, if the surfaces are not perfectly flat, the bands are distorted. The interference of light provides an extremely sensitive method for testing the flatness of surfaces. Surfaces that produce uniform fringes are said to be optically flat—this means that surface irregularities are small relative to the wavelength of visible light (Figure 29.24).

When a lens that is flat on top and has slight convex curvature on the bottom is placed on an optically flat plate of glass and illuminated from above with monochromatic light, a series of light and dark rings is produced. This pattern is known as *Newton's rings* (Figure 29.25). These light and dark rings

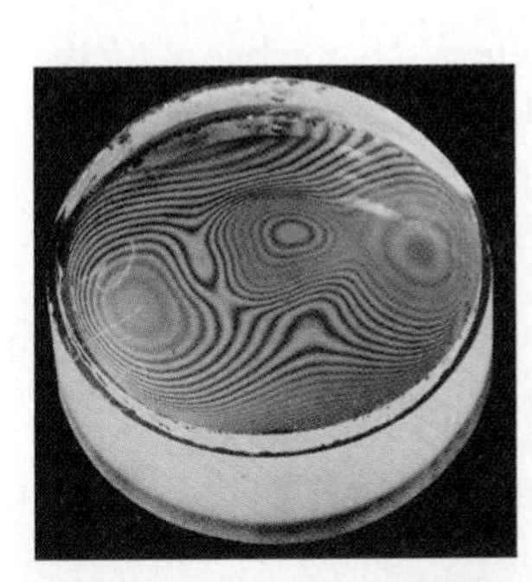

FIGURE 29.24
Optical flats used for testing the flatness of surfaces.

[4]Phase shifts at some reflecting surfaces also contribute to interference. For simplicity and brevity, our concern with this topic will be limited to this footnote. In short, when light in a medium is reflected at the surface of a second medium in which the speed of light is less (when there is a greater index of refraction), there is a 180° phase shift (that is, half a wavelength). However, no phase shift occurs when the second medium is one that transmits light at a higher speed (and has a lower index of refraction). In our air-wedge example, no phase shift occurs for reflection at the upper glass–air surface, and a 180° shift does occur at the lower air–glass surface. So, at the apex of the air wedge where the thickness approaches zero, the phase shift produces cancellation, and the wedge is dark. Likewise with a soap film so thin that its thickness is appreciably smaller than the wavelength of light. This is why parts of a film that are extremely thin appear black. Waves of all frequencies are canceled.

FIGURE 29.25
Newton's rings.

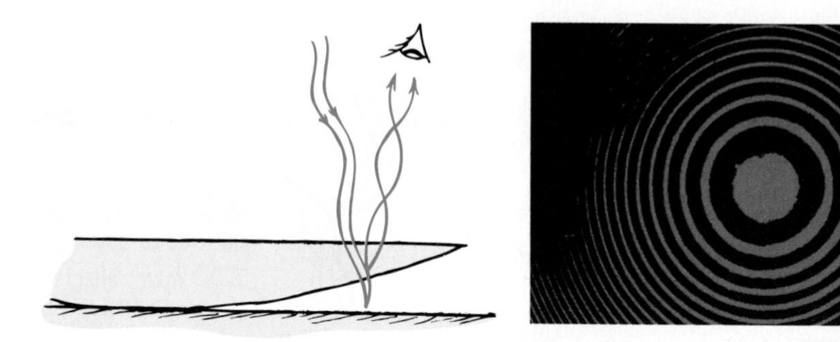

are the same kinds of fringes observed with plane surfaces. This is a useful testing technique in polishing precision lenses.

CHECK YOURSELF

How would the spacings between Newton's rings differ when illuminated by red light and by blue light?

Interference Colors by Reflection from Thin Films

Soap-bubble colors result from the interference of reflected light from the inside and outside surfaces of the soap film. When a color is canceled, what you see is its complementary color.

Insights

We have all noticed the beautiful spectrum of colors reflected from a soap bubble or from gasoline on a wet street. These colors are produced by the *interference* of light waves. This phenomenon is often called *iridescence* and is observed in thin transparent films.

A soap bubble appears iridescent in white light when the thickness of the soap film is about the same as the wavelength of light. Light waves reflected from the outer and inner surfaces of the film travel different distances. When illuminated by white light, the film may be just the right thickness at one place to cause the destructive interference of, say, yellow light. When yellow light is subtracted from white light, the mixture left will appear as the complementary color of yellow, which is blue. At another place, where the film is thinner, a different color may be canceled by interference, and the light seen will be its complementary color. The same thing happens to gasoline on a wet street (Figure 29.26). Light reflects from both the upper gasoline surface and the lower gasoline–water surface. If the thickness of the gasoline is such that it cancels blue, as the figure suggests, then the gasoline surface appears yellow to the eye. This is because the blue is subtracted from the white, leaving the complementary color, yellow. The different colors, then, correspond to different thicknesses of the thin film, providing a vivid "contour map" of microscopic differences in surface "elevations."

Over a wider field of view, different colors can be seen, even if the thickness of the gasoline film is uniform. This has to do with the apparent thickness

CHECK YOUR ANSWER

The rings would be more widely spaced for longer-wavelength red light than for the shorter waves of blue light. Do you see the geometrical reason for this?

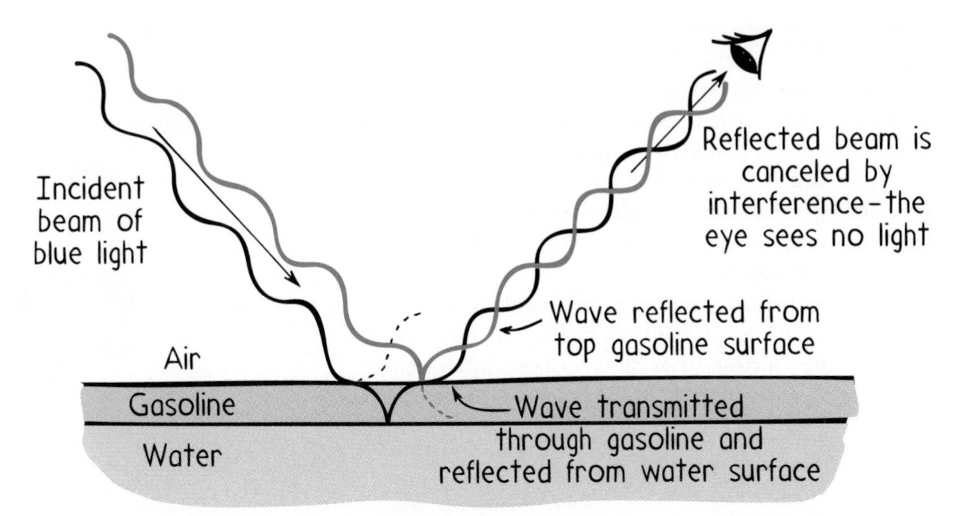

FIGURE 29.26
The thin film of gasoline is just the right thickness to cancel the reflections of blue light from the top and bottom surfaces. If the film were thinner, perhaps shorter-wavelength violet would be canceled. (One wave is drawn in black to show how it is out of phase with the blue wave upon reflection.)

of the film: Light reaching the eye from different parts of the surface is reflected at different angles and traverses different thicknesses. If the light is incident at a grazing angle, for example, the ray transmitted to the gasoline's lower surface travels a longer distance. Longer waves are canceled in this case, and different colors appear.

Soap Bubble Interference

Dishes washed in soapy water and poorly rinsed have a thin film of soap on them. Hold such a dish up to a light source so that *interference colors* can be seen. Then turn the dish to a new position, keeping your eye on the same part of the dish, and the color will change. Light reflecting from the bottom surface of the transparent soap film is canceling light reflecting from the top surface of the soap film. Light waves of different wavelengths are canceled for different angles. Interference colors are best seen in soap bubbles (Figure 29.27). You'll notice that these colors are predominantly cyan, magenta, and yellow, due to the subtraction of primary red, green, and blue.

Interference provides a way to measure the wavelength of light and other electromagnetic radiation. It also makes it possible to measure extremely small distances with great accuracy. Instruments called *interferometers,* which use the principle of interference, are the most accurate instruments known for measuring small distances.

FIGURE 29.27
Physicist author Bob Greenler shows interference colors with *big* bubbles. Why are the bubble colors subtractive primaries?

PRACTICING PHYSICS

Do this physics experiment at your kitchen sink. Dip a dark-colored coffee cup (dark colors make the best background for viewing interference colors) in dishwashing detergent and then hold it sideways and look at the reflected light from the soap film that covers its mouth. Swirling colors appear as the soap flows down to form a wedge that grows thicker at the bottom. The top becomes thinner, so thin that it appears black. This tells you that its thickness is less than one-fourth the wavelength of the shortest waves of visible light. Whatever its wavelength, light reflecting from the inner surface reverses phase, rejoins light reflecting from the outer surface, and cancels. The film soon becomes so thin it pops.

CHECK YOURSELF

In the left column are the colors of certain objects. In the right column are various ways in which colors are produced. Match the right column to the left.

1. yellow daffodil	**a.** interference
2. blue sky	**b.** selective reflection
3. rainbow	**c.** refraction
4. soap bubble	**d.** scattering

Polarization

FIGURE 29.28
A vertically plane-polarized plane wave and a horizontally plane-polarized plane wave.

Interference and diffraction provide the best evidence that light is wavelike. As we learned in Chapter 19, waves can be either longitudinal or transverse. Sound waves are longitudinal, which means the vibratory motion is *along* the direction of wave travel. But when we shake a taut rope, the vibratory motion traveling along the rope is perpendicular, or *transverse,* to the rope. Both longitudinal and transverse waves exhibit interference and diffraction effects. Are light waves, then, longitudinal or transverse? **Polarization** of the light waves demonstrates that they are transverse.

If we shake a taut rope up and down as in Figure 29.28, a transverse wave travels along the rope in one plane. We say that such a wave is *plane-polarized,*[5] meaning the waves traveling along the rope are confined to a single plane. If we shake the rope up and down, we produce a vertically plane-polarized wave. If we shake it from side to side, we produce a horizontally plane-polarized wave.

CHECK YOUR ANSWERS

1-b; 2-d; 3-c; 4-a.

[5]Light may also be circularly polarized and elliptically polarized, which are combinations of transverse polarizations. But we will not study these cases.

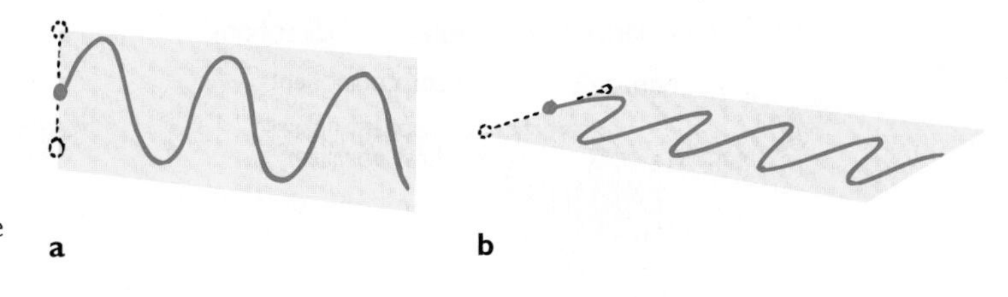

FIGURE 29.29
(a) A vertically plane-polarized wave from a charge vibrating vertically. (b) A horizontally plane-polarized wave from a charge vibrating horizontally.

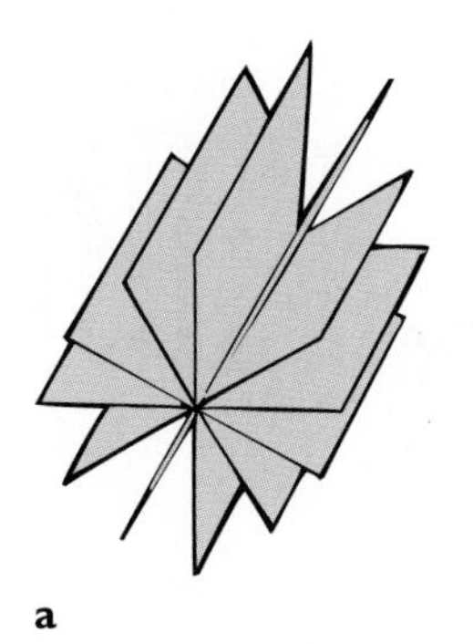

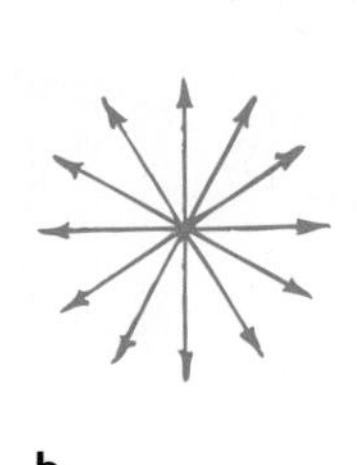

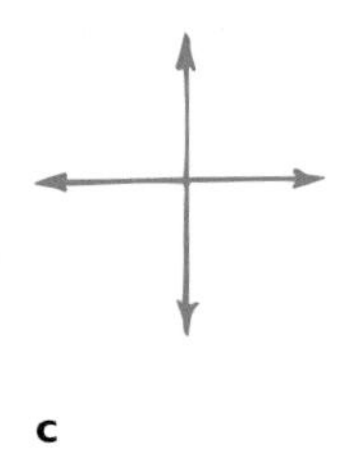

FIGURE 29.30
Representations of plane-polarized waves. Electric vectors, (a) and (b), show the electric part of the electromagnetic wave.

A single vibrating electron can emit an electromagnetic wave that is plane-polarized. The plane of polarization will match the vibrational direction of the electron. A vertically accelerating electron, then, emits light that is vertically polarized, while a horizontally accelerating electron emits light that is horizontally polarized (Figure 29.29).

A common light source—such as an incandescent lamp, a fluorescent lamp, a candle flame, or an arc lamp—emits light that is unpolarized. This is because there is no preferred vibrational direction for the accelerating electrons emitting the light. The planes of vibration might be as numerous as the accelerating electrons producing them. A few planes are represented in Figure 29.30*a*. We can represent all these planes by radial lines (Figure 29.30*b*) or, more simply, by vectors in two mutually perpendicular directions (Figure 29.30*c*), as if we had resolved all the vectors of Figure 29.30*b* into horizontal and vertical components. This simpler schematic represents unpolarized light. Polarized light would be represented by a single vector.

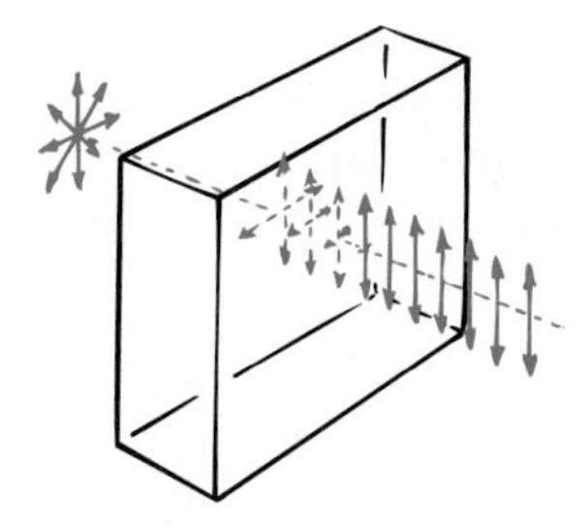

FIGURE 29.31
One component of the incident unpolarized light is absorbed, resulting in emerging polarized light.

All transparent crystals of a noncubic natural shape have the property of transmitting light of one polarization differently from light of another polarization. Certain crystals[6] not only divide unpolarized light into two internal beams polarized at right angles to each other but also strongly absorb one beam while transmitting the other (Figure 29.31). Tourmaline is one such common crystal, but, unfortunately, the transmitted light is colored. Herapathite, however, does the job without discoloration. Microscopic crystals of herapathite are embedded between cellulose sheets in uniform alignment and are used in making Polaroid filters. Some Polaroid sheets consist of certain aligned molecules rather than tiny crystals.[7]

---

[6]Called *dichroic*.

[7]The molecules are polymeric iodine in a sheet of polyvinyl alcohol or polyvinylene.

FIGURE 29.32
A rope analogy illustrates the effect of crossed Polaroids.

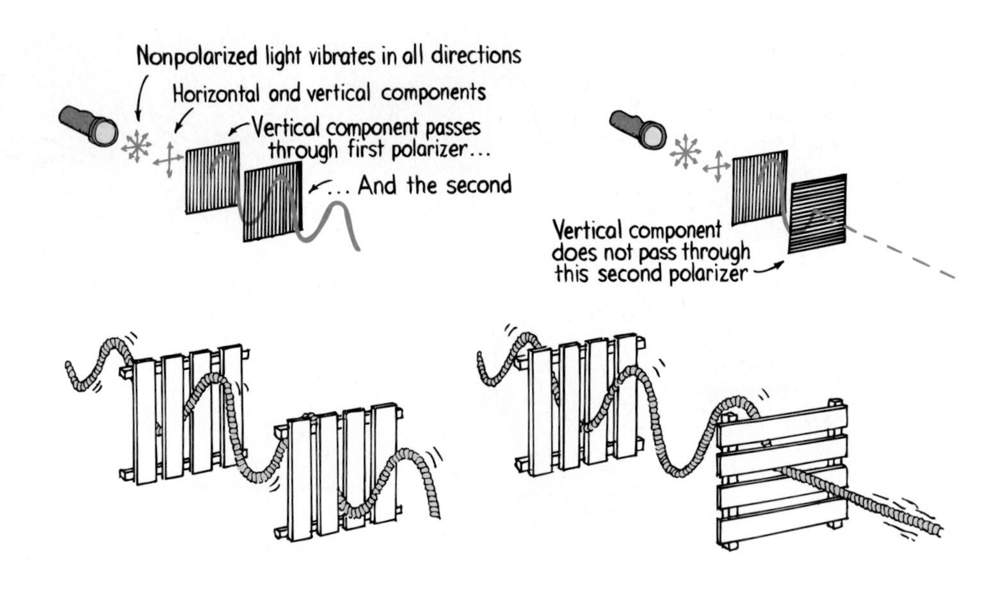
Nonpolarized light vibrates in all directions
Horizontal and vertical components
Vertical component passes through first polarizer...
... And the second
Vertical component does not pass through this second polarizer

FIGURE 29.33
Polaroid sunglasses block out horizontally vibrating light. When the lenses overlap at right angles, no light gets through.

If you look at unpolarized light through a Polaroid filter, you can rotate the filter in any direction, and the light will appear unchanged. But if you are looking at polarized light and you rotate the filter, you can progressively cut off more and more of the light until it is entirely blocked out. An ideal Polaroid will transmit 50% of incident unpolarized light. That 50% is, of course, polarized. When two Polaroids are arranged so that their polarization axes are aligned, light will be transmitted through both (Figure 29.32). If their axes are at right angles to each other (in this case, we say the filters are *crossed*), no light penetrates the pair. (Actually, some light of the shorter wavelengths does get through, but not to any significant degree.) When Polaroids are used in pairs like this, the first one is called the *polarizer* and the second one is called the *analyzer.*

Much of the light reflected from nonmetallic surfaces is polarized. The glare from glass or water is a good example. Except for perpendicular incidence, the reflected ray contains more vibrations parallel to the reflecting surface, whereas the transmitted beam contains more vibrations at right angles to the vibrations of reflected light (Figure 29.34). This is analogous to skipping flat rocks off the surface of a pond. When the rocks hit with their faces parallel to the surface,

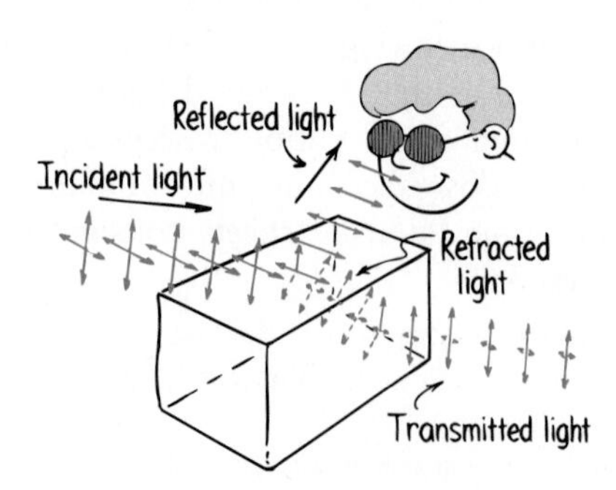
Reflected light
Incident light
Refracted light
Transmitted light

FIGURE 29.34
Most glare from nonmetallic surfaces is polarized. Here we see that the components of incident light parallel to the surface are reflected and that the components perpendicular to the surface pass through the surface into the medium. Because most of the glare we encounter is from horizontal surfaces, the polarization axes of Polaroid sunglasses are vertical.

a

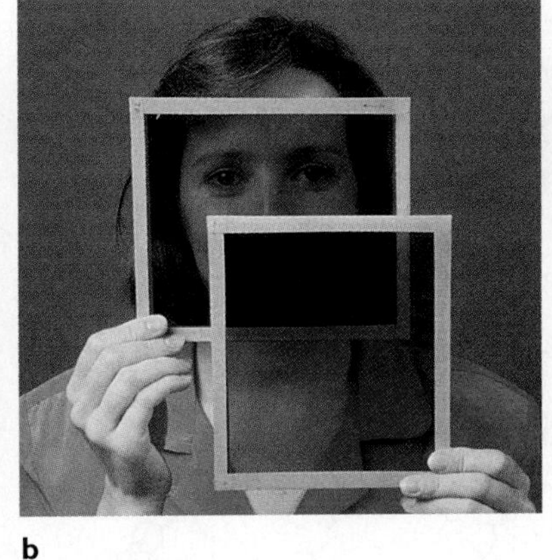
b

c

FIGURE 29.35

Interactive Figure

Light is transmitted when the axes of the Polaroids are aligned (a), but absorbed when Ludmila rotates one so that the axes are at right angles to each other (b). When she inserts a third Polaroid at an angle between the crossed Polaroids, light is again transmitted (c). Why? (For the answer—after you have given this some thought—see Appendix D, "More About Vectors.")

they easily reflect; but, if they hit with their faces tilted to the surface, they "refract" into the water. The glare from reflecting surfaces can be appreciably diminished with the use of Polaroid sunglasses. The polarization axes of the lenses are vertical because most glare reflects from horizontal surfaces. Properly aligned Polaroid eyeglasses enable us to see the projection of stereoscopic movies or slides on a flat screen in three dimensions.

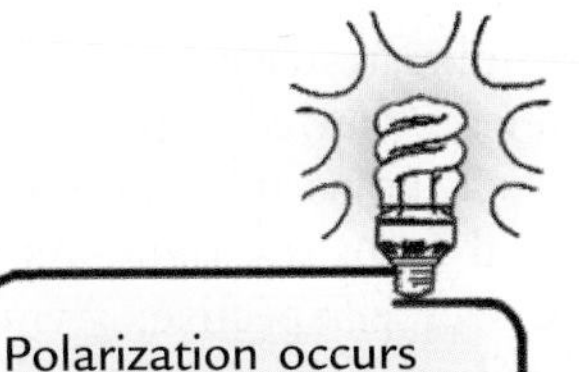

Polarization occurs only for transverse waves. In fact, it is an important way of telling whether a wave is transverse or longitudinal.

Insights

CHECK YOURSELF

Which pair of glasses is best suited for automobile drivers? (The polarization axes are shown by the straight lines.)

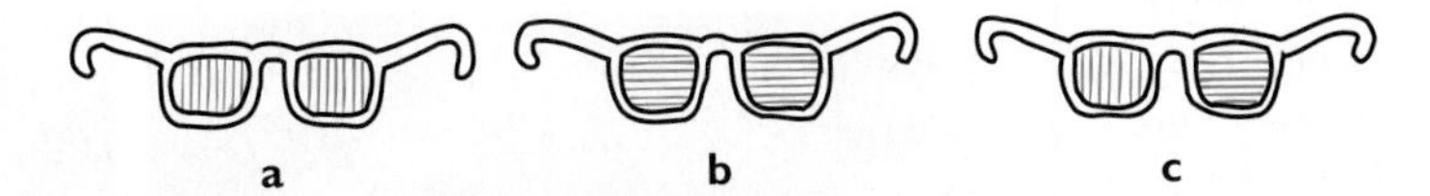

Three-Dimensional Viewing

Polarized Light and 3-D Viewing

Vision in three dimensions depends primarily on the fact that both eyes give their impressions simultaneously (or nearly so), each eye viewing the scene from a slightly different angle. To convince yourself that each eye sees a different view, hold an upright finger at arm's length and see how it appears to shift position from left to right in front of the background as you alternately close each

CHECK YOUR ANSWER

Glasses *a* are best suited because the vertical axis blocks horizontally polarized light, which constitutes much of the glare from horizontal surfaces. Glasses *c* are suited for viewing 3-D movies.

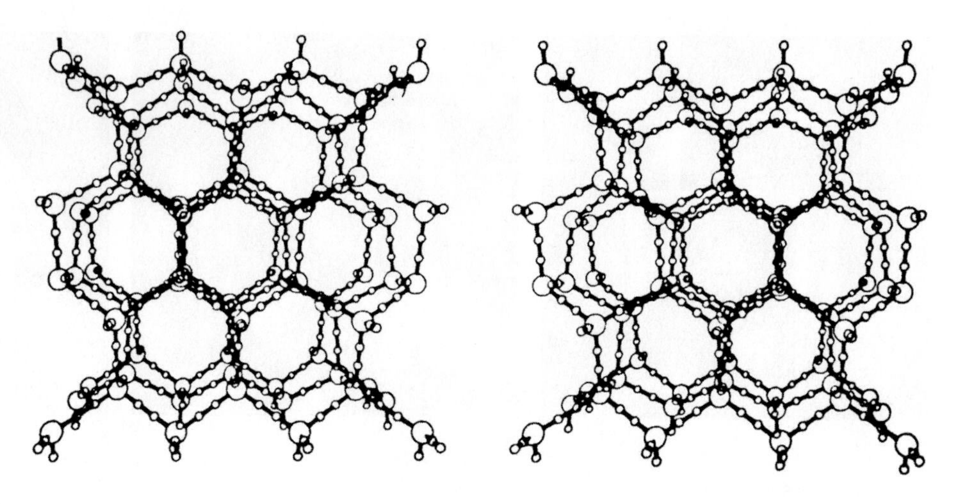

FIGURE 29.36
The crystal structure of ice in stereo. You'll see depth when your brain combines the views of your left eye looking at the left figure and your right eye looking at the right figure. To accomplish this, focus your eyes for distant viewing before looking at this page. Without changing your focus, look at the page, and each figure will appear double. Then adjust your focus so that the two inside images overlap to form a central composite image. Practice makes perfect. (If you instead *cross* your eyes to overlap the figures, near and far are reversed!)

eye. The drawings of Figure 29.36 illustrate a stereo view of the crystal structure of ice.

The familiar handheld stereoscopic viewer (Figure 29.39) simulates the effect of depth. In this device, there are two photographic transparencies (or slides) taken from slightly different positions. When they are viewed at the same time, the arrangement is such that the left eye sees the scene as photographed from the left, and the right eye sees it as photographed from the right. As a result, the objects in the scene sink into relief in correct perspective, giving apparent depth to the picture. The device is constructed so that each eye sees only the proper view. There is no chance for one eye to see both views. If you remove the slides from the hand viewer and project each view on a screen by slide projector (so that the views are superimposed), a blurry picture results.

Cosmic microwave background (CMB) fills all of space and comes at us from every direction. It is an echo of the Big Bang that got our universe started some 14 billion years ago. Recent findings show this radiation to be polarized. Polarization observations are unaffected by gravity and provide a clear and detailed look at the early cosmos.

FIGURE 29.37
A stereo view of snowflakes. View these in the same way as Figure 29.36.

The test of all knowledge
is experiment.
Experiment is the *sole judge*
of scientific "truth."
Richard P. Feynman

The test of all knowledge
is experiment.
Experiment is the *sole judge*
of scientific "truth."
Richard P. Feynman

FIGURE 29.38
With your eyes focused for distant viewing, the second and fourth lines appear to be farther away; if you cross your eyes, the second and fourth lines appear closer.

This is because each eye sees both views simultaneously. This is where Polaroid filters come in. If you place the Polaroids in front of the projectors so that one is horizontal and the other is vertical, and you view the polarized image with polarized glasses of the same orientation, each eye will see the proper view as with the stereoscopic viewer (Figure 29.40). You then will see an image in three dimensions.

FIGURE 29.39
A stereoscopic viewer.

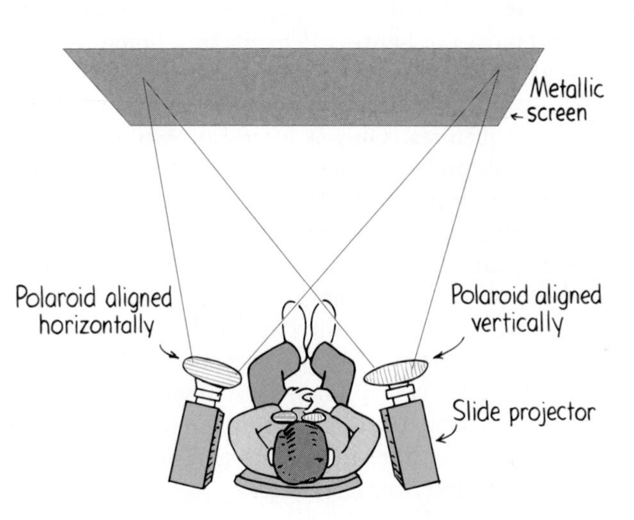

FIGURE 29.40
A 3-D slide show using Polaroids. The left eye sees only polarized light from the left projector, the right eye sees only polarized light from the right projector, and both views merge in the brain to produce depth.

Depth is also seen in computer-generated stereograms, as in Figure 29.41. Here the slightly different patterns are not obvious in a casual view. Use the procedure for viewing the previous stereo figures. Once you've mastered the viewing technique, head for the local mall and check the variety of stereograms in posters and books.

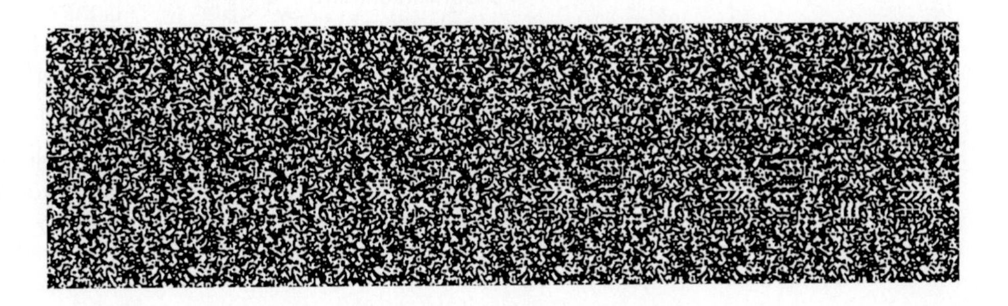

FIGURE 29.41
A computer-generated stereogram.

Holography

Perhaps the most exciting illustration of interference is the **hologram,** a two-dimensional photographic plate illuminated with laser light that allows you to see a faithful reproduction of a scene in three dimensions. The hologram was invented and named by Dennis Gabor in 1947, 10 years before lasers were invented. *Holo* in Greek means "whole," and *gram* in Greek means "message" or "information." A hologram contains the whole message or entire picture. When illuminated with laser light, the image is so realistic that you can actually look around the corners of objects in the image and see the sides.

In ordinary photography, a lens is used to form an image of an object on photographic film. Light reflected from each point on the object is directed by the lens only to a corresponding point on the film, and all the light that reaches the film comes from the object being photographed. In the case of holography, however, no image-forming lens is used. Instead, each point of the object being "photographed" reflects light to the *entire* photographic plate, so every part of the plate is exposed with light reflected from every part of the object. Most importantly, the light used to make a hologram must be of a single frequency and all parts exactly in phase: It must be *coherent.* If, for example, white light were used, the diffraction fringes for one frequency would be washed out by other frequencies. Only a laser can easily produce such light. (We will treat lasers in detail in the next chapter.) Holograms are made with laser light.

A conventional photograph is a recording of an image, but a hologram is a recording of the interference pattern resulting from the combination of two sets of wave fronts. One set of wave fronts is from light reflected by the object, and the other set is from a *reference beam,* which is diverted from the illuminating beam and sent directly to the photographic plate (Figure 29.42). The developed photograph has no recognizable image. The hologram is simply a hodgepodge of whirly lines, tiny fringed areas of varying density—dark where wave fronts from the object and reference beam arrived in phase and light where the wave fronts were out of phase. The hologram is a photographic pattern of microscopic interference fringes.

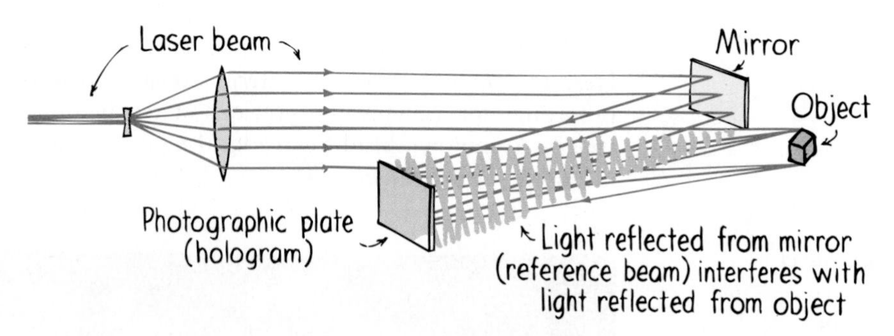

FIGURE 29.42

A simplified arrangement for making a hologram. The laser light that exposes the photographic plate consists of two parts: the reference beam reflected from the mirror and light reflected from the object. The wave fronts of these two parts interfere to produce microscopic fringes on the photographic plate. The exposed and developed plate is then a hologram.

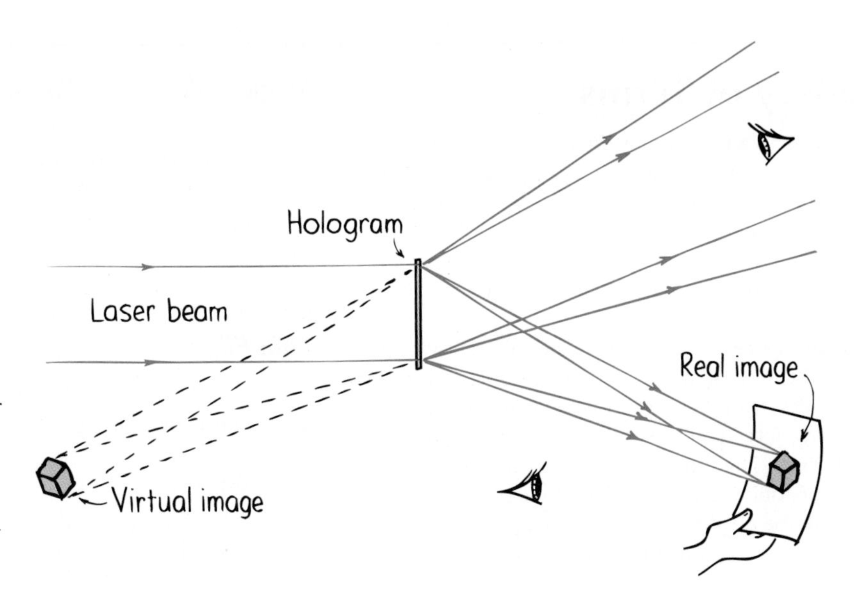

FIGURE 29.43
When laser light is transmitted through the hologram, the divergence of diffracted light produces a three-dimensional image that can be seen when looking *through* the hologram, like looking through a window. This is a *virtual image,* for it appears to be only in back of the hologram, like your virtual image in a mirror. By refocusing your eyes, you can see all parts of the virtual image, near and far, in sharp focus. Converging diffracted light produces a *real image* in front of the hologram that can be projected on a screen. Since the image is three-dimensional, you cannot see the entire image in sharp focus for any single position on a flat screen.

When a hologram is placed in a beam of coherent light, the light is diffracted by the microscopic fringes to produce wave fronts identical in form to the original wave fronts reflected by the object. When viewed with the eye or any other optical instrument, the diffracted wave fronts produce the same effect as the original reflected wave fronts. You look through the hologram and see a full, realistic three-dimensional image as if you were viewing the original object through a window. Depth is evident when you move your head and see down the sides of the object or when you lower your head and look underneath the object. Holographic pictures are extremely realistic.

Interestingly enough, if the hologram is made on film, you can cut it in half and still see the entire image on each half. And you can cut one of the pieces in half again, and yet again. This is because every part of the hologram has received and recorded light from the entire object. Similarly, light outside an open window fills the entire window, so you can see an outside view from every part of this open window. Because vast amounts of information are recorded in a tiny area, the film used for holograms must be much finer-grained than ordinary fine-grain photographic film. The optical storage of information via holograms is finding wide application in computers.

Even more interesting is holographic magnification. If holograms are made using a short-wavelength light and viewed with a longer-wavelength light, the resulting image is magnified in the same proportion as the wavelengths. Holograms made with X-rays would be magnified thousands of times when viewed with visible light and appropriate geometric viewing arrangements. Because holograms require no lenses, the possibilities of an X-ray microscope are particularly attractive.

And as for television, today's two-dimensional screens may appear as quaint to your children as your grandparents' early radio receivers appear to you now.

Light is fascinating—especially when it is diffracted by the interference fringes of a hologram.

Summary of Terms

Huygens' principle Every point on a wave front can be regarded as a new source of wavelets, which combine to produce the next wave front, whose points are sources of further wavelets, and so on.

Diffraction The bending of light that passes around an obstacle or through a narrow slit, causing the light to spread.

Interference The result of superposing different waves, usually of the same wavelength. Constructive interference results from crest-to-crest reinforcement; destructive interference results from crest-to-trough cancellation. The interference of selected wavelengths of light produces colors known as *interference colors.*

Polarization The alignment of the transverse electric vibrations of electromagnetic radiation. Such waves of aligned vibrations are said to be *polarized.*

Hologram A two-dimensional microscopic interference pattern that shows three-dimensional optical images.

Suggested Reading

Falk, D. S., D. R. Brill, and D. Stork, *Seeing the Light: Optics in Nature.* New York: Harper & Row, 1985.

Review Questions

Huygens' Principle

1. According to Huygens, how does every point on a wave front behave?
2. Will plane waves incident upon a small opening in a barrier fan out on the other side or continue as plane waves?

Diffraction

3. Is diffraction more pronounced through a small opening or through a large opening?
4. For an opening of a given size, is diffraction more pronounced for a longer wavelength or for a shorter wavelength?
5. Which is more easily diffracted around buildings, AM or FM radio waves? Why?

Interference

6. Is interference restricted to only some types of waves or does it occur for all types of waves?
7. What exactly did Thomas Young demonstrate in his famous experiment with light?

Single-Color Thin-Film Interference

8. What accounts for the light and dark bands produced when monochromatic light reflects from a glass pane atop another glass pane?
9. What is meant by saying a surface is *optically flat*?
10. What is the cause of Newton's rings?

Interference Colors by Reflection from Thin Films

11. What produces iridescence?
12. What causes the spectrum of colors seen in gasoline splotches on a wet street? Why are these not seen on a dry street?
13. What accounts for the different colors in either a soap bubble or a layer of gasoline on water?
14. Why are interference colors primarily cyan, magenta, and yellow?

Polarization

15. What phenomenon distinguishes longitudinal waves from transverse waves?
16. Is polarization characteristic of all types of waves?
17. How does the direction of polarization of light compare with the direction of vibration of the electron that produces it?
18. Why will light pass through a pair of Polaroids when the axes are aligned but not when the axes are at right angles to each other?
19. How much ordinary light will an ideal Polaroid transmit?
20. When *ordinary* light is incident at an oblique angle upon water, what can you say about the *reflected* light?

Three-Dimensional Viewing

21. Why would depth not be perceived if you viewed duplicates of ordinary slides in a stereo viewer (Figure 29.39), rather than the pairs of slides taken with a stereo camera?
22. What role do polarization filters play in a 3-D slide show?

Holography

23. How does a hologram differ from a conventional photograph?
24. How is *coherent* light different from ordinary light?
25. How can holographic magnification be accomplished?

Projects

1. With a razor blade, cut a slit in a card and look at a light source through it. You can vary the size of the opening by bending the card slightly. See the interference fringes? Try it with two closely spaced slits.
2. Next time you're in the bathtub, froth up the soapsuds and notice the colors of highlights from the illuminating light overhead on each tiny bubble. Notice that different bubbles reflect different colors, due to the different thicknesses of soap film. If a friend is bathing with you, compare the different colors that each of you see reflected from the same bubbles. You'll see that they're different—for what you see depends on your point of view!
3. When you're wearing Polaroid sunglasses, look at the glare from a nonmetallic surface, such as a road or body of water. Tip your head from side to side and see how the glare intensity changes as you vary the magnitude of the electric vector component aligned with the polarization axis of the glasses. Also notice the polarization of different parts of the sky when you hold the sunglasses in your hand and rotate them.
4. Place a source of white light on a table in front of you. Then place a sheet of Polaroid in front of the source, a bottle of corn syrup in front of the sheet, and a second sheet of Polaroid in front of the bottle. Look through the Polaroid sheets that sandwich the syrup and view spectacular colors as you rotate one of the sheets.

5. See spectacular interference colors with a polarized-light microscope. Any microscope, including an inexpensive toy microscope, can be converted into a polarized-light microscope by fitting a piece of Polaroid inside the eyepiece and taping another onto the stage of the microscope. Stretch various pieces of plastic wrap across a slide, rotate the eyepiece, and view impressive changes of colors.

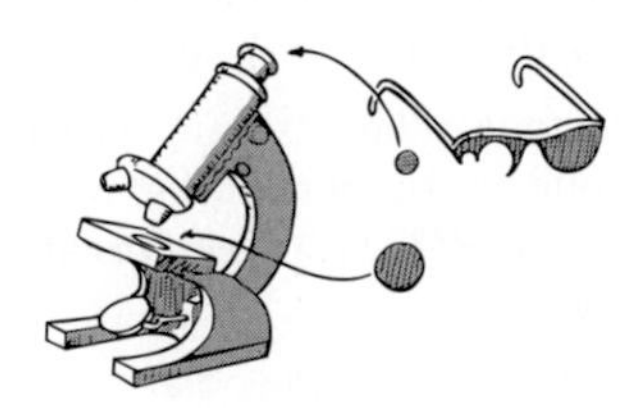

6. Make some slides for a slide projector by sticking some crumpled cellophane onto pieces of slide-sized Polaroid. (Also try strips of cellophane tape or plastic wrap overlapped at different angles.) Project the slides onto a large screen or white wall and rotate a second, slightly larger piece of Polaroid in front of the projector lens in rhythm with your favorite music. You'll have your own light show.

Exercises

1. Why can sunlight that illuminates the Earth be approximated by plane waves, whereas the light from a nearby lamp cannot?
2. In our everyday environment, diffraction is much more evident for sound waves than for light waves. Why is this so?
3. Why do *radio waves* diffract around buildings, while *light waves* do not?
4. Why are TV broadcasts in the VHF (very high frequency) range easier to receive in areas of marginal reception than broadcasts in the UHF (ultrahigh frequency) range? (Hint: UHF has higher frequencies than VHF.)
5. Can you think of a reason why TV channels of lower numbers might give clearer pictures in regions of poor TV reception? (*Hint:* Lower channel numbers represent lower carrier frequencies.)
6. The wavelengths of TV signals for the standard VHF Channels 2 through 13 range from about 5.6 down to about 1.4 meters. New high-definition TV signals are in the UHF band, with wavelengths considerably less than 1 meter. Will this shorter wavelength increase or decrease reception in "shadow areas" (assuming no cable)?
7. Two loudspeakers a meter or so apart emit pure tones of the same frequency and loudness. When a listener walks along a path parallel to the line that joins the loudspeakers, the sound is heard alternating from loud to soft. What is going on?
8. In the preceding exercise, suggest a path along which the listener could walk so as not to hear alternate loud and soft sounds.
9. How are interference fringes of light analogous to the varying intensity that you hear as you walk past a pair of speakers emitting the same sound?
10. By how much should a pair of light rays from a common source differ in length to produce destructive interference?
11. Light illuminates two closely spaced thin slits and produces an interference pattern on a screen behind.

How will the distance between the fringes of the pattern differ for yellow light and green light?

12. A double-slit arrangement produces interference fringes for yellow sodium light. To produce narrower-spaced fringes, should red light or blue light be used?
13. When white light diffracts upon passing through a thin slit, as in Figure 29.8*b*, different color components diffract by different amounts, so that a rainbow of colors appears at the edge of the pattern. Which color is diffracted through the greatest angle? Which color through the smallest angle?
14. Which will give wider-spaced fringes in a double-slit experiment, red light or violet light? (Let Figure 29.19 guide your thinking.)
15. Which will give wider-spaced fringes, a double-slit experiment in air or in water? (Let Figure 29.19 guide your thinking.)
16. If the path-length difference between two identical and coherent beams is two wavelengths when they arrive on a screen, will they produce a dark or a bright spot?
17. Which will produce wider fringes of light when passed through a diffraction grating—light from a red laser or light from a green laser?
18. When the reflected path from one surface of a thin film is one full wavelength different in length from the reflected path from the other surface and no phase change occurs, will the result be destructive interference or constructive interference?
19. When the reflected path from one surface of a thin film is one-half wavelength different in length from the reflected path from the other surface and no phase change occurs, will the result be destructive interference or constructive interference?
20. When the reflected path from one surface of a thin film is one-half wavelength different in length from the reflected path from the other surface and a 180° phase change occurs, why will the film appear black?
21. A pattern of fringes is produced when monochromatic light passes through a pair of thin slits. Would such a pattern be produced by three parallel thin slits? By thousands of such slits? Give an example to support your answers.
22. Suppose you place a diffraction grating in front of a camera lens and take a picture of illuminated streetlights. What will you expect to see in your photograph?
23. What happens to the distance between interference fringes if the separation between two slits is increased?
24. Why is Young's experiment more effective with slits than with the pinholes he first used?
25. In which of these is color formed by refraction—flower petals, rainbow, soap bubbles? By selective reflection? By thin-film interference?
26. The colors of peacocks and hummingbirds are the result not of pigments but of ridges in the surface layers of their feathers. By what physical principle do these ridges produce colors?
27. The colored wings of many butterflies are due to pigmentation, but in others, such as the Morpho butterfly, the colors do not result from any pigmentation. When the wing is viewed from different angles, the colors change. How are these colors produced?
28. Why do the iridescent colors seen in some seashells (such as abalone shells) change as the shells are viewed from various positions?
29. When dishes are not properly rinsed after washing, different colors are reflected from their surfaces. Explain.
30. Why are interference colors more apparent for thin films than for thick films?
31. Will the light from two very close stars produce an interference pattern? Explain.
32. If you notice the interference patterns of a thin film of oil or gasoline on water, you'll note that the colors form complete rings. How are these rings similar to the lines of equal elevation on a contour map?
33. Because of wave interference, a film of oil on water is seen to be yellow to observers directly above in an airplane. What color does it appear to a scuba diver directly below?
34. For the Hubble Space Telescope, which light—red, green, blue, or ultraviolet—is better for seeing fine detail of distant astronomical objects?
35. Polarized light is a part of nature, but polarized sound is not. Why?
36. The digital displays of watches and other devices are normally polarized. What related problem can occur when wearing polarized sunglasses?
37. Why will an ideal Polaroid filter transmit 50% of incident nonpolarized light?
38. Why may an ideal Polaroid filter transmit anything from zero to 100% of incident polarized light?
39. What percentage of light is transmitted by two ideal Polaroids, one on top of the other with their polarization axes aligned? With their axes at right angles to each other?
40. How can you determine the polarization axis for a single sheet of Polaroid?

41. Why do Polaroid sunglasses reduce glare, whereas nonpolarized sunglasses simply cut down the total amount of light reaching the eyes?
42. To remove the glare of light from a polished floor, should the axis of a Polaroid filter be horizontal or vertical?
43. Most of the glare from nonmetallic surfaces is polarized, the axis of polarization being parallel to that of the reflecting surface. Would you expect the polarization axis of Polaroid sunglasses to be horizontal or vertical? Why?
44. How can a single sheet of Polaroid film be used to show that the sky is partially polarized? (Interestingly enough, unlike humans, bees and many insects can discern polarized light and use this ability for navigation.)
45. Light will not pass through a pair of Polaroid sheets when they are aligned perpendicularly. However, if a third Polaroid is sandwiched between the two with its alignment halfway between the alignments of the other two (that is, with its axis making a 45° angle with each of the other two alignment axes), some light does get through. Why?
46. Why is it that, when you stand close to a painting, you get a greater sense of depth looking at it with one eye than with both eyes? (If you're not aware of this, look at paintings close up with one eye and see the difference.)
47. Why did practical holography have to await the advent of the laser?
48. How is magnification accomplished with holograms?
49. Which of these is most central to holography—interference, selective reflection, refraction, or all of these?
50. If you are viewing a hologram and you close one eye, will you still perceive depth? Explain.

Light Emission

George Curtis separates light from an argon source into its component frequencies with a spectroscope.

FIGURE 30.1
Simplified view of electrons orbiting in discrete shells about the nucleus of an atom.

If energy is pumped into a metal antenna in such a way that it causes free electrons to vibrate to and fro a few hundred thousand times per second, a radio wave is emitted. If the free electrons could be made to vibrate to and fro on the order of a million billion times per second, a visible light wave would be emitted. But light is not produced from metallic antennae, nor is it exclusively produced by atomic antennae via oscillations of electrons in atoms, as discussed in previous chapters. We now distinguish between light reflected, refracted, scattered, and diffracted by objects and light emitted by objects. In this chapter, we discuss the physics of light sources—of light *emission*.

The details of light emission from atoms involve the transitions of electrons from higher to lower energy states within the atom. This emission process can be understood in terms of the familiar planetary model of the atom that we discussed in Chapter 11. Just as each element is characterized by the number of electrons that occupy the shells surrounding its atomic nucleus, each element also possesses its own characteristic pattern of electron shells, or energy states. These states are found only at certain radii and energies. Because these states can have only certain energies, we say they are *discrete*. We call these discrete states *quantum states,* and we'll return to them in detail in the next two chapters. For now, we'll concern ourselves only with their role in light emission.

Excitation

An electron farther from the nucleus has a greater electric potential energy with respect to the nucleus than an electron nearer the nucleus. We say that the more distant electron is in a higher energy state, or, equivalently, at a higher energy level. In a sense, this is similar to the energy of a spring door or a pile driver. The wider the door is pulled open, the greater its spring potential energy; the higher the ram of a pile driver is raised, the greater its gravitational potential energy.

When an electron is in any way raised to a higher energy level, the atom is said to be *excited*. The electron's higher position is only momentary: Like the pushed-open spring door, it soon returns to its lowest energy state. The atom loses its temporarily acquired energy when the electron returns to a lower level

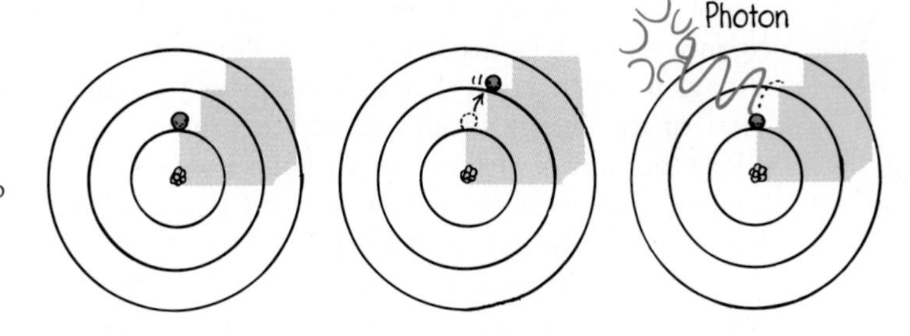

FIGURE 30.2
Interactive Figure

When an electron in an atom is boosted to a higher orbit, the atom is excited. When the electron returns to its original orbit, the atom de-excites and gives off a photon of light.

Exciting an atom is like trying to kick a ball out of a ditch. Many short kicks won't do the job, because the ball keeps falling back. A kick of just the right energy is enough to get the ball out of the ditch. Likewise with the excitation of atoms.

Insights

and emits radiant energy. The atom has undergone the process of **excitation** and *de-excitation.*

Just as each electrically neutral element has its own number of electrons, each element also has its own characteristic set of energy levels. Electrons dropping from higher to lower energy levels in an excited atom emit with each jump a throbbing pulse of electromagnetic radiation called a *photon,* the frequency of which is related to the energy transition of the jump. We think of this photon as a localized corpuscle of pure energy—a "particle" of light—which is ejected from the atom. The frequency of the photon is directly proportional to its energy. In shorthand notation,

$$E \sim f$$

When the proportionality constant h is introduced, this becomes the exact equation

$$E = hf$$

where h is Planck's constant (more about this in the next chapter). A photon in a beam of red light, for example, carries an amount of energy that corresponds to its frequency. Another photon of twice the frequency has twice as much energy and is found in the ultraviolet part of the spectrum. If many atoms in a material are excited, many photons with many frequencies are emitted that correspond to the many different levels excited. These frequencies correspond to characteristic colors of light from each chemical element.

The light emitted in the glass tubes in an advertising sign is a familiar consequence of excitation. The different colors in the sign correspond to the excitation of different gases, although it is common to refer to any of these as "neon." Only the red light is that of neon. At the ends of the glass tube that contains the neon gas are electrodes. Electrons are boiled off these electrodes and are jostled back and forth at high speeds by a high ac voltage. Millions of high-speed electrons vibrate back and forth inside the glass tube and smash into millions of target atoms, boosting orbital electrons into higher energy levels by an amount of energy equal to the decrease in kinetic energy of the bombarding electron. This energy is then radiated as the characteristic red light of neon when the electrons fall back to their stable orbits. The process occurs and recurs many times, as neon atoms continually undergo a cycle of excitation and de-excitation. The overall result of this process is the transformation of electrical energy into radiant energy.

The colors of various flames are due to excitation. Different atoms in the flame emit colors characteristic of their energy-level spacings. Common table salt placed in a flame, for example, produces the characteristic yellow of sodium. Every element, excited in a flame or otherwise, emits its own characteristic color or colors.

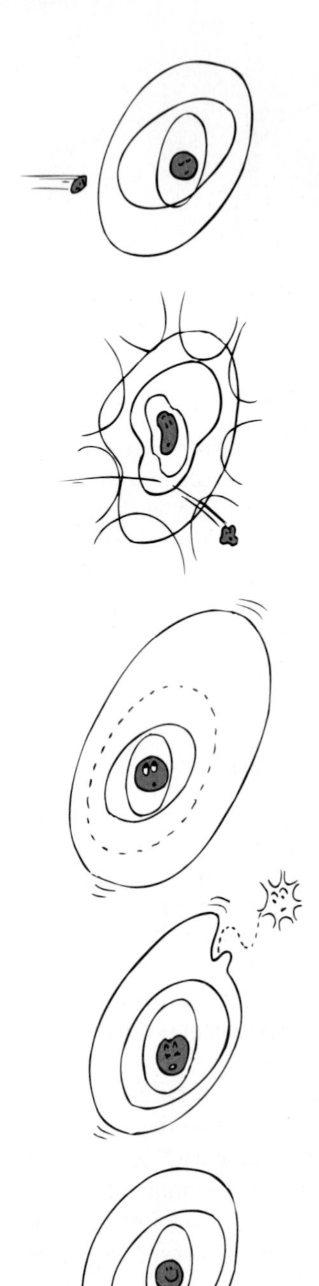

FIGURE 30.3
Excitation and de-excitation.

Street lamps provide another example. City streets are no longer illuminated by incandescent lamps but are now illuminated with the light emitted by gases such as mercury vapor. Not only is the light brighter, it is less expensive. Whereas most of the energy in an incandescent lamp is converted to heat, most of the energy put into a mercury-vapor lamp is converted to light. The light from these lamps is rich in blues and violets and therefore is a different "white" than the light from an incandescent lamp. See if your instructor has a spare prism or diffraction grating you can borrow. Look through the prism or grating at the light from a street lamp and see the discreteness of the colors, which indicates the discreteness of the atomic levels. Also note that the colors from different mercury-vapor lamps are identical, showing that the atoms of mercury are identical.

Excitation is illustrated in the aurora borealis. High-speed electrons that originate in the solar wind strike atoms and molecules in the upper atmosphere. They emit light exactly as occurs in a neon tube. The different colors in the aurora correspond to the excitation of different gases—oxygen atoms produce a greenish-white color, nitrogen molecules produce red-violet, and nitrogen ions produce a blue-violet color. Auroral emissions are not restricted to visible light; they also include infrared, ultraviolet, and X-ray radiation.

The excitation/de-excitation process can be accurately described only by quantum mechanics. An attempt to view the process in terms of classical physics runs into contradictions. Classically, an accelerating electric charge produces electromagnetic radiation. Does this explain light emission by excited atoms? An electron does accelerate in a transition from a higher to a lower energy level. Just as the innermost planets of the solar system have greater orbital speeds than those in the outermost orbits, the electrons in the innermost orbits of the atom have greater speeds. An electron gains speed in dropping to lower energy levels. Fine—the accelerating electron radiates a photon! But not so fine—the electron is continually undergoing acceleration (centripetal acceleration) in any orbit, whether or not it changes energy levels. According to classical physics, it should continually radiate energy. But it doesn't. All attempts to explain the emission of light by an excited atom in terms of a classical model have been unsuccessful. We shall simply say that light is emitted when an electron in an atom makes a "quantum jump" from a higher to a lower energy level and that the energy and frequency of the emitted photon are described by $E = hf$.

CHECK YOURSELF

Suppose a friend suggests that, for a first-rate operation, the gaseous neon atoms in a neon tube should be periodically replaced with fresh atoms because the energy of the atoms tends to be used up with continued excitation, producing dimmer and dimmer light. What do you say to this?

CHECK YOUR ANSWER

The neon atoms don't release any energy that is not given to them by the electric current in the tube and therefore don't get "used up." Any single atom may be excited and reexcited without limit. If the light is, in fact, becoming dimmer and dimmer, it is probably because a leak exists. Otherwise, there is no advantage whatsoever to changing the gas in the tube, because a "fresh" atom is indistinguishable from a "used" one. Both are ageless and older than the solar system.

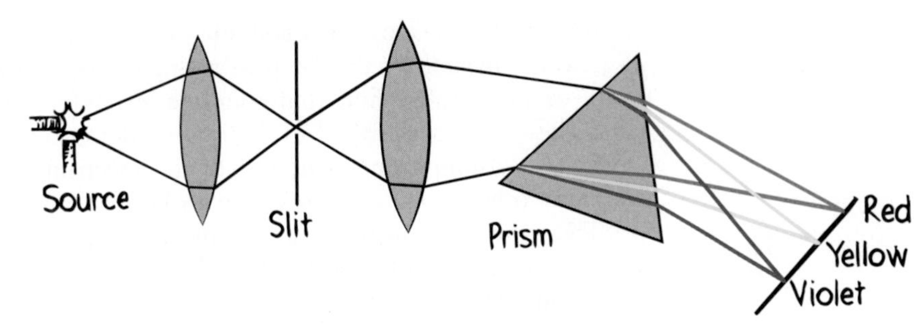

FIGURE 30.4
A simple spectroscope. Images of the illuminated slit are cast on a screen and make up a pattern of lines. The spectral pattern is characteristic of the light used to illuminate the slit.

Emission Spectra

Every element has its own characteristic pattern of electron energy levels and therefore emits light with its own characteristic pattern of frequencies, its **emission spectrum,** when excited. This pattern can be seen when light is passed through a prism—or, better, when it is first passed through a thin slit and then focused through a prism onto a viewing screen behind. Such an arrangement of slit, focusing lenses, and prism (or diffraction grating) is called a **spectroscope,** one of the most useful instruments of modern science (Figure 30.4).

Each component color is focused at a definite position, according to its frequency, and forms an image of the slit on the screen, photographic film, or appropriate detector. The different-colored images of the slit are called *spectral lines.* Some typical spectral patterns labeled by wavelengths are shown in Figure 30.5. It is customary to refer to colors in terms of their wavelengths rather than their frequencies. A given frequency corresponds to a definite wavelength.[1]

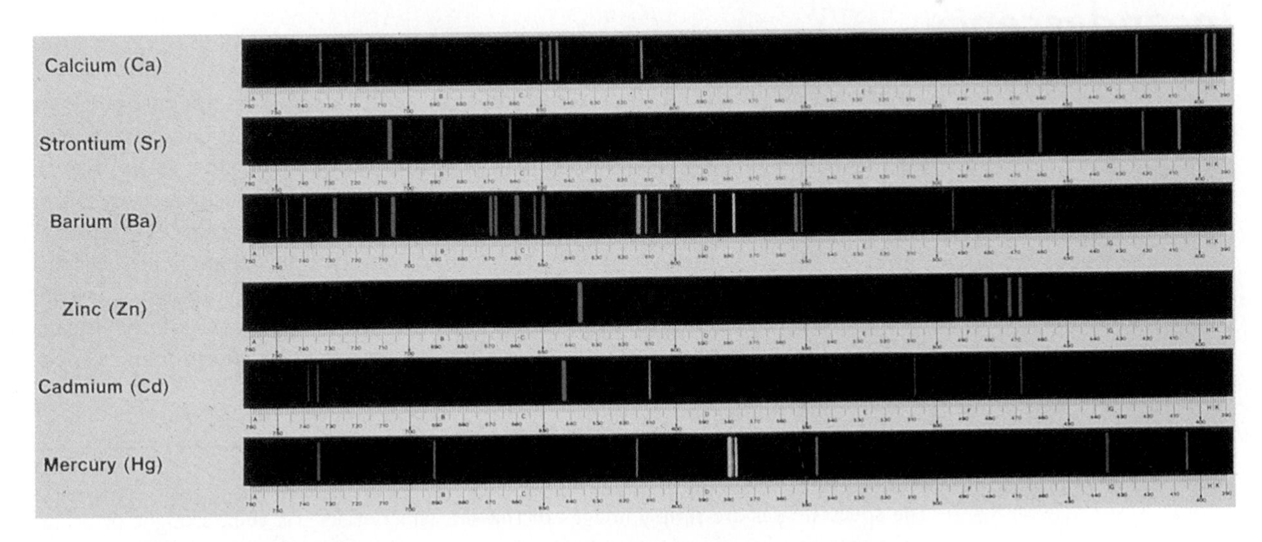

FIGURE 30.5 Interactive Figure
Spectral patterns of some elements.

[1]Recall, from Chapter 19, that $v = f\lambda$, where v is the wave speed, f is the wave frequency, and λ (lambda) is the wavelength. For light, v is the constant c, so we see from $c = f\lambda$ the relationship between frequency and wavelength—namely, $f = c/\lambda$ and $\lambda = c/f$.

If the light given off by a sodium-vapor lamp is analyzed in a spectroscope, a single yellow line predominates—a single image of the slit. If we narrow the width of the slit, we find that this line is really composed of two very close lines. These lines correspond to the two predominant frequencies of light emitted by excited sodium atoms. The rest of the spectrum looks dark. (Actually, there are many other lines, often too dim to be seen with the naked eye.)

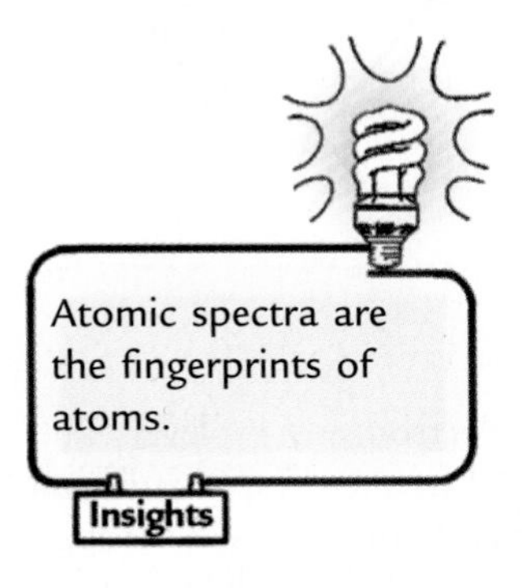

Atomic spectra are the fingerprints of atoms.
Insights

The same happens with all glowing vapors. The light from a mercury-vapor lamp shows a pair of bright yellow lines close together (but in different positions from those of sodium), a very intense green line, and several blue and violet lines. A neon tube produces a more complicated pattern of lines. We find that the light emitted by each element in the vapor phase produces its own characteristic pattern of lines. These lines correspond to the electron transitions between atomic energy levels and are as characteristic of each element as are fingerprints of people. The spectroscope, therefore, is widely used in chemical analysis.

The next time you see evidence of atomic excitation, perhaps the green flame produced when a piece of copper is placed in a fire, squint your eyes and see if you can imagine electrons jumping from one energy level to another in a pattern characteristic of the atom being excited—a pattern that displays a color unique to that atom. That's what's happening!

CHECK YOURSELF

Spectral patterns are not shapeless smears of light but, instead, consist of fine and distinct straight lines. Why is this so?

Incandescence

Light that is produced as a result of high temperature has the property of **incandescence** (from a Latin word meaning "to grow hot"). It can have a reddish tint, as from the heating element of a toaster, or a bluish tint, as from a particularly hot star. Or it can be white, as from the familiar incandescent lamp. What sets incandescent light apart from the light of a neon tube or mercury-vapor lamp is that it contains an infinite number of frequencies, spread smoothly across the spectrum. Does this mean that an infinite number of energy levels characterizes the tungsten atoms making up the filament of the incandescent lamp? The answer is no; if the filament were vaporized and then excited,

CHECK YOUR ANSWER

The spectral lines are simply images of the slit, which is itself a thin, straight opening through which light is admitted before being spread by the prism (or diffraction grating). When the slit is adjusted to make its most narrow opening, closely spaced lines can be resolved (distinguished from one another). A wider slit admits more light, which permits easier detection of dimmer radiant energy. But width is at the expense of resolution when closely spaced lines blur together.

FIGURE 30.6
The sound of an isolated bell rings with a clear and distinct frequency, whereas the sound emanating from a box of bells crowded together is discordant. Likewise with the difference between the light emitted from atoms in the gaseous state and that from atoms in the solid state.

the tungsten gas would emit light with a finite number of frequencies and produce an overall bluish color. Light emitted by atoms far from one another in the gaseous phase is quite different from the light emitted by the same atoms closely packed in the solid phase. This is analogous to the differences in sound from an isolated ringing bell and from a box crammed with ringing bells (Figure 30.6). In a gas, the atoms are far apart. Electrons undergo transitions between energy levels within the atom quite unaffected by the presence of neighboring atoms. But, when the atoms are closely packed, as in a solid, electrons of the outer orbits make transitions not only within the energy levels of their "parent" atoms but also between the levels of neighboring atoms. They bounce around over dimensions larger than a single atom, resulting in an infinite variety of transitions—hence the infinite number of radiant-energy frequencies.

As might be expected, incandescent light depends on temperature, because it is a form of thermal radiation. A plot of radiated energy over a wide range of frequencies for two different temperatures is shown in Figure 30.7. (Recall that we treated the radiation curve for sunlight back in Chapter 27 and discussed blackbody radiation in Chapter 16.) As the solid is heated further, more high-energy transitions occur, and higher-frequency radiation is emitted. In the brightest part of the spectrum, the predominant frequency of emitted radiation, the *peak frequency*, is directly proportional to the absolute temperature of the emitter:

$$\bar{f} \sim T$$

We use the bar above the *f* to indicate the peak frequency, for radiations of many frequencies are emitted from the incandescent source. If the temperature of an object (in kelvins) is doubled, the peak frequency of emitted radiation is doubled. The electromagnetic waves of violet light have nearly twice the frequency of red light waves. A violet-hot star, therefore, has nearly twice the surface temperature of a red-hot star.[2] The temperature of incandescent bodies, whether they be stars or blast-furnace interiors, can be determined by measuring the peak frequency (or color) of the radiant energy they emit.

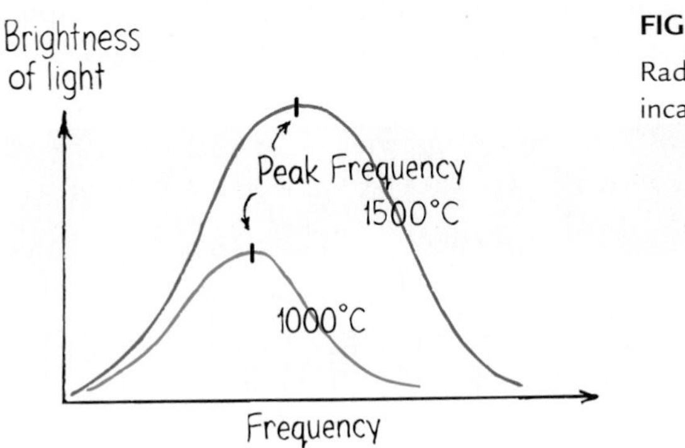

FIGURE 30.7 Interactive Figure
Radiation curves for an incandescent solid.

[2]If you study this topic further, you will find that the time rate at which an object radiates energy (the radiated power) is proportional to the fourth power of its Kelvin temperature. So a doubling of temperature corresponds to a doubling of the frequency of radiant energy but a sixteenfold increase in the rate of emission of radiant energy.

CHECK YOURSELF

From the radiation curves shown in Figure 30.7, which emits the higher average frequency of radiant energy—the 1000°C source or the 1500°C source? Which emits more radiant energy?

Absorption Spectra

When we view white light from an incandescent source with a spectroscope, we see a continuous spectrum over the whole rainbow of colors. If a gas is placed between the source and the spectroscope, however, careful inspection will show that the spectrum is not quite continuous. This is an **absorption spectrum,** and there are dark lines distributed throughout it; these dark lines against a rainbow-colored background are like emission lines in reverse. These are *absorption lines.*

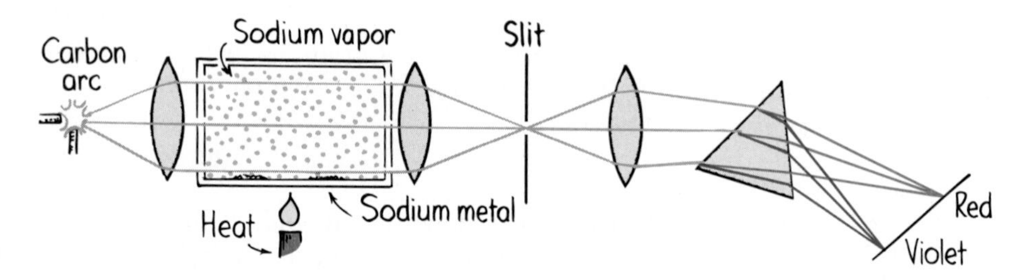

FIGURE 30.8
Interactive Figure
Experimental arrangement for demonstrating the absorption spectrum of a gas.

Atoms absorb light as well as emit light. An atom will most strongly absorb light having the frequencies to which it is tuned—some of the same frequencies it emits. When a beam of white light passes through a gas, the atoms of the gas absorb light of selected frequencies from the beam. This absorbed light is reradiated, but in *all* directions instead of only in the direction of the incident beam. When the light remaining in the beam spreads out into a spectrum, the frequencies that were absorbed show up as dark lines in the otherwise continuous spectrum. The positions of these dark lines correspond exactly to the positions of lines in an emission spectrum of the same gas (Figure 30.9).

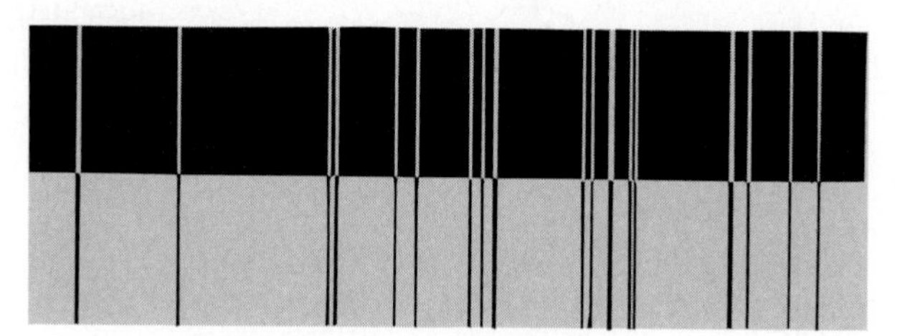

FIGURE 30.9
Emission and absorption spectra.

CHECK YOUR ANSWERS

The 1500°C radiating source emits the higher average frequencies, as noted by the extension of the curve to the right. The 1500°C source is brighter and also emits more radiant energy, as noted by its greater vertical displacement.

The Andromeda galaxy is approaching us, emitting light toward Earth that is blue-shifted.
Insights

Although the Sun is a source of incandescent light, the spectrum it produces, upon close examination, is not continuous. There are many absorption lines, called *Fraunhofer lines* in honor of the Bavarian optician and physicist Joseph von Fraunhofer, who first observed and mapped them accurately. Similar lines are found in the spectra produced by the stars. These lines indicate that the Sun and stars are each surrounded by an atmosphere of cooler gases that absorb some of the frequencies of light coming from the main body. Analysis of these lines reveals the chemical composition of the atmospheres of such sources. We find from these analyses that the stellar elements are the same elements that exist on Earth. An interesting sidelight is that, in 1868, spectroscopic analysis of sunlight showed some spectral lines different from any known on Earth. These lines identified a new element, which was named *helium,* after Helios, the Greek god of the Sun. Helium was discovered in the Sun before it was discovered on Earth. How about that!

We can determine the speed of stars by studying the spectra they emit. Just as a moving sound source produces a Doppler shift in its pitch (Chapter 19), a moving light source produces a Doppler shift in its light frequency. The frequency (not the speed!) of light emitted by an approaching source is higher, while the frequency of light from a receding source is lower, than the frequency of light from a stationary source. The corresponding spectral lines are displaced toward the red end of the spectrum for receding sources. Since the universe is expanding, almost all the galaxies show a red shift in their spectra.

We shall see, in Chapter 31, how the spectra of elements enable us to determine atomic structure.

Fluorescence

So we see that thermal agitation and bombardment by particles, such as high-speed electrons, are not the only means of imparting excitation energy to an atom. An atom may be excited by absorbing a photon of light. From the relationship $E = hf$, we see that high-frequency light, such as ultraviolet, which lies beyond the visible spectrum, delivers more energy per photon than lower-frequency light. Many substances undergo excitation when illuminated with ultraviolet light.

Many materials that are excited by ultraviolet light emit visible light upon de-excitation. The action of these materials is called **fluorescence.** In these materials, a photon of ultraviolet light excites the atom, boosting an electron to a higher energy state. In this upward quantum jump, the atom is likely to

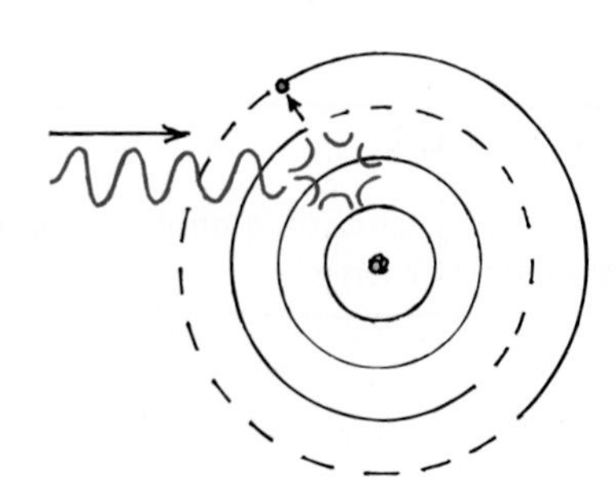

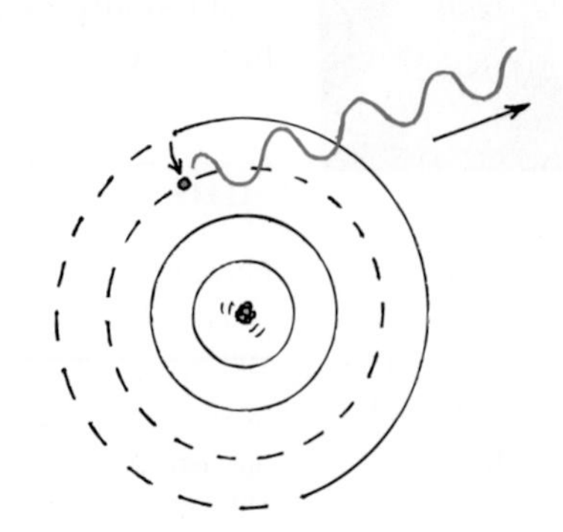

FIGURE 30.10

Interactive Figure

In fluorescence, the energy of the absorbed ultraviolet photon boosts the electron in an atom to a higher energy state. When the electron then returns to an intermediate state, the photon emitted is less energetic and therefore of a lower frequency than the ultraviolet photon.

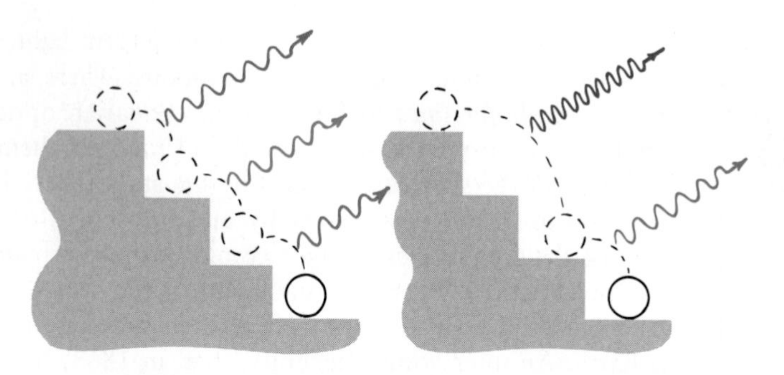

FIGURE 30.11
An excited atom may de-excite in several combinations of jumps.

FIGURE 30.12
Crayons fluorescing in various colors under ultraviolet light.

leapfrog over several intermediate energy states. So when the atom de-excites, it may make smaller jumps, emitting photons with less energy.

This excitation and de-excitation process is like leaping up a small staircase in a single bound, and then descending one or two steps at a time rather than leaping all the way down in a single bound. Since the photon energy released at each step is less than the total energy originally in the ultraviolet photon, lower-frequency photons are emitted. Hence, ultraviolet light shining on the material causes it to glow an overall red, yellow, or whatever color is characteristic of the material. Fluorescent dyes are used in paints and fabrics to make them glow when bombarded with ultraviolet photons in sunlight. These are the Day-Glo colors, which are spectacular when illuminated with an ultraviolet lamp.

CHECK YOURSELF

Why would it be impossible for a fluorescent material to emit ultraviolet light when illuminated by infrared light?

Detergents that make the claim of cleaning your clothes "whiter than white" use the principle of fluorescence. Such detergents contain a fluorescent dye that converts the ultraviolet light in sunlight into blue visible light, so clothes dyed in this way appear to reflect more blue light than they otherwise would. This makes the clothes appear whiter.[3]

The next time you visit a natural-science museum, go to the geology section and take in the exhibit of minerals illuminated with ultraviolet light (Figure 30.13). You'll notice that different minerals radiate various colors. This is to be expected because different minerals are composed of different elements, which, in turn, have

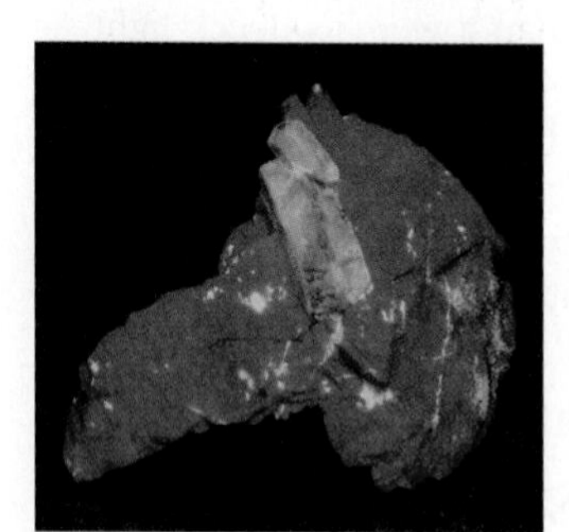

FIGURE 30.13
The rock contains the fluorescing minerals calcite and willemite, which, under ultraviolet light, are clearly seen as red and green, respectively.

CHECK YOUR ANSWER

Photon energy output would be greater than photon energy input, which would violate the law of conservation of energy.

[3]Interestingly enough, the same detergents marketed in Mexico and some other countries are adjusted for a rosier, warmer effect.

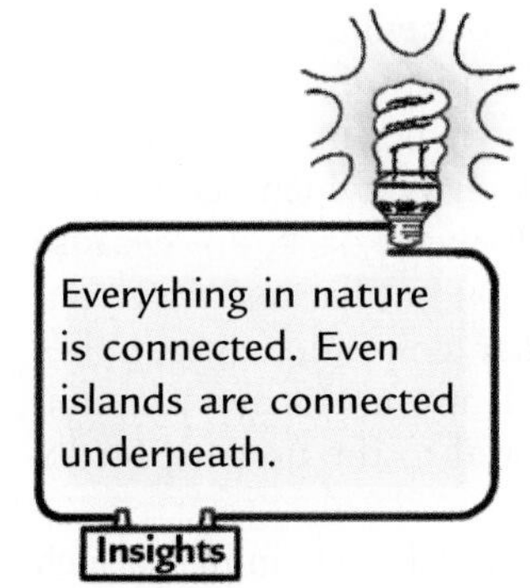

different sets of electron energy levels. Seeing the radiating minerals is a beautiful visual experience, which is even more fascinating when integrated with your knowledge of nature's submicroscopic happenings. High-energy ultraviolet photons strike the minerals, causing the excitation of atoms in the mineral structure. The frequencies of light that you see correspond to the tiny energy-level spacings as the energy cascades down. Every excited atom emits its characteristic frequencies, with no two different minerals giving off light of exactly the same color. Beauty is in both the eye and the mind of the beholder.

Fluorescent Lamps

The common fluorescent lamp consists of a cylindrical glass tube with electrodes at each end (Figure 30.14). In the lamp, as in the tube of a neon sign, electrons are boiled from one of the electrodes and forced to vibrate to and fro at high speeds within the tube by the ac voltage. The tube is filled with very low-pressure mercury vapor, which is excited by the impact of the high-speed electrons. Much of the emitted light is in the ultraviolet region. This is the primary excitation process. The secondary process occurs when the ultraviolet light strikes *phosphors,* a powdery material on the inner surface of the tube. The phosphors are excited by the absorption of the ultraviolet photons and fluoresce, giving off a multitude of lower-frequency photons that combine to produce white light. Different phosphors can be used to produce different colors or "textures" of light.

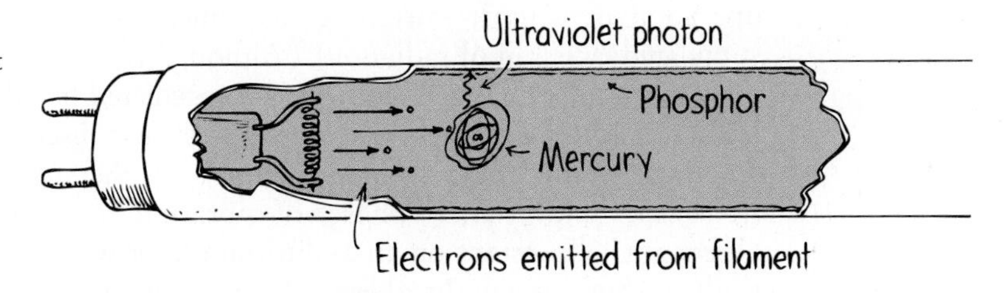

FIGURE 30.14
A fluorescent tube. Ultraviolet (UV) light is emitted by gas in the tube excited by an alternating electric current. The UV light, in turn, excites phosphors on the inner surface of the glass tube, which emit white light.

Phosphorescence

When excited, certain crystals as well as some large organic molecules remain in a state of excitation for a prolonged period of time. Unlike what occurs in fluorescent materials, their electrons are boosted into higher orbits and become "stuck." As a result, there is a time delay between the processes of excitation and de-excitation. Materials that exhibit this peculiar property are said to have **phosphorescence.** The element phosphorus, which is used in luminous clock dials and in other objects that are made to glow in the dark, is a good example. Atoms or molecules in these materials are excited by incident visible light. Rather than de-exciting immediately, as fluorescent materials do, many of the atoms remain in a *metastable state*—a prolonged state of excitation—sometimes as long as several hours, although most de-excite rather quickly. If the source of excitation is removed—for instance, if the lights are put out—an afterglow occurs while millions of atoms spontaneously undergo gradual de-excitation.

A TV screen is slightly phosphorescent, the glow decaying rather quickly, but just slow enough so that successive scans of the picture blend into one another. The afterglow of some phosphorescent light switches in the home may last more than an hour. Likewise for luminous clock dials, which are excited by visible light. Some older clock dials glow indefinitely in the dark, not because of a long time delay between excitation and de-excitation but because they contain radium or some other radioactive material that continuously supplies energy to keep the excitation process going. Such dials are no longer common because of the potential harm of the radioactive material to the user, especially if in a wristwatch or pocket watch.[4]

Many living creatures—from bacteria to fireflies and larger animals, such as jellyfish—chemically excite molecules in their bodies that give off light. We say that such living things are *bioluminescent*. Under some conditions, certain fish become luminescent when they swim but remain dark when still. Schools of these fish hang motionless and are not seen, but, when they are alarmed, they streak the depths with sudden light, creating a sort of deep-sea fireworks. The mechanism of bioluminescence is not well understood and is currently being researched.

Lasers

The phenomena of excitation, fluorescence, and phosphorescence underlie the operation of a most intriguing instrument, the **laser** (light amplification by stimulated emission of radiation).[5] Although the first laser was invented in 1958, the concept of stimulated emission was predicted by Albert Einstein in 1917. To understand how a laser operates, we must first discuss coherent light.

Light emitted by a common lamp is incoherent; that is, photons of many frequencies and in many phases of vibration are emitted. The light is as incoherent as the footsteps on an auditorium floor when a mob of people are chaotically rushing about. Incoherent light is chaotic. A beam of incoherent light spreads out after a short distance, becoming wider and wider and less intense with increased distance.

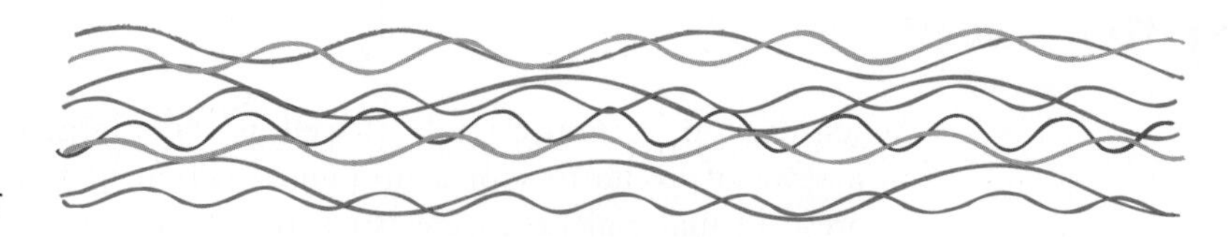

FIGURE 30.15
Incoherent white light contains waves of many frequencies (and of many wavelengths) that are out of phase with one another.

Even if the beam is filtered so that it consists of single-frequency waves (monochromatic light), it is still incoherent, because the waves are out of phase with one another. The beam spreads and becomes weaker with increased distance.

[4] A radioactive form of hydrogen called *tritium*, however, can serve to keep watch dials illuminated harmlessly. This is because its radiation doesn't have enough energy to penetrate the metal or plastic of the watch case.

[5] A word constructed from the initials of a phrase is called an *acronym*.

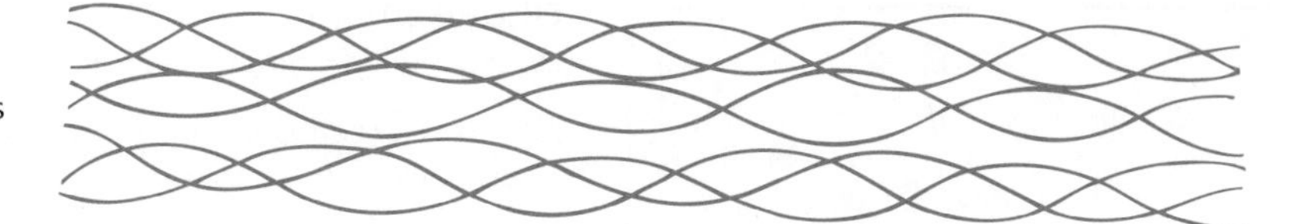

FIGURE 30.16
Light of a single frequency and wavelength still contains a mixture of phases.

A beam of photons having the same frequency, phase, and direction—that is, a beam of photons that are identical copies of one another—is said to be *coherent*. A beam of coherent light spreads and weakens very little.[6]

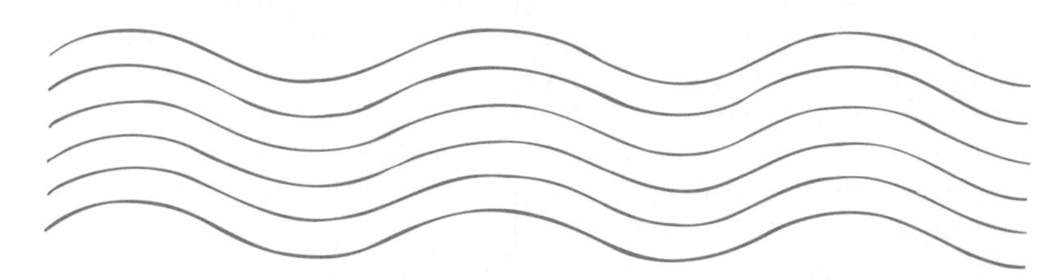

FIGURE 30.17
Coherent light: All the waves are identical and in phase.

A laser beam is not seen unless it scatters off something in the air. Like sunbeams or moonbeams, what you see are the particles in the scattering medium, not the beam itself. When the beam hits a diffuse surface, part of it is scattered toward your eye as a dot.

Insights

A laser is a device that produces a beam of coherent light. Every laser has a source of atoms called an active medium, which can be a gas, liquid, or solid (the first laser was a ruby crystal). The atoms in the medium are excited to metastable states by an external source of energy. When most of the atoms in the medium are excited, a single photon from an atom that undergoes de-excitation can start a chain reaction. This photon strikes another atom, stimulating it into emission, and so on, producing coherent light. Most of this light is initially moving in random directions. Light traveling along the laser axis, however, is reflected from mirrors coated to reflect light of the desired wavelength selectively. One mirror is totally reflecting, while the other is partially reflecting. The reflected waves reinforce each other after each round-trip reflection between the mirrors, thereby setting up a to-and-fro resonance condition wherein the light builds up to an appreciable intensity. The light that escapes through the more transparent-mirrored end makes up the laser beam.

In addition to gas and crystal lasers, other types have joined the laser family: glass, chemical, liquid, and semiconductor lasers. Present models produce beams ranging from infrared through ultraviolet. Some models can be tuned to various frequency ranges. Most exciting is the prospect of an X-ray laser beam.

The laser is not a source of energy. It is simply a converter of energy that takes advantage of the process of stimulated emission to concentrate a certain fraction of its energy (commonly 1%) into radiant energy of a single frequency moving in a single direction. Like all devices, a laser can put out no more energy than is put into it.

Lasers have found wide use in surgery. They cut cleanly. Laser light can be intense and concentrated enough to enable eye surgeons to "weld" detached retinas back into place without making an incision. The light is simply brought to focus in the region where the welding is to take place.

Whereas radio wavelengths span hundreds of meters and television waves span many centimeters, wavelengths of laser light are measured in millionths of a centimeter. Correspondingly, laser-light frequencies are vastly greater than radio or television frequencies. As a result, laser light can carry an enormous number

[6]The narrowness of a laser beam is evident when you see a lecturer produce a tiny red spot on a screen using a laser "pointer." Light from an intense laser pointed at the Moon has been reflected and detected back on Earth.

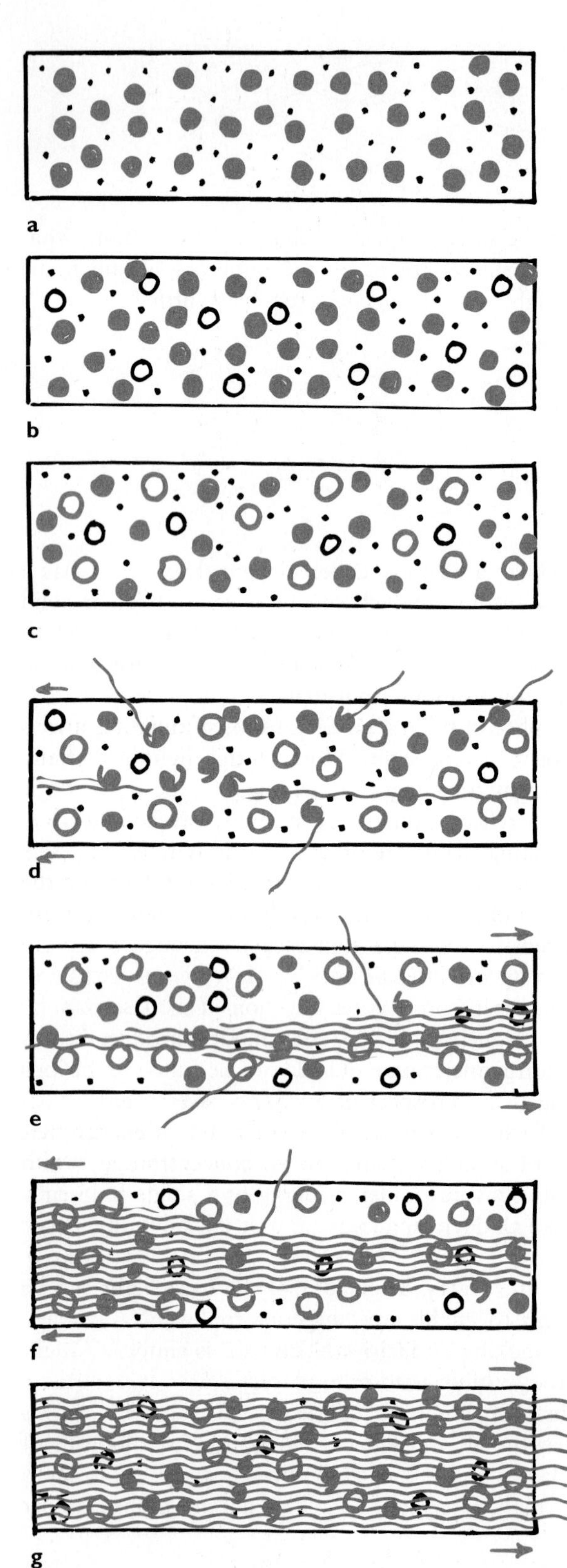

FIGURE 30.18
Laser action in a helium–neon laser.

(a) The laser consists of a narrow Pyrex tube that contains a low-pressure gas mixture consisting of 85% helium (small black dots) and 15% neon (large colored dots).

(b) When a high-voltage current zaps through the tube, it excites both helium and neon atoms to their usual higher states, and they immediately undergo de-excitation, except for one state in the helium that is characterized by a prolonged delay before de-excitation—a *metastable state*. Since this state is relatively stable, a sizable population of excited helium atoms (black open circles) is built up. These atoms wander about in the tube and act as an energy source for neon, which has an otherwise hard-to-come-by metastable state very close to the energy of the excited helium.

(c) When excited helium atoms collide with neon atoms in their lowest energy state (ground state), the helium gives up its energy to the neon, which is boosted to its metastable state (red open circles). The process continues, and the population of excited neon atoms soon outnumbers neon atoms in a lower-energy excited state. This inverted population is, in effect, waiting to radiate its energy.

(d) Some neon atoms eventually de-excite and radiate red photons in the tube. When this radiant energy passes other excited neon atoms, the latter are stimulated into emitting photons exactly in phase with the radiant energy that stimulated the emission. Photons pass out of the tube in irregular directions, giving it a red glow.

(e) Photons moving parallel to the axis of the tube are reflected from specially coated parallel mirrors at the ends of the tube. The reflected photons stimulate the emission of photons from other neon atoms, thereby producing an avalanche of photons having the same frequency, phase, and direction.

(f) The photons flash to and fro between the mirrors, becoming amplified with each pass.

(g) Some photons "leak" out of one of the mirrors, which is only partially reflecting. These make up the laser beam.

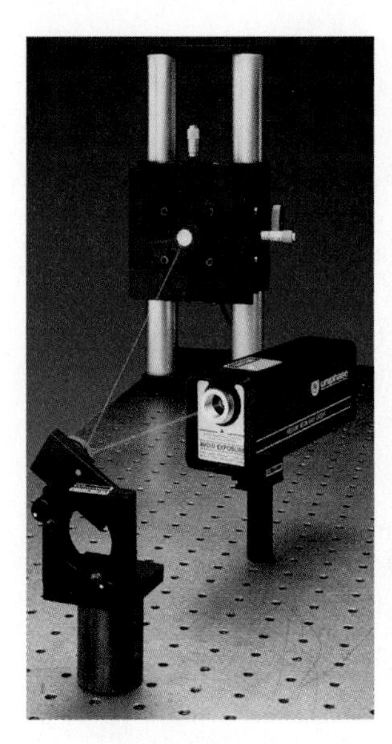

FIGURE 30.19
(left) A helium-neon laser. (right) Lasers are common tools in most school laboratories.

FIGURE 30.20
A textbook's unique identification is contained in the bar code that appears on the back cover.

of messages bunched into a very narrow band of frequencies. Communications can be carried in a laser beam directed through space, through the atmosphere, or through optical fibers (light pipes) that can be bent like cables.

The laser is at work at supermarket checkout counters, where code-reading machines scan the universal product code (UPC) symbol printed on packages and on the back cover of this book (Figure 30.20). Laser light is reflected from the bars and spaces and converted to an electric signal as the symbol is scanned. The signal rises to a high value when reflected from a bright space and falls to a low value when reflected from a dark bar. The information on the thickness and spacing of the bars is "digitized" (converted into the ones and zeros of binary code) and processed by a computer.

Surveyors use reflected laser light to measure distance, and home hobbyists and carpenters use them for levels. Environmental scientists use lasers to measure and detect pollutants in exhaust gases. Different gases absorb light at characteristic wavelengths and leave their "fingerprints" on a reflected beam of laser light. The specific wavelength and amount of light absorbed are analyzed by a computer, which produces an immediate tabulation of the pollutants.

Lasers have ushered in a whole new technology, the promise of which we continually tap. The future for laser applications seems unlimited.

Summary of Terms

Excitation The process of boosting one or more electrons in an atom or molecule from a lower to a higher energy level. An atom in an excited state will usually decay (de-excite) rapidly to a lower state by the emission of a photon. The energy of the photon is proportional to its frequency: $E = hf$.

Emission spectrum The distribution of wavelengths in the light from a luminous source.

Spectroscope An optical instrument that separates light into its constituent wavelengths in the form of spectral lines.

Incandescence The state of glowing while at a high temperature, caused by electrons bouncing around over dimensions larger than the size of an atom, emitting radiant energy in the process. The peak frequency of radiant energy is proportional to the absolute temperature of the heated substance:

$$\bar{f} \sim T$$

Absorption spectrum A continuous spectrum, like that of white light, interrupted by dark lines or bands that result from the absorption of light of certain frequencies by a substance through which the radiant energy passes.

Fluorescence The property of certain substances to absorb radiation of one frequency and to reemit radiation of lower frequency. It occurs when an atom is boosted up to an excited state and loses its energy in two or more downward jumps to a lower energy state.

Phosphorescence A type of light emission that is the same as fluorescence except for a delay between excitation and de-excitation, which provides an afterglow. The delay is caused by atoms being excited to energy states that do not decay rapidly. The afterglow may last from fractions of a second to hours or even days, depending on the type of material, temperature, and other factors.

Laser (**l**ight **a**mplification by **s**timulated **e**mission of **r**adiation) An optical instrument that produces a beam of coherent monochromatic light.

Review Questions

1. If electrons are made to vibrate to and fro at a few hundred thousand hertz, radio waves are emitted. What class of waves are emitted when electrons are made to vibrate to and fro at a few million billion hertz?
2. What does it mean to say an energy state is *discrete*?

Excitation

3. Which has the greater potential energy with respect to the atomic nucleus, electrons in inner electron shells or electrons in outer electron shells?
4. What does it mean to say an atom is *excited*?
5. What is the name given to a single throbbing corpuscle of electromagnetic radiation?
6. What is the relationship between the *difference in energy* between energy levels and the *energy of the photon* that is emitted by a transition between those levels?
7. How is the *energy* of a photon related to its vibrational frequency?
8. Which has the higher *frequency*, red or blue light? Which has the greater *energy* per photon, red or blue light?
9. An electron loses some of its kinetic energy when it bombards a neon atom in a glass tube. What becomes of this energy?
10. Can a neon atom in a glass tube be excited more than once?
11. What do the various colors displayed in the flame of a burning log represent?
12. Which converts the greater percentage of its energy into heat, an incandescent lamp or a mercury-vapor lamp?

Emission Spectra

13. What pattern does every element have that enables each to emit its own characteristic colors of light?
14. What is a *spectroscope*, and what does it accomplish?
15. Why do the colors of light in a spectroscope appear as lines?

Incandescence

16. What is the "color" of all the visible light frequencies mixed together equally?
17. When a gas glows, discrete colors are emitted. When a solid glows, the colors are smudged. Why?
18. How is the peak frequency of emitted light related to the temperature of its incandescent source?

Absorption Spectra

19. How does an absorption spectrum differ in appearance from an emission spectrum?
20. What are Fraunhofer lines?
21. How was helium discovered?
22. How can astrophysicists tell whether a star is receding or approaching Earth?

Fluorescence

23. Name three ways in which atoms can be excited.
24. Why is ultraviolet light, but not infrared light, effective in making certain materials fluoresce?

Fluorescent Lamps

25. Distinguish between the primary and secondary excitation processes that occur in a fluorescent lamp.
26. What enables a fluorescent lamp to give off white light?

Phosphorescence

27. Distinguish between *fluorescence* and *phosphorescence*.
28. What is responsible for the afterglow of phosphorescent materials?
29. What is a *metastable state*?

Lasers

30. Distinguish between *monochromatic light* and *coherent light*.

31. How does the avalanche of photons in a laser beam differ from the hordes of photons emitted by an incandescent lamp?

32. A friend speculates that scientists in a certain country have developed a laser that puts out far more energy than is put into it. Your friend asks for your response to this speculation. What is your response?

Projects

1. Write a letter to Grandma to explain how light is emitted from lamps, flames, and lasers. Tell her why fluorescent dyes and paints are so impressively vivid when illuminated with an ultraviolet lamp. Go on to tell her about the similarities and differences between fluorescence and phosphorescence.

2. Borrow a diffraction grating from your physics instructor. The common kind looks like a photographic slide, and light passing through it or reflecting from it is diffracted into its component colors by thousands of finely ruled lines. Look through the grating at the light from a sodium-vapor street lamp. If it's a low-pressure lamp, you'll see the nice yellow spectral "line" that dominates sodium light (actually, it's two closely spaced lines). If the street lamp is round, you'll see circles instead of lines; if you look through a slit cut in cardboard or some similar material, you'll see lines. What happens with the now-common high-pressure sodium lamps is more interesting. Because of the collisions of excited atoms, you'll see a smeared-out spectrum that is nearly continuous, almost like that of an incandescent lamp. Right at the yellow location, where you'd expect to see the sodium line, is a dark area. This is the sodium absorption band. It is due to the cooler sodium, which surrounds the high-pressure emission region. You should view this a block or so away so that the line, or circle, is small enough to allow the resolution to be maintained. Try this. It is very easy to see!

Exercises

1. Why is a gamma-ray photon more energetic than an X-ray photon?

2. Have you ever watched a fire and noticed that the burning of various materials often produces flames of different colors? Why is this so?

3. Green light is emitted when electrons in a substance make a particular energy-level transition. If blue light were instead emitted from the same substance, would it correspond to a greater or lesser change of energy in the atom?

4. Ultraviolet light causes sunburns, whereas visible light, even of greater intensity, does not. Why is this so?

5. If we double the frequency of light, we double the energy of each of its photons. If we instead double the wavelength of light, what happens to the photon energy?

6. Why doesn't a neon sign finally "run out" of excited atoms and produce dimmer and dimmer light?

7. An investigator wishes spectral lines in a spectrum to be thin crescents. What change in the spectroscope will accomplish this?

8. If light were passed through a round hole instead of a thin slit in a spectroscope, how would the spectral "lines" appear? What is the drawback of a hole in comparison with a slit?

9. If we use a prism or a diffraction grating to compare the red light from a common neon tube and the red light from a helium–neon laser, what striking difference do we see?

10. What is the evidence for the claim that iron exists in the relatively cool outer layer of the Sun?

11. How might the Fraunhofer lines in the spectrum of sunlight that are due to absorption in the Sun's atmosphere be distinguished from those due to absorption by gases in the Earth's atmosphere?

12. In what specific way does light from distant stars and galaxies tell astronomers that atoms throughout the universe have the same properties as those on Earth?

13. What difference does an astronomer see between the emission spectrum of an element in a receding star and a spectrum of the same element in the lab? (*Hint:* This relates to information in Chapter 19.)

14. A blue-hot star is about twice as hot as a red-hot star. But the temperatures of the gases in advertising signs are about the same, whether they emit red or blue light. What is your explanation?

15. Which has the greatest energy—a photon of infrared light, of visible light, or of ultraviolet light?

16. Does atomic excitation occur in solids as well as in gases? How does the radiant energy from an incandescent solid differ from the radiant energy emitted by an excited gas?

17. Low-pressure sodium-vapor lamps emit line spectra with well-defined wavelengths, but high-pressure sodium-vapor lamps emit light whose lines are more spread out. Relate this to the continuous smear of wavelengths emitted by solids.

18. A lamp filament is made of tungsten. Why do we get a continuous spectrum rather than a tungsten line

spectrum when light from an incandescent lamp is viewed with a spectroscope?

19. How can a hydrogen atom, which has only one electron, have so many spectral lines?

20. Since an absorbing gas reemits the light it absorbs, why are there dark lines in an absorption spectrum? That is, why doesn't the reemitted light simply fill in the dark places?

21. If atoms of a substance absorb ultraviolet light and emit red light, what becomes of the "missing" energy?

22. (a) Light from an incandescent source is passed through sodium vapor and then examined with a spectroscope. What is the appearance of the spectrum? (b) The incandescent source is switched off and the sodium is heated until it glows. How does the spectrum of the glowing sodium compare with the previously observed spectrum?

23. Your friend reasons that, if ultraviolet light can activate the process of *fluorescence*, infrared light ought to also. Your friend looks to you for approval or disapproval of this idea. What is your position?

24. When ultraviolet light falls on certain dyes, visible light is emitted. Why does this not happen when infrared light falls on these dyes?

25. Why are fabrics that fluoresce when exposed to ultraviolet light so bright in sunlight?

26. Why do different fluorescent minerals emit different colors when illuminated with ultraviolet light?

27. How is pushing open a screen door with a spring similar to the process of phosphorescence?

28. When a certain material is illuminated with visible light, electrons jump from lower to higher energy states in atoms of the material. When illuminated by ultraviolet light, atoms are ionized as some of them eject electrons. Why do the two kinds of illumination produce such different results?

29. The forerunner to the laser involved microwaves rather than visible light. What does *maser* mean?

30. The first laser consisted of a red ruby rod activated by a photoflash tube that emitted green light. Why would a laser composed of a green crystal rod and a photoflash tube that emits red light not work?

31. A laboratory laser has a power of only 0.8 mW (8×10^{-4} W). Why does it seem more powerful than light from a 100-W lamp?

32. How do the avalanches of photons in a laser beam differ from the hordes of photons emitted by an incandescent lamp?

33. In the operation of a helium–neon laser, why is it important that the metastable state of helium be relatively long-lived? (What would be the effect of this state de-exciting too rapidly?) (Refer to Figure 30.18.)

34. In the operation of a helium–neon laser, why is it important that the metastable state in the helium atom closely match the energy level of a more-difficult-to-come-by metastable state in neon?

35. A friend speculates that scientists in a certain country have developed a laser that produces far more energy than is put into it. Your friend asks for your response to this speculation. What is your response?

36. A laser cannot produce more energy than is put into it. A laser can, however, produce pulses of light with more power output than the power input required to run the laser. Explain.

37. In the equation $\bar{f} \sim T$, what do the symbols $\bar{f}$ and T represent?

38. We know that a lamp filament at 2500 K radiates white light. Does the lamp filament also radiate energy when it is at room temperature?

39. We know that the Sun radiates energy. Does the Earth similarly radiate energy? If so, what is different about their radiations?

40. Since every object has some temperature, every object radiates energy. Why, then, can't we see objects in the dark?

41. If we continue heating a piece of initially room-temperature metal in a dark room, it will begin to glow visibly. What will be its first visible color, and why?

42. We can heat a piece of metal to red hot and then to white hot. Can we heat it until the metal glows blue hot?

43. How do the surface temperatures of reddish, bluish, and whitish stars compare?

44. If you see a red-hot star, you can be certain that its peak intensity is in the infrared region. Why is this? And if you see a "violet-hot" star, you can be certain its peak intensity is in the ultraviolet range. Why is this?

45. We perceive a "green-hot" star not as green but as white. Why? (*Hint:* Consider the radiation curve back in Figure 27.7.)

46. Part *a* in the sketch in the next column shows a radiation curve of an incandescent solid and its spectral pattern as produced with a spectroscope. Part *b* shows the "radiation curve" of an excited gas and its emission spectral pattern. Part *c* shows the curve produced when a cool gas is between an incandescent source and the viewer; the corresponding spectral pattern is left as an exercise for you to construct. Part *d* shows the spectral pattern of an incandescent source as seen through a piece of green glass; you are to sketch in the corresponding radiation curve.

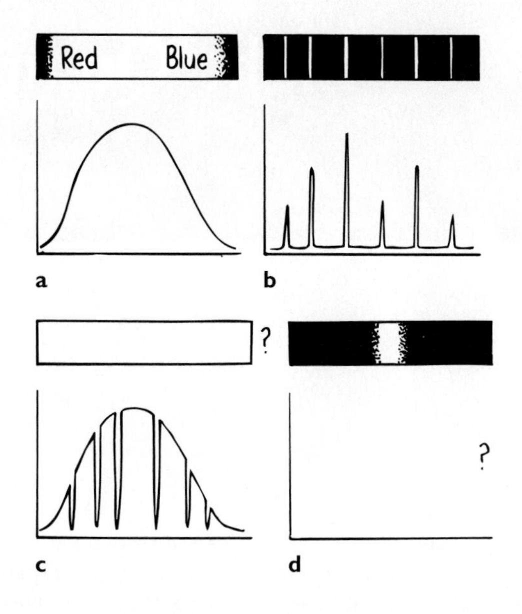

47. Consider just four of the energy levels in a certain atom, as shown in the diagram. How many spectral lines will result from all possible transitions among these levels? Which transition corresponds to the highest-frequency light emitted? To the lowest-frequency light emitted?

n = 4
n = 3
n = 2

n = 1

48. An electron de-excites from the fourth quantum level in the diagram above to the third and then directly to the ground state. Two photons are emitted. How does the sum of their frequencies compare with the frequency of the single photon that would be emitted by de-excitation from the fourth level directly to the ground state?

49. For the transitions described in the previous exercise, is there any relationship among the wavelengths of the emitted photons?

50. Suppose the four energy levels in Exercise 47 were somehow evenly spaced. How many spectral lines would result?

Problem

In the diagram, the energy difference between states A and B is twice the energy difference between states B and C. In a transition (quantum jump) from C to B, an electron emits a photon of wavelength 600 nm. (a) What is the wavelength emitted when the photon jumps from B to A? (b) When it jumps from C to A?

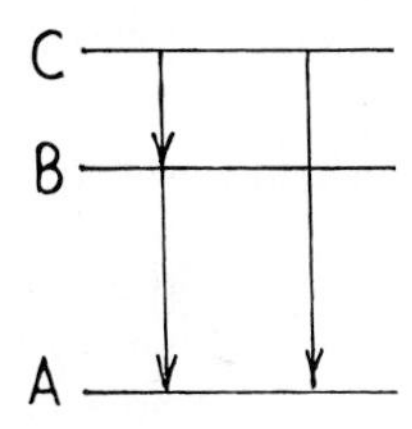

CHAPTER 31

Light Quanta

Phil Wolf, co-author of Problem Solving in Conceptual Physics, *demonstrates the photoelectric effect by directing light of different frequencies onto a photocell and measuring the energies of ejected electrons.*

The classical physics that we have so far studied deals with two categories of phenomena: particles and waves. According to our everyday experience, "particles" are tiny objects like bullets. They have mass and they obey Newton's laws—they *travel* through space in straight lines unless a force acts upon them. Likewise, according to our everyday experience, "waves," like waves in the ocean, are phenomena that *extend* in space. When a wave travels through an opening or around a barrier, the wave diffracts and different parts of the wave interfere. Therefore, particles and waves are easy to distinguish from each other. In fact, they have properties that are mutually exclusive. Nonetheless, the question of how to classify light was a mystery for centuries.

One of the early theories about the nature of light is that of Plato, who lived in the fifth and fourth centuries BC. Plato thought that light consisted of streamers emitted by the eye. Euclid, who lived roughly a century later, also held this view. On the other hand, the Pythagoreans believed that light emanated from luminous bodies in the form of very fine particles, while Empedocles, a predecessor of Plato, taught that light is composed of high-speed waves of some sort. For more than 2000 years, the questions remained unanswered. Does light consist of waves or particles?

In 1704, Isaac Newton described light as a stream of particles or corpuscles. He held this view despite his knowledge of what we now call polarization and despite his experiment with light reflecting from glass plates, in which he noticed fringes of brightness and darkness (Newton's rings). He knew that his particles of light had to have certain wave properties too. Christian Huygens, a contemporary of Newton, advocated a wave theory of light.

With all this history as background, Thomas Young, in 1801, performed the "double-slit experiment," which seemed to prove, finally, that light is a wave phenomenon. This view was reinforced in 1862 by Maxwell's prediction that light carries energy in oscillating electric and magnetic fields. Twenty-five years later, Heinrich Hertz used sparking electric circuits to demonstrate the reality of electromagnetic waves (of radio frequency). In 1905, however, Albert Einstein published a Nobel Prize–winning paper that challenged the wave theory of light by arguing that light interacts with matter, not in continuous waves, as Maxwell envisioned, but in tiny packets of energy that we now call *photons*. This discovery didn't wipe out light waves. It revealed, instead, that light is both wave *and* particle. In this chapter, we shall visit the world of the very small and look into some of the strange and exciting aspects of quantum reality.

Birth of the Quantum Theory

Max Planck (1858–1947)

As the twentieth century arrived, new technologies had reached a level that enabled scientists to design experiments to explore the behavior of very small particles. With the discovery of the electron in 1897 and the investigation of radioactivity at about the same time, experimenters began to probe the atomic structure of matter. In 1900, the German theoretical physicist Max Planck hypothesized that warm bodies emit radiant energy in discrete bundles. Each of these bundles he called a **quantum.** According to Planck, the energy in each energy bundle is proportional to the frequency of radiation. His hypothesis began a revolution of ideas that has completely changed the way we think about the physical world. We will see that the rules we apply to the everyday macroworld, the Newtonian laws that work so well for large objects, such as baseballs and planets, simply don't apply to events in the microworld of the atom. In the macroworld, the study of motion is called *mechanics;* in the microworld, where different rules hold sway, the study of motion is *quantum mechanics.* More broadly, the body of laws developed from 1900 to the late 1920s that describe all quantum phenomena of the microworld is known as **quantum physics.**

Quantization and Planck's Constant

Quantization, the idea that the natural world is granular rather than smoothly continuous, is certainly not a new idea to physics. Matter is quantized; the mass of a brick of gold, for example, is equal to some whole-number multiple of the mass of a single gold atom. Electricity is quantized, as electric charge is always some whole-number multiple of the charge of a single electron.

Quantum physics states that, in the microworld of the atom, the amount of energy in any system is quantized—not all values of energy are possible. This is analogous to saying a campfire can only be so hot. It might burn at 450°C or it might burn at 451°C, but in no way can it burn at 450.5°C. Believe it? Well, you shouldn't, for as far as our macroscopic thermometers can measure, a campfire can burn at any temperature as long as it's above the minimum temperature that is required for combustion. But the energy of the campfire, interestingly enough, is the composite energy of a great number and a great variety of elemental units of energy. A simpler example is the energy in a beam of laser light, which is a whole-number multiple of a single lowest value of energy—one quantum. The quanta of light, and of electromagnetic radiation in general, are the photons. (The plural of quantum is *quanta,* just as *momenta* is the plural form of momentum.)

Recall, from the previous chapter, that the energy of a photon is given by $E = hf$, where h is **Planck's constant** (the single number that results when the energy of a photon is divided by its frequency).[1] We shall see that Planck's constant is a fundamental constant of nature that serves to set a lower limit on the

[1] Planck's constant, h, has the numerical value 6.6×10^{-34} J·s.

smallness of things. It ranks with the velocity of light and Newton's gravitational constant as a basic constant of nature, and it appears again and again in quantum physics. The equation $E = hf$ gives the smallest amount of energy that can be converted to light with frequency f. The radiation of light is not emitted continuously but is emitted as a stream of photons, with each photon throbbing at a frequency f and carrying an energy hf.

The equation $E = hf$ tells us why microwave radiation can't do the damage to molecules in living cells that ultraviolet light and X-rays can. Electromagnetic radiation interacts with matter only in discrete bundles of photons. So the relatively low frequency of microwaves insures low energy per photon. Ultraviolet radiation, on the other hand, can deliver about a million times more energy to molecules because the frequency of ultraviolet radiation is about a million times greater than the frequency of microwaves. X-rays, with even higher frequencies, can deliver even more.

Light quanta, electrons, and other particles all behave as if they were lumps in some respects and waves in others.

Insights

Quantum physics tells us that the physical world is a coarse, grainy place rather than the smooth, continuous place with which we are familiar. The "common-sense" world described by classical physics seems smooth and continuous because quantum graininess is on a very small scale compared with the sizes of things in the familiar world. Planck's constant is small in terms of familiar units. But you don't have to enter the quantum world to encounter graininess underlying apparent smoothness. For example, the blending areas of black, white, and gray in the photograph of Max Planck on the previous page and other photographs in this book do not look smooth at all when viewed through a magnifying glass. With magnification, you can see that a printed photograph consists of many tiny dots. In a similar way, we live in a world that is a blurred image of the grainy world of atoms.

Physicists were reluctant to adopt Planck's revolutionary quantum notion. Before it could be taken seriously, the quantum idea would have to be verified by something besides electromagnetic energy given off by warm bodies. A verification was supplied five years later by Einstein, who extended Planck's ideas to explain the photoelectric effect in the Nobel Prize–winning paper we mentioned earlier. (Even then, scientists were slow to accept so revolutionary an idea. Only after Niels Bohr's work on atomic structure in 1913—covered in the next chapter—was the quantum generally accepted. Einstein's Nobel Prize was delayed until 1921.)

CHECK YOURSELF

1. What does the term *quantum* mean?
2. How much total energy is in a monochromatic beam composed of n photons of frequency f?

CHECK YOUR ANSWERS

1. A *quantum* is the smallest elemental unit of a quantity. Radiant energy, for example, is composed of many quanta, each of which is called a *photon*. So the more photons in a beam of light, the more energy in that beam.
2. The energy in a monochromatic beam of light containing n quanta is $E = nhf$.

Photoelectric Effect

In the latter part of the nineteenth century, several investigators noticed that light was capable of ejecting electrons from various metal surfaces. This is the **photoelectric effect,** now used in electric eyes, in the photographer's light meter, and in "reading" the sound tracks of motion pictures.

An arrangement for observing the photoelectric effect is shown in Figure 31.1. Light shining on the negatively charged, photosensitive metal surface liberates electrons. The liberated electrons are attracted to the positive plate and produce a measurable current. If we instead charge this plate with just enough negative charge that it repels electrons, the current can be stopped. We can then calculate the energies of the ejected electrons from the easily measured potential difference between the plates.

The photoelectric effect was not particularly surprising to early investigators. The ejection of electrons could be accounted for by classical physics, which pictures the incident light waves building an electron's vibration up to greater and greater amplitudes until it finally breaks loose from the metal surface, just as water molecules break loose from the surface of heated water. It should take considerable time for a weak source of light to give electrons in a metal enough energy to make them boil off the surface. Instead, it was found that electrons are ejected as soon as the light is turned on—but not as many are ejected as with a strong light source. Careful examination of the photoelectric effect led to several observations that were quite contrary to the classical wave picture:

1. The time lag between turning on the light and the ejection of the first electrons was unaffected by the brightness or frequency of the light.
2. The effect was easy to observe with violet or ultraviolet light but not with red light.
3. The rate at which electrons were ejected was proportional to the brightness of the light.
4. The maximum energy of the ejected electrons was unaffected by the brightness of the light. However, there were indications that the electron's energy did depend on the frequency of the light.

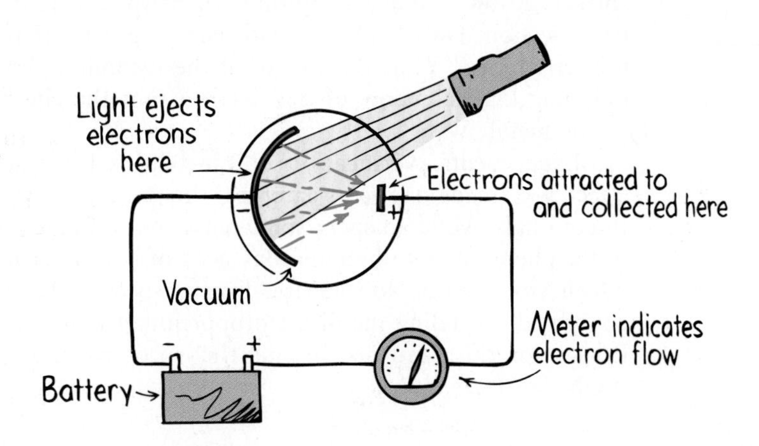

FIGURE 31.1
Interactive Figure
An apparatus used for observing the photoelectric effect. Reversing the polarity and stopping the electron flow provides a way to measure the energy of the electrons.

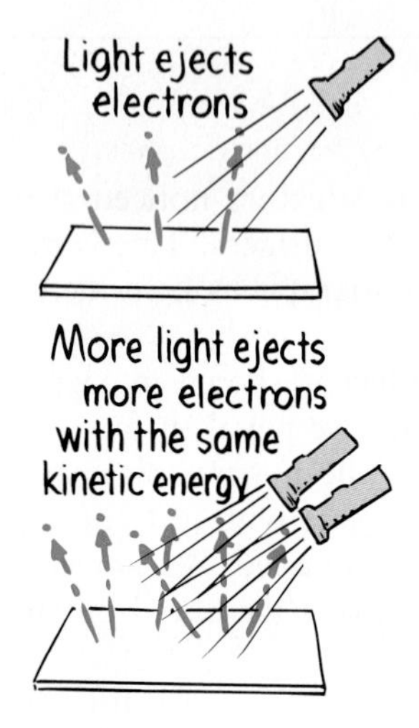

FIGURE 31.2
The photoelectric effect depends on intensity.

The lack of any appreciable time lag was especially difficult to understand in terms of the wave picture. According to the wave theory, an electron in dim light should, after some delay, accumulate sufficient vibrational energy to fly out, while an electron in bright light should be ejected almost immediately. However, this didn't occur. It was not unusual to observe an electron being ejected immediately, even under the dimmest light. The observation that the brightness of light in no way affected the energies of ejected electrons was also perplexing. The stronger electric fields of brighter light did not cause electrons to be ejected at greater speeds. More electrons were ejected in brighter light, but not at greater speeds. A weak beam of ultraviolet light, on the other hand, produced a smaller number of ejected electrons but at much higher speeds. This was most puzzling.

Einstein produced the answer in 1905, the same year he explained Brownian motion and set forth his theory of special relativity. His clue was Planck's quantum theory of radiation. Planck had assumed that the emission of light in quanta was due to restrictions on the vibrating atoms that produced the light. That is, he assumed that energy in *matter* is quantized, but that radiant energy is continuous. Einstein, on the other hand, attributed quantum properties to light itself and viewed radiation as a hail of particles. To emphasize this particle aspect, we speak of photons (by analogy with electrons, protons, and neutrons) whenever we are thinking of the particle nature of light. One photon is completely absorbed by each electron ejected from the metal. The absorption is an all-or-nothing process and is immediate, so there is no delay as "wave energies" build up.

A light wave has a broad front, and its energy is spread out along this front. For the light wave to eject a single electron from a metal surface, all its energy would somehow have to be concentrated on that one electron. But this is as improbable as an ocean wave hurling a boulder far inland with an energy equal to that of the whole wave. Therefore, instead of thinking of light encountering a surface as a continuous train of waves, the photoelectric effect suggests we conceive of light encountering a surface or any detector as a succession of corpuscles, or photons. The number of photons in a light beam controls the brightness of the *whole beam,* whereas the frequency of the light controls the energy of each *individual photon.*

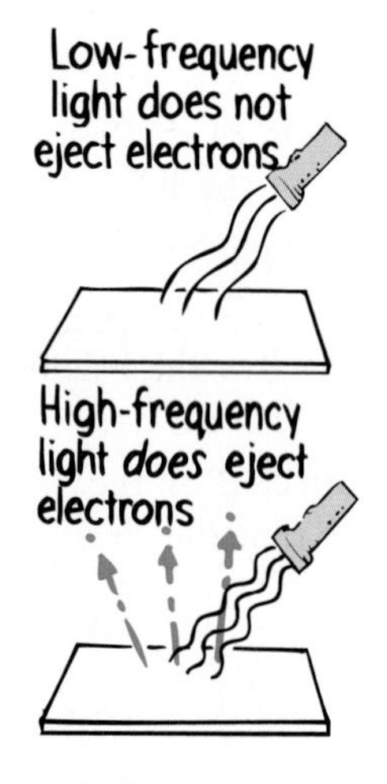

FIGURE 31.3
The photoelectric effect depends on frequency.

Electrons are held in a metal by attractive electrical forces. A minimum energy, called the *work function,* W_0, is required for an electron to leave the surface. A low-frequency photon with energy less than W_0 won't produce electron ejection. Only a photon with energy greater than W_0 results in the photoelectric effect. Thus the energy of the incoming photon will be equal to the outgoing kinetic energy of the electron plus the energy required to get it out of the metal, W_0.

Experimental verification of Einstein's explanation of the photoelectric effect was made 11 years later by the American physicist Robert Millikan. Interestingly, Millikan spent some ten years trying to disprove Einstein's theory of the photon only to become convinced of it as a result of his own experiments, which won him a Nobel Prize. Every aspect of Einstein's interpretation was confirmed, including the direct proportionality of photon energy to frequency. It was for this (and not for his theory of relativity) that Einstein received his Nobel Prize.

The photoelectric effect proves conclusively that light has particle properties. We cannot conceive of the photoelectric effect on the basis of waves. On the other hand, we have seen that the phenomenon of interference demonstrates convincingly that light has wave properties. We cannot conceive of interference in terms of particles. In classical physics, this appears to be and is contradictory. From the point of view of quantum physics, light has properties resembling both. It is "just like a wave" or "just like a particle," depending on the particular experiment. So we think of light as both, as a wave–particle. How about "wavicle"? Quantum physics calls for a new way of thinking.

CHECK YOURSELF

1. Will brighter light eject more electrons from a photosensitive surface than dimmer light of the same frequency?
2. Will high-frequency light eject a greater number of electrons than low-frequency light?

Wave–Particle Duality

The wave and particle nature of light is evident in the formation of optical images. We understand the photographic image produced by a camera in terms of light waves, which spread from each point of the object, refract as they pass through the lens system, and converge to focus on the photographic film. The path of light from the object through the lens system and to the focal plane can be calculated using methods developed from the wave theory of light.

But now consider carefully the way in which the photographic image is formed. The photographic film consists of an emulsion that contains grains of silver halide crystal, each grain containing about 10^{10} silver atoms. Each photon that is absorbed gives up its energy hf to a single grain in the emulsion. This energy activates surrounding crystals in the entire grain and is used in development to complete the photochemical process. Many photons activating many grains produce the usual photographic exposure. When a photograph is taken with exceedingly feeble light, we find that the image is built up by individual photons that arrive independently and are seemingly random in their distribution. We see this strikingly illustrated in Figure 31.4, which shows how an exposure progresses photon by photon.

CHECK YOUR ANSWERS

1. Yes. The number of ejected electrons depends on the number of incident photons.
2. Not necessarily. The energy (not the number) of ejected electrons depends on the frequency of the illuminating photons. A bright source of blue light, for example, may eject more electrons at a lower energy than a dim violet source.

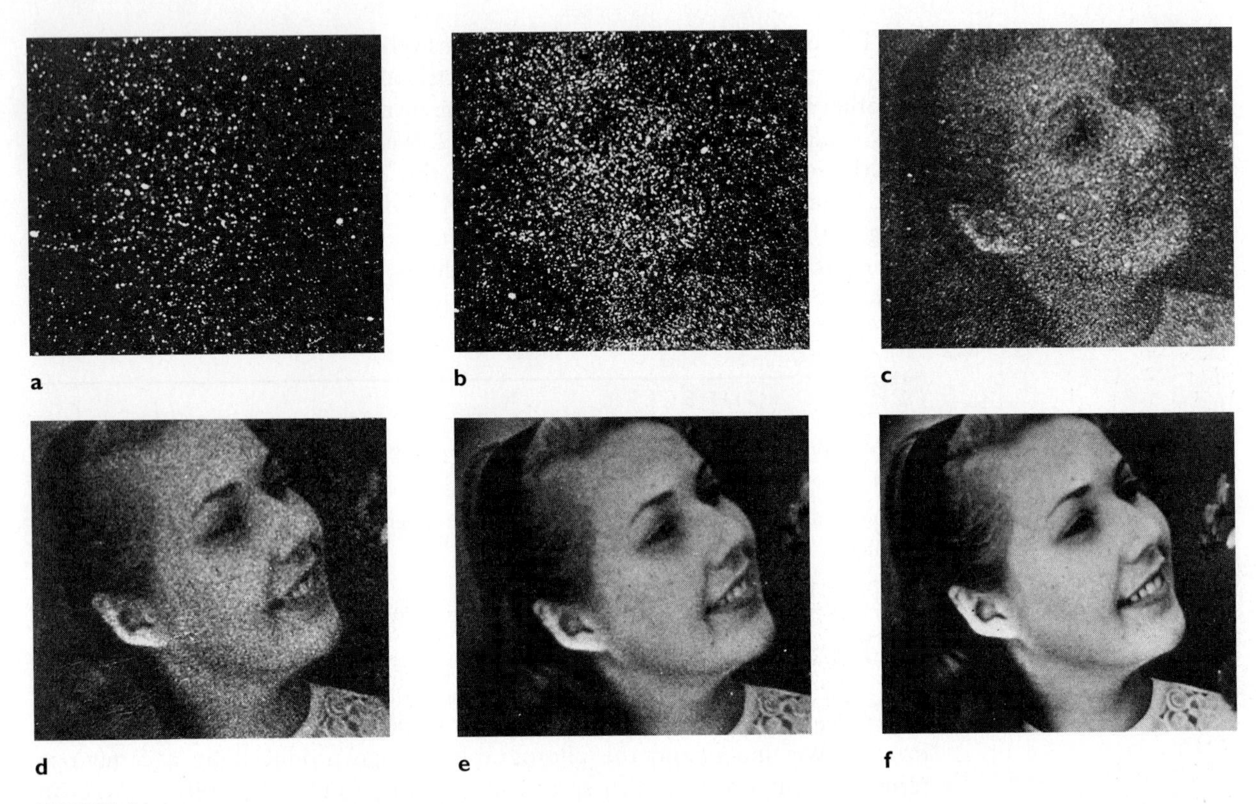

FIGURE 31.4
Stages of film exposure reveal the photon-by-photon production of a photograph. The approximate numbers of photons at each stage are (a) 3×10^3, (b) 1.2×10^4, (c) 9.3×10^4, (d) 7.6×10^5, (e) 3.6×10^6, and (f) 2.8×10^7.

Double-Slit Experiment

Let's return to Thomas Young's double-slit experiment, which we discussed in terms of waves in Chapter 29. Recall that, when we pass monochromatic light through a pair of closely spaced thin slits, we produce an interference pattern (Figure 31.5). Now let's consider the experiment in terms of photons. Suppose we dim our light source so that, in effect, only one photon at a time reaches

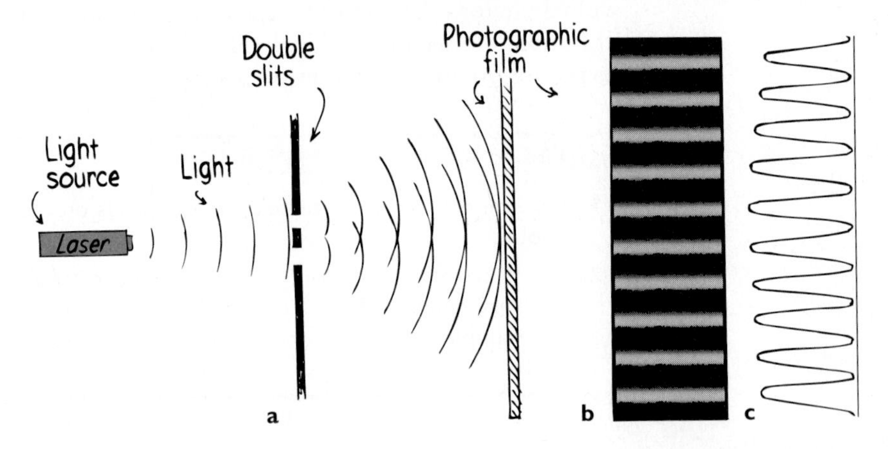

FIGURE 31.5
(a) Arrangement for double-slit experiment. (b) Photograph of interference pattern. (c) Graphic representation of pattern.

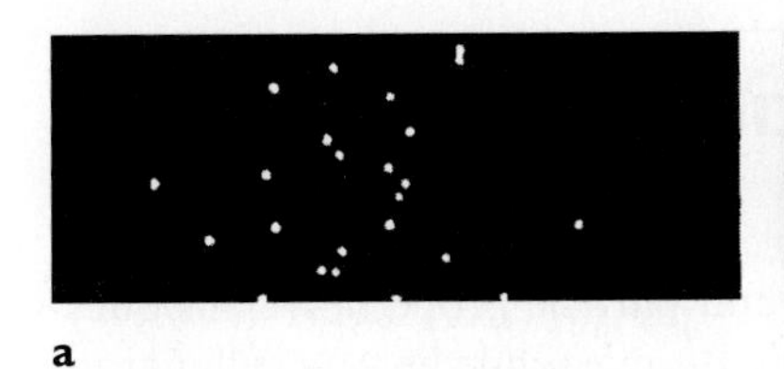
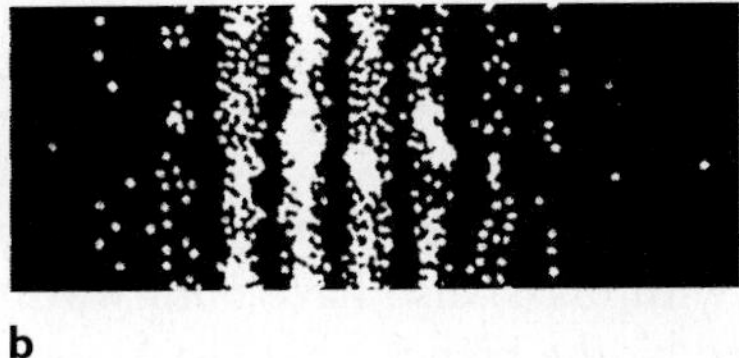
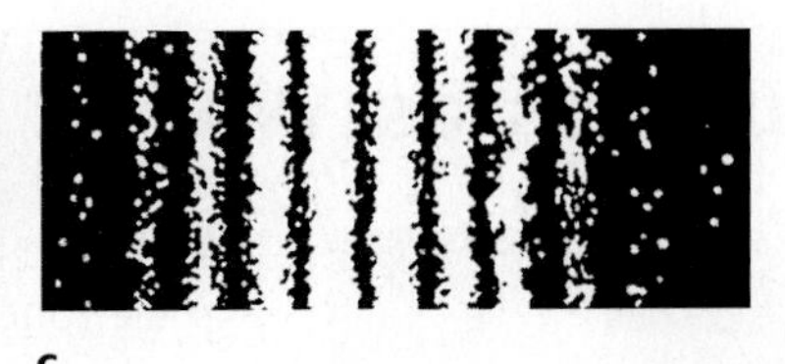

a b c

FIGURE 31.6 Interactive Figure

Stages of two-slit interference pattern. The pattern of individually exposed grains progresses from (a) 28 photons to (b) 1000 photons to (c) 10,000 photons. As more photons hit the screen, a pattern of interference fringes appears.

the barrier with the thin slits. If film behind the barrier is exposed to the light for a very short time, the film gets exposed as simulated in Figure 31.6*a*. Each spot represents the place where the film has been exposed by a photon. If the light is allowed to expose the film for a longer time, a pattern of fringes begins to emerge, as in Figures 31.6*b* and 31.6*c*. This is quite amazing! Spots on the film are seen to progress, photon by photon, to form the same interference pattern characterized by waves!

If we cover one slit so that photons striking the photographic film can pass only through a single slit, the tiny spots on the film accumulate to form a single-slit diffraction pattern (Figure 31.7). We find that photons hit the film at places they would not hit if both slits were open! If we think about this classically, we are perplexed and may ask how photons passing through the single slit "know" that the other slit is covered and therefore fan out to produce the wide single-slit diffraction pattern. Or, if both slits are open, how do photons traveling through one slit "know" that the other slit is open and avoid certain regions, proceeding only to areas that will ultimately fill to form the fringed double-slit interference pattern?[2] The modern answer is that the wave nature of light is not some average property that shows up only when many photons act together. Each single photon has wave properties as well as particle properties. But the photon displays different aspects at different times. *A photon behaves as a particle when it is being emitted by an atom or absorbed by photographic film or other detectors and behaves as a wave in traveling from a source to the place where it is detected.* So the photon strikes the film as a particle but travels to its position as a wave that interferes constructively. The fact that light exhibits both wave and particle behavior was one of the interesting surprises of the early twentieth century. Even more surprising was the discovery that objects with mass also exhibit a dual wave–particle behavior.

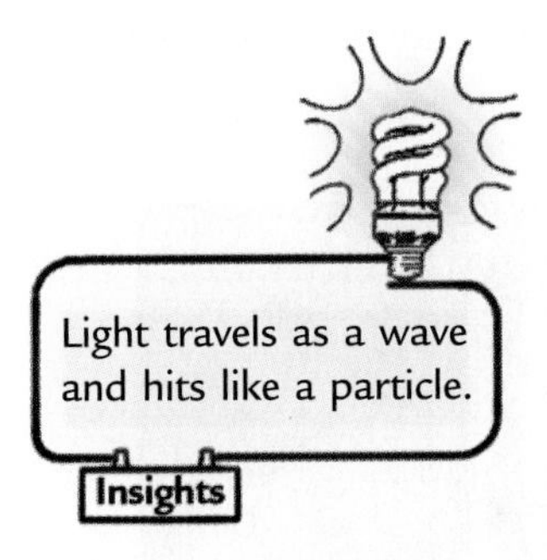

Light travels as a wave and hits like a particle.
Insights

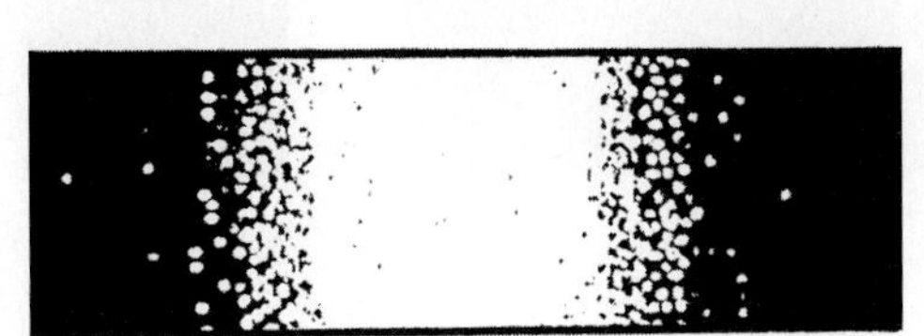

FIGURE 31.7
Single-slit diffraction pattern.

[2]From a prequantum point of view, this wave–particle duality is indeed mysterious. This leads some people to believe that quanta have some sort of consciousness, with each photon or electron having "a mind of its own." The mystery, however, is like beauty. It is in the mind of the beholder rather than in nature itself. We conjure models to understand nature and, when inconsistencies arise, we sharpen or change our models. The wave–particle duality of light doesn't fit a model built on classical ideas. An alternate model is that quanta have minds of their own. Another model is quantum physics. In this book, we subscribe to the latter.

Particles as Waves: Electron Diffraction

Louis de Broglie (1892–1987)

If a photon of light has both wave and particle properties, why can't a material particle (one with mass) also have both wave and particle properties? This question was posed by the French physicist Louis de Broglie while he was still a graduate student in 1924. His answer constituted his doctoral thesis in physics and later earned him the Nobel Prize in physics. According to de Broglie, every particle of matter is somehow endowed with a wave to guide it as it travels. Under the proper conditions, then, every particle will produce an interference or diffraction pattern. Each body—whether an electron, a proton, an atom, a mouse, you, a planet, a star—has a wavelength that is related to its momentum by

$$\text{Wavelength} = \frac{h}{\text{momentum}}$$

where h is Planck's constant. A body of large mass and ordinary speed has such a small wavelength that interference and diffraction are negligible; rifle bullets fly straight and do not pepper their targets far and wide with detectable interference patches.[3] But, for smaller particles, such as electrons, diffraction can be appreciable.

A beam of electrons can be diffracted in the same way a beam of photons can be diffracted, as is evident in Figure 31.8. Beams of electrons directed through double slits exhibit interference patterns, just as photons do. The double-slit experiment discussed in the previous section can be performed with electrons as well as with photons. For electrons, the apparatus is more complex, but the procedure is essentially the same. The intensity of the source can be reduced to direct electrons one at a time through a double-slit arrangement, producing the same remarkable results as with photons. Like photons, electrons

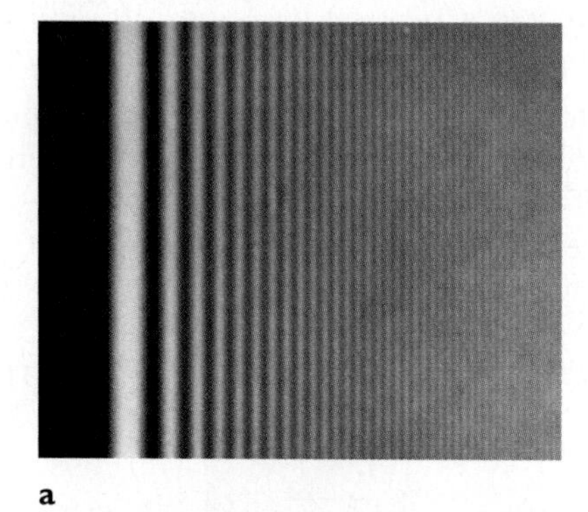

a

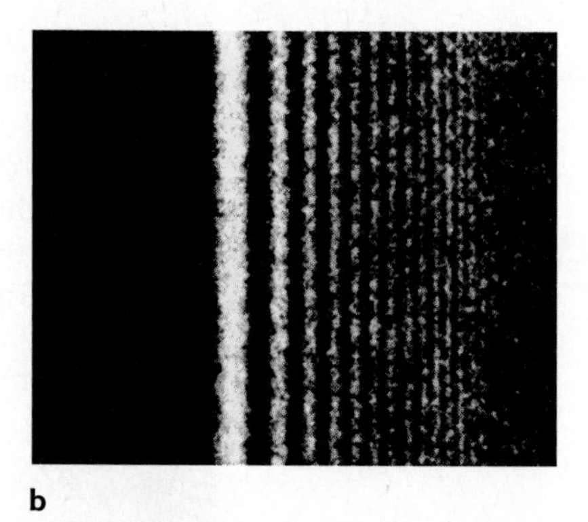

b

FIGURE 31.8
Fringes produced by the diffraction of (a) light and (b) an electron beam.

[3]A bullet of mass 0.02 kg traveling at 330 m/s, for example, has a de Broglie wavelength of

$$\frac{h}{mv} = \frac{6.6 \times 10^{-34}\,\text{J}\cdot\text{s}}{(0.02\text{ kg})(330\text{ m/s})} = 10^{-34}\text{ m},$$

an incredibly small size a million million million millionth the size of a hydrogen atom. An electron traveling at 2% the speed of light, on the other hand, has a wavelength 10^{-10} m, which is equal to the diameter of the hydrogen atom. Diffraction effects for electrons are measurable, whereas diffraction effects for bullets are not.

FIGURE 31.9
An electron microscope makes practical use of the wave nature of electrons. The wavelength of electron beams is typically thousands of times shorter than the wavelength of visible light, so the electron microscope is able to distinguish detail not visible with optical microscopes.

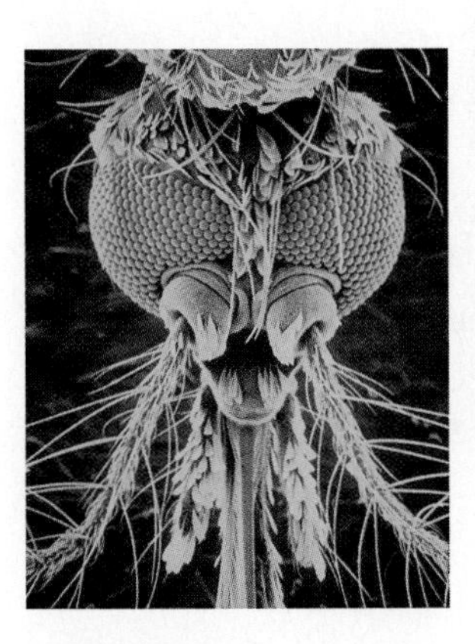

FIGURE 31.10
Detail of the head of a female mosquito as seen with a scanning electron microscope at a "low" magnification of 200 times.

strike the screen as particles, but the *pattern* of arrival is wavelike. The angular deflection of electrons to form the interference pattern agrees perfectly with calculations using de Broglie's equation for the wavelength of an electron.

This wave–particle duality is not restricted to photons and electrons. In Figure 31.11, we see the results of a similar procedure that uses a standard electron microscope. The electron beam of very low current density is directed through an electrostatic biprism that diffracts the beam. A pattern of fringes produced by individual electrons builds up step by step and is displayed on a

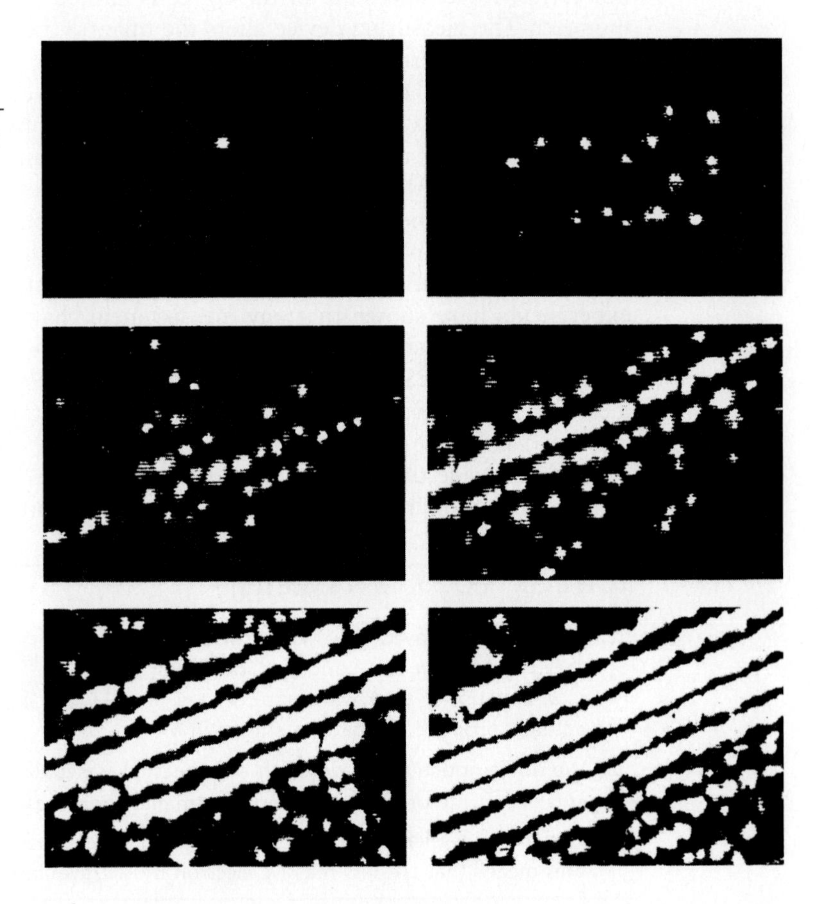

FIGURE 31.11
Electron interference patterns filmed from a TV monitor, showing the diffraction of a very low intensity electron-microscope beam through an electrostatic biprism.

TV monitor. The image is gradually filled by electrons to produce the interference pattern customarily associated with waves. Neutrons, protons, whole atoms, and, to an immeasurable degree, even high-speed rifle bullets exhibit a duality of particle and wave behavior.

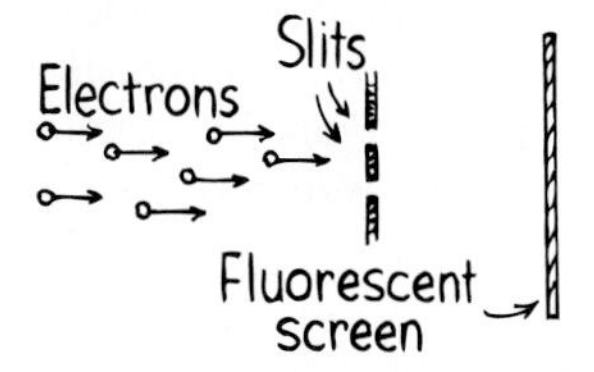

CHECK YOURSELF

1. If electrons behaved only like particles, what pattern would you expect on the screen after the electron passed through the double slits?
2. We don't notice the de Broglie wavelength for a pitched baseball. Is this because the wavelength is very large or because it is very small?
3. If an electron and a proton have the same de Broglie wavelength, which particle has the greater speed?

Uncertainty Principle

The wave–particle duality of quanta has inspired interesting discussions about the limits of our ability to accurately measure the properties of small objects. The discussions center on the idea that the act of measuring something affects the quantity being measured.

For example, we know that, if we place a cool thermometer in a cup of hot coffee, the temperature of the coffee is altered as it gives heat to the thermometer. The measuring device alters the quantity being measured. But we can correct for these errors in measurement if we know the initial temperature of the thermometer, the masses and specific heats involved, and so forth. Such corrections fall well within the domain of classical physics—these are *not* the uncertainties of quantum physics. Quantum uncertainties stem from the wave nature of matter. A wave, by its very nature, occupies some space and lasts for some time. It cannot be squeezed to a point in space or limited to a single instant of time, for then it would not be a wave. This inherent "fuzziness" of a wave gives a fuzziness to measurement at the quantum level. Innumerable experiments have shown that any measurement that in any way probes a system necessarily disturbs the system by at least one quantum of action, h—Planck's constant. So any measurement that involves interaction between the measurer and what is being measured is subject to this minimum inaccuracy.

We distinguish between observing and probing. Consider a cup of coffee on the other side of a room. If you passively glance at it and see steam rising from it, this act of "measuring" involves no physical interaction between your eyes

CHECK YOUR ANSWERS

1. If electrons behaved only like particles, they would form two bands, as indicated in *a*. Because of their wave nature, they actually produce the pattern shown in *b*.

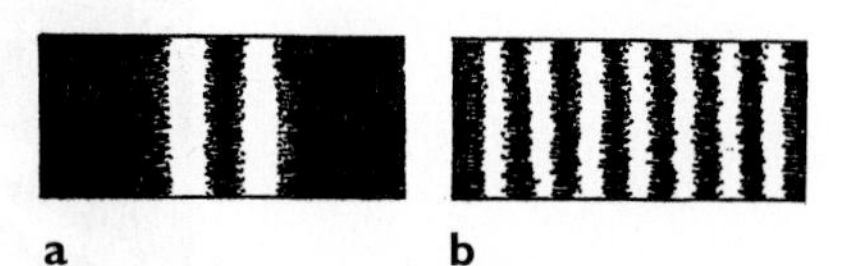

2. We don't notice the wavelength of a pitched baseball because it is extremely small—on the order of 10^{-20} times smaller than the atomic nucleus.
3. The same wavelength means that the two particles have the same momentum. This means that the less massive electron must travel faster than the heavier proton.

Werner Heisenberg
(1901–1976)

and the coffee. Your glance neither adds nor subtracts energy from the coffee. You can assert that it's hot with no *probing*. Placing a thermometer in it is a different story. You physically interact with the coffee and thereby subject it to alteration. The quantum contribution to this alteration, however, is completely dwarfed by classical uncertainties and is negligible. Quantum uncertainties are significant only in the atomic and subatomic realm.

Compare the acts of making measurements of a pitched baseball and of an electron. You can measure the speed of a pitched baseball by having it fly through a pair of photogates that are a known distance apart (Figure 31.12). The ball is timed as it interrupts beams of light in the gates. The accuracy of the ball's measured speed has to do with uncertainties in the measured distance between the gates and in the timing mechanisms. Interactions between the macroscopic ball and the photons it encounters are insignificant. But not so in the case of measuring submicroscopic things like electrons. Even a single photon bouncing off an electron appreciably alters the motion of the electron—and in an unpredictable way. If you wish to observe an electron and determine its whereabouts with light, the wavelength of the light would have to be very short. You fall into a dilemma. Light of a short wavelength, which can "see" the tiny electron better, corresponds to a large quantum of energy, which, in turn, greatly alters the electron's state of motion. If, on the other hand, you use a long wavelength that corresponds to a smaller quantum of energy, the change you induce to the electron's state of motion will be smaller, but the determination of its position by means of the coarser wave will be less accurate. The act of observing something as tiny as an electron probes the electron and, in so doing, produces a considerable uncertainty in either its position or its motion. Although this uncertainty is completely negligible for measurements of position and motion regarding everyday (macroscopic) objects, it is a predominant fact of life in the atomic domain.

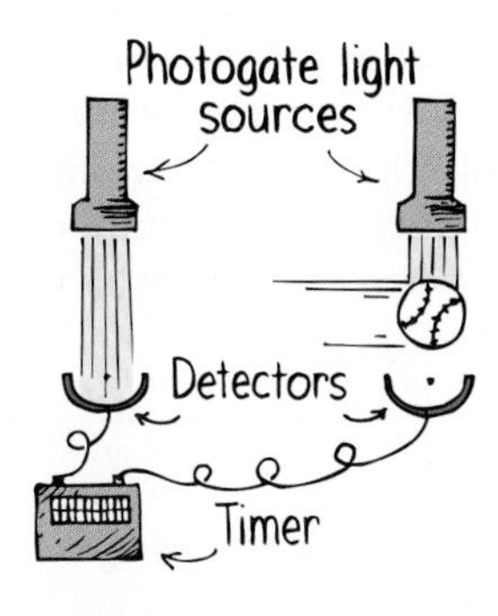

FIGURE 31.12
The ball's speed is measured by dividing the distance between the photogates by the time difference between crossing the two light paths. Photons hitting the ball alter its motion much less than the motion of an oil supertanker would be altered by a few fleas bumping into it.

The uncertainty of measurement in the atomic domain, which was first stated mathematically by the German physicist Werner Heisenberg, is called the **uncertainty principle.** It is a fundamental principle in quantum mechanics. Heisenberg found that, when the uncertainties in the measurement of momentum and position for a particle are multiplied together, the product must be equal to or greater than Planck's constant, h, divided by 2π, which is represented as $\hbar$ (called *h-bar*). We can state the uncertainty principle in a simple formula:

$$\Delta p \Delta x \geq \hbar$$

The Δ here means "uncertainty of": Δp is the uncertainty of momentum (the symbol for momentum is conventionally p), and Δx is the uncertainty of position. The product of these two uncertainties must be equal to or greater than ($\geq$) the size of $\hbar$. For minimum uncertainties, the product will equal $\hbar$; the product of larger uncertainties will be greater than $\hbar$. But in no case can the product of the uncertainties be less than $\hbar$. The significance of the uncertainty principle is that, even in the best of conditions, the lower limit of uncertainty is $\hbar$. This means that if we wish to know the momentum of an electron with great accuracy (small Δp), the corresponding uncertainty in position will be large. Or if we wish to know the position with great accuracy (small Δx), the corresponding uncertainty in momentum will be large. The sharper one of these quantities is, the less sharp is the other.[4]

[4]Only in the classical limit where $\hbar$ becomes zero could the uncertainties of both position and momentum be arbitrarily small. Planck's constant is greater than zero, and we cannot, in principle, simultaneously know both quantities with absolute certainty.

You can never change only one thing! Every equation reminds us of this—you can't change a term on one side without affecting the other side.

Insights

The uncertainty principle operates similarly with energy and time. We cannot measure a particle's energy with complete precision in an infinitesimally short span of time. The uncertainty in our knowledge of energy, ΔE, and the duration taken to measure the energy, Δt, are related by the expression[5]

$$\Delta E \Delta t \geq \hbar$$

The greatest accuracy we can ever hope to attain is that case in which the product of the energy and time uncertainties equals $\hbar$. The more accurately we determine the energy of a photon, an electron, or a particle of whatever kind, the more uncertain we will be of the time during which it has that energy.

The uncertainty principle is relevant only to quantum phenomena. The inaccuracies in measuring the position and momentum of a baseball due to the interactions of observation, for example, are completely negligible. But the inaccuracies in measuring the position and momentum of an electron are far from negligible. This is because the uncertainties in the measurements of these subatomic quantities are comparable to the magnitudes of the quantities themselves.[6]

There is a danger in applying the uncertainty principle to areas outside of quantum mechanics. Some people conclude from statements about the interaction between the observer and the observed that the universe does not exist "out there," independent of all acts of observation, and that reality is created by the observer. Others interpret the uncertainty principle as nature's shield of forbidden secrets. Some critics of science use the uncertainty principle as evidence that science itself is uncertain. The reality of the universe (whether observed or not), nature's secrets, and the uncertainties of science have very little to do with Heisenberg's uncertainty principle. The profundity of the uncertainty principle has to do with the unavoidable interaction between nature at the atomic level and the means by which we probe it.

CHECK YOURSELF

1. Is Heisenberg's uncertainty principle applicable to the practical case of using a thermometer to measure the temperature of a glass of water?
2. A Geiger counter measures radioactive decay by registering the electrical pulses produced in a gas tube when high-energy particles pass through it. The particles emanate from a radioactive source—say, radium. Does the act of measuring the decay rate of radium alter the radium or its decay rate?
3. Can the quantum principle that we cannot observe something without changing it be reasonably extrapolated to support the claim that you can make a stranger turn around and look at you by staring intently at his back?

---

[5]We can see that this is consistent with the uncertainty in momentum and position. Recall that Δmomentum = force × Δtime and that Δenergy = force × Δdistance. Then

$$\begin{aligned} \hbar &= \Delta\text{momentum}\ \Delta\text{distance} \\ &= (\text{force} \times \Delta\text{distance})\Delta\text{time} \\ &= \Delta\text{energy}\ \Delta\text{time} \end{aligned}$$

[6]The uncertainties in measurements of momentum, position, energy, or time that are related to the uncertainty principle for a pitched baseball are only 1 part in about 10 million billion billion billion (10^{-34}). Quantum effects are negligible even for the swiftest bacterium, where the uncertainties are about 1 part in a billion (10^{-9}). Quantum effects become evident for atoms, where the uncertainties can be as large as 100%. For electrons moving in an atom, quantum uncertainties dominate, and we are in the full-scale quantum realm.

Complementarity

The realm of quantum physics seems confusing. Light waves that interfere and diffract deliver their energy in particle packages of quanta. Electrons that move through space in straight lines and experience collisions as if they were particles distribute themselves spatially in interference patterns as if they were waves. In this confusion, there is an underlying order. The behavior of light and electrons is confusing in the same way! Light and electrons both exhibit wave and particle characteristics.

The Danish physicist Niels Bohr, one of the founders of quantum physics, formulated an explicit expression of the wholeness inherent in this dualism. He called his expression of this wholeness **complementarity**. As Bohr expressed it, quantum phenomena exhibit complementary (mutually exclusive) properties—appearing either as particles or as waves—depending on the type of experiment conducted. Experiments designed to examine individual exchanges of energy and momentum bring out particle-like properties, while experiments designed to examine spatial distribution of energy bring out wavelike properties. The wave-like properties of light and the particle-like properties of light complement one another—both are necessary for the understanding of "light." Which part is emphasized depends on what question one puts to nature.

CHECK YOUR ANSWERS

1. No. Although we probably subject the temperature of water to a change by the act of probing it with a thermometer, especially one appreciably colder or hotter than the water, the uncertainties that relate to the precision of the thermometer are quite within the domain of classical physics. The role of uncertainties at the subatomic level is inapplicable here.
2. Not at all, because the interaction involved is between the Geiger counter and the particles, not between the Geiger counter and the radium. It is the behavior of the particles that is altered by measurement, not the radium from which they emanate. See how this ties into the next question.
3. No. Here we must be careful in defining what we mean by *observing*. If our observation involves probing (giving or extracting energy), we indeed change to some degree that which we observe. For example, if we shine a light source onto the person's back, our observation consists of probing, which, however slight, physically alters the configuration of atoms on his back. If he senses this, he may turn around. But simply staring intently at his back is observing in the passive sense. The light you receive (or block by blinking, for example) has already left his back. So whether you stare, squint, or close your eyes completely, you in no physical way alter the atomic configuration on his back. Shining a light or otherwise probing something is not the same thing as passively looking at something. A failure to make the simple distinction between *probing* and *passive observation* is at the root of much nonsense that is said to be supported by quantum physics. Better support for the above claim would be positive results from a simple and practical test, rather than the assertion that it rides on the hard-earned reputation of quantum theory.

PREDICTABILITY AND CHAOS

We can make predictions about an orderly system when we know the initial conditions. For example, we can state precisely where a launched rocket will land, where a given planet will be at a particular time, or when an eclipse will occur. These are examples of events in the Newtonian macroworld. Similarly, in the quantum microworld, we can predict where an electron is likely to be in an atom and the probability that a radioactive particle will decay in a given time interval. Predictability in orderly systems, both Newtonian and quantum, depends on knowledge of initial conditions.

Some systems, however, whether Newtonian or quantum, are not orderly—they are inherently unpredictable. These are called "chaotic systems." Turbulent water flow is an example. No matter how precisely we know the initial conditions of a piece of floating wood as it flows downstream, we cannot predict its location later downstream. A feature of chaotic systems is that slight differences in initial conditions result in wildly different outcomes later. Two identical pieces of wood just slightly apart at one time may be vastly far apart soon thereafter.

Weather is chaotic. Small changes in one day's weather can produce big (and largely unpredictable) changes a week later. Meteorologists try their best, but they are bucking the hard fact of chaos in nature. This barrier to accurate prediction first led the scientist Edward Lorenz to ask, "Does the flap of a butterfly's wings in Brazil set off a tornado in Texas?" We now talk about the *butterfly effect* when we are dealing with situations in which very small effects can amplify into very big effects.

Interestingly, chaos is not all hopeless unpredictability. Even in a chaotic system there can be patterns of regularity. There is *order in chaos.* Scientists have learned how to treat chaos mathematically and how to find the parts of it that are orderly. Artists seek patterns in nature in a different way. Both scientists and artists look for the connections in nature that were always there but are not yet put together in our thinking.

FIGURE 31.13
Opposites are seen to complement one another in the yin–yang symbol of Eastern cultures.

Complementarity is not a compromise, and it doesn't mean that the whole truth about light lies somewhere in between particles and waves. It's rather like viewing the sides of a crystal. What you see depends on what facet you look at, which is why light, energy, and matter appear to be behaving as quanta in some experiments and as waves in others.

The idea that opposites are components of a wholeness is not new. Ancient Eastern cultures incorporated it as an integral part of their worldview. This is demonstrated in the yin–yang diagram of T'ai Chi Tu (Figure 31.13). One side of the circle is called *yin,* and the other side is called *yang*. Where there is yin, there is yang. Only the union of yin and yang forms a whole. Where there is low, there is also high. Where there is night, there is also day. Where there is birth, there is also death. A whole person integrates yin (feminine traits, right brain, emotion, intuition, darkness, cold, wetness) with yang (masculine traits, left brain, reason, logic, light, heat, dryness). Each has aspects of the other. For Niels Bohr, the yin–yang diagram symbolized the principle of complementarity. In later life, Bohr wrote broadly on the implications of complementarity. In 1947, when he was knighted for his contributions to physics, he chose for his coat of arms the yin–yang symbol.

Summary of Terms

Quantum (*pl.* quanta) From the Latin word *quantus*, meaning "how much," a quantum is the smallest elemental unit of a quantity, the smallest discrete amount of something. One quantum of electromagnetic energy is called a photon.

Quantum physics The physics that describes the microworld, where many quantities are granular (in units called *quanta*), not continuous, and where particles of light (*photons*) and particles of matter (such as electrons) exhibit wave as well as particle properties.

Planck's constant A fundamental constant, h, that relates the energy of light quanta to their frequency:

$$h = 6.6 \times 10^{-34} \text{ joule}\cdot\text{second}$$

Photoelectric effect The emission of electrons from a metal surface when light shines upon it.

Uncertainty principle The principle, formulated by Werner Heisenberg, stating that Planck's constant, h, sets a limit on the accuracy of measurement. According to the uncertainty principle, it is not possible to measure exactly both the position and the momentum of a particle at the same time, nor the energy and the time during which the particle has that energy.

Complementarity The principle, enunciated by Niels Bohr, stating that the wave and particle aspects of both matter and radiation are necessary, complementary parts of the whole. Which part is emphasized depends on what experiment is conducted (i.e., on what question one puts to nature).

Suggested Reading

Cole, K. C. *The Hole in the Universe: How Scientists Peered over the Edge of Emptiness and Found Everything*. New York: Harcourt, 2001.

Ford, K. W. *The Quantum World: Quantum Physics for Everyone*. Cambridge, Mass.: Harvard University Press, 2004. An intriguing account of the development of quantum physics, with emphasis on the participating physicists.

Trefil, J. *Atoms to Quarks*. New York: Scribner's, 1980. A nice development of quantum theory in the early chapters, which lead on to particle physics.

Review Questions

1. Which theory of light, the wave theory or the particle theory, did the findings of Young, Maxwell, and Hertz support?
2. Did Einstein's photon explanation of the photoelectric effect support the wave theory or the particle theory of light?

Birth of the Quantum Theory

3. What exactly did Max Planck consider quantized, the energy of vibrating atoms or the energy of light itself?
4. Distinguish between the study of *mechanics* and the study of *quantum mechanics*.

Quantization and Planck's Constant

5. Why is the energy of a campfire not a whole-number multiple of a single quantum, although the energy of a laser beam is?
6. What is a quantum of light called?
7. In the formula $E = hf$, does f stand for wave frequency, as defined in Chapter 19?
8. Which has the lower energy quanta—red light or blue light? Radio waves or X-rays?

Photoelectric Effect

9. What evidence can you cite for the particle nature of light?
10. Which are more successful in dislodging electrons from a metal surface—photons of violet light or photons of red light? Why?
11. Why won't a very bright beam of red light impart more energy to an ejected electron than a feeble beam of violet light?
12. In considering the interaction of light with matter, how did Einstein extend the quantum idea of Planck?
13. Einstein proposed his explanation of the photoelectric effect in 1905. When were his views on this verified?

Wave–Particle Duality

14. Why do photographs in a book or magazine look grainy when magnified?

15. Does light behave primarily as a wave or as a particle when it interacts with the crystals of matter in photographic film?

Double-Slit Experiment

16. Does light travel from one place to another in a wavelike way or in a particle-like way?
17. Does light interact with a detector in a wavelike way or in a particle-like way?
18. When does light behave as a wave? When does it behave as a particle?

Particles as Waves: Electron Diffraction

19. What evidence can you cite for the wave nature of particles?
20. When electrons are diffracted through a double slit, do they arrive at the screen in a wavelike way or in a particle-like way? Is the pattern of hits wavelike or particle-like?

Uncertainty Principle

21. In which of the following are quantum uncertainties significant? Measuring simultaneously the speed and location of a baseball; of a spitball; of an electron.
22. What is the uncertainty principle with respect to motion and position?
23. If measurements show a precise position for an electron, can those measurements show precise momentum also? Explain.
24. If measurement shows a precise value for the energy radiated by an electron, can that measurement show a precise time for this event as well? Explain.
25. Is there a distinction between passively observing an event and actively probing it?

Complementarity

26. What is the principle of complementarity?
27. Cite evidence that the idea of opposites as components of a wholeness preceded Bohr's principle of complementarity.

Exercises

1. Distinguish between *classical physics* and *quantum physics*.
2. What does it mean to say that something is quantized?
3. In the previous chapter, we learned the formula $E \sim f$. In this chapter, we learned the formula $E = hf$. Explain the difference between these two formulas. What is h?
4. The frequency of violet light is about twice that of red light. How does the energy of a violet photon compare with the energy of a red photon?
5. Which has more energy—a photon of visible light or a photon of ultraviolet light?
6. We speak of photons of red light and photons of green light. Can we speak of photons of white light? Why or why not?
7. Which laser beam carries more energy per photon—a red beam or a green beam?
8. If a beam of red light and a beam of blue light have exactly the same energy, which beam contains the greater number of photons?
9. One of the technical challenges facing the original developers of color television was the design of an image tube (camera) for the red portion of the image. Why was finding a material that would respond to red light more difficult than finding materials to respond to green and blue light?
10. Phosphors on the inside of fluorescent lamps convert ultraviolet light to visible light. Why are there no substances that convert visible light to ultraviolet light?
11. Silver bromide ($AgBr$) is a light-sensitive substance used in some types of photographic film. To cause exposure of the film, it must be illuminated with light having sufficient energy to break apart the molecules. Why do you suppose this film may be handled without exposure in a darkroom illuminated with red light? How about blue light? How about very bright red light relative to very dim blue light?
12. Sunburn produces cell damage in the skin. Why is ultraviolet radiation capable of producing this damage, while visible radiation, even if more intense, is not?
13. In the photoelectric effect, does brightness or frequency determine the kinetic energy of the ejected electrons? Which determines the number of the ejected electrons?
14. A very bright source of red light has much more energy than a dim source of blue light, but the red light has no effect in ejecting electrons from a certain photosensitive surface. Why is this so?
15. Why are ultraviolet photons more effective at inducing the photoelectric effect than photons of visible light?
16. Why does light striking a metal surface eject only electrons, not protons?
17. Does the photoelectric effect depend on the wave nature or the particle nature of light?
18. Explain how the photoelectric effect is used to open automatic doors when someone approaches.

19. Explain briefly how the photoelectric effect is used in the operation of at least two of the following: an electric eye, a photographer's light meter, the sound track of a motion picture.
20. If you shine an ultraviolet light on the metal ball of a negatively charged electroscope (shown in Exercise 12 in Chapter 22), it will discharge. But, if the electroscope is positively charged, it won't discharge. Can you venture an explanation?
21. Discuss how the reading of the meter in Figure 31.1 will vary as the photosensitive plate is illuminated by light of various colors at a given intensity and by light of various intensities of a given color.
22. Does the photoelectric effect *prove* that light is made of particles? Do interference experiments *prove* that light is composed of waves? (Is there a distinction between what something *is* and how it *behaves*?)
23. Does Einstein's explanation of the photoelectric effect invalidate Young's explanation of the double-slit experiment? Explain.
24. The camera that took the photograph of the woman's face (Figure 31.4) used ordinary lenses that are well known to refract waves. Yet the step-by-step formation of the image is evidence of photons. How can this be? What is your explanation?
25. What evidence can you cite for the wave nature of light? For the particle nature of light?
26. When does a photon behave like a wave? When does it behave like a particle?
27. Light has been argued to be a wave and then a particle, and then back again. Does this indicate that light's true nature probably lies somewhere between these two models?
28. What laboratory device utilizes the wave nature of electrons?
29. How might an atom obtain enough energy to become ionized?
30. When a photon hits an electron and gives it energy, what happens to the frequency of the photon after it bounces from the electron? (This phenomenon is called the *Compton effect.*)
31. A hydrogen atom and a uranium atom move at the same speed. Which possesses more momentum? Which has the longer wavelength?
32. If a cannonball and a BB have the same speed, which has the longer wavelength?
33. An electron and a proton travel at the same speed. Which has more momentum? Which has the longer wavelength?
34. One electron travels twice as fast as another. Which has the longer wavelength?
35. Does the de Broglie wavelength of a proton become longer or shorter as its velocity increases?
36. We don't notice the wavelength of moving matter in our common experience. Is this because the wavelength is extraordinarily large or extraordinarily small?
37. What principal advantage does an electron microscope have over an optical microscope?
38. Would a beam of protons in a "proton microscope" exhibit greater or less diffraction than electrons of the same speed in an electron microscope? Defend your answer.
39. Suppose nature were entirely different so that an infinite number of photons would be needed to make up even the tiniest amount of radiant energy, the wavelength of material particles was zero, light had no particle properties, and matter had no wave properties. This would be the classical world described by the mechanics of Newton and the electricity and magnetism of Maxwell. What would be the value of Planck's constant for such a world with no quantum effects?
40. Suppose that you lived in a hypothetical world in which you'd be knocked down by a single photon, in which matter would be so wavelike that it would be fuzzy and hard to grasp, and in which the uncertainty principle would impinge on simple measurements of position and speed in a laboratory, making results irreproducible. In such a world, how would Planck's constant compare with the accepted value?
41. Comment on the idea that the theory one accepts determines the meaning of one's observations and not vice versa.
42. A friend says, "If an electron is not a particle, then it must be a wave." What is your response? (Do you hear "either-or" statements like this often?)
43. Consider one of the many electrons on the tip of your nose. If somebody looks at it, will its motion be altered? How about if it is looked at with one eye closed? With two eyes open, but crossed? Does Heisenberg's uncertainty principle apply here?
44. Does the uncertainty principle tell us that we can never know anything for certain?
45. Do we inadvertently alter the realities that we attempt to measure in a public opinion survey? Does Heisenberg's uncertainty principle apply here?
46. If the behavior of a system is measured exactly for some period of time and is understood, does it follow that the future behavior of that system can be exactly predicted? (Is there a distinction between properties that are *measurable* and properties that are *predictable*?)

47. When checking the pressure in tires, some air escapes. Why does Heisenberg's uncertainty principle not apply here?
48. If a butterfly causes a tornado, does it make sense to eradicate butterflies? Defend your answer.
49. We hear the expression "taking a quantum leap" to describe large changes. Is the expression appropriate? Defend your answer.
50. To measure the exact age of Old Methuselah, the oldest living tree in the world, a Nevada professor of dendrology, aided by an employee of the U.S. Bureau of Land Management, cut the tree down in 1965 and counted its rings. Is this an extreme example of changing that which you measure or an example of arrogant and criminal stupidity?

Problems

1. A typical wavelength of infrared radiation emitted by your body is 25 μm (2.5×10^{-5} m). What is the energy per photon of such radiation?
2. What is the de Broglie wavelength of an electron that strikes the back of the face of a TV screen at 1/10 the speed of light?
3. You decide to roll a 0.1-kg ball across the floor so slowly that it will have a small momentum and a large de Broglie wavelength. If you roll it at 0.001 m/s, what is its wavelength? How does this compare with the de Broglie wavelength of the high-speed electron in the previous problem?

PART SEVEN

Atomic and Nuclear Physics

"Know nukes!" The Earth's natural heat that warms this natural hot spring, or that powers a geyser or a volcano, comes from nuclear power—the radioactivity of minerals in the Earth's interior. Power from atomic nuclei is as old as the Earth itself, and isn't restricted to today's nuclear reactors, or "nukes," as they are called. How about that!

CHAPTER 32

The Atom and the Quantum

David Kagan models an orbiting electron with a strip of corrugated plastic, and he models energy levels with stacked wood blocks.

We discussed the atom as a building block of matter in Chapter 11 and as an emitter of light in the preceding chapters. We know that the atom is composed of a central nucleus surrounded by a complex arrangement of electrons; the study of this atomic structure is called *atomic physics*. In this chapter, we will outline some of the developments that led to our present understanding of the atom. We will trace developments in atomic physics from classical to quantum physics. In the two following chapters, we will learn about *nuclear physics*—the study of the structure of the atomic nucleus. This knowledge of the atom and its implications has had a profound impact on human society.

We begin this chapter with a brief look at some events early in the twentieth century that led to our current understanding of the atom.

Discovery of the Atomic Nucleus

Ernest Rutherford (1871–1937)

Half a dozen years after Einstein announced the photoelectric effect, the New Zealand–born British physicist Ernest Rutherford oversaw his now famous gold-foil experiment.[1] This significant experiment showed that the atom is mostly empty space, with most of its mass concentrated in the central region—the *atomic nucleus*.

In Rutherford's experiment, a beam of positively charged particles (alpha particles) from a radioactive source was directed through a sheet of extremely thin gold foil. Because alpha particles are thousands of times more massive than electrons, it was expected that the stream of alpha particles would not be impeded as it passed through the "atomic pudding." This was indeed observed—for the most part. Nearly all alpha particles passed through the gold foil with little or no deflection and produced a spot of light when they hit a fluorescent screen beyond the foil. But some particles were deflected from their

[1]Why "oversaw"? To indicate that more investigators than Rutherford were involved in this experiment. The widespread practice of elevating a single scientist to the position of sole investigator, which seldom is the case, too often denies the involvement of other investigators. There's substance to the saying, "There are two things more important to people than sex and money—*recognition and appreciation*."

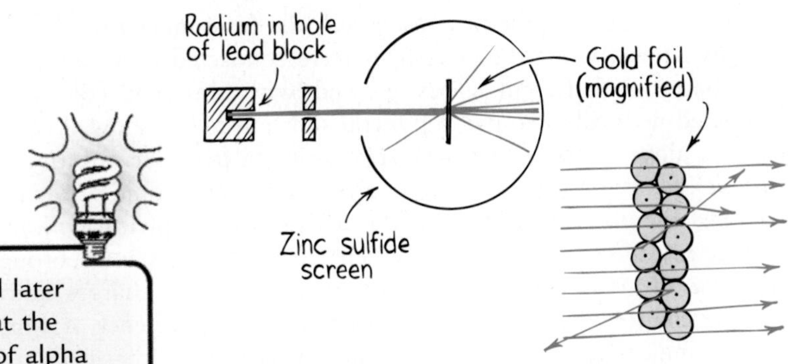

FIGURE 32.1
The occasional large-angle scattering of alpha particles from gold atoms led Rutherford to the discovery of the small, very massive nuclei at their centers.

Rutherford later related that the discovery of alpha particles rebounding backward was the most incredible event of his life—as incredible as if a 15-inch shell rebounded from a piece of tissue paper.

Insights

straight-line paths as they emerged. A few alpha particles were widely deflected, and a small number were even scattered backward! These alpha particles must have hit something relatively massive—but what? Rutherford reasoned that the undeflected particles traveled through regions of the gold foil that were empty space, while the small number of deflected particles were repelled from extremely dense, positively charged central cores. Each atom, he concluded, must contain one of these cores, which he named the **atomic nucleus.**

Discovery of the Electron

FIGURE 32.2
Franklin's kite-flying experiment.

Surrounding the atomic nucleus are electrons. The name *electron* comes from the Greek word for amber, a brownish-yellow fossil resin studied by the early Greeks. They found that, when amber was rubbed by a piece of cloth, it attracted such things as bits of straw. This phenomenon, known as the amber effect, remained a mystery for almost 2000 years. In the late 1500s, William Gilbert, Queen Elizabeth's physician, found other materials that behaved like amber, and he called them "electrics." The concept of electric charge awaited experiments by the American scientist-statesman Benjamin Franklin nearly two centuries later. Franklin experimented with electricity and postulated the existence of an electric fluid that could flow from place to place. An object with an excess of this fluid he called electrically positive, and one with a deficiency of the fluid he called electrically negative. The fluid was thought to attract ordinary matter but to repel itself. Although we no longer talk about electric fluid, we still follow Franklin's lead in how we define positive and negative electricity. Most of us know about Franklin's 1752 experiment with the kite in the lightning storm that showed that lightning is an electrical discharge between clouds and the ground. This discovery told him that electricity is not restricted to solid or liquid objects and that it can travel through a gas.

Franklin's experiments later inspired other scientists to produce electric currents through various dilute gases in sealed glass tubes. Among these, in the 1870s, was Sir William Crookes, an unorthodox English scientist who believed he could communicate with the dead. He is better remembered for his "Crooke's tube," a sealed glass tube containing gas under very low pressure and with

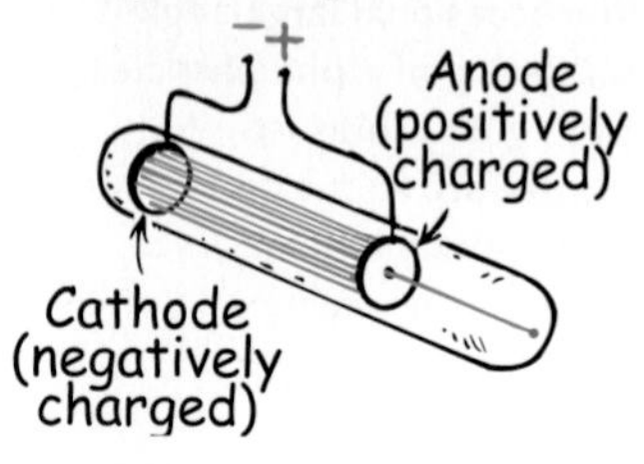

FIGURE 32.3
A simple cathode-ray tube. An electric current is produced in the gas when a high voltage is imposed across the electrodes inside the tube.

electrodes inside the tube near each end (the forerunner to today's neon signs). The gas glowed when the electrodes were connected to a voltage source (such as a battery). Different gases glowed with different colors. Experiments conducted with tubes containing metal slits and plates showed that the gas was made to glow by some sort of a "ray" coming from the negative terminal (the *cathode*). Slits could make the ray narrow and plates could prevent the ray from reaching the positive terminal (the *anode*). The apparatus was named the *cathode-ray tube* (Figure 32.3). When electric charges were brought near the tube, the ray was deflected. It bent toward positive charges and away from negative charges. The ray was also deflected by the presence of a magnet. These findings indicated that the ray consisted of negatively charged particles.

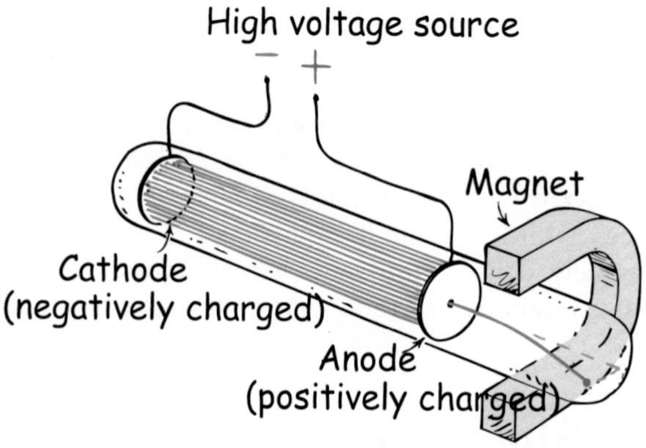

FIGURE 32.4
A cathode ray (electron beam) is deflected by a magnetic field. (This is the forerunner of the television tubes and computer monitors that followed.)

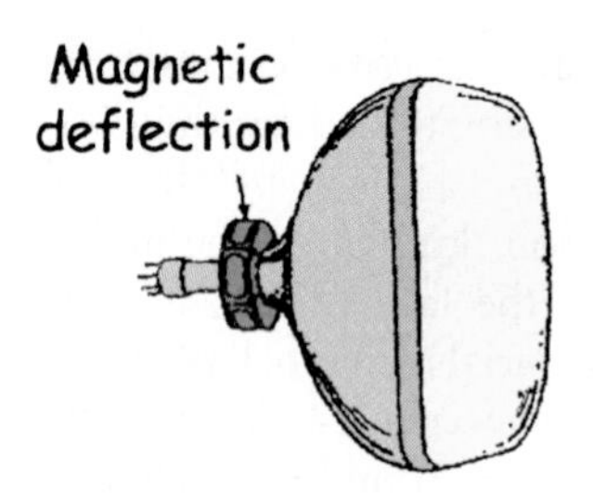

FIGURE 32.5
A familiar CRT (cathode-ray tube).

In 1897, the English physicist Joseph John Thomson ("J. J.", as his friends called him) showed that the cathode rays were indeed particles, smaller and lighter than atoms and apparently all identical. He created narrow beams of cathode rays and measured their deflection in electric and magnetic fields. Thomson reasoned that the amount of the beams' deflection depended on the mass of the particles and their electrical charge. How? The greater each particle's mass, the greater the inertia and the less the deflection. The greater each particle's charge, the greater the force and the greater the deflection. The greater the speed, the less the deflection.

From careful measurements of the deflection of the beam, Thomson succeeded in calculating the mass-to-charge ratio of the cathode-ray particle, which was soon thereafter named the **electron.** All electrons are identical; they are copies of one another. For establishing the existence of the electron, J. J. Thomson was awarded the Nobel Prize in physics in 1906.

Next to investigate the properties of electrons was the American physicist Robert Millikan. He calculated the numerical value of a single unit of electric charge on the basis of an experiment he performed in 1909. In his experiment, Millikan sprayed tiny oil droplets into a chamber between electrically charged plates—into an *electric field*. When the field was strong, some of the droplets moved upward, indicating that they carried a very slight negative charge. Millikan adjusted the field so that droplets hovered motionless. He knew that the downward force of gravity on the droplets was exactly balanced by the upward electrical force. Investigation showed that the charge on each drop was always some multiple of a single very small value, which he proposed to be the fundamental unit of charge carried by each electron. Using this value and the

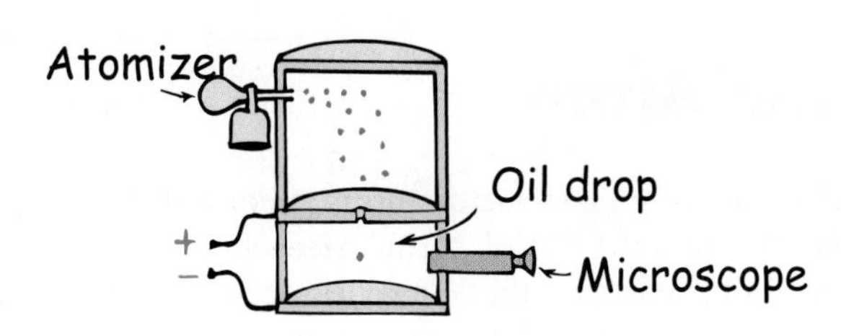

FIGURE 32.6
Millikan's oil-drop experiment for determining the charge on the electron. The pull of gravity on a particular drop can be balanced by an upward electrical force.

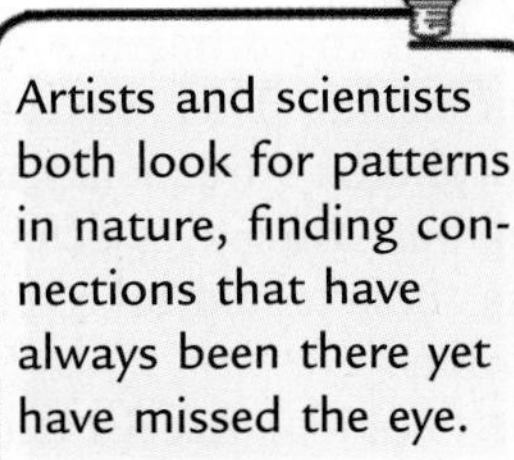

Artists and scientists both look for patterns in nature, finding connections that have always been there yet have missed the eye.

Insights

ratio discovered by Thomson, he calculated the mass of an electron to be about 1/2000 the mass of the lightest known atom, hydrogen. This confirmed that the atom was no longer the least massive particle of matter. For his work in physics, Millikan received the 1923 Nobel Prize.

If atoms contained negatively charged electrons, it stood to reason that atoms must also contain some balancing positively charged matter. J. J. Thomson put forth what he called a "plum-pudding" model of the atom in which electrons were like plums in a sea of positively charged pudding. The experimentation of Rutherford and the gold-foil experiment, previously mentioned, proved this model wrong.

Atomic Spectra: Clues to Atomic Structure

During the period of Rutherford's experiments, chemists were using the spectroscope (discussed in Chapter 30) for chemical analysis, while physicists were busy trying to find order in the confusing arrays of spectral lines. It had long been known that the lightest element, hydrogen, has a far more orderly spectrum than the other elements (Figure 32.7). An important sequence of lines in the hydrogen spectrum starts with a line in the red region, followed by one in the blue, then by several lines in the violet, and many in the ultraviolet. Spacing between successive lines becomes smaller and smaller from the first in the red to the last in the ultraviolet, until the lines become so close that they seem to merge. A Swiss schoolteacher, Johann Jakob Balmer, first expressed the wavelengths of these lines in a single mathematical formula in 1884. Balmer, however, could give no reason why his formula worked so successfully. His guess that the series for other elements might follow a similar formula proved to be correct, leading to the prediction of lines that had not yet been measured.

FIGURE 32.7
A portion of the hydrogen spectrum. Each line, an image of the slit in the spectroscope, represents light of a specific frequency emitted by hydrogen gas when excited (higher frequency is to the right).

Another regularity in atomic spectra was found by the Swedish physicist and mathematician Johannes Rydberg. He noticed that the sum of the frequencies of two lines in the spectrum of hydrogen often equals the frequency of a third line. This relationship was later advanced as a general principle by the Swiss physicist Walter Ritz and is called the **Ritz combination principle.** It states that the spectral lines of any element include frequencies that are either the sum or the difference of the frequencies of two other lines. Like Balmer, Ritz was unable to offer an explanation for this regularity. These regularities were the clues that the Danish physicist Niels Bohr used to understand the structure of the atom itself.

Bohr Model of the Atom

Niels Bohr (1885–1962)

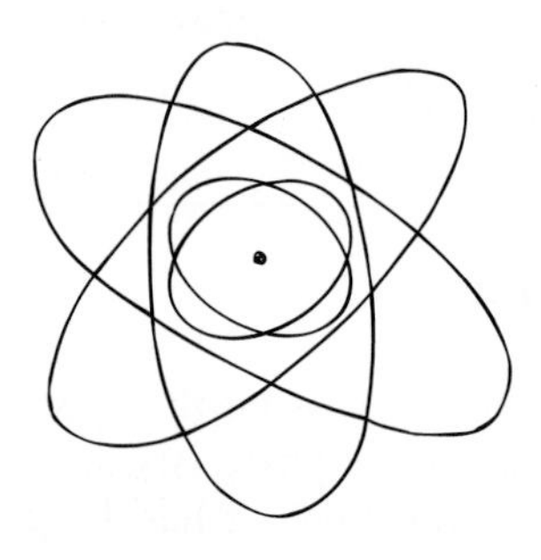

FIGURE 32.8
The Bohr model of the atom. Although this model is very oversimplified, it is still useful in understanding light emission.

In 1913, Bohr applied the quantum theory of Planck and Einstein to the nuclear atom of Rutherford and formulated the well-known planetary model of the atom.[2] Bohr reasoned that electrons occupy "stationary" states (of fixed energy, not fixed position) at different distances from the nucleus and that the electrons can make "quantum jumps" from one energy state to another. He reasoned that light is emitted when such a quantum jump occurs (from a higher to a lower energy state). Furthermore, Bohr realized that the frequency of emitted radiation is determined by $E = hf$ (actually $f = E/h$), where E is the difference in the atom's energy when the electron is in the different orbits. This was an important breakthrough, because it said that the emitted photon's frequency is not the classic frequency at which an electron is vibrating but, instead, is determined by the energy *differences* in the atom. From there, Bohr could take the next step and determine the energies of the individual orbits.

Bohr's planetary model of the atom begged a major question. Accelerated electrons, according to Maxwell's theory, radiate energy in the form of electromagnetic waves. So an electron accelerating around a nucleus should radiate energy continuously. This radiating away of energy should cause the electron to spiral into the nucleus (Figure 32.9). Bohr boldly deviated from classical physics by stating that the electron doesn't radiate light while it accelerates around the nucleus in a single orbit, but that radiation of light occurs only when the electron makes a transition from a higher energy level to a lower energy level. The energy of the emitted photon is equal to the *difference* in energy between the two energy levels, $E = hf$. Color depends on the size of the jump. So the quantization of light energy neatly corresponds to the quantization of electron energy.

Bohr's views, as outlandish as they seemed at the time, explained the regularities found in atomic spectra. Bohr's explanation of the Ritz combination principle is shown in Figure 32.10. If an electron is raised to the third energy level, it can return to its initial level either by a single jump from the third to the first level or by a double jump, first to the second level and then to the first level. These two return paths will produce three spectral lines. Note that the sum of the energy jumps along paths A and B is equal to the energy jump C.

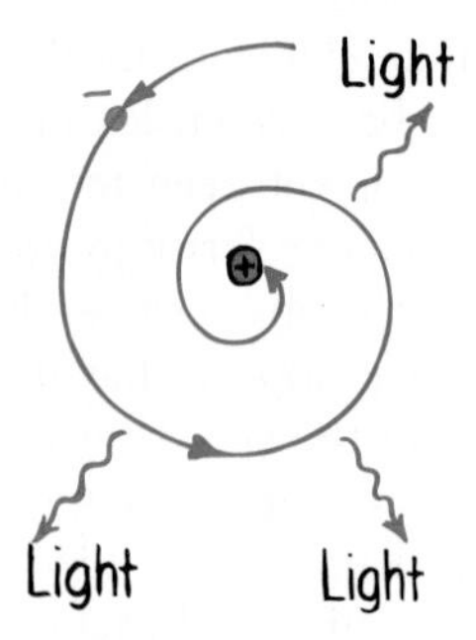

FIGURE 32.9
According to classical theory, an electron accelerating around its orbit should continuously emit radiation. This loss of energy should cause it to spiral rapidly into the nucleus. But this does not happen.

[2]This model, like most models, has major defects because the electrons do not revolve in planes as planets do. Later, the model was revised; "orbits" became "shells" and "clouds." We use *orbit* because it was, and still is, commonly used. Electrons are not just bodies, like planets, but rather behave like waves concentrated in certain parts of the atom.

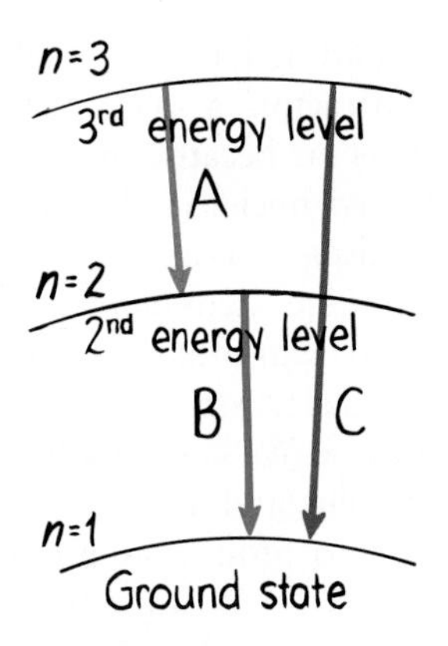

FIGURE 32.10
Three of many energy levels in an atom. An electron jumping from the third level to the second level is shown in red, and one jumping from the second level to the ground state is shown in green. The sum of the energies (and the frequencies) for these two jumps equals the energy (and the frequency) of the single jump from the third level to the ground state, shown in blue.

Since frequency is proportional to energy, the frequencies of light emitted along path A and path B when added equal the frequency of light emitted when the transition is along path C. Now we can see why the sum of two frequencies in the spectrum is equal to a third frequency in the spectrum.

Bohr was able to account for X-rays in heavier elements, showing that they are emitted when electrons jump from outer to innermost orbits. He predicted X-ray frequencies that were later experimentally confirmed. Bohr was also able to calculate the "ionization energy" of a hydrogen atom—the energy needed to knock the electron out of the atom completely. This also was verified by experiment.

Using measured frequencies of X-rays as well as visible, infrared, and ultraviolet light, scientists could map energy levels of all the atomic elements. Bohr's model had electrons orbiting in neat circles (or ellipses) arranged in groups or shells. This model of the atom accounted for the general chemical properties of the elements. It also predicted a missing element, which led to the discovery of hafnium.

Bohr solved the mystery of atomic spectra while providing an extremely useful model of the atom. He was quick to point out that his model was to be interpreted as a crude beginning, and the picture of electrons whirling about the nucleus like planets about the Sun was not to be taken literally (to which popularizers of science paid no heed). His sharply defined orbits were conceptual representations of an atom whose later description involved waves—quantum mechanics. His ideas of quantum jumps and frequencies being proportional to energy differences remain parts of today's modern theory.

CHECK YOURSELF

1. What is the maximum number of paths for de-excitation available to a hydrogen atom excited to level number 3 in changing to the ground state?
2. Two predominant spectral lines in the hydrogen spectrum, an infrared one and a red one, have frequencies 2.7×10^{14} Hz and 4.6×10^{14} Hz, respectively. Can you predict a higher-frequency line in the hydrogen spectrum?

Relative Sizes of Atoms

The diameters of the electron orbits in the Bohr model of the atom are determined by the amount of electrical charge in the nucleus. For example, the positive proton in the hydrogen atom holds one electron in an orbit at a certain radius. If we double the positive charge in the nucleus, the orbiting electron will be pulled into

CHECK YOUR ANSWERS

1. Two (a single jump and a double jump), as shown in Figure 32.10.
2. The sum of the frequencies is $2.7 \times 10^{14} + 4.6 \times 10^{14} = 7.3 \times 10^{14}$ Hz, which happens to be the frequency of a violet line in the hydrogen spectrum. Using Figure 32.10 as a model, can you see that, if the infrared line is produced by a transition similar to path A and the red line corresponds to path B, then the violet line corresponds to path C?

a tighter orbit with half its former radius, since the electrical attraction is doubled. This occurs for a helium ion—a doubly charged nucleus attracting a single electron. Interestingly, an added second electron isn't pulled in as far because the first electron partially offsets the attraction of the doubly charged nucleus. This is a neutral helium atom, which is somewhat smaller than a hydrogen atom.

So two electrons around a doubly charged nucleus assume a configuration characteristic of helium. A third proton found in the nucleus can pull two electrons into an even closer orbit and, furthermore, hold a third electron in a somewhat larger orbit. This is the lithium atom, atomic number 3. We can continue with this process, increasing the positive charge of the nucleus and adding successively more electrons and more orbits all the way up to atomic numbers above 100, to the "synthetic" radioactive elements.[3]

We find that, as the nuclear charge increases and additional electrons are added in outer orbits, the inner orbits shrink in size because of the stronger nuclear attraction. This means that the heavier elements are not much larger in diameter than the lighter elements. The diameter of the xenon atom, for example, is only about four helium diameters, even though it is nearly 33 times as massive. The relative sizes of atoms in Figure 32.11 are approximately to the same scale.

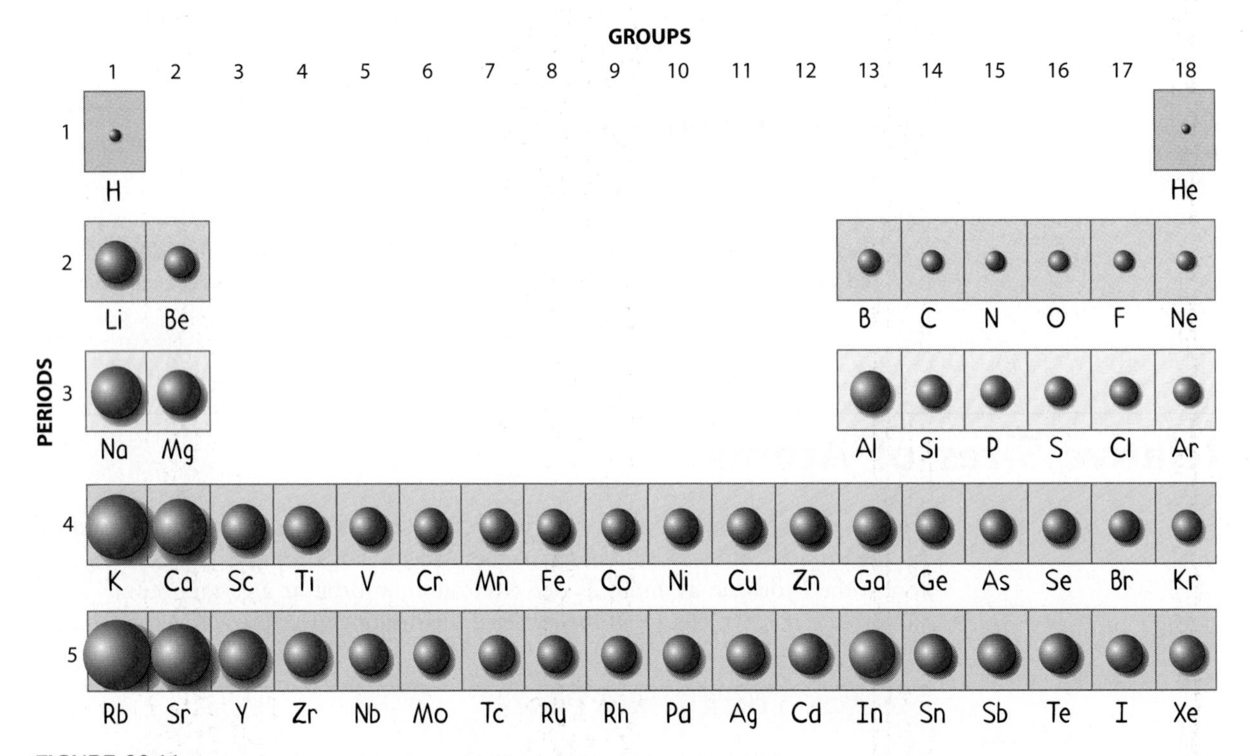

FIGURE 32.11
The sizes of atoms gradually decrease from left to right across the periodic table (only the first 5 periods are shown here).

[3]Each orbit will hold only so many electrons. A rule of quantum mechanics is that an orbit is filled when it contains a number of electrons given by $2n^2$, where n is 1 for the first orbit, 2 for the second orbit, 3 for the third orbit, and so on. For $n = 1$, there are 2 electrons; for $n = 2$, there are $2(2^2)$ or 8 electrons; for $n = 3$, there are a maximum of $2(3^2)$ or 18 electrons, etc. The number n is called the *principal quantum number*.

CHECK YOURSELF

What fundamental force dictates the size of an atom?

Explanation of Quantized Energy Levels: Electron Waves

So we see that a photon is emitted when an electron makes a transition from a higher to a lower energy level, and the frequency of the photon is equal to the energy-level difference divided by Planck's constant, h. If an electron transits a large energy-level difference, the emitted photon has a high frequency—perhaps ultraviolet. If an electron makes a transition through a lesser energy difference, the emitted photon is lower in frequency—perhaps it is a photon of red light. Each element has its own characteristic energy levels; thus, transitions of electrons between these levels result in each element emitting its own characteristic colors. Each of the elements emits its own unique pattern of spectral lines.

The idea that electrons may occupy only certain levels was very perplexing to early investigators and to Bohr himself. It was perplexing because the electron was considered to be a particle, a tiny BB whirling around the nucleus like a planet whirling around the Sun. Just as a satellite can orbit at any distance from the Sun, it would seem that an electron should be able to orbit around the nucleus at any radial distance—depending, of course, like the satellite, on its speed. Moving among all orbits would enable the electrons to emit all energies of light. But this doesn't happen. It can't. Why the electron occupies only discrete levels is understood by considering the electron to be not a particle but a *wave*.

Louis de Broglie introduced the concept of matter waves in 1924. He hypothesized that a wave is associated with every particle and that the wavelength of a matter wave is inversely related to a particle's momentum. These *matter waves* behave just like other waves; they can be reflected, refracted, diffracted, and caused to interfere. Using the idea of interference, de Broglie showed that the discrete values of radii of Bohr's orbits are a natural consequence of standing electron waves. A Bohr orbit exists where an electron wave closes on itself constructively. The electron wave becomes a standing wave, like a wave on a musical string. In this view, the electron is thought of not as a particle located at some point in the atom but as if its mass and charge were spread out into a standing wave surrounding the atomic nucleus—with an integral number of wavelengths fitting evenly into the circumferences of the

CHECK YOUR ANSWER

The electrical force.

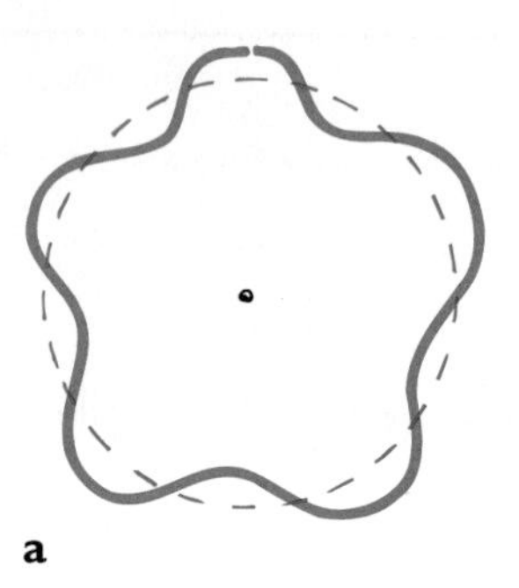

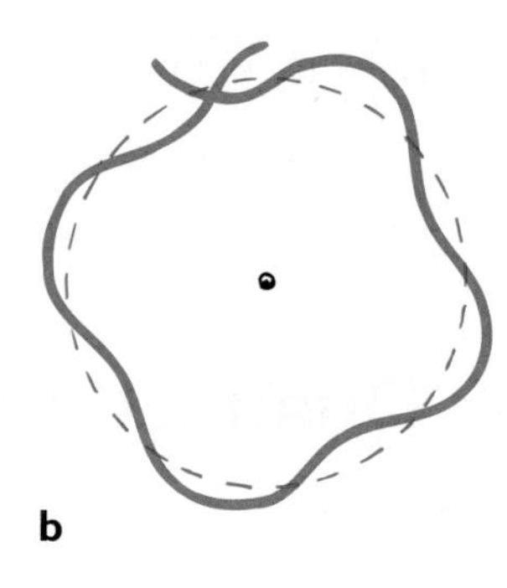

FIGURE 32.12
(a) An orbiting electron forms a standing wave only when the circumference of its orbit is equal to a whole-number multiple of the wavelength. (b) When the wave does not close in on itself in phase, it undergoes destructive interference. Hence, orbits exist only where waves close in on themselves in phase.

orbits (Figure 32.12). The circumference of the innermost orbit, according to this picture, is equal to one wavelength. The second orbit has a circumference of two electron wavelengths, the third has three, and so forth (Figure 32.13). This is similar to a chain necklace made of paper clips. No matter what size necklace is made, its circumference is equal to some multiple of the length of a single paper clip.[4] Since the circumferences of electron orbits are discrete, it follows that the radii of these orbits, and hence the energy levels, are also discrete.

Electron Waves

This model explains why electrons don't spiral closer and closer to the nucleus, causing atoms to shrink to the size of the tiny nucleus. If each electron orbit is described by a standing wave, the circumference of the smallest orbit can be no smaller than one wavelength—no fraction of a wavelength is possible in a circular (or elliptical) standing wave. As long as an electron carries the momentum necessary for wave behavior, atoms don't shrink in on themselves.

In the still more modern wave model of the atom, electron waves move not only around the nucleus but also in and out, toward and away from the nucleus. The electron wave is spread out in three dimensions, leading to the picture of an electron "cloud." As we shall see, this is a cloud of *probability,* not a cloud made up of a pulverized electron scattered over space. The electron, when detected, remains a point particle.

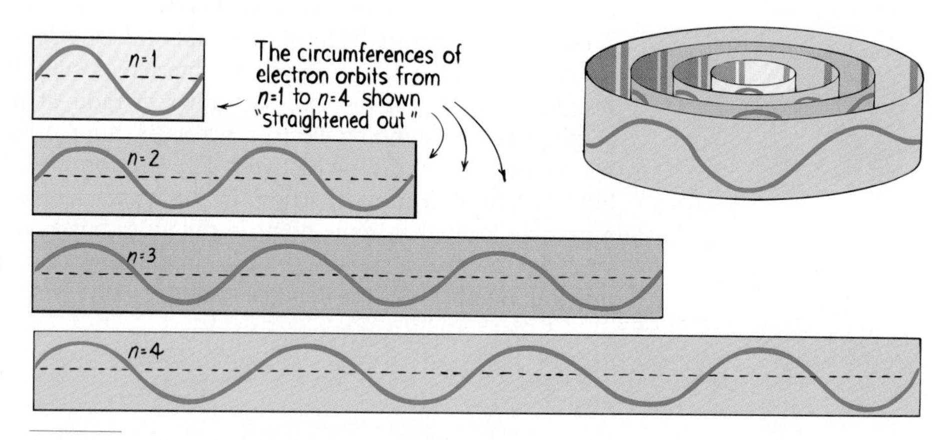

FIGURE 32.13
Interactive Figure
The electron orbits in an atom have discrete radii because the circumferences of the orbits are whole-number multiples of the electron wavelength. This results in a discrete energy state for each orbit. (The figure is greatly oversimplified, as the standing waves make up spherical and ellipsoidal shells rather than flat, circular ones.)

[4]For each orbit, the electron has a unique speed, which determines its wavelength. Electron speeds are less, and wavelengths are longer, for orbits of increasing radii; so, for our analogy to be accurate, we'd have to use not only more paper clips to make increasingly longer necklaces but increasingly larger paper clips as well.

Quantum Mechanics

"I think it is safe to say that no one understands quantum mechanics."
—*Richard P. Feynman*

Insights

Many changes in physics occurred in the mid-1920s. Not only was the particle nature of light established experimentally, but particles of matter were found to have wave properties. Starting with de Broglie's matter waves, the Austrian-German physicist Erwin Schrödinger formulated an equation that describes how matter waves change under the influence of external forces. Schrödinger's equation plays the same role in **quantum mechanics** that Newton's equation (acceleration = force/mass) plays in classical physics.[5] The matter waves in Schrödinger's equation are mathematical entities that are not directly observable, so the equation provides us with a purely mathematical rather than a visual model of the atom—which places it beyond the scope of this book. So our discussion of it will be brief.[6]

Erwin Schrödinger (1887–1961)

In **Schrödinger's wave equation,** the thing that "waves" is the nonmaterial *matter wave amplitude*—a mathematical entity called a *wave function,* represented by the symbol ψ (the Greek letter psi). The wave function given by Schrödinger's equation represents the possibilities that can occur for a system. For example, the location of the electron in a hydrogen atom may be anywhere from the center of the nucleus to a radial distance far away. An electron's possible position and its probable position at a particular time are not the same. A physicist can calculate its probable position by multiplying the wave function by itself ($|\psi|^2$). This produces a second mathematical entity called a *probability density function,* which at a given time indicates the probability per unit volume for each of the possibilities represented by ψ.

Experimentally, there is a finite probability (chance) of finding an electron in a particular region at any instant. The value of this probability lies between the limits 0 and 1, where 0 indicates never and 1 indicates always. For example, if the probability is 0.4 for finding the electron within a certain radius, this signifies a 40% chance that the electron will be there. So the Schrödinger equation cannot tell a physicist where an electron can be found in an atom at any moment but only the *likelihood* of finding it there—or, for a large number of measurements, what fraction of measurements will find the electron in each region. When an electron's position in its Bohr energy level (state) is repeatedly measured and each of its locations is plotted as a dot, the resulting pattern resembles a sort of electron cloud (Figure 32.14). An individual electron may, at various times, be detected anywhere in this probability cloud; it even has an extremely small but finite probability of momentarily existing inside the nucleus. It is detected most of the time, however, close to an average distance from the nucleus, which fits the orbital radius described by Niels Bohr.

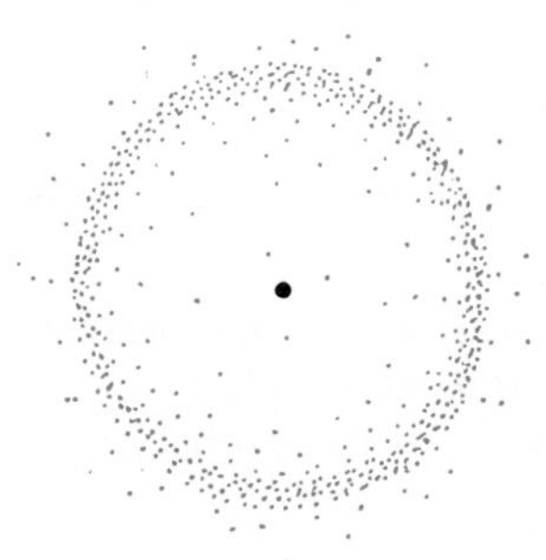

FIGURE 32.14
Probability distribution of an electron cloud.

[5]Schrödinger's wave equation, strictly for math types, is $\left(-\frac{\hbar^2}{2m}\nabla^2 + U\right)\psi = i\hbar\frac{\partial\psi}{\partial t}$

[6]Our short treatment of this complex subject is hardly conducive to any real understanding of quantum mechanics. At best, it serves as a brief overview and possible introduction to further study. The reading suggested at the end of the chapter may be quite useful.

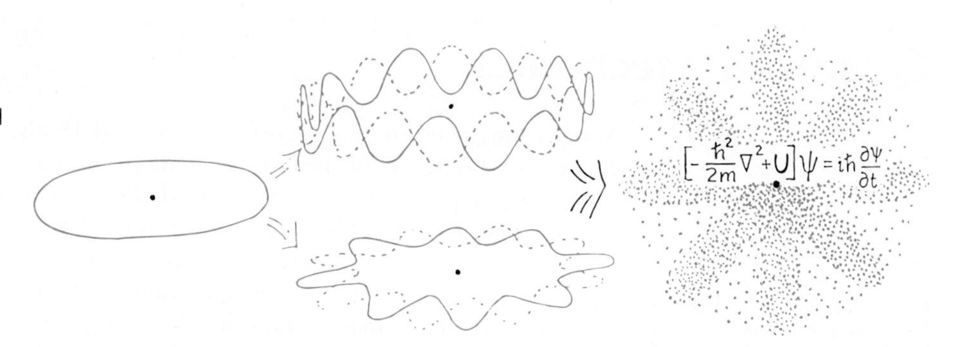

FIGURE 32.15
Interactive Figure
From the Bohr model of the atom, to the modified model with de Broglie waves, to a wave model with the electrons distributed in a "cloud" throughout the atomic volume.

CHECK YOURSELF

1. Consider 100 photons diffracting through a thin slit to form a diffraction pattern. If we detect five photons in a certain region in the pattern, what is the probability (between 0 and 1) of detecting a photon in this region?
2. Suppose that you open a second identical slit and that the diffraction pattern is one of bright and dark bands. Suppose the region where 5 photons hit before now has none. A wave theory says that waves that hit before are now cancelled by waves from the other slit—that crests and troughs combine to zero. But our measurement is of photons that either make a hit or don't. How does quantum mechanics reconcile this?

Considering something to be impossible may reflect a *lack* of understanding, as when scientists thought a single atom could never be seen. Or it may represent a *deep* understanding, as when scientists (and the patent office!) reject perpetual motion machines.

Insights

Most physicists, but not all, view quantum mechanics as a fundamental theory of nature. Interestingly enough, Albert Einstein, one of the founders of quantum physics, never accepted it as fundamental; he considered the probabilistic nature of quantum phenomena as the outcome of a deeper, as yet undiscovered, physics. He stated, "Quantum mechanics is certainly imposing. But an inner voice tells me it is not yet the real thing. The theory says a lot, but does not really bring us closer to the secret of 'the Old One.'"[7]

CHECK YOUR ANSWERS

1. We have approximately a 0.05 probability of detecting a photon at this location. In quantum mechanics we say $|\psi|^2 \approx 0.05$. The true probability could be somewhat more or less than 0.05. Put the other way around, if the true probability is 0.05, the number detected could be somewhat more or less than 5.
2. Quantum mechanics says that photons propagate as waves and are absorbed as particles, with the probability of absorption governed by the maxima and minima of wave interference. Where the combined wave from the two slits has zero amplitude, the probability of a particle being absorbed is zero.

[7]Although Einstein practiced no religion, he often invoked God as the "Old One" in his statements about the mysteries of nature.

Correspondence Principle

If a new theory is valid, it must account for the verified results of the old theory. This is the **correspondence principle,** first articulated by Bohr. New theory and old must correspond; that is, they must overlap and agree in the region where the results of the old theory have been fully verified.

When the techniques of quantum mechanics are applied to macroscopic systems rather than atomic systems, the results are essentially identical with those of classical mechanics. For a large system, such as the solar system, where classical physics is successful, the Schrödinger equation leads to results that differ from classical theory only by infinitesimal amounts. The two domains blend when the de Broglie wavelength is small compared with the dimensions of the system or of the pieces of matter in the system. It is, in fact, impractical to use quantum mechanics in the domains in which classical physics is successful; but, at the atomic level, quantum physics reigns and is the only theory that gives results consistent with what is observed.

The correspondence principle is a general rule not only for good science but for all good theory—even in areas as far removed from science as government, religion, and ethics.

Summary of Terms

Atomic nucleus Positively charged center of an atom, containing protons and neutrons and almost the entire mass of the atom, but only a tiny fraction of its volume.

Electron Negative particle in the shell of an atom.

Ritz combination principle The statement that the frequencies of some spectral lines of the elements are either the sums or the differences of the frequencies of two other lines.

Quantum mechanics The theory of the microworld based on wave functions and probabilities developed especially by Werner Heisenberg (1925) and Erwin Schrödinger (1926).

Schrödinger's wave equation A fundamental equation of quantum mechanics, which relates probability wave amplitudes to the forces acting on a system. It is as basic to quantum mechanics as Newton's laws of motion are to classical mechanics.

Correspondence principle The rule that a new theory must give the same results as the old theory where the old theory is known to be valid.

Suggested Reading

Ford, K. W. *The Quantum World: Quantum Physics for Everyone*. Cambridge, Mass.: Harvard University Press, 2004. A fascinating account of the development of quantum physics with emphasis on the participating physicists.

Gamow, George. *Thirty Years That Shook Physics*. New York: Dover, 1985. A historical tracing of quantum theory by someone who was part of it.

Hey, A. J., and P. Walters. *The Quantum Universe*. New York: Cambridge University Press, 1987. A broad view of modern physics with many illustrations.

Pagels, H. R. *The Cosmic Code: Quantum Physics as the Language of Nature*. New York: Simon & Schuster, 1982. An elegant and highly recommended book for the general reader.

Gleick, James. *Genius: The Life and Science of Richard Feynman*. New York: Vintage, 1993. An inspiring book about an intriguing person.

Review Questions

1. Distinguish between *atomic physics* and *nuclear physics*.

Discovery of the Atomic Nucleus

2. Why do most alpha particles fired through a piece of gold foil emerge almost undeflected?
3. Why do a few alpha particles fired at a piece of gold foil bounce backward?

Discovery of the Electron

4. What did Benjamin Franklin postulate about electricity?
5. What is a cathode ray?
6. What property of a cathode ray is indicated when a magnet is brought near the tube?
7. What did J. J. Thomson discover about the cathode ray?
8. What did Robert Millikan discover about the electron?

Atomic Spectra: Clues to Atomic Structure

9. What did Johann Jakob Balmer discover about the spectrum of hydrogen?

10. What did Johannes Rydberg and Walter Ritz discover about atomic spectra?

Bohr Model of the Atom

11. What relationship between electron orbits and light emission did Bohr postulate?
12. According to Niels Bohr, can a single electron in one excited state give off more than one photon when it jumps to a lower energy state?
13. What is the relationship between the energy differences of orbits in an atom and the light emitted by the atom?

Relative Sizes of Atoms

14. Why is a helium atom smaller than a hydrogen atom?
15. Why are heavy atoms not much larger than the hydrogen atom?

Explanation of Quantized Energy Levels: Electron Waves

16. Why does each element have its own pattern of spectral lines?
17. How does treating the electron as a wave rather than as a particle solve the riddle of why electron orbits are discrete?
18. According to the simple de Broglie model, how many wavelengths are there in an electron wave in the first orbit? In the second orbit? In the *n*th orbit?
19. How can we explain why electrons don't spiral into the attracting nucleus?

Quantum Mechanics

20. What does the wave function ψ represent?
21. Distinguish between a *wave function* and a *probability density function*.
22. How does the probability cloud of the electron in a hydrogen atom relate to the orbit described by Niels Bohr?

Correspondence Principle

23. Exactly what is it that "corresponds" in the correspondence principle?
24. How does Schrödinger's equation fare when applied to the solar system?

Exercises

1. Consider photons emitted from an ultraviolet lamp and a TV transmitter. Which has the greater (a) wavelength, (b) energy, (c) frequency, and (d) momentum?
2. Which color light is the result of a greater energy transition, red or blue?
3. In what way did Rutherford's gold-foil scattering experiment show that the atomic nucleus is both small and very massive?
4. How does Rutherford's model of the atom account for the back-scattering of alpha particles directed at the gold foil?
5. At the time of Rutherford's gold-foil experiment, scientists knew that negatively charged electrons exist within the atom, but they did not know where the positive charge resides. What information about the positive charge was provided by Rutherford's experiment?
6. Uranium-238 is 238 times more massive than hydrogen. Why, then, isn't the diameter of the uranium atom 238 times that of the hydrogen atom?
7. Why does classical physics predict that atoms should collapse?
8. If the electron in a hydrogen atom obeyed classical mechanics instead of quantum mechanics, would it emit a continuous spectrum or a line spectrum? Explain.
9. Why are spectral lines often referred to as "atomic fingerprints"?

10. When an electron makes a transition from its first quantum level to ground level, the energy difference is carried by the emitted photon. In comparison, how much energy is needed to return an electron at ground level to the first quantum level?
11. Figure 32.10 shows three transitions among three energy levels that would produce three spectral lines in a spectroscope. If the energy spacing between the levels were equal, would this affect the number of spectral lines?
12. How can elements with low atomic numbers have so many spectral lines?

13. In terms of wavelength, what is the smallest orbit that an electron can have about the atomic nucleus?

14. Which best explains the photoelectric effect—the particle nature or the wave nature of the electron? Which best explains the discrete levels in the Bohr model of the atom? Explain.
15. How does the wave model of electrons orbiting the nucleus account for discrete energy values rather than arbitrary energy values?
16. Why are the electrons of a helium atom closer to the nucleus than the electron in the hydrogen atom?
17. Why do atoms that have the same number of electron shells decrease in size with increasing atomic number?
18. Would you expect the inner shell of electrons in a uranium atom to be closer to the nucleus than those in an iron atom? Why or why not?
19. Why are atoms with many electrons not appreciably larger than atoms with fewer electrons? (And why are they sometimes smaller than atoms with more electrons?)
20. Why do helium and lithium exhibit very different chemical behavior, even though they differ by only one electron?
21. The Ritz combination principle can be considered to be a statement of energy conservation. Explain.
22. Does the de Broglie model assert that an electron must be moving in order to have wave properties? Defend your answer.
23. Why does no stable electron orbit with a circumference of 2.5 de Broglie wavelengths exist in any atom?
24. An orbit is a distinct path followed by an object in its revolution around another object. An atomic orbital is a *volume of space* in which an electron of a given energy is most likely to be found. What do orbits and orbitals have in common?
25. Can a particle be diffracted? Can it exhibit interference?
26. How does the amplitude of a matter wave relate to probability?
27. If Planck's constant, h, were larger, would atoms be larger also? Defend your answer.
28. What is it that waves in the Schrödinger wave equation?
29. If the world of the atom is so uncertain and subject to the laws of probabilities, how can we accurately measure such things as light intensity, electric current, and temperature?
30. What evidence supports the notion that light has wave properties? What evidence supports the view that light has particle properties?
31. When we say that electrons have particle properties and then continue to say that electrons have wave properties, aren't we contradicting ourselves? Explain.
32. Did Einstein support quantum mechanics as being fundamental physics, or did he think quantum mechanics was inconclusive?
33. When only a few photons are observed, classical physics fails. When many are observed, classical physics is valid. Which of these two facts is consistent with the correspondence principle?
34. When and where do Newton's laws of motion and quantum mechanics overlap?
35. What does Bohr's correspondence principle say about quantum mechanics versus classical mechanics?
36. Does the correspondence principle have application to macroscopic events in the everyday macroworld?
37. Richard Feynman, in his book *The Character of Physical Law,* states: "A philosopher once said, 'It is necessary for the very existence of science that the same conditions always produce the same results.' Well, they don't!" Who was speaking of classical physics, and who was speaking of quantum physics?
38. What does the wave nature of matter have to do with the fact that we can't walk through solid walls, as Hollywood movies often show using special effects?
39. Largeness or smallness has meaning only relative to something else. Why do we usually call the speed of light "large" and Planck's constant "small"?
40. Make up a multiple-choice question that would check a classmate's understanding of the difference between the domains of classical mechanics and quantum mechanics.

Problems

1. The higher the energy level occupied by an electron in the hydrogen atom, the larger the atom. The size of the atom is proportional to n^2, where $n = 1$ designates the lowest (or "ground") state, $n = 2$ is the second state, $n = 3$ is the third state, and so on. If the atom's diameter is 1×10^{-10} m in its lowest energy state, what is its diameter in state number 50? How many unexcited atoms could fit within this one giant atom?
2. We can define zero energy in the hydrogen atom to be the energy of the lowest (or "ground") state. Energies of successive excited states above the ground state are proportional to $100 - (100/n^2)$, for the quantum numbers $n = 1, 2, 3$, etc. So, on this scale, the energy of level $n = 2$ is $[100 - (100/4)] = 75.0$, that of $n = 3$ is $[100 - (100/9)] = 88.9$, and that of $n = 4$, $[100 - (100/16)] = 93.8$, and so on. (a) Sketch, approximately to scale, a diagram that includes the ground state and the lowest four excited states ($n = 1$ to 5). (b) The most prominent red line in the spectrum of hydrogen is caused by an electron transition from state 3 to state 2. Will the transition from 4 to 3 produce a higher-frequency, or a lower-frequency, spectral line? (c) What about the 2-to-1 transition?

Walter Steiger, first pioneer of telescopes in Hawaii, examines vapor trails in a small cloud chamber.

CHAPTER 33

The Atomic Nucleus and Radioactivity

Nuclear Physics

Radioactivity has been around since Earth's beginning.

Insights

In the previous chapter, we were concerned with *atomic physics*—the study of the clouds of electrons that make up the atom. In this chapter, we will burrow beneath the electrons and go deeper into the atom—to the atomic nucleus. Now we study *nuclear physics*, where available energies dwarf those available among electrons. Nuclear physics is a topic of great public interest—and public fear.

Public phobia about anything *nuclear* or anything *radioactive* is much like the phobia about electricity that was common more than a century ago. The fears of electricity in homes stemmed from ignorance. Indeed, electricity can be quite dangerous and even lethal when improperly handled. But, with safeguards and well-informed consumers, society has determined that the benefits of electricity outweigh its risks. Today, we are making similar decisions about nuclear technology's risks versus its benefits. These decisions should be made with an adequate understanding of the atomic nucleus and its inner properties.

Knowledge of the atomic nucleus began with the chance discovery of radioactivity in 1896, which, in turn, was based on the discovery of X-rays two months earlier. So, to begin our study of nuclear physics, we'll consider X-rays.

X-Rays and Radioactivity

Before the turn of the twentieth century, the German physicist Wilhelm Roentgen discovered a "new kind of ray" produced by a beam of "cathode rays" (later found to be electrons) striking the glass surface of a gas-discharge tube. He named these **X-rays**—rays of an unknown nature. Roentgen found that X-rays could pass through solid materials, could ionize the air, showed no refraction in glass, and were undeflected by magnetic fields. Today we know that X-rays are high-frequency electromagnetic waves, usually emitted by the de-excitation of the innermost orbital electrons of atoms. Whereas the electron current in a fluorescent lamp excites the outer electrons of atoms and produces ultraviolet and visible photons, a more energetic beam of electrons striking a solid surface excites the innermost electrons and produces higher-frequency photons of X-radiation.

X-ray photons have high energy and can penetrate many layers of atoms before being absorbed or scattered. X-rays do this when they pass through your soft tissues to produce images of the bones inside of your body (Figure 33.1).

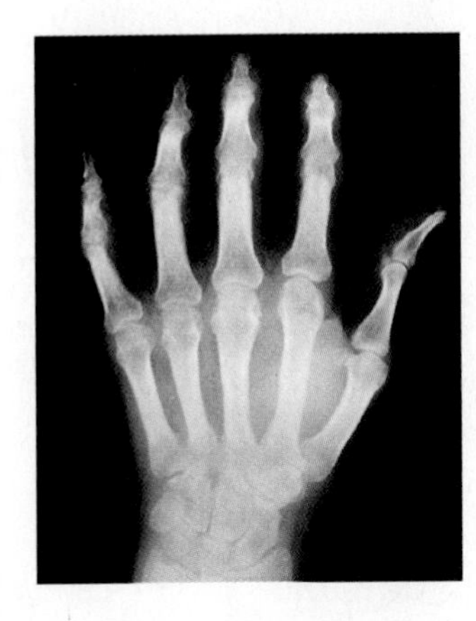

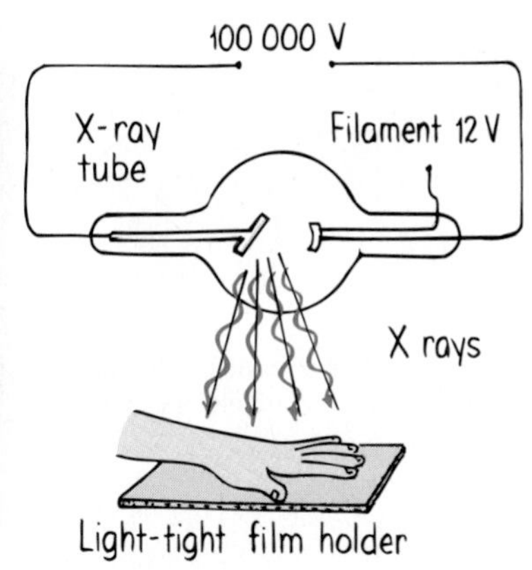

FIGURE 33.1
X-rays emitted by excited metallic atoms in the electrode penetrate flesh more readily than bone and produce an image on the film.

Marie Curie (1867–1934)

In a modern X-ray tube, the target of the electron beam is a metal plate rather than the glass wall of the tube.

Two months after Roentgen announced his discovery of X-rays, the French physicist Antoine Henri Becquerel tried to find out whether any elements spontaneously emitted X-rays. To do this, he wrapped a photographic plate in black paper to keep out the light and then put pieces of various elements against the wrapped plate. From Roentgen's work, Becquerel knew that, if these materials emitted X-rays, the rays would go through the paper and blacken the plate. He found that, although most elements produced no effect, uranium did produce rays. It was soon discovered that similar rays are emitted by other elements, such as thorium, actinium, and two new elements discovered by Marie and Pierre Curie—polonium and radium. The emission of these rays was evidence of much more drastic changes in the atom than atomic excitation. These rays were the result not of changes in the electron energy states of the atom but of changes occurring within the central atomic core—the nucleus. These rays were the result of a spontaneous chipping apart of the atomic nucleus—**radioactivity.**

Alpha, Beta, and Gamma Rays

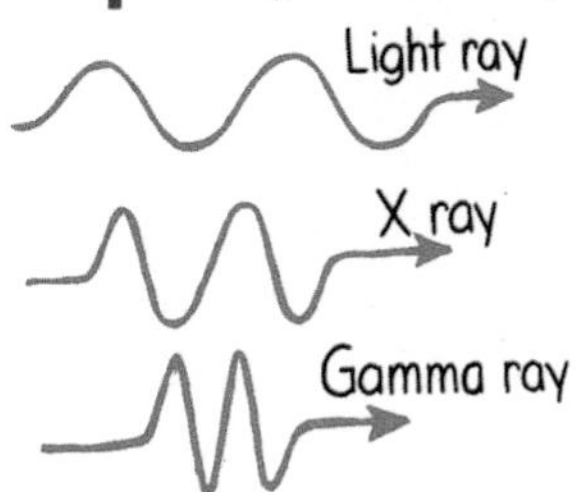

FIGURE 33.2
Interactive Figure
A gamma ray is part of the electromagnetic spectrum. It is simply electromagnetic radiation that is much higher in frequency and energy than light and X-rays.

More than 99.9% of the atoms in our everyday environment are stable. The nuclei in those atoms will likely not change over the lifetime of the universe. But some kinds of atoms are not stable. All elements having an atomic number greater than 82 (lead) are radioactive. These elements, and others, emit three distinct types of radiation, named by the first three letters of the Greek alphabet, α, β, γ—*alpha, beta,* and *gamma.* **Alpha rays** have a positive electrical charge, **beta rays** have a negative charge, and **gamma rays** have no charge at all. The three rays can be separated by placing a magnetic field across their paths (Figure 33.3). Further investigation has shown that an alpha ray is a stream of helium nuclei, and a beta ray is a stream of electrons. Hence, we often call these *alpha particles* and *beta particles*. A gamma ray is electromagnetic radiation (a stream of photons) whose frequency is even higher than that of X-rays. Whereas X-rays originate in the electron cloud outside the atomic nucleus, gamma rays originate in the nucleus. Gamma photons provide information about nuclear structure, much as visible and X-ray photons provide information about atomic electron structure.

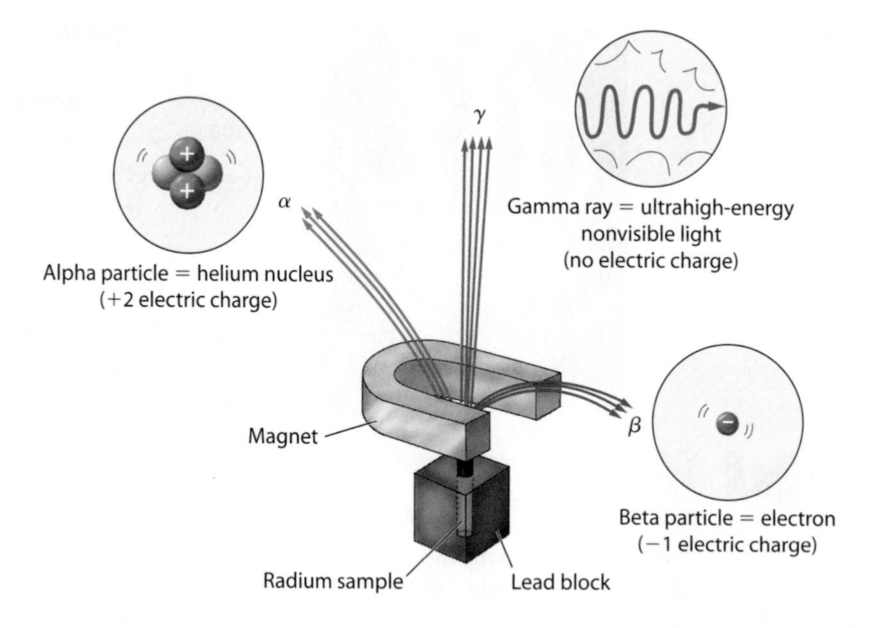

FIGURE 33.3 Interactive Figure

In a magnetic field, alpha rays bend one way, beta rays bend the other way, and gamma rays don't bend at all. The combined beam comes from a radioactive source placed at the bottom of a hole drilled in a lead block.

Once alpha and beta particles are slowed by collisions, they become harmless. They combine to become helium atoms.

Insights

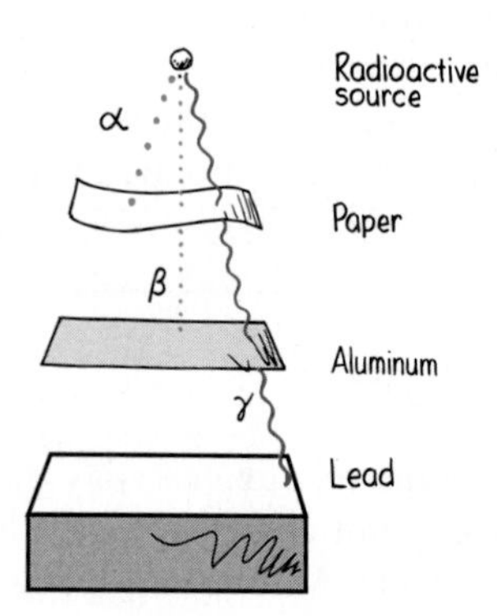

FIGURE 33.4 Interactive Figure

Alpha particles are the least penetrating and can be stopped by a few sheets of paper. Beta particles will readily pass through paper, but not through a sheet of aluminum. Gamma rays penetrate several centimeters into solid lead.

PRACTICING PHYSICS

Some watches and clocks have luminous hands that glow continuously. In some of these, what causes the glow is traces of radioactive radium bromide mixed with zinc sulfide. (Safer clock faces use light rather than radioactive decay as a means of excitation and, as a result, become progressively dimmer in the dark.) If you have a glow-all-the-time type of clock available, take it into a completely dark room and, after your eyes have adjusted to the dark, examine the hands of the clock with a very strong magnifying glass or the eyepiece of a microscope or telescope. You should be able to see individual tiny flashes, which together seem to be a steady source of light to the unaided eye. Each flash occurs when an alpha particle ejected by a radium nucleus strikes a molecule of zinc sulfide.

The Nucleus

Light is emitted by energy-level transitions in atoms; gamma rays are emitted by similar energy transitions within the atomic nucleus.

Insights

As described in earlier chapters, the atomic nucleus occupies only a few quadrillionths the volume of the atom, leaving most of the atom as empty space. Particles occupying the nucleus are called **nucleons.** If they are electrically charged, they are protons; if they are electrically neutral, they are neutrons. The positive charge of the proton is the same in magnitude as the negative charge of the electron. Nucleons have nearly 2000 times the mass of electrons, so the mass of an atom is practically equal to the mass of its nucleus. The neutron's mass is slightly greater than the proton's. We'll see that, when an electron is ejected from a neutron (beta emission), the neutron becomes a proton.

Nuclear radii range from about 10^{-15} meter for hydrogen to about seven times larger for uranium. Some nuclei are spherical, but most deviate from that shape in the "football" way and a few in the "doorknob" way. Protons and neutrons within the nucleus move relatively freely, yet provide a "skin" that gives the nucleus some properties of a liquid drop.

The emission of alpha particles is a quantum phenomenon that can be understood in terms of waves and probability. Just as orbital electrons form a probability cloud about the nucleus, inside the radioactive nucleus there is a similar probability cloud for the clustering of the two protons and two neutrons that constitute an alpha particle. A tiny part of the alpha particle's probability wave extends outside the nucleus, meaning that there is a small chance that the alpha particle will be outside. Once outside, it is hurled violently away by electric repulsion. The electron emitted in beta decay, on the other hand, is not "there" before it is emitted. It is created at the moment of radioactive decay when a neutron is transformed into a proton.

In addition to alpha, beta, and gamma rays, more than 200 various other particles have been detected coming from the nucleus when it is clobbered by energetic particles. We do not think of these so-called elementary particles as being buried within the nucleus and then popping out, just as we do not think of a spark as being buried in a match before it is struck. These particles, like the electrons in beta decay, come into being when the nucleus is disrupted. There are regularities in the masses of these particles as well as the particular characteristics of their creation. Almost all of the new particles created in nuclear collisions can be understood as combinations of just six subnuclear particles—the **quarks.**

Two of the six quarks are the fundamental building blocks of all nucleons. An unusual property of quarks is that they carry fractional electrical charges. One kind, the *up* quark, carries +2/3 the proton charge, and another kind, the *down* quark, has −1/3 the proton charge. (The name *quark,* inspired by a quotation from *Finnegans Wake* by James Joyce, was chosen in 1963 by Murray Gell-Mann, who first proposed their existence.) Each quark has an antiquark with opposite electric charge. The proton consists of the combination *up up down,* and the neutron of *up down down.* The other four quarks bear the whimsical names *strange, charm, top,* and *bottom.* All of the several hundred particles that feel the strong nuclear force appear to be composed of some combination of the six quarks. As with magnetic poles, no quarks have been isolated and experimentally observed—most theorists think quarks, by their nature, cannot be isolated.

Particles lighter than protons and neutrons, like electrons and muons, and still lighter particles called *neutrinos,* are members of a class of six particles called *leptons*. Leptons are not composed of quarks. The six quarks and six leptons are thought to be the truly *elementary particles,* particles not composed of more basic entities. Investigation of elementary particles is at the frontier of our present knowledge and the area of much current excitement and research.

Isotopes

Recall, from Chapter 11, that the nucleus of a hydrogen atom contains a single proton, a helium nucleus contains two protons, a lithium nucleus has three, and so forth. Every succeeding element in the list of elements has one more proton than the preceding element. Recall, also from Chapter 11, that the number of protons in the nucleus is the same as the **atomic number.** Hydrogen has atomic number 1; helium, atomic number 2; lithium, 3; and so on.

The number of neutrons in the nucleus of a given element may vary. Although the nucleus of every hydrogen atom contains one proton, some hydrogen nuclei contain a neutron in addition to the proton. In rare instances, a hydrogen nucleus may contain *two* neutrons in addition to the proton. Recall that atoms containing like numbers of protons but unlike numbers of neutrons are **isotopes** of a given element.

Tritium is used in various self-luminescent devices, such as exit signs in buildings, aircraft dials, gauges, luminous paints, and wristwatches. Tritium is also used in life science research, and in studies investigating the safety of potential new drugs.

Insights

The most common isotope of hydrogen is ${}^1_1H$. The subscript (lower) number refers to the atomic number and the superscript (upper) number refers to the **atomic mass.** The double-mass hydrogen isotope ${}^2_1H$ is called *deuterium.* "Heavy water" is the name usually given to H_2O in which one or both of the H atoms have been replaced by deuterium atoms. In all naturally occurring hydrogen compounds, such as hydrogen gas and water, there is 1 atom of deuterium to about 6000 atoms of hydrogen. The triple-mass hydrogen isotope ${}^3_1H$, which is radioactive and lives long enough to be a known constituent of atmospheric water, is called *tritium.* Tritium is present only in extremely minute amounts—less than 1 per 10^{17} atoms of ordinary hydrogen. The tritium used for practical purposes is made in nuclear reactors or accelerators, and not extracted from natural sources. (Interestingly, tritium occurs in abundance on the Moon's surface.)

All elements have a variety of isotopes. For instance, three isotopes of uranium naturally occur in the Earth's crust; the most common is ${}^{238}_{92}U$. In briefer notation, we can drop the atomic number and simply say uranium-238, or, more concisely, U-238. Of the 83 elements present in significant amounts on Earth, 20 have a single stable (nonradioactive) isotope. The others have from 2 to 10 stable isotopes. More than 2000 distinct isotopes, radioactive and stable, are known.

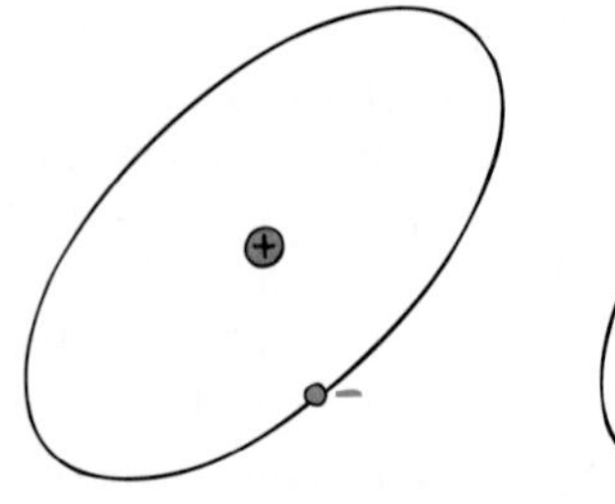
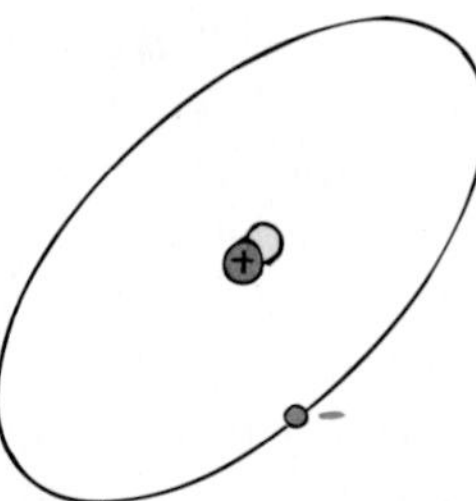
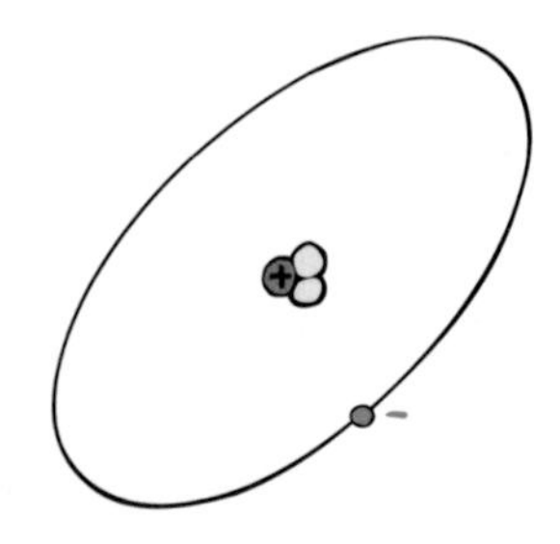

FIGURE 33.5
Three isotopes of hydrogen. Each nucleus has a single proton, which holds a single orbital electron, which, in turn, determines the chemical properties of the atom. The different number of neutrons (yellow) changes the mass of the atom but not its chemical properties.

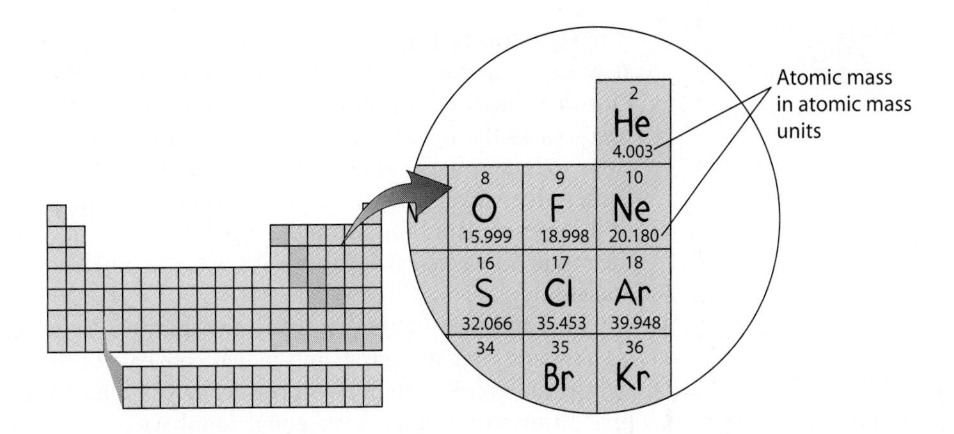

FIGURE 33.6
Helium, He, has an atomic mass of 4.003 amu, and neon, Ne, has an atomic mass of 20.180 amu. These values are averaged by isotopic abundances in Earth's surface.

Recall, from Chapter 11, that a special unit for atomic masses is the **atomic mass unit (amu),** which is based on the mass of the common carbon atom—which, by common agreement, is arbitrarily given the value of exactly 12. An amu of value 1 would be one-twelfth the mass of common carbon-12, equal to 1.661×10^{-27} kg, slightly less than the mass of a single proton. As shown in Figure 33.6, the atomic masses listed in the periodic table are in atomic mass units. The value listed for each element is the average atomic mass of its various isotopes.

CHECK YOURSELF

State the numbers of protons and neutrons in ${}^{1}_{1}H$, ${}^{14}_{6}C$, and ${}^{235}_{92}U$.

Why Atoms Are Radioactive

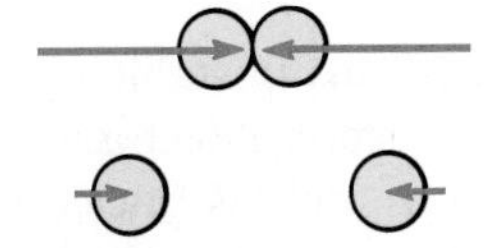

FIGURE 33.7
The nuclear strong interaction is a short-range force. For nucleons very close or in contact, it is very strong; but, a few nucleon diameters away, it is nearly zero.

The positively charged and closely spaced protons in a nucleus have huge electrical forces of repulsion between them. Why don't they fly apart in response to this huge repulsive force? Because there is an even more formidable force within the nucleus—the nuclear force. Both neutrons and protons are bound to each other by this attractive force. The nuclear force is much more complicated than the electrical force. The principal part of the nuclear force, the part that holds the nucleus together, is called the *strong interaction.*[1] The strong interaction is an attractive force that acts between protons, neutrons, and particles called *mesons,* all of which are called *hadrons.* This force acts over only a very short distance (Figure 33.7).

CHECK YOUR ANSWERS

The atomic number gives the number of protons. The number of neutrons is the atomic mass minus the atomic number. So we see 1 proton and no neutrons in ${}^{1}_{1}H$; 6 protons and 8 neutrons in ${}^{14}_{6}C$; and 92 protons and 143 neutrons in ${}^{235}_{92}U$.

[1]Fundamental to the strong interaction is the *color force* (which has nothing to do with visible color). This color force interacts between quarks and holds them together by the exchange of "gluons." Read more about this in H. R. Pagels, *The Cosmic Code: Quantum Physics as the Language of Nature.* New York: Simon & Schuster, 1982.

It is very strong between nucleons about 10^{-15} meter apart, but close to zero at greater separations. So the strong nuclear interaction is a short-range force. Electrical interaction, on the other hand, is a relatively long-range force, for it weakens as the inverse square of separation distance. So as long as protons are close together, as in small nuclei, the nuclear force easily overcomes the electrical force of repulsion. But for distant protons, like those on opposite edges of a large nucleus, the attractive nuclear force may be small in comparison to the repulsive electrical force. Hence, a larger nucleus is not as stable as a smaller nucleus.

The presence of the neutrons also plays a large role in nuclear stability. It so happens that a proton and a neutron can be bound together a little more tightly, on average, than two protons or two neutrons. As a result, many of the first 20 or so elements have equal numbers of neutrons and protons.

Without the nuclear strong force—strong interaction—there would be no atoms beyond hydrogen.

Insights

For heavier elements, it is a different story, because protons repel each other electrically and neutrons do not. If you have a nucleus with 28 protons and 28 neutrons, for example, it can be made more stable by replacing two of the protons with neutrons, resulting in Fe-56, the iron isotope with 26 protons and 30 neutrons. The inequality of neutron and proton numbers becomes more pronounced for elements that are heavier still. For example, in U-238, which has 92 protons, there are 146 (238 − 92) neutrons. If the uranium nucleus were to have equal numbers of protons and neutrons, 92 protons and 92 neutrons, it would fly apart at once because of the electrical repulsion forces. The extra 54 neutrons are needed for relative stability. Even so, the U-238 nucleus is still fairly unstable because of the electrical forces.

To put the matter in another way: There is an electrical repulsion between *every pair* of protons in the nucleus, but there is not a substantial nuclear attractive force between every pair (Figure 33.8). Every proton in the uranium nucleus exerts a repulsion on each of the other 91 protons—those near and those far. However, each proton (and neutron) exerts an appreciable nuclear attraction only on those nucleons that happen to be near it.

One ton of ordinary granite contains about 9 grams of uranium and 20 grams of thorium. One ton of basalt contains 3.5 grams of uranium and 7.7 grams of thorium.

Insights

All nuclei having more than 82 protons are unstable. In this unstable environment, alpha and beta emissions take place. The force responsible for beta emission is called the *weak interaction:* It acts on leptons as well as nucleons. When an electron is created in beta decay, another lighter particle called an *antineutrino* is also created and also shoots out of the nucleus.

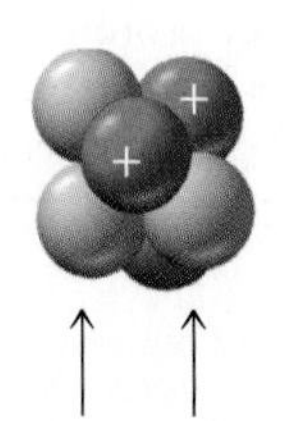

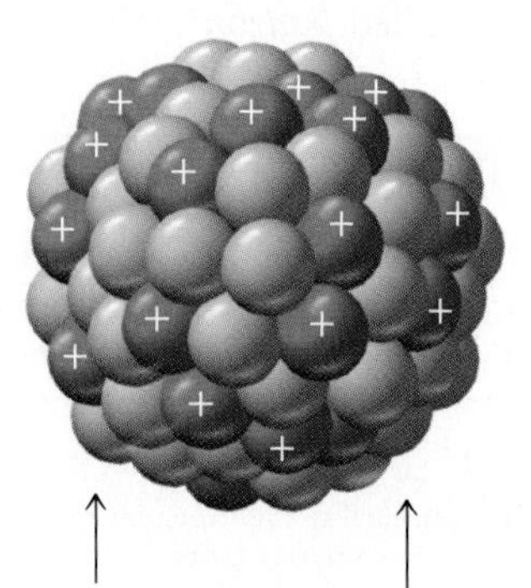

FIGURE 33.8
(a) All nucleons in a small nucleus are close to one another; hence, they experience an attractive strong nuclear force. (b) Nucleons on opposite sides of a larger nucleus are not as close to one another, and so the attractive strong nuclear forces holding them together are much weaker. The result is that the large nucleus is less stable.

Half-Life

The radioactive decay rate of an element is measured in terms of a characteristic time, the **half-life.** This is the time it takes for half of an original quantity of a radioactive isotope to decay. Radium-226, for example, has a half-life of 1620 years. This means that half of any given specimen of radium-226 will be converted into other elements by the end of 1620 years. In the following 1620 years, half of the remaining radium will decay, leaving only one-fourth the original amount of radium (after 20 half-lives, the initial quantity radium-226 will be diminished by a factor of about one million). Cobalt-60, a standard source for radiotherapy, has a half-life of 5.27 years. The isotopes of some elements have a half-life of less than a millionth of a second, while uranium-238, for example, has a half-life of 4.5 billion years. Every isotope of every radioactive element has its own characteristic half-life.

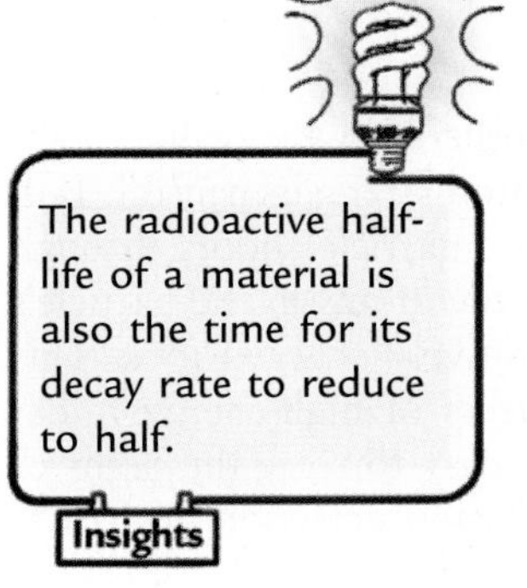

The radioactive half-life of a material is also the time for its decay rate to reduce to half.

Insights

Many elementary particles have very short half-lives. The muon (a close relative of the electron), which is produced when cosmic rays bombard atomic nuclei in the upper atmosphere, has a half-life of 2 millionths of a second (2×10^{-6}s)—actually a very long time on the subnuclear scale. The shortest half-lives of elementary particles are on the order of 10^{-23} second, the time light takes to travel a distance equal to the diameter of a nucleus.

The half-lives of radioactive elements and elementary particles appear to be absolutely constant, unaffected by any external conditions, however drastic. Wide temperature and pressure extremes, strong electric and magnetic fields, and even violent chemical reactions have no detectable effect on the rate of decay of a given element. Any of these stresses, although severe by ordinary standards, is far too mild to affect the nucleus in the deep interior of the atom.

It is not necessary to wait through the duration of a half-life in order to measure it. The half-life of an element can be calculated at any given moment by measuring the rate of decay of a known quantity. This is easily done using a radiation detector. In general, the shorter the half-life of a substance, the faster it disintegrates and the greater its decay rate.

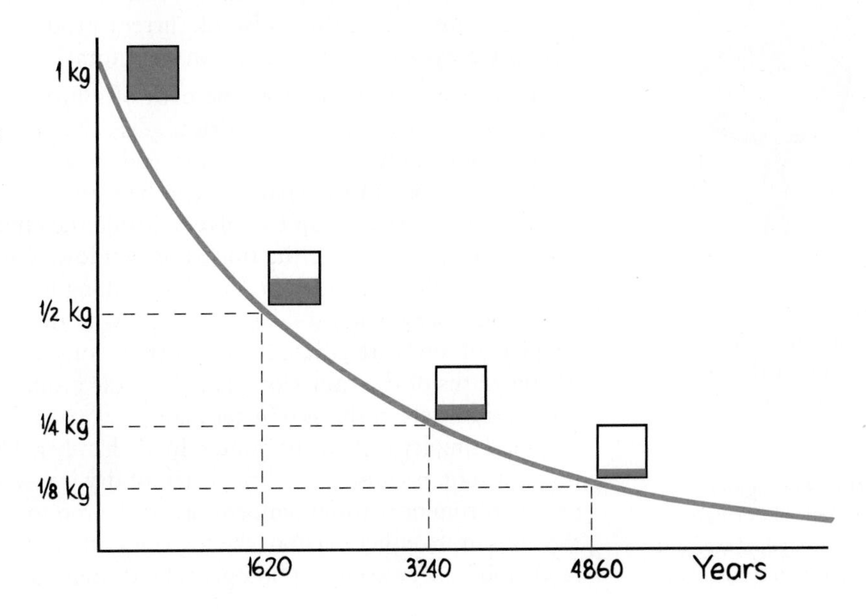

FIGURE 33.9
Interactive Figure
Every 1620 years, the amount of radium decreases by half.

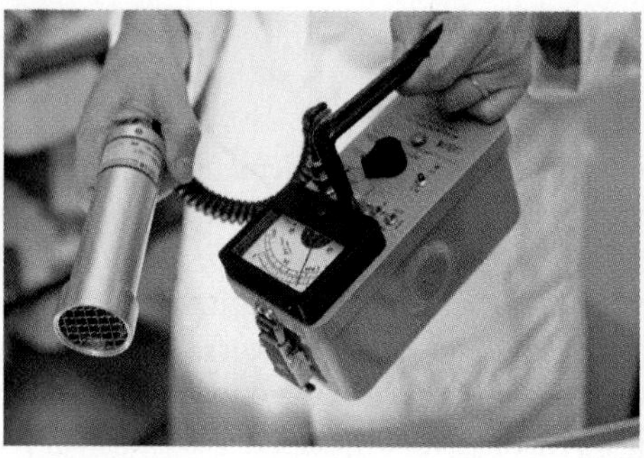

a

b

FIGURE 33.10
Radiation detectors. (a) A Geiger counter detects incoming radiation by the way the radiation ionizes a gas enclosed in the tube. (b) A scintillation counter indicates incoming radiation by flashes of light produced when charged particles or gamma rays pass through the counter.

Radiation Detectors

Ordinary thermal motions of atoms bumping one another in a gas or liquid are not energetic enough to dislodge electrons, and the atoms remain neutral. But, when an energetic particle such as an alpha or a beta particle shoots through matter, electrons one after another are knocked from the atoms in the particle's path. The result is a trail of freed electrons and positively charged ions. This ionization process is responsible for the harmful effects of high-energy radiation in living cells. Ionization also makes it relatively easy to trace the paths of high-energy particles. We will briefly discuss five radiation detection devices.

1. A *Geiger counter* consists of a central wire in a hollow metal cylinder filled with low-pressure gas. An electrical voltage is applied across the cylinder and wire so that the wire is more positive than the cylinder. If radiation enters the tube and ionizes an atom in the gas, the freed electron is attracted to the positively charged central wire. As this electron is accelerated toward the wire, it collides with other atoms and knocks out more electrons, which, in turn, produce more electrons, and so on, resulting in a cascade of electrons moving toward the wire. This makes a short pulse of electric current, which activates a counting device connected to the tube. Amplified, this pulse of current produces the familiar clicking sound we associate with radiation detectors.
2. A *cloud chamber* shows a visible path of ionizing radiation in the form of fog trails. It consists of a cylindrical glass chamber closed at the upper end by a glass window and at the lower end by a movable piston. Water vapor or alcohol vapor in the chamber can be saturated by adjusting the piston.

   The radioactive sample is placed inside the chamber, as shown in Figure 33.11, or outside the thin glass window. When radiation passes through the chamber, ions are produced along its path. If the saturated air in the chamber is then suddenly cooled by motion of the piston, tiny droplets of moisture condense about these ions and form vapor trails, showing the paths of the radiation. These are the atomic versions of the ice-crystal trails left in the sky by jet planes.

   Even simpler is the continuous cloud chamber. This has a steady supersaturated vapor, because it rests on a slab of dry ice, so there is a temperature gradient from near room temperature at the top to very low temperature at the bottom. In either version, the fog tracks that form are illuminated with a lamp and may be seen or photographed through the glass top. The chamber

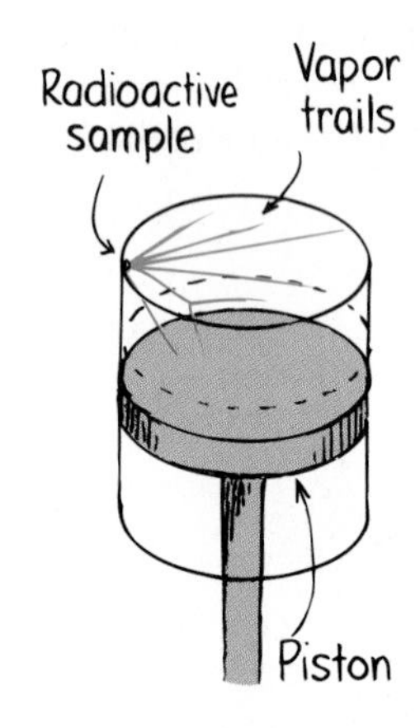

FIGURE 33.11
A cloud chamber. Charged particles moving through supersaturated vapor leave trails. When the chamber is in a strong electric or magnetic field, bending of the tracks provides information about the charge, mass, and momentum of the particles.

may be placed in a strong electric or magnetic field, which will bend the paths in a manner that provides information about the charge, mass, and momentum of the radiation particles. Positively and negatively charged particles will bend in opposite directions.

Cloud chambers, which were critically important tools in early cosmic-ray research, are now used principally for demonstration. Perhaps your instructor will show you one, as does Walter Steiger on page 634.

3. The particle trails seen in a *bubble chamber* are minute bubbles of gas in liquid hydrogen (Figure 33.12). The liquid hydrogen is heated under pressure in a glass and stainless steel chamber to a point just short of boiling. If the pressure in the chamber is suddenly released at the moment an ion-producing particle enters, a thin trail of bubbles is left along the particle's path. All the liquid erupts to a boil, but, in the few thousandths of a second before this happens, photographs are taken of the particle's short-lived trail. As with the cloud chamber, a magnetic field in the bubble chamber reveals the charge and relative mass of the particles being studied. Bubble chambers have been widely used by researchers in past decades, but presently there is greater interest in spark chambers.
4. A *spark chamber* is a counting device that consists of an array of closely spaced parallel plates; every other plate is grounded, and the plates in between are maintained at a high voltage (about 10 kV). Ions are produced in the gas between the plates as charged particles pass through the chamber. Discharge along the ionic path produces a visible spark between pairs of plates. A trail of many sparks reveals the path of the particle. A different design, called a *streamer chamber,* consists of only two widely spaced plates, between which an electric discharge, or "streamer," closely follows the path of the incident charged particle. The principal advantage of spark and streamer chambers over the bubble chamber is that more events can be monitored in a given time.
5. A *scintillation counter* uses the fact that certain substances are easily excited and emit light when charged particles or gamma rays pass through them. Tiny flashes of light, or scintillations, are converted into electric signals by special photomultiplier tubes. A scintillation counter is

FIGURE 33.12
Tracks of elementary particles in a bubble chamber. (The trained eye notes that two particles were destroyed at the point where the spirals emanate, with four others created in the collision.) Note that this image graces the cover of this book, illustrating both the micro and macro physics about us.

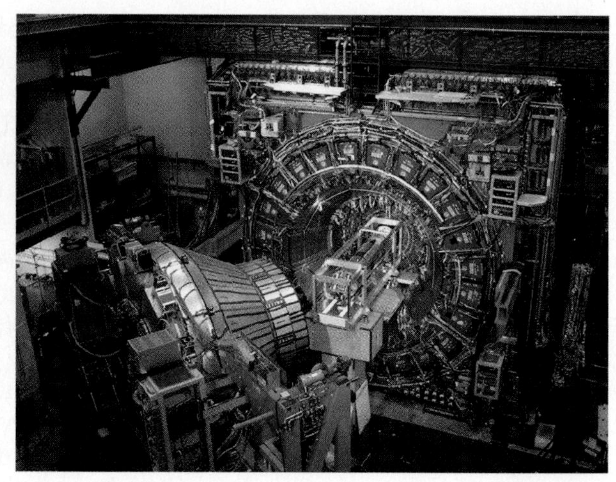

FIGURE 33.13
(a) Installation of the Big European Bubble Chamber (BEBC) at CERN, near Geneva, typical of the large bubble chambers used in the 1970s to study particles produced by high-energy accelerators. The 3.7-m cylinder contained liquid hydrogen at −173°C. (b) The collider detector at Fermilab, which detects and records myriad events when particle beams collide. The detector stands two stories tall, weighs 4500 tons, and was built by a collaboration of more than 170 physicists from the U.S., Japan, and Italy and was built by an international collaboration of about 500 physicists from various countries.

much more sensitive to gamma rays than a Geiger counter, and, in addition, it can measure the energy of charged particles or gamma rays absorbed in the detector. Ordinary water, when highly purified, can serve as a scintillator.

CHECK YOURSELF

1. If a sample of a radioactive isotope has a half-life of 1 day, how much remains at the end of the second day? At the end of the third day?
2. What becomes of isotopes that undergo alpha decay?
3. Which produces a higher counting rate on a radiation detector—a gram of radioactive material having a short half-life, or a gram having a long half-life?

Transmutation of Elements

When a radioactive nucleus emits an alpha or a beta particle, there is a change in atomic number—a different element is formed. This changing of one chemical element to another is called **transmutation.** Transmutation occurs in natural events and is also initiated artificially in the laboratory.

CHECK YOUR ANSWERS

1. At the end of the first day, it decays to one-half. At the end of the second day, it decays to half of this half. Half of one-half is one-fourth. So it decays to one-fourth, and, for a pure sample, one-fourth of the original sample will remain. Can you see that, at the end of three days, one-eighth of the original isotope remains?
2. They become altogether different elements, elements two steps down in atomic number.
3. The material with the shorter half-life decays more quickly and registers a higher counting rate on a radiation detector.

Natural Transmutation

Consider uranium-238, the nucleus of which contains 92 protons and 146 neutrons. When an alpha particle is ejected, the nucleus is reduced by two protons and two neutrons (an alpha particle is a helium nucleus consisting of two protons and two neutrons). An element is defined by the number of protons in its nucleus, so the 90 protons and 144 neutrons left behind are no longer uranium but the nucleus of a different element—*thorium*. This reaction is expressed as

$$^{238}_{92}\mathrm{U} \rightarrow {}^{234}_{90}\mathrm{Th} + {}^{4}_{2}\mathrm{He}$$

The arrow shows that $^{238}_{92}\mathrm{U}$ changes into the two elements to the right of the arrow. When this transmutation occurs, energy is released in three forms; partly as gamma radiation, partly as kinetic energy of the recoiling thorium atom, and mostly as kinetic energy of the alpha particle ($^{4}_{2}\mathrm{He}$). In equations such as this, the mass numbers at the top ($238 = 234 + 4$) and the atomic numbers at the bottom ($92 = 90 + 2$) balance.

Thorium-234, the product of this reaction, is also radioactive. When it decays, it emits a beta particle. Recall that a beta particle is an electron—not an orbital electron but one created within the nucleus. You may find it useful to think of a neutron as a combined proton and electron (although it's not really the case) because, when the electron is emitted, a neutron becomes a proton.[2] A neutron is ordinarily stable when it is locked in a nucleus, but a free neutron is radioactive and has a half-life of 12 minutes. It decays into a proton by beta emission. So, in the case of thorium, which has 90 protons, beta emission leaves the nucleus with one less neutron and one more proton. The new nucleus then has 91 protons and is no longer thorium but the element *protactinium*. Although the atomic number has increased by 1 in this process, the mass number (protons + neutrons) remains the same. The nuclear equation is

$$^{234}_{90}\mathrm{Th} \rightarrow {}^{234}_{91}\mathrm{Pa} + {}^{0}_{-1}e$$

We write an electron as $^{0}_{-1}e$. The 0 indicates that the electron mass is closer to zero than to the 1 of protons and neutrons, which alone contribute to the mass number. The -1 is the charge of the electron. Remember that this electron is

[2]Beta emission is always accompanied by the emission of a neutrino (actually an antineutrino), a neutral particle that travels at about the speed of light. The neutrino ("little neutral one," named by Enrico Fermi) was postulated to retain conservation laws by Wolfgang Pauli in 1930 and detected in 1956. Detecting neutrinos is difficult because they interact very weakly with matter. Millions of them fly through you every second of every day because the universe is filled with them. Only one or two times a year does a neutrino or two interact with the matter of your body.

a beta particle from the nucleus and not an electron from the electron cloud that surrounds the nucleus.

We can see that, when an element ejects an alpha particle from its nucleus, the mass number of the resulting atom decreases by 4 and its atomic number decreases by 2. The resulting atom belongs to an element two places back in the periodic table. When an element ejects a beta particle (electron) from its nucleus, the mass of the atom is practically unaffected, so there is no change in mass number, but its atomic number *increases* by 1. The resulting atom belongs to an element one place forward in the periodic table. Gamma emission results in no change in either the mass number or the atomic number. So we see that the emission of an alpha or beta particle by an atom produces a different atom in the periodic table. Alpha emission lowers the atomic number and beta emission increases it. Radioactive elements can decay backward or forward in the periodic table.[3]

The radioactive decay of $^{238}_{92}\text{U}$ to $^{206}_{82}\text{Pb}$, an isotope of lead, is shown in the decay-scheme chart in Figure 33.14. Each nucleus that plays a role in the decay scheme is shown by a burst. The vertical column containing the burst shows its atomic number, and the horizontal column shows its mass number. Each green arrow shows an alpha decay, and each red arrow shows a beta decay. Notice that some of the nuclei in the series can decay in both ways. This is one of several similar radioactive series that occur in nature.

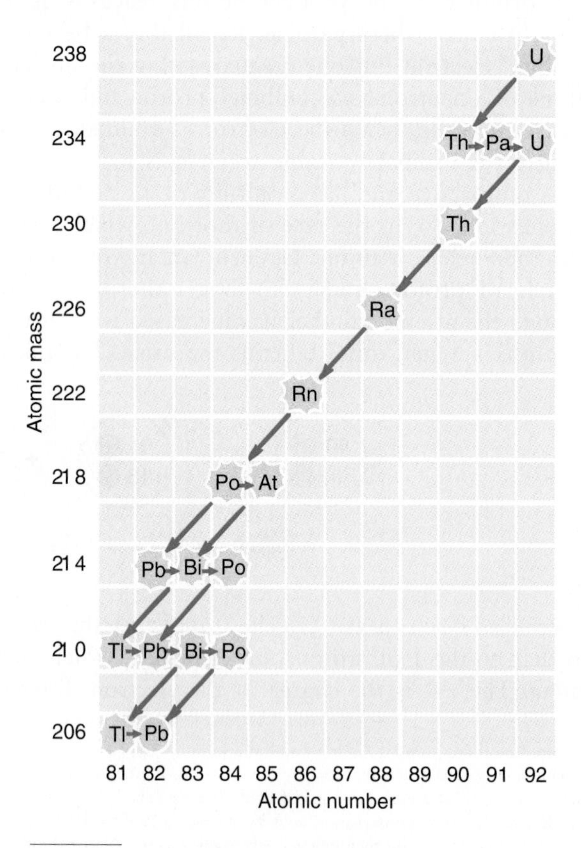

FIGURE 33.14
U-238 decays to Pb-206 through a series of alpha (green) and beta (red) decays.

---

[3]Sometimes a nucleus emits a positron, which is the "antiparticle" of an electron. In this case, a proton becomes a neutron, and the atomic number is decreased.

CHECK YOURSELF

1. Complete the following nuclear reactions.
   a. $^{228}_{88}\text{Ra} \rightarrow ^{?}_{?}? + ^{0}_{-1}e$
   b. $^{209}_{84}\text{Po} \rightarrow ^{205}_{82}\text{Pb} + ^{?}_{?}?$
2. What is the end result of all the uranium-238 that undergoes radioactive decay?

Artificial Transmutation

The alchemists of old tried vainly for more than 2000 years to cause the transmutation of one element into another. Enormous efforts were expended and elaborate rituals performed in the quest to change lead into gold. They never succeeded. Lead, in fact, can be transformed into gold, but not by the chemical means employed by the alchemists. Chemical reactions involve alterations of the outermost shells of the electron clouds of atoms and molecules. To change an element from one kind to another, one must go deep within the electron clouds to the central nucleus, which is immune to the most violent chemical reactions. To change lead into gold, three positive charges must be extracted from the nucleus. Ironically enough, transmutations of atomic nuclei were constantly occurring all around the alchemists, as they are around us today. Radioactive decay of minerals in rocks has been occurring since their formation. But this was unknown to the alchemists, who lacked a model of matter that could have led to the discovery of these radiations. Had alchemists used the high-speed particles ejected from radioactive ores as bullets, they would have succeeded in transmuting some of the atoms in a substance. But the atoms so transmuted would most likely have escaped their notice.

Ernest Rutherford, in 1919, was the first of many investigators to succeed in deliberately transmuting a chemical element. He bombarded nitrogen nuclei with alpha particles and succeeded in transmuting nitrogen into oxygen:

$$^{14}_{7}\text{N} + ^{4}_{2}\text{He} \rightarrow ^{17}_{8}\text{O} + ^{1}_{1}\text{H}$$

CHECK YOUR ANSWERS

1. **a.** $^{228}_{88}\text{Ra} \rightarrow ^{228}_{89}\text{Ac} + ^{0}_{-1}e$
   b. $^{209}_{84}\text{Po} \rightarrow ^{205}_{82}\text{Pb} + ^{4}_{2}\text{He}$
2. All uranium-238 will ultimately become lead-206. Along the way to becoming lead, it will exist as various isotopes of various elements, as indicated in Figure 33.14.

TABLE 33.1
Transuranic Elements

Atomic Number	*Mass Number*	*Name*	*Symbol*	*Discovery Date*
93	237	Neptunium	Np	1940
94	244	Plutonium	Pu	1940
95	243	Americium	Am	1944
96	247	Curium	Cm	1944
97	247	Berkelium	Bk	1949
98	251	Californium	Cf	1950
99	252	Einsteinium	Es	1952
100	257	Fermium	Fm	1952
101	258	Mendelevium	Md	1955
102	259	Nobelium	No	1958
103	262	Lawrencium	Lr	1961
104	261	Rutherfordium	Rf	1964
105	262	Dubnium	Db	1967
106	266	Seaborgium	Sg	1974
107	264	Bohrium	Bh	1981
108	269	Hassium*	Hs	1984
109	268	Meitnerium	Mt	1982
110	271	Darmstadtium*	Ds	1994
111	272	Roentgenium	Rg	1994
112	285	Unnamed		1996
114	289	Unnamed		1998
116	292	Unnamed		2000

*Hassium is named for Hesse (in its Latin spelling), the German state in which the Darmstadt laboratory is located. Other locations recognized by element names are America, Berkeley, California, and Dubna. People honored by heavy-element names are Marie Curie, Albert Einstein, Enrico Fermi, Dmitri Mendele'ev, Alfred Nobel, Ernest Lawrence, Ernest Rutherford, Glenn Seaborg, Niels Bohr, Lise Meitner, and Wilhelm Roentgen, representing nine countries.

Rutherford's source of alpha particles was a radioactive piece of ore. From a quarter-of-a-million cloud-chamber tracks photographed on movie film, he showed seven examples of atomic transmutation. Analysis of tracks bent by a strong external magnetic field showed that, when an alpha particle collided with a nitrogen atom, a proton bounced out and the heavy atom recoiled a short distance. The alpha particle disappeared. The alpha particle was absorbed by the nitrogen nucleus, transforming nitrogen to oxygen.

Following Rutherford's success with transmutation, experimenters produced many such nuclear reactions, first with natural bombarding projectiles from radioactive ores and then with still more energetic projectiles—protons and electrons hurled by huge particle accelerators. Artificial transmutation is what produces the hitherto unknown elements from atomic number 93 to 116 (with the odd-numbered elements 113, 115, and beyond 117 yet to be discovered). Table 33.1 lists the known elements beyond uranium as of 2005. All these artificially made elements have short half-lives. Whatever transuranic elements existed naturally when Earth was formed have long since decayed.

Radioactive Isotopes

All the elements have been made radioactive by bombardment with neutrons and other particles. Radioactive materials are extremely useful in scientific research and industry. In order to check the action of a fertilizer, for example, researchers combine a small amount of radioactive material with the fertilizer and then apply the combination to a few plants. The amount of radioactive fertilizer absorbed by the plants can be easily measured with radiation detectors. From such measurements, scientists can inform farmers about proper usages of fertilizer. When used in this way, radioactive isotopes are called *tracers*.

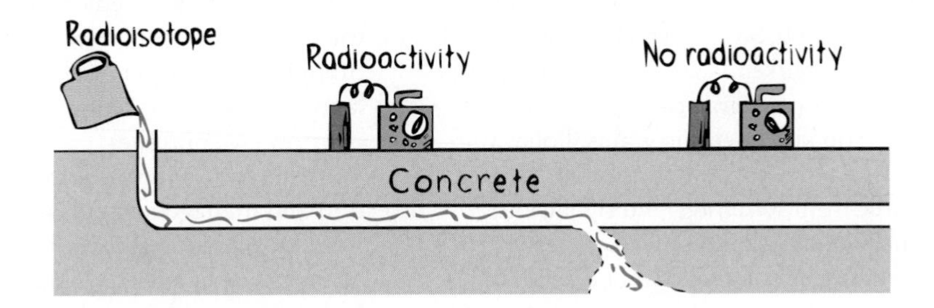

FIGURE 33.15
Tracking pipe leaks with radioactive isotopes.

FIGURE 33.16
Radioisotopes are used to check the action of fertilizers in plants and the progression of food in digestion.

Tracers are widely used in medicine to diagnose disease. Small quantities of particular radioactive isotopes, after being injected into the bloodstream, concentrate at trouble spots, such as at bone fractures or tumors. Using radiation detectors, medical staff can locate isotope concentrations.

Engineers can study how the parts of an automobile engine wear away during use by making the cylinder walls radioactive. While the engine is running, the piston rings rub against the cylinder walls. The tiny particles of radioactive metal that are worn away fall into the lubricating oil, where they can be measured with a radiation detector. By repeating this test with different oils, the engineer can determine which oil provides the least wear and longest life to the engine.

Tire manufacturers also employ radioactive isotopes. If a known fraction of the carbon atoms used in an automobile tire is radioactive, the amount of rubber left on the road when the car is braked can be estimated through a count of the radioactive atoms.

FIGURE 33.17
The shelf-life of fresh strawberries and other perishables is markedly increased when the food is subjected to gamma rays from a radioactive source. The strawberries on the right were treated with gamma radiation, which kills the microorganisms that normally lead to spoilage. The food is only a receiver of radiation and is in no way transformed into an emitter of radiation, as can be confirmed with a radiation detector.

FOOD IRRADIATION

Each week in the United States about 100 people, most of them children or elderly, die from illnesses they contract from food. People stricken ill each week from food-borne diseases number in the millions, according to the Centers for Disease Control and Prevention in Atlanta, Georgia. But never astronauts. Why? Because diarrhea in orbit is a no-no, and food taken on space missions is irradiated with high-energy gamma rays from a radioactive-cobalt source (Co-60). Astronauts, as well as patients in many hospitals and nursing homes, don't have to contend with salmonella, *E. coli*, microbes, or parasites in food irradiated by Co-60. So why isn't more irradiated food available in the marketplace? The answer is public phobia about the *r* word—*radiation*.

Food irradiation kills insects in grains, flour, fruits, and vegetables. Small doses prevent stored potatoes, onions, and garlic from sprouting, and significantly increase the shelf life of soft fruits, such as strawberries. Larger doses kill microbes, insects, and parasites in spices, pork, and poultry. Irradiation can penetrate through sealed cans and packages. What irradiation does *not* do is leave the irradiated food radioactive. No radioactive material touches the food. Gamma rays pass through the food like light passing through glass, destroying most bacteria that can cause disease. No food becomes radioactive, for the gamma rays lack the energy needed to knock neutrons from atomic nuclei.

Irradiation does, however, leave behind traces of broken compounds—identical to those resulting from pyrolysis when charbroiling foods we already eat. Compared with canning and cold storage, irradiation has less effect on nutrition and taste. It's been around for most of the 1900s, and it has been tested for more than 40 years, with no evidence of danger to consumers. Irradiation of foods is endorsed by all major scientific societies, the United Nations' World Health Organization, the U.S. Food and Drug Administration, and the American Medical Association. Irradiation is the method of choice for 37 countries worldwide. Although widely used in Belgium, France, and the Netherlands, its use in the United States is presently small, as controversy continues.

This controversy is another example of risk evaluation and management. Shouldn't risks of injury or death from irradiated food be judged rationally and weighed against the benefits it would bring? Shouldn't the choice be based upon the number of people who *might* die of irradiated food versus those who in fact *do* die because food is not irradiated?

Perhaps what is needed is a name change—expunging the *r* word, as was done with the *n* word when the resisted medical procedure once known as NMRI (nuclear magnetic resonance imaging) was given a more acceptable name, MRI (magnetic resonance imaging).

There are hundreds more examples of the use of trace amounts of radioactive isotopes. The important thing is that this technique provides a way to detect and count atoms in samples of materials too small to be seen with a microscope.

When radioactivity is used in medicine for treatment rather than diagnosis, intense radiation is needed. Strong, short-lived sources of radiation can be used to destroy cancer cells, such as in the thyroid gland or the prostate gland.

CHECK YOURSELF

Suppose that you want to find out how much gasoline is in an underground storage tank. You pour in one gallon of gasoline that contains some radioactive material with a long half-life that gives off 5000 counts per minute. The next day, you remove a gallon from the underground tank and measure its radioactivity to be 10 counts per minute. How much gasoline is in the tank?

CHECK YOUR ANSWER

There are 500 gallons in the tank since, after mixing, the gallon you withdraw has $10/5000 = 1/500$ of the original radioactive particles in it.

Radiometric Dating

Radioactive decay provides scientists who study Earth history and human history a remarkable method for determining the ages of materials. The method depends on a knowledge of the half-lives of radioactive materials. For materials that were once alive, age can be found by a comparison of carbon isotopes in the material. For nonorganic substances, isotopes of uranium, potassium, and other elements are examined.

Carbon Dating

Earth's atmosphere is continuously bombarded by cosmic rays that produce transmutation of many atoms in the upper atmosphere. These transmutations result in many protons and neutrons being "sprayed out" into the environment. Most of the protons quickly capture electrons and become hydrogen atoms in the upper atmosphere. The neutrons, however, continue for longer distances because they have no charge and therefore do not interact electrically with matter. Eventually, many of them collide with atomic nuclei in the denser lower atmosphere. When nitrogen captures a neutron, it becomes an isotope of carbon by emitting a proton:

$$^{14}_{7}\text{N} + ^{1}_{0}n \rightarrow ^{14}_{6}\text{C} + ^{1}_{1}\text{H}$$

This carbon-14 isotope is radioactive and has 8 neutrons (the most common stable isotope, carbon-12, has 6 neutrons). Less than one-millionth of 1% of the carbon in the atmosphere is carbon-14. Both carbon-12 and carbon-14 join with oxygen to become carbon dioxide, which is absorbed by plants. This means that all plants contain a tiny bit of radioactive carbon-14. All animals eat either plants or plant-eating animals and therefore have a little carbon-14 in them. In short, all living things on Earth contain some carbon-14.

Carbon-14 is a beta emitter and decays back to nitrogen:

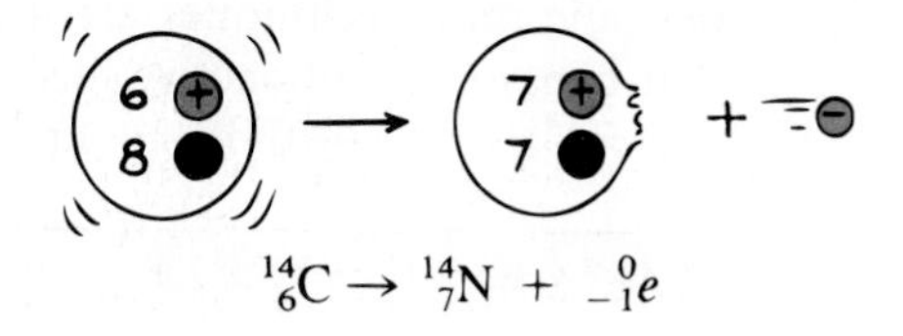

$$^{14}_{6}\text{C} \rightarrow ^{14}_{7}\text{N} + ^{0}_{-1}e$$

Because plants take in carbon dioxide as long as they live, any carbon-14 lost to decay is immediately replenished with fresh carbon-14 from the atmosphere. In this way, a radioactive equilibrium is reached where there is a ratio of about one carbon-14 atom to every 100 billion carbon-12 atoms. When a plant dies, replenishment stops. Then the percentage of carbon-14 decreases at a constant rate given by its radioactive half-life. The longer a plant or animal is dead, the less carbon-14 it contains.

The half-life of carbon-14 is about 5730 years. This means that half of the carbon-14 atoms now present in a plant or animal that dies today will decay

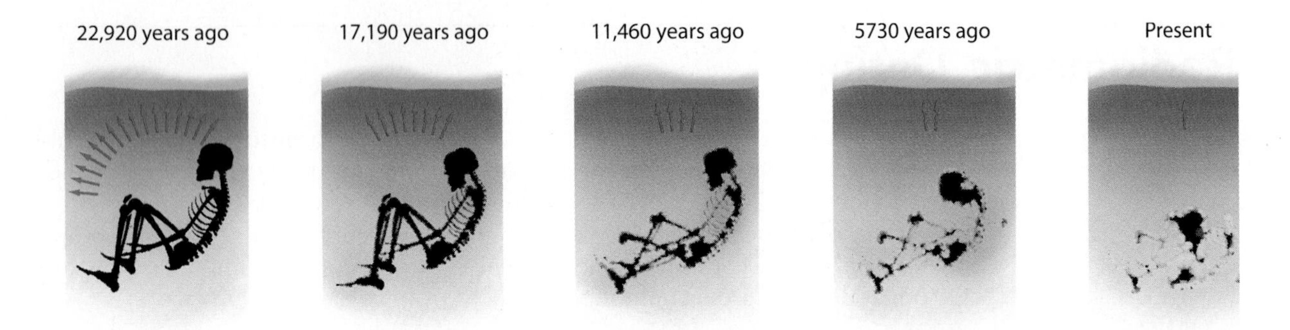

FIGURE 33.18
The radioactive carbon isotopes in the skeleton diminish by one-half every 5730 years. A skeleton today contains only a fraction of the carbon-14 it originally had. The red arrows symbolize relative amounts of carbon-14.

in the next 5730 years. Half of the remaining carbon-14 atoms will then decay in the following 5730 years, and so forth. The radioactivity of living things gradually decreases at a steady rate after they die.

With this knowledge, archeologists are able to calculate the age of carbon-containing artifacts, such as wooden tools or skeletons, by measuring their current level of radioactivity. This process, known as *carbon-14 dating,* enables us to probe as much as 50,000 years into the past.

Carbon dating would be an extremely simple and accurate dating method if the amount of radioactive carbon in the atmosphere had been constant over the ages. But it hasn't been. Fluctuations in the Sun's and Earth's magnetic fields affect cosmic-ray intensities in Earth's atmosphere, which, in turn, produce fluctuations in the amount of carbon-14 in the atmosphere at any given time. In addition, changes in Earth's climate affect the amount of carbon dioxide in the atmosphere. The oceans are great reservoirs of carbon dioxide. When the oceans are cold, they release less carbon dioxide into the atmosphere than when they are warm. Because of all these fluctuations in the production of carbon-14 through the centuries, carbon dating has an uncertainty of about 15%. This means, for example, that the straw of an old adobe brick that is dated to be 500 years old may really be only 425 years old on the low side, or 575 years old on the high side. For many purposes, this is an acceptable level of uncertainty. Laser-enrichment techniques using milligrams of carbon produce smaller uncertainties, and these techniques are used for dating more ancient relics. A technique that bypasses a radioactive measure altogether uses a mass spectrometer device that makes a C-14/C-12 count directly.

CHECK YOURSELF

Suppose an archeologist extracts 1 gram of carbon from an ancient ax handle and finds it to be one-fourth as radioactive as 1 gram of carbon extracted from a freshly cut tree branch. About how old is the ax handle?

CHECK YOUR ANSWER

Assuming the ratio of C-14/C-12 was the same when the ax was made, the ax handle is as old as two half-lives of C-14, or about 11,460 years old.

Uranium Dating

The dating of older, but nonliving, things is accomplished with radioactive minerals such as uranium. The naturally occurring isotopes U-238 and U-235 decay very slowly and ultimately become isotopes of lead—but not the common lead isotope Pb-208. For example, U-238 decays through several stages to finally become Pb-206. U-235, on the other hand, decays to become the isotope Pb-207. Thus, any Pb-206 and Pb-207 that now exist in a uranium-bearing rock were, at one time, uranium. The older the rock, the higher the percentage of these remnant isotopes.

From the half-lives of uranium isotopes and the percentage of lead isotopes in uranium-bearing rock, it is possible to calculate the date at which the rock was formed. Rocks dated in this way have been found to be as much as 3.7 *billion* years old. Samples from the Moon, where there has been an absence of erosion, have been dated at 4.2 billion years, an age that agrees closely with the estimated 4.6-billion-year age of Earth and the rest of the solar system.

Other widely used isotopes include potassium-40 (with a half-life of 1.25 billion years) and rubidium-87 (with a half-life of 49 billion years). As with uranium dating, age is determined by measuring the relative percentage of a given isotope in the material in question.

Effects of Radiation on Humans

A common misconception is that radioactivity is something new in the environment. But radioactivity has been around far longer than the human race. It is as much a part of our surroundings as the Sun and the rain. It is what warms the interior of the Earth and makes it molten. In fact, radioactive decay inside the Earth is what heats the water that spurts from a geyser or that wells up from a natural hot spring. Even the helium in a child's balloon is the offspring of radioactivity. Its nuclei are nothing more than the alpha particles that were once ejected by radioactive nuclei.

As Figure 33.19 shows, more than 98% of our annual exposure to radiation comes from food and water, natural background radiation, and medical and dental X-rays. Fallout from nuclear testing and the coal and nuclear power industries are minor contributors in comparison. Amazingly, the coal industry far outranks the nuclear power industry as a source of radiation. The global combustion of coal annually releases about 16,000 tons of radioactive thorium and about 7000 tons of radioactive uranium into the atmosphere (with nearly 50 tons being fissionable U-235!). Worldwide, the nuclear power industries

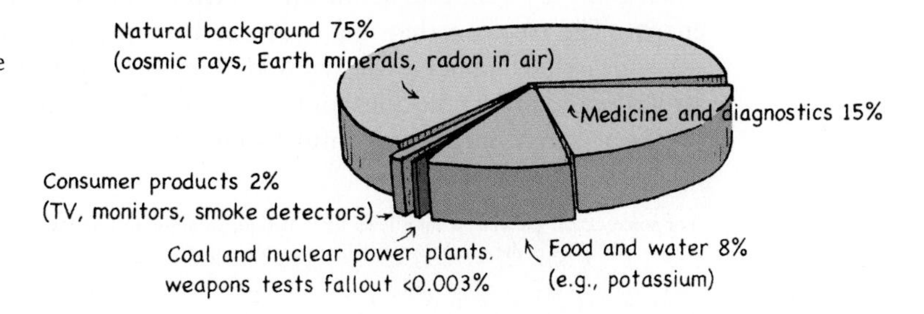

FIGURE 33.19
Origins of radiation exposure for an average individual in the United States.

generate about 10,000 tons of radioactive waste each year. Most all of this waste, however, is contained and *not* released into the environment.

Most of the radiation we encounter originates in the natural environment. It is in the ground we stand on and in the bricks and stones of surrounding buildings. Every ton of ordinary granite contains, on average, some 20 grams of thorium and 9 grams of uranium. Because of the traces of radioactive elements in most all rocks, people who live in brick, concrete, or stone buildings are exposed to greater amounts of radiation than people who live in wooden buildings. This natural background radiation was present before humans emerged in the world. If our bodies couldn't tolerate it, we wouldn't be here. Apart from radioactivity, we are bombarded by cosmic rays. At sea level, the protective blanket of the atmosphere reduces cosmic-ray intensity; at higher altitudes, radiation is more intense. In Denver, the "mile-high city," a person receives more than twice as much radiation from cosmic rays as at sea level. Frequent flyers receive significant radiation exposure. (Is the air time of airline personnel limited because of this extra radiation?)

An average ton of coal contains 1.3 ppm of uranium and 3.2 ppm of thorium. That's why the average coal-burning power plant is a far greater source of airborne radioactive material than a nuclear power plant.

Insights

Even the human body is a source of natural radiation, primarily from the potassium in the food we eat. Our bodies contain about 200 grams of potassium. Of this quantity, about 20 milligrams is the radioactive isotope potassium-40. Between every heartbeat, about 5,000 potassium-40 atoms undergo spontaneous radioactive decay. Added to this are some 3,000 beta particles per second emitted by the carbon-14 in your body. We and all living creatures are to some degree radioactive.

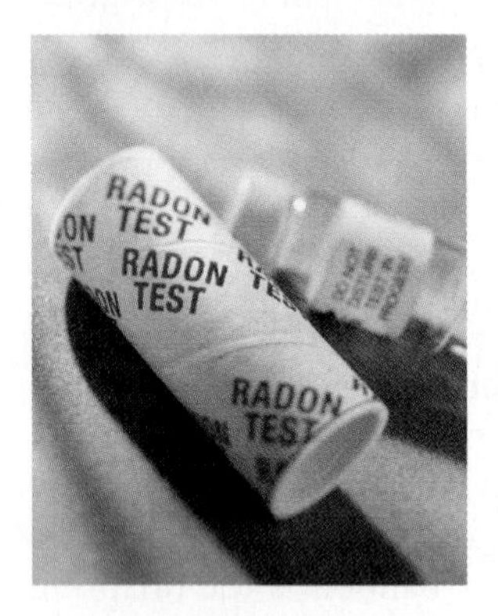

FIGURE 33.20
A commercially available radon test kit for the home.

The leading source of naturally occurring external radiation is radon-222, an inert gas arising from uranium deposits. Radon is a heavy gas that tends to accumulate in basements after it seeps up through cracks in the basement floor. Levels of radon vary from region to region, depending upon local geology. You can check the radon level in your home with a radon detector kit. If levels are abnormally high, corrective measures, such as sealing the basement foundation and maintaining adequate ventilation, are recommended.

Exposure to radiation greater than normal background should be avoided because of the damage it can do.[4] The cells of living tissue are composed of intricately structured molecules in a watery, ion-rich brine. When X-radiation or nuclear radiation encounters this highly ordered soup, it produces chaos on the atomic scale. A beta particle, for example, passing through living matter collides with a small percentage of the molecules and leaves a randomly dotted trail of altered or broken molecules along with newly formed, chemically active ions and free radicals. Free radicals are unbonded, electrically neutral, very chemically active atoms or molecular fragments. The ions and free radicals may dissociate more molecular bonds or they may quickly form strong new bonds, forming molecules that may be useless or harmful to the cell. Gamma radiation produces a similar effect. As a high-energy gamma-ray photon moves through matter, it may rebound from an electron and give the electron a high kinetic energy. The electron then may careen through the tissue, creating havoc in the ways described above. All types of high-energy radiation break or alter the structure of some molecules and create conditions in which other molecules will be formed that may be harmful to life processes.

[4]For some cancer patients, a high level of radiation, carefully directed, can be beneficial by selectively killing cancer cells. This is the province of radiation oncology.

Cells are able to repair most kinds of molecular damage, if the radiation is not too intense. A cell can survive an otherwise lethal dose of radiation if the dose is spread over a long period of time to allow intervals for healing. When radiation is sufficient to kill cells, the dead cells can be replaced by new ones. An important exception to this is most nerve cells, which are irreplaceable. Sometimes a radiated cell will survive with a damaged DNA molecule. Defective genetic information will be transmitted to offspring cells when the damaged cell reproduces, and a cell *mutation* will occur. Mutations are usually insignificant, but, if they are significant, they will probably result in cells that do not function as well as undamaged ones. A genetic change of this type could also be part of the cause of a cancer that will develop later. In rare cases, a mutation may be an improvement.

The concentration of disorder produced along the trajectory of a particle depends upon its energy, charge, and mass. Gamma-ray photons and very energetic beta particles spread their damage out over a long track. They penetrate deeply with widely separated interactions, like a very fast BB fired through a hailstorm. Slow, massive, highly charged particles, such as low-energy alpha particles, do their damage in the shortest distance. They have collisions that are close together, more like a bull charging through a flock of sleepy sheep. They do not penetrate deeply because their energy is absorbed by many closely spaced collisions. Particles that produce especially concentrated damage are the assorted nuclei (called *heavy primaries*) flung outward by the Sun in solar flares and existing in small percentages of cosmic radiation. These include all the elements found on Earth. Some of them are captured in Earth's magnetic field and some are stopped by collisions in the atmosphere, so practically none reaches Earth's surface. We are shielded from most of these dangerous particles by the very property that makes them a threat—their tendency to have many collisions close together.

Astronauts do not have this protection, and they absorb large doses of radiation during their time in space. Every few decades there is an exceptionally powerful solar flare that would almost certainly kill any conventionally protected astronaut who is unprotected by Earth's atmosphere and magnetic field.

We are bombarded most by what harms us least—neutrinos. Neutrinos are the most weakly interacting particles. They have near-zero mass, no charge, and are produced frequently in radioactive decays. They are the most common high-speed particles known, zapping the universe and passing unhindered through our bodies by many millions every second. They pass completely through the Earth with only occasional encounters. It would take a "piece" of lead 6 light years in thickness to absorb half the neutrinos incident upon it. Only about once per year, on the average, a neutrino triggers a nuclear reaction in your body. We don't hear much about neutrinos because they fail to interact with us.

FIGURE 33.21
International symbol that indicates an area where radioactive material is being handled or produced.

Of the radiations we have focused upon in this chapter, gamma radiation is the most penetrating and therefore the hardest to shield against. This, combined with its ability to interact with the matter in our bodies, makes it potentially the most dangerous radiation. It emanates from radioactive materials, and it makes up a substantial part of the normal background radiation. Exposure to it should be minimized.

CHECK YOURSELF

1. People working around radioactivity wear film badges to monitor radiation levels reaching their bodies. These badges consist of a small piece of photographic film enclosed in a light-proof wrapper. What type of radiation do these devices monitor, and how can they determine the amount of radiation a body receives?
2. Suppose you are given three radioactive cookies—one an alpha emitter, one a beta emitter, and one a gamma emitter. You must eat one, hold one in you hand, and put the third in your pocket. What can you do to minimize your exposure to radiation?

Radiation Dosage

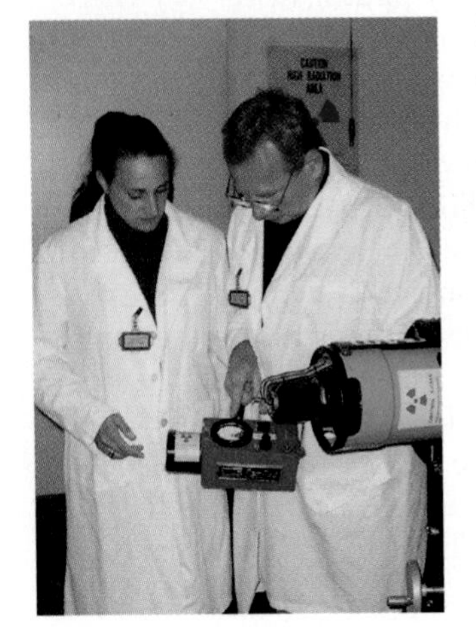

FIGURE 33.22
The film badges attached to the lapels of Tammy's and Larry's lab coats contain audible alerts for both radiation surge and accumulated exposure. The badges are individualized and information from them is periodically downloaded to a database and a complete picture of these lab workers' exposure levels and those of others in the same radioactive environment are analyzed.

Radiation dosage is measured in *rads* (short for radiation), a unit of absorbed energy of ionizing radiation. The number of rads indicates the amount of radiation energy absorbed per gram of exposed material. However, in contexts in which we are concerned with the potential ability of radiation to affect human beings, dosage is measured in *rems* (*r*oentgen *e*quivalent *m*an). In calculating the dosage in rems, the number of rads is multiplied by a factor allowing for the different health effects of different types of radiation. For example, 1 rad of slow alpha particles has the same biological effect as 10 rads of fast electrons. Both of these dosages are 10 rems.

The average person in the United States is exposed to about 0.2 rem per year. This comes from within the body itself, from the ground, buildings, cosmic rays, diagnostic X-rays, television, and so on. It varies widely from place to place on the planet, but it is stronger at higher elevations, where cosmic radiation is more intense, and strongest near the poles, where the Earth's magnetic field doesn't shield against cosmic rays.

A lethal dose of radiation is on the order of 500 rems; that is, a person has about a 50% chance of surviving a dose of this magnitude if it is received over a short period of time. Under radiotherapy—the use of radiation to kill cancer cells—a patient may receive localized doses in excess of 200 rems each day for a period of weeks. A typical diagnostic chest X-ray exposes a person to 5 to 30 millirems, less than one ten-thousandth of the lethal dose. However, even small doses of radiation can produce long-term effects due to mutations within

CHECK YOUR ANSWERS

1. Alpha rays and most beta rays don't penetrate the film wrapper, so the type of radiation that makes it to the film is mostly gamma radiation. Like light on photographic film, greater intensity means greater exposure, as noted by how black the film becomes.
2. Ideally, you should get far away from all the cookies. But, if you must eat one, hold one, and put one in your pocket, hold the alpha because the skin on your hand will shield you. Put the beta in your pocket because your clothing will likely shield you. Eat the gamma because it will penetrate your body in any of these cases anyway.

the body's tissues. And, because a small fraction of any X-ray dose reaches the gonads, some mutations occasionally occur that are passed on to the next generation. Medical X-rays for diagnosis and therapy have a far larger effect on the human genetic heritage than any other artificial source of radiation. Perspective dictates that we keep in mind the fact that we normally receive significantly more radiation from natural minerals in the Earth than from all artificial sources of radiation combined.

Taking all causes into account, most of us will receive a lifetime exposure of less than 20 rems, distributed over several decades. This makes us a little more susceptible to cancer and other disorders. But more significant is the fact that all living beings have always absorbed natural radiation and that the radiation received in the reproductive cells has produced genetic changes in all species for generation after generation. Small mutations selected by nature for their contributions to survival can, over billions of years, gradually produce some interesting organisms—*us,* for example!

Summary of Terms

X-ray Electromagnetic radiation of higher frequencies than ultraviolet; emitted by electron transitions to the lowest energy states in atoms.

Radioactivity Process of the atomic nucleus that results in the emission of energetic subatomic particles.

Alpha ray A stream of alpha particles (helium nuclei) ejected by certain radioactive elements.

Beta ray A stream of electrons (or positrons) emitted during the radioactive decay of certain nuclei.

Gamma ray High-frequency electromagnetic radiation emitted by the nuclei of radioactive atoms.

Nucleon A nuclear proton or neutron; the collective name for either or both.

Quarks The elementary constituent particles or building blocks of nuclear matter.

Atomic number A number associated with an atom, equal to the number of protons in the nucleus or, equivalently, to the number of electrons in the electron cloud of a neutral atom.

Isotopes Atoms whose nuclei have the same number of protons but different numbers of neutrons.

Atomic mass A number associated with an atom, equal to the number of nucleons in the nucleus.

Atomic mass unit (amu) Standard unit of atomic mass, based on the mass of the common carbon nucleus, which is arbitrarily given the value of exactly 12. An amu of value 1 is one-twelfth the mass of this common carbon nucleus.

Half-life The time required for half the atoms in a sample of a radioactive isotope to decay.

Transmutation The conversion of an atomic nucleus of one element into an atomic nucleus of another element through a loss or gain in the number of protons.

Review Questions

X-Rays and Radioactivity

1. What did the physicist Roentgen discover about a cathode-ray beam striking a glass surface?
2. What is the similarity between a beam of X-rays and a beam of light? What is the principal difference between the two?
3. What did the physicist Becquerel discover about uranium?
4. What two elements did Pierre and Marie Curie discover?

Alpha, Beta, and Gamma Rays

5. Why are gamma rays not deflected in a magnetic field?
6. What is the origin of a beam of gamma rays? A beam of X-rays?

The Nucleus

7. Why is the mass of an atom practically equal to the mass of its nucleus?
8. What are *quarks*?
9. Name three different leptons.

Isotopes

10. Give the atomic number for deuterium and the atomic number for tritium.
11. Give the atomic mass number for deuterium and the atomic mass number for tritium.
12. Distinguish between an *isotope* and an *ion*.

Why Atoms Are Radioactive

13. What prevents protons in the nucleus from flying apart due to electrical repulsion?
14. Why do protons in a very large nucleus have a greater chance of flying apart?
15. Why is a larger nucleus generally less stable than a smaller nucleus?

Half-Life

16. What is meant by *radioactive half-life*?
17. What is the half-life of Ra-226? Of a muon?

Radiation Detectors

18. What kind of trail is left when an energetic particle shoots through matter?
19. Which two radiation detectors operate primarily by sensing the trails left by energetic particles that shoot through matter?

Transmutation of Elements

20. What is transmutation?

Natural Transmutation

21. When thorium (atomic number 90) decays by emitting an alpha particle, what is the atomic number of the resulting nucleus?
22. When thorium decays by emitting a beta particle, what is the atomic number of the resulting nucleus?
23. What is the change in atomic mass for each of the above two reactions?
24. What change in atomic number occurs when a nucleus emits an alpha particle? A beta particle? A gamma ray?
25. What is the long-range fate of all the uranium that exists in the world?

Artificial Transmutation

26. The alchemists of old believed that elements could be changed into other elements. Were they correct? Were they effective? Why or why not?
27. When, and by whom, did the first successful intentional transmutation of an element occur?
28. Why are the elements beyond uranium not common in the Earth's crust?

Radioactive Isotopes

29. How are radioactive isotopes produced?
30. What is a radioactive *tracer*?

Radiometric Dating

31. What specific knowledge is needed to enable the dating of historic objects?

Carbon Dating

32. What occurs when a nitrogen nucleus captures an extra neutron?
33. How many carbon-14 atoms are present in nature compared with carbon-12?
34. Why is the quantity of C-14 in new bones greater than in old bones of the same mass?

Uranium Dating

35. Why is lead found in all deposits of uranium ores?
36. What does the proportion of lead and uranium in a rock tell us about the age of the rock?

Effects of Radiation on Humans

37. Where does most of the radiation you encounter originate?
38. Is the human body radioactive?

Radiation Dosage

39. What is the average annual radiation dosage for the average person in the U.S.? What is an average dose from medical X-rays? What is the lethal dose?
40. Which gives humans the greatest radiation dose, radiation from natural minerals in Earth or from artificial sources?

Project

Write a letter to Grandma to dispel any notion she or her friends might have about radioactivity being something new in the world. Tie this to the idea that many people have the strongest views on that which they understand the least.

Exercises

1. In the nineteenth century, the famous physicist Lord Kelvin estimated the age of the Earth to be much much less than the present estimate. What information that Kelvin did not have might have allowed him to avoid making his erroneous estimate?
2. X-rays are most similar to which of the following—alpha, beta, or gamma rays?
3. Gamma radiation is fundamentally different from alpha and beta radiation. What is this basic difference?
4. Why is a sample of radioactive material always a little warmer than its surroundings?
5. Some people say that all things are possible. Is it at all possible for a common hydrogen nucleus to emit an alpha particle? Defend your answer.

6. Why are alpha and beta rays deflected in opposite directions in a magnetic field? Why are gamma rays not deflected?

7. The alpha particle has twice the electric charge of the beta particle but, for the same kinetic energy, deflects less than the beta in a magnetic field. Why is this so?

8. How do the paths of alpha, beta, and gamma rays compare in an electric field?

9. Which type of radiation—alpha, beta, or gamma—produces the greatest change in *mass number* when emitted by an atomic nucleus? Which produces the greatest change in *atomic number*?

10. Which type of radiation—alpha, beta, or gamma—produces the least change in mass number? In atomic number?

11. Which type of radiation—alpha, beta, or gamma—predominates within an enclosed elevator descending into a uranium mine?

12. In bombarding atomic nuclei with proton "bullets," why must the protons be accelerated to high energies if they are to make contact with the target nuclei?

13. Just after an alpha particle leaves the nucleus, would you expect it to speed up? Defend your answer.

14. What do all isotopes of the same element have in common? How do they differ?

15. Why would you expect alpha particles, with their greater charge, to be less able to penetrate into materials than beta particles of the same energy?

16. Two protons in an atomic nucleus repel each other, but they are also attracted to each other. Explain.

17. Which interaction tends to hold the particles in an atomic nucleus together and which interaction tends to push them apart?

18. What evidence supports the contention that the strong nuclear interaction can dominate over the electrical interaction at short distances within the nucleus?

19. Can it be truthfully said that, whenever a nucleus emits an alpha or beta particle, it necessarily becomes the nucleus of another element?

20. Exactly what is a positively charged hydrogen atom?

21. Why do different isotopes of the same element have the same chemical properties?

22. If you make an account of 1000 people born in the year 2000 and find that half of them are still living in 2060, does this mean that one-quarter of them will be alive in 2120 and one-eighth of them alive in 2180? What is different about the death rates of people and the "death rates" of radioactive atoms?

23. Radiation from a point source obeys the inverse-square law. If a Geiger counter 1 m from a small sample registers 360 counts per minute, what will be its counting rate 2 m from the source? What will it be 3 m from the source?

24. Why do the charged particles flying through bubble chambers travel in spiral paths rather than in the circular or helical paths they might ideally follow?

25. What two quantities are always conserved in all nuclear equations?

26. Judging from Figure 33.14, how many alpha and beta particles are emitted in the series of radioactive decay events from a U-238 nucleus to a Pb-206 nucleus?

27. If an atom has 104 electrons, 157 neutrons, and 104 protons, what is its approximate atomic mass? What is the name of this element?

28. When a $^{226}_{88}Ra$ nucleus decays by emitting an alpha particle, what is the atomic number of the resulting nucleus? What is the resulting atomic mass?

29. When a nucleus of $^{218}_{84}Po$ emits a beta particle, it transforms into the nucleus of a new element. What is the atomic number and the atomic mass of this new element?

30. When a nucleus of $^{218}_{84}Po$ emits an alpha particle, what is the atomic number and the atomic mass of the resulting element?

31. Which has the greater number of protons, U-235 or U-238? Which has the greater number of neutrons?

32. State the number of neutrons and protons in each of the following nuclei: $^{2}_{1}H$, $^{12}_{6}C$, $^{56}_{26}Fe$, $^{197}_{79}Au$, $^{90}_{38}Sr$, and $^{238}_{92}U$.

33. How is it possible for an element to decay "forward in the periodic table"—that is, to decay to an element of higher atomic number?

34. How could an element emit alpha and beta particles and result in the same element?

35. When radioactive phosphorus (P) decays, it emits a positron. Will the resulting nucleus be another isotope of phosphorus? If not, what will it be?

36. "Strontium-90 is a pure beta source." How could a physicist test this statement?

37. A friend suggests that nuclei are composed of equal numbers of protons and electrons, and not neutrons. What evidence can you cite to show that your friend is mistaken?

38. Radium-226 is a common isotope on Earth, but it has a half-life of about 1600 years. Given that the Earth is some 5 billion years old, why is there any radium left at all?

39. Elements above uranium in the periodic table do not exist in any appreciable amounts in nature because they have short half-lives. Yet there are several elements below uranium in atomic number with equally short half-lives that do exist in appreciable amounts in nature. How can you account for this?

40. Your friend says that the helium used to inflate balloons is a product of radioactive decay. Another friend disagrees. With whom do you agree?

41. Another friend, fretful about living near a fission power plant, wishes to get away from radiation by traveling to the high mountains and sleeping out at night on granite outcroppings. What comment do you have about this?

42. Still another friend has journeyed to the mountain foothills to escape the effects of radioactivity altogether. While bathing in the warmth of a natural hot spring, she wonders aloud how the spring gets its heat. What do you tell her?

43. Although coal contains only minute quantities of radioactive materials, there is more radiation emitted by a coal-fired power plant than a fission power plant simply because of the vast amount of coal that is burned in coal-fired plants. What does this indicate about methods of preventing the release of radioactivity that are typically implemented at the two kinds of power plants?

44. A friend produces a Geiger counter to check the local normal background radiation. It clicks, randomly but repeatedly. Another friend, whose tendency is to fear most that which is least understood, makes an effort to avoid Geiger counters and looks to you for advice. What do you say?

45. When food is irradiated with gamma rays from a cobalt-60 source, does the food become radioactive? Defend your answer.

46. When the author attended high school some 50 years ago, his teacher showed a piece of uranium ore and measured its radioactivity with a Geiger counter. Would that reading for the same piece of ore be different today?

47. Why is carbon dating ineffective in finding the ages of dinosaur bones?

48. Is carbon dating appropriate for measuring the age of materials that are a few years old? A few thousand years old? A few million years old?

49. The age of the Dead Sea Scrolls was found by carbon dating. Could this technique apply if they were carved in stone tablets? Explain.

50. Make up two multiple-choice questions that would check a classmate's understanding of radioactive dating.

Problems

1. If a sample of a radioactive isotope has a half-life of 1 year, how much of the original sample will be left at the end of the second year? At the end of the third year? At the end of the fourth year?

2. A sample of a particular radioisotope is placed near a Geiger counter, which is observed to register 160 counts per minute. Eight hours later, the detector counts at a rate of 10 counts per minute. What is the half-life of the material?

3. The isotope cesium-137, which has a half-life of 30 years, is a product of nuclear power plants. How long will it take for this isotope to decay to about one-sixteenth its original amount?

4. A certain radioactive isotope has a half-life of one hour. If you start with 1 g of the material at noon, how much of the original isotope will the sample contain at 3:00 p.m.? At 6:00 p.m.? At 10:00 p.m.?

5. Suppose that you measure the intensity of radiation from carbon-14 in an ancient piece of wood to be 6% of what it would be in a freshly cut piece of wood. How old is this artifact?

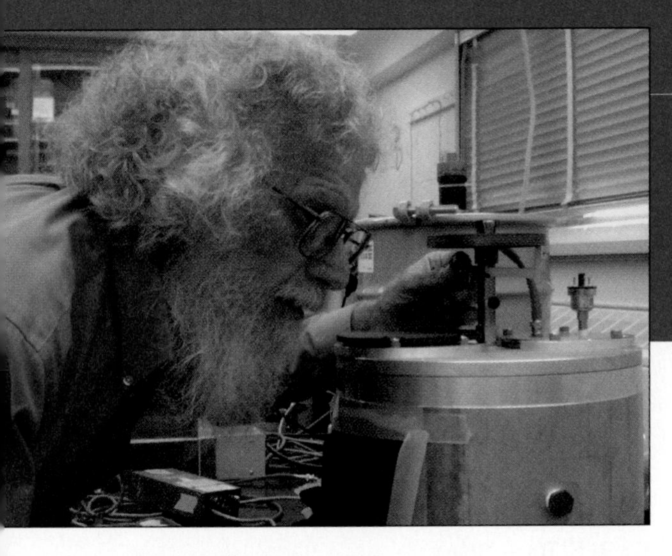

CHAPTER 34

Nuclear Fission and Fusion

Dean Zollman investigates nuclear properties with a modern version of Rutherford's scattering experiment.

In December 1938, two German scientists, Otto Hahn and Fritz Strassmann, made an accidental discovery that was to change the world. While bombarding a sample of uranium with neutrons in the hope of creating new heavier elements, they were astonished to find chemical evidence for the production of barium, an element about half the mass of uranium. They were reluctant to believe their own results. Hahn sent news of this discovery to his former colleague Lise Meitner, a refugee from Nazism working in Sweden. Over the Christmas holidays, she discussed it with her nephew Otto Frisch, also a German refugee, who was visiting her from Denmark, where he worked with Niels Bohr. Together, they came up with the explanation: The uranium nucleus, activated by neutron bombardment, had split in two. Meitner and Frisch named the process *fission*, after the similar process of cell division in biology.[1]

Nuclear Fission

Nuclear fission involves a delicate balance within the nucleus between nuclear attraction and the electrical repulsion between protons. In all known nuclei, the nuclear forces dominate. In uranium, however, this domination is tenuous. If the uranium nucleus is stretched into an elongated shape (Figure 34.1), the electrical forces may push it into an even more elongated shape. If the elongation passes a critical point, nuclear forces yield to electrical ones, and the nucleus separates. This is fission.[2] The absorption of a neutron by a uranium nucleus

[1]Similarly, Ernest Rutherford used a biological term when he chose the word *nucleus* for the center of an atom.

[2]Fission resulting from neutron absorption is called *induced fission*. In rare instances nuclei can also undergo *spontaneous fission* without initial neutron absorption. There is evidence that at least one such major spontaneous fission event occurred in Africa almost two billion years ago when the percentage of U-235 in uranium deposits was greater (see *Scientific American*, July 1976). Interestingly, when U-235 absorbs a neutron it momentarily becomes U-236, which splits in half almost instantaneously. So, strictly speaking, it is U-236, not U-235, that undergoes fission. It's common, however, to speak of the fission of U-235.

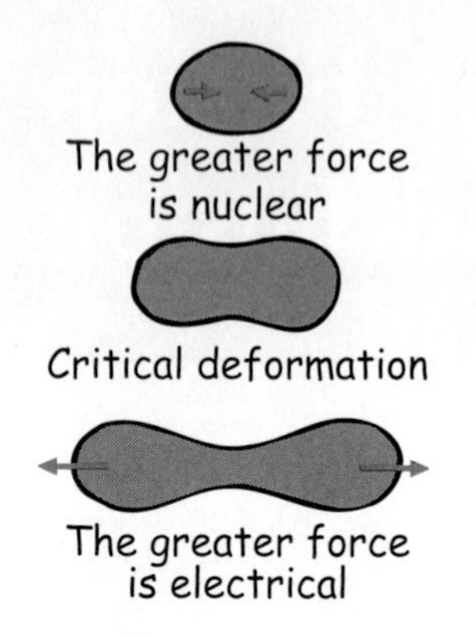

FIGURE 34.1
Nuclear deformation may result when repulsive electrical forces overcome attractive nuclear forces, in which case fission occurs.

supplies enough energy to cause such an elongation. The resultant fission process may produce many different combinations of smaller nuclei. A typical example is

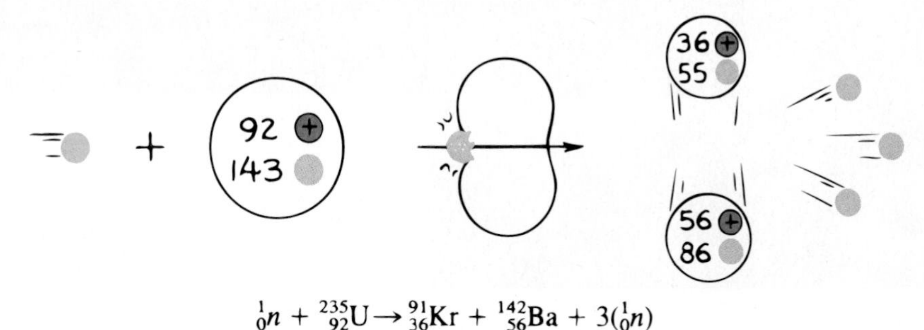

$$^{1}_{0}n + ^{235}_{92}U \rightarrow ^{91}_{36}Kr + ^{142}_{56}Ba + 3(^{1}_{0}n)$$

In this reaction, note that one neutron starts the fission of the uranium nucleus and that the fission produces three neutrons (yellow).[3] Because neutrons have no charge and are not repelled by atomic nuclei, they make good "nuclear bullets" and can cause the fissioning of three other uranium atoms, releasing a total of nine more neutrons. If each of these neutrons succeeds in splitting a uranium atom, the next step in the reaction can produce 27 neutrons, and so on. Such a sequence is called a **chain reaction** (Figure 34.2).

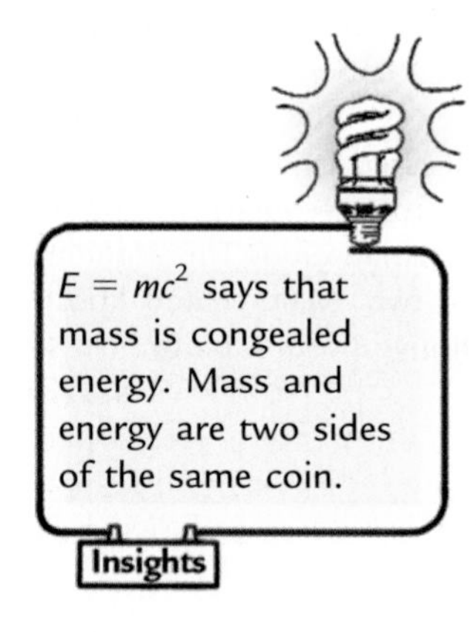

$E = mc^2$ says that mass is congealed energy. Mass and energy are two sides of the same coin.
Insights

A typical fission reaction releases energy of about 200,000,000 electron volts.[4] (By comparison, the explosion of a TNT molecule releases only 30 electron volts.) The combined mass of the fission fragments and neutrons produced in fission is less than the mass of the original uranium nucleus. The tiny amount of missing mass converted to this awesome amount of energy is in accord with Einstein's equation $E = mc^2$. Quite remarkably, the energy of fission is mainly in the form of kinetic energy of the fission fragments that fly apart from one another and of the ejected neutrons. Interestingly, a smaller amount of energy is that of gamma radiation.

The scientific world was jolted by the news of nuclear fission—not only because of the enormous energy release but also because of the extra neutrons liberated in the process. A typical fission reaction releases an average of about two or three neutrons. These new neutrons can, in turn, cause the fissioning of two or three other atomic nuclei, releasing more energy and a total of from four to nine more neutrons. If each of these neutrons splits just one nucleus, the next step in the reaction will produce between 8 and 27 neutrons, and so on. Thus, a whole chain reaction can proceed at an exponential rate.

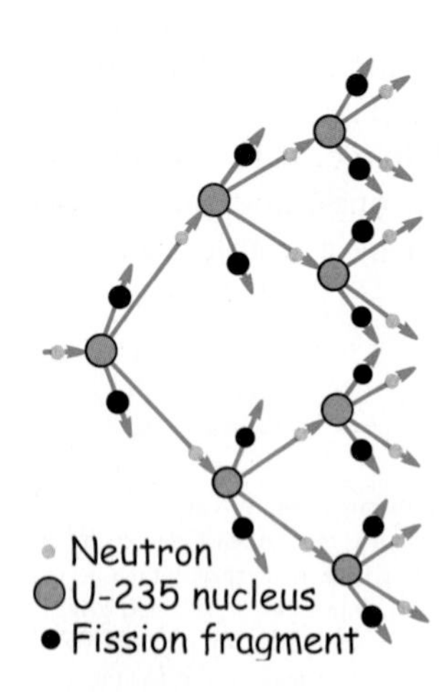

FIGURE 34.2
A chain reaction.

Why doesn't a chain reaction start in naturally occurring uranium deposits? Chain reactions don't ordinarily happen because fission occurs mainly for the rare isotope U-235, which makes up only 0.7% of the uranium in the Earth. Fissionable U-235 is very dilute in natural uranium deposits. The prevalent isotope U-238 absorbs neutrons, but it does not ordinarily undergo fission, so a chain reaction can be quickly snuffed out by the neutron-absorbing U-238 nuclei. The prevalence

[3]In this reaction, three neutrons are ejected when fission occurs. In some other reactions, two neutrons may be ejected—or, occasionally, one or four. On average, fission produces 2.5 neutrons per reaction.

[4]The electron volt (eV) is defined as the amount of kinetic energy an electron acquires in accelerating through a potential difference of 1 V.

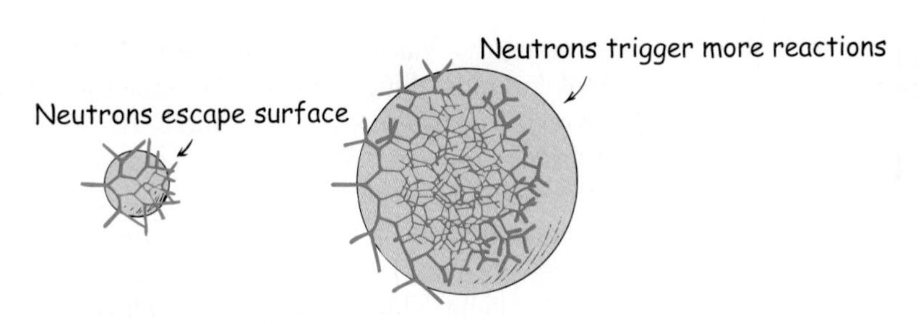

FIGURE 34.3
A chain reaction in a small piece of pure U-235 dies out because neutrons leak from the surface too readily. The small piece has a lot of surface area relative to its mass. In a larger piece, more uranium atoms and less surface area are presented to the neutrons.

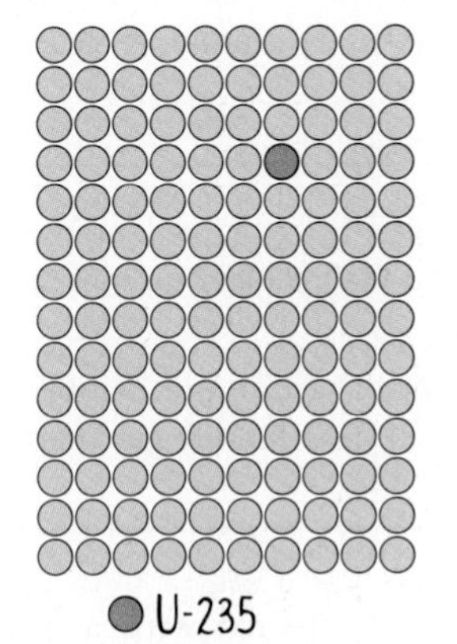

FIGURE 34.4
Only 1 part in 140 (0.7%) of naturally occurring uranium is U-235.

of U-238 lessens the chances of fission. With rare exceptions, naturally occurring uranium is too "impure" to undergo a chain reaction spontaneously.

If a chain reaction occurred in a baseball-size chunk of pure U-235, an enormous explosion would likely result. If the chain reaction were started in a smaller chunk of pure U-235, however, no explosion would occur. This is because a neutron ejected by a fission event travels a certain average distance through the material before it encounters another uranium nucleus and triggers another fission event. If the piece of uranium is too small, a neutron is likely to escape through the surface before it "finds" another nucleus. On the average, fewer than one neutron per fission will be available to trigger more fission, and the chain reaction will die out. In a bigger piece, a neutron can move farther through the material before reaching a surface. Then more than one neutron from each fission event, on the average, will be available to trigger more fission. The chain reaction will build up to enormous energy. We can also understand this geometrically. Recall the concept of scaling in Chapter 12. Small pieces of material have more surface relative to volume than larger pieces (there is more skin on 1 kilogram of small potatoes than on a single 1-kilogram large potato). The larger the piece of fission fuel, the less surface area it has relative to its volume.

The **critical mass** is the amount of mass for which each fission event produces, on the average, one additional fission event. It is just enough to "hold even." A *subcritical* mass is one in which the chain reaction dies out. A *supercritical* mass is one in which the chain reaction builds up explosively.

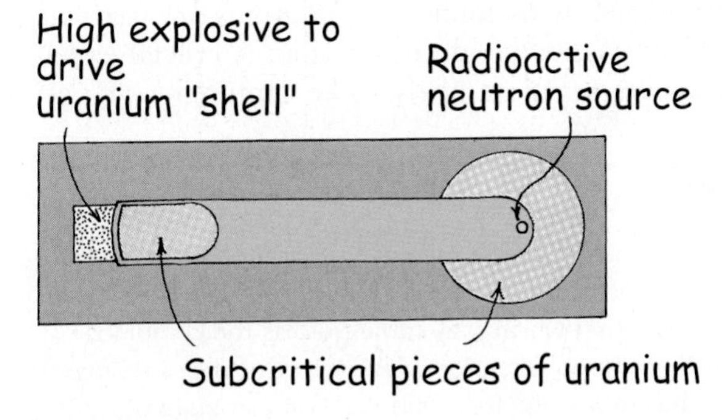

FIGURE 34.5
Simplified diagram of an idealized uranium fission bomb of the "gun type."

Consider a quantity of pure U-235 divided into two units, each having subcritical mass. Neutrons readily reach a surface and escape before a sizable chain reaction builds up. But, if one piece is suddenly driven into the other to form a single piece, the average distance a neutron can move within the material is increased, and fewer neutrons escape through the surface. The total surface area decreases. If the timing is right and the combined mass is greater than critical, a violent explosion takes place. This device is a nuclear fission bomb of the "gun type." Figure 34.5 shows such an idealized uranium fission bomb. Other designs are somewhat more complex.

A chunk of U-235 probably a little larger than a softball was used in the historic Hiroshima blast in

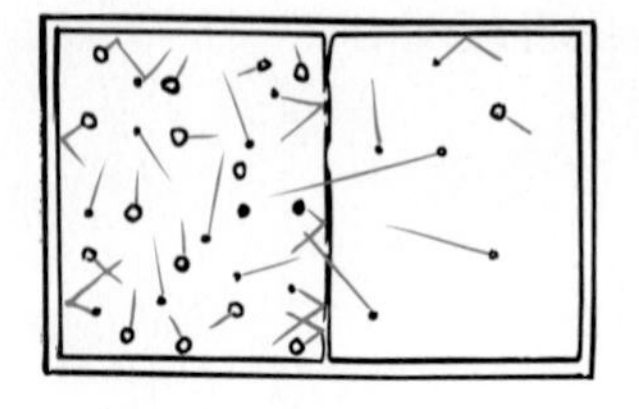

FIGURE 34.6
Lighter molecules move faster than heavier ones at the same temperature and diffuse more readily through a thin membrane.

1945. Separating this much fissionable material from natural uranium was one of the principal and most difficult tasks of the secret Manhattan Project during World War II. Project scientists used two methods of isotope separation. One method employed diffusion, where lighter U-235 has a slightly greater average speed than U-238 at the same temperature. Combined with fluorine to make the gas uranium hexafluoride, the faster isotope has a higher rate of diffusion through a thin membrane or small opening, resulting in a slightly enriched gas containing U-235 on the other side (Figure 34.6). Diffusion through thousands of chambers ultimately produced a sufficiently enriched sample of U-235. The other method, used only for partial enrichment, employed magnetic separation of uranium ions shot into a magnetic field. The smaller-mass U-235 ions were deflected more by the magnetic field than the U-238 ions and were collected atom by atom through a slit positioned to catch them (look ahead to Figure 34.14). After a couple of years, the two methods together netted a few tens of kilograms of U-235.

CHECK YOURSELF

1. A 10-kilogram ball of U-235 is supercritical, but the same ball broken up into small chunks isn't. Explain.
2. Why will molecules of uranium hexafluoride gas made with U-235 move slightly faster at the same temperature than molecules of uranium hexafluoride gas made with U-238?

Uranium-isotope separation today is more easily accomplished with a gas centrifuge. Uranium hexafluoride gas is whirled in a drum at tremendously high rim speeds (on the order of 1500 kilometers per hour). Gas molecules containing the heavier U-238 gravitate to the outside like milk in a dairy separator, and gas containing the lighter U-235 is extracted from the center. Engineering difficulties, overcome only in recent years, prevented the use of this method during the Manhattan Project.

Nuclear Fission Reactors

A chain reaction cannot ordinarily take place in *pure* natural uranium, since it is mostly U-238. The neutrons released by fissioning U-235 atoms are fast neutrons, readily captured by U-238 atoms, which do not fission. A crucial experimental fact is that *slow* neutrons are far more likely to be captured by U-235

CHECK YOUR ANSWERS

1. In smaller pieces, neutrons exit the material before sustaining a chain reaction that won't die out. Put another way, geometrically, the small chunks of U-235 have more combined surface area than the ball from which they came (just as the combined surface area of gravel is greater than the surface area of a boulder of the same mass). Neutrons escape via the surface before a sustained chain reaction can build up.
2. At the same temperature, the molecules of both compounds have the same kinetic energy ($1/2mv^2$). So the molecule made with the less massive U-235 must have a correspondingly higher speed.

Enrico Fermi. It was said jokingly that, when Fermi left Stockholm to return to his native Italy after receiving the Nobel Prize in December, 1938, he got lost and ended up in New York. In fact, he and his Jewish wife Laura carefully planned this escape from Fascist Italy. Fermi became an American citizen in 1945.

FIGURE 34.7
An artist's depiction of the setting in the squash court beneath the stands at the University of Chicago's Stagg Field, where Enrico Fermi and his colleagues constructed the first nuclear reactor.

than by U-238.[5] If neutrons can be slowed down, there is an increased chance that a neutron released by fission will cause fission in another U-235 atom, even amid the more plentiful and otherwise neutron-absorbing U-238 atoms. This increase may be enough to allow a chain reaction to take place.

Within less than a year after the discovery of fission, scientists realized that a chain reaction with ordinary uranium metal might be possible if the uranium were broken up into small lumps and separated by a material that would slow down the neutrons released by nuclear fission. Enrico Fermi, who came to America from Italy at the beginning of 1939, led the construction of the first nuclear reactor—or *atomic pile,* as it was called—in a squash court underneath the grandstands of the University of Chicago's Stagg Field. He and his group used graphite, a common form of carbon, to slow the neutrons. They achieved the first self-sustaining controlled release of nuclear energy on December 2, 1942.

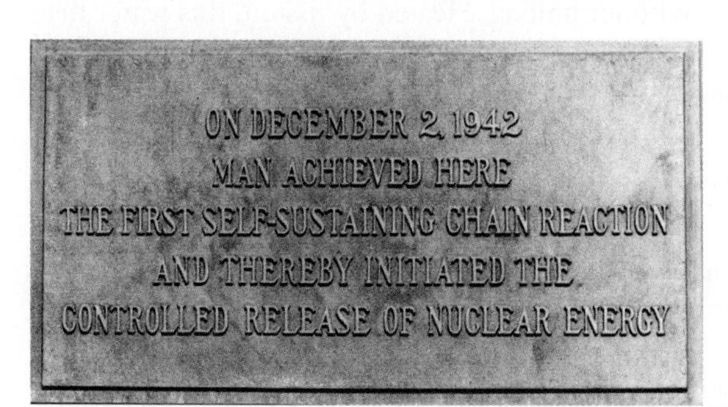

FIGURE 34.8
The bronze plaque at Chicago's Stagg Field commemorates Enrico Fermi's historic fission chain reaction.

Three fates are possible for a neutron in ordinary uranium metal. It may (1) cause fission of a U-235 atom, (2) escape from the metal into nonfissionable surroundings, or (3) be absorbed by U-238 without causing fission. Graphite was used to make the first fate more probable. Uranium was divided into discrete parcels and buried at regular intervals in nearly 400 tons of graphite. A simple analogy clarifies the function of the graphite: If a golf ball rebounds from a massive wall, it loses hardly any speed; but, if it rebounds from a baseball, it loses considerable speed. The case of the neutron is similar. If a neutron rebounds from a heavy nucleus, it loses hardly any speed; but, if it rebounds

[5]This is similar to the selective absorption of various frequencies of light. Just as atoms of various elements absorb light differently, various isotopes of the same element, though chemically almost identical, can have quite different nuclear properties and absorb neutrons differently.

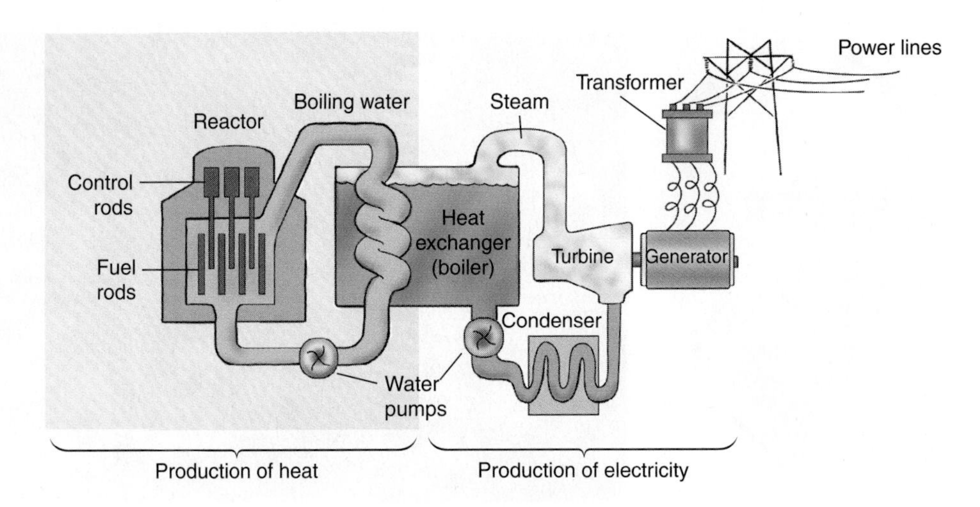

FIGURE 34.9
Diagram of a nuclear fission power plant.

from a lighter carbon nucleus, it loses considerable speed. The graphite was said to "moderate" the neutrons.[6] The whole apparatus is called a *reactor*.

Today's fission reactors contain three components: nuclear fuel, control rods, and a fluid (usually water) to extract heat from the reactor. The nuclear fuel is primarily U-238 plus about 3% U-235. Because the U-235 is so highly diluted with U-238, an explosion like that of a nuclear bomb is not possible.[7] The reaction rate, which depends on the number of neutrons available to initiate fission of other U-235 nuclei, is controlled by rods inserted into the reactor. The control rods are made of a neutron-absorbing material, usually cadmium or boron. Water surrounding the nuclear fuel is kept under high pressure to keep it at a high temperature without boiling. Heated by fission, this water then transfers heat to a second, lower-pressure water system, which operates a turbine and electric generator. Two separate water systems are used so that no radioactivity reaches the turbine.

CHECK YOURSELF

What is the function of a *moderator* in a nuclear reactor? Of *control rods*?

CHECK YOUR ANSWERS

A moderator slows down neutrons that are normally too fast to be absorbed readily by fissionable isotopes, such as U-235. Control rods absorb more neutrons when they are pushed into the reactor and fewer neutrons when they are pulled out of the reactor. They thereby control the number of neutrons that participate in the chain reaction.

[6] *Heavy water,* which contains the heavy hydrogen isotope deuterium, is an even more effective moderator. This is because, in an elastic collision, a neutron transfers a greater portion of its energy to the deuterium nucleus than it would to the heavier carbon nucleus, and a deuteron never absorbs a neutron, as a carbon nucleus occasionally does.

[7] In a worst-case accident, however, heat sufficient to melt the reactor core is possible—and, if the reactor building is not strong enough, to scatter radioactivity into the environment. One such accident occurred at the Chernobyl reactor in 1986 in Ukraine, which was then a constituent republic of the Soviet Union.

Plutonium

Plutonium

When a U-238 nucleus absorbs a neutron, no fission occurs. The nucleus that is created, U-239, is radioactive. With a half-life of 24 minutes, it emits a beta particle and becomes an isotope of the first synthetic element beyond uranium—the transuranic element called *neptunium* (Np, named after the first planet discovered via Newton's law of gravitation). This isotope of Neptunium, Np-239, is also radioactive, with a half-life of 2.3 days. It soon emits a beta particle and becomes an isotope of *plutonium* (Pu, named after Pluto, the second planet to be discovered via Newton's law of gravitation). The half-life of this isotope, Pu-239, is about 24,000 years. Like U-235, Pu-239 will undergo fission when it captures a neutron. Interestingly, Pu-239 is even more fissionable than U-235.

Even before Fermi's atomic pile went critical, physicists realized that reactors could be used to make plutonium, and they set about designing large reactors for that purpose. The reactors constructed at Hanford, Washington, for plutonium production during World War II were 200 million times more powerful than Fermi's atomic pile. By mid-1945, they had manufactured kilograms of this element, which was not found in nature and unknown a few years earlier. Since plutonium is an element distinct from uranium, it can be separated from uranium by ordinary chemical methods when "fuel slugs" are removed from the reactor for processing. Consequently, the reactor provides a process for making fissionable material more easily than by separating the U-235 from natural uranium. The atomic bombs tested in New Mexico and detonated over Nagasaki were plutonium bombs.

Although the process of separating plutonium from uranium is simple in theory, it is very difficult in practice. This is because large quantities of radioactive fission products are formed in addition to plutonium. All chemical processing must be done by remote control to protect the staff against radiation. Also, the element plutonium is chemically toxic in the same sense that lead and arsenic are toxic. It attacks the nervous system and can cause paralysis; death can follow if the dose is sufficiently large. Fortunately, plutonium doesn't remain in its elemental form for long but rapidly combines with oxygen to form three compounds, PuO, PuO_2, and Pu_2O_3, all of which are chemically inert. They will not dissolve in water or in biological systems. These plutonium compounds do not attack the nervous system and have been found to be biologically inert.

Plutonium in any form, however, is radioactively toxic to humans and to other animals. It is more toxic than uranium, although less toxic than radium. Pu-239 emits high-energy alpha particles, which kill cells rather than simply disrupting them. Because damaged cells rather than dead cells produce

92 146 → 92 147 → 93 146 → 94 145

Uranium-238 → Uranium-239 → Neptunium-239 → Plutonium-239

FIGURE 34.10 Interactive Figure

When a nucleus of U-238 absorbs a neutron, it becomes a nucleus of U-239. Within about half an hour, this nucleus emits a beta particle, resulting in a nucleus of about the same mass but with one more unit of charge. This is no longer uranium; it's a new element—*neptunium*. After the neptunium, in turn, emits a beta particle, it becomes plutonium. (In both events, an antineutrino, not shown, is also emitted.)

mutations and lead to cancer, plutonium ranks relatively low as a carcinogen. The greatest danger that plutonium presents to humans is its use in nuclear fission bombs. Its greatest potential benefit is in breeder reactors.

CHECK YOURSELF

Why does plutonium not occur in appreciable amounts in natural ore deposits?

The Breeder Reactor

A remarkable feature of fission power is the *breeding* of plutonium from nonfissionable U-238. Breeding occurs when small amounts of fissionable U-235 are mixed with U-238 in a reactor. Fissioning liberates neutrons that convert the relatively abundant nonfissionable U-238 to U-239, which beta-decays to become Np-239, which, in turn, beta-decays to fissionable plutonium—Pu-239. So, in addition to the abundant energy produced, fission fuel is bred from relatively abundant U-238 in the process.

Some breeding occurs in all fission reactors, but a **breeder reactor** is specifically designed to breed more fissionable fuel than is put into it. Using a breeder reactor is analogous to filling your car's gas tank with water, adding some gasoline, then driving the car and having more gasoline at the end of the trip than at the beginning! The basic principle of the breeder reactor is very attractive: After a few years of operation, a breeder-reactor power plant can produce vast amounts of power while breeding twice as much fuel as it had in the beginning.

The downside of breeder reactors is the enormous complexity of successful and safe operation. The United States gave up on breeders in the 1980s, and only France, Germany, India, and China are still investing in them. Officials in these countries point out that supplies of naturally occurring U-235 are limited. At present rates of consumption, all natural sources of U-235 may be depleted within a century. Countries then deciding to use breeder reactors may well find themselves digging up radioactive wastes they once buried.[8]

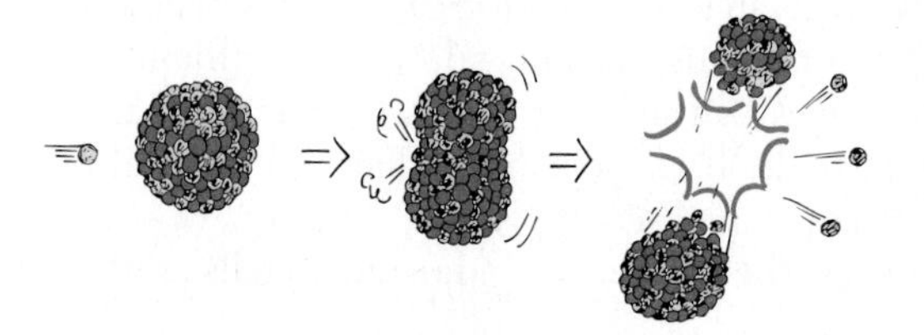

FIGURE 34.11 Interactive Figure

Pu-239 or U-233, like U-235, undergoes fission when it captures a neutron.

CHECK YOUR ANSWER

On a geological time scale, plutonium has a relatively short half-life, so any that exists is produced by very recent transmutations of uranium isotopes.

[8]Many nuclear scientists do not think that deep burial is a desirable solution to the problem of nuclear waste. Devices are presently being studied that could, in principle, convert long-lived radioactive atoms of spent reactor fuel into short-lived or nonradioactive atoms. (See "Will New Technology Solve the Nuclear Waste Problem?" in *The Physics Teacher,* Vol. 35, Feb. 1997.) Nuclear wastes may not plague future generations indefinitely, as has been commonly thought.

CHECK YOURSELF

Complete these reactions, which occur in a breeder reactor:

$^{239}_{92}U \rightarrow$ ________ $+ ^{0}_{-1}e$

$^{239}_{93}Np \rightarrow$ ________ $+ ^{0}_{-1}e$

Fission Power

Energy available from nuclear fission was introduced to the world in the form of nuclear bombs. This violent image still impacts our thinking about nuclear power. Add to this the fearsome 1986 Chernobyl disaster in the Soviet Union, and we find many people viewing nuclear power as evil technology. Nevertheless, about 20% of electric energy in the United States is generated by nuclear-fission reactors. These reactors, sometimes called *nukes,* are simply nuclear furnaces. Like fossil-fuel furnaces, they do nothing more elegant than boil water to produce steam for a turbine. The greatest practical difference is the amount of fuel involved. One kilogram of uranium fuel, a chunk smaller than a baseball, yields more energy than 30 freightcar loads of coal.

FIGURE 34.12
A typical nuclear fission power plant.

One disadvantage of fission power is the generation of radioactive waste products. Light atomic nuclei are most stable when composed of equal numbers of protons and neutrons, and it is mainly heavy nuclei that need more neutrons than protons for stability. For example, there are 143 neutrons but only 92 protons in U-235. When uranium fissions into two medium-weight elements, the extra neutrons in their nuclei make them unstable. These fragments are therefore radioactive, and most of them have very short half-lives. Some of them, however, have half-lives of thousands of years. Safely disposing of these waste products as well as materials made radioactive in the production of nuclear fuels requires special storage casks and procedures. Although fission power goes back a half century, the technology of radioactive waste disposal remains in the developmental stage.

The energy value of radioactive materials released in coal-burning power plants is some 1.5 times more than the energy provided by coal itself.

Insights

The benefits of fission power are (1) plentiful electricity; (2) the conservation of the many billions of tons of coal, oil, and natural gas that every year are literally turned to heat and smoke and that in the long run may be far more precious as sources of organic molecules than as sources of heat; and (3) the elimination of the megatons of sulfur oxides and other poisons, as well as the greenhouse gas carbon dioxide, that are put into the air each year by the burning of these fuels.

CHECK YOUR ANSWERS

$^{239}_{93}Np$; $^{239}_{94}Pu$. (Antineutrinos are also emitted in these beta-decay processes, and they escape unobserved.)

The drawbacks include (1) the problem of storing radioactive wastes; (2) the production of plutonium and the danger of nuclear weapons proliferation; (3) the low-level release of radioactive materials into the air and groundwater; and, most importantly, (4) the risk of an accidental release of large amounts of radioactivity.

Reasoned judgment requires not only that we examine the benefits and drawbacks of fission power but also that we compare its benefits and drawbacks with those of other power sources. For a variety of reasons, public opinion in the United States and much of Europe is now against fission power plants. Although there are growing signs of fission power acceptance, for the most part fission power is currently on the decline, and fossil-fuel power is on the upswing.[9]

CHECK YOURSELF

Coal contains tiny quantities of radioactive materials, enough so there is more environmental radiation surrounding a typical coal-fired power plant than surrounds a fission power plant. What does this indicate about the shielding typically surrounding the two types of power plants?

Mass–Energy Equivalence

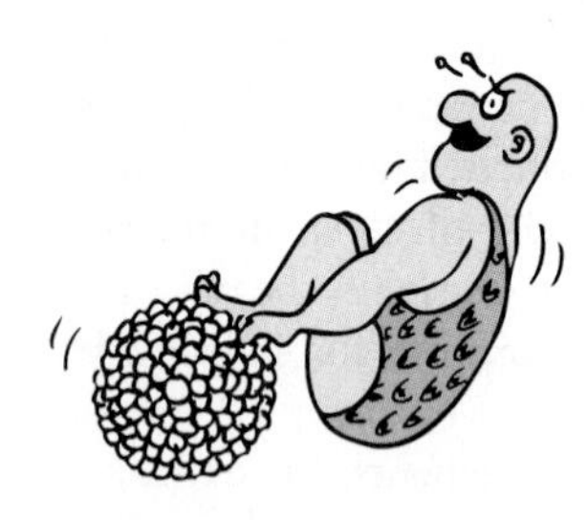

FIGURE 34.13
Work is required to pull a nucleon from an atomic nucleus. This work goes into mass energy.

From Einstein's mass–energy equivalence, $E = mc^2$, mass can be thought of as congealed energy. Mass is like a super storage battery. It stores energy—vast quantities of energy—which can be released if and when the mass decreases. If you were to stack up 238 bricks, the mass of the stack would be equal to the sum of the masses of the individual bricks. At the nuclear level, things are different. The mass of a nucleus is not simply the masses of the individual nucleons that compose it. Consider the work that would be required to separate nucleons from an atomic nucleus.

Recall that work, which is a way of transferring energy, is equal to the product of force and distance. Imagine that you could reach into a U-238 nucleus and, pulling with a force even greater than the attractive nuclear force, remove one nucleon. That would require a lot of work. Then keep repeating the process until you end up with 238 nucleons, stationary and well separated. What

CHECK YOUR ANSWER

Coal-fired power plants are, seemingly, as American as apple pie, with no required (and expensive) shielding to restrict the emissions of radioactive particles. Fission power plants, on the other hand, are required to have shielding to ensure strictly low levels of radioactive emissions.

[9]Public outcry against electricity was common in the nineteenth century. Those with the loudest voices likely knew the least about electricity. Today there is public outcry against nuclear power—"No Nukes!" The position of this book, in contrast, is "Know Nukes!"—first know something about the promises of, as well as the drawbacks to, nuclear power before saying *yes* or *no* to nukes.

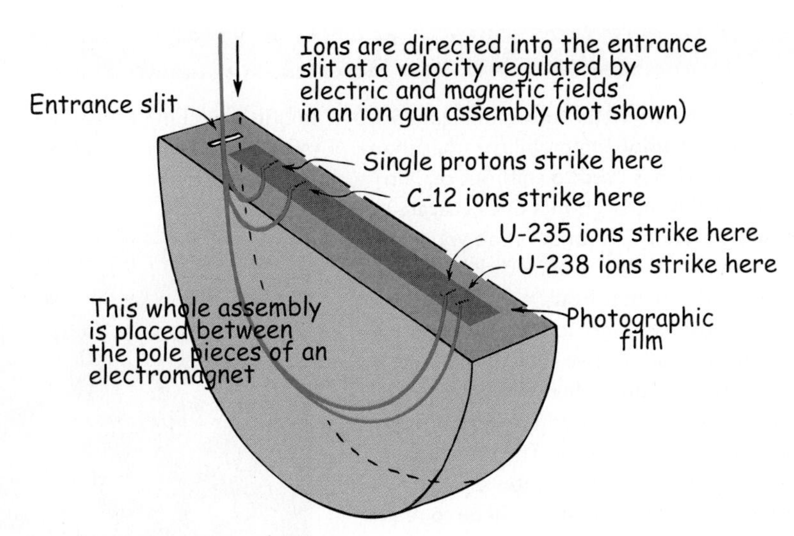
Ions are directed into the entrance slit at a velocity regulated by electric and magnetic fields in an ion gun assembly (not shown)
Entrance slit
Single protons strike here
C-12 ions strike here
U-235 ions strike here
U-238 ions strike here
Photographic film
This whole assembly is placed between the pole pieces of an electromagnet

FIGURE 34.14
The mass spectrometer. Ions of a fixed speed are directed into the semicircular "drum," where they are swept into semicircular paths by a strong magnetic field. Because of different inertia, heavier ions are swept into curves of larger radii and lighter ions are swept into curves of smaller radii. The radius of the curve is directly proportional to the mass of the ion. Using C-12 as a standard, the masses of the isotopes of all the elements are easily determined.

happened to all the work that you have done? You started with one stationary nucleus containing 238 particles and ended with 238 stationary particles. The work that you did has to show up somewhere as additional energy. It shows up as *mass* energy. The separated nucleons have a total mass greater than the mass of the original nucleus, and the extra mass, multiplied by the square of the speed of light, is exactly equal to your energy input: $\Delta E = \Delta mc^2$.

One way to interpret this mass change is to say that an average nucleon inside a nucleus has less mass than a nucleon outside the nucleus. How much less depends on which nucleus. The mass difference is related to the "binding energy" of the nucleus. For uranium, the mass difference is about 0.7%, or 7 parts in a thousand. The 0.7% reduction in the nucleon mass in a uranium atom indicates the binding energy of the nucleus—how much work would be required to disassemble the nucleus.

The experimental verification of this conclusion is one of the triumphs of modern physics. The masses of nucleons and the isotopes of the various elements can be measured with an accuracy of 1 part per million or better. One means of doing this is with a *mass spectrometer* (Figure 34.14).

In a mass spectrometer, charged ions are directed into a magnetic field, where they deflect into circular arcs. The greater the inertia of the ion, the more it resists deflection and the greater the radius of its curved path. All the ions entering this device have the same speed. The magnetic force sweeps the heavier ions into larger arcs and the lighter ions into smaller arcs. The ions pass through exit slits, where they may be collected, or they strike a detector, such as photographic film. An isotope is chosen as a standard, and its position on the film of the mass spectrometer is used as a reference point. The standard is the common isotope of carbon, C-12. The mass of the C-12 nucleus is assigned the value of 12.00000 atomic mass units. As mentioned earlier, the atomic mass unit (amu) is defined to be precisely one-twelfth the mass of the common carbon-12 nucleus. With this reference, the amu values of the other atomic nuclei are measured. The masses of the proton and neutron are found to be greater when they are isolated than when they are in a nucleus. They are 1.00728 and 1.00867 amu, respectively.

PHYSICS AT AIRPORT SECURITY

A version of the mass spectrometer shown in Figure 34.14 is employed in airport security. Ion mobility rather than electromagnetic separation is used to sniff out certain molecules, mainly the few nitrogen-rich ones characteristic of explosives. Security personnel swab your luggage or other belongings with a small disk of paper, which they place in a device that heats it to expel vapors from it. Molecules in the vapor are ionized by exposure to beta radiation from a radioactive source. Most molecules become positive ions, whereas nitrogen-rich molecules become negative, which drift against a flow of air to a positively charged detector. The time for a negative ion to reach the detector indicates the ion's mass—the heavier the ion, the slower it will be to reach the detector.

The same process occurs in body scans, in which a person stands momentarily in an enclosed region the size of a telephone booth where upward puffs of air impinge on the body. The air is then "sniffed" by the same technique, searching for some forty types of explosives and sixty types of drug residues. Presto, green light means none were detected, and red light means—uh-oh!

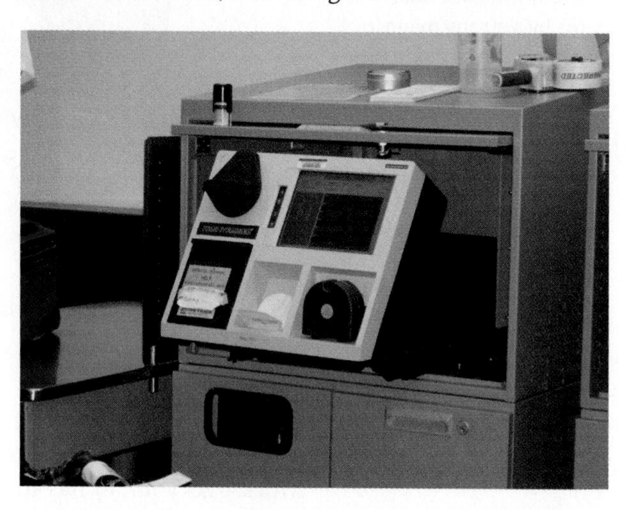

CHECK YOURSELF

Wait a minute! If isolated protons and neutrons have masses greater than 1.0000 amu, why don't 12 of them in a carbon nucleus have a combined mass greater than 12.0000 amu?

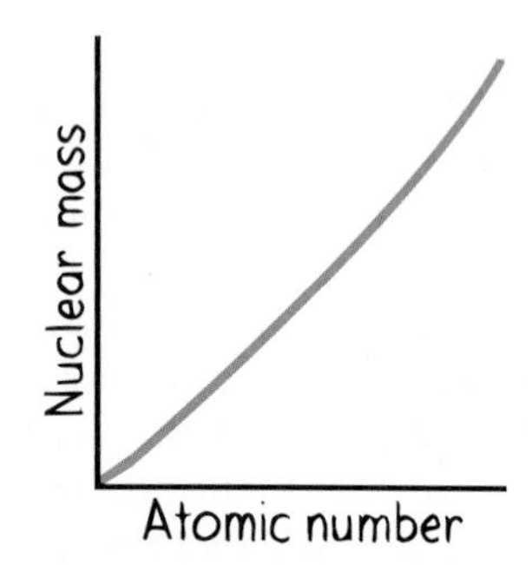

FIGURE 34.15
The plot shows how nuclear mass increases with increasing atomic number.

A graph of nuclear mass as a function of atomic number is shown in Figure 34.15. The graph slopes upward with increasing atomic number, as expected, telling us that elements are more massive as atomic number increases. (The slope curves because there are proportionally more neutrons in the more massive atoms.)

A more important graph results from a plot of average mass *per nucleon* for the elements hydrogen through uranium (Figure 34.16). This is perhaps the most important graph in this book, for it is the key to understanding the energy associated with nuclear processes—fission as well as fusion. To obtain the average mass per nucleon, you divide the total mass of a nucleus by the number of nucleons in the nucleus. (Similarly, if you divide the total mass of a roomful of people by the number of people in the room, you get the average mass per

CHECK YOUR ANSWER

When you pull a nucleon from the nucleus, you do work on it and it gains energy. When that nucleon falls back into the nucleus, it does work on its surroundings and *loses* energy. Losing energy means losing mass. It's as if each nucleon, on average, slims down to a mass of exactly 1.0000 amu when it joins with 11 other nucleons to form C-12. If you pull them back out, you'll get back the original mass. Indeed, $E = mc^2$.

FIGURE 34.16

Interactive Figure

The graph shows that the mass of a nucleon depends on which nucleus it is in. Individual nucleons have the most mass in the lightest (hydrogen) nuclei, the least mass in iron nuclei, and intermediate mass in the heaviest (uranium) nuclei. (The vertical scale is exaggerated.)

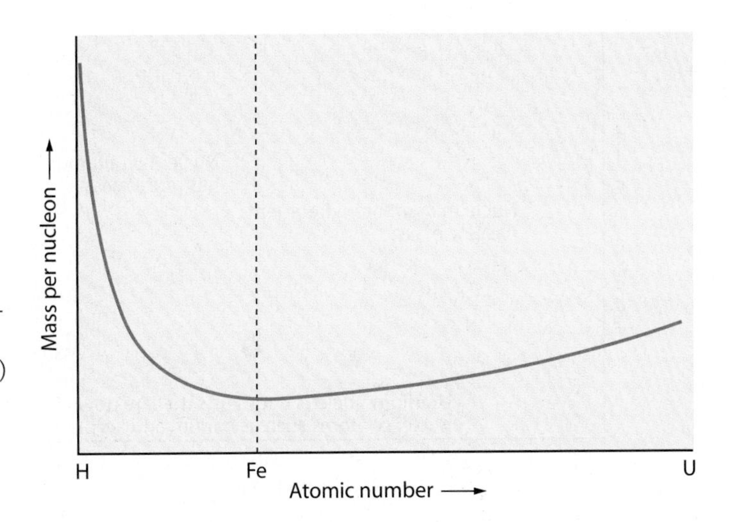

The graph of Figure 34.16 reveals the energy of the atomic nucleus, likely the primary source of energy in the universe—which is why it can be considered the most important graph in this book.

Insights

person.) The major fact we learn from Figure 34.16 is that the average mass per nucleon varies from one nucleus to another.

The greatest mass per nucleon occurs for a proton when it occurs alone as a hydrogen nucleus because then it has no binding energy to pull its mass down. As we progress to elements beyond hydrogen, Figure 34.16 tells us that the mass per nucleon decreases and is least for a nucleon in an iron nucleus. Iron holds its nucleons more tightly than any other nucleus does. Beyond iron, the trend reverses itself as protons (and neutrons) have progressively more and more mass in atoms of increasing atomic number. This continues all the way through the list of elements.

From this graph, we can see why energy is released when a uranium nucleus is split into two nuclei of lower atomic number. When the uranium nucleus splits, the masses of the two fission fragments lie about halfway between the masses of uranium and hydrogen on the horizontal scale of the graph. It is most important to note that the mass per nucleon in the fission fragments is *less than* the mass per nucleon when the same set of nucleons are combined in the uranium nucleus.

TABLE 34.1
Relative Masses and Masses/Nucleon of Some Isotopes

Isotope	*Symbol*	*Mass (amu)*	*Mass/Nucleon (amu)*
Neutron	n	1.008665	1.008665
Hydrogen	$^{1}_{1}H$	1.007825	1.007825
Deuterium	$^{2}_{1}H$	2.01410	1.00705
Tritium	$^{3}_{1}H$	3.01605	1.00535
Helium-4	$^{4}_{2}He$	4.00260	1.00065
Carbon-12	$^{12}_{6}C$	12.00000	1.000000
Iron-58	$^{58}_{26}Fe$	57.93328	0.99885
Copper-63	$^{63}_{29}Cu$	62.92960	0.99888
Krypton-90	$^{90}_{36}Kr$	89.91959	0.99911
Barium-143	$^{143}_{56}Ba$	142.92054	0.99944
Uranium-235	$^{235}_{92}U$	235.04395	1.00019

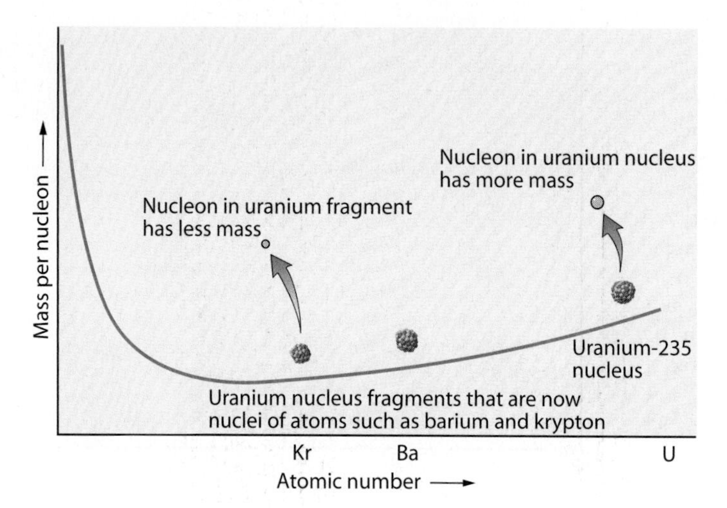

FIGURE 34.17
The mass of each nucleon in a uranium nucleus is greater than the mass of each nucleon in any one of its nuclear-fission fragments. This decrease in mass is mass that has been transformed into energy. Hence, nuclear fission is an energy-releasing process.

When this decrease in mass is multiplied by the speed of light squared, it equals 200,000,000 electron volts, the energy yielded by each uranium nucleus that undergoes fission. As mentioned earlier, most of this enormous energy is the kinetic energy of the fission fragments.

We can think of the mass-per-nucleon curve as an energy valley that starts at the highest point (hydrogen), slopes steeply down to the lowest point (iron), and then slopes more gradually up to uranium. Iron, which is at the bottom of the energy valley, has the most stable nucleus. It also has the most tightly bound nucleus: More energy per nucleon is required to separate nucleons from its nucleus than from any other nucleus. Any nuclear transformation that moves lighter nuclei toward iron by combining them or moves heavier nuclei toward iron by dividing them releases energy.

So the decrease in mass is detectable in the form of energy—much energy—when heavy nuclei undergo fission. As mentioned, a drawback to this process involves the radioactive fission fragments. A more promising long-range source of energy is to be found on the left side of the energy valley.

CHECK YOURSELF

1. Consider a graph like those of Figures 34.16 and 34.17—not for microscopic nucleons, but for a house or other structures made of regular bricks. Would the graph dip, or would it be a straight horizontal line?
2. Correct the following incorrect statement: When a heavy element, such as uranium, undergoes fission, there are fewer nucleons after the reaction than before.

TABLE 34.2
Energy Gain from Fission of Uranium

Reaction:	$^{235}U + n \rightarrow {}^{143}Ba + {}^{90}Kr + 3n + \Delta m$
Mass balance:	$235.04395 + 1.008665 = 142.92054 + 89.91959 + 3(1.008665) + \Delta m$
Mass defect:	$\Delta m = 0.186$ amu
Energy gain:	$\Delta E = \Delta mc^2 = 0.186 \times 931$ MeV $= 173.6$ MeV
Energy gain/nucleon:	$\Delta E/236 = 173.6$ MeV$/236 = 0.74$ MeV/nucleon

(When m is expressed in amu, c^2 is equivalent to 931 MeV.)

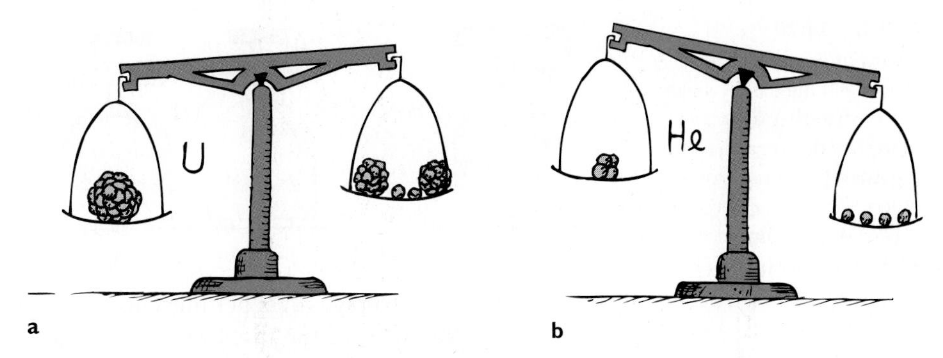

FIGURE 34.18 Interactive Figure

The mass of a nucleus is *not* equal to the sum of the masses of its parts. (a) The fission fragments of a heavy nucleus like uranium are less massive than the uranium nucleus. (b) Two protons and two neutrons are more massive in their free states than when they are combined to form a helium nucleus.

Nuclear Fusion

Inspection of the mass-per-nucleon versus atomic-number graph will show that the steepest part of the energy hill is from hydrogen to iron. Energy is gained as light nuclei *fuse* (which means that they combine). This process is **nuclear fusion**—the opposite of nuclear fission. We can see from Figure 34.19 that, as we move along the list of elements from hydrogen to iron (the left side of the

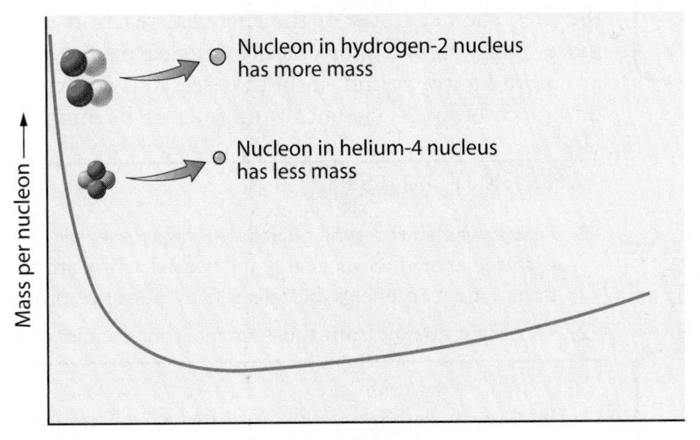

FIGURE 34.19
The average mass of a nucleon in hydrogen is greater than its average mass when fused with another to become helium. The decreased mass is mass that is converted to energy, which is why nuclear fusion of light elements is an energy-releasing process.

CHECK YOUR ANSWERS

1. It would be a straight horizontal line. The mass per brick would be the same for every structure. (Theoretically, however, it would not be *exactly* a straight horizontal line because, even for bricks, binding energy has some effect on mass, although the effect would be far too small to measure.)
2. When a heavy element, such as uranium, undergoes fission, there aren't fewer nucleons after the reaction. Instead, there's *less mass* in the same number of nucleons.

FIGURE 34.20
Fictitious example: The "hydrogen magnets" weigh more when they are apart than they do when they are together. (Adapted from Albert V. Baez, *The New College Physics: A Spiral Approach*. W. H. Freeman and Company, 1967. An oldie but a goodie.)

energy valley), the average mass per nucleon decreases. Thus, if two small nuclei were to fuse, the mass of the fused nucleus would be less than the mass of the two single nuclei before fusion. Energy is gained as light nuclei fuse.

Consider hydrogen fusion. For a fusion reaction to occur, the nuclei must collide at a very high speed in order to overcome their mutual electric repulsion. The required speeds correspond to the extremely high temperatures found in the Sun and other stars. Fusion brought about by high temperatures is called **thermonuclear fusion.** In the high temperatures of the Sun, approximately 657 million tons of hydrogen are fused to 653 million tons of helium each second. The "missing" 4 million tons of mass convert to energy. Such reactions are, quite literally, nuclear burning.

In a sense, nucleons in the heavy elements wish to lose mass and be like nucleons in iron. And nucleons in the light elements also wish to lose mass and become more like those in iron.

Insights

Interestingly, most of the energy of nuclear fusion is in the kinetic energy of fragments, mainly neutrons. When the neutrons are stopped and captured, the energy of fusion turns into heat. In fusion reactions of the future, part of this heat will be transformed to electricity.

Thermonuclear fusion is analogous to ordinary chemical combustion. In both chemical and nuclear burning, a high temperature starts the reaction; the release of energy by the reaction maintains a sufficient temperature to spread the fire. The net result of the chemical reaction is a combination of atoms into more tightly bound molecules. In nuclear reactions, the net result is nuclei that are more tightly bound. In both cases, mass decreases as energy is released. The difference between chemical and nuclear burning is essentially one of scale.

CHECK YOURSELF

1. First it was stated that nuclear energy is released when atoms split apart. Now it is stated that nuclear energy is released when atoms combine. Is this a contradiction? How can energy be released by opposite processes?
2. To obtain energy from the element iron, should iron nuclei be fissioned or fused?

Know nukes before you say "No nukes"!
Insights

In fission reactions, the amount of matter that is converted to energy is about 0.1%; in fusion, it can be as much as 0.7%. These numbers apply whether the process takes place in bombs, in reactors, or in stars.

Some typical fusion reactions are shown in Figure 34.21. Note that all reactions produce at least a pair of particles—for example, a pair of deuterium nuclei that fuse produce a tritium nucleus and a neutron rather than a lone helium nucleus. Either reaction is okay as far as adding the nucleons and charges is concerned, but the lone-nucleus case is not okay with conservation of momentum and energy. If a lone helium nucleus flies away after the reaction, it adds momentum that wasn't there initially. Or if it remains motionless, there's no mechanism for energy release. So, because a single product particle

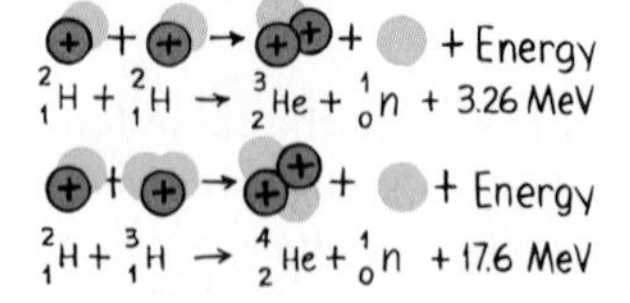

FIGURE 34.21
Two of many fusion reactions.

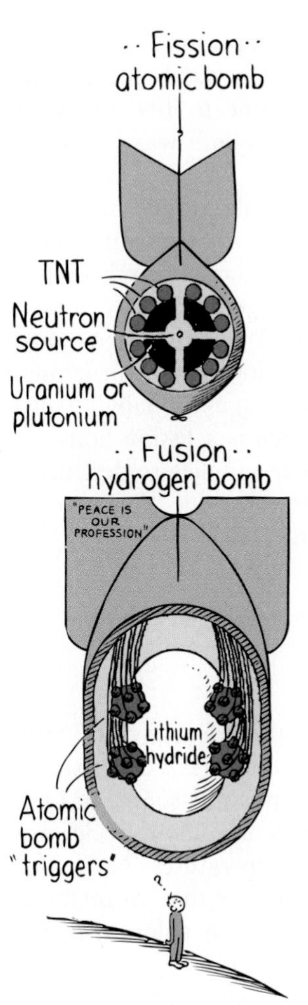

FIGURE 34.22
Fission and fusion bombs.

TABLE 34.3
Energy Gain from Fusion of Hydrogen

Reaction:	${}^{2}\text{H} + {}^{3}\text{H} \rightarrow {}^{4}_{2}\text{H} + n + \Delta m$
Mass balance:	$2.01410 + 3.01605 = 4.00260 + 1.008665 + \Delta m$
Mass defect:	$\Delta m = 0.01888$ amu
Energy gain:	$\Delta E = \Delta mc^2 = 0.018888 \times 931$ MeV $= 17.6$ MeV
Energy gain/nucleon:	$\Delta E/5 = 17.6$ MeV$/5 = 3.5$ MeV/nucleon

can't move and it can't sit still, it isn't formed. Fusion normally requires the creation of at least two particles to share the released energy.[10]

Table 34.3 shows the energy gain from the fusion of hydrogen isotopes deuterium and tritium. This is the reaction proposed for plasma fusion power plants of the future. The high-energy neutrons, according to plan, will escape from the plasma in the reactor vessel and heat a surrounding blanket of material to provide useful energy. The helium nuclei remaining behind will help to keep the plasma hot.

Elements somewhat heavier than hydrogen release energy when fused. But they release much less energy per fusion reaction than hydrogen. The fusion of still heavier elements occurs in the advanced stages of a star's evolution. The energy released per gram during the various fusion stages from helium to iron amounts only to about one-fifth of the energy released in the fusion of hydrogen to helium. Hydrogen, most notably in the form of deuterium, is the choicest fuel for fusion.

CHECK YOUR ANSWERS

1. Energy is released in any nuclear reaction in which the mass of the nuclei after the reaction is less than the mass of the nuclei before. When light nuclei, such as those of hydrogen, fuse to form heavier nuclei, total nuclear mass decreases. The fusion of light nuclei therefore releases energy. When heavy nuclei, such as those of uranium, split to become lighter nuclei, total nuclear mass also decreases. The splitting of heavy nuclei, therefore, releases energy. For energy release, "decrease mass" is the name of the game—any game, chemical or nuclear.
2. Neither, because iron is at the very bottom of the curve (energy valley). If you fuse two iron nuclei, the product lies somewhere on the upper right of iron on the curve, which means the product has a higher mass per nucleon. If you split an iron nucleus, the products lie on the upper left of iron on the curve, which again means a higher mass per nucleon. Because no mass decrease occurs in either reaction, no energy is ever released.

[10]One of the reactions in the Sun's proton–proton fusion cycle does have a one-particle final state. It is proton + deuteron → He-3. This happens because the density in the center of the Sun is great enough that "spectator" particles share in the energy release. So, even in this case, the energy released goes to two or more particles. Fusion in the Sun involves more complicated (and slower!) reactions in which a small part of the energy also appears in the form of gamma rays and neutrinos. The neutrinos escape unhindered from the center of the Sun and bathe the solar system. Interestingly, the fusion of nuclei in the Sun is an occasional process, for the mean spacing between nuclei is vast, even at the high pressures in its center. That's why it takes some 10 billion years for the Sun to consume its hydrogen fuel.

Prior to the development of the atomic bomb, the temperatures required to initiate nuclear fusion on Earth were unattainable. When it was found that the temperatures inside an exploding atomic bomb are four to five times the temperature at the center of the Sun, the thermonuclear bomb was but a step away. This first hydrogen bomb was detonated in 1952. Whereas the critical mass of fissionable material limits the size of a fission bomb (atomic bomb), no such limit is imposed on a fusion bomb (thermonuclear, or hydrogen, bomb). Just as there is no limit to the size of an oil-storage depot, there is no theoretical limit to the size of a fusion bomb. Like the oil in a storage depot, any amount of fusion fuel can be stored with safety until it is ignited. Although a mere match can ignite an oil depot, nothing less energetic than a fission bomb can ignite a thermonuclear bomb. We can see that there is no such thing as a "baby" hydrogen bomb. It cannot be less energetic than its fuse, which is a fission bomb.

The hydrogen bomb is another example of a discovery applied to destructive rather than constructive purposes. The potential constructive side of the picture is the controlled release of vast amounts of clean energy.

Controlling Fusion

The energy released in fusing a pair of hydrogen nuclei is less than that of fissioning a uranium nucleus. But because there are more atoms in a gram of hydrogen than in a gram of uranium, gram for gram, fusion releases several times as much energy as uranium.

Insights

The world's oceans contain deuterium with the potential to release vastly more energy than all the world's fossil fuels, and much more than the world's supply of uranium. Therefore, fusion has to be considered as a possible source of long-term energy needs. Fusion reactions require temperatures of millions of degrees. There are a variety of techniques for attaining high temperatures. No matter how the temperature is produced, a technological problem is that all materials melt and vaporize at the temperatures required for fusion. A solution to this problem is to confine the reaction in a *nonmaterial container*.

One type of nonmaterial container is a magnetic field, which can exist at any temperature and can exert powerful forces on charged particles in motion. "Magnetic walls" provide a kind of magnetic straightjacket for hot plasmas. Magnetic compression further heats the plasma to fusion temperatures. At about a million degrees, some nuclei are moving fast enough to overcome electrical repulsion and to slam together and fuse. The energy output, however, is small relative to the energy used to heat the plasma. Even at 100 million degrees, more energy must be put into the plasma than is given off by fusion. At about 350 million degrees, the fusion reactions produce enough energy to be self-sustaining. At this ignition temperature, all that is needed to produce continuous power is a steady feed of nuclei. This is the sought-after condition called *break-even*.

Although break-even has nearly been achieved for very short times in several fusion devices, instabilities in the plasma have thus far prevented a sustained reaction. A big problem has been devising a field system that can hold the plasma in a stable and sustained position while an ample number of nuclei fuse. A variety of magnetic confinement devices are the subject of much present-day research.

Controlling Nuclear Fusion

Another approach uses high-energy lasers. One proposed technique is to aim an array of laser beams at a common point and drop pellets of frozen hydrogen isotopes through the synchronous crossfire (Figure 34.23). The energy of the multiple beams should crush the pellets to densities 20 times that of lead

Watch for the forthcoming International Thermonuclear Experimental Reactor (ITER), a joint fusion project by the European Union, Japan, and the United States.

Insights

and heat them to the required temperatures. Such laser-induced fusion could produce several hundred times more energy than is delivered by the laser beams that compress and ignite the pellets. Like the succession of fuel/air explosions in an automobile engine's cylinders that convert into a smooth flow of mechanical power, the successive ignition of pellets in a fusion power plant may similarly produce a steady stream of electric power.[11] The success of this technique requires precise timing, for the necessary compression must occur before a shock wave causes the pellets to disperse. High-power lasers that work reliably are yet to be developed. Break-even has not yet been achieved with laser fusion.

Still other approaches involve the bombardment of fuel pellets not by laser light but by beams of electrons and ions. Whatever the method, we are still looking forward to the great day in this twenty-first century when fusion power becomes a reality.

Fusion power, if it can be achieved, will be nearly ideal. Fusion reactors cannot become "supercritical" and go out of control because fusion requires no critical mass. Furthermore, there is no air pollution because the only product of the thermonuclear combustion is helium (good for children's balloons). Except for some radioactivity in the inner chamber of the fusion device because of high-energy neutrons, the by-products of fusion are not radioactive. Disposal of radioactive waste is not a major problem. Furthermore, there is no air pollution because there is no combustion. The problem of thermal pollution, characteristic of conventional steam-turbine plants, can be avoided by direct generation of electricity with MHD generators or by similar techniques using charged-particle fuel cycles that employ a direct energy conversion.

The fuel for nuclear fusion is hydrogen, the most plentiful element in the universe. The reaction that works best at "moderate" temperature is the fusion of the hydrogen isotopes deuterium ($^{2}_{1}H$) and tritium ($^{3}_{1}H$). Deuterium is found in ordinary water and tritium can be produced by the fusion reactor. Thirty

FIGURE 34.23
How laser fusion might work. Pellets of frozen deuterium are rhythmically dropped into synchronized laser crossfire. The resulting heat is carried off by molten lithium to produce steam.

FIGURE 34.24
Pellet chamber at Lawrence Livermore National Laboratory. The laser source is Nova, one of the most powerful lasers in the world, which directs 10 beams into the target region.

[11]The rate of pellet fusion is 5 per second at the National Ignition Facility at Lawrence Livermore National Laboratory. (For comparison, approximately 20 explosions per second occur in each automobile engine cylinder in a car that travels at highway speed.) Such a plant could produce some 1000 MW of electric power, enough to supply a city of about 600,000 people. Five fusion burns per second will provide about the same power as 60 L of fuel oil or 70 kg of coal per second from conventional power plants.

FUSION TORCH AND RECYCLING

A fascinating application for the abundant energy that fusion of whatever kind may provide is the *fusion torch,* a star-hot flame or high-temperature plasma into which all waste materials—whether liquid sewage or solid industrial refuse—could be dumped. In the high-temperature region, the materials would be reduced to their constituent ionized atoms and separated by a mass-spectrometer-type device into various bins ranging from hydrogen to uranium. In this way, a single fusion plant could, in principle, not only dispose of thousands of tons of solid wastes daily but also provide a continuous supply of fresh raw material—thereby closing the cycle from use to reuse.

This would be a major turning point in materials economy (Figure 34.25). Our present concern for recycling materials would reach a grand fruition with this or a comparable achievement, for it would be recycling with a capital *R*! Rather than gut our planet further for raw materials, we'd be able to recycle existing stock over and over again, adding new material only to replace the relatively small amounts that are lost.

Fusion power holds the potential to produce abundant electrical power, to desalinate water, to help to cleanse our world of pollution and wastes, to recycle our materials, and, in so doing, to provide the setting for a better world—not necessarily in the far-off future but perhaps in this century. If and when fusion power plants become a reality, they are likely to have an even more profound impact upon almost every aspect of human society than did the harnessing of electromagnetic energy at the end of the nineteenth century.

When we think of our continuing evolution, we can see that the universe is well suited to those who will live in the distant future. If people are one day to dart about the universe in the same way we are able to jet about the world today, their supply of fuel is assured. The fuel for fusion is found in every part of the universe—not only in the stars but also in the space between them. About 91% of the atoms in the universe are estimated to be hydrogen. For people of this imagined future, the supply of raw materials is also assured; all the elements known to exist result from the fusing of more and more hydrogen nuclei. Simply put, if you fuse 8 deuterium nuclei, you have oxygen; if you fuse 26, you have iron; and so forth. Future humans might synthesize their own elements and produce energy in the process, just as the stars have always done. Humans may one day travel to the stars in ships fueled by the same energy that makes the stars shine.

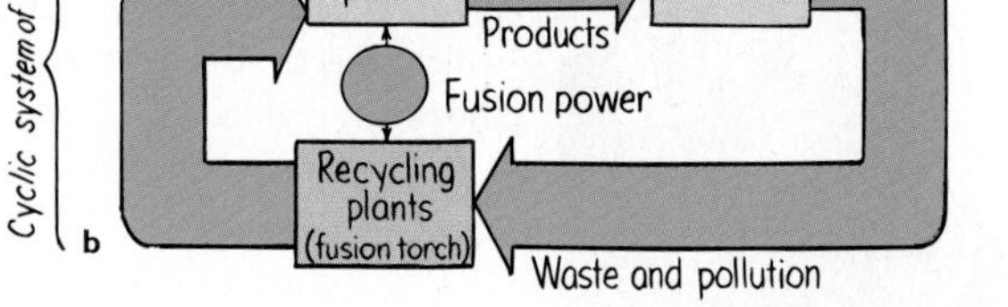

FIGURE 34.25

A closed materials economy could be achieved with the aid of the fusion torch. In contrast to present systems, (a) which are based on inherently wasteful linear material economies, a stationary-state system (b) would be able to recycle the limited supply of material resources, thus alleviating most of the environmental pollution associated with present methods of energy utilization. (Redrawn from "The Prospects of Fusion Power," by William C. Gough and Bernard J. Eastlund. *Scientific American,* Feb. 1971.) This idea remains as visionary and as enticing as it was more than thirty years ago.

liters of seawater contain 1 gram of deuterium, which, when fused, release as much energy as 10,000 liters of gasoline or 80 tons of TNT. Natural tritium is much scarcer, but, given enough to get started, a controlled thermonuclear reactor will breed it from deuterium in ample quantities.

The development of fusion power has been slow and difficult, already extending over fifty years. It is one of the biggest scientific and engineering challenges that we face. Yet there is justified hope to believe that it will be achieved and will be a primary energy source for future generations.

CHECK YOURSELF

1. One last time: Fission and fusion are opposite processes, yet each releases energy. Isn't this contradictory?
2. Would you expect the temperature of the core of a star to increase or decrease as a result of the fusion of intermediate elements to manufacture elements heavier than iron?

CHECK YOUR ANSWERS

1. No, no, no! This is contradictory only if the same element is said to release energy by both the processes of fission and fusion. Only the fusion of light elements and the fission of heavy elements result in a decrease in nucleon mass and a release of energy.
2. Energy is absorbed, not released, when heavier elements fuse, so the star core tends to cool at this late stage of its evolution. Interestingly, however, this allows the star to collapse, which produces an even greater temperature. Nuclear cooling is then more than offset by gravitational heating.

Summary of Terms

Nuclear fission The splitting of the nucleus of a heavy atom, such as uranium-235, into two smaller atoms, accompanied by the release of much energy.

Chain reaction A self-sustaining reaction in which the products of one reaction event stimulate further reaction events.

Critical mass The minimum mass of fissionable material in a reactor or nuclear bomb that will sustain a chain reaction.

Breeder reactor A fission reactor that is designed to breed more fissionable fuel than is put into it by converting nonfissionable isotopes to fissionable isotopes.

Nuclear fusion The combination of light atomic nuclei to form heavier nuclei, with the release of much energy.

Thermonuclear fusion Nuclear fusion produced by high temperature.

Suggested Reading

Bodanis, David. *E* = *mc*²: *A Biography of the World's Most Famous Equation*. New York: Berkley Publishing Group, 2002. Insightful reading that includes some of the adventure in the early quest for nuclear power.

Review Questions

Nuclear Fission

1. What is the role of electrical forces in nuclear fission?
2. When a nucleus undergoes fission, what role can the ejected neutrons play?
3. Why does a chain reaction not occur in uranium mines?
4. Why is a chain reaction more likely in a big piece of uranium than it is in a small piece?
5. What is meant by the idea of a critical mass?
6. Which will leak more neutrons, two separate pieces of uranium or the same pieces stuck together?
7. What were the two methods used to separate U-235 from U-238 in the Manhattan Project during World War II?

Nuclear Fission Reactors

8. What was the function of graphite in the first atomic reactor?
9. What are the three possible fates of neutrons in uranium metal?
10. What are the three main components of a fission reactor?
11. Why can a reactor not explode like a fission bomb?

Plutonium

12. What isotope is produced when U-238 absorbs a neutron?
13. What isotope is produced when U-239 emits a beta particle?
14. What isotope is produced when Np-239 emits a beta particle?

15. What do U-235 and Pu-239 have in common?
16. Why is plutonium more easily separated from uranium metal than particular isotopes of uranium? What makes it difficult?
17. When is plutonium chemically toxic and when is it not?

The Breeder Reactor

18. What is the effect of placing small amounts of fissionable isotopes with large amounts of U-238?
19. Name three isotopes that undergo nuclear fission.
20. How does a breeder reactor breed nuclear fuel?

Fission Power

21. How is a nuclear reactor similar to a conventional fossil-fuel plant? How is it different?
22. Why are the fragments of fission radioactive?
23. What is the main advantage of fission power? What is the main drawback?

Mass–Energy Equivalence

24. What celebrated equation shows the equivalence of mass and energy?
25. Is work required to pull a nucleon out of an atomic nucleus? Does the nucleon, once outside, then have more energy than it did when it was inside the nucleus? In what form is this energy?
26. Which ions are least deflected in a mass spectrometer?
27. What is the basic difference between the graphs of Figure 34.15 and Figure 34.16?
28. In which atomic nucleus do nucleons have the greatest mass? In which nucleus do they have the least mass?
29. What becomes of the missing mass when a uranium nucleus undergoes fission?
30. If the graph in Figure 34.16 is seen as an energy valley, what can be said about the energy of nuclear transformations that progress toward iron?

Nuclear Fusion

31. When two hydrogen nuclei are fused, is the mass of the product nucleus more or less than the sum of the masses of the two hydrogen nuclei?
32. For helium to release energy, should it be fissioned or fused?

Controlling Fusion

33. What kind of containers are used to contain multimillion-degree plasmas?
34. In what form is energy initially released in nuclear fusion?

Project

Write a letter to Grandpa discussing nuclear power. Cite both the ups and downs of it, and explain how the comparison affects your personal view of nuclear power. Also explain to him how nuclear fission and nuclear fusion differ.

Exercises

1. Why doesn't uranium ore spontaneously undergo a chain reaction?
2. Do today's nuclear power plants use fission, fusion, or both?
3. Some heavy nuclei, containing even more protons than the uranium nucleus, undergo "spontaneous fission," splitting apart without absorbing a neutron. Why is spontaneous fission observed only in the heaviest nuclei?
4. Why will nuclear fission probably not be used directly for powering automobiles? How could it be used indirectly to power automobiles?
5. Why does a neutron make a better nuclear bullet than a proton or an electron?
6. Why will the escape of neutrons be proportionally less in a large piece of fissionable material than in a smaller piece?
7. Which shape is likely to need more material for a critical mass, a cube or a sphere? Explain.
8. A 56-kg sphere of U-235 constitutes a critical mass. If the sphere were flattened into a pancake shape, would it still be critical? Explain.
9. Does the average distance that a neutron travels through fissionable material before escaping increase or decrease when two pieces of fissionable material are assembled into one piece? Does this assembly increase or decrease the probability of an explosion?
10. U-235 releases an average of 2.5 neutrons per fission, while Pu-239 releases an average of 2.7 neutrons per fission. Which of these elements might you therefore expect to have the smaller critical mass?
11. Uranium and thorium occur abundantly in various ore deposits. However, plutonium could occur only in exceedingly tiny amounts in such deposits. What is your explanation?
12. Why, after a uranium fuel rod reaches the end of its fuel cycle (typically 3 years), does most of its energy come from the fissioning of plutonium?
13. If a nucleus of $^{232}_{90}\text{Th}$ absorbs a neutron and the resulting nucleus undergoes two successive beta decays (emitting electrons), what nucleus results?

14. The water that passes through a reactor core does not pass into the turbine. Instead, heat is transferred to a separate water cycle that is entirely outside the reactor. Why is this done?

15. Why is carbon better than lead as a moderator in nuclear reactors?

16. Is the mass of an atomic nucleus greater or less than the sum of the masses of the nucleons composing it? Why don't the nucleon masses add up to the total nuclear mass?

17. The energy release of nuclear fission is tied to the fact that the heaviest nuclei have about 0.1% more mass per nucleon than nuclei near the middle of the periodic table of the elements. What would be the effect on energy release if the 0.1% figure were instead 1%?

18. In what way are fission and fusion reactions similar? What are the main differences in these reactions?

19. How is chemical burning similar to nuclear fusion?

20. To predict the approximate energy release of either a fission reaction or a fusion reaction, explain how a physicist makes use of the curve of Figure 34.16 or a table of nuclear masses and the equation $E = mc^2$.

21. What nuclei will result if a U-235 nucleus, after absorbing a neutron and becoming U-236, splits into two identical fragments?

22. If U-238 splits into two even pieces, and each piece emits an alpha particle, what elements are produced?

23. Fermi's original reactor was just "barely" critical because the natural uranium that he used contained less than 1% of the fissionable isotope U-235 (half-life 713 million years). What if, in 1942, the Earth had been 9 billion years old instead of 4.5 billion years old? Would Fermi have been able to make a reactor go critical with natural uranium?

24. The energy of fission is the kinetic energy of its products. What becomes of this energy in a commercial power reactor?

25. U-235 has a half-life of about 700 million years. What does this say about the likelihood of fission power on the Earth 1 billion years from now?

26. Heavy nuclei can be made to fuse—for instance, by firing one gold nucleus at another one. Does such a process yield energy or cost energy? Explain.

27. Light nuclei can be split. For example, a deuteron, which is a proton–neutron combination, can split into a separate proton and separate neutron. Does such a process yield energy or cost energy? Explain.

28. Which process would release energy from gold, fission or fusion? Which would release energy from carbon? From iron?

29. If uranium were to split into three segments of equal size instead of two, would more energy or less energy be released? Defend your answer in terms of Figure 34.16.

30. Mixing copper and zinc atoms produces the alloy brass. What would be produced with the fusion of copper and zinc nuclei?

31. Oxygen and hydrogen atoms combine to form water. If the nuclei in a water molecule were fused, what element would be produced?

32. If a pair of carbon atoms were fused, and the product were to emit a beta particle, what element would be produced?

33. Suppose the curve of Figure 34.16 for mass per nucleon versus atomic number had the shape of the curve shown in Figure 34.15. Then would nuclear fission reactions produce energy? Would nuclear fusion reactions produce energy? Defend your answers.

34. The "hydrogen magnets" in Figure 34.20 weigh more when apart than when combined. What would be the basic difference if the fictitious example instead consisted of "nuclear magnets" half as heavy as uranium?

35. In a nuclear fission reaction, which has more mass, the initial uranium or its products?

36. In a nuclear fusion reaction, which has more mass, the initial hydrogen isotopes or the fusion products?

37. Which produces more energy, the fissioning of a single uranium nucleus or the fusing of a pair of deuterium nuclei? The fissioning of a gram of uranium or the fusing of a gram of deuterium? (Why do your answers differ?)

38. Why is there, unlike fission fuel, no limit to the amount of fusion fuel that can be safely stored in one locality?

39. If a fusion reaction produces no appreciable radioactive isotopes, why does a hydrogen bomb produce significant radioactive fallout?

40. List at least two major potential advantages of power production by fusion rather than by fission.

41. Sustained nuclear fusion has yet to be achieved and remains a hope for abundant future energy. Yet the energy that has always sustained us has been the energy of nuclear fusion. Explain.

42. Explain how radioactive decay has always warmed the Earth from the inside and how nuclear fusion has always warmed the Earth from the outside.

43. What effect on the mining industry can you foresee in the disposal of urban waste by means of a fusion torch coupled with a mass spectrometer?

44. The world has never been the same since the discovery of electromagnetic induction and its applications to electric motors and generators. Speculate and list some of the worldwide changes that are likely to follow the advent of successful fusion reactors.

45. Discuss, and make a comparison of, pollution by conventional fossil-fuel power plants and nuclear-fission power plants. Consider thermal pollution, chemical pollution, and radioactive pollution.

46. Ordinary hydrogen is sometimes called a perfect fuel, both because of its almost unlimited supply on Earth and because, when it burns, harmless water is the product of the combustion. So why don't we abandon fission energy and fusion energy, not to mention fossil-fuel energy, and just use hydrogen?

47. In reference to the fusion torch, if a star-hot flame is positioned between a pair of large, electrically charged plates, one positive and the other negative, and materials dumped into the flame are dissociated into bare nuclei and electrons, in which direction will the nuclei move? In which direction will the electrons move?

48. Suppose that the negative plate of a fusion torch has a hole in it so that atomic nuclei that move toward it and pass through it constitute a beam. Further suppose that the beam is then directed between the pole pieces of a powerful electromagnet. Will the beam of charged nuclei continue in a straight line or will the beam be deflected?

49. Supposing that the beam in the preceding exercise is deflected, will all nuclei, both light and heavy, be deflected by the same amount? How is this device similar to a mass spectrometer?

50. In a bucketful of seawater are minute amounts of gold. You can't separate them from the water with an ordinary magnet, but, if the bucket were dumped into the fusion torch speculated about in the chapter, a magnet could indeed separate them. If hydrogen atoms were collected in bin #1 and uranium atoms were collected in bin #92, what would be the bin number for gold?

Problems

1. The kiloton, which is used to measure the energy released in an atomic explosion, is equal to 4.2×10^{12} J (approximately the energy released in the explosion of 1000 tons of TNT). Recalling that 1 kilocalorie of energy raises the temperature of 1 kg of water by 1°C and that 4,184 joules is equal to 1 kilocalorie, calculate how many kilograms of water can be heated through 50°C by a 20-kiloton atomic bomb.

2. The isotope of lithium used in a hydrogen bomb is Li-6, whose nucleus contains 3 protons and 3 neutrons. When a Li-6 nucleus absorbs a neutron, a nucleus of the heaviest hydrogen isotope, tritium, is produced. What is the other product of this reaction? Which of these two products fuels the explosive reaction?

3. An important fusion reaction in both hydrogen bombs and controlled-fusion reactors is the "DT reaction," in which a deuteron and a triton (nuclei of heavy hydrogen isotopes) combine to form an alpha particle and a neutron with the release of much energy. Use momentum conservation to explain why the neutron resulting from this reaction receives about 80% of the energy, while the alpha particle gets only about 20%.

PART EIGHT

Relativity

Before the advent of Special Relativity, people thought the stars were beyond human reach. But distance is relative—it depends on motion. In a frame of reference moving almost as fast as light, distance contracts and time stretches enough to allow future astronauts access to the stars and beyond! We *are* like Evan's chickie on opening page 1, at the verge of a whole new beginning. Newton's physics got us to the moon; Einstein's physics points us to the stars. We live at an exciting time!

CHAPTER 35

Special Theory of Relativity

Ken Ford, former CEO of the American Institute of Physics, brings the beauty of relativity to his high-school students.

Einstein was 26 in 1905 when he published three major papers that ushered in a new era of physics. One was on the quantum theory of light and the photoelectric effect, the second on an explanation of Brownian motion, and the third on special relativity. He won the Nobel Prize for his quantum explanation of the photoelectric effect—not for relativity.

Insights

As an eager young student of physics in the 1890s, Albert Einstein was troubled by a difference between Newton's laws of mechanics and Maxwell's laws of electromagnetism. Newton's laws were independent of the state of motion of an observer; Maxwell's laws were not—or so it seemed. Someone at rest and someone in motion would find that the *same* laws of mechanics apply to a moving object being studied, but they would find that *different* laws of electricity and magnetism apply to a moving charge being studied. Newton's laws suggest that there is no such thing as absolute motion; only relative motion matters. But Maxwell's laws seemed to suggest that motion is absolute.

In a celebrated 1905 paper titled "On the Electrodynamics of Moving Bodies," written when he was 26, Einstein showed that Maxwell's laws can, after all, like Newton's laws, be interpreted as being independent of the state of motion of an observer—but at a cost! The cost of achieving this unified view of nature's laws is a total revolution in how we understand space and time.

Einstein showed that, as the forces between electric charges are affected by motion, the very measurements of space and time are also affected by motion. All measurements of space and time depend on relative motion.

For example, the length of a rocket ship poised on its launching pad and the ticks of clocks within are found to change when the ship is set into motion at high speed. It has always been common sense that we change our position in space when we move, but Einstein flouted common sense and stated that, in moving, we also change our rate of proceeding into the future—time itself is altered. Einstein went on to show that a consequence of the interrelationship between space and time is an interrelationship between mass and energy, given by the famous equation $E = mc^2$.

These are the ideas that make up this chapter—the ideas of special relativity—ideas so remote from your everyday experience that understanding them requires stretching your mind. It will be enough to become acquainted with these ideas, so be patient with yourself if you don't understand them right away. Perhaps in some future era, when high-speed interstellar space travel is commonplace, your descendants will find that relativity makes common sense.

Motion Is Relative

Recall from Chapter 3 that, whenever we talk about motion, we must always specify the vantage point from which motion is being observed and measured. For example, a person who walks along the aisle of a moving train may be walking at a speed of 1 kilometer per hour relative to his seat but at 60 kilometers per hour relative to the railroad station. We call the place from which motion is observed and measured a **frame of reference**. An object may have different velocities relative to different frames of reference.

To measure the speed of an object, we first choose a frame of reference and pretend that we are in that frame of reference standing still. Then we measure the speed with which the object moves relative to us—that is, relative to the frame of reference. In the foregoing example, if we measure from a position of rest within the train, the speed of the walking person is 1 kilometer per hour. If we measure from a position of rest on the ground, the speed of the walking person is 60 kilometers per hour. But the ground is not really still, for Earth spins like a top about its polar axis. Depending on how near the train is to the equator, the speed of the walking person may be as much as 1600 kilometers per hour relative to a frame of reference at the center of the Earth. And the center of the Earth is moving relative to the Sun. If we place our frame of reference on the Sun, the speed of the person walking in the train, which is on the orbiting Earth, is nearly 110,000 kilometers per hour. And the Sun is not at rest, for it orbits the center of our galaxy, which moves with respect to other galaxies.

Michelson–Morley Experiment

Isn't there some reference frame that is still? Isn't space itself still, and can't measurements be made relative to still space? In 1887, the American physicists A. A. Michelson and E. W. Morley attempted to answer these questions by performing an experiment that was designed to measure the motion of the Earth through space. Because light travels in waves, it was then assumed that something in space vibrates—a mysterious something called *ether*, thought to fill all space and to serve as a frame of reference attached to space itself. These physicists used a very sensitive apparatus called an *interferometer* to make their observations (Figure 35.1). In this instrument, a beam of light from a monochromatic source was separated into two beams with paths at right angles to each other; these were reflected and recombined to show whether there was any difference in average speed over the two back-and-forth paths. The interferometer was set with one path parallel to the motion of the Earth in its orbit; then either Michelson or Morley carefully watched for any changes in average speed as the apparatus was rotated to put the other path parallel to the motion of the Earth. The interferometer was sensitive enough to measure the difference in the round-trip times of light going with and against Earth's orbital velocity of 30 kilometers per second and going back and forth across Earth's path through space. But no changes were observed. None. Something was wrong with the sensible idea that the speed of light measured by a moving receiver should be its usual speed in a vacuum, c, plus or minus the contribution from

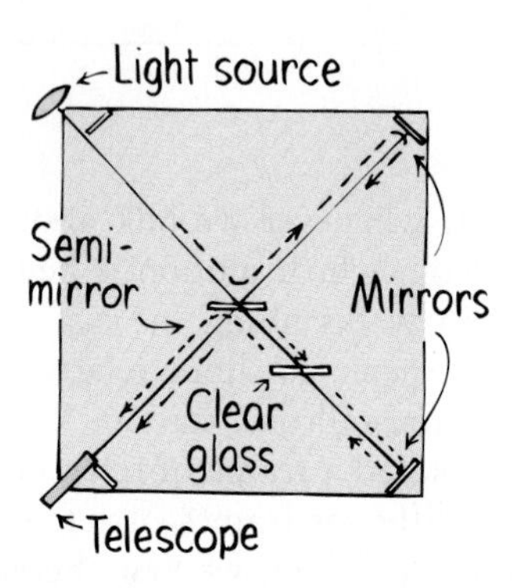

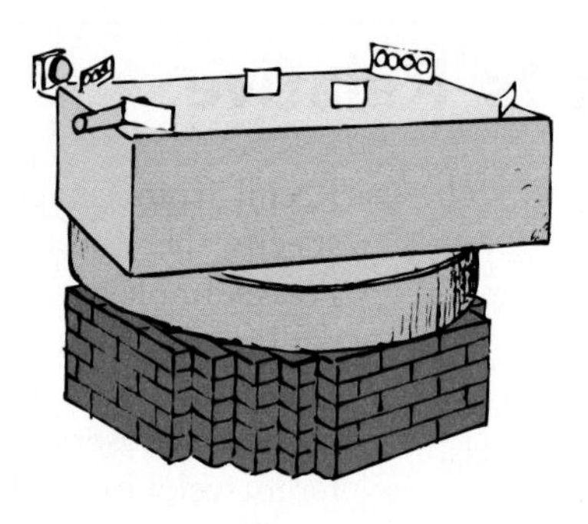

FIGURE 35.1
The Michelson–Morley interferometer, which splits a light beam into two parts and then recombines them to form an interference pattern after they have traveled different paths. Rotation was accomplished in their experiment by floating a massive sandstone slab in mercury. This schematic diagram shows how the half-silvered mirror splits the beam into two rays. The clear glass assured that both rays traverse the same amount of glass. Four mirrors at each end were used to lengthen the paths.

the motion of the source or receiver. Many repetitions and variations of the Michelson–Morley experiment by many investigators showed the same null result. This was one of the puzzling facts of physics when the twentieth century opened.

One interpretation of the bewildering result was suggested by the Irish physicist G. F. FitzGerald, who proposed that the length of the experimental apparatus shrank in the direction in which it was moving by just the amount required to counteract the presumed variation in the speed of light. The needed "shrinkage factor," $\sqrt{1 - v^2/c^2}$, was worked out by the Dutch physicist Hendrik A. Lorentz. This arithmetical factor accounted for the discrepancy, but neither FitzGerald nor Lorentz had a suitable theory for why this was so. Interestingly, the same factor was derived by Einstein in his 1905 paper, where he showed it to be the shrinkage factor of space itself, not just of matter in space.

How much the Michelson–Morley experiment influenced Einstein, if at all, is unclear. In any event, Einstein advanced the idea that the speed of light in free space is the same in all reference frames, an idea that was contrary to the classical ideas of space and time. Speed is a ratio of distance through space to a corresponding interval of time. For the speed of light to be a constant, the classical idea that space and time are independent of each other had to be rejected. Einstein saw that space and time are linked, and, with simple postulates, he developed a profound relationship between the two.

Postulates of the Special Theory of Relativity

Einstein saw no need for the ether. Gone with the stationary ether was the notion of an absolute frame of reference. All motion is relative, not to any stationary hitching post in the universe, but to arbitrary frames of reference. A rocket ship cannot measure its speed with respect to empty space but only with respect to other objects. If, for example, rocket ship A drifts past rocket ship B in empty space, spaceman A and spacewoman B will each observe the relative motion, and, from this observation, each will be unable to determine who is moving and who is at rest, if either.

This is a familiar experience to a passenger on a train who looks out his window and sees the train on the next track moving by his window. He is aware only of the relative motion between his train and the other train and cannot tell which train is moving. He may be at rest relative to the ground and the other train may be moving, or he may be moving relative to the ground and the other train may be at rest, or they both may be moving relative to the ground. The important point here is that, if you were in a train with no windows, there would be no way to determine whether the train was moving with uniform velocity or was at rest. This is the first of Einstein's postulates of the special theory of relativity:

> **All laws of nature are the same in all uniformly moving frames of reference.**

On a jet airplane going 700 kilometers per hour, for example, coffee pours as it does when the plane is at rest; if we swing a pendulum in the moving plane, it swings as it would if the plane were at rest on the runway. There is no physical experiment that we can perform, even with light, to determine our state of uniform motion. The laws of physics within the uniformly moving cabin are the same as those in a stationary laboratory.

FIGURE 35.2
The speed of light is measured to be the same in all frames of reference.

Any number of experiments can be devised to detect accelerated motion, but none can be devised, according to Einstein, to detect a state of uniform motion. Therefore, absolute motion has no meaning.

It would be very peculiar if the laws of mechanics varied for observers moving at different speeds. It would mean, for example, that a pool player on a smoothly moving ocean liner would have to adjust her style of play to the speed of the ship, or even to the season as the Earth varies in its orbital speed about the Sun. It is our common experience that no such adjustment is necessary. And, according to Einstein, this same insensitivity to motion extends to electromagnetism. No experiment, mechanical or electrical or optical, has ever revealed absolute motion. That is what the first postulate of relativity means.

One of the questions that Einstein, as a youth, asked his schoolteacher was, "What would a light beam look like if you traveled along beside it?" According to classical physics, the beam would be at rest to such an observer. The more Einstein thought about this, the more convinced he became that one could not move with a light beam. He finally came to the conclusion that, no matter how fast two observers might be moving relative to each other, each of them would measure the speed of a light beam passing them to be 300,000 kilometers per second. This was the second postulate in his special theory of relativity:

> **The speed of light in free space has the same measured value for all observers, regardless of the motion of the source or the motion of the observer; that is, the speed of light is a constant.**

FIGURE 35.3
The speed of a light flash emitted by the space station is measured to be *c* by observers on both the space station and the rocket ship.

To illustrate this statement, consider a rocket ship departing from the space station shown in Figure 35.3. A flash of light traveling at 300,000 kilometers per second, or *c*, is emitted from the station. Regardless of the velocity of the rocket, an observer in the rocket sees the flash of light pass her at the same speed *c*. If a flash is sent to the station from the moving rocket, observers on the station will measure the speed of the flash to be *c*. The speed of light is measured to be the same regardless of the speed of the source or receiver. *All* observers who measure the speed of light will find it has the same value *c*. The more you think about this, the more you think it doesn't make sense. We will see that the explanation has to do with the relationship between space and time.

Simultaneity

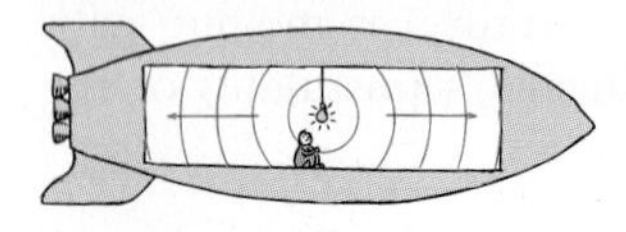

FIGURE 35.4
Interactive Figure

From the point of view of the observer who travels with the compartment, light from the source travels equal distances to both ends of the compartment and therefore strikes both ends simultaneously.

An interesting consequence of Einstein's second postulate occurs with the concept of **simultaneity**. We say that two events are simultaneous if they occur at the same time. Consider, for example, a light source in the exact center of the compartment of a rocket ship (Figure 35.4). When the light source is switched on, light spreads out in all directions at speed c. Because the light source is equidistant from the front and back ends of the compartment, an observer inside the compartment finds that light reaches the front end at the same instant it reaches the back end. This occurs whether the ship is at rest or moving at constant velocity. The events of hitting the back end and hitting the front end occur *simultaneously* for this observer within the rocket ship.

But what about an outside observer who views the same two events in another frame of reference—say, from a planet not moving with the ship? For that observer, these same two events are *not* simultaneous. As light travels out from the source, this observer sees the ship move forward, so the back of the compartment moves toward the beam while the front moves away from it. The beam going to the back of the compartment, therefore, has a shorter distance to travel than the beam going forward (Figure 35.5). Since the speed of light is the same in both directions, this outside observer sees the event of light hitting the back of the compartment *before* seeing the event of light hitting the front of the compartment. (Of course, we are making the assumption that the observer can discern these slight differences.) A little thought will show that an observer in another rocket ship that passes the ship in the opposite direction would report that the light reaches the front of the compartment first.

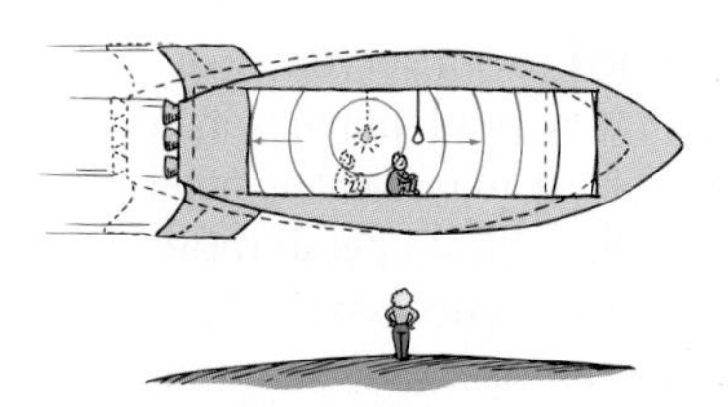

FIGURE 35.5
Interactive Figure

The events of light striking the front and back of the compartment are not simultaneous from the point of view of an observer in a different frame of reference. Because of the ship's motion, light that strikes the back of the compartment doesn't have as far to go and strikes sooner than light that strikes the front of the compartment.

Two events that are simultaneous in one frame of reference need not be simultaneous in a frame moving relative to the first frame.

This nonsimultaneity of events in one frame that are simultaneous in another is a purely relativistic result—a consequence of light always having the same speed for all observers.

CHECK YOURSELF

1. How is the nonsimultaneity of hearing thunder *after* seeing lightning similar to relativistic nonsimultaneity?
2. Suppose that the observer standing on a planet in Figure 35.5 sees a pair of lightning bolts simultaneously strike the front and rear ends of the compartment in the high-speed rocket ship. Will the lightning strikes be simultaneous to an observer in the middle of the compartment in the rocket ship? (We assume here that an observer can detect any slight differences in time for light to travel from the ends to the middle of the compartment.)

Spacetime

When we look up at the stars, we realize that we are actually looking backward in time. The stars we see farthest away are the stars we are seeing longest ago. The more we think about this, the more apparent it becomes that space and time must be intimately tied together.

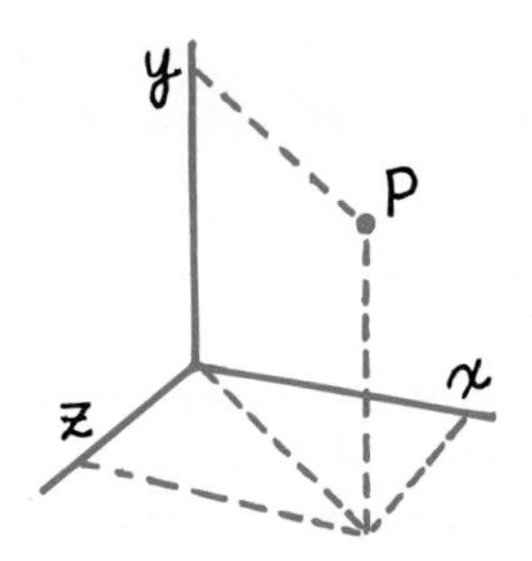

FIGURE 35.6
Point P can be specified with three numbers: the distances along the *x* axis, the *y* axis, and the *z* axis.

The space we live in is three-dimensional; that is, we can specify the position of any location in space with three dimensions. For example, these dimensions could be north–south, east–west, and up–down. If we are at the corner of a rectangular room and wish to specify the position of any point in the room, we can do this with three numbers. The first would be the number of meters the point is along a line joining the side wall and the floor; the second would be the number of meters the point is along a line joining the adjacent back wall and the floor; and the third would be the number of meters the point lies above the floor or along the vertical line joining the walls at the corner. Physicists speak of these three lines as the *coordinate axes* of a reference frame (Figure 35.6). Three numbers—the distances along the x axis, the y axis, and the z axis—will specify the position of a point in space.

We also use three dimensions to specify the size of objects. A box, for example, is described by its length, width, and height. But the three dimensions do not give a complete picture. There is a fourth dimension—time. The box was not always a box of given length, width, and height. It began as a box only at a certain point in time, on the day it was made. Nor will it always be a box. At any moment, it may be crushed, burned, or otherwise destroyed. So the three dimensions of space are a valid description of the box only during a certain specified period of time. We cannot speak meaningfully about space without implying time. Things exist in **spacetime**. Each object, each person, each planet, each star, each galaxy exists in what physicists call "the spacetime continuum."

Two side-by-side observers at rest relative to each other share the same reference frame. Both would agree on measurements of space and time intervals between given events, so we say they share the same realm of spacetime. If there is relative motion between them, however, the observers will not agree on these measurements of space and time. At ordinary speeds, differences in their measurements are imperceptible, but, at speeds near the speed of light—so-called relativistic speeds—the differences are appreciable. Each observer is in a different realm of spacetime, and her measurements of space and time differ from the measurements of an observer in some other realm of spacetime. The measurements differ not haphazardly but in such a way that each observer will always

CHECK YOUR ANSWERS

1. It isn't! The duration between hearing thunder and seeing lightning has nothing to do with moving observers or relativity. In such a case, you simply make corrections for the time the signals (sound and light) take to reach you. The relativity of simultaneity is a genuine discrepancy between observations made by observers in relative motion, and not simply a disparity between different travel times for different signals.

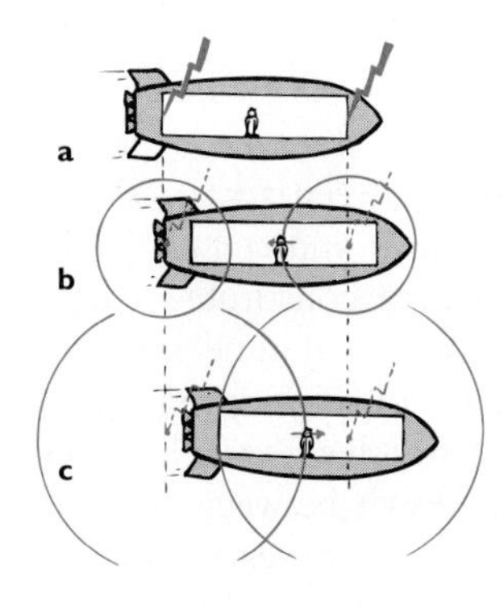

2. No; an observer in the middle of the compartment will see the lightning that hits the front end of the compartment before seeing the lightning that hits the rear end. This is shown in positions (a), (b), and (c) to the left. In (a), we see both lightning bolts striking the ends of the compartment simultaneously, according to the outside observer. In position (b), light from the front lightning bolt reaches the observer within the rocket ship. Slightly later, in (c), light from the rear lightning bolt reaches this observer.

CLOCKWATCHING ON A TROLLEY CAR RIDE

Pretend you are Einstein at the turn of the twentieth century, riding in a trolley car moving away from a huge clock in the village square. The clock reads 12 noon. To say it reads 12 noon is to say that light that carries the information "12 noon" is reflected by the clock and travels toward you along your line of sight. If you suddenly move your head to the side, the light carrying the information, instead of meeting your eye, continues past, presumably out into space. Out there, an observer who *later* receives the light says, "Oh, it's 12 noon on Earth now." But, from your point of view, it's now later than that. You and the distant observer see 12 noon at different times. You wonder more about this idea. If the trolley car traveled as fast as the light, then the trolley car would keep up with the light's information that says "12 noon." Traveling at the speed of light, then, tells you it's always 12 noon at the village square. In other words, time at the village square is frozen!

If the trolley car is not moving, you see the village-square clock move into the future at the rate of 60 seconds per minute; if you move at the speed of light, you see seconds on the clock taking infinite time. These are the two extremes. What's in between? How would the advance of the clock's hands be viewed as you move at speeds less than the speed of light?

A little thought will show that you will receive the message "1 o'clock" anywhere from 60 minutes to an infinity of time after you receive the message "12 noon," depending on what your speed is between the extremes of zero and the speed of light. From your high-speed (but less than c) frame of reference, you see all events taking place in the reference frame of the clock (which is the Earth) as happening in slow motion. If you reverse direction and travel at high speed back toward the clock, you'll see all events taking place in the clock's reference frame as being speeded up. When you return and are once again sitting in the square, will the effects of going and coming compensate each other? Amazingly, no! Time will be stretched. The wristwatch you were wearing the whole time and the village clock will disagree. This is time dilation.

FIGURE 35.7
All space and time measurements of light are unified by c.

measure the same ratio of space and time for light: the greater the measured distance in space, the greater the measured interval of time. This constant ratio of space and time for light, c, is the unifying factor between different realms of spacetime and is the essence of Einstein's second postulate.

Time Dilation

Let's examine the notion that time can be stretched. Imagine that we are somehow able to observe a flash of light bouncing to and fro between a pair of parallel mirrors, like a ball bouncing to and fro between a floor and ceiling. If the distance between the mirrors is fixed, then the arrangement constitutes a *light clock*, because the back-and-forth trips of the flash take equal time intervals (Figure 35.8). Suppose this light clock is inside a transparent, high-speed spaceship. An observer who travels along with the ship and watches the light clock (Figure 35.9a) sees the flash reflecting straight up and down between the two mirrors, just as it would if the spaceship were at rest. This observer sees no

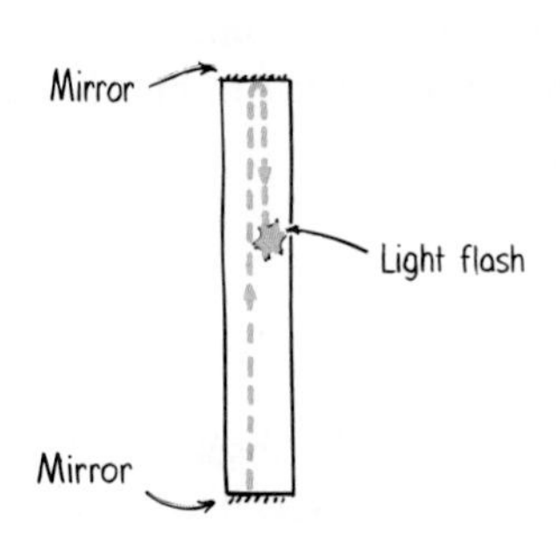

FIGURE 35.8
A light clock. A flash of light will bounce up and down between parallel mirrors and "tick off" equal intervals of time.

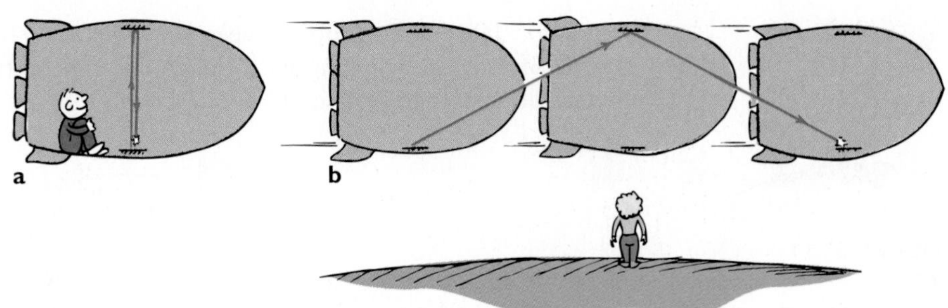

FIGURE 35.9 Interactive Figure
(a) An observer moving with the spaceship observes the light flash moving vertically between the mirrors of the light clock. (b) An observer who sees the moving ship pass by observes the flash moving along a diagonal path.

unusual effects. Note that, because the observer is in the ship moving along with it, there is no relative motion between the observer and the light clock; we say that the observer and the clock share the same reference frame in spacetime.

Suppose now that we are standing on the ground as the spaceship whizzes by us at a high speed—say, half the speed of light. Things are quite different from our reference frame, for we do not see the light path as being simple up-and-down motion. Because each flash moves horizontally while it moves vertically between the two mirrors, we see the flash follow a diagonal path. Notice in Figure 35.9b that, from our earthbound frame of reference, the flash travels a *longer distance* as it makes one round trip between the mirrors, considerably longer than the distance it travels in the reference frame of the observer riding along with the ship. Because the speed of light is the same in all reference frames (Einstein's second postulate), the flash must travel for a corresponding longer time between the mirrors in our frame than in the reference frame of the on-board observer. This follows from the definition of speed—distance divided by time. *The longer diagonal distance must be divided by a correspondingly longer time interval to yield an unvarying value for the speed of light.* This stretching out of time is called **time dilation.**

We have considered a light clock in our example, but the same is true for any kind of clock. All clocks run more slowly when moving than when at rest. Time dilation has to do not with the mechanics of clocks but with the nature of time itself.

The relationship of time dilation for different frames of reference in spacetime can be derived from Figure 35.10 with simple geometry and

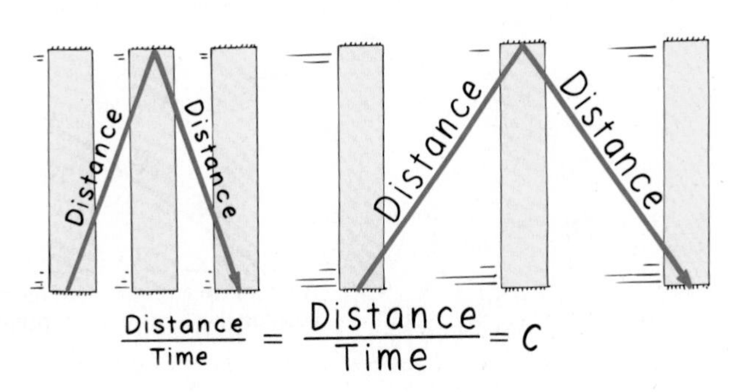

FIGURE 35.10
Interactive Figure
The longer distance covered by the light flash in following the longer diagonal path on the right must be divided by a correspondingly longer time interval to yield an unvarying value for the speed of light.

FIGURE 35.11
When we see the rocket at rest, we see it traveling at the maximum rate in time: 24 hours per day. When we see the rocket traveling at the maximum rate through space (the speed of light), we see its time standing still.

algebra.[1] The relationship between the time t_0 (call it the *proper time*) in the frame of reference moving with the clock and the time t measured in another frame of reference (call it the *relative time*) is

$$t = \frac{t_0}{\sqrt{1 - \dfrac{v^2}{c^2}}}$$

where v represents the speed of the clock relative to the outside observer (the same as the relative speed of the two observers) and c is the speed of light. The quantity

$$\sqrt{1 - \frac{v^2}{c^2}}$$

is the same factor used by Lorentz to explain length contraction. We call the inverse of this quantity the *Lorentz factor* γ (gamma). That is,

$$\gamma = \frac{1}{\sqrt{1 - \dfrac{v^2}{c^2}}}$$

Then we can express the time dilation equation more simply as

$$t = \gamma t_0$$

Let's look at the terms in γ. Some mental tinkering will show that γ is always greater than 1 for any speed v greater than zero. Note that, since speed v is always less than c, the ratio v/c is always less than 1; likewise for v^2/c^2. Can you see it follows that γ is greater than 1? Now consider the case where $v = 0$. This ratio v^2/c^2 is zero, and for everyday speeds, where v is negligibly small compared to c,

[1]The light clock is shown in three successive positions in the figure below. The diagonal lines represent the path of the light flash as it starts from the lower mirror at position 1, moves to the upper mirror at position 2, and then back to the lower mirror at position 3. Distances on the diagram are marked ct, vt, and ct_0, which follows from the fact that the distance traveled by a uniformly moving object is equal to its speed multiplied by the time.

The symbol t_0 represents the time it takes for the flash to move between the mirrors, as measured from a frame of reference fixed to the light clock. This is the time for straight up or down motion. The speed of light is c, and the path of light is seen to move a vertical distance ct_0. This distance between mirrors is at right angles to the motion of the light clock and is the same in both reference frames.

The symbol t represents the time it takes the flash to move from one mirror to the other, as measured from a frame of reference in which the light clock moves with speed v. Because the speed of the flash is c and the time it takes to go from position 1 to position 2 is t, the diagonal distance traveled is ct. During this time t, the clock (which travels horizontally at speed v) moves a horizontal distance vt from position 1 to position 2.

As the figure shows, these three distances make up a right triangle in which ct is the hypotenuse and ct_0 and vt are legs. A well-known theorem of geometry, the Pythagorean theorem, states that the square of the hypotenuse is equal to the sum of the squares of the two legs. If we apply this formula to the figure, we obtain:

$$c^2t^2 = c^2t_0^2 + v^2t^2$$

$$c^2t^2 - v^2t^2 = c^2t_0^2$$

$$t^2[1 - (v^2/c^2)] = t_0^2$$

$$t^2 = \frac{t_0^2}{1 - (v^2/c^2)}$$

$$t = \frac{t_0}{\sqrt{1 - (v^2/c^2)}}$$

ct
ct₀
vt
Path of light as seen from a position of rest
Mirrors at position 1
Mirrors at position 2
Mirrors at position 3

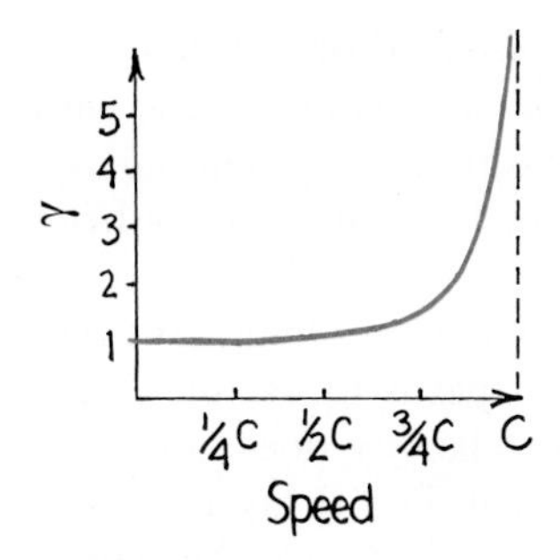

FIGURE 35.12
A plot of the Lorentz factor γ as a function of speed.

it's practically zero. Then $1 - v^2/c^2$ has a value of 1, as has $\sqrt{1 - v^2/c^2}$, which makes $\gamma = 1$. Then we find $t = t_0$—time intervals appear the same in both reference frames. For higher speeds, v/c is between zero and 1, and $1 - v^2/c^2$ is less than 1; likewise, $\sqrt{1 - v^2/c^2}$. This makes γ greater than 1, so t_0 multiplied by a factor greater than 1 produces a value greater than t_0—an elongation—a dilation of time.

To consider some numerical values, assume that v is 50% the speed of light. Then we substitute $0.5c$ for v in the time-dilation equation and, after some arithmetic, find that $\gamma = 1.15$; so $t = 1.15\,t_0$. This means that, if we viewed a clock on a spaceship traveling at half the speed of light, we would see the second hand take 1.15 minutes to make a revolution, whereas an observer riding with the clock would see it take 1 minute. If the spaceship passes us at 87% the speed of light, $\gamma = 2$; and $t = 2\,t_0$. We would measure time events on the spaceship taking twice the usual intervals, for the hands of a clock on the ship would turn only half as fast as those on our own clock. Events on the ship would seem to take place in slow motion. At 99.5% the speed of light, $\gamma = 10$ and $t = 10\,t_0$; we would see the second hand of the spaceship's clock take 10 minutes to sweep through a revolution requiring 1 minute on our clock.

To put these figures another way, at $0.995c$, the moving clock would appear to run a tenth of our rate; it would tick only 6 seconds while our clock ticks 60 seconds. At $0.87c$, the moving clock ticks at half rate and shows 30 seconds to our 60 seconds; at $0.50c$, the moving clock ticks 1/1.15 as fast and ticks 52 seconds to our 60 seconds. Moving clocks run slow.

Nothing is unusual about a moving clock itself; it is simply ticking to the rhythm of a different time. The faster a clock moves, the slower it appears to run as viewed by an observer not moving with the clock. If it were possible to make a clock fly by us at the speed of light, the clock would not appear to be running at all. We would measure the interval between ticks to be infinite. The clock would be ageless! If we could move with such an imaginary clock, however, the clock would not show any slowing down of time. To us, the clock would be operating normally. This is because there would be no motion of the clock relative to us. The v in γ would then be zero, and $t = t_0$; we and the clock would share the same frame in spacetime.

If a person whizzing past us were to check a clock in our reference frame, he would find our clock to be running as slowly as we would find his to be. Each would see the other's clock running slowly. There is really no contradiction here, for it is physically impossible for two observers in relative motion to refer to one and the same realm of spacetime. The measurements made in one realm of spacetime need not agree with the measurements made in another realm of spacetime. The measurement that all observers always agree on, however, is the speed of light.

The Global Positioning System (GPS) takes account of the time dilation of orbiting atomic clocks. Otherwise, your GPS receiver would badly miss your location.

Time dilation has been confirmed in the laboratory innumerable times with particle accelerators. The lifetimes of fast-moving radioactive particles increase as the speed goes up, and the amount of increase is just what Einstein's equation predicts.

Time dilation has been confirmed also for not-so-fast motion. In 1971, to test Einstein's theory, four cesium-beam atomic clocks were twice flown on regularly scheduled commercial jet flights around the world, once eastward and once westward, to test Einstein's theory of relativity with macroscopic clocks. The clocks indicated different times after their round trips. Relative to the atomic time scale of the U.S. Naval Observatory, the observed time differences, in billionths of a second, were in accord with Einstein's prediction. Now, with atomic clocks orbiting Earth as part of the Global Positioning System, adjustments for the effects of time dilation are essential in order to use signals from the clocks to pinpoint locations on Earth.

This all seems very strange to us only because it is not our common experience to deal with measurements made at relativistic speeds or atomic-clock-type measurements at ordinary speeds. The theory of relativity does not make common sense. But common sense, according to Einstein, is that layer of prejudices laid down in the mind prior to the age of 18. If we had spent our youth zapping through the universe in high-speed spaceships, we would probably be quite comfortable with the results of relativity.

CHECK YOURSELF

1. If you were moving in a spaceship at a high speed relative to Earth, would you notice a difference in your pulse rate? In the pulse rate of the people back on Earth?
2. Will observers A and B agree on measurements of time if A moves at half the speed of light relative to B? If both A and B move together at half the speed of light relative to Earth?
3. Does time dilation mean that time really passes more slowly in moving systems or only that it seems to pass more slowly?

The Twin Trip

The Twin Trip Animation

A dramatic illustration of time dilation is provided by identical twins, one an astronaut who takes a high-speed round-trip journey in the galaxy while the other stays home on Earth. When the traveling twin returns, he is younger than the stay-at-home twin. How much younger depends on the relative speeds involved.

CHECK YOUR ANSWERS

1. There would be no relative speed between you and your pulse because the two share the same frame of reference. Therefore, you would notice no relativistic effects in your pulse. There would be, however, a relativistic effect between you and people back on Earth. You would find their pulse rate to be slower than normal (and they would find your pulse rate to be slower than normal). Relativity effects are always attributed to the other guy.
2. When A and B move relative to each other, each observes a slowing of time in the other's frame of reference. So they do not agree on measurements of time. When they are moving in unison, they share the same frame of reference and agree on measurements of time. They see each other's time as passing normally, and they each see events on Earth in the same slow motion.
3. The slowing of time in moving systems is not merely an illusion resulting from motion. Time really does pass more slowly in a moving system relative to one at relative rest, as we shall see in the next section. Read on!

FIGURE 35.13
The traveling twin does not age as fast as the stay-at-home twin.

Cosmonaut Sergei Avdeyev spent more than two years orbiting Earth in the *Mir* spacecraft, and, due to time dilation, he is today two-hundredths of a second younger than he would be if he'd never been in space!

Insights

If the traveling twin maintains a speed of 50% the speed of light for 1 year (according to clocks aboard the spaceship), 1.15 years will have elapsed on Earth. If the traveling twin maintains a speed of 87% the speed of light for a year, then 2 years will have elapsed on Earth. At 99.5% the speed of light, 10 Earth years would pass in one spaceship year. At this speed, the traveling twin would age a single year while the stay-at-home twin would age 10 years.

One question often arises: Since motion is relative, why doesn't the effect work equally well the other way around? Why wouldn't the traveling twin return to find his stay-at-home twin younger than himself? We will show that, from the frames of reference of both the earthbound twin and traveling twin, it is the earthbound twin who ages more.

First, consider a spaceship hovering at rest relative to Earth. Suppose the spaceship sends brief, regularly spaced flashes of light to the planet (Figure 35.14). Some time will elapse before the flashes get to the planet, just as 8 minutes elapse before sunlight gets to Earth. The light flashes will encounter the receiver on the planet at speed c. Since there is no relative motion between the sender and receiver, successive flashes will be received as frequently as they are sent. For example, if a flash is sent from the ship every 6 minutes, then, after some initial delay, the receiver will receive a flash every 6 minutes. With no motion involved, there is nothing unusual about this.

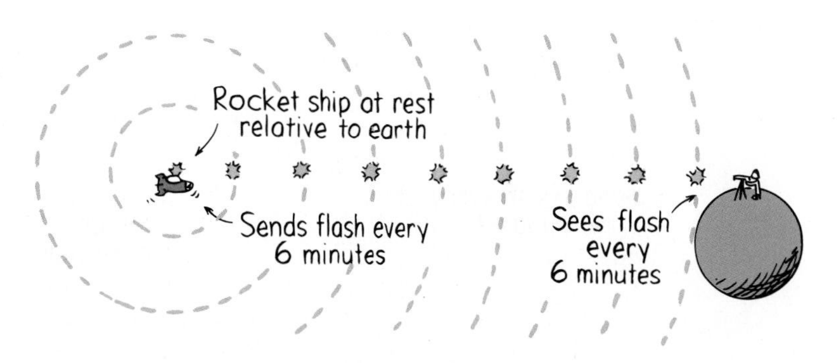

FIGURE 35.14
When no motion is involved, the light flashes are received as frequently as the spaceship sends them.

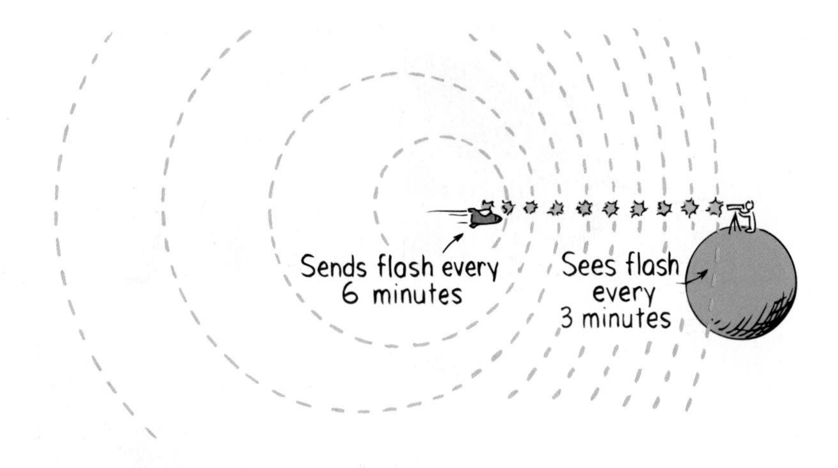

FIGURE 35.15
When the sender moves toward the receiver, the flashes are seen more frequently.

When motion is involved, the situation is quite different. It is important to note that the speed of the flashes will still be *c*, no matter how the ship or receiver may move. How frequently the flashes are seen, however, very much depends on the relative motion involved. When the ship travels toward the receiver, the receiver sees the flashes more frequently. This happens not only because time is altered due to motion, but mainly because each succeeding flash has less distance to travel as the ship gets closer to the receiver. If the spaceship emits a flash every 6 minutes, the flashes will be seen at intervals of less than 6 minutes. Suppose the ship is traveling fast enough for the flashes to be seen twice as frequently. Then they are seen at intervals of 3 minutes (Figure 35.15).

If the ship recedes from the receiver at the same speed and still emits flashes at 6-minute intervals, these flashes will be seen half as frequently by the receiver—that is, at 12-minute intervals (Figure 35.16). This is mainly because each succeeding flash has a longer distance to travel as the ship gets farther away from the receiver.

The effect of moving away is just the opposite of moving closer to the receiver. So if the flashes are received twice as frequently when the spaceship is approaching (6-minute flash intervals are seen every 3 minutes), they are

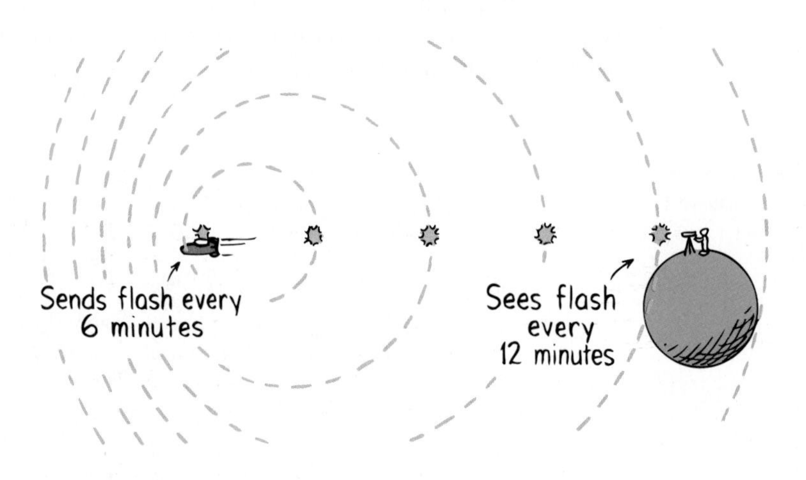

FIGURE 35.16
When the sender moves away from the receiver, the flashes are spaced farther apart and are seen less frequently.

received half as frequently when it is receding (6-minute flash intervals are seen every 12 minutes).[2]

This means that, if two events are separated by 6 minutes according to the spaceship clock, they will be seen to be separated by 12 minutes when the spaceship recedes and by only 3 minutes when the ship is approaching.

CHECK YOURSELF

1. If the spaceship emits an initial "starting gun" signal followed by a flash every 6 min for an hour, how many flashes will be emitted?
2. The ship sends equally spaced flashes every 6 min while approaching the receiver at constant speed. Will these flashes be equally spaced when they encounter the receiver?
3. If the receiver sees these flashes at 3-min intervals, how much time will elapse between the initial signal and the last flash (in the frame of reference of the receiver)?

Let's apply this doubling and halving of flash intervals to the twins. Suppose the traveling twin recedes from the earthbound twin at the same high speed for 1 hour and then quickly turns around and returns in 1 hour. Follow this line of reasoning with the help of Figure 35.17. The traveling twin takes a round trip of 2 hours, according to all clocks aboard the spaceship. This trip will not be seen to take 2 hours from the Earth frame of reference, however. We can see this with the help of the flashes from the ship's light clock.

CHECK YOUR ANSWERS

1. The ship will emit a total of ten flashes in 1 h, since (60 min)/(6 min) = 10 (11 if the initial signal is counted).
2. Yes; as long as the ship moves at constant speed, the equally spaced flashes will be seen equally spaced but more frequently. (If the ship accelerated while sending flashes, then they would not be seen at equally spaced intervals.)
3. Thirty minutes, since the ten flashes are coming every 3 min.

---

[2]This reciprocal relationship (halving and doubling of frequencies) is a consequence of the constancy of the speed of light and can be illustrated with the following example: Suppose that a sender on Earth emits flashes 3 min apart to a distant observer on a planet that is at rest relative to Earth. The observer, then, sees a flash every 3 min. Now suppose a second observer travels in a spaceship between Earth and the planet at a speed great enough to allow him to see the flashes half as frequently—6 min apart. This halving of frequency occurs for a speed of recession of $0.6c$. We can see that the frequency will double for a speed of approach of $0.6c$ by supposing that the spaceship emits its own flash every time it sees an Earth flash—that is, every 6 min. How does the observer on the distant planet see these flashes? Since Earth flashes and the spaceship flashes travel together at the same speed c, the observer will see not only Earth flashes every 3 min but the spaceship flashes every 3 min as well. So, although a person on the spaceship emits flashes every 6 min, the observer sees them every 3 min at twice the emitting frequency. So, for a speed of recession where frequency appears halved, frequency appears doubled for the same speed of approach. If the ship were traveling faster so that the frequency of recession were 1/3 or 1/4 as much, then the frequency of approach would be threefold or fourfold, respectively. This reciprocal relationship does not hold for waves that require a medium. In the case of sound waves, for example, a speed that results in a doubling of emitting frequency for approach produces 2/3 (not 1/2) the emitting frequency for recession.

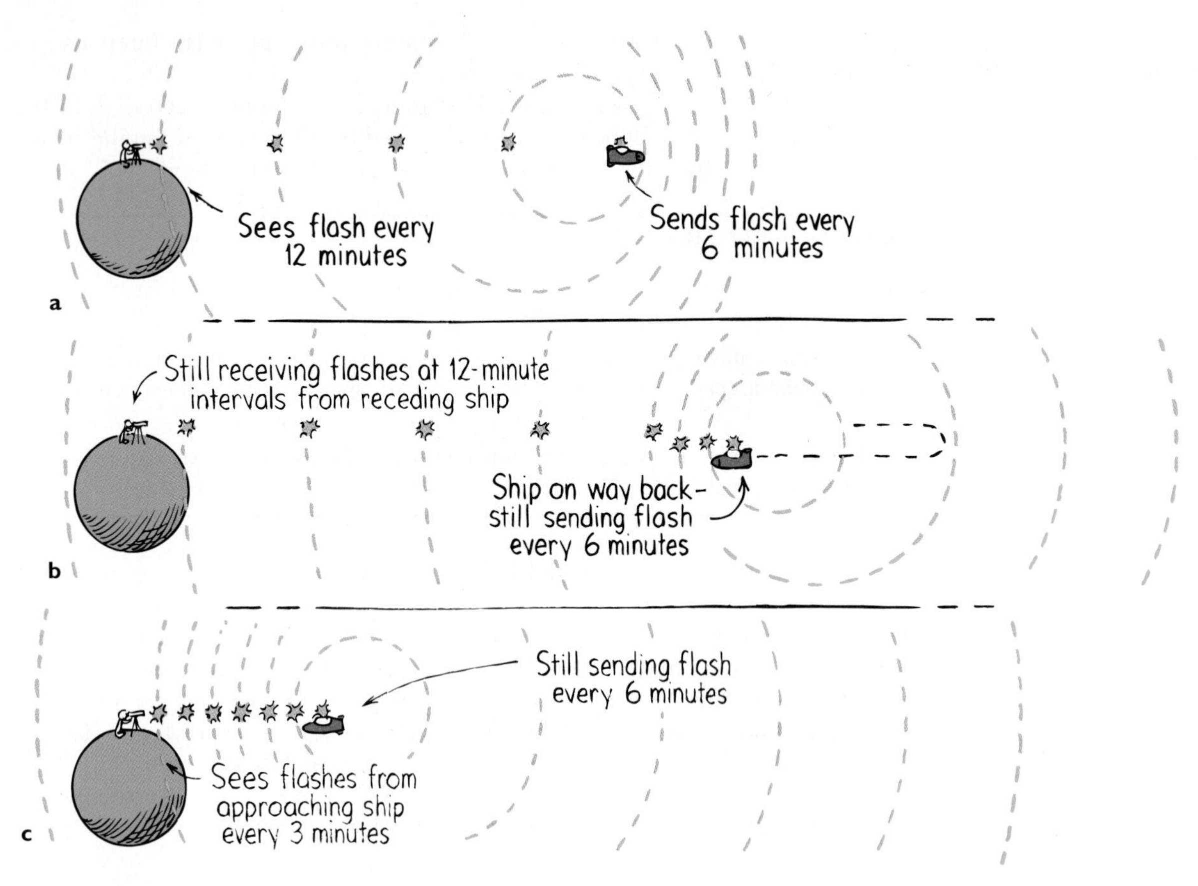

FIGURE 35.17
The spaceship emits flashes every 6 min during a 2-hour trip. During the first hour, it recedes from the Earth. During the second hour, it approaches the Earth.

As the ship recedes from Earth, it emits a flash of light every 6 minutes. These flashes are received on Earth every 12 minutes. During the hour of going away from Earth, a total of ten flashes are emitted (after the "starting gun" signal). If the ship departs from Earth at noon, clocks aboard the ship read 1 p.m. when the tenth flash is emitted. What time will it be on Earth when this tenth flash reaches Earth? The answer is 2 p.m. Why? Because the time it takes Earth to receive 10 flashes at 12-minute intervals is 10 × (12 minutes), or 120 minutes (= 2 hours).

Suppose the spaceship is somehow able to turn around "on a dime" (in a negligibly short time) and return at the same high speed. During the hour of return, it emits ten more flashes at 6-minute intervals. These flashes are received every 3 minutes on Earth, so all ten flashes come in 30 minutes. A clock on Earth will read 2:30 p.m. when the spaceship completes its 2-hour trip. We see that the earthbound twin has aged half an hour more than the twin aboard the spaceship!

The result is the same from either frame of reference. Consider the same trip again, only this time with flashes emitted from Earth at regularly spaced 6-minute intervals in Earth time. From the frame of reference of the receding spaceship, these flashes are received at 12-minute intervals (Figure 35.19*a*). This

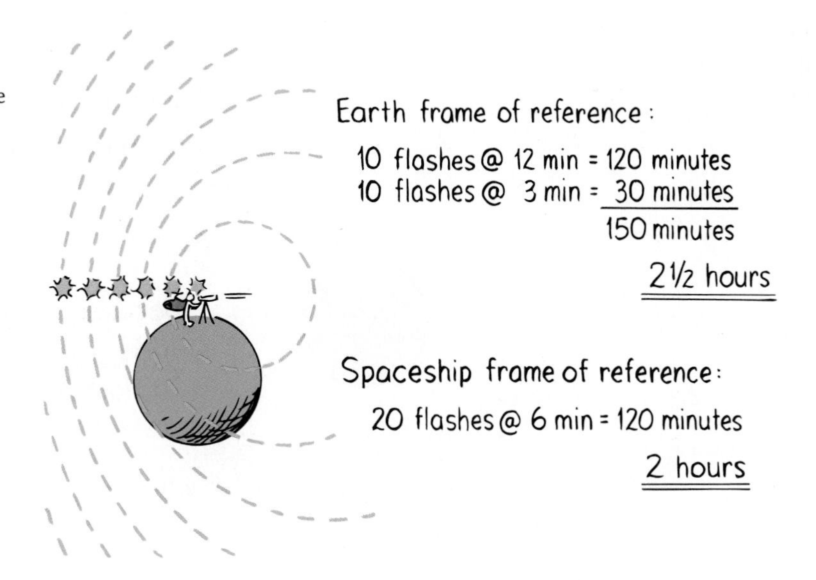

Earth frame of reference:
10 flashes @ 12 min = 120 minutes
10 flashes @ 3 min = 30 minutes
150 minutes
2 1/2 hours
Spaceship frame of reference:
20 flashes @ 6 min = 120 minutes
2 hours

FIGURE 35.18
The trip that takes 2 h in the frame of reference of the spaceship takes 2 1/2 h in the Earth's frame of reference.

means that five flashes are seen by the spaceship during the hour of receding from Earth. During the spaceship's hour of approaching, the light flashes are seen at 3-minute intervals (Figure 35.19*b*), so twenty flashes will be seen.

So we see that the spaceship receives a total of twenty-five flashes during its 2-hour trip. According to clocks on Earth, however, the time it took to emit the twenty-five flashes at 6-minute intervals was $25 \times (6 \text{ minutes})$, or 150 minutes$(= 2.5 \text{ hours})$. This is shown in Figure 35.20.

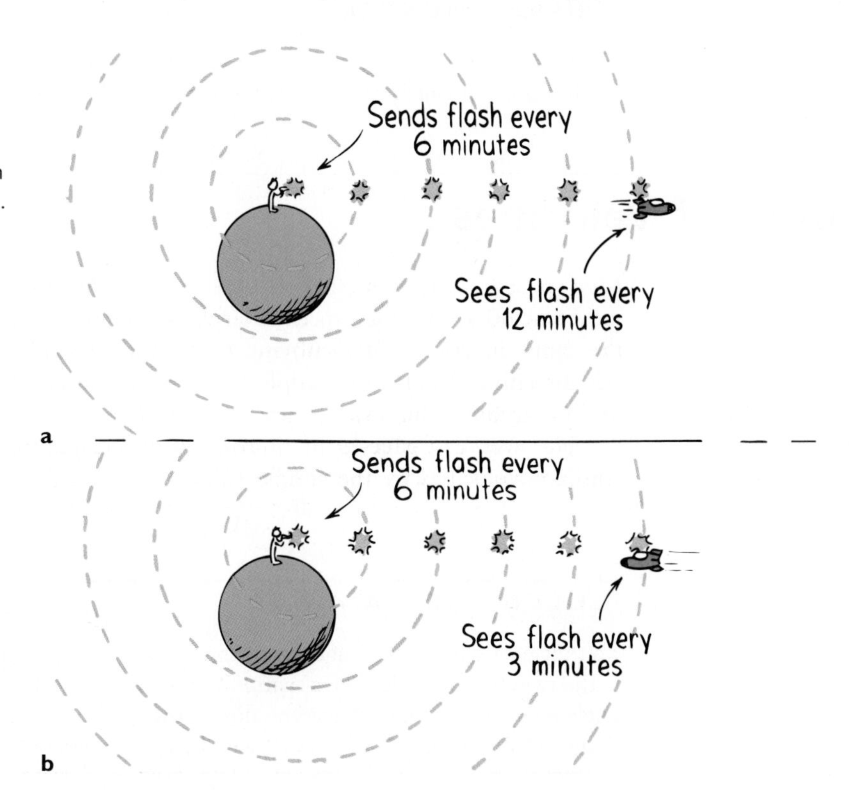

FIGURE 35.19
Flashes sent from Earth at 6-min intervals are seen at 12-min intervals by the ship when it recedes and at 3-min intervals when it approaches.

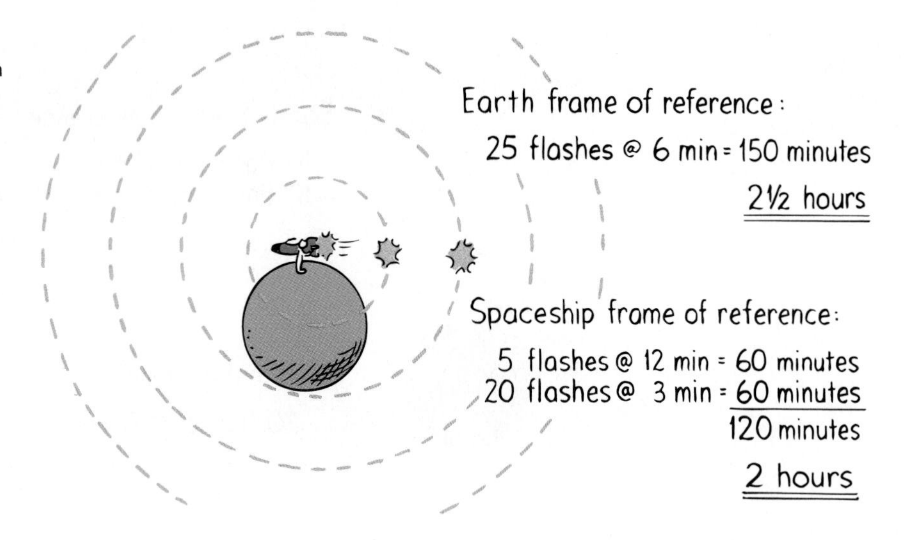

FIGURE 35.20
A time interval of 2 1/2 h on Earth is seen to take 2 h in the spaceship's frame of reference.

So both twins agree on the same results, with no dispute as to who ages more. While the stay-at-home twin remains in a single reference frame, the traveling twin has experienced two different frames of reference, separated by the acceleration of the spaceship in turning around. The spaceship has in effect experienced two different realms of time, while Earth has experienced a still different but single realm of time. The twins can meet again at the same place in space only at the expense of time.

CHECK YOURSELF

Since motion is relative, can't we say as well that the spaceship is at rest and the Earth moves, in which case the twin on the spaceship ages more?

Addition of Velocities

Most people know that, if you walk at 1 km/h along the aisle of a train that moves at 60 km/h, your speed relative to the ground is 61 km/h if you walk in the same direction as the moving train and 59 km/h if you walk in the opposite direction. What most people know is *almost* correct. Taking special relativity into account, one's speeds are *very nearly* 61 km/h and 59 km/h, respectively.

For everyday objects in uniform (nonaccelerating) motion, we ordinarily combine velocities by the simple rule

$$V = v_1 + v_2$$

CHECK YOUR ANSWER

No, not unless Earth then undergoes the turnaround and returns, as our spaceship did in the twin-trip example. The situation is not symmetrical, for one twin remains in a single reference frame in spacetime during the trip while the other makes a distinct change of reference frame, as evidenced by the acceleration in turning around.

But this rule does not apply to light, which always has the same velocity c. Strictly speaking, the above rule is an approximation of the relativistic rule for adding velocities. We'll not treat the long derivation but simply state the rule:

$$V = \frac{v_1 + v_2}{1 + \frac{v_1 v_2}{c^2}}$$

The numerator of this formula makes common sense. But this simple sum of two velocities is altered by the second term in the denominator, which is significant only when both v_1 and v_2 are nearly c.

As an example, consider a spaceship moving away from you at a velocity of $0.5c$. It fires a rocket that thrusts in the same direction, also away from you, at a speed of $0.5c$ relative to itself. How fast does the rocket move relative to you? The nonrelativistic rule would say that the rocket moves at the speed of light in your reference frame. But, in fact,

$$V = \frac{0.5c + 0.5c}{1 + \frac{0.25c^2}{c^2}} = \frac{c}{1.25} = 0.8c$$

which illustrates another consequence of relativity: No material object can travel as fast as, or faster than, light.

Suppose that the spaceship instead fires a pulse of laser light in its direction of travel. How fast does the pulse move in your frame of reference?

$$V = \frac{0.5c + c}{1 + \frac{0.5c^2}{c^2}} = \frac{1.5c}{1.5} = c$$

No matter what the relative velocities between two frames, light moving at c in one frame will be seen to be moving at c in any other frame. If you try chasing light, you can never catch it.

Space Travel

One of the old arguments against the possibility of human interstellar travel was that our life span is too short. It was argued, for example, that the star nearest Earth (after the Sun), Alpha Centauri, is 4 light-years away, and a round trip even at the speed of light would require 8 years.[3] And even a speed-of-light voyage to the center of our galaxy, 25,000 light-years distant, would require 25,000 years. But these arguments fail to take into account time dilation. Time for a person on Earth and time for a person in a high-speed spaceship are not the same.

A person's heart beats to the rhythm of the realm of spacetime in which it finds itself. And one realm of spacetime seems the same as any other to the

[3]A light-year is the distance light travels in 1 year, 9.46×10^{12} km.

FIGURE 35.21
From the Earth frame of reference, light takes 25,000 years to travel from the center of our Milky Way galaxy to our solar system. From the frame of reference of a high-speed spaceship, the trip takes less time. From the frame of reference of light itself, the trip takes no time. There is no time in a speed-of-light frame of reference.

heart, but not to an observer who stands outside the heart's frame of reference. For example, astronauts traveling at 99% of *c* could go to the star Procyon (10.4 light-years distant) and back in 21 Earth years. Because of time dilation, however, only three years would pass for the astronauts. This is what all their clocks would tell them—and, biologically, they would be only 3 years older. It would be the space officials greeting them on their return who would be 21 years older!

At higher speeds, the results are even more impressive. At a speed of 99.99% of *c*, travelers could travel a distance of slightly more than 70 light-years in a single year of their own time; at 99.999% of *c*, this distance would be pushed appreciably farther than 200 light-years. A 5-year trip for them would take them farther than light travels in 1000 Earth-time years!

Present technology does not permit such journeys. Getting enough propulsive energy and shielding against radiation are both prohibitive problems. Spaceships traveling at relativistic speeds would require billions of times the energy used to put a space shuttle into orbit. Even some kind of interstellar ramjet that scooped up interstellar hydrogen gas for burning in a fusion reactor would have to overcome the enormous retarding effect of scooping up the hydrogen at high speeds. And the space travelers would encounter interstellar particles just as if they had a large particle accelerator pointed at them. No way of shielding such intense particle bombardment for prolonged periods of time is presently known. For the present, interstellar space travel must be relegated to science fiction. Not because of scientific fantasy, but simply because of the impracticality of space travel. Traveling close to the speed of light in order to take advantage of time dilation is completely consistent with the laws of physics.

We can see into the past, but we cannot go into the past. For example, we experience the past when we look at the night skies. The starlight impinging on our eyes left those stars dozens, hundreds, even millions of years ago. What we see is the stars as they were long ago. We are thus eyewitnesses to ancient history—and can only speculate about what may have happened to the stars in the interim.

If we are looking at light that left a star, say, 100 years ago, then it follows that any sighted beings in that solar system are seeing us by light that left

CENTURY HOPPING

Let's push our science fiction to a possible time in the future when the prohibitive problems of energy supplies and of radiation have been overcome and space travel is a routine experience. People will have the option of taking a trip and returning to any future century of their choosing. For example, one might depart from Earth in a high-speed spaceship in the year 2100, travel for 5 years or so, and return in the year 2500. One could live among the earthlings of that period for a while and depart again to try out the year 3000 for style.

People could keep jumping into the future with some expense of their own time—but they could not trip into the past. They could never return to the same era on Earth to which they had bade farewell. Time, as we know it, travels one way—forward. Here on Earth, we move constantly into the future at the steady rate of 24 hours per day. An astronaut leaving on a deep-space voyage must live with the fact that, upon return, much more time will have elapsed on Earth than the astronaut has subjectively and physically experienced during the voyage. The credo of all star travelers, whatever their physiological condition, will be permanent farewell.

here 100 years ago and that, further, if they possessed super telescopes, they might very well be able to eyewitness earthly events of a century ago—the aftermath of the American Civil War, for instance. They would see our past, but they would still see events in a forward direction; they would see our clocks running clockwise.

We can speculate about the possibility that time might just as well move counterclockwise into the past as clockwise into the future. Why is it, we might ask, that in space we can move forward or back, left or right, up or down, but we can move only in one direction through time? Quite interestingly, the mathematics of elementary-particle interactions permits "time reversal," although there are some particle interactions that favor only one direction in time. Hypothetical particles that move backward in time are called *tachyons*. In any case, for the complex organism called a human being, time has only one direction.[4]

This conclusion is blithely ignored in a limerick that is a favorite with scientist types:

> There was a young lady named Bright
> Who traveled much faster than light.
> She departed one day
> In a relative way
> And returned on the previous night.

Even with our heads fairly well into relativity, we may still unconsciously cling to the idea that there is an absolute time and compare all these relativistic effects to it—recognizing that time changes this way and that way for this speed and that speed, yet feeling that there still is some basic or absolute time. We may tend to think that the time we experience on Earth is fundamental and that other times are wrong. This is understandable: We're earthlings. But the idea is confining. From the point of view of observers elsewhere in the universe, we may be moving at relativistic speeds; they see us living in slow motion. They may see us living lifetimes a hundred times as long as theirs, just as with super

[4]It has been speculated that, if we moved backward through time, we wouldn't know it, for then we would remember our future and would think it was our past!

telescopes we would see them living lifetimes a hundredfold longer than ours. There is no universally standard time—none.

We think of time and then we think of the universe. We think of the universe and we wonder about what went on before the universe began. We wonder about what will happen if the universe ceases to exist in time. But the concept of time applies to events and entities within the universe, not to the universe as a whole. Time is "in" the universe; the universe is not "in" time. Without the universe, there is no time; no before, no after. Likewise, space is "in" the universe; the universe is not "in" a region of space. There is no space "outside" the universe. Spacetime exists within the universe. Think about that!

Length Contraction

As objects move through spacetime, space as well as time changes. In a nutshell, space is contracted, making the objects look shorter when they move by us at relativistic speeds. This **length contraction** was first proposed by the physicist George F. FitzGerald and mathematically expressed by another physicist, Hendrick A. Lorentz (mentioned earlier). Whereas these physicists hypothesized that matter contracts, Einstein saw that what contracts is space itself. Nevertheless, because Einstein's formula is the same as Lorentz's, we call the effect the *Lorentz contraction*:

$$L = L_0\sqrt{1 - \frac{v^2}{c^2}}$$

where v is the relative velocity between the observed object and the observer, c is the speed of light, L is the measured length of the moving object, and L_0 is the measured length of the object at rest.[5]

Suppose that an object is at rest so that $v = 0$. When we substitute $v = 0$ in the Lorentz equation, we find $L = L_0$, as we would expect. When we substitute various large values of v in the Lorentz equation, we begin to see the calculated L get smaller and smaller. At 87% of c, an object would be contracted to half its original length. At 99.5% of c, it would contract to one-tenth its original length. If the object were somehow able to move at c, its length would be zero. This is one of the reasons we say that the speed of light is the upper limit for the speed of any moving object. Another limerick popular with the science heads is this one:

FIGURE 35.22
The Lorentz contraction. The meterstick is measured to be half as long when traveling at 87% of the speed of light relative to the observer.

There was a young fencer named Fisk,
Whose thrust was exceedingly brisk.
So fast was his action
The Lorentz contraction
Reduced his rapier to a disk.

As Figure 35.23 indicates, contraction takes place only in the direction of motion. If an object is moving horizontally, no contraction takes place vertically.

[5]We can express this as $L = \frac{1}{\gamma}L_0$. Where $\frac{1}{\gamma}$ is always 1 or less (because γ is always 1 or greater).

Note that we do not explain how the length-contraction equation or other equations come about. We simply state equations as "guides to thinking" about the ideas of special relativity.

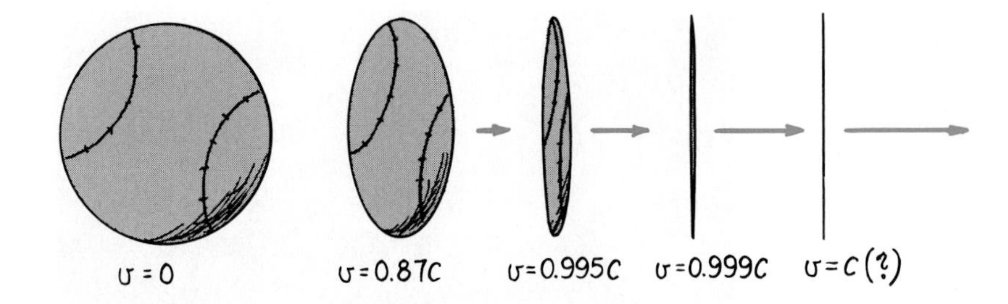

FIGURE 35.23
As speed increases, length in the direction of motion decreases. Lengths in the perpendicular direction do not change.

Time dilation: Moving clocks run slowly. Length contraction: Moving objects are shorter (in the direction of motion).

Insights

Length contraction should be of considerable interest to space voyagers. The center of our Milky Way galaxy is 25,000 light-years away. Does this mean that, if we traveled in that direction at the speed of light, it would take 25,000 years to get there? From an Earth frame of reference, yes; but to the space voyagers, decidedly not! At the speed of light, the 25,000 light-year distance would be contracted to no distance at all. Space voyagers would arrive there instantly!

For hypothetical travel near the speed of light, length contraction and time dilation are just two faces of the same phenomenon. If astronauts go so fast that they find the distance to the nearest star to be just one light-year instead of the four light-years measured from Earth, they make the trip in a little more than one year. But observers back on Earth say that the clocks aboard the spaceship have slowed so much that they tick off only one year in four years of Earth time. Both agree on what happens: The astronauts are only a little more than a year older when they reach the star. One set of observers say it's because of length contraction, the other set say it's because of time dilation. Both are right.

If space voyagers are ever able to boost themselves to relativistic speeds, they will find distant parts of the universe drawn closer by space contraction, while observers back on Earth will see the astronauts covering more distance because they age more slowly.

CHECK YOURSELF

A rectangular billboard in space has the dimensions 10 m × 20 m. How fast, and in what direction with respect to the billboard, would a space traveler have to pass for the billboard to appear square?

FIGURE 35.24
Interactive Figure
In the frame of reference of our meterstick, its length is 1 meter. Observers in a moving frame see *our* metersticks contracted, while we see *their* metersticks contracted. The effects of relativity are always attributed to "the other guy."

CHECK YOUR ANSWER

The space traveler would have to travel at 0.87c in a direction parallel to the longer side of the board.

FIGURE 35.25
The Stanford Linear Accelerator is 3.2 km (2 miles) long. But to electrons moving through it at 0.9999995*c*, the accelerator is only 3.2 meters long. The electrons start their journey in the foreground, and they smash into targets, or are otherwise studied, in the experimental areas beyond the freeway (near the top of the photo).

Relativistic Momentum

Recall our study of momentum in Chapter 6. We learned that the change of momentum mv of an object is equal to the impulse Ft applied to it: $Ft = \Delta mv$, or, $Ft = \Delta p$, where $p = mv$. If you apply more impulse to an object that is free to move, the object acquires more momentum. Double the impulse, and the momentum doubles. Apply ten times the impulse, and the object gains ten times as much momentum. Does this mean that momentum can increase without any limit? The answer is *yes*. Does this mean that speed can also increase without any limit? The answer is *no*! Nature's speed limit for material objects is c.

To Newton, infinite momentum would mean infinite mass or infinite speed. But not so in relativity. Einstein showed that a new definition of momentum is required. It is

$$p = \gamma mv$$

where γ is the Lorentz factor (recall that γ is always 1 or greater). This generalized definition of momentum is valid in all uniformly moving reference frames. *Relativistic momentum* is larger than mv by a factor of γ. For everyday speeds much less than c, γ is nearly equal to 1, so p is nearly equal to mv. Newton's definition of momentum is valid at low speeds.

At higher speeds, γ grows dramatically, and so does relativistic momentum. As speed approaches c, γ approaches infinity! No matter how close to c an object is pushed, it would still require infinite impulse to give it the last bit of speed needed to reach c—clearly impossible. Hence we see that no body with mass can be pushed to the speed of light, much less beyond it.

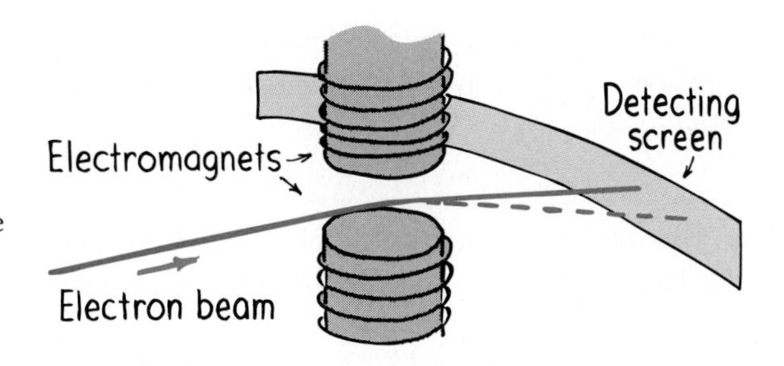

FIGURE 35.26
If the momentum of the electrons were equal to the Newtonian value *mv*, the beam would follow the dashed line. But, because the relativistic momentum γmv is greater, the beam follows the "stiffer" trajectory shown by the solid line.

Subatomic particles are routinely pushed to nearly the speed of light. The momenta of such particles may be thousands of times more than the Newtonian expression *mv* predicts. Classically, the particles behave as if their masses increase with speed. Einstein initially favored this interpretation, and later changed his mind to keep mass a constant, a property of matter that is the same in all frames of reference. So it is γ that changes with speed, not mass. The increased momentum of a high-speed particle is evident in the increased "stiffness" of its trajectory. The more momentum it has, the "stiffer" is its trajectory and the harder it is to deflect.

We see this when a beam of electrons is directed into a magnetic field. Charged particles moving in a magnetic field experience a force that deflects them from their normal paths. For small momentum, the path curves sharply. For large momentum, there is greater stiffness and the path curves only a little (Figure 35.26). Even though one particle may be moving only a little faster than another one—say, 99.9% of the speed of light instead of 99% of the speed of light—its momentum will be considerably greater and it will follow a straighter path in the magnetic field. This stiffness must be compensated for in circular accelerators like cyclotrons and synchrotrons, where momentum dictates the radius of curvature. In the linear accelerator shown in Figure 35.25, the particle beam travels in a straight-line path and momentum changes don't produce deviations from a straight-line path. Deviations occur when the beam of electrons is bent at the exit port by magnets, as indicated in Figure 35.26. Whatever the type of particle accelerator, physicists working with subatomic particles every day verify the correctness of the relativistic definition of momentum and the speed limit imposed by nature.

To summarize, we see that, as the speed of an object approaches the speed of light, its momentum approaches infinity—which means there is no way that the speed of light can be reached. There is, however, at least one thing that reaches the speed of light—light itself! But the photons of light are massless, and the equations that apply to them are different. Light travels always at the same speed. So, interestingly, a material particle can never be brought to the speed of light; and light can never be brought to rest.

Mass, Energy, and $E = mc^2$

Einstein linked not only space and time but also mass and energy. A piece of matter, even at rest and not interacting with anything else, has an "energy of being." This is called its *rest energy*. Einstein concluded that it takes energy to make mass and that energy is released if mass disappears. The amount of energy E is related to the amount of mass m by the most celebrated equation of the twentieth century:

$$E = mc^2$$

The c^2 is the conversion factor between energy units and mass units. Because of the large magnitude of c, a small mass corresponds to an enormous quantity of energy.[6]

Recall, from Chapter 11, that tiny decreases of nuclear mass in both nuclear fission and nuclear fusion produced enormous releases of energy, all in accord with $E = mc^2$. To the general public, $E = mc^2$ is synonymous with nuclear energy. If we were to weigh a fully fueled nuclear power plant, then weigh it again a week later, we'd find it weighs slightly less. Part of the fuel's mass, about 1 part in a thousand, has been converted to energy. Now, interestingly enough, if we were to weigh a coal-burning power plant and all the coal and oxygen it consumes in a week, and then weigh it again with all the carbon dioxide and other combustion products that came out during the week, we'd also find it all weighs slightly less. Again, mass has been converted to energy. About 1 part in a billion has been converted. Get this: If both plants produce the same amount of energy, the mass change will be the same for both—whether energy is released by nuclear or chemical mass conversion makes no difference. The chief difference lies in the amount of energy released in each individual reaction and the amount of mass involved. Fissioning of a single uranium nucleus releases 10 million times as much energy as the combustion of carbon to produce a single carbon dioxide molecule. Hence a few truckloads of uranium fuel will power a fission plant while a coal-burning plant consumes many hundred car trainloads of coal.

FIGURE 35.27
Saying that a power plant delivers 90 million megajoules of energy to its consumers is equivalent to saying that it delivers 1 gram of energy to its consumers, because mass and energy are equivalent.

[6]When c is in meters per second and m is in kilograms, then E will be in joules. If the equivalence of mass and energy had been understood long ago when physics concepts were first being formulated, there would probably be no separate units for mass and energy. Furthermore, with a redefinition of space and time units, c could equal 1, and $E = mc^2$ would simply be $E = m$.

FIGURE 35.28
In 1 second, 4.5 million tons of mass are converted to radiant energy in the Sun. The Sun is so massive, however, that in 1 million years only one ten-millionth of the Sun's mass will have been converted to radiant energy.

When we strike a match, phosphorus atoms in the match head rearrange themselves and combine with oxygen in the air to form new molecules. The resulting molecules have very slightly less mass than the separate phosphorus and oxygen molecules. From a mass standpoint, the whole is slightly less than the sum of its parts, by amounts that escape our notice. For all chemical reactions that give off energy, there is a corresponding decrease in mass of about one part in a billion.

For nuclear reactions, a decrease in mass by one part in a thousand can be directly measured by a variety of devices. This decrease of mass in the Sun by the process of thermonuclear fusion bathes the solar system with radiant energy and nourishes life. The present stage of thermonuclear fusion in the Sun has been going on for the past 5 billion years, and there is sufficient hydrogen fuel for fusion to last another 5 billion years. It is nice to have such a big Sun!

The equation $E = mc^2$ is not restricted to chemical and nuclear reactions. A change in energy of any object at rest is accompanied by a change in its mass. The filament of a lightbulb energized with electricity has more mass than when it is turned off. A hot cup of tea has more mass than the same cup of tea when cold. A wound-up spring clock has more mass than the same clock when unwound. But these examples involve incredibly small changes in mass—too small to be measured. Even the much larger changes of mass in radioactive change were not measured until after Einstein predicted the mass–energy equivalence. Now, however, mass-to-energy and energy-to-mass conversions are measured routinely.

Consider a coin with a mass of 1 g. You'd expect two of the same coins to have a mass of 2 g, ten coins to have a mass of 10 g, and 1,000 coins piled in a box to have a mass of 1 kg. Not so if the coins attract or repel each other. Suppose, for example, that each coin carries a negative electric charge so that each coin repels all other coins. Then forcing them together in the box takes work. This work adds to the mass of the collection. So a box containing 1,000 negatively charged coins has more than 1 kg of mass. If, on the other hand, the coins all attracted one another (as nucleons in the nucleus attract one another), it takes work to separate them; then a box of 1,000 coins would have a mass less than 1 kg. So the mass of an object is not necessarily equal to the sum of the masses of its parts, as we know from measuring the masses of nuclei. The effect would be dramatically enormous if we could deal with bare charged particles. If we could force together a number of electrons whose masses add separately to 1 g into a 10-cm diameter sphere, the collection

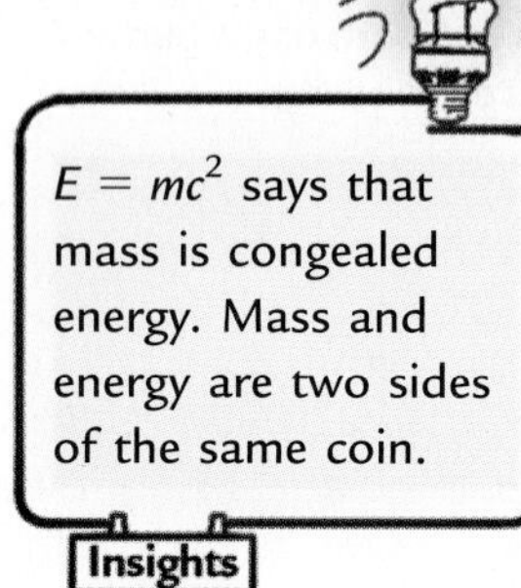

$E = mc^2$ says that mass is congealed energy. Mass and energy are two sides of the same coin.

Insights

would have a mass of 10 trillion kilograms! The equivalence of mass and energy is indeed profound.

Some physicists speculate that the mass of a single electron is merely the energy equivalent of the work required to compress its charge, assuming that, if its charge were spread out, it would have no mass at all.[7]

In ordinary units of measurement, the speed of light c is a large quantity and its square is even larger—hence a small amount of mass stores a large amount of energy. The quantity c^2 is a "conversion factor." It converts the measurement of mass to the measurement of equivalent energy. Or it is the ratio of rest energy to mass; $E/m = c^2$. Its appearance in either form of this equation has nothing to do with light and nothing to do with motion. The magnitude of c^2 is 90 quadrillion (9×10^{16}) joules per kilogram. One kilogram of matter has an "energy of being" equal to 90 quadrillion joules. Even a speck of matter with a mass of only 1 milligram has a rest energy of 90 billion joules.

The equation $E = mc^2$ is more than a formula for the conversion of mass into other kinds of energy, or vice versa. It states even more: that energy and mass are the *same thing*. Mass is congealed energy. If you want to know how much energy is in a system, measure its mass. For an object at rest, its energy *is* its mass. Energy, like mass, exhibits inertia. Shake a massive object back and forth; it is energy itself that is hard to shake.

CHECK YOURSELF

Can we look at the equation $E = mc^2$ in another way and say that matter transforms into pure energy when it is traveling at the speed of light squared?

The first evidence for the conversion of radiant energy to mass was provided in 1932 by the American physicist Carl Anderson. He and a colleague at Caltech discovered the *positron* by the track it left in a cloud chamber. The positron is the *antiparticle* of the electron, equal in mass and spin to the electron but opposite in charge. When a high-frequency photon comes close to an atomic nucleus, it can create an electron and a positron together as a pair,

CHECK YOUR ANSWER

No, no, no! Matter cannot be made to move at the speed of light, let alone the speed of light squared (which is not a speed!). The equation $E = mc^2$ simply means that energy and mass are "two sides of the same coin."

[7]San Francisco Sidewalk Astronomer John Dobson speculates that, just as a clock becomes more massive when we do work on it by winding it against the resistance of its spring, the mass of the entire universe is nothing more than the energy that has gone into winding it up against mutual gravitation. In this view, the mass of the universe is equivalent to the work done in spreading it out. So, perhaps each electron has mass because its charge is confined, and the atoms that make up the universe have mass because they are dispersed.

thus creating mass. The created particles fly apart. The positron is not part of normal matter because it lives such a short time. As soon as it encounters an electron, the pair is annihilated, sending out two gamma rays in the process. Then mass is converted back to radiant energy.[8]

The Correspondence Principle

We introduced the correspondence principle in Chapter 32. Recall that it states that any new theory or any new description of nature must agree with the old where the old gives correct results. If the equations of special relativity are valid, they must correspond to those of classical mechanics when speeds much less than the speed of light are considered.

The relativity equations for time, length, and momentum are:

$$t = \frac{t_0}{\sqrt{1 - \frac{v^2}{c^2}}} = \gamma t_0$$

$$L = L_0\sqrt{1 - \frac{v^2}{c^2}} = L_0/\gamma$$

$$p = \frac{mv}{\sqrt{1 - \frac{v^2}{c^2}}} = \gamma mv$$

Note that these equations each reduce to Newtonian values for speeds that are very small compared with c. Then the ratio v^2/c^2 is very small and, for everyday speeds, may be taken to be zero. The relativity equations become

$$t = \frac{t_0}{\sqrt{1 - 0}} = t_0 \cdot$$

$$L = L_0\sqrt{1 - 0} = L_0$$

$$p = \frac{mv}{\sqrt{1 - 0}} = mv$$

So, for everyday speeds, the momentum, length, and time of moving objects are essentially unchanged. The equations of special relativity hold for all speeds, although they differ appreciably from classical equations only for speeds near the speed of light.

Einstein's theory of relativity has raised many philosophical questions. What, exactly, is time? Can we say that it is nature's way of seeing to it

[8]Recall that the energy of a photon is $E = hf$ and that the mass energy of a particle is $E = mc^2$. High-frequency photons routinely convert their energy to mass when they produce pairs of particles in nature—and in accelerators, where the processes can be observed. Why pairs? Mainly because that's the only way the conservation of charge is not violated. So, when an electron is created, an antiparticle positron is created also. Equating the two equations, $hf = 2mc^2$, where m is the mass of a particle (or antiparticle), we see the minimum frequency of a gamma ray for the production of a particle pair is $f = 2mc^2/h$. Producing heavier particles means more energy—hence the high energies of particle accelerators.

BIOGRAPHY: ALBERT EINSTEIN (1879–1955)

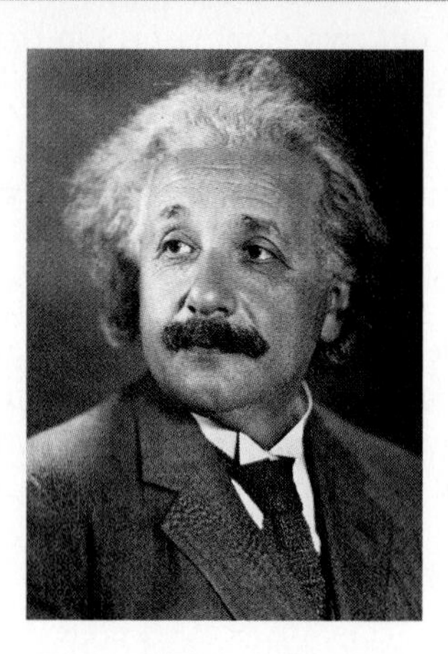

Albert Einstein was born in Ulm, Germany, on March 14, 1879. According to popular legend, he was a slow child and learned to speak at a much later age than average; his parents feared for a while that he might be mentally retarded. Yet his elementary-school records show that he was remarkably gifted in mathematics, physics, and playing the violin. He rebelled, however, at the practice of education by regimentation and rote, and he was expelled just as he was preparing to drop out of school at the age of 15. Largely because of business reasons, his family moved to Italy. Young Einstein renounced his German citizenship and went to live with family friends in Switzerland. There he was allowed to take the entrance examinations for the renowned Swiss Federal Institute of Technology in Zurich, although he was two years younger than the normal age for enrollment. But, because of difficulties with the French language, he did not pass the examination. He spent a year at a Swiss preparatory school in Aarau, where he was "promoted, with protest, in French." He tried the entrance exam again at Zurich and passed. As a student, he cut many lectures, preferring to study on his own, and, in 1900, he succeeded in passing his examinations by cramming with the help of a friend's meticulous notes. He said later of this, ". . . after I had passed the final examination, I found the consideration of any scientific problem distasteful to me for an entire year." During this year, he became a citizen of Switzerland; he accepted a temporary summer teaching position and tutored two young high-school students. He advised their father, a high-school teacher himself, to remove the boys from school, where, he maintained, their natural curiosity was being destroyed. Einstein's job as a tutor was short-lived.

It was not until two years after graduation that he got a steady job, as a patent examiner at the Swiss Patent Office in Berne. Einstein held this position for more than seven years. He found the work rather interesting, sometimes stimulating his scientific imagination, but mainly freeing him of financial worries while providing time to ponder the problems in physics that puzzled him.

With no academic connections whatsoever, and with essentially no contact with other physicists, he laid out the main lines along which twentieth-century theoretical physics has developed. In 1905, at the age of 26, he earned his Ph.D. in physics and published three major papers. The first was on the quantum theory of light, including an explanation of the photoelectric effect, for which he won the 1921 Nobel Prize in physics. The second paper was on the statistical aspects of molecular theory and Brownian motion, a proof for the existence of atoms. His third and most famous paper was on special relativity. In 1915, he published a paper on the theory of general relativity, which presented a new theory of gravitation that included Newton's theory as a special case. These trailblazing papers have greatly affected the course of modern physics.

Einstein's concerns were not limited to physics. He lived in Berlin during World War I and denounced the

that everything does not all happen at once? And why does time seem to move in one direction? Has it always moved *forward*? Are there other parts of the universe in which time moves *backward*? Is it likely that our three-dimensional perception of a four-dimensional world is only a beginning? Could there be a fifth dimension? A sixth dimension? A seventh dimension? And, if so, what would the nature of these dimensions be? Perhaps these unanswered questions will be answered by the physicists of tomorrow. How exciting!

German militarism of his time. He publicly expressed his deeply felt conviction that warfare should be abolished and that an international organization should be founded to govern disputes between nations. In 1933, while Einstein was visiting the United States, Hitler came to power. Einstein spoke out against Hitler's racial and political policies and resigned his position at the University of Berlin. No longer safe in Germany, Einstein came to the United States and accepted a research position at the Institute for Advanced Study in Princeton, New Jersey.

In 1939, one year before Einstein became an American citizen, and after German scientists fissioned the uranium atom, he was urged by several prominent Hungarian-American scientists to write the famous letter to President Roosevelt pointing out the scientific possibilities of a nuclear bomb. Einstein was a pacifist, but the thought of Hitler developing such a bomb prompted his action. The outcome was the development of the first nuclear bomb, which, ironically, was detonated on Japan after the fall of Germany.

Einstein believed that the universe is indifferent to the human condition, and he stated that, if humanity were to continue, it must create a moral order. He intensely advocated world peace through nuclear disarmament. Nuclear bombs, Einstein remarked, had changed everything but our way of thinking.

C. P. Snow, who was acquainted with Einstein, in a review of *The Born–Einstein Letters, 1916–1955,* says this of him: "Einstein was the most powerful mind of the twentieth century, and one of the most powerful that ever lived. He was more than that. He was a man of enormous weight of personality, and perhaps most of all, of normal stature. . . . I have met a number of people whom the world calls great; of these, he was by far, by an order of magnitude, the most impressive. He was—despite the warmth, the humanity, the touch of the comedian—the most different from other men."

Einstein was more than a great scientist; he was a man of unpretentious disposition with a deep concern for the welfare of his fellow beings. The choice of Einstein as the person of the century by *Time* magazine at the end of the 1900s was most appropriate—and noncontroversial.

Summary of Terms

Frame of reference A vantage point (usually a set of coordinate axes) with respect to which position and motion may be described.

Postulates of the special theory of relativity (1) All laws of nature are the same in all uniformly moving frames of reference. (2) The speed of light in free space has the same measured value regardless of the motion of the source or the motion of the observer; that is, the speed of light is a constant.

Simultaneity Occurring at the same time. Two events that are simultaneous in one frame of reference need not be simultaneous in a frame moving relative to the first frame.

Spacetime The four-dimensional continuum in which all events take place and all things exist: Three dimensions are the coordinates of space, and the fourth is time.

Time dilation The slowing of time as a result of speed.

Length contraction The contraction of space in an observer's direction of motion as a result of speed.

Suggested Reading

Einstein, Albert. *The Meaning of Relativity*. Princeton, N.J.: Princeton University Press, 1950. Written for the average person by Einstein himself.

Epstein, Lewis C. *Relativity Visualized*. San Francisco: Insight Press, 1983.

Gamow, George. *Mr. Tompkins in Wonderland*. New York: Macmillan, 1940. An excellent and very interesting little book.

Gardner, Martin. *The Relativity Explosion*. New York: Vintage Books, 1976.

Taylor, Edwin F., and John A. Wheeler. *Spacetime Physics*. San Francisco: W. H. Freeman, 1966.

Review Questions

Motion Is Relative

1. If you walk at 1 km/h down the aisle of a train that moves at 60 km/h, what is your speed relative to the ground?
2. In the previous question, is your approximate speed relative to the Sun as you walk down the aisle of the bus slightly more or very much more?

Michelson–Morley Experiment

3. What hypothesis did G. F. FitzGerald make to explain the findings of Michelson and Morley?
4. What classical idea about space and time was rejected by Einstein?

Postulates of the Special Theory of Relativity

5. Cite two examples of Einstein's first postulate.
6. Cite one example of Einstein's second postulate.

Simultaneity

7. Inside the moving compartment of Figure 35.4, light travels a certain distance to the front end and a certain distance to the back end of the compartment. How do these distances compare as seen in the frame of reference of the moving rocket?
8. How do the distances in question 7 compare as seen in the frame of reference of an observer on a stationary planet?

Spacetime

9. How many coordinate axes are usually used to describe three-dimensional space? What does the fourth dimension measure?
10. Under what condition will you and a friend share the same realm of spacetime? When will you not share the same realm?
11. What is special about the ratio of the distance traveled by a flash of light and the time the light takes to travel this distance?

Time Dilation

12. Time is required for light to travel along a path from one point to another. If this path is seen to be longer because of motion, what happens to the time it takes for light to travel this longer path?
13. What do we call the "stretching out" of time?
14. What is the value of the Lorentz factor γ (gamma)?
15. How do measurements of time differ for events in a frame of reference that moves at 50% the speed of light relative to us? At 99.5% the speed of light relative to us?
16. What is the evidence for time dilation?

The Twin Trip

17. When a flashing light approaches you, each flash that reaches you has a shorter distance to travel. What effect does this have on how frequently you receive the flashes?
18. When a flashing light source approaches you, does the speed of light or the frequency of light—or both—increase?
19. If a flashing light source moves toward you fast enough so that the duration between flashes seems half as long, how will the duration between flashes seem if the source is moving away from you at the same speed?
20. How many frames of reference does the stay-at-home twin experience in the twin trip? How many frames of reference does the traveling twin experience?

Addition of Velocities

21. What is the maximum value of v_1v_2/c^2 in an extreme situation? What is the smallest value?
22. Is the relativistic rule

$$V = \frac{v_1 + v_2}{1 + \frac{v_1v_2}{c^2}}$$

consistent with the fact that light can have only one speed in all uniformly moving reference frames?

Space Travel

23. What two main obstacles prevent us from traveling today throughout the galaxy at relativistic speeds?
24. What is the universal standard of time?

Length Contraction

25. How long would a meterstick appear to be if it were traveling like a properly thrown spear at 99.5% the speed of light?

26. How long would the meterstick in the previous question appear to be if it were traveling with its length perpendicular to its direction of motion? (Why is your answer different from your answer to the previous question?)
27. If you were traveling in a high-speed rocket ship, would metersticks on board appear to you to be contracted? Defend your answer.

Relativistic Momentum

28. What would be the momentum of an object pushed to the speed of light?
29. When a beam of charged particles moves through a magnetic field, what is the evidence that their momentum is greater than the value *mv*?

Mass, Energy, and $E = mc^2$

30. Compare the amount of mass converted to energy in nuclear reactions and in chemical reactions.
31. How does the energy from the fissioning of a single uranium nucleus compare with the energy from the combustion of a single carbon atom?
32. Does the equation $E = mc^2$ apply only to nuclear and chemical reactions?
33. What is the evidence for $E = mc^2$ in cosmic-ray investigations?

The Correspondence Principle

34. How does the correspondence principle relate to special relativity?
35. Do the relativity equations for time, length, and momentum hold true for everyday speeds? Explain.

Project

Write a letter to Grandma and explain how Einstein's theories of relativity concern the fast and the big—that relativity is not only "out there," but that it affects this world. Tell her how these ideas stimulate your quest for more knowledge about the universe.

Exercises

1. The idea that force causes acceleration doesn't seem strange. This and other ideas of Newtonian mechanics are consistent with our everyday experience. But the ideas of relativity do seem odd and more difficult to grasp. Why is this?
2. If you were in a smooth-riding train with no windows, could you sense the difference between uniform motion and rest? Between accelerated motion and rest? Explain how you could make such a distinction with a bowl filled with water.
3. A person riding on the roof of a freight train fires a gun pointed forward. (a) Relative to the ground, is the bullet moving faster or slower when the train is moving than when it is standing still? (b) Relative to the freight car, is the bullet moving faster or slower when the train is moving than when the train is standing still?
4. Suppose instead that the person riding on top of the freight car shines a searchlight beam in the direction in which the train is traveling. Compare the speed of the light beam relative to the ground when the train is at rest and when it is in motion. How does the behavior of the light beam differ from the behavior of the bullet in Exercise 3?
5. Why did Michelson and Morley at first consider their experiment a failure? (Have you ever encountered other examples where failure has to do not with the lack of ability but with the impossibility of the task?)
6. When you drive down the highway, you are moving through space. What else are you moving through?
7. In Chapter 26, we learned that light travels more slowly in glass than in air. Does this contradict Einstein's second postulate?
8. Astronomers view light coming from distant galaxies moving away from the Earth at speeds greater than 10% the speed of light. How fast does this light meet the telescopes of the astronomers?
9. Does special relativity allow *anything* to travel faster than light? Explain.
10. When a light beam approaches you, its frequency is greater and its wavelength less. Does this contradict the postulate that the speed of light cannot change? Defend your answer.
11. The beam of light from a laser on a rotating turntable casts into space. At some distance, the beam moves across space faster than *c*. Why does this not contradict relativity?
12. Can an electron beam sweep across the face of a cathode-ray tube at a speed greater than the speed of light? Explain.
13. Consider the speed of the point where scissors blades meet when the scissors are closed. The closer the blades are to being closed, the faster the point moves. The point could, in principle, move faster than light. Likewise for the speed of the point where an ax meets wood when the ax blade meets the wood not quite horizontally. The contact point travels faster than the ax. Similarly, a pair of laser beams that are crossed and moved toward being parallel produce a point of intersection that can move faster than light. Why do these examples not contradict special relativity?
14. If two lightning bolts hit exactly the same place at exactly the same time in one frame of reference, is it possible that observers in other frames will see the bolts hitting at different times or at different places?

14. If two lightning bolts hit exactly the same place at exactly the same time in one frame of reference, is it possible that observers in other frames will see the bolts hitting at different times or at different places?
15. Event A occurs before event B in a certain frame of reference. How could event B occur before event A in some other frame of reference?
16. Suppose that the lightbulb in the rocket ship in Figures 35.4 and Figure 35.5 is closer to the front than to the rear of the compartment, so that the observer in the ship sees the light reaching the front before it reaches the back. Is it still possible that the outside observer will see the light reaching the back first?
17. The speed of light is a speed limit in the universe—at least for the four-dimensional universe we comprehend. No material particle can attain or surpass this limit even when a continuous, unremitting force is exerted on it. What evidence supports this?
18. Since there is an upper limit on the speed of a particle, does it follow that there is also an upper limit on its momentum? On its kinetic energy? Explain.
19. Light travels a certain distance in, say, 20,000 years. How is it possible that an astronaut, traveling slower than light, could go as far in 20 years of her life as light travels in 20,000 years?
20. Is it possible for a human being who has a life expectancy of 70 years to make a round-trip journey to a part of the universe thousands of light-years distant? Explain.
21. A twin who makes a long trip at relativistic speeds returns younger than his stay-at-home twin sister. Could he return before his twin sister was born? Defend your answer.
22. Is it possible for a son or daughter to be biologically older than his or her parents? Explain.
23. If you were in a rocket ship traveling away from Earth at a speed close to the speed of light, what changes would you note in your pulse? In your volume? Explain.
24. If you were on Earth monitoring a person in a rocket ship traveling away from Earth at a speed close to the speed of light, what changes would you note in his pulse? In his volume? Explain.
25. Due to length contraction, you see people in a spaceship passing by you as being slightly narrower than they normally appear. How do these people view you?
26. Because of time dilation, you observe the hands of your friend's watch to be moving slowly. How does your friend view your watch—as running slowly, running rapidly, or neither?
27. Does the equation for time dilation show dilation occurring for all speeds, whether slow or fast? Explain.
28. If you lived in a world where people regularly traveled at speeds near the speed of light, why would it be risky to make a dental appointment for 10:00 a.m. next Thursday?
29. How do the measured densities of a body compare at rest and in motion?
30. If stationary observers measure the shape of a passing object to be exactly circular, what is the shape of the object according to observers on board the object, traveling with it?
31. The formula relating speed, frequency, and wavelength of electromagnetic waves, $v = f\lambda$, was known before relativity was developed. Relativity has not changed this equation, but it has added a new feature to it. What is that feature?
32. Light is reflected from a moving mirror. How is the reflected light different from the incident light, and how is it the same?
33. As a meterstick moves past you, your measurements show its momentum to be twice its classical momentum and its length to be 1 m. In what direction is the stick pointing?
34. In the preceding exercise, if the stick is moving in a direction along its length (like a properly thrown spear), how long will you measure its length to be?
35. If a high-speed spaceship appears shrunken to half its normal length, how does its momentum compare with the classical formula $p = mv$?
36. How can the momentum of a particle increase by 5%, with only a 1% increase in speed?
37. The two-mile linear accelerator at Stanford University in California "appears" to be less than a meter long to the electrons that travel in it. Explain.
38. Electrons end their trip in the Stanford accelerator with an energy thousands of times greater than their initial rest energy. In theory, if you could travel with them, would you notice an increase in their energy? In their momentum? In your moving frame of reference, what would be the approximate speed of the target they are about to hit?
39. Two safety pins, identical except that one is latched and one is unlatched, are placed in identical acid baths. After the pins are dissolved, what, if anything, is different about the two acid baths?
40. A chunk of radioactive material encased in an idealized, perfectly insulating blanket gets warmer as its nuclei decay and release energy. Does the mass of the radioactive material and the blanket change? If so, does it increase or decrease?
41. The electrons that illuminate the screen in a typical television picture tube travel at nearly one-fourth the

speed of light and possess nearly 3% more energy than hypothetical nonrelativistic electrons traveling at the same speed. Does this relativistic effect tend to increase or decrease your electric bill?

42. Muons are elementary particles that are formed high in the atmosphere by the interactions of cosmic rays with atomic nuclei in the upper atmosphere. Muons are radioactive and have average lifetimes of about two-millionths of a second. Even though they travel at almost the speed of light, very few should be detected at sea level after traveling through the atmosphere—at least according to classical physics. Laboratory measurements, however, show that muons in great number do reach the Earth's surface. What is the explanation?

43. How might the idea of the correspondence principle be applied outside the field of physics?

44. What does the equation $E = mc^2$ mean?

45. According to $E = mc^2$, how does the amount of energy in a kilogram of feathers compare with the amount of energy in a kilogram of iron?

46. Does a fully charged flashlight battery weigh more than the same battery when dead? Defend your answer.

47. When we look out into the universe, we see into the past. John Dobson, founder of the San Francisco Sidewalk Astronomers, says that we cannot even see the backs of our own hands *now*—in fact, we can't see anything *now*. Do you agree? Explain.

48. One of the fads of the future might be "century hopping," where occupants of high-speed spaceships would depart from the Earth for several years and return centuries later. What are the present-day obstacles to such a practice?

49. Is the statement by the philosopher Kierkegaard that "Life can only be understood backwards; but it must be lived forwards" consistent with the theory of special relativity?

50. Make up four multiple-choice questions, one each that would check a classmate's understanding of (a) time dilation, (b) length contraction, (c) relativistic momentum, and (d) $E = mc^2$.

Problems

Recall, from this chapter, that the factor gamma (γ) governs both time dilation and length contracion, where

$$\gamma = \frac{1}{\sqrt{1 - (v^2/c^2)}}$$

When you multiply the time in a moving frame by γ, you get the longer (dilated) time in your fixed frame. When you divide the length in a moving frame by γ, you get the shorter (contracted) length in your fixed frame.

1. Consider a high-speed rocket ship equipped with a flashing light source. If the frequency of flashes seen on an approaching ship is twice what it was when the ship was a fixed distance away, by how much is the period (time interval between flashes) changed? Is this period constant for a constant relative speed? For accelerated motion? Defend your answer.

2. The starship *Enterprise,* passing Earth at 80% of the speed of light, sends a drone ship forward at half the speed of light relative to the *Enterprise*. What is the drone's speed relative to Earth?

3. Pretend that the starship *Enterprise* in the previous problem is somehow traveling at c with respect to the Earth, and it fires a drone forward at speed c with respect to itself. Use the equation for the relativistic addition of velocities to show that the speed of the drone with respect to the Earth is still c.

4. A passenger on an interplanetary express bus traveling at $v = 0.99c$ takes a five-minute catnap, according to his watch. How long does the nap last from your vantage point on a fixed planet?

5. According to Newtonian mechanics, the momentum of the bus in the preceding problem is $p = mv$. According to relativity, it is $p = \gamma mv$. How does the actual momentum of the bus moving at $0.99c$ compare with the momentum it would have if classical mechanics were valid? How does the momentum of an electron traveling at $0.99c$ compare with its classical momentum?

6. The bus in the previous problems is 70 feet long, according to its passengers and driver. What is its length from your vantage point on a fixed planet?

7. If the bus in Problem 4 were to slow to a "mere" 10% of the speed of light, how long would you measure the passenger's catnap to be?

8. If the bus driver in Problem 4 decided to drive at 99.99% of the speed of light in order to gain some time, what would you measure the length of the bus to be?

9. Assume that rocket taxis of the future move about the solar system at half the speed of light. For a one-hour trip as measured by a clock in the taxi, a driver is paid 10 stellars. The taxi-driver's union demands that pay be based on Earth time instead of taxi time. If their demand is met, what will be the new payment for the same trip?

10. The fractional change of mass to energy in a fission reactor is about 0.1%, or 1 part in a thousand. For each kilogram of uranium that undergoes fission, how much energy is released? If energy costs three cents per megajoule, how much is this energy worth in dollars?

CHAPTER 36

General Theory of Relativity

Richard Crowe kicks off a lecture on general relatitivity with a celestial sphere.

The special theory of relativity is "special" in the sense that it deals with uniformly moving reference frames—that is, reference frames that are not accelerated. The **general theory of relativity,** a new theory of gravitation, incorporates accelerating reference frames. Underlying it is the idea that the effects of gravitation and acceleration cannot be distinguished from one another. Einstein presented a new theory of gravitation.

Recall that Einstein postulated, in 1905, that no observation made inside an enclosed chamber could determine whether the chamber is at rest or moving with constant velocity; that is, no mechanical, electrical, optical, or any other physical measurement that one could perform inside a closed compartment in a smoothly riding train traveling along a straight track (or in an airplane flying through still air with the window curtains drawn) could possibly give any information as to whether the train was moving or at rest (or whether the plane was airborne or at rest on the runway). But, if the track were not smooth and straight (or if the air were turbulent), the situation would be entirely different: Uniform motion would give way to accelerated motion, which would be easily noticed. Einstein's conviction that the laws of nature should be expressed in the same form in every frame of reference, accelerated as well as nonaccelerated, was the primary motivation that led him to the general theory of relativity.

Principle of Equivalence

Long before there were real spaceships, Einstein could imagine himself in a vehicle far away from gravitational influences. In such a spaceship at rest or in uniform motion relative to the distant stars, he and everything within the ship would float freely; there would be no "up" and no "down." But, when the rocket motors were turned on and the ship accelerated, things would be different; phenomena similar to gravity would be observed. The wall adjacent to the rocket motors would push up against any occupants and become the floor, while the opposite wall would become the ceiling. Occupants in the ship would be able to stand on the floor and even jump up and down. If the acceleration of the spaceship were equal to g, the occupants could well

FIGURE 36.1
Everything is weightless on the inside of a nonaccelerating spaceship far away from gravitational influences.

be convinced the ship was not accelerating but was at rest on the surface of the Earth.

To examine this new "gravity" in an accelerating spaceship, let's consider the consequence of dropping two balls inside the spaceship, one ball of wood and the other of lead. When the balls are released, they continue to move upward side by side with the velocity the ship had at the moment of release. If the ship were moving at *constant velocity* (zero acceleration), the balls would remain suspended in the same place because they and the ship move the same distance in any given time interval. But, because the ship is accelerating, the floor moves upward faster than the balls, with the result that the floor soon catches up with the balls (Figure 36.3). Both balls, regardless of their mass, meet the floor at the same time. Remembering Galileo's demonstration at the Leaning Tower of Pisa, occupants of the ship might be prone to attribute their observations to the force of gravity.

The two interpretations of the falling balls are equally valid, and Einstein incorporated this equivalence, or impossibility of distinguishing between gravitation and acceleration, in the foundation of his general theory of relativity. The **principle of equivalence** states that observations made in an accelerated reference frame are indistinguishable from observations made in a Newtonian gravitational field. This equivalence would be interesting but not revolutionary if it could be applied only to mechanical phenomena, but Einstein went further and stated that the principle holds for all natural phenomena; it holds for optical and all electromagnetic phenomena as well.

FIGURE 36.2
When the spaceship accelerates, an occupant inside feels "gravity."

FIGURE 36.3 Interactive Figure
To an observer inside the accelerating ship, a lead ball and a wood ball appear to fall together when released.

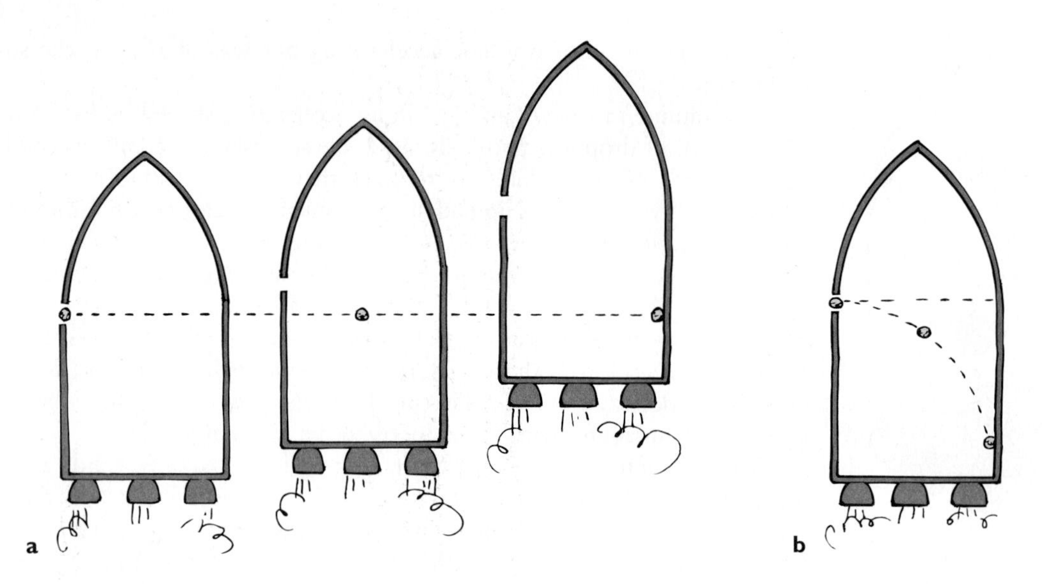

FIGURE 36.4 Interactive Figure

(a) An outside observer sees a horizontally thrown ball travel in a straight line. Because the ship is moving upward while the ball travels horizontally, the ball strikes the wall below a point opposite the window. (b) To an inside observer, the ball bends as if in a gravitational field.

Bending of Light by Gravity

A ball thrown sideways in a stationary spaceship in a gravity-free region will follow a straight-line path relative both to an observer inside the ship and to a stationary observer outside the spaceship. But, if the ship is accelerating, the floor overtakes the ball just as in our previous example. An observer outside the ship still sees a straight-line path, but, to an observer in the accelerating ship, the path is curved; it is a parabola (Figure 36.4). The same holds true for a beam of light.

Imagine that a light ray enters the spaceship horizontally through a side window, passes through a sheet of glass in the middle of the cabin, leaving a visible trace, and then reaches the opposite wall, all in a very short time. The outside observer sees that the light ray enters the window and moves horizontally along a straight line with constant velocity toward the opposite wall. But the spaceship is accelerating upward. During the time it takes for the light to reach the glass sheet, the spaceship moves up some distance, and, during the equal time for the light to continue to the far wall, the spaceship moves up a greater distance. So, to observers in the spaceship, the light has followed a downward curving path (Figure 36.5). In this accelerating frame of reference, the light ray is deflected downward toward the floor, just as the thrown ball in Figure 36.4 is deflected. The curvature of the slow-moving ball is very pronounced; but, if the ball were somehow thrown horizontally across the spaceship cabin at a velocity equal to that of light, its curvature would match the light ray's curvature.

An observer inside the ship feels "gravity" because of the ship's acceleration. The observer is not surprised by the deflection of the thrown ball, but might be quite surprised by the deflection of light. According to the principle of equivalence, if light is deflected by acceleration, it must be deflected by gravity. Yet how

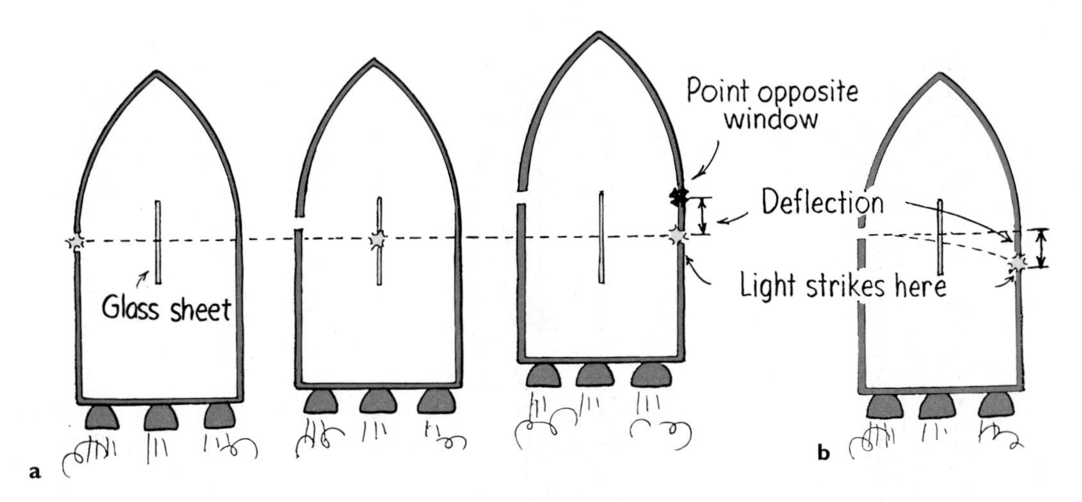

FIGURE 36.5
(a) An outside observer sees light travel horizontally in a straight line, but, like the ball in the previous figure, it strikes the wall slightly below a point opposite the window. (b) To an inside observer, the light bends as if responding to a gravitational field.

FIGURE 36.6
The trajectory of a flashlight beam is identical to the trajectory that a baseball would have if it could be "thrown" at the speed of light. Both paths curve equally in a uniform gravitational field.

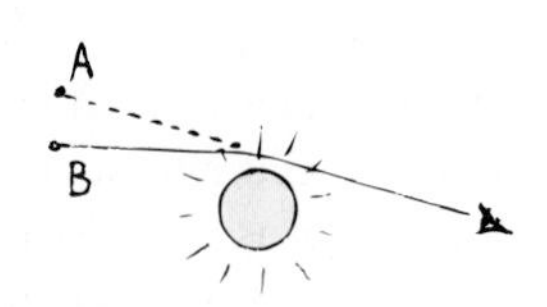

FIGURE 36.7
Starlight bends as it grazes the Sun. Point A shows the apparent position; point B shows the true position.

can gravity bend light? According to Newton's physics, gravitation is an interaction between masses; a moving ball curves because of the interaction between its mass and the mass of Earth. But what of light, which is pure energy and is massless? Einstein's answer was that light may be massless, but it's not "energyless." Gravity pulls on the energy of light because energy is equivalent to mass.

This was Einstein's first answer, before he fully developed the general theory of relativity. Later, he gave a deeper explanation—that light bends when it travels in a spacetime geometry that is bent. We shall see later in this chapter that the presence of mass results in the bending or warping of spacetime. The mass of Earth is too small to appreciably warp the surrounding spacetime, which is practically flat, so any such bending of light in our immediate environment is not ordinarily detected. Close to bodies of mass much greater than Earth's, however, the bending of light is large enough to detect.

Einstein predicted that starlight passing close to the Sun would be deflected by an angle of 1.75 seconds of arc—large enough to be measured. Although stars are not visible when the Sun is in the sky, the deflection of starlight can be observed during an eclipse of the Sun. (Measuring this deflection has become a standard practice at every total eclipse since the first measurements were made during the total eclipse of 1919.) A photograph taken of the darkened sky around the eclipsed Sun reveals the presence of the nearby bright stars. The positions of the stars are compared with those in other photographs of the same area taken at other times in the night with the same telescope. In every instance, the deflection of starlight has supported Einstein's prediction (Figure 36.7).

Light bends in Earth's gravitational field also—but not as much. We don't notice it because the effect is so tiny. For example, in a constant gravitational field of 1 *g*, a beam of horizontally directed light will "fall" a vertical distance of 4.9 meters in 1 second (just as a baseball would), but it will travel a horizontal distance of 300,000 kilometers in that time. Its curve would hardly be noticeable when you're this far from the beginning point. But, if the light traveled 300,000 kilometers in multiple reflections between idealized parallel mirrors,

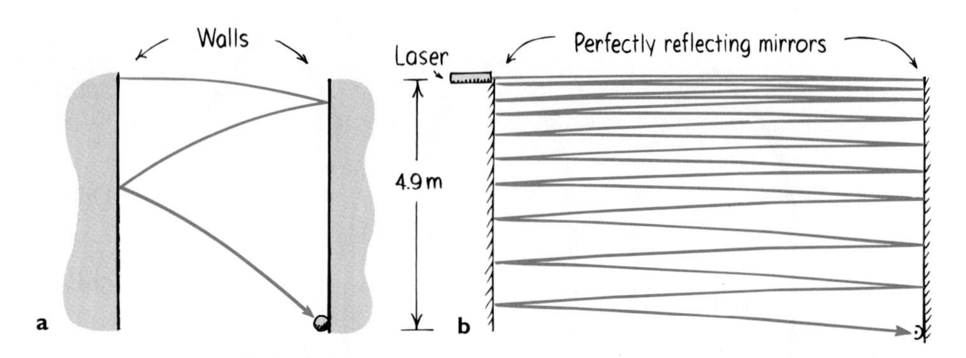

FIGURE 36.8
(a) If a ball is horizontally projected between a vertical pair of parallel walls, it will bounce back and forth and fall a vertical distance of 4.9 m in 1 s. (b) If a horizontal beam of light is directed between a vertical pair of perfectly parallel ideal mirrors in a uniform gravitational field, it will reflect back and forth and fall a vertical distance of 4.9 m in 1 s. The number of back-and-forth reflections is overly simplified in the diagram; if the mirrors were 300 km apart, for example, 1000 reflections would occur in 1 s.

the effect would be quite noticeable (Figure 36.8). (Doing this would make a dandy home project for extra credit—like earning credit for a Ph.D.)

CHECK YOURSELF

1. Whoa! We learned previously that the pull of gravity is an interaction between masses. And we learned that light has no mass. Now we say that light can be bent by gravity. Isn't this a contradiction?
2. Why do we not notice the bending of light in our everyday environment?

Gravity and Time: Gravitational Red Shift

According to Einstein's general theory of relativity, gravitation causes time to slow down. If you move in the direction in which the gravitational force acts—from the top of a skyscraper to the ground floor, for instance, or from the surface of the Earth to the bottom of a well—time will run slower at the point you reach than at the point you left behind. We can understand the slowing of clocks by gravity by applying the principle of equivalence and time dilation to an accelerating frame of reference.

Imagine our accelerating reference frame to be a large rotating disk. Suppose we measure time with three identical clocks, one placed on the disk at its center, a second placed on the rim of the disk, and the third at rest on the ground nearby (Figure 36.9). From the laws of special relativity, we know that the clock attached to the center, since it is not moving with respect to the ground, should run at the same rate as the clock on the ground—but not at the same rate as the clock attached to the rim of the disk. The clock at the rim is in motion with respect to the ground and should therefore be observed to be running more slowly than the

CHECK YOUR ANSWERS

1. There is no contradiction when the mass–energy equivalence is understood. It's true that light has no mass, but it is not "energyless." The fact that gravity deflects light is evidence that gravity pulls on the energy of light. Energy indeed is equivalent to mass!
2. Only because light travels so fast; just as, over a short distance, we do not notice the curved path of a high-speed bullet, we do not notice the curving of a light beam.

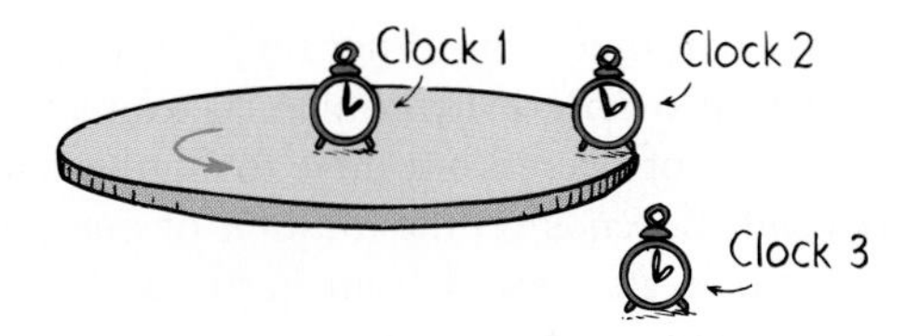

FIGURE 36.9
Clocks 1 and 2 are on an accelerating disk, and clock 3 is at rest in an inertial frame. Clocks 1 and 3 run at the same rate, while clock 2 runs slower. From the point of view of an observer at clock 3, clock 2 runs slow because it is moving. From the point of view of an observer at clock 1, clock 2 runs slow because it is in a stronger centrifugal force field.

ground clock—and therefore more slowly than the clock at the center of the disk. Although the clocks on the disk are attached to the same frame of reference, they do not run synchronously; the outer clock runs slower than the inner clock.

An observer on the rotating disk and an observer at rest on the ground both see the same difference in clock rates between themselves and the clock on the rim. Interpretations of the difference for the two observers are not the same, however. To the observer on the ground, the slower rate of the clock on the rim is due to its motion. But, to an observer on the rotating disk, the disk clocks are not in motion with respect to each other; instead, a centrifugal force acts on the clock at the rim, while no such force acts on the clock at the center. The observer on the disk is likely to conclude that the centrifugal force has something to do with the slowing of time. He notices that, as he moves in the direction of the centrifugal force, outward from the center to the edge of the disk, time is slowed. By applying the principle of equivalence, which says that any effect of acceleration can be duplicated by gravity, we must conclude that, as we move in the direction in which a gravitational force acts, time will also be slowed.

Astrophysics goes beyond describing how the sky looks by explaining how it got to be as it is.

Insights

This slowing down will apply to all "clocks," whether physical, chemical, or biological. An executive working on the ground floor of a tall city skyscraper will age more slowly than her twin sister working on the top floor. The difference is very small, only a few millionths of a second per decade, because, by cosmic standards, the distance is small and the gravitation is weak. For larger differences in gravitation, such as between the surface of the Sun and the surface of the Earth, the differences in time are larger (although still tiny). A clock at the surface of the Sun should run measurably slower than a clock at the surface of the Earth. Years before he completed his general relativity theory, Einstein suggested a way to measure this when he formulated the principle of equivalence in 1907.

All atoms emit light at specific frequencies characteristic of the vibrational rate of electrons within the atom. Every atom is therefore a "clock," and a slowing down of atomic vibration indicates the slowing down of such clocks. An atom on the Sun should emit light of a lower frequency (slower vibration) than light emitted by the same element on Earth. Since red light is at the low-frequency end of the visible spectrum, a lowering of frequency shifts the color toward the red. This effect is called the **gravitational red shift.** The gravitational red shift is observed in light from the Sun, but various disturbing influences prevent accurate measurements of this tiny effect. It wasn't until 1960 that an entirely new technique, using gamma rays from radioactive atoms, permitted incredibly precise and confirming measurements of the gravitational slowing of time between the top and bottom floors of a laboratory building at Harvard University.[1]

[1]In the late 1950s, shortly after Einstein's death, the German physicist Rudolph Mössbauer discovered an important effect in nuclear physics that provides an extremely accurate method of using atomic nuclei as atomic clocks. The *Mössbauer effect,* for which its discoverer was awarded the Nobel Prize, has many practical applications. In late 1959, Robert Pound and Glen Rebka at Harvard University conceived an application that was a test for general relativity and performed the confirming experiment.

FIGURE 36.10
Interactive Figure

If you move from a distant point down to the surface of the Earth, you move in the direction in which the gravitational force acts—toward a location where clocks run more slowly. A clock at the surface of the Earth runs slower than a clock farther away.

So measurements of time depend not only on relative motion, as we learned in the last chapter, but also on gravity. In special relativity, time dilation depends on the *speed* of one frame of reference relative to another one. In general relativity, the gravitational red shift depends on the *location* of one point in a gravitational field relative to another one. As viewed from Earth, a clock will be measured to tick more slowly on the surface of a star than on Earth. If the star shrinks, its surface moves inward to ever-stronger gravity, which causes time on its surface to slow down more and more. We would measure longer intervals between the ticks of the star clock. But, if we made our measurements of the star clock from the star itself, we would notice nothing unusual about the clock's ticking.

Suppose, for example, that an indestructible volunteer stands on the surface of a giant star that begins collapsing. We, as outside observers, will note a progressive slowing of time on the clock of our volunteer as the star surface recedes to regions of stronger gravity. The volunteer himself, however, does not notice any differences in his own time. He is viewing events within his own frame of reference, and he notices nothing unusual. As the collapsing star proceeds toward becoming a black hole and time proceeds normally from the viewpoint of the volunteer, we on the outside perceive time for the volunteer as approaching a complete stop; we see him frozen in time with an infinite duration between the ticks of his clock or the beats of his heart. From our view, his time stops completely. The gravitational red shift, instead of being a tiny effect, is dominating.

We can understand the gravitational red shift from another point of view—in terms of the gravitational force acting on photons. As a photon flies from the surface of a star, it is "retarded" by the star's gravity. It loses energy (but not speed). Since a photon's frequency is proportional to its energy, its frequency decreases as its energy decreases. When we observe the photon, we see that it has lower frequency than that it would have had if it had been emitted by a less-massive source. Its time has been slowed, just like the ticking of a clock is slowed. In the case of a black hole, a photon is unable to escape at all. It loses all its energy and all its frequency in the attempt. Its frequency is gravitationally red shifted to zero, consistent with our observation that the rate at which time passes on a collapsing star approaches zero.

It is important to note the relativistic nature of time both in special relativity and in general relativity. In both theories, there is no way that you can extend the duration of your own existence. Others moving at different speeds or in different gravitational fields may attribute a great longevity to you, but your longevity is seen from *their* frame of reference, never from your own. Changes in time are always attributed to "the other guy."

The Global Positioning System (GPS) must take account of the effect of gravity as well as speed on orbiting atomic clocks. Because of gravity, clocks run faster in orbit. Because of speed, they run slower. The effects vary during each elliptical orbit and they don't cancel. When your GPS unit tells you exactly where you are, thank Einstein.

Insights

CHECK YOURSELF

Will a person at the top of a skyscraper age more than or less than a person at ground level?

CHECK YOUR ANSWER

More—going from the top of the skyscraper to the ground is going in the direction of the gravitational force, so it is going to a place where time runs more slowly.

Gravity and Space: Motion of Mercury

From the special theory of relativity, we know that measurements of space and time undergo transformations when motion is involved. Likewise with the general theory: Measurements of space differ in different gravitational fields—for example, close to and far away from the Sun.

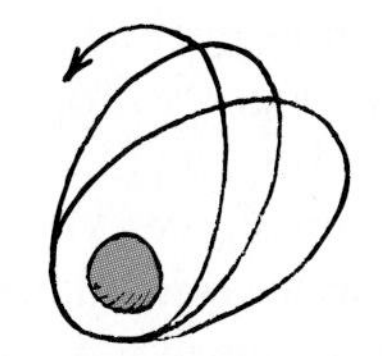

FIGURE 36.11
A precessing elliptical orbit.

Planets orbit the Sun and stars in elliptical orbits and move periodically into regions farther from the Sun and closer to the Sun. Einstein directed his attention to the varying gravitational fields experienced by the planets orbiting the Sun and found that the elliptical orbits of the planets should *precess* (Figure 36.11)—independently of the Newtonian influence of other planets. Near the Sun, where the effect of gravity on time is the greatest, the rate of precession should be the greatest; and far from the Sun, where time is less affected, any deviations from Newtonian mechanics should be virtually unnoticeable.

An incorrect hypothesis, rightly treated, can sometimes produce more new useful information than unguided observation.

Insights

Mercury, the planet nearest the Sun, is in the strongest part of the Sun's gravitational field. If the orbit of any planet exhibits a measurable precession, it should be Mercury, and the fact that the orbit of Mercury does precess—above and beyond effects attributable to the other planets—had been a mystery to astronomers since the early 1800s. Careful measurements showed that Mercury's orbit precesses about 574 seconds of arc per century. Perturbations by the other planets were found to account for all but 43 seconds of arc per century. Even after all known corrections due to possible perturbations by other planets had been applied, the calculations of physicists and astronomers failed to account for the extra 43 seconds of arc. Either Venus was extra massive or a never-discovered other planet (called Vulcan) was pulling on Mercury. And then came the explanation of Einstein, whose general relativity field equations applied to Mercury's orbit predict an extra 43 seconds of arc per century!

The mystery of Mercury's orbit was solved, and a new theory of gravity was recognized. Newton's law of gravitation, which had stood as an unshakable pillar of science for more than two centuries, was found to be a special limiting case of Einstein's more general theory. If the gravitational fields are comparatively weak, Newton's law turns out to be a good approximation of the new law—enough so that Newton's law, which is easier to work with mathematically, is the law that today's scientists use most of the time, except for cases involving enormous gravitational fields.

Gravity, Space, and a New Geometry

We can begin to understand that measurements of space are altered in a gravitational field by again considering the accelerated frame of reference of our rotating disk. Suppose that we measure the circumference of the outer rim with a measuring stick. Recall the Lorentz contraction from special relativity: The measuring stick will appear contracted to any observer not moving along with the stick, while the dimensions of an identical measuring stick moving much more slowly near the center will be nearly unaffected (Figure 36.12). All distance measurements along a *radius* of the rotating disk should be completely unaffected by motion, because motion is perpendicular to the radius. Since only distance measurements parallel to and around the circumference are affected,

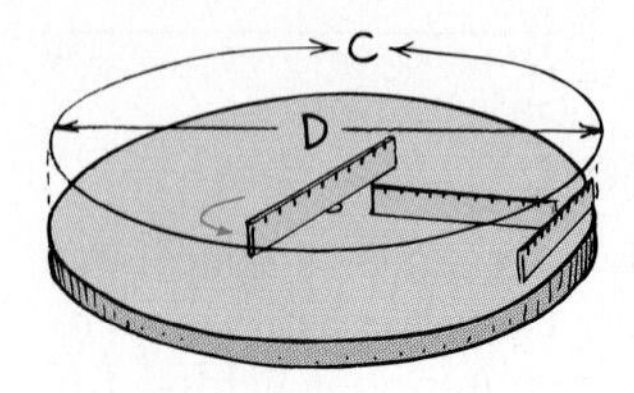

FIGURE 36.12
A measuring stick along the edge of the rotating disk appears contracted, while a measuring stick farther in and moving more slowly is not contracted as much. A measuring stick along a radius is not contracted at all. When the disk is not rotating, $C/D = \pi$; but, when the disk is rotating, C/D is not equal to π and Euclidean geometry is no longer valid. Likewise in a gravitational field.

the ratio of circumference to diameter when the disk is rotating is no longer the fixed constant π (3.14159 . . .) but is a variable depending on angular speed and the diameter of the disk. According to the principle of equivalence, the rotating disk is equivalent to a stationary disk with a strong gravitational field near its edge and a progressively weaker gravitational field toward its center. Measurements of distance, then, will depend on the strength of gravitational field (or, more exactly, for relativity buffs, on gravitational potential), even if no relative motion is involved. Gravity causes space to be non-Euclidean; the laws of Euclidean geometry taught in high school are no longer valid when applied to objects in the presence of strong gravitational fields.

The familiar rules of Euclidean geometry pertain to various figures you can draw on a flat surface. The ratio of the circumference of a circle to its diameter is equal to π; all the angles in a triangle add up to 180°; the shortest distance between two points is a straight line. The rules of Euclidean geometry are valid in flat space; but, if you draw these figures on a curved surface, like a sphere or a saddle-shaped object, the Euclidean rules no longer hold (Figure 36.13). If you measure the sum of the angles for a triangle in space, you call the space flat if the sum is equal to 180°, spherelike or positively curved if the sum is larger than 180°, and saddlelike or negatively curved if it is less than 180°.

Of course, the lines forming the triangles in Figure 36.13 are not all "straight" from a three-dimensional view, but they are the "straightest," or *shortest,* distances between two points if we are confined to the curved surface. These lines of shortest distance are called *geodesic lines* or simply **geodesics.**

The path of a light beam follows a geodesic. Suppose three experimenters on Earth, Venus, and Mars measure the angles of a triangle formed by light beams traveling between these three planets. The light beams bend when passing the Sun, resulting in the sum of the three angles being larger than 180° (Figure 36.14). So the space around the Sun is positively curved. The planets that orbit the Sun travel along four-dimensional geodesics in this positively curved spacetime. Freely falling objects, satellites, and light rays all travel along geodesics in four-dimensional spacetime.

The standard model of cosmology assumes a flat universe dominated by dark matter and dark energy that formed by rapid inflation from its hot, dense origins.

Insights

"Small" parts of the universe are certainly curved. What about the universe as a whole? Recent study of the low-temperature radiation in space that is a remnant of the Big Bang suggests that the universe is flat. If it were open-ended like the saddle in Figure 36.13*c*, it would extend forever and beams of light that started out parallel would diverge. If it were closed like the spherical surface in Figure 36.13*b*, beams of light that started out parallel would eventually cross

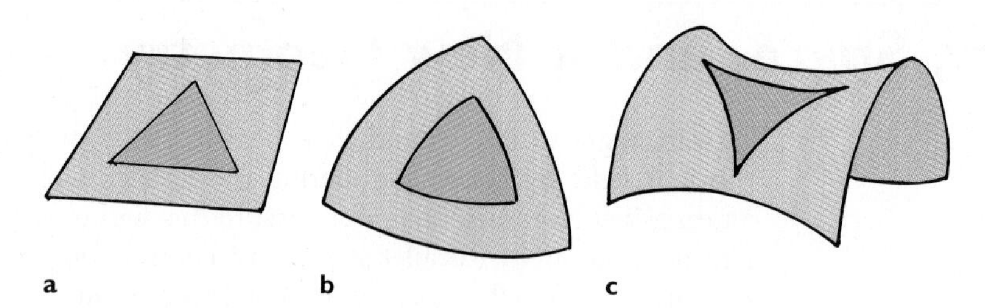

FIGURE 36.13
The sum of the angles of a triangle depends on which kind of surface the triangle is drawn on. (a) On a flat surface, the sum is 180°. (b) On a spherical surface, the sum is greater than 180°. (c) On a saddle-shaped surface, the sum is less than 180°.

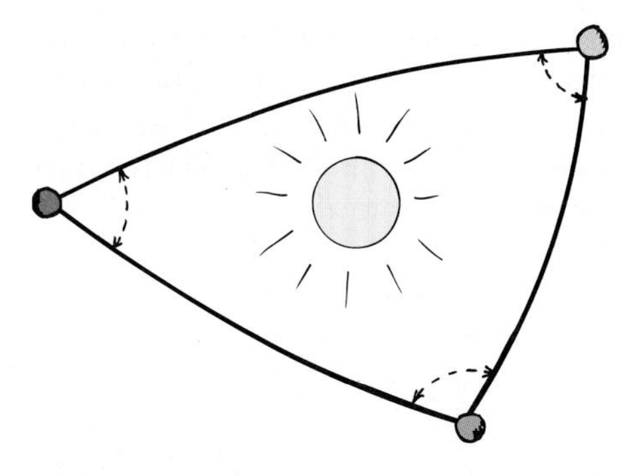

FIGURE 36.14
The light rays joining the three planets form a triangle. Since light passing near the Sun bends, the sum of the angles of the resulting triangle is greater than 180°.

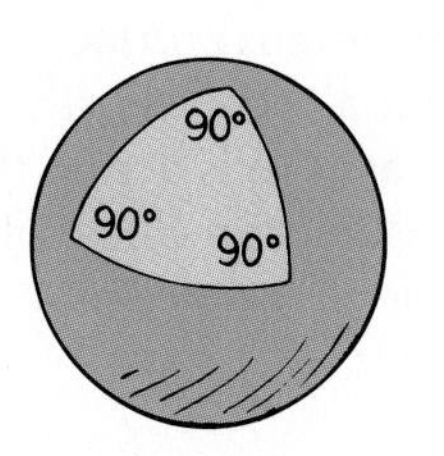

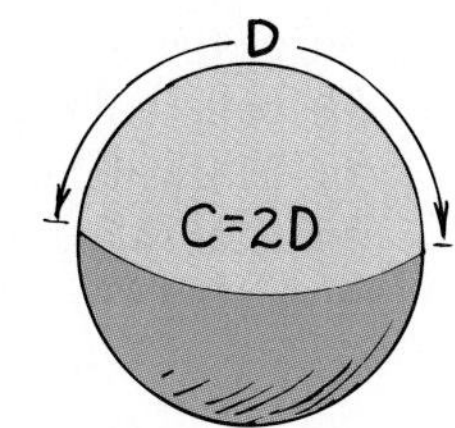

FIGURE 36.15
The geometry of the curved surface of the Earth differs from the Euclidean geometry of flat space. Note, in the globe on the left, that the sum of the angles for an equilateral triangle in which each side equals 1/4 the Earth's circumference is clearly greater than 180°. The globe on the right shows the Earth's circumference is only twice its diameter instead of 3.14 times its diameter. Euclidean geometry is also invalid in curved space.

and circle back to their starting point. In such a universe, if you could look infinitely into space through an ideal telescope, you would see the back of your own head (after waiting patiently for enough billions of years)! In our actual flat universe, parallel beams of light remain parallel and will never return.

General relativity calls for a new geometry: Rather than space simply being a region of nothingness, space is a flexible medium that can bend and twist. How it bends and twists describes a gravitational field. General relativity is a geometry of curved four-dimensional spacetime.[2] The mathematics of this geometry is too formidable to present here. The essence, however, is that the presence of mass produces the curvature, or *warping*, of spacetime. By the same token, a curvature of spacetime reveals itself as mass. Instead of visualizing gravitational forces between masses, we abandon altogether the notion of force and instead think of masses responding in their motion to the warping of the spacetime they inhabit. It is the bumps, depressions, and warpings of geometrical spacetime that *are* the phenomena of gravity.

We cannot visualize the four-dimensional bumps and depressions in spacetime because we are three-dimensional beings. We can get a glimpse of this warping by considering a simplified analogy in two dimensions: a heavy ball resting on the middle of a waterbed. The more massive the ball, the greater it dents or

One of general relativity's predictions was a subtle twisting of spacetime around a massive spinning object. A test for this "frame-dragging" effect would be predictable tiny changes in the orientations of satellite orbits and orbiting gyroscopes. Researchers in 2004 found such confirming evidence.

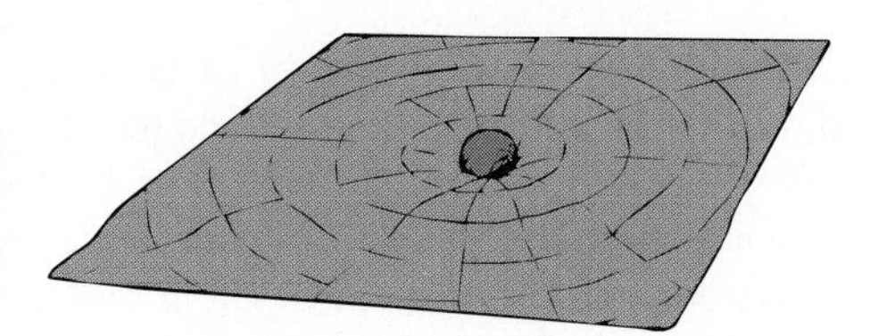

FIGURE 36.16
A two-dimensional analogy of four-dimensional warped spacetime. Spacetime near a star is curved in a way similar to the surface of a waterbed when a heavy ball rests on it.

[2]Don't be discouraged if you cannot visualize four-dimensional spacetime. Einstein himself often told his friends, "Don't try. I can't do it either." Perhaps we are not too different from the great thinkers of Galileo's time who couldn't think of a moving Earth!

warps the two-dimensional surface. A marble rolled across the bed, but far from the ball, will roll in a relatively straight-line path, whereas a marble rolled near the ball will curve as it rolls across the indented surface. If the curve closes upon itself, its shape resembles an ellipse. The planets that orbit the Sun similarly travel along four-dimensional geodesics in the warped spacetime about the Sun.

Gravitational Waves

Every object has mass and therefore warps the surrounding spacetime. When an object undergoes a change in motion, the surrounding warp moves in order to readjust to the new position. These readjustments produce ripples in the overall geometry of spacetime. This is similar to moving a ball that rests on the surface of a waterbed. A disturbance ripples across the waterbed surface in waves; if we move a more massive ball, then we get a greater disturbance and the production of even stronger waves. Similarly for spacetime in the universe. Similar ripples travel outward from a gravitational source at the speed of light and are **gravitational waves.**

Space is stretching out, carrying the galaxies with it. Visible light in the early universe has stretched out now to be relatively long-wavelength microwave radiation.

Insights

Any accelerating object produces a gravitational wave. In general, the more massive the object and the greater its acceleration, the stronger the resulting gravitational wave. But even the strongest waves produced by ordinary astronomical events are extremely weak—the weakest known in nature. For example, the gravitational waves emitted by a vibrating electric charge are a trillion trillion trillion times weaker than the electromagnetic waves emitted by the same charge. Detecting gravitational waves is enormously difficult, and no confirmed detection has occurred to date. Recently completed wave detectors are expected to detect gravitational waves from supernovae, which may radiate away as much as 0.1% of their mass as gravitational waves, and perhaps from even more cataclysmic events, such as colliding black holes.

As weak as they are, gravitational waves are everywhere. Shake your hand back and forth: You have just produced a gravitational wave. It is not very strong, but it exists.

Newtonian and Einsteinian Gravitation

When Einstein formulated his new theory of gravitation, he realized that, if his theory is valid, his field equations must reduce to Newtonian equations for gravitation in the weak-field limit. He showed that Newton's law of gravitation is a special case of the broader theory of relativity. Newton's law of gravitation is still an accurate description of most of the interactions between bodies in the solar system and beyond. From Newton's law, one can calculate the orbits of comets and asteroids and even predict the existence of undiscovered planets. Even today, when computing the trajectories of space probes to the Moon and planets, only ordinary Newtonian theory is used. This is because the gravitational field of these bodies is very weak, and, from the viewpoint of general relativity, the surrounding spacetime is essentially flat. But for regions of more intense gravitation, where spacetime is more appreciably curved, Newtonian theory cannot adequately account for various phenomena—such as the precession of Mercury's orbit close to the Sun and, in the case of stronger fields, the gravitational red shift and other apparent distortions in measurements of space and time. These distortions reach their limit in the case of a star that collapses to a black hole, where spacetime completely folds over on itself. Only Einsteinian gravitation reaches into this domain.

We saw, in Chapter 32, that Newtonian physics is linked at one end with quantum theory, whose domain is the very light and very small—tiny particles and atoms. And now we have seen that Newtonian physics is linked at the other end with relativity theory, whose domain is the very massive and very large.

We do not see the world the way the ancient Egyptians, Greeks, or Chinese did. It is unlikely that people in the future will see the universe as we do. Our view of the universe may be quite limited, and perhaps filled with misconceptions, but it is most likely clearer than the views of others before us. Our view today stems from the findings of Copernicus, Galileo, Newton, and, more recently, Einstein—findings that were often opposed on the grounds that they diminished the importance of humans in the universe. In the past, being important meant having risen above nature—being apart from nature. We have expanded our vision since then by enormous effort, painstaking observation, and an unrelenting desire to comprehend our surroundings. Seen from today's understanding of the universe, we find our importance in being very much a part of nature, not apart from it. We are the part of nature that is becoming more and more conscious of itself.

If a learner's first course in physics is enjoyable, the rigor of a second course will be welcome and meaningful.

Insights

Summary of Terms

General theory of relativity The second of Einstein's theories of relativity, which treats the effects of gravity on space and time.

Principle of equivalence Because observations made in an accelerated frame of reference are indistinguishable from observations made in a gravitational field, any effect produced by gravity can be duplicated by accelerating a frame of reference.

Gravitational red shift The lengthening of the waves of electromagnetic radiation escaping from a massive object.

Geodesic The shortest path between two points in various models of space.

Gravitational wave A gravitational disturbance generated by an accelerating mass that propagates through spacetime.

Suggested Reading

Einstein, Albert. *Relativity: The Special and General Theory*. New York: Crown, 1961. (Originally published in 1916.)

Hawking, Stephen W. *A Brief History of Time: From the Big Bang to Black Holes*. New York: Bantam Books, 1988.

Thorne, Kip S., *Black Holes and Time Warps: Einstein's Outrageous Legacy*. New York: Norton, 1994. An expert's readable account of black holes, neutron stars, gravitational waves, time machines, and more.

Review Questions

1. What is the principal difference between the theory of special relativity and the theory of general relativity?

Principle of Equivalence

2. In a spaceship accelerating at g, far from Earth's gravity, how does the motion of a dropped ball compare with dropping one on Earth?
3. Exactly what is *equivalent* in the principle of equivalence?

Bending of Light by Gravity

4. Compare the bending of the paths of baseballs and photons by a gravitational field.
5. Why must the Sun be eclipsed to measure the deflection of starlight passing near the Sun?

Gravity and Time: Gravitational Red Shift

6. What is the effect of strong gravitation on measurements of time?
7. Which runs slower, a clock at the top of the Sears Tower in Chicago or a clock on the shore of Lake Michigan?
8. How does the frequency of a particular spectral line observed in sunlight compare with the frequency of that line observed from a source on Earth?
9. If we view events occurring on a star that is collapsing to become a black hole, do we see time speeded up or slowed down?

Gravity and Space: Motion of Mercury

10. Of all the planets, why is Mercury the best candidate for finding evidence of the relationship between gravitation and space?
11. In what kind of gravitational field are Newton's laws valid?

Gravity, Space, and a New Geometry

12. A measuring stick placed along the circumference of a rotating disk will appear contracted, but, if it is oriented along a radius, it will not. Explain.
13. The ratio of circumference to diameter for measured circles on a disk equals π when the disk is at rest, but not when the disk is rotating. Explain.
14. What effect does mass have on spacetime?

Gravitational Waves

15. What occurs in the surrounding space when a massive object undergoes a change in its motion?
16. A star 10 light-years away explodes and produces gravitational waves. How long will it take these waves to reach the Earth?
17. Why are gravitational waves so difficult to detect?

Newtonian and Einsteinian Gravitation

18. Does Einstein's theory of gravitation invalidate Newton's theory of gravitation? Explain.
19. Is Newtonian physics adequate to get a rocket to the Moon?
20. How does Newtonian physics link with quantum theory and relativity theory?

Exercises

1. What is different about the reference frames that apply to special relativity and to general relativity?
2. An astronaut awakes in her closed capsule, which actually sits on the Moon. Can she tell whether her weight is the result of gravitation or of accelerated motion? Explain.
3. You wake up at night in your berth on a train to find yourself being "pulled" to one side of the train. You naturally assume that the train is rounding a curve, but you are puzzled that you don't hear any sounds of motion. Offer another possible explanation that involves only gravity, not acceleration of your frame of reference.
4. Since gravity can duplicate the effects of acceleration, it can also balance the effects of acceleration. Cite how and when an astronaut can experience no net force (as measured by a scale) because of the cancelling effects of gravity and acceleration.
5. An astronaut is provided a "gravity" when the ship's engines are activated to accelerate the ship. This requires the use of fuel. Is there a way to accelerate and provide "gravity" without the sustained use of fuel? Explain, perhaps using ideas from Chapter 8.
6. In a spaceship far from the reaches of gravity, under what conditions would you feel as if the spaceship were stationary on Earth's surface?
7. In his famous novel *Journey to the Moon,* Jules Verne stated that occupants in a spaceship would shift their orientation from up to down when the ship crossed the point where the Moon's gravitation became greater than the Earth's. Is this correct? Defend your answer.
8. What happens to the separation distance between two people if they both walk north at the same rate from two locations on the Earth's equator? And just for fun, where in the world is a step in every direction a step south?
9. We readily note the bending of light by reflection and refraction, but why are we not aware of the bending of light by gravity?
10. Light *does* bend in a gravitational field. Why is this bending not taken into consideration by surveyors who use laser beams as straight lines?
11. Why do we say that light travels in straight lines? Is it strictly accurate to say that a laser beam provides a perfectly straight line for purposes of surveying? Explain.
12. Your friend says that light passing the Sun is bent whether or not Earth experiences a solar eclipse. Do you agree or disagree, and why?
13. In 2004, when Mercury passed between the Sun and the Earth, light was not appreciably bent as it passed Mercury. Why?

14. At the end of 1 s, a horizontally fired bullet drops a vertical distance of 4.9 m from its otherwise straight-line path in a gravitational field of 1 *g*. By what distance would a beam of light drop from its otherwise straight-line path if it traveled in a uniform field of 1 *g* for 1 s? For 2 s?

15. Light changes its energy when it "falls" in a gravitational field. This change in energy is not evidenced by a change in speed, however. What is the evidence for this change in energy?

16. Would we notice a slowing down or speeding up of a clock if we carried it to the bottom of a very deep well?

17. If we witnessed events occurring on the Moon, where gravitation is weaker than on Earth, would we expect to see a gravitational red shift or a gravitational blue shift? Explain.

18. Armed with highly sensitive detection equipment, you are in the front of a railroad car that is accelerating forward. Your friend at the rear of the car shines green light toward you. Do you find the light to be red-shifted (lowered in frequency), blue-shifted (increased in frequency), or neither? Explain. (*Hint:* Think in terms of the principle of equivalence. What is your accelerating railroad car equivalent to?)

19. Why will the gravitational field intensity increase on the surface of a shrinking star?

20. Will a clock at the equator run slightly faster or slightly slower than an identical clock at one of the Earth's poles?

21. Do you age faster at the top of a mountain or at sea level?

22. Splitting hairs, should a person who worries about growing old live at the top or at the bottom of a tall apartment building?

23. Time slows in a strong gravitational field. Would time slow in the artificial gravity produced in a rotating space habitat? Why or why not?

24. Prudence and Charity are twins raised at the center of a rotating kingdom. Charity goes to live at the edge of the kingdom for a time and then returns home. Which twin is older when they reunite? (Ignore any time-dilation effects associated with travel to and from the edge.)

25. Splitting hairs, if you shine a beam of colored light to a friend above in a high tower, will the color of light your friend receives be the same color you send? Explain.

26. Is light emitted from the surface of a massive star red-shifted or blue-shifted by gravity?

27. From our frame of reference on Earth, objects slow to a stop as they approach black holes in space because time gets infinitely stretched by the strong gravity near the black hole. If astronauts accidentally falling into a black hole tried to signal back to Earth by flashing a light, what kind of "telescope" would we need to detect the signals?

28. Would an astronaut falling into a black hole see the surrounding universe red-shifted or blue-shifted?

29. How can we "observe" a black hole if neither matter nor radiation can escape from it?

30. Should it be possible in principle for a photon to circle a star?

31. Why does the gravitational attraction between the Sun and Mercury vary? Would it vary if the orbit of Mercury were perfectly circular?

32. Your friend whimsically says that, at the North Pole, a step in any direction is a step south. Do you agree?

33. In the astronomical triangle shown in Figure 36.14, with sides defined by light paths, the sum of the interior angles is more than 180°. Is there any astronomical triangle whose interior angles sum to less than 180°?

34. Do binary stars (double-star systems that orbit about a common center of mass) radiate gravitational waves? Why or why not?

35. Gravitational waves are difficult to detect. Is this due to having long wavelengths or short ones?

36. Based on what you know about the emission and absorption of electromagnetic waves, suggest how gravitational waves are emitted and how they are absorbed. (Scientists seeking to detect gravitational waves must arrange for them to be absorbed.)

37. Comparing Einstein's and Newton's theories of gravitation, how can the correspondence principle be applied?

38. Current findings suggest that the universe is flat. What are the implications of this finding?

39. Make up a multiple-choice question to check a classmate's understanding of the principle of equivalence.

40. Make up a multiple-choice question to check a classmate's understanding of the effect of gravity on time.

Epilogue

I hope you've enjoyed *Conceptual Physics* and will value your knowledge of physics as a worthwhile component of your general education. Viewing physics as a study of the rules of nature contributes to your sense of wonder and will enhance the way you see the physical world—knowing that so much in nature is connected, with seemingly diverse phenomena often following the same basic rules. How intriguing that the rules governing a falling apple also apply to a space station orbiting the Earth, that a sky's redness at sunset is connected to its blueness at midday, that the rules discovered by Faraday and Maxwell show how electricity and magnetism connect to become light.

The value of science is more than its applications to fast cars, DVD players, computers, and other products. Its greatest value lies in its methods of understanding and investigating nature—that hypotheses are framed so that they are capable of being disproved, and experiments are designed so that their results can be reproduced by others. Science is more than a body of knowledge; it is a way of thinking.

And then there are the purveyors of junk science who dress up their claims in the language of science but intentionally ignore its methods. The several boxes on pseudoscience throughout this book are an attempt to expose this. Being able to distinguish between scientific experiments and unsupported claims is particularly important because so much misinformation and hype are used by charlatans to peddle their bogus wares and bogus ideas. Pseudoscience cheapens science. Its purveyors wish to topple the scientific way of viewing the world and disavow skeptical thinking.

Skeptical thinking, in addition to sharpening common sense, is an essential ingredient in formulating a hypothesis that requires a test for wrongness: If I'm wrong, how would I know? This key question can accompany any important idea, scientific or otherwise. When it is applied to social, political, and religious questions, you become stronger for it. Socially, you see others' points of view more clearly. Politically, you see all social movements as experiments. Religiously, you see that the supposed conflicts between science and religion stem mainly from misapplications of one or both of them. Properly applied, science is not only compatible with spirituality but can be a profound source of spirituality.

Contemplating the immensity of the universe and the geologic time scale of our planet evokes a soaring feeling that is surely spiritual. We've learned that four hundred million years ago, long before mammals, there were fish; then came amphibians and then reptiles. In the struggle of species survival, trillions upon trillions of life forms passed their genetic traits on to their offspring, sometimes here and there making adaptive changes. After a long and prodigious ascent, humans emerged. We should not ignore the sacrifices of the innumerable lives that brought us to where we are, and we all should celebrate this long and astounding journey of life—for we are the benefactors.

Science offers a modern-day means of establishing our origins, how we can survive, and even who we can become. We're at a present vantage point where science can proceed from "how" to "why"—ironically at a time when the potential for world calamity has never been greater. Overpopulation denial, energy greed, and other socioeconomic and political problems beset our age. Yet science provides us with the physical and intellectual tools to improve our lives and our relationships with one another and our environment. Our hope lies with those whose open scientific minds understand and can sensibly address the global issues that threaten our survival. Earth is the only home we all share, and it deserves our utmost care. Caring, knowledgeable people applying the methods of science are humanity's best hope.

APPENDIX A

Systems of Measurement

Two major systems of measurement prevail in the world today: the *United States Customary System* (USCS, formerly called the British system of units), used in the United States of America and in Burma, and the *Système International* (SI) (known also as the international system and as the metric system), used everywhere else. Each system has its own standards of length, mass, and time. The units of length, mass, and time are sometimes called the *fundamental units* because, once they are selected, other quantities can be measured in terms of them.

United States Customary System

Based on the British Imperial System, the USCS is familiar to everyone in the United States. It uses the foot as the unit of length, the pound as the unit of weight or force, and the second as the unit of time. The USCS is presently being replaced by the international system—rapidly in science and technology (all 1988 Department of Defense contracts) and some sports (track and swimming), but so slowly in other areas and in some specialties that it seems that the change may never come. For example, we will continue to buy seats on the 50-yard line. Camera film is measured in millimeters, but computer disks are measured in inches.

For measuring time, there is no difference between the two systems except that in pure SI the only unit is the *second* (s, *not* sec) with prefixes; but, in general, minute, hour, day, year, and so on, with two or more lettered abbreviations (h, *not* hr), are accepted in the USCS.

Système International

During the 1960 International Conference on Weights and Measures held in Paris, the SI units were defined and given status. Table A.1 shows SI units and their symbols. SI is based on the *metric system,* originated by French scientists after the French revolution in 1791. The orderliness of this system makes it useful for scientific work, and it is used by scientists all over the world. The metric system branches into two systems of units. In one of these, the unit of length is the meter, the unit of mass is the kilogram, and the unit of time is the second. This is called the *meter-kilogram-second* (mks) system and is preferred in physics. The other branch is the *centimeter-gram-second* (cgs) system, which, because of its smaller values, is favored in chemistry. The cgs and mks units are related to each other as follows: 100 centimeters equal 1 meter; 1000 grams equal 1 kilogram. Table A.2 shows several units of length related to each other.

TABLE A.1
SI Units

Quantity	*Unit*	*Symbol*
Length	meter	m
Mass	kilogram	kg
Time	second	s
Force	newton	N
Energy	joule	J
Current	ampere	A
Temperature	kelvin	K

TABLE A.2
Table Conversions Between Different Units of Length

Unit of Length	*Kilometer*	*Meter*	*Centimeter*	*Inch*	*Foot*	*Mile*
1 kilometer	= 1	1000	100,000	39,370	3280.84	0.62140
1 meter	= 0.00100	1	100	39.370	3.28084	6.21×10^{-4}
1 centimeter	= 1.0×10^{-5}	0.0100	1	0.39370	0.032808	6.21×10^{-6}
1 inch	= 2.54×10^{-5}	0.02540	2.5400	1	0.08333	1.58×10^{-5}
1 foot	= 3.05×10^{-4}	0.30480	30.480	12	1	1.89×10^{-4}
1 mile	= 1.60934	1609.34	160,934	63,360	5280	1

One major advantage of a metric system is that it uses the decimal system, in which all units are related to smaller or larger units by dividing or multiplying by 10. The prefixes shown in Table A.3 are commonly used to show the relationship among units.

TABLE A.3
Some Prefixes

Prefix	*Definition*
micro-	One millionth: a microsecond is 1 millionth of a second
milli-	One thousandth: a milligram is 1 thousandth of a gram
centi-	One hundredth: a centimeter is 1 hundredth of a meter
kilo-	One thousand: a kilogram is 1000 grams
mega-	One million: a megahertz is 1 million hertz

Meter

The standard of length of the metric system orginally was defined in terms of the distance from the north pole to the equator. This distance was thought at the time to be close to 10,000 kilometers. One ten-millionth of this, the meter, was carefully determined and marked off by means of scratches on a bar of platinum-iridium alloy. This bar is kept at the International Bureau of Weights and Measures in France. The standard meter in France has since been calibrated in terms of the wavelength of light—it is 1,650,763.73 times the wavelength of orange light emitted by the atoms of the gas krypton-86. The meter is now defined as being the length of the path traveled by light in a vacuum during a time interval of 1/299,792,458 of a second.

FIGURE A.1
The standard kilogram.

Kilogram

The standard unit of mass, the kilogram, is a block of platinum-iridium alloy (platinum 90%, iridium 10%), also preserved at the International Bureau of Weights and Measures located in France (Figure A.1). The kilogram equals 1000 grams. A gram is the mass of 1 cubic centimeter (cc) of water at a temperature of 4° Celsius. (The standard pound is defined in terms of the standard kilogram; the mass of an object that weighs 1 pound is equal to 0.4536 kilogram.) The kilogram is the only unit of measure still defined by a physical object. Researchers are presently trying to find a way to define the kilogram in terms of constants of nature.

Second

The official unit of time for both the USCS and the SI is the second. Until 1956, it was defined in terms of the mean solar day, which was divided into 24 hours. Each hour was divided into 60 minutes and each minute into 60 seconds. Thus, there were 86,400 seconds per day, and the second was defined as 1/86,400 of the mean solar day. This proved unsatisfactory because the rate of rotation of the Earth is gradually becoming slower. In 1956, the mean solar day of the year 1900 was chosen as the standard on which to base the second. In 1964, the second was officially defined as the duration of 9,192,631,770 periods of radiation corresponding to the transition between the two hyperfine levels of the ground state of the cesium-133 atom.

Newton

One newton is the force required to accelerate 1 kilogram at 1 meter per second per second. This unit is named after Sir Isaac Newton.

Joule

One joule is equal to the amount of work done by a force of 1 newton acting over a distance of 1 meter. In 1948, the joule was adopted as the unit of energy by the International Conference on Weights and Measures. Therefore, the specific heat of water at 15°C is now given as 4185.5 joules per kilogram Celsius degree. This figure is always associated with the mechanical equivalent of heat—4.1855 joules per calorie.

Ampere

The ampere is defined as the intensity of the constant electric current that, when maintained in two parallel conductors of infinite length and negligible cross section and placed 1 meter apart in a vacuum, would produce between them a force equal to 2×10^{-7} newton per meter length. In our treatment of electric current in this text, we have used the not-so-official but easier-to-comprehend definition of the ampere as being the rate of flow of 1 coulomb of charge per second, where 1 coulomb is the charge of 6.25×10^{18} electrons.

Kelvin

The fundamental unit of temperature is named after the scientist William Thomson, First Baron Kelvin. The kelvin is defined to be 1/273.15 the thermodynamic temperature of the triple point of water (the fixed point at which ice, liquid water, and water vapor coexist in equilibrium). This definition was adopted in 1968, when it was decided to change the name *degree Kelvin* (°K) to *kelvin* (K). The temperature of melting ice at atmospheric pressure is 273.15 K. The temperature at which the vapor pressure of pure water is equal to standard atmospheric pressure is 373.15 K (the temperature of boiling water at standard atmospheric pressure).

Area

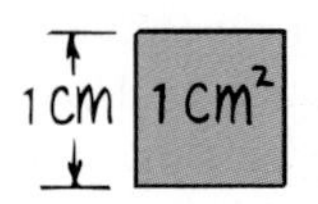

FIGURE A.2
Unit square.

The unit of area is a square that has a standard unit of length as a side. In the USCS, it is a square with sides that are each 1 foot in length, called 1 square foot and written 1 ft^2. In the international system, it is a square with sides that are 1 meter in length, which makes a unit of area of 1 m^2. In the cgs system it is 1 cm^2. The area of a given surface is specified by the number of square feet, square meters, or square centimeters that would fit into it. The area of a rectangle equals the base times the height. The area of a circle is equal to πr^2, where $\pi \approx 3.14$ and r is the radius of the circle. Formulas for the surface areas of other objects can be found in geometry textbooks.

Volume

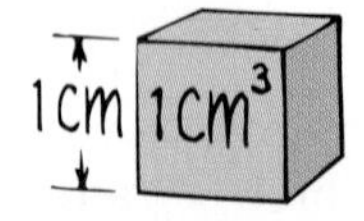

FIGURE A.3
Unit volume.

The volume of an object refers to the space it occupies. The unit of volume is the space taken up by a cube that has a standard unit of length for its edge. In the USCS, one unit of volume is the space occupied by a cube 1 foot on an edge and is called 1 cubic foot, written 1 ft^3. In the metric system, it is the space occupied by a cube with sides of 1 meter (SI) or 1 centimeter (cgs). It is written 1 m^3 or 1 cm^3 (or cc). The volume of a given space is specified by the number of cubic feet, cubic meters, or cubic centimeters that will fill it.

In the USCS, volumes can also be measured in fluid ounces, liquid pints, dry pints, liquid quarts, dry quarts, gallons, pecks, bushels, and cubic inches, as well as in cubic feet. There are 1728 ($12 \times 12 \times 12$) cubic inches in 1 ft^3. A U.S. gallon is a volume of 231 in^3. Four quarts equal 1 gallon. In the SI, volumes are also measured in liters. A liter is equal to 1000 cm^3.

Scientific Notation

It is convenient to use a mathematical abbreviation for large and small numbers. The number 50,000,000 can be obtained by multiplying 5 by 10, and again by 10, and again by 10, and so on until 10 has been used as a multiplier seven times. The short way of showing this is to write the number 5×10^7. The number 0.0005 can be obtained from 5 by using 10 as a divisor four times. The short way of showing this is to write 5×10^{-4} for 0.0005. Thus, 3×10^5 means $3 \times 10 \times 10 \times 10 \times 10 \times 10$, or 300,000; and 6×10^{-3} means $6/(10 \times 10 \times 10)$, or 0.006. Numbers expressed in this shorthand manner are said to be in scientific notation. See the inside back cover.

APPENDIX B

More About Motion

When we describe the motion of something, we say how it moves relative to something else (Chapter 3). In other words, motion requires a reference frame (an observer, origin, and axes). We are free to choose this frame's location and to have it moving relative to another frame. When our frame of motion has zero acceleration, it is called an *inertial frame.* In an inertial frame, force causes an object to accelerate in accord with Newton's laws. When our frame of reference is accelerated, we observe fictitious forces and motions (Chapter 8). Observations from a carousel, for example, are different when it is rotating and when it is at rest. Our description of motion and force depends on our "point of view."

We distinguish between *speed* and *velocity* (Chapter 3). Speed is how fast something moves, or the time rate of change of position (excluding direction): a *scalar* quantity. Velocity includes direction of motion: a *vector* quantity whose magnitude is speed. Objects moving at constant velocity move the same distance in the same time in the same direction.

Another distinction between speed and velocity has to do with the difference between distance and net distance, or *displacement.* Speed is *distance per duration,* while velocity is *displacement per duration.* Displacement differs from distance. For example, a commuter who travels 10 kilometers to work and back travels 20 kilometers, but has "gone" nowhere. The distance traveled is 20 kilometers and the displacement is zero. Although the instantaneous speed and instantaneous velocity have the same value at the same instant, the average speed and average velocity can be very different. The average speed of this commuter's round trip is 20 kilometers divided by the total commute time—a value greater than zero. But the average velocity is zero. In science, displacement is often more important than distance. (To avoid information overload, we have not treated this distinction in the text.)

Acceleration is the rate at which velocity changes. This can be a change in speed only, a change in direction only, or both. Negative acceleration is often called *deceleration.*

In Newtonian space and time, space has three dimensions—length, width, and height—each with two directions. We can go, stop, and return in any of them. Time has one dimension, with two directions—past and future. We cannot stop or return, only go. In Einsteinian space-time, these four dimensions merge (Chapter 35).

Computing Velocity and Distance Traveled on an Inclined Plane

Recall, from Chapter 2, Galileo's experiments with inclined planes. Consider a plane tilted such that the speed of a rolling ball increases at the rate of 2 meters per second each second—an acceleration of 2 m/s^2. At the instant it starts moving, its velocity is zero; and 1 second later, it is rolling at 2 m/s; at the end of the next second, 4 m/s; at the end of the next second, 6 m/s; and so on. The velocity of the ball at any instant is simply

$$\text{Velocity} = \text{acceleration} \times \text{time}$$

or, in shorthand notation,

$$v = at$$

(It is customary to omit the multiplication sign, ×, when expressing relationships in mathematical form. When two symbols are written together, such as the *at* in this case, it is understood that they are multiplied.)

How fast the ball rolls is one thing; how *far* it rolls is another. To understand the relationship between acceleration and distance traveled, we must first investigate the relationship between instantaneous velocity and *average velocity*. If the ball shown in Figure B.1 starts from rest, it will roll a distance of 1 meter in the first second. Question: What will be its average speed? The answer is 1 m/s (because it covered 1 meter in the interval of 1 second). But we have seen that the *instantaneous velocity* at the end of the first second is 2 m/s. Since the acceleration is uniform, the average in any time interval is found the same way we usually find the average of any two numbers: add them and divide by 2. (Be careful not to do this when acceleration is not uniform!) So, if we add the initial speed (zero in this case) and the final speed of 2 m/s and then divide by 2, we get 1 m/s for the average velocity.

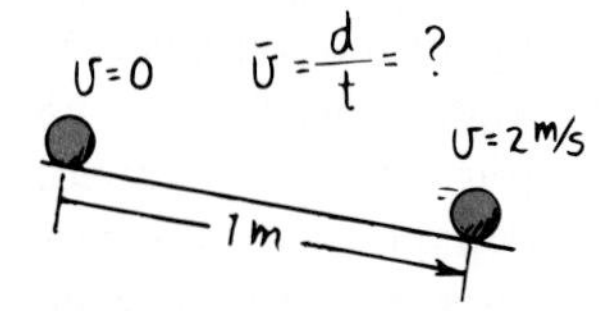

FIGURE B.1
The ball rolls 1 m down the incline in 1 s and reaches a speed of 2 m/s. Its average speed, however, is 1 m/s. Do you see why?

In each succeeding second, we see the ball roll a longer distance down the same slope in Figure B.2. Note the distance covered in the second time interval is 3 meters. This is because the average speed of the ball in this interval is 3 m/s. In the next 1-second interval, the average speed is 5 m/s, so the distance covered is 5 meters. It is interesting to see that successive increments of distance increase as a *sequence of odd numbers*. Nature clearly follows mathematical rules!

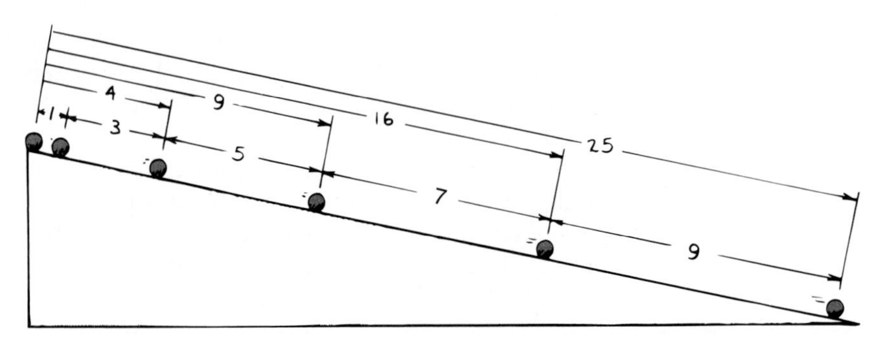

FIGURE B.2
If the ball covers 1 m during its first second, then, in each successive second, it will cover the odd-numbered sequence of 3, 5, 7, 9 m, and so on. Note that the total distance covered increases as the square of the total time.

CHECK YOURSELF

During the span of the second time interval, the ball begins at 2 m/s and ends at 4 m/s. What is the *average speed* of the ball during this 1-s interval? What is its *acceleration?*

Investigate Figure B.2 carefully and note the *total* distance covered as the ball accelerates down the plane. The distances go from zero to 1 meter in 1 second, zero to 4 meters in 2 seconds, zero to 9 meters in 3 seconds, zero to 16 meters in 4 seconds, and so on in succeeding seconds. The sequence for *total distances* covered is of the *squares of the time.* We'll investigate the relationship between distance traveled and the square of the time for constant acceleration more closely in the case of free fall.

Computing Distance When Acceleration Is Constant

How far will an object released from rest fall in a given time? To answer this question, let us consider the case in which it falls freely for 3 seconds, starting at rest. Neglecting air resistance, the object will have a constant acceleration of about 10 meters per second each second (actually more like 9.8 m/s^2, but we want to make the numbers easier to follow).

$$\text{Velocity at the } \textit{beginning} = 0 \text{ m/s}$$

$$\text{Velocity at the } \textit{end} \text{ of 3 seconds} = (10 \times 3) \text{ m/s}$$

$$\begin{aligned} \textit{Average} \text{ velocity} &= \tfrac{1}{2} \text{ the sum of these two speeds} \\ &= \tfrac{1}{2} \times (0 + 10 \times 3) \text{ m/s} \\ &= \tfrac{1}{2} \times 10 \times 3 = 15 \text{ m/s} \end{aligned}$$

$$\begin{aligned} \text{Distance traveled} &= \text{average velocity} \times \text{time} \\ &= (\tfrac{1}{2} \times 10 \times 3) \times 3 \\ &= \tfrac{1}{2} \times 10 \times 3^2 = 45 \text{ m} \end{aligned}$$

We can see from the meanings of these numbers that

$$\text{Distance traveled} = \tfrac{1}{2} \times \text{acceleration} \times \text{square of time}$$

This equation is true for an object falling not only for 3 seconds but for any length of time, as long as the acceleration is constant. If we let d stand for the distance traveled, a for the acceleration, and t for the time, the rule may be written, in shorthand notation,

$$d = \tfrac{1}{2} at^2$$

This relationship was first deduced by Galileo. He reasoned that, if an object falls for, say, twice the time, it will fall with *twice the average speed.* Since it falls for *twice* the time at *twice* the average speed, it will fall *four*

CHECK YOUR ANSWER

$$\text{Average speed} = \frac{\text{beginning + final speed}}{2} = \frac{2 \text{ m/s} + 4 \text{ m/s}}{2} = 3 \text{ m/s}$$

$$\text{Acceleration} = \frac{\text{change in velocity}}{\text{time interval}} = \frac{4 \text{ m/s} - 2 \text{ m/s}}{1 \text{ s}} = \frac{2 \text{ m/s}}{1 \text{ s}} = 2 \text{ m/s}^2$$

times as far. Similarly, if an object falls for *three* times the time, it will have an average speed *three* times as great and will fall *nine* times as far. Galileo reasoned that the total distance fallen should be proportional to the *square* of the time.

In the case of objects in free fall, it is customary to use the letter g to represent the acceleration instead of the letter a (g because acceleration is due to *gravity*). While the value of g varies slightly in different parts of the world, it is approximately equal to 9.8 m/s^2 (32 ft/s^2). If we use g for the acceleration of a freely falling object (negligible air resistance), the equations for falling objects starting from a rest position become

$$v = gt$$

$$d = \frac{1}{2}gt^2$$

Much of the difficulty in learning physics, like learning any discipline, has to do with learning the language—the many terms and definitions. Speed is somewhat different from velocity, and acceleration is vastly different from speed or velocity.

CHECK YOURSELF

1. An auto starting from rest has a constant acceleration of 4 m/s^2. How far will it go in 5 s?
2. How far will an object released from rest fall in 1 s? In this case the acceleration is $g = 9.8$ m/s^2.
3. If it takes 4 s for an object to freely fall to the water when released from the Golden Gate Bridge, how high is the bridge?

Mass and weight are related but are different from each other. Likewise for work, heat, and temperature. Please be patient with yourself as you find that learning the similarities and the differences among physics concepts is not an easy task.

CHECK YOUR ANSWER

1. Distance $= \frac{1}{2} \times 4 \times 5^2 = 50$ m
2. Distance $= \frac{1}{2} \times 9.8 \times 1^2 = 4.9$ m
3. Distance $= \frac{1}{2} \times 9.8 \times 4^2 = 78.4$ m

Notice that the units of measurement when multiplied give the proper units of meters for distance:

$$d = \frac{1}{2} \times 9.8\ \text{m}/\cancel{\text{s}^2} \times 16\cancel{\text{s}^2} = 78.4\ \text{m}$$

Graphing

Graphs: A Way to Express Quantitative Relationships

Graphs, like equations and tables, show how two or more quantities relate to each other. Since investigating relationships between quantities makes up much of the work of physics, equations, tables, and graphs are important physics tools.

Equations are the most concise way to describe quantitative relationships. For example, consider the equation $v = v_0 + gt$. It compactly describes how a freely falling object's velocity depends on its initial velocity, acceleration due to gravity, and time. Equations are nice shorthand expressions for relationships among quantities.

Tables give values of variables in list form. The dependence of v on t in $v = v_0 + gt$ can be shown by a table that lists various values v for corresponding times t. Table 3.2 on page 48 is an example. Tables are especially useful when the mathematical relationship between quantities is not known, or when numerical values must be given to a high degree of accuracy. Also, tables are handy for recording experimental data.

Graphs *visually* represent relationships between quantities. By looking at the shape of a graph, you can quickly tell a lot about how the variables are related. For this reason, graphs can help clarify the meaning of an equation or table of numbers. And, when the equation is not already known, a graph can help reveal the relationship between variables. Experimental data are often graphed for this reason.

Graphs are helpful in another way. If a graph contains enough plotted points, it can be used to estimate values between the points (interpolation) or values following the points (extrapolation).

Cartesian Graphs

The most common and useful graph in science is the *Cartesian* graph. On a Cartesian graph, possible values of one variable are represented on the vertical axis (called the *y-axis*) and possible values of the other variable are plotted on the horizontal axis (*x-axis*).

Figure C.1 shows a graph of two variables, x and y, that are *directly proportional* to each other. A direct proportionality is a type of *linear* relationship. Linear relationships have straight-line graphs—the easiest kinds of graphs to interpret. On the graph shown in Figure C.1, the continuous straight-line rise from left to right tells you that, as x increases, y increases. More specifically, it shows that y increases at a constant rate with respect to x. As x increases, y increases. The graph of a direct proportionality often passes through the "origin"—the point at

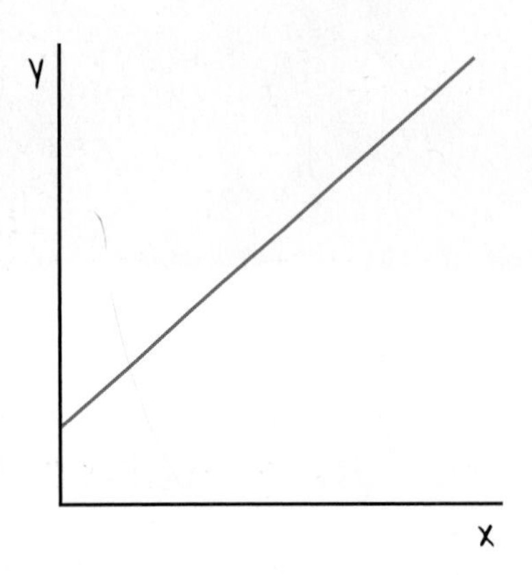

FIGURE C.1

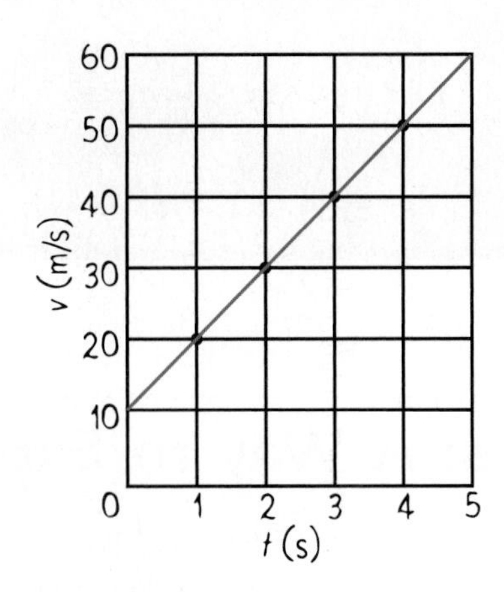

FIGURE C.2

the lower left where $x = 0$ and $y = 0$. In Figure C.1, however, we see the graph begins where y has a nonzero value when $x = 0$. The value y has a "head start."

Figure C.2 shows a graph of the equation $v = v_0 + gt$. Speed v is plotted along the y-axis, and time t along the x-axis. As you can see, there is a linear relationship between v and t. Note that the initial speed is 10 m/s. If the initial speed were 0, as in dropping an object from rest, then the graph would intercept the origin, where both v and t are 0. Note that the graph originates at $v = 10$ m/s when $t = 0$, showing a 10-m/s "head start."

Many physically significant relationships are more complicated than linear relationships, however. If you double the size of a room, the area of the floor increases four times; tripling the size of the room increases the floor area nine times; and so on. This is one example of a *nonlinear* relationship. Figure C.3 shows a graph of another nonlinear relationship: distance versus time in the equation of free fall from rest, $d = \frac{1}{2}gt^2$.

Figure C.4 shows a *radiation curve.* The *curve* (or graph) shows the rather complex nonlinear relationship between intensity I and radiation wavelength λ for a glowing object at 2000 K. The graph shows that radiation is most intense when λ equals about 1.4 μm. Which is brighter, radiation at 0.5 μm or radiation at 4.0 μm? The graph can quickly tell you that radiation at 4.0 μm is appreciably more intense.

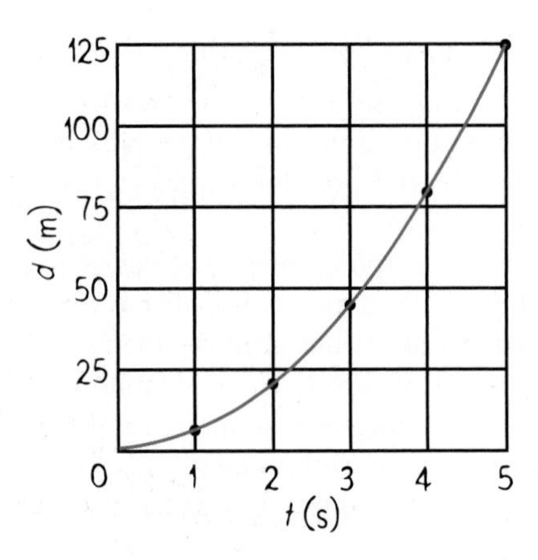

FIGURE C.3

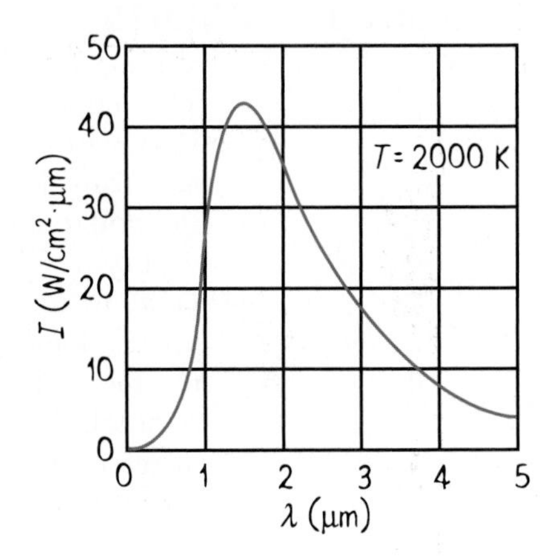

FIGURE C.4

Slope and Area Under the Curve

Quantitative information can be obtained from a graph's *slope* and the *area under the curve.* The slope of the graph in Figure C.2 represents the rate at which v increases relative to t. It can be calculated by dividing a segment Δv along the y-axis by a corresponding segment Δt along the x-axis. For example, dividing Δv of 30 m/s by Δt of 3 s gives $\Delta v/\Delta t$ = 10 m/s·s = 10 m/s^2, the acceleration due to gravity. By contrast, consider the graph in Figure C.5, which is a horizontal straight line. Its slope of zero shows zero acceleration—that is, constant speed. The graph shows that the speed is 30 m/s, acting throughout the entire five-second interval. The rate of change, or slope, of the speed with respect to time is zero—there is no change in speed at all.

The area under the curve is an important feature of a graph because it often has a physical interpretation. For example, consider the area under the graph of v versus t shown in Figure C.6. The shaded region is a rectangle with sides 30 m/s and 5 s. Its area is 30 m/s × 5 s = 150 m. In this example, the area is the distance covered by an object moving at constant speed of 30 m/s for 5 s ($d = vt$).

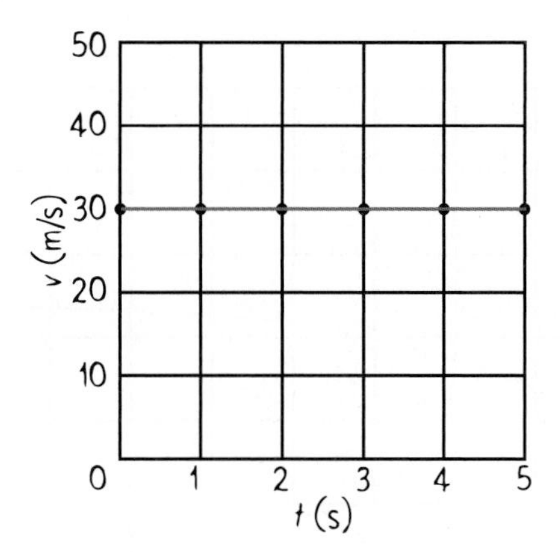

FIGURE C.5

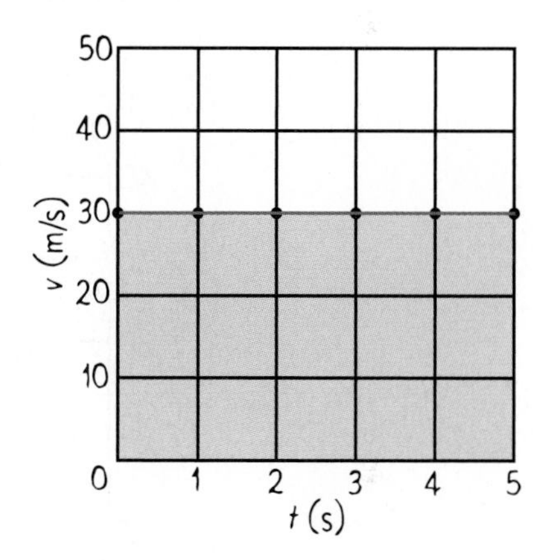

FIGURE C.6

The area need not be rectangular. The area beneath any curve of v versus t represents the distance traveled in a given time interval. Similarly, the area beneath a curve of acceleration versus time gives the change of velocity in a time interval. The area beneath a force-versus-time curve gives the change of momentum. (What does the area beneath a force-versus-distance curve give?) The nonrectangular area under various curves, including rather complicated ones, can be found by way of an important branch of mathematics—*integral calculus.*

Graphing with Conceptual Physics

You may develop basic graphing skills in the laboratory part of this course. The lab "Blind as a Bat" introduces you to graphing concepts. It also gives you a chance to work with a computer and sonic-ranging device. The lab "Trial and Error" will show you the useful technique of converting a nonlinear graph to a linear one to discover a direct proportionality. The area under the curve is the basis of the lab experiment "Impact Speed." You will learn more about graphing in other labs as well.

You may also learn in the lab part of your *Conceptual Physics* course that computers can graph data for you. You are not being lazy when you graph your data with a software program. Instead of investing time and energy scaling the axes and plotting points, you spend your time and energy investigating the meaning of the graph, a high level of thinking!

CHECK YOURSELF

Figure C.7 is a graphical representation of a ball dropped into a mine.

1. How long did the ball take to hit the bottom?
2. What was the ball's speed when it struck bottom?
3. What does the decreasing slope of the graph tell you about the acceleration of the ball with increasing speed?
4. Did the ball reach terminal speed before hitting the bottom of the shaft? If so, about how many seconds did it take to reach its terminal speed?
5. What is the approximate depth of the mine shaft?

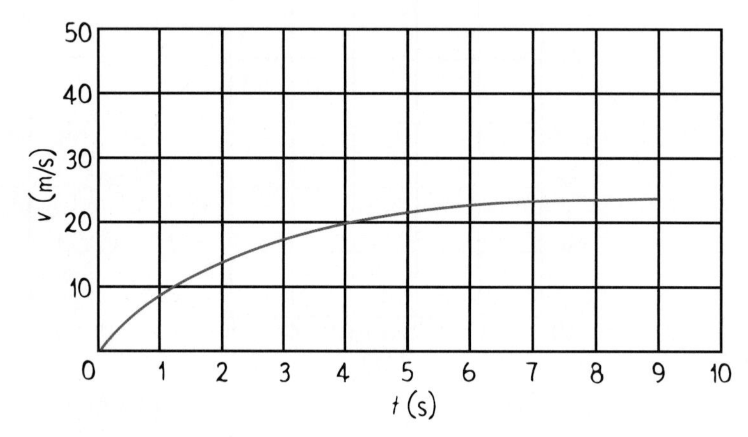

FIGURE C.7

CHECK YOUR ANSWERS

1. 9 s
2. 25 m/s
3. Acceleration decreases as speed increases (due to air resistance).
4. Yes (since slope curves to zero), about 7 s.
5. Depth is about 170 m. (The area under the curve is about 17 squares, each of which represents 10 m.)

APPENDIX D

More About Vectors

Vectors and Scalars

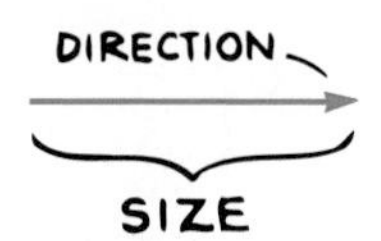

FIGURE D.1

A *vector* quantity is a directed quantity—one that must be specified not only by magnitude (size) but by direction as well. Recall, from Chapter 5, that velocity is a vector quantity. Other examples are force, acceleration, and momentum. In contrast, a *scalar* quantity can be specified by magnitude alone. Some examples of scalar quantities are speed, time, temperature, and energy.

Vector quantities may be represented by arrows. The length of the arrow tells you the magnitude of the vector quantity, and the arrowhead tells you the direction of the vector quantity. Such an arrow drawn to scale and pointing appropriately is called a *vector.*

Adding Vectors

Vectors that add together are called *component vectors.* Recall, from Chapter 5, that the sum of component vectors is called a *resultant.*

To add two vectors, make a parallelogram with two component vectors acting as two of the adjacent sides (Figure D.2). (Here our parallelogram is a rectangle.) Then draw a diagonal from the origin of the vector pair; this is the resultant (Figure D.3).

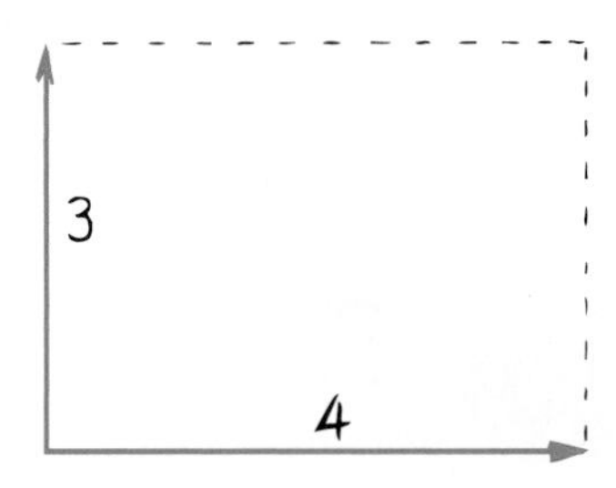

FIGURE D.2

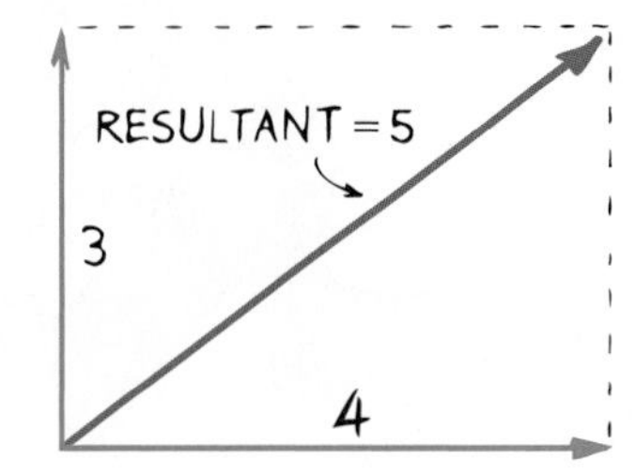

FIGURE D.3

Caution: Do not try to mix vectors! We cannot add apples and oranges, so velocity vector combines only with velocity vector, force vector combines only with force vector, and acceleration vector combines only with acceleration vector—each on its own vector diagram. If you ever show different kinds of vectors on the same diagram, use different colors or some other method of distinguishing the different kinds of vectors.

Finding Components of Vectors

Recall, from Chapter 5, that to find a pair of perpendicular components for a vector, first draw a dotted line through the tail end of the vector in the direction of one of the desired components. Second, draw another dotted line through the tail end of the vector at right angles to the first dotted line. Third, make a rectangle whose diagonal is the given vector. Draw in the two components. Here we let **F** stand for "total force," **U** stand for "upward force," and **S** stand for "sideways force."

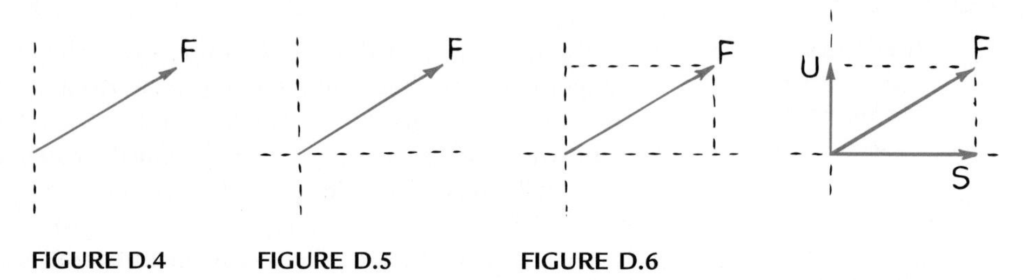

FIGURE D.4 FIGURE D.5 FIGURE D.6

Examples

1. Ernie Brown pushing a lawn mower applies a force that pushes the machine forward and also against the ground. In Figure D.7, **F** represents the force applied by Ernie. We can separate this force into two components. The vector **D** represents the downward component, and **S** is the sideways component, the force that moves the lawnmower forward. If we know the magnitude and direction of the vector **F**, we can estimate the magnitude of the components from the vector diagram.

FIGURE D.7

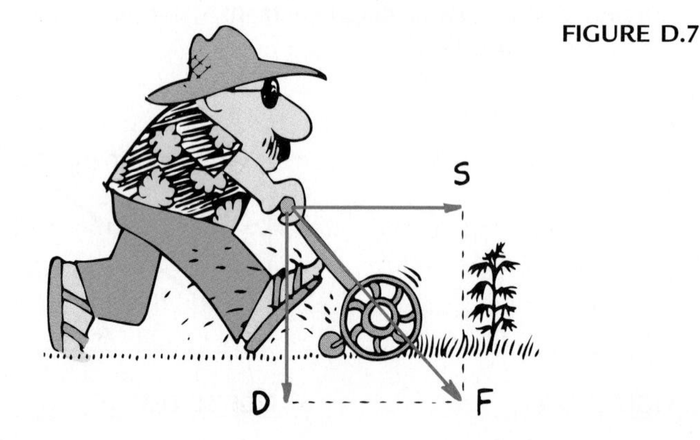

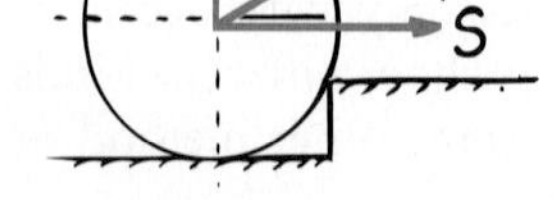

FIGURE D.8

2. Would it be easier to push or pull a wheelbarrow over a step? Figure D.8 shows a vector diagram for each case. When you push a wheelbarrow, part of the force is directed downward, which makes it harder to get over the step. When you pull, however, part of the pulling force is directed upward, which helps to lift the wheel over the step. Note that the vector diagram suggests that pushing the wheelbarrow may not get it over the step at all. Do you see that the height of the step, the

radius of the wheel, and the angle of the applied force determine whether the wheelbarrow can be pushed over the step? We see how vectors help us analyze a situation so that we can see just what the problem is!

3. If we consider the components of the weight of an object rolling down an incline, we can see why its speed depends on the angle (Figure D.9). Note that the steeper the incline, the greater the component **S** becomes and the faster the object rolls. When the incline is vertical, **S** becomes equal to the weight, and the object attains maximum acceleration, 9.8 meters per second squared.

FIGURE D.9

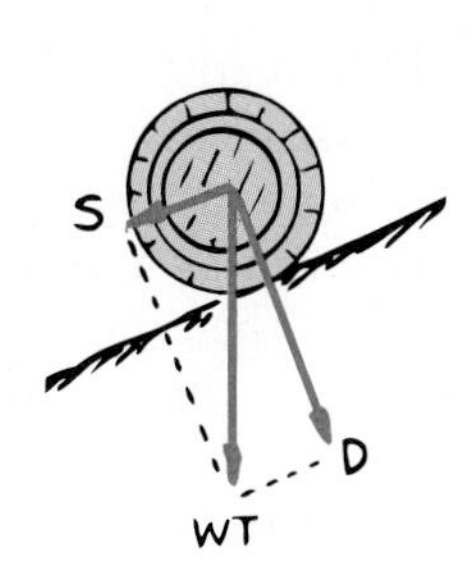

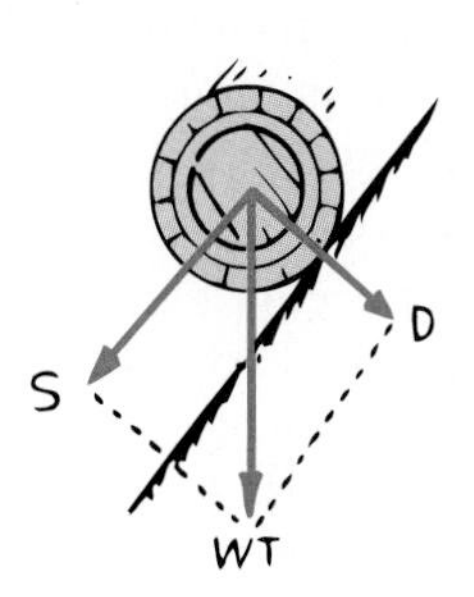

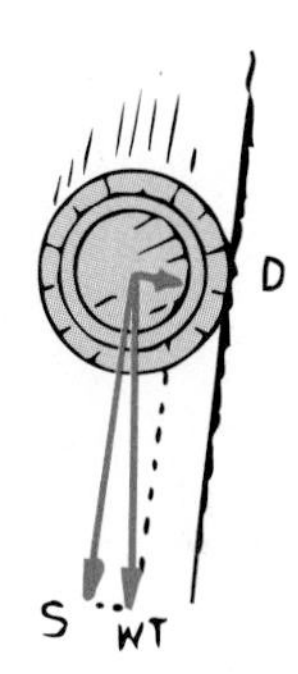

There are two more force vectors that are not shown: the normal force **N**, which is equal and oppositely directed to **D**, and the friction force **f**, acting at the barrel-plane contact.

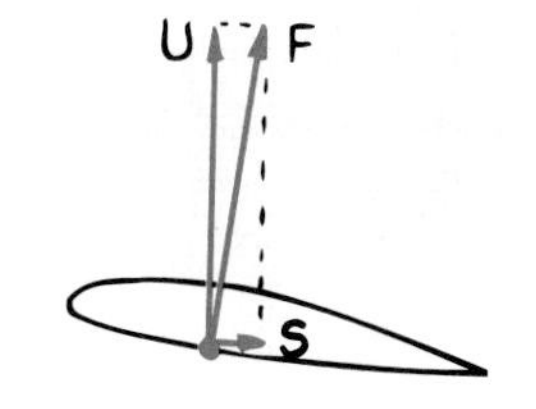

FIGURE D.10

4. When moving air strikes the underside of an airplane wing, the force of air impact against the wing may be represented by a single vector perpendicular to the plane of the wing (Figure D.10). We represent the force vector as acting midway along the lower wing surface, where the dot is, and pointing above the wing to show the direction of the resulting wind impact force. This force can be broken up into two components, one sideways and the other up. The upward component, **U**, is called *lift.* The sideways component, **S**, is called *drag.* If the aircraft is to fly at constant velocity at constant altitude, then lift must equal the weight of the aircraft and the thrust of the plane's engines must equal drag. The magnitude of lift (and drag) can be altered by changing the speed of the airplane or by changing the angle (called *angle of attack*) between the wing and the horizontal.

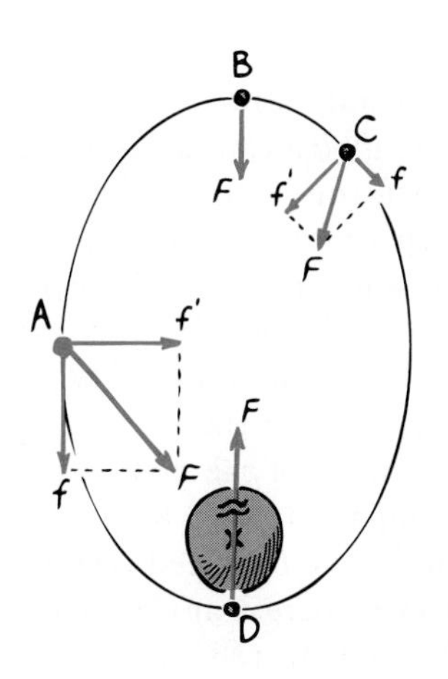

FIGURE D.11

5. Consider the satellite moving clockwise in Figure D.11. Everywhere in its orbital path, gravitational force *F* pulls it toward the center of the host planet. At position A we see *F* separated into two components: *f,* which is tangent to the path of the projectile, and *f′*, which is perpendicular to the path. The relative magnitudes of these components in comparison to the magnitude of *F* can be seen in the imaginary rectangle they compose; *f* and *f′* are the sides, and *F* is the diagonal. We see that component *f* is along the orbital path but against the direction of motion of the satellite. This force component reduces the speed of the satellite. The other component, *f′*, changes the direction of the satellite's motion and pulls it away from its tendency to go in a straight line. So the path of the satellite curves. The satellite loses speed until it reaches position B. At this farthest point from the planet (apogee), the gravitational force is somewhat

weaker but perpendicular to the satellite's motion, and component f has reduced to zero. Component f', on the other hand, has increased and is now fully merged to become F. Speed at this point is not enough for circular orbit, and the satellite begins to fall toward the planet. It picks up speed because the component f reappears and is in the direction of motion as shown in position C. The satellite picks up speed until it whips around to position D (perigee), where once again the direction of motion is perpendicular to the gravitational force, f' blends to full F, and f is nonexistent. The speed is in excess of that needed for circular orbit at this distance, and it overshoots to repeat the cycle. Its loss in speed in going from D to B equals its gain in speed from B to D. Kepler discovered that planetary paths are elliptical, but never knew why. Do you?

6. Refer to the Polaroid filters held by Ludmila back in Chapter 29, in Figure 29.35 (page 573). In the first picture (a), we see that light is transmitted through the pair of Polaroids because their axes are aligned. The emerging light can be represented as a vector aligned with the polarization axes of the Polaroids. When the Polaroids are crossed (b), no light emerges because light passing through the first Polaroid is perpendicular to the polarization axes of the second Polaroid, with no components along its axis. In the third picture (c), we see that light is transmitted when a third Polaroid is sandwiched at an angle between the crossed Polaroids. The explanation for this is shown in Figure D.12.

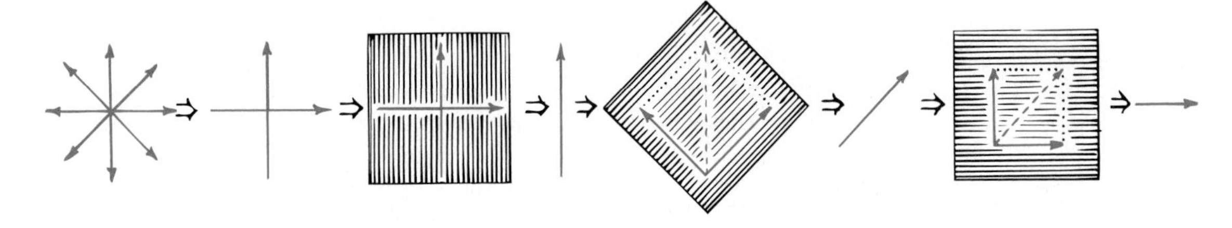

FIGURE D.12

Sailboats

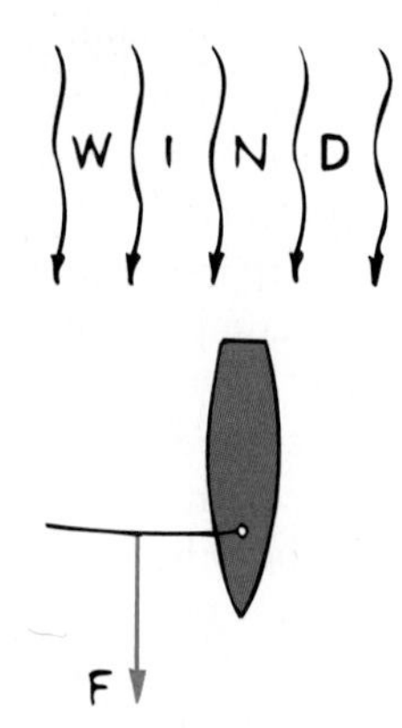

FIGURE D.13

Sailors have always known that a sailboat can sail downwind, in the direction of the wind. Sailors have not always known, however, that a sailboat can sail upwind, against the wind. One reason for this has to do with a feature that is common only to recent sailboats—a finlike keel that extends deep beneath the bottom of the boat to ensure that the boat will knife through the water only in a forward (or backward) direction. Without a keel, a sailboat could be blown sideways.

Figure D.13 shows a sailboat sailing directly downwind. The force of wind impact against the sail accelerates the boat. Even if the drag of the water and all other resistance forces are negligible, the maximum speed of the boat is the wind speed. This is because the wind will not make impact against the sail if the boat is moving as fast as the wind. The wind would have no speed relative to the boat and the sail would simply sag. With no force, there is no acceleration. The force vector in Figure D.13 *decreases* as the boat travels faster. The force vector is maximum when the boat is at rest and the full impact of the

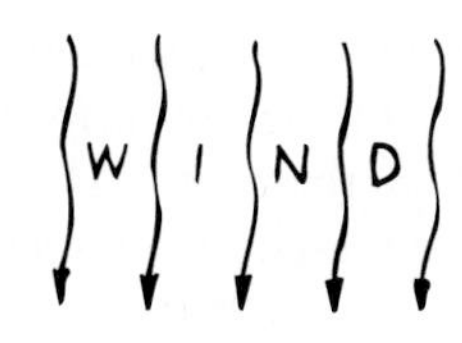

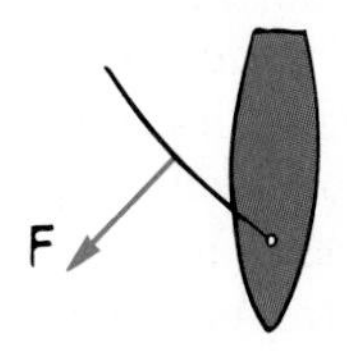

FIGURE D.14

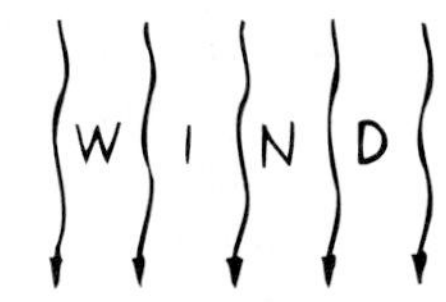

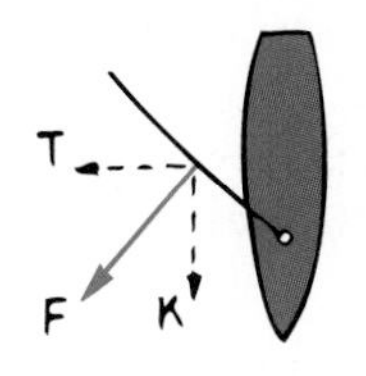

FIGURE D.15

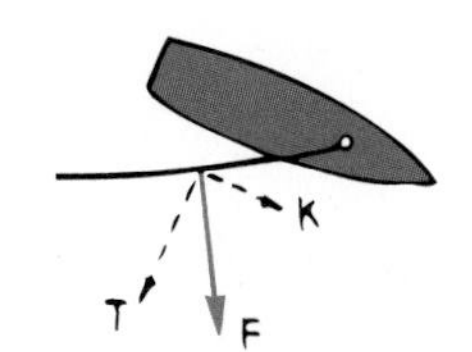

FIGURE D.16

wind fills the sail, and is minimum when the boat travels as fast as the wind. If the boat is somehow propelled to a speed faster than the wind (by way of a motor, for example), then air resistance against the front side of the sail will produce an oppositely directed force vector. This will slow the boat down. Hence, the boat when driven only by the wind cannot exceed wind speed.

If the sail is oriented at an angle, as shown in Figure D.14, the boat will move forward, but with less acceleration. There are two reasons for this:

1. The force on the sail is less because the sail does not intercept as much wind in this angular position.
2. The direction of the wind impact force on the sail is not in the direction of the boat's motion but is perpendicular to the surface of the sail. Generally speaking, whenever any fluid (liquid or gas) interacts with a smooth surface, the force of interaction is perpendicular to the smooth surface.* The boat does not move in the same direction as the perpendicular force on the sail, but is constrained to move in a forward (or backward) direction by its keel.

We can better understand the motion of the boat by resolving the force of wind impact, ***F***, into perpendicular components. The important component is that which is parallel to the keel, which we label ***K***, and the other component is perpendicular to the keel, which we label ***T***. It is the component ***K***, as shown in Figure D.15, that is responsible for the forward motion of the boat. Component ***T*** is a useless force that tends to tip the boat over and move it sideways. This component force is offset by the deep keel. Again, maximum speed of the boat can be no greater than wind speed.

Many sailboats sailing in directions other than exactly downwind (Figure D.16) with their sails properly oriented can exceed wind speed. In the case of a sailboat cutting across the wind, the wind may continue to make impact with the sail even after the boat exceeds wind speed. A surfer, in a similar way, exceeds the velocity of the propelling wave by angling his surfboard across the wave. Greater angles to the propelling medium (wind for the boat, water wave for the surfboard) result in greater speeds. A sailcraft can sail faster cutting across the wind than it can sailing downwind.

As strange as it may seem, maximum speed for most sailcraft is attained by cutting into (against) the wind, that is, by angling the sailcraft in a direction upwind! Although a sailboat cannot sail directly upwind, it can reach a destination upwind by angling back and forth in a zigzag fashion. This is called

*You can do a simple exercise to see that this is so. Try bouncing one coin off another on a smooth surface, as shown. Note that the struck coin moves at right angles (perpendicular) to the contact edge. Note also that it makes no difference whether the projected coin moves along path A or path B. See your instructor for a more rigorous explanation, which involves momentum conservation.

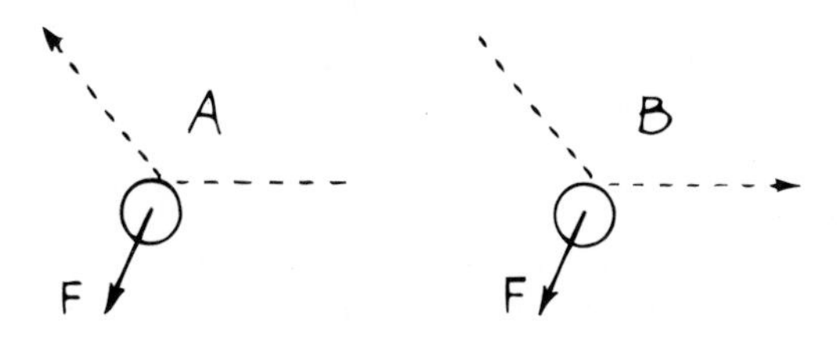

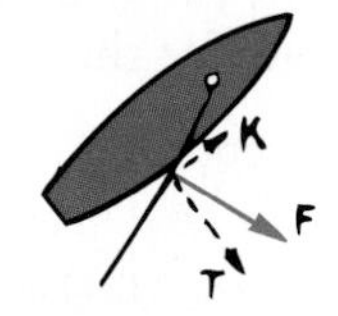

FIGURE D.17

tacking. Suppose the boat and sail are as shown in Figure D.17. Component ***K*** will push the boat along in a forward direction, angling into the wind. In the position shown, the boat can sail faster than the speed of the wind. This is because as the boat travels faster, the impact of wind is increased. This is similar to running in a rain that comes down at an angle. When you run into the direction of the downpour, the drops strike you harder and more frequently; but when you run away from the direction of the downpour, the drops don't strike you as hard or as frequently. In the same way, a boat sailing upwind experiences greater wind impact force, while a boat sailing downwind experiences a decreased wind impact force. In any case, the boat reaches its terminal speed when opposing forces cancel the force of wind impact. The opposing forces consist mainly of water resistance against the hull of the boat. The hulls of racing boats are shaped to minimize this resistive force, which is the principal deterrent to high speeds.

Iceboats (sailcraft equipped with runners for traveling on ice) encounter no water resistance and can travel at several times the speed of the wind when they tack upwind. Although ice friction is nearly absent, an iceboat does not accelerate without limits. The terminal velocity of a sailcraft is determined not only by opposing friction forces but also by the change in relative wind direction. When the boat's orientation and speed are such that the wind seems to shift in direction, so the wind moves parallel to the sail rather than into it, forward acceleration ceases—at least in the case of a flat sail. In practice, sails are curved and produce an airfoil that is as important to sailcraft as it is to aircraft. The effects are discussed in Chapter 14.

APPENDIX E

Exponential Growth and Doubling Time*

Try folding a piece of paper in half, then folding it again upon itself successively nine times. You'll soon see that it gets too thick to keep folding. And, if you could fold a fine piece of tissue paper upon itself 50 times, it would be more than 20 million kilometers thick! The continual doubling of a quantity builds up astronomically. If you give a child a penny on his first birthday, two pennies on his second birthday, four pennies on his third birthday, and so on, doubling the number of pennies every birthday, when this child reaches age 30, he will have accumulated $10,737,418.23! One of the most important things we have trouble perceiving is the process of exponential growth and why it proliferates out of control.

When a quantity of something, such as money in the bank, population, or the rate of consumption of a resource, steadily grows at a fixed percentage per year, the growth is said to be *exponential.* Money in the bank may grow at 5% or 6% per year; world population is presently growing at about 2% per year; the electric-power–generating capacity in the United States grew at about 7% per year for the first three quarters of the twentieth century. The important thing about exponential growth is that the time required for the growing quantity to double in size (to increase by 100%) is constant. For example, if the population of a growing city takes 10 years to double from 10,000 to 20,000 people and it continues with exponential growth, in the next 10 years the population will double to 40,000, and, in the next 10 years, to 80,000, and so on.

There is an important relationship between the percentage growth rate and its *doubling time,* the time it takes to double a quantity:**

$$\text{doubling time} = \frac{69.2\%}{\text{percentage growth per unit time}} = \frac{70\%}{\text{percentage growth rate}}$$

This means that, to estimate the doubling time for a steadily growing quantity, we simply divide 70% by the percentage growth rate. For example, when electric-power–generating capacity in the United States was growing at 7% per year, the capacity doubled every 10 years (since $[70\%]/[70\%\text{year}] = 10$ years). If world population grew steadily at 2% per year, the world population would double in 35 years (since $[70\%]/[2\%\text{year}] = 35$ years). A city planning commission that

*This appendix is adapted from material written by University of Colorado physics professor Albert A. Bartlett, who asserts, "The greatest shortcoming of the human race is our inability to understand the exponential function."

**For exponential decay, we speak about *half-life,* the time for a quantity to reduce to half its value. An example of this case is radioactive decay, which is treated in Chapter 33.

accepts what seems like a modest growth rate of 3.5% per year may not realize that this means that doubling will occur in 20 years (since [70%]/[3.5%year] = 20 years). That means doubling the capacity for such things as water supply, sewage-treatment plants, and other municipal services every 20 years.

Continued growth and continued doubling lead to enormous numbers. In two doubling times, a quantity will double twice ($2^2 = 4$), or quadruple in size; in three doubling times, its size will increase eightfold ($2^3 = 8$); in four doubling times, it will increase sixteenfold ($2^4 = 16$); and so on.

FIGURE E.1
A single grain of wheat placed on the first square of the chessboard is doubled on the second square, and this number is doubled on the third square, and so on. Note that each square contains one more grain than all the preceding squares combined. Does enough wheat exist in the world to fill all 64 squares in this manner?

This is best illustrated by the story of the court mathematician in India who years ago invented the game of chess for his king. The king was so pleased with the game that he offered to repay the mathematician, whose request seemed modest enough. The mathematician requested a single grain of wheat on the first square of the chessboard, two grains on the second square, four on the third square, and so on, doubling the number of grains on each succeeding square until all squares had been used. At this rate there would be 2^{63} grains of wheat on the sixty-fourth square alone. The king soon saw that he could not fill this "modest" request, which amounted to more wheat than had been harvested in the entire history of humankind.

As Table E.1 shows, the number of grains on any square is one grain more than the total of all grains on the preceding squares. This is true anywhere on the board. For example, when four grains are placed on the third square, that number of grains is one more than the total of three grains already on the board. The number of grains (eight) on the fourth square is one more than the total of seven grains already on the board. The same pattern occurs everywhere on the board. In any case of exponential growth, a greater quantity is represented in one doubling time than in all the preceding growth. This is important enough to be repeated in different words: Whenever steady growth occurs, the numerical count of a quantity that exists after a single doubling time is one greater than the total count of that quantity in the entire history of growth.

Steady growth in a steadily expanding environment is one thing, but what happens when steady growth occurs in a finite environment? Consider the

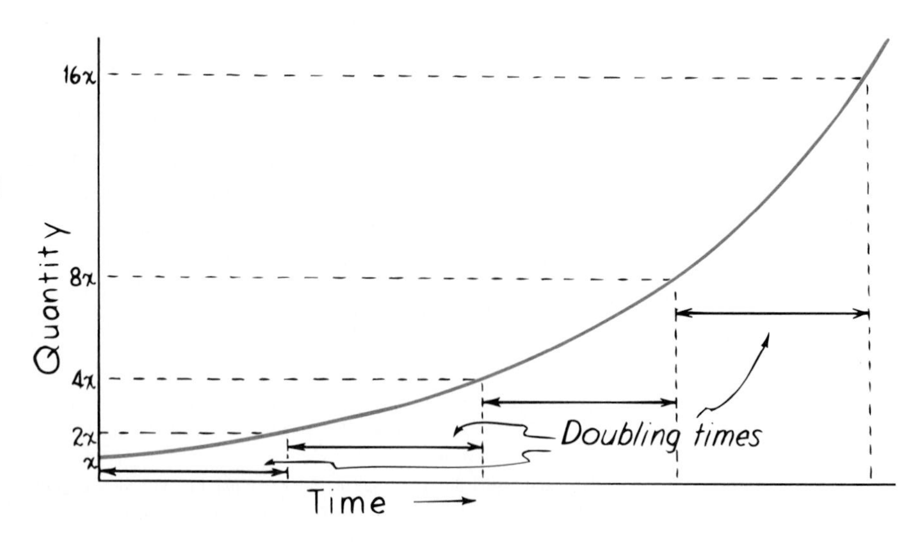

FIGURE E.2
Graph of a quantity that grows at an exponential rate. Notice that the quantity doubles during each of the successive equal time intervals marked on the horizontal scale. Each of these time intervals represents the doubling time.

TABLE E.1
Filling the Squares on the Chessboard

Square Number	*Grains on a Square*	*Total Grains Thus Far*
1	1	1
2	2	3
3	$4 = 2^2$	7
4	$8 = 2^3$	15
5	$16 = 2^4$	31
6	$32 = 2^5$	63
7	$64 = 2^6$	127
⋮	⋮	⋮
10	2^9	about 1,000
⋮	⋮	⋮
20	2^{19}	about 1,000,000
⋮	⋮	⋮
30	2^{29}	about 1,000,000,000
⋮	⋮	⋮
40	2^{39}	about 1,000,000,000,000
⋮	⋮	⋮
64	2^{63}	$2^{64} - 1$ (more than 10 billion!)

growth of bacteria that grow by division, so that one bacterium becomes two, the two divide to become four, the four divide to become eight, and so on. Suppose that the division time for a certain kind of bacteria is one minute. This is then steady percentage growth—the number of bacteria grows exponentially, with a doubling time of one minute. Further, suppose that one bacterium is put in a bottle at 11:00 a.m. and that growth continues steadily until the bottle becomes full of bacteria at 12 noon.

CHECK YOURSELF

When was the bottle half full?

CHECK YOUR ANSWER

At 11:59 a.m., since the bacteria will double in number every minute!

FIGURE E.3

TABLE E.2
The Last Minutes in the Bottle

Time	*Portion Full*	*Portion Empty*
11:54 a.m.	$\frac{1}{64}$ (1.5%)	$\frac{63}{64}$ (98.5%)
11:55 a.m.	$\frac{1}{32}$ (3%)	$\frac{31}{32}$ (97%)
11:56 a.m.	$\frac{1}{16}$ (6%)	$\frac{15}{16}$ (94%)
11:57 a.m.	$\frac{1}{8}$ (12%)	$\frac{7}{8}$ (88%)
11:58 a.m.	$\frac{1}{4}$ (25%)	$\frac{3}{4}$ (75%)
11:59 a.m.	$\frac{1}{2}$ (50%)	$\frac{1}{2}$ (50%)
12:00 noon	Full (100%)	None (0%)

It is startling to note that, at 2 minutes before noon, the bottle was only 1/4 full, and, at 3 minutes before noon, only 1/8 full. Table E.2 summarizes the amount of space left in the bottle in the last few minutes before noon. If bacteria could think, and if they were concerned about their future, at what time do you suppose they would sense they were running out of space? Would a serious problem have been evident at, say, 11:55 a.m., when the bottle was only 3% full (1/32) and had 97% open space just yearning for development? The point here is that there isn't much time between the moment the effects of growth become noticeable and the time when they become overwhelming.

Suppose that, at 11:58 a.m., some farsighted bacteria see that they are running out of space and launch a full-scale search for new bottles. And further suppose that they consider themselves lucky, because they find three new empty bottles. This is three times as much space as they have ever known. It may seem to the bacteria that their problems are solved—and just in time.

CHECK YOURSELF

If the bacteria are able to migrate to the new bottles and their growth continues at the same rate, what time will it be when the three new bottles are filled to capacity?

Table E.3 illustrates that the discovery of the new bottles extends the resource by only two doubling times. In this example, the resource is space—such as

TABLE E.3
Effects of the Discovery of Three New Bottles

Time	*Effect*
11:58 a.m.	Bottle 1 is $\frac{1}{4}$ full; bacteria divide into four bottles, each $\frac{1}{16}$ full
11:59 a.m.	Bottles 1, 2, 3, and 4 are each $\frac{1}{8}$ full
12:00 noon	Bottles 1, 2, 3, and 4 are each $\frac{1}{4}$ full
12:01 p.m.	Bottles 1, 2, 3, and 4 are each $\frac{1}{2}$ full
12:02 p.m.	Bottles 1, 2, 3, and 4 are each all full

CHECK YOUR ANSWER

All four bottles will be filled to capacity at 12:02 p.m.!

land area for a growing population. But it could be coal, oil, uranium, or any nonrenewable resource.

CHECK YOURSELF

1. According to a French riddle, a lily pond starts with a single leaf. Each day the number of leaves doubles, until the pond is completely full on the thirtieth day. On what day was the pond half covered? One-quarter covered?
2. In 2000, the population grew to 6 billion (and will likely grow to 7 billion in 2013, and to 8 billion in 2027). At the 2000 world growth rate of 1.2% per year, how long will it take for the world population to reach 12 billion?
3. What annual percentage increase in world population would be required to double the world population in 100 years?

Growth in the empty bottles discovered by the bacteria can proceed unrestricted (until the bottles are full)—very untypical of nature. Although bacteria and other organisms have the potential to multiply exponentially, limiting factors usually restrict the growth. The number of mice in a field, for example, depends not only on birthrate and food supply but also on the number of hawks and other predators in the vicinity. A "natural balance" of competing factors is struck. If the predators are removed, exponential growth of the population of mice can proceed for a while. If you remove certain plants from a region, other plant populations may tend to grow exponentially. All plants, animals, and creatures that inhabit the Earth are in states of balance—states that change with changing conditions. Hence the environmental adage, "You never change only one thing."

The consumption of a nonrenewable resource cannot grow exponentially for an indefinite period because the resource is finite and its supply finally expires. This is shown in Figure E.4*a*, in which the rate of consumption, such as the number of barrels of oil produced per year, is plotted against time. In such a graph, the shaded area under the curve represents the supply of the resource. When the supply is exhausted, consumption ceases altogether. Such a sudden change is rarely the case, because the rate of extracting the supply falls as the resource becomes more scarce. This is shown in Figure E.4*b*. Note that the area under the curve is equal to the area under the curve in Figure E.4*a*. Why? Because the total supply is the same in both cases. The difference is the time taken to extract that supply. History shows that the rate of production of a nonrenewable resource rises and falls in a nearly symmetric manner, as shown in Figure 4*c*. The time during which production rates rise is approximately equal to the time during which these rates fall to zero (or near zero).

CHECK YOUR ANSWERS

1. The pond was half covered on the twenty-ninth day, and was one-quarter covered on the twenty-eighth day!
2. 2058, since the doubling time (70%)/(1.2%/year) ≈ 58 years.
3. 0.7%, since (70%)/(0.7%/year) = 100 years. You can rearrange the equation so that it reads percentage growth rate = (70%)/(doubling time). Using the rearranged equation gives (70%)/(100 years) = 0.7%/year.

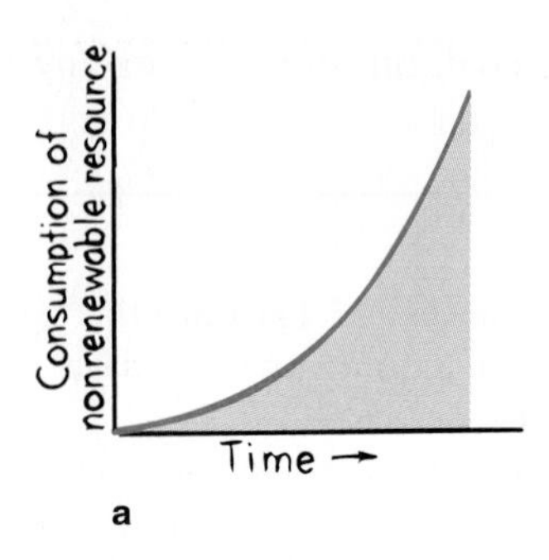

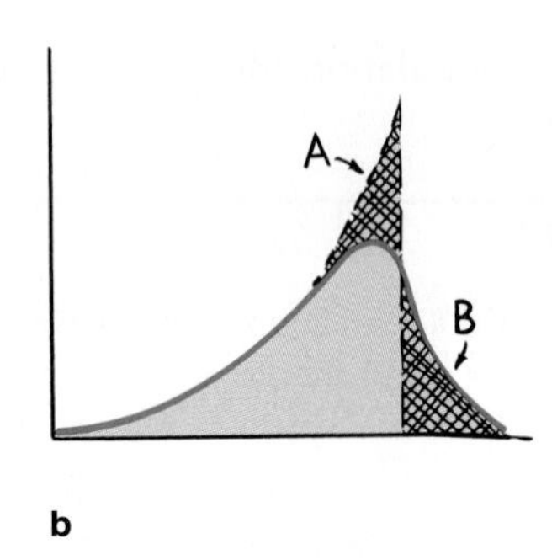

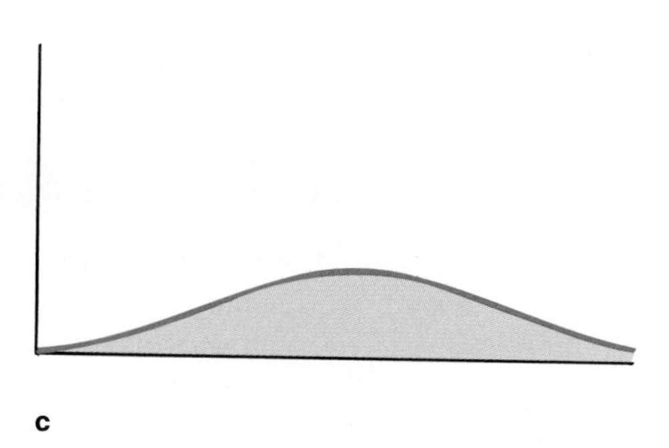

FIGURE E.4
(*a*) If the exponential rate of consumption for a nonrenewable resource continues until it is depleted, consumption falls abruptly to zero. The colored area under this curve represents the total supply of the resource. (*b*) In practice, the rate of consumption levels off and then falls less abruptly to zero. Note that the crosshatched area A is equal to the crosshatched area B. Why? (*c*) At lower consumption rates, the same resource lasts a longer time.

The consequences of unchecked exponential growth are staggering. It is very important to ask: Is growth really good? In answering this question, bear in mind that human growth is an early phase of life that continues normally through adolescence. Physical growth stops when physical maturity is reached. What do we say of growth that continues in the period of physical maturity? We say that such growth is obesity—or worse: cancer.

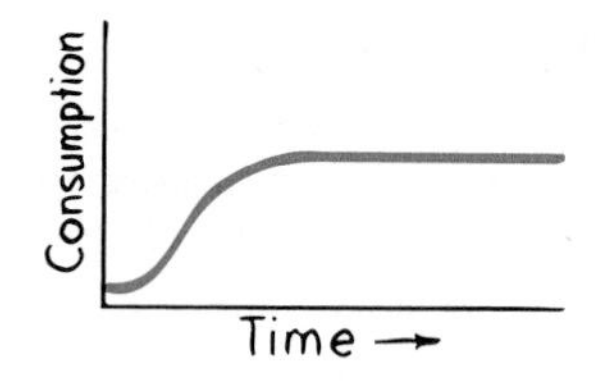

FIGURE E.5
A curve showing the rate of consumption of a renewable resource, such as agricultural or forest products, where a steady rate of production and consumption can be maintained for a long period, provided that this production is not dependent upon the use of a nonrenewable resource that is waning in supply.

Questions to Ponder

1. In an economy that has a steady inflation rate of 7% per year, in how many years does a dollar lose half its value?
2. At a steady inflation rate of 7%, what will be the price every 10 years for the next 50 years for a theater ticket that now costs $20? For a coat that now costs $200? For a car that now costs $20,000? For a home that now costs $200,000?
3. If the population of a city with one overloaded sewage-treatment plant grows steadily at 5% annually, how many overloaded sewage-treatment plants will be necessary 42 years later?
4. If world population doubles in 40 years and world food production also doubles in 40 years, how many people then will be starving each year compared with the number who are starving now?
5. Suppose that you get a prospective employer to agree to hire your services for a wage of a single penny for the first day, two pennies the second day, and doubling each day thereafter. If the employer keeps to the agreement for a month, what will be your total wages for the month?
6. In the previous question, how will your wages for only the thirtieth day compare with your total wages for the previous 29 days?
7. If fusion power were harnessed today, the abundant energy resulting would likely sustain and even further encourage our present appetite for continued growth in energy use and, in a relatively few doubling times, would produce an appreciable fraction of the solar power input to the Earth. Make an argument that the current delay in harnessing fusion is a blessing for the human race.

Glossary

A (a) Abbreviation for *ampere*. (b) When in lowercase italic *a*, the symbol for *acceleration*.

aberration Distortion in an image produced by a lens or mirror, caused by limitations inherent to some degree in all optical systems. See *spherical aberration* and *chromatic aberration*.

absolute zero Lowest possible temperature that any substance can have; the temperature at which the atoms of a substance have their minimum kinetic energy. The temperature of absolute zero is −273.15°C, which is −459.7°F and 0 kelvin.

absorption lines Dark lines that appear in an absorption spectrum. The pattern of lines is unique for each element.

absorption spectrum Continuous spectrum, like that generated by white light, interrupted by dark lines or bands that result from the absorption of light of certain frequencies by a substance through which the light passes.

ac Abbreviation for *alternating current*.

acceleration (a) Rate at which an object's velocity changes with time; the change in velocity may be in magnitude (speed), or direction, or both.

$$\text{acceleration} = \frac{\text{change of velocity}}{\text{time interval}}$$

acceleration due to gravity (*g*) Acceleration of a freely falling object. Its value near the Earth's surface is about 9.8 meters per second each second.

achromatic lenses See *chromatic aberration*.

acoustics Study of the properties of sound, especially its transmission.

action force One of the pair of forces described in Newton's third law. See also *Newton's laws of motion, Law 3*.

additive primary colors Three colors of light—red, blue, and green—that when added in certain proportions will produce any color of the spectrum.

adhesion Molecular attraction between two surfaces making contact.

adiabatic Term applied to expansion or compression of a gas occurring without gain or loss of heat.

adiabatic process Process, often of fast expansion or compression, wherein no heat enters or leaves a system. As a result, a liquid or gas undergoing an expansion will cool, or undergoing a compression will warm.

air resistance Friction, or drag, that acts on something moving through air.

alchemist Practitioner of the early form of chemistry called alchemy, which was associated with magic. The goal of alchemy was to change base metals to gold and to discover a potion that could produce eternal youth.

alloy Solid mixture composed of two or more metals or of a metal and a nonmetal.

alpha particle Nucleus of a helium atom, which consists of two neutrons and two protons, ejected by certain radioactive nuclei.

alpha ray Stream of alpha particles (helium nuclei) ejected by certain radioactive nuclei.

alternating current (ac) Electric current that rapidly reverses in direction. The electric charges vibrate about relatively fixed positions, usually at the rate of 60 hertz.

AM Abbreviation for *amplitude modulation*.

ammeter A device that measures current. See *galvanometer*.

ampere (A) SI unit of electric current. One ampere is a flow of one coulomb of charge per second—6.25×10^{18} electrons (or protons) per second.

amplitude For a wave or vibration, the maximum displacement on either side of the equilibrium (midpoint) position.

amplitude modulation (AM) Type of modulation in which the amplitude of the carrier wave is varied above and below its normal value by an amount proportional to the amplitude of the impressed signal.

amu Abbreviation for *atomic mass unit*.

analog signal Signal based on a continuous variable, as opposed to a digital signal made up of discrete quantities.

aneroid barometer Instrument used to measure atmospheric pressure; based on the movement of the lid of a metal box, rather than on the movement of a liquid.

angle of incidence Angle between an incident ray and the normal to the surface it encounters.

angle of reflection Angle between a reflected ray and the normal to the surface of reflection.

angle of refraction Angle between a refracted ray and the normal to the surface at which it is refracted.

angular momentum Product of a body's rotational inertia and rotational velocity about a particular axis. For an object that is small compared with the radial distance, it is the product of mass, speed, and radial distance of rotation.

$$\text{angular momentum} = mvr$$

antimatter Matter composed of atoms with negative nuclei and positive electrons.

antinode Any part of a standing wave with maximum displacement and maximum energy.

antiparticle Particle having the same mass as a normal particle, but a charge of the opposite sign. The antiparticle of an electron is a positron.

antiproton Antiparticle of a proton; a negatively charged proton.

apogee Point in an elliptical orbit farthest from the focus around which orbiting takes place. See also *perigee*.

Archimedes' principle Relationship between buoyancy and displaced fluid: An immersed object is buoyed up by a force equal to the weight of the fluid it displaces.

armature Part of an electric motor or generator where an electromotive force is produced. Usually the rotating part.

astigmatism Defect of the eye caused when the cornea is curved more in one direction than in another.

atmospheric pressure Pressure exerted against bodies immersed in the atmosphere resulting from the weight of air pressing down from above. At sea level, atmospheric pressure is about 101 kPa.

atom Smallest particle of an element that has all the element's chemical properties. Consists of protons and neutrons in a nucleus surrounded by electrons.

atomic bonding Linking together of atoms to form larger structures, such as molecules and solids.

atomic mass number Number associated with an atom, equal to the number of nucleons (protons plus neutrons) in the nucleus.

atomic mass unit (amu) Standard unit of atomic mass. It is based on the mass of the common carbon atom, which is arbitrarily given the value of exactly 12. An amu of one is one-twelfth the mass of this common carbon atom.

atomic number Number associated with an atom, equal to the number of protons in the nucleus, or, equivalently, to the number of electrons in the electron cloud of a neutral atom.

aurora borealis Glowing of the atmosphere caused by ions from above the atmosphere that dip into the atmosphere; also called northern lights. In the southern hemisphere, they are called aurora australis.

average speed Path distance divided by time interval.

$$\text{average speed} = \frac{\text{total distance covered}}{\text{time interval}}$$

Avogadro's number 6.02×10^{23} molecules.

Avogadro's principle Equal volumes of all gases at the same temperature and pressure contain the same number of molecules, 6.02×10^{23} in one mole (a mass in grams equal to the molecular mass of the substance in atomic mass units).

axis (pl. axes) (a) Straight line about which rotation takes place. (b) Straight lines for reference in a graph, usually the x-axis for measuring horizontal displacement and the y-axis for measuring vertical displacement.

barometer Device used to measure the pressure of the atmosphere.

beats Sequence of alternating reinforcement and cancellation of two sets of superimposed waves differing in frequency, heard as a throbbing sound.

bel Unit of intensity of sound, named after Alexander Graham Bell. The threshold of hearing is 0 bel (10^{-12} watts per square meter). Often measured in decibels (dB, one-tenth of a bel).

Bernoulli's principle Pressure in a fluid decreases as the speed of the fluid increases.

beta particle Electron (or positron) emitted during the radioactive decay of certain nuclei.

beta ray Stream of beta particles (electrons or positrons) emitted by certain radioactive nuclei.

Big Bang Primordial explosion that is thought to have resulted in the creation of our expanding universe.

bimetallic strip Two strips of different metals welded or riveted together. Because the two substances expand at different rates when heated or cooled, the strip bends; used in thermostats.

binary code Code based on the binary number system (which uses a base of 2). In binary code any number can be expressed as a succession of ones and zeros. For example, the number 1 is 1, 2 is 10, 3 is 11, 4 is 100, 5 is 101, 17 is 10001, etc. These ones and zeros can then be interpreted and transmitted electronically as a series of "on" and "off" pulses, the basis for all computers and other digital equipment.

bioluminescence Light emitted from certain living things that have the ability to chemically excite molecules in their bodies; these excited molecules then give off visible light.

biomagnetism Magnetic material located in living organisms that may help them navigate, locate food, and affect other behaviors.

black hole Concentration of mass resulting from gravitational collapse, near which gravity is so intense that not even light can escape.

blind spot Area of the retina where all the nerves carrying visual information exit the eye and go to the brain; this is a region of no vision.

blue shift Increase in the measured frequency of light from an approaching source; called the blue shift because the apparent increase is toward the high-frequency, or blue, end of the color spectrum. Also occurs when an observer approaches a source. See also *Doppler effect.*

boiling Change from liquid to gas occurring beneath the surface of the liquid; rapid vaporization. The liquid loses energy, the gas gains it.

bow wave V-shaped wave produced by an object moving on a liquid surface faster than the wave speed.

Boyle's law The product of pressure and volume is a constant for a given mass of confined gas regardless of changes in either pressure or volume individually, as long as temperature remains unchanged.

$$P_1V_1 = P_2V_2$$

breeder reactor Nuclear fission reactor that not only produces power but produces more nuclear fuel than it consumes by converting a nonfissionable uranium isotope into a fissionable plutonium isotope. See also *nuclear reactor.*

British thermal unit (BTU) Amount of heat required to change the temperature of 1 pound of water by 1 Fahrenheit degree.

Brownian motion Haphazard movement of tiny particles suspended in a gas or liquid resulting from bombardment by the fast-moving molecules of the gas or liquid.

BTU Abbreviation for *British thermal unit.*

buoyancy Apparent loss of weight of an object immersed or submerged in a fluid.

buoyant force Net upward force exerted by a fluid on a submerged or immersed object.

butterfly effect Situation in which a very small change in one place can amplify into a large change somewhere else.

C Abbreviation for *coulomb*.

cal Abbreviation for *calorie*.

calorie (cal) Unit of heat. One calorie is the heat required to raise the temperature of one gram of water 1 Celsius degree. One Calorie (with a capital C) is equal to one thousand calories and is the unit used in describing the energy available from food; also called a kilocalorie (kcal).

$$1 \text{ cal} = 4.184 \text{ J} \quad \text{or} \quad 1 \text{ J} = 0.24 \text{ cal}$$

capacitor Device used to store charge in a circuit.

capillarity Rise of a liquid in a fine, hollow tube or in a narrow space.

carbon dating Process of determining the time that has elapsed since death by measuring the radioactivity of the remaining carbon-14 isotopes.

Carnot efficiency Ideal maximum percentage of input energy that can be converted to work in a heat engine.

carrier wave High-frequency radio wave modified by a lower-frequency wave.

Celsius scale Temperature scale that assigns 0 to the melt-freeze point for water and 100 to the boil-condense point of water at standard pressure (one atmosphere at sea level).

center of gravity (CG) Point at the center of an object's weight distribution, where the force of gravity can be considered to act.

center of mass Point at the center of an object's mass distribution, where all its mass can be considered to be concentrated. For everyday conditions, it is the same as the center of gravity.

centrifugal force Apparent outward force on a rotating or revolving body.

centripetal force Center-directed force that causes an object to follow a curved or circular path.

CG Abbreviation for *center of gravity*.

chain reaction Self-sustaining reaction that, once started, steadily provides the energy and matter necessary to continue the reaction.

charge See *electric charge*.

charging by contact Transfer of electric charge between objects by rubbing or simple touching.

charging by induction Redistribution of electric charges in and on objects caused by the electrical influence of a charged object close by but not in contact.

chemical formula Description that uses numbers and symbols of elements to describe the proportions of elements in a compound or reaction.

chemical reaction Process of rearrangement of atoms that transforms one molecule into another.

chinook Warm, dry wind that blows down from the eastern side of the Rocky Mountains across the Great Plains.

chromatic aberration Distortion of an image caused when light of different colors (and thus different speeds and refractions) focuses at different points when passing through a lens. Achromatic lenses correct this defect by combining simple lenses made of different kinds of glass.

circuit Any complete path along which electric charge can flow. See also *series circuit* and *parallel circuit*.

circuit breaker Device in an electric circuit that breaks the circuit when the current gets high enough to risk causing a fire.

coherent light Light of a single frequency with all photons exactly in phase and moving in the same direction. Lasers produce coherent light. See also *incoherent light* and *laser*.

complementarity Principle enunciated by Niels Bohr stating that the wave and particle aspects of both matter and radiation are necessary, complementary parts of the whole. Which part is emphasized depends on what experiment is conducted (i.e., on what questions one puts to nature.)

complementary colors Any two colors of light that, when added, produce white light.

component Parts into which a vector can be separated and that act in different directions from the vector. See *resultant*.

compound Chemical substance made of atoms of two or more different elements combined in a fixed proportion.

compression (a) In mechanics, the act of squeezing material and reducing its volume. (b) In sound, the region of increased pressure in a longitudinal wave.

concave mirror Mirror that curves inward like a "cave."

condensation Change of phase of a gas into a liquid; the opposite of evaporation.

conduction (a) In heat, energy transfer from particle to particle within certain materials, or from one material to another when the two are in direct contact. (b) In electricity, the flow of electric charge through a conductor.

conduction electrons Electrons in a metal that move freely and carry electric charge.

conductor (a) Material through which heat can be transferred. (b) Material, usually a metal, through which electric charge can flow. Good conductors of heat are generally good electric charge conductors.

cones See *retina*.

conservation of angular momentum When no external torque acts on an object or a system of objects, no change of angular momentum takes place. Hence, the angular momentum before an event involving only internal torques is equal to the angular momentum after the event.

conservation of charge Principle that net electric charge is neither created nor destroyed but is transferable from one material to another.

conservation of energy Principle that energy cannot be created or destroyed. It may be transformed from one form into another, but the total amount of energy never changes.

conservation of energy for machines Work output of any machine cannot exceed the work input.

conservation of momentum In the absence of a net external force, the momentum of an object or system of objects is unchanged.

$$mv_{(\text{before event})} = mv_{(\text{after event})}$$

conserved Term applied to a physical quantity, such as momentum, energy, or electric charge, that remains unchanged during interactions.

constructive interference Combination of waves so that two or more waves overlap to produce a resulting wave of increased amplitude. See also *interference.*

convection Means of heat transfer by movement of the heated substance itself, such as by currents in a fluid.

converging lens Lens that is thicker in the middle than at the edges and refracts parallel rays of light passing through it to a focus. See also *diverging lens.*

convex mirror Mirror that curves outward. The virtual image formed is smaller and closer to the mirror than the object. See also *concave mirror.*

cornea Transparent covering over the eyeball, which helps focus the incoming light.

correspondence principle If a new theory is valid, it must account for the verified results of the old theory in the region where both theories apply.

cosmic ray One of various high-speed particles that travel throughout the universe and originate in violent events in stars.

cosmology Study of the origin and development of the entire universe.

coulomb (C) SI unit of electrical charge. One coulomb is equal to the total charge of 6.25×10^{18} electrons.

Coulomb's law Relationship among electrical force, charges, and distance: The electrical force between two charges varies directly as the product of the charges (q) and inversely as the square of the distance between them. (k is the proportionality constant $9 \times 10^9 \text{ N} \cdot \text{m}^2/\text{C}^2$) If the charges are alike in sign, the force is repulsive; if the charges are unlike, the force is attractive.

$$F = k\frac{q_1 q_2}{d^2}$$

crest One of the places in a wave where the wave is highest or the disturbance is greatest in the opposite direction from a trough. See also *trough.*

critical angle Minimum angle of incidence for which a light ray is totally reflected within a medium.

critical mass Minimum mass of fissionable material in a nuclear reactor or nuclear bomb that will sustain a chain reaction. A subcritical mass is one in which the chain reaction dies out. A supercritical mass is one in which the chain reaction builds up explosively.

crystal Regular geometric shape found in a solid in which the component particles are arranged in an orderly, three-dimensional, repeating pattern.

current See *electric current.*

cyclotron Particle accelerator that imparts high energy to charged particles such as protons, deuterons, and helium ions.

dark matter Unseen and unidentified matter that is evident by its gravitational pull on stars in the galaxies—comprising perhaps 90% of the matter of the universe.

dB Abbreviation for decibel. See *bel.*

dc Abbreviation for *direct current.*

DDT Abbreviation for the chemical pesticide dichloro diphenyl trichloroethane.

de Broglie matter waves All particles have wave properties; in de Broglie's equation, the product of momentum and wavelength equals Planck's constant.

decibel (dB) One-tenth of a *bel.*

de-excitation See *excitation.*

density Mass of a substance per unit volume. Weight density is weight per unit volume. In general, any item per space element (e.g., number of dots per area).

$$\text{density} = \frac{\text{mass}}{\text{volume}}$$

$$\text{weight density} = \frac{\text{weight}}{\text{volume}}$$

destructive interference Combination of waves so that crest parts of one wave overlap trough parts of another, resulting in a wave of decreased amplitude. See also *interference.*

deuterium Isotope of hydrogen whose atom has a proton, a neutron, and an electron. The common isotope of hydrogen has only a proton and an electron; therefore, deuterium has more mass.

deuteron Nucleus of a deuterium atom; it has one proton and one neutron.

dichroic crystal Crystal that divides unpolarized light into two internal beams polarized at right angles and strongly absorbs one beam while transmitting the other.

diffraction Bending of light that passes around an obstacle or through a narrow slit, causing the light to spread and to produce light and dark fringes.

diffraction grating Series of closely spaced parallel slits or grooves that are used to separate colors of light by interference.

diffuse reflection Reflection of waves in many directions from a rough surface. See also *polished.*

digital audio Audio reproduction system that uses binary code to record and reproduce sound.

digital signal Signal made up of discrete quantities or signals, as opposed to an analog signal which is based on a continuous signal.

diode Electronic device that restricts current to a single direction in an electric circuit; a device that changes alternating current to direct current.

dipole See *electric dipole.*

direct current (dc) Electric current whose flow of charge is always in one direction.

dispersion Separation of light into colors arranged according to their frequency, for example by interaction with a prism or a diffraction grating.

displaced Term applied to the fluid that is moved out of the way when an object is placed in fluid. A submerged object always displaces a volume of fluid equal to its own volume.

diverging lens Lens that is thinner in the middle than at the edges, causing parallel rays of light passing through it to diverge as if from a point. See also *converging lens.*

Doppler effect Change in frequency of a wave of sound or light due to the motion of the source or the receiver. See also *red shift* and *blue shift.*

echo Reflection of sound.

eddy Changing, curling paths in turbulent flow of a fluid.

efficiency In a machine, the ratio of useful energy output to total energy input, or the percentage of the work input that is converted to work output.

$$\text{Efficiency} = \frac{\text{useful energy output}}{\text{total energy input}}$$

elastic collision Collision in which colliding objects rebound without lasting deformation or heat generation.

elastic limit Distance of stretching or compressing beyond which an elastic material will not return to its original state.

elasticity Property of a solid wherein a change in shape is experienced when a deforming force acts on it, with a return to its original shape when the deforming force is removed.

electric charge Fundamental electrical property to which the mutual attractions or repulsions between electrons or protons is attributed.

electric current Flow of electric charge that transports energy from one place to another. Measured in amperes, where one ampere is the flow of 6.25×10^{18} electrons (or protons) per second.

electric dipole Molecule in which the distribution of charge is uneven, resulting in slightly opposite charges on opposite sides of the molecule.

electric field Force field that fills the space around every electric charge or group of charges. Measured by force per charge (newtons/coulomb).

electric potential Electric potential energy (in joules) per unit of charge (in coulombs) at a location in an electric field; measured in volts and often called voltage.

$$\text{voltage} = \frac{\text{electrical energy}}{\text{charge}} = \frac{\text{joules}}{\text{coulomb}}$$

electric potential energy Energy a charge has due to its location in an electric field.

electric power Rate of electrical energy transfer or the rate of doing work, which can be measured by the product of current and voltage.

$$\text{power} = \text{current} \times \text{voltage}$$

electrical force Force that one charge exerts on another. When the charges are the same sign, they repel; when the charges are opposite, they attract.

electrical resistance Resistance of a material to the flow of electric charge through it; measured in ohms (symbol Ω).

electrically polarized Term applied to an atom or molecule in which the charges are aligned so that one side is slightly more positive or negative than the opposite side.

electricity General term for electrical phenomena, much like gravity has to do with gravitational phenomena, or sociology with social phenomena.

electrode Terminal, for example of a battery, through which electric current can pass.

electrodynamics Study of moving electric charge, as opposed to electrostatics.

electromagnet Magnet whose magnetic properties are produced by electric current.

electromagnetic induction Phenomenon of inducing a voltage in a conductor by changing the magnetic field near the conductor. If the magnetic field within a closed loop changes in any way, a voltage is induced in the loop. The induction of voltage is actually the result of a more fundamental phenomenon: the induction of an electric field. See also *Faraday's law.*

electromagnetic radiation Transfer of energy by the rapid oscillations of electromagnetic fields, which travel in the form of waves called electromagnetic waves.

electromagnetic spectrum Range of frequencies over which electromagnetic radiation can be propagated. The lowest frequencies are associated with radio waves; microwaves have a higher frequency, and then infrared waves, light, ultraviolet radiation, X rays, and gamma rays in sequence.

electromagnetic wave Energy-carrying wave emitted by vibrating charges (often electrons) that is composed of oscillating electric and magnetic fields that regenerate one another. Radio waves, microwaves, infrared radiation, visible light, ultraviolet radiation, X rays, and gamma rays are all composed of electromagnetic waves.

electromotive force (emf) Any voltage that gives rise to an electric current. A battery or a generator is a source of emf.

electron Negative particle in the shell of an atom.

electron volt (eV) Amount of energy equal to that an electron acquires in accelerating through a potential difference of 1 volt.

electrostatics Study of electric charges at rest, as opposed to electrodynamics.

element Substance composed of atoms that all have the same atomic number and, therefore, the same chemical properties.

elementary particles Subatomic particles. The basic building blocks of all matter, consisting of two classes of particles, the quarks and the leptons.

ellipse Closed curve of oval shape wherein the sum of the distances from any point on the curve to two internal focal points is a constant.

emf Abbreviation for *electromotive force.*

emission spectrum Distribution of wavelengths in the light from a luminous source.

energy That which that can change the condition of matter. Commonly defined as the ability to do work; actually only describable by examples. It is *not* a material substance.

entropy A measure of the disorder of a system. Whenever energy freely transforms from one form to another, the direction of transformation is toward a state of greater disorder and therefore toward one of greater entropy.

equilibrium In general, a state of balance. For mechanical equilibrium, the state in which no net forces and no net torques act. In liquids, the state in which evaporation equals condensation. More generally, the state in which no net change of energy occurs.

equilibrium rule $\Sigma F = 0$. On an object or system of objects in mechanical equilibrium, the sum of forces equal zero. Also, $\Sigma\tau = 0$; the sum of the torques equal zero.

escape velocity Velocity that a projectile, space probe, etc., must reach to escape the gravitational influence of the Earth or celestial body to which it is attracted.

ether Hypothetical invisible medium that was formerly thought to be required for the propagation of electromagnetic waves, and thought to fill space throughout the universe.

eV Abbreviation for *electron volt.*

evaporation Change of phase from liquid to gas that takes place at the surface of a liquid. The opposite of condensation.

excitation Process of boosting one or more electrons in an atom or molecule from a lower to a higher energy level. An atom in an excited state will usually decay (de-excite) rapidly to a lower state by the emission of radiation. The frequency and energy of emitted radiation are related by

$$E = hf$$

excited See *excitation.*

eyepiece Lens of a telescope closest to the eye; which enlarges the real image formed by the first lens.

fact Close agreement by competent observers of a series of observations of the same phenomena.

Fahrenheit scale Temperature scale in common use in the United States. The number 32 is assigned to the melt-freeze point of water, and the number 212 to the boil-condense point of water at standard pressure (one atmosphere, at sea level).

Faraday's law Induced voltage in a coil is proportional to the product of the number of loops and the rate at which the magnetic field changes within those loops. In general, an electric field is induced in any region of space in which a magnetic field is changing with time. The magnitude of the induced electric field is proportional to the rate at which the magnetic field changes. See also *Maxwell's counterpart to Faraday's law.*

$$\text{voltage induced} \sim \text{number of loops} \times \frac{\text{magnetic field change}}{\text{change in time}}$$

Fermat's principle of least time Light takes the path that requires the least time when it goes from one place to another.

field See *force field.*

field lines See *magnetic field lines.*

flotation See *principle of flotation.*

fluid Anything that flows; in particular, any liquid or gas.

fluorescence Property of certain substances to absorb radiation of one frequency and to re-emit radiation of a lower frequency.

FM Abbreviation for *frequency modulation.*

focal length Distance between the center of a lens and either focal point; the distance from a mirror to its focal point.

focal plane Plane, perpendicular to the principal axis, that passes through a focal point of a lens or mirror. For a converging lens or a concave mirror, any incident parallel rays of light converge to a point somewhere on a focal plane. For a diverging lens or a convex mirror, the rays appear to come from a point on a focal plane.

focal point For a converging lens or a concave mirror, the point at which rays of light parallel to the principal axis converge. For a diverging lens or a convex mirror, the point from which such rays appear to come.

focus (pl. foci) (a) For an ellipse, one of the two points for which the sum of the distances to any point on the ellipse is a constant. A satellite orbiting the Earth moves in an ellipse that has the Earth at one focus. (b) For optics, a focal point.

force Any influence that tends to accelerate an object; a push or pull; measured in newtons. Force is a vector quantity.

force field That which exists in the space surrounding a mass, electric charge, or magnet, so that another mass, electric charge, or magnet introduced into this region will experience a force. Examples of force fields are gravitational fields, electric fields, and magnetic fields.

forced vibration Vibration of an object caused by the vibrations of a nearby object. The sounding board in a musical instrument amplifies the sound through forced vibration.

Fourier analysis Mathematical method that disassembles any periodic wave form into a combination of simple sine waves.

fovea Area of the retina that is in the center of the field of view; region of most distinct vision.

frame of reference Vantage point (usually a set of coordinate axes) with respect to which position and motion may be described.

Fraunhofer lines Dark lines visible in the spectrum of the sun or a star.

free fall Motion under the influence of gravity only.

free radical Unbonded, electrically neutral, very chemically active atom or molecular fragment.

freezing Change in phase from liquid to solid; the opposite of melting.

frequency For a vibrating body or medium, the number of vibrations per unit time. For a wave, the number of crests that pass a particular point per unit time. Frequency is measured in hertz.

frequency modulation (FM) Type of modulation in which the frequency of the carrier wave is varied above and below its normal frequency by an amount that is proportional to the amplitude of the impressed signal. In this case, the amplitude of the modulated carrier wave remains constant.

friction Force that acts to resist the relative motion (or attempted motion) of objects or materials that are in contact.

fuel cell A device that converts chemical energy to electrical energy, but unlike a battery is continually fed with fuel, usually hydrogen.

fulcrum Pivot point of a lever.

fundamental frequency See *partial tone.*

fuse Device in an electric circuit that breaks the circuit when the current gets high enough to risk causing a fire.

g (a) Abbreviation for *gram.* (b) When in lower-case italic g, the symbol for the acceleration due to gravity (at the Earth's surface, 9.8 m/s^2). (c) When in lower case bold **g** the gravitational field vector (at the Earth's surface, 9.8 N/kg). (d) When in upper-case italic G, the symbol for the *universal gravitation constant* (6.67×10^{-11} N·m^2/kg^2).

galvanometer Instrument used to detect electric current. With the proper combination of resistors, it can be converted to an ammeter or a voltmeter. An ammeter is calibrated to measure electric current. A voltmeter is calibrated to measure electric potential.

gamma ray High-frequency electromagnetic radiation emitted by atomic nuclei.

gas Phase of matter beyond the liquid phase, wherein molecules fill whatever space is available to them, taking no definite shape.

general theory of relativity Einstein's generalization of special relativity, which deals with accelerated motion and features a geometric theory of gravitation.

generator Machine that produces electric current, usually by rotating a coil within a stationary magnetic field.

geodesic Shortest path between points on any surface.

geosynchronous orbit A satellite orbit in which the satellite orbits the Earth once each day. When moving westward, the satellite remains at a fixed point (about 42,000 km) above the Earth's surface.

global warming See *greenhouse effect.*

gram (g) A metric unit of mass. It is one thousandth of a kilogram.

gravitation Attraction between objects due to mass. See also *law of universal gravitation and universal gravitational constant.*

gravitational field Force field that exists in the space around every mass or group of masses; measured in newtons per kilogram.

gravitational potential energy Energy that a body possesses because of its position in a gravitational field. On Earth, potential energy (PE) equals mass (m) times the acceleration due to gravity (g) times height (h) from a reference level such as the Earth's surface.

$$\text{PE} = mgh$$

gravitational red shift Shift in wavelength toward the red end of the spectrum experienced by light leaving the surface of a massive object, as predicted by the general theory of relativity.

gravitational wave Gravitational disturbance that propagates through spacetime made by a moving mass. (Undetected at this writing.)

graviton Quantum of gravity, similar in concept to the photon as a quantum of light. (Undetected at this writing.)

greenhouse effect Warming effect caused by short-wavelength radiant energy from the sun that easily enters the atmosphere and is absorbed by the Earth, but when radiated at longer wavelengths cannot easily escape the Earth's atmosphere.

grounding Allowing charges to move freely along a connection from a conductor to the ground.

group Elements in the same column of the periodic table.

h (a) Abbreviation for hour (though hr. is often used). (b) When in italic h, the symbol for *Planck's constant.*

hadron Elementary particle that can participate in strong nuclear force interactions.

half-life Time required for half the atoms of a radioactive isotope of an element to decay. This term is also used to describe decay processes in general.

harmonic See *partial tone.*

heat The energy that flows from one object to another by virtue of a difference in temperature. Measured in *calories* or *joules.*

heat capacity See *specific heat capacity.*

heat engine A device that uses heat as input and supplies mechanical work as output, or that uses work as input and moves heat "uphill" from a cooler to a warmer place.

heat of fusion Amount of energy that must be added to a kilogram of a solid (already at its melting point) to melt it.

heat of vaporization Amount of energy that must be added to a kilogram of a liquid (already at its boiling point) to vaporize it.

heat pump A device that transfers heat out of a cool environment and into a warm environment.

heat waves See *infrared waves.*

heavy water Water (H_2O) that contains the heavy hydrogen isotope deuterium.

hertz (Hz) SI unit of frequency. One hertz is one vibration per second.

hologram Two-dimensional microscopic interference pattern that shows three-dimensional optical images.

Hooke's law Distance of stretch or squeeze (extension or compression) of an elastic material is directly proportional to the applied force. Where Δx is the change in length and k is the spring constant,

$$F = k\Delta x$$

humidity Measure of the amount of water vapor in the air. Absolute humidity is the mass of water per volume of air. Relative humidity is absolute humidity at that temperature divided by the maximum possible, usually given as a percent.

Huygens' principle Light waves spreading out from a light source can be regarded as a superposition of tiny secondary wavelets.

hypothesis Educated guess; a reasonable explanation of an observation or experimental result that is not fully accepted as factual until tested over and over again by experiment.

Hz Abbreviation for *hertz.*

ideal efficiency Upper limit of efficiency for all heat engines; it depends on the temperature difference between input and exhaust.

$$\text{ideal efficiency} = \frac{T_{\text{hot}} - T_{\text{cold}}}{T_{\text{hot}}}$$

impulse Product of force and the time interval during which the force acts. Impulse produces change in momentum.

$$\text{impulse} = Ft = \Delta\ (mv)$$

incandescence State of glowing while at a high temperature, caused by electrons bouncing around over dimensions larger than the size of an atom, emitting radiant energy in the process. The peak frequency of radiant energy is proportional to the absolute temperature of the heated substance:

$$\bar{f} \sim T$$

incoherent light Light containing waves with a jumble of frequencies, phases, and possibly directions. See also *coherent light* and *laser.*

index of refraction (n) Ratio of the speed of light in a vacuum to the speed of light in another material.

$$n = \frac{\text{speed of light in vacuum}}{\text{speed of light in material}}$$

induced (a) Term applied to electric charge that has been redistributed on an object due to the presence of a charged object nearby. (b) Term applied to a voltage, electric field, or magnetic field that is created due to a change in or motion through a magnetic field or electric field.

induction Charging of an object without direct contact. See also *electromagnetic induction.*

inelastic Term applied to a material that does not return to its original shape after it has been stretched or compressed.

inelastic collision Collision in which the colliding objects become distorted and/or generate heat during the collision, and possibly stick together.

inertia Sluggishness or apparent resistance of an object to change its state of motion. Mass is the measure of inertia.

inertial frame of reference Unaccelerated vantage point in which Newton's laws hold exactly.

infrared Electromagnetic waves of frequencies lower than the red of visible light.

infrared waves Electromagnetic waves that have a lower frequency than visible red light.

infrasonic Term applied to sound frequencies below 20 hertz, the normal lower limit of human hearing.

in parallel Term applied to portions of an electric circuit that are connected at two points and provide alternative paths for the current between those two points.

in phase Term applied to two or more waves whose crests (and troughs) arrive at a place at the same time, so that their effects reinforce each other.

in series Term applied to portions of an electric circuit that are connected in a row so that the current that goes through one must go through all of them.

instantaneous speed Speed at any instant.

insulator (a) Material that is a poor conductor of heat and that delays the transfer of heat. (b) Material that is a poor conductor of electricity.

intensity Power per square meter carried by a sound wave, often measured in decibels.

interaction Mutual action between objects where each object exerts an equal and opposite force on the other.

interference Result of superposing different waves, often of the same wavelength. Constructive interference results from crest-to-crest reinforcement; destructive interference results from crest-to-trough cancellation. The interference of selected wavelengths of light produces colors known as interference colors. See also *constructive interference, destructive interference, interference pattern,* and *standing wave.*

interference pattern Pattern formed by the overlapping of two or more waves that arrive in a region at the same time.

interferometer Device that uses the interference of light waves to measure very small distances with high accuracy. Michelson and Morley used an interferometer in their famous experiments with light.

internal energy The total energy stored in the atoms and molecules within a substance. Changes in internal energy are of principal concern in thermodynamics.

inverse-square law Law relating the intensity of an effect to the inverse square of the distance from the cause. Gravity, electric, magnetic, light, sound, and radiation phenomena follow the inverse-square law.

$$\text{intensity} \sim \frac{1}{\text{distance}^2}$$

inversely When two values change in opposite directions, so that if one increases and the other decreases by the same amount, they are said to be inversely proportional to each other.

ion Atom (or group of atoms bound together) with a net electric charge, which is due to the loss or gain of electrons. A positive ion has a net positive charge. A negative ion has a net negative charge.

ionization Process of adding or removing electrons to or from the atomic nucleus.

iridescence Phenomenon whereby interference of light waves of mixed frequencies reflected from the top and bottom of thin films produces an assortment of colors.

iris Colored part of the eye that surrounds the black opening through which light passes. The iris regulates the amount of light entering the eye.

isotopes Atoms whose nuclei have the same number of protons but different numbers of neutrons.

J Abbreviation for *joule.*

joule (J) SI unit of work and of all other forms of energy. One joule of work is done when a force of one newton is exerted on an object moved one meter in the direction of the force.

K (a) Abbreviation for *kelvin.* (b) When in lower case k, the abbreviation for the prefix *kilo-*. (c) When in lower case italics *k*, the symbol for the electrical proportionality constant in *Coulomb's law.* It is approximately $9 \times 10^9\ \text{N} \cdot \text{m}^2/\text{C}^2$.(d) When in lower case italics *k*, the symbol for the spring constant in *Hooke's law.*

kcal Abbreviation for *kilocalorie.*

KE Abbreviation for *kinetic energy.*

kelvin SI unit of temperature. A temperature measured in kelvins (symbol K) indicates the number of units above absolute zero. Divisions on the Kelvin scale and Celsius scale are the same size, so a change in temperature of one kelvin equals a change in temperature of one Celsius degree.

Kelvin scale Temperature scale, measured in kelvins K, whose zero (called absolute zero) is the temperature at which it is impossible to extract any more internal energy from a material. 0 K = −273.15°C. There are no negative temperatures on the Kelvin scale.

Kepler's laws

Law 1: The path of each planet around the Sun is an ellipse with the Sun at one focus.

Law 2: The line from the Sun to any planet sweeps out equal areas of space in equal time intervals.

Law 3: The square of the orbital period of a planet is directly proportional to the cube of the average distance of the planet from the Sun ($T^2 \sim r^3$ for all planets).

kg Abbreviation for *kilogram.*

kilo- Prefix that means thousand, as in kilowatt or kilogram.

kilocalorie (kcal) Unit of heat. One kilocalorie equals 1000 calories, or the amount of heat required to raise the temperature of one kilogram of water by 1°C. Equal to one food Calorie.

kilogram (kg) Fundamental SI unit of mass. It is equal to 1000 grams. One kilogram is very nearly the amount of mass in one liter of water at 4°C.

kilometer (km) One thousand *meters.*

kilowatt (kW) One thousand *watts.*

kilowatt-hour (kWh) Amount of energy consumed in 1 hour at the rate of 1 kilowatt.

kinetic energy (KE) Energy of motion, equal (nonrelativistically) to half the mass multiplied by the speed squared.

$$\text{KE} = \frac{1}{2}mv^2$$

km Abbreviation for *kilometer.*

kPa Abbreviation for *kilopascal.* See *pascal.*

kWh Abbreviation for *kilowatt-hour.*

L Abbreviation for *liter.* (In some textbooks, lowercase l is used.)

laser Optical instrument that produces a beam of coherent light—that is, light with all waves of the same frequency, phase, and direction. The word is an acronym for light amplification by stimulated emission of radiation.

latent heat of fusion The amount of energy required to change a unit mass of a substance from solid to liquid (and vice versa).

latent heat of vaporization The amount of energy required to change a unit mass of a substance from liquid to gas (and vice versa).

law General hypothesis or statement about the relationship of natural quantities that has been tested over and over again and has not been contradicted. Also known as a *principle.*

law of conservation of momentum In the absence of an external force, the momentum of a system remains unchanged. Hence, the momentum before an event involving only internal forces is equal to the momentum after the event:

$$mv_{(\text{before event})} = mv_{(\text{after event})}$$

law of inertia See *Newton's laws of motion, Law 1.*

law of reflection The angle of incidence for a wave that strikes a surface is equal to the angle of reflection. This is true for both partially and totally reflected waves. See also *angle of incidence* and *angle of reflection.*

law of universal gravitation For any pair of particles, each particle attracts the other particle with a force that is directly proportional to the product of the masses of the particles, and inversely proportional to the square of the distance between them (or their centers of mass if spherical objects), where F is the force, m is the mass, d is distance, and G is the gravitation constant:

$$F \sim \frac{m_1 m_2}{d^2} \quad \text{or} \quad F = G\frac{m_1 m_2}{d^2}$$

least time See *Fermat's principle of least time.*

length contraction Shrinkage of space, and therefore of matter, in a frame of reference moving at relativistic speeds.

lens Piece of glass or other transparent material that can bring light to a focus.

lepton Class of elementary particles that are not involved with the nuclear force. It includes the electron and its neutrino, the muon and its neutrino, and the tau and its neutrino.

lever Simple machine made of a bar that turns about a fixed point called the fulcrum.

lever arm Perpendicular distance between an axis and the line of action of a force that tends to produce rotation about that axis.

lift In application of Bernoulli's principle, the net upward force produced by the difference between upward and downward pressures. When lift equals weight, horizontal flight is possible.

light Visible part of the electromagnetic spectrum.

light-year The distance light travels in a vacuum in one year: 9.46×10^{12} km.

line spectrum Pattern of distinct lines of color, corresponding to particular wavelengths, that are seen in a spectroscope when a hot gas is viewed. Each element has a unique pattern of lines.

linear momentum Product of the mass and the velocity of an object. Also called momentum. (This definition applies at speeds much less than the speed of light.)

linear motion Motion along a straight-line path.

linear speed Path distance moved per unit of time. Also called simply speed.

liquid Phase of matter between the solid and gaseous phases in which the matter possesses a definite volume but no definite shape: it takes on the shape of its container.

liter (L) Metric unit of volume. A liter is equal to 1000 cm^3.

logarithmic Exponential.

longitudinal wave Wave in which the individual particles of a medium vibrate back and forth in the direction in which the wave travels—for example, sound.

Lorentz contraction See *length contraction.*

loudness Physiological sensation directly related to sound intensity or volume. Relative loudness, or sound level, is measured in decibels.

lunar eclipse Event wherein the full moon passes into the shadow of the Earth.

m (a) Abbreviation for *meter.* (b) When in italic *m,* the abbreviation for *mass.*

Mach number Ratio of the speed of an object to the speed of sound. For example, an aircraft traveling *at* the speed of sound is rated Mach 1.0; traveling at *twice* the speed of sound, Mach 2.0.

machine Device for increasing (or decreasing) a force or simply changing the direction of a force.

magnet Any object that has magnetic properties, that is the ability to attract objects made of iron or other magnetic substances. See also *electromagnetism* and *magnetic force.*

magnetic declination Discrepancy between the orientation of a compass pointing toward magnetic north and the true geographic north.

magnetic domain Microscopic cluster of atoms with their magnetic fields aligned.

magnetic field Region of magnetic influence around a magnetic pole or a moving charged particle.

magnetic field lines Lines showing the shape of a magnetic field. A compass placed on such a line will turn so that the needle is aligned with it.

magnetic force (a) Between magnets, it is the attraction of unlike magnetic poles for each other and the repulsion between like magnetic poles. (b) Between a magnetic field and a moving charged particle, it is a deflecting force due to the motion of the particle: The deflecting force is perpendicular to the magnetic field lines and the direction of motion. This force is greatest when the charged particle moves perpendicular to the field lines and is smallest (zero) when it moves parallel to the field lines.

magnetic monopole Hypothetical particle having a single north or a single south magnetic pole, analogous to a positive or negative electric charge.

magnetic pole One of the regions on a magnet that produces magnetic forces.

magnetic pole reversal When the magnetic field of an astronomical body reverses its poles, that is, the location where the north magnetic pole existed becomes the south magnetic pole, and the south magnetic pole becomes the north magnetic pole.

magnetism Property of being able to attract objects made of iron, steel, or magnetite. See also *electromagnetism* and *magnetic force.*

magnetohydrodynamic (MDH) generator Device generating electric power by interaction of a plasma and a magnetic field.

maser Instrument that produces a beam of microwaves. The word is an acronym for **m**icrowave **a**mplification by **s**timulated **e**mission of **r**adiation.

mass (m) Quantity of matter in an object; the measurement of the inertia or sluggishness that an object exhibits in response to any effort made to start it, stop it, or change in any way its state of motion; a form of energy.

mass spectrometer Device that magnetically separates charged ions according to their mass.

mass-energy equivalence Relationship between mass and energy as given by the equation

$$E = mc^2$$

where c is the speed of light.

matter waves See *de Broglie matter waves.*

Maxwell's counterpart to Faraday's law A magnetic field is created in any region of space in which an electric field is changing with time. The magnitude of the induced magnetic field is proportional to the rate at which the electric field changes. The direction of the induced magnetic field is at right angles to the changing electric field.

mechanical advantage Ratio of output force to input force for a machine.

mechanical energy Energy due to the position or the movement of something; potential or kinetic energy (or a combination of both).

mechanical equilibrium State of an object or system of objects for which any impressed forces cancel to zero and no acceleration occurs and when no net torque exists. That is, $\Sigma F = 0$, and $\Sigma\tau = 0$.

mega- Prefix that means million, as in megahertz or megajoule.

melting Change in phase from solid to liquid; the opposite of freezing. Melting is a different process from dissolving, in which an added solid mixes with a liquid and the solid dissociates.

meson Elementary particle with an atomic weight of zero; can participate in the strong interaction.

metastable state State of excitation of an atom that is characterized by a prolonged delay before de-excitation.

meter (m) Standard SI unit of length (3.28 feet).

MeV Abbreviation for million *electron volts,* a unit of energy, or equivalently a unit of mass.

MHD Abbreviation for *magnetohydrodynamic.*

mi Abbreviation for mile.

microscope Optical instrument that forms enlarged images of very small objects.

microwaves Electromagnetic waves with frequencies greater than radio waves but less than infrared waves.

min Abbreviation for minute.

mirage False image that appears in the distance and is due to the refraction of light in the Earth's atmosphere.

mixture Substances mixed together without combining chemically.

MJ Abbreviation for megajoules, million *joules.*

model Representation of an idea created to make the idea more understandable.

modulation Impressing a signal wave system on a higher frequency carrier wave, amplitude modulation (AM) for amplitude signals and frequency modulation (FM) for frequency signals.

molecule Two or more atoms of the same or different elements bonded to form a larger particle.

momentum Inertia in motion. The product of the mass and the velocity of an object (provided the speed is much less than the speed of light). Has magnitude and direction and therefore is a vector quantity. Also called linear momentum, and abbreviated p.

$$p = mv$$

monochromatic light Light made of only one color and therefore waves of only one wavelength and frequency.

muon Elementary particle in the class of elementary particles called leptons. It is short-lived with a mass that is 207 times that of an electron; may be positively or negatively charged.

music Scientifically speaking, sound with periodic tones, which appear on an oscilloscope as a regular pattern.

N Abbreviation for *newton.*

nanometer Metric unit of length that is 10^{-9} meter (one billionth of a meter).

natural frequency Frequency at which an elastic object naturally tends to vibrate if it is disturbed and the disturbing force is removed.

neap tide Tide that occurs when the moon is halfway between a new moon and a full moon, in either direction. The tides due to the sun and the moon partly cancel, so that the high tides are lower than average and the low tides are not as low as average. See also *spring tide.*

net force Combination of all the forces that act on an object.

neutrino Elementary particle in the class of elementary particles called leptons. It is uncharged and almost massless; three kinds—electron, muon, and tau neutrinos, are the most common high-speed particles in the universe; more than a billion pass unhindered through each person every second.

neutron Electrically neutral particle that is one of the two kinds of nucleons that compose an atomic nucleus.

neutron star Star that has undergone a gravitational collapse in which electrons are compressed into protons to form neutrons.

newton (N) SI unit of force. One newton is the force applied to a one-kilogram mass that will produce an acceleration of one meter per second per second.

Newton's law of cooling The rate of cooling of an object—whether by conduction, convection, or radiation—is approximately proportional to the temperature difference between the object and its surroundings.

Newton's laws of motion

Law 1: Every body continues in its state of rest, or of motion in a straight line at constant speed, unless it is compelled to change that state by a net force exerted upon it. Also known as the law of inertia.

Law 2: The acceleration produced by a net force on a body is directly proportional to the magnitude of the net force, is in the same direction as the net force, and is inversely proportional to the mass of the body.

Law 3: Whenever one body exerts a force on a second body, the second body exerts an equal and opposite force on the first.

node Any part of a standing wave that remains stationary; a region of minimal or zero energy.

noise Scientifically speaking, sound that corresponds to an irregular vibration of the eardrum produced by some irregular vibration, which appears on an oscilloscope as an irregular pattern.

nonlinear motion Any motion not along a straight-line path.

normal At right angles to, or perpendicular to. A normal force acts at right angles to the surface on which it acts. In optics, a normal defines the line perpendicular to a surface about which angles of light rays are measured.

normal force Component of support force perpendicular to a supporting surface. For an object resting on a horizontal surface, it is the upward force that balances the weight of the object.

northern lights See *aurora borealis.*

nuclear fission Splitting of an atomic nucleus, particularly that of a heavy element such as uranium-235, into two lighter elements, accompanied by the release of much energy.

nuclear force Attractive force within a nucleus that holds neutrons and protons together. Part of the nuclear force is called the strong interaction. The strong interaction is an attractive force that acts between protons, neutrons, and mesons (another nuclear particle); however, it acts only over very short distances (10^{-15} meter). The weak interaction is the nuclear force responsible for beta (electron) emission.

nuclear fusion Combining of nuclei of light atoms, such as hydrogen, into heavier nuclei, accompanied by the release of much energy. See also *thermonuclear fusion.*

nuclear reactor Apparatus in which controlled nuclear fission or nuclear fusion reactions take place.

nucleon Principal building block of the nucleus. A neutron or a proton; the collective name for either or both.

nucleus (pl. nuclei) Positively charged center of an atom, which contains protons and neutrons and has almost all the mass of the entire atom but only a tiny fraction of the volume.

objective lens In an optical device using compound lenses, the lens closest to the object observed.

octave In music, the eighth full tone above or below a given tone. The tone an octave above has twice as many vibrations per second as the original tone; the tone an octave below has half as many vibrations per second.

ohm (Ω) SI unit of electrical resistance. One ohm is the resistance of a device that draws a current of one ampere when a voltage of one volt is impressed across it.

Ohm's law Current in a circuit is directly proportional to the voltage impressed across the circuit, and is inversely proportional to the resistance of the circuit.

$$\text{Current} = \frac{\text{voltage}}{\text{resistance}}$$

opaque Term applied to materials that absorb light without re-emission, and consequently do not allow light through them.

optical fiber Transparent fiber, usually of glass or plastic, that can transmit light down its length by means of total internal reflection.

oscillation Same as vibration: a repeating to-and-fro motion about an equilibrium position. Both oscillation and vibration refer to periodic motion, that is, motion that repeats.

oscillatory motion To-and-fro vibratory motion, such as that of a pendulum.

out of phase Term applied to two waves for which the crest of one wave arrives coincident with a trough of the second wave. Their effects tend to cancel each other.

overtone Musical term where the first overtone is the second harmnic. See also *partial tone.*

oxidize Chemical process in which an element or molecule loses one or more electrons.

ozone Gas, found in a thin layer in the upper atmosphere, composed of molecules of three oxygen atoms. Atmospheric oxygen gas is composed of molecules of two oxygen atoms.

Pa Abbreviation for the SI unit *pascal.*

parabola Curved path followed by a projectile acting under the influence of gravity only.

parallax Apparent displacement of an object when viewed by an observer from two different positions; often used to calculate the distance of stars.

parallel circuit Electric circuit with two or more devices connected in such a way that the same voltage acts across each one and any single one completes the circuit independently of the others. See also *in parallel.*

partial tone One of the many tones that make up one musical sound. Each partial tone (or partial) has only one frequency. The lowest partial of a musical sound is called the fundamental frequency. Any partial whose frequency is a multiple of the fundamental frequency is called a harmonic. The fundamental frequency is also called the first harmonic. The second harmonic has twice the frequency of the fundamental; the third harmonic, three times the frequency, and so on.

pascal (Pa) SI unit of pressure. One pascal of pressure exerts a normal force of one newton per square meter. A kilopascal (kPa) is 1000 pascals.

Pascal's principle Changes in pressure at any point in an enclosed fluid at rest are transmitted undiminished to all points in the fluid and act in all directions.

PE Abbreviation for *potential energy.*

penumbra Partial shadow that appears where some of the light is blocked and other light can fall. See also *umbra.*

percussion In musical instruments, the striking of one object against another.

perigee Point in an elliptical orbit closest to the focus about which orbiting takes place. See also *apogee.*

period In general, the time required to complete a single cycle. (a) For orbital motion, the time required for a complete orbit. (b) For vibrations or waves, the time required for one complete cycle, equal to 1/frequency.

periodic table Chart that lists elements by atomic number and by electron arrangements, so that elements with similar chemical properties are in the same column (group). See Figure 11.15, pages 214–215.

perturbation Deviation of an orbiting object (e.g., a planet) from its path around a center of force (e.g., the sun) caused by the action of an additional center of force (e.g., another planet).

phase (a) One of the four main forms of matter: solid, liquid, gas, and plasma. Often called *state.* (b) The part of a cycle that a wave has advanced at any moment. See also *in phase* and *out of phase.*

phosphor Powdery material, such as that used on the inner surface of a fluorescent light tube, that absorbs ultraviolet photons, then gives off visible light.

phosphorescence Type of light emission that is the same as fluorescence except for a delay between excitation and de-excitation, which provides an afterglow. The delay is caused by atoms being excited to energy levels that do not decay rapidly. The afterglow may last from fractions of a second to hours, or even days, depending on such factors as the type of material and temperature.

photoelectric effect Ejection of electrons from certain metals when exposed to certain frequencies of light.

photon Localized corpuscle of electromagnetic radiation whose energy is proportional to its radiation frequency: $E \sim f$, or $E = hf$, where h is Planck's constant.

pigment Fine particles that selectively absorb light of certain frequencies and selectively transmit others.

pitch Term that refers to our subjective impression about the "highness" or "lowness" of a tone, which is related to the frequency of the tone. A high-frequency vibrating source produces a sound of high pitch; a low-frequency vibrating source produces a sound of low pitch.

Planck's constant (*h*) Fundamental constant of quantum theory that determines the scale of the small-scale world. Planck's constant multiplied by the frequency of radiation gives the energy of a photon of that radiation.

$$E = hf, \quad h = 6.6 \times 10^{-34} \text{ joule-second}$$

plane mirror Flat-surfaced mirror.

plane-polarized wave A wave confined to a single plane.

plasma Fourth phase of matter, in addition to solid, liquid, and gas. In the plasma phase, existing mainly at high temperatures, matter consists of positively charged ions and free electrons.

polarization Aligning of vibrations in a transverse wave, usually by filtering out waves of other directions. See also *plane-polarized wave* and *dichroic crystal.*

polished Describes a surface that is so smooth that the distances between successive elevations of the surface are less than about one-eighth the wavelength of the light or other incident wave of interest. The result is very little diffuse reflection.

positron Antiparticle of an electron; a positively charged electron.

postulates of special relativity

First: All laws of nature are the same in all uniformly moving frames of reference.

Second: The speed of light in free space has the same measured value regardless of the motion of the source or the motion of the observer; that is, the speed of light is invariant.

potential difference Difference in electric potential (voltage) between two points. Free charge flows when there is a difference, and will continue until both points reach a common potential.

potential energy (PE) Energy of position, usually related to the relative position of two things, such as a stone and the Earth (gravitational PE), or an electron and a nucleus (electric PE).

power Rate at which work is done or energy is transformed, equal to the work done or energy transformed divided by time; measured in watts.

$$\text{power} = \frac{\text{work}}{\text{time}}$$

precession Wavering of a spinning object, such that its axis of rotation traces out a cone.

pressure Force per surface area where the force is normal (perpendicular) to the surface; measured in pascals. See also *atmospheric pressure.*

$$\text{pressure} = \frac{\text{force}}{\text{area}}$$

primary colors See *additive primary colors* and *subtractive primary colors.*

principal axis Line joining the centers of curvature of the surfaces of a lens. Line joining the center of curvature and the focus of a mirror.

principle General hypothesis or statement about the relationship of natural quantities that has been tested over and over again and has not been contradicted; also known as a law.

principle of equivalence Observations made in an accelerating frame of reference are indistinguishable from observations made in a gravitational field.

principle of flotation A floating object displaces a weight of fluid equal to its own weight.

prism Triangular solid of a transparent material such as glass, that separates incident light by refraction into its component colors. These component colors are often called the spectrum.

projectile Any object that moves through the air or through space, acted on only by gravity (and air resistance, if any).

proton Positively charged particle that is one of the two kinds of nucleons in the nucleus of an atom.

pseudoscience Fake science that pretends to be real science.

pulley Wheel that acts as a lever used to change the direction of a force. A pulley or system of pulleys can also multiply forces.

pupil Opening in the eyeball through which light passes.

quality Characteristic timbre of a musical sound, governed by the number and relative intensities of partial tones.

quantum (pl. quanta) From the Latin word *quantus,* meaning "how much," a quantum is the smallest elemental unit of a quantity, the smallest discrete amount of something. One quantum of electromagnetic energy is called a photon. See also *quantum mechanics* and *quantum theory.*

quantum mechanics Branch of physics concerned with the atomic microworld based on wave functions and probabilities, introduced by Max Planck (1900) and developed by Werner Heisenberg (1925), Erwin Schrödinger (1926), and others.

quantum physics Branch of physics that is the general study of the microworld of photons, atoms, and nuclei.

quantum theory Theory that describes the microworld, where many quantities are granular (in units called quanta), rather than continuous, and where particles of light (photons) and particles of matter (such as electrons) exhibit wave as well as particle properties.

quark One of the two classes of elementary particles. (The other is the lepton.) Two of the six quarks (up and down) are the fundamental building blocks of nucleons (protons and neutrons).

rad Unit used to measure a dose of radiation; the amount of energy (in centijoules) absorbed from ionizing radiation per kilogram of exposed material.

radiant energy Any energy, including heat, light, and X rays, that is transmitted by radiation. It occurs in the form of electromagnetic waves.

radiation (a) Energy transmitted by electromagnetic waves. (b) The particles given off by radioactive atoms such as uranium. Do not confuse radiation with radioactivity.

radiation curve of sunlight See *solar radiation curve.*

radio waves Electromagnetic waves of the longest frequency.

radioactive Term applied to an atom having an unstable nucleus that can spontaneously emit a particle and become the nucleus of another element.

radioactivity Process of the atomic nucleus that results in the emission of energetic particles. See *radiation.*

radiotherapy Use of radiation as a treatment to kill cancer cells.

rarefaction Region of reduced pressure in a longitudinal wave.

rate How fast something happens or how much something changes per unit of time; a change in a quantity divided by the time it takes for the change to occur.

ray Thin beam of light. Also lines drawn to show light paths in optical ray diagrams.

reaction force Force that is equal in strength and opposite in direction to the action force, and one that acts simultaneously on whatever is exerting the action force. See also *Newton's third law.*

real image Image formed by light rays that converge at the location of the image. A real image, unlike a virtual image, can be displayed on a screen.

red shift Decrease in the measured frequency of light (or other radiation) from a receding source; called the *red shift* because the decrease is toward the low-frequency, or red, end of the color spectrum. See also *Doppler effect.*

reflection Return of light rays from a surface in such a way that the angle at which a given ray is returned is equal to the angle at which it strikes the surface. When the reflecting surface is irregular, the light is returned in irregular directions; this is *diffuse reflection.* In general, the bouncing back of a particle or wave that strikes the boundary between two media.

refraction Bending of an oblique ray of light when it passes from one transparent medium to another. This is caused by a difference in the speed of light in the transparent media. In general, the change in direction of a wave as it crosses the boundary between two media in which the wave travels at different speeds.

regelation Process of melting under pressure and the subsequent refreezing when the pressure is removed.

relationship of impulse and momentum Impulse is equal to the change in the momentum of the object that the impulse acts upon. In symbol notation,

$$Ft = \Delta mv$$

relative Regarded in relation to something else, depending on point of view, or frame of reference. Sometimes referred to as "with respect to."

relative humidity Ratio between how much water vapor is in the air and the maximum amount of water vapor that could be in the air at the same temperature.

relativistic Pertaining to the theory of relativity; or approaching the speed of light.

relativity See *special theory of relativity, postulates of the special theory of relativity,* and *general theory of relativity.*

rem Acronym of **r**oentgen **e**quivalent **m**an, it is a unit used to measure the effect of ionizing radiation on human beings.

resistance See *electrical resistance.*

resistor Device in an electric circuit designed to resist the flow of charge.

resolution (a) Method of separating a vector into its component parts. (b) Ability of an optical system to make clear or to separate the components of an object viewed.

resonance Phenomenon that occurs when the frequency of forced vibrations on an object matches the object's natural frequency, producing a dramatic increase in amplitude.

rest energy The "energy of being" given by the equation $E = mc^2$.

resultant Net result of a combination of two or more vectors.

retina Layer of light-sensitive tissue at the back of the eye, composed of tiny light-sensitive antennae called rods and cones. Rods sense light and darkness. Cones sense color.

reverberation Persistence of a sound, as in an echo, due to multiple reflections.

revolution Motion of an object turning around an axis that lies outside the object.

Ritz combination principle For an element, the frequencies of some spectral lines are either the sum or the difference of the frequencies of two other lines in that element's spectrum.

rods See *retina.*

rotation Spinning motion that occurs when an object rotates about an axis located within the object (usually an axis through its center of mass).

rotational inertia Reluctance or apparent resistance of an object to change its state of rotation, determined by the distribution of the mass of the object and the location of the axis of rotation or revolution.

rotational speed Number of rotations or revolutions per unit of time; often measured in rotations or revolutions per second or minute.

rotational velocity Rotational speed together with a direction for the axis of rotation or revolution.

RPM Abbreviation for rotations or revolutions per minute.

s Abbreviation for second.

satellite Projectile or smaller celestial body that orbits a larger celestial body.

saturated Term applied to a substance, such as air, that contains the maximum amount of another substance, such as water vapor, at a given temperature and pressure.

scalar quantity Quantity in physics, such as mass, volume, and time, that can be completely specified by its magnitude, and has no direction.

scale In music, a succession of notes of frequencies that are in simple ratios to one another.

scaling Study of how size affects the relationship among weight, strength, and surface area.

scatter To absorb sound or light and re-emit it in all directions.

scattering Emission in random directions of light that encounters particles that are small compared to the wavelength of light; more often at short wavelengths (blue) than at long wavelengths (red).

Schrödinger's wave equation Fundamental equation of quantum mechanics, which interprets the wave nature of material particles in terms of probability wave amplitudes. It is as basic to quantum mechanics as Newton's laws of motion are to classical mechanics.

scientific method Orderly method for gaining, organizing, and applying new knowledge.

self-induction Induction of an electric field within a single coil, caused by the interaction of the loops within the same coil. This self-induced voltage is always in a direction opposing the changing voltage that produces it, and is commonly called back electromotive force or back emf.

semiconductor Device made of material not only with properties that fall between a conductor and an insulator but with resistance that changes abruptly when other conditions change, such as temperature, voltage, and electric or magnetic field.

series circuit Electric circuit with devices connected in such a way that the electric current through each of them is the same. See also *in series.*

shadow Shaded region that appears where light rays are blocked by an object.

shell model of the atom Model in which the electrons of an atom are pictured as grouped in concentric shells around the nucleus.

shock wave Cone-shaped wave produced by an object moving at supersonic speed through a fluid.

short circuit Disruption in an electric circuit, caused by the flow of charge along a low-resistance path between two points that should not be directly connected, thus deflecting the current from its proper path; an effective "shortening of the circuit."

SI Abbreviation for Système International, an international system of units of metric measure accepted and used by scientists throughout the world. See Appendix A for more details.

simple harmonic motion Vibratory or periodic motion, like that of a pendulum, in which the force acting on the vibrating body is proportional to its displacement from its central equilibrium position and acts toward that position.

simultaneity Occurring at the same time. Two events that are simultaneous in one frame of reference need not be simultaneous in a frame moving relative to the first frame.

sine curve Curve whose shape represents the crests and troughs of a wave, as traced out by a pendulum that drops a trail of sand while swinging at right angles to and over a moving conveyor belt.

sine wave The simplest of waves with only one frequency and the shape of a sine curve.

sliding friction Contact force produced by the rubbing together of the surface of a moving object with the material over which it slides.

solar constant 1400 J/m^2 received from the sun each second at the top of the Earth's atmosphere, expressed in terms of power, 1.4 kW/m^2.

solar eclipse Event wherein the moon blocks light from the sun and the moon's shadow falls on part of the Earth.

solar power Energy per unit time derived from the sun. See also *solar constant.*

solar radiation curve Graph of brightness versus frequency (or wavelength) of sunlight.

solid Phase of matter characterized by definite volume and shape.

solidify To become solid, as in freezing or the setting of concrete.

sonic boom Loud sound resulting from the incidence of a shock wave.

sound Longitudinal wave phenomenon that consists of successive compressions and rarefactions of the medium through which the wave travels.

sound barrier The pile up of sound waves in front of an aircraft approaching or reaching the speed of sound, believed in the early days of jet aircraft to create a barrier of sound that a plane would have to break through in order to go faster than the speed of sound. The sound barrier does not exist.

spacetime Four-dimensional continuum in which all events take place and all things exist: Three dimensions are the coordinates of space and the fourth is of time.

special theory of relativity Comprehensive theory of space and time that replaces Newtonian mechanics when velocities are very large. Introduced in 1905 by Albert Einstein. See also *postulates of the special theory of relativity.*

specific heat capacity Quantity of heat required to raise the temperature of a unit mass of a substance by one degree Celsius (or equivalently, by one kelvin). Often simply called specific heat.

spectral lines Colored lines that form when light is passed through a slit and then through a prism or diffraction grating, usually in a spectroscope. The pattern of lines is unique for each element.

spectrometer See *spectroscope.*

spectroscope An optical instrument that separates light into its constituent frequencies or wavelengths in the form of spectral lines. A *spectrometer* is an instrument that can also measure the frequencies or wavelengths.

spectrum (pl. spectra) For sunlight and other white light, the spread of colors seen when the light is passed through a prism or diffraction grating. The colors of the spectrum, in order from lowest frequency (longest wavelength) to highest frequency (shortest wavelength) are red, orange, yellow, green, blue, indigo, violet. See also *absorption spectrum, electromagnetic spectrum, emission spectrum,* and *prism.*

speed How fast something moves; the distance an object travels per unit of time; the magnitude of velocity. See also *average speed, linear speed, rotational speed,* and *tangential speed.*

$$\text{speed} = \frac{\text{distance}}{\text{time}}$$

spherical aberration Distortion of an image caused when the light that passes through the edges of a lens focuses at slightly different points from the point where the light passing through the center of the lens focuses. It also occurs with spherical mirrors.

spring tide High or low tide that occurs when the sun, Earth, and moon are all lined up so that the tides due to the sun and moon coincide, making the high tides higher than average and the low tides lower than average. See also *neap tide.*

stable equilibrium State of an object balanced so that any small displacement or rotation raises its center of gravity.

standing wave Stationary wave pattern formed in a medium when two sets of identical waves pass through the medium in opposite directions. The wave appears not to be traveling.

static friction Force between two objects at relative rest by virtue of contact that tends to oppose sliding.

streamline Smooth path of a small region of fluid in steady flow.

strong force Force that attracts nucleons to each other within the nucleus; a force that is very strong at close distances but decreases rapidly as the distance increases. Also called strong interaction. See also *nuclear force.*

strong interaction See *strong force.*

subcritical mass See *critical mass.*

sublimation Direct conversion of a substance from the solid to the vapor phase, or vice versa, without passing through the liquid phase.

subtractive primary colors The three colors of light-absorbing pigments—magenta, yellow, and cyan—that when mixed in certain proportions will reflect any color in the spectrum.

superconductor Material that is a perfect conductor with zero resistance to the flow of electric charge.

supercritical mass See *critical mass.*

superposition principle In a situation where more than one wave occupies the same space at the same time, the displacements add at every point.

supersonic Traveling faster than the speed of sound.

support force Upward force that balances the weight of an object on a surface.

surface tension Tendency of the surface of a liquid to contract in area and thus behave like a stretched elastic membrane.

tachyon Hypothetical particle that can travel faster than light and thus move backward in time.

tangent Line that touches a curve in one place only and is parallel to the curve at that point.

tangential speed Linear speed along a curved path.

tangential velocity Component of velocity tangent to the trajectory of a projectile.

tau The heaviest elementary particle in the class of elementary particles called leptons.

technology Method and means of solving practical problems by implementing the findings of science.

telescope Optical instrument that forms images of very distant objects.

temperature Measure of the average translational kinetic energy per molecule of a substance, measured in degrees Celsius or Fahrenheit, or in kelvins.

temperature inversion Condition wherein upward convection of air is stopped, sometimes because an upper region of the atmosphere is warmer than the region below it.

terminal speed Speed attained by an object wherein the resistive forces, often air resistance, counterbalance the driving forces, so motion is without acceleration.

terminal velocity Terminal speed together with the direction of motion (down for falling objects).

terrestrial radiation Radiant energy emitted from the Earth.

theory Synthesis of a large body of information that encompasses well-tested and verified hypotheses about aspects of the natural world.

thermal contact State of two or more objects or substances in contact such that heat can flow from one object or substance to the other.

thermal equilibrium State of two or more objects or substances in thermal contact when they have reached a common temperature.

thermal pollution Undesirable heat expelled from a heat engine or other source.

thermodynamics Study of heat and its transformation to mechanical energy, characterized by two principal laws:

First Law: A restatement of the law of conservation of energy as it applies to systems involving changes in temperature: Whenever heat is added to a system, it transforms to an equal amount of some other form of energy.

Second Law: Heat cannot be transferred from a colder body to a hotter body without work being done by an outside agent.

thermometer Device used to measure temperature, usually in degrees Celsius, degrees Fahrenheit, or kelvins.

thermonuclear fusion Nuclear fusion brought about by extremely high temperatures; in other words, the welding together of atomic nuclei by high temperature.

thermostat Type of valve or switch that responds to changes in temperature and that is used to control the temperature of something.

time dilation Slowing down of time for an object moving at relativistic speeds.

torque Product of force and lever-arm distance, which tends to produce rotational acceleration.

$$\text{torque} = \text{lever-arm distance} \times \text{force}$$

total internal reflection The 100% reflection (with no transmission) of light that strikes the boundary between two media at an angle greater than the critical angle.

transformer Device for increasing or decreasing voltage or transferring electric power from one coil of wire to another by means of electromagnetic induction.

transistor See *semiconductor.*

transmutation Conversion of an atomic nucleus of one element into an atomic nucleus of another element through a loss or gain in the number of protons.

transparent Term applied to materials that allow light to pass through them in straight lines.

transuranic element Element with an atomic number above 92, which is the atomic number for uranium.

transverse wave Wave with vibration at right angles to the direction the wave is traveling. Light consists of transverse waves.

tritium Unstable, radioactive isotope of hydrogen whose atom has a proton, two neutrons, and an electron.

trough One of the places in a wave where the wave is lowest or the disturbance is greatest in the opposite direction from a crest. See also *crest.*

turbine Paddle wheel driven by steam, water, etc., that is used to do work.

turbogenerator Generator that is powered by a turbine.

ultrasonic Term applied to sound frequencies above 20,000 hertz, the normal upper limit of human hearing.

ultraviolet (UV) Electromagnetic waves of frequencies higher than those of violet light.

umbra Darker part of a shadow where all the light is blocked. See also *penumbra.*

uncertainty principle The principle formulated by Heisenberg, stating that Planck's constant, h, sets a limit on the accuracy of measurement at the atomic level. According to the uncertainty principle, it is not possible to measure exactly both the position and the momentum of a particle at the same time, nor the energy and the time associated with a particle simultaneously.

universal gravitational constant The proportionality constant G that measures the strength of gravity in the equation for Newton's law of universal gravitation.

$$F = G\frac{m_1 m_2}{d^2}$$

unstable equilibrium State of an object balanced so that any small displacement or rotation lowers its center of gravity.

UV Abbreviation for *ultraviolet.*

V (a) In lower-case italic v; the symbol for *speed* or *velocity.* (b) In uppercase V, the abbreviation for *voltage.*

vacuum Absence of matter; void.

Van Allen radiation belts Two donut-shaped belts of radiation that surround the Earth.

vaporization The process of a phase change from liquid to vapor; evaporation.

vector Arrow whose length represents the magnitude of a quantity and whose direction represents the direction of the quantity.

vector quantity Quantity in physics that has both magnitude and direction. Examples are force, velocity, acceleration, torque, and electric and magnetic fields.

velocity Speed of an object and its direction of motion; a vector quantity.

vibration Oscillation; a repeating to-and-fro motion about an equilibrium position—a "wiggle in time."

virtual image Image formed by light rays that do not converge at the location of the image. Mirrors, converging lenses used as magnifying glasses, and diverging lenses all produce virtual images. The image can be seen by an observer, but cannot be projected onto a screen.

visible light Part of the electromagnetic spectrum that the human eye can see.

visible spectrum See *electromagnetic spectrum.*

volt (V) SI unit of electric potential. One volt is the electric potential difference across which one coulomb of charge gains or loses one joule of energy. 1 V = 1 J/C

voltage Electrical "pressure" or a measure of electrical potential difference.

$$\text{voltage} = \frac{\text{electric potential energy}}{\text{unit of charge}}$$

voltage source Device, such as a dry cell, battery, or generator, that provides a potential difference.

voltmeter See *galvanometer.*

volume Quantity of space an object occupies.

W (a) Abbreviation for *watt*. (b) When in italic *W*, the abbreviation for *work*.

watt SI unit of power. One watt is expended when one joule of work is done in one second. 1 W = 1 J/s

wave A "wiggle in space and time"; a disturbance that repeats regularly in space and time and that is transmitted progressively from one place to the next with no net transport of matter.

wave front Crest, trough, or any continuous portion of a two-dimensional or three-dimensional wave in which the vibrations are all the same way at the same time.

wave speed Speed with which waves pass a particular point.

$$\text{wave speed} = \text{wavelength} \times \text{frequency}$$

wave velocity Wave speed stated with the direction of travel.

wavelength Distance between successive crests, troughs, or identical parts of a wave.

weak force Also called weak interaction. The force within a nucleus that is responsible for beta (electron) emission. See *nuclear force*.

weak interaction See *nuclear force* and *weak force*.

weight The force that an object exerts on a supporting surface (or, if suspended, in a supporting string)—often, but not always, due to the force of gravity.

weight density See *density*.

weightlessness Condition of free fall toward or around the Earth, in which an object experiences no support force (and exerts no force on a scale).

white light Light, such as sunlight, that is a combination of all the colors. Under white light, white objects appear white and colored objects appear in their individual colors.

work (W) Product of the force on an object and the distance through which the object is moved (when force is constant and motion is in a straight line in the direction of the force); measured in joules.

$$\text{work} = \text{force} \times \text{distance}$$

work-energy theorem Work done on an object is equal to the kinetic energy gained by the object.

$$\text{work} = \text{change in energy} \quad \text{or} \quad W = \Delta KE$$

wormhole Hypothetical enormous distortion of space and time, similar to a black hole, but opens out again in some other part of the universe.

X ray Electromagnetic radiation, higher in frequency than ultraviolet, emitted by atoms when the innermost orbital electrons undergo excitation.

zero-point energy Extremely small amount of kinetic energy that molecules or atoms have even at absolute zero.

Photo Credits

Frontmatter: p xv Lillian Lee Hewitt

Part Openers: p 1 John Suchocki; **Part 1 p 21** David Vasquez, Collection of Paul G. Hewitt; **Part 2 p 209** David Riveros; **Part 3 p 289** Olan Mills; **Part 4 p 361** Paul G. Hewitt; **Part 5 p 409** Bob Abrams; **Part 6 p 495** Victor Lee; **Part 7 p 619** Paul G. Hewitt; **Part 8 p 685** Reprinted with permission from *Physics Today,* 54(8), Copyright 2001, American Institute of Physics

Chapter 1: Opening photo Paul G. Hewitt; **1.7** CORBIS; **1.8** Jay Pasachoff

Chapter 2: Opening photo Paul G. Hewitt; **p 23** Bettmann/CORBIS; **p 24** Erich Lessing/Art Resource. Artist: German School; **p 25** Alinari/Art Resource. Artist: Justus Sustermans; **p 29** Giraudon/Art Resource. Artist: Godfrey Kneller

Chapter 3: Opening photo Keith Bardin, Collection of Paul G. Hewitt; **3.2** Alan Schein Photography/CORBIS; **p 53** NBAE/Getty Images

Chapter 4: Opening photo Paul G. Hewitt; **4.5** Pearson Education; **4.10** Caterpillar; **4.16** Fundamental Photographs

Chapter 5: Opening photo Judith Brand, Exploratorium; **5.19** Henry R. Fox/Animals Animals; **5.20** Lillian Lee Hewitt; **p 88** Paul G. Hewitt

Chapter 6: Opening photo Benjamin Alexander, Collection of Paul G. Hewitt; **6.1** Craig Aurness/Westlight; **6.4** The Harold E. Edgerton Trust/Palm Press; **6.7** Brand X Pictures/Getty Images; **6.9** Paul G. Hewitt; **6.10** Paul G. Hewitt; **6.16** Paul G. Hewitt; **6.20** DC Heath with Educational Development Center

Chapter 7: Opening photo Burl Grey, Collection of Paul G. Hewitt; **7.1** Charles Krupa/AP; **7.3** NASA; **7.7** Paul G. Hewitt; **7.8** Paul G. Hewitt; **7.11** Michael Vollmer; **7.20** Jack Hancock; **p 129** Collection of Paul G. Hewitt

Chapter 8: Opening photo Anna-Marie Darden; **8.11** Silvestre Machado/Photolibrary.com; **8.17** Meidor Hu; **8.23** Richard Megna/Fundamental Photographs; **8.28** Getty Images; **8.35** Paul G. Hewitt; **8.36** Paul G. Hewitt; **p 146** Collection of Paul G. Hewitt; **8.50** NASA, Ames Research Center; **8.51** Paul G. Hewitt; **8.55** Gerard Lacz/Peter Arnold; **p 155** Lillian Lee Hewitt; **p 159** Paul G. Hewitt

Chapter 9: Opening photo Lillian Lee Hewitt; **9.11** NASA

Chapter 10: Opening photo Ted Lobosky, Collection of Paul G. Hewitt; **10.5** Chuck Stone; **10.12** Getty Images; **p 191** NBAE/Getty Images; **10.19** NASA, Johnson Space Center; **10.25** Pearson Education; **10.26** Diane Schiumo/Fundamental Photographs; **p 198** NASA, Goddard Space Flight Center; **p 199 top** Royal Observatory, Edinburgh/Photo Researchers; **p 199 bottom** Erich Lessing/Art Resource. Artist: Anonymous; **10.34** NASA, Goddard Space Flight Center; **10.35** Steven Hobbs, NASA, JPL-Caltech; **p 205** Fred R. Myers Jr.

Chapter 11: Opening photo Collection of Paul G. Hewitt; **11.4** University of Chicago; **11.5** Almaden Research Center, IBM; **11.8** Paul G. Hewitt

Chapter 12: Opening photo Manfred Kage/Peter Arnold; **12.5** Paul G. Hewitt; **12.6** The Harold E. Edgerton Trust/Palm Press; **12.11** Lillian Lee Hewitt; **12.12** Raymond V. Schoder; **12.13** Paul G. Hewitt; **12.14** Paul G. Hewitt; **12.15** The Stock Market/CORBIS; **12.18** W. Geirsperger/Camerique/H. Armstrong Roberts; **12.19** James P. Rowan; **p 246** Paul G. Hewitt

Chapter 13: Opening photo Keith Bardin, Collection of Paul G. Hewitt; **13.2** Paul Ryan; **13.3** The Stock Market/CORBIS; **13.6** The Stock Market/CORBIS; **13.19** Milo Patterson; **13.24** Adrian Greeman/constructionphotography.com; **13.27** Brand X Pictures/Alamy; **13.29** Olivier Blaise, Collection of Paul G. Hewitt; **p 265** Margaret Ellenstein

Chapter 14: Opening photo Paul G. Hewitt; **14.2** The Granger Collection; **14.9** David Hewitt; **14.19** Lillian Lee Hewitt; **14.27** Carey B. Van Loon; **14.28** David Hewitt; **p 285** Lillian Lee Hewitt

Chapter 15: Opening photo Paul G. Hewitt; **15.2** Kasia Werel, Collection of Paul G. Hewitt; **15.12** LU Engineers; **15.13** Meidor Hu; **15.14** AP; **15.17** Paul G. Hewitt; **p 304** Ed Young/Photo Researchers

Chapter 16: Opening photo Tracy Suchocki; **16.3** Don Hynek, Wisconsin Division of Energy; **16.5** Nancy Rogers; **16.6** Paul G. Hewitt; **16.8** Collection of Paul G. Hewitt; **16.17** Paul G. Hewitt; **16.18** David Cavagnaro; **16.19** Paul G. Hewitt; **16.23** Stephen W. Hewitt; **16.24** Inga Spence/Visuals Unlimited; **p 324** Paul G. Hewitt

Chapter 17: Opening photo Dean Baird, Collection of Paul G. Hewitt; **17.1** Paul G. Hewitt; **17.2** Paul G. Hewitt; **17.3** Ralph A. Reinhold/Animals Animals; **17.4** Paul G. Hewitt; **17.19** Paul G. Hewitt; **17.20** Rick Povich Photography

Chapter 18: Opening photo Nicole Minor, Exploratorium; **18.4** Paul G. Hewitt; **18.7** The Stock Market/CORBIS; **18.14** Dennis Wong; **18.15** Michael Howell/Robert Harding; **18.18** Paul G. Hewitt

Chapter 19: Opening photo Dave Eddy; **19.2** Collection of Paul G. Hewitt; **19.5** The Image Bank/Getty Images; **19.12** Richard Megna/Fundamental Photographs; **19.21** U.S. Navy; **19.22** The Harold E. Edgerton Trust/Palm Press

Chapter 20: Opening photo Andrew Morrison; **20.4** Paul G. Hewitt; **20.7** Terrence McCarthy, San Francisco Symphony; **20.9** Stone/Getty Images; **20.10** Laura Pike & Steve Eggen; **20.12** Paul G. Hewitt; **20.14 left and right** AP; **20.14 center** Bettmann/CORBIS; **20.19** Norman Synnestvedt; **20.21** Pearson Education

Chapter 21: Opening photo David Yee; **p 399 top** Ivan Nikitin; **p 399 bottom** Bernd Kammerer/AP; **21.2** Paul G. Hewitt; **p 400** Time Life Pictures/Getty Images; **p 401** Mark J. Terrill/AP; **p 402** Richard Drew/AP; **21.6** Ronald B. Fitzgerald; **p 403** Globe Photos; **21.10** Meidor Hu; **21.12** Andrew Syred/Photo Researchers

Chapter 22: Opening photo Jim Stith, Collection of Paul G. Hewitt; **22.4** Camerique; **22.5a** Pearson Education; **22.5b** Paul G. Hewitt; **22.10** Grant W. Goodge, National Climatic Data Center; **22.19** Palmer Physical Laboratory, Princeton University; **22.22** BBC; **22.28** Ahmed Eid; **22.29** Dorling Kindersley; **22.31** Paul G. Hewitt

Chapter 23: Opening photo Keith Bardin, Collection of Paul G. Hewitt; **23.2** Dorling Kindersley; **23.3** Zig Leszczynski/Animals Animals; **23.6** Dorling Kindersley; **23.10** Dorling Kindersley; **23.16** Pearson Education; **23.17** Pearson Education; **23.18** Pearson Education; **23.21** David Hewitt

Chapter 24: Opening photo Gary Kuwahara, Collection of Paul G. Hewitt; **24.2** George Haling/Photo Researchers; **24.4** Richard Megna/Fundamental Photographs; **24.5** Duane Ackerman; **24.10** Richard Megna/Fundamental Photographs; **24.11** John Suchocki; **24.12** Str/AP; **24.18** Pearson Education; **24.23** Pekka Parvianinen/Polar Image; **24.24** Corby Waste, NASA, JPL-Caltech; **24.25** John Downer/naturepl.com; **p 473** Pete Saloutos/CORBIS

Chapter 25: Opening photo Lillian Lee Hewitt; **25.5** John Suchocki; **p 481** The Granger Collection; **25.13** Paul G. Hewitt; **25.17** Lillian Lee Hewitt

Chapter 26: Opening photo Sasha Sokolova, Collection of Paul G. Hewitt; **p 497** Dorling Kindersley; **26.9** Paul G. Hewitt; **26.10** Diane Schiumo/Fundamental Photographs; **26.11** Diane Schiumo/Fundamental Photographs; **26.16** Linnart Nilsson; **26.19** Pearson Education

Chapter 27: Opening photo Paul G. Hewitt; **27.3** Cordelia Molloy/SPL/Photo Researchers; **27.4** Meidor Hu; **27.10** Dave Vasquez; **27.11** Paul G. Hewitt III; **27.12** Dave Vasquez; **27.13** Paul G. Hewitt; **27.16** Meidor Hu; **27.17** J.H. Robinson/Animals Animals; **27.19** H. Armstrong Roberts; **27.20** The Image Bank/Getty Images; **27.21** Dean Baird; **p 528** Collection of Paul G. Hewitt

Chapter 28: Opening photo Launch McKenzie, Collection of Paul G. Hewitt; **28.8** Paul G. Hewitt; **28.11** David Nunuk/Photo Researchers; **28.12** Dennis Kunkel/Visuals Unlimited; **28.20** Ted Mahiew; **28.22** Robert Greenler; **28.34** Paul G. Hewitt; **28.41** Will & Deni McIntyre/Photo

Researchers; **28.45** Ron Fitzgerald; **28.50** Paul G. Hewitt; **p 554** Fred R. Myers Jr.; **p 555 top** Barbara Thomas, Collection of Paul G. Hewitt; **p 555 bottom** Camerique/H. Armstrong Roberts; **p 556** Milo Patterson

Chapter 29: Opening photo Udo von Mulert; **29.1** Martin Dohrn/SPL/Photo Researchers; **29.7** Education Development Center; **29.10** Ken Kay/Fundamental Photographs; **29.12** C.K. Menka; **29.14** Paul G. Hewitt; **29.15** Education Development Center; **29.24** Bausch & Lomb; **29.25** Bausch & Lomb; **29.27** Paul G. Hewitt; **29.33** Diane Schiumo/Fundamental Photographs; **29.35** Paul G. Hewitt

Chapter 30: Opening photo Lillian Lee Hewitt; **30.5** Sargent-Welch; **30.12** Aaron Haupt/SPL/Photo Researchers; **30.13** Mark A. Schneider/Visuals Unlimited; **30.19 left** JDS Uniphase; **30.19 right** Collection of Paul G. Hewitt; **30.20** Pearson Education

Chapter 31: Opening photo Neil Chapman, Collection of Paul G. Hewitt; **p 601** Bettmann/CORBIS; **31.4** Albert Rose; **31.6** Elisha Huggins; **31.7** AIP Niels Bohr Library; **p 608** Meggers Gallery/American Institute of Physics/SPL/Photo Researchers; **31.8** H. Raether, Elecktroninterferenzen, Handbuck der Physik, Vol. 32, 1957/Springer-Verlag, Berline & Heidelberg, NY; **31.9** Lawrence Migdale/Photo Researchers; **31.10** Tony Brain/SPL/Photo Researchers; **31.11** P.G. Merli, G.F. Missiroli, and G. Pozzo. "On the Statistical Aspect of Electron Interference Phenomena," *American Journal of Physics*, Vol. 44, No. 3, March 1976. Copyright 1976 by the American Association of Physics Teachers; **p 611** Archives for the history of quantum, AIP Niels Bohr Library; **p 614 left** SuperStock; **p 614 right** National Oceanic and Atmospheric Administration, National Severe Storms Laboratory

Chapter 32: Opening photo Mary Murphy-Waldorf; **p 620** Bettmann/CORBIS; **p 624** Margethe Bohr Collection, AIP Niels Bohr Library; **p 629** Bettmann/CORBIS

Chapter 33: Opening photo Lillian Lee Hewitt; **33.1** New York Hospital; **p 635** AP; **33.10a** Hank Morgan/Photo Researchers; **33.10b** Kevin Schaefer/Peter Arnold; **33.12** Lawrence Berkeley Lab, University of California; **33.13a** CERN/SPL/Photo Researchers; **33.13b** Fermilab; **33.17** International Atomic Energy Agency; **33.20** Photodisc Green/Getty Images; **33.22** Jerry Nulk and Joshua Baker

Chapter 34: Opening photo Dean Zollman; **p 665** Fermi Film Collection, AIP Niels Bohr Library; **34.7** Chicago Historical Society; **34.8** National Argonne Library; **34.12** Roger Ressmeyer/Starlight; **p 672** Lillian Lee Hewitt; **34.24** Lawrence Livermore National Laboratory

Chapter 35: Opening photo Paul G. Hewitt; **35.21** Mark Paternostro, NASA, Kennedy Space Center; **35.25** SLAC/Photo Researchers; **35.27** Stone/Getty Images; **p 714** Library of Congress; **p 715** California Institute of Technology Archives

Chapter 36: Opening photo Lillian Lee Hewitt

Appendix A: A.1 Collection of Paul G. Hewitt

Index